Mineral Spectroscopy: A Tribute to Roger G. Burns

SPECIAL PUBLICATION SERIES

No. 1. *Magmatic Processes: Physicochemical Principles.* B. O. Mysen (Editor), 1987.

No. 2. *Fluid-Mineral Interactions: A Tribute to H. P. Eugster.* R. J. Spencer and I-Ming Chou (Editors), 1990.

No. 3. *Stable Isotope Geochemistry: A Tribute to Samuel Epstein.* H. P. Taylor, Jr., J. R. O'Neil and I. R. Kaplan (Editors), 1991.

No. 4. *Victor Moritz Goldschmidt: Father of Modern Geochemistry.* Brian Mason, 1992.

No. 5. *Mineral Spectroscopy: A Tribute to Roger G. Burns.* M. D. Dyar, C. McCammon and M. W. Schaefer (Editors), 1996.

Special Publication Series Editor: Scott M. McLennan
Department of Earth and Space Sciences
State University of New York at Stony Brook
Stony Brook, NY, 11794-2100, U.S.A.

Mineral Spectroscopy: A Tribute to Roger G. Burns

Edited by:

M. Darby Dyar

Department of Geology and Astronomy, West Chester University,
West Chester, PA 19383, U.S.A.

Catherine McCammon

Bayerisches Geoinstitut, Universität Bayreuth, D-95440 Bayreuth, Germany

Martha W. Schaefer

Department of Geology and Geophysics, Yale University,
New Haven, CT 06520, U.S.A.

Special Publication No. 5

THE GEOCHEMICAL SOCIETY

Published by The Geochemical Society
Department of Chemistry
University of Houston
Houston, TX 77204-5641, U.S.A.

Library of Congress Catalogue Number 96-077227

ISBN 0-941809-04-8

Printed in the United States of America
by Lancaster Press

TABLE OF CONTENTS

Part B. Spectroscopy and Mineralogy of the Earth's Mantle

Part C. Remote Sensing of Planetary Surfaces

PREFACE

Mineral Spectroscopy: A Tribute to Roger G. Burns is a collection of 29 papers written by former students, postdoctoral fellows, colleagues, friends, and admirers of ROGER G. BURNS. The idea for this volume was conceived following a memorial service in Roger's honor that was held at MIT in March, 1994 and attended by many of his former students. Contributions for the volume were solicited from participants in spectroscopic presentations at the 16th Meeting of the International Mineralogical Association, and in special sessions organized in Roger's memory at the 25th Annual Lunar and Planetary Science Conference and the 1994 Fall Meeting of the American Geophysical Union.

ROGER BURNS was widely recognized as a pioneer in the development of the field of mineral spectroscopy. He began his career studying chemistry and geology at Victoria University of Wellington in his native New Zealand, and was drawn to the full-time study of minerals with BILL FYFE at Berkeley in the 1960's. Four postdoctoral years in New Zealand and England led to the development with colleagues of a new research approach that used spectroscopy to understand the structure and color of minerals, site occupancies of elements in minerals, and the factors influencing partitioning of elements between melt and crystals. After arriving at MIT in 1970, Roger continued to make important advances in the field of crystal chemistry of rock-forming minerals. His book on the *Mineralogical Applications of Crystal Field Theory* (1970 and 1993) remains the classic, influential text in the field.

This volume is divided into three sections. The first section (Part A) is composed largely of papers submitted by former students and associates of ROGER BURNS. An introduction on the future of geochemistry and the Earth is followed by fourteen papers on the theory and application of spectroscopy to problems in geology and mineralogy. Included in this section are papers utilizing optical absorption, infrared, and Mössbauer spectroscopy, three techniques with which ROGER BURNS did much of his pioneering work. The remaining two sections of the book each focus on a single topic. The first is spectroscopy of mantle minerals (Part B), an area that is currently the subject of much focus at scientific meetings and in journals. The theme is an appropriate one, because a large portion of ROGER BURNS' research focused on the properties of mantle minerals and implications for physical and chemical properties of the mantle. With this section, we bring together a compilation of work on mantle mineralogy that provides a summary of current knowledge in that field. The second topic (Part C) includes results from the planetary remote sensing community, where much of the current research utilizes skills and concepts that can be directly traced to Roger's research. This section begins with a review of methodology for quantitative mineral analysis of planetary surfaces using reflectance spectroscopy, and continues with a group of papers discussing the mineralogy of surface materials on Mars.

It is fair to say that essentially all modern workers in the field of mineral spectroscopy have in some way been touched or guided by ROGER BURNS, either by personal contact or through the fundamental advances presented in his books and papers. Those who were fortunate enough to know him will long remember Roger for his genuine kindness, generous intellect, and infectious enthusiasm for our science. With this book, we pay tribute to the memory of his brilliant career.

MELINDA DARBY DYAR
CATHERINE A. MCCAMMON
MARTHA W. SCHAEFER
April, 1996

ACKNOWLEDGMENTS

Any volume of this size and complexity must by necessity be the product of many individuals' efforts, and this Special Publication is no exception. We have been touched by the willingness of the scientific community to assist with this volume in every way, most notably in the form of [largely] prompt, thorough, and constructive manuscript reviews. We are very grateful to the following individuals who devoted their time to be reviewers for this volume: David Agresti, Georg Amthauer, Orson Anderson, Hu Barnes, Olof Beckman, James F. Bell III, Diana Blaney, Lawrence Bowen, D.W. Bullett, John Cashion, Tore Ericsson, B.J. Evans, Yingwei Fei, Embaie Ferrow, Robert Finkelman, John Freeman, Fari Goodarzi, Colin Graham, Grant Henderson, Anne Hofmeister, Demelza Hugh-Jones, Phillip Ihinger, David Jenkins, Ralf Krupp, Thomas B. McCord, Richard V. Morris, Scott L. Murchie, James Murowchick, John Mustard, Jane Nielson, James Penner-Hahn, Robert K. Popp, Mati Raudsepp, Shiv Sharma, John Tossell, and Roger Weppling. We also acknowledge the efforts of several reviewers who chose to remain anonymous.

M.W.S. acknowledges the support of NASA grant NAGW-3588 through the University of Maryland. M.D.D. wishes to thank the College of Arts and Sciences (David Buchanan, Dean) and the Department of Geology and Astronomy (Gil Wiswall, Chairperson) at West Chester University for their support during the leave of absence over which this book was produced. Other support during this period came from grants from the National Science Foundation, EAR-9304304, EAR-9303958, EAR-9405310, and EAR-9527020. She is also grateful to the Departments of Geology at both Amherst College and Smith College for providing her with logistical support during the leave, and to her husband and son, Peter and Duncan Crowley, for their moral support and patience with her during the long duration of this project.

Finally, we thank the Geochemical Society for its willingness to fund this endeavor, and Special Publication Series Editor Scott McLennan for all his help with the production aspects of this book.

Roger George Burns

Roger G. Burns
Scholar and a Gentleman

Roger George Burns was born in Wellington, New Zealand on December 28, 1937. His B.Sc. (1959, Chemistry and Geology) and M.Sc. (1961, with 1st class Honors in Chemistry) were from Victoria University of Wellington. Inspired by work on the Skaergaard intrusion by the Oxford chemist R. J. P. Williams that explained fractionation patterns during basalt crystallization using crystal field theory, Roger traveled to the United States to continue his graduate studies at the University of California at Berkeley under the supervision of fellow New Zealander, William S. Fyfe.

The fall of 1964 and especially the spring of 1965 were exciting times on the Berkeley campus. Every day there was a huge protest rally going on at noon in front of the administration building. Many photos were taken from the roof of that building looking down on the mass of attending students. In many of them, a group of eight geology graduate students can be observed enjoying the lunchtime antics of the student leaders while consuming their brown bag lunches (B. Crawford, personal communication, 1995). Close inspection of these photos might identify them as part of an extremely talented cohort of students who were at Berkeley at that time, whose collective contributions have changed the geosciences over the course of their careers. This group included petrologists Ned Brown, Weecha and Bill Crawford, Eric Essene, Jay Hodgson, Mike Holdaway, Derrill Kerrick, Peter Reed, and Dan Weill as well as Roger Burns; other classmates included Stuart Blusson, Brent Dalrymple, Greg Davis, Priscilla Grew, Sherman Gromme, Hal Gluskoter, John Obradovich, Bill Romey, and Roger's future wife, Virginia Mee.

Roger's graduate work at Berkeley pioneered the use of absorption spectra to study the location and atomic coordination of transition metals in minerals. His Ph.D. thesis (1965, Geochemistry), ''Electronic Spectra of Silicate Minerals: Application of Crystal-Field Theory to Aspects of Geochemistry,'' was a breakthrough in using chemical principles to understand the bonding environment of transition metals in silicates.

Roger is remembered from his time at Berkeley as an intense yet remarkably friendly colleague. He always made the time for any person who needed help. Despite his chemical background, Roger's performance as a field geologist was remarkable (E. Essene, personal communication, 1995). His skills with bowling at informal cricket matches during Departmental picnics were as impressive as his font of chemical and mineralogical knowledge.

Following his Ph.D., Roger spent four very productive years in England and New Zealand, first as a Senior Research Visitor at Cambridge University (1966), then as a Senior Lecturer in Geochemistry at Victoria University of Wellington (1967), and finally as Lecturer in Geochemistry at Oxford University (1968–1970). At Cambridge, Roger Burns (along with Mike Bancroft, Roger Strens and Eric Essene) kept the Department of Mineralogy and Petrology lively. At this time he began a fruitful collaboration and long-standing friendship with G. Michael Bancroft. Michael and Roger met at a Chemical Society meeting in Nottingham in 1965, shortly after Michael had written to Roger in Berkeley for a sample of the unusual mineral gillespite. Coincidentally, in Nottingham Roger gave a paper on that ''attractive pink silicate'' (as labeled by Michael Clark, another long-standing collaborator and friend). Together Burns and Bancroft developed a new research approach for understanding the structure of minerals by applying a variety of newly-developing spectroscopic techniques, notably Mössbauer spectroscopy, to rock-forming minerals. Nine joint papers appeared in 1967 and 1968 from one year of intense, exciting collaboration. Important advances were made, not only in the specialized field of crystal chemistry, but also in understanding the causes of color in minerals and gems, an aspect of their research that has significance beyond the realm of science. Roger Burns rapidly gained international recognition with publications resulting from this early research, particularly the 1970 publication of his book ''Mineralogical Applications of Crystal Field Theory,'' which was based on a series of lectures given at Oxford and Cambridge Universities. The first edition of his book was soon translated into Japanese, Chinese, and Russian, and a revised second edition was published in 1993 following a Guggenheim-sponsored sabbatical leave spent with David Vaughan at Manchester University, England.

Roger joined the faculty of the M.I.T. Department of Earth and Planetary Sciences in 1970 as Associate Professor of Geochemistry. Within two years he was promoted to Professor of Mineralogy and Geochemistry. At M.I.T. his expertise in mineralogy and geochemistry and rapport with students immediately led to innovative research. In addition to his energetic pursuit of the application of spectroscopic techniques to mineralogical problems, Roger broke new ground in several areas of earth and planetary science. In the 1970's, Roger and his wife Virginia were instrumental in characterizing and understanding the ubiquitous formation of manganese-rich minerals on the sea floor. His efforts were in part motivated by a desire to pursue research that would have a positive impact on society. In recent years, Roger pursued this goal through evaluations of the reactivity of zeolites in proposed repositories for high-level nuclear waste. Also, throughout his career he was an active consultant in diverse specialities ranging from the chemistry of batteries to the manufacture of synthetic gems and the Nevada nuclear waste isolation program.

Also in the 1970's, Roger collaborated with experimental mineralogists to obtain spectroscopic data while minerals were subjected to very high pressure in a diamond anvil cell. He and his students were among the first to observe atomic coordination changes in silicate minerals at high pressure. Many of the papers in the mantle mineralogy section of this volume represent various outgrowths of this pioneering line of research. Roger was also interested in the oxidation states of planetary interiors and how they might be recorded by transition metals in magmas generated at great depths. He was among the first to report Mössbauer and electronic absorption spectra measurements of lunar samples, and he was instrumental in applying many spectroscopic techniques to study the oxidation state of the lunar interior as recorded in glasses from the lunar regolith.

Roger's involvement with the lunar program quickly expanded into more broadly-based studies of remotely-sensed spectra of many types of planetary surfaces. Working with planetary scientists at M.I.T. and Brown University, he used measured laboratory spectra of likely planetary surface minerals to identify the phases present on planetary surfaces and the presence of specific transition metals. Beginning in the 1980's he made major contributions to understanding the composition and mineralogy of the Martian surface. Most recently, Roger and his students were using Mössbauer spectroscopy to characterize the Fe^{3+}-bearing minerals in meteorites with the objective of understanding how meteorites are affected by the terrestrial environment.

As a teacher, Roger was popular with both undergraduate and graduate students, and his courses in "Chemistry and Physics of Minerals and Rocks" and "Geochemistry of the Transition Elements" attracted not only M.I.T. students, but generations of Wellesley and Harvard students as well. Both courses were memorable for a variety of reasons, as fondly recalled by Dan Orange (personal communication, 1996):

"First of all, Roger had the most amazingly impeccable handwriting of anybody I've ever seen, before or since. In addition, he could write in a straight line all the way across the board (no mean feat). He could write and talk at the same time, without seeming to stare at the board, and if all you did was copy the notes, by the time the course was over you practically had a textbook in your notebook! Of course, there was more than just mere penmanship. Roger could draw the most elegant yet straightforward figures on the board at the drop of a hat. I still marvel at how he was able to convey crystal symmetry, lattice structure, etc. and *make it seem simple*."

"All who knew Roger could immediately tell that he was a caring person . . . he genuinely cared about his students both inside the classroom and out. Roger was always available for a question or a comment, and he never made you feel like you were stupid (even if you were asking a pretty basic question). Thus, in the classroom Roger would belabor a point and explain it umpteen different ways until a student got it. If a student cared to visit him in his office he would take whatever time was needed out of what he was doing to help a student along."

"The amazing thing about Roger is that the approach he brought to teaching was evident in all other aspects of him. I took a higher level analytical geochemistry course a few years later, . . . and once again, Roger would painstakingly explain things and clearly illustrate them (on the back of a napkin or in a notebook) until I got it. At professional meetings years later we would discuss the state of our science, career paths, and research, and Roger would offer thoughtful suggestions and insightful comments. There are cliches about "nice guys," but I was always struck by the fact that with Roger, he had risen high up on the heap without losing his "nice guy"-ness: his humanity, his caring, and his devotion to his students."

During his career, Roger Burns supervised 23 Ph.D., 10 M.S., and numerous B.S. and B.A. theses; his students have gone on to productive careers in both industry and academia. Roger was especially committed to promoting the careers of women in science, and he shared his lab and his mentorship with a succession of M.I.T. and Wellesley College women through ongoing contacts with his former student James Besancon, a professor at Wellesley College. He would be pleased that this memorial volume is edited by three of his former women students.

Roger G. Burns published more than 140 research papers, served as an editor for several journals and books, and received additional degrees and many honors, awards, and fellowships. Among these, he was proudest of his M.A. and D.Sc. in mineralogy from Oxford University and of his Guggenheim Fellowship received in 1991. The 1975 presentation of the Mineralogical Society of America award to Roger Burns showed that his leadership and impact in mineralogy were recognized early in his career. He was a Life Fellow of the Mineralogical Society of America, a member of the Mineralogical Society of Great Britain (and its Hallimond Lecturer in 1987), and of the American Geophysical Union, the Geochemical Society, and the New Zealand Geochemical Group.

Roger's many non-scientific interests included a strong commitment to his family of wife Virginia Mee Burns and sons Kirk and Jonathan. Himself the son of a music teacher, Roger was known to time his other commitments around scheduled broadcasts of his favorite operas. He was a dedicated distance runner who commuted daily on foot from his home in north Cambridge to M.I.T. He was proud of the many marathons (including the Boston Marathon) he completed, and he often scheduled professional trips around local races! He frequently found solace, peace, and quiet at Virginia's family home on Peaks' Island off the coast of Portland, Maine, where he said the rocky coast reminded him of New Zealand.

Roger G. Burns died on January 7, 1994. All who knew Roger sorely miss his keen intellect, gentlemanly charm, and unselfish enthusiasm for the earth and planetary sciences. With this book, we celebrate his memory through publications on many aspects of the applications of chemistry and spectroscopy to geologic problems.

Part A.
Overview, Theory, and Application of Spectroscopic Methods

INTRODUCTION

A retrospective study of the complete works of ROGER G. BURNS would reveal the great diversity of his contributions to the field of spectroscopy. In spanning the gamut of applications of spectroscopy to scientific problems, Roger pursued the study of disparate topics ranging from splat-quenched ferrite glasses to manganese nodules to Martian surfaces. In this section, we showcase applications of different types of spectroscopy to diverse problems in geology and mineralogy.

We begin with a plenary paper by WILLIAM FYFE, Roger's Ph.D. supervisor at Berkeley, that gives an overview of the prognosis for sustainable development of human life on Earth and its relationship to the study of earth materials. FYFE calls upon earth scientists to become involved with the fundamental earth science problems that will affect and be affected by the demographic facts of the world of the next century, including energy, water, and soil resources as well as materials needs, waste management, and knowledge of geofluctuations.

ROGER BURNS' interests in quantum mechanics are represented in a paper by TOSSELL presenting calculations that reproduce many of the structural and energetic properties of As and Sb sulfide molecules. The calculated structures and vibrational spectra are in good agreement with scaled experimental results, and they demonstrate the tremendous power of molecular quantum mechanical techniques in understanding structures, stabilities, and vibrational spectra of minerals.

The next three papers present data from optical absorption studies of minerals, the area of expertise in which ROGER BURNS first began his career. ROSSMAN discusses the correlation of optical absorption intensities with magnetic moments of Fe^{3+}-bearing minerals, focussing on the magnetic interactions responsible for the intense red color of the mineral hematite. He invokes crystal field theory to explain the differences in intensity of color observed for Fe^{3+}-bearing silicates, phosphates, sulfates, and oxides.

LANGER AND ANDRUT report on another trivalent transition metal, Cr^{3+}, and discuss the competing effects of crystal field stabilization energies vs. steric considerations on its intercrystalline partitioning behavior. They show that Cr^{3+} concentrations are a function of crystal field stabilization energies for the majority of minerals, except where crystallochemical peculiarities exist. In a related study, PLATONOV *et al.* investigate the influence of coordinating ligands and next nearest cation neighbors on the octahedral crystal field parameter Δ_o. Their work on Cr^{3+}-bearing phyllosilicates shows that a negative correlation exists between Δ_o and the strength of the ligand field surrounding the Cr^{3+} as represented by the Racah parameter B.

Infrared spectroscopy was another of ROGER BURNS' ongoing interests, and it is represented in this book by two papers on infrared spectroscopy of amphiboles. HAWTHORNE *et al.* present a general discussion of the problems of short-range order and long-range order in amphiboles, providing a careful theoretical treatment of the factors affecting the relative intensities of the four fine-structure bands in the principle OH-stretching region of amphibole. DELLA VENTURA *et al.* apply this understanding to the study of the synthetic (Ni, Mg, Co)-potassium richterite series. They demonstrate that significant short-range order involving MgMgMg, NiNiNi, and/or CoCoCo clustering may be found within this series.

Raman spectroscopy is represented in this volume by a paper from ORANGE *et al.*, who present spectra of crude oils and hydrocarbon fluid inclusions. They show that Raman spectroscopy can provide useful constraints on the structure and maturity of organic liquids, providing a powerful tool for fingerprinting hydrocarbons and for identifying the major functional groups within each sample.

The next group of three papers includes studies involving the Mössbauer spectrometer, the tool that became the focus of ROGER BURNS' research in the latter part of his career. ERICSSON *et al.* combine Mössbauer studies with X-ray diffraction and magnetization measurements of pyrrhotite-related solid solutions. They conclude that substitution of anions with identical charges in Fe_7Se_8 affects ordering processes in cation positions. DA COSTA *et al.* use Mössbauer spectroscopy to study highly-substituted aluminum maghemite. Their results show that Al is distributed equally between octahedral and tetrahedral sites in the structure. DEGRAVE *et al.* apply the Mössbauer technique to the study of the mineral franklinite, in conjunction with X-ray diffraction and atomic absorption spectroscopy. Their results suggest that Fe^{3+} is distributed between the tetrahedral A sites and the octahedral B sites of the structure.

The final four papers in this section utilize various types of X-ray absorption (XAS) spectroscopy and the analysis of fine structure in spectra at or near the absorption edge (XANES) and in the energy region extending well above the edge (EXAFS). LENNIE and VAUGHAN examine the iron monosulfide mackinawite and the amorphous FeS with which it forms in aqueous systems at low temperatures. Their model facilitates understanding of the phase relationships in the Fe-S system at low temperatures. HUGGINS and HUFFMAN review the application of XAFS spectroscopy to coal geochemistry, where it is used to understand geochemical and utilization behavior of elements in coal. Their work establishes that most elements occur in coal in forms that are either mineralogical or organic in nature. LI *et al.* use XANES to provide direct information on the nearest-neighbor coordination and intermediate-range structure of silicate and aluminosilicate glasses. They present spectra and interpretation of glasses along the $CaMgSi_2O_6$—$NaAlSi_3O_8$ and $NaAlSi_2O_6$—$NaAlSi_3O_8$ joins, as well as data on $Na_2Si_4O_9$ and $K_2Si_4O_9$ compositions. Finally, DELANEY *et al.* show that micro-XANES can be used for *in situ* microanalysis of $Fe^{3+}/\Sigma Fe$ in amphibole with spatial resolution comparable to that of other microbeam analytical techniques. Their XANES results correlate well with macroscopic measurements (by wet chemistry and Mössbauer spectroscopy) of the same samples, with the added benefits of discrimination of zoning and microinclusion effects.

Mineral Spectroscopy: A Tribute to Roger G. Burns

Editors: M. D. Dyar, C. McCammon and M. W. Schaefer

The science of earth materials—New priorities in a changing world: A tribute to Roger Burns

W. S. FYFE

Department of Earth Sciences, University of Western Ontario, London, Ontario, N6A 5B7, Canada

PROLOGUE

IT IS WITH great sadness that I write these notes devoted to the memory of my former student and colleague in research, Roger Burns. Roger came from New Zealand to work at Berkeley in the early 60's. His main background preparation was in chemistry (but he had to take the rigorous Berkeley field course!) It was a time when various groups of chemists around the world had realized that crystal field theory (ORGEL, 1960) could explain much of the behavior of transition metals and their compounds. I had been fortunate, in that the National Science Foundation had given me the support to acquire a then state of the art visible - uv spectrophotometer, with a sample chamber that allowed us to insert a standard petrographic microscope (with a u-stage), or an optical cell that could hold fluids up to 10 kilobars pressure. Roger and I could see a host of problems that looked interesting, and he had the fresh young mind, the energy, and the background preparation to attack the difficult experiments. A steady stream of papers was to follow (BURNS *et al.*, 1964). After Berkeley, Roger spent time at Cambridge, U.K., and there had the great fortune to meet many in their excellent chemistry department, including Mike Bancroft, and this encounter was to produce many seminal developments in mineral spectroscopy. It is interesting that many of the most productive workers in Earth System Science came with diverse backgrounds, a fact that our geology departments, which have tended to become over-specialized, should remember.

EARTH MATERIALS: EARTH HISTORY

As many have stressed, earth materials hold the record of events on this planet. The exact study of such materials is at the heart of our understanding of time, P-T stress history, and processes influencing all parts of the planet, from outer space, to atmosphere, to core.

From the earliest recorded history, humankind has been concerned with earth materials. The discovery of many of the elements came from those who worked with minerals. Those involved in improving our understanding of the nature of the chemical bond and inorganic structures, such as PAULING and BRAGG, were very familiar with naturally occurring substances. The discovery of radioactivity, the early developments in the understanding of the atomic nucleus, and the appreciation that these nuclear processes must hold the record of time, led to the technologies and the theories needed to advance the science of natural isotopes in all planetary systems and to understand the thermal histories of the planets. I often think that we forget the contributions of earth science to science in general.

At the present time, those who live in the rich (developed) worlds use something like 20 tons of rock materials per person per year to satisfy their various needs (from building materials to gold). If our world is to achieve some form of social equality and tranquility in the coming decades, given the present rate of population growth, in the next century, we may use something like 100 km^3 of earth materials per year—we will move and modify more rock than all natural processes on our planet (FYFE, 1981).

At this time, there are great and growing concerns with the state of our planet. But, if one examines many of the most fundamental aspects of our environmental and resource problems, almost all involve our knowledge of earth materials and their processes of formation and modification. At a time when the geosciences are under threat (KERR, 1995), this is also a time of great opportunity. As the papers in this volume show, we have powerful techniques available today to improve our precise quantification and understanding of the behavior of earth materials.

EARTH MATERIALS AND THE LIFE SUPPORT SYSTEM

Consider some of the most basic problems facing sustainable development, and the relationship between these problems and the study of earth materials. A list of such mega-problems must also recognize the new demographic facts of the world of the next century. First, over 80% of the future population will live in the so-called developing countries. Second, the future population will be urban; it is already 85% in the developed world.

Some fundamental problems with which all earth scientists must be involved (not in order of importance) include:

- Energy resources
- Water resources
- Soil resources
- Materials for 10 billion
- Waste management
- Understanding and predicting fundamental geo-fluctuations

ENERGY

At the present time, the bulk of world energy comes from the combustion of coal, oil and gas; all these resources are non-sustainable, natural capital. The thoughtless waste of such valuable resources is a global disaster. Of the fossil carbon sources, only coal and certain types of carbon rich sediments have reserves of interest for more than a few decades. In a general way, there has been little change in burning technologies - add air - burn - and exhaust to the atmosphere.

There is no need here to discuss the potential future impacts of the climate changes related to the fact that we have rapidly changed the chemistry of the atmosphere. Discussion is normally concerned with CO_2, CH_4, and acid compounds, but, as stressed previously (FYFE and POWELL, 1995), many coals contain significant quantities of all halogens: F, Cl, Br, I, and the steadily increasing ozone depletion catastrophe may be influenced by combustion of these fuels. Also, time after time, the detailed chemistry of coal and coal ash is not well known, and many coals have significant quantities of elements like uranium and arsenic, and an array of heavy metals immobilized in the reducing, sulfur-rich, medium of coal. It is amazing how little is known about the detailed chemistry and phase chemistry of this major world fuel.

It is certain that nations like China and India will depend on coal for decades to come. Can the technology be changed at reasonable cost to reduce the environmental impact of coal combustion? I think the answer is positive. We have been studying the fixation of CO_2 and organics in the cracked, permeable basalts in the caves of Kauai, Hawaii, deep beneath a very heavy forest cover. Every crack is covered with white materials (silica, clays, carbonates) formed by the action of organics with the basalt, a process mediated by ubiquitous bacterial biofilms. Bacteria can live to depths of over 4 km, up to 110°C, in favorable locations (PEDERSEN, 1994). Can such processes be used to fix the exhaust gases of coal combustion? Certainly, some rock types will be better than others, and volcanics with Ca-feldspars and rich in Fe-Mg phases should be ideal, as in Hawaii. Also it is interesting to note that adding H_2O-CO_2 to appropriate rocks can be a highly exothermic process and the gas disposal could lead to a geothermal energy bonus! Recently, on a field trip in China (East of Beijing), we discussed the possibility of using their rapidly exploited oil-gas fields for disposal of wastes of many types. If a basin can isolate oil-gas for millions of years, it undoubtedly has capacity to isolate wastes (DESSEAULT, 1995). And, generally, oil field structures are well known.

The growing knowledge of the deep biosphere also raises the possibility, with certain types of carbonaceous sediments, of using microorganisms for *in-situ* methane production. In place of opening deep mines, with all the related water pollution problems, could it be possible to produce bio-gas?

There is no shortage of energy sources on this planet. Ultimately, the world must move to solar energy of all types (photovoltaics, wind, tidal) and geothermal energy. Wind energy use is increasing across the world, and photovoltaic devices are becoming more efficient and cheaper (NEW SCIENTIST, 1995). Geothermal sources are normally associated with regions of high heat flow (volcanic systems) but, for some purposes (city heating, greenhouses and aquaculture systems), the normal geothermal gradient can provide background heating. There are many regions of the ocean floor with impressive potential for geothermal energy. All such potential use requires exact knowledge of deep geologic structures, porosity, permeability and geochemistry.

WATER

According to POSTAL (1992), today, forty nations have a crisis of water supply. In many places, uncontrolled extraction of ground water (mining) is being used to promote non-sustainable increases in food production. In vast areas, bad water-management has led to salinization of soils and massive pollution by agri-chemical residues. Examples of pollution abound as in Calcutta, and their major problem of groundwater arsenic pollution, which has led to serious health problems with very large numbers of people who live from ground waters. Where does the arsenic come from? This problem is not solved, but arsenic compounds have been used for a long time in rodent control.

There are many examples of the use of mineral surfaces in controlling heavy metal pollution (as with mercury adsorption on pyrite, BROWN *et al.*, 1979). In general, by using our mineralogical-geochemical knowledge, most metal pollution problems can be solved. And, increasingly, living microorganisms can be used to secrete metals and degrade organics (BEVERIDGE and FYFE, 1985). Possibilities exist of using reactions like:

$$\text{Organo-Cl} + \text{Na(silicates)} \rightarrow \text{graphite} + SiO_2 + NaCl.$$

to dispose of certain compounds subsurface. Most of these processes have very favorable thermodynamics.

In the developing world, there are few places where the total water cycle is adequately described. While sea level is slowly rising in many sensitive coastal regions, land subsidence and apparent sea level rise are often associated with subsidence, compaction, following the mining of groundwater. Again, exact geoscience is required to describe the water resource potential of any region, and to live with the natural fluctuations in precipitation.

SOIL

At this time, at least one billion humans do not have an adequate supply of food of well balanced nutritional values (SADIK, 1989). Across the world, wood is becoming an expensive and declining commodity. Despite the electronic revolution, the use of paper products is increasing (per capita, 3x in the last 40 years). In addition, the world's marine resources are declining at an alarming rate. Again the rich-poor gap is dramatically increasing the nutritional difference in the world's population.

Sustainable food-fiber production depends on climate, climate fluctuations, soil quality and water resources, and knowledge from the geosciences is involved in all these parameters. Given that we are not adequately providing nutrition for the present human population, what are the prospects for the next 5 billion?

All organisms require a large array and balance of the chemical elements (about 50) for efficient production of the organics needed for life (MERTZ, 1981). The geochemistry and the mineralogy of soils are critical in estimating the capacity of a soil for sustainable organic productivity. According to the Worldwatch Institute, topsoil loss globally is approaching 1% per year, while natural remediation can take 100's of years. The technologies exist now for erosion and salinization control, but such technologies are not adequately used, and there is great need for new soil maps which clearly show good soils, soils for forests only, and ones we should leave alone (FYFE, 1989)!

Given the chemical and physical properties of a soil, additives may greatly enhance bio-productivity. Often, such additives require the addition of simple mineral materials containing species like K, Mg, Ca, P . . . . , and appropriate trace metals like Co, Mo, etc., that may be critical in biofunctions like nitrogen fixation. The types of additives may be closely linked to soil type and climate. For many situations as with the laterite soils of the humid tropics, slow release mineral fertilizers (K in feldspars, rock phosphates . . . .) may be more effective and less wasteful than soluble chemical fertilizers.

Soil contains a complex array of ultra-fine inorganic and bio-mineral materials with vast surface areas which control key soil-bio functions. Today, with the modern techniques of surface chemistry (Auger, ESCA . . . .) and the power of modern high resolution transmission electron microscopy (TAZAKI *et al.*, 1987; TAZAKI and FYFE, 1986), we can precisely examine the inorganic-bio-gas-liquid interactions, which was not possible a decade ago. In soils, many of the mineral-forming processes involve reactions with living cells. There is a new world in the science of biomineralization which, as H. LOWENSTAM showed decades ago, is of vast importance in all aquatic environments to at least 100°C (LOWENSTAM, 1981).

MATERIALS-MINERALS-MINING

Advanced societies use about 20 ton of rock per person per year for their needs. Most is for various forms of construction, highways, buildings, etc., and giant mining operations include those for fertilizers, ores like iron, and coal. The average "student" gold ring weighs about 10g and represents the processing of about 3 ton of rock. It is amazing, but humans in general do not understand where their resources come from, and the impacts on the ecosphere, geosphere, hydrosphere made by mining operations. With modern understanding of Earth convection, our understanding of the resource base and prospecting strategies has dramatically increased. If the world needs more copper, we know where to look, and we find it.

Since mining technology has historically been careless, many large companies have increasingly moved to developing countries with less stringent environmental laws. For example, I was amazed that, when the NAFTA agreement was signed, it was clearly stated that only local environmental regulations applied. What of the future of the use of mineral resources?

First, the world must move to more efficient recycling and, for this to be successful, it requires careful quality control at all stages of the use—mining to fabrication. In mining, there must first be careful 3D mapping of structures, permeability, porosity, faulting, etc., to assess accurately the environmental impacts, and there is great need for extreme quality control in the extraction of ore materials. The total geochemistry must be known: the desired and the undesired elements. Such data must be available to plan the mining technology, and to assess the environmental impact of the operation, and waste products from mining must be studied for potential uses in construction, soil remineralization, etc. The growing knowledge of microorganisms at depths (now over 4 km) opens a host of new technological opportunities. Silica secreting organisms might be used for permeability control; metal-secreting organisms can be used for removal of heavy metals and secondary recovery, as has been well demonstrated. Also a host of new possibilities must be considered for "*in situ*" metal extraction via sulfide oxidizing species. The same is true for "*in situ*" methane production from carbonaceous sediments. To develop such technologies, there is need for cooperation between geologists, geo-microbiologists, hydrogeologists, engineers and economists. Far too little thought has been given to the end use of mines. In some cases, by careful planning, these could become waste disposal sites for urban areas, a growing world problem.

There is no doubt that, with the correct team for planning *from the start*, the environmental impact of extracting resources from the crust can be vastly reduced and I am sure that, in many cases, the long term economics of the operation will be improved. The development of new materials is producing new opportunities. According to the British journal "The Economist", there is a world shortage of high purity silicon for modern electronic devices, and soon there will be vastly increased use of photovoltaics. For such purposes, there is a giant difference between 99.99% SiO_2 and 99.9999% SiO_2.

I have always been intrigued by the possibilities for use of near ocean ridge sites for metal extraction and energy production (*cf* the Salton Sea Thermal fields of California). On Canada's well-described Juan de Fuca system, by drilling through the impermeable sediment cover, with thermal gradients of up to 300°C/km, one might simultaneously extract both metals and energy.

WASTE MANAGEMENT

This century will go down in history as that of careless technologies and waste production. There is no doubt that the long term costs can be staggering (as with the arsenic pollution in India, mentioned above). The complexity of modern wastes is enormous, from organics to radionuclides, urban garbage, etc. It is amazing that the nuclear industries developed before any serious attention was given to wastes—the "we-will-do-it-when-necessary" philosophy. In Ontario, Canada, where nuclear electricity dominates the system, the current estimates are that it will cost something like 15 billion dollars for nuclear waste disposal over the next decades and, at this time, no "best" site has been proposed. This problem exists in many nations, and it is strange that only recently has a combustion gas like CO_2 been considered a waste product. Combustion has changed the concentration of many critical components of the Earth's atmosphere.

The time has come to change drastically our philosophy on wastes. First, we must precisely describe the nature of waste, chemistry, compounds, etc. Then, we must search for uses for the waste, and recognize that time after time it can be a resource. (Denmark recycles 97% of its paper, Canada 17%.) The secret of domestic waste management is the five minutes a day spent in separating the components. As mentioned above, we have shown that, by using careful geochemistry, many types of coal ash (but not all) can be a valuable soil additive. In the past few years, a group from The University of Western Ontario, Bombay and Orissa State in India have been studying the use of coal ash in soil remineralization for forestry. The results have been spectacular and widely recognized (Young, 1994; Geoskop, 1994). But the success has been related to very detailed geochemistry and phase studies of the coal ash (Sahu, 1991). The same is true for most urban sewage, as long as it is not mixed with other toxic wastes from, for example, the chemical industries. Again, we need teams of the appropriate scientists, not just engineers! One case I would like to emphasize again is that of the gas products from combustion, CO_2, NO_x, SO_x and, as we have shown with halogen bearing coals that are common, possible halogen-organics. Are the latter partly responsible for the growing problems of stratospheric ozone destruction? As mentioned above, is disposal of these gases below ground possible and not simply vent to the atmosphere? I was recently in China, in the Beijing region, which has catastrophic atmospheric pollution from combustion and other industries. The cost reduction on public health of reducing all the lung problems might well cover the additional engineering costs of gas disposal. Japan is seriously considering the marine dumping of CO_2.

Every waste product requires a unique approach. Many wastes can be resources and, for most, secure disposal is possible. For geological disposal, we require a new precision in the total description of the subsurface environment.

GEO-FLUCTUATIONS

Given the future population of our planet, and the fact that most of this population will be urban, geo-fluctuations will have increasingly serious social and economic impacts, if they are not fully considered in planning all forms of development. Imagine the cost of careless site selection and waste packaging in nuclear waste disposal. But, for earth system scientists, certain problems present major challenges. Some, in which we require vastly increased knowledge include:

- prediction of the sites, magnitude and character of future volcanic events—events which can cause a serious climate perturbation. We have not had a mega-eruption this century (STOMMEL and STOMMEL, 1994).
- prediction of the sites and magnitude of major seismic events. One need only reflect on Kobe, Japan, and some recent events in southern California, to appreciate the economic impact.
- the general problems of slope stability.
- the monitoring of the arrival of significantly large objects from space.

If one considers the first three of these processes listed above, and the fundamental science behind the process, it is obvious that they involve complex aspects of mineral-fluid interactions in various depth regimes. Their understanding requires integrating knowledge and techniques from almost all Earth science areas of specialization. The new and developing International Program of Continental Scientific Drilling (ZOBACK and EMMERMAN, 1993), provides an example of the types of cooperation urgently needed to address critical global problems. Data from the ODP program has made massive contributions.

CONCLUSION

As world human population continues to grow, to move to over 10 billion next century, the need for exact geoscience must be a priority in planning the needed future development of the support systems. There is urgent need to improve the communication and effective cooperation between all the experts in modern science and technology, economists, engineers, politicians, and all educated citizens. At this time, Europe leads in demonstrating that sound economic and environmental policies are not in conflict, but must form a working partnership. We must and can reduce pollution and wastes; must recognize the limits of the Earth system; and we must develop holistic natural science. Given the future numbers of humans on Earth, the cost of errors will become intolerable, and it is teams of people with spirit and talent, like Roger Burns, who are needed.

REFERENCES

BEVERIDGE T. J. and FYFE W. S. (1985) Metal fixation by bacterial cell walls. *Can. J. Earth Sci.* **22**, 1893–1898.

BROWN J. R., BANCROFT G. M., FYFE W. S. and MCLEAN R. A. (1979) Mercury removal from water by iron sulfide minerals. An electron spectroscopy for chemical analysis (ESCA) study. *Env. Sci. & Tech.* 1142–1144.

BURNS R. G., FYFE W. S. and CLARK R. H. (1964) Applications of crystal-field theory in geology. In *Chemistry of the Earth's Crust,* Vol.2, Vernadsky memorial volumes, Acad. Sci. U.S.S.R., 88–106.

DESSEAULT M. B. (1995) Radioactive waste disposal. *Nature* **375**, 625.

FYFE W. S. (1981) The environmental crisis: quantifying geosphere interactions. *Science* **213**, 105–110.

FYFE W. S. (1989) Soil and global change. *Episodes* **12,** 249–254.

FYFE W. S. and POWELL M. A. (1995) Halogens in coal and the ozone hole, Letter to the editor. *Chemical Engineering News*, (American Chem. Soc.) April 24, 6.

GEOSKOP (1994) Umwwelt. Müll Hilft Kranken boden. *GEOSKOP*, October 1994, 168.

KERR R. A. (1995) Downsizing squeezes basic research at the U.S.G.S. *Science* **213**, 105–110.

LOWENSTAM H. A. (1981) Minerals formed by organisms. *Science* **211**, 1126–1131.

MERTZ W. (1981) The essential trace elements. *Science* **213**, 1332–1338.

NEW SCIENTIST, 1995. Cheap solar power, January 14:11.

ORGEL L. E. (1960) *An introduction to Transition-metal Chemistry: ligand-field theory*. Methuen & Co. Ltd.

PEDERSEN K. (1994) The deep subterranean biosphere. *Earth Science Reviews* **34**, 243–260.

POSTAL S. (1992) *Last oasis - facing water scarcity*. W. W. Norton and Co., New York).

SADIK N. (1989) *The state of world population, 1989.* United Nations Population Fund, U.N., New York.

SAHU K. C. (1991) Environmental impact of coal utilization. *Proceedings, International Conference, IIT, Bombay*.

STOMMEL H. and STOMMEL E. (1994) In the Enigma of weather, *Scientific American Inc.* 45–51.

TAZAKI K. and FYFE W. S. (1986) Observations of primitive clay precursors during microcline weathering. *Contrib. Min. Pet.* **92**, 86–90.

TAZAKI K. FYFE W. S. and IWATZAKI M. (1987) Clues to Arctic soil erosion from cryoelectron microscopy of smectite, *Nature* **333**, 245–247.

YOUNG M. (1994) Waste does wonders for India's forests. *New Scientist* **1934**, 19.

ZOBACK M. D. and EMMERMAN R. (1993) *Scientific Rationale for establishment of an international program of continental scientific drilling*. GeoForschungsZentrum, Potsdam, Germany.

Mineral Spectroscopy: A Tribute to Roger G. Burns
© The Geochemical Society, Special Publication No. 5, 1996
Editors: M. D. Dyar, C. McCammon and M. W. Schaefer

Theoretical studies on As and Sb sulfide molecules

J. A. TOSSELL
Department of Chemistry and Biochemistry
University of Maryland, College Park, MD 20742, U.S.A.

Abstract—Dimorphite (As_4S_3) and realgar and pararealgar (As_4S_4) occur as crystalline solids containing As_4S_3 and As_4S_4 molecules, respectively. Crystalline As_2S_3 (orpiment) has a layered structure composed of rings of AsS_3 triangles, rather than one composed of discrete As_4S_6 molecules. When orpiment dissolves in concentrated sulfidic solutions the species produced, as characterized by IR and EXAFS, are mononuclear, *e.g.* $AsS_3H_2^{-1}$, but solubility studies suggest trimeric species in some concentration regimes. Of the antimony sulfides only Sb_2S_3 (stibnite) has been characterized and its crystal structure does not contain Sb_4S_6 molecular units. We have used molecular quantum mechanical techniques to calculate the structures, stabilities, vibrational spectra and other properties of As_4S_3, As_4S_4, As_4S_6, As_4S_{10}, Sb_4S_3, Sb_4S_4, Sb_4S_6 and Sb_4S_{10} (as well as S_8 and P_4S_3, for comparison with previous calculations). The calculated structures and vibrational spectra are in good agreement with experiment (after scaling the vibrational frequencies by the standard correction factor of 0.893 for polarized split valence Hartree-Fock self-consistent-field calculations). The calculated geometry of the As_4S_4 isomer recently characterized in pararealgar crystals also agrees well with experiment and is calculated to be about 2.9 kcal/mole less stable than the As_4S_4 isomer found in realgar. The calculated heats of formation of the arsenic sulfide gas-phase molecules, compared to the elemental cluster molecules As_4, Sb_4 and S_8, are smaller than the experimental heats of formation for the solid arsenic sulfides, but shown the same trend with oxidation state. As(III) sulfides, like As_4S_6, are calculated to be stable toward disproportionation to a mixture of As(II) and As(V) sulfides and As_4S_4 does not disproportionate to As(III) and elemental As. As_4S_6 reacts exothermically with H_2S to give $As_3S_3(SH)_3$. Calculated electric field gradients at the As atoms in the arsenic sulfides are found to be determined by the As coordination number and by the number of As—As bonds within the first coordination sphere. UV spectral energies and intensities and polarizabilities are calculated for the isomers of As_4S_4 using the random-phase-approximation. Based on a comparison with the spectral properties of S_8 we project an excitation energy of about 3.3 eV to the lowest energy singlet and 2.2 eV to the lowest energy triplet state of realgar. The optimized geometry of the triplet state shows breaking of one of the As—As bonds, allowing a facile rearrangement to the pararealgar form.

INTRODUCTION

ALTHOUGH MOST minerals crystallize in ionic or covalent structures without discrete "molecules", elemental S and the sulfides of As and Sb sometimes adopt "molecular" structures, in which ring or cage molecules of moderate size are held together in the solid by weak van der Waals forces. Such molecular units may also exist in amorphous forms of such materials or in solution. We have recently begun a program of study of sulfide species existing in solution and have characterized monomeric and small oligomeric sulfide species of Sb (TOSSELL, 1994) and As (HELZ *et al.*, 1995). It is important to know whether larger oligomeric species, *e.g.* As_4S_4, which exist in both realgar (MULLEN and NOWACKI, 1972) and pararealgar (BONAZZI, 1995), in amorphous films of As_2S_3 (NEMANICH *et al.*, 1978) and in the gas phase (MARTIN, 1984) also exist in solution and/or react to produce other solution species. As a first step in such a study we here present calculations of the structural, energetic and spectral properties of a number of As and Sb sulfide molecules.

Studies of speciation in solution typically employ solubility methods in which the variation of total metal concentration with pH and ligand concentration is used to develop a model of the species present (*e.g.*, KRUPP, 1988). In most cases it is not possible to significantly vary the activities of the metal sulfide component, making it difficult to identify oligomeric species. In the Cu-S system THOMPSON and HELZ (1994) were able to vary CuS activity to some extent and to develop a model with multinuclear Cu species. A powerful tool for determining speciation is vibrational spectroscopy, particularly Raman spectroscopy, (*e.g.*, WOOD, 1989) which yields spectral signatures for the different species (although identifying a particular peak with a given species can be quite difficult). EXAFS can also be used to determine local geometry about metal centers, which can help in identifying the species present (*e.g.*, HELZ, *et al.*, 1993) but for highly insoluble sulfides obtaining high enough solution concentrations to yield usable EXAFS spectra can be difficult.

To assist in determining speciation quantum mechanical calculations can be very valuable (TOS-

SELL and VAUGHAN, 1993; TOSSELL, 1994; HELZ *et al.*, 1995). For small gas-phase molecules quantum mechanical techniques exist which can determine heats of formation to within a few kcal/mol of experiment (CURTISS *et al.*, 1991). Other properties, *e.g.* structures and vibrational spectra, can also be calculated to high accuracy for small molecules. Even for relatively large gas-phase molecules Hartree-Fock level quantum calculations combined with standard correction factors can give good agreement with experiment. For condensed phase species such calculations are more difficult although enormous progress has been made in the calculation of the properties of crystalline materials using both Hartree-Fock (PISANI *et al.*, 1988) and density-functional based theories. Calculating the electronic strucutre of crystalline realgar at the ab initio Hartree-Fock level would presently be a very formidable task, since the unit cell contains 4 As_4S_4 molecules. In addition, a correct description of the van der Waals forces which hold such a solid together requires a method which incorporates electron correlation, making the computational problem even more difficult. For the present we will apply molecular quantum mechanical methods to a study of the discrete molecules, such as As_4S_4, which exist in such phases.

A considerable number of quantum mechanical computations have been performed on various neutral (HOHL *et al.*, 1988; RAGHAVACHARI *et al.*, 1990) and anionic (GANTEFOR *et al.*, 1995) oligomers of elemental sulfur. Fewer calculations have been done on As species, although YSTENES *et al.* (1993, 1994) have carried very complete studies of the vibrational spectra of P_4S_3 and P_4S_4 and have used their calculated force fields, after rescaling, to interpret the vibrational spectrum of As_4S_3 and As_4S_4 (realgar). SUNDARARAJAN and KUMAR (1995) have recently used the ab initio molecular dynamics density functional approach to study Sb clusters, up to Sb_{12}. As_4S_4 and As_2S_3 crystals have been studied using a non-self-consistent LCAO approach by BULLETT (1976), who concentrated on their density of states and photoemission spectra and BABIC *et al.* (1989) have treated the As_4S_4 and As_4S_6 molecules using density functional theory. They employed an s,p orbital only basis and did only a partial geometry optimization, although they obtained equilibrium As—As and As—S distances fairly close to experiment.

COMPUTATIONAL METHODS

We employ the conventional methods of *ab initio* Hartree-Fock self-consistent-field molecular orbital theory as described in quantum chemistry texts (SZABO and OSTLUND, 1989) and monographs (HEHRE *et al.*, 1986) and as implemented in the program GAMESS (SCHMIDT *et al.*, 1993). All the calculations were done on a cluster of DEC Alpha workstations and were done in direct SCF mode, *i.e.* without disk storage of two-electron repulsion integrals. TOSSELL and VAUGHAN (1992) have reviewed the applications of quantum chemical calculations to a number of mineralogical and geochemical problems and SIMONS (1991) has written a brief experimentalist's guide to the use of *ab initio* quantum chemical techniques and the interpretation of the results. SAUER (1989) has reviewed the application of quantum mechanical methods to a number of condensed phase systems.

Although the *ab initio* Hartree-Fock MO techniques commonly applied to large molecules do not give exact agreement with experiment, the errors due to the use of a finite basis set for expansion of the MO's and the neglect of electron correlation are often quite systematic. For example, when studying neutral molecules at the polarized valence double zeta basis set self-consistent-field (SCF) level (*i.e.* Hartree-Fock *ab initio* MO theory with two functions to describe each valence atomic orbital plus additional functions with higher azimuthal quantum numbers, *e.g.* 3d on oxygen, to describe the polarization of the electron density which occurs upon bond formation), we expect calculated bond distances to be too short by a few hundredths of an angstrom and vibrational frequences to be too high by about 10% (HEHRE *et al.*, 1986). The root-mean-square error in the vibrational frequencies for an extensive series of neutral gas-phase main-group molecules was found to be minimized when the calculated vibrational frequencies were scaled down by the factor 0.893 (POPLE *et al.*, 1993). This scaling factor appeared to be appropriate for all types of vibrations. The scaling factor, however, is somewhat larger for highly charged anions, as discussed in TOSSELL (1994) and YSTENES *et al.* (1993).

The basis sets used to expand the molecular orbitals in our studies of geometry, stability and vibrational spectra are of the relativistic compact effective core-potential valence double zeta type (STEVENS *et al.* 1992), with polarization functions of d type added on each of the non-H atoms. We shall refer to these as polarized SBK basis sets. For Sb the relativistic effective core potential replaces 46 core electrons so that only the $5s^25p^3$ electrons are explicitly treated, with the 5s and 5p shells each described by two contracted Gaussian functions. Although these basis functions are quite new they have already been successfully used to study a number of different problems, including Zn complexes as models for the enzyme carbonic anhydrase (GARMER and KRAUSS, 1992) and the Sb sulfides (TOSSELL, 1994). In all cases so far examined the calculated geometries and other properties were very similar to those obtained at the all-electron polarized valence double zeta SCF level. For these effective core potential basis set calculations we can then use the standard correction factors appropriate for polarized valence double zeta SCF calculations. The elimination of core electrons in the Sb sulfide calculations is very important since it reduces the number of orbitals in the basis set and consequently the computer time required, which scales as approximately the third to fourth power of the number of basis functions.

For calculation of the electric field gradients at the As nuclei we use all-electron basis sets of the 6–31G* type (HEHRE *et al.*, 1986), which are basically of single zeta quality for the core orbitals and double zeta quality for the valence orbitals, with appropriate polarization functions added. We have also evaluated optical excitation energies and intensities as well as frequency dependent molecular dipole polarizabilities for the two isomers of As_4S_4 using the program RPAC (BOUMAN and HANSEN, 1991), interfaced to GAMESS. This approach evaluates excitation properties in terms of the random-phase approximation, or RPA (HANSEN and BOUMAN, 1985), which is a method that includes those first-order electron correlation effects that are important for electronic intensities and excitation energies.

For each molecule we have determined the minimum energy or equilibrium geometry and have calculated the vibrational spectrum, making sure that this spectrum has no imaginary frequencies and that the geometry thus corresponds to (at least a local) minimum on the potential energy surface. For each of molecules studied we have also calculated the vibrational energy spectrum and have displayed the normal modes of vibration using the capabilities of GAMESS and its auxiliary graphics programs.

RESULTS

Calculated bond distances and angles for the As sulfides are collected in Table 1 and compared with experiment (the crystal structures of the Sb sulfides do not contain molecular units). Bond distances are

Table 1. Calculated bond distances and angles in As_4S_3, the two isomers of As_4S_4, As_4S_6 and As_4S_{10} compared with experimental data for alpha-dimorphite, realgar, pararealgar, gas-phase As_4S_4 and gas phase As_4S_6 (bond distances in Å and bond angles in degrees)

Molecule	Calc. geom.	Exp. geom.
As_4S_3 (assumed C_{3v})	R(As—S) = 2.217 R(As—As) = 2.442 <S—As—S = 109.5 <As—S—As = 91.6 <As—As—As = 60	R(As—S) = 2.200–2.220 R(As—As) = 2.445–2.450 <S—As—S = 97.8–98.0 <As—S—As = 105.7–106.1 <As—As—As = 60.2 (WHITFIELD, 1970)
As_4S_4, realgar (assumed C2v)	R(As—S) = 2.238 R(As—As) = 2.563 <S—As—S = 95.3 <As—S—As = 100.4	R(As—S) = 2.228–2.247 R(As—As) = 2.566–2.571 <S—As—S = 94.5–95.1 <As—S—As = 100.8–101.2 (MULLEN and NOWACKI (1972), crystalline realgar) R(As—S) = 2.23 ± 0.02 R(As—As) = 2.49 ± 0.04 <S—As—S = 93 <As—S—As = 101 (LU and DONOHUE (1994), gas phase electron diffraction)
As_4S_4, pararealgar (assumed C_S)	R(As—S) As1—S 2.248, 2.251 As2, As3—S 2.249, 2.263 As4—S4 2.212 As—As 2.492	R(As—S) As1—S1 2.254 As1—S3 2.238 As1—S4 2.261 As2—S1 2.251 As2—S2 2.252 As3—S2 2.244 As3—S3 2.228 As4—S4 2.190 As2—As4 2.484 As3—As4 2.534 (BONAZZI *et al.* (1995), crystalline solid)
As_4S_6 (assumed C_{3v})	R(As—S) = 2.250 <As—S—As = 116.5 <S—As—S = 105.7	R(As—S) = 2.25 ± 0.02 <As—S—As = 100 ± 2 <S—As—S = 114 ± 2 (LU and DONOHUE (1994), gas-phase electron diffraction)
As_4S_{10} (assumed C_{3V})	R(As—S) = 2.040 (terminal) 2.230 (bridging)	

satisfactorily reproduced in all cases although some bond angles deviate substantially from experiment. The apparent disagreement between gas-phase and solid state experimental results for As_4S_4 may not be real since the gas-phase results were from an early electron-diffraction study in which only the shorter As—S distance could be directly determined from the data. Note the considerably different structures for the realgar and pararealgar forms of As_4S_4 (Fig. 1). Each molecule can be described as the result of adding four S atoms to tetrahedral As_4 and has 8 As—S and 2 As—As bonds but they are arranged very differently. In realgar there are two As—As groups on opposite sides of the molecule while in pararealgar there is a group of three As atoms on one side. Both structures are minima on the potential energy surface for the singlet ground state with the realgar structure calculated to be more stable by 2.9 kcal/mole. Notice also that although As_4S_3, realgar As_4S_4 and pararealgar As_4S_4 exhibit As—As single bonds the bond lengths are quite different, with the longest bond (2.563Å) in realgar and the shortest (2.442 Å) in the As_3 grouping of As_4S_3.

For the Sb compounds the calculated structures are very much like those for the As analogues. For example, a realgar-like structure of Sb_4S_4 shows Sb—Sb distances of 2.93 Å and Sb—S distances of 2.42 Å. The pararealgar-like isomer is 4.2 kcal/mol higher in energy and shows a structure very much like the corresponding sulfide. Detailed geometries for the other Sb sulfides are available from the author. Although Sb_4S_4 and Sb_4S_6 molecules have not been characterized, the $Sb_4S_6^{-2}$ anion is known (MARTIN *et al.*, 1995), and shows an Sb—Sb bond distance of 2.86 Å and Sb—$S_{bridging}$ distances of 2.44 – 2.48 Å.

In Table 2 we collect calculated formation energies for the As and Sb sulfides. These energies are obtained from SCF calculations of internal energies for these gas-phase molecules and for the most stable gas-phase forms of the elements, S_8, As_4 and Sb_4, at their optimized geometries. We can compare these with conventional heats of formation, *i.e.* enthalpy differences between the solid crystalline compounds and the stable (solid) forms of the elements. It should be noted that the newest values of the formation enthalpy of As_4S_4 and As_2S_3 (BRYNDZIA and KLEPPA, 1988 and JOHNSON *et al.* 1980) are much smaller (in absolute value) than earlier results such as those tabulated in MILLS (1974). One might suspect that the Sb sulfide for-

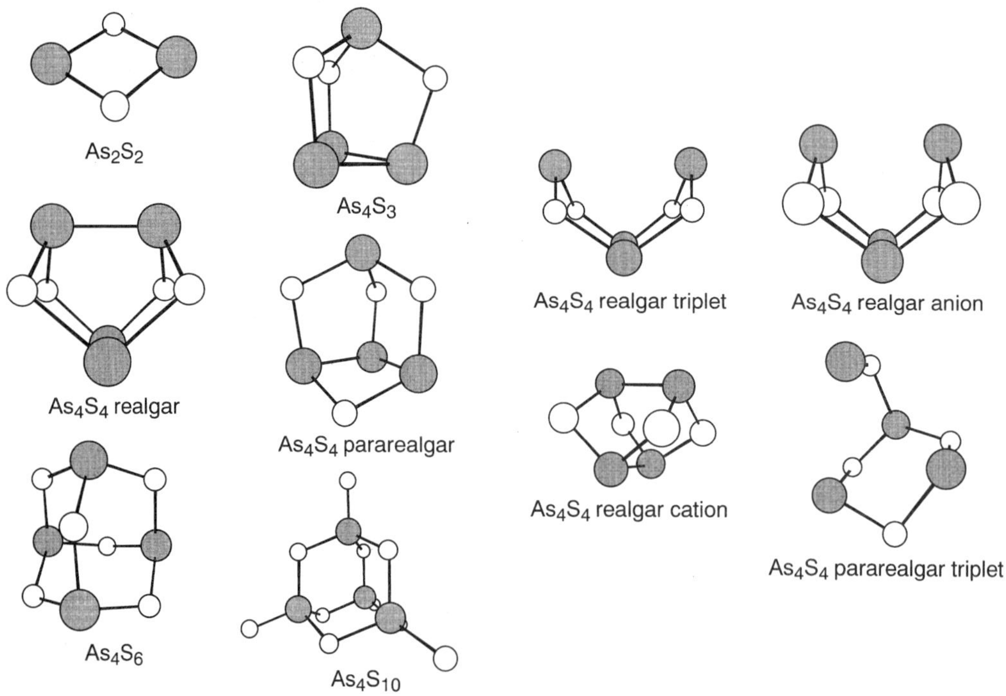

FIG. 1. Equilibrium geometries for various As sulfide species. As are shown as larger, solid spheres, S as smaller, open spheres.

Table 2. Calculated internal energies of formation of As and Sb sulfides from As_4, Sb_4 and S_8 molecules, in kcal/mole (of formula shown)

Molecule	Calculated SCF formation energy
As_2S_2	+8
As_4S_3	−48
As_4S_4	−61 (realgar), −58 (pararealgar)
As_4S_6	−69
As_4S_{10}	−15
Sb_4S_3	−60
Sb_4S_4	−80
Sb_4S_6	−99
Sb_4S_{10}	−37

mation enthalpies reported in MILLS (1974) are also in error. Based on a As_4S_4 formation enthalpy of −26.9 kcal/mol (of As_4S_4) from BRYNDZIA and KLEPPA (1988) and sublimation energies of As(s) to As_4 and S(s) to S_8 reported in MILLS (1974) we obtain an experimental value of −74.5 kcal/mol. Our calculated values are −61 kcal/mol for the realgar form of As_4S_4 and −58 kcal/mol for the pararealgar form. A similar calculation for As_4S_6 using data from BRYNDZIA and KLEPPA (1988) and MILLS (1974) gives −93.2 kcal/mol (of As_4S_6) for the formation enthalpy of crystalline orpiment, As_2S_3, while our calculated SCF value for the formation enthaly of gas-phase As_4S_6 is −69 kcal/mol (of As_4S_6). We have also studied the formation of As_4S_4 and As_4S_6 from As_4 and S_8 (but not the analogous Sb sulfides) at a somewhat higher quantum mechanical level. Incorporating zero-point vibrational energy corrections makes the formation energy less negative by 0.7 kcal/mol for As_4S_4 and 0.6 kcal/mol for As_4S_6 while incorporating electron correlation at the level of second-order Moller-Plesset perturbation theory (MP2, see HEHRE *et al.*, 1986) makes the formation energy less negative by 4.8 kcal/mol for As_4S_4 and 9.8 kcal/mol for As_4S_6. These corrections change the calculated formation energies to −56 kcal/mol for As_4S_4 and −59 kcal/mol for As_4S_6, a substantial change from the SCF energies in Table 2. The difference between our best calculated value for the formation enthalpy of the As_4S_4 realgar molecule and the experimental value for As_4S_4 in the realgar crystal is about 18 kcal/mol, fairly close to the 24.5 kcal/mol sublimation enthalpy of S_8 (MILLS, 1974). It seems reasonable that since As_4S_4 and S_8 both form molecular solids and their basic structural units in both the solid and gas phases are very similar, they would have similar sublimation energies. For As_4S_6 the difference between the calculated gas-phase formation enthalpy and the experimental formation enthalpy for orpiment is 34 kcal/mol, consistent with a greater stability for the non-molecular crystal structure of orpiment, compared to a hypothetical molecular solid with As_4S_6 units.

Based on the SCF formation energies we can establish that the most stable oxidation state for As and/or Sb in combination with S is the trivalent state, *i.e.* As_4S_6 (or As_2S_3). As(III) and Sb(III) sulfides would not be expected to disproportionate to mixtures of divalent and quintavalent states and the stability of the quintavalent sulfides will be very limited. It is not immediately clear how to reconcile this with the mass spectrometric data of MARTIN (1984) which showed As_4S_4 to be more abundant than As_4S_6 in the gas phase over either solid AsS or As_2S_3. As noted by Martin it may be that some of the $As_4S_4^+$ cluster ions observed are actually decomposition products of As_4S_6 or other molecules. The highest energy occupied molecular orbitals (HOMO's) of As_4S_6 are As—S bonding while the lowest energy unoccupied orbitals (LUMO's) are As—S antibonding. Exciting electrons from HOMO to LUMO could then lead to fragmentation. It could also be that the As_4S_4 molecule is the first formed, even over (amorphous) As_2S_3, and that the reaction of As_4S_4 with elemental S to give As_4S_6 is slow under the conditions of the experiment.

Based on our energies for As, S, As_4, S_8 and As_4S_6 we can also evaluate energies of atomization and compare with experimental atomization enthalpies (see *e.g.* HUHEEY *et al.*, 1993, Table E.1). We calculate an atomization energy of 30.4 kcal/per As—As bond for As_4 (exp. 35 kcal) and 46.1 kcal for the atomization energy per S—S bond in S_8 (exp. 54 kcal). The calculated atomization energy of the As—S bond in As_4S_6 is 43.1 kcal.

We can qualitatively understand the transformations from As_4 (or Sb_4) to the various sulfides by recognizing that the HOMO's of twenty valence electron tetrahedral molecules like As_4 are p lone-pair orbitals directed out of the tetrahedron while the LUMO's are p sigma antibonding orbitals directed along the edges of the tetrahedra (BURDETT, 1980, p. 240). Attack by a nucleophile occurs through intereaction with these LUMO's. For example in As_4^{-2} one As—As bond will be broken to give a rhombus structure (TOSSELL, 1983). When the nucleophile is sulfur, a S atom can be inserted into the broken As—As bond. Insertion of three S atoms along three of the six As—As edges of As_4 gives the molecule As_4S_3 while four S insertions

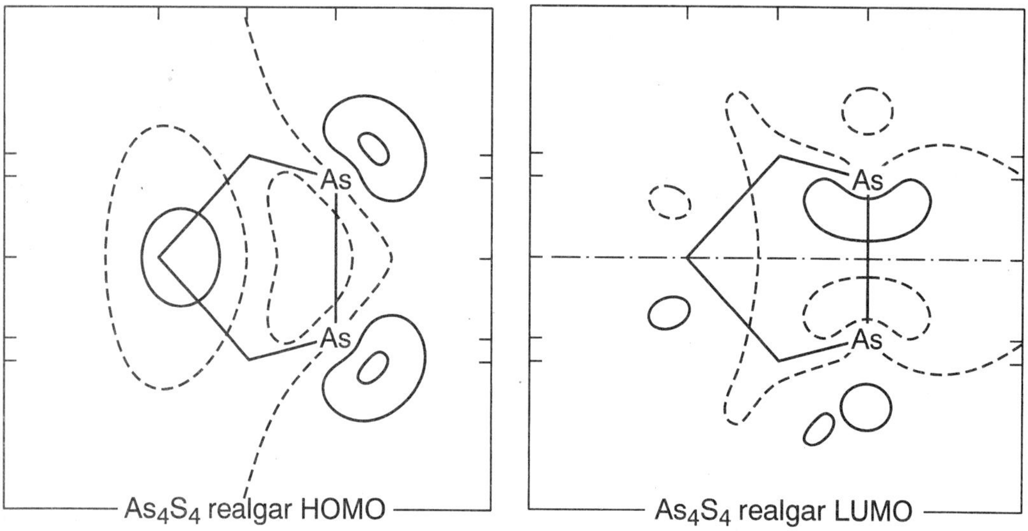

FIG. 2 Contour plots of the highest energy occupied MO and the lowest energy unoccupied MO for the ground state of As_4S_4 realgar. The plane shown contains one As—As bond and bisects the molecule. The positions of the As atoms are marked. Positive contours are shown as solid lines, negative contours as broken lines.

give As_4S_4. The two different isomers of As_4S_4 seen in realgar and pararealgar result from addition of four S atoms to a different set of four As—As edges of a As_4 tetrahedron. In the realgar form of As_4S_4 the HOMO is a As 5p lone pair distributed over the atoms in the As—As bonds while the LUMO is an As—As sigma antibonding orbital, as shown in Fig. 2. Further addition of S breaks the final two As—As bonds to give As_4S_6. To convert As_4S_6 to As_4S_{10} we must oxidize As by transferring its tangential lone-pair electrons to S.

In Table 3 we collect calculated vibrational energies for P_4S_3, As_4S_3 and Sb_4S_3 and compare them with the calculations of YSTENES *et al.* (1993) for P_4S_3 and with experiment for P_4S_3 and As_4S_3. We have scaled the calculated vibrational frequencies for P_4S_3 and As_4S_3 in Table 3 by the factor of 0.86 favored by YSTENES *et al.* (1993) (they report a scaling factor of 0.74 for the *force constants* which is equivalent to a scaling factor of 0.86 for the *frequencies*). If we had scaled the frequencies by the more standard factor of 0.893 the overall degree of agreement would have been very similar. Our results for P_4S_3 are very similar to those of YSTENES *et al.* (1993), with an identical order of frequencies and assignment of bands, and our calculations on

Table 3. Calculated vibrational frequencies for P_4S_3, As_4S_3, scaled by 0.86 and compared to experimental data from BUES *et al.* (1980) and the calculations of YSTENES *et al.* (1993) (×2 signifies a doubly degenerate vibration)

P_4S_3 6-31G* (YSTENES *et al.*, 1993)	Pol. SBK	Pol SBK, scaled by 0.86	Exp	As_4S_3 pol. SBK	Pol. SBK scaled by 0.86	Exp.	Sb_4S_3 pol. SBK	Pol. SBK scaled by 0.86
567	582	501	483	427	367	375	395	292
533 × 2	540 × 2	464	(none)	413 × 2	355	370	362 × 2	311
513 × 2	524 × 2	451	440	406	349	354	349	299
493	523	450	420	397 × 2	341	340	340 × 2	292
484	485	417	417	314	270	268	219	188
406 × 2	411 × 2	353	351	248	213	217	176	151
394	392	337	339	230 × 2	198	197	162 × 2	139
332 × 2	330 × 2	284	287	207 × 2	178	181	158 × 2	136
252 × 2	251 × 2	216	218	198 × 2	170	175	140 × 2	120
219	216	186	187	137	118	142	102	88

Table 4. Calculated vibrational frequencies in cm^{-1} (scaled by 0.893) for the two isomers of As_4S_4 and for As_4S_6 (with calculated relative IR intensities in parentheses), compared with experiment (FORNERIS, 1969) for realgar (×2 indicates a doubly degenerate vibration)

As_4S_4 realgar calculated frequencies scaled by 0.893	As_4S_4 realgar exp. frequencies	As_4S_4 pararealgar calculated frequencies scaled by 0.893	As_4S_6 calculated frequencies scaled by 0.893
104 (0)	47, 60	113 (.001)	106 (0) × 3
153 (0)	141, 143	114 (.023)	121 (0) × 2
168 (.088) × 2	170	158 (.001)	151 (.073) × 3
190 (0)	183	158 (.031)	214 (.253) × 3
196 (.113)	194	178 (.085)	233 (.0004)
219 (.066) × 2	225 (e)	185 (.134)	316 (.104) × 3
225 (0)	222	208 (.031)	328 (.0004) × 2
234 (.100)	—	232 (.038)	342 (0)
343 (0)	330	243 (.094)	355 (0) × 3
359 (0)	343	244 (.0004)	408 (2.67) × 3
360 (.881) × 2	358	334 (.084)	
370 (0)	367 (e)	344 (.075)	
372 (.009) × 2		345 (.181)	
382 (1.107)	373 (e)	353 (.357)	
		362 (.128)	
		375 (.271)	
		377 (1.25)	
		393 (1.92)	

As_4S_3 also agree well with their scaled force field results and with experiment. As expected, the calculated spectrum of Sb_4S_3 is qualitatively similar to that of P_4S_3 or As_4S_3, with all the vibrations displaced to lower frequency.

The calculated frequences for the two isomers of As_4S_4 and for As_4S_6, scaled by the commonly used 0.893 factor, are listed in Table 4, along with calculated relative IR intensities, and are compared with the data of FORNERIS (1969) for realgar (WARD (1968) and others have published similar realgar spectra). The calculated vibrational normal modes for the two As_4S_4 isomers are shown in Fig. 3. The agreement of scaled calculated and experimental frequencies is quite good, except in the very low frequency region, in which the vibrational motions of one As_4S_4 molecule may be strongly coupled with those of others. Our results for As_4S_4 are also in good agreement with the scaled force field results of YSTENES *et al.* (1994). As noted by FORNERIS (1969), the As_4S_4 molecule will have 18 vibrational normal modes, of which 8 are expected to involve As—S stretching, 2 As—As stretching and 8 mainly bending motions. The As—S stretches are expected to occur at highest frequency and we indeed find them at scaled calculated frequencies of 343–382 cm^{-1} (Table 4 and Fig. 3), with corresponding experimental values from 330 to 373 cm^{-1}. In agreement with the analysis of YANG *et al.* (1987), based on the temperature dependence of EXAFS, the calculated normal modes shown in Fig. 3 indicate that the Raman feature at a scaled calculated frequency of 225 cm^{-1} is mainly an As—As stretching mode. An IR active mode at a scaled calculated frequency of 234 cm^{-1} is partly As—As and partly As—S stretching. The modes calculated from 104–219 cm^{-1} are predominantly bending modes.

Our predicted spectrum for pararealgar is significantly different in detail from that of realgar. The stretching vibrations of the As_3 group in pararealgar occur at about 5–15 cm^{-1} higher frequency than the As—As stretch in realgar and the highest energy vibration in pararealgar is mostly a stretch of the S atoms vs. the ''terminal''-type As and is about 10 cm^{-1} higher than the highest energy As—S stretch in realgar. Note that our calculated vibrational spectrum for As_4S_6 is not directly comparable to that measured for solid As_2S_3 since the crystalline material does not contain such molecular units. The calculated vibrational spectrum of As_4S_{10} is quite similar to that of As_4S_6, the main difference being the presence of high energy As–$S_{terminal}$ stretches in As_4S_{10}, at scaled energies of 472 cm^{-1} for the three-fold degenerate vibration and 481 cm^{-1} for the totally symmetric one.

We have also calculated a number of the electronic properties of the As_4S_4 isomers. The calculated ionization potentials (IP's) and electron affinities (EA's), both vertical and adiabatic, are given

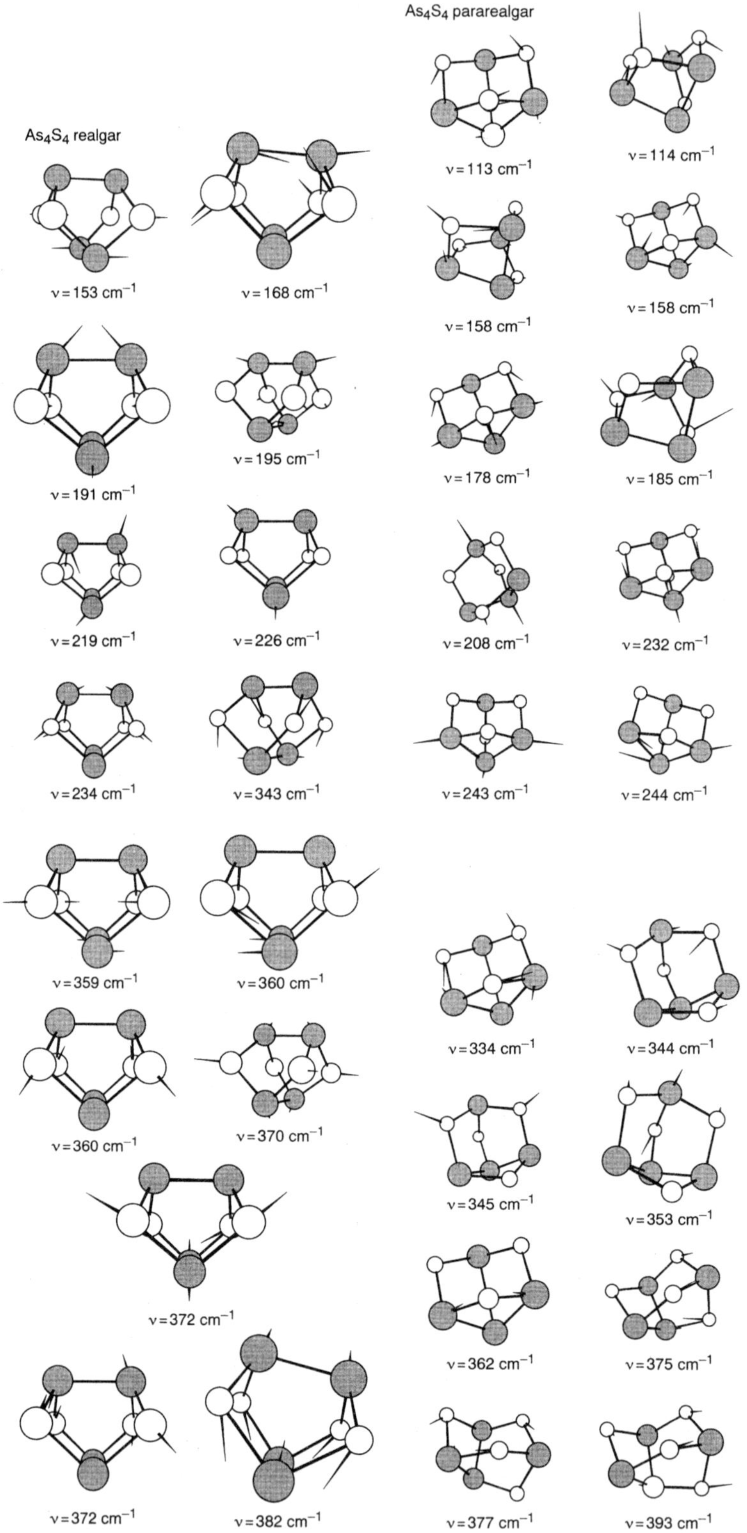

FIG. 3. Vibrational normal modes in the two isomers of As_4S_4. As are shown as larger, solid spheres, S as smaller, open spheres. The arrows are proportional to the nuclear displacements. The normal modes are identified by their scaled calculated frequencies.

Table 5. Calculated properties of different electronic and charge states of As_4S_4, realgar isomer, in eV (IP = ionization potential, EA = electron affinity)

Vertical IP (ΔE_{SCF})	8.62
Adiabatic IP (reopt. geom.)	8.35
Vertical EA (ΔE_{SCF})	−0.45
Adiabatic EA (reopt. geom.)	0.73
Transition energy to triplet state:	
At ground singlet equilibrium geom.	3.69
At triplet equilibrium geom.	0.80
Transition energy to lowest singlet at ground state singlet geom.	4.83
Negatives of orbital eigenvalues for highest occupied orbitals in neutral singlet: 9.14, 9.68, 3 × 10.46, 10.83, 11.25, 2 × 11.54, 2 × 13.43, 13.82, 14.02, 14.90	

for the realgar molecule in Table 5. We also list the negatives of the orbital eigenvalues, which according to Koopmans' theorem approximate the vertical ionization energies, for the highest energy occupied orbitals of the realgar isomer. These relative ionization energies match reasonably well against the UV photoemission data of TAKANASHI and HARADA (1980). Although the geometry of the relaxed As_4S_4 cation is not much different from that of the neutral, the relaxed anion shows a broken As—As bond. This change in geometry is consistent with the As—As antibonding character in the LUMO of neutral As_4S_4. The optimized geometry of the triplet state (corresponding to a HOMO-LUMO excitation) is very close to that in the anion. The geometries calculated for the anion and the triplet state are like that observed for $As_4S_6^{-2}$ (PORTER and SHELDRICK, 1971), which may be seen as the product of nucleophilic attack of S_2^{-2} on one of the As—As bonds of As_4S_4.

In Table 6 we collect the calculated electric field gradients at the As nucleus in the various molecules. Since the electric field gradient is a property depending strongly on the electron density near the nucleus we utilize an all-electron basis, the polarized double zeta 6−31G* basis (HEHRE *et al.*, 1986), to calculate it. All the As sulfides except the As(V) compound As_4S_{10} show an electric field gradient slightly smaller than that for gas-phase $AsCl_3$. To calculate quadrupole coupling constants from the electric field gradients we have used the value of 0.27 barns for the nuclear quadrupole moment of ^{75}As (SEMIN *et al.*, 1975), which is probably uncertain to at least 10%. Given this uncertainty our calculated coupling constant for $AsCl_3$ is in reasonable agreement with experiment. For crystalline and amorphous As_2S_3 quadrupole resonance frequencies of about 72−73 MHz are observed (RUBINSTEIN and TAYLOR, 1972). Assuming an asymmetry parameter reasonably close to zero (as must be exactly the case for a molecule with a three-fold rotation axis and consistent with the later results of SZEFTEL and ALLOUL (1979) for vitreous As_2S_3) this translates to a nuclear quadrupole coupling constant of 144−146 MHz, in reasonable agreement with our calculations for $As(SH)_3$ or As_4S_6. An interesting feature of the calculated electric field gradients is the small variation within the lower oxidation state As sulfides. Only for As2 of As_4S_3, which is bonded to two As atoms within a triangular geometry and only one S (see Table 1) is the field gradient significantly different from the average. Even As4 in As_4S_4 pararealgar, which is bonded to two As atoms and only one S, has an

Table 6. Calculated principal axis components of the electric field gradient (in atomic units) and quadrupole coupling constant (in MHz) at the As nuclei in $AsCl_3$, $As(SH)_3$, As_4S_3, As_4S_4 (realgar and pararealgar), As_4S_6 and As_4S_{10} (using the 6-31G* basis at the polarized SBK equilibrium geometry)

Molecule	q_{11}	q_{22}	q_{33}	e^2qQ/h
$AsCl_3$	−1.32	−1.32	2.63	168 (exp. value 173[a])
$As(SH)_3$	−1.23	−1.23	2.46	156
As_4S_3 As1	−1.21	−1.21	2.42	154
As2	−1.67	0.28	1.39	106
As_4S_4, realgar	−1.63	−0.84	2.47	157
As_4S_4, pararealgar				
As1	−1.48	−1.02	2.50	159
As2, As3	−1.49	−1.02	2.50	159
As4	−1.52	−0.86	2.38	151
As_4S_6	−1.14	−1.14	2.29	146
As_4S_{10}	1.47	1.47	−2.94	187

[a] SEMIN *et al.* (1975).

electric field gradient very much like that in the other sulfides. For the As_4S_{10} molecule, containing four-coordinate As(V), the electric field gradient components are larger (and of opposite sign), indicating a higer quadrupole coupling constant.

Calculated HOMO-LUMO energy gaps for all the molecules studied are shown in Table 7. It is clear that As and S containing molecules have similar valence orbital energetics while Sb containing species have somewhat smaller energy gaps. For the As_4S_y and Sb_4S_y series the trends in HOMO-LUMO energy gaps are similar, with the gaps increasing from y = 3 through y = 6 and then decreasing for y = 10, in the same order as the heats of formation. The HOMO-LUMO gap is also smaller in the less stable pararealgar isomer of As_4S_4. According to the principle recently formulated by PARR and CHATTARAJ (1991), the most stable molecules are those with a maximum "hardness". Although this "hardness" is actually defined in terms of second derivatives of the total energy with respect to the number of electrons, it can be approximated by the difference of LUMO and HOMO eigenvalues, or the energy gap. Consistent with this principle, realgar has a larger HOMO-LUMO gap than pararealgar and the largest gaps appear for the As(III) and Sb(III) sulfides, which have the most negative calculated formation energies.

For S_8 we calculate the lowest energy allowed dipole transition at 5.81 eV using the polarized SBK basis and the RPA method, compared to experimental values of 4.3 eV (COOK and SPEAR, 1969) and 4.4–4.7 eV (STEUDEL *et al.*, 1988). Calculations of the spectrum of S_3^- by JOHANSEN (1995) showed that basis set expansion and correlation decreased the calculated energy of the transitions near 4.5 eV. Modest expansion of our basis set for the As_4S_4 calculation (using two 3d polarization functions on each atom instead of one) indeed reduces the calculated excitation energy from 4.83 to 4.63 eV. This suggests that we are overestimating the excitation energies by about 1.5 eV, leading to an estimated excitation energy to the lowest singlet state of about 3.3 eV in realgar and about 2.2 eV for excitation to the lowest triplet. Because of the very high absorption of realgar in the UV its full spectrum has not been characterized. Studies of its photoconductivity by STREET and GILL (1966) yielded an optical gap of about 2.4 eV and a photoconductivity peak at about 2.9 eV.

For the isomers of As_4S_4 molecular dipole polarizabilities have also been calculated for energies from 0 to 4.5 eV and are presented as well in Table 8. Based on an average dielectric constant of about 2.64 for realgar in sodium D light (E = 2.1 eV) from WINCHELL and WINCHELL (1964), the formulas of HARTSHORNE and STUART (1960) and a volume of 199.8 $Å^3$ per molecule (MULLEN and NOWACKI, 1972) we obtain an experimental polarizability per molecule of 31.8 $Å^3$, compared to a calculated value of 25.1 $Å^3$ with our best basis set. It is not immediately clear whether all the discrepancy with experiment for the excitation energies and polarizability can be attributed to basis set truncation and correlation effects or whether solid-state effects within the realgar crystal are also of some importance. Of course we are also using an effective core potential basis set which excludes any contribution from the core electrons of As and S.

Table 7. Calculated HOMO-LUMO gaps (in eV) for molecules studied

S_8	11.63
As_4	10.65
Sb_4	8.85
As_4S_3	9.06
As_4S_4	
realgar	9.93
pararealgar	9.64
As_4S_6	10.19
As_4S_{10}	7.71
Sb_4S_3	8.70
Sb_4S_4	8.85
Sb_4S_6	9.36
Sb_4S_{10}	9.20

CONCLUSIONS

It is apparent that quantum mechanical calculations at this level can satisfactorily reproduce many of the structural and energetic properties of As and Sb sulfide molecules. Using standard scaling factors we can also reproduce the vibrational spectra of these molecules and assign the normal modes. The vibrational spectra of the realgar and pararealgar forms of As_4S_4 are predicted to differ in significant and understandable ways. Calculating the UV spectra and polarizabilities of the arsenic sulfide molecules would seem to be a greater challenge, with both compromises in the quantum mechanical treatment and solid state effects potentially important in explaining the discrepancies with experiment. There is also evidence that the UV spectrum of realgar may depend significantly upon temperature, since the mineral's color changes from bright red at room temperature to green-yellow at −180°C (BERAN *et al.*, 1994).

Table 8. Calculated singlet and triplet excitation energies, ΔE_1 and ΔE_3 (in eV), oscillator strengths (in parentheses) and dynamic polarizabilities, $\alpha(E)$ (in Å^3) for realgar and pararealgar (basis set a only) As_4S_4 molecules, using the random-phase-approximation and the specified basis sets (a: polarized SBK; b: polarized SBK + diffuse s,p; c: SBK + double 3d)

Realgar ΔE_1 a	Realgar ΔE_1 b	Realgar ΔE_1 c	Realgar ΔE_3 a	Realgar ΔE_3 b	Realgar ΔE_3 c
4.83 (0.002)	4.84 (0.002)	4.63 (0.005)	3.69	3.93	3.52
5.38 (0.0)	5.53 (0.0)	5.17 (0.0)	4.56	4.66	4.28
5.54 (0.0)	5.57 (0.0)	5.31 (0.0)	4.62	4.72	4.35
5.63 (0.082)	5.82 (0.138)	5.40 (0.049)	4.74	4.78	4.53
5.65 (0.00)	5.83 (0.00)	5.42 (0.00)	4.98	5.22	4.74
5.78 (0.123)	5.91 (0.0)	5.60 (0.097)	5.06	5.34	4.81
5.84 (0.00)			5.29	5.41	4.97

Realgar $\alpha(E)$ E	a	b	c
0.0	20.5	20.6	24.2
0.5	20.6	20.7	24.3
1.0	20.8	20.8	24.4
1.5	21.0	21.1	24.7
2.0	21.4	21.6	25.1
2.5	22.0	22.2	25.7
3.0	22.8	23.0	26.5
3.5	23.8	24.1	27.6
4.0	25.1	25.6	29.1
4.5	27.2	28.0	32.1

Pararealgar ΔE_1	pol. SBK ΔE_3
4.306 (0.001)	3.385
5.014 (0.005)	3.910
5.042 (0.007)	4.311
5.217 (0.011)	4.395
5.273 (0.011)	4.580
5.312 (0.003)	4.664
5.335 (0.000)	4.740
5.456 (0.024)	4.780
5.560 (0.002)	4.848
5.814 (0.034)	4.900

Studies of the light-induced alteration of realgar to pararealgar (Douglas *et al.*, 1992) indicate that light with energy around 2 eV is most effective in effecting the transformation and that there is an intermediate phase (labeled c) which gradually transforms to pararealgar. Based on the comparison of calculated S_8 and As_4S_4 spectra this may be a triplet state of the As_4S_4 molecule, which relaxes to break one of the As—As bonds. Formation of pararealgar from this triplet state can be accomplished by motion of one of the S atoms, without significant energy penalty. Alternatively it is possible that more than one As_4S_4 molecule is involved in the transformation step.

The next step in studying the solution chemistry of these compounds will be to evaluate the interaction of these molecules with species present in sulfidic solution. At the SCF level As_4S_6 reacts exothermically with H_2S to give solution species such as the trimer $As_3S_3(SH)_3$ (see Helz *et al.*, 1995 for a discussion of this species), but the reaction energy per mole of $As_3S_3(SH)_3$ formed is only -3.3 kcal. As_4S_4 reacts exothermically with H_2S and S_8 to give the same product with a reaction energy of -9.4 kcal per mole of $As_3S_3(SH)_3$ formed.

It is also clear that more work remains to be done in elucidating the molecular mechanism of the realgar-pararealgar conversion as well as the mechanism of thermal annealing of vapor-deposited As_2S_3 films. Also many questions remain concerning the effect of light on various amorphous As sulfides (Lee *et al.*, 1990).

Acknowledgements—This work was suppoprted by NSF grant EAR 9403521 and by DOE grant DE-FG02-94ER14467.

REFERENCES

Babic D., Rabi S. and Bernholc J. (1989) Structural and ele ctronic properties of arsenic chalcogenide molecules. *Phys. Rev. B* **39,** 10831–10838.

Beran A., Gotzinger M. A. and Rieck B. (1994) A fluid inclusion study of realgar from the Allchar deposit, FYR Macedonia. *N. Jb. Miner. Abh.* **167,** 345–348.

Bouman, T. D. and Hansen A. E. (1991) *RPAC Molecular Properties Package, Version 9.0,* Southern Illinois University at Edwardsville.

Bonazzi P., Menchetti S. and Pratesi, G. (1995) The crystal structure of pararealgar, As_4S_4. *Amer. Mineral.* **80,** 400–403.

Bryndzia L. T. and Kleppa O. J. (1988) Standard molar enthalpies of formation of realgar (aAsS) and orpiment

(As_2S_3) by high-temperature direct-synthesis calorimetry. *J. Chem. Thermo.* **20,** 755–764.
BULLETT D. W. (1976) Electronic strucutre of arsenic chalcogenides. *Phys. Rev. B* **14,** 1683–1692.
BURDETT J. K. (1980) *Molecular Shapes: Theoretical Models of Inorganic Stereochemistry,* Wiley.
COOK B. E. and SPEAR W. E. (1969) The optical properties of orthorhombic sulphur crystals in the vacuum ultraviolet. *J. Phys. Chem. Solids* **30,** 1125–1134.
CURTISS L. A., RAGHAVACHARI K., TRUCKS G. W. and POPLE J. A. (1991) Gaussian-2 theory for molecular energies of first- and second-row compounds. *J. Chem. Phys.* **94,** 7221–7230.
DOUGLASS D. L., SHING C. and WANG G. (1992) The light-induced alteration of realgar to pararealgar. *Amer. Mineral.* **77,** 1266–1274.
FORNERIS, R. (1969) The infrared and Raman spectra of realgar and orpiment. *Amer. Mineral.* **54,** 1062–1074.
GANTEFOR G., HUNSICKER S. and JONES R. O. (1995) Prediction and observation of ring and chain isomers in S_n^- ions. *Chem. Phys. Lett.,* **236,** 43–49.
GARMER D. R. and KRAUSS M. (1992) Metal substitution and the active site of carbonic anhydrase. *J. Amer. Chem. Soc.* **114,** 6487–6493.
HANSEN A. E. and BOUMAN T. D. (1985) Optical activity of monoolefins: RPA calculations and extraction of the mechanisms in Kirkwood's theory. Application to (−)-trans-cyclooctene and (3R)-3-methylcyclopentene. *J. Amer. Chem. Soc.,* **107,** 4828–4839.
HARTSHORNE N. H. and STUART M. (1960) *Crystals and the Polarizing Microscope: A Handbook for Chemists and Others.* Arnold.
HEHRE W. J., RADOM L. SCHLEYER P. v. R. and POPLE J. A. (1986) *Ab Initio Molecular Orbital Theory.* Wiley.
HELZ G. R., CHARNOCK J. M., VAUGHAN D. J. and GARNER C. D. (1993) Multinuclearity of aqueous copper and zinc bisulfide complexes: an EXAFS investigation. *Geochim. Cosmochim. Acta* **57,** 15–25.
HELZ G. R., TOSSELL J. A., CHARNOCK J. M., PATTRICK R. A. D., VAUGHAN D. J. and GARNER C. D. (1996) Oligomerization in arsenic(III) sulfide solutions: theoretical constraints and spectroscopic evidence. *Geochim. Cosmochim. Acta,* in press.
HOHL D., JONES R. O., CAR R. and PARRINELLO M. (1988) Structure of sulfur clusters using simulated annealing: S_2 to S_{12}. *J. Chem. Phys.* **89,** 6823–6835.
HUHEEY J. E., KEITER E. A. and KEITER R. L. (1993) *Inorganic Chemistry: Principles of Structure and Reactivity.* Harper Collins.
JOHNSON G. K., PAPATHEODOROU G. N. and JOHNSON C. E. (1980) The enthalpes of formation and high-temperature thermodynamic functions of As_4S_4 and As_2S_3. *J. Chem. Thermo.* **12,** 545–557.
JOHANSEN H. (1995) A theoretical investigation of the electronic structure and spectrum for S_3^-. *Acta Chem. Scand.* **49,** 79–84.
KRUPP R. E. (1988) Solubility of stibnite in hydrogen sulfide solutions, speciation, and equilibrium constants, from 25 to 350°C. *Geochim. Cosmochim. Acta* **52,** 3005–3015.
LEE J. M., PAESLER M. A., SAYERS D. E. and FONTAINE A. (1990) Kinetic x-ray absorption studies and computer structural modelling of photo-darkening in amorphous arsenic sulfide. *J. Non-Cryst. Solids* **123,** 295–309.
LU C. and DONOHUE J. (1944) An electron diffraction investigation of sulfur nitride, arsenic disulfide (realgar), arsenic trisulfide (orpiment) and sulfur. *J. Amer. Chem. Soc.* **66,** 818–827.
MARTIN T. M., SCHIMEK G. L., PENNINGTON W. T. and KOLIS J. W. (1995) Synthesis of two new antimony sulfide clusters: structures of $[PPh_4]_2[Sb_6S_6]$ and $[PPh_4]_2[Sb_4S_6]$. *J. Chem. Soc. Dalton Trans.* 501–502.
MARTIN T. P. (1984) Stability of As—S clusters. *J. Chem. Phys.* **80,** 170–175.
MILLS K. C. (1974) *Thermodynamic Data for Inorganic Sulphides, Selenides and Tellurides.* Butterworths
MULLEN D. J. E. and NOWACKI W. (1972) Refinement of the crystal structures of realgar, AsS, and orpiment, As_2S_3. *Zeit. Krist.* **136,** 48–65.
NEMANICH R. J., CONNELL G. A. N., HAYES T. M. and STREET R. A. (1978) Thermally induced effects in evaporated chalcogenide films. I. Structure. *Phys. Rev. B,* **18,** 6900–6914.
PARR R. G. and CHATTARAJ P. K. (1991) Principle of maximum hardness. J. *Amer. Chem. Soc.* **113,** 1854–1855.
PISANI C., DOVESI R. and ROETTI C. (1988) *Hartree-Fock Ab Initio Treatment of Crystalline Systems.* Springer-Verlag.
POPLE J. A., SCOTT A. P., WONG M. W. and RADOM L. (1993) Scaling factors for obtaining fundamental vibrational frequncies and zero-point energies from HF/6–31G* and MP2/6–31G* harmonic frequencies. *Isr. J.Chem.* **33,** 345–350.
PORTER E. J. and SHELDRICK G. M. (1971) Crystal structure of piperidinium hexathiotetra-arsenate. *J. Chem. Soc. A,* 3130–3132.
RAGHAVACHARI K., ROHLFING C. McM. and BINKLEY J. S. (1990) Structures and stabilities of sulfur clusters. *J. Chem. Phys.* **93,** 5862–5874.
RUBENSTEIN M. and TAYLOR P. C. (1972) Determination of local structural order in vitreous As_2S_3 by nuclear quadrupole resonance. *Phys. Rev. Lett.* **29,** 119–121.
SAUER J. (1989) Molecular models in ab initio studies of solids and surfaces: from ionic crystals and semiconductors to catalysts. *Chem. Rev.* **89,** 199–255.
SCHMIDT M. W., BALDRIDGE K. K., BOATZ J. A., ELBERT S. T., GORDON M. S., JENSEN J. H., KOSEKI S., MATSUNAG N., NGUYEN K. A., SU S. J., WINDUS T. L., DUPUIS M. and MONTGOMERY J. A. (1993) General Atomic and Molecular Electronic Structure System. *J. Comput. Chem.* **14,** 1347–1363.
SEMIN G. K., BABUSHKINA T. A. and YAKOBSON G. G. (1975) *Nuclear Quadrupole Resonance in Chemistry,* Wiley.
SIMONS J. (1991) An experimental chemist's guide to ab initio quantum chemistry. *J. Phys. Chem.* **95,** 1017–1029.
STEUDEL R., JENSEN D., GOBEL P. and HUGO P. (1988) Optical absorption spectra of the monocyclic sulfur molecules S_n (n = 6,7,8,9,10,12,15,20) in solution. *Ber Bunsenges. Phys. Chem.* **92,** 118–122.
STEVENS W. J., KRAUSS M., BASCH H. and JANSEN P. G. (1992) Relativistic compact effective potentials and efficient, shared-exponent basis sets for the third-, fourth- and fifth-row atoms. *Canad. J. Chem.* **70,** 612–630.
STREET G. B. and GILL W. D. (1966) Photoconductivity and drift mobilities in single crystal realgar (As_4S_4). *Phys. Stat. Sol.* **18,** 601–607.
SUNDARARAJAN V. and KUMAR V. (1995) Ab initio molecular dynamics study of antimony clusters. *J. Chem. Phys.* **102,** 9631–9637.

SZABO A. and OSTLUND N. S. (1989) *Modern Quantum Chemistry: Introduction to Advancced Electronic Structure Theory,* revised 1st. ed., McGraw-Hill.

SZEFTEL J. and ALLOUL H. (1979) Study of the local structure in vitreous chalcogenides via measurement of asymmetry of the electric-field-gradient tensor. *Phys. Rev. Lett.* **42,** 1691–1694.

TAKSNASHI T. and HARADA Y. (1980) Thermally and photo-induced changes in the valence states of vapour-deposited As_2S_3 films: a comparative photoemission study of As_2S_3 and As_4S_4. *Solid State Comm.* **35,** 191–194.

THOMPSON R. A. and HELZ G. R. (1994) Copper speciation in sulfidic solutions at low sulfur activity: further evidence for cluster complexes? *Geochim. Cosmochim. Acta* **58,** 2971–2983.

TOSSELL J. A. (1983) A qualitative molecular orbital study of the stability of polyanions in mineral structures. *Phys. Chem. Mineral.* **9,** 115–123.

TOSSELL J. A. (1994) The speciation of antimony in sulfidic solutions: a theoretical study. *Geochim. Cosmochim. Acta* **58,** 5093–5104.

TOSSELL J. A. and VAUGHAN D. J. (1992) *Theoretical Geochemistry: Applications of Quantum Mechanics in the Earth and Mineral Sciences,* Oxford.

TOSSELL J. A. and VAUGHAN D. J. (1993) Bisulfide complexes of zinc and cadmium in aqueous solution: calculation of structure, stability, vibrational, and NMR spectra, and of speciation on sulfide mineral surfaces. *Geochim. Cosmochim. Acta* **57,** 1935–1945.

WARD A. T. (1968) Raman spectroscopy of sulfur, sulfur-selenium, and sulfur-arsenic mixtures. *J. Phys. Chem.* **72,** 4133–4139.

WINCHELL A. N. and WINCHELL H. (1964) *The Microscopical Characters of Artificial Inorganic Solid Substances: Optical Properties of Artificial Minerals,* Academic.

WOOD S. A. (1989) Raman spectroscopic dtermination of the speciation of ore metals in hydrothermal solutions: I. Speciation of antimony in alkaline sulfide solutions at 25°C. *Geochim. Cosmochim. Acta* **53,** 237–244.

WUENSCH B. J. (1974) Sulfide crystal chemistry, In *Min. Soc. Amer. Short Course Notes, Vol. 1 Sulfide Mineralogy,* (ed. P. H. Ribbe), Ch. 2, pp. W21-W34, Mineralogical Society of America.

YANG C. Y., PAESLER M. A. and SAYES D. E. (1987) Determination of bond strengths of arsenic and arsenic chalcogen compounds using the temperature dependence of extended x-ray-absorption fine structure. *Phys. Rev. B,* **36,** 980–988.

YSTENES M., BROCKNER W. and MENZEL F. (1993) Scaled quantum mechanical (SQM) calculations and vibrational analyses of the cage-like molecules P_4S_3, As_4Se_3, P_4Se_3, As_4S_3, and PAs_3S_3. *Vibrat. Spectrosc.* **5,** 195–204.

YSTENES M., MENZEL F. and BROCKNER W. (1994) *Ab initio* quntum mechanical calculations of energy, geometry, vibrational frequencies and IR intensities of tetraphosphorus tetrasulphide, α-P_4S_4 (D_{2d}), and vibrational analysis of As_4S_4 and As_4Se_4. *Spectrochim. Acta* **50A,** 225–231.

Mineral Spectroscopy: A Tribute to Roger G. Burns
© The Geochemical Society, Special Publication No. 5, 1996
Editors: M. D. Dyar, C. McCammon and M. W. Schaefer

Why hematite is red: Correlation of optical absorption intensities and magnetic moments of Fe^{3+} minerals

GEORGE R. ROSSMAN
Division of Geological and Planetary Sciences, California Institute of Technology, Pasadena, California 91125, U.S.A.

Abstract—Structures with Fe^{3+} shared through oxo- or hydroxyl-groups have antiferromagnetic interactions. Such interactions result in enhanced intensity of the Fe^{3+} optical absorption bands which in some systems can be as great as a factor of 100 compared to isolated, octahedrally coordinated Fe^{3+} ions. A comparison is presented between the intensity of the lowest energy crystal-field band of Fe^{3+} minerals and their magnetic moments which demonstrates the dependence of optical absorption intensity and antiferromagnetic interactions in the host phase. Hematite, which is usually responsible for the red color of geological materials, owes its intense color to these magnetic interactions.

INTRODUCTION

OXIDIZED IRON (Fe^{3+}) is associated with the red color which is commonly observed in many soils, sedimentary rocks and weathering products (BLODGETT *et al.,* 1993). In these materials, Fe^{3+}, primarily in the form of finely disseminated hematite (Fe_2O_3), is an intense pigment. The prevalence of hematite as a red pigment in geological materials leads to the common association of red color with oxidized iron in general.

On the other hand, a variety of other Fe^{3+} minerals and chemical compounds are pale colored. Many examples exist of pale colored Fe^{3+} minerals including light greenish-yellow silicates such as andradite garnet, and lavender phosphates and sulfates such as strengite and coquimbite. A problem has been to reconcile the intense color of hematite and related hydrous iron oxides with the pale color of many other Fe^{3+} minerals and compounds.

ROSSMAN (1975, 1976a,b) noted that the intensity of color per unit of Fe^{3+} ions in iron sulfates increases dramatically when the iron ions are joined through shared oxide and hydroxide ions (shared edges or vertices of coordination polyhedra). He observed that such systems are antiferromagnetic and suggested that, in these systems, the intensity of color was related to the extent of magnetic interaction. Furthermore, ferric sulfate systems joined through shared oxide ions often show much greater magnetic interaction and stronger color than those joined through hydroxide ions.

SHERMAN (1985) and SHERMAN and WAITE (1985) discussed the results of molecular orbital calculations on clusters with Fe^{3+}-O and Fe^{3+}-OH units. They found that the Fe^{3+}-OH bond was more ionic and had a smaller spin-polarization than the Fe^{3+}-O bond. This gave rise to much weaker magnetic exchange (superexchange) between hydroxyl-bridged Fe^{3+} cations compared to oxo-bridged Fe^{3+} cations. In addition to the effects discussed by Sherman on the internal magnetic hyperfine fields observed in the Mössbauer spectra, superexchange interactions produce the major intensifications of the Fe^{3+} ligand field transitions in the optical spectra (ROSSMAN, 1975, 1976a,b).

This paper presents empirical correlations between the intensities of optical absorption bands and the magnetic susceptibility of Fe^{3+} minerals. These observations provide experimental support for the concepts developed by SHERMAN (1985). As part of these correlations, the special case of the color of hematite and other hydrous iron oxides is considered.

EXPERIMENTAL

Magnetic and optical data were taken from the literature or were determined by the methods described in ROSSMAN (1975). Both new and reviewed data are presented in Table 1.

RESULTS AND DISCUSSION

The optical absorption spectrum of Fe^{3+} in an octahedral site isolated from other Fe^{3+} octahedra consists of two broad bands at lower energies (labeled $^4T_{1g}$ and $^4T_{2g}$ in order of increasing energy) and a pair of bands, often overlapping near 440 nm, labeled ($^4A_{1g}$,^4E_g). The light not absorbed by these bands determines the color of Fe^{3+} minerals.

Table 1. Spectral and magnetic data

Sample		Formula	$\lambda(^4T_{1g})$	ε	$\lambda(^4A_{1g},^4E_g)$	ε	μ
Regular Octahedra							
1	andradite	$Ca_3Fe_2(SiO_4)_3$	854	0.08	440	1.5	-
Distorted Octahedra							
2	coquimbite	$Fe_2(SO_4)_3{\cdot}9H_2O$	778	0.18	427	2.0	5.87
3	phosposiderite	$FePO_4{\cdot}2H_2O$	746	0.34	423	6.5	-
Dimer							
4	magnesiocopiapite	$MgFe_4(SO_4)_6(OH)_2{\cdot}20H_2O$	864	1.7	430	36	4.70
Chains							
5	butlerite	$Fe(SO_4)(OH){\cdot}2H_2O$	920	2.4	424	33	3.8
6	parabutlerite	$Fe(SO_4)(OH){\cdot}2H_2O$	912	2.5	426	28	3.3
7	stewartite	$MnFe_2(PO_4)_2(OH)_2{\cdot}8H_2O$	880	2.3	428	55	-
8	fibroferrite	$Fe(SO_4)(OH){\cdot}5H_2O$	840	1.9	423	11	3.72
9	botryogen	$MgFe(SO_4)_2(OH){\cdot}7H_2O$	939	3.4	432	96	3.97
Clusters							
10	metavoltine	$K_2Na_6FeFe_6(SO_4)_{12}O_2{\cdot}18H_2O$	855	5.5	464	66	3.42
11	amarantite	$Fe(SO_4)(OH){\cdot}3H_2O$	866	12.7	442	90	2.53
12	leucophosphite	$KFe_2(PO_4)_2(OH){\cdot}2H_2O$	800	0.78	441	8.4	5.13
13	pharmocosiderite	$KFe_4(AsO_4)_3(OH)_4{\cdot}7H_2O$	820	0.5	446	19	4.71
Extended Structures							
14	hematite	Fe_2O_3	855	15.3	-	-	2.06
15	goethite	FeO(OH)	915	7.0	-	-	3.23
16	lepidocrocite	FeO(OH)	918	13.5	-	-	3.01
17	bernalite	$Fe(OH)_3$	885	0.12	431	1.95	-

Wavelengths of absorption bands (λ) in nm; molar absorption coefficients (ε) in l/mol· cm^{-1}; effective magnetic moments (μ) in Bohr magnetons at approximately 295°C.

Sources of data: 1 (MAO and BELL, 1974); 2-9 (ROSSMAN, 1975, 1976a,b); 10 (previously unpublished); 11,12 (ROSSMAN, 1976b); 13 (previously unpublished); 14 (BAILEY, 1960; HOFER *et al.*, 1946); 15,16 (MAO *et al.* 1974, HOFER *et al.*, 1946); 17 (MCCAMMON *et al.*, 1995)

Andradite garnet, $Ca_3Fe_2(SiO_4)_3$, is a mineral containing such isolated octahedral Fe^{3+} sites. The general features of the spectrum of andradite (Fig. 1) are typical for Fe^{3+} in octahedral coordination in a variety of silicate, sulfate and phosphate minerals. The two broad bands at longer wavelengths ($^4T_{1g}$ and $^4T_{2g}$) have low intensity and do not produce strong colors in a thickness of a few millimeters and the more intense, sharp band near 440 nm is so far in the violet that it contributes little to the color of the mineral.

In hematite, all Fe^{3+} sites are adjacent to other Fe^{3+} sites. In Fig. 1 the hematite spectrum (in the (0001) plane) is compared to the spectrum of andradite. The spectra are normalized for density and iron concentration such that they are presented for equal amounts of Fe^{3+} in the sample path. From this comparison, it can be seen that hematite has a much greater absorption intensity than andradite. The intense, deep red color of hematite is determined by the narrow transmission window near 750 nm.

The absolute intensity of the bands is measured in terms of the molar absorption coefficient, ϵ, defined by absorbance = ϵ (l/mol × cm^{-1}) × path (cm) × concentration (moles/liter). The ϵ value for

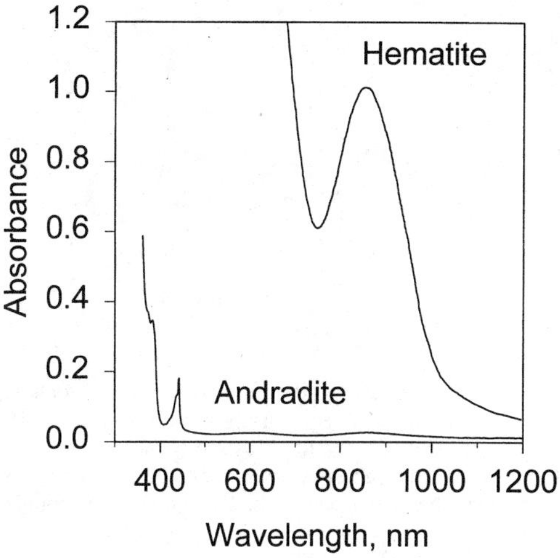

FIG. 1. Comparison of the optical absorption spectra of hematite and andradite garnet. The spectra have been normalized for density and Fe-concentration so that they are presented for identical amounts of Fe in the sample path.

hematite is 15.3 in the 850 nm region compared to 0.08 for andradite. This increase of intensity of absorption by over two orders of magnitude coupled with increased intensity of the higher energy bands is responsible for the intense color of hematite.

A general observation which follows from these studies is that intense color and high absorption band intensity is associated with minerals which have chains, clusters or extended networks of Fe^{3+} cations. This observation is reinforced by Fig. 2 which illustrates that the intensity of the $^4T_{1g}$ band is generally much greater for various clusterings of Fe^{3+} cations than for Fe^{3+} in isolated, symmetrical or moderately distorted octahedra. Similar behavior is observed for the often sharp ($^4A_{1g}$,^4E_g) band near 440 nm (Table 1). In the case of anisotropic absorption, the data are presented for the polarization direction with the greatest intensity. Comparable intensification is not observed for the $^4T_{2g}$ band in many of these systems, so this band is not further considered. The apparent exceptions to this generalization which involve clusters and extended structures based on hydroxyl units as the shared ion are rationalized on the basis of the magnetic interactions in such units.

Magnetic interactions usually accompany clustering or polymerization of the Fe^{3+} cations. Thus it is necessary to consider the correlation between intensity of absorption and the strength of the magnetic interaction. For many minerals, the intensity of the ($^4A_{1g}$,^4E_g) band is often difficult to measure due to overlap with the tail of intense absorption bands in the ultraviolet. In nearly all cases, the $^4T_{1g}$ band near 800 nm is unaffected by this tail. Its intensity is therefore more amenable to quantification. The strength of the magnetic interaction can be measured through the magnetic moment, derived from measurements of the mineral's bulk magnetic susceptibility. The effective magnetic moment of Fe^{3+} ions isolated from anti-ferromagnetic interactions is about 5.9 Bohr magnetons. Strong antiferromagnetic interactions decrease this value to about 2.0 in the case of pairs of Fe^{3+} bridged by a nearly linear oxo-bridge (SCHUGAR *et al.*, 1972). The intensity of the $^4T_{1g}$ band increases markedly as the magnetic moment per Fe^{3+} decreases due to magnetic exchange interactions (Fig. 3). Although the concept is well established that absorption band intensification is related to magnetic interactions, Fig. 3 provides a quantitative demonstration of this relationship and illustrates that there is some variation from a smooth trend, undoubtedly caused by variations in the structural details of the interacting Fe^{3+} sites.

A visible manifestation of the importance of these magnetic interactions is the intensity of color of the host phases. If Fe^{3+} in the host mineral is free of magnetic interactions the mineral is usually pale yellow-green (typically silicates) or pale lavender (phosphates and sulfates). The reds and browns usually associated with Fe^{3+} are observed

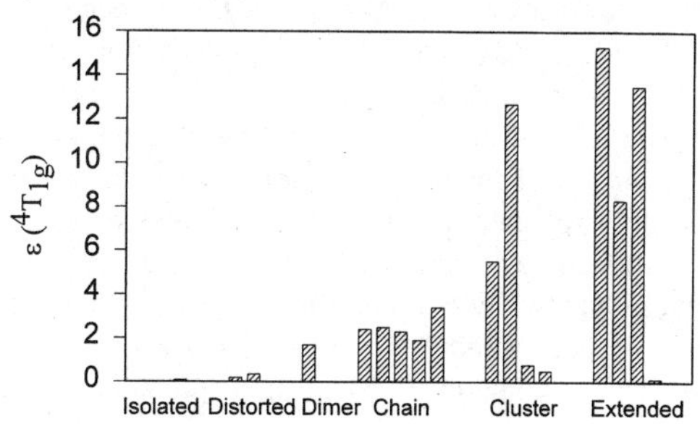

FIG. 2. Bar graph illustrating the range of molar absorption coefficients for the $^4T_{1g}$ band in the 750 - 950 nm range in a variety of minerals with Fe^{3+} in different distortions and types of polyhedral linkages. In all cases, Fe^{3+} absorption intensity is much greater when the Fe^{3+} ions are magnetically coupled through shared polyhedra.

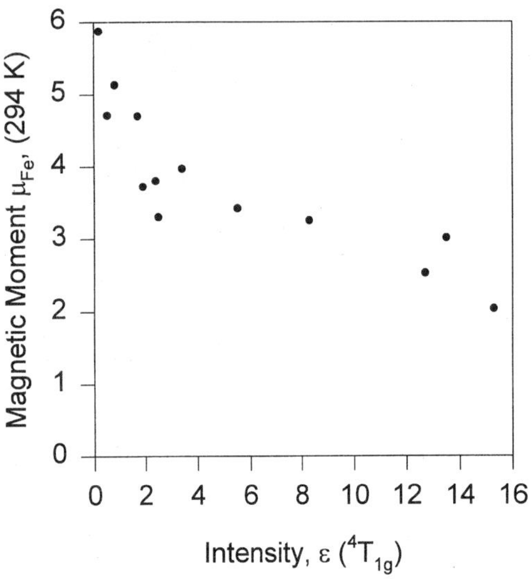

FIG. 3. Comparison of the effective room temperature magnetic moment per Fe^{3+} and the intensity of the first ligand field band ($^4T_{1g}$). Absorption intensity is greatest for minerals with strong antiferromagnetic coupling between adjacent Fe^{3+} cations.

only in minerals with strong magnetic interactions due to extensive polyhedral sharing.

The importance of O-linkages between Fe^{3+} rather than OH-linkages in providing the pathway for absorption band enhancement has previously been noted (ROSSMAN, 1976b) in the tetrameric clusters of amarantite (deep red, oxo-bridged cluster) and leucophosphite (pale amber, hydroxyl-bridged cluster). The importance of oxo-linkages is further demonstrated by the contrast between hematite, Fe_2O_3, which has oxo-linkages and a deep red color with $\epsilon_{855\ nm}$ = 15.3, and bernalite, $Fe(OH)_3$, which has hydroxyl-linkages (BIRCH *et al.,* 1993), a green color, and $\epsilon_{885\ nm}$ = 0.12 calculated from the data presented in MCCAMMON *et al.* (1995).

The relationship between magnetic interactions and enhanced intensity of light absorption can be simply explained using crystal field theory by considering the electronic states of isolated Fe^{3+} ions and magnetically coupled ions. In isolated octahedrally coordinated Fe^{3+}, all electronic transitions require a change in the electronic spin state of the ion. As such, to first order, they are forbidden by the quantum mechanical laws which govern such transitions and will happen, in practice, with low probability. Consequently, the intensity of light absorption will be low as will be the corresponding intensity of color.

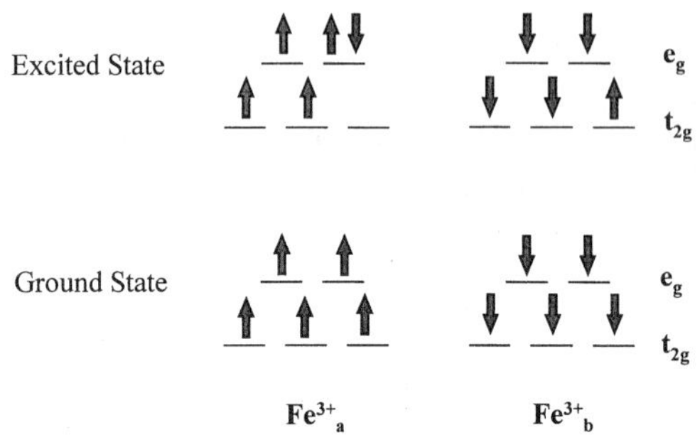

FIG. 4. Orbital energy diagram for a pair of adjacent magnetically coupled Fe^{3+} ions before (bottom) and after (top) an electronic transition illustrating the coupled electronic promotion and spin orientation changes which results in no net electronic spin change for the coupled system.

In a magnetically coupled system, the electrons in adjacent Fe^{3+} ions interact to circumvent the spin selection rules by acting collectively as a system. Fig. 4 illustrates the ground state of a pair of Fe^{3+} ions with strong magnetic interaction. When an electronic excitation (promotion of an electron to an orbital of higher energy) occurs on Fe^{3+}_a, the promoted electron must change its spin state when it pairs with the higher energy electron. If simultaneously an electron on the adjacent Fe^{3+}_b changes its spin state without a corresponding promotion to a higher energy orbital, the net spin state of the system remains unchanged. Thus, the simultaneous electronic and magnetic state transition of the paired system does not represent a spin-forbidden process. The transition is spin-allowed; the intensity of light absorption will be much higher, as will be the corresponding intensity of color. This concept, developed for a pair of interacting Fe^{3+} ions, is also applicable to more extended interacting clusters and networks of Fe^{3+} ions such as hematite.

Acknowledgments—Support for optical spectroscopy of minerals has been provided by the White Rose Foundation and the National Science Foundation (USA). Contribution number 5624 of the Division of Geological and Planetary Sciences.

REFERENCES

BAILEY P. C. (1960) Absorption and reflectivity measurements on some rare earth iron garnets and α-Fe_2O_3. *J. Appl. Phys.* **31**, 39s-40s.

BIRCH W. D., PRING A., RELLER A. and SCHMALLE H. W. (1993) Bernalite, $Fe(OH)_3$, a new mineral from

Broken Hill, New South Wales—description and structure. *Amer. Mineral.* **78**, 827–834.

Blodgett R. H., Crabaugh J. P. and McBride E. F. (1993) The color of red beds; a geologic perspective. In: *Soil Color*, Soil Sci. Soc. Amer., Madison, WI, USA, Spec. Pub. **31**, pp. 127–159.

Hofer L. J. E., Peebles W. C. and Dieter W. E. (1946) X-ray diffraction and magnetic studies of unreduced ferric oxide Fischer-Tropsch catalysts. *J. Amer. Chem. Soc.* **68**, 1953–1956.

Mao H. K. and Bell P. M. (1974) Crystal-field effects of ferric iron in goethite and lepidocrocite: band assignment and geochemical applications at high pressure. *Carnegie Inst. Wash. Year Book* **73**, 502–506.

McCammon C. A., Pring A., Keppler H. and Sharp T. (1995) A study of bernalite, $Fe(OH)_3$, using Mössbauer spectroscopy, optical spectroscopy and transmission electron microscopy. *Phys. Chem. Mineral.* **22**, 11–20.

Rossman G. R. (1975) Spectroscopic and magnetic studies of ferric iron hydroxy sulfates: intensification of color in ferric iron clusters bridged by a single hydroxide ion. *Amer. Mineral.* **60,** 698–704.

Rossman G. R. (1976a) Spectroscopic and magnetic studies of ferric iron hydroxy sulfates: the series $Fe(OH)SO_4 \times nH_2O$ and the jarosite. *Amer. Mineral.* **61**, 398–404.

Rossman G. R. (1976b) The optical spectroscopic comparison of ferric iron tetrameric clusters in amarantite and leucophosphite. *Amer. Mineral.* **61**, 933–938.

Schugar H. J, Rossman G. R., Barraclough C. G. and Gray H. B. (1972) Electronic structure of oxo-bridged iron(III) dimers. *J. Amer. Chem. Soc.* **94**, 2683–2690.

Sherman D. M. (1985) The electronic structures of Fe^{3+} coordination sites in iron oxides; Applications to spectra, bonding and magnetism. *Phys. Chem. Mineral.* **12**, 161–175.

Sherman D. M. and Waite T. D. (1985) Electronic spectra of Fe^{3+} oxides and oxide hydroxides in the near IR to near UV. *Amer. Mineral.* **70**, 1262–1269.

Mineral Spectroscopy: A Tribute to Roger G. Burns
© The Geochemical Society, Special Publication No. 5, 1996
Editors: M. D. Dyar, C. McCammon and M. W. Schaefer

The crystal field concept (CFC) in geosciences: Does the crystal field stabilization energy of Cr^{3+} rule its intercrystalline partitioning behavior?

K. LANGER and M. ANDRUT*
Institute of Mineralogy and Crystallography, Technical University, D-10623 Berlin, Germany

Abstract—According to crystal field concepts, CFC (BURNS, 1970a and 1993), $3d^N$-ion partitioning between different phases in geochemical systems or between different structural sites in multisite phases is governed by the crystal field stabilization energies of the $3d^N$-ions within the coexisting phases or the different sites, respectively. Although CFC was qualitatively shown to be valid in many systems, other authors claim that ionic sizes and distances within the polyhedra incorporating the $3d^N$-ions play a more important role. A decision over CFC and this latter predominantly geometrical concept may be taken when quantitative relations of the type

$$c_{3d^N} = f(CFSE_{3d^N}) \text{ or } K_{D,3d^N,(p,T)}^{(Ph1/Ph2)} = f(\Delta CFSE_{3d^N}^{(Ph1-Ph2)})$$

with positive f are found. Such quantitative relations may be seen when influences of different bulk chemistry and p,T-conditions during formation on c_{3d^N} are ruled out. Therefore, we determine c_{3d^N} and $CFSE_{3d^N}$, for the case of Cr^{3+}, on coexisting minerals in the following parageneses, which contain Cr^{3+} above the trace level (>1000 ppm): 1 and 2: Gt/Cpx/Ky, grosspydite Zagadochnaya/Yakutia; 3: Gt/Cpx/Amph, eclogite Urals/Russia; 4: Cor/Parag/ Phlog, Raj Iz/Antarctica; 5: Cor/Ky/Fuchs, Mysore/India; 6: Cor/Amph, Longido/Tanzania; 7: Opx/Cpx, Cpx exsolution in Opx from an eclogite Urals/Russia. The results quantitatively confirm CFC for parageneses 1 through 6. In case of paragenesis 7, a negative correlation between c_{Cr3+} and $CFSE_{Cr3+}$ was observed. This deviation from CFC can be explained on the basis of crystallochemical arguments.

INTRODUCTION

Outline of the Crystal Field Concept, CFC

THE d-ELECTRONS of $3d^N$-ions in the gaseous state are subject to the spherical fields of their nuclei and, therefore, are degenerate with respect to both energy and symmetry. When incorporated into nonspherical fields, *e.g.*, ligand fields of regular or distorted octahedra in a mineral structure, the degeneracies are at least partly lifted. The respective spectroscopic states of the $3d^N$-ions are split to produce an electronic ground state and concomitant excited state or states. This effect is called crystal field splitting. It is characterized by the crystal field parameter 10Dq, or in the case of octahedral fields, $10Dq_{oct.}$. These relations are outlined in Fig. 1, which presents the octahedral crystal field splittings of geochemically important $3d^N$-ions with total d-electron numbers N between 1 and 9.

For the same $3d^N$-ion ligand-distances, the crystal field parameters behave as

$$10Dq_{oct}:10Dq_{cub}:10Dq_{tetr} = 1:\left(-\frac{8}{9}\right):\left(-\frac{4}{9}\right) \quad (1)$$

in cubic and tetrahedral fields[1].

**Present adress:* GeoForschungszentrum Potdsdam, Telegraphenberg, D-144407 Potsdam, Germany.

[1]The − sign is a consequence of the reversal of electronic states compared to their sequence in octahedral fields.

In the splitting schemes of the lower part of Fig. 1, the spectroscopic ground states of the free ions are represented by symbols derived from Russell-Saunders- or LS-coupling, $^{(2S+1)}L_J$, of the d-electrons. The crystal field split states are characterized by the symbols of the irreducible representations of their symmetry (*cf. e.g.* MCCLURE, 1958; DUNN *et al.*, 1965; SCHLÄFER and GLIEMANN, 1980).

One essential aspect of all the splitting schemes shown in Fig. 1 is that the electronic ground states in the crystal field are energetically lower than the spectroscopic states of the free ions from which they originate. This energy gain of the $3d^N$-ions on incorporation into a crystal field is called *crystal field stabilization energy, CFSE*, and is indicated for the case of an octahedral field in Fig. 1 by multiples of Dq for all $3d^N$-ions $1 \leqslant N \leqslant 9$.

The crystal field stabilization energy in octahedral fields, point group O_h, is obtained from

$$-CFSE_{3d^N,oct} = [4n - 6(N - n)] \cdot Dq \quad (2)$$

or from

$$-CFSE_{3d^N,tetr,cub} = [6n - 4(N - n)] \cdot Dq \quad (3)$$

for tetrahedal or cubic fields of point group T_d or O_h, respectivly. N is the total number of d-electrons and n is the d-electrons in the crystal field-derived ground state (t_{2g} in octahedral fields

d^1	d^2	d^3	d^4	d^5	d^6	d^7	d^8	d^9
Ti^{3+}	V^{3+}	Cr^{3+}	Mn^{3+}	Fe^{3+}	Co^{3+}			
		V^{2+}	Cr^{2+}	Mn^{2+}	Fe^{2+}	Co^{2+}	Ni^{2+}	Cu^{2+}
^2D	^3F	^4F	^5D	^6S	^5D	^4F	^3F	^2D

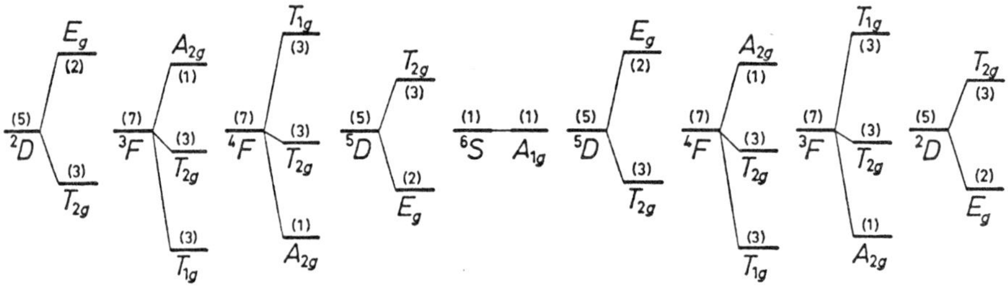

$-CFSE_{oct}$ $4Dq$ $8Dq$ $12Dq$ $6Dq$ 0 $4Dq$ $8Dq$ $12Dq$ $6Dq$

FIG. 1. d-electron configurations, spectroscopic ground states, $^{(2S+1)}L_J$, of the free ion case, crystal field-splitting of these free ion ground states as well as crystal field stabilization energies of the whole series of $3d^N$-ions in an octahedral environment, point group O_h (after SCHLÄFER and GLIEMANN, 1980; in case of d^2 and d^7, 6Dq was corrected to 8Dq). The ordinate in the split level diagrams is energy in arbitrary units. For further explanations *cf.* text.

or e_g in tetrahedral and cubal fields for the one electron case.)

In addition to the type and symmetry of the coordination polyhedra, two peculiarities of the crystal field determine the crystal field splitting parameter 10Dq of a given $3d^N$-ion,

(i) *The nature of the ligands:* In the case of rock forming-minerals and related geochemical systems, these are predominantly O^{2-}, OH^-, H_2O, F^-, and S^{2-}. Generally, 10Dq increases within the "spectrochemical series" (TSUCHIDA, 1938a, b, and c; SHIMURA and TSUCHIDA, 1956) which for the just mentioned ligands is $F^- < OH^- < O^{2-} < H_2O < S^{2-}$ (BURNS, 1993).

(ii) *The mean $3d^N$-ion ligand-distance:* In the case of octahedral fields of ligands with effective charge Z_L, crystal field theory yields

$$10Dq_{oct} = \frac{5}{3}\frac{(Z_L \times e^2)}{R^5} \times \overline{r^4} \qquad (4)$$

(*e.g.*, DUNN *et al.*, 1965; LEVER, 1968), wherein $\overline{r^4}$ = mean fourth power radius of the d-electrons of the central ion, a quantity that is thought to be approximately constant for $3d^N$-ions of the same valence, and R = the mean cation-ligand distance.

On decendent symmetry from O_h or T_d, the remnant degeneracies of states are gradually removed, leading to further splitting of the E- and T-states, shown in Fig. 1. When ground states are subject to such low symmetry splitting, as is the case for all $3d^N$-ions except for $3d^3$, $3d^5$, and $3d^8$ (*cf.* Fig. 1), then the resulting ground state is lowered again compared to the high symmetry case. Consequently, $CFSE_{3d^N}$ is higher at such sites for $3d^N$-ions with degenerate crystal field ground state.

The energy gaps between the ground and excited crystal field states of Fig. 1, eventually split by low symmetry, are in the range of 25000 to 4000 cm^{-1}, corresponding to 3.1 to 0.5 eV. This means that they occur in the UV to NIR spectral region with wavelengths of 400 to 2500 nm. Electronic resonant absorption spectroscopy within the aforementioned region is used to measure such energies and to extract 10Dq and, from this $CFSE_{3d^N}$. Polarized radiation and crystallographically oriented crystal slabs are necessary in the case of anisotropic crystals where symmetry related selection rules for the electronic transitions are active.

The importance of these crystal field considerations and their experimental evaluation is obvious from the following: in the early fifties, it was shown that the heat of hydration of $3d^N$-ions as well as the lattice energy of $3d^N$-metal sulfides can only be correctly calculated when the crystal field stabilization energy of the $3d^N$-ions was included into the calculations that were based on the electrostatic approach, valid for spherical ions with $3d^0$-configuration.

This result stimulated many geoscientists to suggest that the crystal field stabilization energy, $CFSE_{3d^N}$, should also affect or even rule the thermodynamic properties of $3d^N$-ion bearing minerals and the behavior of $3d^N$-ions in homogeneous or inhomogeneous geochemical systems, *e.g.*, their intra- and intercrystalline distribution behavior and their partitioning between crystals and melt (WILLIAMS, 1959; BURNS and FYFE, 1964, 1966, 1967a and b; BURNS *et al.*, 1964; CURTIS, 1964; MERLINO, 1965; SCHWARCZ, 1967; STRENS, 1968; BURNS, 1968, 1970a, and 1976; BURNS and SUNG, 1978). This approach was further stimulated by the pioneering book of BURNS (1970b, 1993).

Considering $3d^N$-ion distributions, different $CFSE_{3d^N}$ in phases with different cation-ligand distances and different site symmetries may force the $3d^N$-ions predominantly into one of the phases in multiphase assemblages. This concept to interpret interphase distribution is called crystal field concept, CFC, throughout the paper. It predicts that $3d^N$-ions fractionate into those sites and/or phases wherein $CFSE_{3d^N}$ is highest (BURNS, 1970b, 1993).

THE PROBLEM AND AN APPROACH TO SOLVE IT

CFC proved capable of qualitatively interpreting geochemical observations on $3d^N$-ion partioning between melt and crystals, between coexisting minerals in magmatic and metamorphic rocks from the earth's crust and mantle between different structural sites in multisite mineral structures, or solid solution thermodynamics and phase transitions of $3d^N$-ion bearing minerals. This was comprehensively presented 1970 and 1993 by Roger BURNS.

Despite the abundant qualitative evidence for the validity of CFC, concepts based on ionic radii and site geometry in mineral structures are often thought to be superior in the interpretation of partitioning patterns of not only the spherical $3d^0$-ions but also of nonspherical $3d^N$- or $4f^N$-ions (*e.g.*, JENSEN, 1973; BEATTIE, personal comm.). The argument for this is that the Madelung part of the lattice energy of crystals is much higher than any crystal field effects therein. However, this may be true only for trace amounts of $3d^N$-ions but not for solid solutions with concentrations of $3d^N$-ion end members higher than 1 to 2 mole%, because $CFSE_{3d^N}$ typically has values between 50 and 260 [KJ/g-ion] depending on the type of $3d^N$-ion and of its structural site.

The problem is, thus, to decide between CFC and the just mentioned geometrical approach in the case of $3d^N$-ion partioning in a concentration range above the trace level. This problem can only be solved by checking for *quantitative* relations between $3d^N$-ion concentrations and $CFSE_{3d^N}$

$$C_{Cr^{3+}} = f(CFSE_{Cr^{3+}}) \quad (5)$$

or between distribution coefficients and differences in the crystal field stabilization energies

$$K^{Ph1/Ph2}_{D,Cr^{3+}(p,T)} = f(\Delta CFSE^{Ph1-Ph2}_{Cr^{3+}}) \quad (6)$$

Such quantitative relations may be found only when any effects of the bulk chemistry and of temperature and/or pressure changes on the partitioning of $3d^N$-ions in the respective geochemical system, *e.g.*, between the minerals of a rock, can be ruled out. Therefore, our approach is to obtain $CFSE_{3d^N}$ and c_{3d^N} for coexisting, homogeneous minerals. Results of such work are presented here for the case of Cr^{3+} in paragenetic minerals (*cf.* ANDRUT, 1995). The behavior of this $3d^3$-ion was studied because its $CFSE_{Cr^{3+}}$ is highest among the geochemically significant $3d^N$-ions and because its ground state in octahedral crystal fields, strongly preferred by this ion (BURNS, 1970b and 1993), is nondegenerate (*cf.* Fig. 1). Furthermore, the energy of the spin-allowed transition $^4A_{2g} - {}^4T_{2g}$ of $Cr^{3+[6]}$ (Fig. 1) is equal to $10Dq_{oct}$ (TANABE and SUGANO, 1954 and b).

SAMPLES AND METHODS

Crystal fragments of coexisting minerals with suitable grain size were drilled out of a series of chromium containing rocks, compiled in Table 1.

For measurement of polarized single crystal spectra, anisotropic mineral crystals were oriented by means of spindle stage techniques (BLOSS, 1981) such that one of the main directions of their indicatrix, either X, Y, or Z (notation of TRÖGER, 1952), became the rotation axis. Two crystal fragments of each of the coexisting minerals were prepared to facilitate measurements with E parallel to X, Y and Z. The oriented crystals were embedded and ground to obtain the suitable thickness, then polished from both sides. Quartz crystals oriented parallel to *c*, placed aside the crystals to be studied, served to determine the thickness of the mounts by evaluating the birefringence of quartz. In this way, single crystal slabs with thicknesses between 50 and 500 μm and containing the main optical directions (Y,Z), (X,Z), or (X,Y), respectively, were obtained. The procedures of preparation and thickness measurements are described elsewhere (LANGER, 1988). The slabs had lateral dimensions between about 200 and 900 μm. Their orientation was checked by conoscopic observations, and is thought to be correct to within 3° to 5° for

Table 1. Parageneses studied and their sources. Abbreviations: Gt = garnet, Kpx = clinopyroxene, Opx = orthopyroxene, Ky = kyanite, Amph = clinoamphibole, Ru = corundum (ruby), Parag = paragonite, Phlo = phlogopite, Fu = fuchsite.

Pargeneses studied	Symbols used for minerals studied	Source	Obtained from
Garnet Clinopyroxene Kyanite	Zag C1-Gt, Zag G1-Gt[1] Zag C1-Cpx, Zag G1-Cpx Zag C1-Ky, Zag G1-Ky	Grosspydite xenolith Z13, Zagadochnaya, Yakutia	N.V. Sobolev, Novosibirsk, Russia
Garnet Clinopyroxene Clinoamphibole	MC1-A Gt MC-1A Cpx MC1-A Amph	Eclogite MC1, Urals, Russia	A.N. Platonov, Kiev, Ukraine
Corundum Paragonite Phlogopite	RI Ru RI Parag RI Phlo	Raj Iz, Antarctica	A.N.Platonov, Kiev, Ukraine
Corundum Kyanite Fuchsite	My Ru My Ky My Fu	Mysore, India	S. Herting-Agthe, Techn. Univ. Berlin
Corundum Clinoamphibole	Lon Ru Lon Amph	Amphibolite Longido, Tanzania	S. Herting-Agthe, Techn. Univ. Berlin
Clinopyroxene Orthopyroxene	OPXCPX-Cpx OPXCPX-Opx	Exsolution of clinopyroxene in orthopyroxene from an eclogite, Urals	A.N. Platonov, Kiev, Ukraine

[1]Two sets of the minerals garnet, clinopyroxene, and kyanite were extracted from the grosspydite xenolith, C1 and G1.

observation in the green part of the spectrum. In the case of OPXCPX-Kpx (Table 1), *i.e.,* the clinopyroxene lamellae exsolved from orthopyroxene at an orientation independent from that of orthopyroxene could not be achieved. The orthopyroxene slab was oriented parallel to Y, hence E//X and E//Z spectra could be measured on Opx. The indicatrix of the clinopyroxene lamellae is tilted against that of Opx. Therefore, the clinopyroxene spectra were, in this case, polarized E// ca.X and E// ca.Y. However, this will not greatly influence 10Dq, as the baricenter of the $^4A_{2g} - {}^4T_{2g}$ transition does not strongly depend on the polarization.

Polarized single crystal spectra, $\log(I_o/I) = f(\tilde{\nu})$, in the spectral range 40000 to 4000 cm^{-1} were scanned in a single-beam microscope spectrometer (Zeiss UMSP 80) using UV-transparent 10× objectives (Zeiss ULTRAFLUAR) as condenser and objective. Details of the procedure are given elsewhere (Langer and Frentrup, 1979; Langer, 1988). Luminous field and measuring diaphragms were 80 and 32 μm, respectively, in the UV/VIS or 100 and 60 μm, respectively, in the NIR. Spectral band widths and step widths were 1 nm each in the UV/VIS or 8 and 10 nm, respectively, in the NIR. Spectra were obtained as averages of up to 200 scans, the reference $I_o = f(\tilde{\nu})$ being obtained in the quartz plate (see above). Special care was used to select optically clear measuring spots on the crystal slabs studied. Curve deconvolution and band positions were achieved by means of a peak fit program, assuming Gaussian functions to represent the component bands in complex spectra.

Microprobe analyses on a variety of different points on each crystal were performed on a Cameca Camebax in the ZELMI laboratory of the Technical University Berlin subsequent to the spectral measurements. To check for the homogeneity of the crystals studied, N repeated analyses on M various points on a grain were performed. From these, standard deviations were calculated for the means of N_i analyses of one individual point i as well as of all N analyses on M points. When the standard deviation of the mean of all N measurements on M points on a grain does not exceed 2s of N_i measurements on one individual point i, there is no significant compositional inhomogeneity with respect to the elements measured in the crystal grain.

X-ray diffraction. For data evaluation, the molar volumes, V_M in [dm^3], of the minerals under study were needed (see below). These data were calculated from the volumes of the unit cells, V_e in [A^3], by

$$V_M = \frac{10^{-27} \times V_e \times N_L}{Z} \tag{7}$$

($N_L = 6.022*10^{23}$ $mole^{-1}$, Z = number of formula units per unit cell). Cell volumes were obtained from the lattice constants determined from powders prepared from a number of additional crystals extracted from the rock samples in addition to the coexisting ones, which had been prepared for spectroscopy and subsequent microprobe. The powder diffractograms were scanned on a Huber Guinier diffractometer G 645 (courtesy W. Depmeier, University of Kiel), using silicon as external standard.

EXPERIMENTAL RESULTS AND THEIR EVALUATION

Composition of the coexisting minerals studied

The crystallochemical formulae of the coexisting minerals studied as well as their lattice constants are presented in Table 2. The crystallochemical formulae were calculated from the wt% values of the oxide components. These values were calculated as the averages of repeated measurements on several spots on the respective crystal slab, including the spot on which the spectral measurements had been performed. From a comparison of the standard deviations of mean values of repeated analyses on one spot with those obtained by averaging the data measured on several points on one slab, there was no indication of significant zoning or other chemical inhomogeneities of the minerals studied. Typical examples of such standard deviations are those of wt% Cr_2O_3 in kyanite from the Zagadochnaya parageneses C1 and G1 (*cf.* Table 1):

ZagC1-Ky:

n = 3 measurements on the same spot :
2.17%, s=0.06%

measurements on M = 7 different spots :
2.20%, s=0.09%

ZagG1-Ky:

n = 3 measurements on the same spot :
1.78% s=0.05%

measurements on M = 7 different spots :
1.83%, s=0.09%

From this, it may be estimated that the uncertainty in $n_{Cr^{3+}}$, *i.e.* the number of Cr^{3+}-ions per formula unit, of the formulae in Table 2 occurs in the third decimal point. This is important to note, as $n_{Cr^{3+}}$ will be used to calculate the volume-normalized chromium concentrations held by the coexisting phases (see below).

The spectra and the extraction of $10Dq_{cr3+}$ from them

As shown in the introduction, the crystal field parameter $10Dq_{Cr^{3+}}$ is equal to the lowest-energy spin-allowed electronic transition of Cr^{3+}, $A_{2g} - {}^4T_{2g}(F)$. This transition occurs in the absorption spectra of Cr^{3+}-bearing, oxygen-based structures as strong, broad band, centered between about 18500 and 15500 cm^{-1}, depending on the properties of the crystal field in the respective structural sites as described by eqn. (4). The half widths of this band range up to about 2500 cm^{-1}.

Examples of such spectra, measured during this study on coexisting Cr^{3+}-bearing garnet, clinopyroxene and kyanite, extracted as paragenesis G1 from the Zagadochnaya grosspydite xenolith, are shown in Figs. 2 to 5. It would go beyond the scope of this paper to present the spectra measured on all the crystals studied (Table 1 and 2) and all parameters extracted. This will be the aim of a forthcoming paper (ANDRUT and LANGER, in preparation). Here, we shall concentrate on the problems of the extraction of 10Dq, using Figs. 2 to 5 as examples.

All the minerals studied contain Cr^{3+} in coordination octahedra with site symmetries lower than O_h – m3m. For example, in garnet, the octahedra have site symmetry $\bar{3}$, in corundum 3, in clinopyroxene M(1) 2 and in kyanite 1. As pointed out in the introduction, fields of lower symmetry than O_h – m3m give in principle rise to two effects, important in the extraction of correct $10Dq_{Cr^{3+}}$-values:

(i) splitting of the excited, triply degenerate T(F) states of Cr^{3+} and

(ii) symmetry-related selection rules for the respective transitions between the nondegenerate crystal-field ground state of Cr^{3+} and the now nondegenerate excited states, originating from the aforementioned T-states by low-symmetry splitting. Such symmetry-related selection rules make allowance for a transition only, when the electric field vector of the exciting radiation has a special orientation with respect to the electric dipole moment vector of the transition considered (*cf. e.g.* WILSON *et al.*, 1955; COTTON, 1971).

The latter effect causes the "pleochroism" of crystals of low-symmetry minerals and other crystalline material. Other effects may contribute in the case of chemically complex natural minerals. For example, charge-transfer phenomena, may also give rise to strongly polarized absorption bands (SMITH and STRENS, 1976).

A third effect that eventually makes the extraction of $10Dq_{Cr^{3+}}$ from the spectra difficult is:

(iii) band-overlap of the (${}^4A_{2g} - {}^4T_{2g}$)-derived Cr^{3+}-transition, or transitions respectively, by other excitations occurring in chemically complex natural minerals.

An example of (i) is the splitting of the two spin-allowed Cr^{3+}-bands in the clinopyroxene spectra of Fig. 4. This splitting, as well as the band polarizations (ii) that are obvious in Fig. 4, was interpreted to be caused by the nonregular symmetry of the Cr^{3+}-containing M(1)-octahedra of the clinopyroxene structure, the effects (i) and (ii) being most consistent with a D_{2d} – $\bar{4}$2m pseudosymmetry (ABS-WURMBACH *et al.*, 1985). Another example of effect (ii) is the polarization of band intensities in the kyanite spectra of Fig. 3. An example of effect (iii) is the overlap between the (${}^4A_{2g} - {}^4T_{2g}$)-derived Cr^{3+}-band in the kyanite spectra of Fig. 3 with a band originating from $Fe^{2+} - Ti^{4+}$ charge-

Table 2. Chemical composition, recalculated to crystallochemical formulae, and lattice constants of the paragenetic minerals studied. Compositions are averages of data calculated from microprobe analyses on a variety of spots on the respective crystal slab.

Crystals	Analyzed composition	a [A]	b [A]	c [A]	[°]	[°]	[°]
ZagC1-Gt	$(Ca_{1.65}Mg_{0.75}Mn_{0.01}Fe_{0.62})^{[8]}$ $(Al_{1.92}Cr_{0.09}Ti_{0.01})^{[6]}$ $[SiO_4]_3$	11.649(4)					
ZagC1-Cpx	$(Ca_{0.60}Na_{0.35})^{[8]}$ $(Mg_{0.52}Fe_{0.06}Al_{0.43}Cr_{0.03})^{[6]}$ $[Si_{1.93}Al_{0.07}O_6]$	9.87(5)	8.61(5)	5.24(2)		106.0(4)	
ZagC1-Ky	$(Al_{1.95}Cr_{0.04}Fe_{0.01})^{[6]}$ $[O/SiO_4]$	7.113(6)	7.860(5)	5.570(5)	90.11(8)	101.96(8)	105.89(7)
ZagG1-Gt	$(Ca_{1.62}Mg_{0.80}Mn_{0.01}Fe_{0.60})^{[8]}$ $(Al_{1.92}Cr_{0.08}Ti_{0.02})^{[6]}$ $[SiO_4]_3$	11.649(4)					
ZagG1-Cpx	$(Ca_{0.60}Na_{0.35})^{[8]}$ $(Mg_{0.48}Fe_{0.06}Al_{0.44}Cr_{0.02})^{[6]}$ $[Si_{1.95}Al_{0.05}O_6]$	9.87(5)	8.61(5)	5.24(2)		106.0(4)	
ZagG1-Ky	$(Al_{1.90}Cr_{0.04}Fe_{0.01})^{[6]}$ $([O/SiO_4]$	7.113(6)	7.860(5)	5.570(5)	90.11(8)	101.96(8)	105.89(7)
MC1-A Gt	$(Ca_{0.65}Mg_{1.25}Mn_{0.03}Fe_{1.07})^{[8]}$ $Al_{2.00}{}^{[6]}$ $[SiO_4]_3$	11.5583(14)					
MC1-A Cpx	$(Ca_{0.68}Na_{0.30})^{[8]}$ $(Mg_{0.60}Fe_{0.08}Al_{0.33}Cr_{0.01})^{[6]}$ $[Si_{1.94}Al_{0.06}O_6]$	9.9960(18)	8.52(2)	5.207(9)		106.1(2)	
MC1-A Amph	$(K_{0.14}Na_{0.47})^{[12]}$ $(Na_{0.31}Ca_{1.58}Fe_{0.11})^{M4[8]}$ $(Mg_{3.43}Fe_{0.52}Al_{0.98}Cr_{0.02}Ti_{0.05})^{M1,2,3[6]}$ $[Al_{1.39}Si_{6.61}O_{22}\ (OH)_2]$	9.797(10)	17.93(2)	5.297(6)		104.85(9)	
RI Ru	$(Al_{1.97}Cr_{0.03})^{[6]}$ O_3	4.7631(19)		12.999(14)			
RI Parag	$(K_{0.03}Na_{0.79})^{[12]}$ $(Fe_{0.03}Mg_{0.06}Al_{1.92}Cr_{0.09}Ti_{0.01})^{[6]}$ $[Al_{1.21}Si_{2.79}O_{10}\ (OH)_2]$	5.177(11)	9.03(3)	20.53(7)		95.0(2)	
RI Phlo	$(K_{0.62}Na_{0.28})^{[12]}$ $(Fe_{0.30}Mg_{2.00}Al_{0.43}Cr_{0.11}Ti_{0.06})^{[6]}$ $[Al_{1.33}Si_{2.67}O_{10}\ (OH)_2]$	5.310(12)	9.29(2)	20.44(2)		95.9(2)	
My Ru	$(Al_{1.99}Cr_{0.01})^{[6]}$ O_3	4.7537(15)		12.980(5)			
My Ky	$(Al_{1.99}Cr_{0.01})^{[6]}$ $[O/SiO_4]$	7.112(5)	7.842(6)	5.574(5)	90.08(7)	101.08(8)	105.88(6)
My Fu	$(K_{0.91}Na_{0.09})^{[12]}$ $(Fe_{0.02}Mg_{0.02}Al_{1.80}Cr_{0.01}Ti_{0.09})^{[6]}$ $[Al_{0.96}Si_{3.04}O_{10}\ (OH)_2]$	5.180(3)	8.992(6)	20.034(11)		95.80(5)	
Lon Ru	$(Al_{1.97}Cr_{0.03})^{[6]}$ O_3	4.7613(19)		12.995(4)			
Lon Amph	$(K_{0.01}Na_{0.84})^{[12]}$ $(Na_{0.02}Ca_{1.94}Fe_{0.04})^{M4[8]}$ $(Mg_{2.80}Fe_{0.84}Al_{1.11}Cr_{0.22})^{M1,2,3[6]}$ $[Al_{2.22}Si_{5.78}O_{22}\ (OH)_2]$	9.843(8)	17.970(8)	5.298(5)		105.22(8)	
OPXCPX Cpx	$(Ca_{0.86}Na_{0.11})^{[8]}$ $(Mg_{0.82}Fe_{0.05}Al_{0.11}Cr_{0.05})^{[6]}$ $[Si_{1.93}Al_{0.07}O_6]$						
OPXCPX Opx	$(Ca_{0.02}Mg_{1.73}Fe_{0.18}Al_{0.06}Cr_{0.02})^{[6]}$ $[Si_{1.92}Al_{0.08}O_6]$						

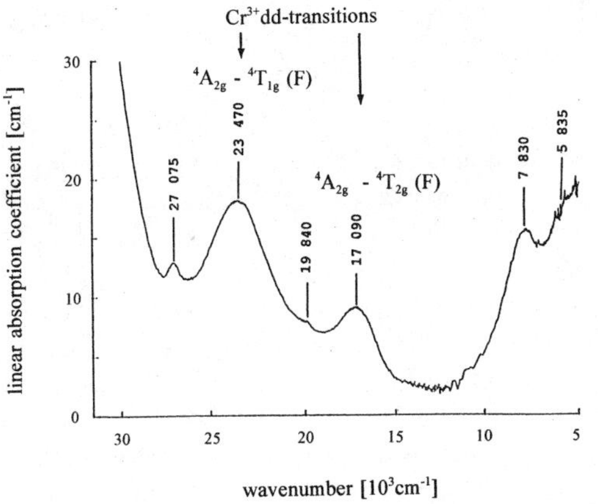

FIG. 2. Single crystal spectrum of the Cr^{3+}-bearing ($Pyr_{26}Gross_{49}Alm_{20}$)-garnet ZagG1-Gt (*cf.* Tables 1and 2) and assignment of the spin-allowed dd-transitions of $Cr^{3+[6]}$. Other bands in the UV/VIS and NIR ranges are dd-transitions of $Fe^{2+[8]}$. The crystal slab was 247 μm thick.

transfer in edge-connected octahedra in the structure of this mineral (SMITH and STRENS, 1976).

As a consequence, the extraction of $10Dq_{Cr^{3+}}$-values necessary for our purpose of quantitatively checking CFC is a nontrivial problem and, hence, we ought to shortly comment on it for the Cr^{3+}-containing, coexisting minerals studied here.

Corundum—The trigonal field of the Cr^{3+}-centered octahedra yields two bands in the 10Dq-region at around 17900 and 18300 cm^{-1}, the first occurring in $E\|n_e$ and the second in $E\|n_o$ (*cf. e.g.* MCCLURE, 1958 and 1962). It is possible to base the interpretation of the spectra on a trigonal field approach (MCCLURE 1958; MACFARLANE, 1963) and to extract a trigonal field parameter $10Dq_{trig}$. In the case of all other minerals, the interpretation is based on the approximation of a regular octahedral field parameter $10Dq_{cub}$, because an exact treatment is impossible and, in any case, will not significantly enhance the accuracy of the crystal field parameter obtained. In case of corundum, this is obvious from the following consideration: We can recalculate $10Dq_{trig}$, obtained from the exact treatment for the point symmetry 3 of the corundum octahedra to the cubic field parameter by

$$Dq_{cub} = Dq_{trig} + {}^{7}\!/_{18} \times D\tau \qquad (8)$$

(KÖNIG and KREMER, 1977), wherein $D\tau$ is a parameter taking into account the trigonal distortion of the octahedra along C_3. $D\tau$ can be determined from the energies of all the transitions of Cr^{3+} in the corundum-spectra (KÖNIG and KREMER, 1977).

The other possible interpretation is to determine the baricenters of the two polarized components, derived from ($^4A_{2g} - {}^4T_{2g}$), in the $E\|n_e$ and $E\|n_o$ spectra and to calculate the mean value. This then yields $10Dq_{cub}$ for the regular approach. In the case of corundum, the latter procedure yields $10Dq_{cub}$-values which are only 1.3% lower than those obtained in the first mentioned and, with respect to the symmetry, correct procedure.

Because a strict symmetry treatment is not possible in the case of the other minerals studied and as the deviation between the crystal field parameter as obtained in the strict vs. the cubic approach is small, the $10Dq_{cub}$-approximation is adopted for the evaluation of all the spectra measured in this study. If so, why are polarized and not unpolarized spectra used, which are more easily obtained? Such a procedure might produce erroneous $10Dq_{cub}$-values

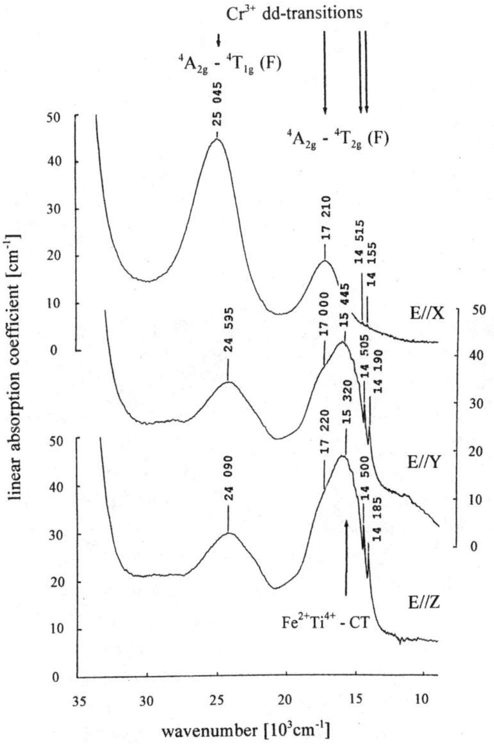

FIG. 3. Polarized single crystal spectra of the Cr^{3+}-bearing kyanite ZagG1-Ky (*cf.* Tables 1 and 2) and assignment of dd-transitions of Cr^{3+} as well as Fe^{2+}-Ti^{4+} charge-transfer. The two narrow low-energy bands of Cr^{3+} are due to spin forbidden transitions of this ion. Wavenumber positions of component bands are obtained by the curve fitting process (*cf.* text). Spectra E//X and E//Z are measured on a slab $\perp$Y (−Ky−3), 237 μm thick, the spectrum E//Y on a slab $\perp$Z (−Ky−1), 227 μm thick.

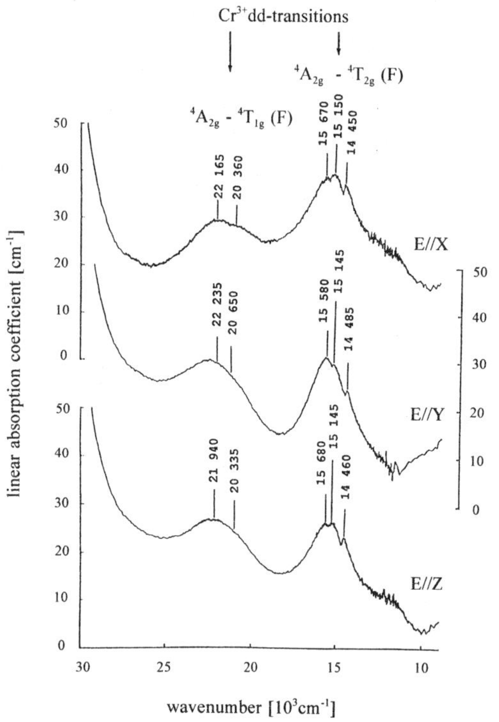

FIG. 4. Polarized single crystal spectra of the Cr^{3+}-bearing ($Jd_{45}Di_{47}$)-clinopyroxene ZagG1-Kpx (*cf.* Tables 1 and 2). Baricenters of Cr^{3+} dd-transitions are indicated. Wavenumber positions of component bands are obtained by the curve fitting process. Spectra E//X and E//Y were measured on a slab ⊥Z (−Cpx−3), 110 μm thick, the E//Z spectrum on a slab X (−Cpx−5), 68 μm thick.

because the position of the baricenters depend on the polarization. Therefore, all three polarizations are needed, whereas an unpolarized spectrum of a crystal plate is a mixture of only two. Also, effect (iii) can only be identified, if present, on the basis of polarized spectra.

Garnet—The trigonal site distortion of the garnet octahedra incorporating Cr^{3+} is so small (NOVAK and GIBBS, 1971; ARNI *et al.,* 1985) that any band splitting that might result from effect (i) is indicated only by a slight asymmetry of the strong spin-allowed dd-bands of Cr^{3+}, even at low temperatures (TARAN *et al.,* 1994). Symmetry-related selection rules for the site symmetry $\bar{3}$ of the garnet octahedra are destroyed by the Ia3d symmetry of the overall structure. Hence, an evaluation based on the baricenters and yielding $10Dq_{cub}$ is in this case the only possible way to obtain the required data from the spectra, an example of which is shown in Fig. 2.

Kyanite—The spectra obtained (*e.g.* Fig. 3) exhibit in the spectral region around 16000 cm^{-1}, where the ($^4A_{2g}$ − $^4T_{2g}$)-derived bands occur, a complex band system. This contains at least two strong and broad components, a high energy component at around 17200 cm^{-1}, polarized $\alpha_x < \alpha_y < \alpha_z$, and a low energy component at around 15500 cm^{-1}, which is not observed in the E‖X-spectrum and, in E‖Y and E‖Z, with $\alpha_Y < \alpha_Z$ polarization (Fig. 3). In addition, there occur two narrow bands between 14000 14500 cm^{-1}, caused by spin-forbidden transitions of Cr^{3+}, which show up with remarkably high intensity in kyanite (Fig. 3). These bands do not play a role in the extraction of $10Dq_{cub}$ and, therefore, will not be further commented upon here. The assignment of the strong component bands is based on the following results published so far:

(a) Pure, synthetic Cr^{3+}-bearing kyanites (LANGER and SEIFERT, 1971), with chromium-contents comparable to those of the natural kyanites studied here, show the ($^4A_{2g}$ − $^4T_{2g}$)-derived transition of Cr^{3+} at 16850 cm^{-1} in unpolarized powder remission spectra (LANGER, 1976). Single crystal spectra, of crystals with $n_{Cr^{3+}} = 0.3$ pfu, measured with E‖c and E⊥c which is close to the E‖Z and E‖X measurements of the present study, show the baricenter of this band at 16480 cm^{-1} with $\alpha_{\perp c} < \alpha_{\|c}$ polarization (LANGER, 1984 and 1988). Hence, the strong, high-energy component observed here is identified as the 10Dq-band of Cr^{3+}. We determined $10Dq_{cub}$ by averaging the baricenters of this component band as obtained by curve resolution of the three polarized spectra (Fig. 3).

(b) Most kyanites from metamorphic rocks, containing traces of chromium only, exhibit a typical blue color, the intensity of which is often undulous or zoned within the crystals. The polarized spectra of such kyanites were first obtained by WHITE and WHITE (1967) and are dominated by a strongly polarized, intense band system with maximum at 16000 cm^{-1}, absent in E‖X and with $\alpha_Y > \alpha_Z$. After an intense debate about the origin of these spectral features, it became obvious (SMITH and STRENS, 1976; PARKIN *et al.,* 1977) that the band system originates from Fe^{2+}-Ti^{4+} charge-transfer transitions in the c-parallel chains of edge-connected M(1)-, M(2)-octahedra of the structure (BURNHAM, 1963). Contributions of Fe^{2+}-Fe^{3+} charge-transfer in the low-energy wing of the band system do also contribute when chemistry and conditions of the kyanite formation allows for it. Because all features of the low-energy component band under discussion in our kyanite spectra correspond to those of the Fe^{2+}-Ti^{4+} charge-transfer band identified in blue kyanites (*l.c.*), its assignment to charge-transfer of this type is obvious.

Clinopyroxene—The identification of the 10Dq-band in the spectra (*cf.* Fig. 4) and the extraction of $10Dq_{cub}$ of Cr^{3+} in the M(1)-octahedra of the clinopyroxene structure (CLARK *et al.*, 1969) pose no problems thanks to the spectral studies of ABS-WUMBACH *et al.* (1985) and of KHOMENKO and PLATONOV (1985). The baricenters of the ($^4A_{2g}$ – $^4T_{2g}$)-derived band system in E‖X, E‖Y and E‖Z polarization (Fig. 4) were determined and averaged to obtain $10Dq_{cub}$ in all clinopyroxenes of Table 1.

The only difficulty arose with the clinopyroxene exsolution from orthopyroxene, *i.e.* the spectra of the OPXCPX-Cpx sample in Table 1. Due to the (100) intergrowth of the clinopyroxene exsolution lamellae in the orthopyroxene matrix, the indicatrix of clinopyroxene is tilted against the face of the slab ‖ Y of orthopyroxene, corresponding to (100) within the error of the orientation. Therefore, the E-vector of the measuring radiation could only be approximately oriented parallel to X or Y of clinopyroxene in this case.

Orthopyroxene—Considering the radius of Cr^{3+} and the mean M-O distances within the octahedral M(1)- and M(2)-sites of the structure (CAMERON and PAPIKE, 1981), as well as the fairly weak distortion from regular geometry of M(1) as compared to M(2), it is to be expected that Cr^{3+} enters the former site, charge balance being achieved *e.g.* by $Al^{3+[4]}$ for $Si^{4+[4]}$ substitution. Entrance of chromium into the M(1)-sites of the orthopyroxene structure is confirmed by the evaluation of the polarized spectra of a synthetic Cr^{3+}-bearing enstatite (ROSSMAN, 1980). Thus, our evaluation of the spectra of OPXCPX-Opx (Table 1) is based on these results, $10Dq_{cub}$ being obtained as the average of the baricenters of the 10Dq-band in E‖X and E‖Z polarization. E‖Z could not be measured on this sample (see above), and, in this polarization, the baricenter of the 10Dq-band system occurs at slightly higher wavelengths, *i.e.* lower energies compared to the two former polarizations (ROSSMAN, 1980). Therefore, the $10Dq_{cub}$-value obtained here may be too high by about 100 cm^{-1}, which would correspond, in $CFSE_{Cr^{3+}}$, to +1.4 kcal/g-atom$_{Cr^{3+}}$.

Clinoamphiboles, dioctahedral and trioctahedral micas—The 10Dq-band system in the spectra obtained on the clinoamphiboles of Tables 1 and 2 exhibits, if any, only a very weak low-symmetry splitting and also weak polarization with $\alpha_Y > \alpha_X \cong \alpha_Z$. Hence, Cr^{3+} is incorporated in octahedral sites that are only slightly distorted from regular geometry. This is the case for the M(1,2,3) sites of the structure (*e.g.*, PAPIKE *et al.*, 1969), which are, therefore, assumed to incorporate chromium.

There is no information in the literature about the distribution of Cr^{3+} in trioctahedral biotite or dioctahedral muscovite and paragonite that can help in the interpretation of the spectra of our samples RI Phlo, My Fu and RI Parag (Tables 1 and 2). Effects (i) and (ii) are not observed in the case of RI Phlog; low band intensity may be a consequence of the low Cr^{3+}-concentration in this sample (Table 2). My Fu and RI Parag, with pleochroic schemes $\alpha_Y \cong \alpha_Z < \alpha_X$ and $\alpha_Y < \alpha_Z \ll \alpha_X$, respectively, of the 10Dq-band system, do not show any low symmetry splitting although band intensities are strong enough to detect it. In any case, Cr^{3+} will preferentially enter the octahedral sites with lowest distortion from regular geometry, *i.e.* in phlogopite the M(1)-sites with trans-configuration of the two OH-groups. FAYE (1968), in his interpretation of the spectra of a fuchsite from Madagaskar, considers the M(2)-position with cis-configuration of OH to be "relatively distorted," whereas the M(1)-position, normally vacant in dioctahedral micas, is close to regular geometry (*e.g.*, BAILEY, 1984).

Due to very weak band intensity and a possible polarization $\alpha_Y \cong \alpha_Z \ll \alpha_x$, the 10Dq-band in RI Phlo could only be obtained from the E‖X spectrum, which bears on the accuracy of $CFSE_{Cr^{3+}}$ for this coexisting mineral. In case of the dioctahedral micas, $10Dq_{cub}$ was calculated from the baricenters for all three polarizations.

In summary—The spectral results on Cr^{3+} in the various coexisting minerals studied here, as well as the considerations about their evaluation, lead to values of $10Dq_{cub}$ and these to values of $CFSE_{Cr^{3+}}$ as listed in columns 2 to 4 of Table 3. $10Dq_{cub}$ values are estimated to be correct to within about ±30 cm^{-1}, corresponding to ±0.43 KJ/g-atom for the values of $CFSE_{Cr^{3+}}$, except for OPXCPX and RI Phlo as discussed before.

The choice of concentration unit

When Cr^{3+} partitions between different phases of a system, a common concentration unit, valid for all the phases, must be chosen. Thus, wt% Cr_2O_3 or mol% Cr^{3+}-end member in the respective solid solution are inappropriate as they relate the Cr^{3+}-concentration to the very phase just under consideration. Therefore, the concentrations must be recalculated to a unit volume, *e.g.* 1dm^3 or l, of all the phases studied. This is achieved by

$$C_{Cr^{3+}[g-atom/l]} = n_{Cr^{3+}}/V_M \qquad (9)$$

wherein $n_{Cr^{3+}}$ are Cr^{3+}-ions per formula unit of the respective crystallochemical formulae of the minerals and V_M their molar volumes in dm^3 as obtained from eqn. (7). The values of V_{el}, as obtained from the X-ray diffraction data, of V_M, $n_{Cr^{3+}}$ and, finally, of $c_{Cr^{3+}}$ in g-atom/l are compiled in columns 5 to 8 of Table 3. The error in $c_{Cr^{3+}}$, resulting from the uncertainties in the determination of V_{el} and of Cr^{3+}-analyses by microprobe, is given in parentheses.

CRYSTAL FIELD STABILIZATION AND INTERCRYSTALLINE PARTITIONING OF CR^{3+}—THE CONCLUSIONS

Figure 5 shows a plot of the uniformly normalized Cr^{3+}-concentrations, $C_{Cr^{3+}}$[g-atom/l], as a function of the crystal field stabilization energy, $CFSE_{Cr^{3+}}$, that the chromium ions gain in their respective structural matrix. From this figure, it is obvious that there exists a positive, quantitative correlation between the two quantities, as postulated by eqn. (5), the function f being of the same type for all the parageneses studied. The only exception, OPXCPX, may be explained by a value of CFSE that is too high in the case of orthopyroxene (see above). Crystallochemical peculiarities involved in the Cr^{3+}-substitutions based on $Ca_{M2}Mg_{M1} \ll Na_{M2}Cr_{M1}$ in clinopyroxene and on $M_{M1}Si_T \ll Cr_{M1}Al_T$ in orthopyroxene, whereby the additional presence of significant amounts of Fe^{2+} in the latter phase may also be of influence for the exceptional sample.

Calculating $K_{D,Cr^{+3}}$-values and plotting them as a function of the difference of the crystal field stabilization energies of the respective two phases Ph1 and Ph2 (Fig. 6), also shows positive correlations for the various parageneses, except for OPXCPX with a slightly negative K_D.

From this it is obvious that the data obtained show a quantitative, positive correlation between Cr^{3+} [g-atom/l] and $CFSE_{Cr^{3+}}$, as predicted by CFC, except for a clinopyroxene exsolution in orthopyroxene, a deviation that may be explained by possible errors of the $10Dq_{cub}$-determination and crystallochemical peculiarities of the substitutions in this case.

This proves the CFC concept to be valid, at least for Cr^{3+}-concentrations above the trace level.

Table 3. Crystal field splitting, $10Dq_{Cr3+}$, crystal field stabilization enrgy, $CFSE_{Cr3+}$, and concentration of chromium, c_{Cr3+} in [g-atom/l], in the parageneticminerals studied. V_{el}, V_M and n_{Cr3+} are the unit cell volume, molar volume and atoms per formula unit, respectively. [a] the extraction of 10Dq is discussed in the text. [b] The M(1,2,3) positions are taken into account (cf. text).

Crystals (cf. Tab 1)	$10Dq_{Cr3+}$[1] [cm^{-1}]	$CFSE_{Cr3+}$ [cm^{-1}]	$CFSE_{Cr3+}$ [KJ/g-atom]	V_{el} [A^3]	V_M [dm^3/mole]	n_{Cr3+}	c_{Cr3+} [g-atom/l]
Zag C1-Gt	17 030	20 435	244.5	1600.5(1.6)	0.1190(1)	0.093	0.77 (3)
Zag C1-Cpx	15 340	18 408	220.2	428 (4)	0.0644(6)	0.027	0.418(18)
Zag-C1.Ky	17 100	20 520	245.5	293.6 (4)	0.0442(1)	0.045	1.09 (9)
Zag G1-Gt	17 090	20 510	245.3	1600.5(1.6)	0.1190(1)	0.076	0.63 (3)
Zag G1-Cpx	15 265	18 318	219.1	428 (4)	0.0644(6)	0.023	0.36 (2)
Zag G1-Ky	17 145	20 579	246.1	293.6 (4)	0.0442(1)	0.042	0.94 (7)
MC1-A Gt				1544.1 (5)	0.1162(1)	0.001	
MC1-ACpx	15 435	18 522	221.6	424.2(1.3)	0.0639(2)	0.019	0.14 (2)
MC1-A Amph	15 200	18 240	218.2	899.4(1.5)	0.2708(5)	0.024	0.089(15)[b]
RI Ru	18 130	21 756	260.3	255.4	0.0256(1)	0.029	1.12 (6)
RI Parag	16 605	19 926	238.4	957 (5)	0.1441(8)	0.088	0.61 (4)
RI Phlo	16 695	20 034	239.3	1002 (4)	0.1509(6)	0.113	0.75 (4)
My Ru	18 140	21 768	260.4	254.0(2)	0.0255(1)	0.011	0.44 (2)
My Ky	17 150	20 580	246.6	293.0(4)	0.0441(3)	0.009	0.204(17)
My Fu	16 255	19 506	233.3	928.3(9)	0.1398(1)	0.014	0.100(4)
Lon Ru	18 100	21 720	259.8	255.1(1)	0.0265(1)	0.025	0.98 (6)
Lon Amph	15 550	18 636	222.9	904.2(1.2)	0.2723(3)	0.217	0.80 (3)
OPXCPX-Cpx	15 640	18 768	224.5	440 (10)	0.062(15)	0.046	0.70 (4)
OPXCPX-Opx	15 900	19 085	228.3	830 (10)	0.0625(7)	0.023	0.37 (5)

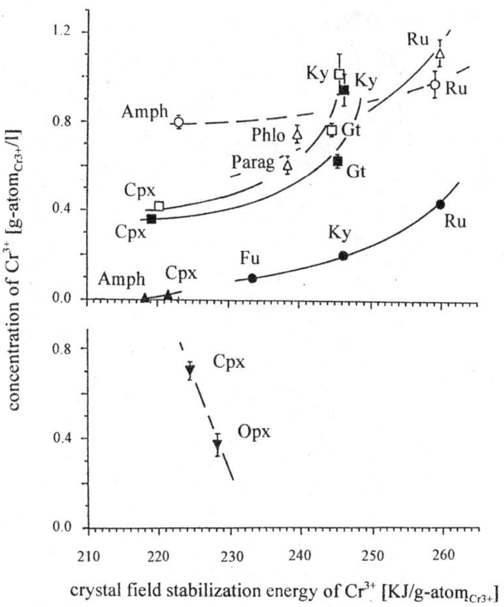

FIG. 5. Plot of the Cr^{3+}-concentration, normalized to the volume unit for all minerals studied (*cf.* text), $c_{Cr^{3+}}$ versus the $CFSE_{Cr^{3+}}$ in their structure. The different symbols represent the various parageneses studied: C1, G1, MC1-A, RI, My, Lon, OPXCPX (*cf.* Table 1); abbreviations as in Table 1.

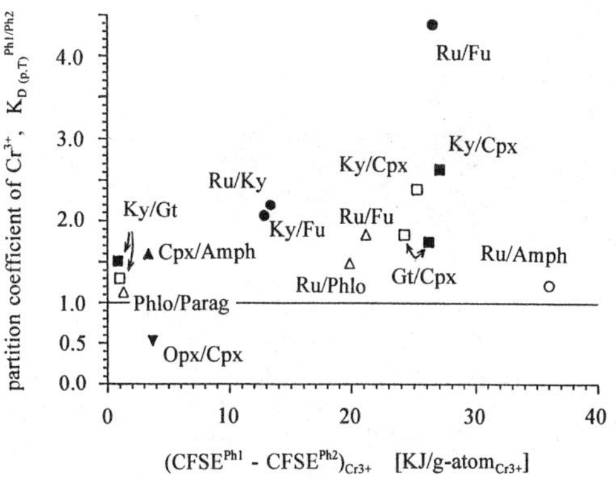

FIG. 6. Partition coefficient $K_{D,Cr3+,(p,T)}^{Ph1/Ph2}$ as a function of the difference of the crystal field stabilization energy of chromium in the respective two phases $CFSE^{(Ph1-Ph2)}$. Symbols and abbreviations as in Fig. 5.

Acknowledgments—Prof. A. N. Platonov, Kiev, kindly provided samples MC-1, RI, and OPXCPX; Prof. N. V. Sobolev, Novosibirsk, sample Z13 from Zagadochnaya; and S. Herting-Agthe, Berlin, samples My and Lo. Drs. Helfmeier and Galbert, ZELMI-laboratory of the Technische Universität Berlin helped in the microprobe analyses; H. Reuff, Berlin, carefully prepared slabs from oriented crystal grains extracted from the rock samples; and Prof. W. Depmeier, Kiel, made the Guinier-diffractometer available for the lattice constants determinations. Dr. M. Wildner, Vienna, helped by commenting on the energy level calculations and Profs. G. Franz and D. Lattard helped through discussions of geochemical and petrological problems. The Deutsche Forschungsgemeinschaft, Bonn-Bad-Godesberg, supported this study within the major research program ''Elementverteilung'' under grant no. La 324/31. To all of them, the authors' thanks are due.

REFERENCES

ABS-WURMBACH I., LANGER K. and OBERHÄNSLI R. (1985) Polarized absorption spectra of single crystals of the chromium bearing clinopyroxenes kosmochlore and Cr-aegirine-augite. *N. Jb. Miner. Abh.* **152,** 293–319.

ANDRUT M. (1995) Kristallfeldstabilisierungsenergie und interkristalline Verteilung von Übergangsmetallionen in silikatischen Paragenesen. Dr.-Dissertation, Technische Universität Berlin, 123 pp.

ARNI R., LANGER K. and TILLMANNS E. (1985) Mn^{3+} in garnets. III. Absence of Jahn-Teller distortion in synthetic Mn^{3+}-bearing garnet. *Phys. Chem. Minerals* **12,** 279–282.

BAILEY S. W. (1984) Crystal chemistry of the true micas. In *Micas* (ed. S.W. BAILEY). *Rev. in Mineral.* **13,** pp. 13–60, Mineralogical Society of America.

BLOSS F. D. (1981) *The spindle stage: Principles and practice.* Cambridge University Press, Cambridge.

BURNHAM C. W. (1963) Refinement of the crystal structure of kyanite. *Z. Kristallogr.* **118,** 337–360.

BURNS R. G. (1968) Enrichments of transition metal ions in silicate structures. In *Symposion on the origin and distribution of elements, Section 5: Terrestrial abundances* (ed. T.H. AHRENS). Pergamon Press, New York, 1151–1164.

BURNS R. G. (1970a) Crystal field spectra and evidence for cation ordering in olivine minerals. *Amer. Mineral.* **55,** 1608–1632.

BURNS R. G. (1970b) *Mineralogical applications of crystal field theory.* Cambridge University Press, Cambridge.

BURNS R. G. (1976) Partioning of transition metals in mineral structures in the mantle. In *The physics and chemistry of minerals and rocks* (ed. R. G. J. STRENS). Wiley, New York, 555–572.

BURNS R. G. (1993) *Mineralogical applications of crystal field theory.* Cambridge University Press, Cambridge, 2nd ed.

BURNS R. G., CLARK J. R. and FYFE W. S. (1964) Crystal field theory and applications to problems in geochemistry. In *Chemistry of the earthś crust* (ed. V. VINOGRADOV). *Proc. Vernadsky Centen. Symp.* **2,** 88–106.

BURNS R. G. and FYFE W. S. (1964) Site preference energy and selective uptake of transition metal ions during magmatic recrystallization. *Science* **144,** 1001–1003.

BURNS R. G. and FYFE W. S. (1966) Distribution of elements in geological processes. *Chem. Geol.* **1,** 49–56.

BURNS R. G. and FYFE W. S. (1967a) Trace element distribution rules and their significance. *Chem. Geol.* **2,** 89–104.

BURNS R. G. and FYFE W. S. (1967b) Crystal field theory and the geochemistry of transition elements. In *Researches in Geochemistry* (ed. P.H. ABELSON), Wiley, New York.

BURNS, R. G. and SUNG C. M. (1978) The effect of crystal field stabilization on the olivine-spinel transition in the system Mg_2SiO_4-Fe_2SiO_4. *Phys. Chem. Minerals* **2,** 349–364.
CAMERON M. and PAPIKE J. J. (1981) Structural and chemical variations in pyroxenes. *Amer. Mineral.* **66,** 1–50.
CLARK J. R., APPLEMAN D. E. and PAPIKE J. J. (1969) Crystal chemical characterization of clinopyroxenes based on eight new structure refinements. *Min. Soc. Amer., Spec. Pap.* **2,** 31–50.
COTTON A. F. (1971) *Chemical applications of group theory*. Wiley, New York.
CURTIS D. C. (1964) Applications of the crystal field theory to the inclusions of trace transition elements in minerals during magmatic differentiation. *Geochim. Cosmochim. Acta* **28,** 389–402.
DUNN T. M., MCCLURE D. S. and PEARSON R. G. (1965) *Some aspects of crystal field theory*. Harper & Row, New York.
FAYE G. H. (1968) Optical spectra of certain transition metal ions in muscovite, lepidocrocite and fuchsite. *Can. J. Earth Sci.* **5,** 31–38.
JENSEN B. B. (1973) Patterns of trace element partitioning. *Geochim. Cosmochim. Acta* **37,** 2227–2242.
KHOMENKO V. M. and PLATONOV A. N. (1985) Electronic absorption spectra of Cr^{3+}-ions in natural pyroxenes. *Phys. Chem. Minerals* **11,** 261–265.
KÖNIG E. and KRÄMER S. (1977) *Ligand field energy diagrams*. Plenum Press, New York.
LANGER K. (1976) Synthetic 3d-transition metal bearing kyanites $(Al_{2-x}M_x)SiO_5$. In *The physics and chemistry of minerals and rocks* (ed. R. G. J. STRENS). Wiley, New York, 389–402.
LANGER K. (1984) Die Farbe von Mineralen und ihre Aussagefähigkeit für die Kristallchemie. *Rhein. Westfäl. Akad. Wiss., Vortr.* N**332,** 7–60.
LANGER K. (1988) UV to NIR spectra of silicate minerals obtained by microscope spectrometry and their use in mineral thermodynamics and kinetics. In *Physical properties and thermodynamic behaviour of minerals* (ed. E. K. H. SALJE). Reidel, Dordrecht, 639–685.
LANGER K. and FRENTRUP K.R. (1979) Automated microscope-absorption-spectrophotometry of rock-forming minerals in the range 40,000–5,000 cm^{-1} (250–2,000 nm). *J. Microscopy* **116,** 311–320.
LANGER K. and SEIFERT F. (1971) High pressure-high temperature synthesis and properties of chromium kyanite, $(Al,Cr)_2SiO_5$. *Z. anorg. allg. Chem.* **383,** 29–39.
LEVER A. B. P. (1968) *Inorganic electronic spectroscopy*. Elsevier, Amsterdam.
MACFARLANE R. M. (1963) Analysis of the spectrum of $3d^3$ ions in trigonal crystal field. *J. Chem. Phys.* **39,** 3118–3126.
MCCLURE D. (1958) Electronic spectra of molecules and ions in crystals. *Solid State Phys.* **8,** 1–47.
MCCLURE D. (1962) Optical spectra of transition metal ions in corundum. *J. Chem. Phys.* **36,** 2757–2768.
MERLINO S. (1965) Applicazione delle teoria del campo cristallino allo studio della ripartizione di elementi in tracce. *Atti della Soc. Tosc. Sci. Nat., Ser. A,* **72,** 14 pp.
NOVAK G. A. and GIBBS G. V. (1971) The crystal chemistry of the silicate garnets. *Amer. Mineral.* **56,** 791–825.
PAPIKE J. J., ROSS M. and CLARK J. R. (1969) Crystal chemical characterization of clinoamphiboles based on five new structure refinements. *Min. Soc. Amer., Spec. Pap.* **2,** 117–136.
PARKIN K. M., LOEFFLER B. M. and BURNS R. G. (1977) Mössbauer spectra of kyanite, aquamarine and cordierite showing intervalence charge transfer. *Phys. Chem. Minerals* **1,** 301–311.
ROSSMAN G. R. (1980) Pyroxene spectroscopy. In: *Pyroxenes* (ed. C. T. PREWITT) *Rev. in Mineral.* **7,** pp. 91–115, Mineralogical Society of America.
SCHLÄFER H. L. and GLIEMANN G. (1980) *Einführung in die Ligandenfeldtheorie*. Akademische Verlagsgesellschaft, Frankfurt, 2nd ed.
SCHWARTZ H. P. (1976) The effect of crysatl field stabilization on the distribution of transition metals between metamorphic minerals. *Geochim. Cosmochim. Acta* **31,** 503–517.
SHIMURA Y. and TSUCHIDA R. (1956) Absorption spectra of Co(III) complexes. II. Redetermination of the spectrochemical series. *Bull. Chem. Soc. Japan* **31,** 311–316.
SMITH G. and STRENS R. G. J. (1976) Intervalence-transfer absorption in some silicate, oxide and phosphate minerals. In *The physics and chemistry of minerals and rocks* (ed. R. G. J. STRENS), Wiley, New York, 583–612.
STRENS R. G. J. (1968) Stability of the Al_2 SiO_5 solid solutions. *Mineral. Mag.* **36,** 839–849.
TANABE Y. and SUGANO S. (1954a) On the absorption spectra of complex ions. I. *J. Phys. Soc. Japan* **9,** 753–766.
TANABE Y. and SUGANO S. (1954b) On the absorption spectra of complex ions. II. *J. Phys. Soc. Japan* **9,** 767–779.
TARAN M. N., LANGER K., PLATONOV A. N. and INDUTNY V. V. (1994) Optical absorption investigation of Cr^{3+}-bearing minerals in the temperature range 77–797 K. *Phys. Chem. Minerals* **21,** 360–372.
TRÖGER W. E. (1952) *Tabellen der gesteinsbildenden Minerale*. Schweizerbart, Stuttgart.
TSUCHIDA R. (1938a) Absorption spectra of coordination compounds. I. *Bull. Chem. Soc. Japan* **13,** 388–400.
TSUCHIDA R. (1938b) Absorption spectra of coordination compounds. II. *Bull. Chem. Soc. Japan* **13,** 436–450.
TSUCHIDA R. (1938c) Absorption spectra of coordination compounds. III. Special bands of Cr^{3+}-complexes. *Bull. Chem. Soc. Japan* **13,** 471–480.
WHITE E. W. and WHITE W. B (1967) Electron microprobe and optical absorption study of colored kyanites. *Science* **158,** 915–917.
WILLIAMS R. J. P. (1959) Deposition of trace elements in basic magma. *Nature* **184,** 44.
WILSON E. B., DECIUS J. C. and CROSS P. C. (1955) *Molecular vibrations - The theory of infrared and Raman spectra*. McGraw-Hill, New York.

Mineral Spectroscopy: A Tribute to Roger G. Burns
© The Geochemical Society, Special Publication No. 5, 1996
Editors: M. D. Dyar, C. McCammon and M. W. Schaefer

Cr^{3+} in phyllosilicates: Influence of the nature of coordinating ligands and their next cationic neighbors on the crystal field parameters

A. N. PLATONOV[1], K. LANGER[2], M. ANDRUT[2,3], G. CALAS[4]

[1]Institute of Geochemistry, Mineralogy and Ore Formation, Academy of Science of Ukraine, 252680 Kiev, Ukraine
[2]Institute of Mineralogy and Crystallography, Technical University, D-10623 Berlin, Germany
[3]GeoForschungszentrum Potsdam, D-14473 Potsdam, Deutschland
[4]Laboratoire de Minéralogie et de Cristallographie, Universitè de Paris 6 et 7, F-75251 Paris, France

Abstract—The electronic absorption spectra of Cr^{3+}-bearing clinochlore (I, kämmererite), amesite (II), muscovite (III, fuchsite), dickite (IV), and montmorillonite (V, volkonskite) analysed by electron microprobe were obtained on single crystals. Microscope-spectrometric techniques and polarized radiation in the spectral range 10000-38000 cm^{-1} (I, II, III) or (on fine grained material) diffuse reflectance spectrometry in the spectral range 8000-50000 cm^{-1} (IV, V) were used. The ligand field theoretical evaluation of the spectra showed the following: (i) The $\Delta_o = 10Dq = f(1/R^5)$ relation, wherein Δ_o is the octahedral crystal field parameter and R the mean cation ligand distance, is valid within each series of layer silicates containing octahedral Cr^{3+} either in a trioctahedral layer (I, II and phlogopite) or in a dioctahedral layer (III, IV, V). Between the two functions, $\Delta_{o,trioct} = f(1/R^5_{trioct})$ and $\Delta_{o,dioct} = f(1/R^5_{dioct})$, there exists an energy difference of about 2200 cm^{-1}. (ii) There is no support for the sequence of the effective charges of $OH^- < O^{2-}$ as suggested by the spectrochemical series. Quite the contrary, Δ_o decreases in the sequence $Cr(OH)_6$ (I), $CrO_2(OH)_4$ (II), $CrO_4(OH)_2$ (phlogopite). (iii) A negative higher-order correlation between Racah parameter B and crystal field parameter Δ_o exists, indicating an increase in covalency of the Cr-(O,OH) bond with increasing strength of the crystal field, wherein the dioctahedral t-o-t phases represent lowest Δ_o and highest B. This is interpreted to result from next-neighbor effects on the $(Si_{1-x}Al_x)$-coordinated apical oxygens forming, together with 2 OH^-, the octahedral coordination in the o-layers of these phases.

INTRODUCTION

THE LIGAND FIELD theoretical evaluation of electronic absorption spectra of $3d^N$ transition metal ion-bearing, oxygen based minerals (*e.g.*, LEVER, 1984; SCHLÄFER and GLIEMANN, 1980) reveals the octahedral crystal field parameter $\Delta_o = 10Dq$, possibly parameters describing the distortion of the field from cubic symmetry (KÖNIG and KREMER, 1977), and the Racah parameters B and C, taking into account the interelectronic repulsion of the d-electrons (TANABE and SUGANO, 1954a,b).

The octahedral crystal field parameter $\Delta_o = 10Dq$, which allows for the calculation of the crystal field stabilization energy of the respective $3d^N$-ion in the respective structural site or sites (*e.g.*, BURNS, 1970, 1993), is theoretically derived from the point charge model of the crystal field theory as

$$10_{Dq} = \frac{5}{3}\frac{Z_L \times e^2}{R^5}\langle r^4\rangle \qquad (1)$$

(*e.g.*, DUNN *et al.*, 1965; LEVER, 1968). Herein, Z_L is the effective charge of the ligands, R is the mean central ion-ligand distance and $\langle r^4\rangle$, the mean value of the forth power of the radial distance of the 3d-orbitals from the nucleus.

Equation 1 is appropriate to determine, from spectroscopically measured 10Dq-values, mean distances between central $3d^N$-ion and its ligands in various oxygen-based mineral structures. Such distances are local values, which are valid for the $3d^N$-ion containing site or, in the case of substitutive Cr^{3+}, of that part of the respective crystallographic site where the substitution inside the structure really occurs. This procedure presupposes that the same transition metal ion is considered and that the product $(Z_L \cdot e^2)\cdot\langle r^4\rangle$ is constant. The latter assumption is, at least in a first approximation, fulfilled for $3d^N$-ion containing solid solutions, *i.e.* in one and the same structural type. Indeed, it has been shown that mean local octahedral $3d^N$-oxygen distances can be obtained from spectra of intermediate members of solid solution series (LANGER, 1988).

On the other hand, the above approach is likely to fail when comparing different structure types, because the product $(Z_L \cdot e^2)\cdot\langle r^4\rangle$ is not constant. This is indicated by observations on the Racah B vs. Cr—O distances of various minerals (ABU-EID and BURNS, 1976) as well as by a study of some chromium-bearing minerals from alterated rocks by EXAFS and electronic spectroscopy (CALAS *et al.*, 1984). Recently, the influence of next nearest cations on the effective charge of the ligands, *i.e.* on $Z_L \cdot e^2$, was qualitatively demonstrated for the case of chromium-bearing amesite (PLATONOV *et al.*, 1995).

It is the aim of the present paper to study the influence of changing effective charge of the ligands on the crystal field parameters and the degree of covalency of the ligand-central ion bonds. For this purpose we obtained and evaluated the electronic absorption spectra of a series of Cr^{3+}-bearing sheet silicates with structures that are especially appropriate for the following reasons:

(i) The phyllosilicates, whether they are of the 2:1-(T—O—T)-type or of the 1:1-(T—O)-type, provide octahedral layers isolated from each other and interconnected with T-layers only. This allows for the study of influences of the cationic neighbours of the second coordination sphere on the coordinating ligands of an octahedron under consideration.

(ii) The coordination octahedra within the O-layers of different phyllosicates provide $[O_{6-n}(OH)_n]$ ligands where n_{OH} may be 2 in 2:1-minerals, 4 in 1:1-minerals, and 6 in the interlayer sheets of chlorites.

(iii) The site occupancies in the O-layers may be different. Thus there exist a variety of different next neighbor relations for the ligands of an octahedron under consideration.

SAMPLES AND METHODS

The following samples were studied:

(1) *Chromian clinochlore (kämmererite)*, reddish-purple, mm-sized, clear and inclusion-free single crystals approximately 100 μm thick, from the Erzincan deposit, Turkey (BAILEY, 1986) with perfect cleavage parallel (001) and weak pleochroism in purple hues.

(2) *Chromian amesite:* Small, clear, lilac crystals from Sobótka deposit, Poland (PLATONOV *et al.*, 1995).

(3) *Chromian muscovite (fuchsite),* clear, emerald green, platy crystals, similar in size and thickness as the kämmererite sample, from Murun-keu eclogites, Arctic Urals, Russia (UDOVKINA, 1971).

(4) *Chromian dickite,* fine-grained aggregate of semitransparent quartz crystallites and powdery bluish-green dickite in 3.5:1 proportion (pseudo-jade) from Chinlon deposit, China (CAO YUNGCHENG, 1983).

(5) *Volkonskoite (chromium montmorillonite),* deep-green, fine-grained aggregate from Urals, Russia (YUSHKIN *et al.,* 1986).

Microprobe analyses

Microprobe data are presented in Table 1. They were obtained on the Camebax Microbeam in the ZELMI-laboratory at the Technical University of Berlin using the following standards: natural quartz (Si), wollastonite (Ca), rutile (Ti), spinel (Al), olivine or spinel (Mg), albite (Na) and metallic iron or chromium (Fe or Cr). The microprobe beam diameter was 2 μm with acceleration voltage of 15 kV and beam current of 20 nA. The chromium-bearing pseudo-jade was analyzed by OES-ICP. Because this material is not single phase, the data obtained are not included in Table 1. Its essential feature for the present purpose is the chromium content, which is calculated to be near 0.09 wt% Cr from the analytical data and the quartz to dickite proportion (see above).

Preparation of samples and spectroscopic measurements

Where grain sizes were large enough, *i.e.* in case of clinochlore, amesite and muscovite, individual crystals could be oriented and prepared for polarized single crystal spectroscopy. In these cases, the crystals were oriented by means of spindle stage methods, embedded in the oriented position and subsequently ground and polished on two opposite sides to produce slabs on which up to two polarizations could be measured. Slabs parallel to (001) were prepared for all three minerals mentioned and additional slabs parallel to (001) were prepared in case of clinochlore and muscovite. Hence, for the two latter minerals spectra with E‖X (α-spectra), E‖Y (β-spectra) and E‖Z (γ-spectra) could be recorded. In case of amesite, only spectra parallel and perpendicular to the c-axis were obtained.

The single crystal spectra in the spectral range 38000 to 12000 cm^{-1} were scanned at room temperature in a microscope-spectrometer (Zeiss UMSP-80) using UV-transparent optics (Zeiss Ultrafluars 10×) and a Glan-Thompson-type calcite prism polarizer. Entrance and measuring apertures had effective diameters of 30 and 21 μm, respectively. The spectral slit width was 1 nm and the step width was also 1 nm. The reference I_o-spectrum was taken in air. Sample spectra were averaged from 20 scans.

In the case of dickite and volkonskoite, where no single crystal spectra could be obtained, powder reflectance spectra were scanned in the range 50000 to 10000 cm^{-1} using a Perkin-Elmer Lambda 19 spectrometer equipped with an integrating sphere covered with teflon as the reflecting material. The compact natural sample was ground and polished on one side before measurement. Teflon powder served as a white standard. Measuring steps were 0.1 nm, the spectral resolution was between 0.05 an 5 nm and the wavelength accuracy was 0.15 and 0.6 nm in the UV/VIS and NIR, respectively.

Evaluation of the spectra

The band characteristics in the polarized single crystal spectra, especially the position of the band maxima, were obtained by applying peak fitting procedures (program PeakFit, Jandel Scientific, 1991) assuming a Gaussian shape for the component bands. The energy values for the band maxima of the various polarization directions were averaged. The accuracy of the average values obtained is better than 50 cm^{-1}. Energy positions of band maxima in powder reflectance spectra were obtained by the same fitting procedures, resulting in the accuracy of band positions mentioned above.

RESULTS AND DISCUSSION

The crystal chemistry of the minerals under study (Table 1) is typically characterized by their chromium content, while other substituting transition elements are low in concentration. The polar-

Table 1. Composition of chromian phyllosilicates studied as obtained by electronmicroprobe analyses given in wt% of the oxide components and in atoms per formula units. n.d. = below the limit of detection. t = tetrahedral layer, o = octahedral layer, i.l. interlayer

Oxide components [wt.%]	Clinochlore (kämmererite)		Amesite		Muscovite (fuchsite)		Montmorillonite (volkonskoite)	
SiO_2	33.81		21.70		49.49		47.09	
TiO_2	n.d.		n.d.		0.89		n.d.	
Al_2O_3	8.08		32.15		30.10		3.10	
Fe_2O_3 tot.	1.37		0.46		1.18		0.97	
Cr_2O_3	7.47		3.37		0.55		21.05	
MgO	35.02		27.03		3.31		8.55	
CaO	n.d.		n.d.		n.d.		4.21	
Na_2O	n.d.		n.d.		0.85		n.d.	
K_2O	n.d.		n.d.		10.06		0.19	
Total	85.60		84.67		96.27		85.05	
Basis	28 neg. charges		14 neg. charges		22 neg. charges		22 neg. charges	
Si	3.29	} 4.00^t	1.05	} 2.00^t	3.26	} 4.00^t	4.04^t	
Al – $Al^{[4]}$	0.71		0.95		0.74			
$Mg^{[6]}$	3.00^o							
$Al^{[6]}$	0.21	} $2.99^{i.l.}$	0.88	} 2.98^o	1.60	} 2.06^o	0.31	} 2.67^o
Ti	<0.01		<0.01		0.04		<0.01	
Fe^{3+}	0.12		0.02		0.06		0.07	
Cr^{3+}	0.58		0.13		0.03		1.43	
Mg	2.08		1.95		0.33		0.86	
Ca	<0.01		<0.01		<0.01	} $0.95^{i.l.}$	0.39	} $0.60^{i.l.}$
Na	<0.01		<0.01		0.11		<0.01	
K	<0.01		<0.01		0.84		0.21	

ized electronic absorption spectra of Fig. 1 as well as the unpolarized spectra in Fig. 2 show the two strong bands, v_1 and v_2, in the visible range. They are typical of dd-transitions of octahedral Cr^{3+} and are derived from the transitions $^4A_{2g}$-$^4T_{2g}$ (F) and $^4A_{2g}$-$^4T_{1g}$ (F), respectively, under point symmetry O_h. The unpolarized reflectance spectra of Fig. 2 show an additional broad band in the UV, attached to the UV-absorption edge. This band is displayed in the polarized spectra of amesite only. The reason is that the UV above about 33000 cm^{-1} is masked in the other spectra by absorption in the embedding epoxy resin. A peculiarity of the muscovite spectra is the position of the UV edge at relatively low energies compared to the spectra of the other minerals. This may be due to the Fe^{3+} content of this sample (Table 1).

The polarization of the two bands, v_1 and v_2, displayed in Fig. 1 is caused by symmetry related selection rules of the respective transitions of Cr^{3+} in the octahedra of low site symmetry in the structures of the minerals under study. Because a true low-symmetry ligand-field treatment is currently possible only in special sases (KÖNIG and KREMER, 1977; *cf.* also WILDNER and LANGER, 1994) and because the actual or effective octahedral site symmetries in the mineral structures involved here are lower than orthorhombic, our interpretation is based on the commonly used approximation of the cubic crystal field parameter Δ_o = 10Dq. In this procedure, the averages of band maxima in the different polarizations are determined (see above). Crystal field parameters obtained by this approximation are slightly lower than the correct ones (1.3% in corundum, *cf.* LANGER and ANDRUT, 1996). This will not influence the conclusions to be drawn in this paper, as the slight error introduced is expected to be nearly the same in all the phases studied here. However, for these reasons, the band polarizations observed in Fig. 1 will not be evaluated in detail.

The values obtained are listed in Table 2, along with the relevant crystallochemical characteristics. The band positions listed are used to calculate the parameters quoted in the right part of the table, which characterize octahedral Cr^{3+} in the respective structural matrix. The cubic crystal field parameter Dq is equal to $^1/_{10}$ of the energy of ν_1. The Racah parameter B is derived from ν_1 and ν_2 using the respective solutions of the secular determinant

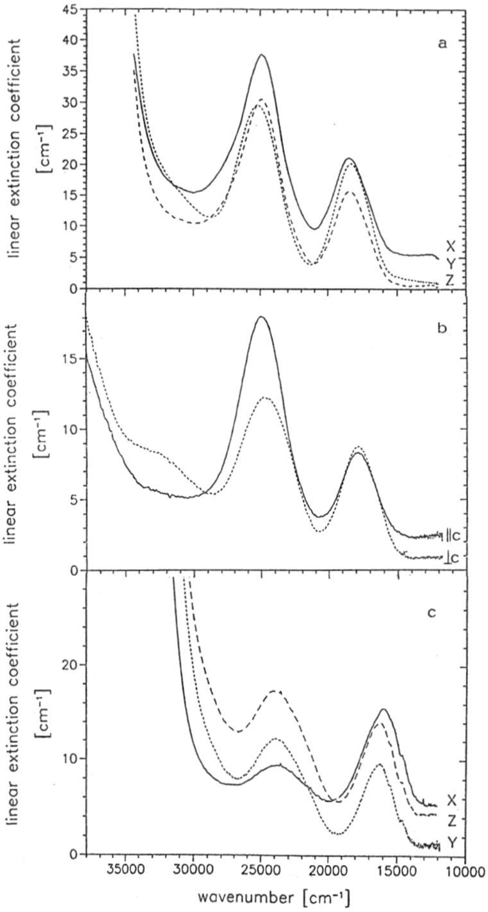

FIG. 1. Electronic single crystal absorption spectra of chromium-bearing clinochlore (a), amesite (b) and muscovite (c). The orientation of the electrical vector of the measuring radiation is given at the respective absorption spectrum. Measurements are performed at ambient conditions. For compositions of samples see Table 1.

(*cf., e.g.,* LEVER, 1984). The quantity of B, taking into account the interelectronic repulsion of the d-electrons of Cr^{3+}, is a measure of the degree of covalency of the chromium-oxygen bond. To characterize differences in covalency, it is general practice to calculate the nephelauxetic ratio $\beta = B_{ion,cryst}/B_{ion,free}$. The free ion value, $B_{ion,free}$, for Cr^{3+} adopted here is 918 cm^{-1} (TANABE and SUGANO, 1954a). Finally, the table compiles the crystal field stabilization energy of Cr^{3+} in octahedral coordination in the minerals listed.

Using the values of the crystal field parameters Dq and B quoted in Table 2, it is possible to calculate the energy of a third spin-allowed transition of octahedral Cr^{3+}, $^4A_{2g}$-$^4T_{1g}$ (P), which is expected in the UV and possibly causes the aforementioned broad band in this spectral region. The calculation is done by using the respective solution of the secular determinant for the excited state (*cf., eg.,* LEVER, 1984). The band positions as observed in the spectra of amesite (32400 cm−1) of dickite (37300 cm^{-1}) and of volkonskoite (38300 cm−1) compare well with the values calculated for the above transitions to the excited crystal field state derived from ^4P, which are 33900, 37200 and 36200 cm^{-1}, respectively. The fair agreement between calculated and observed values indicates that the above assignment is correct. It might be mentioned here that this transition has been to date observed in only a few chromium-bearing minerals.

The octahedral first coordination sphere of the central Cr^{3+}-ions is different in three aspects in the structures of the various minerals studied: (i) the mean Cr^{3+}-(O,OH) distances, R, in the structures, (ii) the ratio of oxygen and hydroxyl per octahedron, (iii) the next neighbor situation of the ligands as created by the dioctahedral or trioctahedral nature and the cationic population of the octahedral sites by cations other than Cr^{3+}.

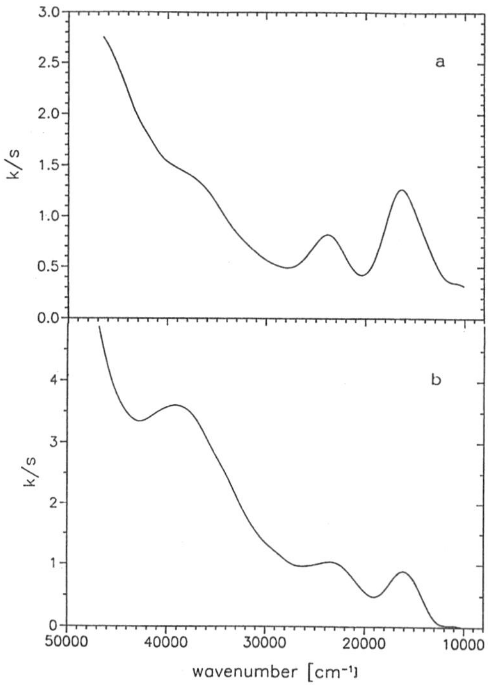

FIG. 2. Electronic absorption spectra, scanned as diffuse reflectance spectra, of chromium-bearing dickite (a) and volkonskoite (b). k/s is the solution of the KUBELKA-MUNK-function, $F(R_\infty)$, wherein k = absorption coefficient and s = scattereing coefficient. For compositions of samples see Table 1.

Table 2. Crystallochemical aspects of phyllosilicates and of Cr^{3+}-accomodation therein, as well as related spectroscopic parameters of this $3d^N$-ion accomodated in octahedra of phyllosilicate structures. Abbreviations: t = tetrahedral layer, o = octahedral layer attached to t, $o^{i.l.}$ = octahedral interlayer (''brucite type'' layer $[M(OH)_6]$). The uncertainty in ν_1, 10Dq and CFSE is ca. + 40 cm^{-1} or +0.5 kJ/g-atom, resp., in v_2 ca. + 80 cm^{-1} and in B ca. + 20 cm^{-1}

Mineral	Type of layering	Cr^{3+} in octahedra of	Type of Cr^{3+}-containing coordination octahedra	Mean octahedral distance, R_{M-O} [A], R_{Cr-O}, EXAFS R_{M-O}, XRD	Cr_2O_3 content, wt%	Spectroscopic parameters of Cr^{3+}[6] ν_1 cm^{-1}	ν_2 cm^{-1}	10Dq cm^{-1}	B cm^{-1}	β	CFSE cm^{-1}	kJ/g-atom
Chromian clinochlore (kämmererite)	t-o-t, $o^{i.l.}$	$o^{i.l.}$, trioctahedral $[M_2^{2+}M^{3+}]$	$[M(OH)_6]$	1.97[1] 1.963[2]	7.47	18340	25170	18340	668	0.728	22010	263.8
Chromian amesite	t-o	o, trioctahedral $[M_2^{2+}M^{3+}]$	$[MO_2(OH)_4]$	— 2.03[3]	3.90	17820	24810	17820	694	0.756	21380	256.3
Chromium dickite	t-o	o, dioctahedral $[M_2^{3+}\square]$	$[MO_2(OH)_4]$	— 1.904[4]	0.09[8]	16480	23870	16480	772	0.841	19780	237.1
Chromian smectite	t-o-t	o, dioctahedral $[M_2^{3+}\square]$	$[MO_4(OH)_2]$	1.96[1] —	17.5[1]	16800[1]	24000[1]	16800	738	0.804	20160[1]	241.6
Chromian muscovite (fuchsite)	t-o-t	o, dioctahedral $[M_2^{3+}\square]$	$[MO_4(OH)_2]$	1.97[1] 1.95[5]	0.55	16200	23880	16200	826	0.899	19440	233.0
Chromian montmorillonite (volkonskoite)	t-o-t	o, di-/trioctahedral $[(M^{3+}, M^{2+})_{3-x}\square_x]$	$[MO_4(OH)_2]$	— 2.00[6]	21.05	16100	23200	16100	737	0.802	19320	231.6
Chromian phlogopite	t-o-t	o, trioctahedral $[M_3^{2+}]$	$[MO_4(OH)_2]$	— 2.07[3]	1.24[9]	16900[9]		16900[9]			20280[9]	243.1

[1]CALAS et al. (1984), [2]PHILLIPS et al. (1980), [3]WIEWIORA et al. (1991), [4]JOSWIG and DRITS (1986), [5]RULE and BAILEY in: BAILEY (1984), [6]this mean distance was estimated from $Cr^{[6]}$-O distances obtained by XRD refinements of a series of Cr-minerals near to end member compositions (e.g. eskolaite, magnesiochromite, uvarovite) on the basis of the fact that R_{M-O} in the respective (CrO_6)-octahedra themselves does not change much in series of solid solutions (LANGER, 1988; LANGER and GARSCHE, in prep.), [7]BAILEY (1984, Tab. 1), [8]CAO YUNGCHENG (1983), [9]KHOMENKO (personal communication, 1992).

Table 2 includes information on these characteristics for the minerals studied here. With respect to mean distances R, it should be pointed out that the values obtained in the refinement of X-ray diffraction data are crystal averaged and do not necessarily equal the local Cr—O distances. This is also indicated by the larger Cr—O EXAFS distances compared to the X-ray values in case of clinochlore and muscovite.

In Fig. 3, the values of the octahedral crystal field parameter Δ_o = 10Dq (Table 2) are plotted as a function of $1/R^5$ of eqn. (1). Where available, X-ray spectroscopic data on octahedral chromium-ligand distances were used to obtain $1/R^5$ (*cf.* Table 2). Where such local octahedral distances were not available, mean distances from X-ray diffraction refinements were used. We are aware that the local Cr-(O,OH) distances may be larger, in these cases, than quoted, because X-ray diffraction averages the distances in all the octahedra of one crystallographic type of site (*cf.* LANGER, 1988). It is obvious from Fig. 3 that there is indeed a $1/R^5$-proportionality of Δ_o. However, this is only valid when we are within one series of layer silicates, either the trioctahedral or the dioctahedral series. Between these two, there exists an extended gap of Δ_o. Thus, clinochlore and muscovite, with the same mean octahedral M^{3+}-O EXAFS distances of 1.97 A, show a difference in Δ_o of 2140 cm^{-1}, *i.e.* of more than 10%! In smectite only, the Δ_o-value of Cr^{3+} (CALAS *et al.*, 1984) is anomalously high. The reason for this is not known.

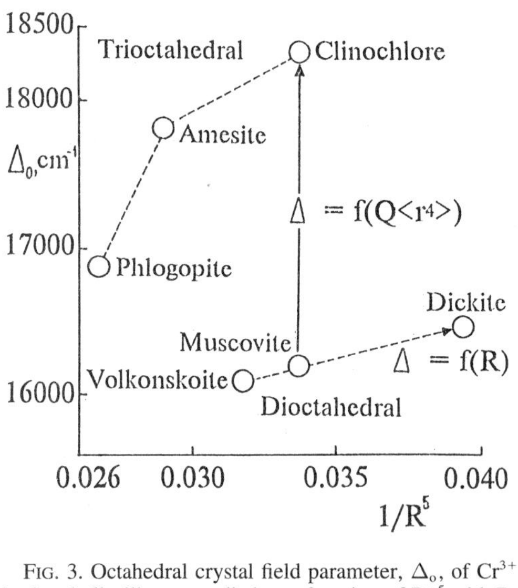

FIG. 3. Octahedral crystal field parameter, Δ_o, of Cr^{3+} in the phyllosilicates studied as a function of R^{-5} with R, the mean octahedral distance (see text). Note the large energy gap Δ between crystal field parameters in trioctahedral vs. dioctahedral phases. This gap is suggested to be caused by differences in the mean effective charge in case of trioctahedral members as compared to dioctahedral ones.

This gap is obviously due to differences in the effective charge of the ligands, $Z_L \cdot e^2$ (*cf.* eqn. 1), coordinating the Cr^{3+}-ions in the structure of the minerals of the two series. The effective charges of hydroxyl and oxygen are different and, thus, OH is expected to create a weaker field than O^{2-} according to the spectrochemical series (BURNS, 1993), such that the Δ_o-values are expected to decrease in the sequence $CrO_4(OH)_2 > CrO_2(OH)_4 > Cr(OH)_6$ in accordance with point (ii) above. No such correlation is clearly demonstrated by the data in Table 2 and Fig. 3. Obviously, point (iii) predominantly accounts for the observed gap. This is consistent with the fact that Δ_o is generally lower in the dioctahedral minerals compared to the trioctahedral ones. Furthermore, the same ratio of hydroxyl to oxygen in the coordination sphere of chromium produces higher crystal field strength in trioctahedral than in dioctahedral minerals as is obvious from the pairs phlogopite/muscovite and amesite/dickite (*cf.* Tab. 2 and Fig. 3).

The second parameter dependent on the effective charge of the coordinating oxygens is the Racah parameter B and the nephelauxetic ratio, β, derived from it (see above). It was proposed (ABU-EID and BURNS, 1976; BURNS, 1993) that the value of B is larger in hydroxyl-bearing minerals than in those with O^{2-} as the coordinating ligands. The argument for this ist that the electron density at OH^- is small compared to O^{2-}, such that the orbital overlap with the metal d-orbitals is smaller in the first case. This is also consistent with the nephelauxetic series (BURNS, 1993), from which a decrease in B and β is expected in the sequence $Cr(OH)_6 > CrO_2(OH)_4 > CrO_4(OH)_2$. However, the values of B and β derived from the spectra of the present study and listed in Table 3 show that the degree of covalency of the chromium-oxygen bond is higher in $Cr(OH)_6$ octahedra of clinochlore than in $CrO_4(OH)_2$ octahedra in muscovite. This shows that the effective charge on the oxygens attached to Cr^{3+} in the octahedral layer of muscovite is even smaller than that of hydroxyl ligands in the clinochlore interlayer. The reason may be that the four oxygen ligands per coordination polyhedron in the octahedral layer of muscovite and of the other 2:1-phyllosilicates, being the apices of the tetrahedral layers, are coordinated by tetrahedral $(Si_{4-x}Al_x)$. On the other hand, the hydroxyl ligands of the octahedral interlayer in clinochlore are not subject to such influence. That this consideration is likely to explain the unexpected behavior of B and β, is obvious from the fact that B and β in the 1:1 layer silicates, containing only two such oxygen ligands per coordination octahedron, have values between those in clinochlore and muscovite (Table 2).

Increasing covalency of the chromium-oxygen bond, as reflected by decreasing B- and β-values of $Cr^{3+[6]}$, is expected to be the result of increasing strength of the ligand field, *i.e.* of increasing crystal field parameter Δ_o = 10Dq. Hence, there should exist a negative correlation between Δ_o and B. It is obvious that this is really the case in the minerals studied here. Fig. 4 quite clearly shows that the strength of the field, as reflected in Δ_o = 10Dq, greatly determines the degree of covalency of the Cr^{3+}-(O,OH)-bonds as reflected in the values of Racah-parameter B. Only the data of volkonskite do not fit into the correlation of Fig. 4. Here, Δ_o is too low possibly as a result of the exceptionally high chromium concentration (Table 1) and the relatively large chromium-oxygen distance resulting from that.

CONCLUSION

The $1/R^5$-proportionality of the octahedral crystal field parameter Δ_o = 10Dq is confirmed for octahedral chromium in phyllosilicates only when either trioctahedral or dioctahedral layers are considered. Between the $\Delta_o = f(1/R^5)$-relations for the tri- or di-octahedral series, respectively, there occurs a broad energy gap which is probably due to next-neighbor influences on the donor properties of the coordinating ligands. The nature of O^{2-} or OH^-, seems to play a minor role, an unexpected result when considering the position of the two types of ligands within the spectrochemical series (BURNS, 1993). Racah-parameter B was found to be negatively correlated with Δ_o = 10Dq, a result that clearly demonstrates an increase of the covalency of the Cr^{3+}-(O,OH)-bonds with increasing strength of the crystal field.

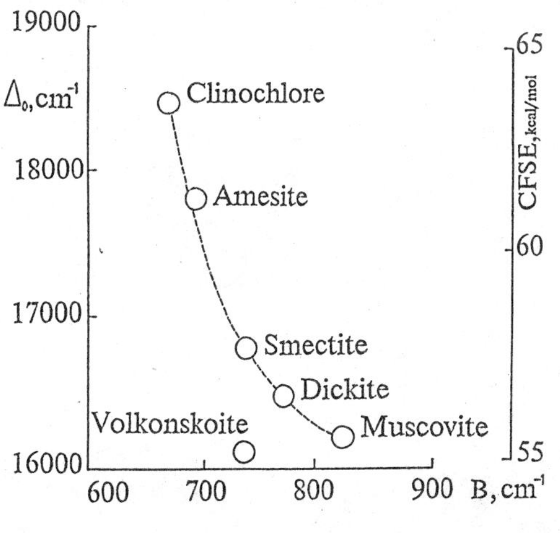

FIG. 4. Relationship between the octahedral crystal field parameter, as well as the interrelated crystal field stabilization energy, and the Racah parameter B of Cr^{3+} in the minerals studied. Remember that decreasing B indicates increasing covalency of the Cr-(O,OH) bond.

Acknowledgements—This work was made possible by financial support provided by INTAS, Brussels, project no. 94-2428, and by travel grants provided by the Deutsche Forschungsgemeinschaft and the Alexander von Humboldt-Stiftung, both Bonn-Bad Godesberg. Prof. I. Abs-Wurmbach, Berlin, introduced and helped in scanning the reflectance spectra and Dr. H. Galbert, ZELMI laboratory Technische Universität Berlin, helped with the microprobe analyses. V. M. Khomenko, Kiev, made available unpublished data on chromium-bearing phlogopite. To all these persons and institutions our thanks are due.

REFERENCES

ABU-EID R. M. and BURNS R. G. (1976) The effect of pressure on the dgree of covalency of the cation oxygen bond in minerals. *Am. Mineral.* **61,** 391–397.

BAILEY S. W. (1984) Crystal chemistry of the true micas. In *Micas* (ed. S. W. BAILEY) *Reviews in Minerology* **13,** 13–59.

BAILEY S. W. (1986) Reevaluation of ordering and local charge balance of Ia chlorite. *Can. Mineral.* **24,** 649–654.

BURNS R. G. (1970) Mineralogical Applications of Crystal Field Theory. Cambridge University Press, 1st ed.

BURNS R. G. (1993) Mineralogical Applications of Crystal Field Theory. Cambridge University Press, 2nd ed.

CALAS G., MANCEAU A., NOVIKOFF A. and BOUKILI H. (1984) Compartment du chrome dans les mineraux d'altération du gisement de Campo Formoso (Bahia, Brésil). *Bull. Minéral.* **107,** 755–766.

CAO YUNGCHENG (1983) Acta Mineral. *Sinica* **3,** 183–191.

DUNN T., MCCLURE D. S. and PEARSON R. G. (1965) Some Aspects of Crystal Field Theory. Harper and Row, New York.

JOSWIG W. and DRITS V. A. (1986) The orientation of the hydroxyl groups in dickite by X-ray diffraction. N. Jb. Mineral., Mh. 19–22.

KÖNIG E. and KREMER K. (1977) Ligand Field Energy Diagrams. Plenum Press, New York and London.

KORTÜM G. (1969) Reflexionsspektroskopie. Springer. Berlin, Heidelberg and New York.

LANGER K. (1988) UV to NIR spectra of silicate minerals obtained by miccroscope spectrometry and their use in mineral thermodynamics and kinetics. In: Salje E. K. H. ed.: Physical Properties and Thermodynamic behaviour of Minerals. Reidel, Dordrecht 639–685.

LANGER K. and ANDRUT M. (1996) The crystal field concept, CFC, in geosciences: Does the crystal field stabilization energy of Cr^{3+} rule its intercrystalline partitioning behaviour? In *Mineral Spectroscopy: A tribute to* ROGER *G.* BURNS. (eds. M. D. DYAR, M. W.

SCHAEFER and C. MCCAMMON) Geochem. Soc. Spec. Pub. No. 5, 29–40.

LEVER A. B. P. (1984) Inorganic Electronic Spectroscopy. Elsevier, Amsterdam, 2nd ed.

PHILLIPS T. L., LOVELESS J. K. and BAILEY S. W. (1980) Cr^{3+} coordination in chlorites: a structural study of ten chromian chlorites. *Am. Mineral.* **65,** 112–122.

PLATONOV A. N., LANGER K., WIEWIORA A. and ANDRUT M. (1995) Electronic absorption spectra of chromium-bearing amesite. *Eur. J. Mineral.* **7,** 961–965.

SCHLÄFER H. L. and GLIEMANN G. (1980) Einführung in die Ligandenfeldtheorie. Akademische Verlagsgesellschaft, Frankfurt/M, 2nd ed.

TANABE Y. and SUGANO S. (1954a) On the absorption spectra of complex ions. I. *J. Phys. Soc. Japan* **9,** 753–779.

TANABE Y. and SUGANO S. (1954b) On the absorption spectra of complex ions. II. *J. Phys. Soc. Japan* **9,** 767–779.

UDOVKINA N. G. (1971) Eclogites of the Arctic Urals. Nauka, Moscow (in Russian)

WIEWIORA A., RAUSELL-COLOM J. A. and GARCIÁ-GONZÁLES T. (1991) The crystal structure of amesite from Mount Sobótka: a nonstandard polytype. *Am. Mineral.* **76,** 647–652.

WILDNER M. and LANGER K. (1994) Co^{2+} in trigonal fields of oxygen based structures: Electronicabsorption spectra of buetschliite-type $K_2Co(SeO_3)_2$, $K_2Co_2(SeO_3)_3$ and Zemannite-type $K_2Co_2(SeO_3)_3$*$2H_2O$. *Phys. Chem. Minerals.* **20,** 460–468.

YUSHKIN N. P., IVANOV O. K. and POPOV V. A. (1986) Introduction to the topomineralogy of Urals. Nauka, Moscow (in Russian)

Mineral Spectroscopy: A Tribute to Roger G. Burns
Editors: M. D. Dyar, C. McCammon and M. W. Schaefer

Short-range order and long-range order in amphiboles: A model for the interpretation of infrared spectra in the principal OH-stretching region

FRANK C. HAWTHORNE[1], GIANCARLO DELLA VENTURA[2], and JEAN-LOUIS ROBERT[3]

[1]Department of Geological Sciences, University of Manitoba, Winnipeg, Manitoba, R3T 2N2, Canada
[2]Dipartimento di Scienze Geologiche, Terza Università di Roma, via Ostiense 169, I-00154 Roma, Italy
[3]CRSC-CNRS, 1A Rue de la Férollerie, F-45071, Orléans Cedex 2, France

Abstract—The factors affecting the relative intensities of the four fine-structure bands in the principal OH-stretching region of the infrared spectrum of binary amphibole solid-solutions of the type $^{C}M^{2+}$ [M=(Mg,Fe^{2+}), (Mg,Co^{2+})] are examined. The band intensities are affected by both LRO (Long-Range Order) and SRO (Short-Range Order). If the LRO is characterized by a different method, (*e.g.*, Rietveld structure refinement, single-crystal structure refinement), the expected band intensities can be calculated for this state of LRO; any deviation from the observed band intensities (corrected for variation in transition moment with absorption frequency) is due to SRO. The possible variations in band intensities are derived as a function of the different possible types of SRO, subject to the constraint that the state of LRO is fixed for a specific crystal. There are 22 different types of SRO that give rise to 16 different ways in which the observed band intensities can depart from the band intensities calculated for no SRO and a known state of LRO. In principle, SRO can be detected in this fashion provided the observed band-intensities can be corrected for variation in integrated molar absorptivity (transition moment) as a function of band energy.

INTRODUCTION

IN AMPHIBOLES, the fundamental band of the OH-stretching vibration in the infrared occurs in the range 3750–3600 cm^{-1}. Ordered end-member amphiboles have a single sharp band in this region. However, in intermediate-composition amphiboles, this band shows considerable fine-structure that may be attributed primarily to cation substitution effects at those cation sites coordinated by the hydroxyl ion. This fine-structure was first used by STRENS (1966) and BURNS and STRENS (1966) to characterize cation order over the M(1,2,3) sites in amphiboles, and extensive work on amphiboles was published in the following five years (see HAWTHORNE, 1981, 1983). However, several problems with the method gradually became apparent:

(1) The spectra are sensitive to all substituents at the M(1) and M(3) sites, and in chemically complicated amphiboles, the spectra can be too complicated to be adequately resolved.

(2) Substitution at cation *and* anion sites not coordinated directly to OH may broaden bands or even introduce new bands.

(3) Infrared spectroscopy is sensitive to both long-range order *and* short-range order, and it is still not clear whether or not these effects can be adequately distinguished by this method.

As a result of these problems, the method was little used in site-occupancy studies of amphiboles for ~15 years. However, the last six years have seen a revival of the method, specifically to characterize ordering in synthetic amphiboles. In such work, problems (1) and (2) are not relevant because the composition of the amphibole is controlled as part of the experiment, and the spectral effects of specific components can be examined one at a time. This approach is the underlying methodology of the work by DELLA VENTURA and ROBERT (*e.g.*, DELLA VENTURA, 1992; DELLA VENTURA and ROBERT, 1990; DELLA VENTURA *et al.*, 1991, 1993a, b; ROBERT *et al.*, 1989). However, because of problem (3), the infrared method has only been used in a qualitative fashion in this work (see also RAUDSEPP *et al.*, 1987a,b, 1991). In this paper, we consider the spectral effects of long-range order and short-range order on the infrared spectra of amphiboles in the principal OH-stretching region, and use some results from well-characterized synthetic amphiboles to test our model.

SHORT-RANGE ORDER (SRO) AND LONG-RANGE ORDER (LRO)

Previous investigators have used the terms 'segregation' and 'clustering' to designate various aspects of ordering in amphiboles that pertain to infrared spectroscopy. Both of these terms are superfluous. 'Segregation' is equivalent to long-range order (LRO) and 'clustering' is equivalent to short-range order (SRO); consequently, we will use the latter terms when referring to ordering in general. However, the terms 'clustering' and 'anti-clustering' are very useful to describe short-range arrangements with pair-correlations respectively greater and less than that expected for no SRO, and hence we will use these terms here.

THE O(3) SITE

The O(3) site and its local environment is illustrated in Fig. 1; it can be occupied by OH^-, F^-, Cl^- (OBERTI *et al.*, 1993) and O^{2-} (HAWTHORNE *et al.*, 1995). When O(3) = OH^-, the O(3)–H distance is ~0.96 Å (HAWTHORNE and GRUNDY, 1976) and the O–H bond is approximately orthogonal to (100) (HANISCH, 1966). The O(3) anion bonds directly to two M(1) and one M(3) cations (Fig. 1); although the point symmetry of the O(3) site is *m*, it is also pseudo-trigonal, a very important factor that affects the resolution of the infrared spectra (STRENS, 1966).

PREVIOUS WORK

Most work has focused on the situation where there are two cation species (*e.g.*, Mg, Fe^{2+}) distributed over the M(1) and M(3) sites. As O(3) bonds to these M sites, there are 2 × 2 × 2 = 8 different possibilities for the local environment of an OH^- anion at the O(3) site (Table 1); however, the corresponding spectra show only four bands (*e.g.*, Fig. 2). This is a result of the pseudo-trigonal nature of the local arrangement; the M(1) and M(3) cations are pseudo-symmetrically related, introducing an accidental degeneracy into the spectrum (Table 1). In spectra such as that shown in Fig. 2, the peaks are labelled A, B, C and D, and the assignments are shown in Table 1.

STRENS (1966, 1974) derived criteria for identifying both LRO and SRO, but LAW (1976) showed that these were not correct. LAW (1976) developed a model for LRO assuming that no SRO is present, but application to two amphiboles characterized by other means (Mössbauer spectroscopy, crystal-structure refinement) showed poor agreement between the various sets of results. WHITTAKER (1979) explored the spectral effects of possible models of SRO in amphiboles, and his results gave the impression that the resolution of SRO and LRO in amphiboles by infrared spectroscopy is not feasible.

Table 1. Possible local arrangements around the O(3) site in amphiboles with (Mg, M^{2+}) at M(1) and M(3)

	M(1)	M(1)	M(3)	Band*	Band⁺
(1)	Mg	Mg	Mg	A	A
(2)	Mg	Mg	M^{2+}	B′	
(3)	M^{2+}	Mg	Mg	B′′	B
(4)	Mg	M^{2+}	Mg		
(5)	M^{2+}	M^{2+}	Mg	C′	
(6)	M^{2+}	Mg	M^{2+}	C′′	C
(7)	Mg	M^{2+}	M^{2+}		
(8)	M^{2+}	M^{2+}	M^{2+}	D	D

* ideal band-structure; (3) and (4) are symmetry equivalent and give rise to only one band: B′′; similarly (5) and (6) are symmetry equivalent and give rise to only one band: C″.

⁺ band structure usually observed; (2) and [(3)+(4)] are pseudo-symmetrically degenerate, and only one band (B) is observed experimentally; similarly, (5) and [(6)+(7)] are pseudo-symmetrically degenerate and only one band (C) is observed experimentally.

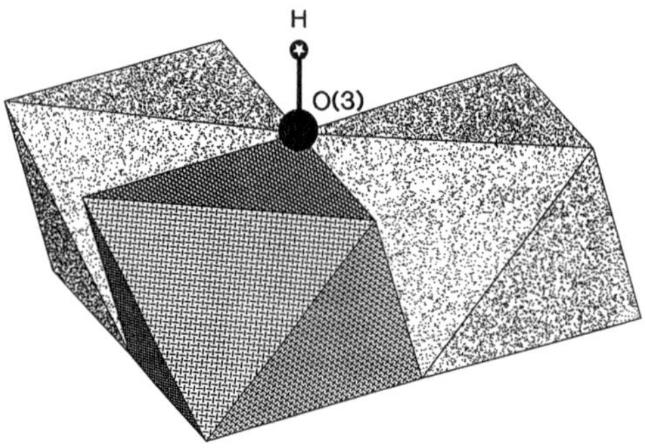

FIG. 1. The (OH) group in the amphibole structure, associated with the M(1) and M(3) coordination polyhedra. The M(1) octahedra are stippled, the M(3) octahedron is hatched, the oxygen atom at the O(8) site is shown as a black circle, and the H atom is shown as a starred circle.

THE CURRENT MODEL

We will consider binary occupancy of M(1) and M(3) by Mg and a transition metal M^{2+}. Site occupancies and C-group compositions are denoted in the following way:

$$
\begin{aligned}
y_{M(1)} &= M^{2+} \text{ at } M(1) \\
y_{M(3)} &= M^{2+} \text{ at } M(3) \\
x_{M(1)} &= \text{Mg at } M(1) \\
x_{M(3)} &= \text{Mg at } M(3) \\
X &= \text{total Mg at } M(1,3) = 2x_{M(1)} + x_{M(3)} \\
Y &= \text{total } M^{2+} \text{ at } M(1,3) = 2y_{M(1)} + y_{M(3)}
\end{aligned} \tag{1}
$$

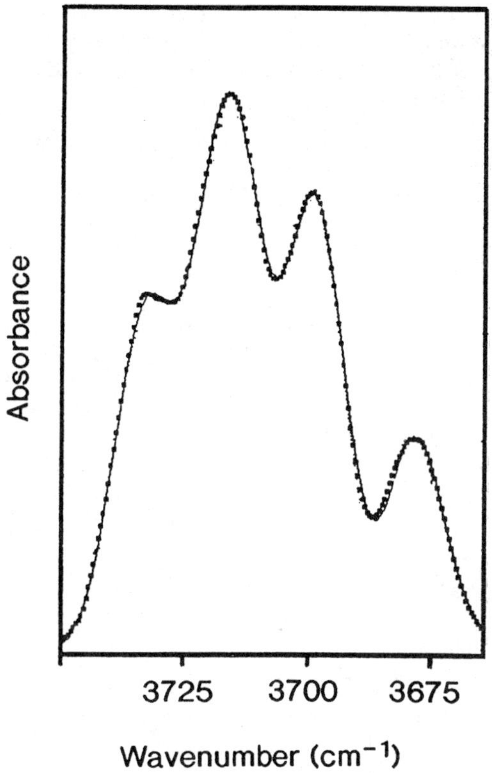

FIG. 2. The infrared spectrum in the principal OH-stretching region of a synthetic (Mg,Co)-potassium-richterite of composition $K(CaNa)(Mg_3Co_2)Si_8O_{22}(OH)_2$.

We can express *all* occupancies in terms of x and y by manipulating the individual terms of equation set (1):

$$x_{M(1)} = 1 - y_{M(1)} = 1 - \tfrac{1}{2}[Y - y_{M(3)}]$$
$$y_{M(1)} = \tfrac{1}{2}[Y - y_{M(3)}]$$
$$x_{M(3)} = 1 - y_{M(3)} \qquad (2)$$

The case of no SRO

Following STRENS (1966, 1974) and LAW (1976), the (degenerate) relative band intensities, $I^C_{A,B,C,D}$, are given by

$$I^C_A = x^2_{M(1)}x_{M(3)}$$
$$I^C_B = x^2_{M(1)}y_{M(3)} + 2x_{M(1)}y_{M(1)}x_{M(3)}$$
$$I^C_C = y^2_{M(1)}x_{M(3)} + 2x_{M(1)}y_{M(1)}y_{M(3)}$$
$$I^C_D = y^2_{M(1)}y_{M(3)} \qquad (3)$$

where the superscript *C* on the band intensities denotes that they are *calculated* values. Substituting (2) into (3) gives

$$I^C_A = (1 - \tfrac{1}{2}[Y - y_{M(3)}])^2(1 - y_{M(3)})$$
$$I^C_B = (1 - \tfrac{1}{2}[Y - y_{M(3)}])^2 \times y_{M(3)}$$
$$+ 2(1 - \tfrac{1}{2}[Y - y_{M(3)}]) \times \tfrac{1}{2}[Y - y_{M(3)}](1 - y_{M(3)})$$
$$I^C_C = \tfrac{1}{4}[Y - y_{M(3)}]^2(1 - y_{M(3)})$$
$$+ 2(1 - \tfrac{1}{2}[Y - y_{M(3)}]) \times \tfrac{1}{2}[Y - y_{M(3)}] \times y_{M(3)}$$
$$I^C_D = \tfrac{1}{4}[Y - y_{M(3)}]^2 \times y_{M(3)} \qquad (4)$$

These equations (set 4) give the calculated band intensities as a function of cation ordering (*i.e.*, $y_{M(3)}$) for a specific composition Y when there is no SRO present. Following LAW (1976), we may assume that there is no SRO and use equation (4) to derive the state of LRO in the amphibole from the *observed* values of the band intensities (I^A_O, I^B_O, I^C_O, I^D_O). Alternatively, we may measure the LRO (*i.e.*, $y_{M(3)}$) by some other method (*e.g.*, Rietveld structure refinement: RAUDSEPP *et al.*, 1987a,b; DELLA VENTURA *et al.*, 1993a,b) and use

Table 2. Behavior of band intensities* with short-range order

MgMgMg		$M^{2+}M^{2+}M^{2+}$	I^O_A	I^O_B	I^O_C	I^O_D
c^A	>	$2c^D$	*U*	*D*	*u*	*u*
c^A	=	$2c^D$	*U*	*D*	*E*	*u*
c^A	< >	$2c^D$ c^D	*u*	*D*	*d*	*u*
c^A	=	c^D	*U*	*D*	*D*	*U*
$2c^A$ c^A	> <	c^D	*u*	*d*	*D*	*u*
$2c^A$	=	c^D	*u*	*E*	*D*	*u*
$2c^A$	<	c^D	*u*	*u*	*D*	*U*
c^A		n^D	*U*	*D*	*u*	*E*
n^A		c^D	*E*	*u*	*D*	*u*
a^A	>	$2a^D$	*D*	*U*	*d*	*d*
a^A	=	$2a^D$	*d*	*U*	*E*	*d*
a^A	< >	$2a^D$ a^D	*d*	*U*	*u*	*d*
a^A	=	a^D	*D*	*U*	*U*	*D*
$2a^A$ a^A	> <	a^D	*d*	*u*	*U*	*d*
$2a^A$	=	a^D	*d*	*E*	*U*	*d*
$2a^A$	<	a^D	*d*	*d*	*U*	*d*
c^A		a^D	*U*	*D*	*U*	*D*
a^A		c^D	*D*	*U*	*D*	*U*
a^A		n^D	*d*	*U*	*d*	*E*
n^A		a^D	*E*	*d*	*U*	*d*

* the symbols *u*/*U* (up), *d*/*D* (down) and *E* (equal) denote the *observed* band intensity (I^O_A...) relative to the band intensity (I^C_A...) calculated for the same degree of LRO and zero SRO; upper and lower case qualitatively denote the relative sizes of the differences.

** *c* denotes clustering, *n* denotes no clustering, *a* denotes anticlustering; >, = and < denote the relative degree of clustering or anticlustering.

the observed ($I^O_A \cdots$) and calculated ($I^C_A \cdots$) band intensities to derive information on SRO in amphiboles.

SPECTRAL EXPRESSION OF SRO IN AMPHIBOLES

Consider the band assignments as given in Table 1. It should be noted that the aggregate intensity of certain combinations of bands are related to the site occupancies in the following way:

$$I^O_A + I^O_{B''} + I^O_{C'} = x_{M(3)}$$

$$I^O_{B'} + I^O_{C''} + I^O_D = y_{M(3)}$$

$$2I^O_A + 2I^O_{B'} + I^O_{B''} + I^O_{C''} = 2x_{M(1)}$$

$$I^O_{B''} + 2I^O_{C'} + I^O_{C''} + 2I^O_D = 2y_{M(1)} \quad (5)$$

Note that this equation holds only if there is no variation in molar absorptivity with band frequency *within a single sample.* As shown by DELLA VENTURA *et al.* (this volume), this condition holds for binary (Mg,Ni)- and (Mg,Co)-potassium-richterite solid solutions, and hence we can use equation (5) for these amphiboles. For a given state of LRO, SRO can only find its spectral expression within the constraints of equation (5); the band intensities will vary with different SRO, but the band-intensity *sums* in equation (5) are fixed by the state of LRO. With these constraints, we can now examine the effects of different sorts of SRO on the spectrum for a given state of LRO.

Let us first consider the total possible number of ways in which the band intensities can deviate from the band intensities expected for complete short-range disorder. There are four bands, and for each band there are three possibilities: (1) $I^O > I^C$; (2) $I^O = I^C$; (3) $I^O < I^C$; we will designate these as *U*, *E* and *D*, respectively. There are 3^4=81 possible arrangements of the symbols {*U,E,D*} taken four at a time; are all of these possible deviations from short-range disorder? The answer to this is no; there are various constraints on allowed combinations of these symbols. First, as the bulk composition of the crystal is fixed, any combination of symbols that would require a change in the composition of the crystal is forbidden. Thus any combination of symbols that does *not* include both *U* and *D* is forbidden (except for *EEEE*); of the 81 possible arrangements, 30 fall within this forbidden category. For the same reason, any combination of symbols that does *not* include both *U* and *D* in the first and in the last triplets of symbols is also forbidden; 26 arrangements fall within this category. This leaves 25 arrangements to consider.

Let each band differ by $^{\Delta}I_i$ (i = A $\cdots$ D) from the values calculated for no short-range order (*i.e.*, $^{\Delta}I_i = I^O_i - I^C_i$). The Mg and M^{2+} contents of the crystal are fixed, and hence

$$\text{for Mg: } 3^{\Delta}I_A + 2^{\Delta}I_B + {}^{\Delta}I_C = 0 \quad (6.1)$$

$$\text{for } M^{2+}\text{: } {}^{\Delta}I_B + 2^{\Delta}I_C + 3^{\Delta}I_D = 0 \quad (6.2)$$

Rearranging gives the following relationships:

$$^{\Delta}I_B = {}^{\Delta}I_D - 2^{\Delta}I_A \quad (7.1)$$

$$^{\Delta}I_C = {}^{\Delta}I_A - 2^{\Delta}I_D \quad (7.2)$$

Thus the variations in I^O_i relative to I^C_i must conform to the constraints of equation (7); this will be very important in associating specific band intensities with specific local arrangements. Inspection of the remaining 25 combinations of symbols shows that 8 are incompatible with equation (7), leaving 17 combinations that are possible (within the constraint of constant chemical composition).

MgMgMg and/or $M^{2+}M^{2+}M^{2+}$ clustering

Let us first consider the case of MgMgMg clustering. This will increase the relative intensity of the A band above the value of I^C_A given in equations (3) and (4). From equation (5)

$$^{\Delta}I_A + {}^{\Delta}I_{B'} + \tfrac{1}{2}({}^{\Delta}I_{B''} + {}^{\Delta}I_{C''}) = 0 \quad (8.1)$$

$$^{\Delta}I_A + {}^{\Delta}I_{B''} + {}^{\Delta}I_{C'} = 0 \quad (8.2)$$

With MgMgMg clustering, $^{\Delta}I_A$ is positive, and hence equation (8) reduces to

$$^{\Delta}I_{B'} + \tfrac{1}{2}({}^{\Delta}I_{B''} + {}^{\Delta}I_{C''}) < 0 \quad (9.1)$$

$$^{\Delta}I_{B''} + {}^{\Delta}I_{C'} < 0 \quad (9.2)$$

However, there cannot be MgMgMg clustering ($^{\Delta}I_A > 0$) without a compensating difference in the number of other configurations. We will examine this aspect of the problem next.

No $M^{2+}M^{2+}M^{2+}$ clustering. This condition means that $^{\Delta}I_D = 0$; also, the state of LRO for M^{2+} is fixed, and hence

$$\tfrac{1}{2}({}^{\Delta}I_{B''} + {}^{\Delta}I_{C''}) + {}^{\Delta}I_{C'} + {}^{\Delta}I_D = \tfrac{1}{2}({}^{\Delta}I_{B''} + {}^{\Delta}I_{C''}) + {}^{\Delta}I_{C'} = 0 \quad (10.1)$$

$$^{\Delta}I_{B'} + {}^{\Delta}I_{C''} + {}^{\Delta}I_D = {}^{\Delta}I_{B'} + {}^{\Delta}I_{C''} = 0 \quad (10.2)$$

Combining (9) and (10) gives $^{\Delta}I_B < 0$ and $^{\Delta}I_C > 0$. Thus for MgMgMg clustering with no $M^{2+}M^{2+}M^{2+}$ clustering (or anticlustering), $I^O_B < I^C_B$ and $I^O_C > I^C_C$. We may summarize this behaviour by the symbol *uDuE* (the size of the *U/u* and *D/d* symbols are a crude measure of the relative values of $^{\Delta}I_i$).

$M^{2+}M^{2+}M^{2+}$ clustering. This condition means that $^{\Delta}I_D > 0$. As we are dealing with MgMgMg clustering, equation (9) holds. As we are dealing with $M^{2+}M^{2+}M^{2+}$ clustering,

$$\tfrac{1}{2}(^{\Delta}I_{B''} + {}^{\Delta}I_{C''}) + {}^{\Delta}I_{C'} < 0 \qquad (11.1)$$

$$^{\Delta}I_{B'} + {}^{\Delta}I_{C''} < 0 \qquad (11.2)$$

There are three distinct cases for MgMgMg clustering in combination with $M^{2+}M^{2+}M^{2+}$ clustering: MgMgMg clustering greater than ($^{\Delta}I_A > {}^{\Delta}I_D$), equal to ($^{\Delta}I_A = {}^{\Delta}I_D$) or less than ($^{\Delta}I_A < {}^{\Delta}I_D$) $M^{2+}M^{2+}M^{2+}$ clustering.

First we consider the case for MgMgMg clustering greater than $M^{2+}M^{2+}M^{2+}$ clustering: ($^{\Delta}I_A > {}^{\Delta}I_D$). The relations between $^{\Delta}I_A$ and $^{\Delta}I_D$ indicate that

$$|^{\Delta}I_{B'} + \tfrac{1}{2}(^{\Delta}I_{B''} + {}^{\Delta}I_{C''})| > |\tfrac{1}{2}(^{\Delta}I_{B''} + {}^{\Delta}I_{C''}) + {}^{\Delta}I_{C'}| \qquad (12.1)$$

$$|^{\Delta}I_{B''} + {}^{\Delta}I_{C'}| > |^{\Delta}I_{B'} + {}^{\Delta}I_{C''}| \qquad (12.2)$$

In this case, we may subtract equation (11) from equation (9) [2.(9.1)–(11.2)] to show that $^{\Delta}I_B < 0$. Similarly, we can show that $^{\Delta}I_B + {}^{\Delta}I_C < 0$ but we cannot explicitly determine $^{\Delta}I_C$ in this case [relationship (12) does not allow us to subtract (9.2) from 2.(11.1) as we have no knowledge of the relative magnitudes of the terms in each equation]. So we may temporarily represent this behaviour by the symbol *UD*u*. To determine the * symbol, consider equation (7.2): $^{\Delta}I_C = {}^{\Delta}I_A - 2{}^{\Delta}I_D$. As $^{\Delta}I_A > {}^{\Delta}I_D$, $^{\Delta}I_C > 0$ if $^{\Delta}I_A > 2{}^{\Delta}I_D$; $^{\Delta}I_C = 0$ if $^{\Delta}I_A = 2{}^{\Delta}I_D$; $^{\Delta}I_C < 0$ if $^{\Delta}I_D < {}^{\Delta}I_A < 2{}^{\Delta}I_D$. These conditions correspond to the symbols *UDuu*, *uDEu* and *uDdu*, respectively. Thus MgMgMg clustering greater than $M^{2+}M^{2+}M^{2+}$ clustering has three different spectral expressions, depending on the relative amount of clustering in each case.

Next consider the case for MgMgMg clustering equal to $M^{2+}M^{2+}M^{2+}$ clustering: ($^{\Delta}I_A = {}^{\Delta}I_D$), and hence the > signs in relation (12) become = signs. Expressions (9) and (11) reduce to $^{\Delta}I_B = {}^{\Delta}I_C$. Combination of this relation with equation (7) indicates that $^{\Delta}I_B = {}^{\Delta}I_C < 0$. These conditions correspond with the symbol *UDDU*.

Last, consider the case of MgMgMg clustering less than $M^{2+}M^{2+}M^{2+}$ clustering. This case is symmetrically related to the case of MgMgMg clustering greater than $M^{2+}M^{2+}M^{2+}$ clustering, and the solutions follow directly: *uuDU*, *uEDu* and *udDu*. All of these configurations and their spectral expression as quartets of symbols are listed in Table 2.

MgMgMg and $M^{2+}M^{2+}M^{2+}$ anticlustering

These arrangements are symmetrically related to the cases involving clustering derived in detail above. Hence the arrangements may be derived directly from the algebraic symmetry of the equations. All distinct configurations and their spectral expression are listed in Table 2.

Clustering in combination with anticlustering

The combination of clustering of one component (*e.g.*, Mg) with anticlustering of another (*e.g.*, M^{2+}) is somewhat different from combinations involving only clustering or only anticlustering. This arises because $^{\Delta}I_A$ and $^{\Delta}I_D$ have different signs in equations (8) and (10), and hence the relative complexities involving whether or not $^{\Delta}I_A$ is less than, equal to or greater than $2{}^{\Delta}I_D$ [equation (7)] do not arise.

Consider first the case where MgMgMg clustering is greater than or equal to $M^{2+}M^{2+}M^{2+}$ anticlustering: $^{\Delta}I_A > 0$, $^{\Delta}I_D < 0$, $|^{\Delta}I_A| \geq |^{\Delta}I_D|$. Equation (11) becomes

$$\tfrac{1}{2}(^{\Delta}I_{B''} + {}^{\Delta}I_{C''}) + {}^{\Delta}I_{C'} > 0 \qquad (13.1)$$

$$^{\Delta}I_{B'} + {}^{\Delta}I_{C''} > 0 \qquad (13.2)$$

Relation (12) holds (as ≥ rather than >), and we may combine expressions (9) and (11) to show that $^{\Delta}I_B < 0$ [2.(9.1) − 11.2], giving rise to the symbol *UD*D*; we cannot derive an analogous expression for $^{\Delta}I_C$ under the conditions of relation (12). From equation (7.2), $^{\Delta}I_C > 0$ as $^{\Delta}I_A > 0$ and $^{\Delta}I_D < 0$, and hence the general symbol is *UDUD*.

For $^{\Delta}I_A > 0$, $^{\Delta}I_D < 0$, $|^{\Delta}I_A| < |^{\Delta}I_D|$, expressions (9) and (11) combine to show that $^{\Delta}I_C > 0$ [2.(11.1) − (9.2)], and equation (7.1) gives $^{\Delta}I_B < 0$ for the same general symbol *UDUD*.

The arguments for MgMgMg anticlustering combined with $M^{2+}M^{2+}M^{2+}$ clustering are symmetrical with regard to those given above, and the general symbol for this behaviour is thus *DUDU*.

VARIATION IN TRANSITION PROBABILITY WITH ABSORPTION ENERGY

All earlier work implicitly assumed that the transition moment was the same for all of the different bands in the OH-stretching spectra of complex amphiboles. It now seems that this may not be the case. In a polarized single-crystal absorption study of amphiboles, SKOGBY and ROSSMAN (1991) showed that the integrated molar absorptivity for the principal OH-stretching band increases with decreasing stretching frequency. This is also the same for vesuvianite (GROAT *et al.*, 1995) in which the range in OH-stretching frequency is much greater than in amphibole. Also, BURNS and HAWTHORNE (1994) have shown the analogous relation for normalized OH-stretching-band intensities in powder infrared spectra of borate minerals. Thus there is not a 1:1 correspondence between the intensity of OH-stretching absorption bands and the relative abundance of short-range configurations with which they are associated. This relationship must be characterized and the measured intensities of the absorption bands must be corrected for this phenomenon before any effects of SRO can be examined.

Della Ventura *et al.* (this volume) consider this problem for synthetic (Ni,Mg,Co)-potassium-richterite and show that the curve of Skogby and Rossman (1991) is *not* compatible with the variation in relative band intensities as a function of composition [at the M(1) and M(3) sites] as measured by Rietveld refinement. Preliminary spectra on other synthetic amphibole series indicate that a variety of behaviour may occur with regard to variation in molar absorptivity as a function of band frequency. However, the arguments of Della Ventura *et al.* (this volume) suggest that this problem is susceptible to solution for many amphibole series of interest. This being the case, the equations given here allow characterization of SRO over the M(1) and M(3) sites.

Acknowledgements—We thank M. R. Raudsepp and an anonymous referee for their comments. FCH was supported by a Killam Fellowship and by Operating, Infrastructure and Equipment grants from the Natural Sciences and Engineering Research Council of Canada.

REFERENCES

Burns P. C. and Hawthorne F. C. (1994) Structure and hydrogen bonding in preobrazhenskite, a complex heteropolyhedral borate. *Can. Mineral.* **32**, 387–396.

Burns R. G. and Strens R. G. J. (1966) Infrared study of the hydroxyl bonds in clinoamphiboles. *Science* **153**, 890–892.

Della Ventura G. (1992) Recent developments in the synthesis and characterization of amphiboles. Synthesis and crystal chemistry of richterites. *Trends in Mineral.* **1**, 153–192.

Della Ventura G. and Robert J.-L. (1990) Synthesis, XRD and FTIR studies of strontium richterites. *Eur. J. Mineral.* **2**, 171–175.

Della Ventura G., Robert J. L. and Bény J.-M. (1991) Tetrahedrally coordinated Ti^{4+} in synthetic Ti-rich potassic richterite: Evidence from XRD, FTIR and Raman studies. *Amer. Mineral.* **76**, 1134–1140.

Della Ventura G., Robert J. L., Raudsepp M. and Hawthorne F. C. (1993a) Site occupancies in monoclinic amphiboles: Rietveld structure refinement of synthetic nickel magnesium cobalt potassium richterite. *Amer. Mineral.* **78**, 633–640.

Della Ventura G., Robert J. L., Bény J.-M., Raudsepp M. and Hawthorne F. C. (1993b) The OH—F substitution in Ti-rich potassium-richterites: Rietveld structure refinement and FTIR and micro-Raman spectroscopic studies of synthetic amphiboles in the system K_2O—Na_2O—CaO—MgO—SiO_2—TiO_2—H_2O—HF. *Amer. Mineral.* **78**, 980–987.

Della Ventura G., Robert J. L. and Hawthorne F. C. (1996) Infrared spectroscopy of synthetic (Ni—Mg—Co)-potassium richterite. In *Mineral Spectroscopy: A Tribute to Roger G. Burns* (eds. Dyar M. D., McCammon C. A. and Schaefer M. S.), pp. 55–63. The Geochemical Society.

Groat L. A., Hawthorne F. C., Rossman G. R. and Ercit T. S. (1995) The infrared spectroscopy of vesuvianite in the OH region. *Can. Mineral.* **33**, 609–626.

Hanisch K. (1966) Messung des Ultrarot-Pleochroismus von Mineralen. VI. Der Pleochroismus der OH-Streckfrequenz in Riebeckit. *Neues. Jahrb. Mineral. Monatsh.* **1966**, 109–112.

Hawthorne F. C. (1981) Amphibole Spectroscopy. In *Amphiboles and Other Hydrous Pyriboles—Mineralogy.* (ed. D. R. Veblen), *Rev. Mineral.* 9a, pp. 103–140. Mineralogical Society of America.

Hawthorne F.C. (1983) The crystal chemistry of the amphiboles. *Can. Mineral.* **21**, 173–480.

Hawthorne F. C. and Grundy H. D. (1976) The crystal chemistry of the amphiboles. IV. X-ray and neutron refinements of the crystal structure of tremolite. *Can. Mineral.* **14**, 334–345.

Hawthorne F. C., Oberti R., Cannillo E., Sardone N., Zanetti A., Grice J. D. and Ashley P. M. (1995) A new anhydrous amphibole from the Hoskins mine, Grenfell, New South Wales, Australia: Description and crystal structure of ungarettiite, $NaNa_2(Mn_2^{2+}Mn_3^{3+})Si_8O_{22}O_2$. *Amer. Mineral.* **80**, 165–172.

Law A. D. (1976) A model for the investigation of hydroxyl spectra of amphiboles. In *The Physics and Chemistry of Minerals and Rocks* (ed. R. G. J. Strens), pp. 677–686. Wiley.

Oberti R., Ungaretti L., Cannillo E. and Hawthorne F. C. (1993) The mechanism of Cl incorporation in amphibole. *Amer. Minieral.* **78**, 746–752.

Raudsepp M., Turnock A. C. and Hawthorne F. C. (1987a) Characterization of cation ordering in synthetic scandium-fluor-eckermannite, indium-fluor-eckermannite and scandium-fluor-nyböite by Rietveld structure refinement. *Amer. Mineral.* **72**, 959–964.

Raudsepp M., Turnock A. C. and Hawthorne F. C. (1991) Amphiboles synthesis at low-pressure: what grows and what doesn't. *Eur. J. Mineral.* **3**, 983–1004.

Raudsepp M., Turnock A. C., Hawthorne F. C., Sherriff B. L. and Hartman J. S. (1987b) Characterization of synthetic pargasitic amphiboles ($NaCa_2Mg_4M^{3+}Si_6Al_2O_{22}(OH,F)$; M^{3+} = Al, Cr, Ga, Sc, In) by infrared spectroscopy, Rietveld structure refinement and ^{27}Al, ^{29}Si, and ^{19}F MAS NMR spectroscopy. *Amer. Mineral.* **72**, 580–593.

Robert J.-L., Della Ventura G. and Thauvin J.-L. (1989) The infrared OH-stretching region of synthetic richterites in the system Na_2O—K_2O—CaO—MgO—SiO_2—H_2O—HF. *Eur. J. Mineral.* **1**, 203–211.

Skogby H. and Rossman G. R. (1991) The intensity of amphibole OH bands in the infrared absorption spectrum. *Phys. Chem. Mineral.* **18**, 64–68.

Strens R. G. J. (1966) Infrared study of cation ordering and clustering in some (Fe,Mg) amphibole solid solutions. *Chem. Commun.* **519**.

Strens R. G. J. (1974) The common chain, ribbon and ring silicates. In *The Infrared Spectra of Minerals* (ed. V. C. Farmer), pp. 305–330. Mineralogical Society, London.

Whittaker E. J. W. (1979) Clustering of cations in amphiboles. *Phys. Chem. Mineral.* **4**, 1–10.

Mineral Spectroscopy: A Tribute to Roger G. Burns
© The Geochemical Society, Special Publication No. 5, 1996
Editors: M. D. Dyar, C. McCammon and M. W. Schaefer

Infrared spectroscopy of synthetic (Ni,Mg,Co)-potassium-richterite

GIANCARLO DELLA VENTURA,[1] JEAN-LOUIS ROBERT,[2] FRANK C. HAWTHORNE[3]
[1]Dipartimento di Scienze Geologiche, Terza Università di Roma, via Ostiense 169, I-00154 Roma, Italy
[2]CRSC-CNRS, 1A Rue de la Ferollerie, F-45071 Orléans Cedex 2, France
[3]Department of Geological Sciences, University of Manitoba, Winnipeg, Manitoba, R3T 2N2, Canada

Abstract—The two series (Mg,Ni)- and (Mg,Co)-potassium-richterite were synthesized at 750°C, 1 kbar $P(H_2O)$ and characterized by FTIR spectroscopy. The infrared spectra of nominal end-member compositions show a single main band assigned to an OH anion bonded to three identical octahedral M^{2+} (M^{2+} = Ni, Mg or Co) cations and directed toward K at the A-site, and a minor band at lower wavenumbers assigned to an OH anion bonded to three identical octahedral cations and directed toward a □ (vacancy) at the A-site. Intermediate binary compositions show fine-structure caused by ordering of cations over the M(1) and M(3) sites in the structure; this fine structure constitutes a quartet of bands for each single band in the nominal end-member compositions. Precise band-intensities were derived by non-linear least-squares fitting of Gaussian band-shapes to the observed spectra. The relative intensities of the combinations of bands $3I^{o}_{A} + 2I^{o}_{B} + I^{o}_{C}$ and $I^{o}_{B} + 2I^{o}_{C} + 3I^{o}_{D}$ are in accord with the equations of BURNS and STRENS (1966) and seem to contradict the experimental results of SKOGBY and ROSSMAN (1991) on variation in molar absorptivity with principal-stretching frequency in amphiboles. Using the equations of HAWTHORNE *et al.* (1996) and the long-range site-populations of DELLA VENTURA *et al.* (1993) for the same series indicates that there is significant short-range order in these amphiboles involving MgMgMg and NiNiNi (or CoCoCo) clustering.

INTRODUCTION

ORDERED END-MEMBER amphiboles have a single absorption band in the principal OH-stretching region of the infrared. In intermediate amphibole compositions, this band has a prominent fine-structure that can be attributed to cation substitution effects within the structure. STRENS (1966) and BURNS and STRENS (1966) used the relative band-intensities of this fine structure to derive information on LRO (long-range order) and SRO (short-range order) of divalent cations, specifically Mg and Fe^{2+}, over the M(1) and M(3) sites in amphiboles (see surveys by STRENS, 1974 and HAWTHORNE, 1981, 1983). However, work by LAW (1976), WHITTAKER (1979) and SKOGBY and ROSSMAN (1991) indicated significant problems with the models used for the quantitative interpretation of these spectra, and the method fell into disuse for such work. In contrast to natural amphiboles, infrared spectroscopy in the principal OH-stretching region has played an increasingly important role in the characterization of ordering in synthetic amphiboles (SEMET, 1973; RAUDSEPP *et al.*, 1987, 1991; ROBERT *et al.*, 1989; DELLA VENTURA and ROBERT, 1990; DELLA VENTURA, 1992; DELLA VENTURA *et al.*, 1991, 1993a). However, none of these applications has addressed the question of ordering of divalent cations over the M(1) and M(3) sites, the original problem to which the method was applied in natural amphiboles. It is apparent from all of the original work done on natural amphiboles that the fine structure of the absorption in the principal OH-stretching region contains significant information on ordering of divalent cations over the M(1) and M(3) sites. This being the case, it seems wasteful not to extract this information and characterize ordering in amphiboles. With this end in mind, here we examine the infrared spectra of the two solid-solution series (Mg, Ni)-potassium-richterite and (Mg,Co)-potassium-richterite, $K(CaNa)(Mg,M^{2+})_5Si_8O_{22}(OH)_2$, M^{2+} = Ni or Co. DELLA VENTURA *et al.* (1993b) described the synthesis and characterization of these amphiboles, and reported site-populations derived by Rietveld structure-refinement. We note that these latter data provide significant quantitative constraints on our interpretation of band-intensity variations in the infrared spectra of these amphiboles.

EXPERIMENTAL

Samples were prepared as KBr pellets using the procedure of ROBERT *et al.* (1989). Spectra in the range 4000–3000 cm^{-1} were recorded on a Perkin Elmer model 1760 spectrometer equipped with a DTGS detector and a KBr beamsplitter, and operating at a nominal resolution of 1 cm^{-1}.

Digitized spectra (average of 64 scans) were fitted by interactive optimization followed by least-squares refinement. The background was modelled as linear and all peaks were modelled as symmetric Gaussians (STRENS, 1974). The distribution of absorption, y, as a function of energy (wavenumber, x) was described by the relationship $y = A\, exp[-0.5(x-P/W)^2]$ where A is the amplitude, P is the peak centroid and W is the full-width at half-maximum height (FWHM). The spectra were fitted to the smallest number of peaks needed to get an accurate description of the spectral profile. For intermediate compositions, all parameters were well defined by the envelope of the spectrum and all peak parameters could be refined unconstrained. For compositions close to the end members, the weak bands tended to merge with the background if all their parameters were refined unconstrained; consequently, for the weak bands, the positions were fixed at the values determined in those spectra in which the corresponding peaks are intense, and the band widths, W, were constrained to be equal to the width of the same band.

RESULTS

The raw spectra and band nomenclature are shown in Fig. 1 and the fitted spectra are shown in Fig. 2. Band positions, widths and relative intensities ($=I/I^{TOTAL}$) are given in Table 1. The spectrum of end-member potassium-richterite shows an intense band centered at 3735 cm^{-1}, assigned to an MgMgMg—OH → K configuration and designated the A band, and a minor band at 3670 cm^{-1} due to the presence of some vacant A-sites in the amphibole (band A* in Fig. 1). We have tried to synthesize both richterite and potassium-richterite under a variety of conditions in order to get rid of the A* band and produce an amphibole with a fully occupied A-site. However, the A* band is always present, even in richterites synthesized in concentrated NaCl or KCl solutions, and we conclude that a significant amount of vacancy at the A site is characteristic of the richterite structure. The spectrum of Ni(20) shows four reasonably intense bands designated A, B, C and D, and these bands are present in the spectra of all intermediate synthetic amphiboles examined here. These bands are due to the local configurations around the OH site (Table 2), as first established by STRENS (1966, 1974) and BURNS and STRENS (1966).

Examination of the spectra of the intermediate-composition amphiboles (Fig. 1) shows that the A* band is also associated with a triplet of additional bands that are most easily seen in the spectra of Ni(40), Co(40) and Co(60) (Fig. 1). These bands are designated as B*, C* and D*, and their assignment to specific local configurations is given in Table 2.

DISCUSSION

Variation in molar absorptivity with frequency

The primary interest in using infrared spectroscopy in the principal OH-stretching region to look at amphiboles is to derive information on LRO (Long-Range Order) and SRO (Short-Range Order) of cations in the amphibole structure. In the approach developed by STRENS (1966), BURNS and STRENS (1966) and LAW (1976), the relative intensities of the A, B, C and D bands are directly related to cation ordering over the M(1) and M(3) sites. However, this approach implicitly assumed that the molar absorptivity is the same for each of these bands (*i.e*, the transition moment is the same for all local cation configurations in the structure). In a polarized single-crystal infrared absorption study of amphiboles, SKOGBY and ROSSMAN (1991) showed that the integrated molar absorptivity for the principal OH-stretching-band increases with decreasing stretching frequency. The analogous relation was also observed in polarized single-crystal spectra of vesuvianite (GROAT *et al.*, 1995). BURNS and HAWTHORNE (1994) showed the analogous relation for normalized OH-stretching-band intensities in powder infrared spectra of borate minerals. Thus the implicit assumption of previous work on powder spectra of amphiboles, that there is no variation in molar absorptivity with frequency, seems invalid. This relation must be experimentally characterized if we wish to relate the intensities of the OH-stretching bands to the concentrations of specific local configurations in the amphibole structure.

HAWTHORNE *et al.* (1996) show that the binary site occupancies x ($=Mg/[Mg + M^{2+}]$) and y ($=M^{2+}/[Mg + M^{2+}]$) at the M(1) and M(3) sites in the amphibole structure are related to the observed relative intensities, I^o, in the following way, provided that there is no variation in molar absorptivity with band frequency *within a single sample*.

$$
\begin{aligned}
I^o_A + I^o_{B''} + I^o_{C'} &= x_{M(3)} \\
I^o_{B'} + I^o_{C''} + I^o_D &= y_{M(3)} \\
2I^o_A + 2I^o_{B'} + I^o_{B''} + I^o_{C''} &= 2x_{M(1)} \\
I^o_{B''} + 2I^o_{C'} + I^o_{C''} + 2I^o_D &= 2y_{M(1)} \qquad (1)
\end{aligned}
$$

If we sum the band intensities in equation (1) relating to the occupancies of M(1) and M(3), we get the following relations:

$$Mg_{M(1,3)} = 3I^o_A + 2I^o_B + I^o_C \qquad (2.1)$$

$$M^{2+}_{M(1,3)} = I^o_B + 2I^o_C + 3I^o_D \qquad (2.2)$$

These are the original equations of BURNS and STRENS (1966) who used them for amphiboles in which the M(2) site is occupied by Al and Fe^{3+}, and for which the unit formulae give directly the cations occupying the M(1,3) sites. For the (Ni,Mg,Co)-potassium-richterites of this work, DELLA VENTURA *et al.* (1993b) derived the populations of the M(1), M(2) and M(3) sites by Rietveld structure-refinement, and hence the contents of Mg and M^{2+} at the M(1,3) sites are known for these amphiboles. In equation (2), the band intensities must be corrected for variation in molar absorptivity as a function of absorption fre-

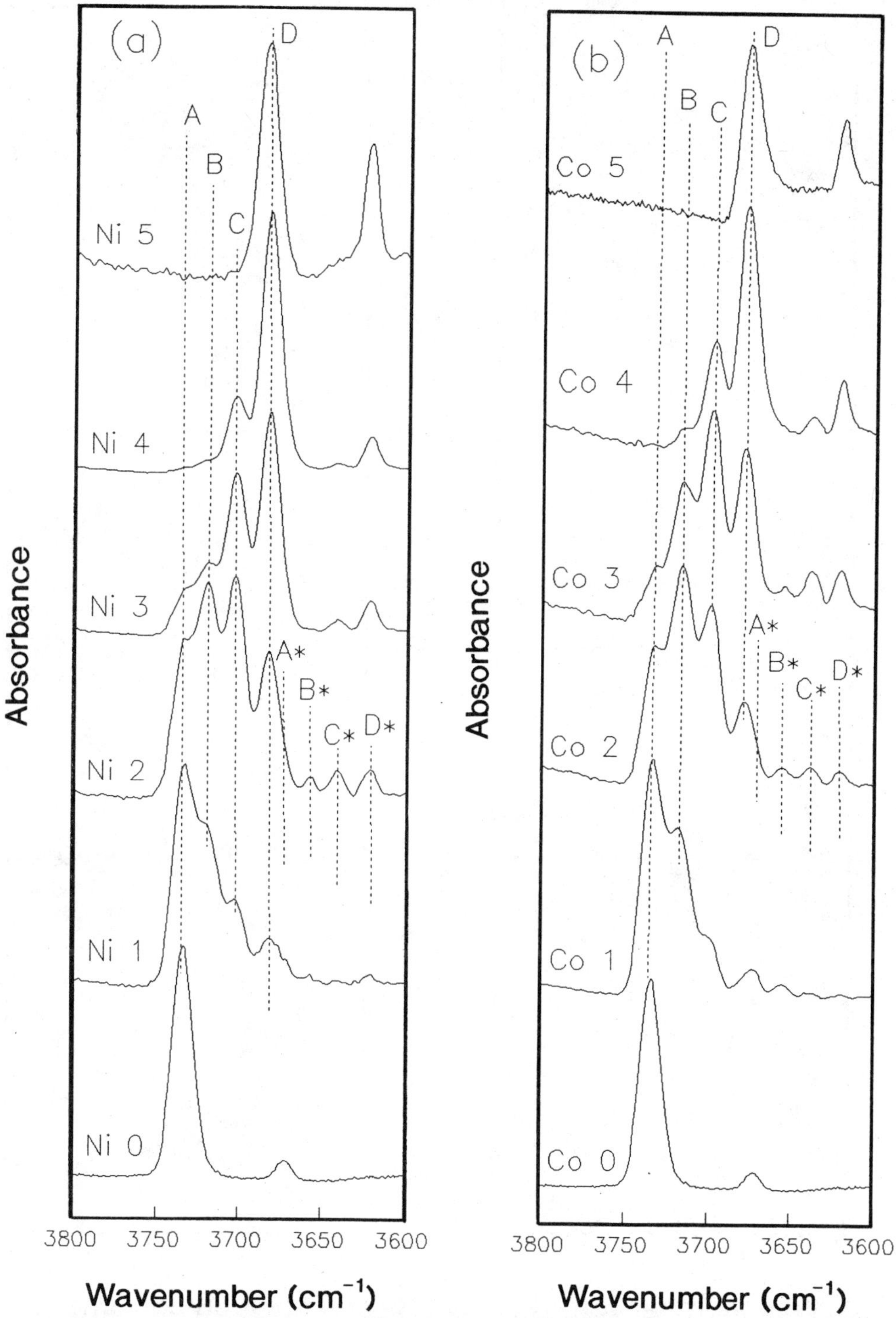

FIG. 1. FTIR spectra in the OH-stretching region for amphiboles synthesized along (a) the Ni-potassium-richterite—potassium-richterite join; (b) the potassium-richterite—Co-potassium-richterite join. The band nomenclature is also given.

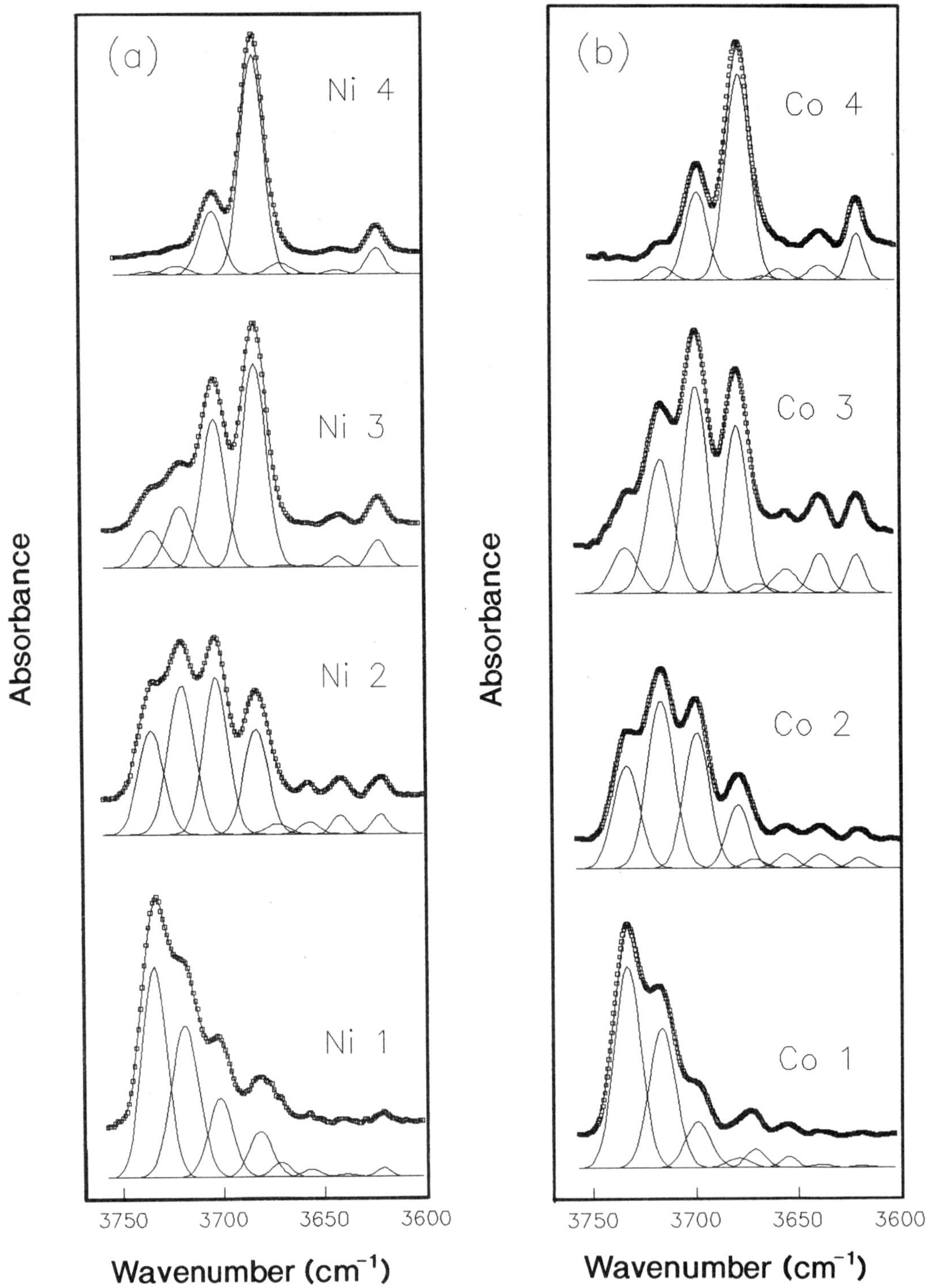

FIG. 2. Resolved spectra for the intermediate amphiboles synthesized along (a) the (Mg,Ni)-potassium-richterite join; (b) the (Mg,Co)-potassium-richterite join. Resolved bands are vertically displaced for clarity; observed intensities are shown by the hollow squares, and the line following the observed intensities is the envelope of the sum of the fitted component bands.

Table 1. Refined positions (cm^{-1}), widths (cm^{-1}) and normalized integrated intensities of fine-structure bands in the principal OH-stretching region of the infrared spectra of synthetic (Ni,Mg,Co)-potassium-richterites

		A	B	C	D	A′	B′	C′	D′
	Position[1]	3734.4	3719.5	3701.6	3682.1	3671.7	3656.8	3639.5	3622.0
Ni(20)	Width[2]	14.0	16.1	14.8	12.1	10.7	10.4	11.9	10.8
	Intensity	0.390	0.324	0.155	0.075	0.026	0.011	0.006	0.013
	Position	3735.0	3719.9	3702.7	3682.6	3672.4	3656.6	3641.1	3621.5
Ni(40)	Width	14.3	15.3	14.4	13.8	12.7	10.5	9.1	9.7
	Intensity	0.181	0.282	0.280	0.179	0.016	0.016	0.022	0.024
	Position	3735.2	3720.1	3702.7	3682.5	3670.5	3656.5	3641.3	3621.8
Ni(60)	Width	13.3	16.1	13.8	14.0	11.6	9.7	8.9	9.3
	Intensity	0.067	0.149	0.303	0.425	0.004	0.003	0.013	0.036
	Position	3734.5	3720.7	3703.2	3682.7	3669.2	3656.8	3641.2	3621.8
Ni(80)	Width	11.2	13.6	13.5	13.3	12.5	11.2	10.5	10.6
	Intensity	0.005	0.026	0.192	0.663	0.034	0.003	0.012	0.065
	Position	3733.3	3717.0	3699.2	3678.4	3671.3	3654.9	3638.5	3618.7
Co(20)	Width	14.6	15.7	13.9	13.2	11.3	8.7	10.4	11.7
	Intensity	0.470	0.351	0.104	0.020	0.032	0.015	0.005	0.003
	Position	3733.3	3716.5	3698.9	3678.8	3671.0	3655.4	3638.7	3619.7
Co(40)	Width	14.8	15.4	14.6	12.5	11.8	11.8	11.6	10.3
	Intensity	0.203	0.344	0.265	0.106	0.017	0.024	0.024	0.017
	Position	3733.4	3716.2	3698.6	3678.7	3668.3	3654.8	3638.0	3620.36
Co(60)	Width	13.4	15.3	13.9	13.1	12.6	13.8	10.5	9.5
	Intensity	0.065	0.228	0.322	0.248	0.012	0.036	0.047	0.042
	Position	3734.7	3714.8	3698.1	3677.4	3666.4	3657.4	3638.9	3619.6
Co(80)	Width	12.2	11.1	12.8	13.9	10.0	11.7	9.0	8.7
	Intensity	0.005	0.030	0.231	0.583	0.009	0.028	0.031	0.083

[1] standard deviations are ≤0.5 cm^{-1}; [2] standard deviations are ≤0.4 cm^{-1}.

quency. Although the relation of SKOGBY and ROSSMAN (1991) indicates a significant variation in molar absorptivity as a function of frequency, we cannot rigorously apply their relationship to our powder data as they did not measure the $E//\beta$ spectrum. However, we will quantitatively apply their relation to our data in order to evaluate qualitatively the order of magnitude of the effect. We emphasize here that we are examining this effect on the relative band intensities within a single spectrum (and hence avoid grain-size effects, etc.); we are not comparing band intensities between samples. The inclined line in Fig. 3 shows the variation in integrated molar absorptivity, ϵ, as a function of wavenumber, and the position of the four bands, A, B, C and D are marked, together with the corresponding values of integrated molar absorptivity ϵ_A, ϵ_B, ϵ_C and ϵ_D, respectively. Let us normalize the band intensities to the wavelength of the A band by multiplying the observed B, C and D band intensities by ϵ_A/ϵ_B, ϵ_A/ϵ_C and ϵ_A/ϵ_D, respectively. From Fig. 3,

$$\epsilon_A/\epsilon_B = \Delta/(\Delta + \delta) \qquad (3.1)$$

$$\epsilon_A/\epsilon_C = \Delta/(\Delta + 2\delta) \qquad (3.2)$$

$$\epsilon_A/\epsilon_D = \Delta/(\Delta + 3\delta) \qquad (3.3)$$

where Δ is the distance (in cm^{-1}) of the A band from the projected position of zero molar absorptivity, and δ is the separation of the A, B, C and D bands. We may write the corrected band intensities, I^T, as

$$I_A^T = I_A^o$$

$$I_B^T = I_B^o \times \Delta/(\Delta + \delta)$$

$$I_C^T = I_C^o \times \Delta/(\Delta + 2\delta)$$

$$I_D^T = I_D^o \times \Delta/(\Delta + 3\delta) \qquad (4)$$

In equation (2), we need to replace the observed intensities, I^o, by the corrected intensities, I^T, as equation (2) implicity assumes that the integrated

Table 2. Possible local arrangements around the O(3) site in with (Mg, M^{2+}) at M(1) and M(3)

	M(1)	M(1)	M(3)	Band⁺	Band (A=Na)⁺⁺	Band (A=□)
(1)	Mg	Mg	Mg	A	A	A*
(2)	Mg	Mg	M^{2+}	B′		
(3)	M^{2+}	Mg	Mg	B″	B	B*
(4)	Mg	M^{2+}	Mg			
(5)	M^{2+}	M^{2+}	Mg	C′		
(6)	M^{2+}	Mg	M^{2+}	C″	C	C*
(7)	Mg	M^{2+}	M^{2+}			
(8)	M^{2+}	M^{2+}	M^{2+}	D	D	D*

⁺ ideal band-structure; (3) and (4) are symmetry equivalent and give rise to only one band: B″; similarly (5) and (6) are symmetry equivalent and give rise to only one band: C″.

⁺⁺ band structure usually observed; (2) and [(3)+(4)] are pseudo-symmetrically degenerate, and only one band (B) is observed experimentally; similarly, (5) and [(6)+(7)] are pseudo-symmetrically degenerate and only one band (C) is observed experimentally.

molar absorptivity is the same for all bands in the spectrum. Thus for the M^{2+} content of the M(1,3) sites, we may write

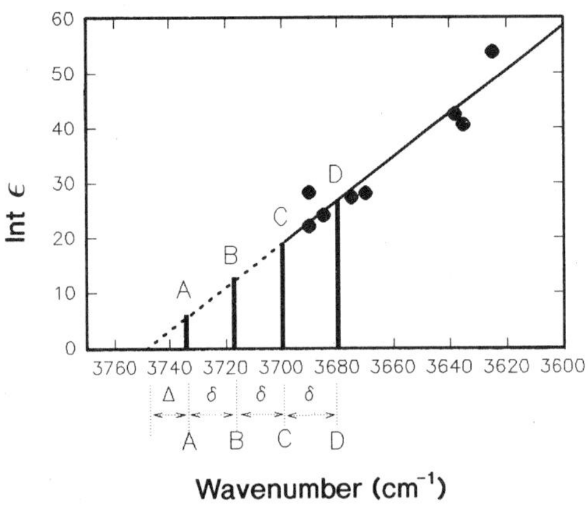

FIG. 3. The variation in integrated molar absorptivity vs. absorption frequency (data points for miscellaneous single-crystal amphibole spectra taken from SKOGBY and ROSSMAN, 1991) projected to zero absorptivity (at 3750 cm^{-1}). The band frequencies for the synthetic (Ni, Mg,Co)-potassium-richterites of this work are marked by the letters A, B, C, D corresponding to the band nomenclature of Table 1; the lengths of the vertical lines at the band frequencies for A, B, C and D are relative measures of the integrated molar absorptivities at the corresponding absorption frequencies.

$$M^{2+}_{M(1,3)} = I^T_B + 2I^T_C + 3I^T_D \qquad (5.1)$$

$$= I^o_B \times [\Delta/(\Delta + \delta)] + 2I^o_C \times [\Delta/(\Delta + 2\delta)] + 3I^o_D \times [\Delta/(\Delta + 3\delta)] \qquad (5.2)$$

The curve of SKOGBY and ROSSMAN (1991) intersects the wavenumber axis at 3750 cm^{-1}. For M^{2+} = Ni and Co, the A band occurs at ~3735 cm^{-1} and the separation between the individual A, B, C and D bands is 15 cm^{-1}; thus $\Delta = \delta = 15$ cm^{-1}. Insertion of these values into equation (5.2) allows calculation of the amount of Ni and Co at the M(1,3) sites from the relative band intensities.

There is a problem in directly applying this relationship to the synthetic amphiboles of this work due to the presence of a second set of absorptions, the A*. . . bands. However, we may correct the band intensities for this effect using the curve of Fig. 3 and the same algebraic argument given above. Insertion of these values into equation (5), together with the values $\Delta = \delta = 15$ cm^{-1} allows calculation of the amount of Ni or Co at the M(1,3) sites. The resultant values are shown in Fig. 4 (hollow symbols). Although there is a linear relationship between the two sets of values, there is also a complete lack of correspondence between the magnitudes of the two sets of values. For the most Ni- and Co-rich amphiboles, the predicted Ni and Co contents of the M(1,3) sites are incompatible with the bulk composition of the amphiboles.

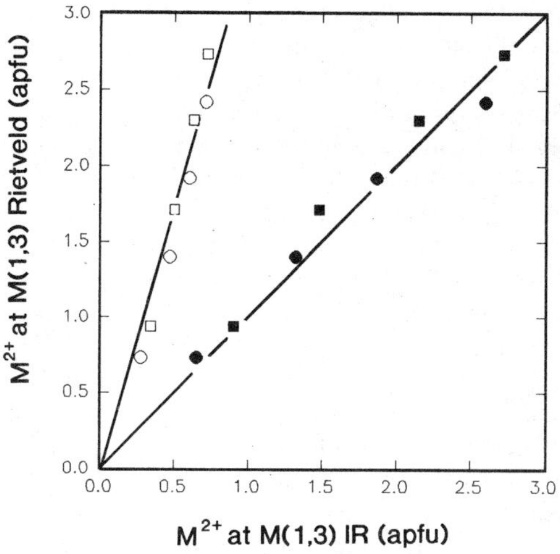

FIG. 4. Variation in Ni (squares) and Co (circles) content of the M(1,3) sites in (Ni,Mg,Co)-potassium-richterites calculated from the IR band intensities vs. the corresponding values measured by Rietveld structure-refinement; hollow symbols: calculated using the integrated molar-absorptivity curve of SKOGBY and ROSSMAN (1991); shaded symbols: calculated assuming the same relative integrated molar absorptivity (=1) for all bands in the spectrum.

If we assume that the integrated molar absorptivity does *not* vary as a function of absorption frequency, then we can sum the A. . . and A*. . . band-intensities and use the resulting values in equation (2). The predicted Ni and Co contents of the M(1,3) sites show very close agreement with the values measured by Rietveld structure-refinement (Fig. 4, shaded symbols). This provides us with somewhat of a conundrum, as this result seems totally incompatible with the results of SKOGBY and ROSSMAN (1991) and BURNS and HAWTHORNE (1994). Unfortunately, we *absolutely* cannot compare band intensities *between* samples, as SKOGBY and ROSSMAN (1991) were able to do with their single-crystal spectra, because of differential particle effects between samples, and hence we are limited to examining *relative* band intensities *within* a single spectrum. Nevertheless, the argument given above indicates that there is *not* any significant variation in integrated molar absorptivity with absorption frequency in the synthetic (Ni,Mg,Co)-potassium-richterites. Perhaps this is due to some form of coupling within a specific sample; this matter is currently being examined.

The A, B*, C*, D* bands*

Previous work on nominal end-member richterite and potassium-richterite has assigned the A* band to an empty A-site configuration (DELLA VENTURA *et al.*, 1991, 1993a; ROBERT *et al.*, 1989). In accord with this assignment, the quartet of bands A, B, C, D are accompanied by a corresponding set of A*, B*, C*, D* bands (Fig. 1). It is of interest to compare the relative intensities of the individual corresponding bands, as there is no intrinsic reason why the two sets of bands should show the same distribution of relative intensities. Fig. 5 compares the *normalized* relative intensities of all the bands calculated as $^{NORM}I^o_A = I^o_A/(I^o_A + I^o_B + I^o_C + I^o_D)$ and $^{NORM}I^o_{A*} = I^o_{A*}/(I^o_{A*} + I^o_{B*} + I^o_{C*} + I^o_{D*})$. There is a general correspondence of the individual bands, as they scatter about a 1:1 relationship. However, there are systematic patterns in the distribution of the intensities. The A* and D* bands tend to be relatively more intense than the A and D bands [except for the D band of Ni(80) and Co(80)], and the B* and C* bands tend to be relatively less intense than the B and C bands [except for the C* band in Co(80)]. This suggests that the ordering over M(1) and M(3) associated with a vacant A-site differs from that associated with an occupied A-site.

Short-range disorder

HAWTHORNE *et al.* (1996) show that the relative band-intensities of amphiboles involving binary solid-solution of C-group cations are dependent on both LRO and SRO. If the state of LRO is known, then the state of SRO may be derived from the

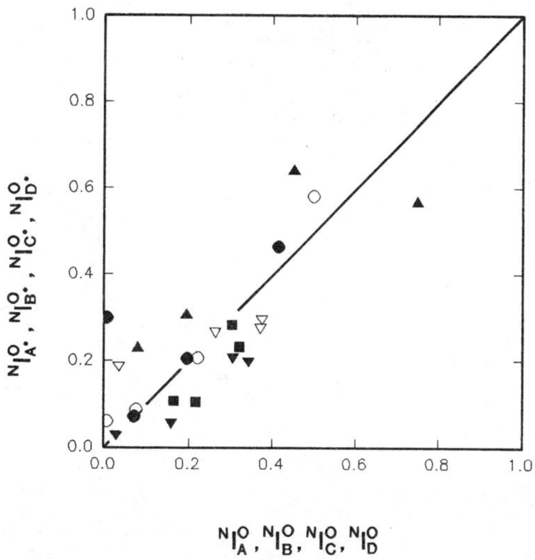

FIG. 5. The relative intensities of the A* (○), B* (▽), C* (□) and D* (△) bands vs. the relative intensities of the A, B, C and D bands in (Mg,Ni)-potassium-richterite (full symbols) and (Mg,Co)-potassium-richterite (hollow symbols).

observed band intensities. Referring to equation (1), although SRO can change the values of $I^o_A \ldots$, the intensity sums of equation (1) are fixed by the state of LRO over the M(1) and M(3) sites. Assuming no SRO, the relative band intensities are given by the following relations (LAW, 1976; HAWTHORNE *et al.*, 1996):

$$I^C_A = (1 - \tfrac{1}{2}[Y - y_{M(3)}])^2(1 - y_{M(3)})$$

$$I^C_B = (1 - \tfrac{1}{2}[Y - y_{M(3)}])^2 \times y_{M(3)} + 2(1 - \tfrac{1}{2}[Y - y_{M(3)}]) \times \tfrac{1}{2}[Y - y_{M(3)}](1 - y_{M(3)})$$

$$I^C_C = \tfrac{1}{4}[Y - y_{M(3)}]^2(1 - y_{M(3)}) + 2(1 - \tfrac{1}{2}[Y - y_{M(3)}]) \times \tfrac{1}{2}[Y - y_{M(3)}]y_{M(3)}$$

$$I^C_D = \tfrac{1}{4}[Y - y_{M(3)}]^2 \times y_{M(3)} \qquad (6)$$

where Y is the total amount of M^{2+} cations at the M(1) and M(3) sites. HAWTHORNE *et al.* (1996) show that specific patterns of deviation of the observed band intensities, $I^o_A \ldots$, from the calculated band intensities, $I^C_A \ldots$, are characteristic of particular patterns of SRO. In accord with the relation of Fig. 4 (squares and circles), we sum corresponding A... and A*... band intensities to use in this calculation. Although Fig. 5 indicates that there are systematic differences in the two sets of intensities, the A... set is sufficiently more intense than the A*... set that any error introduced by this summation procedure should be negligible. The observed and calculated values are shown in Fig. 6. The general pattern of values does deviate from that expected for complete short-range disorder (EEEE), following the general scheme UEDu in the nomenclature of HAWTHORNE *et al.* (1996). This corresponds to there being significant MgMgMg clustering and very slight NiNiNi (or CoCoCo) clustering.

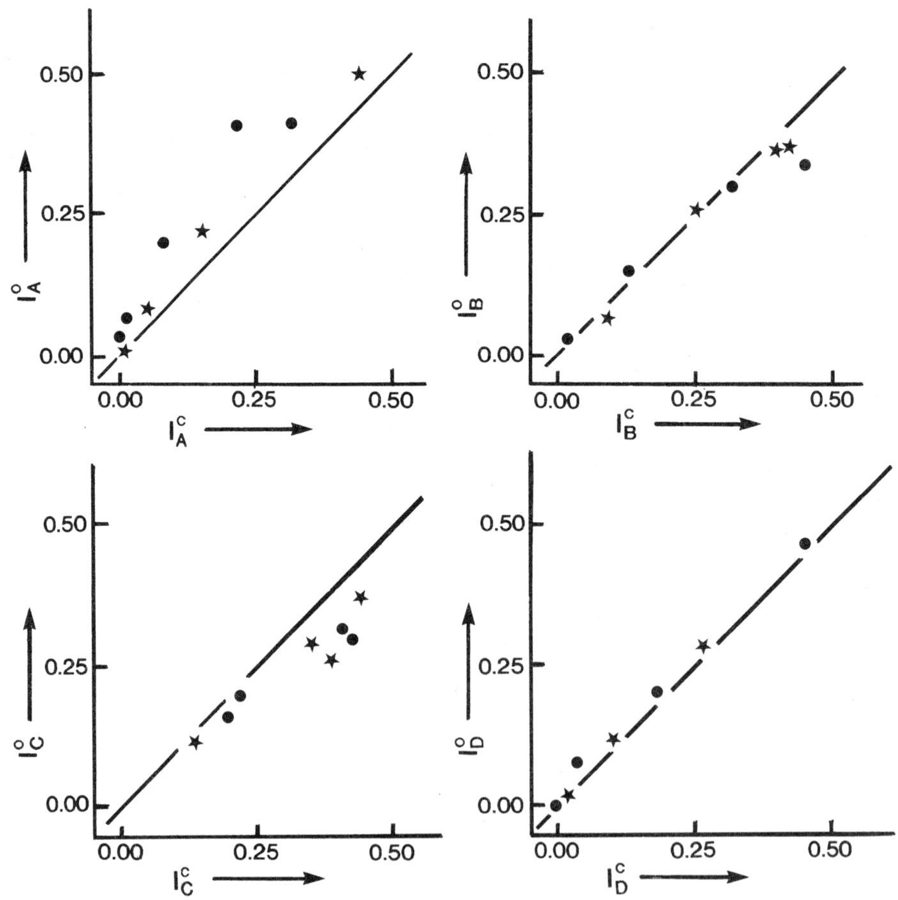

FIG. 6. The observed relative intensities of the A, B, C and D bands ($I^o_A \cdots$) versus the relative intensities of these bands calculated from equation (6), assuming no SRO in the amphibole; ●: (Mg,Ni)-potassium-richterite; ○: (Mg,Co)-potassium-richterite.

Fig. 6 shows the behaviour for two separate richterite series, (Mg,Ni)-potassium-richterite and (Mg,Co)-potassium-richterite; we see the same type of behaviour for both series. This suggests that the deviation from an EEEE pattern of band-intensity differences is a real effect. This deviation cannot be due to variation in band intensity as a function of band frequency, as discussed above. Variation in molar absorptivity with band frequency as described by SKOGBY and ROSSMAN (1991) would *decrease* the relative intensity of I_A^o, whereas I_A^o is larger than I_A^C. Thus, if the relative intensities are affected by such variations in molar absorptivity, the real deviation from an EEEE pattern of band intensities will be even greater than that currently observed.

SUMMARY

(1) There seems to be no effect of principal OH-stretching band frequency on band intensity (molar absorptivity) in these amphiboles.

(2) Satellite bands (denoted as A*. . .) show a different pattern of intensities than the corresponding A. . . bands. This indicates that (Mg,Ni) and (Mg,Co) ordering differs around an empty A-site from that around an occupied A-site.

(3) Comparison of the observed band-intensities with ideal values calculated for no SRO indicate that there is significant MgMgMg clustering and slight NiNiNi/CoCoCo clustering in these amphiboles.

Acknowledgements—We thank David M. Jenkins and an anonymous reviewer for their comments which helped improve the clarity of this paper. Financial assistance was provided by a CNR grant to GDV and by an Operating grant from the Natural Sciences and Engineering Research Council of Canada to FCH.

REFERENCES

BURNS P. C. and HAWTHORNE F. C. (1994) Structure and hydrogen bonding in preobrazhenskite, a complex heteropolyhedral borate. *Can. Mineral.* **32,** 387–396.

BURNS *R. G. and S*TRENS R. G. J. (1966) Infrared study of the hydroxyl bonds in clinoamphiboles. *Science* **153,** 890–892.

DELLA VENTURA G. (1992) Recent developments in the synthesis and characterization of amphiboles. Synthesis and crystal chemistry of richterites. *Trends in Mineral.* **1,** 153–192.

DELLA VENTURA G. and ROBERT J.-L. (1990) Synthesis, XRD and FTIR studies of strontium richterites. *Eur. Jour. Mineral.* **2,** 171–175.

DELLA VENTURA G., ROBERT J.-L. and BÉNY J.-M. (1991) Tetrahedrally coordinated Ti^{4+} in synthetic Ti-rich potassic richterite: Evidence from XRD, FTIR and Raman studies. *Amer. Mineral.* **76,** 1134–1140.

DELLA VENTURA G., ROBERT J.-L., RAUDSEPP M., and HAWTHORNE F. C. (1993a): Site occupancies in monoclinic amphiboles: Rietveld structure refinement of synthetic nickel magnesium cobalt potassium richterite. *Amer. Mineral.* **78,** 633–640.

DELLA VENTURA G., ROBERT J.-L., BÉNY J.-M., RAUDSEPP M., and HAWTHORNE F.C. (1993b) The OH—F substitution in Ti-rich potassium-richterites: Rietveld structure refinement and FTIR and micro-Raman spectroscopic studies of synthetic amphiboles in the system $K_2O—Na_2O—CaO—MgO—SiO_2—TiO_2—H_2O—HF$. *Amer. Mineral.* **78,** 980–987.

GROAT L. A., HAWTHORNE F. C., ROSSMAN G. R. and ERCIT, T. S. (1995): The infrared spectroscopy of vesuvianite in the OH region. *Can. Mineral.* **33,** 609–626.

HAWTHORNE F. C. (1981) Amphibole Spectroscopy. In *Amphiboles and Other Hydrous Pyriboles–Mineralogy.* (ed. D. R. VEBLEN). *Rev. Mineral.* 9a, pp. 103–140. Mineralogical Society of America.

HAWTHORNE F. C. (1983) The crystal chemistry of the amphiboles. *Can. Mineral.* **21,** 173–480.

HAWTHORNE F. C., DELLA VENTURA G. and ROBERT J.-L. (1996) Short-range order and long-range order in amphiboles: a model for the interpretation of infrared spectra in the principal OH-stretching region. *Geochim. Cosmochim. Acta* (in press).

LAW A. D. (1976) A model for the investigation of hydroxyl spectra of amphiboles. In *The Physics and Chemistry of Minerals and Rocks.* (ed. R. G. J. STRENS), pp. 677–686. Wiley.

RAUDSEPP M., TURNOCK A. C. and HAWTHORNE F. C. (1991) Amphiboles synthesis at low-pressure: what grows and what doesn't. *Eur. Jour. Mineral.* **3,** 983–1004.

RAUDSEPP M., TURNOCK A. C., HAWTHORNE F. C, SHERRIFF B. L. and HARTMAN J. S. (1987) Characterization of synthetic pargasitic amphiboles ($NaCa_2Mg_4M^{3+}Si_6Al_2O_{22}(OH,F)$; M^{3+} = Al, Cr, Ga, Sc, In) by infrared spectroscopy, Rietveld structure refinement and ^{27}Al, ^{29}Si, and ^{19}F MAS NMR spectroscopy. *Amer. Mineral.* **72,** 580–593.

ROBERT J.-L., DELLA VENTURA G. C. and THAUVIN J.-L. (1989) The infrared OH-stretching region of synthetic richterites in the system $Na_2O—K_2O—CaO—MgO—SiO_2—H_2O—HF$. *Eur. Jour. Mineral.* **1,** 203–211.

SEMET M. P. (1973) A crystal-chemical study of synthetic magnesiohastingsite. *Amer. Mineral.* **58,** 480–494.

SKOGBY H. and ROSSMAN G. R. (1991) The intensity of amphibole OH bands in the infrared absorption spectrum. *Phys. Chem. Mineral.* **18,** 64–68.

STRENS R. G. J. (1966) Infrared study of cation ordering and clustering in some (Fe,Mg) amphibole solid solutions. *Chem. Comm.* **15,** 519–520.

WHITTAKER E. J. W. (1979) Clustering of cations in amphiboles. *Phys. Chem. Mineral.* **4,** 1–10.

Mineral Spectroscopy: A Tribute to Roger G. Burns
Editors: M. D. Dyar, C. McCammon and M. W. Schaefer

Raman spectroscopy of crude oils and hydrocarbon fluid inclusions: A feasibility study

DANIEL ORANGE,* ELISE KNITTLE, DANIEL FARBER,‡ and QUENTIN WILLIAMS
Department of Earth Sciences and Institute of Tectonics, University of California, Santa Cruz, CA 95064, U.S.A.

Abstract—Visible Raman spectroscopy provides useful constraints on the structure and maturity of both hydrocarbon fluid inclusions and crude oils. In particular, the presence or absence of hydrocarbon functional groups may be qualitatively identified, as well as relative degrees of maturation. The primary constraints on the utility of this technique are: 1) spectral contamination by fluorescence of either the hydrocarbon inclusion or the matrix in which it sits; and 2) photodegradation of hydrocarbons under laser excitation. Both of these difficulties can be averted through a combination of appropriate selection of the laser excitation line and minimization of the excitation volume of fluorescence. We present and interpret spectra of synthetic hydrocarbon inclusions, natural hydrocarbon inclusions contained in a matrix fluorescent under blue-green laser excitation ($CaCO_3$:Mn) from the Hoh Accretionary Complex (Olympic Peninsula, Washington), and of crude oils collected from the same region. The latter two types of samples demonstrate the utility of this technique in providing constraints on the structural differences between natural hydrocarbons.

INTRODUCTION

ORGANIC MATTER is present in all sedimentary material, and when such sediment undergoes diagenesis, catagenesis, and metamorphism, this material can be transformed into organic liquids and gases (TISSOT and WELTE, 1984; GAUTIER *et al.*, 1985). Organic fluids present in rock include hydrocarbons ranging from methane to high molecular weight hydrocarbons (KVENVOLDEN and ROEDDER, 1971), and these fluids can undergo complex chemical changes depending on the conditions present during their migration and maturation. Hydrocarbon inclusions are direct samples of the fluid present at that time, and can provide information on the composition of as well as the conditions under which the fluids were emplaced (ROEDDER, 1984). Hydrocarbon inclusions are thus of interest because their compositions are sensitive to the complex interplay between source, transport, and deposition, while the variation in composition between present-day crude oils and inclusions can provide information on the genesis of the fluid (*e.g.*, JENSENIUS and BURRUSS, 1990; MCLIMANS, 1987; ROEDDER, 1984; BURRUSS *et al.*, 1985; NOONER *et al.*, 1973; PERING, 1973). Numerous studies, principally optical, fluorescence, and gas chromatographic measurements, have been conducted on hydrocarbon inclusions to elucidate the evolution of organic material under a range of geological conditions (PIRONON and BARRES, 1990; JENSENIUS and BURRUSS, 1990; ETMINAN and HOFFMAN, 1989; MCLIMANS and VIDETICH, 1987; MCLIMANS, 1987; PAGEL *et al.*, 1986; HORSFIELD and MCLIMANS, 1984; BURRUSS *et al.*, 1985; KVENVOLDEN and ROEDDER, 1971; MURRAY, 1957).

Distinct advantages exist for techniques able to probe individual hydrocarbon fluid inclusions: their composition (and all the information contained therein) may be placed in a relative temporal sequence based on cross-cutting relationships with microfabrics and with inclusions in other veins and cements (MCLIMANS, 1987). Temporal variations in the chemical composition of hydrocarbons within a region can thus provide information on the evolution of these fluids (TISSOT and WELTE, 1984). By analyzing inclusions on a grain by grain basis, it may be possible to differentiate inclusions from radically different reservoirs, as well as inclusions formed by mixing of different sources. When placed in a diagenetic/structural/temporal framework, the compositions of crude oils and hydrocarbon inclusions potentially allow constraints to be placed on the timing of fluid transport events, the reservoirs involved in each event, and on the maximum temperatures attained by the material present in the inclusion (MCLIMANS, 1987; ROEDDER, 1984; TISSOT and WELTE, 1984).

Typically, compositional studies of individual inclusions have been conducted using vibrational spectroscopic techniques. Analyses of hydrocarbon inclusions have been conducted using infrared spectroscopic techniques (PASTERIS *et al.* 1983;

Present address: *Monterey Bay Aquarium Research Institute, P.O. Box 628, Moss Landing, CA 95039, U.S.A. and ‡Dept. of Earth and Space Sciences, University of California, Los Angeles, CA 90024, U.S.A.

BARRES *et al.*, 1987; GUILHAUMOU and SZYDLOWSKI, 1989; O'GRADY *et al.*, 1989; WOPENKA *et al.*, 1990; PIRONON and BARRES, 1990), yet such measurements are limited by diffraction effects to inclusions larger than ~15 μm in diameter (WOPENKA *et al.*, 1990), and a number of hydrocarbon vibrations are weak or inactive in the infrared (DOLLISH *et al.*, 1974). Raman spectroscopic techniques, although frequently applied to inorganic inclusions, have been underutilized in the study of hydrocarbon inclusions. Indeed, studies on natural organic inclusions have been primarily conducted on methane-bearing assemblages, particularly those in the N_2-CO_2-CH_4 system (CHOU *et al.*, 1990; PASTERIS *et al.*, 1985, 1988; SEITZ *et al.*, 1987, 1989). Exceptions to this include the study of DELE-DUBOIS *et al.* (1980) on hydrocarbon (possibly pentane) inclusions in emerald, and that of GUILHAUMOU *et al.* (1988) who, although unable to spectroscopically characterize fluorescent primary and secondary inclusions, were able to analyze for methane, ethane, and propane in vapor-dominated inclusions. PIRONON and BARRES (1990) obtained spectra of a single hydrocarbon inclusion with exceptionally low fluorescence, and interpreted their results as indicating the presence of n-heptane and a minor amount of aromatics.

Raman spectra of more complicated hydrocarbon inclusions are considerably less common: this originates in the perception that these studies are difficult or impossible (ROEDDER, 1984; TOURAY *et al.*, 1985; BURRUSS *et al.*, 1989). The usual reasons cited for the lack of feasibility of such measurements are several: first, the presence of laser-induced fluorescence from both natural hydrocarbons and many host minerals can overwhelm the Raman scattering signal (ROSASCO and ROEDDER, 1979; GUILHAUMOU, 1982; TOURAY *et al.*, 1985; GUILHAUMOU *et al.*, 1988; WOPENKA *et al.*, 1990); second, laser-induced photodegradation may alter the chemical composition of the inclusion (TOURAY *et al.*, 1985; GUILHAUMOU, 1982); and finally, the low laser power per area required in studies of these inclusions would make the Raman scattering intensity correspondingly low (TOURAY *et al.*, 1985; ROSASCO and ROEDDER, 1979).

Although each of these criticisms has merit for Raman experiments on natural inclusions using either blue or green laser lines of the argon-ion lasers used in many Raman laboratories, one of the intents of this paper is to demonstrate that a judicious choice of laser excitation, coupled with photon counting detection and a low incident laser intensity, render Raman spectra of highly fluorescent natural inclusions in a fluorescent matrix feasible, if not routine. In particular, the utilization of a laser which is not absorbed by either the inclusion or the host mineral is critical for the success of such measurements.

Recently, PIRONON *et al.* (1991) have reported Raman spectra in the C—H stretching region of several hydrocarbon inclusions using Fourier Transform-Raman spectroscopy with excitation at 1064 nm. In this study we will first demonstrate that the interferometer and specialized infrared laser and optical system necessary for FT-Raman measurements is not mandatory for conducting Raman spectroscopy on complex hydrocarbon fluid inclusions (indeed, they may be conducted with equipment available in many Raman labs). Secondly, we will demonstrate the ability of this non-destructive technique to identify different functional units within (and thus, on the thermal and geochemical history of) hydrocarbon inclusions.

In addition to providing information on the chemistry of the inclusions, Raman spectroscopy also has the advantage of having extremely small spatial resolution (as small as 2 μm: PASTERIS *et al.*, 1988). Thus, it is possible to carry out studies on an inclusion by inclusion basis in a sample, thereby characterizing multiple phases of fluid interaction. These individual inclusions can also be compared with present-day crude oils to examine differences in hydrocarbon composition, information that may relate to the evolutionary history of both the host rock and the oil.

In this paper, we present a study of a suite of synthetic hydrocarbon fluid inclusions, n-pentadecane, bromobenzene and cyclohexane, to illustrate the feasibility of studying fluorescent hydrocarbons using Raman spectroscopy. We then apply the technique to two fluorescent and chemically complex crude oils from the Hoh accretionary complex, Olympic Peninsula, Washington. Previous gas chromatographic studies of these oils indicate that they are similar in composition: our results are in accord with these analyses (KVENVOLDEN *et al.*, 1989). Finally, we present an analysis of a representative natural hydrocarbon fluid inclusion. This inclusion occurs in rocks located near the crude oils, although the inclusions were trapped in the Miocene (ORANGE *et al.*, 1993). The inclusion is qualitatively similar in structure to the crude oils, yet differs in several important respects that may reflect a different migration history.

EXPERIMENTAL METHODS

Raman system

The Raman spectra of the synthetic fluid inclusions were collected using a Coherent Innova 70–4 argon-ion laser tuned at 488 nm. An Optometrics laser monochromator was used to remove plasma lines, and the laser beam at the sample was attenuated to between 1 and 5 mW. A modified Leitz Orthoplan microscope is used to image the fluid inclusions and to focus the laser into the sample. The microscope is fitted with a Leitz UM 32X objective to focus the laser and collect the scattered light in a 0 degree backscattering geometry. An aperture in the microscope system is used to isolate the sample area containing the fluid inclusion. The scattered light is then focused with an Olympus camera lens into the front slit of a SPEX Triplemate spectrograph equipped with a 600 gr/mm holographic grating in the filter stage and a 1200 gr/mm holographic grating in the spectrograph stage. Spectra are collected using a thermoelectrically cooled ITT Model F4146 two-dimensional micro-channel plate controlled by a Surface Sciences Laboratory Position Computer (Quantar Technologies). This detector couples the photon-counting resolution, high quantum efficiency and high signal to noise of a conventional photomultiplier tube with the two dimensionality of CCD detectors (VEIRS *et al.*, 1987). Because of the photon-counting ability of this detector and the ability to collect spatially resolved data in a spectrographic mode (that is, without scanning a monochromater), we are able to collect Raman data using extremely low incident light levels. Typical exposure times for these inclusions were 1 to 6 hours.

For the natural fluid inclusion and the crude oils studied, a Uniphase 18 mW He—Ne laser (632.8 nm) was used for sample excitation: both the natural fluid inclusions and the crude oils, as well as the host calcite matrix of the natural inclusions, strongly fluoresce at argon-ion wavelengths (488.0 and 514.5 nm), so a red laser was used to minimize sample absorbance and fluorescence. Oil samples were contained in fused silica microtubes. Laser plasma lines were removed with a standard Oriel laser-line filter, and the laser power at the sample was again between 1 and 5 mW. For the natural fluid inclusions, a Leitz 100X microscope objective was used to focus the laser beam into the inclusion and collect the scattered light. The exposure times for the Raman scattering were 12 to 24 hours. The low levels of Raman excitation used did not cause detectable photodegradation in either the oils or fluid inclusions, as manifested by either a change in visual appearance of the inclusion, or by any time dependence of the spectra.

Sample preparation

J. Pironon of CREGU (France) provided synthetic n-pentadecane, bromobenzene and cyclohexane fluid inclusions. These inclusions are made using a technique described elsewhere (PIRONON, 1990), and are typically 10–50 mm in diameter. In contrast to most natural inclusions, these synthetic inclusions were contained in high purity KCl, which not only does not fluoresce, but also has no first-order Raman spectrum. Moreover, these samples are thick and monocrystalline, reducing the amount of stray scattered light from the sample. Thus, the solid matrix containing the synthetic inclusions is comparatively ideal for Raman spectroscopic measurements. We limited the light intensity incident on the sample to less than 5 mW in all cases, and utilized highly attenuated Ar-ion laser light to obtain spectra of these inclusions (~1 mW).

Natural hydrocarbon fluid inclusions were obtained from a suite of rocks collected from the Hoh rock assemblage of the western Olympic peninsula, Washington (RAU, 1973, 1975, 1979). The natural hydrocarbon fluid inclusions occur most frequently in syn-deformational sparitic calcite veins. These inclusions range from 2 to 60 μm in long dimension, and are commonly in the 5 to 15 μm range. Inclusions vary in color from translucent to light brown to dark brown, and fluoresce strongly (under incident UV) in colors ranging from blue to greenish yellow (~485–535 nm). Highly polished standard thin sections (30 mm thick) were prepared from the Hoh rock samples for petrographic and Raman analysis. These were separated from their supporting glass slides, and the rock chip was mounted on a block of NaCl, as halite has no first order Raman spectrum. The use of kerosene was avoided in all parts of the sample preparation process because of possible interference of hydrocarbon contamination with the Raman analyses of the fluid inclusions. Extensive polish helped prevent excess light scattering off surface topography of the sample.

We initially examined thin sections of samples using optical and cathodoluminescence petrography in order to clarify the relative cross-cutting relationships between the fluid inclusions and the microstructures of the sample (ORANGE *et al.*, 1993). We then carried out detailed optical and ultraviolet fluorescence petrography to identify potential hydrocarbon inclusions. Inclusions that appeared light tan to dark brown in visible light and fluoresced under ultraviolet light were selected for further study using Raman spectroscopy. Both the manganese-bearing calcite matrix and the complex hydrocarbons present within these inclusions necessitated utilization of He—Ne laser excitation to produce Raman spectra.

BACKGROUND

The three primary experimental difficulties in Raman spectroscopy of fluid inclusions are: 1) laser-induced fluorescence from the solid matrix holding the inclusion; 2) fluorescence from the material of the inclusion itself; and 3) laser-induced photodegradation within the inclusion. The first of these problems is endemic to all inclusions within certain types of minerals: the example of primary relevance to this study is that of calcite with Mn^{2+} as a trace substituent. The latter two problems, inclusion fluorescence and photodegradation, are each primarily associated with hydrocarbon inclusions, and are generated by the range of energies of anti-bonding excited electronic states available to complex hydrocarbons. In the following discussion, we emphasize the fluorescence behavior of hydrocarbons present in moderately mature inclusions (falling generally toward the blue end of the visible spectrum under UV excitation). In particular, we focus on the fluorescence and photodegradation (or lack thereof) undergone by alkanes (paraffins), cycloalkanes (naphthenes), and aromatics (particularly benzenoids) under various types of laser irradiation. We explicitly do not treat more complex organic compounds of limited abundances and thermal stabilities, such as porphyrin-based materials. Our intent is to document that for a range of hydrocarbon inclusions, both laser-induced sample photodegradation and sample fluorescence can be reduced to noninterferring levels, and that significant qualitative insight into the molecular species present within these inclusions (and thus insight into their thermal and chemical histories) may be gained from appropriately collected Raman spectra.

Notably, both photodegradation and fluorescence are critically dependent upon the absorption spectrum of the inclusion or host mineral: as a result, these problems may be avoided by an appropriate choice of the Raman excitation wavelength. That is, avoiding the absorption bands of a hydrocarbon is crucial when choosing a Raman excitation line. We utilize this relatively straightforward approach to determine spectra of both our natural and synthetic inclusion samples. Our guiding principle is to use the highest frequency (lowest wavelength) laser line feasible for each mineral/inclusion pair: this choice is motivated by the (generally) higher detector sensitivity for higher frequency photons and by the enhanced Raman scattering efficiency obtainable at shorter excitation wavelengths.

Solid matrix fluorescence

In this discussion, we deal principally with the example of Mn^{2+} impurity fluorescence within the calcite lattice. Although the fluorescent characteristics of each host mineral (or its associated impurities) with inclusions of interest should be evaluated on an individual basis prior to Raman spectral study, we consider the case of $CaCO_3:Mn^{2+}$ to represent a particularly severe test for visible laser-induced Raman scattering measurements. The difficulties associated with this mineral (and specifically our samples of this mineral) are several-fold. First, the fluorescence of manganese in these samples of calcite is intense and, under blue excitation, overwhelms the Raman signal from not only the inclusion, but also the host calcite. This fluorescence intensity is particularly large in calcite partially because the quenching ability of other ions, particularly Fe^{2+}, in calcite is low (MEDLIN, 1968; TEN HAVE and HEIJNEN, 1985). Second, the Mn^{2+} $T_{1g} \rightarrow A_{1g}$ emission band lies between 520 and 730 nm, a region of primary interest for Raman scattering experiments using blue laser excitation (MEDLIN, 1968; WALKER *et al.*, 1989). Finally, the absorption bands which pump the emission band of manganese in calcite span much of the visible region of the spectrum from 450 to 630 nm (MEDLIN, 1968; WALKER *et al.*, 1989).

Despite its fluorescent characteristics, we note the importance of being able to conduct Raman spectral measurements on hydrocarbon fluid inclusions in a calcite matrix containing transition-metal ion impurities. Because liquid hydrocarbons form in a comparatively low pressure, low temperature environment, inclusions containing such fluids are frequently found in calcite veins.

In order to measure Raman spectra in this matrix, we utilized the 632.8 nm line of a He-Ne laser to produce Raman spectra of the natural inclusions, instead of the more commonly used Ar-ion emission lines at 457.9, 488.0 and 514.5 nm. The 632.8 nm line of He-Ne, although significantly less powerful (for lasers of comparable expense) than those of the argon-ion laser, has the advantage that the resultant fluorescence of calcite is extremely low, and thus the line is able to produce Raman spectra of acceptable quality. The physical mechanism behind this improvement is simple: the incident laser wavelength is chosen so that the sample absorption (and thus emission) is minimized. A spectra produced under these conditions of a non-inclusion bearing monocrystalline region of this calcite from the Hoh accretionary prism is shown in Fig. 1. No significant fluorescence is observed, and the primary Raman lines of the calcite spectrum are well-resolved, as is the isotopically shifted component of the strong symmetric stretching vibration of the carbonate unit (inset, Fig. 1). Because this spectrum is of a non-oriented single crystal, the relative peak intensities differ somewhat from those of calcite powder, although their location is in excellent agreement with previous studies (*e.g.*, WHITE, 1974). Notably, either a Kr-ion laser or red-tuned dye laser could be used with similar effect (and with potentially greater power). Thus, judicious selection of an appropriate visible (or infrared) laser line can minimize or eliminate the effects of unwanted mineral fluorescence in Raman spectra of fluid inclusions in calcite. We note that the alternative technique of time lapse spectroscopy, which takes advantage of the long lifetime of fluorescence relative to vibrational measurements, can also be utilized to eliminate fluorescence effects (*e.g.*, FRANTZ *et al.*, 1982).

Petroleum fluorescence and laser-induced photodegradation

The relatively high absorption of many hydrocarbon inclusions in the ultraviolet and resultant

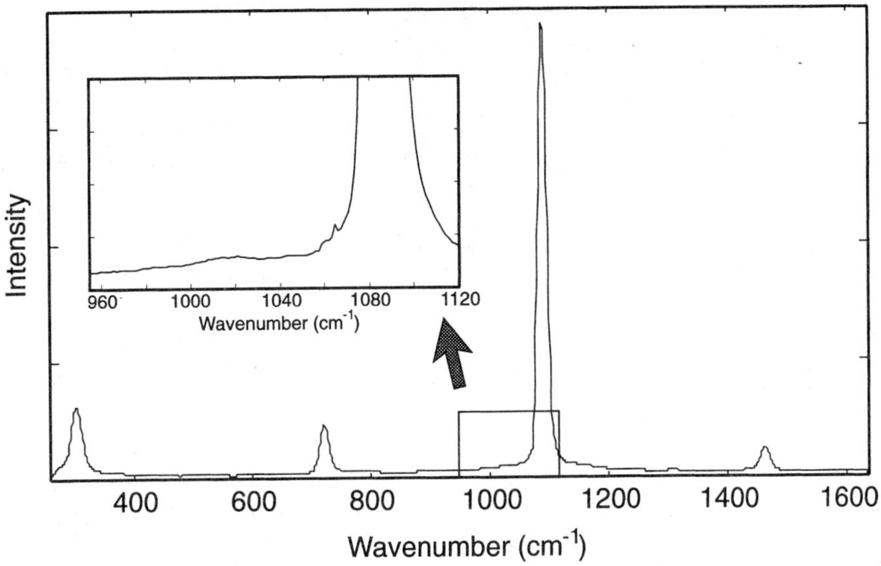

FIG. 1. Raman spectrum of manganese-bearing calcite from the Hoh accretionary peninsula: the spectrum was produced using 632.8 nm excitation to avoid manganese-induced sample fluorescence. The inset shows the spectral feature associated with the symmetric stretching vibration of carbon-13 bearing carbonate groups.

visible fluorescence (note that for simplicity we do not distinguish between fluorescence and phosphorescence of hydrocarbons, but use the former term to describe all light emission associated with relaxation of electronic excited states) has been widely used to verify the presence of hydrocarbons within fluid inclusions (BURRUSS, 1981; BURRUSS *et al.*, 1983, 1985; HASZELDINE *et al.*, 1984; HORSFIELD and MCLIMANS, 1984; PAGEL *et al.*, 1986; GUILHAUMOU *et al.*, 1988; JENSENIUS and BURRUSS, 1990). Because of the high flux of photons associated with laser Raman measurements, as well as the relative weakness of the spontaneous Raman effect (efficiencies typically of order 10^{-6}: NAKAMOTO (1982)), even a small amount of sample absorption (and reemission) may induce a sufficient amount of fluorescence to interfere with the Raman spectrum. The most efficient mechanisms by which hydrocarbons fluoresce is by transitions from bonding or non-bonding states to anti-bonding states, principally either $\pi \rightarrow \pi^*$ or $n \rightarrow \pi^*$ transitions, followed by subsequent radiative relaxation to the ground state (*e.g.*, BECKER, 1969). The lower energy of the resultant fluorescent photon from that of the initial excitation photon is primarily produced by the environment of the excited molecule shifting to a lower energy configuration following the excitation, but prior to emission. This phenomenon is described by the Franck-Condon effect (*e.g.*, ENGLMAN, 1979).

Notably, in contrast to the Raman effect, characteristic quantum efficiencies of the fluorescence of organic compounds (particularly aromatics) are high: typically between 0.05 and 0.99 (DAWSON and WINDSOR, 1968). It is these high efficiencies, when coupled with the comparative weakness of the Raman effect, which have rendered characterization of the Raman spectra of hydrocarbon fluid inclusions difficult: even a trace quantity of hydrocarbons which fluoresce under the incident laser excitation can generate a background larger than that of the Raman scattering from the sample.

Fortunately, the primary absorption bands of simple alkanes and cycloalkanes fall below 200 nm in wavelength. The major simple ($C_n < \sim 30$) naturally-occurring carbon compounds with absorption bands which can either be present in, or tail into, the blue region of the visible are either aromatic compounds (particularly those with multiple rings) or polyolefins with $C_n > \sim 10$. For most hydrocarbons of this type, absorption seldom extends to wavelengths higher than about 550 nm (BERLMAN, 1965; DIAS, 1987; BURDETT and TAYLOR, 1955). We note in this context that extremely immature hydrocarbons with significantly more complex molecules might require infrared excitation to avoid sample absorption/fluorescence (*e.g.*, PIR-

ONON *et al.*, 1991). However, for inclusions varying in net complexity from methane, ethane, and other alkanes (*e.g.*, SEITZ *et al.*, 1987; GUILHAUMOU *et al.*, 1988; PIRONON and BARRES, 1990) to those with hydrocarbons of intermediate complexity, most complex hydrocarbon fluorescence and/or photodegradation can again be avoided by utilization of an appropriate visible (in this case, red) laser line.

In this context, it is of critical importance to examine any hydrocarbon inclusions for visible absorption prior to Raman spectral characterization. In the case of each of the synthetic hydrocarbons we have examined, the absorption spectra of the hydrocarbons are well-known (*e.g.*, BERLMAN, 1965; *Sadtler Handbook of UV Spectra,* 1982), and each is amenable to examination using either the 488.0 or 514.5 nm lines of argon-ion lasers. This is again in contrast to the case of the natural inclusions examined, which not only fluoresce under excitation at these frequencies, but also appear to undergo photodegradation. The 632.8 nm line of the He-Ne laser avoids not only the problem of fluorescence but also, by implication, that of photodegradation.

Photodegradation of hydrocarbons results from the enhanced reactivities of molecules within their excited states: in this sense, photodegradation and fluorescence represent competitive processes of non-radiative and radiative decay of excited states (*e.g.*, COWAN and DRISKO, 1976). Although small amounts of photodegradation are difficult to preclude from having occurred within our natural inclusions, we are confident that photodegradation from 632.8 nm excitation does not result in any major changes in the chemical species present within these inclusions. Our rationales for this are several-fold. First, no time-dependence of the spectra is observed (other than the normal increase in signal to noise for longer collection times). Neither are there any observable changes in the physical appearance of the inclusion during or following laser-irradiation (for reference, under argon-ion excitation, these inclusions were observed to darken in color and seep into cleavage planes in the calcite, indicating that the material in the inclusion had undergone both a chemical change and a decrease in viscosity). Finally, the lack of fluorescence from these samples, as well as their apparent transparency to red light, implies that any absorption in the sample at this wavelength must be weak.

RESULTS AND DISCUSSION

Synthetic fluid inclusions

n-Pentadecane. The n-pentadecane ($C_{15}H_{32}$) inclusion studied is a colorless, pseudospherical, one-phase inclusion, 35 mm in diameter. Under long-wavelength ultraviolet light (366 nm Hg) the n-pentadecane inclusion exhibits medium intensity impurity fluorescence in a dull green color (~530 nm). This fluorescence was manifested as a broad, low amplitude signal centered at 1800 cm^{-1} away from the incident 488 nm laser line, with a half-wavelength of ~300 cm^{-1}. We note that the impurity level in these samples is likely to be one part in 10^3-10^6. The fluorescence peak was small and centered away from any peaks of interest, and thus presented no difficulties in terms of collecting Raman spectra. The spectra for this inclusion with peak assignments (*e.g.*, DOLLISH *et al.*, 1974) are shown in Fig. 2a.

PIRONON and BARRES (1990) published infrared and Raman spectra of these pentadecane inclusions in the C—H stretching spectral region (between 2850 and 2975 cm^{-1}) using both synthetic inclusions and a liquid standard. Their peak identifications for both the standard and the inclusion are identical to those of this study; our spectra (Fig. 2) however, show a stronger methyl C—H symmetric stretch (2848 cm^{-1}), and resolve a number of side peaks (possibly combination bands). PIRONON and BARRES (1990) also identify a small peak at 2730 cm^{-1} which we do not observe in our spectra. Many of the bands of these synthetic hydrocarbons (such as the C—H stretching vibrations) are endemic to hydrocarbons, and are accordingly not particularly useful in determining the presence or absence of different structural groups. Instead, we focus on bands which allow us to discriminate between functional groups, in order to produce particular regions of interest for unknown inclusions. For example, the vibrations centered at 1440 cm^{-1} are due to the methylene scissors vibration, and therefore are dependent on the presence of $—CH_2—$ groups; peaks near this vibrational frequency are indicative of alkanes and cycloalkanes.

Cyclohexane. The cyclohexane (C_6H_{12}) occurs as numerous pseudospherical inclusions ranging from 25 to 80 mm in diameter. The vast majority of these inclusions are one phase, although several contain a small bubble (probably aqueous) as well. Under long wavelength ultraviolet light (366 nm Hg) these inclusions exhibit an extremely subtle impurity fluorescence in a very dark red color (~635 nm). Using the Argon-ion 488 nm laser line for the Raman spectra, this fluorescence will appear at a wave number shift much higher than the C—H stretching vibration; we thus expect no interference from this fluorescence with the Raman spectra. The spectra and peak assignments are shown in Fig. 2b.

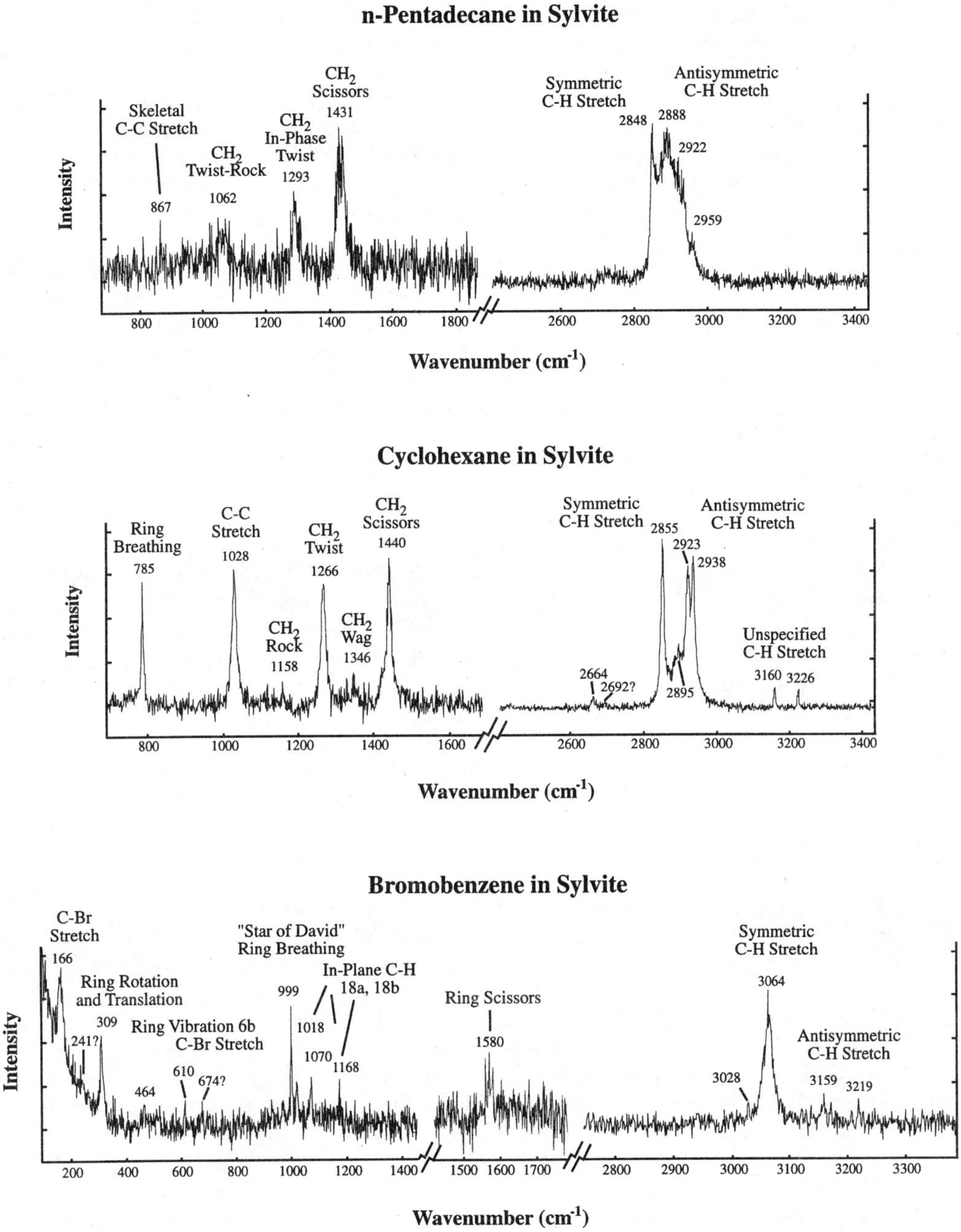

FIG. 2. Raman spectrum of synthetic a) n-pentadecane, b) cyclohexane, and c) bromobenzene inclusions in sylvite. Note that the collection times, and thus the intensity scales, are different for each spectral region.

Our Raman spectrum of the cyclohexane inclusion matches in both peak location and relative intensity the cyclohexane Raman bands found in PIRONON's (1990) spectra of a cyclohexane standard or a synthetic inclusion containing 50% cyclohexane and 50% bromobenzene. The spectra shown in Fig. 2b, however, better resolve the details of the manifold (in particular, the 2895 cm^{-1} methyl symmetric

C—H stretch) due to the high signal to noise of our system. Note also that the small C—H stretch peak at 2664 cm^{-1} is resolved, illustrating the sensitivity of this system.

The strong peaks at 785 cm^{-1} and 1028 cm^{-1} (Fig. 2b) are diagnostic of cyclohexane and represent the vibrations for ring-breathing and C—C stretching vibrations, respectively. Although the position of the ring breathing peak can change slightly as a function of substituted groups, strong peaks in the 700 cm^{-1} to 1000 cm^{-1} range can signify the presence of cycloalkanes.

Bromobenzene. Bromobenzene (C_6H_5Br) inclusions occur as pseudospherical, one phase (and rare two-phase) inclusions ranging in diameter from 15 to 30 mm. Under long wavelength ultraviolet light (366 nm Hg) the impurities in the bromobenzene inclusions fluoresce slightly in a light blue color (~505 nm). The inclusion fluorescence in this sample was again low and centered away from peaks of interest; we were therefore able to collect quality spectra using the Argon ion 488 nm laser line. The Raman spectra with identified peaks (DOLLISH *et al.*, 1974) are illustrated in Fig. 2c. Our Raman spectrum of the bromobenzene inclusion matches in both peak location and intensity, over the regions of overlap (i.e. the mid- and hi-frequency regions), the bromobenzene Raman bands found in PIRONON's (1990) published spectra of a bromobenzene standard and a synthetic inclusion with 50% cyclohexane and 50% benzene.

The very strong peak at 999 cm^{-1} is due to the breathing vibration of the benzene ring; within endmember benzene, this band has A_{1g} symmetry, and involves simultaneous inward and outward motion of all the carbon atoms (DOLLISH *et al.*, 1974; LIN-VIEN *et al.*, 1991). Within substituted benzenes, the trigonal "Star of David" vibration, derived from the infrared-active B_{1u} symmetry mode of endmember benzene, is Raman active. This mode involves alternate inward and outward motions of the carbon atoms within the plane of the ring, and lies at nearly the same energy as the A_{1g} breathing vibration. In some cases, these vibrations will mix. Thus, intense vibrations are present in benzene and all benzene derivatives at 990–1010 cm^{-1} (DOLLISH *et al.*, 1974), and are therefore diagnostic of the presence of this molecule or its derivatives. A second, weaker ring vibration is observed at 610 cm^{-1}. In benzene derivatives, there are no methylene (H—C—H) vibrations but only bands due to C—H stretches. This leads to the single sharp peak at 3064 cm^{-1}, due to the C—H symmetric stretch. Note that the alkane C—H stretching vibrations occurred as a broad manifold in the region 2850–2975 cm^{-1} (Fig. 2a); the intense single peak at 3064 cm^{-1} is therefore distinct from the alkane peaks, and is, with the 990–1010 cm^{-1} peak, diagnostic of aromatic compounds (GUILHAUMOU, 1982).

Crude oils

Two oil samples from the Hoh accretionary complex were examined to determine both the feasibility of conducting visible Raman spectroscopy on fluorescent, complex hydrocarbon mixtures, as well as to provide a basis for comparison between modern crude oils and the mid-Miocene fluid inclusion described in the following section. The two oil samples were produced by the Jefferson and Medina wells. The Jefferson oil is from the Jefferson Oil Company Hoh Head No. 2 well, located just north of the mouth of the Hoh River in the Olympic National Park. This well was spudded at ~350 m depth in sheared rocks of the Hoh mélange, and produced significant oil production in the 1930's, but was not economic. The other oil in this study comes from the Sunshine Mining Company Medina No. 1 well, located 80 km south of the Jefferson well in the Ocean City area north of Greys Harbor (for the exact location of both wells, see Fig. 1 of KVENVOLDEN *et al.* (1989)). This well was drilled in 1957 and produced 12,000 barrels of 38° gravity oil from sheared Hoh mélange rocks at the 1200 m depth level. Crude oil analyses yielded 35.6% gasoline by volume, 20.5% kerosene, and 42.9% reduced crude. Sulfur contents of the samples were between 0.014 and 0.15 wt.% (KVENVOLDEN *et al.*, 1989). Previous gas chromatographic studies by KVENVOLDEN *et al.* (1989) indicated that the Jefferson and Medina oils had similar molecular weight distributions, which were interpreted to indicate similar source regions and migration histories for these two widely separated samples.

Because of the complexity of the spectra of these hydrocarbons, we will focus principally on the dominant features in each spectrum. Given the wide range of possible vibrations, we view the strongest peaks as being indicative of the dominant chemical species within each sample, while the remainder are likely to be produced by less abundant functional groups. For reference, spectra of mixtures of as few as five to ten hydrocarbon species would be expected to show dozens of peaks in all of the frequency ranges examined here (DOLLISH *et al.*, 1974); this degree of complexity is, not surprisingly, what we observe.

Raman spectra from the Jefferson and Medina oils are similar, but not identical (Fig. 3): this is consistent with gas chromatographic studies (KVENVOLDEN *et al.*, 1989). The spectra from the natural oils include a number of the peaks identified in the synthetic inclusions as discussed in the previous section, and include both chain and cyclic alkanes, unsaturated carbon, and aromatic rings, as well as sulfur complexes. We focus initially on the lowest frequency region of these spectra.

Both oils show a number of peaks in the region between 670 and 950 cm^{-1} (Figs. 3a and 3b). Peaks in this region are probably correlated with either chain or cyclic aliphatic hydrocarbons (DOLLISH *et al.*, 1974). C—C skeletal stretching vibrations are a function of chain length and occur over the range between 850–900 cm^{-1}; in spectra of both oils, a complex manifold is observed in this region. Cycloalkanes exhibit a distinctive vibration due to ring ''breathing'' that varies as a function of ring size, from 703 (cyclooctane) to 1188 cm^{-1} (cyclopropane). The cyclohexane ring (a common constituent of hydrocarbons) has a ring breathing vibration that occurs at 802 cm^{-1}. Notably, the frequencies of both alkane and cycloalkane vibrations in this region are somewhat altered by any substitutions for hydrogen. Vibrations in this region could also be produced by internal carbon atoms, such as tert-butyl, isopropyl or quaternary geometries (DOLLISH *et al.*, 1974).

The Jefferson oil also has a number of peaks at 960–1020 cm^{-1} and at 1071 cm^{-1} (Fig. 3a, 3c), whereas the Medina exhibits a smaller manifold centered at 1023 cm^{-1} and a distinctive band at ~1077 cm^{-1} (Figs. 3b, 3d). The peaks near 1000 cm^{-1} are in a region which cannot be attributed to vibrations of cycloalkanes or internal carbons, although some substituted C—C stretching vibrations can occur in this region. Benzene, the prototypical aromatic hydrocarbon, has an extremely intense vibration due to the A_{1g}-symmetry ring breathing mode at 990–1010 cm^{-1}, which shifts on substitution of hydrogen by halogens in benzenes to 1060–1100 cm^{-1} for heavy substituents (>25 amu) and 830–620 cm^{-1} for lighter substituents (DOLLISH *et al.*, 1974). Accordingly, the bands near 1000 cm^{-1} are probably produced by aromatic hydrocarbons.

Both the Jefferson and Medina oils have complex manifolds in the 1250 and 1300 cm^{-1} region (Figs. 3c, 3d), while a weak manifold in the Medina that occurs at 1135 cm^{-1} may have a poorly resolved counterpart in the Jefferson oil. The former set of these bands are produced by CH_2 rocking motions (1175–1310 cm^{-1}), and are thus diagnostic of the presence of CH_2 units (CH_2 twisting vibrations also occur in the range between 720–1060 cm^{-1}). Other carbon configurations also lead to weak peaks in this region of the spectra.

The stretching vibrations of multiply bonded carbons occur in the 1400–2100 cm^{-1} region of the spectra, as well as the CH_2 scissors vibration at 1430–1470 cm^{-1} (Figs. 3e, 3f). The CH_2 is seen in both spectra as a distinctive manifold (due to both symmetric and antisymmetric scissoring motions). For alkanes, these vibrations occur at 1446 to 1473 cm^{-1}, and vary in frequency primarily as a function of the length of the carbon chain and the type of substitutions for hydrogen (DOLLISH *et al.*, 1974), while in alkenes they range from 1389 to 1426 cm^{-1}. Thus, the breadth of this manifold plausibly reflects a range of CH_2 environments being present within these oils.

The peak present near 1500 cm^{-1} is more difficult to uniquely assign, as it could be due to either carbon-nitrogen linkages, C=C stretching within cyclopentadienes, or to vibrations of doubly-bonded carbon-oxygen (although the latter two species are unexpected in natural oils [HUNT, 1979]). In contrast to the 1500 cm^{-1} peak, the strong vibrations at 1604 cm^{-1} (Medina) and 1614 cm^{-1} (Jefferson) are almost certainly produced by the relatively strong ring stretching vibrations of aromatic hydrocarbons. These characteristically occur within 30 cm^{-1} of 1600 cm^{-1} (DOLLISH *et al.*, 1974); thus, this assignment is consistent with our observation of the ring breathing vibration of such species at frequencies near 1000 cm^{-1}. The manifold peak at 1849 cm^{-1} in the Jefferson oil may be due to an additional C=O vibration. We have no simple explanation for the weak band near 2000 cm^{-1}, but believe that it could be associated with any number of triple bonded organic species, consecutive double bonds or possibly an overtone vibration.

In the highest frequency region, both oils show three large peaks at ~2420, 2465, and 2570 cm^{-1} (Figs. 3g, 3h). These are in addition to features at 2848 and 2929 cm^{-1}, which are the expected peaks produced by symmetric and antisymmetric C—H stretching vibrations. The only common functional group giving rise to vibrational bands in the 2400–2600 cm^{-1} region is the thiol (SH) group. Analyses of the Medina oil indicated the presence of sulfur in small concentrations, so some vibrations due to these species are expected, yet these peaks are comparably intense to the C—H stretching vibrations. A possible means through which this en-

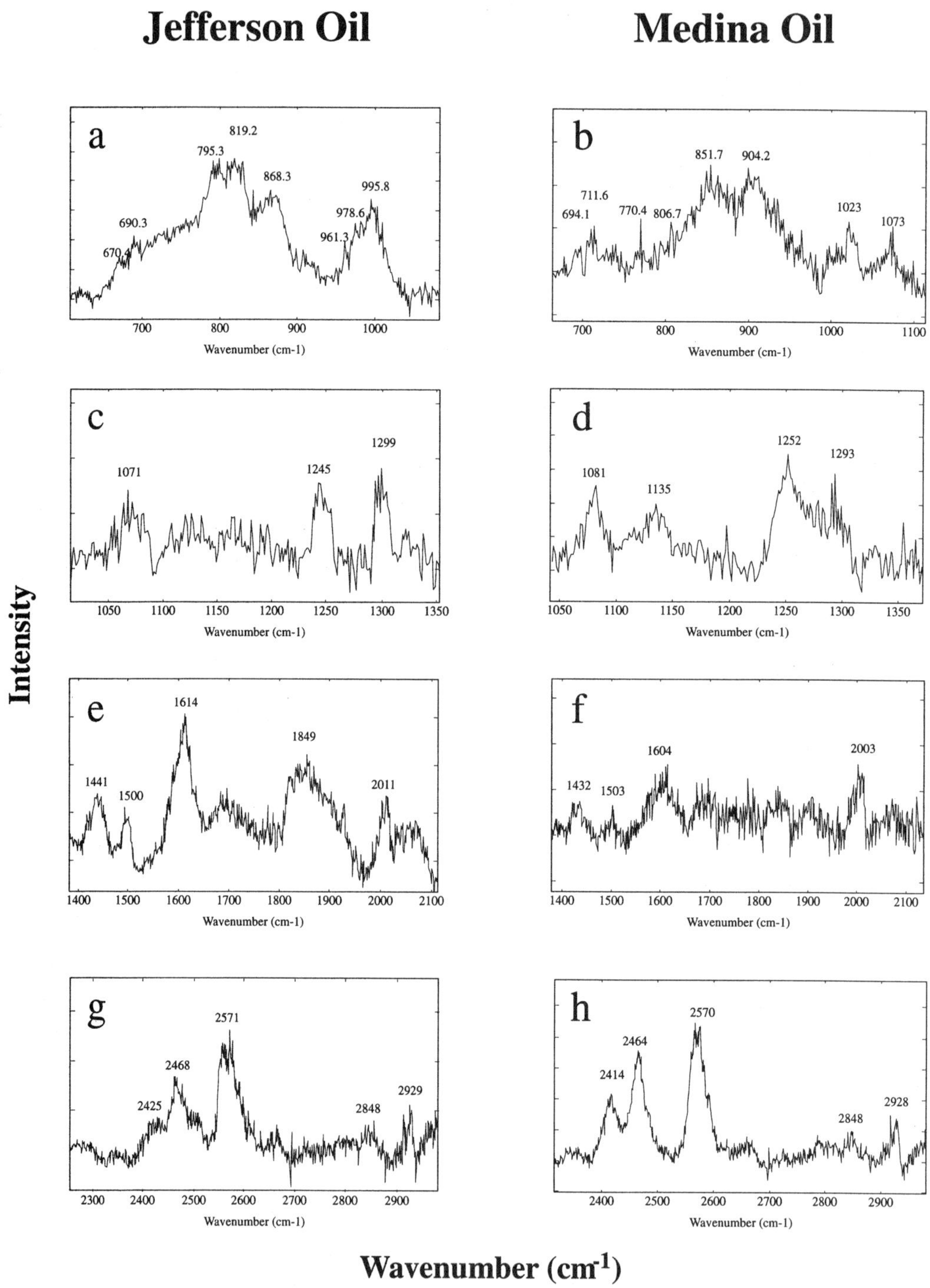

FIG. 3. Raman spectra of complex fluorescent crude oils from the western Olympic Peninsula, Washington. The spectra are notably similar, supporting the gas chromatograph studies of KVENVOLDEN *et al.* (1989) that indicated similar source regions and migration pathways for these oils, despite the fact that the wells are 80 km apart. Peaks are related to the species that are present, and include alkanes, cycloalkanes, aromatic hydrocarbons (benzene) and sulfur. See text for discussion.

hancement in intensity could be generated is through a resonance Raman effect (*e.g.*, MELVEGER, 1978). Such resonance effects have been extensively documented to occur in sulfonated species, although typically under blue-green laser excitation: this resonance is induced by an electronic transition associated with sulfur within hydrocarbons (*e.g.*, DELLA VEDOVA, 1992). Because of the weakness of our non-resonant hydrocarbon spectra, it is possible that even a weak resonance enhancement of the Raman intensity of these bands could produce a significant augmentation of their intensity beyond that expected from the abundance of sulfonated species in the sample. In natural oils, sulfur can occur either as H_2S or as a substituent in higher order hydrocarbons (HUNT, 1979). If sulfur is present as part of a hydrocarbon, then the S=C stretching vibration might be expected to be observed in the 800–1065 cm^{-1} region, with the S—C stretch in the 620–715 cm^{-1} region. Both vibrations occur in the same region as the C—C stretch and cycloalkane ring breathing frequencies. Thus, we cannot preclude that some of the peaks in this region may be produced by organosulfur species.

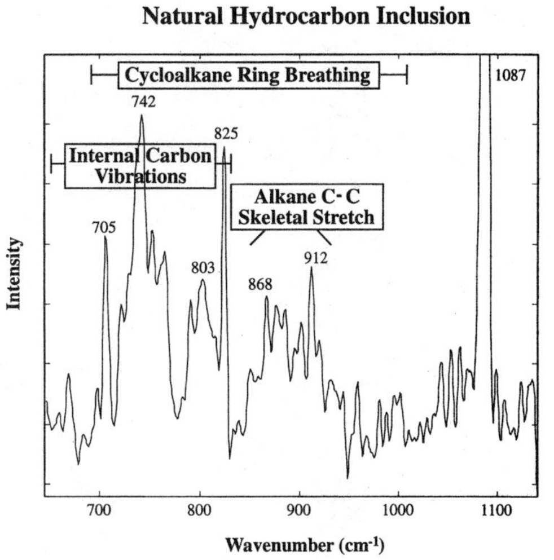

FIG. 4. Raman spectrum of a natural hydrocarbon inclusion from the Hoh accretionary complex, Olympic Peninsula, Washington. Peak assignments are listed in Table 1. Note the symmetric stretching peak of the carbonate unit at 1087 cm^{-1}, and the manifolds at 700 to 750 cm^{-1}, 803 cm^{-1}, and 850–920 cm^{-1}. These vibrations can be assigned to alkanes or cycloalkanes, with a component produced by the symmetric bending vibration of the carbonate group present near 720 cm^{-1}. No major peaks are visible near 1000 cm^{-1}, where a strong vibration of benzene and its derivatives occurs.

Natural fluid inclusions

The particular sample analyzed here comes from a 40 m thick fault zone related to the main phase of compressional deformation in the Hoh accretionary complex. The family of inclusions analyzed occur as ~20 μm diameter secondary inclusions in a late syn-deformational sparitic calcite vein.

In the low frequency region (below 1100 cm^{-1}), our Raman spectra contain peaks from both the host calcite mineral and the hydrocarbon inclusions (Fig. 4; for peak assignments see Table 1). Vibrations assignable to the calcite include the CO_3 symmetric stretching vibration at 1087 cm^{-1}, as well as the CO_3 symmetric bending vibration at 722 cm^{-1} (WHITE, 1974). In addition to the carbonate peaks, this spectral region is characterized by a multitude of peaks and manifolds in the 700 to 950 cm^{-1} region, which are produced by the hydrocarbons and are similar to the peaks seen in the oil samples. Most, if not all, of these peaks can be assigned to the C—C stretching vibration of alkanes, the breathing vibrations of cycloalkanes, or vibrations related to internal carbon configurations (Table 1). The 990 to 1010 cm^{-1} region, however, is of particular interest. As observed in the bromobenzene synthetic inclusion and both oil samples, peaks in this spectral region are diagnostic of the presence of aromatic rings. These vibrations (because of their highly symmetric character) are quite intense, and we would expect to observe them if benzene (or its derivatives) were present in abundance in the sample. Because such peaks are absent in this inclusion, benzene or substituted benzenes (if present) must be in concentrations significantly lower than within the oil samples.

In the mid-frequency range, there are two manifolds observed near 1440 cm^{-1} and 1650 cm^{-1} (Fig. 5; see peak assignments listed in Table 2). Although one of the peaks in the 1440 cm^{-1} manifold could be due to the antisymmetric stretching vibration of the carbonate group (WHITE, 1974) or the symmetric stretch of the carboxylate ion group (LIN-VIEN *et al.*, 1991). The other peaks in this region are plausibly assigned to the methylene (CH_2) scissors vibrations seen in the synthetic alkanes, cycloalkanes and oils (Table 2). The other high intensity region in this spectrum is centered at 1650 cm^{-1}, which could be associated with the presence of carbonyl groups; such groups are relatively common constituents of crude oils (HUNT, 1979). We note that because

Table 1. Natural hydrocarbon inclusion, 650-1150 cm^{-1}.

n-alkane vibrations:	C-C stretch (DOLLISH *et al.*)	identified peak
n-butane	837	--
n-pentane	869	868
n-hexane	898	900
n-heptane	905	in manifold?
n-octane	899	900
n-nonane	888	886
n-decane	886	886
n-dodecane	892	894
n-hexadecane	888	886

Strong peaks in the 800-1000 cm^{-1} range are very common for all alkanes, and (although not diagnostic) are due to various combinations of C-C stretching and CH_3 (methyl) twisting-rocking (DOLLISH *et al.*, 1974).

Additional strong peaks due to the configuration of carbon atoms:	
tert-butyl	680-750
isopropyl	740-830
internal quaternary carbon atom	670-710
internal tertiary carbon atom	800-860
two adjacent tertiary carbon atoms	730-760

(Note: these latter peaks are not diagnostic, as they vary as a function of the base chain of which they are a part.)

cycloalkane vibrations	ring breathing (DOLLISH *et al.*)	identified peak
cyclobutane	1001	1002
cyclopentane	886	886
cyclohexane	802	803
cycloheptane	733	735 (shoulder on low frequency side of 742 peak?)
cyclooctane	703	705
benzene and its derivatives	990-1010	--

there is no prominent peak in the 1000 cm^{-1} region (diagnostic of aromatic hydrocarbons), we cannot attribute the 1600 cm^{-1} peak to the weaker band expected in substituted benzene rings. A single band is also present near 1540 cm^{-1} in the spectrum of the natural fluid inclusion (Fig. 5). We have no simple assignment for the origin of this feature, but note that peaks in this region of the spectrum could be generated by vibrations associated with carbon-nitrogen linkages (we do not view the assignment of this band to C=C species as likely, due to the scarcity of such species in natural oils [HUNT, 1979]). Alternatively, this peak could be generated by either a combination band or a resonance enhanced vibration (or overtone) of a sulfonated species.

Vibrations related to C—H stretching occur as a manifold centered at 2919 cm^{-1}, with numerous other weaker peaks occurring throughout the frequency range between 2825 and 3095 cm^{-1} (Fig. 6). This manifold is due to the juxtaposition of symmetric and antisymmetric vibrations of hydrocarbon compounds containing single bonded carbons. For example, the peak at 2911 cm^{-1} may plausibly be associated with either methane or the methylene antisymmetric stretch, the 2924 cm^{-1} band with ethane or the methyl antisymmetric stretch, and the 2945 cm^{-1} due to the methyl and methylene antisymmetric and symmetric stretches (GUILHAUMOU, pers. comm.). Almost all of these peaks can in fact be assigned to alkanes or cycloalkanes (Table 3). Higher frequency C—H vibrations (3000 to 3100 cm^{-1}) can be assigned to C—H stretching from either alkenes or aromatic compounds. The lack of a significant peak at 3062 cm^{-1} argues against there being a significant amount of benzene in this sample, consistent with the lack of a ring breathing peak observed in the lower frequency spectral region.

Thus, for this natural fluid inclusion from the Hoh accretionary complex, we find compelling evidence

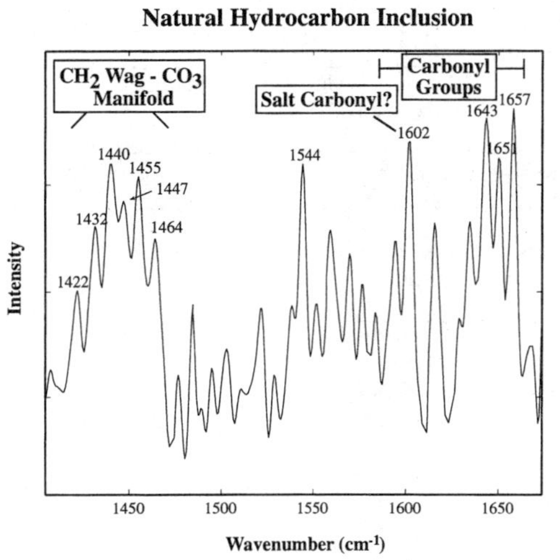

FIG. 5. Raman spectrum of a natural hydrocarbon inclusion from the Hoh accretionary complex, Olympic Peninsula, Washington. Peak assignments are listed in Table 2. The two manifolds seen in this spectra are due to the H-C-H methylene scissors vibrations (1420 to 1465 cm^{-1}) and carboxylic acid. The vibration near 1540 cm^{-1} may be associated with carbon-nitrogen bonds; see text for discussion.

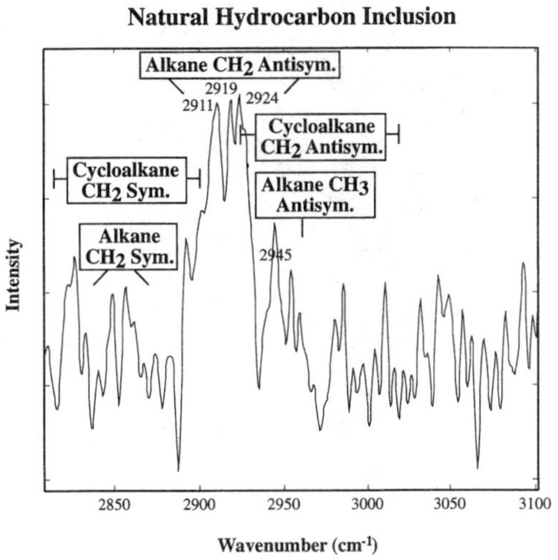

FIG. 6. Raman spectrum of a natural hydrocarbon inclusion from the Hoh accretionary complex, Olympic Peninsula, Washington. Peak assignments are listed in Table 3. The large manifold peak near 2900 cm^{-1} is indicative of alkane and cycloalkane methyl and methylene symmetric and antisymmetric stretching vibrations. Note the lack of a significant peak at 3062 cm^{-1}, indicating the absence of appreciable quantities of benzene or its derivatives.

for the presence of alkanes and cycloalkanes. Evidence for the presence of aromatic hydrocarbons (benzene and its derivatives) within these inclusions is notably lacking. For comparison, the Raman spectra of the oils document the presence of alkanes, cycloalkanes and aromatic rings, as well as indications of the presence of sulfur.

Comparison between oils and inclusion with possible causes of chemical differences

Whereas spectra from the oil samples are similar to the natural hydrocarbon inclusion, there is one highly significant difference. Like the oil samples, the inclusion contains alkanes and cycloalkanes.

Table 2. Natural hydrocarbon inclusion, 1400-1675 cm^{-1}.

compound	methylene scissors (DOLLISH *et al.*)	identified peak
alkane, in-plane	1446-1473	large manifold peak, 1447-1464
alkane, out-of-plane	1465	1464
cyclobutane	1443	1440
cyclopentane	1455	1455
cyclohexane	1452	1455
cycloheptane	1450	1447
cyclooctane	1467	1464
alkenes, and substituted alkenes (?)	1389-1426	1422, 1432

compound	expected peak	identified peak
carboxylic acid*	~1650	manifold at 1643-1657
alkenes, and substituded aklenes (??)	1575-1693	1602, 1616, etc.
cycloalkenes (??)	1614-1672	1626-1657

*Note: alkenes and substituted alkenes have C=C vibrations in this region, yet these unsaturated hydrocarbon constituents are extremely rare in naturally occurring oils [HUNT, 1979]; we thus attribute vibrations in this region to carboxylic acid.

Table 3. Natural hydrocarbon inclusion, 2800-3100 cm^{-1}. This zone is entirely due to the various C-H stretching modes, and is indicative of the compounds present.

compound	CH_3 antisym. stretch (DOLLISH *et al.*)	identifed peak	CH_3 sym. stretch (DOLLISH *et al.*)	identifed peak
alkanes and derivatives	2965-2969	2959 and shoulder?	2884	2892?

compound	CH_2 antisym. stretch (DOLLISH *et al.*)	identifed peak	CH_2 sym. stretch (DOLLISH *et al.*)	identifed peak
cyclopropane	3101-3090	3093	3038-3019	3032
cyclobutane	2987-2975	2985	2895-2887	2892
cyclopentane	2959-2952	2959, 2954	2866-2853	2861, 2857
cyclohexane	2933-2915	in 2924 manifold	2897-2852	numerous peaks
cycloheptane	2935-2917	in 2924 manifold	2862-2851	2861, 2857
cyclooctane	2925	2924	2855	2857
alkanes	2929-2912	in 2919 manifold	2861-2849	2861, 2857, 2849

The Raman spectrum of the inclusion, however, contains little, if any, evidence for the existence of benzene or substituted benzenes. We suggest that the different histories of the oil and the inclusion led to their spectroscopic differences. Both the oil and the inclusion come from sheared rocks within the Hoh rock assemblage. The oils were recently extracted from pore spaces in near-surface rocks, while the inclusion is fundamentally different. Since the inclusion is part of a vein, the hydrocarbons (unless they are locally derived) must have traveled in the dominantly aqueous system that deposited the carbonate vein. The vein the inclusion occurs in was emplaced during thrusting in Miocene time (ORANGE *et al.*, 1991, 1993) at conditions in the range of 50–100 MPa and 80–100°C (ORANGE *et al.*, 1992).

Differences between the oils and the hydrocarbon inclusion could be due to differences in age, source region, migration, and alteration (TISSOT and WELTE, 1984). Since both the oils and the inclusion occur in sheared Miocene Hoh rocks related to large scale thrusting on the Olympic Peninsula we believe that they are likely to have similar ages, migration histories, and source regions. Gas chromatograph studies of petroliferous samples throughout the Hoh accretionary complex support this hypothesis (KVENVOLDEN *et al.*, 1989). Thus, the chemical differences between the oil and the inclusion may be due to the alteration history. The oil within the inclusion may have experienced higher temperatures than the oils, leading to thermal degradation. Benzene is stable at relatively high temperatures, and so the lack of benzene can not be attributed to high temperature alteration. Benzene is, however, quite soluble in aqueous solutions. The high surface area to volume ratio in the fluid inclusion could have led to preferential removal of benzene by water washing, which selectively removes lighter hydrocarbon molecules from their source as these compounds are much more soluble in aqueous solution (TISSOT and WELTE, 1984; WELTE *et al.*, 1982; LAFARGUE and BARKER, 1988).

CONCLUSIONS

Visible Raman spectroscopy can successfully identify the major vibrational modes within synthetic hydrocarbon inclusions, complex fluorescent oils, and fluorescent natural hydrocarbon inclusions in calcite. The synthetic inclusions analyzed represent alkanes, cycloalkanes, and benzene derivatives. Problems related to fluorescence and photodegradation in natural inclusions and their matrix can be avoided by using judiciously selected, longer laser excitation wavelengths, low laser power and long integration times. Oil samples provide similar problems with fluorescence that can be overcome again by selecting longer laser wavelengths and increasing the detector integration times.

This technique is capable of determining what type of hydrocarbon species are present in abundance within such inclusions, based on the presence or absence of characteristic vibrational bands of organic functional groups. We note that at present, because of the complications associated with quantification of the relative Raman scattering cross-sections of the range of different hydrocarbons likely to be present in such fluid inclusions, this technique remains a qualitative tool capable of characterizing which types of organic functional groups are present in abundance within inclusions (*e.g.*, WOPENKA and PASTERIS, 1986).

Determining the geochemical history of hydrocarbon evolution in a complex geologic region such as the Hoh accretionary complex clearly requires more detailed data on the temporal evolution of inclusion chemistry. However, we have used variations in Raman spectra to derive some qualitative (and geologically consistent) insights into possible differences in migration histories between a mid-Miocene inclusion and present-day crude oils from the same complex. Although the regional applicability of these results remains uncertain, they demonstrate the utility of Raman spectroscopy as a powerful tool for fingerprinting hydrocarbons and for identifying the major functional groups within each sample.

Acknowledgments—We would like to thank J. Pironon for his generous contribution of the synthetic hydrocarbon inclusion samples. Keith Kvenvolden provided valuable insight into the organic geochemistry of hydrocarbon species. We thank J. J. Freeman, R. J. Hemley, K. Kvenvolden, J. C. Moore, and J. Gieskes for helpful reviews which improved this paper. This research was partially supported by the W. M. Keck Foundation, Weyerhaeuser Corporation, the National Science Foundation and the American Chemical Society-Petroleum Research Fund. E. K. is an A. P. Sloan Foundation Fellow. Institute of Tectonics (Mineral Physics Lab) Contribution No. 111.

REFERENCES

BARRES O., BURNEAU A., DUBESSY J. and PAGEL M. (1987) Application of micro-FT-IR spectroscopy to individual hydrocarbon fluid inclusion analysis. *Appl. Spectr.* **41**, 1000–1008.

BECKER R. S. (1969) *Theory and interpretation of fluorescence and phosphorescence.* Wiley-Interscience.

BERLMAN B. (1965) *Handbook of Fluorescence Spectra of Aromatic Molecules.* Academic Press, New York.

BURDETT R. A. and TAYLOR L. W. (1955) Determination of aromatic hydrocarbons in lubricating oil fractions by far ultra-violet absorption spectroscopy. In *Molecular Spectroscopy, report of a conference organized by The Spectroscopic Panel of the Hydrocarbon Research Group of the Institute of Petroleum, London, 28–29 October 1954* (ed. G. SELL), 30–41.

BURRUSS R. C. (1981) Hydrocarbon fluid inclusions in studies of sedimentary diagenesis. In *Mineralogical Association of Canada Short Course on Fluid Inclusions: Applications to Petrology* (eds. L.S. HOLLISTER and M.L. CRAWFORD), vol. 6, pp. 138–156. Mineralogical Association of Canada.

BURRUSS R. C., CERCONE K. R. and HARRIS P. M. (1983) Fluid inclusion petrography and tectonic-burial historu of the Al Ali No. 2 well: Evidence for the timing of diagenesis and oil migration, northern Oman Foredeep. *Geology* **11**, 567–570.

BURRUSS R. C., CERCONE K. R. and HARRIS P. M. (1985) Timing of hydrocarbon migration: evidence from fluid inclusions in calcite cements, tectonics, and burial history: in *Carbonate Cements* (eds. N. SCHNEIDERMANN and P. M. HARRIS), Special Publication No. 36, pp. 277–289. Society of Economic Paleontologists and Mineralogists.

BURRUSS R. C., GING T. and CARLSON, D. (1989) Raman microprobe observations of carbon and oxygen stable isotopes in geologic materials. In *Microbeam Analysis - 1989* (ed. P. E. RUSSELL), pp. 173–175. San Francisco Press.

CHOU I.-M., PASTERIS J. D. and SEITZ J. C. (1990) High-density volatiles in the system C-O-H-N for the calibration of a laser Raman microprobe. *Geochim. Cosmochim. Acta* **54,** 535–543.

COWAN D. O. and DRISKO R.L. (1976) *Elements of Organic Photochemistry.* Plenum.

DAWSON W. R. and WINDSOR M. W. (1968) Fluorescence yields of aromatic compounds. *J. Phys. Chem.* **72,** 3251.

DELE-DUBOIS M.-L., DHAMELINCOURT P. and SCHNUBEL H.-J. (1980) Studies by Raman spectroscopy of inclusions in diamonds, sapphires, and emeralds, parts 1 and 2. *Rev. Gemmologie,* **63,** 11–14 (part 1), and **64,** 13–16 (part 2) (in French).

DELLA VEDOVA C. O. (1992) Towards the absolute determination of the conformaton in thioester compounds through the first pre-resonance Raman study in X-NSO molecules, *J. Molec. Struc.* **274,** 9–23.

DIAS J. R. (1987) *Handbook of Polycyclic Hydrocarbons: Part A Benzenoid Hydrocarbons.* Elsevier, Amsterdam.

DOLLISH F. R., FATELEY W. G. and BENTLEY F. F. (1974) *Characteristic Raman frequencies of organic compounds,* John Wiley and Sons, New York.

ENGLMAN, R. (1979) *Non-radiative decay of ions and molecules in solids,* 336 pp., North-Holland Publishing Co., New York.

ETMINAN H. and HOFFMAN C.F. (1989) Biomarkers in fluid inclusions: A new tool in constraining source regimes and its implications for the genesis of Mississippi Valley-type deposits. *Geology* **17,** 19–22.

FRANTZ J. D., VIRGO D. and MYSEN B. O. (1982) Time-lapse spectroscopy for fluorescence radiation rejection, *Carnegie Institute of Washington Year Book 81,* 437–440.

GAUTIER D. L., KHARAKA Y. K. and SURDAM R. C. (1985) Relationship of Organic Matter and Mineral Diagenesis, SEPM Short Course No. 17, Tulsa, OK.

GUILHAUMOU N. (1982) Accurate analysis of fluid inclusions by the laser molecular microprobe (MOLE) and by microthermometry. *Travaux Lab. Géol. Ecole Normale Supérieure,* **14,** 68 p. (in French).

GUILHAUMOU N. and SZYDLOWSKI N. (1989) Analyse ponctuelle a micro FTIR des inclusions d'hydrocarbures de H.Zriba-Guebli. Mise en évidence de modifications par surchauffe sous-pression de confinement (300 C - 40 Mpa). *C. R. Acad. Sci. Paris*v **309,** series II, 1171–1176.

GUILHAUMOU N., JOUAFFRE D., VELDE B. and BÉNY C. (1988) Raman microprobe analysis on gaseous inclusions from diagenetically altered Terres Noures (S-E, France). *Bull. Mineral.* **111,** 577–585.

HUNT J. M. (1979), Petroleum Geochemistry and Geology, 617 pp., W.H. Freeman, San Francisco.

HASZELDINE R. S., SAMSON I. M. and COMFORD C. (1984) Dating diagenesis in a petroleum basin, a new fluid inclusion method. *Nature* **307,** 354–357.

HORSFIELD B. and MCLIMANS R. K. (1984) Geothemometry and geochemistry of aqueous and oil-bearing fluid inclusions from Fateh Field, Dubai. *Org. Geochem.* **6,** 733–740.

JENSENIUS J. and BURRUSS R. C. (1990) Hydrocarbon-water interactions during brine migration: Evidence from hydrocarbon inclusions in calcite cements from Danish North Sea oil fields. *Geochim. Cosmochim. Acta* **54,** 705–713.

KVENVOLDEN K. A. and ROEDDER E. (1971) Fluid inclusions in quartz crystals from South-West Africa. *Geochim. Cosmochim. Acta* **35,** 1209–1229.

KVENVOLDEN K. A., RAPP J. B., HOSTETTLER F. D. and SNAVELY P. D. Jr. (1989) Preliminary Evaluation of the Petroleum Potential of the Tertiary Accretionary Terrane, West Side of the Olympic Peninsula, Washington: Part 2 - Comparison of Molecular Markers in Oil and Rock Extracts. *U.S. Geol. Survey Bull. 1892,* p. 19–36.

LAFARGUE E. and BARKER C. (1988) Effect of water washing on crude oil composition. *Bull. Amer. Assoc. Petrol. Geol.* **72,** 263–276.

LIN-VEIN D., COLTHUP N.B., FATELEY W.G. and GRASSELLI J.G. (1991) *The Handbook of Infrared and Raman Characteristic Frequencies of Organic Molecules,* Academic Press, New York.

MCLIMANS R. K. (1987) The application of fluid inclusions to migration of oil and diagenesis in petroleum reservoirs. *Appl. Geochem.* **2,** 585–603.

MCLIMANS R. K. and VIDETICH P.E. (1987) Reservoir diagenesis and oil migration, Middle Jurassic Great Oolite Limestone, Wealdon Basin, Southern England. In *Petroleum Geology of North West Europe* (ed. J. BROOKS and K. GLENNIE), pp. 419–429. Graham and Trodman.

MEDLIN W. L. (1968) The nature of traps and emission centers in thermoluminescent rock materials. In *Thermoluminescence of Geologic Materials* (ed. D. J. MCDOUGALL), pp. 193–223. Academic Press, New York.

MELVEGER, A.J. (1978) *Resonance Raman Spectroscopy as an Analytic Tool,* Franklin Inst. Press, Philadelphia.

MURRAY R. C. (1957) Hydrocarbon fluid inclusions in quartz. *Bull. Amer. Assoc. Petrol. Geol.* **41,** 950–956.

NAKAMOTO K. (1982) *Infrared and Raman Spectra of Inorganic and Coordination Compounds*. 4th Ed., Wiley-Interscience, New York.

NOONER D. W., UPDEGROVE D. A., FLORY D. A., ORO J. and MUELLER G. (1973) Isotopic and chemical data of bitumens associated with hydrothermal veins from Windy Knoll, Derbyshire, England. *Chem. Geol.* **11,** 189–202.

O'GRADY M. R., BODNAR R. J., HELLGETH J. W., CONROY C. M., TAYLOR L. T. and KNIGHT C. L. (1989) Fourier transform infrared microprobe (FTIRM) analysis of individual petroleum fluid inclusions in geologic samples. In *Microbeam Analysis* (ed. P. E. RUSSELL), pp. 579–582. San Francisco Press.

ORANGE D. L., GEDDES D., COX R. and UNDERWOOD M. (1991) Out-of-Sequence Thrust Faulting in a Young Accretionary Complex: Evidence from Fluid Inclusion and Vitrinite Reflectance Data. *EOS (Trans., Amer. Geophys. Union),* **44,** 549.

ORANGE D. L., GEDDES D., MOORE J. C., COX R. and UNDERWOOD M. (1992) Constraining the temperature, pressure, and fluid compositional evolution of an accretionary complex by an integrated study of cement stratigraphy and tectonics, *7th International Conference on Water-Rock Interaction, Park City, Utah,* (eds. Y. KHARAKA and A. MAEST), Balkema, Rotterdam, p. 1537–1541.

ORANGE D. L., GEDDES D., and MOORE J. C. (1993) Structural and fluid evolution of a young accretionary complex: The Hoh rock assemblage of the western Olympic Peninsula, Washington. *Geol. Soc. of Amer. Bull.* **105,** 1053–1075.

PAGEL M., WALGENWITZ F. and DUBESSY J. (1986) Fluid inclusions in oil and gas-bearing sedimentary formations. In *Thermal Modeling in Sedimentary Basins,* pp. 565–583. Ed. Technip., Paris.

PASTERIS J. D., PATEL R., BERGMAN S. C. and ADAR F. (1983) Comparative spectroscopy and microthermometry on fluid inclusions in a mantle xenolith. *EOS (Trans., Amer. Geophys. Union),* **64,** 340.

PASTERIS J. D., SEITZ J. C. and WOPENKA B. (1985) Compositional interpretation of synthetic C—O—H fluid inclusions by micro-Raman analysis and microthermometry. In *Microbeam Analysis—1985* (ed. J. T. ARMSTRONG), pp. 25–28. San Francisco Press.

PASTERIS J. D., WOPENKA B. and SEITZ J. C.(1988) Practical aspects of quantitative laser Raman microprobe spectroscopy for the study of fluid inclusions. *Geochim. Cosmochim. Acta* **52,** 979–988.

PERING K.L. (1973) Bitumens associated with lead, zinc and fluorite ore minerals in North Derbyshire, England. *Geochim. Cosmochim. Acta* **37,** 401–417.

PIRONON J. (1990) Synthesis of hydrocarbon fluid inclusions at low temperature. *Amer. Mineral.* **75,** 226–229.

PIRONON J. and BARRES O. (1990) Semi-quantitative FT-IR micronanalysis limits: Evidence from synthetic hydrocarbon fluid inclusions in sylvite. *Geochim. Cosmochim. Acta* **54,** 509–518.

PIRONON J., SAWATZKI J. and DUBESSY J. (1991) NIR FT-Raman microspectroscopy of fluid inclusions - Comparisons with Vis Raman and FT-IR microspectroscopies, *Geochim. Cosmochim. Acta* **55,** 3885–3891.

RAU W. W. (1973) Geology of the Washington Coast between Point Grenville and the Hoh River. *Washington Department of Natural Resources, Geology and Earth Resources Division, Bulletin* no. 66, 58 p.

RAU W. W. (1975) Geologic map of the Destruction Island and Taholah Quadrangles, Washington. *State of Washington Department of Natural Resources, Division of Geology and Earth Resources,* scale 1:62,500, Geologic Map GM-13.

RAU W. W. (1979) Geologic map in the vicinity of the Lower Bogachiel and Hoh rivers, and the Washington Coast. *Washington Department of Natural Resources, Division of Geology and Earth Resources, Geologic Map GM-24.*

ROEDDER E. (1984) *Fluid Inclusions. Reviews in Mineral-*

ogy, 12, Mineralogical Society of America, Washington, D.C.

ROSASCO G. J. and ROEDDER E. (1979) Application of a new Raman microprobe spectrometer to nondestructive analysis of sulfate and other ions in individual phases in fluid inclusions in minerals. *Geochim. Cosmochim. Acta* **43,** 1907–1915.

Sadtler Handbook of UV Spectra (1982) (ed W. W. Simmons) Sadtler Research Laboratories. Philadelphia, PA.

SEITZ J. C., PASTERIS J. D. and WOPENKA B. (1987) Characterization of CO_2-CH_4-H_2O fluid inclusions by microthermometry and laser Raman microprobe spectroscopy: Inferences for clathrate and fluid equilibria. *Geochim. Cosmochim. Acta* **51,** 1651–1664.

SEITZ J. C., PASTERIS J. D. and CHOU I.-M. (1989) Quantitative composition and density determinations of CH_4-N_2-bearing fluid inclusions by Raman micro-spectroscopy. *Geol. Soc. Amer. Abstr. Prog.* **21,** A358.

TEN HAVE T. and HEIJNEN W. (1985) Cathodoluminescence activation and zonation in carbonate rocks: an experimental approach. *Geol. Mijnbouw* **64,** 297–310.

TISSOT B. P. and WELTE D. H. (1984) *Petroleum Formation and Occurrence,* Springer-Verlag, New York.

TOURAY J. C., BENY-BASSEZ C., DUBESSY J. and GUILHAUMOU N. (1985) Microcharacterization of fluid inclusions in minerals by Raman microprobe. *Scanning Electron Microscopy, Vol. I, SEM Inc.,* pp. 103–118. AMF O'Hare.

VEIRS D.K., CHIA V.K.F. and ROSENBLATT G.M. (1987) Raman spectroscopy applications of an imaging P.M.T., *Appl. Optics* **26,** 3530–3535.

WALKER G., ABUMERE O. E. and KAMALUDDIN B. (1989) Luminescence spectroscopy of Mn^{2+} centres in rock-forming carbonates. *Mineral Mag.* **53,** 201–211.

WELTE D. H., KRATOCHIVIL H., RULLKOTTER J., LADWEIN J. and SCHAEFER R. G. (1982) Organic geochemistry of crude oils from the Vienna basin and an assessment of their origin. *Chem. Geol.* **35,** 33–68.

WHITE, W.B. (1974) The carbonate minerals. In *The Infrared Spectra of Minerals* (ed. V.C. FARMER), pp. 227–279. Mineralogical Society, London.

WOPENKA B. and PASTERIS J.D. (1986) Limitations to quantitative analysis of fluid inclusions in geological samples by laser Raman microprobe spectroscopy. *Appl. Spectroscopy* **40,** 144–151.

WOPENKA B., PASTERIS J. D. and FREEMAN J. J. (1990) Analysis of individual fluid inclusions by Fourier transform infrared and Raman microspectroscopy. *Geochim. Cosmochim. Acta* **54,** 519–533.

Mineral Spectroscopy: A Tribute to Roger G. Burns
© The Geochemical Society, Special Publication No. 5, 1996
Editors: M. D. Dyar, C. McCammon and M. W. Schaefer

Pyrrhotite-related $Fe_x(Se_{1-y}Te_y)_8$; $x \approx 7$, $y \leq 0.15$ solid solutions, studied by X-ray diffraction, magnetization and Mössbauer spectroscopy techniques

TORE ERICSSON[1], ÖRJAN AMCOFF[1] and PER NORDBLAD[2]
[1]University of Uppsala, Institute of Earth Sciences, Mineralogy-Petrology, Norbyvägen 18 B, S-752 36 Uppsala, Sweden
[2]University of Uppsala, Department of Technology, Box 534, S-751 21 Uppsala, Sweden

Abstract—Pyrrhotite-related ~FeX; X = Se,Te, S compounds with compositions in the range of Fe_7Se_8 have been synthesized. At 873 K the solubility of Te in Fe_7Se_8 is between 15 and 20 mole%, decreasing to less than one mole% at 423 K. The corresponding solubility of Te in hexagonal Fe_7S_8 is less than one mole% at 873 K. Vacancy ordering in Fe_7Se_8 (*1c*), giving rise to a hexagonal *3c*-superstructure in rapidly cooled samples, is more strongly hindered by Te-substitution compared with S-substitution; however, Mössbauer spectroscopy indicates local ordering of vacancies. These phenomena can be seen as a change in the macroscopic magnetization. They can be modelled by a continuous decrease in the size of the vacancy-ordered domains with Te-substitution (and to a lesser degree with S-substitution), as a result of domain-walls being arrested by the foreign atoms. Also, a characteristic spin re-orientation takes place at 130 K in Fe_7Se_8 (*3c*), with spins turning from the *ab* plane to the c-axis on lowering the temperature. The temperature for this re-orientation is increased when Se is substituted by Te (and decreased when Se is substituted by S), reflecting the corresponding change in relative magnitude of the distances between neighbouring Fe-atoms.

INTRODUCTION

IRON SULFIDES are quantitatively the most important sulfide minerals in nature. In mineralizations formed in near surface environments both pyrite and pyrrhotite are common, while the latter phase dominates in mafic and ultramafic igneous rocks because of a low oxygen fugacity. As a consequence, pyrrhotite can be expected to be of widespread occurrence in the mantle.

Pyrrhotite-related compounds show a considerable complexity, especially at low and intermediate temperatures, as seen both in nature and in the synthetic Fe-S system; thus, the mineral has been much studied (*e.g.*, THIEL, 1969, VAUGHAN and CRAIG, 1978, KISSIN and SCOTT, 1982, BARKER and PARKS, 1986 and KULLERUD, 1986).

Studies of compositions along the pseudobinary Fe_7S_8 $-Fe_7Se_8$ join (ERICSSON *et al.*, 1994) show that the low field magnetic permeability decreases gradually inwards from both end compositions. This indicates a change from ordered to disordered structure with an increase in (S,Se)−mixing, a finding that was supported by X-ray diffraction (XRD). However, Mössbauer spectroscopy (MS) indicated local ordering throughout the solid solution series. We explain these superficially contradictory results by assuming a model where the size of the ordered regions in the mix crystals decreases continuously as the middle composition along the Fe_7S_8 − Fe_7Se_8 join is approached. This yields a corresponding decrease in the low field permeability inwards if it is assumed that the ferromagnetic coupling within an *ab* layer is intact over the phase − antiphase boundaries.

In the present paper we have pursued our studies on pyrrhotite-like compounds further and investigated the effect of anion-mixing by introducing Te into the system. Of special interest is the question of how a spin re-orientation, from the *ab*-plane to the *c*-axis with decreasing temperature, is affected by Te/Se-substitution. This substitution is assumed to be completely random, as in the case of compositions along the Fe_7S_8 − Fe_7Se_8 join (ERICSSON *et al.*, 1994).

CRYSTAL STRUCTURE AND MAGNETIC PROPERTIES

In the hexagonal structure, which is NiAs-related, the iron deficiency is compensated by vacancies, which at elevated temperatures are statistically distributed among the cations giving a primitive *1c*-structure. Because of vacancy avoidance (BERTAUT, 1953) superstructures form on lowering the temperature, and the lattice multiplicity of the supercell depends both on composition and temperature (PUTNIS and MCCONELL, 1980).

Compositions near Fe_7S_8 yield superstructures where the *a*-axis is repeated twice and the *c*-axis three or four times (*3c* or *4c*), depending on heat treatment. In the analogous Fe-Se system a similar situation is present for compositions around Fe_7Se_8. In the primitive *1c*-structure, the magnetic moments are ferromagnetically coupled (FM) within a cation *ab*-layer, but antiferromagnetically coupled (AFM) between adjacent *ab*-layers, giving zero net moment for the crystal. On the other hand, the superstructures *3c* and *4c* show a ferrimagnetic moment (FiM), as the vacancies are concentrated in alternating *ab*-layers.

In the hexagonal unit-cell of Fe_7Se_8 composition, an Fe atom is surrounded by six Se atoms in octahedral configuration at 2.55 Å, and by two Fe atoms in the *c*-direction at 2.94 Å. In the same *ab*-plane as the studied Fe atom, there are six Fe atoms at 3.61 Å, and in the fourth shell there are 12 Fe atoms at 4.65 Å (AMCOFF *et al.*, 1994). The situation is described in Fig. 1. In the *3c* and *4c* superstructures, where the vacancies are concentrated in every second metal *ab*-layer, the nearest surrounding to an Fe-atom can be described as being of A-, B- or C-type, as shown in the bottom part of Fig. 1. In the figure the numbers denote vacancies as well as heights in the *c*-direction (in *1c* units). The anion-layers have been excluded for clarity.

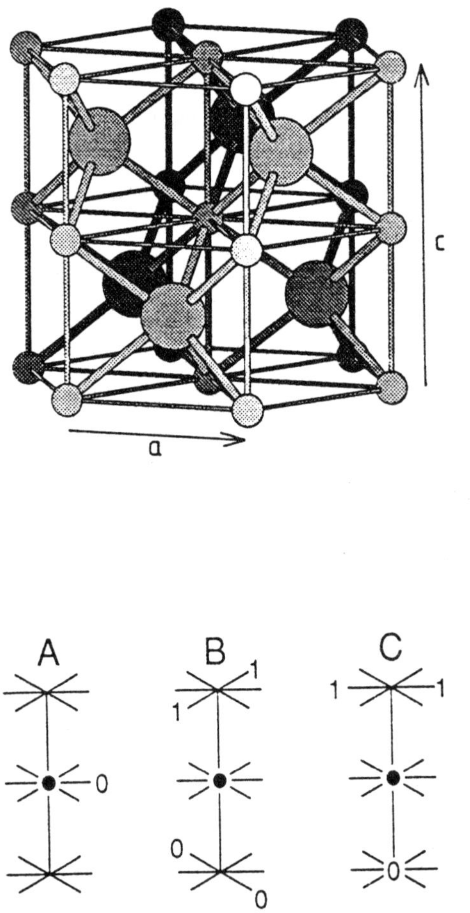

FIG. 1. Crystal structure of "NiAs,1c" and A, B, C vacancy configurations. In the upper figure (drawn by use of the program "ATOMS" by Shape Software), three primitive unit cells are shown. In the lower figure only the cation layers are sketched. The Mössbauer atom is black, and the numbers denote vacancies as well as heights in c-axis direction (primitive *1c*-units).

EXPERIMENTAL

Sample preparation and XRD

All preparations were undertaken using elements of 99.999% purity and employing the technique of dry synthesis in evacuated silica glass capsules. After weighing, loading and prereacting at 770 K for three days, each sample (0.5–1g) was ground under acetone and homogenized at 1073 K for one day. Subsequently one series was annealed at 873 K for one day and quenched, while a second series was annealed at 423 K for ten days and quenched.

Compositions yielding one-phase products after quenching in cold water from 873 K were: y = 0.01, 0.05, 0.1 and 0.15, and x = 7.08, 7, and 6.92 in $Fe_x(Se_{1-y} \times Te_y)_8$. Prior to SQUID and MS measurements all samples were analysed using XRD with CuK_a radiation and Si as an internal standard (a = 5.43088 Å). Standard deviations of the obtained lattice parameters were less than 0.001 Å. Including a possible weighing error of 0.0005 g for each element, the errors in *c*- and *a*-axis data were calculated to be less than 0.004 Å and 0.002 Å, respectively. To avoid oxidation, all samples were stored under acetone between measurements.

SQUID

The magnetization measurements were performed in a Quantum Design MPMS5 SQUID magnetometer. The magnetization of each sample was measured in an applied field of 1 kOe versus temperature (10 – 300 K) and magnetization versus magnetic field curves were recorded at some different temperatures in fields 0–50 kOe.

Mössbauer spectroscopy

Transmission ^{57}Fe Mössbauer spectra (MS) were recorded at room and low temperatures, using a computerized spectrometer in constant acceleration mode. The establishment of the velocity scale was done using natural iron at room temperature as reference. The centroid shift CS is thus given relative to metallic Fe at 295 K. The source at room temperature was of CoRh-type. The powder absorbers contained less than 5 mg/cm^2 of natural Fe; thus we saw no need for thickness corrections when fitting the spectra using a locally developed Mössbauer fitting program (JERNBERG and SUNDQVIST, 1983). An Oxford Cryogenic helium flow cryostat, with a temperature stability better than ± 1 K, was used when measuring at low temperature.

ANALYSES AND RESULTS

Phase relations and lattice parameters

A NiAs-type iron selenide exists at low temperature with a composition around Fe_7Se_8

(SCHUSTER *et al.*, 1979; AMCOFF *et al.*, 1994), which shows complete solubility with pyrrhotite of Fe_7S_8 composition (ERICSSON *et al.*, 1994). On the other hand, NiAs-type phases in the Fe-Te system have Fe:Te ratios below 7:8, and are only stable above ~800 K (IPSER *et al.*, 1974). From this it can be deduced that the solubility of Te in Fe_7Se_8 must be limited.

Indeed, our results show a maximum solubility of between 15 and 20 mole% Te in $Fe_x(Se_{1-y}Te_y)_8$ for samples quenched from 873 K. Samples richer in Te decompose into a two-phase assemblage, with a Fe-rich PbO-type phase (composition around $Fe_{1.04}(Se_{1-y}Te_y)$ (AMCOFF *et al.*, 1994), plus a vacancy-rich phase, probably equivalent to a second NiAs-type phase which is distorted and present on the Se rich side of Fe_7Se_8 in the Fe-Se system (SCHUSTER *et al.*, 1979). Furthermore, employing an identical experimental procedure along an imaginary Fe_7S_8-Fe_7Te_8 join, our results show a solubility of less than one mole% Te in pyrrhotite of Fe_7S_8 composition.

For pure Fe_7Se_8, quenching in cold water from a high temperature (*e.g.*, 873 K) results in the formation of a well-ordered hexagonal superstructure (*3c*); while vacancy ordering is confined to smaller and smaller regions when S is substituted into the Se substructure. Slow cooling of the pure end-member phase, or annealing at *e.g.* 423 K, leads to formation of a triclinically distorted superstructure (*4c*). This transition fails in S-doped samples. Samples of $Fe_x(Se_{1-y}Te_y)_8$ compositions annealed at 423 K, with x = 6.92, 7, 7.08 and y = 0.01, 0.05, 0.1, 0.15, decomposed into the two-phase assemblage described above, except for one sample of composition $Fe_{6.92}(Se_{0.99}Te_{0.01})_8$. For this reason the discussion below is limited to "873 K samples."

When x is decreased at constant y in $Fe_x(Se_{1-y} \times Te_y)_8$, the *c*-axis in the basic unit cell contracts considerably more than the *a*-axis (Fig. 2). This finding is comparable with observed changes in pyrrhotites (KRUSE, 1990) and in Fe_xSe_8 (AMCOFF *et al.*, 1994). When y is increased at constant x in $Fe_x(Se_{1-y}Te_y)_8$ the *a*-axis expands in a regular way with increasing Te, while there is a smaller change in the *c*-axis. For the most vacancy-rich samples, $Fe_{6.92}(Se_{1-y}Te_y)_8$, the *c*-axis even contracts with increase in Te; possibly there is also such a tendency for Te-rich samples with x = 7 and 7.08 (Fig. 2).

Comparing the observed changes in lattice parameters with results from the pseudobinary Fe_7S_8 - Fe_7Se_8 join (ERICSSON *et al.*, 1994), it is obvious that the change in *a*-axis when Te is substituted into the Se-sublattice is more pronounced than with S-Se substitution (Fig. 3). This is not unexpected, since Se and Te differ more, with regard to both ionic radius and electronegativity, than do Se and S. However, the decrease in the *c*-axis when Te is substituted for Se is unexpected. The rapid flattening of the unit cell (increase in *a/c* ratio) with increasing Te is probably related to the approach of the decomposition point, where the hexagonal phase is split into a PbO phase + a distorted NiAs phase.

Using XRD, *3c* of Fe_7Se_8 composition can be distinguished from *1c* by a set of well-defined superreflections (AMCOFF *et al.*, 1994). Two observations deserve to be mentioned in this connection: (*i*) With increasing Te, the superreflections decrease in intensity and disappear somewhere between y = 0.05 and y = 0.1 in $Fe_x(Se_{1-y}Te_y)_8$. We have

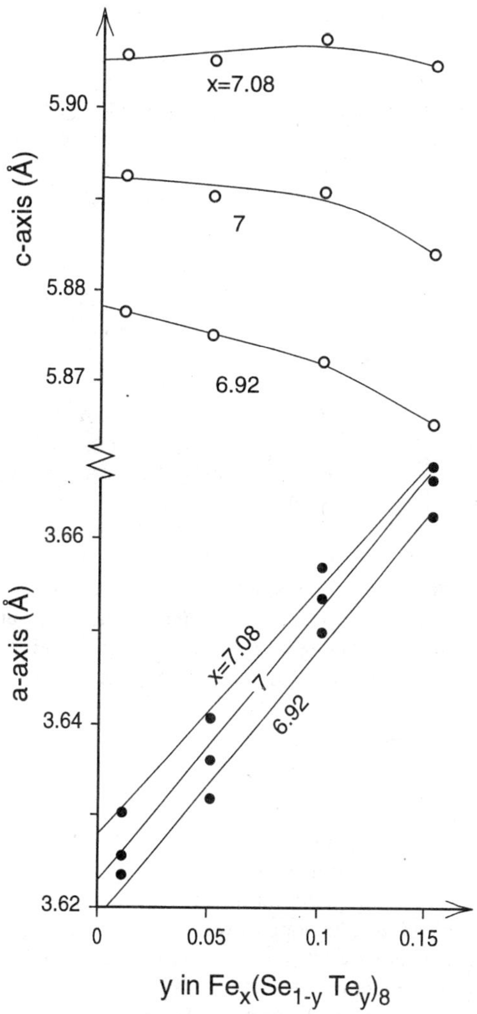

FIG. 2. Lattice parameter changes in $Fe_x(Se_{1-y}Te)_8$ mix crystals. The curves have been drawn by free-hand.

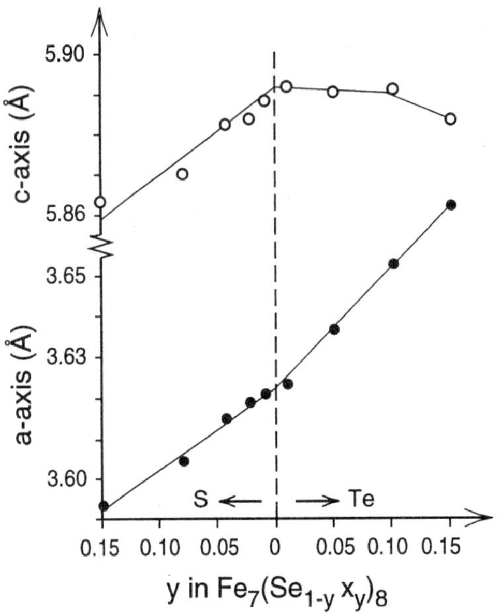

FIG. 3. Lattice parameter changes in $Fe_7(Se,X)_8$ around the pure end-composition, where X corresponds to S (to the left) and Te (to the right), respectively. Data for S/Se substitution have been taken from ERICSSON *et al.* (1994).

also noticed this phenomenon for compositions along the pseudobinary Fe_7S_8-Fe_7Se_8 join; but it is more accentuated for Se-Te substitution. (*ii*) With decreasing x (from 7), in $Fe_x(Se_{1-y}Te_y)_8$, the *3c*-super-reflections become stronger, a finding which was also observed for compositions along the Fe-Se join (AMCOFF *et al.*, 1994). This can be explained partly by kinetics and partly by the fact that ideal *3c* can only form in vacancy-rich samples (Fe_7X_8 according to *e.g.* (BERTAUT, 1953)).

Below, the lattice parameter changes as well as the changes in *3c*-superreflection intensities are put in relation to the results from the magnetisation- and Mössbauer measurements.

SQUID

The magnetic response of $Fe_7(Se_{1-x}Te_x)_8$ is governed by the ferrimagnetic character of the sequential vacancy occupation in adjacent *ab*-planes. As in the case of substituting S for Se (ERICSSON *et al.*, 1994), the substitution of Te for Se hinders the occurrence of long range vacancy ordering, a vacancy ordering that characterises both the *3c* and the *4c* structures. However, a local ferrimagnetic order is evident on short length scales. The local magnetic structure remains essentially unchanged with Te substitution, whereas the macroscopic magnetic response changes drastically, *e.g.,* the permeability decreases rapidly. Additionally, the temperature for the spin reorientation T_{sr} increases with increasing Te substitution. At T_{sr}, the spins flip from aligning parallel to the *c*-axis at low temperatures to being confined to the *ab*-plane at higher temperatures. We illustrate the macroscopic magnetic behavior by examining two details of the magnetic response that are highly sensitive to small compositional changes: (*i*) the temperature for the spin reorientation (occurring at T $\approx$ 130 K for the pure selenium compound) and (*ii*) the low field permeability.

In Fig. 4, the magnetization at an applied field of 1 kOe is plotted versus temperature for some different samples. The magnetization is recorded after cooling the sample in zero field, applying the field at 10 K and then recording the magnetization with increasing temperature (zero field cooled procedure). The most striking feature of the curves is a marked increase of the magnetization with increasing temperature and an abrupt levelling off above a well defined temperature. We identify this knee in the curves as the spin reorientation temperature. It is seen from Fig. 4 that the magnetization (and consequently also the permeability) decreases rapidly with increasing Te substitution. In Fig. 5 we illustrate the permeability behaviour for the different samples by magnetization curves measured at 200 K. The macroscopic response changes from typical ferrimagnetic behaviour for the low Te contents to a linear response at higher Te content. This behaviour reflects the hindered vacancy ordering on increasing anion substitution.

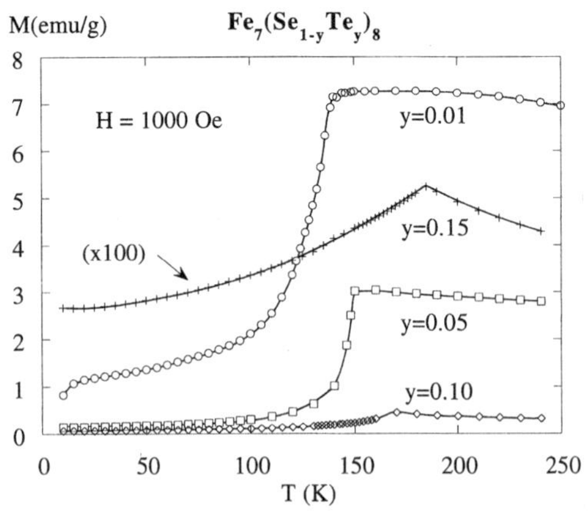

FIG. 4. Magnetization versus temperature at an applied field of 1 kOe for $Fe_7(Se_{1-y}Te_y)_8$ samples of different Te concentration. Note that the magnetisation for y = 0.15 has been multiplied by a factor of 100.

Fig. 6 shows M versus T and Fig. 7 M versus H curves for the nonstoichiometric compositions $Fe_{7.08}(Se_{0.95}Te_{0.05})_8$ and $Fe_{6.92}(Se_{0.95}Te_{0.05})_8$, together with the corresponding stoichiometric sample $Fe_7(Se_{0.95}Te_{0.05})_8$. As is seen from Fig. 6, the spin reorientation temperature decreases with an increasing number of vacancies and from Fig. 7 it is clearly seen that the saturation magnetisation increases with an increasing number of vacancies. The estimated saturation magnetisation scales reasonably well with the number of vacancies. At temperatures below T_{sr}, the material becomes magnetically softer with an increasing number of vacancies.

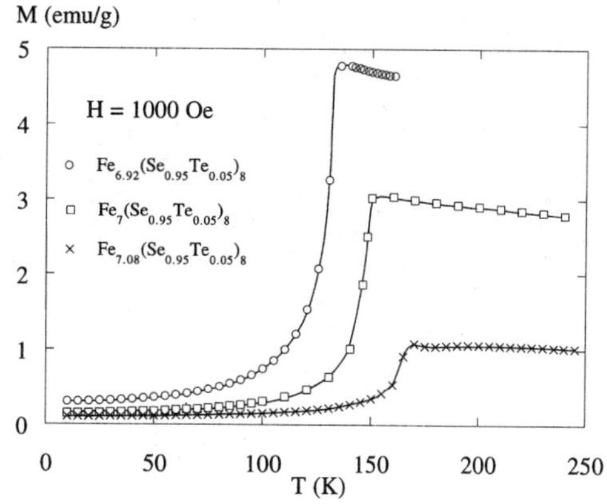

FIG. 6. Magnetization versus temperature at an applied field of 1 kOe for $Fe_x(Se_{0.95}Te_{0.05})_8$, three samples with x = 6.92, 7 and 7.08.

Mössbauer spectroscopy

Room temperature spectra. Spectra of Fe_7S_8 and Fe_7Se_8 recorded at room temperature are very similar in shape (ERICSSON *et al.*, 1994). LEVINSON and TREVES, (1968) presented a model for Fe_7S_8 in which the hyperfine field depends mainly on the number of Fe—S—Fe super-exchange bonds between iron in adjacent *ab*-planes. Fe atoms at A, B and C sites (see Fig. 1) have 18, 14, and 13 bonds (note that two near by Fe atoms along the same *c*-axis are connected via three Fe—S—Fe bonds), thus the hyperfine field for Fe at an A site should be larger than at a B site and smallest at a C site. The ferromagnetic coupling between the Fe atoms within the same *ab*-plane leads to a further division into A_1 (moment directed against a vacancy) and A_2 sites (moment directed against an Fe atom). The occurrence of the different configurations in the structure is $A_1{:}A_2{:}B{:}C = 1{:}2{:}2{:}2$.

We have used the four sextet model to fit all spectra recorded at room temperature. It fits nicely for Fe_7Se_8 (AMCOFF *et al.*, 1994) and $Fe_7(Se_{0.99} \times Te_{0.01})_8$, and can still be used for Te-richer samples (Fig. 8). IGAKI *et al.*(1981; 1982) studied $Fe_{1-x}S$ ($0.083 \leq x \leq 0.125$) by MS at room temperature and found that a vacancy in the same *ab*-plane increased the hyperfine field by 0.75 T (ΔB_a), a vacancy obliquely above (or below) decreased it by 0.875 T (ΔB_b) and a vacancy straight above (or below) along the same *c*-axis decreased it by 4.75 T (ΔB_c). The actual hyperfine field for a special configuration could then be calculated using a linear combination of DB_a, DB_b and DB_c, depending on the number of near-by vacancies. Transforming these values to Fe_7Se_8 leads to a reduction of $\approx 21\%$, thus $\Delta B_a = +0.59$ T, ΔB_b

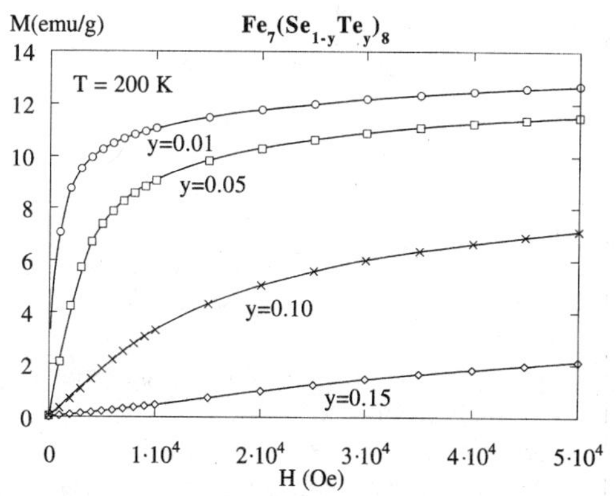

FIG. 5. Magnetization versus applied field H for four different samples of $Fe_7(Se_{1-y}Te_y)_8$ at 200 K.

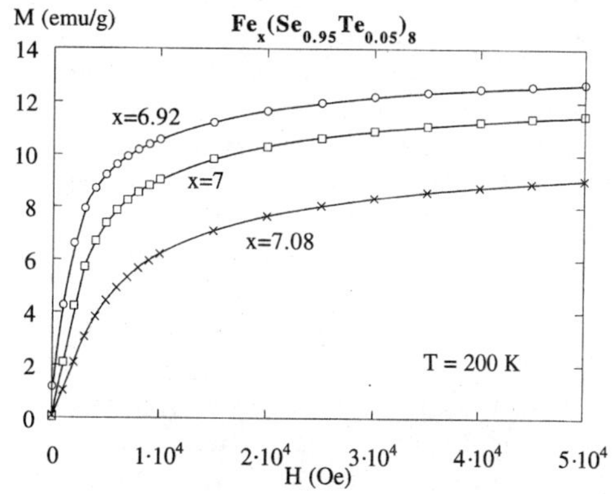

FIG. 7. Magnetization of $Fe_x(Se_{0.95}Te_{0.05})_8$ for three samples with x = 6.92, 7 and 7.08 versus applied field at T = 200 K.

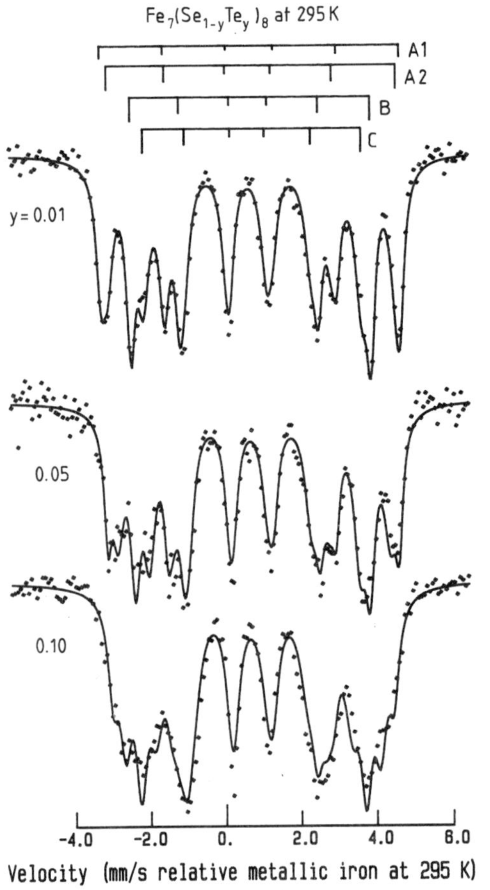

FIG. 8. Mössbauer spectra, recorded at room temperature of $Fe_7(Se_{1-y}Te_y)_8$ with y = 0.01, 0.05 and 0.10. The spectra are fitted using four sextets A_1, A_2, A and C with intensities 1:2:2:2. The bar diagram above shows the peak-positions for y = 0.01.

= −0.69 T and ΔB_c = −3.75 T (ERICSSON *et al.*, 1994). Assuming a random distribution of vacancies over the metal positions in Fe_7Se_8 should then result in a spectrum described by a broad sextet with a hyperfine field of ≈23 T and weak shoulders on the inner sides, coming from a broad sextet with a field of ≈19 T. None of our spectra can be fitted to such a model, and as a consequence we assume that vacancy-vacancy repulsion leads to short range order with A, B, and C like surroundings even for Te-rich compositions. The broad spectra obtained for Te ≥ 0.05 can be understood as a result of the random Se-Te mixing in the anion shells surrounding the central iron atom. In the fittings the centroid shifts for A_1 and A_2 were constrained to equal values, and intensities of A_1, A_2, B and C patterns were fixed to 1:2:2:2. With increasing Te-mixing, the half widths increase from 0.2 mm/s to 0.4 mm/s. CS(A_1) = CS(A_2) are a few hundreds of a mm/s bigger than CS(B) ≈ CS(C), but the centroid shifts did not show any significant variation with composition, on average CS = 0.715(10) mm/s in the whole interval. The quadrupole splittings (defined as DQ = 0.5 ($v_6 - v_5 - v_2 + v_1$), where $v_1, \ldots . v_6$ are peak positions in the sextet with increasing velocity) are small but positive for all sites with no significant variation with composition: DQ(A_1) = 0.05(5), DQ(A_2) = 0.05(5), DQ(B) = 0.05(5) and DQ(C) = 0.13(3) mm/s.

The magnetic field is 28.7, 28.6, 27.4 and 25.9 tesla for the A_1, A_2, B and C sites in Fe_7Se_8 and decreases linearly to 21.7, 19.8, 17.4 and 14,9 tesla in $Fe_7(Se_{0.85}Te_{0.15})_8$ for the respective sites. Thus, the hyperfine field is the only Mössbauer parameter that shows a clear composition dependence in the studied interval.

Mössbauer spectra recorded at low temperatures. In Fe_7Se_8 *3c*, the room temperature quadrupole splittings are in the range 0–0.15 mm/s for the four sites: A_1, A_2, B and C. However, at temperatures around 100 K, the A_1 and A_2 patterns coincide and DQ is smaller (0, −0.3 and 0 mm/s for the A, B and C sites; (AMCOFF *et al.*, 1994; OK *et al.*, 1993). This can be understood as a result of the spin reorientation from the *ab*-plane to the *c*-axis with decreasing temperature. The same tendency is observed for the compositions studied here. Although the spectra are broadened due to the Se-Te mixing. DQ is close to or below zero at temperatures around 100 K, and positive, showing values some tenths of a mm/s higher, around 200 K. DQ is related to the electric field gradient (EFG) and the direction of the hyperfine field relative EFG, defined through the polar and azimuthal angles Θ and Φ:

$$DQ = \frac{V_{zz}eQ}{2}\left(\frac{3\cos^2\Theta - 1 + \eta\sin^2\Theta\cos 2\Phi}{2}\right)$$

Here η is the asymmetry parameter and V_{zz} is the principal component of the EFG. In Fe_7Se_8 *3c*, on decreasing temperature, there is a decrease in DQ at 130 K of ≈ 0.3 mm/s (OK *et al.*, 1993). Thus, it is obvious that the substitution of Se by Te does not lead to a significant randomization of EFG directions relative to the magnetization direction. (The quadrupole splitting should not be sensitive to a spin rotation in such a case.)

Spectra recorded in the vicinity of liquid helium temperature are broadened compared to spectra at 100 K and above. To get reasonable fits we have used 8 sextets (with some constraints in order to keep the number of free parameters low), three coming from A sites (thus having equal CS), two from B and C sites each, and a low field ($\approx$12 tesla) sextet with 5% intensity at 128 K, increasing to 10% at 15 K. A component having an extraordinary low field has earlier been observed in Fe_7Se_8 (AMCOFF *et al.*, 1994). FATSEAS *et al.*, (1977) and BOUMFORD and MORRISH (1974) also noticed that Fe_7Se_8 spectra recorded at 4.2 K were more complex compared to the spectra recorded at 77 K. To get comparable parameter values, these eight sextets have been also used in fitting $Fe_7(Se_{.90} \times Te_{0.10})_8$ at 200 K,128 K and 15 K (see Fig. 9). The intensity-weighted averaged centroid shift, $CS_{aver.}$ = 0.86, 0.82 and 0.77 mm/s at 15 K, 128 K and 200 K respectively, which gives a slope of -7.8×10^{-4} mm/(s K) above 128 K. This is comparable with the high temperature slope of the second order dopplershift (SOD) of -7.3×10^{-4} mm/(s K) (COHEN, 1976). The Debye temperature of Fe_7Se_8 is low, 248(5) K (OK *et al.*, 1993), giving an increase in SOD of 0.04 mm/s from 128 K to 15 K, in full agreement with the experimental values above. $DQ_{aver.} = -0.08(2)$ mm/s below 128 K, and the hyperfine field $B_{aver.}$ increases from 22 T at 128 K to 23 T at 15 K. Fe_7Se_8 at 7 K and $Fe_7(Se_{0.85}Te_{0.15})_8$ at 15 K gave average hyperfine fields of 23.9 and 22.9 tesla, respectively. Thus, the clear difference in hyperfine fields with composition, as measured at room temperature, is probably a result of a decrease in magnetic ordering temperature T_N with increase in Te/Se substitution. (Our SQUID magnetometer cannot be used at temperatures above 300 K, so we have not yet studied how T_N varics with composition). Contrary to the observations above, OK *et al.* (1993) claimed that they had successfully fitted Mössbauer spectra of Fe_7Se_8 *3c*, recorded down to 12 K, using only three sextets.

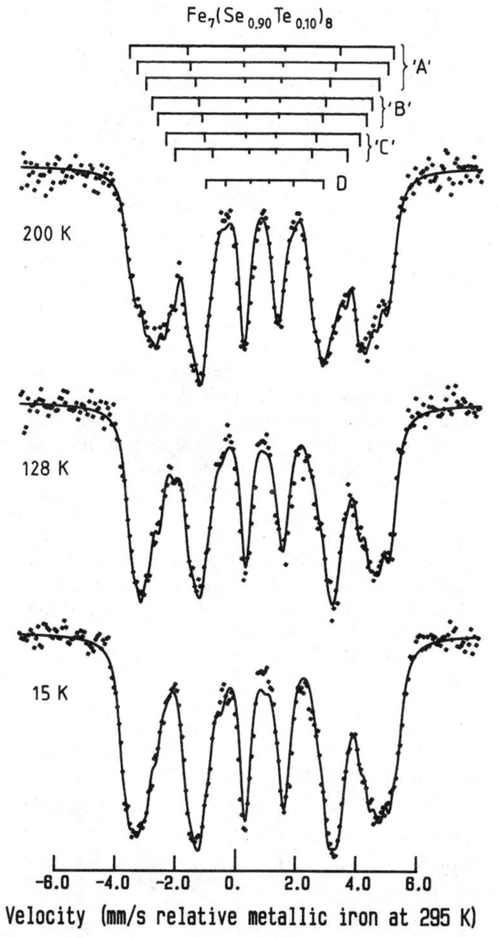

FIG. 9. Mössbauer spectra, recorded at 200 K, 128 K and 15 K of $Fe_7(Se_{0.90}Te_{0.10})_8$. All spectra are fitted using 8 sextets, assigned to 'A', 'B', 'C' and D sites. For further details, see text.

DISCUSSION

The low field permeability

The magnitude of the magnetisation in an external field of 1 kOe decreases strongly with concentration of Te in $Fe_7(Se_{1-y}Te_y)_8$, at temperatures below T_{sr} more than 100 times and at temperatures above T_{sr} more than 50 times from y = 0.01 to y = 0.15, a change of order 25 times stronger than the corresponding decrease in $Fe_7(Se_{1-y}S_y)_8$. The magnetization increases also with concentration of vacancies in $Fe_x(Se_{1-y}Te_y)_8$. The fraction of vacant Fe-sites is given by (8 $-$x)/8, *i. e.* 13.5%, 12.5% and 11.5% for x = 6.92, 7 and 7.08, respectively. Using Fig. 7 (values for $Fe_x(Se_{0.95}Te_{0.05})_8$), and assuming that the magnetization values at an external field of 5 tesla are proportional to the saturation value, one gets the following ratios between saturation magnetization and concentration of vacancies: 12.8 (emu/g)/13.5% = 0.95, 11.5 (emu/g)/12.5% = 0.92 and 9 (emu/g)/11.5% = 0.78. The nearly equal value of the ratio for the two more vacancy-rich samples can be interpreted to reflect a similar degree of vacancy ordering in alternate *ab* layers in those samples. The significantly lower ratio between saturation magnetization and number of vacancies for the 11.5% sample may be understood as a result of a less ideal *3c* structure in this sample. In this respect it is interesting to note that the homogeneity range for the *3c* structure can be expected to move to the vacancy-rich side of a nominal $Fe_7(Se, Te)_8$

composition as the concentration of Te increases (IPSER *et al.*, 1974). The low ratio between saturation magnetization and concentration of vacancies for x = 7.08 is also in agreement with results for pure Fe selenides, showing a rapid decrease in this ratio with decrease in number of vacancies (AMCOFF *et al.*, 1994).

A simple model

Our XRD, SQUID and MS results can all be described in the following schematic way. In Fe_7Se_8 *3c*, the vacancies are confined to every second metal *ab*-layer. The antiferromagnetic coupling between the layers and the ferromagnetic coupling within a layer give ferrimagnetism, and the net magnetization of a domain is given by the difference in magnetization of vacancy-free layers and vacancy layers (see Fig. 10). When S, and especially Te, is substituted into the system, the size of the ordered regions decreases and phase and antiphase regions start to coexist within a magnetic domain. The randomly distributed Te (or S) anions hinder the diffusion of vacancies. The macroscopic magnetization will now depend on the difference in "phase volume" (PV) and "anti phase volume" (APV) within a magnetic domain. With increasing S and Te substitutions, the number of phase and antiphase domains within a magnetic domain increases rapidly (Fig. 10). This leads to a comparably low value of the permeablity. The results (see *e.g.* Fig. 4) also indicate a significant increase in anisotropy when the spins are aligned along the *c* axis as compared to in-plane alignment. The smaller individual PVs and APVs result in a lack of supercell reflections in XRD and also give a magnetically harder material, as the magnetization of a magnetic domain decreases significantly with S/Se or Te/Se substitution. Still, there is local ordering of vacancies around an individual iron atom, leading to A, B and C like surroundings as shown by the Mössbauer spectra above. The broadening in MS for Te-rich samples could both be due to the Se/Te mixing and to the increasing numbers of phase-antiphase boundaries.

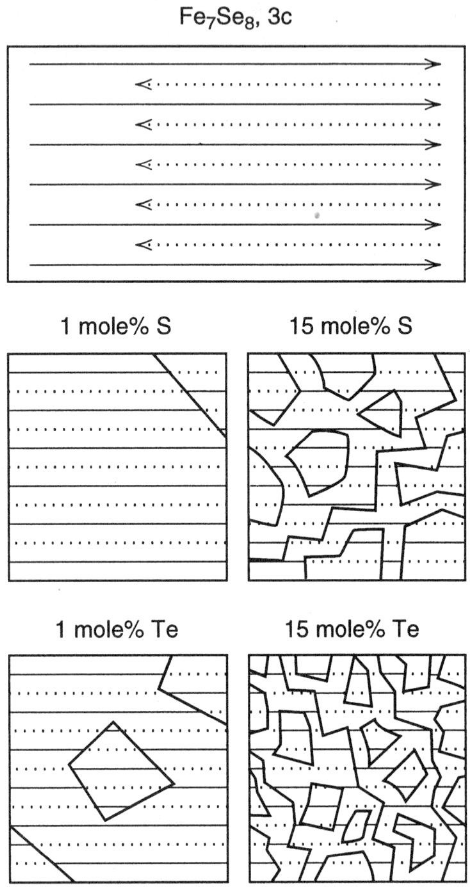

FIG. 10. A schematic model for magnetization of $Fe_7(Se_{1-x}X_x)_8$, with X = S or Te above the spin reorientation temperature. The rectangles denote a magnetic domain with a ferromagnetic coupling within an *ab* layer passing the phase-antiphase border. The situation in ideal *3c* is visualized at top. Due to 25 % vacancies in every second cation layer (dotted lines) there is a ferrimagnetic moment and a net macroscopic magnetisation. The number of phase-antiphase borderlines within the magnetic domain will increase with concentration of S and especially with Te, giving a reduced macroscopic magnetisation (middle and bottom pictures).

The spin reorientation

The spin reorientation temperature T_{sr} increases with decreasing number of vacancies (Fig. 6) and with increasing Te/Se substitution (Fig. 4). The covariation of *a/c* and T_{sr} with anion substitution (Fig. 11), as well as with concentration of vacancies, indicates a geometrical factor for the dependence of T_{sr} with composition. Low temperature favors spins parallel with the *c*-axis and high temperatures spins in the *ab*-plane. Apparently, short distances within the *ab*-plane or in the *c* direction favor spin orientation in the *ab*-plane or parallel to the *c*-axis, respectively.

A spin reorientation also appears in related systems. T_{sr} is suppressed to $\approx$0 K for $Fe_7(Se_{0.4}S_{0.6})_8$, thus there is no spin reorientation in ordinary pyrrhotite. However, there is a similar spin transition in FeS at 445 K, with a decrease of 40 K for each

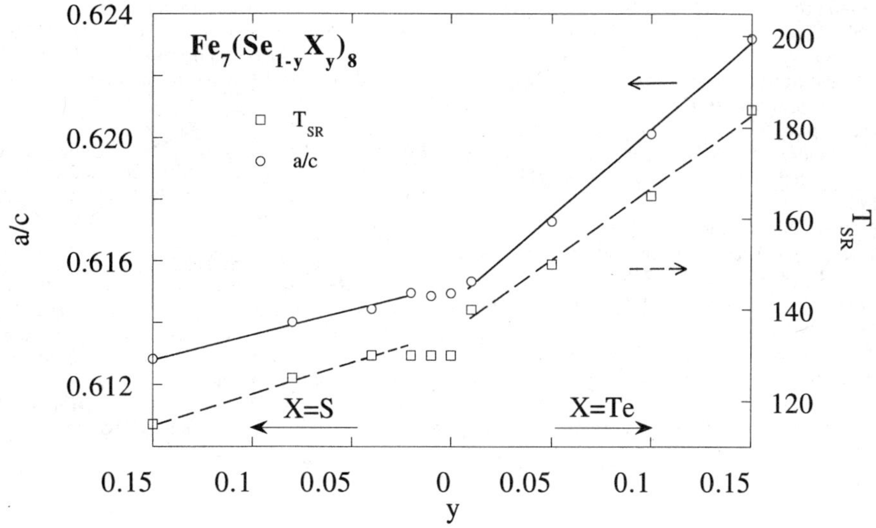

FIG. 11. Ratio of the lattice parameters *a/c* (left axis) and spin reorientation temperature, T_{sr} (right axis), plotted vs concentration of anion substitution y for $Fe_7(Se_{1-y}X_y)_8$. Data are shown for X = S, (left part of the x axis) and X = Te, (right part of the x axis).

% of metal vacancies on the sulfur-rich side (HORWOOD *et al.*, 1976). Thus, increases in concentration of vacancies decrease T_{sr} in FeS as well as in the system $Fe_x(Se_{1-y}Te_y)_8$. In $(Fe_{1-y}Co_y)_7Se_8$ the spin transition temperature increases with y, in $(Fe_{0.92}V_{0.08})_7Se_8$ *4c,* the spin reorientations occurs abruptly at 123 K, while there is a continuous change below 220 K in Fe_7Se_8 *4c* (SATO *et al.*, 1985 and OK *et al.*, 1994). XRD data at low temperatures might improve our understanding of the spin rotation mechanism, and both cation as well as anion substitutions may give hints to explain the spin reorientation in ordinary pyrrhotites as well.

Mineralogical considerations

The present results show that substitution of anions with identical charges in Fe_7Se_8 affects ordering processes in cation positions and that the magnitude of this effect, among other factors, depends on chemical differences between the anions. Indirectly, it can be inferred that substitution of ions in *any* variable position in a crystal structure affects other variable positions in the structure in ways that are not always obvious. The idea presented here of ''induced disorder'' may be of importance, especially in relation to order-disorder geothermometry and order-disorder geospeedometry, as natural minerals are invariably ''contaminated'' by foreign atoms which may act as ''nuclei of disorder''.

Acknowledgements—This investigation was supported by the Swedish Natural Science Research Council.

REFERENCES

AMCOFF Ö., ERICSSON T. and GISMELSEED A. (1994) An X-ray, Mössbauer and magnetization investigation of hexagonal FeSe. *Zeits. Krist.* **209**, 197–205.

BARKER W. W. and PARKS, T. C. (1986) The thermodynamic properties of pyrrhotite and pyrite. A re-evaluation. *Geochim. Cosmochim. Acta* **50**, 2185–2194.

BERTAUT E. F. (1953) Contribution a létude des structures lacunaires: La pyrrhotine. *Acta Crystallo.* **6**, 557–561.

BOUMFORD C. and MORRISH A. H. (1974) Magnetic properties of the iron selenide Fe_7Se_8. *Phys. Stat. Sol. (a)* **22**, 435–444.

COHEN R. L. (1976) *Application of Mössbauer spectroscopy, volume I.* Academic Press, New York San Fransisco London.

ERICSSON T., AMCOFF Ö. and NORDBLAD P. (1994) Vacancy ordering in Fe_7Se_8 -Fe_7S_8 solid solutions studied by Mössbauer, X-ray and magnetization techniques. *Hyper. Interact.* **90**, 515–520.

FATSEAS G. A., DORMANN J. L., DRUILHE R., BROSSARD L. and GIBART P. (1977) Test of the ''3c'' and ''4c'' Fe_7Se_8 Okazaki's superstructures by Mössbauer effect. *Physica* **86–88B**, 887–888.

HORWOOD J. L., TOWNSEND M. G. and WEBSTER A. H. (1976) Magnetic susceptibility of single crystal $Fe_{1-x}S$. *J. Solid State Chem.* **17**, 35–42.

IGAKI K., SATO M. and SHINOHARA T. (1981) Mössbauer study on the iron vacancy distribution in iron sulfide $Fe_{1-x}S$ ($0.083 \leq x \leq 0.125$). *Trans. JIM* **22**(9), 627–632.

IGAKI K., SATO M. and SHINOHARA T. (1982) Mössbauer study on the distribution of iron vacancies in iron sulfide $Fe_{1-x}S$. *Trans. JIM* **23**(5), 221–228.

IPSER H., KOMAREK K. L. and MIKLER H. (1974) Transition metal—chalcogen systems, V.: The iron-tellurium phase diagram. *Mh. Chem.* **105**, 1322–1334.

JERNBERG P. and SUNDQVIST T. (1983) A versitile Mössbauer analysis program. *Inst. Phys. Reports* 1090, University of Uppsala.

KISSIN S. A. and SCOTT S. D. (1982) Phase relations involving pyrrhotite below 350°C. *Econ. Geol.* **77**, 1739–1754.

KRUSE O. (1990) Mössbauer and X-ray study of the effects of vacancy concentration in synthetic hexagonal pyrrhotites. *Amer. Mineral.* **75**, 755–763.

KULLERUD G. (1986) Monoclinic pyrrhotite. *Bull. Geol. Soc. of Finland* **58**, 293–305.

LEVINSON L. M. and TREVES D. (1968) Mössbauer study of the magnetic structure of Fe_7S_8. *J. Phys. Chem. Solids* **29**, 2227–2231.

OK H. N., BAEK K. S. and KIM E. C. (1993) Mössbauer study of *"3c"* superstructure of the ferrimagnetic Fe_7Se_8. *Solid State Commun.* **87**, 1169–1172.

OK H. N., BAEK K. S. and PARK E. B. (1994) Effects of vanadium impurities on the spin-rotation transition of iron selenide *"4c"* Fe_7Se_8. *Solid State Commun.* **92**, 459–461.

PUTNIS A. and MCCONELL J. D. (1980) *Principles of Mineral Behaviour.* Blackwell Scientific Publications.

SATO M., KAMIMURA T. and IWATA T. (1985) Magnetic properties of $(Fe_{1-x}Co_x)_7Se_8$. *J. Appl. Phys.* **57**(1), 3244–3246.

SCHUSTER W., MIKLER H. and KOMAREK K. L. (1979) Transition metal-chalcogen systems. VII.: The iron-selenium phase diagram. *Mh. Chem.* **110**, 1153–1170.

THIEL R. C. (1969) A Mössbauer study in the troilite-pyrrhotite system. *Ph. D.-thesis*, Rijksuniversiteit te Groningen, Netherlands.

VAUGHAN D. J. and CRAIG J. R. (1978) *Mineral chemistry of metal sulfides.* Cambridge University Press.

Mineral Spectroscopy: A Tribute to Roger G. Burns
© The Geochemical Society, Special Publication No. 5, 1996
Editors: M. D. Dyar, C. McCammon and M. W. Schaefer

A comprehensive Mössbauer study of highly-substituted aluminum maghemite*

G. M. DA COSTA,[1,2] CH. LAURENT,[3] E. DE GRAVE,[1] and R. E. VANDENBERGHE[1]

[1]Laboratory of Magnetism, Department of Subatomic and Radiation Physics, University of Gent, Proeftuinstraat 86, B-9000, Belgium

[2]On leave from Departamento de Química, Universidade Federal de Ouro Preto, Ouro Preto, MG, Brazil

[3]Laboratoire de Chimie des Materiaux Inorganiques, Universite Paul Sabatier, Toulouse Cedex, France

Abstract—Synthetic Al-substituted maghemite samples, γ-$(Al_yFe_{1-y})_2O_3$, with y = 0, 0.10, 0.15, 0.22, 0.37, 0.55 and 0.66, were prepared from an oxalate precursor. The samples were studied by Mössbauer spectroscopy (MS) at 4.2 K in an applied magnetic field of 60 kOe, as well as at 8 K and in the range 80 K to 475 K at steps of 30 K in zero-field. Powder X-ray diffraction (XRD) and transmission electron microscopy (TEM) were applied as well. The particles are more or less spherical and the sizes determined from the XRD experiments decreased from approximately 80 nm for the Al-free sample to about 8 nm in the highly substituted samples. At 4.2 K the dependence of the hyperfine fields was found to be $H_{hf,A}$ = (520 − 30y) kOe and $H_{hf,B}$ = (531 − 26y) kOe, suggesting that the Al ions are located on both sites. Further information on the Al distribution was obtained from the relative areas, which are in agreement with a 50/50 distribution, *i.e.,* half of the Al on the tetrahedral sites and half on the octahedral ones. The canting angles were found to be Al-dependent, and for high substitutions the moments seem to be no longer collinear. The temperature dependence of the A- and B-site hyperfine fields could be reproduced using molecular field theory assuming an antiparallel spin configuration. The exchange integrals were found to be $J_{AA} = -18$ K, $J_{AB} = -27$ K and $J_{BB} = -4$ K. The Curie temperature (T_C) decreased with Al as T_C = (948 − 990y) K for y < 0.22. The A-site hyperfine field becomes larger than that of the B site in the vicinity of 300 K, and this crossing temperature depends on the Al content. The variation of the center shifts with T was interpreted in terms of the Debye model, but the derived characteristic Mössbauer temperature and intrinsic isomer shift bear no relation to Al.

INTRODUCTION

THE DISCOVERY of the Mössbauer effect, and in particular of the existence of that effect in ^{57}Fe, has created a new incentive for the study of both synthetic and natural Fe-containing minerals. The use of this technique for the study of soils and sediments has been reviewed on a regular basis by Bowen and co-authors (BOWEN, 1979; BOWEN and WEED 1984; BOWEN *et al.*, 1993). Since such systems are normally composed of mixtures of various and mostly poorly crystalline compounds, emphasis has been given to the correct characterization of the pure constituents. Another fact that has caught the attention of many researchers is the possibility that some of these natural compounds contain foreign cations isomorphically substituting for iron. Thus, the effects of both poor crystallinity and isomorphic substitution have been extensively studied, especially in the case of hematite (α-Fe_2O_3) and goethite (α-FeOOH), which are known to be present in almost all kinds of soils. Nevertheless, it was realised in recent years that another iron oxide, maghemite (γ-Fe_2O_3), also occurs in many tropical soils (SCHWERTMANN and FECHTER, 1984). This mineral, occurring with or without magnetite (Fe_3O_4) is believed to be the main one responsible for the magnetic character of such soils.

Maghemite is a cation-deficient spinel with 21 1/3 1/3 Fe^{3+} ions distributed among eight tetrahedral (A) and sixteen octahedral (B) sites of the unit cell; $2\frac{2}{3}$ sites per unit cell are vacant. It is commonly represented by the formula $(Fe)[Fe_{5/3}\square_{1/3}]O_4$, where ( ) denotes tetrahedral sites, [ ] octahedral sites and □ vacancies. However, it was reported that the vacancies can occur on both A and B sites simultaneously (ARMSTRONG, 1966; ANNERSTEN and HAFNER, 1973; HANEDA and MORRISH, 1977a). Well crystallized maghemite further exhibits a 1 : 5 ordering of □ and Fe within the octahedral sublattice, giving rise to X-ray reflections which are normally not observed for a cubic spinel structure (VAN OOSTERHOUT and ROOIJMANS, 1958). This ordering of the vacancies is size-dependent, and for particles smaller than approximately 20 nm it ceases to exist (HANEDA and MORRISH, 1977b).

Little is known about the influence of isomorphic substitution on the properties of maghemite, and

* Contribution No SSF95-03-02 from the Department of Subatomic and Radiation Physics, University of Gent, Belgium.

only very recently a systematic study of the effect of both Al and poor crystallinity was reported by this research group. Nano-sized maghemite and Al-maghemites prepared by thermal treatment of lepidocrocite (γ-FeOOH) containing up to 18 mole % Al, and hereafter referred to as ML samples, were studied by DA COSTA *et al.* (1994a, 1995a). The first paper emphasizes the difficulties in analysing the zero-field Mössbauer spectra (ZFMS) of such samples. In addition, the results of the spectra collected at 4.2 K under an applied field of 60 kOe (AFMS) were presented and from them it was concluded that 10 mole % may be a solubility limit for Al in the maghemite structure. This proposition had been suggested earlier by WOLSKA and SCHWERTMANN (1989) and BOWEN *et al.* (1994), but GILLOT and ROUSSET (1990) claimed that up to 66 mole % could be incorporated in maghemite samples prepared from an oxalate precursor. The second paper (DA COSTA *et al.*, 1995a) dealt with the results derived from the spectra collected at 8 K, and between 80 K and 475 K (steps of 25 K) without an applied field. Several important parameters such as the blocking temperatures, average hyperfine fields at 8 K, A- and B-site quadrupole splittings, and the so-called *Characteristic Mössbauer temperature,* have been obtained.

Relatively well-crystallized Al-maghemites prepared from Al-hematites, hereafter referred to as M samples, were also examined by BOWEN *et al.* (1994) and DA COSTA *et al.* (1994b, 1995b). One of the main objectives of these latter studies was to clarify an unacceptably wide scatter in the Mössbauer data concerning non-substituted γ-Fe_2O_3 as reported in literature. It was shown from the AFMS, both at 4.2 K as well as at 275 K, that the A-site center shift is on the average 0.12 mm/s smaller than that of the B site, regardless of temperature and Al content. This important finding enabled the authors to correctly fit the ZFMS collected at 8 K, and between 80K and 475 K at steps of 25 K (DA COSTA *et al,* 1995b). Thus, both the *Characteristic Mössbauer temperature* and the intrinsic isomer shift for the B site showed an increase with increasing the Al content, at least up to 10 mole %. The temperature dependence of the A- and B-sites hyperfine fields could be satisfactorily reproduced using the molecular-field theory assuming an antiparallel spin configuration. The exchange integrals were found to be $J_{AB} = -25$ K, $J_{AA} = -18$ K, and $J_{BB} = -3$ K. The Curie temperature for the Al-free sample was determined to be 930 K, and it decreased with increasing Al content.

To the best of the authors' current knowledge only two other papers dealing with the MS of Al-maghemites have been published. In one of them, four samples containing between zero and 11 mole % Al were prepared by oxidation of magnetite (DE JESUS FILHO *et al.,* 1993). The cell parameter was reported to decrease from 0.8337 nm to 0.8292 nm for the 11 mole % sample. Magnetization measurements gave a lower than expected value for the spontaneous magnetization, and to explain this result the authors proposed the existence of spin canting on B sites only, and that the vacancies are distributed over the two sites. The second paper was published by BOWEN *et al.* (1994), and concerns Al-maghemites prepared by several thermal treatments of a mixture of sucrose and Al-hematites containing up to 15 mole % Al. Both vacancies and Al were found to occupy the B sites exclusively.

As mentioned above, the solubility limit of Al in maghemite is still a matter of some dispute. In an attempt to solve this question, Mössbauer spectra of several highly-substituted samples have been collected at 4.2 K in an applied-field of 60 kOe, and also at 8 K and between 80 K and 475 K at steps of 30 K in zero field. These samples were prepared by the method reported by ROUSSET and PARIS (1972). In what follows it will be demonstrated that up to 66 mole % of Al can indeed be incorporated into the maghemite lattice.

EXPERIMENTAL PROCEDURES

All samples were prepared according to the method originally proposed by ROUSSET and PARIS (1972). It consists of several thermal treatments of co-precipitated Fe-Al oxalate precursors. Initially the mixed oxalates $(NH_4)_3[Fe_{1-x}Al_x(C_2O_4)_3] \times 3H_2O$, with $0 < x < \frac{2}{3}$, are decomposed in air at 400°C. The materials thus obtained are subjected to reduction in a H_2-H_2O atmosphere in order to produce magnetite. Finally, oxidation was carried-out in air at 210°C. Samples are code-named DG00, DG10, DG15, DG22, DG37, DG55 and DG66, where the numbers refer to the mole % Al in γ-$(Fe_{1-y}Al_y)_2O_3$, as determined by atomic absorption.

Powder X-ray diffraction (XRD) patterns were obtained using a Philips diffractometer with Co$K\alpha$ radiation and a graphite monochromator. The scans were done in the range of 15–80° (2θ) at a speed of $\frac{1}{4}$° min^{-1}, and the reflected intensities were recorded in a multichannel analyser. The diffractograms were fitted with a sum of pseudo-Lorentzians, two for each peak in order to account for the $K\alpha1$ and $K\alpha2$ radiation. Mean crystallite dimensions (MCD) along the [110] and [311] lattice directions were estimated from the full widths at half maximum using Scherrer's equation with $K = 0.9$. Lattice parameters were determined by the Nelson-Riley extrapolation method (KLUG and ALEXANDER, 1974). An internal standard (Si) was used in some runs.

Zero-field Mössbauer spectra were collected at 8 K and between 80–475 K in steps of 30 K. An external field of 60 kOe parallel to the direction of the incident γ-rays was applied at 4.2 K for all samples, and at intermediate temperatures for sample DG66. All center shifts are

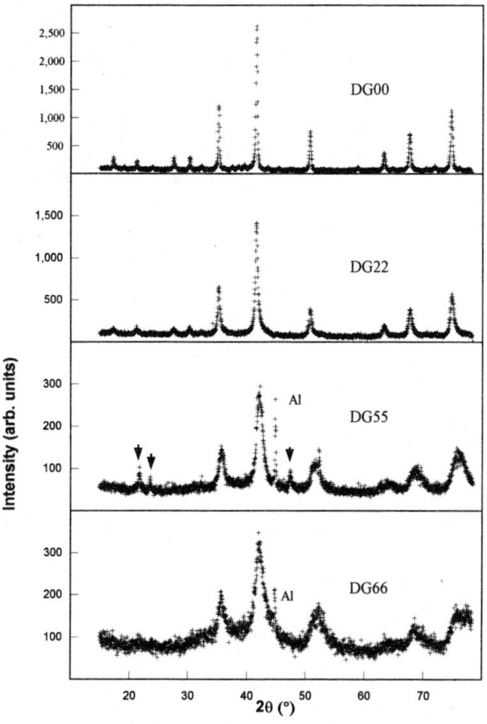

FIG. 1. X-ray diffraction patterns of maghemite (DG00) and Al-substituted maghemites (DG22, DG55 and DG66). Some superstructure lines are clearly visible in samples DG00 and DG22. Peaks marked with an arrow in sample DG55 are due to bayerite, while peaks marked as Al are due to the sample holder.

quoted relative to metallic iron. More details about the experimental set-up can be found elsewhere (DA COSTA *et al.*, 1994b). The overall number of spectra exceeds 100 and not all data will be presented here, but are available upon request.

Transmission electron microscopy (TEM) was performed with a JEOL 200CX operating at 200 kV. The powders were ultrasonically dispersed in ethanol and a drop of the dispersion was deposited onto a copper grid coated with a collodion film.

RESULTS AND DISCUSSION

X-ray diffraction

The XRD of samples DG00, DG22, DG55 and DG66 are shown in Fig. 1. Some superstructure lines are evident in the diffractograms of samples DG00, DG10, DG15 and DG22, indicating that for these compositions the vacancies are ordered on the B sites. For higher Al contents the lines become increasingly broader, probably due to the combined effects of small-particle sizes and high amounts of aluminum. There are three reflections for DG55 which do not correspond to a spinel phase. They were identified as belonging to bayerite ($Al_2O_3 \times 3H_2O$). Furthermore, it is to be noted that for DG66 the background in the region of 20–40° seems to be higher than outside that region, perhaps indicating the presence of another phase. One possible candidate is ferrihydrite (or most probably Al-ferrihydrite), which has the main peak around 40°. This substance is known to be very poorly crystallized, giving rise to a characteristic XRD pattern showing only 2 or 6 extremely broad lines (EGGLETON and FITZPATRICK, 1988). Ferrihydrite transforms to hematite by heating at around 300°C, but the stability of Al-ferrihydrite is much higher. For a 16 mole % Al the transformation to hematite occurs at 500°C (CHADWICK *et al.*, 1986). Thus, it is not impossible that sample DG66 contains Al-ferrihydrite as an impurity.

The cell parameters and the particle sizes along the [110] and [311] lattice directions are given in Table 1. Both quantities decrease with increasing Al content. In particular the particle size drastically lowers from the Al-free sample to the 66 mole % sample. The cell parameters listed in Table 1 differ considerably from those reported by GILLOT and ROUSSET (1990), especially for higher substitutions. In order to check for any possible systematic errors, the diffractograms of some samples were also obtained with a Siemens D5000 diffractometer, and the cell parameters derived from these runs are essentially the same as those in Table 1. Thus, the origin of the discrepancy between the present results and those reported in Gillot's paper could not be traced. A plot of cell parameters vs. Al content does not give a straight line, and a possible reason for that might be the smaller

Table 1. Lattice constants (a) and mean crystallite dimensions (MCD) along the [110] and [311] lattice directions. The estimated error on the lattice parameters is 0.0005 nm.

Sample	a (nm)	MCD110 (nm)	MCD311 (nm)
DG00	0.8348	80	60
DG10	0.8337	50	40
DG15	0.8326	35	35
DG22	0.8329	40	30
DG37	0.8341	20	20
DG55	0.8231	10	10
DG66	0.8146	10	5

MCD of the highly substituted samples. This effect was reported to become important in Co ferrites smaller than 20 nm (VANDENBERGHE *et al.*, 1980), and the samples with y > 0.37 have sizes of that order.

TRANSMISSION ELECTRON MICROSCOPY

TEM observations revealed that the DG00, DG10, DG15 and DG22 specimens are very similar to one another. A representative micrograph of these powders is shown in Fig. 2 for sample DG22. The grains are isotropic in shape and their sizes are in the range 25–200 nm. Indexing of the electron microdiffraction patterns (EMD) revealed that all rings belong to a maghemite-type phase. However, some grains form larger, elongated particles as if they were sintered, but the diffraction spots on the EMD patterns of these grains also belong to a maghemite phase. A few agglomerates composed of small black dots in a grey matrix were also observed for DG15. Their EMD pattern shows diffraction rings and spots that could be ascribed to a mixture of maghemite and some unreacted magnetite-type oxide. As for DG55 and DG66, their smaller crystallite sizes are obvious from the TEM micrographs (Fig. 2). These powders are also more heterogeneous than those containing less alumi-

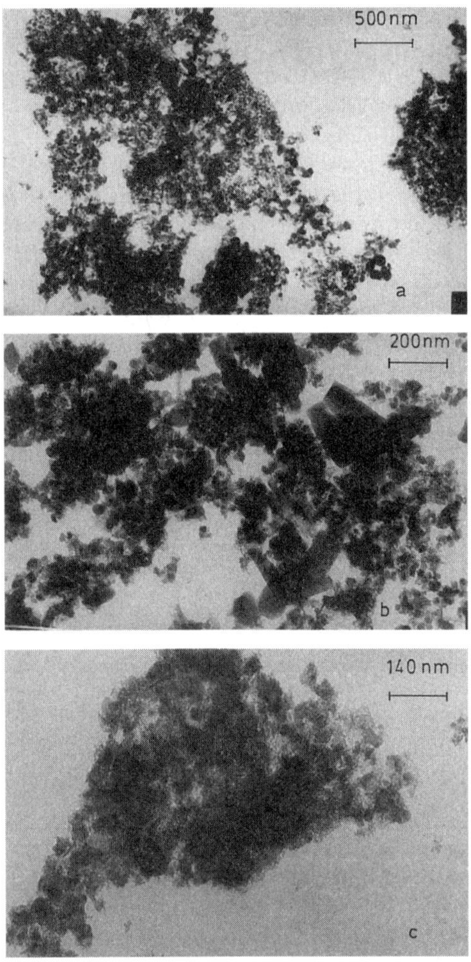

FIG. 2. TEM photographs of samples DG22 (a), DG55 (b) and DG66 (c).

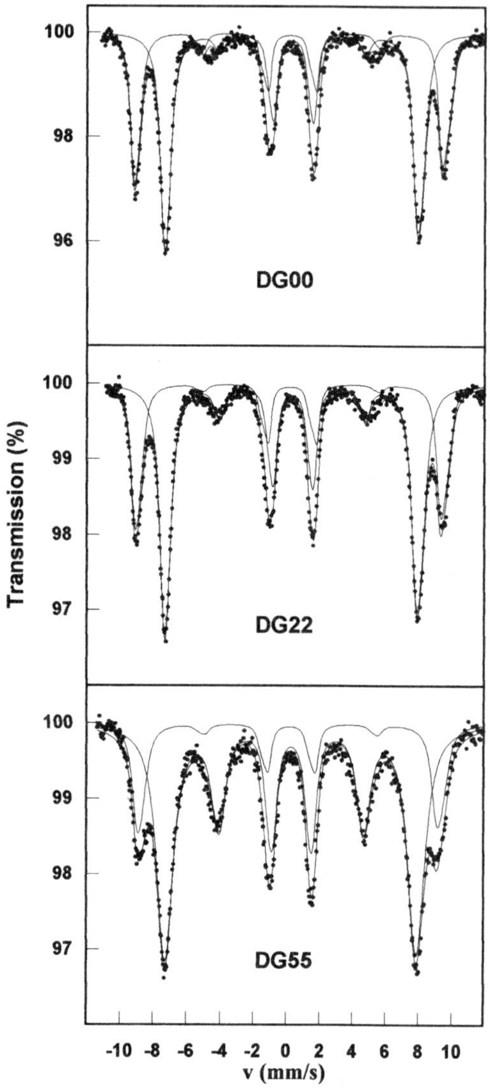

FIG. 3. Mössbauer spectra collected at 4.2 K in an applied field of 60 kOe. Dots represent the experimental data, and full lines represent the calculated subspectra and their sum.

num. For instance, small needle-shaped grains could be observed in some micrographs, especially in DG37 (not shown in Fig. 2). These needles are probably due to maghemite grains that did not yet sinter to a more isotropic shape. Indeed, it is known that there is a threshold temperature at which such a sintering phenomenon occurs, which notably depends on the formation temperature and kinetics of the maghemite phase and on its grain size (ROUSSET *et al.,* 1979). In addition, the micrograph of sample DG55 (Fig. 2) shows the presence of "bottle-shaped" grains. The EMD pattern of such a "bottle" is fairly poor, but the measured distances could match those of $FeAl_2O_4$.

MÖSSBAUER SPECTROSCOPY

Applied-field spectra

Figure 3 shows the Mössbauer spectra at 4.2 K in an applied-field of 60 kOe for samples DG00, DG22 and DG55. These spectra were fitted with two main asymmetric sextet components. The asymmetry is due to the combined effects of random orientation of the hyperfine field with respect to the electric-field-gradient's principal axis and a non-zero dipolar contribution to the hyperfine field (DE GRAVE *et al.,* 1993). For samples DG15 and DG37 a weak hematite component (~3 mole %) had to be included to account for the large broadening of lines 2 and 5 (BOWEN *et al.,* 1994). In AFMS the observed hyperfine field (H_{eff}) is given by the vectorial summation of the applied field (H_{ext}) and the internal field (H_{hf}). The parameters derived from this fit procedure are given in Table 2, and several important conclusions can be reached from them. The averaged difference between the B- and A-site center shifts is 0.12 mm/s, in excellent agreement with the value reported earlier for the M samples. As discussed in detail by DA COSTA *et al.* (1995b), the correct interpretation of the ZFMS of maghemite requires that the difference between the A- and B-site center shifts remains fixed at 0.12 mm/s. This same constraint was applied to analyse the spectra of the present samples.

The presence of lines 2 and 5, which are expected to vanish if the hyperfine fields are completely aligned along the external field (H_{ext}), is a well-documented aspect of the AFMS of maghemite and is due to a canting of the spins with respect to $\vec{H}_{ext}$. There are several propositions in literature to explain the origin of this canting (COEY, 1971; MORRISH *et al.,* 1976; HENDRIKSEN *et al.,* 1994a) and the interested reader is referred to these original papers. In addition, it is still a matter of controversy as to whether or not the A- and B-site moments remain collinear with one another. Some investigators have suggested that the A-site moments are completely aligned with $\vec{H}_{ext}$, while those from the B site are canted (POLLARD, 1990; DE JESUS FILHO *et al.,* 1993). On the other hand it has also been proposed that the moments remain collinear, although canted with respect to H_{ext} (POLLARD and MORRISH, 1987; DE BAKKER *et al.,* 1991). From the relative areas $A_{2,5} : A_{1,6}$ of the middle to outer lines, one can determine the canting angle θ, defined as the angle between $\vec{H}_{hf}$ and H_{ext}, (VANDENBERGHE and DE GRAVE, 1989). The results given in Table 2 for θ_A and θ_B show that for low substitutions ($y < 0.15$) the summation of θ_A and θ_B is close to 180°, indicating that the net moments are likely to be collinear. For higher substitutions there seems to exist a systematic decrease of θ_B with increasing Al content, but not for θ_A. Nevertheless, it cannot be ruled out that the lower intensity of the A-site absorption prevents an accurate determination of the area of lines 2 and 5. On the other hand it is clear that the summation of θ_A and θ_B deviates substantially from 180°, suggesting that the moments are no longer collinear in these highly substituted maghemites.

The inclusion of Al in the structure causes the loss of some exchange paths, hence reducing the

Table 2. Effective hyperfine fields H_{eff} (kOe), hyperfine fields H_{hf} (kOe), canting angles θ (°), center shifts δ (mm/s), full-widths at half maximum of the third and fourth absorption lines (Γ, mm/s) and relative area RA (%) obtained from the spectra collected at 4.2 K in an applied field of 60 kOe. A and B refer to tetrahedral and octahedral sites respectively. Estimated errors are 1 kOe for H_{eff}, 4 kOe for H_{hf}, 5° for θ, and 0.005 mm/s for δ.

Sample	$H_{eff.A}$	$H_{hf.A}$	θ_A	δ_A	Γ_A	RA	$H_{eff.B}$	$H_{hf.B}$	θ_B	δ_B	Γ_B	RA
DG00	577	521	22	0.33	0.34	40	473	530	161	0.45	0.44	60
DG10	575	517	12	0.34	0.42	36	473	530	160	0.45	0.47	64
DG15	572	514	16	0.36	0.31	37	470	527	159	0.49	0.34	63
DG22	572	514	15	0.33	0.37	34	472	527	156	0.45	0.48	66
DG37	567	508	8	0.36	0.44	27	469	521	149	0.48	0.44	73
DG55	560	504	20	0.32	0.43	22	470	517	140	0.42	0.63	78

values of the hyperfine fields. The observed steady decrease in fields (Table 2) can be fitted according to the eqns:

$$H_{hf,A} = (520 - 30y) \text{ kOe}$$

$$H_{hf,B} = (531 - 26y) \text{ kOe}$$

BOWEN *et al.* (1994) collected external-field spectra for the M samples ($y \leq 0.15$) and reported a decrease of about 40 kOe per mole of Al for the A-site hyperfine field. In addition, DA COSTA *et al.* (1995b) also indicated a decrease of the same magnitude based on the ZFMS collected at 8 K for the same M series. As for the B sites, any change in H_{hf} remained inconclusive. The fact that for the present maghemites both A- and B-site hyperfine fields decrease with increasing Al content is a strong indication that the Al ions are located on both sites. If there is no substitution on the A sites, the B-site hyperfine field should remain fairly constant because no reduction of supertransfer of spin density from A- to B-site ferric ion would occur.

The above conclusion with respect to the Al distribution is corroborated by the relative A- and B-site area ratios (RA) derived from the AFMS (Table 2). This parameter for the A-site absorption is plotted as a function of Al substitution in Fig. 4, together with values expected on the basis of some specific assumptions. Surprisingly the best overall agreement is not with a random substitution but instead with a 50/50 distribution, *i.e.*, half of the Al entering the tetrahedral sites and half the octahedral sites. For low y values, a random distribution could not be excluded, but for higher substitutions it is clear that a random substitution is grossly in disagreement with the experimental data. As the synthesis of maghemite was carried out at fairly low temperatures, *i.e.*, 210°C, it can be expected that the same cation distribution should exist in the magnetite precursor. Thus, if we represent magnetite as $Fe^{3+}(Fe^{2+}Fe^{3+})O_4$, and consider that the Al ions would replace Fe^{3+} cations, it seems that indeed a random inclusion of Al results in a ratio of 50/50, and not of 37.5/62.5 as predicted by considering the maghemite formula. This Al distribution differs from that derived for the ML samples, in which a random distribution of both Al and vacancies was proposed. However, it is conceivable that the cation distribution depends upon the formation mechanism.

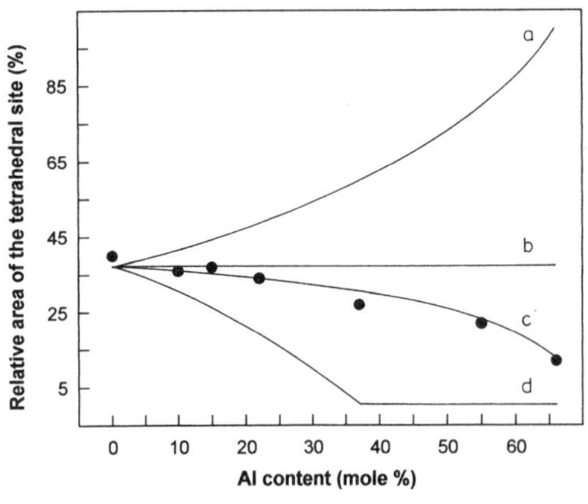

FIG. 4. Theoretical (solid lines) and experimental (dots) tetrahedral area as a function of Al content. The theoretical curves were calculated assuming that the Al ions are located: on B sites (a), distributed at random (b), distributed in a 50/50 ratio (c), and on A sites (d). The experimental data were obtained from the in-field spectra at 4.2 K.

The spectra of sample DG66 at 4.2 K and 55 K, both with and without an external field clearly indicate the existence of an impurity phase in addition to the maghemite (DA COSTA, 1995c). As this impurity tremendously complicates the analysis of the MS, no detailed attention will be paid to this composition.

Zero-field spectra

The ZFMS of samples DG00, DG22 and DG55 at 8 K and 300 K are reproduced in Fig. 5. The spectra of the other samples, as well as those at other temperatures, are similar to those displayed. At 8 K one broad and asymmetric sextet, characteristic for maghemite is obtained. At a given temperature T, depending on y, the spectra of all samples, except DG00, contain an additional doublet, which grows at the expense of the sextet component. This feature is typical for superparamagnetic particles (MØRUP and TOPSØE, 1976), and the appearance of a doublet is expected considering the estimated particle sizes of samples DG37, DG55 and DG66.

Two different approaches to fit the experimental data have been followed. For samples DG00, DG10, DG15 and DG22 the spectra were fitted as a superposition of two asymmetric Lorentzian-sextet profiles (VANDENBERGHE *et al.*, 1994) with three sets of width parameters, equal for A- and B-site subspectra. Thus, lines one and six, two and five, and three and four of each subspectrum were assumed to have the same shape with a high-field (Γ_{HF}) and a low-field (Γ_{LF}) line width. A very important restriction concerns the A-site center shift, which was constrained as $\delta_A = \delta_B - 0.12$ mm/s. Due to the strong overlap of the two patterns some other constraints were imposed: quadrupole shift $\epsilon_Q = 0$ fixed for both sextets, line-

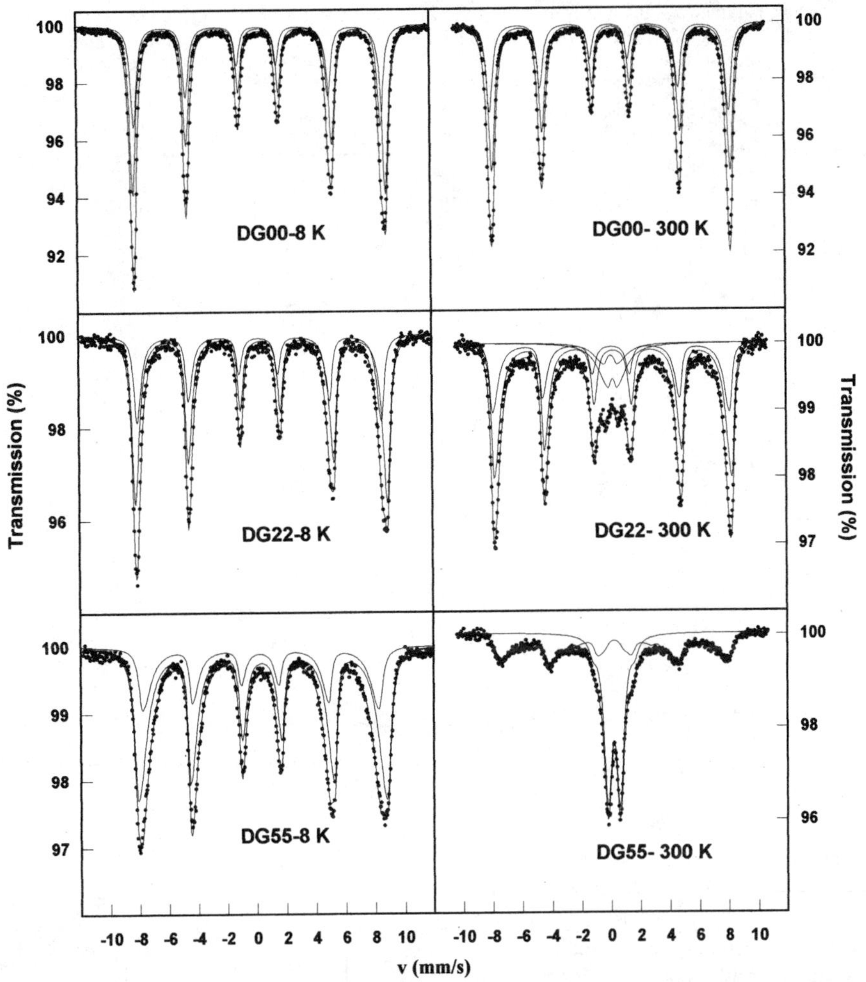

FIG. 5. Zero-field Mössbauer spectra of samples DG00, DG22 and DG55 at 8 K and 300 K. Dots represent the experimental data, and full lines represent the calculated subspectra and their sum.

area ratios fixed at 3 : x : 1, and fixed B- to A-site area fraction as obtained from the external field measurements. Typical iterated values for x were in the range 1.9–2.1. The doublet component was fitted as a superposition of two contributions with different quadrupole splittings, using the same constraints for the isomer shifts and fractional area ratios as listed above for the sextet components (DA COSTA *et al.*, 1995a).

The solid lines in Fig. 5 are the adjusted curves based on the methodology described above (except sample DG55, see below). The agreement between the experimental and fitted spectra is satisfactory. Some numerical results at selected temperatures are presented in Table 3. The ZFMS recorded at 8 K for DG37 and DG55 could be fit in the same manner, and the data are included as well. The agreement between the A- and B-site hyperfine fields at 8 K with the values obtained from the AFMS (4.2 K) is reasonable, and this finding supports the adopted fitting procedure. However, the differences for the B site show more disagreement at high Al contents than for the A site. This is probably related to the fact that as the Al content increases the lines become more asymmetric, and not necessarily in the same amount for the two sites. Some authors have found a much poorer agreement between the hyperfine-field values as obtained with and without an external field (DE BAKKER *et al.*, 1991; HENDRIKSEN *et al.*, 1994b). However, DA COSTA *et al.* (1995b) have demonstrated that this disagreement is an artifact of the fitting procedure the authors applied to analyse

Table 3. Mössbauer parameters obtained from fitting the zero-field spectra of Al-maghemites with two asymmetric-Lorentzian profiles. Γ_{HF} and Γ_{LF} are the high-field and low-field widths at half maximum of the first and sixth absorption lines. Estimated errors at 8 K are 2 kOe for H_{hf} and 0.01 mm/s for δ.

Sample	T (K)	$H_{hf,A}$ (kOe)	$H_{hf,B}$ (kOe)	δ_B (mm/s)	Γ_{HF} (mm/s)	Γ_{LF} (mm/s)
DG00	8	519	530	0.48	0.49	0.50
	80	519	529	0.47	0.48	0.57
	300	501	500	0.36	0.49	0.56
	450	476	465	0.27	0.52	0.68
DG10	8	516	529	0.48	0.56	0.62
	80	516	528	0.47	0.45	0.64
	300	498	499	0.36	0.46	0.70
	450	477	464	0.26	0.45	0.91
DG15	8	515	530	0.48	0.47	0.66
	80	516	529	0.47	0.47	0.77
	300	498	499	0.36	0.44	0.89
	450	473	462	0.27	0.52	1.1
DG22	8	515	531	0.47	0.46	0.70
	80	515	528	0.47	0.41	0.80
	300	498	500	0.36	0.41	0.94
	450	472	464	0.26	0.45	1.2
DG37	8	512	529	0.47	0.47	1.0
DG55	8	498	524	0.45	0.59	1.4

the data. The present results corroborate that conclusion.

As seen from Table 3 the A-site hyperfine field becomes higher than that of the B site in the vicinity of 300 K. This crossing of fields was also observed for the M samples (DA COSTA *et al.*, 1995b), as well as for $Li_{0.5}Fe_{2.5}O_4$ (YASUOKA *et al.*, 1962) and $MnFe_2O4$ (SAWATZKY *et al.*, 1969). A quantitative treatment of the temperature dependence of the hyperfine fields was carried out using the molecular field theory with the assumption of a collinear ferrimagnetic spin configuration (VANDENBERGHE and DE GRAVE, 1989). A spin of $^5/_2$ was attributed to Fe^{3+}, and the vacancies were restricted to B sites. The Al ions were distributed in a ratio 50/50 among the tetrahedral and octahedral sites. Example fitted curves are plotted in Fig. 6 for DG00 and DG22, and they are very similar for all four compositions. The calculated exchange integrals J_{AA}, J_{BB} and J_{AB} (see Table 4) are in remarkably good agreement with the values earlier found for the M samples. The calculated crossing temperatures, T_{inv1}, increases with increasing Al content, at least up to 15 mole %.

The Curie temperatures T_C, calculated from the exchange integrals, together with the data for the M samples, are plotted as a function of the Al content in Fig. 7. The observed decrease in T_C is due to the loss of exchange paths caused by the incorporation of diamagnetic Al ions. It is seen from Fig. 7 that the present results are in agreement

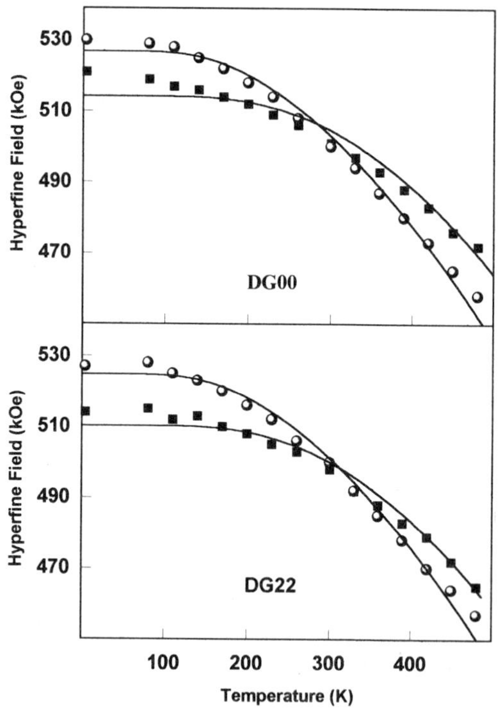

FIG. 6. Temperature dependence of the A- and B-sites hyperfine fields (squares and circles, respectively) of samples DG00 and DG22. Solid lines represent the theoretical dependence calculated using the molecular-field theory assuming a ferrimagnetic spin configuration.

Table 4. Exchange integrals J_{AA}, J_{AB}, and J_{BB}, and temperature at which $H_{hf,A}$ becomes higher than $H_{hf,B}$ (T_{invl}). All quantities are expressed in K.

Sample	J_{AA}	J_{AB}	J_{BB}	T_{invl}
DG00	−18	−27	−4	285
DG10	−16	−26	−3	298
DG15	−16	−26	−3	316
DG22	−18	−26	−4	316

with those obtained for the M samples, and a linear fit to both sets of data results in the following relations:

$$\text{DG samples: } T_C = 948 - 990y$$

$$\text{M samples: } T_C = 913 - 980y$$

The somewhat higher values for the DG samples as compared to the M samples might well be within the uncertainties (an error of 15 K would not be unreasonable) but an influence of the particle size, which is on average larger for the DG samples, cannot be ruled out.

The temperature dependences of the B-site center shifts are similar for all four DG samples, and two examples are shown in Fig. 8. The data for the A-site center shift are displaced by −0.12 mm/s. The solid curves represent the calculated variations assuming the Debye model for the lattice vibrations. The two important parameters derived from this analysis are the *characteristic Mössbauer temperature* (Θ_M) and the intrinsic isomer shift (δ_I). A conclusive variation of these two quantities with Al content could not be detected. Θ_M was found to be scattered within the range (460 ± 20) K, while a range (0.598 ± 0.006) mm/s was obtained for δ_I. These figures can be compared with the results previously reported for Al-maghemites, viz., Θ_M ~ 480 K and δ_I ~ 0.603 mm/s (M samples), and Θ_M ~ 530 K and δ_I ~ 0.588 mm/s (ML samples). These values are all typical for Fe^{3+} in oxides and oxyhydroxides (DE GRAVE and VAN ALBOOM, 1991).

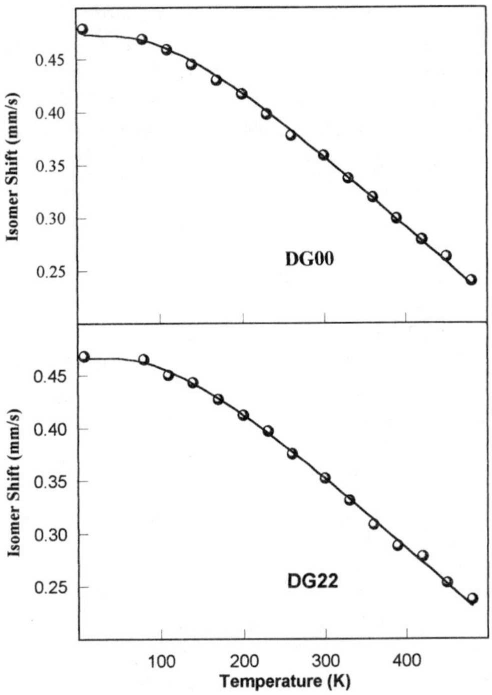

FIG. 8. Temperature dependence of the B-site isomer shift of samples DG00 and DG22. Solid lines are the calculated behavior assuming a Debye model for the lattice vibrations.

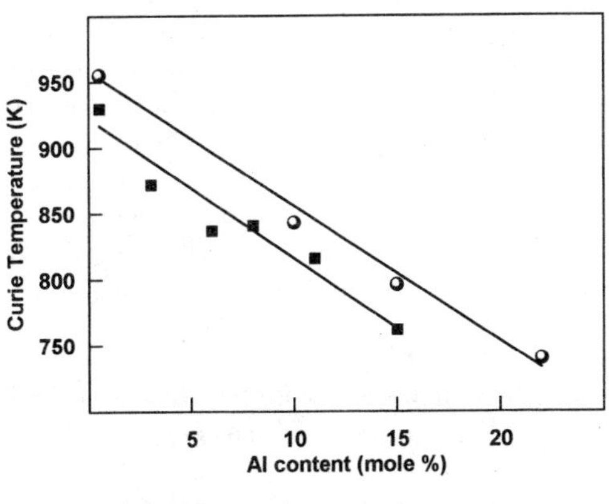

FIG. 7. Curie Temperature versus aluminum content for the DG samples (circles) and M samples (squares). Solid lines are the best linear fits to the experimental data.

Finally, concerning the line widths, Table 3 shows that the HF widths do not change significantly with temperature or Al substitution. On the other hand, a huge broadening is noticed for the LF widths with increasing temperature, and at a given temperature, with increasing Al content. Several possible reasons for this asymmetric broadening can be enumerated (collective magnetic excitations, surface effects, inter-particle interactions, etc.) and most probably they are all operating simultaneously, thus producing highly asymmetric line shapes. For a more specific discussion in this respect, the reader is referred to an earlier paper of DA COSTA *et al.* (1995b).

The ZFMS of samples DG37, DG55 and DG66 at 80 K and higher display a very broad sextet component with a superimposed central doublet arising from Fe^{3+} species. For DG66 a second quadrupole doublet due to Fe^{2+} (5%, against 46% for the ferric doublet, at RT) had to be included. For obvious reasons no further attention is given to this ferrous component. It was found that a superposition of two (*i.e.* A and B sites) asymmetric sextets and doublets could not describe the experimental spectra in a proper way. Therefore, these ZFMS have been analysed with one single model-independent magnetic hyperfine field distribution (MHFD) plus a quadrupole-splitting distribution (QSD). The hyperfine field was generally chosen to vary within the range 200–550 kOe at steps of 5 kOe, and the quadrupole splitting in the range of 0–3 mm/s at steps of 0.05 mm/s. Both center shifts and line widths of the elementary sextet/doublet were adjustable. Within reasonable uncertainty limits these parameters at a given temperature were iterated to the same values for the doublet and sextet, and not significantly affected by the Al content. For the sake of comparison, the spectra of samples DG00, DG10, DG15 and DG22 at 80 K and 300 K were also fitted with this distribution approach. Some relevant parameter values for 80 K and 300 K are given in Table 5. The characteristics of the QSD profiles were not found to be affected in a relevant way by the presence of Al. The generally weak fraction of the doublet component undoubtedly prevents uncovering any significant correlation in that respect. In contrast, both the maximum-probability hyperfine field (H_{hf}^{max}) and the average hyperfine field ($\bar{H}_{hf}$) are affected by the incorporation of Al into the lattice. However, it is known that $\bar{H}_{hf}$ is strongly influenced by the fitting conditions, especially in cases where high quadrupole splitting coexist with a very broad MHFD (LE CAER *et al.*, 1984). Hence, the maximum probability hyperfine field is considered to be a more characteristic parameter, and the relations between the Al content and H_{hf}^{max} were found to be:

$$H_{hf}^{max}\ (80\ K) = (525 - 21y)\ kOe$$

$$H_{hf}^{max}\ (300\ K) = (500 - 34y)\ kOe$$

It is beyond any doubt that the external field results are the most reliable ones for the purpose of studying the effect of Al substitution. However, considering that this equipment is not available in the vast majority of the Mössbauer labs, the above equations can be used to estimate the Al content in a unknown maghemite sample, provided Al content is sufficiently high and the morphological characteristics are similar to the samples investigated in this work.

As mentioned above, the doublet contribution to the spectra increases with increasing temperature, and this behavior is due to the existence of particles in a superparamagnetic state. The relative doublet area is plotted as a function of temperature in Fig. 9, and the temperature corresponding to a 50/50 doublet-sextet spectrum ($T_{1/2}$) can be defined from these experimental curves. The data for DG00 (no doublet at any temperature), and for DG10 and DG15 (doublets are too weak to be properly fitted) are not included in the plots. The results for sam-

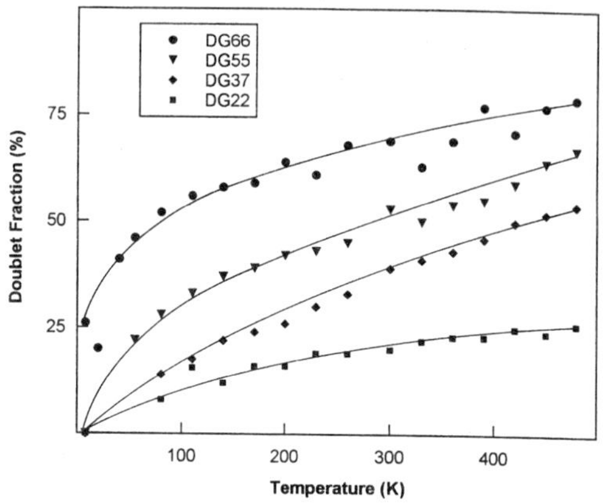

FIG. 9. Variation of the fractional area of the superparamagnetic doublet component with temperature for samples with different Al contents. Solid lines serve as a guide for the eye.

Table 5. Hyperfine parameters at 80 K and 300 K obtained from the fits with distributions of hyperfine fields and quadrupole splittings. H_{hf}^{max} and $\bar{H}_{hf}$ are the maximum probability and average hyperfine fields; σ is the square-root of the second-order moment of the distribution and δ is the center shift.

Sample	T (K)	H_{hf}^{max} (kOe)	$\bar{H}_{hf}$ (kOe)	σ (kOe)	δ (mm/s)
DG00	80	525	522	14	0.43
	300	500	496	21	0.32
DG10	80	523	518	14	0.44
	300	495	487	28	0.32
DG15	80	522	513	28	0.41
	300	494	480	33	0.32
DG22	80	519	507	34	0.42
	300	494	478	35	0.32
DG37	80	516	474	78	0.43
	300	488	407	96	0.31
DG55	80	515	444	98	0.43
	300	476	399	89	0.35
DG66	80	510	446	92	0.42
	300	480	392	88	0.30

ples DG37, DG55 and DG66 were obtained from the fits with a superposition of a magnetic hyperfine field distribution and a quadrupole splitting distribution. As expected, there is a steady increase of the doublet fraction with increasing Al content at any given T. The values of $T_{1/2}$ for samples DG37, DG55 and DG66 were found to be 420 K, 330 K and 70 K, respectively. It is suggested that the variations with Al are due to the concomitant effects of particle size and diamagnetic substitution. The value of $T_{1/2}$ for sample DG66 has to be considered with some caution because, as discussed previously, this sample has an impurity phase.

Another intriguing aspect reflected in Fig. 9 is the extremely broad temperature range over which mixed sextet-doublet spectra are observed. According to some authors this would indicate the absence of inter-particle magnetic interactions (KOCH *et al.*, 1986; PRENE *et al.*, 1993). The same phenomenon was found for the Al-free ML sample, but not for the Al-substituted members of the ML series. This remarkable behavior has remained unexplained so far, and certainly deserves further investigation.

CONCLUSIONS

The parameters derived from the external-field Mössbauer spectra collected at 4.2 K showed that at least 37 mole % Al can be incorporated into the maghemite lattice. Although an impurity in DG55 and DG66 prevents an accurate determination of the various Mössbauer parameters and of the exact Al content, there are strong indications that the maghemite phase present in those samples indeed contains more than 37 mole % Al. Thus, the apparent controversy with respect to the amount of Al that can be incorporated in maghemite seems to be related to the formation process of the investigated samples. The huge amount of data collected over the years in this laboratory for differently-prepared Al-maghemite series indeed have demonstrated that the preparation method plays a crucial role, both in the amount of Al^{3+} replacing Fe^{3+} as well as in the final characteristics of the obtained products.

Acknowledgements—This work was supported in part by the Fund for Joint Basic Research (Belgium, Grant Nr 2.0014.93), and by CNPq (Brazil).

References

ARMSTRONG J. R., MORRISH A. H. and SAWATZKY G. A. (1966) Mössbauer study of ferric ions in the tetrahedral and octahedral sites of a spinel. *Phys. Lett.* **23,** 414–416.

ANNERSTEN H. and HAFNER S. S. (1973) Vacancy distribution in synthetic spinel of the series Fe_3O_4-γ-Fe_2O_3. *Z. Kristallogr.* **137,** 321–340.

BOWEN L. H. (1979) Mössbauer spectroscopy of ferric oxides and hydroxides. *Mössbauer Effect Data J.* **2,** 76–94.

BOWEN L. H. and WEED S. B. (1984) Mössbauer spectroscopy of soils and sediments. In *Chemical Mössbauer Spectroscopy* (ed. R.H. HERBER) Chap. IX, pp. 217–242. Plenum Press.

BOWEN L. H., DE GRAVE E. and VANDENBERGHE R. E. (1993) Mössbauer effect studies of magnetic soils and sediments. In *Mössbauer Spectroscopy Applied to Magnetism and Materials Science* (ed. G. J. LONG and F. GRANDJEAN). Vol. 1. Chap. 4. pp. 115–159. Plenum Press.

BOWEN L. H., DE GRAVE E., and BRYAN A. M. (1994) Mössbauer studies in external field of well crystallized Al-maghemites made from hematite. *Hyperfine Interact.* **94,** 1977–1982.

CHADWICK J., JONES D. H., THOMAS M. F., TATLOCK G. J. and DEVENISH R. W. (1986) A Mössbauer study of ferrihydrite and aluminium substituted ferrihydrites. *J. Magn. Magn. Mater.* **61,** 88–100.

COEY J. M. D. (1971) Noncollinear spin arrangement in ultrafine ferrimagnetic crystallites. *Phys. Rev. Lett.* **27,** 1140–1142.

DA COSTA G. M., DE GRAVE E., BRYAN A. M. and BOWEN L. H. (1994a) Mössbauer studies of nano-sized aluminum-substituted maghemites. *Hyperfine Interact.* **94,** 1983–1988.

DA COSTA G. M., DE GRAVE E., BOWEN L. H., DE BAKKER P. M. A. and VANDENBERGHE R. E. (1994b) The center shift in Mössbauer spectra of maghemite and aluminum maghemites. *Clays Clay Minerals* **42,** 628–633.

DA COSTA G. M., DE GRAVE E., BOWEN L. H., DE BAKKER P. M. A. and VANDENBERGHE R. E. (1995a) Variable-temperature Mössbauer spectroscopy of nano-sized maghemite and Al-substituted maghemites. *Clays Clay Minerals* **43,** 562–568.

DA COSTA G. M., DE GRAVE E., BOWEN L. H., DE BAKKER P. M. A. and VANDENBERGHE R. E. (1995b) Temperature dependence of the hyperfine parameters of maghemite and Al-substituted maghemites. *Phys. Chem. Minerals* **22,** 178–185.

DA COSTA G. M. (1995c) Mössbauer spectroscopy and X-ray diffraction studies of maghemite (γ-Fe_2O_3) and aluminum-substituted maghemites [γ-$(Fe_{1-y}Al_y)_2O_3$] with $0 < y \leq 0.66$. Ph.D. dissertation, University of Gent.

DE BAKKER P. M. A., DE GRAVE E., VANDENBERGHE R. E., BOWEN L. H., POLLARD R. J. and PERSOONS R. M. (1991) Mössbauer study of the thermal decomposition of lepidocrocite and characterisation of the decomposition products. *Phys. Chem. Minerals* **18,** 131–143.

DE GRAVE E. and VAN ALBOOM A. (1991) Evaluation of ferrous and ferric Mössbauer fractions. *Phys. Chem. Minerals* **18,** 337–342.

DE GRAVE E., PERSOONS R. M., VANDENBERGHE R. E. and DE BAKKER P. M. A. (1993) Mössbauer study of the high-temperature phase of Co-substituted magnetites, $Co_xFe_{3-x}O_4$. I. $x \leq 0.04$. *Phys. Rev. B* **47,** 5881–5893.

DE JESUS FILHO M. F., DA NOVA MUSSEL W., QI Q. and COEY J. M. D. (1993) Magnetic properties of aluminum-doped γ-Fe_2O_3. *Proc. 6th Int. Conf. Ferrites, Tokyo 1992,* 126–128.

EGGLETON R. A. and FITZPATRICK R. W. (1988) New data

and a revised structural model for ferrihydrite. *Clays Clay Minerals* **36,** 111–124.

GILLOT B. and ROUSSET A. (1990) On the limit of aluminum substitution in Fe_3O_4 and γ-Fe_2O_3. *Phys. Status Solidi (a)* **118,** K5-K8.

HANEDA K. and MORRISH A. H. (1977a) Vacancy ordering in γ-Fe_2O_3 small particles. *Solid State Commun.* **22,** 779–782.

HANEDA K. and MORRISH A. H. (1977b) On the hyperfine field of γ-Fe_2O_3 small particles. *Phys. Lett.* **64A,** 259–262.

HENDRIKSEN P. V., LINDEROTH S., OXBORROW C. A. and MØRUP S. (1994a) Ultrafine maghemite particles: II. The spin-canting effect revisited. *J. Phys.Condens. Matter* **6,** 3091–3100.

HENDRIKSEN P. V., BØDKER F., LINDEROTH S., WELLS S. and MØRUP S. (1994b) Ultrafine maghemite particles: I. Studies of induced magnetic texture. *J. Phys.Condens. Matter* **6,** 3081–3090.

KOCH J. W. C., MADSEN M. B. and MØRUP S. (1986) Decoupling of magnetically interacting crystallites of goethite. *Hyperfine Interact.* **28,** 549–552.

KLUG H. P. and ALEXANDER L. E. (1974) *X-ray Diffraction Procedures for Polycrystalline and Amorphous Materials.* John Wiley and Sons.

LE CAER G., DUBOIS J. M., FISCHER H., GONSER U. and WAGNER H. G. (1984) On the validity of ^{57}Fe hyperfine field distribution calculation from Mössbauer spectra of magnetic amorphous alloys. *Nucl. Instr. Methods* **B5,** 25–33.

MORRISH A. H., HANEDA K. and SCHURER P. J. (1976) Surface magnetic structure of small γ-Fe_2O_3 particles. *J. de Phys.* **37,** C6301–305.

MØRUP S. and TOPSØE H. (1976) Mössbauer studies of thermal excitations in magnetically ordered microcrystals. *Appl. Phys.* **11,** 63–66.

POLLARD R. J. (1990) The spin canting anomaly in ferrimagnetic particles. *J. Phys. Condens. Matter.* **2,** 983–991.

POLLARD R. J. and MORRISH A. H. (1987) High-field magnetism in non-polar γ-Fe_2O_3 recording particles. *IEEE Trans. Magn.* **23,** 42–44.

PRENE P., TRONC E., JOLIVET J. P., LIVAGE J., CHERKAOUI R., NOGUES M., DORMANN J. L. and FIORANI D. (1993) Magnetic properties of isolated γ-Fe_2O_3 particles. *IEEE Trans. Magn.* **29,** 2658–2660.

ROUSSET A. and PARIS J. (1972) Formation des solutions solides binaires et ternaires des sesquioxydes de chrome, d'aluminium et de fer III- Le systeme Fe_2O_3-Al_2O_3. *Bull. Soc. Chim.* **10,** 3729–3733.

ROUSSET A., and PARIS J. and CHASSAGNEUX F. (1979) Interet de la reactivite des solides pour l'obtention de phases nouvelles. *Ann. Chim. Fr.* **4,** 115–122.

SAWATZKY G. A., VAN DER WOUDE F. and MORRISH A. H. (1969) Mössbauer study of several ferrimagnetic spinels. *Phys. Rev.* **187,** 747–757.

SCHWERTMANN U., and FECHTER H. (1984) The influence of aluminum on iron oxides: XI. Aluminum-substituted maghemite in soils and its formation. *Soil Sci. Soc. Am. J.* **48,** 1462–1463.

VANDENBERGHE R. E., VANLEERBERGHE R., DE GRAVE E. and ROBBRECHT G. (1980) Preparation and magnetic properties of ultra-fine cobalt ferrites. *J. Magn. Magn. Mater.* **15–18,** 1117–1118.

VANDENBERGHE R. E. and DE GRAVE E. (1989) Mössbauer Studies of Oxidic Spinels. In *Mössbauer Spectroscopy Applied to Inorganic Chemistry* (ed. G. J. LONG and F. GRANDJEAN). Vol. 3. Chap. 3, pp 59–182. Plenum.

VANDENBERGHE R. E., DE GRAVE E. and DE BAKKER P. M. A. (1994) On the methodology of the analysis of Mössbauer spectra. *Hyperfine Interact.* **83,** 29–49.

VAN OOSTERHOUT G. W. and ROOIJMANS C. J. M (1958) A new superstructure in gamma-ferric oxide. *Nature* **181,** 44.

WOLSKA E. and SCHWERTMANN U. (1989) The vacancy ordering and distribution of aluminium ions in γ-$(Fe,Al)_2O_3$. *Solid State Ionics* **32/33,** 214–218.

YASUOKA H., HIRAI A., MATSUURA M. and HASHI T. (1962) Hyperfine field at ^{57}Fe in lithium ferrite measured by NMR. *J. Phys. Soc. Japan* **17,** 1071–1072.

Mineral Spectroscopy: A Tribute to Roger G. Burns
© The Geochemical Society, Special Publication No. 5, 1996
Editors: M. D. Dyar, C. McCammon and M. W. Schaefer

A Mössbauer effect study of franklinite from Sterling Hill, New Jersey*

E. DE GRAVE[1], R. VOCHTEN[2], and R. E. VANDENBERGHE[1]

[1]Department of Subatomic and Radiation Physics, University of Gent, B-9000 Gent, Belgium
[2]Laboratory of Physical and Chemical Mineralogy, Department of Chemistry, University of Antwerp, B-2020 Antwerp, Belgium

Abstract—Franklinite crystals from the Sterling Hill deposit (Ogdenburg, New Jersey, U.S.A.) have been characterized by X-ray diffraction and atomic absorption spectrometry. The composition of this ferrite was determined to be $Zn_{0.68}Mn_{0.36}Fe_{1.96}O_4$. Mössbauer spectra were recorded at temperatures between 4.3 and 480 K, and at selected low temperatures in the presence of external magnetic fields up to 60 kOe. Both the paramagnetic and the magnetically split spectra consistently reveal that $\approx$0.06 Fe atoms per formula unit are located on the A sites. The doublet spectra are adequately described as a superposition of an Fe^{3+} A-site doublet and two Fe^{3+} B-site doublets, the latter having quadrupole splittings close to the values reported for $ZnFe_2O_4$ and $MnFe_2O_4$, respectively. This finding could be related to the previously suggested presence of Mn- and Zn-rich clusters. Slightly but consistently better fits were obtained using an A-site doublet and a quasi-continuous distribution of quadrupole splittings for the B sites. At temperatures below 245 K and down to $\approx$20 K, the spectral envelope is complex and not entirely interpretable by either static or dynamic (relaxation) effects. The spectra recorded at 4.3 and 9 K exhibit significant asymmetry in peak depths for corresponding pairs of lines. Applied-field Mössbauer spectra reveal the existence of locally canted spin states and reversed B-site spins. Both phenomena cease to exist at a temperature of 30 K and indicate a re-entrant ferrimagnetic state. The predominant B-site sub-pattern is asymmetrical and was reproduced adequately assuming a random angle α between the hyperfine field and the principal axis of the electric field gradient. Additional distributions on the effective hyperfine fields were introduced to interpret the applied-field spectra at 14 and 30 K. The models permitted the determination of the B-site dipolar-field contribution, *i.e.* -8 ± 2 kOe. The quadrupole splitting was found to be negative, and its magnitude is in line with the average value of the paramagnetic quadrupole-splitting distributions. The existence of locally canted spin states implies that in zero applied field the angle α is also fluctuating, and the observed asymmetry in the spectra at 4.3 and 9 K could be explained accordingly. It was further experienced that annealing of the sample at 600°C in vacuum did not produce substantially different Mössbauer spectra.

INTRODUCTION

FRANKLINITE IS a naturally occurring Mn-Zn ferrite with a cubic spinel structure (space group *Fd*3m). Franklinite crystals from the Sterling Hill (Ogdenburg, New Jersey, U.S.A.) deposit are known to have been crystallized at extremely low cooling rates (METSGER *et al.*, 1958). According to VOGEL *et al.* (1976), this slow cooling process could have resulted in a clustering of the tetrahedral Mn and Zn cations. Such a clustering effect could explain why the room-temperature (RT) Mössbauer spectra (MS), measured for five franklinite samples, could be interpreted in terms of two quadrupole doublets with hyperfine parameters fixed at the values observed for synthetic $ZnFe_2O_4$ (quadrupole splitting ΔE_Q = 0.35 mm/s) and $MnFe_2O_4$ (ΔE_Q = 0.58 mm/s) respectively. A MS recorded at 50 K in a longitudinal external magnetic field of 50 kOe was also presented. Two components, both arising from octahedral Fe^{3+}, were obvious and reportedly provide direct evidence that some sort of ordering is present in the as-collected samples.

To the best of the authors' knowledge, a detailed temperature-dependent Mössbauer study of franklinite has so far not been published. Such a study is reported in the present paper. An additional objective was to investigate in more detail the effect of clustering, if present in this particular sample, upon the microscopic magnetic properties of the Fe^{3+} cations as reflected in the temperature variations of the Mössbauer parameters.

MATERIAL AND METHODS

The franklinite sample was collected by one of the authors (RV) from the aforementioned deposit. The crystals were embedded in a manganoan-calcite matrix and were recovered from it by treatment in diluted HCl. A qualitative analysis with energy-dispersive X-rays showed Fe, Mn and Zn to be the major, practically sole observable constituents. Al was observed as a minor element. Some crystals were crushed and dissolved in a Na_2O melt to determine the elements Fe, Mn, and Zn by atomic absorp-

* Contribution No SSF95-03-01, Department of Subatomic and Radiation Physics, University of Gent, Belgium.

tion spectrometry. The MS (see next section) did not exhibit any obvious Fe^{2+} component (detection limit estimated as 1% of total iron), and the composition was found to be close to $Zn_{0.68}Mn_{0.36}Fe_{1.96}O_4$.

X-ray diffraction (XRD) photographs were obtained with a Guinier-Hägg camera (diameter 100 mm), using silicon powder as an internal standard. All the observed reflections could be attributed to the spinel phase. The lattice parameter, calculated as the arithmetic mean from the eight most prominent reflections, was found as $a = 0.8450 \pm 0.0005$ nm, which is in close agreement with relevant literature data for franklinite (METSGER *et al.*, 1958; VOGEL *et al.,* 1976) and for a synthetic Mn-Zn ferrite with similar composition (MORRISH and CLARK, 1975). Powder XRD patterns, recorded with a Co tube, show relatively sharp spinel reflections, with α_1 and α_2 components well separated at higher diffraction angles. Considering the high sensitivity of the lattice parameters towards the Mn/Zn ratio (MORRISH and CLARK, 1975), this observation implies that the cation distribution is fairly homogeneous.

Mössbauer spectra at variable temperatures between 4.2 and 500 K, and at selected low temperatures ($T \leq 60$ K) in a magnetic field of 60 kOe applied parallel to the γ-ray beam, were obtained with a conventional, constant-acceleration spectrometer with triangular reference signal. The data were recorded in 1024 channels (unfolded spectrum). The spectrometer was periodically calibrated using the spectra of either metallic iron or hematite. The velocity increment was approximately 0.01 mm/s per channel for the paramagnetic MS, and 0.045 mm/s for the magnetically split ones. The source was ^{57}Co in a Rh matrix; however all isomer-shift data quoted hereafter refer to metallic iron at RT. The absorber consisted of powdered crystals and had an effective thickness of $\approx$10 mg Fe/cm^2. Additional measurements at selected temperatures were performed on a sample annealed for 24 hrs. at 600°C in a vacuum of $\approx 10^{-4}$ mbar. Spectra were recorded until an off-resonance count rate of at least 2.10^6 was reached. Unless otherwise stated, all spectra were fitted with superpositions of symmetrical Lorentzian-shaped doublets or sextets. For each doublet the center shift δ, the quadrupole splitting ΔE_Q, and one width parameter Γ were adjusted. For the sextets, only two width parameters were iterated, *viz.* Γ and $\Delta\Gamma$, so that the widths of the outer, middle, and inner absorption lines are given as $\Gamma_1 = \Gamma + 2\Delta\Gamma$, $\Gamma_2 = \Gamma + \Delta\Gamma$, and $\Gamma_3 = \Gamma$ respectively. Center shifts, quadrupole shifts, $2\epsilon_Q$, and magnetic hyperfine fields, H_{hf}, were iterated by non-linear least-squares fitting. Intensity parameters were adjusted by linear regression. However, in many instances it was necessary to employ an asymmetrical line shape for the sextet spectra and some of the variables were constrained in order to obtain reasonable and consistently varying parameter values. More details about this line shape and the constraints will be presented where relevant.

RESULTS AND DISCUSSION

Paramagnetic spectra

The MS recorded at temperatures above 250 K consist of a slightly asymmetric doublet; this asymmetry can be characteristic of a superposition of at least two distinct components with different values

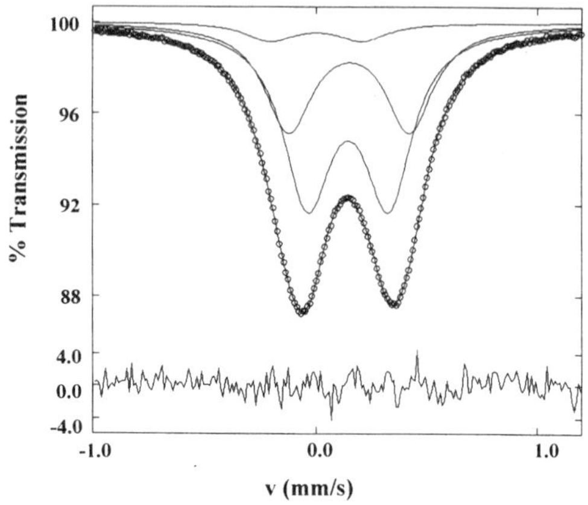

FIG. 1. Mössbauer spectrum (Rh source) of franklinite sample recorded at 470 K. Solid curves are the three quadrupole doublets and their superposition adjusted to the experimental data. The lack of significant structure in the residual (bottom) supports the adequacy of the applied fitting model.

for the hyperfine parameters δ and ΔE_Q. A sample spectrum is shown in Fig.1 (T = 470 K) and is representative of spectra for all temperatures in the paramagnetic regime. The full line through the data points is the fitted sum of three symmetrical quadrupole doublets, all assumed to have equal line widths, which are also reproduced in the drawing. The lower part of Fig.1 represents the residual function which obviously lacks any fine structure. This is not the case if only the two predominant quadrupole-split components are employed in the fit. The residual is then not featureless and consistently suggests the presence of a third spectral component. The goodness-of-fit, χ^2, is typically 25 to 30% lower for a three-doublet fit than for a two-doublet fit. This improvement is statistically significant and is believed to have a physical significance as well.

The quadrupole splittings $\Delta E_{Q,i}$ and the fractional areas A_i (i = 1,2,3) of the three doublets, henceforth denoted as D1, D2, and D3, do not vary monotonically with temperature. Their fitted values in the temperature range 260–485 K are scattered between the lower and upper limits listed in Table 1. The line width Γ gradually decreases with increasing T from 0.30 mm/s at 260 K to 0.26 mm/s at 485 K. Such a variation can be explained by the lowering of the Mössbauer fraction, and hence of the effective thickness of the absorber, as the temperature increases (DE GRAVE and VOCHTEN, 1985). Compared to the 260 K spectrum, the line

Table 1. Intervals within which the iterated values of the quadrupole splitting ΔE_Q and of the fractional area A are scattered for the tetrahedral (D3) and two octahedral (D1, D2) doublets fitted to the Mössbauer spectra recorded at temperatures between 260 and 485 K (15 spectra).

	D1	D2	D3
ΔE_Q (mm/s)	0.35-0.38	0.53-0.55	0.30-0.40
A	0.58-0.65	0.31-0.36	0.03-0.08

width of the doublets at 250 K is markedly higher. This broadening implies that the Curie temperature T_C, as probed by the Mössbauer effect, is in the vicinity of 250 K, which is in fair agreement with values reported in the literature for synthetic (Mn, Zn) ferrites with comparable Mn/Zn ratios (see *e.g.* KÖNIG, 1972).

The center shifts of D1 and D2 are consistently equal within experimental error limits (0.358 and 0.363 mm/s respectively at RT), and their values are characteristic of six-fold oxygen coordination in spinel oxides. The quadrupole splittings of the two B-site doublets are indeed very close to those found for Zn and Mn ferrite, respectively. Moreover, the fractional-area ratio A_2/A_1 is comparable with the analytically determined Mn/Zn ratio. These findings, together with the narrow line width of both doublets, suggest that the sample is composed of clusters with fairly uniform compositions close to $ZnFe_2O_4$ and $MnFe_2O_4$. The third doublet has $\delta = 0.24$ mm/s at RT which is typical for a tetrahedral coordination in a spinel ferrite (VANDENBERGHE and DE GRAVE, 1989). This finding implies that the cation distribution in the franklinite studied here exhibits a small degree of inversion with respect to Fe, *i.e.* 0.055 ± 0.025 (see Table 1), in reasonable agreement with the conclusions in that respect put forward by MORRISH and CLARK (1975) for synthetic (Mn, Zn) ferrites. The temperature variation of the B-site center shifts follows normal behavior as expected on the basis of the second-order Doppler shift (DE GRAVE and VAN ALBOOM, 1991). At each temperature, the fitted δ values for the tetrahedral sites are consistently lower by 0.10 to 0.15 mm/s than the corresponding B-site values. The average difference is 0.13 mm/s, which is typical for spinel ferrites (DA COSTA *et al.*, 1994).

All doublet MS have additionally been analyzed in terms of a model-independent quadrupole-splitting distribution (QSD) with an adjustable linear correlation between δ and ΔE_Q. The latter quantity was allowed to vary in the range 0.1 to 2.0 mm/s in steps of 0.05 mm/s. The value of the smoothing parameter was chosen according to the "first rule of fist" as described by VANDENBERGHE *et al.* (1994). In order to be consistent with the three-doublet model, a non-distributed A-site doublet, with ΔE_Q and Γ fixed at the values iterated in this latter model, was included as well. Its center-shift value was forced to be 0.13 mm/s lower than that of the B-site doublets. These severe restrictions for the A-site component were required because of its weak contribution and the strong overlap with the dominant B-site component. Some relevant numerical results from this QSD procedure are presented in Table 2. For comparison, the goodness-of-fit values produced by the three-doublet fits, χ^2_{3d}, are listed as well. It is clear that the QSD approach consistently produces better fits than the three-doublet model, but not substantially so. With the exclusion of the center shift, all other parameters are not affected by the temperature, and the range of scatter in the numerical values is significantly smaller than that which is indicated in Table 1. The calculated distribution profiles are smooth and do not provide any indication whatsoever that two basically distinct B sites would be present in the structure. Instead, as quantitatively reflected in the small differences between the respective average quadrupole splittings $\langle \Delta E_Q \rangle$ and the maximum-probability values ΔE_Q^m, the profiles are nearly symmetrically shaped. Finally, the correlation between δ and ΔE_Q is weak (see parameter c in Table 2) and believed to be of little physical significance.

At this point, it remains inconclusive which of the two approaches, *i.e.* the three-doublet fit or the QSD fit, is physically more justified. Considering the average composition of the compound being investigated, it is difficult to imagine that two structurally more or less distinct B sites can have formed, even in the event that clustering has occurred. The relatively large difference in ionic radii of Mn^{2+} and Zn^{2+} is instead expected to result in a wide range of local distortions of the oxygen octahedra containing the Fe probes. In that respect, a QSD seems to be more realistic than just two isolated doublets (see also RANCOURT, 1994). The observation that the two doublets adjusted to the franklinite MS exhibit quadrupole splittings close to the values found for the respective end-members would then be merely fortuitous.

Table 2. Some relevant Mössbauer parameters derived from fitting a superposition of an A-site quadrupole doublet with a model-independent, B-site quadrupole-splitting distribution to the paramagnetic spectra of franklinite at selected temperatures *T:* average and maximum-probability splittings $\langle \Delta E_Q \rangle$ and ΔE^m_Q, center shift δ^m corresponding to maximum probability, slope *c* of the linear correlation between center shift and quadrupole splitting, and relative spectral area A_t of the A-site contribution. The quantity χ^2 is the goodness-of-fit and χ^2_{3d} refers to the numerical analysis using one A-site and two B-site doublets.

T (K)	χ^2	χ^2_{3d}	$\langle \Delta E_Q \rangle$ (mm/s)	ΔE^m_Q (mm/s)	δ^m (mm/s)	c	A_t
265	446	465	0.435	0.407	0.378	-0.016	0.05
280	587	605	0.440	0.405	0.368	-0.017	0.05
300	507	537	0.436	0.410	0.356	-0.019	0.05
325	559	611	0.435	0.412	0.342	-0.018	0.06
350	578	609	0.434	0.416	0.327	-0.019	0.05
380	514	526	0.438	0.412	0.309	-0.017	0.05
410	569	617	0.439	0.418	0.291	-0.013	0.06
440	518	541	0.439	0.416	0.272	-0.010	0.07
470	497	557	0.442	0.418	0.253	-0.011	0.07

Low-temperature Mössbauer spectra: external field absent

A set of representative low-temperature MS is reproduced in Fig. 2, showing the dramatic changes in line shape with increasing temperature. Synthetic (Me, Zn) ferrites, with Me as divalent cation, are known to undergo electronic relaxation effects that give rise to characteristic Mössbauer patterns at temperatures that are comparable to those covered in this study (see references cited in the review by VANDENBERGHE and DE GRAVE, 1989). It is clear that the line shapes observed here are markedly different from these reported patterns, implying that a simple collective relaxation process cannot explain these franklinite MS. A notable distinction from the relaxation spectra measured for the synthetic ferrites concerns the apparent splitting of the outer absorption lines (see Fig. 2 for $T = 80$ and 115 K) into two, well-separated components. Another peculiar feature is the appearance of seemingly extraneous peaks as clearly observed in the 80 K spectrum (see Fig. 2, at velocities of approximately -1.8 mm/s and 2.2 mm/s). Attempts to interpret the low-temperature MS ($30 \leq T \leq 230$ K) using a two-component, Blume-Tjon relaxation model (for a review and relevant references, see HOY, 1984) have remained unsuccessful so far. Hence, it cannot be concluded whether the appearance of the two spectral components could be an indication for the existence of two kinds of clusters.

The spectra recorded at 4.2 K (see Fig. 2) and 9 K seem to present fewer challenges to their interpretation. However, they exhibit substantially asymmetric peak depths for corresponding pairs of lines, pointing at first glance to the presence of at least two distinct subspectra with different Mössbauer parameters. As it was experienced, the numerical interpretation of these MS is not straightforward, and several different physical approaches, described in the following paragraphs, have been considered in fitting the MS. The goodness-of-fit (χ^2) values, obtained under the various assumptions, are indicated in Table 3.

In a first attempt, two symmetrical sextets with equal line-width parameters were used to fit both spectra (model 2SS in Table 3). This model is the simplest one that could possibly produce results consistent with those of the three-doublet fits to the high-temperature MS, and hence with the suggestion of clustering. The values for the hyperfine fields, H_{hf}, and the quadrupole shifts, $2\epsilon_Q$, were iterated to be $H_{hf,1} = 501 \pm 2$ kOe and $2\epsilon_{Q,1} = -0.19 \pm 0.04$ mm/s, and $H_{hf,2} = 509 \pm 2$ kOe and $2\epsilon_{Q,2} = 0.13 \pm 0.03$ mm/s for the first and second components respectively. For both subspec-

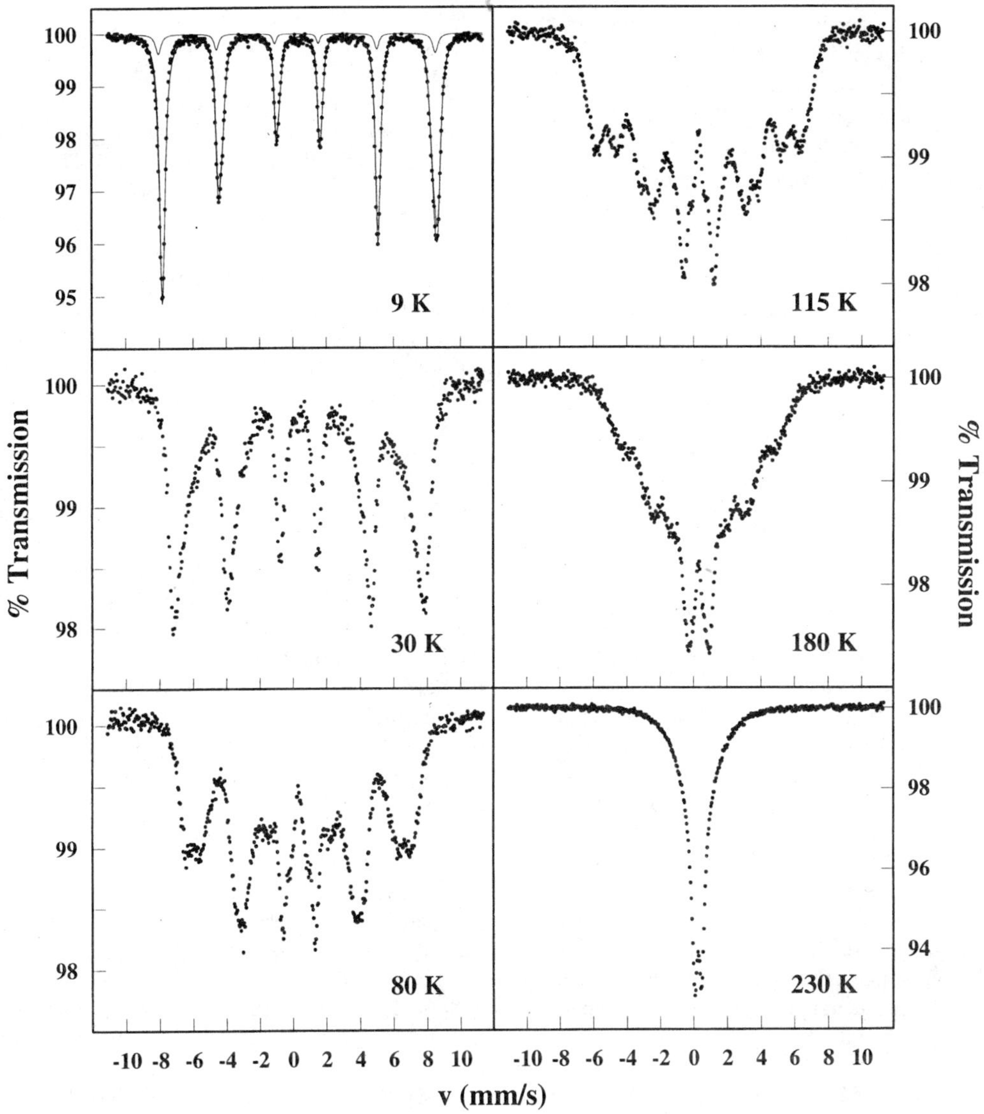

FIG. 2. Mössbauer spectra of franklinite at some selected low temperatures as indicated. The full line through the data points of the 9 K spectrum is the calculated spectrum using the asymmetrical line shape for the B-site pattern. The weak A-site sextet, assumed to be Lorentzian, is shown. The appearance of two components at intermediate temperatures (80 and 115 K) is obvious.

tra at 4.2 and 9 K, a value of 0.480 ± 0.006 mm/s for δ was obtained from the data, again typical for a six-fold coordination. In contrast to the paramagnetic spectra, the presence of a weak third component arising from tetrahedral Fe^{3+} ions could not be resolved straightforwardly from the magnetically-split MS. At this point, no further attempts were made to fit such a component with constraint parameter values.

If the clustering model is assumed to be real and responsible for the appearance of two magnetic components, then on the basis of the literature values of the hyperfine fields at 4.2 K for $ZnFe_2O_4$ and $MnFe_2O_4$, *viz.* respectively 502 (VARRET *et al.*, 1971) and 513 to 517 kOe (MORRISH and CLARK, 1975), it would be logical to assign the component with the highest field to the Mn-rich clusters. However, this assignment is not consistent with the re-

Table 3. The goodness-of-fit χ^2 obtained for the various fitting models applied to interpret the B-site subpattern in the franklinite Mössbauer spectra recorded at 4.2 and 9 K. 2SS: two unconstrained symmetric sextets; 2SS1/1 and 2SS1/3: two symmetric sextets with area ratios 1/1and 1/3 respectively; MHFD: a model-independent distribution of hyperfine fields; 1AS: one asymmetric sextet.

Model	$T = 4.2$ K	$T = 9$ K
2SS	1165	1252
2SS1/1	1193	1345
2SS1/3	1182	1671
MHFD	1047	1115
1AS	1040	948

sults derived from the paramagnetic spectra in the sense that the area ratio of the respective sextet components is found to be 60:40, while approximately the reverse number is obtained for the corresponding doublet components. Fixing the area ratio of the sextets to this latter number results in an increase of χ^2 by 35% (4.3 K) to 75% (9 K), with significant deviations between experimental and calculated line shapes.

Even if in the above 2SS model the assignment of the sextets is reversed, there remains an inconsistency, in particular with respect to the iterated $2\epsilon_Q$ values, that renders the model doubtful. Depending on the sign and the relative magnitudes of the anisotropy constants, the domains' spin direction, and hence that of the hyperfine fields, in cubic collinear ferrimagnets is along one of three crystallographic axes [100], [110], and [111] (SMIT and WIJN, 1959). On the other hand, the principal axis of the electric field gradient (EFG) in spinel ferrites is generally along one of the four local [111] axes. Therefore, for a spin direction coinciding with [100], one sextet component with a shift $2\epsilon_Q = 0$ mm/s is expected. In contrast, a [111] orientation, as in $MnFe_2O_4$ (KÖNIG, 1971), leads to two sextets with $2\epsilon_Q$ values of respectively ΔE_Q and $-\Delta E_Q/3$, and with relative abundance of 1:3, while a [110] direction yields two sextets as well, however with shifts of $2\epsilon_Q = -\Delta E_Q/2$ or $\Delta E_Q/2$ and with equal area (DE GRAVE *et al.*, 1993). If in these two latter cases the observed spectrum is fitted as a single symmetrical component, the obtained quadrupole shift will be close to zero. The two components resolved from the low-temperature MS irrefutably have non-zero quadrupole shifts. Hence, it may be concluded that the 2SS model as defined above produces results that are not in line with those derived from the doublet MS. Therefore it does not confirm the suggested clustering mechanism.

In a subsequent stage, and following the above reasoning with respect to the possible spin directions, only one physical component was considered, using two different hyperfine-field values as a result of the mentioned orientation effects. This difference is due to the dipolar-field contribution to H_{hf}, which shows a θ dependence similar to that referring to $2\epsilon_Q$ (DE GRAVE *et al.*, 1993). Both sextets were forced to have the same center shift and line-width parameters. Their relative areas were constrained according to the spin state: 1:1 (spins along [110], model 2SS1/1 in Table 3) or 1:3 (spins along [111], model 2SS1/3 in Table 3). An additional sextet, with fixed relative area of 0.06, and referring to A-site Fe^{3+} species, could be included in the fit; however its hyperfine parameters are highly unprecise. The resulting values for the hyperfine fields and quadrupole shifts of the two B-site subspectra, B1 and B2, at 4.3 and 9 K are listed in Table 4. The center shift was in the range 0.47–0.48 mm/s, and the outer line width was in the range 0.40–0.45 mm/s. As seen from Table 3, the agreement between experimental and calculated spectra has not improved compared to the 2SS approach. Consistent quadrupole splittings ΔE_Q (averaged for the 4.3 and 9 K runs) of -0.34 and -0.25 mm/s are calculated from the adjusted $2\epsilon_Q$ values (see Table 4) for respectively the 2SS1/1 and 2SS1/3 models. The magnitude of these splittings, in particular of the latter one, is rather low in view of the results derived from the quadrupole doublets, using for example a model-independent quadrupole-splitting distribution to fit these MS (note that the sign of ΔE_Q is undetermined from a quadrupole doublet). Otherwise, the calculated dipolar fields seem to possess reasonable values regardless of the applied approach. From the numerical data reported for $MnFe_2O_4$ by KÖNIG (1971), one can deduce $H_{dip} = -7.5$ kOe, which compares excellently with the present results. In conclusion, although both the 2SS1/1 and 2SS1/3 models are believed to have a more solid, physical background than the 2SS model, it remains unfeasible to express any preference towards either of the three models merely on the basis of the obtained fitting results.

Table 4. Hyperfine fields H_{hf} and quadrupole shifts $2\varepsilon_Q$ at 4.3 K and 9 K for the two symmetrical sextets B1 and B2 fitted to the experimental spectra in case the spins are assumed to be aligned along a domain's [110] crystallographic direction (model 2SS1/1) or along a [111] axis (model 2SS1/3). The relative area ratios were fixed at 1:1 and 1:3 respectively. H_{dip} is the dipolar field. In the 1AS model, only one B-site pattern, with asymmetric line shape, has been considered and the data contained in the fourth column refer to the quadrupole splitting ΔE_Q. The two bottom rows refer to the annealed franklinite sample.

Model	T (K)	H_{hf}(B1) (kOe)	$2\varepsilon_Q$(B1) (mm/s)	H_{hf}(B2) (kOe)	$2\varepsilon_Q$(B2) (mm/s)	H_{dip} (kOe)
2SS1/1	4.3	502	-0.17	508	0.17	-6
	9	503	-0.12	512	0.14	-8
2SS1/3	4.3	497	-0.24	508	0.07	-8
	9	501	-0.26	510	0.10	-7
1AS	4.3	505	-0.44			-8
	9	507	-0.42			-9
1AS	4.3	506	-0.42			-9
	9	507	-0.40			-10

For the matter of completeness, the 4.2 and 9 K spectra have also been analysed in a next step by a model-independent magnetic hyperfine field distribution (model MHFD), reflecting the non-uniqueness of the field as a result of fluctuations in the chemical environment of the probe iron nuclei (VANDENBERGHE *et al.*, 1994). The observed line shape was reasonably well reproduced, however not substantially better than in the case of the previous models (Table 3). Within the experimental uncertainties, the maximum-probability and the average field values were found to be equal, *i.e.* 506 and 505 kOe for T = 4.3, and 509 and 508 kOe for T = 9 K. Hence, the field distributions seem to be symmetrical. This finding is to some extent unexpected considering the fact that chemical disorder in spinel ferrites is known to commonly give rise to asymmetric distributions, even at 4.3 K and especially on the octahedral sites (SAWATZKY *et al.*, 1969). Finally, the two remaining models mentioned in Table 3 were inspired by the results of the applied-field Mössbauer measurements, and will be discussed in the next section.

Low-temperature Mössbauer spectra: applied field

A collection of applied-field Mössbauer spectra (AFMS) is presented in Fig.3. In contrast to what one would expect for a collinear ferrimagnet in a strong, longitudinal external field H_{ext}, the middle lines (Δm_I = 0 nuclear transitions) have not vanished, meaning that the spins are not aligned along the direction of the applied field. The average canting clearly decreases with increasing field strength and, at a given field, with increasing temperature T. At 30 K in H_{ext} = 60 kOe, the canting has almost completely vanished. Since Mn^{2+} and Fe^{3+} are both S-state cations, and hence only poorly anisotropic, this canting is almost certainly not a result of the applied fields being too weak to dominate any preferred orientation of the net magnetic moment. Instead, the presence of the middle lines, and the changes of their intensities with changing T and H_{ext}, can be explained by the occurrence of so-called *locally canted spin states* (LCSS) (ROSENCWAIG, 1970). These LCSS arise from the selective magnetic dilution of the tetrahedral sublattice. B-site Fe^{3+} species with none or only a small number of magnetic A-site nearest-neighbours are subjected to a weak antiferromagnetic A-B superexchange interaction, while the B-B interaction, which opposes the ferrimagnetic alignment, is not affected. As a result, the corresponding B-site spins take on an orientation that does not coincide with the orientation of the domain's net magnetic moment, *i.e.* with the direction of H_{ext}. Higher temperatures and stronger fields reduce these LCSS. Such a magnetic structure is often termed *re-entrant ferrimagnetism* (BRAND *et al.*, 1985).

A second notable feature about the AFMS is the presence of a weak high-field component. As the numerical analyses of the spectra show (see later), this component at 4.3 K is partly due to tetrahedral Fe^{3+} ions , and partly to octahedral ones that have their spins reversed. Since the net B-site magnetic moment is higher than that of the A sites, the latter one is antiparallel to the external field and hence this field adds to the hyperfine field acting on the A sites and on the spin-reversed B sites. At 30 K, spin reversal has ceased to exist.

Finally, a third observation should be stressed. This concerns the asymmetric peak depths of corresponding pairs of absorption lines. This asymmetry is also obvious in the AFMS of synthetic (Mn, Zn) ferrites and was accounted for by MORRISH and CLARK (1975) by introducing in their fitting procedure two B-site patterns according to a different number of magnetic A-site neighbours (Mn^{2+}) to the probe iron nuclei. One can mention at least two arguments that question the full correctness of these authors' interpretation. First, for the composition range comparable with the presently studied franklinite, configurations consisting of Zn_iMn_{6-i} with i = 5, 4, 3, and 2 all have a significant probability, so that three or four subpatterns would be required to reproduce the experimental line shape. Secondly, the fitted B-site center-shift values were found to differ by approximately 0.08 mm/s, and such a large difference is unlikely to be realistic.

Asymmetric B-site peak depths in AFMS of spinel ferrites are not unusual and occur even for members in which the B-site Fe atoms exhibit a unique nearest-neighbour A-site cation configuration such as magnetite and Co-substituted magnetites (DE GRAVE *et al.*, 1993; PERSOONS *et al.*, 1993) and maghemite (BOWEN *et al.*, 1994). The appearance of asymmetry is a consequence of the spin alignment along $\vec{H}_{ext}$. As a result, the angle θ between the spin direction and the EFG's principal axis becomes random, and due to the θ-dependence of the dipolar-field contribution, the hyperfine field becomes anisotropic. This anisotropy leads to an asymmetrical line shape which can be expressed analytically in the case of axial symmetry (DE GRAVE *et al.*, 1993). The full lines in the 4.3 K spectra as shown in Fig. 3 are the result of fitting such a line function to the B-site absorption (AS component). Because the weak-field component was featureless, a simple Lorentzian shape was assumed (LS component). The relative area of the middle lines was allowed to vary, but it turned out to be small. It is therefore unclear whether or not all A-site spins are aligned along the direction opposite to that of the external field. A similar fitting procedure could successfully be applied for the (9 K, 60 kOe) spectrum (not presented in Fig. 3). In all cases, the observed AFMS are reasonably well reproduced by the adjusted ones, with minor deviations especially as far as the middle lines are concerned. These deviations are due to fluctuations on the magnitude H_{hf} of the hyperfine field and on the orientation, with respect to the propagation of the γ-rays, of the effective field $\vec{H}_{eff}$ (*i.e.* the vectorial sum of the hyperfine and external fields) that determines the intensity of the ($\Delta m_I = 0$) transitions. This latter effect cannot be modelled analytically in the fitting routine. The non-unique strength of H_{hf} is reflected in the rather broad width obtained for the outer lines, *viz.* 0.45 mm/s.

The AFMS recorded at 14 K and 30 K could not be described adequately in the way mentioned in the preceding paragraph. In order to quantify some of their properties, a superposition of two, model-independent HFD was tentatively used, with effective-field values varying in the ranges 330–500 kOe and 550–600 kOe respectively, at steps of 3 kOe, for T = 14K, and 230–500 kOe and 500–600 kOe, at steps of 5 kOe, for T = 30 K. For the most prominent component, asymmetrical elementary sextets, with adjustable line width Γ, were chosen. Center shift δ, dipolar-field contribution H_{dip}, quadrupole splitting ΔE_Q, and area ratios A_2/A_1 (second to first line) were all fitted. Lorentzian elementary quadruplets were considered for the weak high-field pattern (*i.e.* assuming zero canting), with adjustable δ value and quadrupole shift ϵ_Q fixed at zero.

Numerical data obtained from the fitting procedures applied to the AFMS are collected in Table 5. The dipolar-field value, being on the average -8 ± 2 kOe, and the center shift of the AS component are typical for octahedral Fe^{3+} in a spinel ferrite. The quadrupole splitting is found to be scattered within the range $\Delta E_Q = -0.44 \pm 0.03$ mm/s and is in excellent agreement with the average $\langle \Delta E_Q \rangle$ calculated from the paramagnetic spectra (see Table 2). Its sign is consistent with that found for $ZnFe_2O_4$ (EVANS *et al.*, 1971). The behavior of the relative line area parameter A_1/A_2 quantitatively confirms the reduction in spin canting with increasing applied-field strength and temperature, and hence supports the suggested LCSS mechanism. The variation of the fractional contribution A_{LS} reflects the gradual decrease of the number of re-

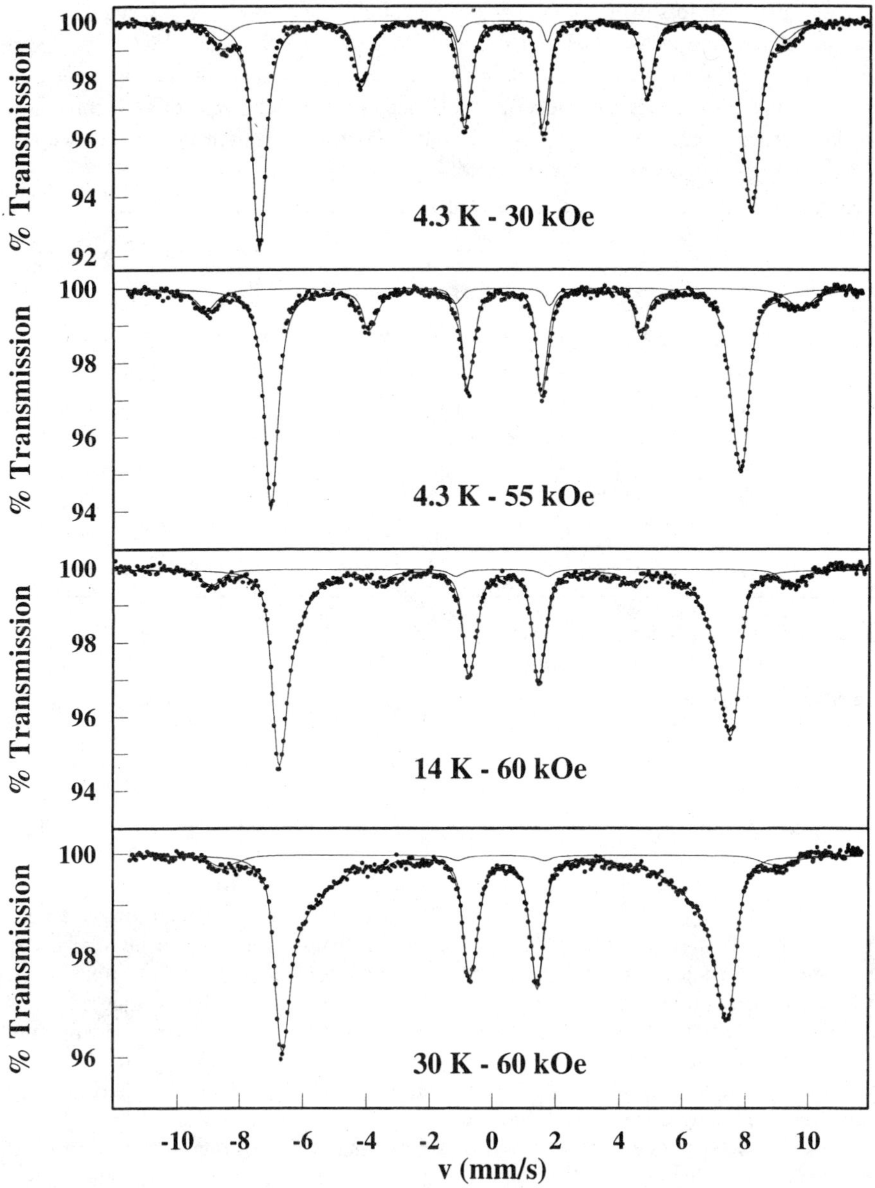

FIG. 3. Applied-field Mössbauer spectra of franklinite at field strengths and temperatures as indicated. The full lines are the calculated spectra using either a superposition of a symmetrical (high-field component) and an asymmetrical subpattern (4.3 K), or a superposition of two hyperfine-field distributions with an asymmetrical, elementary sextet for the B-site component (14 K and 30 K).

versed B-site spins as T increases, and the δ value for the LS component at 30 K indicates that this component can exclusively be attributed to tetrahedral Fe^{3+}. It proves that Fe is indeed present on the A sites and the A_{LS} value is in excellent accordance with the results derived from the high-temperature MS (see Table 2, parameter A_t). In summary, the AFMS are consistent with the LCSS picture and their numerical interpretation has provided reasonable results for the various physical quantities,

Table 5. Some relevant Mössbauer parameters obtained from the external-field spectra of as-collected franklinite and of an annealed sample. A superposition of an asymmetrical sextet (AS) for the main component, and a Lorentzian sextet (LS) for the weak high-field component was used for the spectra corresponding to $T = 4.3$ and 9 K. For higher T values, the effective fields H_{eff} of both components were given a model-independent distribution. A_2/A_1 is the relative spectral area of the second line with respect to the first one, and A_{LS} is the fractional contribution of the LS component.

		AS					LS			
T (K)	H_{ext} (kOe)	H_{eff} (kOe)	H_{dip} (kOe)	δ (mm/s)	ΔE_Q (mm/s)	A_1/A_2	H_{eff} (kOe)	δ (mm/s)	A_1/A_2	A_{LS}
4.3	30	481	-6	0.47	-0.47	0.28	554	0.40	0.13	0.12
4.3	55	459	-6	0.48	-0.45	0.18	583	0.42	0.06	0.12
4.3*	60	451	-8	0.48	-0.46	0.10	577	0.41	0.04	0.14
9	60	443	-10	0.47	-0.41	0.10	578	0.41	0.11	0.09
14	60	442†	-7	0.48	-0.46	0.07	557†	0.38	0.‡	0.08
30	60	436†	-8	0.48	-0.43	0.05	551†	0.34	0.‡	0.07

†value corresponding to maximum probability
‡fixed in the iteration
*data referring to the annealed sample.

which are moreover completely in line with the data obtained from fitting the paramagnetic MS as a superposition of an A-site quadrupole doublet and a B-site QSD.

At this point, the zero-field MS recorded at 4.3 and 9 K need to be clarified. The interpretation is proposed to be as follows. Due to the LCSS in the octahedral sublattice, the angle θ between the hyperfine field and the EFG's principal axis varies from site to site, and a situation similar to that encountered when the absorber is subjected to a strong field can be suggested. Therefore, the B-site component is again fitted with an asymmetrical pattern with all parameters adjustable (model 1AS in Table 3). A Lorentzian shape is assumed for the A-sites, with all parameters, except the hyperfine field, fixed: $\delta = 0.34$ mm/s, $2\epsilon_Q = 0$ mm/s, and $A_t = 0.07$. The fits are reasonable (see Table 3) and an example is shown in Fig. 2 for the 9 K spectrum. The relevant B-site parameters are included in Table 4, and are, as far as H_{dip} and ΔE_Q are concerned, in excellent agreement with the results derived from the AFMS. For the A-site hyperfine field, values of 511 and 514 kOe were iterated for $T = 4.3$ and 9 K respectively. This would mean that the A-site hyperfine field is larger than the B-site one, which at first glance seems rather unusual for spinel ferrites (VANDENBERGHE and DE GRAVE, 1989). However, one can understand the observed feature for the present ferrite compound on the basis of the weaker supertransfer from Mn_A to Fe_B as compared to that from *e.g.* Fe_A to Fe_B, and the high number of nearest-neighbour B-site Fe^{3+} cations to any of the A-site probe nuclei, making the total supertransferred spin density from B to A sites relatively large.

A final point of discussion concerns the magnitudes of the B-site hyperfine fields H_{hf} as obtained from the zero-field MS and of the effective fields H_{eff} derived from the AFMS. The relation between these two quantities is generally given by:

$$H_{eff}^2 = H_{ext}^2 + H_{hf}^2 - 2H_{hf}H_{ext}\cos\alpha \qquad (1)$$

α being the angle between the γ-ray direction (*i.e.* the direction of H_{ext}) and the spin direction. The (average) canting α can be determined from the area ratio A_1/A_2 as defined earlier:

$$\alpha = \arcsin\left[\frac{1.5\,A_2/A_1}{1 + 0.75A_2/A_1}\right]^{1/2} \qquad (2)$$

Using the data for A_1/A_2 as listed in Table 5 and referring to the 4.3 K spectra, canting angles of 36° and 29° are calculated for applied fields of 30 and 55 kOe respectively. Subsequently, from the H_{hf} values of Table 4, eqn.(1) yields for H_{eff} 481 and 458 kOe respectively, in agreement with the experimental results (see Table 5). Finally, it should be stressed that similar calculations for the A-site components cannot be performed, firstly because this component in the AFMS cannot be distinguished from the B-site reversed-spin component, and secondly because no information is available about the A-site canting and dipolar-field contribution.

Annealed sample

According to VOGEL *et al.* (1976), the changes in the RT Mössbauer line shape upon annealing their sample (indicated composition $Zn_{0.730}Mn_{0.346}Fe_{1.924}O_4$) provide an indication that some kind of ordering is present in the as-collected franklinite. For the present sample, annealing was not found to produce any significant visible effect on the MS recorded at RT, 9 K and 4.3 K, without an externally applied field and at 4.3 K in a 60 kOe field. The RT doublet, when fitted with a QSD for the B-site component, has an average quadrupole splitting of 0.48 mm/s, and a maximum-probability value of 0.44 mm/s. Both quantities are about 10% larger than the results for the parent material (see Table 2), as is the width of the distribution (1xσ of 0.23 mm/s as compared to 0.20 mm/s for the non-annealed mineral at RT). These differences might be considered significant, suggesting increased deformations of the structural and/or surrounding-charge symmetries of the Fe-containing octahedra. A slight rearrangement of the cation distribution could indeed explain such a behavior. The relative spectral area of the A-site component, *i.e.* 0.055, was found to have remained unchanged upon annealing. Concerning the low-temperature MS of the annealed sample, the same trends in the goodness-of-fit χ^2 were obtained when fitting the various models as discussed in the preceding sections. Some results of the numerical analyses based on a asymmetrical B-site line shape are included in Tables 4 and 5. It is obvious that none of the involved parameters has markedly changed, implying that the annealing, and the possible alteration of the cation distribution resulting from it, has but a little effect on the low-temperature magnetic structure.

CONCLUSIONS

Both the paramagnetic doublet spectra and the low-temperature, magnetically split spectra (in particular those recorded with the absorber subjected to the external fields) consistently reveal that in the investigated franklinite mineral about 0.06 Fe atoms per formula unit are located in the tetrahedral sublattice. The high-temperature spectra could not detect any significant amount of Fe^{2+} and are adequately described as a superposition of an A-site doublet and two B-site doublets. The fitted quadrupole-splitting values of these latter two are close to the literature values reported for Zn and Mn ferrite respectively. This finding could be related to the suggested presence of Mn-rich and Zn-rich clusters. However, slightly but consistently better fits were obtained using an A-site doublet and a quasi-continuous quadrupole-splitting distribution for the B sites. All parameters except the center shift remain remarkably constant as the temperature varies. The average quadrupole splitting is found to be within the range 0.438 $\pm$ 0.004 mm/s. This fitting model does not provide any support in favor of the cluster formation.

At temperatures below 245 K and down to $\approx$20 K, the spectral shape is governed by relaxation effects. The relaxation process seems to be unusual in the sense that it is not a collective one. Instead, two clearly separated subpatterns emerge. A two-component Blume-Tjon model is unsuccessful at describing the observed line shapes in all details. The spectra recorded at 4.3 and 9 K seem to be free from relaxation effects, but clearly exhibit significant asymmetry in peak depths for corresponding pairs of lines. In view of the suggested clustering, two symmetrical B-site sextets were fitted to these spectra. The results were found to be inconsistent with those obtained in the paramagnetic regime. Two other models, assuming the spins to be aligned along a domain's [111] or [110] crystallographic axis respectively, were unsuccessful as well. A similar conclusion was arrived at with respect to a model-independent magnetic hyperfine field distribution.

Applied-field Mössbauer spectra recorded at 4.3 and 9 K reveal the existence of locally canted spin states and reversed B-site spins. Both phenomena ceased to exist at a temperature in the vicinity of 30 K and are characteristic for a re-entrant ferrimagnetic structure. The predominant B-site subpattern again is asymmetrical and could be reproduced adequately by a line shape calculated on the basis of a random angle between the hyperfine field and the EFG's principal axis. An additional distribution on the effective hyperfine fields was introduced to interpret the AFMS at 14 and 30 K. The models allowed the determination of the dipolar contribution to the B-site magnetic hyperfine field which was found to be -8 ± 2 kOe. The B-site quadrupole splitting was found to be negative and its magnitude is in excellent agreement with the average value of the QSD fitted to the paramagnetic spectra. The existence of LCSS at low temperatures implies that in zero applied field the angle between the B-site hyperfine field and the EFG's principal axis is also fluctuating, so that the asymmetry in the corresponding spectra can be explained accordingly. Using one asymmetrical B-site pattern and a constrained Lorentzian A-site sextet, the MS at 4.3 and 9 K were reproduced adequately, with parameter values that are in line with the results derived from the AFMS. As a final conclusion, it may be inferred that there appears to be no need to assume the formation of Mn- and Zn-rich clusters in order to explain the MS observed for the investigated franklinite mineral. The present results furthermore raise some questions regarding the correctness of earlier reported interpretations of MS, obtained for synthetic (Mn,Zn) ferrites, in terms of possible nearest-neighbor A-site cation configurations.

Acknowledgements—The authors wish to thank Dr. R. P. Kruk (Jagiellonian University, Krakow) for his assistance with the interpretation of the relaxational Mössbauer spectra. This work was supported by the Fund for Joint Basic Research, Belgium (Grants No. 32.0014.93 and No. 32.0055.93).

REFERENCES

BOWEN L. H., DE GRAVE E. and BRYAN A. M. (1994) Mössbauer studies in an external field of well-crystallized Al-maghemites made from hematite. *Hyperfine Interact.* **94**, 1977–1982.

BRAND R. A., GEORGES-GIBERT H., HUBSCH J. and HELLER J.A. (1985) Ferrimagnetic to spin glass transition in the mixed spinel $Mg_{1+t}Fe_{2-2t}Ti_tO_4$: a Mössbauer and DC susceptibility study. *J. Phys. F: Met. Phys.* **15**, 1987–2007.

DA COSTA G. M., DE GRAVE E., BOWEN L. H., VANDENBERGHE R. E. and DE BAKKER P. M. A. (1994) The center shift in Mössbauer spectra of maghemite and aluminum maghemites. *Clays Clay Minerals* **42**, 628–633.

DE GRAVE E. and VAN ALBOOM A. (1991) Evaluation of ferrous and ferric Mössbauer fractions. *Phys. Chem. Minerals* **18**, 337–342.

DE GRAVE E. and VOCHTEN R. (1985) An ^{57}Fe Mössbauer effect study of ankerite. *Phys. Chem. Minerals* **12**, 108–113.

DE GRAVE E., PERSOONS R. M., VANDENBERGHE R. E. and DE BAKKER P. M. A. (1993) Mössbauer study of the high-temperature phase of Co-substituted magnetites, $Co_xFe_{3-x}O_4$. I. $x \leq 0.04$. *Phys. Rev. B* **47**, 5881–5893. (with Erratum, *ibid.* **48**, 3581.)

EVANS B. J., HAFNER S. S. and WEBER H. P. (1971) Electric field gradients at ^{57}Fe in $ZnFe_2O_4$ and $CdFe_2O_4$. *J. Chem. Phys.* **55**, 5282–5288.

HOY G. R. (1984) Relaxation phenomena for chemists. In *Mössbauer Spectroscopy Applied to Inorganic Chemistry* (ed.G. J. LONG), Vol. 1, Chap. 8, pp. 195–226. Plenum Press.

KÖNIG U. (1971) Note on the Mössbauer effect in $MnFe_2O_4$. *Solid State Commun.* **9**, 425-427.

KÖNIG U. (1972) Die magnetische Ordnung in Mangan-Zink-Ferriten. *Techn. Mitt. Krupp, Forsch.-Ber.* **30,** 1–14.

METSGER R. W., TENNANT C. B. and RODDA J. L. (1958) Geochemistry of the Sterling Hill zinc deposit, Sussex County, New Jersey. *Bull. Geol. Soc. Am.* **69,** 775–788.

MORRISH A. H. and CLARK P. E. (1975) High-field Mössbauer study of manganese-zinc ferrites. *Phys. Rev.* **B 11,** 278–286.

PERSOONS R. M., DE GRAVE E., DE BAKKER P. M. A. and VANDENBERGHE R. E. (1993) Mössbauer study of the high-temperature phase of Co-substituted magnetites, $Co_xFe_{3-x}O_4$. II. $x \geq 0.1$. *Phys. Rev. B* **47**, 5894–5905.

RANCOURT D. G. (1994) Mössbauer spectroscopy of minerals. I. Inadequacy of Lorentzian-line doublets in fitting spectra arising from quadrupole splitting distributions. *Phys. Chem. Minerals* **21**, 244–249.

ROSENCWAIG A. (1970) Localized canting model for substituted ferrimagnets. I. Singly substituted YIG systems. *Can. J. Phys.* **48**, 2857–2867.

SAWATZKY G. A., VAN DER WOUDE F. and MORRISH A. H. (1969) Mössbauer study of several ferrimagnetic spinels. *Phys. Rev.* **187**, 747–757.

SMIT J. and WIJN H. P. J. (1959) *Ferrites.* Philips' Technical Library.

VANDENBERGHE R. E. and DE GRAVE E. (1989) Mössbauer studies of oxidic spinels. In *Mössbauer Spectroscopy Applied to Inorganic Chemistry* (eds. G. J. LONG and F. GRANDJEAN), Vol. 3, Chap. 3, pp. 59–182. Plenum Press.

VANDENBERGHE R. E., DE GRAVE E. and DE BAKKER P. M. A. (1994) On the methodology of the analysis of Mössbauer spectra. *Hyperfine Interact.* **83**, 29–49.

VARRET F., GERARD A. and IMBERT P. (1971) Magnetic field distribution analysis of the broadened Mössbauer spectra of zinc ferrite. *Phys. Status Solidi (b)* **43**, 723–730.

VOGEL R. H., EVANS B. J. and SWARTZENDRUBER L. J. (1976) Intra site cation ordering and clustering in natural Mn-Zn ferrites. *In AIP Conf. Proc. No 34—Magnetism and Magnetic Materials* (eds. J. J. BECKER and G. H. LANDER), pp. 125–127. American Institute of Physics.

Mineral Spectroscopy: A Tribute to Roger G. Burns
Editors: M. D. Dyar, C. McCammon and M. W. Schaefer

Spectroscopic studies of iron sulfide formation and phase relations at low temperatures

ALISTAIR R. LENNIE and DAVID J. VAUGHAN
Department of Earth Sciences, The University of Manchester, Oxford Road, Manchester M13 9PL, England

Abstract—The iron monosulfides mackinawite ($Fe_{1+x}S$) and "amorphous FeS," the first formed iron sulfides in aqueous systems at low temperatures, have been studied using X-ray photoelectron and X-ray absorption spectroscopies. The divalent oxidation state of Fe and the local structure environment of Fe and S in mackinawite have been confirmed. Comparisons of the Fe K-edge X-ray absorption spectra of synthetic mackinawite with those of "amorphous FeS" precipitates subjected to ageing for periods ranging from 1 to 1800 seconds, and precipitated under pH conditions between pH 3.9 and pH 7.4, show the precipitates to have local structure consistent with that found in crystalline mackinawite. Combining these data with results in the literature leads to the proposal of a model which accounts for the formation from aqueous solution of Cubic FeS, troilite, and mackinawite. We use this model, which takes account of the effect of electron withdrawal by H^+ at the crystal surface, to propose that these three phases can all be formed from Fe in tetrahedral coordination with S. This model facilitates understanding of the phase relationships in the Fe-S system at low temperatures, whereby initial formation is of iron "monosulfides" with either cubic close-packed (ccp) or hexagonal close-packed (hcp) S sublattices. Subsequent sulfidation of either sub-lattice type will form pyrite (or marcasite). However, the predominant sulfidation sequence at low temperatures is:

[Cubic FeS]/amorphous FeS → mackinawite → greigite → marcasite/pyrite. Both the nucleation of hcp S sub-lattice phases in low temperature aqueous iron sulfide systems, and the formation of hcp phases by transformation from the cubic monosulfides, are inhibited by kinetic factors.

INTRODUCTION

THE MINERALS of the iron-sulfur system are important Earth and planetary materials for a number of reasons. Troilite (FeS) is an important phase in iron meteorites, suggesting that sulfur may coexist with iron in the core of the Earth. Iron sulfides, as well as occurring as accessory minerals in many common rocks, are major phases in the sulfide ore deposits that are still the main suppliers of many base (Cu, Pb, Zn), ferroalloy (Ni, Co, Mn), and rare or precious (Ga, Ge, Ag, Pt) metals. Studies of iron sulfide formation in Recent sediments via sulfate-reducing-bacteria, and in hydrothermal systems presently active at or near mid-ocean ridges, have stimulated new ideas regarding the role of sulfides in the geochemical cycling of the elements. These studies have not only furthered understanding of the processes of formation of metal sulfide ore deposits, but have also suggested that metal sulfides may entrap heavy metal pollutants in sediments. Furthermore, the discovery of life forms adapted to the warm but anoxic environments of sea-floor hydrothermal systems, and the known sulfur-metabolizing ancient bacteria, has fueled a debate that anoxic sulfide environments could have been the sites for the emergence of life on Earth. This debate has centred around iron sulfides which may have played a central role in supplying energy and, perhaps, in catalysing reactions leading to the build-up of complex organic molecules.

At temperatures in excess of ~350°C, the phase relations in the Fe-S system are relatively straightforward, but at lower temperatures the Fe-S system becomes much more complex and less well understood. In part, this arises from ordering of vacancies upon cooling of the pyrrhotite ($Fe_{1-x}S$) solid solution, forming complex superstructures. A further complexity concerns phases found only in relatively low temperature environments in nature, which can be synthesized only in solution. For this reason, these lower temperature phases may not even appear in the (equilibrium) phase diagrams that have generally been constructed on the basis of experiments involving reaction between elemental iron and sulfur in sealed evacuated silica capsules. Several important phases (notably marcasite, FeS_2, and mackinawite $Fe_{1+x}S$) are of this type, along with a number of rarer minerals (greigite Fe_3S_4 and, possibly, smythite ~Fe_9S_{11}) and phases so far found only in synthesis experiments (*e.g.*, Cubic FeS). It seems very likely that all of these phases are metastable; indeed calculations of the free energies of formation of these phases confirm this view (see Fig. 1). Although some of these minerals appear to be rare in their geological occurrence, they are of considerable importance as transient species produced when iron and sulfur react in aqueous solutions, subsequently transforming to the more stable

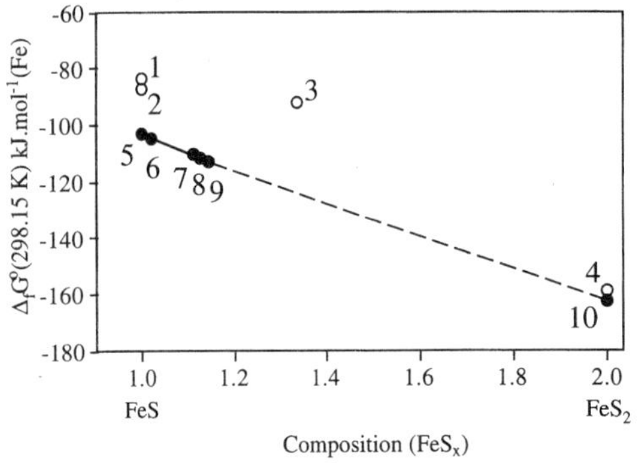

FIG. 1. Schematic free energy-composition diagram for the iron sulfide minerals. The gain in free energy, $\Delta_f G^o$(298.15 K), in kJ.mol^{-1} (of Fe), is that obtained from forming the various sulfides from the elements. Open circles represent Amorphous FeS (1), mackinawite (2), greigite (3) and marcasite (4). Filled circles represent the pyrrhotites: Troilite (5), $Fe_{11}S_{12}$ (6), $Fe_{10}S_{11}$ (7), Fe_9S_{10} (8), Fe_7S_8 (9), and pyrite (10). Data for 1 - 3 are calculated from solubility data of DAVISON (1991); for 4 and 10 from CHASE *et al.* (1985) and for 5 - 9 from GRØNVOLD and STØLEN (1992). The dashed line is the join between monoclinic pyrrhotite and pyrite; the solid line is the join between troilite and monoclinic pyrrhotite. Free energy values for phases 1, 2, 3, and 4 lie above these lines suggesting metastability for these phases at this temperature.

phases (BONEV *et al.*, 1989). The keys to understanding these phases are concerned with their mechanisms of formation in solution, and with relationships to other iron sulfides, as determined by the kinetics and mechanisms of transformations. The known iron sulfide minerals, their structures and stabilities are summarized in Table 1. Detailed accounts of the mineral chemistry of these sulfides are provided by WARD (1970), POWER and FINE (1976), VAUGHAN and CRAIG (1978), VAUGHAN and LENNIE (1991).

Formation of iron-sulfur phases is affected by the d-electron configuration of Fe, resulting in a more complex range of structures than would be otherwise expected. As an example of this, the formation of (metastable) marcasite (FeS_2) from acidic aqueous solution has been explained by the effect of bonding of H^+ at the crystal surface on the Fe 3d electronic structure. We were interested in examining the formation of the reduced phase(s) to determine if electron withdrawing effects might also be responsible for formation of Cubic FeS and troilite, both of which, like marcasite, show pH-dependant formation.

In the present article, we examine the monosulfide mackinawite and explore the properties of "amorphous FeS" as a key to understanding the formation of mackinawite, Cubic FeS and troilite. This work forms part of a series of contributions to studies of the crystal chemistry and phase relations of the iron sulfides (LENNIE, 1994; LENNIE *et al.* 1995a, 1995b).

MACKINAWITE

Mackinawite possesses a layer structure similar to that found in tetragonal lead oxide (see Table 1) with Fe and S occupying the sites of O and Pb, respectively, in PbO. Here we investigate further the oxidation state and local structure of Fe and S in synthetic mackinawite.

Synthesis

Several different methods for synthesis of mackinawite have been described in the literature. BERNER (1964) found that reaction of metallic Fe with a saturated solution of H_2S gave well crystallised tetragonal FeS. Alternatively, mackinawite has been prepared by reacting aqueous solutions of Fe(II) ions and HS^- ions forming an X-ray amorphous precipitate which, even on ageing, gives a poorly crystalline mackinawite.

In the present work, sulfidation of a commercial iron wire (Roncraft, Sheffield) was used to produce crystalline mackinawite. Here, a solution of Na_2S was added slowly to an 0.5 molar acetic acid/acetate buffer (pH = 4.6) in which approximately 4–5g of finely divided iron wire had been previously immersed. Dissolution of the iron wire by reaction with the acetate buffer for 30 minutes or so evolved H_2, providing a reducing solution environment. The Na_2S solution was then added slowly, and exceeded the capacity of the acetate buffer, increasing the pH to a value of approximately pH = 6.5. This solution, containing the iron wire, was allowed to stand for 24 hours. The generated H_2 trapped in the wool matte floated the iron to the top of the solution. Mackinawite forming on the iron surface spalled off and dropped to the bottom of the flask. After reaction, the remaining iron wool was removed, the supernatent solution poured off, and the mackinawite carefully rinsed then dried under vacuum. Samples were sealed under vacuum in glass tubes until required for further analysis.

Crystal structure

The mackinawite synthesised as described above was examined using high resolution X-ray powder diffraction from which a Rietveld structure refinement was obtained. The structural parameters

Table 1. The iron sulfides, their structures, stabilities and natural occurrence*

Composition mineral	Crystal structure type structural data †	Stability ‡	Natural occurrence
FeS_2 pyrite	pyrite-type (cubic) *Pa*3; *a* = 5.42	stable (<742°C)	The most abundant sulfide. Found as an accessory mineral in many igneous, metamorphic and sedimentary rocks including coals. The major sulfide in many massive, bedded or vein type ores.
FeS_2 marcasite	marcasite-type (orthorhombic) *Pnnm*; *a* = 4.44; *b* = 5.41; *c* = 3.38	metastable	Occurs as a primary mineral in certain sediments and lower temperature hydrothermal deposits. Also formed through weathering of other sulfide minerals.
Fe_3S_4 greigite	spinel-type (cubic) ?*Fd3m*; *a* = 9.88	metastable	A very rare mineral reported from Recent sediments and low temperature hydrothermal deposits.
~Fe_9S_{11}(?) smythite	?pseudorhombohedral a = 3.47; c = 34.4	?stable (~ 75°C)	A very rare (?) mineral reported from magmatic and hydrothermal sulfide ore deposits.
~Fe_7S_8 monoclinic pyrrhotite	NiAs-type; superstructure (mono.) *F2/d*; *a* = 11.90; *b* = 6.86; *c* = 22.79; *β* = 90°26'	?stable ~254°C	As an important mineral in sulfide ores of "magmatic' origin and an accessory mineral in ultramafic and mafic rocks. Also in metamorphosed massive sulfides and some hydrothermal ores.
Fe_9S_{10} intermediate pyrrhotite (5C)	NiAs-type; superstructure (hexagonal) *a* = 6.88; *c* = 28.70	?stable (? ~ 100°C)	"
$Fe_{10}S_{11}$ intermediate pyrrhotite (11C)	NiAs-type; superstructure (orthorhombic) *Cmca* or *C2Ca*; *a* = 6.89; *b* = 11.95; *c* = 63.18	?stable (?~ 100°C)	"
$Fe_{11}S_{12}$ intermediate pyrrhotite (6C)	NiAs-type; superstructure (hexagonal) *a* = 6.89; *c* = 34.48	?stable (?~100°C)	"
~Fe_9S_{10}-$Fe_{11}S_{12}$ non-integral pyrrhotite (nC)	NiAs-type; superstructure (orthorhombic or monoclinic); *a* = 6.89; *b* = 11.95; *c* = variable	?stable (?~220°C)	"
FeS troilite	NiAs-type (distorted superstructure) (hexagonal) $P\bar{6}2c$; *a* = 5.97; *c* = 11.75	stable (<138°C)	Found in iron meteorites and lunar rocks. Rare on Earth but found in some serpentinized ultramafic rocks.
$Fe_{1+x}S$ (x = 0.01-0.08) mackinawite	PbO-type (tetragonal) *P4/nmm*; *a* = 3.60; *c* = 5.03	metastable	As the first-formed very fine grained black sulfide precipitated in Recent sediments. Also as small grains and lamellae within other sulfides (pyrrhotite, chalcopyrite etc.) from a variety of environments.
$Fe_{1+x}S$ (x = 0.03-0.10) cubic FeS (synthetic)	Sphalerite-type (cubic) $F\bar{4}3m$; *a* = 5.42	metastable	Known only from synthetic studies at present and formed as a layer on sulfidized iron surfaces.

* Data from Vaughan and Craig (1978), Kissin and Scott (1982)
† Information given is on structure type (crystal class); space group; unit cell parameters (in Å)
‡ Upper temperature of stability in parentheses

obtained from this refinement, reported elsewhere (LENNIE *et al.*, 1995b), are given in Table 2.

X-ray photoelectron spectra

X-ray photoelectron spectra were obtained on a Kratos XSAM 800 surface analysis system operating at a base pressure of less than 1 x 10^{-9} τ. Removing mackinawite from the glass tube and subsequent transfer into the spectrometer vacuum chamber was carried out under O_2-free N_2 conditions. Narrow region XPS spectra were recorded at an analyser pass energy of 20 eV. The spectrometer was calibrated using the peaks: Au ($4f_{7/2}$) 84.00

Table 2. Refined structure parameters for synthetic mackinawite. Errors (σ) are in parentheses (after LENNIE *et al.*, 1995b)

Space group	*P4/nmm* (No. 129) origin at $\bar{4}m2$			
Cell parameters (Rietveld)	*a* (Å)		3.6735(4)	
	b (Å)		3.6735(4)	
	c (Å)		5.0328(7)	
	V(Å^3)		67.914(24)	
Atomic positional parameters	*x*	*y*	*z*	*B*(iso)
Fe in 2(a):	0	0	0	0.41(2)
S in 2(c):	0	0.5	0.2602(3)	0.20(2)
Bond distances (Å)				
Fe-S:	2.2558(9)			
Fe-Fe:	2.5976(3)			

eV, Cu ($2p_{3/2}$) 932.66 eV, Ag ($3d_{5/2}$) 368.27 eV with Mg Kα X-rays (1253.6 eV) as the exciting radiation, and the static charge effect was corrected using adventitious C with the 1s electron binding energy of C taken as 284.6 eV. For comparison with mackinawite, XPS measurements were also made on a natural sample of pyrite cleaved under O_2-free N_2 conditions prior to installation in the spectrometer. A succession of spectra of the cleaved pyrite were taken before, and after, a series of Ar^+ etches of the surface. Fig. 2 shows the Fe, S and O XPS spectra obtained for synthetic mackinawite. These spectra are fitted using a Shirley background (SHIRLEY, 1972).

The Fe ($2p_{3/2}$) peak of mackinawite has a binding energy of 707.25 eV, similar to that of pyrite (707.2 eV), which provides confirmation of the Fe(II) oxidation state predicted for mackinawite. The mackinawite Fe($2p_{3/2}$) peak shows considerable asymmetric broadening at the higher binding energy side of this peak, which may arise from the high conductivity (WERTHEIM and BUCHANAN, 1977) in the (001) plane of mackinawite due to Fe-Fe interactions inferred by KJEKSHUS *et al.* (1972).

Following background subtraction, the S (2p) photoelectron spectrum of mackinawite has been fitted with a doublet in a 2:1 ratio to match the 2:1 intensity ratio of the spin-orbit splitting of S $2(p_{3/2})$ and $2(p_{1/2})$ peaks. These have been fitted with asymmetric curves made up of symmetric 80% Gaussian, 20% Lorentzian components. The binding energies of the fitted doublet are 161.82 eV and 163.02 eV for the $2(p_{3/2})$ and $2(p_{1/2})$ peaks respectively. The O (1s) spectrum in Fig. 2 shows the small fraction of O in the sample of mackinawite. The peak position (532 eV) is consistent with the presence of iron hydroxide on the surface of mackinawite (MCINTYRE and ZETARUK, 1977).

The pyrite XPS Fe ($2p_{3/2}$) spectrum binding energy is 707.2 eV, and the S(2p) spectra were fitted with symmetric Gaussian 50%, Lorentzian 50% curves at 162.5 eV and 163.7 eV respectively. Following Ar^+ etching, the Fe ($2p_{3/2}$) peak showed considerable broadening and increased asymmetry towards the higher binding energy side of this peak. The S(2p) spectrum showed formation of additional peaks at lower binding energies which were fitted with symmetric Gaussian 50%, Lorentzian 50% curves at 161.3 eV and 162.5 eV respectively. These changes in Fe and S photoelectron peaks are consistent with the partial reduction of S_2^{2-} to S^{2-} species, and partial oxidation of Fe(II) which would occur upon reduction to pyrrhotite. The S($2p_{3/2}$) peak of pyrite (S^-) has a higher binding energy (162.5 eV) than that observed for S (S^{2-}) in mackinawite (161.85 eV), consistent with observed trends which show increasing binding energy with increase in formal oxidation state.

The binding energy may also reflect differences in the lattice site environment between two different phases of the same oxidation state. In the case of mackinawite, S (S^{2-}) is coordinated to four Fe atoms in a pyramidal arrangement, with a lone pair of electrons at the apex of this pyramid. In contrast, S in pyrrhotite (S^{2-}) is coordinated to 6 Fe atoms in a trigonal prism arrangement. The binding energy of S in mackinawite (161.85 eV) is greater than the binding energies reported for pyrrhotite phases (161.1 - 161.25). This suggests a slight decrease in the electronic population of the S valence band(s) for mackinawite compared with that of the S valence band(s) for the pyrrhotites.

A compilation of S ($2p_{3/2}$) and Fe ($2p_{3/2}$) binding energies reported in the literature for Fe-S minerals, together with those obtained for pyrite and mackinawite from our experiments, is presented in Table 3.

X-ray absorption spectra

X-ray absorption spectroscopy (XAS) and the analysis of the fine structure in spectra at and near the absorption edge (so called 'XANES'), and in the energy region extending well above the edge (so-called 'EXAFS'), have been very comprehensively described in the literature (see CALAS *et al.* 1984, 1990; BROWN *et al.*, 1988; WAYCHUNAS and BROWN, 1984, for excellent reviews). Here only specific details of our experiments and their interpretation will be discussed.

Mackinawite was examined by X-ray Absorp-

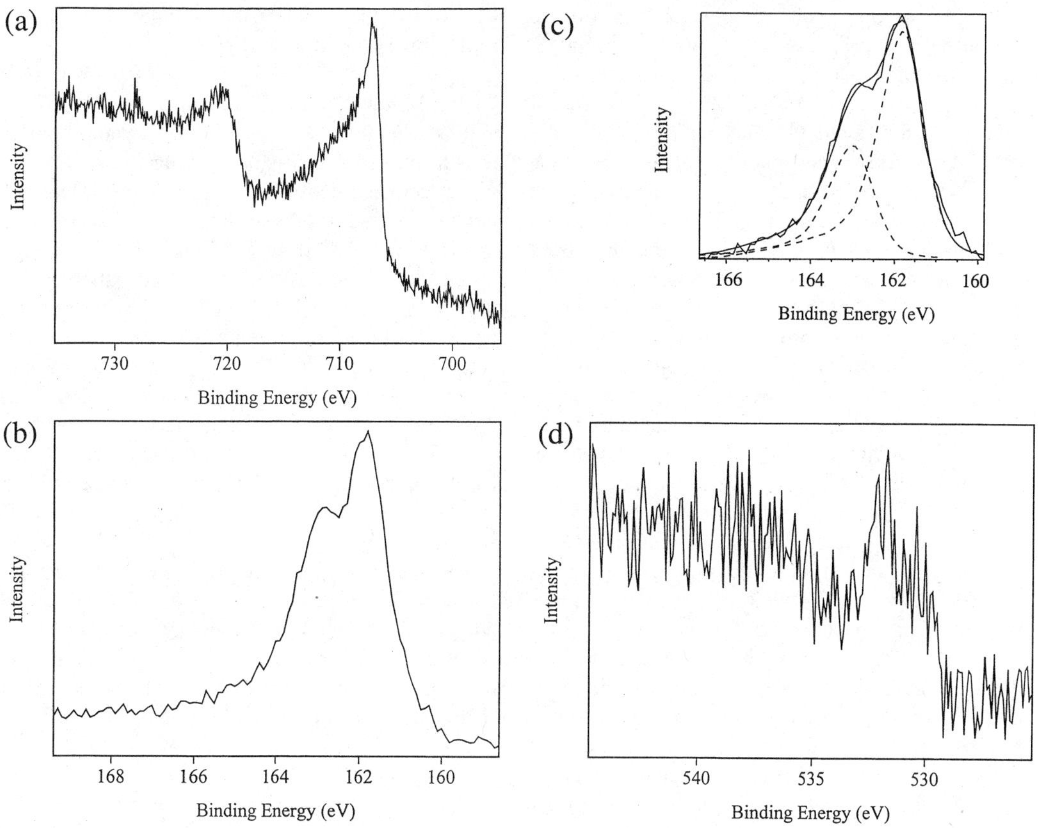

FIG. 2. XPS spectra of synthetic mackinawite: (a) Fe (2p) spectrum and (b) S (2p) spectrum. (c) The S (2p) spectrum following background subtraction and fitted with a doublet of asymmetric curves at 161.82 eV($2p_{3/2}$) and 163.02 eV ($2p_{1/2}$). (d) O (1s) spectrum: the small fraction of oxygen in the synthetic material, due to surface hydroxide species, is indicated by the low signal to noise ratio.

Table 3. XPS binding energies of S ($2p_{3/2}$) and Fe ($2p_{3/2}$) found in iron-sulfur minerals, including results from this study

Species	Mineral	Binding energy (eV)	Reference
Fe ($2p_{3/2}$)	FeS_2 (pyrite)	707	BUCKLEY and WOODS (1987)
		707.2	this study
	$Fe_{1+x}S$ (mackinawite)	707.25	this study
	$Fe_{0.89}S$ (pyrrhotite)	708	BUCKLEY and WOODS (1985)
	Fe_7S_8 (pyrrhotite)	707.45	PRATT *et al.* (1994)
	Ar^+ reduced FeS_2 surface	707.8	this study
S($2p_{3/2}$)	S(0)	163.7	HYLAND and BANCROFT (1989)
	FeS_2 (pyrite)	162.3	BUCKLEY and WOODS (1987)
		162.4	HYLAND and BANCROFT (1989)
		162.5	this study
	$Fe_{1+x}S$ (mackinawite)	161.82	this study
	$Fe_{0.89}S$ (pyrrhotite)	161.1	BUCKLEY and WOODS (1985)
	Fe_7S_8 (pyrrhotite)	161.25	PRATT *et al.* (1994)
	Ar^+ reduced FeS_2 surface	161.24	this study

tion Spectroscopy using synchrotron radiation to establish the local enviroment around Fe and S, and to provide model XAS data for amorphous iron sulfide precipitates. To obtain Fe K-edge absorption data, a solid synthetic mackinawite sample was pressed into Al samples holders with adhesive-tape windows and cooled under vacuum using a liquid N_2 cryostat. The spectrum for Fe was collected in transmission mode on Station 7.1 at the Daresbury Synchrotron Radiation source, operating at 2 GeV with an average current of 150 mA. A Si(111) double-crystal monochromator was used with the crystal faces offset to give 50% rejection of the beam in order to reduce harmonic contamination. Ion chambers to detect the incident and transmitted radiation were filled with Ar-He mixtures to absorb 20% and 80% of the beam, respectively.

The spectrum for S was taken on Station 3.4 using a Ge(111) double crystal monochromator; a focussing mirror is used to remove higher harmonics. The sample of synthetic mackinawite was diluted with graphite and coated as a slurry in acetone onto aluminium stubs. Following evaporation of the acetone, the stubs were mounted in the spectrometer and the chamber evacuated. Spectra over the S K-edge were taken using photo-electron yield as the detection technique. In the case of the S spectrum, four separate scans were summed to improve the signal to noise ratio.

Spectra were prepared for analysis by background subtraction of pre-edge and EXAFS regions of the spectra, followed by normalisation using the Daresbury programs, EXCALIB and EXBACK (DAIKUN *et al.*, 1984). Standard k^3 weighting was used for the EXAFS spectra. The data were analysed using single-scattering curved wave theory in the program EXCURV92 (BINSTED *et al.*, 1991; LEE and PENDRY, 1975; GURMAN *et al.*, 1984, 1986). Phase shifts are derived from *ab initio* calculations within EXCURV92 using the default parameters for the muffin tin radii and exchange terms. The phase shifts were not further refined using model compounds.

A fit was produced by generating a theoretical spectrum from a structure of several shells of backscatterers around the central atom, and allowing parameters refining radial distance from the central scattering atom and Debye-Waller factors within each shell to vary iteratively to give the best agreement with the experimental data. Coordination numbers were kept fixed for analysis of synthetic mackinawite, and only those shells that made a significant improvement to the fit were included in the simulations. The goodness of fit of the theoretical spectrum to the experimental data is given by R(min), the value of the fit index at predicted minimum (JOYNER *et al.*,1987).

Fig. 3 shows k^3 weighted spectra from EXCURV92 for S and Fe of mackinawite plotted against the inverse wave-vector, k, and Fourier transforms of these spectra giving the radial distance from the central scattering atoms. Data obtained from the iterative fitting of these spectra are given in Table 4. Both S K-edge and Fe K-edge spectra have been fitted with three shells. The model set up for the S K-edge data uses an approximation for the second and third shells—the second shell was modelled by 12 S atoms at 3.62Å (instead of 4 S at 3.548Å, 4 S at 3.674Å and 4 S at 3.688Å); the third shell by 12 Fe atoms at 4.26Å (not 4 Fe at 4.153 and 8 Fe at 4.311). Distances and coordination numbers for initial fitting were calculated from the structural refinement of mackinawite presented in Table 2.

The results of these iterative fits to the Fe K-edge and S K-edge EXAFS spectra of mackinawite are consistent with the structure determined by X-ray powder diffraction, and confirm both the tetrahedral coordination of Fe by S, and the short Fe-Fe distances in the (001) plane of mackinawite.

AMORPHOUS IRON SULFIDE (“FeS”)

The material initially formed when dissolved Fe and S react in solution is an extremely fine-grained “X-ray amorphous” black precipitate. Such amorphous iron sulfide precipitates, which are easily prepared in the laboratory, form in nature by reactions taking place in Recent sediments, principally during processes that take place just beneath the surface of certain fine sediments during early diagenesis. Here, the most common interactions involve sulfide S derived from reduction of seawater sulfate by bacteria, and Fe from detrital or other sources.

These fine grained materials are difficult to study using techniques such as X-ray diffraction or electron microprobe analysis. In our work, we have used X-ray Absorption Spectroscopy employing synchrotron radiation to study the structure of these precipitates. This work has involved attempts to study such precipitates by quenching in liquid N_2 immediately following nucleation, after controlled periods of “ageing”, and following buffering in a range of solution pH’s.

FeS Amorphous Precipitates; Ageing Experiments

A series of FeS precipitates were prepared as follows. Equal volumes of an Fe acetate solution

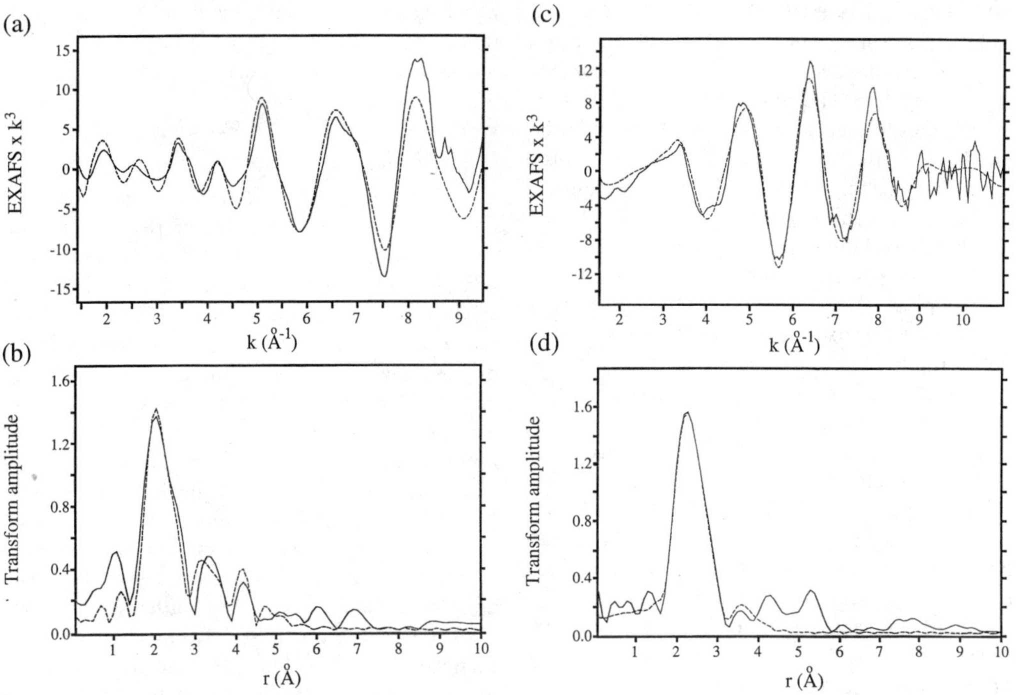

FIG. 3. (a) The k^3-weighted S K-edge EXAFS spectrum of mackinawite (298 K). (b) The Fourier transform of the S K-edge EXAFS spectrum fitted with three shells (as listed in Table 4). (c) The k^3-weighted Fe K-edge EXAFS spectrum of mackinawite (liq. N_2 cooled). (d) The Fourier transform of the Fe K-edge EXAFS spectrum fitted with three shells (as listed in Table 4.). Solid lines, experimental data; dashes, data simulated in EXCURV92.

Table 4. Results of refinements of EXAFS S K-edge and Fe K-edge spectra for mackinawite

Central Atom	Shell	Element	r(Å)	A: $2\sigma^2$(Å^2)	N	r_{XRD}(Å)	N_{XRD}
S (298 K)	1	Fe	2.24	0.01	4	2.255	4(Fe)
[R(min) = 14.9]	2	S	3.62	0.03	12	3.548	4(S)
						3.674	4(S)
						3.688	4(S)
	3	Fe	4.26	0.04	12	4.153	4(Fe)
						4.311	8(Fe)
Fe (298 K)	1	S	2.26	0.03	4	2.255	4(S)
[R(min) = 7.7]	2	Fe	2.56	0.03	4	2.598	4(Fe)
Fe (liq N_2)	1	S	2.24	0.01	4	2.255	4(S)
[R(min) = 6.9]	2	Fe	2.59	0.02	4	2.598	4(Fe)
	3	Fe	3.68	0.03	4	3.674	4(Fe)

r = radial distance of fitted shell from central atom; A = Debye-Waller factor
N = coordination number; R(min) is the value of the fit index at predicted minimum
r_{XRD} = radial distance from central atom calculated from XRD data in Table 2
N_{XRD} = coordination number of element (in brackets) calculated from XRD data in Table 2

(0.05 molar) and a 1:1 mixture of 1N NaOH and a solution 0.21 molar in $Na_2S.9H_2O$ were injected at the same time directly into polyethylene tubes sealed at one end. The solution pH above the precipitates thus formed was 7.5. After measured time periods (1, 5, 10, 20, 60, 300 and 1800 seconds), the tubes containing the precipitates were plunged into liquid nitrogen to quench the reaction. Rapid quenching also limits the formation of ice-crystals which would give Bragg reflections in the absorption experiments. Acetate was selected as the anion in these experiments to prevent confusion with sulfide during EXAFS analysis. Cl^- as an anion was avoided as this has similar X-ray scattering properties to S, being close in atomic mass.

At Daresbury, the tube containing the frozen precipitate was cut to provide a 2.5 cm length of tubing plus precipitate. The frozen precipitate contained within this length was removed and installed in an aluminium holder which was placed in a liquid N_2 cryostat and evacuated to 10^{-4} mbar. Fe K-edge XAS measurements were made in fluorescence mode on Station 8.1 by detecting fluorescence using the Canberra solid state detector.

These spectra, the XANES region for which are shown in Fig. 4, have been analysed using EXCURV92 as described above. The coordination number has also been allowed to vary during iteration. Data obtained from these fits are given in Table 5. These show that a precipitate is rapidly formed which has Fe tetrahedrally coordinated by a first shell of S, and a second shell of Fe atoms with short distances (2.59Å) from the central Fe atom. These radial distances are consistent with formation of a mackinawite-type structured phase as early as 1 second after intial reaction. The quality of data obtained for 10 and 20 seconds ageing time is not sufficient to provide second shell information for these samples.

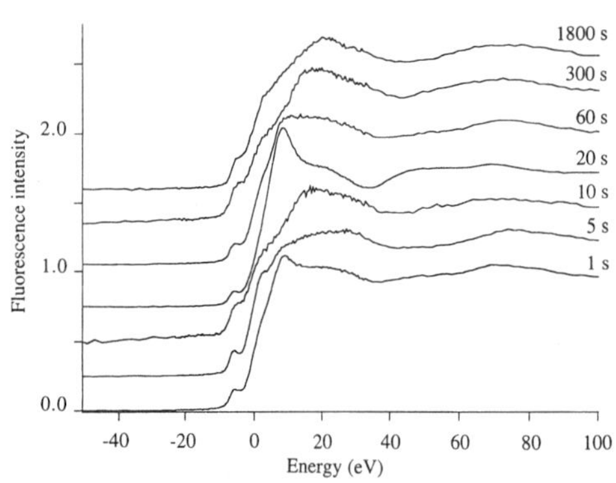

FIG. 4. Edge region Fe K-edge fluorescence spectra of iron sulfide precipitate ageing experiments. The energy scale (eV) is set so that the Fe K adsorption edge is at 0 eV.

FeS Amorphous Precipitates; pH Buffering Experiments

A series of 'FeS' solids were precipitated, then buffered as follows at different pH values. Iron acetate solution (1.6 mls, 0.05 M) was added to 3 mls of a solution 0.21 molar in $Na_2S.9H_2O$. This mixture gave a black amorphous FeS precipitate; the resulting solution pH was 4.5. Buffering to higher pH values, as determined by pH meter, was achieved by adding drops of a 1N NaOH solution. The lower pH (3.9) was achieved by addition of acetic acid.

The precipitate, with buffering solution, was injected into a polyethylene tube sealed at one end, held for 15 mins. at 298 K, then plunged into liquid N_2 to quench the reaction. Spectra over the Fe K-edge were then taken on Station 8.1 as described in the previous section. No S K-edge EXAFS spectra have been taken of the pH buffered or quenched amorphous precipitate experiments as the Station 3.4 spectrometer design prevents analysis of frozen precipitates.

Data obtained from analysis of these Fe K-edge spectra using EXCURV92 are given in Table 6. Again, the coordination number has been allowed to vary during iteration. The XANES region of these spectra are shown in Fig. 5. These precipitates show ordering after 15 minutes reaction sufficient to provide second shell information, and indicate formation of a mackinawite-like structure for precipitates buffered at pH's of 4.5 and above. The results for the precipitate buffered to pH 3.9, which has been fitted with two S shells surrounding the Fe atom, may have been affected by partial oxidation of sulfide during addition of acetic acid to achieve the required pH.

The data obtained from these pH buffered precipitates do not show obvious changes in coordination of Fe by S between ph 4.5 and 6.5; there is no evidence of 6-fold coordination of Fe by sulfur in these amorphous precipitates. We will refer to this result in the following discussion of a possible mechanism of formation for the phases Cubic FeS and troilite in the region of pH 4–5.

METASTABLE Fe-S PHASE RELATIONS

Formation of Cubic FeS and troilite from aqueous solution

Cubic FeS, a metastable synthetic iron sulfide with the sphalerite structure, first reported by

Table 5. Results of refinements of Fe K-edge EXAFS spectra for quenched reaction time iron sulfide precipitates

FeS precipitate	Shell	Element	r (Å)	A: $2\sigma^2$ (Å^2)	N
1 second	1	S	2.24	0.02	3.8
[R(min) = 44.3]	2	Fe	2.59	0.02	2.0
5 seconds	1	S	2.22	0.02	3.8
[R(min) = 38.1]	2	Fe	2.57	0.02	2.8
10 seconds	1	S	2.26	0.01	3.0
[R(min) = 63.5]					
20 seconds	1	S	2.25	0.01	3.4
[R(min) = 46.5]					
60 seconds	1	S	2.20	0.02	3.9
[R(min) = 52.7]	2	Fe	2.59	0.03	3.5
300 seconds	1	S	2.23	0.01	4.0
[R(min) = 54.3]	2	Fe	2.68	0.02	2.5
1800 seconds	1	S	2.21	0.02	4.0
[R(min) = 39.7]	2	Fe	2.62	0.02	2.8

r = radial distance of fitted shell from central Fe atom; A = Debye-Waller factor
N = coordination number; R(min) is the value of the fit index at predicted minimum

DE MEDICIS (1970 a, b), was synthesised by reacting Fe metal at 298 K with aqueous H_2S solutions having pH values between 4.0 and 4.5. TAKENO *et al.* (1970) synthesised Cubic FeS by reacting Fe metal with H_2S in solution at 323 K and report X-ray powder diffraction data for Cubic FeS synthesised in solutions with a pH of 5.7. These authors observed that euhedral troilite (and anhedral mackinawite) are also formed under the acidic reducing conditions required for Cubic FeS formation.

SHOESMITH *et al.* (1980) showed that formation of both troilite and Cubic FeS from stainless steels corroded by aqueous sulfide solutions at 294 K shows similar dependance on solution pH. Cubic FeS formed in greater quantities than troilite with the quantity of troilite and Cubic FeS formed being greatest at pH's between 4 and 5. At higher pH, mackinawite is the predominant iron sulfide formed while at lower pH, all these iron monosulfide phases readily dissolve. MUROWCHICK and BARNES (1986a) found similar pH dependant distri-

Table 6. Results of refinements of Fe K-edge EXAFS spectra for pH buffered iron sulfide amorphous precipitates

FeS precipitate	Shell	Element	r (Å)	A: $2\sigma^2$ (Å^2)	N
pH 3.9	1	S	2.20	0.01	4
[R(min) = 42.6]	2	S	2.37	0.01	2
pH 4.5	1	S	2.25	0.01	4
[R(min) = 46.1	2	Fe	2.67	0.02	3
	3	S	4.24	0.02	8
pH 5.4	1	S	2.19	0.01	4
[R(min) = 48.7]	2	Fe	2.61	0.02	4
pH 6.5	1	S	2.25	0.01	4
[R(min) = 40.3]	2	Fe	2.68	0.03	4
pH 7.4	1	S	2.23	0.01	4
[R(min) = 38.5]	2	Fe	2.65	0.03	2.7

r = radial distance of fitted shell from central Fe atom; A = Debye-Waller factor
N = coordination number; R(min) is the value of the fit index at predicted minimum

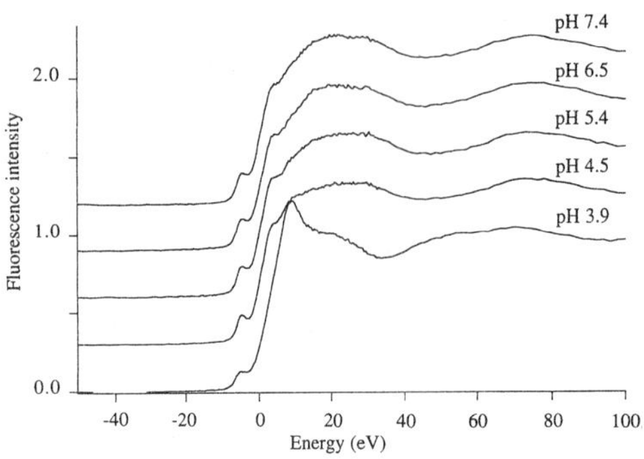

FIG. 5. Edge region Fe K-edge fluorescence spectra of pH buffered iron sulfide precipitates. The energy scale (eV) is set so that the Fe K adsorption edge is at 0 eV.

butions of troilite and Cubic FeS formation when Fe metal was sulfidised by aqueous H_2S in experiments carried out in solutions covering a pH range from 2 to 7, and between 328 K to 366 K.

Although troilite and above 413 K, stoichiometric hexagonal pyrrhotite, are considered stable phases in the Fe-S system, formation of the less S-rich pyrrhotites from aqueous solution under equilibrium conditions (hydrothermal synthesis) is prevented due to the unmanageably high H_2 pressures generated by dissociation of H_2O at low S fugacity (KISSIN and SCOTT, 1982). Troilite is therefore formed under non-equilibrium conditions in the above experimental studies.

Cubic FeS spontaneously transforms to mackinawite at room temperature. SHOESMITH *et al.* (1980) measured the rate of this solid state transformation at 298 K using changing intensities of X-ray diffraction peaks, and proposed a mechanism involving migration of Fe atoms for this transformation. Although no measurements have been made of the temperature dependence of the rate of this transformation, the observation that Cubic FeS does not persist above 366 K (MUROWCHICK and BARNES, 1986a) implies extremely rapid transformation at or above this temperature. There are small deviations from a stoichiometric formula of FeS reported in the literature for both Cubic FeS and mackinawite, which may indicate either iron excess or sulfur deficiency in these phases. However, the crystallographic origin of these compositional variations has not been definitely established.

Proposed mechanism of formation of Cubic FeS and troilite from solution

We now propose a mechanism to explain the similar dependance on pH of both Cubic FeS and troilite formation. We first refer to a study by TOSSELL *et al.* (1981) which accounted for pH dependant formation of marcasite by suggesting that interaction of surface bound H^+ with electrons in one of the $1\pi_g^*$ molecular orbitals of S_2^{2-} would effectively remove 2 electrons from the M3dσ-$A_21\pi_g^*$ system, distorting the S geometry around each Fe. Metastable marcasite would thus be formed in preference to stable pyrite.

We propose that a similar electronic effect may also be occurring in the formation of Cubic FeS and troilite under acidic conditions. At pH's between 4–5, H^+ ions bound to S at the surface of the iron monosulfide would act to withdraw electrons from Fe 3d orbitals. Ferrous ions ($3d^6$) would assume $3d^5$ character, thus making this Fe isoelectronic with Mn^{2+}. It has been observed that in the Mn-S system, two metastable phases, β-MnS and γ-MnS, form by precipitation with H_2S gas from aqueous manganous solutions (FURUSETH and KJEKSHUS, 1965). These phases have sphalerite and wurtzite structures respectively: there are no NiAs-type or tetragonal PbO-type structures found in the MnS system.

If Fe ions at the surface of a FeS crystal growing in acidic solution are assumed to have $3d^5$ character as proposed above, formation of FeS under these conditions may be expected to result in iron monosulfides having sphalerite and wurtzite structures, as found for β-MnS and γ-MnS. Cubic FeS formed under acidic conditions has the sphalerite structure, in agreement with this theory. However, there are no reports in the literature of wurtzite-type FeS formed under acidic conditions. Instead, it is troilite (with a distorted NiAs-type structure) which is observed under these conditions.

We note however that both NiAs and wurtzite-type structures have a hexagonal close-packed anion sublattice, as shown in Fig. 6. The arrangement of Fe (in hexagonal pyrrhotite/troilite) and Zn (in wurtzite) are different; Fe occupies octahedral sites in troilite, whereas in wurtzite, Zn atoms are in tetrahedral sites.

By analogy with Cubic FeS, which transforms spontaneously to mackinawite by rearrangement of Fe, a rapid solid-state transformation at the crystal surface from a (hypothetical) wurtzite-type FeS structure to FeS-troilite, after the removal of electron withdrawing H^+ ions from the surface, would explain both the absence of a wurtzite-type FeS and the presence of troilite in the experiments referred to above. The degree of rearrangement of Fe required for transformation of a wurtzite-type FeS to troilite would be much smaller than that

(a)

(b)

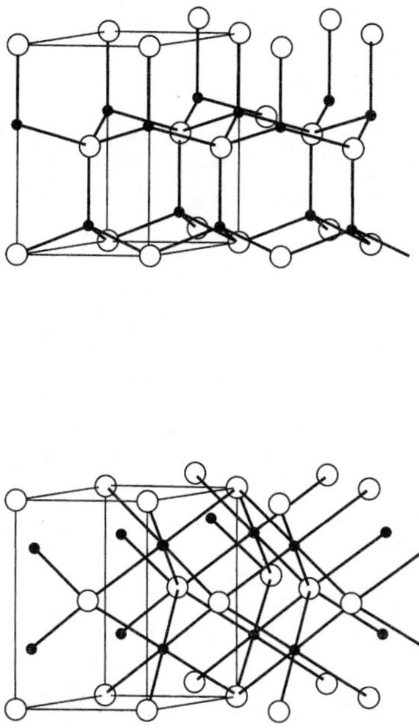

FIG. 6. Crystal structure diagrams for (a) wurtzite, the proposed structure for the metastable hexagonal FeS precursor, and (b) the NiAs structure which forms the basis of troilite and hexagonal pyrrhotite. In (a), open circles represent S while filled circles represent Zn (Fe). In (b), open circles represent As (S), while filled circles represent Ni (Fe).

for transformation of Cubic FeS to mackinawite (see Fig. 6). This proposed transformation would therefore be much more rapid than the observed, spontaneous, transformation of Cubic FeS to mackinawite.

Spontaneous transformation of Cubic FeS to mackinawite (and of a wurtzite-type FeS to troilite) indicates that electrons surplus to eight valence electrons per formula unit are accommodated by structural rearrangements which increase electronic stability. In sphalerite (ZnS), structural stability is achieved because all low-energy molecular orbitals are occupied and a large energy gap is present between highest occupied and lowest unoccupied molecular orbitals (TOSSELL and VAUGHAN, 1992). Electronic structure models for the FeS phases Cubic FeS and mackinawite (WELZ and ROSENBERG, 1987) show that mackinawite has a reduced density of states at the Fermi level when compared with Cubic FeS. This increased band width of mackinawite is due to contributions from both Fe d and S p orbitals. Considerable Fe-Fe interaction between edge sharing FeS_4 tetrahedra in mackinawite causes a rearrangement and splitting of the Fe d-orbitals while the increased band width of S p orbitals is ascribed to strong deviation of the S environment from cubic to pyramidal symmetry, and to direct S-S contact between Fe-carrying sheets. These changes in electronic and crystal structure thus give mackinawite greater stability than Cubic FeS. For the purposes of electronic structure modeling, WELZ and ROSENBERG (1987) assumed stoichiometric compositions for both Cubic FeS and mackinawite.

BURDETT (1980) uses fragment formalism to view the structure of some simple ('AX') solids in terms of the packing of planar hexagonal sheets and their puckered analogues. The wurtzite structure, often regarded as a metastable version of the sphalerite arrangement, is, in general, found for four-coordinate structures with the largest electronegativity difference (Δ_χ) between A and X. Burdett finds that the energy difference between these two AX structures decreases as the electronegativity of X increases. As Δ_χ increases further, the rocksalt structure is produced, as shown by Pearson diagrams (PEARSON, 1962).

Solids having more than 8 valence electrons per AX unit adopt structures that may be understood via bond-breaking processes of the electron-precise wurtzite structure (BURDETT, 1980). Thus, Burdett argues that the NiAs-type structure can be generated by considering the effect of adding electrons to an 'AX' puckered sheet structure. Following these arguments presented by Burdett, we propose that introduction of additional electrons into the 8 electron structural framework of (hypothetical) 'wurtzite-type FeS ' is the driving force for transformation to a NiAs-type FeS.

Thus our model for iron monosulfide formation can be used to explain the origin of all three different FeS crystal structures (troilite, Cubic FeS, mackinawite) from Fe tetrahedrally coordinated by S. This proposed mechanism overcomes several difficulties. Previous studies have shown that the initial amorphous FeS precipitate formed on reaction of aqueous Fe(II) and dissolved sulfide ages to form mackinawite (BERNER, 1964; RICKARD, 1989), and from EXAFS evidence (PARISE and SCHOONEN, 1991; this study), we know that tetrahedral nucleation of Fe by S in reducing solutions develops rapidly. Hence, we consider it unlikely that direct troilite nucleation from solution species with Fe coordinated by 6 S atoms occurs. By showing that troilite could nucleate from solution via Fe tetrahedrally coordinated by S, we avoid the need for solution complexes, such as the trimeric Fe_3S_3 precursor proposed by TAYLOR (1980), as species required for troilite formation.

Our model may also be used to interpret the

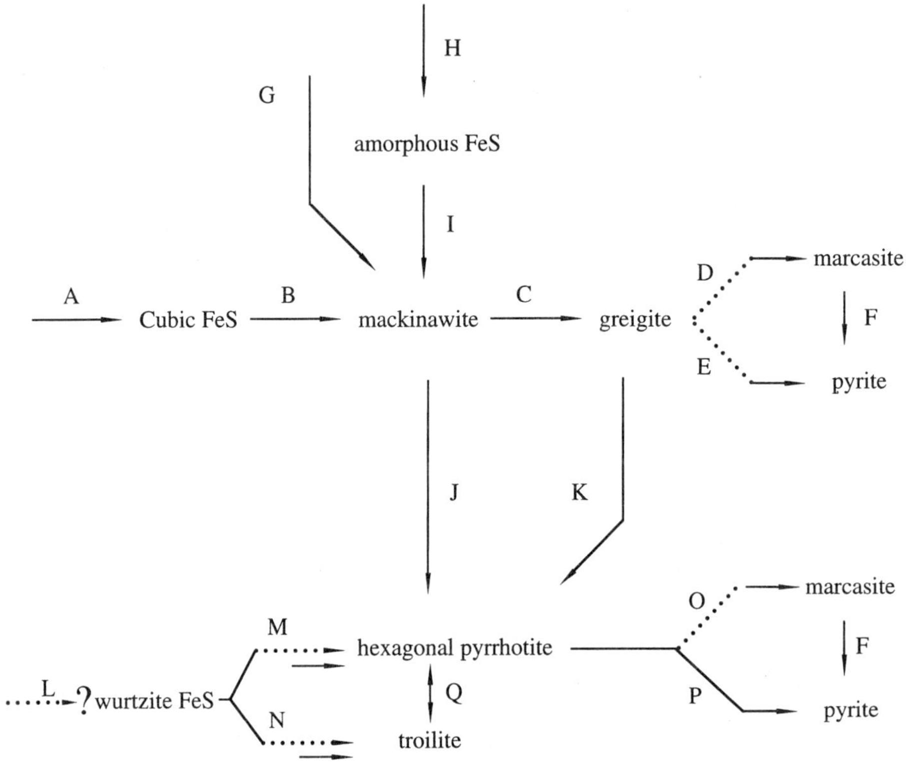

FIG. 7. Summary of known, and speculative phase transformations and formation reactions in the low temperature iron-sulfur system. Transformations that have not been directly demonstrated experimentally are shown by dotted lines. The key to the letters in this figure is given below: A. Formation from Fe^{2+} or Fe, and HS^- or H_2S. pH 4–5 (<368 K); MUROWCHICK and BARNES (1986a). B. Solid state transformation, kinetics only partially established, see SHOESMITH *et al.* (1980). C. Solid state transformation involving oxidation (of Fe), no kinetic data available; see TAYLOR *et al.* (1979b). D. Inferred reaction, by sulfidation of greigite, at pH<5, T<523 K; MUROWCHICK and BARNES (1986b). E. Inferred reaction, by sulfidation of greigite, at pH>5, T<523 K; MUROWCHICK and BARNES (1986b). For kinetics of pyrite formation from iron monosulfide precursors, see RICKARD (1975). F. Solid state transformation, for kinetics of this reactions, see LENNIE and VAUGHAN (1992), structural aspects described by FLEET (1970). G. Sulfidation of Fe, by HS^- or H_2S, pH in range 3–10 (T<523 K). H. Precipitation from Fe^{2+}, by HS^- or H_2S, pH in range 3–10 (T<523 K). I. Ageing of amorphous precipitate, see BERNER (1964). J. Solid state transformation; kinetics described in LENNIE *et al.* (1995a). K. Solid state transformation, kinetics not established. L. Precipitation from Fe^{2+}, by HS^- or H_2S, pH in range 4–6 wide range of temperatures, to above 573 K. Wurtzite-type FeS phase speculated, see this study. M & N. Formation of either hexagonal pyrrhotite (>413 K) M, or troilite (<413 K) N from Fe^{2+} and HS^- or H_2S between pH 4–5. A (speculated) transformation from the proposed wurtzite-type FeS may also form these phases. O. Marcasite after pyrrhotite, see studies by MUROWCHICK (1992), FLEET (1978). P. Pyrite formed by sulfidation of pyrrhotite, see YUND and HALL (1970) for solid state kinetic study. This reaction is rapid above 573 K. Q. Troilite-hexagonal pyrrhotite reversible phase transition, see KRUSE (1992), KING and PREWITT (1982), HORWOOD *et al.* (1976) and PUTNIS (1974) for further details on intermediate structures and transition mechanisms.

observation (SCHOONEN and BARNES, 1991c) that the conversion rate of mackinawite to hexagonal pyrrhotite is relatively slow in alkaline solutions at elevated temperatures (453–473 K) when compared to the conversion rate in acidic and neutral solutions. The rate of the solid-state transformation of mackinawite to pyrrhotite has been demonstrated to be slow below 530K (LENNIE *et al.*, 1995a) and, because it is a solid state transformation, unlikely to be affected by pH. Instead, we suggest that direct nucleation of hexagonal pyrrhotite from solution is enhanced at elevated temperatures under acidic conditions, competing with formation of mackinawite. In addition, under acid conditions, dissolution of mackinawite will be more rapid than that of pyrrhotite allowing pyrrhotite to concentrate relative to mackinawite in acid solutions.

Sulfidation of iron 'monosulfides' forming iron disulfides

The formation of pyrite and marcasite in sedimentary and hydrothermal systems provides an important control on Fe and S in the environment. It has been proposed that at temperatures below 523 K or so, the direct nucleation of iron disulfides from solution is extremely slow compared to formation as a result of sulfidation of the 'monosulfide' precursors (SCHOONEN and BARNES, 1991a,b,c).

Sulfidation of amorphous FeS by H_2S or HS^- proceeds at an insignificantly slow rate at 338 K (SCHOONEN and BARNES, 1991b). However, at or above 373 K, sulfidation by H_2S proceeds at measurable rates (DROBNER *et al.*, 1990; SCHOONEN and BARNES, 1991c; TAYLOR *et al.*, 1979a,b); reduction of H ions forming H_2 balances the redox requirements of this reaction. Sulfidation of iron 'monosulfides' proceeds rapidly in the presence of S as polysulfide and/or zerovalent S (RICKARD, 1975).

It is expected that mackinawite, being the predominant 'FeS' phase formed in reducing solutions, would provide facile topotactic sites for pyrite or marcasite nucleation. Because mackinawite is so easily oxidised to greigite, greigite may also form a suitable precursor for pyrite or marcasite nucleation. To form a small pyrite (or marcasite) nucleus of unit-cell dimensions for further crystal growth from the surface would require relatively few bonds of the greigite structure to be broken and reformed. Although the role of the crystal surface is thought to be of importance in these sulfidation reactions, the site(s) of nucleation of pyrite or marcasite crystallites on the iron monosulfide precursor remains unknown. Further progress in elucidating mechanisms of S addition to Fe monosulfides will depend on carefully designed surface science studies of these reactions.

CONCLUDING REMARKS

Experimental data reported here can be combined with other literature data to obtain a summary of known low temperature phase relations in the iron-sulfur system. This summary is presented in Fig. 7, with footnotes referencing the literature sources of the relevant information.

The sequence of sulfidation of iron 'monosulfides' is seen to follow two main paths, the first being that of sulfidation of iron sulfides with ccp S sub-lattices, leading to formation of pyrite and marcasite (the Cubic FeS-mackinawite-greigite-marcasite/pyrite sequence). The second is that of sulfidation of pyrrhotite, again leading to pyrite, marcasite.

For iron sulfides FeS_{1+x}, with $x<0.33$, the thermodynamic driving force is in the direction of formation of hcp S sub-lattices from ccp S sub-lattices, i.e. mackinawite to hexagonal pyrrhotite, greigite to ?hexagonal pyrrhotite (see Fig. 1). The predominant sulfidation sequence occurring in nature at low temperatures (<523 K), however, is the first sequence shown in Fig. 7. The abundant nucleation of pyrite or marcasite in sedimentary systems thus depends on metastable precursors which, in turn, are prevented by kinetic factors from transforming to stable pyrrhotites. It is noted that additional formation pathways resulting from dissolution or oxidation of iron sulfides, with subsequent precipitation of iron sulfide phases, are not addressed in Fig. 7.

This figure serves to show that the complex phase relations in the low temperature iron-sulfur system are now being clarified although further work is still required to clearly establish the relationship of smythite to other phases in this system. Our study demonstrates again the importance of metastable phases in the iron-sulfur system. Further studies are now required to clarify the role of surface structure and reactivity in formation and transformation of these phases.

Acknowledgements—The support of the Natural Environment Research Council is acknowledged. Thanks are due to Drs. J. M. Charnock, K.E.R. England, J. F. W. Mosselmans and P. F. Schofield for help with various aspects of the experimental work. Discussions with Professors George Helz and Jack Tossell were of great value in developing this work.

This paper is dedicated to the memory of Roger Burns who inspired and encouraged the efforts of both authors in studies of the crystal chemistry and geochemistry of iron sulfides.

REFERENCES

BERNER R. A. (1964) Iron sulfides formed from aqueous solution at low temperatures and atmospheric pressure. *J. Geol.* **72**, 293–306.

BINSTEAD N., CAMPBELL J. W., GURMAN S. J. and STEPHENSON P. C. (1991) 'SERC Daresbury Laboratory EXCURV92 program'.

BONEV I. K., KHRISCHEV KH. G., NEIKOV H. N. and GEORGIEV V. M. (1989) Mackinawite and greigite in iron sulphide concretions from Black Sea sediments. *C. R. Acad. Bulg. Sci.* **42**, 2, 97–100.

BROWN G. E. JR., CALAS G., WAYCHUNAS G. A. and PETIAU J. (1988) X-ray absorption spectroscopy and its applications in mineralogy and geochemistry. In *Spectroscopic Methods in Mineralogy and Geology* (ed. F. C. HAWTHORNE) *Rev. in Mineral. 18*, pp. 431–512. Mineralogical Society of America.

BUCKLEY A. N. and WOODS R. (1985) X-ray photoelectron spectroscopy of oxidised pyrrhotite surfaces. I. Exposure to air. *Appl. Surf. Sci.* **22/23**, 280–287.

BUCKLEY A. N. and WOODS R. (1987) The surface oxidation of pyrite. *Appl. Surf. Sci.* **27**, 437–452.

BURDETT J. K. (1980) How some simple solids hold together. The use of the fragment formalism in crystal chemistry. *J. Am. Chem. Soc.* **102**, 450–460.

CALAS G., BASSETT W. A., PETIAU J., STEINBERG D., TCHOUBAR D. and ZARKA A. (1984) Mineralogical applications of synchrotron radiation. *Phys. Chem. Miner.* **11**, 17–36.

CALAS G., MANCEAU A., COMBES J. M. and FARGES F. (1990) Application of EXAFS in mineralogy. In *Absorption spectroscopy in mineralogy* (eds. A. MOTTANA and F. BURRAGETO) pp. 172–205. Elsevier, Amsterdam.

CHASE M. W., JR., DAVIES C. A., DOWNEY J. R., JR., FRURIP D. J., MCDONALD R. A. and SYVERUD A. N. (1985) JANAF thermochemical tables. *J. Phys. Chem. Ref. Data* **14**, 1198–1199.

DAIKUN G. P., GREAVES G. N., HASNAIN S. S. and QUINN P. D. (1984) Daresbury Laboratory Technical Memorandum, DL/SCI/TM38.

DAVISON W. (1991) The solubility of iron sulphides in synthetic and natural waters at ambient temperature. *Aquatic Sciences* **53/4**, 309–329.

DE MEDICIS R. (1970a) Une nouvelle forme de sulfure de fer. *Rev. Chim. Miner.* **7**, 723–728.

DE MEDICIS R. (1970b) Cubic FeS, a metastable iron sulfide. *Science* **170**, 1191–1192.

DROBNER E., HUBER H., WACHTERHAUSER G., ROSE D. and STRETTER K. O. (1990) Pyrite formation linked with hydrogen evolution under anaerobic conditions. *Nature* **346**, 6286, 742–744.

FLEET M. E. (1970) Structural aspects of the marcasite-pyrite transformation. *Canad. Mineral.* **10**, 225–231.

FLEET M. E. (1978) The pyrrhotite-marcasite transformation. *Canad. Mineral.* **16**, 31–35.

FURUSETH S. and KJEKSHUS A. (1965) On the properties of α-MnS and MnS_2. *Acta Chem. Scand.* **19**, 1405–1410.

GRØNVOLD F. and STØLEN S. (1992) Thermodynamics of iron sulfides II. Heat capacity and thermodynamic properties of FeS and of $Fe_{0.875}S$ at temperatures from 298.15 K to 1000 K, of $Fe_{0.98}S$ from 298.15 K to 800 K, and of $Fe_{0.89}S$ from 298.15 K to about 650 K. Thermodynamics of formation. *J. Chem. Thermodyn.* **24**, 913–936.

GURMAN S. J., BINSTED N. and ROSS I. (1984) A rapid, exact curved-wave theory for EXAFS calculations. *J. Phys. C* **17**, 143–151.

GURMAN S. J., BINSTED N. and ROSS I. (1986) A rapid, exact curved-wave theory for EXAFS calculations: II. The multiple-scattering contributions. *J. Phys. C* **19**, 1845–1861.

HORWOOD J. L., TOWNSEND M. G. and WEBSTER A. H. (1976) Magnetic susceptibility of single-crystal $Fe_{1-x}S$. *J. Sol. St. Chem.* **17**, 35–42.

HYLAND M. H. and BANCROFT G. M. (1989) An XPS study of gold deposition at low temperatures on sulphide minerals: Reducing agents. *Geochim. Cosmochim. Acta* **53**, 367–372.

JOYNER R. W., MARTIN K. J. and MEEHAN P. (1987) Some applications of statistical tests in analysis of EXAFS and SEXAFS data. *J. Phys. Chem.* **20**, 4005–4012.

KING H. E. JR. and PREWITT C. T. (1982) High-pressure and high-temperature polymorphism of iron sulfide (FeS) *Acta Cryst. B* **38**, 1877–1887.

KISSIN S. A. and SCOTT S. D. (1982) Phase relations involving pyrrhotite below 350°C. *Econ. Geol.,* **77**. 1739–1754.

KJEKSHUS, A, NICHOLSON, D. G. and MUKHERJEE, A. D. (1972) On the bonding in tetragonal FeS. *Acta Chem. Scand.* **26**, 1105–1110.

KRUSE O. (1992) Phase transitions and kinetics in natural FeS measured by X-ray diffraction and Mössbauer spectroscopy at elevated temperatures. *Amer. Mineral.* **77**, 391–398.

LEE P. A. and PENDRY J. B. (1975) Theory of the extended X-ray absorption fine structure. *Phys. Rev. B* **11**, 2795–2811.

LENNIE A. R. (1994) Aspects of relations between metastable and stable phases in the iron-sulphur system. Unpublished PhD thesis, University of Manchester.

LENNIE A. R. and VAUGHAN D. J. (1992) Kinetics of the marcasite-pyrite transformation: An infrared spectroscopic study. *Amer. Mineral.* **77,** 1166–1171.

LENNIE A. R., ENGLAND K. E. R. and VAUGHAN D. J. (1995a) Transformation of synthetic mackinawite to hexagonal pyrrhotite: a kinetic study. *Amer. Mineral.* **80,** 960–967.

LENNIE A. R., REDFERN S. A. T., SCHOFIELD P. F. and VAUGHAN D. J. (1995b) Synthesis and Rietveld crystal structure refinement of mackinawite, tetragonal FeS. *Min. Mag.* **59,** 677–683.

MCINTYRE N. S. and ZETARUK D. G. (1977) X-ray photoelectron spectroscopic studies of iron oxides. *Anal. Chem.* **49**, 1521–1529.

MUROWCHICK J. B (1992) Marcasite inversion and the petrographic determination of pyrite ancestry. *Econ. Geol.* **87**, 1141–1452.

MUROWCHICK J. B. and BARNES H. L. (1986a) Formation of Cubic FeS. *Amer. Mineral.* **71**, 1243–1246.

MUROWCHICK J. B. and BARNES H. L. (1986b) Marcasite precipitation from hydrothermal solutions. *Geochim. Cosmochim. Acta* **50**, 2615–2629.

PARISE, J. B. and SCHOONEN, M. A. A. (1990) An extended X-ray absorption fine structure (EXAFS) spectrographic study of amorphous FeS. Geol. Soc. America Abst. Prog. 22, A293.

PEARSON W. B. (1962) Relative atomic size in semiconductor chemistry. *J. Phys Chem. Solids* **23**, 103–108.

POWER L. F. and FINE H. A. (1976) The iron-sulfur system. I. The structure and physical properties of the compounds of the low temperature phase fields. *Miner. Sci. Eng.* **8**, 106–128.

PRATT A. R., MUIR I. J. and NESBITT H. W. (1994) X-ray photoelectron and Auger electron spectroscopic studies of pyrrhotite and mechanism of air oxidation. *Geochim. Cosmochim. Acta* **58**, 827–841.

PUTNIS A. (1974) Electron-optical observations on the α-transformation in troilite. *Science* **186**, 439–440.

RICKARD D. T. (1975) Kinetics and mechanism of pyrite formation at low temperatures. *Amer. J. Sci.* **275**, 636–652.

RICKARD D. T. (1989) Experimental concentration time curves for the iron (II) sulfide precipitation process in aqueous solution and their interpretation. *Chem. Geol.* **78**, 315–324.

SCHOONEN M. A. A. and BARNES H. L. (1991a) Reactions forming pyrite and marcasite from solution: I. Nucleation of FeS_2 below 100°C. *Geochim. Cosmochim. Acta* **55**, 1495–1504.

SCHOONEN M. A. A. and BARNES H. L. (1991b) Reactions forming pyrite and marcasite from solution: II. Via FeS precursors below 100°C. *Geochim. Cosmochim. Acta* **55**, 1505–1514.

SCHOONEN M. A. A. and BARNES H. L. (1991c) Mechanisms of pyrite and marcasite formation from solution: III. Hydrothermal processes. *Geochim. Cosmochim. Acta* **55**, 3491–3504.

SHIRLEY D. A (1972) High-resolution X-ray photoemission spectrum of the valence bands of gold. *Phys. Rev. B* **5**, 4709–4714.

SHOESMITH D. W., TAYLOR P., BAILEY M. G. and OWEN D. G. (1980) The formation of ferrous monosulfide polymorphs during the corrosion of iron by aqueous hydrogen sulfide at 21 C. *J. Electrochem. Soc.* **127**,1007–1015.

TAKENO S., ZOKA H. and NIIHARA T. (1970) Metastable cubic iron sulfide -with special reference to mackinawite. *Amer. Mineral.* **55**, 1639–1649.

TAYLOR P. (1980) The stereochemistry of iron sulfides-a structural rationale for the crystallisation of some metastable phases from aqueous solution. *Amer. Mineral.* **65**, 1026–1030.

TAYLOR P., RUMMERY T. E. and OWEN D. G.(1979a) Reactions of iron monosulphide solids with aqueous hydrogen sulphide up to 160°C. *J. Inorg. Nucl. Chem.* **41**, 1683–1687.

TAYLOR P., RUMMERY T. E. and OWEN D. G.(1979b) On the conversion of mackinawite to greigite. *J. Inorg. Nucl. Chem.* **41**, 595–596.

TOSSELL J. A. and VAUGHAN D. J. (1992) *Theoretical geochemistry: applications of quantum mechanics in the earth and mineral sciences.* Oxford University Press.

TOSSELL J. A., VAUGHAN D. J. and BURDETT J. K. (1981) Pyrite, marcasite, and arsenopyrite type minerals: crystal chemical and structural principles. *Phys. Chem. Minerals* **7**, 177–184.

VAUGHAN D. J. and CRAIG J. R. (1978) *Mineral chemistry of metal sulfides.* Cambridge University Press.

VAUGHAN D. J. and LENNIE A. R. (1991) The iron sulphide minerals: their chemistry and role in nature. *Sci. Prog.* **75**, 371–388.

WARD J. C. (1970) The structure and properties of some iron sulfides. *Rev. Pure Appl. Chem.* **20**, 175–206.

WAYCHUNAS G. A. and BROWN G. E. JR. (1984) Applications of EXAFS and XANES spectroscopy to problems in mineralogy and geochemistry. In *EXAFS and Near Edge Structure III.* (ed. K. O. HODGSON), Springer Proc. Phys. **2**, 336- 342. Springer.

WELZ D. and ROSENBERG M. (1987) Electronic band structure of tetrahedral iron sulphides. *J. Phys. C* **20**, 3911–3924.

WERTHEIM G. K. and BUCHANAN D. N. E. (1977) Core-electron line shapes in x-ray photoemission spectra from semimetals and semiconductors. *Phys. Rev. B* **16**, 2613–2617.

YUND R. A. and HALL H. T. (1970) Kinetics and mechanism of pyrite exsolution from pyrrhotite. *J. Petrol.* **11**, 381–404.

Mineral Spectroscopy: A Tribute to Roger G. Burns

Editors: M. D. Dyar, C. McCammon and M. W. Schaefer

Application of XAFS spectroscopy to coal geochemistry

FRANK E. HUGGINS and GERALD P. HUFFMAN
Department of Chemical and Materials Engineering, University of Kentucky, Lexington, KY 40506, U.S.A.

Abstract—Elements in coal can be found in various geochemical forms (modes of occurrence) that vary from elements covalently bound to the organic matrix to elements that form discrete minerals. Such wide variation in the modes of occurrence presents a significant challenge to a full understanding of the geochemistry and utilization behavior of elements in coal. The application of X-ray absorption fine structure (XAFS) spectroscopy for determining elemental modes of occurrence in U.S. coals is reviewed. The elements discussed are from the first transition-series row of the periodic table (K–Br) and vary in concentration from major elements such as calcium to trace elements such as selenium, as well as varying from alkali metals to halogens in their geochemical behavior. Although more data are needed, this work establishes that many elements occur in coal in two or more distinct forms: one is basically mineralogic; the other is an organic association that involves direct bonding with the macerals, most commonly via oxygen functional groups. This latter mode of occurrence is more prevalent in low-rank coals, but persists for many elements in coals of rank at least as high as high-volatile B bituminous. Formation of discrete minerals appears to be relatively uncommon for first transition-series row elements of trace abundance; they are more likely to be incorporated in illite or pyrite, depending on whether they are lithophilic or chalcophilic.

INTRODUCTION

ROGER BURNS was largely responsible for demonstrating to the geoscience community the potential of spectroscopic techniques for resolving significant problems in mineralogy and geochemistry and for bringing such methods into the main-stream of research in Earth Sciences. The techniques he pioneered in the 1960s, electronic absorption spectroscopy (BURNS, 1965, 1970) and Mössbauer spectroscopy (in conjunction with G. M. Bancroft), provided mineralogists with the capability to focus on the behavior and properties of specific elements in minerals, a capability lacking in the traditional petrographic and crystallographic methods that had dominated mineralogy up to that time. In addition, this ability to focus on individual elements also forged a link to traditional geochemistry, and ever since the distinctions between the disciplines of crystallography, mineralogy, geochemistry, and petrography have been less apparent. In particular, Roger Burns was able to provide much valuable crystallochemical insight into the behavior of the specific elements in the first series of transition elements (Ti–Cu) in not only specific silicate and sulfide structures, but also in broader petrological and geochemical systems (BURNS, 1970).

Now, some twenty-five years later, spectroscopy-based papers make up significant, sometimes major portions of geoscience journals. Although the spectroscopic techniques and problems have changed to some extent, the goal of such investigations remains the same as that in much of Roger Burns' work: to understand better the behavior of specific elements in earth science materials. With this goal in mind, we present a review of results relevant to the geochemistry of first transition-series row (K–Br) elements in coal obtained principally from X-ray absorption fine structure (XAFS) spectroscopy. Application of this method for element speciation in coal was pioneered in the early 1980s (MAYLOTTE *et al.*, 1981a, 1981b; SANDSTROM *et al.*, 1982; WONG *et al.*, 1983; HUGGINS *et al.*, 1983b). More recently, the impetus for such investigations has come from research related to potential environmental issues arising from the behavior of minor and trace elements in coal combustion (SWAINE, 1990; CLARKE and SLOSS, 1992) that may be subject to regulation under Title III of the 1990 Clean Air Act Amendments (CAAA, 1990) in the U.S. and similar legislation in other countries.

Coal is a difficult material with which to investigate the occurrence of minor and trace elements because it is composed of two dissimilar major components: the organic fraction and the mineral fraction. Consequently, there is a variety of possible occurrences for how a given element may be found in coal. Furthermore, elements tend to be dispersed at relatively low concentrations in the organic fraction, but to be much more localized and concentrated in mineralogical occurrences. Hence, determining the mode of occurrence of a specific minor element in coal can be a challenging problem just because of the diversity of the possible modes of occurrence.

As discussed in detail elsewhere (HUGGINS *et al.*, 1994a; HUGGINS and HUFFMAN, 1996), there are a number of methods for determining the mode of occurrence of an element in coal. These vary from indirect methods based on how the element partitions in a particular test on the coal, such as float/sink testing, successive leaching by different solutions, or ashing at different temperatures, to direct microscopic or spectroscopic methods. The electron microscope and various microprobe methods are well suited for mineralogical occurrences in which the element is significantly enriched above its overall concentration in the coal, but are not so good for occurrences in which the trace element is widely distributed at relatively low concentrations throughout the organic matrix. Element-specific spectroscopic techniques, such as the synchrotron-based technique of XAFS spectroscopy, are not limited in this regard.

XAFS spectroscopy can provide information on specific elements in coal regardless of the mode of occurrence because it is more or less equally sensitive to both highly dispersed and discrete mineralogical occurrences of an element. Furthermore, it has excellent sensitivity, and is capable of providing useful information relevant to the mode of occurrence of elements in the middle of the periodic table (K–Mo) at concentrations as low as 5 ppm (HUGGINS *et al.*, 1993a). The main drawbacks of the technique are that (i) only a single spectrum is obtained that represents the weighted sum of all the element occurrences, and (ii) the need for a synchrotron as a high intensity source of X rays. In this review, we will summarize the geochemical implications of results obtained from XAFS spectroscopy applied to the determination of the mode of occurrence of specific elements in coal.

BACKGROUND ON MODES OF OCCURRENCE OF ELEMENTS IN COAL

As mentioned above, coal consists of two distinct components: the combustible organic matter and the noncombustible mineral matter. The organic fraction of coal is derived from plant residues accumulated in peat swamps that have undergone low-grade metamorphism (coalification). During this process, the plant-derived organic material is converted into carbon-rich entities that are known collectively as macerals (STACH *et al.*, 1975). Based on their optical properties, the macerals can be subdivided further into individual types, such as vitrinite, liptinite, fusinite, *etc.*; these components in coal are approximately analogous to the individual minerals in a rock. Elements can be covalently bound in the organic carbon matrix, in heterocyclic or other organic functional groups, or even as porphyrin structures (BONNETT *et al.*, 1987), or they may be present as ions associated with polar organic functional groups. In addition, coal usually contains a significant fraction of minerals, the so-called mineral matter or "ash" fraction, which generally consists of some combination of quartz, clays, carbonates, and sulfides (STACH *et al.*, 1975). A minor or trace element in coal may be associated with one or more of these major minerals or it may form a discrete accessory mineral (defined as a minor mineral which would not exist except for the essential presence of the minor or trace element). Finally, there may also be a significant aqueous phase in coal that is largely associated with the pores and capillaries in the macerals. Depending on rank, the moisture content of coal can vary from as much as 65% in brown coal to less than 1% in anthracites (STACH *et al.*, 1975). Often, such moisture will contain significant amounts of soluble elements. Hence, as summarized in Fig. 1, the forms or modes of occurrence of elements in coal can run the gamut from highly dispersed, covalently-bound elements in the macerals to substitutional replacement in minerals, or to hydrated cations dissolved in the coal moisture.

Not only is coal heterogeneous with respect to the types of elemental occurrences present in coal, it is also heterogeneous with respect to distribution of these entities in the seam in both a vertical and horizontal sense. A coal seam is basically a sedimentary formation, and in a vertical (time) direction, there are clearly distinguishable layers on a variety of scales, both macroscopic and microscopic, reflecting different ratios of the macerals relative to the minerals, variation in sulfur content, and so on (STACH *et al.*, 1975). The chemistry of the individual layers will change in a horizontal direction, reflecting changes in the environment of deposition (*e.g.*, whether freshwater or brackish) and changes in distances to the sources of the different components in the original peat. In addition, these layers may also thin out or thicken as the distances to these sources change. Furthermore, there can be differences across the seam in terms of rank, reflecting regional or sometimes local variation in the geological conditions. The net result of these trends is that coal can be a highly variable material and that each sample of coal tends to be unique. Hence, extrapolation from results on a given sample to broader local or even regional conclusions for the coal seam must be approached conservatively.

The modes of occurrence of an element in different coals can be expected to reflect differences in both the rank and chemistry of the coals. Although probably of lesser significance, factors prevalent at the time of formation of the original peat swamp, such as the type of peat deposit, climate, *etc.*, may also be reflected in the mode of occurrence. Another factor that can be important in determining the mode of occurrence of an element in coal is weathering of the coal as a result of its exposure to air at an outcrop. The weathering zone associated with an outcrop can extend many meters back into the seam and the mineral and maceral constitution of the coal in this zone may become highly altered. Such alteration may promote the formation of unusual elemental forms (HUGGINS and HUFFMAN, 1989). Finally, one further factor that can also influence the mode of occurrence of an element is alteration of the coal sample after collection. Exposure of a pulverized sample to air promotes both drying of the coal and oxidation of principally mineral forms, especially sulfide minerals. Both of these phenomena have been demonstrated to alter the modes of occurrence of certain elements such as arsenic (HUGGINS *et al.*, 1993a) and chlorine (HUGGINS and HUFFMAN, 1995).

EXPERIMENTAL OVERVIEW

XAFS spectroscopy

XAFS spectroscopy is basically a measurement of the variation (or fine structure) of the X-ray absorption coef-

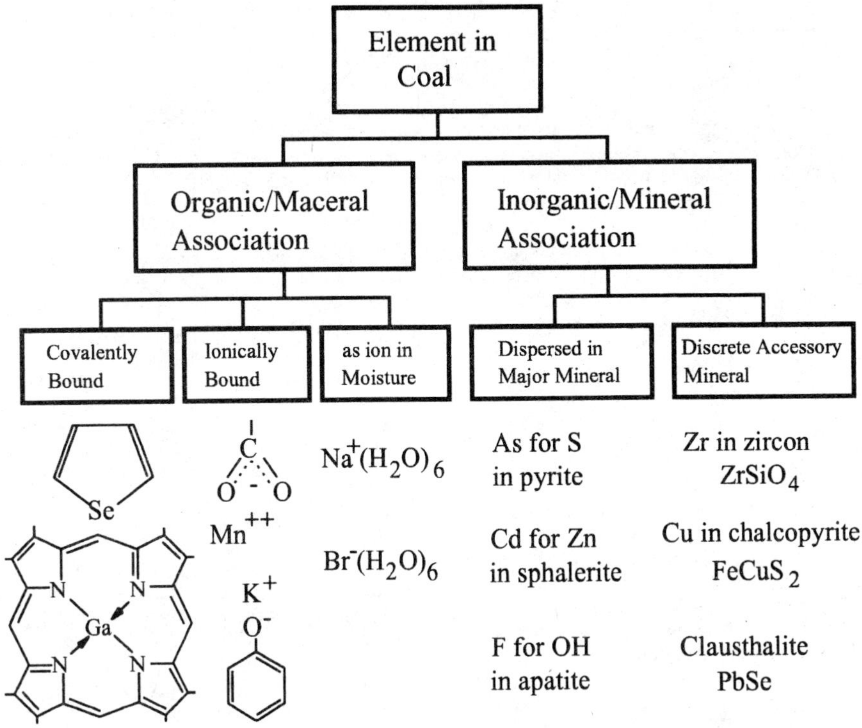

FIG. 1. Classification scheme for modes of occurrence of elements in coal.

ficient with energy associated with one of the characteristic absorption edges of the absorbing element. The XAFS spectrum is normally divided into two distinct regions for analysis: the X-ray absorption near-edge structure (XANES) region and the extended X-ray absorption fine structure (EXAFS) region. As these names imply, these regions incorporate the fine structure in the vicinity of the edge itself and further away from the edge, respectively (Fig. 2). Quite different analysis is done for each

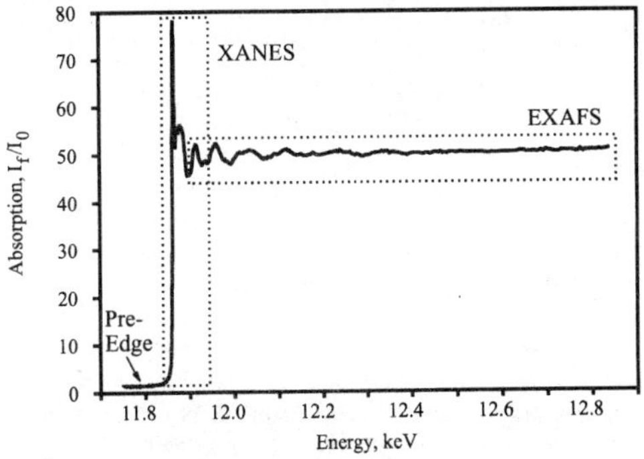

FIG. 2. Arsenic XAFS spectrum of arsenical pyrite, showing arbitrary division of spectrum into separate near-edge (XANES) and extended fine-structure (EXAFS) regions.

region of the spectrum. The XANES spectrum is used, without significant modification, as a ''fingerprint'' for the element in the material under investigation, whereas the EXAFS region can be extensively mathematically manipulated to obtain information on the short-range local structure around the absorbing atom or ion.

Owing to the low abundance of trace and minor elements, XAFS investigations of such elements in coal must be undertaken at a synchrotron radiation facility, where the X-ray flux is some 6 to 9 orders of magnitude greater than that available with a conventional laboratory X-ray tube. In the U.S., the National Synchrotron Light Source (NSLS), Brookhaven National Laboratory, NY, and the Stanford Synchrotron Radiation Laboratory (SSRL), Stanford University, CA are the best facilities currently available for XAFS spectroscopy of dilute species. All of the data discussed in this review were obtained at these two facilities.

Synchrotron radiation is produced by electron (or positron) storage rings in which highly energetic particle beams are ''stored'' in a stable orbit in an ultrahigh vacuum (10^{-9} torr). The particle beam in the storage ring emits synchrotron radiation as a consequence of the principle of conservation of momentum when the charged particles, moving at velocities very near to that of light, are forced to change direction through the application of a magnetic field normal to the path of the particle beam. This synchrotron radiation is a broad ''white'' radiation that is emitted tangentially to the orbit of the charged particles. It is then collimated by a series of slits and/or mirrors and delivered to an experimental beam-line station. For XAFS experiments, a monochromatic beam is required rather than white radiation, so a monochromator or grating is inserted into the white beam to select a specific energy that is determined by the spacing of the reflection planes of the monochromator crystal or grating, the angle of the monochromator with respect to the beam, and application of Bragg's law. There are usually additional slits and focussing mirrors between the monochromator and the exit of the beam in the experimental hutch.

To generate an XAFS spectrum, the absorption by the sample is measured as a function of the monochromator angle or, equivalently, the energy of the incident X-ray beam. Although there exists a quick scanning XAFS method (QEXAFS) that records the spectrum continuously as the monochromator rotates (FRAHM, 1992), most XAFS spectra are still recorded in a discrete manner by stepping across an energy interval that includes an X-ray absorption edge of the element of interest. Typically, the energy interval may start as much as 100–200 eV below the edge and extend as much as 1000–1500 eV above the edge. However, the stepping will not be uniform across the energy interval but will be subdivided into at least three regions (the pre-edge, XANES, and EXAFS regions; *q.v.* Fig. 2) so as to maximize the information quality and thereby facilitate the data analysis. For the pre-edge region (from −200 eV to around −20 eV below the edge), which contains no spectral information but has to be measured in order to determine the pre-edge slope, the stepping is generally coarse with a short time/step. The edge or XANES region, which for trace elements often contains much, if not all, of the useful information, is then measured very carefully and slowly (0.1–0.25 eV/step with a longer time/step) from about −20 eV below the edge to as much as +50 eV above the edge. The EXAFS region is then measured out to at least +250 eV up to as much as +1500 eV above the edge, depending on the strength of the EXAFS oscillations or chi signal. The stepping across the EXAFS region is usually now done in terms of reciprocal-space inverse distance (*e.g.*, 0.05 $Å^{-1}$/step) with the time per step determined by the actual interval in real space. This procedure gives a more uniform appearance to the chi vs. k spectrum, which, after application of a Fourier transform, results in a better quality radial structure function (RSF) spectrum.

The XAFS spectrum can be acquired by one of three different methods, depending on the concentration of the element under investigation. For standard samples or for samples in which the element of interest comprises more than about 5 wt% of a reasonably X-ray transparent material such as coal, the XAFS spectrum is best measured in absorption geometry (Fig. 3) by measurement of the X-ray intensity before and after absorption by the sample. Such measurements are made with simple ion chambers. The relationship between the X-ray intensity and the absorption coefficient ($\mu(E)$) is as follows:

$$I_t = I_0 \exp(-\mu t) \quad (1)$$

where I_0 is the intensity of the X-ray beam incident on the sample, I_t is the intensity of the X-ray beam after transmittance through the sample, and t is the thickness of the sample. The thickness of the sample is invariant; hence, any variation in ln (I_0/I_t) with energy is directly equivalent to variation in the absorption coefficient. For samples with elements of interest with concentrations ranging from 0.1 to 5 wt%, it is better practice to measure the spectrum in fluorescence geometry. In this experiment, the intensity of the X rays fluoresced by the sample in response to the absorption process is measured. Typically, a large solid angle ion chamber fluorescence detector (STERN and HEALD, 1979; LYTLE *et al.*, 1984) is located at 90° to the incident X-ray beam with the sample turned to 45° with respect to both the detector and the incident beam (Fig. 3). In this case, the relationship between the fluorescent X-ray intensity (I_f) and the absorption coefficient is as follows:

$$I_f = I_0\mu(E) \quad (2)$$

It should be noted that this equation is only exact in the limit that the sample is thin and/or the element of interest is sufficiently dilute not to give rise to self-absorption phenomena. However, if the element of interest is of trace abundance only (5–500 ppm), the background fluorescent X-ray intensity from other elements will overwhelm the fluorescence signal from the desired element. At this point, the ion chamber should be replaced by a thirteen-element, solid-state germanium detector that has been optimized for the measurement of the XAFS spectra of dilute elements in materials (CRAMER *et al.*, 1988). This type of detector counts X rays only in a narrow energy window set specifically for the X rays fluoresced by the element of interest and thereby rejects most of the X rays fluoresced by other elements. This signal selection process can greatly enhance the signal/noise ratio. However, care must be taken to avoid count-rate saturation effects with this type of detector when the signal is reasonably strong. This can be achieved by simply moving the detector away from the sample or by increasing the filter thickness. Use of the appropriate filters and Soller slits (STERN and HEALD, 1979) will also enhance the signal/noise ratio with either fluorescent detection method (Fig. 3). Finally, for exceedingly dilute concentrations (<100 ppm), the spectra can be repetitively scanned and added together to obtain a composite spectrum of better quality.

Analysis of experimental XAFS data consists of the following steps: (i) isolation of the EXAFS from the edge step, (ii) conversion of the energy scale to reciprocal space (k-space) inverse dimensions, generating the "chi vs. k" spectrum, and (iii) application of a Fourier transform to the chi spectrum to create a radial structure function (RSF) that describes the position and coordination number of atomic shells that surround the absorbing atom or ion. These techniques and procedures are well described in reviews and textbooks on XAFS spectroscopy (EISENBERGER and KINCAID, 1978; LEE *et al.*, 1981; KONINGSBERGER and PRINS, 1988; BROWN *et al.*, 1988). Unfortunately, analysis of the EXAFS region for trace elements in coal generally becomes uninformative as the element's concentration decreases below about 50 ppm because the relatively weak EXAFS oscillations get lost in the background noise. Hence, much reliance is placed on interpretation of the XANES region for elements in coal that are in the concentration range 5 to 50 ppm. Generally, the interpretation of elemental modes of occurrence from just the XANES spectrum relies primarily upon comparison with standards. However, more creative measures are often needed if the element occurs in two or more distinct forms. Such measures may also have to be applied because of natural-world phenomena such as diadochy (solid solution), polymorphism in mineral systems, small-particle phenomena, *etc.*, that can cause the appearance of the XANES spectrum to vary from that of a particular bulk standard. Specific measures may include (i) comparison of derivative XANES spectra; (ii) comparison with XANES spectra from different but chemically similar elements; (iii) subdivision

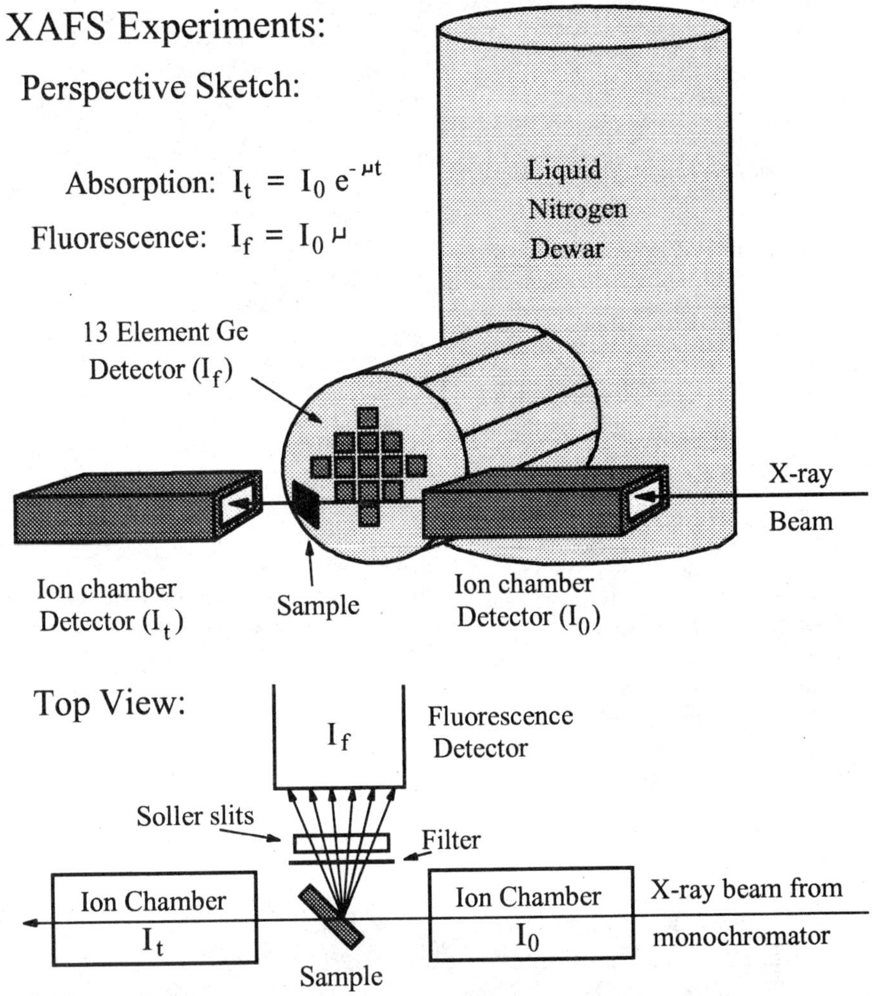

FIG. 3. Simplified schematic of experimental arrangement for X-ray absorption spectroscopy. Both absorption and fluorescent experimental geometries are indicated.

of the coal into different fractions to separate or concentrate different forms of occurrence; (iv) simulation of spectra from weighted additions of XANES spectra of standards and/or coals or coal fractions. Examples of these procedures applied to the interpretation of XANES spectra of minor and trace elements in coal are described elsewhere (HUGGINS and HUFFMAN, 1996).

Coal samples

Preparation of coal samples for XAFS analysis is not complicated except, as in any coal analysis, to ensure that the analyzed sample is representative of the bulk sample by following standard procedures for subdividing and pulverizing the collected sample (ASTM, 1976). At the synchrotron, the volume of coal sampled by the X-ray beam is estimated to be between about 0.01 and 0.1 cm^3, depending on the X-ray energy. To sample at least 10,000 particles in the minimum volume requires that the mean particle size of the coal be reduced to about 100 μm. Hence, size reduction of the divided sample to pass at least a -150 mesh sieve (100 μm top-size) should satisfy this requirement.

To measure the XAFS spectrum of a pulverized coal sample, the coal is generally suspended in the X-ray beam in a plastic bag made of ultrathin ($\approx$6 μm) polypropylene or in the form of a pellet made by pressing the coal into the surface of a boric acid (HBO_3) pellet of 2.5 cm diameter. Generally all XAFS spectra of elements in coal are measured in fluorescence geometry.

XAFS data for a wide variety of coal samples are to be discussed in this review and space only permits the briefest description of the coals to be made. Most of the coals originate from deposits in various U.S. states. Their rank varies from lignite (lig), through subbituminous (sbb), and the various subdivisions of bituminous coal: high volatile C, B, and A (hvCb, hvBb, hvAb), medium volatile (mvb), and low volatile (lvb), to anthracite (anthr). In this paper, only the seam name, state of origin, rank, and appropriate elemental concentration will be supplied. Further details can be found in the original publications. Table 1 lists the elements across the first transition-metal row of the periodic table and data on the range and average concentration of each element extracted from the compilation of over 12,000 U.S. coal analyses (COALQUAL database) recently published by the U.S. Geological Survey (BRAGG *et al.*, 1994).

MODES OF OCCURRENCE OF ELEMENTS IN COAL

Potassium

The concentration of potassium in U.S. coals is generally in the range 120 to 5,000 ppm (Table 1), with a tendency to be higher in high-rank (bituminous, anthracite) coals than in low-rank (lignite, subbituminous) coals. XAFS studies of potassium in bituminous coals (SPIRO *et al.*, 1986, HUFFMAN *et al.*, 1986) indicate that much, if not all, of the potassium in such coals is associated with the clay mineral, illite. Illite and the closely related mica minerals, muscovite and biotite, exhibit similar yet distinctive potassium XAFS spectra that are mirrored by potassium XAFS spectra of bituminous coals (Fig. 4). The spectrum of a Pennsylvania anthracite reported by SPIRO *et al.* (1986) is closely similar to the muscovite spectrum shown in Fig. 4. The small but prominent peak at about 12–15 eV above the potassium K-edge appears to reflect the crystallinity of the layer-silicate mineral and SPIRO *et al.* (1986) suggested that the rank of the coal might be estimated from this particular spectral feature.

The potassium XANES spectra of two low-rank coals from the western U.S. were found (SPIRO *et al.*, 1986) to be quite different from those of the bituminous coals. The reported spectra were very similar to that shown for the K-ion-exchanged lignite in Fig. 4. This particular spectrum was obtained from a lignite after exposure of it to a 0.2M solution of K^+ ions, which promotes exchange of the K^+ ions for carboxyl-bound Ca^{2+} and Na^+ cations originally present in the lignite. Hence, it is clear that potassium in certain low-rank coals can have a quite different mode of occurrence to that in bituminous coals. However, further work needs to be done to document the significance of this form of potassium in low-rank coals as illite and other K-bearing silicate minerals have been routinely noted as being present in coals of all ranks.

Calcium

Calcium is generally the most abundant non-acidic element in low-rank coals and may contrib-

Table 1. Average Concentration, 90% Range, and Maximum Concentration in ppm for First Transition Row Elements in U.S. Coals Extracted from Analyses Compiled in the NCRDS COALQUAL Database.*

Element	5% percentile	Mean (90% range)	95% percentile	Maximum
K	120	1450	5000	14000
Ca	420	3550	15000	71000
Sc	0.9	3.4	8	37
Ti	160	710	1800	9400
V	4.3	20	52	330
Cr	2.0	13	33	200
Mn	3.5	30	150	2500
Fe	1400	12000	40000	140000
Co	0.8	5.2	15	180
Ni	1.7	13	40	280
Cu	3.7	15	39	280
Zn	2.5	17	75	51000
Ga	1.25	5.2	12.5	41
Ge	0.22	4.3	20	220
As	0.7	16	100	2200
Se	0.4	2.6	7.5	150
Br	1.0	16	58	160

*Source of data: BRAGG *et al.* (1994).
Number of coal analyses included in range and average calculations varies from a minimum of 4,855 for bromine to over 7,400.

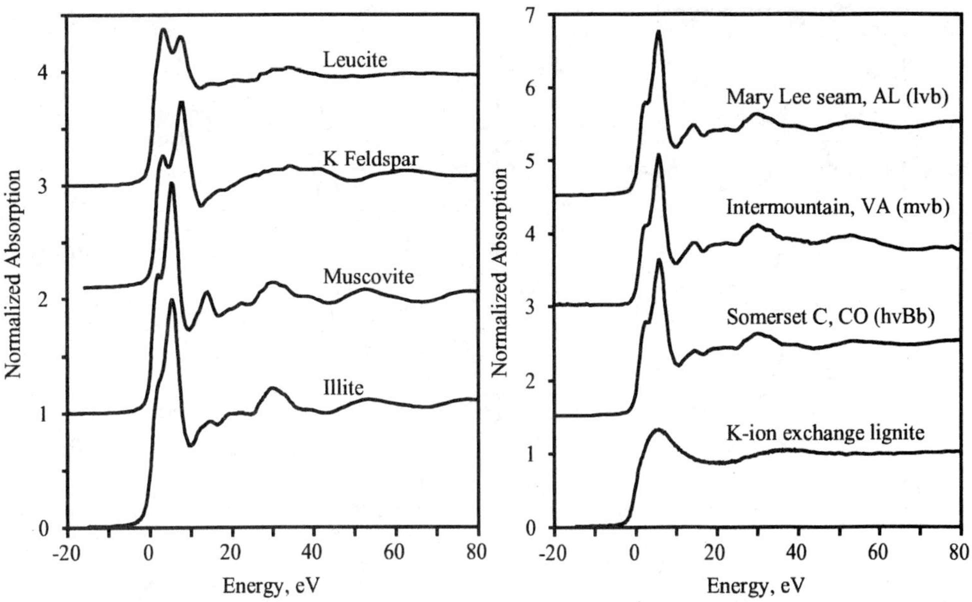

FIG. 4. Potassium XANES spectra of various mineral standards and of coals of different rank. Zero point of energy for potassium K absorption edge is 3,608.4 eV.

ute as much as 40% by weight (as oxide) to the ash of a lignite or a subbituminous coal. In high-rank coals, calcium is often much less significant. Hence, the concentration range for calcium in U.S. coals is broad, varying from as little as 400 ppm to as much as 15,000 ppm. The XAFS spectra of calcium in coal show significant variation with rank. As shown in Fig. 5, low-rank coals exhibit calcium XAFS spectra that are distinctive from those of bituminous coals. Such spectra are similar to the spectra of calcium salts of carboxylic acids, (*q.v.* the spectrum for Ca oxalate in Fig. 5) whereas spectra of high-rank bituminous coals show evidence for calcium as carbonate (HUGGINS *et al.*, 1983b; HUFFMAN and HUGGINS, 1984). Coals of intermediate rank (*e.g.* Illinois #6 in Fig. 5) can often be simulated as the sum of a carbonate component and a carboxyl component (HUFFMAN and HUGGINS, 1984). From analysis of the EXAFS region, the Ca-O bond distance and coordination number for the carboxyl-bound calcium were established as 2.40 ± 0.02 Å and 6, respectively, and it was speculated that at least some of the molecules around carboxyl-bound calcium were water molecules (HUGGINS *et al.*, 1983b).

It has also been demonstrated (HUGGINS *et al.*, 1983a) that extremely weathered bituminous coals can become significantly enriched in calcium, much of which appears to be present as carboxyl-bound calcium, based on the similarity of the Ca XANES spectrum of the outcrop sample from the Lower Kittanning seam, PA, shown in Fig. 6, in comparison to those shown in Fig. 5 for low-rank coals and Ca carboxylate standards (*e.g.*, the spectrum for Ca oxalate). This carboxyl-bound calcium form in weathered bituminous coal may be derived partly from dissolution of calcite in the original coal, but it is also likely that Ca^{2+} ions in groundwater are captured by ion-exchange processes once carboxyl sites are formed in the weathered coal macerals.

Titanium

As a result of concerns expressed in the late 1970s regarding the possible poisoning of catalysts by titanium during coal liquefaction, XAFS investigations at the titanium K-edge (SANDSTROM *et al.* 1982; and WONG *et al.* 1983) were among the first to be carried out on coal. These early studies indicated the presence of two different major titanium forms in coal, one mineralogical, the other associated with the organic fraction of the coal. The more recent work to be discussed here is largely in agreement with such findings and has resolved the mineralogical forms better.

Titanium is a significant minor element in coal and is usually present in the range 150 to 2,000

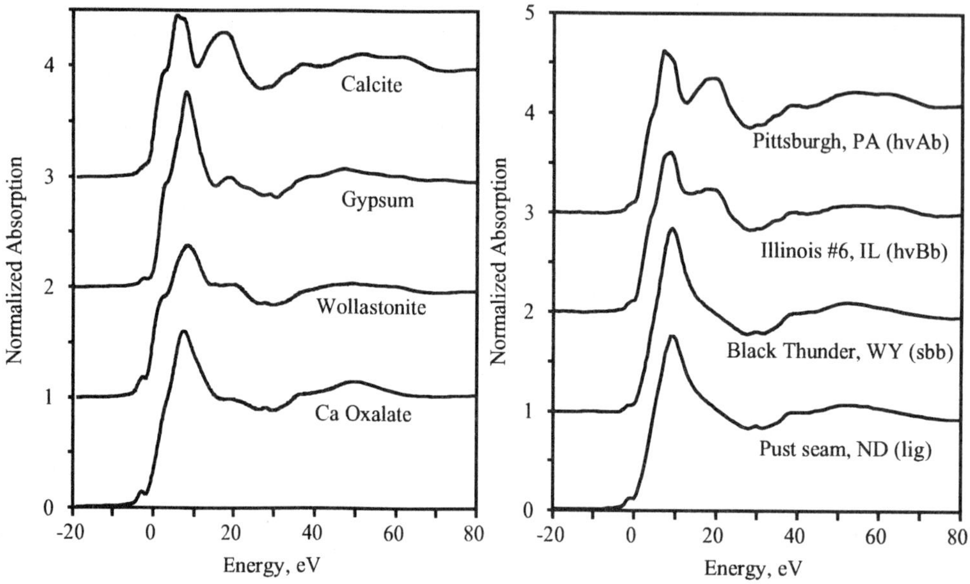

FIG. 5. Calcium XANES spectra of various standards and of four coals of different rank. Zero point of energy for calcium K absorption edge is 4,038.5 eV.

ppm in U.S. coals (SWAINE, 1990; BRAGG *et al.*, 1994). The XANES spectra of titanium in various coals are shown in Fig. 7. As can be seen by comparing these spectra with the spectra of titanium minerals and compounds shown in the same figure and elsewhere (WAYCHUNAS, 1987), there is no convincing match between any one coal and any single standard. However, as shown elsewhere (HUGGINS and HUFFMAN, 1996), it is possible to simulate the spectra of the two highest-rank coals (Lewiston-Stockton, WV, and Pocahontas No. 3, VA) as the weighted sum of Ti in the form of illite, rutile, and/or anatase. Further, the Ti XANES spectra of the lower rank coals are all somewhat similar to each other, and to the spectra of the float (reduced mineral content) fractions from cleaning tests on the Kentucky #9 coal shown in Fig. 8. The spectra of the corresponding tailings (mostly mineral matter) fractions appear to derive principally from titanium in illite. Hence, it can be concluded that there is a significant non-mineralogical mode of occurrence for titanium in many coals in addition to various mineral forms such as Ti in illite and the Ti oxides, rutile and anatase. Such conclusions are consistent with findings from electron microscopy investigations (FINKELMAN and STANTON, 1978; FINKELMAN, 1988), which showed that the occurrence of mineralogical Ti forms could not account for all of the titanium present in coal.

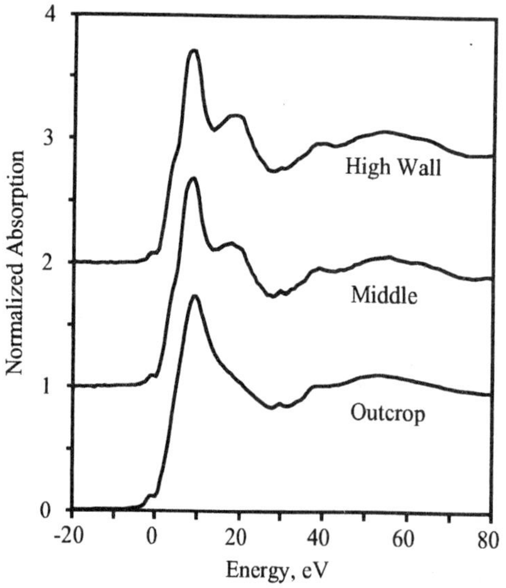

FIG. 6. Calcium XANES spectra of weathered coal samples from an outcrop on the Lower Kittanning seam in western Pennsylvania. Degree of weathering increases from the high wall to the outcrop.

Some information on the nature of the organically associated titanium in coal can be deduced both from the appearance of the XANES spectra and from analysis of the EXAFS regions. SANDSTROM *et al.* (1982) and WONG *et al.* (1983) noted

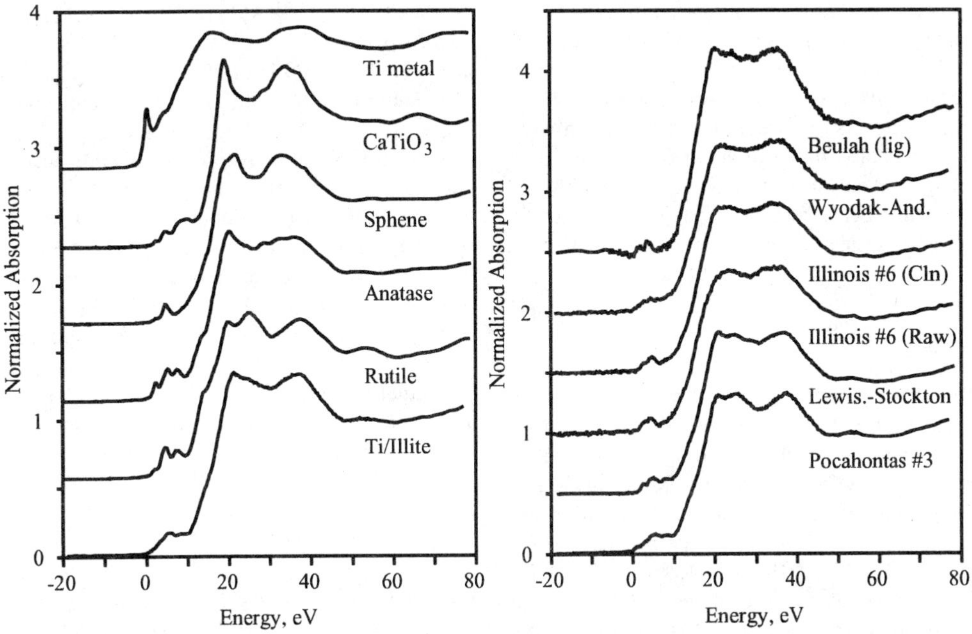

FIG. 7. Titanium XANES spectra of various standards and of five coals of different rank. Rank increases from top to bottom. Zero point of energy for titanium K absorption edge is 4,966 eV.

the close correspondence of the titanium XANES spectrum of the organically associated titanium with that of titanium alkoxides and closely related organo-Ti (''Tyzor'') compounds, respectively. In such compounds, the nearest neighbor shell to the titanium consists of six oxygen atoms. Further-

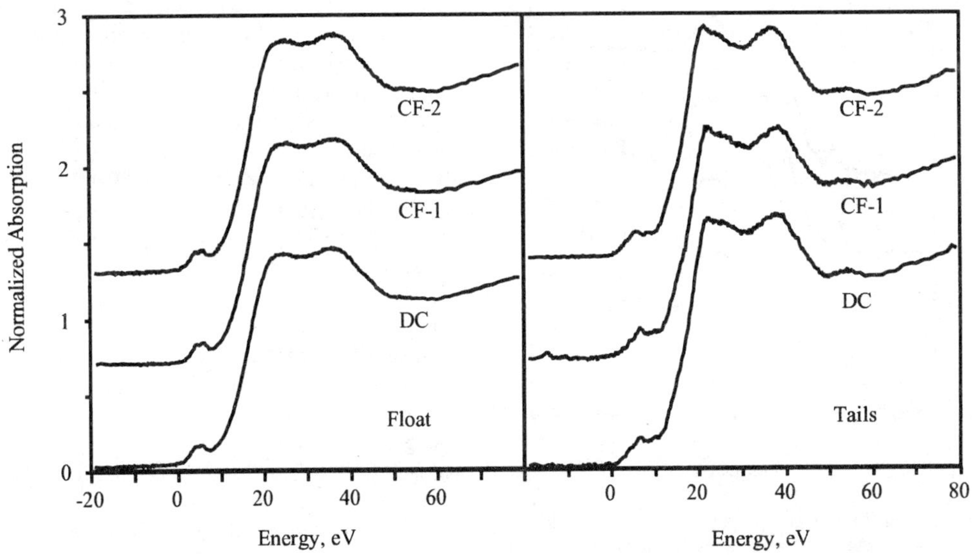

FIG. 8. Titanium XANES spectra of float and tailings fractions from a sample of Kentucky #9 coal subject to cleaning by Denver cell flotation (DC) and by column flotation (CF).

more, the lack of any significant enhancement of the pre-edge peak (*e.g.* spectra of float fractions in Fig. 8) indicates that the titanium site in coal is reasonably symmetric (WAYCHUNAS, 1987). Hence, it appears likely that the organically associated Ti in coal is found in octahedral coordination by oxygen. Analysis (HUGGINS *et al.*, unpublished work) of the EXAFS region of Ti in the Kentucky #9 float fractions is also compatible with such findings.

Vanadium

Vanadium rarely exceeds 50 ppm in U.S. coal samples (Table 1), although local concentrations within seams at specific bedding horizons have been reported to be as high as 2,000 ppm. XAFS investigations of vanadium in one high vanadium bedding horizon in the Kentucky #9 coal (MAYLOTTE *et al.*, 1981a, 1981b; WONG *et al.*, 1983) and more recently from a whole seam sample of the same coal (HUGGINS *et al.*, 1995) indicate that there are two major vanadium forms present in this coal. As was observed for titanium, the vanadium XAFS spectra of float fractions are quite different from those of the tailings fractions (Fig. 9) and indicate the presence of both mineralogical and organically associated vanadium in this coal. Detailed analysis (MAYLOTTE *et al.*, 1981b; WONG *et al.*, 1983) of the EXAFS region of the vanadium XAFS spectrum from a bedding horizon at the top of the Kentucky #9 seam that was highly enriched in vanadium (≈1800 ppm) indicated that the organically associated vanadium was present as V^{4+} in a pyramidal or highly distorted octahedral structure with one short V—O bond at about 1.6 Å and 4 or 5 longer V—O bonds at about 2.0 Å. The XAFS spectrum for the mineralogical form was shown to be similar to that of the vanadium-rich muscovite, roscoelite (MAYLOTTE *et al.*, 1981b; WONG *et al.*, 1983). In this mineral form, the vanadium oxidation state is V^{3+}. However, it is more likely that the actual mineral in coal is illite, as this mineral appears to accumulate minor lithophile elements in Kentucky #9 coal (HUGGINS *et al.*, 1995).

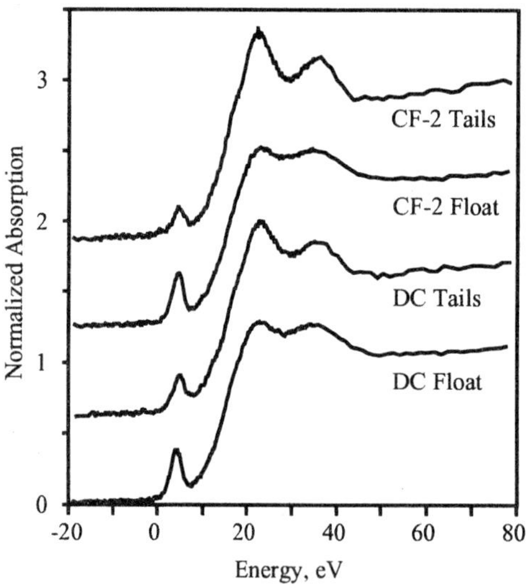

FIG. 9. Vanadium XANES spectra of float and tailings fractions (see Fig. 8 caption for identification) from a sample of Kentucky #9 coal subjected to cleaning by flotation methods. Zero point of energy for vanadium K absorption edge is 5,465 eV.

Chromium

The average concentration of chromium in U.S. coals is less than 15 ppm, with low-rank coals tending to have lower concentrations than higher-rank coals (BRAGG *et al.*, 1994). Compared to many other elements, there appears to be relatively little tendency for chromium to be found in anomalously high concentrations. Hence, as indicated in Table 1, its concentration range in U.S. coals is relatively small: 2 to 33 ppm. Despite this low abundance, chromium is an element of significant interest and concern during coal utilization because it is listed as one of the eleven ''air-toxic'' elements in the 1990 Clean Air Act Amendments (CAAA, 1990). The reason it appears on this list is because of the toxic and carcinogenic properties of the Cr(VI) oxidation state; the Cr(III) oxidation state poses no threat to human health at such low concentrations.

The pre-edge peak in chromium XAFS spectra is highly diagnostic for distinguishing between the Cr^{3+} and Cr(VI) oxidation states, because we need only consider Cr coordinated by oxygen in naturally occurring minerals (BURNS and BURNS, 1975). Owing to the variation of the crystal-field stabilization energy of Cr^{3+} in different coordination symmetries (BURNS, 1970, 1975), Cr^{3+} is found exclusively in octahedral coordination, whereas Cr(VI) is found exclusively in tetrahedral coordination in chromate, CrO_4^{2-}, or dichromate, $Cr_2O_7^{2-}$, oxoanions in terrestrial minerals (BURNS and BURNS, 1975). As shown in Fig. 10, the height of the chromium pre-edge peak amounts to no more than 5% of the edge-step for Cr^{3+}, but as much as 95% for Cr(VI). Hence, often merely by inspection of the Cr XANES spectra, it can be concluded whether or not a significant fraction of Cr(VI) is present. None of the coals examined has shown evidence of Cr(VI).

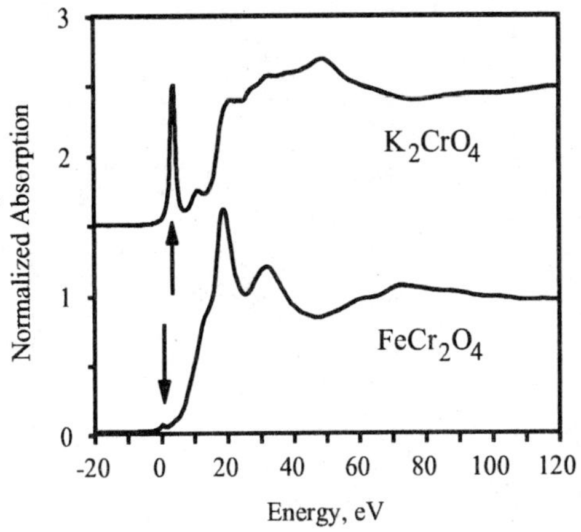

FIG. 10. Chromium XANES spectra of Cr^{3+} in chromite and of Cr(VI) in K_2CrO_4. The arrows indicate the contrast in intensity of the 1s-3d pre-edge feature at about 2–5 eV for the two oxidation states. Zero point of energy for chromium K absorption edge is 5,989 eV.

The chromium XANES spectra of coals tend to show relatively minor variations that are difficult to quantify. However, as discussed elsewhere (HUGGINS and HUFFMAN, 1996), it is easier to draw distinctions from derivative XANES spectra than from the spectra themselves. As was found for titanium and vanadium, there appear to be two major forms of chromium present in coal: one mineral, the other organic in association. The mineral form most often recognized from the XANES spectrum of a coal, especially tailings fractions, is a Cr-bearing illite. The spectra of the organically associated form most closely resemble that of an amorphous or small-particle chromium oxyhydroxide (HUGGINS *et al.*, 1993a), but the match is not exact. Well crystallized oxyhydroxide (*e.g.*, grimaldiite) and oxide (*e.g.*, chromite, eskolaite) minerals and other Cr^{3+} standards, however, do not match at all well. Hence, it is likely that the organically associated chromium consists of small clusters of $Cr(O,OH)_6$ units that also involve bonding to the maceral surface by oxygen functional groups.

The lack of observation of chromite by XAFS spectroscopy deserves comment in view of the fact that this mineral is the only chromium mineral that has been positively identified in some coals (SWAINE, 1990; FINKELMAN, 1994; BROWNFIELD *et al.*, 1995). However, well characterized chromite occurrences have been reported only for coals with anomalously high chromium contents ($\gg$100 ppm), whereas the XAFS measurements have been made largely on coals with typical chromium contents that fall within the 90% range shown in Table 1. Furthermore, as has been well documented by BROWNFIELD *et al.* (1995), the chromite in highly enriched chromium coals appears to be detrital and its origin can be traced to nearby weathered ultramafic (ophiolite) deposits. Hence, for the typical coal, uninfluenced by ultramafics, it would appear that detrital chromite is normally insignificant, based on its lack of observation in XAFS spectra, and that the chromium is introduced by different processes, such as ion-exchange or by incorporation in clay minerals.

Manganese

The concentration of manganese in U.S. coals is generally in the range 5–150 ppm (Table 1). As established by XAFS studies (HUGGINS *et al.*, 1993b; HUGGINS and HUFFMAN, 1996), the geochemical behavior of manganese in coal is similar to that of calcium: it is principally present in low-rank coals as carboxyl-bound Mn^{2+} and in high-rank coals as Mn^{2+} in carbonate minerals. However, the manganese XANES spectra of some bituminous coals (Kentucky #9, Upper Freeport) can not be explained just by these two manganese forms. At least one other major manganese form, (for which there is no corresponding calcium form), must be present in these two coals. From the XANES systematics, this form is not a sulfide form, since oxygen is indicated as the next nearest neighbor to manganese, and our most recent work (HUGGINS *et al.*, unpublished data) suggests that this form arises from Mn in illite.

Iron

Despite the importance of iron minerals in many utilization problems arising from coal, such as acid-mine drainage, slagging and fouling in coal combustion, acid rain, *etc.*, there have been very few investigations of coal using Fe K-edge XAFS spectroscopy. The main reason for this is that ^{57}Fe Mössbauer spectroscopy is able to provide much the same information, but with considerably better precision and certainty of identification in most instances. Mössbauer investigations of coal have shown that iron is present in coal most commonly as pyrite, but if the sulfur content is low, then other minerals, such as iron-bearing clays (illite or chlorite) and siderite ($FeCO_3$) are also present. Furthermore, a wide variety of iron oxyhydroxide and sulfate minerals derived from pyrite can be present in oxidized or weathered coal (HUGGINS and HUFFMAN, 1989). Often these minerals can be identified and determined with a reasonable precision (±3% of the iron) from the parameters and relative areas of quadrupole or magnetic components in Mössbauer spectra, especially if spectra are also collected at cryogenic temperatures to supplement the room-temperature spectrum.

Experience with attempting to extract quantitative information about the relative contributions of individual forms to XANES spectra of coal, however, suggests that two distinct forms can be determined with a precision of typically no better than ±5 to 10%. For three contributions, the precision may degrade to perhaps as much as ±20%, and any attempt to extract more than three contributions is generally an exercise in futility. The reason for this difference in deconvolution of XAFS spectra compared to Mössbauer spectra is that the fitting of Mössbauer spectra is based on a well defined generic peak-shape (*viz.*, lorentzian), whereas the fitting of XANES spectra is based on combining unique standard spectra to match a unique unknown spectrum. The XANES spectrum of a material may show significant variation, due to compositional effects, particle-size effects, sample thickness effects, *etc.*, which contributes to uncertainty in fitting standard spectra to XANES spectra. Furthermore, whereas the Mössbauer spectrum itself will generally indicate the number of discrete contributions to the spectrum by the number of discernable peaks and shoulders, there is generally no way of knowing, *a priori*, how many forms of an element contribute to an XAFS spectrum.

For these reasons, Mössbauer spectroscopy is a much better technique for extracting information about iron minerals from spectra of coal. The only exception might be if iron in a particular coal sample was of very low concentration (<1,000 ppm). In such a situation, the much better sensitivity of XAFS for trace quantities of an element in coal would compensate for the poorer precision of the data reduction.

Nickel

Nickel is one of the more difficult elements in coal to examine by XAFS spectroscopy. The combination of relatively low concentrations (typically less than 40 ppm) and a strong background in the nickel XAFS spectrum because of fluorescence from the abundant iron in coal makes for relatively noisy spectra and ambiguous interpretation. The XANES spectra for nickel standards shown in Fig. 11 indicate that it is possible to discriminate between nickel in an oxygen nearest-neighbor environment and nickel in sulfide environment merely by inspection. Examination of nickel XAFS spectra of bituminous coals suggests that nickel may be found in coal in both oxidic and sulfidic environments. The Ni XANES spectrum for the Illinois #6 coal shown in Fig. 12 appears to be more like that expected from a sulfide, whereas the spectrum for the Kentucky #9 tailings, with its much more prominent white line, is more consistent with oxygen as the nearest neighbor. The float fraction from the Kentucky #9 appears to be intermediate. It should be noted that a number of other elements (iron, zinc, *etc.*) have been found in both sulfidic and oxidic environments in the same coal.

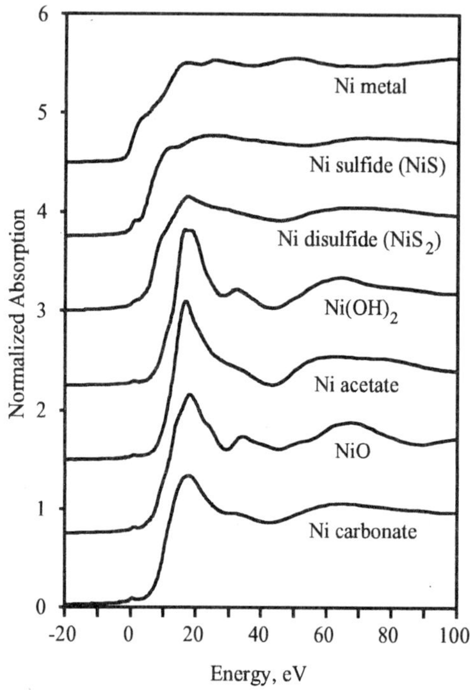

FIG. 11. Nickel XANES spectra of various nickel standard reference materials. Zero point of energy for nickel K absorption edge is 8,333 eV.

The nickel XANES spectrum of the Illinois #6 coal does not match either of the specific nickel sulfide phases shown in Fig. 11, and it has been tentatively proposed that the spectrum arises from nickel substituting for iron in pyrite (HUGGINS *et al.*, 1993b, 1994b). Such an occurrence would be similar to that proposed for nickel in U.K. bituminous coals based on synchrotron microprobe findings (WHITE *et al.*, 1989; SPEARS, 1991). Additional and better data are needed before the specific nickel form or forms present in the Kentucky #9 coal can be identified.

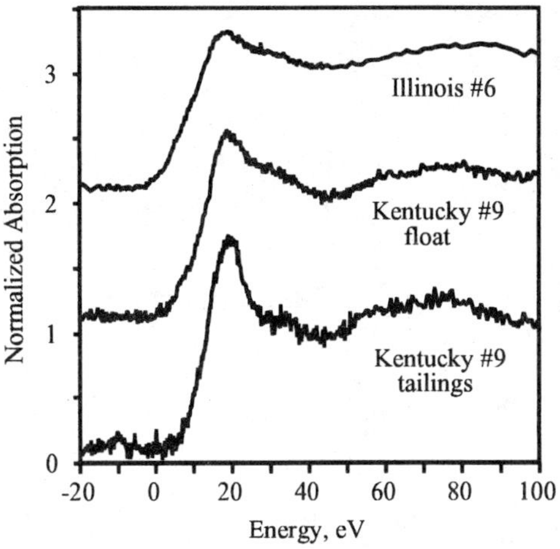

FIG. 12. Nickel XANES spectra for Illinois #6 coal and for Kentucky #9 coal fractions from Denver cell flotation tests.

Zinc

An XAFS spectroscopy survey of zinc occurrences in various U.S. coals (HUGGINS and HUFFMAN, 1996) indicates that Illinois coals that are significantly enriched in zinc (>100 ppm) appear to contain predominantly, if not exclusively, zinc sulfide. This is true even for Illinois basin coals that are relatively low in sulfur (<2.0 wt% total sulfur). Such findings are consistent with previous mineralogical investigations (*e.g.*, HATCH *et al.*, 1976), which have reported significant concentrations of zinc (as much as 5,000 ppm or more on a whole coal basis) occurring as epigenetic sphalerite in Illinois basin coals. However, other bituminous coals from different coal regions in the U.S. with much lower and perhaps more typical zinc abundances (6–20 ppm) appear to exhibit a mixture of both sulfidic and oxidic forms, whereas a subbituminous coal (Wyodak, WY) appears devoid of zinc sulfide and may have most of its zinc present as carboxyl-bound cations, based on the similarity of the spectral profile to that for carboxyl-bound calcium or manganese.

Arsenic

Investigations into the mode of occurrence of arsenic in bituminous coals have generally reported finding this element to be associated with sulfide minerals, especially pyrite (SWAINE, 1990; FINKELMAN, 1994), as might be expected given its chalcophilic nature. XAFS data for arsenic in unoxidized coals (HUGGINS *et al.*, 1993a; HUFFMAN *et al.*, 1994) are generally in agreement with such findings. However, the same XAFS investigations have also shown that significant fractions of arsenic may be present in arsenate mineral form, if the coal has not been protected from oxidation, or if the XAFS examination is not carried out within a *few days* of sample pulverization. In fact, as much as 50% of the arsenic associated with pyrite in a coal may be oxidized to arsenate within six months.

One of the best demonstrations of the power of XAFS spectroscopy for mode of occurrence determinations is the fact that it has proved possible to discriminate between arsenic in the pyrite structure and arsenic in the mineral arsenopyrite (FeAsS), despite the close relationship that exists between the two structures (WYCKOFF, 1965). However, the EXAFS region shows (Fig. 13) clear distinctions between the two structures and reveals the presence of a peak in the XANES region that can also be used to discriminate between the different arsenic forms by inspection (Fig. 14). Virtually all bituminous coals examined by arsenic XAFS appear to contain mostly arsenical pyrite, rather than arsenopyrite.

In North American coals, however, arsenic is generally much less abundant in low rank than in bituminous coals (BRAGG *et al.* 1994). Furthermore, the XANES spectra of arsenic in low-rank coals are quite different from those of arsenic in unoxidized bituminous coals (Fig. 14). Based on the close similarity with the spectrum of arsenic in a lignite in which arsenic had been greatly enriched by ion-exchange procedures, it would appear that the arsenic in low-rank coals may be present principally as As^{3+} at maceral carboxyl sites (HUGGINS *et al.*, 1996).

Selenium

Selenium in U.S. coals rarely exceeds 5 ppm (BRAGG *et al.*, 1994). As a result, relatively little XAFS characterization of this element in coal has been attempted. XAFS spectra at the selenium K-edge have been recorded only from fresh and oxidized samples of two bituminous coals (HUGGINS and HUFFMAN, 1996), both of which contain above average concentrations of selenium. Based on the white-line positions in the selenium XANES spectra of the coals (Fig. 15), which are slightly positive (<1 eV) with respect to elemental selenium, it would appear that the selenium may well be present largely in an organoselenium form. This is consistent with previous deliberations on the mode of occurrence of selenium in coal (GOODARZI, 1988; SWAINE, 1990; FINKELMAN, 1994). There is some

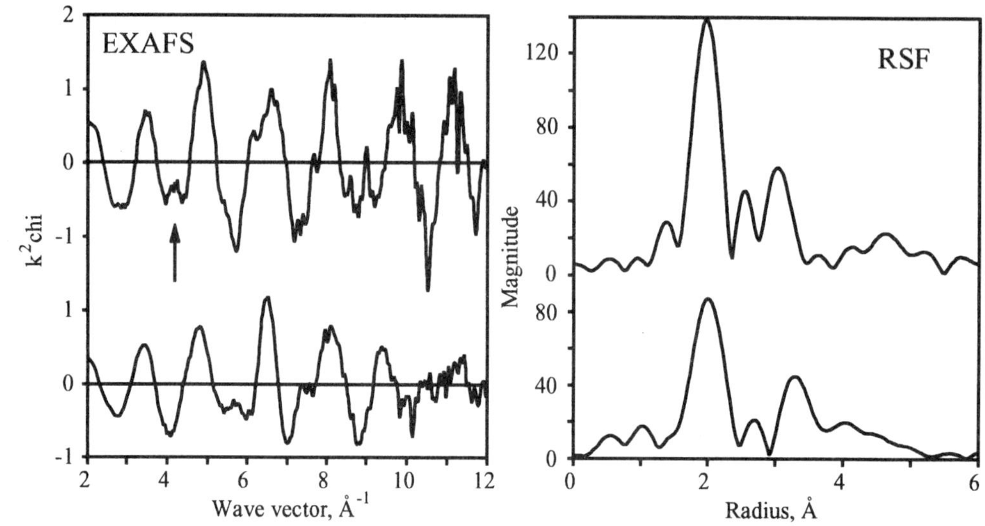

FIG. 13. Comparison of the isolated arsenic EXAFS regions and Fourier transforms (RSF) for arsenical pyrite (top) and arsenopyrite (bottom). The arrow indicates a small peak in the XANES that can be used to discriminate between arsenical pyrite and arsenopyrite by inspection.

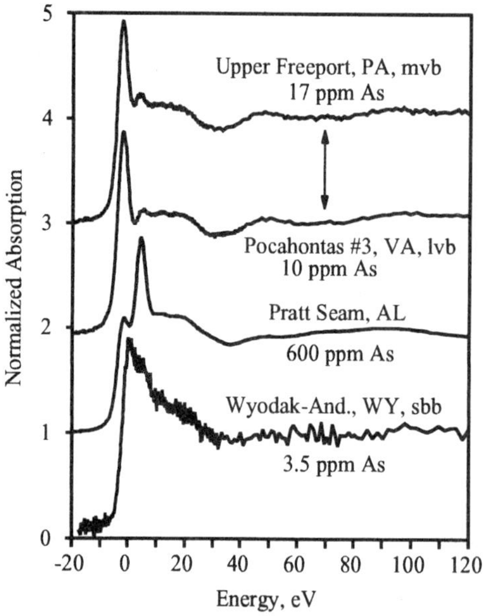

FIG. 14. Arsenic XANES spectra for two unoxidized bituminous coals, an highly oxidized bituminous coal from Alabama, and a subbituminous coal. The arrow denotes the same small peak as indicated in Fig. 13 that discriminates between arsenical pyrite and arsenopyrite. Zero point of energy for arsenic K absorption edge is 11,867 eV.

variation in the spectra between the fresh and oxidized samples, indicating that selenium is relatively susceptible to oxidation, even during storage. In one of the oxidized coal samples (Illinois #6), a second prominent peak was observed at a position characteristic of the selenate (SeO_4^{2-}) oxidation state. A detailed study by DREHER and FINKELMAN (1992) of selenium in a Powder River Basin coal indicates just how complex the geochemical behavior of this element can be.

Bromine

As shown in Fig. 16, the XANES spectra of the two halogen elements, chlorine and bromine, appear very similar for the Upper Freeport coal, implying that the occurrence of bromine in coal is essentially identical to that of chlorine (HUGGINS and HUFFMAN, 1995). Also, such halogen XANES spectra from coals were found to be similar to spectra of the halide anions in aqueous solution and quite different from that of any organohalogen compound or inorganic halide. Hence, it was deduced, in part by elimination of all other postulated possibilities, that the halogen elements in coal exist largely as halide anions (Cl^-, Br^-) associated with the moisture found in the pores and capillaries of the coal macerals. However, the XANES spectra are not exact duplicates of the XANES spectra of halide anions in aqueous solution because the EXAFS oscillations of the halide anions in coal are

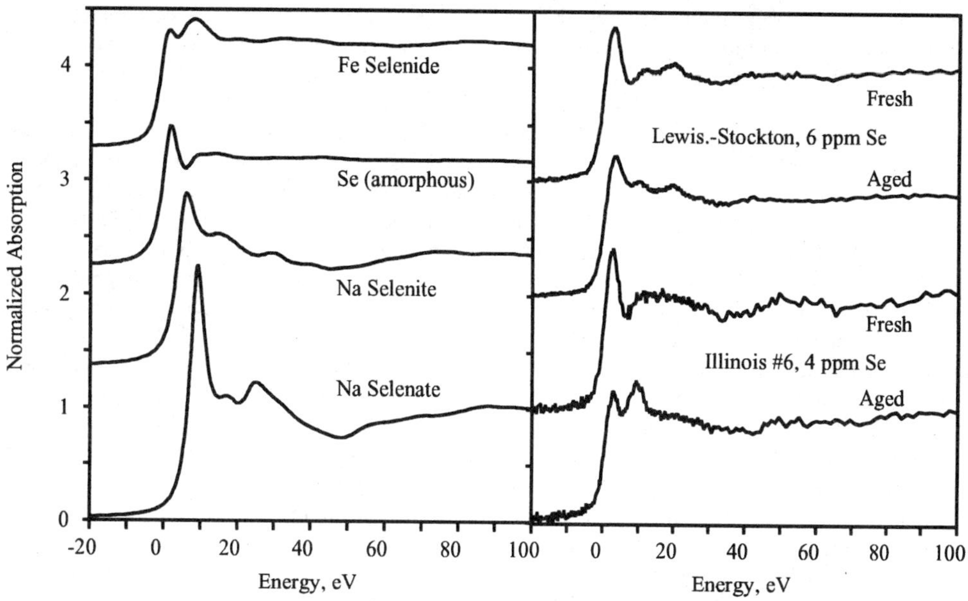

FIG. 15. Comparison of selenium XANES spectra of coals and standards. Note variation of coal spectra between fresh and aged samples. Zero point of energy for selenium K absorption edge is 12,658 eV.

broadened and flattened compared to those for the anions in solution. Furthermore, the halogen XANES profiles from coal persist to temperatures at which most of the free water is driven out from coal. Hence, it was inferred (HUGGINS and HUFFMAN, 1995) that there existed a positive interaction between the halide anions and polar organic functional groups that causes the halide anions to be anchored to the maceral surface and prevents them from behaving as if they were in a true solution. Essentially similar conclusions were reached by CHOU (1991) using less direct evidence.

The chlorine XANES spectrum of the Pocahontas #3 coal appears different from that of bromine in Fig. 16. In particular, there is a prominent small peak at about 12.5 eV in the chlorine spectrum that derives from crystalline sodium chloride. The observation of solid NaCl is relatively uncommon in chlorine XAFS spectra of coals (most spectra are closely similar to that of the Upper Freeport coal), and is thought to be an artefact brought about by crystallization of NaCl from a saturated solution of chloride anions in the moisture of coal as the sample of coal dries during storage after size reduction (HUGGINS and HUFFMAN, 1995). From the chlorine and moisture contents, it was calculated that the moisture in Pocahontas #3 coal is close to a saturated solution of NaCl. However, the much

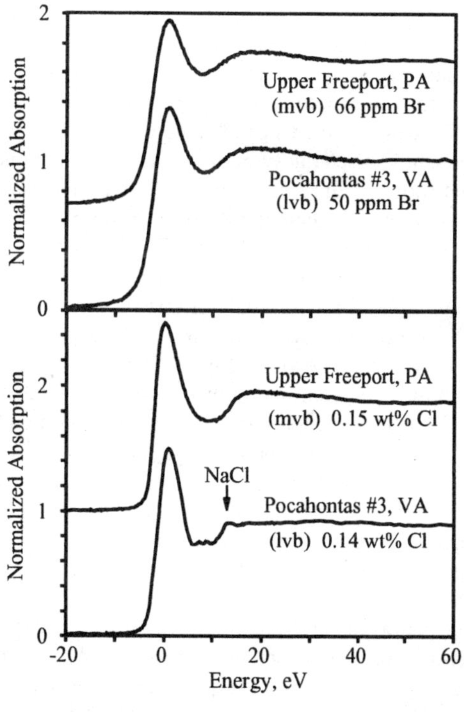

FIG. 16. Comparison of chlorine and bromine XANES spectra of Upper Freeport and Pocahontas #3 coals. Zero points of energy for chlorine and bromine K absorption edges occur at 2,825 and 13,474 eV, respectively.

lower concentration of bromine ensures that it would be entirely in solution as the solubility of NaBr is greater than that of NaCl.

DISCUSSION

As has been demonstrated in this review, XAFS spectroscopy has an unparalleled ability to focus directly on a specific element in a coal sample and extract details about its local structure and chemical bonding, from which the modes of occurrence of the element in the coal may be deduced. Also, the application of this technique is essentially nondestructive and requires a minimum of sample preparation. Furthermore, because of the high X-ray flux of the synchrotron source coupled with the sensitivity of the 13 element Ge detector, measurements can be conducted at concentration levels as low as 5 ppm. In addition, because XAFS spectroscopy is sensitive to dispersed maceral occurrences as well as to discrete mineral occurrences of an element, it complements microscope or microprobe investigations, which provide information primarily on the more concentrated mineral occurrences of an element.

XAFS spectroscopy clearly demonstrates that for many elements there are often two distinct forms present in coal: one associated with minerals, the other associated with the organic matrix. For coals of intermediate rank (subbituminous–high volatile B bituminous), both of these major elemental forms have been observed in the same coal. Hence, the frequently raised question of whether there are distinct forms of an element in float and tailings fractions or whether there is merely an incomplete separation of one elemental form between float and tailings fractions appears to be resolved in favor of multiple forms for coals of this rank.

Hence, of the five different generalized modes of occurrence shown in Fig. 1, it would appear, from the data presented in this review for elements of the first transition row, that the most common modes are ionically-bound forms in organic association and substitutional mineralogical occurrences involving illite or pyrite for the lithophile and chalcophile elements, respectively. Except for abundant elements such as potassium, calcium and iron, which are necessary for forming some of the common major minerals in coal, the only first transition-row elements conclusively identified as forming their own specific minerals are titanium and zinc. The other exceptions include selenium, which appears to be present in organoselenium forms, and bromine, which like chlorine is principally present in anionic form in the moisture associated with the macerals. However, it should be remembered that the data obtained to date for these two elements are quite limited.

Also it should be recalled that this discussion has centered exclusively on the first transition-row elements, which are more compatible with incorporation into illite or pyrite than elements further down the periodic table. From evidence summarized by FINKELMAN (1988) and SWAINE (1990), it would appear that discrete accessory mineral formation involving heavier trace elements is much more common.

Those elements that are found in both an organic as well as a mineralogical association in coal tend to exhibit the organic association more completely in low-rank coals and the mineralogical occurrence more so in higher-rank coals. This observation implies that metamorphic grade plays a role in determining modes of occurrence. Possible evidence for regional variations in the mode of occurrence was demonstrated by zinc in coals of the Illinois basin, which was predominantly, if not exclusively in the form of sphalerite, whereas zinc in other coals was present in an oxidic environment as well. Variation of the mode of occurrence due to oxidation during storage has been noted for arsenic, selenium, and iron (from Mössbauer studies). Most other elements appear to be relatively insensitive to oxidation; however, for many elements, this aspect has not yet been adequately examined. Other factors that might influence the mode of occurrence, such as sulfur content, may also be important, but the XAFS database is not yet sufficient to test this possibility.

As many researchers have speculated in the past, one of the more important processes in coal geochemistry is the decarboxylation step that occurs near the rank boundary between bituminous and subbituminous coals. The loss of carboxyl functionality due to increasing coalification results in many elements undergoing mineralization as coals of bituminous rank start to form. The moisture in coal can be expected to play a major role in helping to transfer elements between the increasingly tenuous carboxyl sites and the more stable mineral forms. As rank increases, the concentration of oxygen functional groups at the coal surface decreases and ions are forced into solution, and thence into mineral formation. Furthermore, the moisture content also decreases and becomes increasingly saturated with respect to dissolved species as coalification increases, which further impels the ionic species to enter into mineralization. The valence state of the cation or, perhaps more appropriately, the ionic potential (Z/r, where Z is the formal valence charge and r is the ionic radius, *q.v.* CARTLEDGE, 1928) is likely to be largely responsible for the order in which cations are released by carboxyl groups, their solubility in aqueous fluids, and their rate of mineralization, either capture by a host mineral or precipitation of a new mineral. These factors basically act together to keep the lowest charge cations (M^+) in solution in the highest concentrations and the highest charge cations the least. In the general case, decarboxylation may be written as:

$$(R{-}COO^-)_nM^{n+} + nH_2O \rightleftharpoons nRH + nCO_2 + [M^{n+}_{aq}] + nOH^-,$$

where the coal is represented as R, and mineralization may be written as:

$$[M^{n+}_{aq}] + nOH^- \rightleftharpoons MO_{n/2} + 0.5nH_2O.$$

If K_1 is the equilibrium constant for the decarboxylation reaction and K_2 is the equilibrium constant for the mineralization reaction, represented by precipitation of the oxide, then, based on valence or ionic potential considerations, it follows that K_1 decreases in the order $M^+ > \cdot\cdot\cdot > M^{4+}$ and K_2 increases in the order $M^+ < \cdot\cdot\cdot\cdot < M^{4+}$. Consequently, $[M^{n+}_{aq}]$, the concentration of the cation in moisture, can be expected to decrease in the order $n = 1 \gg \cdot\cdot\cdot \gg n = 4$.

The observation of many elements with a distinct organic association in coals must also reflect the importance of geochemical processes occurring during peat formation. Hence, it would appear that one of the most important of these processes is the introduction of elements by ion-exchange between cations in the water flowing through the peat and acidic sites in the peat itself. Similar ideas have been expressed by MILLER and GIVEN (1986, 1987) and GIVEN and MILLER (1987).

CONCLUSION

XAFS spectroscopy is capable of revealing information directly and nondestructively about the geochemical properties and behavior of elements in coal. With current synchrotron radiation facilities in the U.S., useful information can be obtained from as little as 5 ppm of an element in coal. With the much brighter Advanced Photon Source to be commissioned at Argonne National Laboratory within the next two years, it should be possible to extend such studies to many of the remaining elements in the periodic table, including most of the elements that are potentially of environmental concern (SWAINE, 1990).

The most important result presented in this review is the demonstration that many elements in coal, even of as high rank as high-volatile B bituminous, appear to have a significant organic association that is distinct from their mineralogical occurrences. This result has important ramifications not only with respect to the geochemistry of coal, but also to various options for the removal or control of trace elements in coal cleaning and utilization.

Acknowledgements—We are grateful to the following personnel from the University of Kentucky who have assisted us with the collection of XAFS spectra at NSLS or SSRL from 1989 to present: Bhaswati Ganguly, Fulong Lu, Sudipa Mitra, Mohammad Najih, Anup Shah, Naresh Shah, Mehdi Taghiei, Shreeniwas Vaidya, and Jianmin Zhao. In addition, we would like to acknowledge Drs. Farrel Lytle and Robert Greegor who introduced us to XAFS spectroscopy in the early 1980s and who have shared with us many insights and data over the years. This work has been supported by contracts with the U.S. Department of Energy and the Electric Power Research Institute. We also acknowledge the U.S. Department of Energy for its support of synchrotron radiation facilities at both NSLS and SSRL.

REFERENCES

ASTM (1976) Coal and coke. In *1976 Annual Book of ASTM Standards. Part 26. Gaseous Fuels; Coal and Coke; Atmospheric Analysis*, pp. 179–426. American Society for Testing and Materials.

BONNETT R., BURKE P. J. and CZECHOWSKI F. (1987) Metalloporphyrins in lignite, coal, and calcite. In *Metal Complexes in Fossil Fuels* (eds. R. H. FILBY and J. F. BRANTHAVER); *ACS Symposium Series 344*, Chapter 11, pp. 173–185. American Chemical Society.

BRAGG L. J., OMAN J. K., TEWALT, S. J., OMAN, C. L., REGA, N. H., WASHINGTON, P. M. and FINKELMAN, R. B. (1994) Coal quality (COALQUAL) database: Version 1.3. *Open-File Report 94–205*, 35Mb CD-ROM. U.S. Geological Survey.

BROWN G. E., JR., CALAS G., WAYCHUNAS G. A. and PETIAU J. (1988) X-ray absorption spectroscopy and its application in mineralogy and geochemistry. In *Spectroscopic Methods in Mineralogy and Geology* (ed. F. C. HAWTHORNE); *Rev. in Mineral. 18*, Chapter 11, pp. 431–512. Mineralogical Society of America.

BROWNFIELD M. E., AFFOLTER R. H., STRICKER G. D. and HILDEBRAND, R. T. (1995) High chromium coals in Tertiary coal deposits of northwestern Washington—a key to their depositional history. *Int. J. Coal Geol.* **27,** 153–169.

BURNS R. G. (1965) Electronic spectra of silicate minerals: applications of crystal field theory to aspects of geochemistry. Ph.D. Thesis, University of California, Berkeley, CA.

BURNS R. G. (1970) *Mineralogical Applications of Crystal Field Theory*. Cambridge University Press.

BURNS R. G. (1975) Crystal field effects in chromium and its partitioning in the mantle. *Geochim. Cosmochim. Acta* **39,** 857–864.

BURNS V. M. and BURNS R. G. (1975) Mineralogy of chromium. *Geochim. Cosmochim. Acta* **39,** 903–910.

CAAA (1990) Amendments to the Clean Air Act. *U.S. Public Law 101–549, Nov. 15, 1990.* 314 pp. U.S. Government Printing Office.

CARTLEDGE G. H. (1928) Studies on the periodic system. I. The ionic potential as a periodic function. *J. Am. Chem. Soc.* **50,** 2855–2863; II. The ionic potential and related properties. *J. Am. Chem. Soc.* **50,** 2863–2872.

CHOU, C-L. (1991) Distribution and forms of chlorine in Illinois Basin coals. In *Chlorine in Coal* (eds. J. STRINGER and D. D. BANERJEE), pp. 11–29. Elsevier Science Publishers.

CLARKE L. B. and SLOSS L. L. (1992) Trace Elements. *IEA Coal Research Report 49,* 111 pp. International Energy Agency.

CRAMER S. P., TENCH O., YOCUM N. and GEORGE G. N. (1988) A 13-element Ge detector for fluorescence EXAFS. *Nucl. Instrum. Meth.* **A266,** 586–591.

DREHER G. B. and FINKELMAN R. B. (1992) Selenium mobiliization in a surface coal mine, Powder River, Wyoming. *Environ. Geol. Water Sci.* **19,** 155–167.

EISENBERGER P. and KINCAID B. M. (1978) EXAFS: New horizons in structure determination. *Science* **200,** 1441–1447.

FINKELMAN R. B. (1988) The inorganic geochemistry of coal: a scanning electron microscopy view. *Scanning Microscopy* **2,** 97–105.

FINKELMAN R. B. (1994) Modes of occurrence of potentially hazardous elements in coal: levels of confidence. *Fuel Proc. Technol.* **39,** 21–34.

FINKELMAN R. B. and STANTON R. W. (1978) Identification and significance of accessory minerals from a bituminous coal. *Fuel* **57,** 763–768.

FRAHM R. (1992) Quick XAFS: potentials and practical applications in materials science. In *X-ray Absorption Fine Structure* (ed. S. S. HASNAIN), Chapter 191, pp. 731–737. Ellis Horwood.

GIVEN P. H. and MILLER R. N. (1987) The association of major, minor, and trace inorganic elements with lignites. III. Trace elements in four lignites and general discussion of all data from this study. *Geochim. Cosmochim. Acta* **51,** 1843–1853.

GOODARZI F. (1988) Elemental distribution in coal seams at the Fording coal mine, British Columbia, Canada. *Chem. Geol.* **68,** 129–154.

HATCH J. R., GLUSKOTER H. J. and LINDAHL P. C. (1976) Sphalerite in coals from the Illinois basin. *Econ. Geol.* **71,** 613–624.

HUFFMAN G. P. and HUGGINS F. E. (1984) Analysis of the inorganic constituents in low-rank coals. In *Chemistry of Low-Rank Coals* (ed. H. H. SCHOBERT); *ACS Symposium Series 264,* Chapter 10, pp. 159–174. American Chemical Society.

HUFFMAN G. P., HUGGINS F. E., SHOENBERGER R. W., WALKER J. S., GREEGOR R. B. and LYTLE F. W. (1986) Investigation of the structural forms of potassium in coke by electron microscopy and X-ray absorption spectroscopy. *Fuel* **65,** 621–632.

HUFFMAN G. P., HUGGINS F. E., SHAH N. and ZHAO J. (1994) Speciation of arsenic and chromium in coal and combustion ash by XAFS spectroscopy. *Fuel Proc. Technol.* **39,** 47–62.

HUGGINS F. E. and HUFFMAN G. P. (1989) Coal weathering and oxidation: the early stages. In *Chemistry of Coal Weathering* (ed. C. R. NELSON); *Coal Science and Technology 14,* Chapter 2, pp. 33–60. Elsevier.

HUGGINS F. E. and HUFFMAN G. P. (1995) Chlorine in coal: an XAFS spectroscopic investigation. *Fuel* **74,** 556–569.

HUGGINS F. E. and HUFFMAN G. P. (1996) Modes of occurrence of trace elements in coal from XAFS spectroscopy. To be published in *Proceedings, 1995 International Chemical Congress of Pacific Basin Societies* (eds. P. C. LYONS, C. A. PALMER and R. M. BUSTIN); *Internat. J. Coal Geol.*

HUGGINS F. E., HUFFMAN G. P. and LIN M. C. (1983a) Observations on low-temperature oxidation of minerals in bituminous coals. *Int. J. Coal Geol.* **3,** 157–182.

HUGGINS F. E., HUFFMAN G. P., LYTLE F. W. and GREEGOR R. B. (1983b) An EXAFS investigation of calcium in coal. In *Proceedings, 1983 International Conference on Coal Science (Pittsburgh, PA),* pp. 679–682. International Energy Agency.

HUGGINS F. E., SHAH N., ZHAO J., LU F. and HUFFMAN G. P. (1993a) Nondestructive determination of trace element speciation in coal and ash by XAFS spectroscopy. *Energy & Fuels* **7,** 482–489.

HUGGINS F. E., ZHAO J., SHAH N. and HUFFMAN G. P. (1993b) Speciation of trace elements in coal from XAFS spectroscopy. *Proceedings, 7th. International Conference on Coal Science (Banff, Canada),* (ed. K. H. MICHAELIAN), Vol. 1, pp. 660–663.

HUGGINS F. E., HUFFMAN G. P. and HELBLE J. J. (1994a) Determination of the form of occurrence of hazardous trace elements in coal and ash. In *Proceedings: Second International Conference on Managing Hazardous Air Pollutants, 1993* (eds. W. CHOW and L. LEVIN); *EPRI Report TR-104295,* pp. II-113 to II-133. Electric Power Research Institute.

HUGGINS F. E., ZHAO J., SHAH N. and HUFFMAN G. P. (1994b) Modes of occurrence of trace elements in coal from XAFS spectroscopy. *Preprints, ACS Div. Fuel Chem.* **39(2),** 504–510. American Chemical Society.

HUGGINS F. E., PAREKH B. K., ROBERTSON J. D. and HUFFMAN G. P. (1995) Modes of occurrence of trace elements in coal: geochemical constraints from XAFS and PIXE spectroscopic analysis of advanced coal cleaning tests. In *Proceedings, 8th. International Conference on Coal Science (Oviedo, Spain)* (eds. J. A. PAJARES and J. M. D. TASCÓN), Vol. 1, 679–682. Elsevier.

HUGGINS F. E., GOODARZI F. and LAFFERTY C. J. (1996) Nature of arsenic in subbituminous coals. *Energy & Fuels* **10,** in press.

KONINGSBERGER D. C. and PRINS R. (1988) *X-ray Absorption. Principles, Applications, Techniques of EXAFS, SEXAFS, and XANES.* J. Wiley & Sons.

LEE P. A., CITRIN P. H., EISENBERGER P. A. and KINCAID B. M. (1981) Extended X-ray absorption fine structure—its strengths and limitations as a structural tool. *Rev. Mod. Phys.* **53,** 769–808.

LYTLE F. W., GREEGOR R. B., SANDSTROM D. R., MARQUES E. C., WONG J., SPIRO C. L., HUFFMAN G. P. and HUGGINS F. E. (1984) Measurement of soft X-ray absorption spectra with a fluorescent ion chamber detector. *Nucl. Instrum. Meth.* **226,** 542–548.

MAYLOTTE D. H., WONG J., ST. PETERS R. L., LYTLE F. W. and GREEGOR R. B. (1981a) Trace vanadium in coal: An EXAFS study. In *Proceedings, International Conference on Coal Science (Düsseldorf, Germany)* (eds. International Energy Agency), pp. 756–761, Verlag Glückauf GmbH.

MAYLOTTE D. H., WONG J., ST. PETERS R. L., LYTLE F. W. and GREEGOR R. B. (1981b) X-ray absorption spectroscopic investigation of trace vanadium sites in coal. *Science* **214,** 554–556.

MILLER R. N. and GIVEN P. H. (1986) The association of major, minor, and trace inorganic elements with lignites. I. Experimental approach and study of a North Dakota lignite. *Geochim. Cosmochim. Acta* **50,** 2033–2043.

MILLER R. N. and GIVEN P. H. (1987) The association of major, minor, and trace inorganic elements with lignites. II. Minerals, and major and minor element profiles, in four seams. *Geochim. Cosmochim. Acta* **51,** 1311–1322.

SANDSTROM D. R., FILBY R. H., LYTLE F. W. and GREEGOR R. B. (1982) Study of Ti in solvent-refined coal by X-ray absorption spectroscopy. *Fuel* **61,** 195–197.

SPEARS D. A. (1991) Pyrite in some U.K. coals. In *Processing and Utilization of High Sulfur Coals IV* (eds. P. R. DUGAN, D. R. QUIGLEY and Y. A. ATTIA); *Coal Science and Technology 18,* pp. 85–93. Elsevier.

SPIRO C., WONG J., MAYLOTTE D. H., LAMSON S., GLOVER B., LYTLE F. W. and GREEGOR R. B. (1986) Nature of potassium impurities in coal. *Fuel* **65,** 327–336.

STACH E., MACKOWSKY M.-TH., TEICHMÜLLER M., TAYLOR G. H., CHANDRA D. and TEICHMÜLLER R. (1975) *Coal Petrology.* Gebrüder Borntraeger.

STERN E. A. and HEALD S. M. (1979) X-ray filter assembly for fluorescence measurements of X-ray absorption fine structure. *Rev. Sci. Instrum.* **50,** 1579–1582.

SWAINE D. J. (1990) *Trace Elements in Coal.* Butterworths.

WAYCHUNAS G. A. (1987) Synchrotron radiation XANES spectroscopy of Ti in minerals: Effects of Ti bonding distances, Ti valence, and site geometry on absorption edge structure. *Amer. Mineral.* **72,** 89–101.

WHITE R. N., SMITH J. V., SPEARS D. A., RIVERS M. L. and SUTTON S. R. (1989) Analysis of iron sulphides from U.K. coal by synchrotron radiation X-ray fluorescence. *Fuel* **68,** 1480–1486.

WONG J., MAYLOTTE D. H., LYTLE F. W., GREEGOR R. B. and ST. PETERS R. L. (1983) EXAFS and XANES investigations of trace V and Ti in coal. In *EXAFS and Near Edge Structure* (eds. A. BIANCONI, L. INCOCCIA and S. STIPCICH); *Springer Series in Chemical Physics 27,* pp. 206–209. Springer-Verlag.

WYCKOFF R. W. G. (1965) *Crystal Structures* (2nd. edition), Vol. 1. Interscience Publishers.

Mineral Spectroscopy: A Tribute to Roger G. Burns
© The Geochemical Society, Special Publication No. 5, 1996
Editors: M. D. Dyar, C. McCammon and M. W. Schaefer

Coordination and local structure of Si and Al in silicate glasses: Si and Al K-edge XANES spectroscopy

DIEN LI,*[,1] G. M. BANCROFT,[1] and M. E. FLEET[2]

[1]Department of Chemistry, University of Western Ontario, London, Ontario N6A 5B7, Canada
[2]Department of Earth Sciences, University of Western Ontario, London, Ontario N6A 5B7, Canada

Abstract—Silicon K-edge XANES spectra of low-pressure glasses along the $CaMgSi_2O_6$ (Di)-$NaAlSi_3O_8$ (Ab) join and high-pressure $Na_2Si_2O_5$, $Na_2Si_4O_9$, and $K_2Si_4O_9$ glasses, and Al K-edge XANES spectra of high-pressure glasses along the $NaAlSi_2O_6$ (Jd)-$NaAlSi_3O_8$ (Ab) join are reported using synchrotron radiation. Ab_{100} glass is very similar to albite in respect to the energy position and FWHM of the Si K-edge peak. Along the $CaMgSi_2O_6$ (Di)-$NaAlSi_3O_8$ (Ab) join, the Si K-edge peak shifts to lower energy and its FWHM increases dramatically with increase in Di content. These results indicate that the glasses become less polymerized and structural entities tend to become more complicated toward the Di end-member composition. The Di_{100} glass contains Q^0, Q^1, Q^2, Q^3 and Q^4 structural entities, although Q^1 (and possibly Q^0 also) entities are predominant.

Aluminum K-edge XANES spectra of high-pressure glasses along the $NaAlSi_2O_6$ (Jd)-$NaAlSi_3O_8$ (Ab) join provide direct experimental evidence for the pressure-induced coordination change of Al from 4 to 6. Five- and six-coordinated Al ($^{[5]}Al$, $^{[6]}Al$) appear in the $Jd_{60}Ab_{40}$ glass and become more prominent in the Jd_{100} glass, indicating that the pressure required for the coordination change of Al in Jd_{100} glass (*e.g.*, ~4 GPa) is lower than that in Ab_{100} glass. The abundances of each of $^{[5]}Al$ and $^{[6]}Al$ in the Jd_{100} glass are estimated to be about 6% of total Al. Silicon K-edge XANES spectra are also used to probe pressure-induced coordination change of Si in high-pressure glasses of composition $Na_2Si_2O_5$, $Na_2Si_4O_9$, and $K_2Si_4O_9$. Although the Si K-edge spectra provide no evidence for five-fold coordinated Si ($^{[5]}Si$) in these glasses, six-fold coordinated Si ($^{[6]}Si$) is present in $Na_2Si_4O_9$ and $K_2Si_4O_9$ composition glasses, confirming earlier ^{29}Si MAS NMR measurements. However, the abundance of $^{[6]}Si$ is higher in $K_2Si_4O_9$ glasses than in $Na_2Si_4O_9$ glasses, and increases with increasing pressure for glasses of the same chemical composition.

INTRODUCTION

SILICON IS four-fold coordinated ($^{[4]}Si$) with oxygen in silicates of the Earth's crust (*e.g.*, SMYTH and BISH, 1988), but becomes six-fold coordinated ($^{[6]}Si$) at high pressure. For example, silicon dioxide (SiO_2) has numerous polymorphic modifications with 4:2 coordinated structures (*e.g.*, quartz, cristobalite and tridymite), and occurs as stishovite with 6:3 coordinated structure at high temperature and pressures beyond about 9 GPa. Materials in the Earth's mantle are believed to be iron-bearing magnesium silicates with $^{[4]}Si$ structures dominant in the upper mantle and $^{[6]}Si$ structures predominant in the lower mantle (*e.g.*, ITO and TAKAHASHI, 1987; JEANLOZ, 1990; FINGER and HAZEN, 1991). Crystalline SiP_2O_7 is one of several compounds in which Si occurs in octahedral coordination with oxygen at atmospheric pressure (*e.g.*, LIEBAU, 1985; FINGER and HAZEN, 1991).

Aluminum is either four- or six-fold coordinated ($^{[4]}Al$ or $^{[6]}Al$) with oxygen in most crystalline aluminosilicate minerals but it also occurs in five-fold coordination ($^{[5]}Al$) in andalusite (*e.g.*, SMYTH and BISH, 1988). However, both Al and Si are four-fold coordinated with oxygen in silicate and aluminosilicate glasses at room temperature and pressure, because the octahedral coordination of Si and Al in glasses would force periodicity and disrupt the vitreous state (ZACHARIASEN, 1932). The only exceptions to date are that six-fold coordinated Si is evident in silicate-phosphate glasses with P_2O_5 above 32 mol% (DUPREE *et al.*, 1987; SEKIYA *et al.*, 1988; LI *et al.*, 1995c), and five-fold coordinated Si ($^{[5]}Si$) has been reported in $K_2Si_4O_9$ glass at atmospheric pressure (STEBBINS, 1991). Aluminum and Si have been reported in five- or six-fold coordination with oxygen in high-pressure silicate glasses and melts (*e.g.*, OHTANI *et al.*, 1985; STEBBINS, 1987; STEBBINS and MCMILLAN, 1989; STEBBINS and SYKES, 1990; STOLPER and AHRENS, 1987; WILLIAMS and JEANLOZ, 1988; XUE *et al.*, 1989, 1991; LI *et al.*, 1994c), although the four-fold coordinated state dominates for both Al and Si in all of the vitreous materials investigated to date.

A knowledge of the structure of silicate melts and its relationship to their physical and chemical properties is necessary in order to understand magmatic processes. However, direct *in situ* spectroscopic study of silicate melt structure at both high temperature and high pressure is difficult experimentally, even though recent *in situ* Raman (HEMLEY *et al.*, 1986; MYSEN, 1990, 1995; MYSEN and

* *Present address:* National Institute of Materials and Chemical Research, Tsukuba, Ibaraki 305, Japan

FRANTZ, 1993, 1994; MCMILLAN *et al.*, 1992; 1994) and MAS NMR studies (FARNAN and STEBBINS, 1990; POE *et al.*, 1993; STEBBINS, 1988a, 1988b) have provided important information on silicate melts at high temperature or at combined high temperature and pressure. It is generally assumed that the structure of a quenched glass represents that of the liquid at its glass transition temperature (*e.g.*, GIBBS and DIMARZIO, 1958), although some local structural changes may occur during quenching (MCMILLAN *et al.*, 1992; MYSEN and FRANTZ, 1993).

The structure of silicate glasses has been extensively studied by X-ray diffraction (TAYLOR and BROWN, 1979), infrared (WONG and ANGELL, 1976) and Raman spectroscopy (MCMILLAN, 1984a; MCMILLAN and PIRIOU, 1983; MYSEN, 1988), MAS NMR spectroscopy (KIRKPATRICK, 1988, STEBBINS, 1988b, STEBBINS and FARNAN, 1989), and metal K-edge X-ray absorption spectroscopy (GREAVES *et al.*, 1981; WAYCHUNAS *et al.*, 1988; JACKSON *et al.*, 1993; GALOISY and CALAS, 1994). The results indicate that the intermediate-range structure of silicate and aluminosilicate glasses may be described by a mixture of different structural units or entities of interconnected TO_4 tetrahedra (T = Si and Al) (MYSEN, 1988; MCMILLAN, 1984a).

Aluminum K-edge X-ray absorption spectra of silicate glasses (MCKEOWN *et al.*, 1985) and Si and O K-edge XANES spectra of densified vitreous silica (DAVOLI *et al.*, 1992) have been reported. In our previous papers, we reported Si K- and L-edge XANES spectra of silica polymorphs (LI *et al.*, 1994a), crystalline SiP_2O_7 (LI *et al.*, 1994b), silicate minerals (LI *et al.*, 1995a) and SiO_2-P_2O_5 and Na_2O-SiO_2-P_2O_5 glasses (LI *et al.*, 1995b, 1995c), and Al K-edge XANES spectra of aluminosilicate minerals (LI *et al.*, 1995d). The Si and Al K-edge XANES spectra have been used to distinguish different coordinations of Si ($^{[4]}$Si and $^{[6]}$Si) and Al ($^{[4]}$Al, $^{[5]}$Al and $^{[6]}$Al) (LI *et al.*, 1994a, 1994b, 1995d). The Si K-edge shifts to higher energy with increase in polymerization of SiO_4 tetrahedra and to lower energy with the substitution of Al for Si in silicate minerals (LI *et al.*, 1995a). Silicon remains four-fold coordinated in SiO_2-P_2O_5 and Na_2O-SiO_2-P_2O_5 glasses with P_2O_5 below 32 mol%, but Na_2O depolymerizes and P_2O_5 copolymerizes these glasses (LI *et al.*, 1995b). However, $^{[6]}$Si occurs in SiO_2-P_2O_5 and Na_2O-SiO_2-P_2O_5 glasses with P_2O_5 above 32 mol%; the proportion of $^{[6]}$Si increases with increasing content of P_2O_5, and is semi-quantitatively estimated by curve-fitting the Si K-edge spectra (LI *et al.*, 1995c).

In this paper, we report Si and Al K-edge XANES spectra of $CaMgSi_2O_6$-$NaAlSi_3O_8$ glasses quenched at low pressure (1 bar), and $NaAlSi_2O_6$-$NaAlSi_3O_8$, $Na_2Si_2O_5$, $Na_2Si_4O_9$ and $K_2Si_4O_9$ glasses quenched at high pressure. We extend applications of Si and Al K-edge XANES spectroscopy at room temperature and pressure to investigate the polymerization of SiO_4 tetrahedra in room-pressure silicate glasses, and pressure-induced coordination changes of Si and Al in high-pressure silicate and aluminosilicate glasses. Preliminary results for the $NaAlSi_2O_6$-$NaAlSi_3O_8$ join are reported elsewhere (LI *et al.*, 1995e).

EXPERIMENTAL METHODS

Sample preparation

The $CaMgSi_2O_6$-$NaAlSi_3O_8$ glass samples were from the study of KEPPLER (1992). Mixtures of glassy powders of albite and diopside starting materials were melted in a platinum crucible using a room-pressure high-temperature gas mixing furnace. Experiments were performed in air, pure O_2, or H_2-CO_2 mixtures. After usually 12–24 h at 1200–1500°C, samples were quenched by dropping the crucible into cold water. Other details on the sample preparation and composition of these glasses are given in KEPPLER (1992).

For study of glasses along the $NaAlSi_2O_6$ (Jd)-$NaAlSi_3O_8$ (Ab) join, the end-member compositions were prepared from stoichiometric proportions of Na_2CO_3, Al_2O_3 and commercial "amorphous" SiO_2, that were ground and mixed, contained in a platinum crucible with a closely fitted lid, and heated initially at 900°C for 1 day and finally at 1000°C for several hours. Intermediate starting compositions were prepared from mechanical mixtures of molar proportions of the end-members. High-pressure glasses were prepared using the 1000-t cubic anvil press at UWO. Samples were contained in BN crucibles, held at 4.4 GPa and 1575°C for 5 to 10 minutes, and quenched by switching off the power to the furnace (see LI *et al.*, 1995e for further details).

The high-pressure alkali silicate glasses were prepared using the MA6/8 superpress at the University of Alberta, Edmonton. Alkali silicate starting materials were prepared from carbonate-oxide mixtures. For the high-pressure experiments, samples were contained in a platinum capsule, and capsules were dried at 300°C for at least 1 hour before welding. Other experimental details are given in Table 1.

Si and Al K-edge XANES spectroscopy

Both Si and Al K-edge XANES spectra of glasses were collected on a double crystal monochromator (DCM) using synchrotron radiation. InSb (111) monochromator crystals were used for Si K-edge measurements, giving an energy resolution of 0.8 eV at 1840 eV (YANG *et al.*, 1992); and α-quartz ($10\bar{1}0$) crystals were used for the Al K-edge measurements with an energy resolution of about 0.7 eV at 1560 eV. For both Si and Al K-edge measurements, very fine powder samples were spread uniformly on electric carbon tape supported on a stainless steel sample holder. The area covered by the sample was about 10 mm × 10 mm and the sample thickness was reasonably constant for the measurement of each sample. Both Si and Al K-edge spectra were recorded by total electron yield (TEY) which measures the sample current from the escape of electrons of different energies from the surface

Table 1. Synthesis of $Na_2Si_2O_5$, $Na_2Si_4O_9$ and $K_2Si_4O_9$ glasses

Sample	Composition	Furnace	P (GPa)	T (°C)	Duration (min.)
1666	$Na_2Si_2O_5$	18M stepped graphite	8	1500	30
1665	$Na_2Si_2O_5$	18M stepped graphite	5	1500	30
1675*	$Na_2Si_2O_5$	18M stepped graphite	4	900	240
1709	$Na_2Si_4O_9$	14M $LaCrO_3$ High-T	12	1900	20
1685	$Na_2Si_4O_9$	18M stepped $LaCrO_3$	8	1900	20
1670	$Na_2Si_4O_9$	18M stepped $LaCrO_3$	6	1900	20
1673*	$Na_2Si_4O_9$	18M low-T stepped graphite	4	900	240
1676	$K_2Si_4O_9$	18M stepped $LaCrO_3$	6	1900	20
1682	$K_2Si_4O_9$	18M high-T stepped graphite	4	1900	20
1669*	$K_2Si_4O_9$	18M low-T stepped graphite	4	900	240

* #1675 is ϵ-$Na_2Si_2O_5$ (FLEET and HENDERSON, 1995a); #1673 appears to be vitreous; #1669 is wadeite-structure $K_2Si_4O_9$ (SWANSON and PREWITT, 1983).

due to excitation by synchrotron X-radiation, and is proportional to the absorption coefficient (μ).

All samples were prepared in a similar manner to minimize the effect of sample thickness and particle size on the XANES features. Three measurements were made for each sample. The spectrum for each measurement was normalized by I/I_0, where I is the intensity of TEY signal and I_0 is the intensity of photon flux. The raw spectrum for each sample investigated was averaged from three normalized measurements, and smoothed. A linear pre-edge background was removed from each spectrum. All Si K-edge XANES spectra were calibrated against the Si K-edge of α-quartz at 1846.8 eV, and all Al K-edge spectra were calibrated by the Al K-edge of Al foil at 1560.0 eV.

RESULTS AND DISCUSSION

Si K-edge XANES spectra of $CaMgSi_2O_6$-$NaAlSi_3O_8$ glasses

Figure 1 shows the Si K-edge XANES spectra of a series of glasses along the $CaMgSi_2O_6$ (Di)-$NaAlSi_3O_8$ (Ab) join, as well as crystalline diopside and albite. The Si K-edge features of the end-member $CaMgSi_2O_6$ (Di_{100}) and $NaAlSi_3O_8$ (Ab_{100}) glasses and an intermediate composition (Ab_{50} Di_{50}) glass are expanded as dot, dash and solid lines, respectingly, in the inset. The spectra of these Di-Ab glasses are generally similar. The Si K-edge peak, designated as C-$^{[4]}$Si, is prominent in the glass spectra, but the post-edge features are more-or-less smeared out compared with spectra for the corresponding minerals. This is anticipated from the disordered structure of glasses. The Si K-edge peak is assigned to transitions of Si 1s electrons to the antibonding 3p-like t_2 states (LI *et al.*, 1994a). The energy position of the Si K-edge peak indicates that Si in the glasses is four-fold coordinated (LI *et al.*, 1994a). However, we note that the position of the Si K-edge peak of the glasses shifts to lower energy, and its full width at half maximum (FWHM) increases progressively, from Ab_{100} glass to $Ab_{50}Di_{50}$ and Di_{100} glasses.

Figure 2 shows the variations in energy position and FWHM of the Si K-edge peak with composition in this series of glasses. The open symbols are data for the glasses, and the filled symbols are data for crystalline diopside and albite. The Si K-edge peak shifts to lower energy by about 0.5 eV and its FWHM increases from about 2.5 to 4.1 eV with increase in molar proportion of Di component through the binary composition series. In addition, the energy position of the Si K-edge peak and its FWHM are very similar for both crystalline albite and Ab_{100} glass. However, the Si K-edge peak of Di_{100} glass shifts to much lower energy and its FWHM increases greatly compared with crystalline diopside.

The Si K-edge XANES spectra of silicate minerals demonstrated that the Si K-edge peak shifts to higher energy with increase in polymerization of SiO_4 tetrahedra and to lower energy with substitution of Al for Si (LI *et al.*, 1995a). Thus, the similarity in both Si K-edge peak position and FWHM for Ab_{100} glass and albite indicates that Ab_{100} glass has fully polymerized (Q^4) structural units, as in crystalline albite. However, the glasses along the

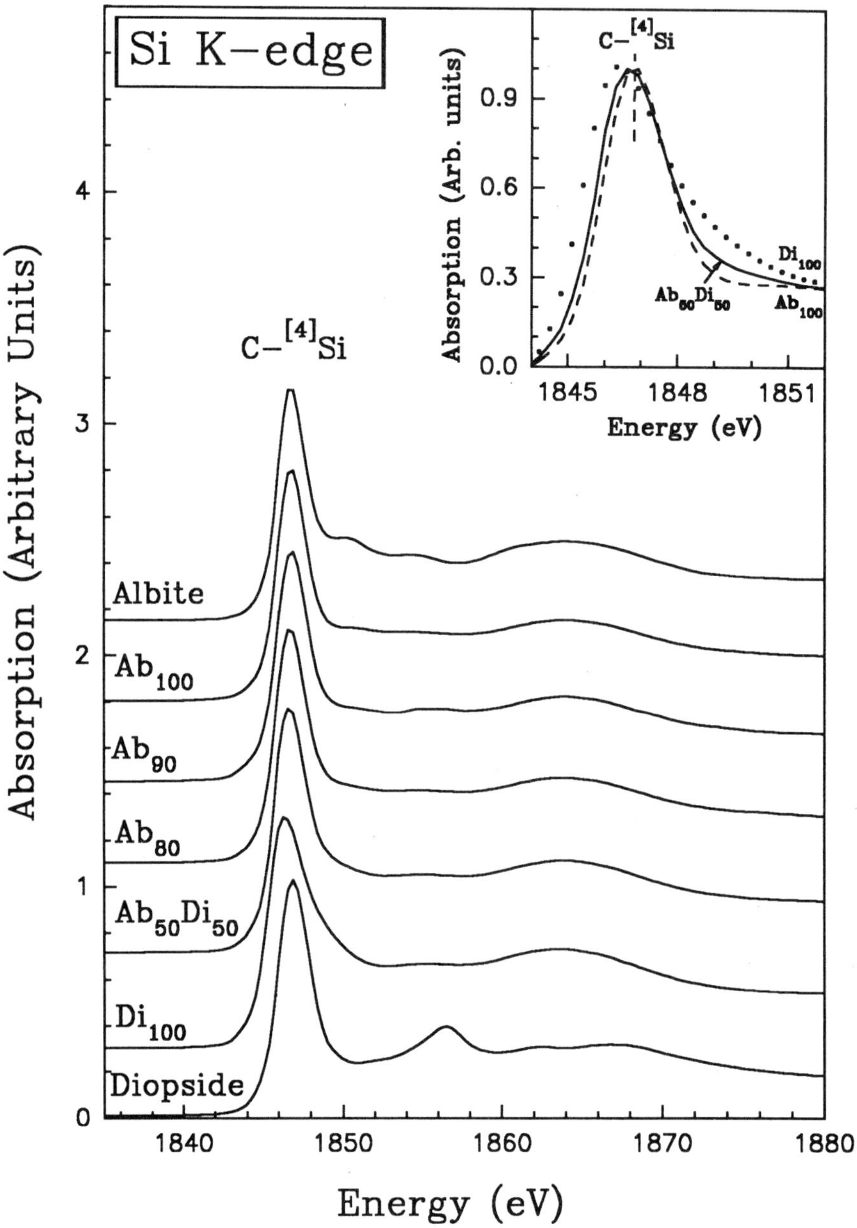

FIG. 1. Silicon K-edge XANES spectra of low-pressure glasses along the $CaMgSi_2O_6$ (Di)-$NaAlSi_3O_8$ (Ab) join. The [4]Si K-edge features of Di_{100}, $Ab_{50}Di_{50}$ and Ab_{100} glasses are expanded in the inset as dot, solid and dash lines, respectively.

Ab-Di join become progressively less polymerized and the Si K-edge peak shifts to lower energy with increasing Di content, paralleling changes in the structure and Si K-edge XANES spectra of the corresponding crystalline minerals (LI *et al.,* 1994a). We have two interpretations for the progressive increase in the FWHM of the Si K-edge peak of glasses along the Di-Ab join. First, five- and/or six-coordinated Si ([5]Si, [6]Si) may occur in the glasses towards the Di end member, because a shoulder at about 1848.5 eV appears in the Si K-edge XANES spectra of $Ab_{50}Di_{50}$ glass and becomes more evident in the spectrum of Di_{100} glass (LI *et al.,* 1994a). However, we were not able to confirm this conclusion by ^{29}Si MAS NMR measurements because of the very small amount of

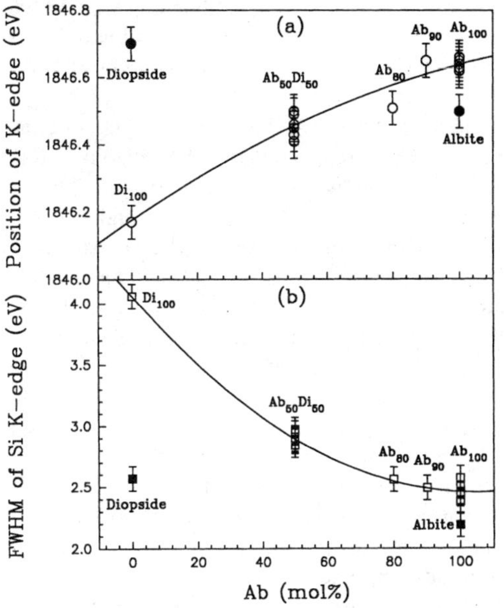

FIG. 2. Variations of energy position (a) of Si K-edge peak and its FWHM with the content of Ab in glasses (open symbols) along the Di-Ab join. The filled symbols represent data for crystalline albite and diopside. The Si K-edge peak shifts to lower energy, and its FWHM increases with decreasing content of Ab, from Ab_{100} to Di_{100}. Note that the structure of Di_{100} glass appears to be significantly different from that of crystalline diopside.

glass samples available for this study. Second, the most likely interpretation is that the intermediate-range structure in these glasses is complex, and becomes progressively more complicated toward the Di end-member composition. Di_{100} glass has an average of two nonbridging oxygens per tetrahedral cation. However, although individual peaks in the Si K-edge region are not resolved, as in corresponding ^{29}Si MAS NMR spectra (KIRKPATRICK *et al.,* 1986), the Di_{100} glass may contain local structural entities with different polymerization (Q^0, Q^1, Q^2, Q^3 and Q^4). These results are in good agreement with ^{29}Si MAS NMR (MURDOCH *et al.,* 1985; KIRKPATRICK *et al.,* 1986) and Raman spectroscopy (MCMILLAN, 1984b).

Al K-edge XANES spectra of $NaAlSi_2O_6$-$NaAlSi_3O_8$ *glasses*

Figure 3 shows the Al K-edge XANES spectra of high-pressure glasses along the $NaAlSi_2O_6$ (Jd)-$NaAlSi_3O_8$ (Ab) join. The Al K-edge spectra for these glasses of different composition are generally similar. The Al K-edge peak is prominent (shown normalized to one unit in intensity), and is assigned

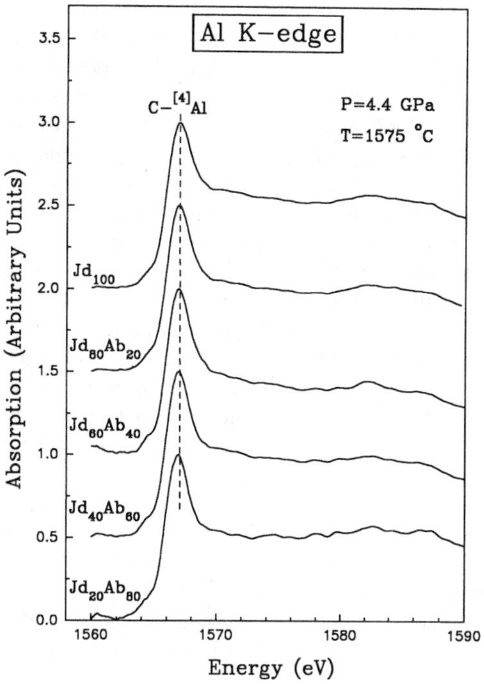

FIG. 3. Aluminum K-edge XANES spectra of high-pressure glasses along the $NaAlSi_2O_6$ (Jd)-$NaAlSi_3O_8$ (Ab) join. The prominent edge peak is attributed to $^{[4]}Al$. A shoulder at about 1568.5 eV becomes more prominent with increasing content of Jd.

to the transitions of Al 1s electrons to antibonding 3p-like t_2 states, by analogy with the Si K-edge peak, as in LI *et al.* (1995d). The energy position of the Al K-edge peak indicates that Al, in general,

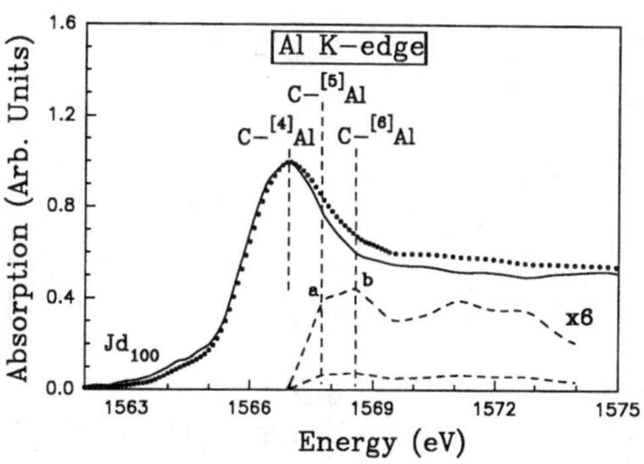

FIG. 4. Comparison of Al K-edge spectra of Jd_{100} glass (dot line) and crystalline albite (solid line). The $^{[4]}Al$ K-edge peak is normalized to one unit in intensity, and the difference spectrum (dash line) between Jd_{100} glass and albite represents, in the immediate post-peak region, non-$^{[4]}Al$ contributions to the glass spectrum.

is four-fold coordinated (Li *et al.,* 1995d). However, a weak shoulder at about 1568.5 eV appears in the $Jd_{60}Ab_{40}$ spectrum and becomes more evident in the Jd_{100} spectrum (Fig. 4).

Figure 4 compares the Al K-edge spectra of the 4.4 GPa Jd_{100} glass (dot line) and crystalline albite (solid line). Aluminum is four-fold coordinated in albite. Therefore, the difference spectrum (dash line) between the Jd_{100} glass and crystalline albite represents some other contribution to the Jd_{100} spectrum, in addition to $^{[4]}$Al.

Figure 5 compares the difference spectra (dot lines) of Jd_{100}, $Jd_{80}Ab_{20}$ and $Jd_{60}Ab_{40}$ glasses, derived as in Fig. 4, with the Al K-edge spectra (solid lines) of crystalline corundum, containing $^{[6]}$Al only, and crystalline andalusite, containing both $^{[5]}$Al and $^{[6]}$Al, as reference spectra (Li *et al.,* 1995d). The interval between adjacent data points in the region of the Al K-edge is 0.1 eV. Peaks a and b in the difference spectra of these glasses evidently correspond to the Al K-edge peaks of $^{[5cb}$Al and $^{[6]}$Al, respectively (Li *et al.,* 1995d). Peak a is attributed to $^{[5]}$Al and further assigned to transitions of Al 1s electrons to antibonding $^{[5]}$Al 3p-like states. Peak b is attributed to $^{[6]}$Al and assigned to transitions of 1s electrons to antibonding $^{[6]}$Al 3p-like t_{1u} states.

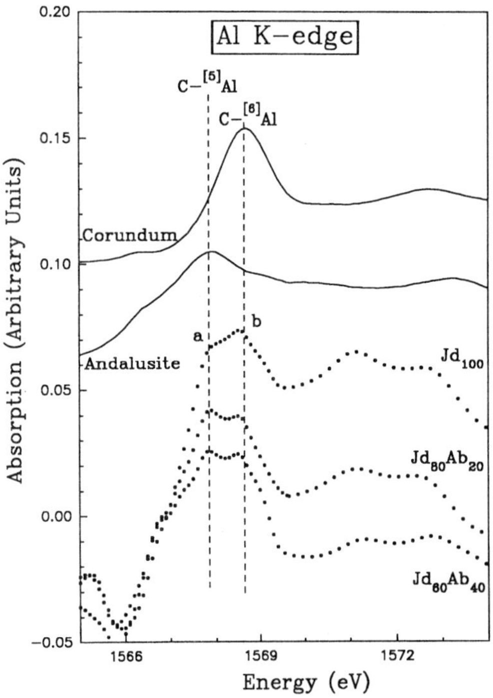

FIG. 5. The Al K-edge difference spectra of high-pressure glasses and crystalline albite (dot lines) compared with Al K-edge spectra (solid lines) of crystalline corundum ($^{[6]}$Al only) and andalusite ($^{[5]}$Al and $^{[6]}$Al). Peaks a and b in the difference spectra of glasses correspond to the Al K-edge peak of $^{[5]}$Al and $^{[6]}$Al, respectively.

The Al K-edge XANES spectra of glasses along the Jd-Ab join indicate that at 4.4 GPa no or little $^{[6]}$Al is present in $Jd_{20}Ab_{80}$ and $Jd_{40}Ab_{60}$ glasses, but $^{[6]}$Al appears in $Jd_{60}Ab_{40}$ glass and becomes more evident with increasing Jd composition, to Jd_{100} glass. These results also indicate that the pressure required for the four- to six-fold coordination change of Al in jadeite composition glass (*e.g.,* 4 GPa) is lower than that for albite glass quenched from the same temperature. The relative abundances of the different Al species in these glasses have been estimated from the relative intensity (simply represented by peak height) of the Al K-edge peak (Table 2). The abundances of both $^{[5]}$Al and $^{[6]}$Al species are estimated to be less than 1% in the $Jd_{20}Ab_{80}$ and $Jd_{40}Ab_{60}$ glasses, about 2.5% in the $Jd_{60}Ab_{40}$ glass, and increase to about 6% in the Jd_{100} glass.

The temperature-pressure-composition conditions for which Al^{3+} may become six-fold coordinated in silicate melts have been debated among geochemists and geophysicists for some time. Computer simulation studies have suggested that pressures beyond 10 GPa may be required for the quantitative conversion of four- to six-fold coordinated Al (Angell *et al.,* 1982). Hochella and Brown (1985) studied the structures of albite and jadeite composition glasses quenched from 1 and 1.5 GPa using X-ray scattering radial distribution functions, and reported no change in coordination of Al, observing only distortion of tetrahedral sites in the high-pressure materials. Ohtani *et al.* (1985) reported $^{[6]}$Al in high-pressure glasses using ^{27}Al MAS NMR, but their results were not reproduced by Stebbins and Sykes (1990). The present result of increasing $^{[5]}$Al and $^{[6]}$Al toward the Di end-member composition is consistent with the temperature-pressure-structure-composition relations along the $NaAlSiO_4$-SiO_2 join. However, the crystal-chemical control favoring $^{[6]}$Al (and $^{[5]}$Al) at the metasilicate ($NaAlSi_2O_6$) composition is not presently evident. The structure of high-pressure glasses beyond this composition (i.e., along the $NaAlSi_2O_6$-$NaAlSiO_4$ join) is now of considerable interest.

Si K-edge XANES spectra of $Na_2Si_2O_5$, $Na_2Si_4O_9$ and $K_2Si_4O_9$ glasses

Figure 6 shows the Si K-edge XANES spectra of the high-pressure phases ϵ-$Na_2Si_2O_5$ and wadeite-structure $K_2Si_4O_9$, and high-pressure glasses of composition $Na_2Si_2O_5$, $Na_2Si_4O_9$ and $K_2Si_4O_9$. The energy position of the Si K-edge peak indicates

Table 2. Abundances (%) of Al species in glasses along Jd-Ab join at 4.4 GPa

	$^{[4]}Al$			$^{[5]}Al$			$^{[6]}Al$		
Sample	Al K-edge*			Al K-edge*			Al K-edge*		
	Position	Height	Abundance	Position	Height	Abundance	Position	Height	Abundance
Jd_{100}	1566.6	1.0	87.7	1567.7	0.066	5.8	1568.6	0.074	6.5
$Jd_{80}Ab_{20}$	1566.6	1.0	92.4	1567.7	0.043	4.0	1568.5	0.040	3.6
$Jd_{60}Ab_{40}$	1566.5	1.0	95.0	1567.7	0.026	2.5	1568.6	0.025	2.5
$Jd_{40}Ab_{60}$	1566.5	1.0	~100.0	-	-	≤1.0	-	-	≤1.0
$Jd_{80}Ab_{80}$	1566.5	1.0	~100.0	-	-	≤1.0	-	-	≤1.0
Andalusite				1567.8					
Corundum							1568.8		

* Errors for the position of Al K-edge are ±0.1 eV; the height of Al K-edge is in arbitrary units. The errors for abundance values of $^{[5]}Al$ and $^{[6]}Al$ are ±30%.

that Si is dominantly four-fold coordinated in the glasses. The post-edge features for the glasses are more or less washed out compared to those of the corresponding crystalline materials, as expected for the disordered structure of glasses. The Si K-edges of ϵ-$Na_2Si_2O_5$ (#1675) and the $Na_2Si_2O_5$ glasses are shifted to lower energy compared with vitreous $Na_2Si_4O_9$, in good agreement with the lower degree of polymerization of SiO_4 tetrahedra expected for the disilicate composition. The structure of epsilon sodium disilicate is based on a distorted disilicate sheet (Q^3) of alternating six-membered rings of UUUUDD and DDDDUU SiO_4 tetrahedra (FLEET and HENDERSON, 1995a). The slight upward shift in the Si K-edge of the disilicate glasses may reflect greater distortion of the disilicate rings at higher pressure. FLEET and HENDERSON (1995a) noted that ϵ-$Na_2Si_2O_5$ may be a model structure for the densification of silicate melts to moderate pressure, with densification accommodated dominantly by decrease in dihedral bond angle through crimping of ring structures.

In recent experiments with $Na_2Si_2O_5$ composition (FLEET and HENDERSON, work in progress), the ϵ-$Na_2Si_2O_5$ phase is stable to about 8.5 GPa, at which point it breaks down to $Na_2Si_3O_7$ (which has a novel structure with silicon in both 4- and 6-fold coordination and [$^{[6]}Si$: $^{[4]}Si$ = 1 : 2]; FLEET and HENDERSON, 1995b), and an unidentified phase. Thus, $^{[4]}Si$ alone persists in the subsolidus to high pressure, and large amounts of $^{[6]}Si$ were not expected in the present sodium disilicate glasses. Similarly, sodium tetrasili-

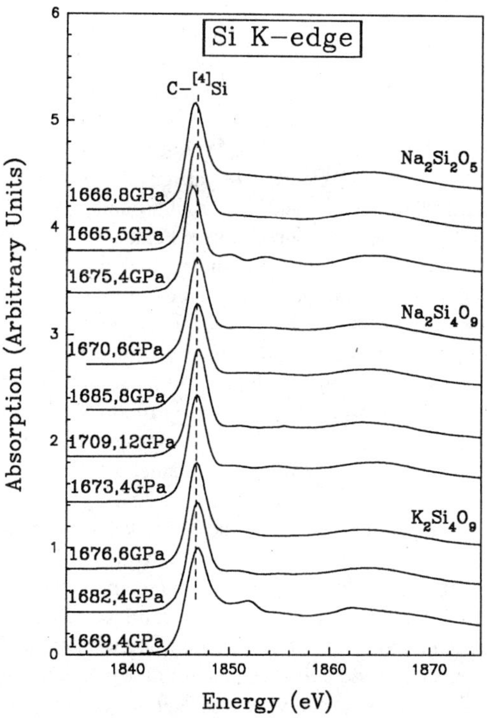

FIG. 6. Silicon K-edge XANES spectra of crystalline ϵ-$Na_2Si_2O_5$ (#1675), wadeite-structure $K_2Si_4O_9$ (#1669) and high-pressure $Na_2Si_2O_5$, $Na_2Si_4O_9$ and $K_2Si_4O_9$ glasses. The preparation conditions of the glasses are given in Table 1.

cate ($Na_6Si_3[Si_9O_{27}]$) (with $^{[6]}Si : ^{[4]}Si = 1 : 3$; FLEET, 1996) appears only above about 6 GPa. For example, experiment #1673 (Table 1, Fig. 6) yielded a product that was amorphous by optical examination, and this is confirmed by the suppressed post-edge features of its Si K-edge XANES spectrum. Therefore, we did not expect to see prominent features of $^{[6]}Si$ in the edge structures of the Si K-edge XANES spectra for the vitreous high-pressure sodium silicates. On the other hand, wadeite-structure $K_2Si_4O_9$ (#1669) is stable to much lower pressures (*e.g.*, SWANSON and PREWITT, 1983), no doubt because of the larger ionic radius of K^+. A high energy shoulder is present at about 1849.0 eV in the Si K-edge spectrum of #1669, and represents the overlapped peak of C-$^{[6]}Si$ (LI *et al.*, 1994a). A similar shoulder is also present in the Si K-edge spectra of the two other high-pressure $K_2Si_4O_9$ glasses.

Figure 7 shows data reduction for the $K_2Si_4O_9$ glass quenched at 6 GPa (#1676). Fig. 7a compares the Si K-edge XANES spectra of the $K_2Si_4O_9$ glass (dot line) and room-pressure SiO_2 glass (solid line). The Si K-edge peak C-$^{[4]}Si$ of $K_2Si_4O_9$ glass and SiO_2 glasss are normalized to one unit in intensity. Because Si is all four-fold coordinated in room-pressure SiO_2 glass, the difference spectrum (dash line) between $K_2Si_4O_9$ glass and SiO_2 glass represents the contribution of $^{[5]}Si$ and/or $^{[6]}Si$ to the Si K-edge spectrum of the $K_2Si_4O_9$ glass. The Si K-edge spectra (dot lines) of $^{[4]}Si$ and $^{[6]}Si$ are curve-fitted into the corresponding Gaussian components (dash lines) in Figs. 7b and 7c, respectively, in which the solid curves are the envelopes of the fitted components and the backgrounds are included in the fitting procedures. Peak a is at about 1848.8 eV, very close to the typical position of C-$^{[6]}Si$ for crystalline materials containing six-fold coordinated Si at 1849.0 eV. Thus, peak a is attributed to $^{[6]}Si$ and assigned to transitions of Si 1s electrons to the antibonding 3p-like t_{1u} states (LI *et al.*, 1994a). The integrated areas of the $^{[4]}Si$ and $^{[6]}Si$ edge peaks are derived and used for the semi-quantitative estimation of abundance of $^{[6]}Si$ species (Table 3).

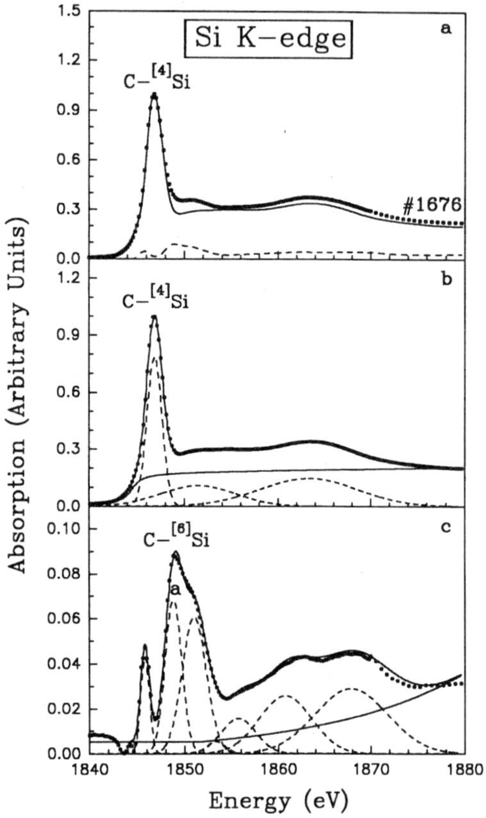

FIG. 7. Data reduction of the Si K-edge spectrum of the $K_2Si_4O_9$ glass (sample 1676) at 6 GPa. (a) Comparison of Si K-edge spectra of the $K_2Si_4O_9$ glass (dot line) and room-pressure SiO_2 glass (solid line), the dash curve is their corresponding difference spectrum. (b) Room-pressure SiO_2 glass spectrum (filled dot line) and (c) the difference spectrum (open dot line) are curve-fitted into the corresponding Gaussian components (dash lines). The solid curves are the corresponding envelopes of the fitted Gaussian components. The backgrounds are included in the fitting procedures.

As noted above, crystalline $K_2Si_4O_9$ has the wadeite structure (SWANSON and PREWITT, 1983), for which $^{[6]}Si : ^{[4]}Si = 1 : 3$. Following the procedure in Fig. 7, the abundance of $^{[6]}Si$ in crystalline $K_2Si_4O_9$ is estimated to be about 22 % based on the integrated areas of $^{[6]}Si$ and $^{[4]}Si$ edge peaks, in good agreement with its X-ray crystal structure. The present analysis reveals little $^{[6]}Si$ is in $Na_2Si_2O_5$ composition glasses up to 8 GPa, and in $Na_2Si_4O_9$ composition glasses up to 8–12 GPa, as anticipated from the foregoing discussion. A trace of $^{[6]}Si$ is present in $Na_2Si_4O_9$ glass at 12 GPa. As expected, the $K_2Si_4O_9$ glasses have the highest abundance of $^{[6]}Si$ among the glasses investigated. For the present series of $Na_2Si_4O_9$ and $K_2Si_4O_9$ glasses, the abundance of $^{[6]}Si$ increases with increasing pressure, in good agreement with ^{29}Si MAS NMR spectroscopy (Table 3; XUE *et al.*, 1991). However, there are two principal discrepancies between the Si K-edge XANES and ^{29}Si MAS NMR measurements. First, little or no $^{[5]}Si$ is observed in the Si K-edge XANES spectra of the three series of glasses studied, but this may be because the $^{[5]}Si$ signal is too weak to be resolved in the Si K-edge spectra. Second, the ^{29}Si MAS NMR spectra showed a higher abundance of $^{[6]}Si$ in $Na_2Si_4O_9$ glasses than in $K_2Si_4O_9$ at the same pressure.

Table 3. Abundance of $^{[6]}Si$ in $Na_2Si_2O_5$, $Na_2Si_4O_9$, and $K_2Si_4O_9$ glasses, ϵ-$Na_2Si_2O_5$ and wadeite-structure $K_2Si_4O_9$ from Si K-edge XANES and ^{29}Si MAS NMR

Sample	P (GPa)	$^{[4]}Si$ edge (± 0.1 eV)	Si K-edge* $^{[6]}Si$	^{29}Si MAS NMR** $^{[5]}Si$	^{29}Si MAS NMR** $^{[6]}Si$
$Na_2Si_2O_5$					
1666	8	1846.6	≤ 1.0	1.5	~ 1.0
1665	5	1846.7	≤ 1.0	≤ 0.3	≤ 0.3
1675	4	1846.3	0.0		
$Na_2Si_4O_9$					
1709	12	1846.9	3.0	8.5	6.3
1685	8	1846.8	≤ 1.0	4.8	4.4
1670	6	1846.8	≤ 1.0	3.0	2.1
1673	4	1846.8	0.0		
$K_2Si_4O_9$					
1676	6	1846.9	8.9	3.9	1.5
1682	4	1846.8	5.0	1.5	0.2
1669	4	1846.9	22.0		

* Errors for the abundance values from Si K-edge XANES are $\pm 30\%$; ideal value is 0 for ϵ-$Na_2Si_2O_5$ (#1675), and 25 for wadeite-structure $K_2Si_4O_9$ (#1669).
** The abundance values from ^{29}Si MAS NMR are cited in XUE *et al.* (1991).

Application of Si and Al K-edge XANES spectroscopy to study of glass structure

Silicon and Al K-edge XANES spectroscopy provide direct information on the nearest-neighbour coordination and intermediate-range structure in silicate and aluminosilicate glasses. The results to date are comparable to those of ^{29}Si and ^{27}Al MAS NMR spectroscopy, respectively, although detailed comparisons between these two techniques vary with the chemical element probed. ^{29}Si MAS NMR has superior resolution of spectral features compared with Si K-edge XANES, but Al K-edge XANES appears rather more useful than ^{27}Al MAS NMR. Also, XANES with synchrotron radiation requires only small (milligram) amounts of sample, and this is a distinct advantage for studying the products of high-pressure experiments.

The chemical and structural features revealed by Si and Al K-edge XANES collected by total electron yield (TEY) are not dominated by surface and near-surface contributions that are unrepresentative of the bulk glasses. Our experience to date is that Si K-edge XANES collected by TEY and fluorescence yield (FY) give comparable results. Fluorescence yield has appreciably greater sampling depth than TEY. This was elegantly demonstrated by KASRAI *et al.* (1996), who used films of amorphous SiO_2 of different thickness on silicon substrates to determine sampling depths of 70 nm for Si K-edge XANES collected by TEY, and several hundred nm for comparable spectra collected by FY.

The pressure-induced changes in coordination of Si and Al presently reported are all close to the minimum detection limit of XANES spectroscopy. The estimated abundances of $^{[6]}Si$ in the alkali silicate glasses and of $^{[5]}Al$ and $^{[6]}Al$ in the $NaAlSi_2O_6$-$NaAlSi_3O_8$ glasses are at best only semi-quantitative. Therefore, it is premature to attach too much significance to the discrepancies in abundance of $^{[6]}Si$ in the high-pressure alkali silicate glasses with the ^{29}Si MAS NMR study of XUE *et al.* (1991), although we have noted that the XANES results are more consistent with the stability of $^{[6]}Si$ in the subsolidus phases.

We also recognize that there is ambiguity in assigning specific structural features to shoulders on the edge peak of XANES spectra. However, the position of the edge peak is determined largely by the effective charge on the absorber atom (LI *et al.*, 1995a), which varies with nearest-neighbour coordination, degree of polymerization, network and network-modifying cation substitution, and so on. In silicate glasses, compression and distortion of TO_4 tetrahedral ring structures will be of subordinate importance, and are manifest largely by shift in position of the whole edge peak (rather than by development of a shoulder), as well as change in FWHM.

In the absence of a quantitative unifying theory for XANES spectra, the use of labelled compounds is essential for their meaningful interpretation. We have used ϵ-$Na_2Si_2O_5$, wadeite-structure $K_2Si_4O_9$ and room-pressure SiO_2 glass to interpret the Si K-edge XANES spectra of the high-pressure alkali silicate glasses. Comparison of the Si K-edge XANES spectra of SiO_2 glass, quartz and stishovite (LI *et al.,* 1994a) with those of wadeite-structure $K_2Si_4O_9$ and the two high-pressure $K_2Si_4O_9$ glasses strongly supports the present structure assignments for the high-pressure alkali silicate glasses. The estimate of 22 % for the proportion of $^{[6]}Si$ in crystalline $K_2Si_4O_9$ (in good agreement with the ideal value of 25 %) was determined directly from the excess absorption of the edge peak after subtraction of the contribution due to $^{[4]}Si$. Unfortunately, for the study of the high-pressure $NaAlSi_2O_6$-$NaAlSi_3O_8$ glasses, we omitted to collect Al K-edge XANES spectra for the corresponding room-pressure glasses, and therefore used crystalline albite as the reference material.

Acknowledgements—We thank G. S. Henderson and J. A. Tossell for helpful reviews of the manuscript, H. Keppler for provision of $CaMgSi_2O_6$-$NaAlSi_3O_8$ glass samples, R. A. Secco for synthesis of $NaAlSi_2O_6$-$NaAlSi_3O_8$ glasses, Y. Thibault and R. W. Luth of C. M. Scarfe Laboratory of Experimental Petrology, University of Alberta, for running the experiments on $Na_2Si_2O_5$, $Na_2Si_4O_9$ and $K_2Si_4O_9$ samples, and M. Kasrai for assistance with graphics. We also acknowledge X. H. Feng at the Canadian Synchrotron Radiation Facility and staff at the Synchrotron Radiation Center (SRC), University of Wisconsin, for their technical assistance, and the National Science Foundation (NSF) for support of the SRC. This work was supported by the Natural Sciences and Engineering Research Council of Canada.

REFERENCES

ANGELL C. A., CHEESEMAN P. A. and TAMADDON S. (1982) Pressure enhancement of ion mobility in liquid silicates from computer simulation studies to 800 kilobars. *Science* **218,** 885–887.

DAVOLI I., PARIS E., STIZZA, S., BENFATO M., FANFONI M., GARGANO A., BIANCONI A. and SEIFERT F. (1992) Structure of densified vitreous silica: Silicon and oxygen XANES spectra and multiple scattering calculations. *Phys. Chem. Minerals* **19,** 171–175.

DUPREE R., HOLLAND D. and MORTUZA M. G. (1987) Six-coordinated silicon in glasses. *Nature* **328,** 416–417.

FARNAN I. and STEBBINS J. F. (1990) A high temperature ^{29}Si investigation of solid and molten silicates. *J. Amer. Chem. Soc.* **112,** 32–39.

FINGER L. W. and HAZEN R. M. (1991) Crystal chemistry of six-coordinated silicon: a key to understanding the Earth's interior. *Acta Cryst.* **B47,** 561–580.

FLEET M. E. (1996) Sodium tetrasilicate: A complex high-pressure framework silicate ($Na_6Si_3[Si_9O_{27}]$). *American Mineralogist,* in press.

FLEET M. E. and HENDERSON G. S. (1995a) Epsilon sodium disilicate: A high pressure layer structure. *J. Solid State Chem.* **119,** 400–404.

FLEET M. E. and HENDERSON G. S. (1995b) Sodium trisilicate: A new high-pressure silicate structure ($Na_2Si[Si_2O_7]$). *Phys. Chem. Minerals* **22,** 383–386

GALOISY L. and CALAS G. (1994) Network-forming Ni in silicate glasses. *Amer. Mineral.* **77,** 677–680.

GIBBS J. H. and DIMARZIO E. A. (1958) Nature of the glass transition and the glassy state. *J. Chem. Phys.* **28,** 373–383.

GREAVES G. N., FONTAINE A., LAGARDE P., RAOUS D. and GURMAN S. J. (1981) Local structure of silicate glasses. *Nature* **293,** 611–616.

HEMLEY R. J., MAO H. K., BELL P. M. and MYSEN B. O. (1986) Raman spectroscopy of SiO_2 glass at high pressure. *Phys. Rev. Lett.* **57,** 747–750.

HOCHELLA M. F. and BROWN G. E. (1985) The structures of albite and jadeite composition glasses quenched from high pressure. *Geochim. Cosmochim. Acta* **49,** 1137–1142.

ITO E. and TAKAHASHI T. (1987) Ultrahigh-pressure phase transformations and theconstitution of the deep mantle. In *High-Pressure Research in Mineral Physics* (eds. M. H. MANGHNANI and Y. SYONO), *Geophys. Mono.* **39,** 221–229, American Geophysical Union, Washington D.C.

JACKSON W. E., MUSTRE DE LEON J., BROWN G. E., WAYCHUNAS G. A., CONRADSON S. D. and COMBES J. M. (1993) High-temperature XAS study of Fe_2SiO_4 liquid: Reduced coordination of ferrous iron. *Science* **262,** 229–233.

JEANLOZ R. (1990) The nature of the Earth's core. *Ann. Rev. Earth Planet. Sci.* **18,** 357–386.

KASRAI M., LENNARD W. N., BRUNNER R. W., BANCROFT G. M., BARDWELL J. A. and TAN K. H. (1966) Sampling depth of total electron and fluorescence measurements in Si L- and K-edge absorption spectroscopy. *Appl. Surface Science,* submitted.

KEPPLER H. (1992) Crystal field spectra and geochemistry of transition metal ions in silicate melts and glasses. *Amer. Mineral.* **77,** 62–77.

KIRKPATRICK R. J. (1988) MAS NMR spectroscopy of minerals and glasses. In *Spectroscopic Methods in Mineralogy and Geochemistry* (ed. F.C. HAWTHORNE), *Rev. Mineral.* **18,** 341–403, Mineralogical Society of America, Washington D.C.

KIRKPATRICK R. J., OESTRIKE R. and WEISS C. A. (1986) High-resolution ^{27}Al and ^{29}Si NMR spectroscopy of glasses and crystals along the join $CaMgSi_2O_6$-$CaAl_2SiO_6$. *Amer. Mineral.* **71,** 705–711.

LI DIEN, BANCROFT G. M., KASRAI M, FLEET M. E., SECCO R. A., FENG X. H., TAN K. H. and YANG B. X. (1994a) X-ray absorption spectroscopy of silicon dioxide (SiO_2) polymorphs: The structural characterization of opal. *Amer. Mineral.* **79,** 622–632.

LI DIEN, BANCROFT G. M., KASRAI M, FLEET M. E., FENG X. H. and TAN, K. H. (1994b) High-resolution Si and P K- and L-edge XANES spectra of crystalline SiP_2O_7 and amorphous SiO_2-P_2O_5. *Amer. Mineral.* **79,** 785–788.

LI DIEN, FLEET M. E., BANCROFT G. M., KASRAI M. and HENDERSON G. S. (1994c) Pressure-induced coordination change of Si in alkali silicate glasses by Si K-edge X-ray absorption spectroscopy (XAS). *EOS* **75,** 371.

LI DIEN, BANCROFT G. M., FLEET M. E. and FENG

X. H. (1995a) Silicon K-edge XANES spectra of silicate minerals. *Phys. Chem. Minerals.*, **22,** 115–122.

LI DIEN, FLEET M. E., BANCROFT G. M., KASRAI M. and PAN Y. (1995b) Local structure of Si and P in SiO_2-P_2O_5 and Na_2O-SiO_2-P_2O_5 glasses: A XANES study. *J. Non-Cryst. Solids.*, **188,** 181–189.

LI DIEN, BANCROFT G. M. and FLEET M. E. (1995c) Coordination of Si in Na_2O-SiO_2-P_2O_5 glasses using Si K- and L-edge XANES. *Amer. Mineral.*, in press.

LI DIEN, BANCROFT G. M. and FLEET M. E. (1995d) Aluminum K-edge XANES spectra of aluminosilicate minerals. *Amer. Mineral.*, **80,** 432–440.

LI DIEN, SECCO R. A., BANCROFT G. M. and FLEET M. E. (1995e) Pressure-induced coordination change of Al in silicate melts from Al K-edge XANES of high-pressure $NaAlSi_2O_6$-$NaAlSi_3O_8$ glasses. *Geophys. Res. Lett.*, in press.

LIEBAU F. (1985) *Structural Chemistry of Silicates.* Springer-Verlag, Berlin.

MCKEOWN, D. A., WAYCHUNAS, G. A. and BROWN G. E. (1985) EXAFS study of the coordination environment of aluminum in a series of silica-rich glasses and selected minerals within the Na_2O-Al_2O_3-SiO_2 system. *J. Non-Cryst. Solids* **74,** 349–371.

MCMILLAN P. F. and PIRIOU B. (1983) Raman spectroscopic studies of silicate and related glass structure: A review. *Bull. Mineral.* **106,** 57–75.

MCMILLAN P. F. (1984a) Structural studies of silicate glasses and melts —— Applications and limitations of Raman spectroscopy. *Amer. Mineral.* **69,** 622–644.

MCMILLAN P. F. (1984b) A Raman spectroscopic study of glasses in the system CaO-MgO-SiO_2. *Amer. Mineral.* **69,** 645–659.

MCMILLAN P. F., WOLF G. H. and POE B. T. (1992) Vibrational spectroscopy of silicate liquids and glasses. *Chem. Geol.* **96,** 351–366.

MCMILLAN P. F., POE B. T., GILLET PH. and REYNARD B. (1994) A study of SiO_2 glass and supercooled liquid to 1950 K via high temperature Raman spectroscopy. *Geochim. Cosmochim. Acta* **58,** 3653–3664.

MURDOCH J. B., STEBBINS J. F. and CARMICHAEL I. S. E. (1985) High-resolution ^{29}Si NMR study of silicate and aluminosilicate glasses. The effect of network-modifying cations. *Amer. Mineral.* **70,** 332–343.

MYSEN B. O. (1988) *Structure and properties of silicate melts.* Elsevier, Amsterdam.

MYSEN B. O. (1990) Role of Al in depolymerized, peralkaline aluminosilicate melts in the systems Li_2O-Al_2O_3-SiO_2, Na_2O-Al_2O_3-SiO_2, and K_2O-Al_2O_3-SiO_2. *Amer. Mineral.* **75,** 120–134.

MYSEN B. O. (1995) Structural behavior of Al^{3+} in silicate melts: In situ, high-temperature measurements as a function of bulk chemical composition. *Geochim. Cosmochim. Acta* **59,** 455–474.

MYSEN B. O. and FRANTZ J. D. (1993) Structure and properties of alkali silicate melts at magmatic temperatures. *Eur. J. Mineral.* **5,** 393–407.

MYSEN B. O. and FRANTZ J. D. (1994) Structure of haplobasaltic liquids at magmatic temperature: In situ, high temperature study of melts on the join $Na_2Si_2O_5$-$Na_2 \times (NaAl)_2O_5$. *Geochim. Cosmochim. Acta* **58,** 1711–1733.

OHTANI E., TAULELLE F. and ANGELL, C. A. (1985) Al^{3+} coordination changes in liquid aluminosilicates under pressure. *Nature* **314,** 78–81.

POE B. T., MCMILLAN P. F., COTE B., MASSIOT D. and COUTURES J. P. (1993) The Al environment in $MgAl_2O_4$ and $CaAl_2O_4$ liquids: An in situ study via high temperature NMR. *Science* **259,** 786–788.

SEKIYA T., MOCHIDA N., OHTSUKA A. and UCHIDA K. (1988) 6-Coordinated Si^{4+} in SiO_2-$PO_{5/2}$ glasses — ^{29}Si MAS NMR method. *Nippon Seramikkusu Kyokai Gakujutsu Ronbunshi* **96,** 571–573.

SMYTH J. R. and BISH D. L. (1988) *Crystal structures and cation sites of the rock-forming minerals.* Allen and Unwin, Boston.

STEBBINS J. F. (1987) Identification of multiple structural species in silicate glasses by ^{29}Si NMR. *Nature* **330,** 465–467.

STEBBINS J. F. (1988a) Effect of temperature and composition on silicate glass structure and dynamics: Si-29 NMR results. *J. Non-Cryst. Solids* **106,** 359–369.

STEBBINS J. F. (1988b) NMR spectroscopy and dynamic process in mineralogy and geochemistry. In *Spectroscopic Methods in Mineralogy and Geochemistry* (ed. F. C. HAWTHORNE), *Rev. Mineral.* **18,** 405–429, Mineralogical Society of America, Washington D.C.

STEBBINS J. F. (1991) NMR evidence for five-coordinated silicon in a silicate glass at atmospheric pressure. *Nature* **351,** 638–639.

STEBBINS J. F. and FARNAN I. (1989) Nuclear magnetic resonance spectroscopy in the earth sciences: Structure and dynamics. *Science* **245,** 257–263.

STEBBINS J. F. and MCMILLAN P. (1989) Five- and six-coordinated Si in $K_2Si_4O_9$ liquid at 1.9 GPa and 1200°C. *Amer. Mineral.* **74,** 965–968.

STEBBINS J. F. and SYKES D. (1990) The structure of $NaAlSi_3O_8$ liquid at high pressure: new constraints from NMR spectroscopy. *Amer. Mineral.* **75,** 943–946.

STOLPER E. M. and AHRENS T. J. (1987) On the nature of pressure-induced coordination changes in silicate melts and glasses. *Geophys. Res. Lett.* **14,** 1231–1233.

SWANSON D. K. and PREWITT C. T. (1983)The crystal structure of $K_2Si^{vi}Si_3^{iv}O_9$. *Amer. Mineral.* **58,** 581–585.

TAYLOR M. and BROWN G. E. (1979) Structure of silicate mineral glasses. I. The feldspar glasses $NaAlSi_3O_8$, $KAlSi_3O_8$, $CaAl_2Si_2O_8$. *Geochim. Cosmochim. Acta* **43,** 1467–1473.

WAYCHUNAS G. A., BROWN G. E., PONADER C. W. and JACKSON W. E. (1988) Evidence from X-ray absorption for network-forming Fe^{2+} in molten silicates. *Nature* **332,** 251–253.

WILLIAMS Q. and JEANLOZ R. (1988) Spectroscopic evidence for pressure-induced coordination changes in silicate glasses and melts. *Science* **239,** 902–905.

WONG J. and ANGELL C. A. (1976) *Glass Structure by Spectroscopy.* Marcel Dekker Inc., New York.

XUE X., STEBBINS J. F., KANZAKI M. and TRØNNES R. G. (1989) Silicon coordination and speciation changes in a silicate liquid at high pressures. *Science* **245,** 962–964.

XUE X., STEBBINS J. F., KANZAKI M., MCMILLAN P. F. and POE B. (1991) Pressure-induced silicon coordination and tetrahedral structural changes in alkali oxide-silica melts up to 12 GPa: NMR, Raman, and infrared spectroscopy. *Amer. Mineral.* **76,** 8–26.

YANG B. X., MIDDLETON F. H., OLSSON B. G., BANCROFT G. M., CHEN J. M., SHAM T. K., TAN K. H. and WALLACE J. D. (1992) The design and performance of a soft X-ray double crystal monochromator beamline at Aladdin. *Nucl. Instr. Methods Phys. Res.* **A316,** 422–436.

ZACHARIASEN W. H. (1932) The atomic arrangement in glass. *J. Amer. Chem. Soc.*, **54,** 3841–3851.

Mineral Spectroscopy: A Tribute to Roger G. Burns
Editors: M. D. Dyar, C. McCammon and M. W. Schaefer

In situ microanalysis of $Fe^{3+}/\Sigma Fe$ in amphibole by x-ray absorption near edge structure (XANES) spectroscopy

J. S. DELANEY,[1] S. BAJT,[2] S. R. SUTTON,[2,3] and M. D. DYAR[4]

[1]Department of Geology, Rutgers University, New Brunswick, NJ 08903 U.S.A.
[2]Consortium for Advanced Radiation Sources, University of Chicago, Chicago, IL 60637 U.S.A.
[3]Department of Geophysical Sciences, University of Chicago, Chicago, IL 60637 U.S.A.
[4]Department of Geology and Astronomy, West Chester University, West Chester, PA 19383 U.S.A.

Abstract—Determination of $Fe^{3+}/\Sigma Fe$ in micrometer scale volumes on individual mineral grains has long been a goal of petrologists and crystallographers. Such measurements can be made using Synchrotron microXANES (SmX) spectroscopy. Comparison of microXANES determinations of the $Fe^{3+}/\Sigma Fe$ in $50 \times 30 \mu m$ areas on polished amphibole samples, with Mössbauer and wet chemical analyses of the bulk material on which the SmX analyses were made, shows generally good matches. Results presented indicate that the $Fe^{3+}/\Sigma Fe$ of amphiboles can be calibrated against data from simple oxides and silicates despite the known influence of the crystal structure on the position and intensity of the measured features of XANES spectra. The measurement technique used successfully minimizes crystal chemical influences and permits direct assessment of iron oxidation state. Measurements of the $Fe^{3+}/\Sigma Fe$ in complex silicate minerals such as amphibole can be made, therefore, with comparable spatial resolution to the microanalysis techniques used for major element abundance. Zoning of $Fe^{3+}/\Sigma Fe$ in amphibole crystals has been detected and can be quantified.

INTRODUCTION

ELEMENTS FREQUENTLY occur in multiple oxidation states in geological and extraterrestrial materials. Fe, Cr, Ti, Mn, V, Eu, Ce and U are all elements whose existence in different oxidation states reflects the conditions under which the host phase crystallized. Of these, the most commonly studied is iron. The variability of $Fe^{3+}/\Sigma Fe$ [$Fe^{3+}/(Fe^{3+} + Fe^{2+})$] in minerals has been recognized almost since the beginning of geochemical research, and its significance as an indicator of oxygen fugacity is widely recognized. However, modern reliance on instrumental microanalytical techniques that are incapable of measuring oxidation state in addition to element abundance has resulted in this fundamental parameter often being ignored.

The influence of oxygen fugacity is fundamental in both phase equilibria and geochemical kinetics involving iron (FROST, 1991). Unfortunately, the oxygen fugacity under which a natural rock or assemblage formed is difficult to measure. One possible approach for estimating oxygen fugacity is to establish the oxidation states of multivalent elements such as iron in diagnostic mineral assemblages and compare these results with calibrations of the influence of f_{O_2} on those oxidation states. The coexistence of minerals containing an element in different oxidation states strongly constrains the environment in which equilibration occurred.

Measurements of the $Fe^{3+}/\Sigma Fe$ in minerals have been made using bulk analytical techniques, as no microanalytical technique was available. The measurement of $Fe^{3+}/\Sigma Fe$ at spatial resolutions comparable with those of modern microanalytical techniques (sampling nanogram instead of milligram masses) is a fundamental technological challenge to geochemists. The use of x-ray spectroscopic techniques that have become possible with the availability of intense synchrotron x-ray sources now permits this challenge to be addressed (CALAS *et al.*, 1984; CHEN *et al.*, 1984; BROWN *et al.*, 1988; WAYCHUNAS, 1988). Synchrotron based spectroscopy can be carried out on the K and L absorption edges of iron with sufficient energy resolution to distinguish between ferrous iron and ferric iron (BAJT *et al.*, 1994, CRESSEY *et al.*, 1993). A synchrotron x-ray beam can be focussed and collimated to produce a true microbeam (1–100 micrometers in diameter) permitting the use of petrographic constraints in the selection of the areas to be analyzed.

Other techniques for analysis of $Fe^{3+}/\Sigma Fe$ in micrometer scale volumes are beginning to be explored by HÖFER *et al.* (1994) and ILTON *et al.* (1993). HÖFER *et al.* (1994) have developed a refined version of the ALBEE and CHODOS (1970) electron microprobe technique of monitoring Fe $L\alpha/L\beta$ x-ray line ratios as an indicator of iron oxidation state. The technique appears to have provided consistent $Fe^{3+}/\Sigma Fe$ for a limited suite of samples but is probably limited by the energy resolution (8–20 eV) of wavelength dispersive spectrometers available on current electron microprobes. Uncertainties about changes in soft x-ray line shapes as a function of site coordination have not been addressed in that study (BASTIN and HEIJLIGERS, 1991). Similarly, ILTON *et al.* (1993) use x-ray photoelectron spectroscopy to determine $Fe^{3+}/\Sigma Fe$ in the near surface region (~5 nm thick) of biotite samples in a 2 × 0.1 mm spot. This technique has potential although the areas analyzed are much larger than those associated with true microbeam techniques. In addition, the strong surface sensitivity of XPS may limit its usefulness in petrographically constrained studies.

The energy of the pre-edge peak (1s–3d transition) associated with the Fe-K x-ray absorption edge in amphibole minerals is correlated with the $Fe^{3+}/\Sigma Fe$ of the amphibole. The pre-edge energy can be determined with good energy resolution (~1–1.5 eV) by using synchrotron microXANES spectroscopy (XANES = x-ray absorption near-edge structure.) Early studies on relatively simple minerals show that the energy of the main K-absorption edge shifts 2–3 eV per valence charge, but high resolution spectra show also that the shape and position, of the main absorption edge, is sensitive to the coordination geometry (WAYCHUNAS *et al.*, 1983; SUTTON *et al.*, 1993a). In contrast, pre-edge peak energies appear to be sensitive mainly to element oxidation states (WAYCHUNAS *et al.*, 1983, MANCEAU *et al.*, 1992 a,b) although its intensity is dependent on coordination. The coordination geometry affects the pre-edge energy much less than the main absorption edge. The 3d final states represented by these peaks are more tightly bound and therefore less sensitive to coordination changes in the local environment and solid state effects than the higher energy states associated with the main edges. In the amphibole minerals the potential for complex variation of both composition and structure requires that the features of the absorption edge be studied in great detail to distinguish between oxidation state and coordination related energy shifts. The presence of several crystallographic sites hosting ferric and ferrous iron implies that the interactions that are characteristic of the XANES region will be very complex. *This preliminary study will not attempt to provide that level of coverage.* The original calibration of the pre-edge energy shifts generated using oxide minerals (BAJT *et al.*, 1994) will be compared here with an independent ''calibration'' based on the available bulk analyses of these amphiboles by wet chemistry and Mössbauer spectroscopy.

WHY MEASURE $Fe^{3+}/\Sigma Fe$ IN AMPHIBOLE?

ROBINSON *et al.* (1982) describe amphibole as ''a mineralogical shark in a sea of unsuspecting elements.'' The amphibole minerals are among the most compositionally complex of the silicate mineral groups and have a variety of sites (M1-M4, A) that can accommodate many different cations. Amphibole is one of the few mineral groups in which there is sufficient 'flexibility' in the structure that it can accommodate ferric and ferrous iron in response to both the crystallization environment ***and*** the crystal chemical constraints. Fe^{2+} is generally located in the M1-M3 sites with Fe^{3+} often in M2 (HAWTHORNE, 1981) although in oxyamphibole Fe^{3+} shows a preference for M1 and M3 (PHILLIPS *et al.*, 1988). The responsiveness of amphibole $Fe^{3+}/\Sigma Fe$ to external conditions, makes them potentially valuable indicators of system oxygen fugacity.

The substitution of Fe^{3+} and Fe^{2+} into amphibole may be diagnostic of oxygen and hydrogen fugacities at the time of equilibration. Two major mechanisms for the substitution of Fe^{3+} into amphibole have been discussed in recent work: (i) Charge balance of the excess charge of Fe^{3+} by oxidation/dehydrogenation of amphibole. The end member substitution of ferric iron corresponds to the formation of an oxyamphibole component (by dehydrogenation) (POPP *et al.*, 1990, 1995; DYAR *et al.*, 1993). (ii) Substitution of Fe^{3+} for Al^{3+} (POPP *et al.*, 1990). The strong anti-correlation of Fe^{3+} with H^+ in a variety of kaersutitic amphibole samples is, however, very suggestive that oxidation/dehydrogenation is the dominant mechanism for stabilizing these Fe^{3+} bearing amphiboles (DYAR *et al.*, 1993; POPP *et al.*, 1990, 1995). The sensitivity of $Fe^{3+}/\Sigma Fe$ in amphibole to f_{H_2} and f_{O_2}, in turn, suggests that the $Fe^{3+}/\Sigma Fe$ be dependent on water fugacity of the source region. Thus study of amphibole $Fe^{3+}/\Sigma Fe$ *in situ* provides information on the volatile components extant during crystal growth and/or recrystallization.

SAMPLES STUDIED

Two suites of amphibole samples were used for this study. Both suites were used previously for comprehensive studies of the significance of $Fe^{3+}/\Sigma Fe$ in amphibole crystal chemistry (COSCA *et al.*, 1991; DYAR *et al.*, 1993) and both were previously analyzed by bulk analytical techniques. These samples provide a test of the ability of synchrotron microXANES spectroscopy to reproduce ''conventional'' measurements of the $Fe^{3+}/\Sigma Fe$ of amphibole. Eight hornblendes (Table 1) from amphibolites and granulites of the Grenville Orogen, were described by COSCA *et al.* (1991), have a range of $Fe^{3+}/\Sigma Fe$ between 0.18 and 0.35 and were analyzed by wet chemistry. These Grenville amphiboles have a range of major element compositions and therefore provide a test of the sensitivity of the SmX technique to crystal chemical variables as well as $Fe^{3+}/\Sigma Fe$. Fourteen samples described by DYAR *et al.* (1993) were studied by Mössbauer spectroscopy and have $Fe^{3+}/\Sigma Fe$ between 0.23 to 1.00 (Table 1). These samples

Table 1. Analytical Results

Sample	$Fe^{3+}/\Sigma Fe$ %	XANES Pre-Edge (eV)	Notes
FA86-1	18	-1.54	a
HL8611	31	-0.86	a
HL862C	35	-0.78	a
MIN864	33	-1.16	a,d
MR865A	22	-1.21	a
SSA-5	29	-1.12	a
SSA-10	23	-1.20	a
SSA-13	35	-0.52	a
AK-M1	28.2	-1.31	b
AK-M2	25	-1.18	b
AK-M3	27.7	-0.24	b,c,d
AK-M4	28.8	-0.85	b
AK-M5	36.1	-0.76	b
BA-5	36.6	-0.89	b
DL-9	98.7	0.95	b
FR-11	72.2	-0.74	b,c,d
FR-12	100	0.78	b,d
84-BR	44.4	-0.5	b,c
H366A	34	-0.56; -1.00	b,d,e
H366-92	92.3	0.64	b
Kakanui	28.3	-0.79	b
TM	77	-0.11	b,d

Notes: (a) Wet chemical analysis from COSCA *et al.*, 1991. (b) Mössbauer analysis from DYAR *et al.*, 1993. (c) Published FeO and microprobe analyses of grains studied show slight mismatch (10-15% difference). (d) Fe-oxide inclusions present in the analyzed grains. (e) Heterogeneity of $Fe^{3+}/\Sigma Fe$ detected.

are kaersutite rich mantle megacrysts that grew in an oxidized metasomatic fluid (DYAR *et al.*, 1993). This suite of mantle megacrysts has a narrow range of major element compositions and varies mainly in the $Fe^{3+}/\Sigma Fe$. The similarity of compositions indicates that crystal chemical differences between the mantle amphiboles are small and that these grains are particularly suitable to test the SmX technique as a measure of $Fe^{3+}/\Sigma Fe$.

Chips (diam. 0.5–2.0 μm) of all the amphiboles were mounted on pure SiO_2 glass disks with an epoxy resin adhesive known to be Fe-free and the samples polished to 1 μm relief as for electron microprobe investigation. The samples were analyzed with a JEOL JXA-8600 electron microprobe at the Rutgers University Microanalysis facility (at 15 kV and 15 nA beam current with analyses corrected using standard BENCE and ALBEE type procedures) to confirm that they were the same amphiboles analyzed by COSCA *et al.* (1991) and DYAR *et al.* (1993). Several grains that had major element compositions significantly different from the published analyses of DYAR *et al.* (1993) were rejected at this point. The polished samples were photographed and used for synchrotron microXANES analysis. The thickness of the samples was not controlled rigorously but was typically ~100 μm.

EXPERIMENTAL TECHNIQUES

Synchrotron microXANES spectroscopy of hornblende

Measurements were made at the synchrotron x-ray microprobe (beamline X26A at the National Synchrotron Light Source, Brookhaven National Laboratory, New York). The basis of XANES is the mapping of the x-ray absorption edge of the element of interest with energy resolution comparable to the natural x-ray line width. The $Fe^{3+}/\Sigma Fe$ in 30 × 50 μm areas on polished sections of amphibole grains are derived from measurements of the energy of a pre-edge peak at the Fe K-absorption edge (Fig. 1). Previous *in situ* measurements of the $Fe^{3+}/\Sigma Fe$ in minerals were used to establish a calibration line (Fig. 2a) based on the pre-edge peak energies of synthetic fayalite (Fe_2SiO_4), and natural magnetite (Fe_3O_4) and hematite (Fe_2O_3) as a function of their known $Fe^{3+}/\Sigma Fe$ (BAJT *et al.*, 1994). Details of the beamline operation are given in BAJT *et al.* (1994) who demonstrated that reproducible $Fe^{3+}/\Sigma Fe$ could be determined for a suite of oxide minerals.

XANES spectra were obtained by measuring the Fe Kα fluorescence intensity from the sample as a function of incident beam energy from 65 eV below the main absorption edge energy (7111 eV for Fe) to about 60 eV above the main edge. The incident beam energy was controlled by a Si(III) channel cut monochromator. The incident X-ray energy was incremented by 0.3 eV intervals over the most critical energy range of −10 to +20 eV relative to the main absorption-edge energy. This provides detailed mapping of the relationship between the pre-edge

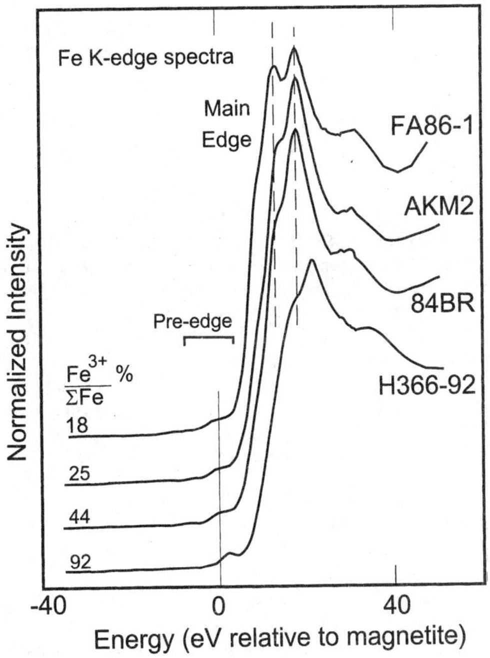

FIG. 1. Synchrotron microXANES spectra of four amphibole samples (FA86–1; AKM2; 84BR and H366–92, with $Fe^{3+}/\Sigma Fe$ of 0.18; 0.25; 0.44; 0.92 respectively) with energies expressed relative to the pre-edge peak energy for Minas Gerais magnetite (USNM standard). Spectra are intensity normalized to the maximum intensity of the main Fe K-absorption edge and to the synchrotron ring current. The individual spectra are offset vertically to facilitate comparison of shapes.

peak and the main absorption edge for comparison with a selected magnetite standard for which the pre-edge position is arbitrarily defined as 0.0 eV. Between −65 and −30 eV, and above +30 eV, the x-ray energy was incremented by larger intervals (2 eV) to reduce data collection times. Each energy interval was counted between 10 and 20 live seconds for a total XANES spectrum acquisition time of ~30 minutes. Counting times were adjusted to obtain at least 10^5 counts per energy step at energies above the absorption edge.

XANES data analysis

The energy of the pre-edge peak that was used as an indicator of Fe oxidation state, was defined to be the centroid of a Gaussian fit to the peak after background subtraction. The centroid of the pre-edge peak for a magnetite standard (USNM114887) from Minas Gerais, Brazil (JAROSEWICH *et al.,* 1980) was used as the reference x-ray energy (zero relative energy) and all pre-edge energies were computed relative to that energy. The peak fitting procedure consisted of a least-squares fit of a second-order polynomial to the background (including the low energy part of the main absorption-edge) followed by a Gaussian curve fit to the pre-edge peak. The Gaussian function was determined empirically to be a better approximation to the shape of the peak than a Lorentzian function (BAJT *et al.,* 1994). The magnetite standard was typically measured between every two or three unknown samples during each experimental session in order to correct for long-term monochromator drifts (about 0.5 eV) caused by crystal heating and/or small vertical drifts in the synchrotron source position. Using the magnetite as a reference, relative energies for all other pre-edge peak measurements could be determined with a reproducibility of about ± 0.1 eV. This reproducibility reflects comparison of multiple (usually >5) Gaussian fits to each standard and unknown pre-edge peak and has been tested by comparison of repeated samples from multiple analytical sessions. These repeated fits used data points that differed from fit to fit to assess the sensitivity of the peak position to background subtraction. In the present study, the apparent centroid position derived from these procedures appears to be fairly insensitive to the details of these fitting procedures, but this is a potential problem when dealing with very weak pre-edges that may be typical of minerals with undistorted octahedral sites. The pre-edge energies of all 'unknowns' are reported relative to that of the pre-edge centroid in a magnetite spectrum acquired within two hours of the unknown.

The siting of iron in amphiboles is exclusively in octahedral M-sites, so the pre-edge peak being measured in these studies will contain only octahedral site contributions. It is certain that the pre-edge studied is a composite peak with Fe^{3+} and Fe^{2+} components (CALAS and PETIAU, 1983). The resolution of the monochromator system used in this study is not, however, sufficient at present to permit recognition of these subpeaks. Future studies using higher energy resolution may well resolve pre-edge features that are presently obscured.

The pre-edge peaks in amphiboles are quite weak, as would be expected for octahedrally coordinated Fe, (WAYCHUNAS *et al.,* 1983) and are similar in this respect to the oxides studied by BAJT *et al.* (1994). A potential spectral interference from Mn Kβ has been identified in Mn rich samples. However, tests of the magnitude of this interference suggest that it is trivial for these amphiboles which mostly contain less than 0.5% MnO. Well defined differences in the shape of the main absorption edge features between oxide standards (BAJT *et al.* 1994) and between different amphiboles (Fig. 1) can be recognized but are not being used for interpretation of oxidation state, or structural information at present. There is a clear sequence of shape changes in the main edge with increasing $Fe^{3+}/\Sigma Fe$ (Fig. 1), but the relative influences of oxidation state and site coordination on the absorption edge shape are unknown. The mantle kaersutites documented by DYAR *et al.* (1993) all have similar compositions and their spectral shapes are similar, despite differences in their $Fe^{3+}/\Sigma Fe$ (Fig. 1). The spectrum of the COSCA *et al.* (1991) sample (FA86−1), which is compositionally distinct, shows a well developed extra peak at the top of the main edge but this feature may be also represented in the Dyar samples by a much weaker peak. The differences between the spectra illustrated do not reflect the differences of their $Fe^{3+}/\Sigma Fe$ in a manner that can be readily quantified at present but may relate instead to the abundance and distribution of iron in the M-sites relative to other cations. The COSCA *et al.* (1991) amphiboles appear to have more Fe^{2+} substituting for Ti and Al in their M-sites, than the DYAR *et al.* (1993) samples. There is clearly great potential for study of these main edge features in the future.

RESULTS

The relative pre-edge energies of 23 amphibole samples with $Fe^{3+}/\Sigma Fe$ between 0.18 and 1.0 have been obtained during three separate analytical sessions. The $Fe^{3+}/\Sigma Fe$ determined by wet chemical techniques (COSCA *et al.,* 1991); and by Mössbauer spectroscopy (DYAR *et al.,* 1993) have been used as the nominal $Fe^{3+}/\Sigma Fe$ of the analyzed grains (Fig. 2a). Most amphiboles fall in a linear array correlating pre-edge position with the $Fe^{3+}/\Sigma Fe$. Fig. 2 also contains the linear fit of the oxide mineral data that were acquired during the last amphibole run and were used to monitor fluctuations in the operating conditions for comparison with the original SmX calibration of BAJT *et al.* (1994). The oxide data provide both a "calibration" of the $Fe^{3+}/\Sigma Fe$ of the amphibole and a test of the stability of the beamline and the x-ray microprobe during measurements. Experimental changes in the configuration of the synchrotron during the period of these measurements (50% changes in maximum ring current available) resulted in changes in the thermal regime at the beamline. The combined data sets show greater apparent scatter than would be seen within a single session. For this reason, although each analytical session produced similar relationships among the samples, only data for the last of the three analytical sessions are shown in Fig. 2.

A strong correlation exists between the nominal $Fe^{3+}/\Sigma Fe$ of the samples and those estimated from the linear fits to the oxide data of BAJT *et al.* (1994)

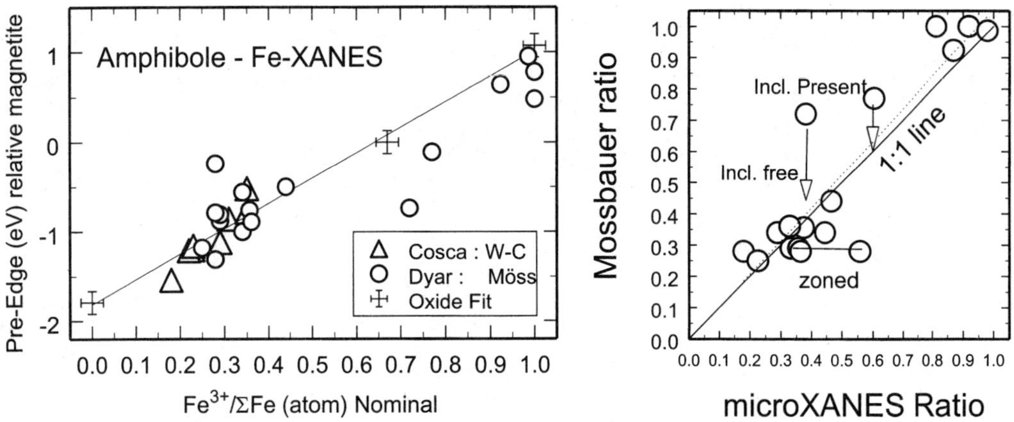

FIG. 2(A) *Left.* Changes of pre-edge energy (relative to the energy of Minas Gerais magnetite) of the Fe K-absorption edge for the amphibole samples, as a function of nominal $Fe^{3+}/\Sigma Fe$ determined by Mössbauer spectroscopy and wet chemical techniques. Calibration curve (solid line) for $Fe^{3+}/\Sigma Fe$ of simple oxides and silicates as a function of pre-edge energy from BAJT *et al.* (1994) is also shown. FIG. 2(B) *Right.* Comparison of $Fe^{3+}/\Sigma Fe$ for amphibole grains determined by Mössbauer spectroscopy and using the microXANES technique of BAJT *et al.* (1994). The dotted line is the linear best fit for the amphibole data using the microXANES calibration, and includes three grains known to have iron oxide inclusions that may have modified the bulk ratios of these samples. Removal of these data results in superposition of the linear best fit on the 1 : 1 line.

within each of the three analytical sessions. Most of the samples have $Fe^{3+}/\Sigma Fe$ between 0.2 and 0.4, and within this region the fit of the microXANES data to the "classical" measurements are very good. Greatest scatter is seen in comparison of the ferric rich amphiboles with the Mössbauer data tabulated by DYAR *et al.* (1993). At least one grain of amphibole H366A has provided reproducibly different spectra from areas separated by about 150 μm (Table 1), indicating the presence of zoning of $Fe^{3+}/\Sigma Fe$. Measurement of $Fe^{3+}/\Sigma Fe$ zoning in amphibole crystals may, therefore, be made quantitatively. Three amphibole grains (TM, FR-11, and one analysis of FR-12) have pre-edge energies suggesting that they are significantly enriched in ferrous iron relative to the wet chemical and Mössbauer analyses. Micrometer sized inclusions of iron oxide minerals were identified in each of these samples. (Linear arrays of tiny inclusions with ~300 μm spacing between the arrays were observed in two of the three samples). Closely spaced inclusions could skew the results from bulk analytical techniques toward ferric iron as a result of incomplete phase separation. This effect may be seen in the comparison of the nominal $Fe^{3+}/\Sigma Fe$ data and the SmX results for these samples. Only one grain (AKM3) appears to be ferric iron enriched relative to the Mössbauer data, but this grain has a slightly different electron microprobe composition (11.6% "FeO" instead of 12%; 2.9% Na_2O instead of 3.3%) than the quoted analysis and the observed differences probably reflect heterogeneity (zoning) within the bulk sample. Comparison of the $Fe^{3+}/\Sigma Fe$ determined by Mössbauer spectroscopy (DYAR *et al.,* 1993) and those obtained using the BAJT *et al.* (1994) microXANES calibration technique indicate very good consistence between the two techniques (Fig. 2b). If the samples known to be contaminated by microinclusions were removed from the data set, the best fit of these result would be indistinquishable from the 1 : 1 correlation.

DISCUSSION

The strong correlation between the energy of the Fe-K pre-edge peak for the amphibole samples with the $Fe^{3+}/\Sigma Fe$ determined by bulk analytical techniques indicates that synchrotron microXANES spectroscopy can be used for *in situ* determination of the $Fe^{3+}/\Sigma Fe$ in complex silicates such as amphibole. This supports the conclusions reached previously from studies of much simpler silicates and oxide minerals with abundant octahedrally coordinated iron (BAJT *et al.,* 1994). The amphibole data fall close to the regression fit for the standard oxide samples determined during the same data collecting sessions and suggest that small differences in the coordination environment of the Fe between simple oxide and complex silicate minerals are *not* influencing the pre-edge measurements made. Consequently the calibration line determined for the oxide mineral study of BAJT *et al.* (1994) can also be applied to the amphibole minerals. Preliminary data for other rock-forming mineral groups suggest that the present calibration can be used for many minerals in which iron occurs at octahedrally coordinated sites. Since that includes many mineral groups, this calibration may be of fairly widespread usefulness. For mineral groups that contain iron in other coordinations the present calibration is not expected to be valid. The present amphibole measurements, although preliminary, suggest that an alternate calibration of the $Fe^{3+}/\Sigma Fe$ may also be made using independently analyzed samples although such a protocol does not permit assessment of errors caused by heterogeneity within individual grains of amphibole.

The present results were produced with a relatively large x-ray beam (30×50 μm) but smaller beams ($\sim 8 \times 30$ μm) have been used for micro-XANES measurements of Fe-rich minerals and with third generation synchrotron facilities becoming available at the Advanced Photon Source (Argonne National Lab.), beams of a few micrometers diameter should be feasible. The ability to measure $Fe^{3+}/\Sigma Fe$ in micrometer scale volumes on thin sections provides a powerful new geochemical technique. Other multivalent elements that occur in natural materials can also be determined by microXANES spectroscopy. Measurements have been made of $Cr^{2+}/\Sigma Cr$ in lunar and terrestrial olivine (SUTTON *et al.,* 1993b), U and Mn oxidation states in soils (BERTSCH *et al.,* 1994; SCHULZE *et al.,* 1995), and Zn speciation in individual fluid inclusions (ANDERSON *et al.,* 1995). Many other elements remain to be studied. The ability to make measurements of the OXIDATION STATE of an element as well as its abundance in micrometer scale volumes removes a fundamental limitation of the numerous microbeam analytical techniques that have become the methods of choice in the last three decades.

The ability to investigate the distribution of elements in terms of both their abundance and their physical states holds enormous potential for future mineralogical, petrological and geo/cosmochemical studies of terrestrial and extraterrestrial materials. With the availability of a microanalytical technique for measuring $Fe^{3+}/\Sigma Fe$, studies of oxygen, hydrogen or water fugacity and their relationships to mineral zoning as well as studies of differences between crystals of different generations of amphibole become feasible. Systematic measurements of other minerals and glasses containing both ferric and ferrous iron are in progress.

Acknowledgements—This research was supported by the following grants: NASA NAG9-106, NAGW-3651 (SRS) and NAGW-3649 (JSD), NSF-EAR 95-08128 (JSD), EAR 93-04304 (MDD), US DOE DE-FG02-92ER14244 (SRS) and DE-AC02-76CH00016 (BNL). Constructive reviews by R. K. Popp and an anonymous reviewer are gratefully acknowledged. We also thank the NSLS staff for providing the synchrotron radiation for this work and H. N. Chapman for software support.

REFERENCES

ALBEE A. L. and CHODOS A. A. (1970) Semiquantitative electron microprobe determination of Fe^{2+}/Fe^{3+} and Mn^{2+}/Mn^{3+} in oxides and silicates and its application to petrologic problems. *Amer. Mineral.* **55,** 491–501.

ANDERSON A. J., MAYANOVIC R. A. and BAJT S. (1995) Determination of the local structure and speciation of zinc complexes in individual fluid inclusions by micro-XAFS. *Can. Mineral.* **33,** 499–508.

BAJT S., SUTTON S. R. and DELANEY J. S. (1994) Microanalysis of iron oxidation states in silicates and oxides using x-ray absorption near edge structure (XANES). *Geochim. Cosmochim. Acta* **58,** 5209–5214.

BASTIN, G. F. and HEIJLIGERS, H. J. M. (1991) Quantitative electron probe microanalysis of ultra-light elements (boron-oxygen). In *Electron Probe Quantitation* (eds. K. F. J. HEINRICH and D. E. NEWBURY) pp. 145–161. Plenum Press, New York.

BERTSCH, P. M., HUNTER, D. B., SUTTON, S. R., BAJT, S. and RIVERS, M. L. (1994) *In situ* chemical speciation of uranium of soils and sediments by micro x-ray absorption spectroscopy. *Environ. Sci. Tech.* **28,** 980–984.

BROWN G. E. Jr., CALAS G., WAYCHUNAS G. A. and PETIAU J. (1988) X-ray absorption spectroscopy and its applications in mineralogy and geochemistry. In *Spectroscopic methods in mineralogy and geology* (ed. F. C. HAWTHORNE), *Rev. Mineral.* **18,** 431–512. Mineral. Soc. America.

CALAS G. and PETIAU J. (1983) Coordination of iron in oxide glasses through high-resolution K-edge spectra: information from the pre-edge. *Solid State Comm.* **48,** 625–629

CALAS G., BASSETT W. A., PETIAU J., STEINBERG M., TCHOUBAR D. and ZARKA A. (1984) Some mineralogical applications of synchrotron radiation. *Phys. Chem. Minerals* **B11,** 17–36.

CHEN J. R., GORDON B. M., HANSON A. L., JONES K. W., KRANER H. W., CHAO E. C. T. and MINKIN J. A. (1984) Synchrotron x-ray fluorescence and extended x-ray absorption fine structure analysis. *Scanning* **IV,** 1483–1500.

COSCA M. A., ESSENE E. J. and BOWMAN J. R. (1991) Complete chemical analyses of metamorphic hornblendes: implications for normalizations, calculated H_2O activities and thermobarometry. *Contrib. Mineral. Petrol.* **108,** 472–484.

CRESSEY G., HENDERSON C. M. B. and VAN DER LAAN G. (1993) Use of L-edge X-ray absorption spectroscopy to characterize multiple oxidation states of 3d elements; a new probe for mineralogical and geochemical research. *Phys. Chem. Minerals* **20,** 111–119.

DYAR M. D., MACKWELL S. J., MCGUIRE A. V., CROSS L. R. and ROBERTSON J. D. (1993) Crystal chemistry of Fe^{3+} and H^+ in mantle kaersutite: Implications for mantle metasomatism. *Amer. Mineral.* **78,** 968–979.

FROST B. R. (1991) Introduction to oxygen fugacity and its petrological importance. In *Oxide Minerals: Petrologic and magnetic significance* (ed. D. H. LINDSLEY), *Rev. Mineral.* **18,** 1–10. Mineral. Soc. America.

HAWTHORNE F. C. (1981) Crystal chemistry of the amphiboles. In *Amphiboles and other hydrous pyriboles—mineralogy* (ed. D. R. VEBLEN), *Rev. Mineral.* **9A,** 1–102. Mineral. Soc. America.

HÖFER H. E., BREY G. P., SCHULZ-DOBRICK B. and OBERHÄNSI R. (1994) The determination of the oxidation state of iron by electron microprobe. *Eur. J. Mineral.* **6,** 407–418.

ILTON E. S., RAEBURN S. P., KERRICK, D. M. and VEBLEN, D. R. (1993) X-ray photoelectron spectroscopy as an in situ microbeam method for determining ferrous/ferric ratios in minerals in thin sections: results for biotites from metapelites. *GSA Programs with Abstracts,* **25,** A100–A101.

JAROSEWICH E., NELEN J. A. and NORBERG J. A. (1980) Reference samples for electron microprobe analysis. *Geostand. Newslett.* **4,** 43–47.

MANCEAU A., GORSHKOV A. I. and DRITS V. A. (1992a) Structural chemistry of Mn, Fe, Co, and Ni in manganese hydrous oxides: Part I. Information from XANES spectroscopy. *Amer. Mineral.* **77,** 1133–1143.

MANCEAU A., GORSHKOV A. I. and DRITS V. A. (1992b) Structural chemistry of Mn, Fe, Co, and Ni in manganese hydrous oxides: Part II. Information from EXAFS spectroscopy and electron and X-ray diffraction. *Amer. Mineral.* **77,** 1144–1157.

PHILLIPS M. W., POPP R. K. and CLOWE C. A. (1988) Structural adjustments accompanying oxidation-dehydrogenation in amphiboles. *Amer. Mineral.* **73,** 500–506

POPP R. K., PHILLIPS M. W. and HARRELL J. A. (1990) Accommodation of Fe^{3+} in natural, Fe^{3+}-rich, calcic and subcalcic amphiboles: evidence from published analyses. *Amer. Mineral.* **75,** 163–169.

POPP R. K., VIRGO D., YODER H. S. J., HOERING T. C. and PHILLIPS M. W. (1995) An experimental study of phase equilibria and Fe oxy-component in kearsutitic amphibole: implications for the f_{H2} and a_{H2O} in the upper mantle. *Amer. Mineral.* **80,** 534–549.

ROBINSON P., SPEAR F. S., SCHUMACHER J. C., LAIRD J., KLEIN C., EVANS B. W. and DOOLAN B. L. (1982) Phase relations of metamorphic amphiboles: natural occurrence and theory. In *Amphiboles: Petrology and experimental phase relations* (eds. D. R. VEBLEN and P. H. RIBBE), *Rev. Mineral.* **9B,** 1–228. Mineral. Soc. America.

SCHULZE D. G., SUTTON S. R. and BAJT S. (1995) Determination of manganese oxidation state in soils using x-ray absorption near edge structure (XANES) spectroscopy. *Soil Sci. Soc. Amer. J.* **59,** 1540–1548.

SUTTON S. R., DELANEY J. S., BAJT S., RIVERS M. L. and SMITH J. V. (1993a) Microanalysis of iron oxidation state in iron oxides using X-ray absorption near edge structure (XANES). In *Lunar and Planetary Science,* Vol. XXIV, pp. 1385–1386. Lunar and Planetary Institute.

SUTTON S. R., JONES K. W., GORDON B., RIVERS M. L., BAJT S. and SMITH J. V. (1993b) Reduced chromium in olivine grains from lunar basalt 15555: x-ray absorption near edge structure (XANES). *Geochim. Cosmochim. Acta* **57,** 461–468.

WAYCHUNAS G. A. (1988) Luminescence, x-ray emission and new spectroscopies. In *Spectroscopic methods in mineralogy and geology* (ed. F. C. HAWTHORNE), *Rev. Mineral.* **18,** 639–698. Mineral. Soc. America.

WAYCHUNAS G. A., APTED M. J. and BROWN G. E. JR. (1983) X-ray K-edge absorption spectra of Fe minerals and model compounds: near-edge structure. *Phys. Chem. Minerals* **10,** 1–9.

Part B.
Spectroscopy and Mineralogy of the Earth's Mantle

INTRODUCTION

A large portion of ROGER BURNS' research focused on the properties of mantle minerals and implications for physical and chemical properties of the mantle, providing an excellent opportunity to showcase current research in this area.

Much of our knowledge of the Earth's interior has come from laboratory measurements of mineral properties and comparison with bulk properties of the Earth measured by geophysical techniques. The success of this approach lies in the strong dependence between microscopic behavior (*e.g.,* at the atomic level) and the macroscopic bulk properties of mineral phases. Spectroscopy provides fundamental information about the microscopic structure that can then be used to determine or constrain macroscopic observables, a common theme in all papers within this section.

Major divisions within the Earth are delineated by seismic discontinuities, which are caused by phase transitions of the dominant mineral phases. Physical and chemical properties of the high-pressure phases are significantly different to their low-pressure modifications; therefore it is important to study the high-pressure phases themselves. Many phases, however, including most of those studied by authors in this section, cannot be quenched to ambient conditions, and must be studied *in situ* at either high pressure, high temperature, or both. The diamond anvil cell is an important tool in such studies, and is ideally suited to spectroscopy since diamonds are essentially transparent to many wavelengths of radiation. ROGER BURNS was one of the first to apply Mössbauer spectroscopy in the diamond anvil cell, providing a foundation for future research. All papers reporting high-pressure experiments within this section are based on work in the diamond anvil cell, a testament to its wide application within the high-pressure community.

An introduction to the wide variety of structures and phases found within the Earth's interior is provided by MCMILLAN, HEMLEY and GILLET. They present a review of vibrational spectroscopy—a compendium of the uses of infrared and Raman spectroscopy with a review of the available vibrational data for phases from the upper mantle, transition zone and lower mantle, as well as studies of OH and C volatile species in mantle phases.

HOFMEISTER gives a detailed illustration of one application of vibrational spectroscopy—to determine the thermodynamic properties of stishovite. She demonstrates that results consistent with all available thermodynamic and elastic measurements can be obtained, providing a method for determining values that cannot be measured, and confidence for extrapolations to mantle pressures and temperatures.

CHOPELAS, REICHMANN and HANG present a different method to obtain high quality elastic properties from vibrational spectra—the sideband fluorescence technique. A crucial ingredient is the doping of a fluorescing transition metal cation with 3d electrons, particularly appropriate for this volume given ROGER BURNS' strong interest in applications of 3d transition metal chemistry.

Techniques for *in situ* experiments at both high pressure and high temperature are an important development in laboratory simulation of the Earth's interior, allowing properties of phases to be studied within their stability field at mantle conditions, removing uncertainties associated with extrapolation and enabling the study of high-pressure phases that cannot be quenched, for example FeO. In an experiment pushing the edge of current technology, FEI presents results on FeO from a study to 100 GPa and 1100 K using synchrotron radiation. The paper clarifies many points that have long been a subject of controversy.

Another development in *in situ* high pressure, high temperature techniques is presented by HÅLENIUS, ANNERSTEN and JÖNSSON. In the first work of this kind, they report Mössbauer data for Fe metal and olivine to 11 GPa and 600 K, demonstrating the viability of obtaining Mössbauer spectra under such conditions.

Another direction for *in situ* high pressure, high temperature studies is reported by BASSETT, WU, CHOU, HASELTON, FRANTZ, MYSEN, HUANG,

SHARMA and SCHIFERL. They describe a new type of diamond anvil cell that can be used for studies at high pressure at temperatures from 83 to 1473 K, and has been used so far for optical microscopy, X-ray diffraction, Raman spectroscopy, luminescence and infrared spectroscopy.

Pyroxene was a favorite topic of study for ROGER BURNS, and the recent discovery of a new polymorph stable at mantle pressures and temperatures has produced a new generation of studies of this phase. ROSS and SOWERBY report electronic absorption spectra of the end member, $FeSiO_3$, at both ambient conditions and high pressure, providing essential data on the crystal field parameters of the new phase as well as changes occurring during the phase transition.

In a complementary study, MCCAMMON and TENNANT present high-pressure Mössbauer data for $FeSiO_3$ at room temperature up to 3.6 GPa. The hyperfine parameters of both the low- and high-pressure phases, as well as changes through the phase transition, are used to extract information about the bonding and crystal field environment of iron atoms in the different structures.

Pyroxene incorporates many minor elements, and one of the most important for mantle studies is hydrogen. When hydrogen diffusivity is fast, loss of hydrogen can occur in pyroxene megacrysts during transport from the upper mantle to the Earth's surface, resulting in increased Fe^{3+}/Fe^{2+} in erupted samples. Thus, both hydrogen and iron redox state must be studied in tandem to understand mantle f_{O_2} and f_{H_2}. To address this situation, DYAR, MARTIN, MACKWELL, CARPENTER, GRANT and MCGUIRE use Mössbauer spectroscopy, wet chemistry, hydrogen isotope and hydrogen diffusivity data to investigate the relation between H_2O and Fe^{3+} in augite and Fe-rich diopside.

Mineral Spectroscopy: A Tribute to Roger G. Burns
© The Geochemical Society, Special Publication No. 5, 1996
Editors: M. D. Dyar, C. McCammon and M. W. Schaefer

Vibrational spectroscopy of mantle minerals

PAUL F. MCMILLAN,[1,2] RUSSELL J. HEMLEY[2,3] and PHILIPPE GILLET[2]

[1]Department of Chemistry and Biochemistry, Arizona State University, Tempe, AZ 85287–1604, U.S.A.
[2]Laboratoire de Sciences de la Terre, Ecole Normale Supérieure des Sciences, 46 allée d'Italie, 69365 Lyon Cedex 07, France
[3]Geophysical Laboratory and Center for High Pressure Research, Carnegie Institution of Washington, 5251 Broad Branch Road, N.W., Washington D.C. 20015–1305, U.S.A.

Abstract—Vibrational spectroscopy has been used extensively to characterize the high pressure minerals important in the Earth's mantle. The results of IR and Raman studies complement structural information obtained from diffraction experiments, for example in providing information on [6]-coordinated silicon in high-pressure phases such as stishovite and $MgSiO_3$ perovskite, on the tetragonal distortion of $MgSiO_3$ garnet, and on the presence of the inter-tetrahedral SiOSi linkage in β-Mg_2SiO_4. Such measurements provide detailed information on phase transitions and tests of *ab initio* and empirical calculations of mineral properties. *In situ* studies at high pressure and/or temperature have been used to study soft-mode behaviour associated with displacive phase transitions, as demonstrated recently for stishovite. The presence of OH in natural minerals and synthetic candidate high-pressure phases has been demonstrated by IR spectroscopy, permitting the identification of potential storage sites for hydrogen in the mantle. *In situ* high *P-T* Raman spectroscopy has demonstrated the stability of magnesite ($MgCO_3$) to mantle temperatures and pressures, important in determining the carbon budget within the deep Earth. Micro-Raman and (more recently) micro-IR spectroscopies have proven very useful for non-destructive, *in situ* analysis of minerals in complex natural and experimental sample assemblages. Experimental vibrational data have been used to construct and constrain models for calculation of thermodynamic parameters such as heat capacity and entropy. These have been useful in establishing *P-T* phase diagrams for mantle mineral assemblages. We summarize these results and collect together vibrational spectra obtained to date for principal mineral phases of the mantle. We also discuss both limitations and future directions of the techniques.

INTRODUCTION

OUR KNOWLEDGE of the internal structure of the Earth has been largely gleaned from analysis of seismic and other geophysical data, bounded by cosmological constraints on its composition (ANDERSON, 1989; POIRIER, 1991; GILLET, 1995; BINA and HEMLEY, 1996). Some direct sampling of mantle minerals has been possible, from examination of ultramafic xenoliths brought to the surface in kimberlites or alkali basalts, or as inclusions in diamonds (SAUTTER and GILLET, 1994). However, most of our present understanding of mantle mineralogy is based on experimentally determined phase relations and thermophysical properties of candidate minerals and assemblages. Vibrational spectroscopy has played an important role in this work (MCMILLAN and HOFMEISTER, 1988; MCMILLAN, 1989). Such measurements provide detailed information on structure and bonding in materials, thereby complementing direct structural studies by diffraction techniques and spectroscopy of electronic transitions. Vibrational spectroscopy provides a straightforward means for identifying phase transitions, as well as the mechanism of such transformations (*e.g.*, soft modes and order parameters). Minerals can be analyzed both as microscopic grains and in thin sections of multiphase natural samples as well as in experimental charges quenched from high pressures and temperatures. Finally, vibrational data have been used extensively for modeling the thermodynamic properties of mantle minerals. This has been an important part of this research because such vibrational data in many cases have provided the first, and in some cases the only, constraints on the thermodynamic properties of such materials at high *P-T* conditions corresponding to the deep Earth.

We review here the uses of Raman and infrared vibrational spectroscopy in the study of mantle minerals. We begin with an overview of the kinds of information that can be obtained from such measurements and their relevance to the Earth's mantle. We discuss in some detail the vibrational properties of the major mantle minerals, summarizing the available vibrational data for each phase. Our motivation has been to collect the spectra obtained to date in a single set of figures, for convenience of comparison and for identifying areas where further work is needed. This then extends and complements earlier and more general reviews of the applications of Raman and IR spectroscopy to Earth and planetary sciences (HEMLEY *et al.*, 1987a; MCMILLAN and HOFMEISTER, 1988; MCMILLAN,

1989; SHARMA, 1989, 1990). We also discuss vibrational spectroscopic studies of the role of water and carbon in the mantle. A brief description of future directions is given in the concluding remarks. The review is limited to crystalline minerals. Vibrational spectroscopy has demonstrated that large structural changes take place in silicate glasses at high pressures (and by inference melts), but this is beyond the scope of the present article and is reviewed elsewhere (MCMILLAN and WOLF, 1995). We also discuss results of theoretical studies of the vibrational dynamics of mantle minerals; excellent reviews of theoretical techniques and the calculation of other properties have been written by COHEN (1994), BUKOWINSKI (1994), and CHIZMESKHYA *et al.*, (1994).

GENERAL CONSIDERATIONS

Overview of mantle mineralogy

The global structure of the mantle is defined by its major seismic discontinuities near 410 km and 660 km, which delineate the regions known as the upper mantle (20–410 km), transition zone (410–660 km), and lower mantle (660–2900 km) (Fig. 1). Based on extensive laboratory studies and theoretical modeling over the past two decades, it is likely that the dominant mineral phase in the lower mantle is a silicate perovskite with composition near $(Mg_{0.9}Fe_{0.1})SiO_3$, probably containing up to several per cent Al_2O_3 component, which co-exists with (Mg,Fe)O (magnesiowüstite) with the rock salt (B1) structure, and a nearly end-member $CaSiO_3$ perovskite (WILLIAMS *et al.*, 1989; ANDERSON, 1989; HEMLEY and COHEN, 1992; IRIFUNE, 1994). Free silica may be present as stishovite in basaltic regions trapped near the top of the lower mantle from subducted oceanic crust (KESSON *et al.*, 1994), which could give rise to aluminous phases (IRIFUNE *et al.*, 1991; YAGI *et al.*, 1994; AHMED-ZAID and MADON, 1995). A free SiO_2 phase near the base of the lower mantle has also been proposed (KNITTLE and JEANLOZ, 1991; JEANLOZ, 1993). The mineralogies of the transition zone and upper mantle are likely to be more complex (ANDERSON, 1989; GASPARIK, 1990; GILLET, 1995). The upper mantle is dominated by ultramafic minerals, principally olivine, pyroxene and garnet phases, but substantial amounts of plagioclase feldspar, kaersutitic amphibole, phlogopite mica, and many other accessory minerals may be found. The petrology is highly dependent on the type of province encountered, and the thermal and compositional regime involved. Many of these phases are volatile-rich, both providing host phases for water, CO_2 and F-, Cl- and S-bearing species degassing from deep within the Earth, or returning via subduction processes (THOMPSON, 1992; BELL and ROSSMAN, 1992a; GASPARIK, 1993; GILLET, 1993a; CARROLL and HOLLOWAY, 1994). In addition, this region is characterized by partial melting, so that the properties of molten silicates are important in the upper mantle, and perhaps even deeper within the Earth (WILLIAMS, 1990; ITO and KATSURA, 1992; WOLF and MCMILLAN, 1995). The transition zone is dominated by the mineralogy of refractory ultramafic phases including the β-(wadsleyite, or "modified spinel") and γ-(ringwoodite, spinel) forms of $(Mg,Fe)_2SiO_4$, along with the garnet and perhaps ilmenite phases of $MgSiO_3$ (GASPARIK, 1990). Both $MgSiO_3$ ilmenite and garnet contain [6]-coordinate silicon, and at least the majorite garnet will exist as a solid solution with Mg^{2+}, Fe^{2+} and Ca^{2+} substituting on the large ([12]-coordinate) site, Al^{3+}, Si^{4+} and Fe^{3+} on the octahedral site, and Si^{4+} and Al^{3+} on tetrahedral sites (IRIFUNE, 1994; KESSON *et al.*, 1994). All of these nominally anhydrous mantle minerals can contain significant amounts of dissolved H as "defect" OH species (BELL and ROSSMAN, 1992a; GILLET, 1993a). In addition, several high pressure hydrous phases, particularly within the system $MgO—SiO_2—H_2O$, may be present in the transition zone (KANZAKI, 1991; PREWITT and FINGER, 1992).

Structure and bonding

Although it is not a technique for direct structure determination, vibrational spectroscopy gives important insight into bonding in materials and structure-property relations, thereby complementing x-ray and neutron diffraction methods. Morever, vibrational spectroscopy has revealed transitions and structural changes that were not apparent from diffraction measurements (KINGMA *et al.*, 1995), although in other cases, such claims have not been supported by subsequent, and more accurate, direct structural investigations (*e.g.*, DOWNS *et al.*, 1996). Measurements as a function of pressure and/or temperature permit detailed identification of response of the structure to these thermodynamic variables (*i.e.*, changes in compression mechanisms in the absence of a thermodynamic phase transition). Vibrational spectroscopy has played an important role in characterization of mantle phases since the earliest days of experimental mantle petrology and mineralogy. Infrared spectroscopy confirmed the octahedral coordination of Si (or Ge in silicate analogues) in several of these high pressure phases (LYON 1962; TARTE and RINGWOOD 1964; WENG

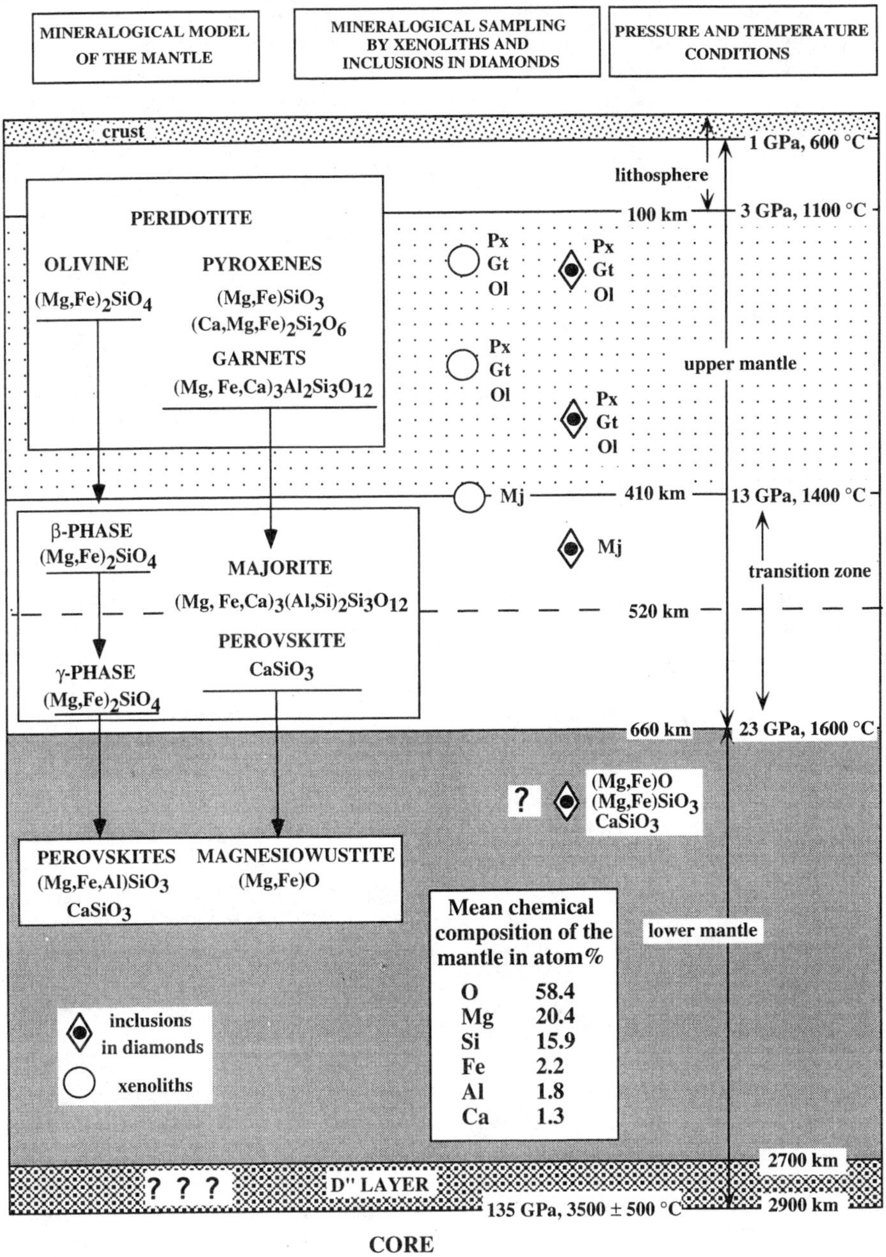

FIG. 1. Schematic (from GILLET, 1995) of the mineralogy of the Earth's mantle, inferred from results of high pressure-high temperature experiments, and natural sampling via xenoliths and inclusions in diamonds.

et al., 1983). In other cases, vibrational spectroscopy was used to reveal unusual structural features, such as the presence of the SiOSi linkage between tetrahedral SiO_4 groups in the β-("modified" spinel, or wadsleyite) polymorph of Mg_2SiO_4 (JEANLOZ, 1980; MCMILLAN and AKAOGI, 1987). Another example is the identification and characterization of ring structures in minerals such as coesite [SHARMA *et al.,* 1981]. This has been important in determining compression mechanisms in glasses as well as amorphization phenomena (HEMLEY *et al.,* 1994; MCMILLAN and WOLF, 1995).

Vibrational spectroscopy has also been essential for characterizing structural components such as hydrogen (bound stoichiometrically or as defects) which are difficult to study by other techniques, as discussed below.

Phase transitions

Because vibrational spectra typically provide a unique signature of crystal (and amorphous) structure, the measurement of such spectra has proven extremely useful for identifying phase transitions, particularly *in situ* at high pressures and/or temperatures. Moreover, such measurements can be used successfully to determine phase transitions mechanisms. A classic example is the study of soft modes and the identification of order parameters associated with displacive phase transitions (SCOTT, 1974; GHOSE, 1985, 1988; MCMILLAN, 1985, 1989; MCMILLAN and HOFMEISTER, 1988; SALJE, 1989). This includes both pressure-induced soft modes ($d\nu/dP < 0$) as well as the more commonly studied temperature-induced soft modes ($d\nu/dT < 0$). Both types of modes are known to occur for materials with the perovskite and rutile structures as functions of pressure and temperature (SAMARA and PEERCY, 1973; SCOTT, 1974), including the $MgSiO_3$ perovskite analog $CaTiO_3$ (GILLET *et al.*, 1993a,b). Early *ab initio* calculations suggested that such critical transitions might be important for $MgSiO_3$ perovskite under mantle *P-T* conditions (WOLF and BUKOWINSKI, 1987; BUKOWINSKI and WOLF, 1988). However, no soft mode behaviour has been observed experimentally in the Raman spectrum of metastable $MgSiO_3$ perovskite at ambient pressure (DURBEN and WOLF, 1992), consistent with more recent theoretical calculations (STIXRUDE and COHEN, 1993), as discussed below. A displacive phase transition associated with pressure-induced mode softening has been identified by Raman spectroscopy in stishovite, which undergoes a transition to a $CaCl_2$ structure at ~50 GPa (COHEN, 1992; KINGMA *et al.*, 1995).

Phase identification and multiphase assemblages

An important recent application is the identification of minerals in experimental samples, in particular, in polyphase assemblages quenched from high pressures and temperature. Micro-Raman has been used in most of these studies because of its high spatial resolution (~1 μm or the diffraction limit of visible light) (HEMLEY *et al.* 1987a). More recently, new micro-IR spectroscopic techniques have been developed (REFFNER *et al.*, 1994), which permit detailed IR mapping of complex samples with spatial resolution approaching the diffraction limit. FEI *et al.* (1991) used the micro-Raman technique to distinguish β- from γ-$(Mg,Fe)SiO_4$ in a study of the phase relations in the $MgO—FeO—SiO_2$ system. ZHANG *et al.* (1993) identified SiO_2 phases in experimental changes quenched from high *P-T* to determine the phase relations in the system to 2800°C at 9–14 GPa. More recently, BERTKA and FEI (1996) used the technique to identify a large number of different silicates in a high-pressure phase equilibrium study of the mineralogy of the martian mantle.

Vibrational spectroscopy, in particular micro-Raman spectroscopy, has been useful for the analysis of mineral specimens in thin section, in some cases providing identification that was not possible by standard petrographic techniques (MAO *et al.*, 1987). These techniques have played a role in studying natural mantle xenoliths as well as deep crustal samples. GILLET *et al.* (1984) used Raman spectroscopy to identify coesite inclusions within pyrope grains in metasedimentary rocks from subducted continental crust (CHOPIN, 1984), placing constraints on the *P-T* history of the assemblage. Similarly, identification of coesite in a clinopyroxene from Norwegian eclogites demonstrated that the rock formation pressure exceeded 3 GPa (SMITH, 1984; BOYER *et al.*, 1985). Micro-Raman spectroscopy has been used to identify coesite in eclogites from central China (WANG *et al.*, 1989). SOBOLEV and SHATSKY (1990) and SHUTONG *et al.* (1992) used the technique to identify diamond crystals included in zircons and garnets, in high grade (>100 km) metamorphic rocks. The technique was also applied to study exsolution of garnet from clinopyroxene in high pressure rocks (MALEZIEUX, 1990), and to investigate eclogitic garnets included in diamonds (LIU *et al.*, 1990). Infrared spectroscopy has been useful in the study of fluid inclusions in diamonds (NAVON *et al.*, 1988), and was used to demonstrate the presence of solid CO_2 in one sample, fixing the entrapment depth at 220–270 km (SCHRAUDER and NAVON, 1993). Vibrational spectroscopy of solid inclusions in diamonds are giving further valuable information on the volatile content of the mantle (D. R. BELL, unpublished).

Characterization of impurities and defects

Vibrational spectroscopy has also been very useful for the identification of impurities, such as dissolved H in nominally anhydrous mantle minerals. Quantitative infrared absorbance measurements on nominally anhydrous minerals such as olivines, pyroxenes and garnets have revealed the presence of often substantial (up to 0.1 wt%) OH contained in defect sites within the structure (see BELL and ROSSMAN, 1992a), with implications for the possible water content of the mantle. The OH sites in hydrous phases in the system MgO—SiO_2—H_2O have been characterized by Raman and infrared spectroscopy (AKAOGI and AKIMOTO, 1986; FINGER *et al.*, 1989; MCMILLAN *et al.*, 1991), and the importance of hydrogen-bonding in stabilizing such high pressure hydrous phases has been discussed (WILLIAMS, 1992). These materials may store large amounts of water within the mantle. SMYTH (1987) has suggested that the wadsleyite (β-) phase of Mg_2SiO_4 might provide an important storage site for H in the mantle. This possibility has been confirmed by vibrational spectroscopy (MCMILLAN and HOFMEISTER, 1988; MCMILLAN *et al.*, 1991; YOUNG *et al.*, 1993). Quantitative IR absorbance measurements have been used recently to investigate the OH content of the α-, β- and γ-phases of Mg_2SiO_4 as a function of composition and synthesis pressure by KOHLSTEDT *et al.* (1996). Finally, the H content and structural position of OH groups contained in defect sites in synthetic stishovite and (Mg,Fe)SiO_3 perovskite have been examined using such techniques (PAWLEY *et al.*, 1993; LU *et al.*, 1994; MEADE *et al.*, 1994).

Thermodynamic properties

Thermodynamic modeling based on vibrational spectroscopic data has played an important role in the development of our present knowledge of the properties of mantle minerals. A systematic and highly successful approach to this problem for minerals was developed by KIEFFER (1979a,b,c, 1980, 1982, 1985). This method permits the heat capacity and vibrational entropy of a mineral phase of interest to be calculated from a model density of states function, constructed from experimental infrared and Raman data, along with acoustic velocities. This application spurred many measurements of the IR and Raman spectra of high pressure silicates (AKAOGI *et al.*, 1984; NAVROTSKY, 1985, 1989; MCMILLAN and ROSS, 1987; HOFMEISTER *et al.*, 1989; FEI *et al.*, 1990; CHOPELAS, 1990a,b, 1991a,b; HOFMEISTER and CHOPELAS, 1991a,b; HOFMEISTER and ITO, 1992; CHOPELAS *et al.*, 1994). The resulting thermodynamic models have proven useful in establishing phase diagrams consistent with direct phase equilibrium data (AKAOGI *et al.*, 1984, 1989; NAVROTSKY, 1985, 1989; FEI *et al.*, 1990; SAXENA *et al.*, 1993).

In these calculations, anharmonic effects have generally been taken into account by using a quasiharmonic $C_p - C_v$ correction ($C_p = C_v + TV\alpha^2 K_T$) from the measured bulk modulus and thermal expansivity (KIEFFER, 1979b; 1985). This approach ignores the effects of intrinsic mode anharmonicity; *i.e.*, variations in vibrational mode frequencies with temperature at constant volume due to phonon-phonon interactions (GILLET *et al.* 1989a, 1990, 1991, 1992, 1993a; HEMLEY, 1991; STIXRUDE and HEMLEY, unpublished). LIU (1993) has in fact suggested that such anharmonic effects are not present for minerals. However, there is ample experimental and theoretical evidence for the existence of intrinsic mode anharmonicity in many minerals (GERVAIS *et al.*, 1972, 1973; GERVAIS and PIRIOU, 1975; GERVAIS, 1983; WOLF and JEANLOZ, 1984; REYNARD *et al.*, 1992; WINKLER and DOVE, 1992), although anharmonic effects do tend to decrease with increasing pressure (*e.g.*, HEMLEY and GORDON, 1985). These anharmonic contributions permit the high temperature heat capacities to exceed the Dulong-Petit limit, and cause the C_p values calculated within the quasiharmonic approximation to be underestimated by several per cent at mantle temperatures (RICHET *et al.*, 1992). The magnitude of anharmonic corrections can be evaluated experimentally by separately measuring vibrational mode frequencies at high temperature and high pressure (SAMARA and PEERCY, 1973; MAMMONE and SHARMA, 1980; GILLET *et al.*, 1989a; 1990). However, it is also found that intrinsic anharmonic corrections to C_p do not substantially affect the calculated phase boundaries over a wide temperature range (AKAOGI *et al.*, 1987; NAVROTSKY, 1989; FEI *et al.*, 1990; YUSA *et al.*, 1993; AKAOGI and ITO, 1993a), and differences between calculations and experiment are likely to be associated with other uncertainties (*e.g.*, free energies of formation, thermal expansivity). The effects of intrinsic mode anharmonicities are important for calculating the high-temperature heat capacity and isotopic fractionation between minerals, properties that depend strongly on vibrational anharmonicity (GILLET *et al.*, 1996).

There have been suggestions to use experimentally determined vibrational frequencies to estimate mineral elastic properties such as the bulk modulus, since K_T is proportional to $\Sigma\nu_i^2/V^{1/3}$ in the quasiharmonic approximation (HOFMEISTER, 1991). The approach involves a large number of approximations for materials beyond highly symmetric structures but has nevertheless has been applied to several mantle phases (HOFMEISTER, 1991; LU *et al.*, 1994). LIU (1992) has suggested estimating the bulk modulus from the pressure shift of an easily observable vibrational mode; however, this approach suffers from an unfortunate circularity because it requires a prior determination of the mineral compressibility. Such applications of vibrational spectroscopy may be useful for exploring possible relationships between (high-frequency) optical modes and elastic properties (low-frequency acoustic modes). However, their predictive value may be questioned in view of the fact that high *P-T* elastic properties can now be measured directly with higher accuracy (ANDERSON *et al.*, 1993; DUFFY *et al.*, 1995). Moreover, first-principles theoretical calculations in many cases are more reliable (*e.g.*, COHEN, 1992).

Density of states and lattice dynamics

Perhaps of most concern in the construction of vibrational models for thermochemical calculation is the very incomplete nature of the vibrational data set available from infrared and Raman spectroscopy. These techniques give information on vibrational modes at the Brillouin zone centre, and generally give little or no insight into dispersion of vibrational frequencies (MCMILLAN, 1985). In addition, many zone center modes may not appear in IR or Raman spectra due to selection rules imposed by crystal symmetry, and others may not be observed experimentally due to low intensity (MCMILLAN and HOFMEISTER, 1988). Because the heat capacity and entropy are obtained as integrals over the entire vibrational density of states $g(\omega)$, the result is a vibrational model for calculating these which may be poorly constrained by experiment. In principle, this problem can be solved by directly determining $g(\omega)$ from inelastic neutron scattering; such measurements gives $S(\omega)$, the vibrational density of states weighted by an atomic participation factor in each vibration, and its neutron scattering cross section (*e.g.*, GHOSE, 1985, 1988). This technique, however, requires large samples (typically a few grams), and is not usually possible for high pressure silicates, especially *in situ* at high pressure. Alternatively, lattice dynamics calculations may be carried out using either *ab initio* (first-principles) or empirically derived interatomic potentials (CATLOW and PRICE, 1990; COHEN, 1994; BUKOWINSKI, 1994). Both approaches have been applied successfully to mantle minerals. The empirical calculations require experimental data as input, which may include vibrational spectroscopic data. It is of particular interest that both IR and Raman spectra of minerals can be obtained readily at high pressures and (most recently) high temperatures (HEMLEY *et al.*, 1987a; MCMILLAN and HOFMEISTER, 1988; CHOPELAS and BOEHLER, 1992), so that not only mode frequencies and relative intensities are available for comparison between theory and experiment, but also their changes with pressure and temperature. Recent developments in *ab initio* techniques permit calculation of mineral lattice dynamics and thermophysical properties with increasing accuracy (COHEN, 1994; BUKOWINSKI, 1994), and it appears that we are entering a new era of sophistication in modeling structure and dynamics of minerals under mantle *P-T* conditions. Infrared and Raman spectra obtained *in situ* at high pressures and temperatures are invaluable for evaluating and constraining the quality of these calculations.

LOWER MANTLE MINERALS

(Mg,Fe)SiO$_3$ perovskite

Extensive studies of (Mg,Fe)SiO$_3$ perovskite have been performed in recent years because of its importance for lower mantle mineralogy (NAVROTSKY and WEIDNER, 1989; HEMLEY and COHEN, 1992). (Mg,Fe)SiO$_3$ perovskite was first synthesized by LIU (1974, 1975) and characterized in detail in subsequent years (see HEMLEY and COHEN, 1992). KNITTLE and JEANLOZ (1987) showed experimental evidence for its stability at *P-T* conditions corresponding to most of the lower mantle, indicating that the phase may be the dominant mineral in the Earth. The structure and thermodynamic properties of this phase are reviewed by WILLIAMS *et al.* (1989) and by HEMLEY and COHEN (1992). The infrared spectrum of a sample of pure $MgSiO_3$ perovskite, synthesized in the diamond anvil cell, was obtained by WENG *et al.* (1983), and the Raman spectrum was first recorded by WILLIAMS *et al.* (1987). The number of IR-active modes, and the appearance of a first-order Raman spectrum, confirmed the distortion from cubic symmetry of this phase (Fig. 2).

Perovskite with the ideal cubic structure (*Fm3m*) would have three IR-active modes and no Raman spectrum: from symmetry analysis,

$$\Gamma_{vib} = 3F_{1u}(IR) + F_{2u}(inactive). \quad (1)$$

Within such a hypothetical cubic structure, the highest frequency mode involves asymmetric stretching of the octahedral groups (SiO_6), the mid-frequency vibration corresponds to a deformation vibration of the octahedra, and the lowest frequency IR mode involves translational motions of the alkaline earth cation in the dodecahedral site. Based on systematic infrared spectroscopy of other

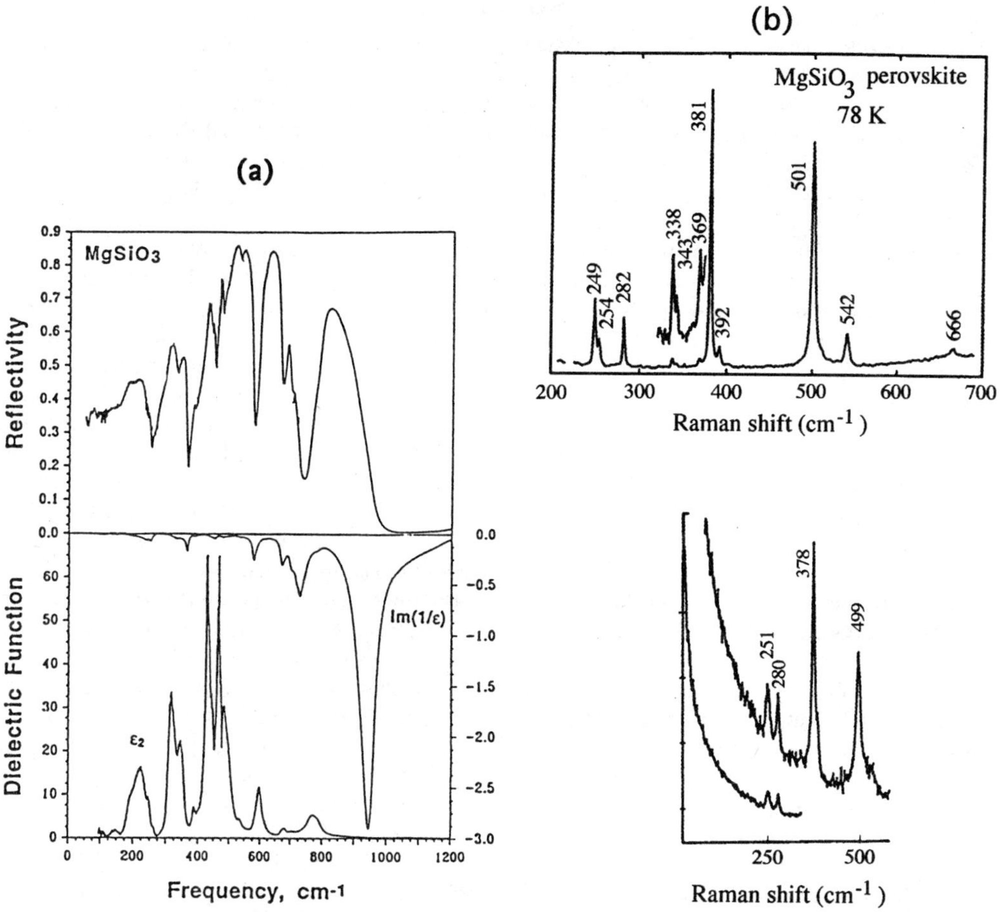

FIG. 2. (a) Infrared and (b) Raman spectra of $MgSiO_3$ perovskite. The infrared reflectivity data are redrawn from LU *et al.* (1994; Fig. 1, p. 11, 798). The peaks and minima in the dielectric functions $\epsilon_2(\nu)$ and $\mathrm{Im}(1/\epsilon(\nu))$ (middle left) indicate the positions of the TO and LO mode frequencies, respectively (LU *et al.*, 1994). The Raman spectrum at top is redrawn from DURBEN and WOLF (1992; Fig. 2, p. 891), for a sample cooled to 78 K. The spectrum at bottom (WILLIAMS *et al.*, 1987) shows the detail of the low-frequency region, to 30 cm^{-1}.

cubic perovskites, and calculations for hypothetical cubic $MgSiO_3$ perovskites, these modes would be expected to occur in the regions 750–1000 cm^{-1}, 500–700 cm^{-1}, and 250–400 cm^{-1}, respectively (WILLIAMS *et al.*, 1987; HEMLEY *et al.*, 1987b; LU *et al.*, 1994). Reduction of the symmetry to orthorhombic *Pbnm* with $Z = 4$ units of $MgSiO_3$ in the unit cell causes splitting of the IR-active modes, the occurrence of additional peaks in the IR spectrum, and the appearance of a first order Raman spectrum:

$$\Gamma_{vib} = 7A_g(R) + 7B_{1g}(R) + 5B_{2g}(R) + 5B_{3g}(R) + 8A_u(inactive) + 7B_{1u}(IR) + 9B_{2u}(IR) + 9B_{3u}(IR) \quad (2)$$

(WILLIAMS *et al.*, 1987; WOLF and BUKOWINSKI, 1987; HEMLEY *et al.*, 1989).

The mid-IR transmission spectra exhibited bands with maxima at 797, 683, 614 and 544 cm^{-1} (WENG *et al.*, 1983; WILLIAMS *et al.*, 1987). MADON and PRICE (1989) observed low frequency peaks at 390, 347, 320 and 282 cm^{-1}, and also suggested splitting in some of the higher frequency bands. The powder transmission minima correspond quite well with TO (transverse optic) mode frequencies obtained by LU *et al.* (1994) from their single crystal and powder reflectivity studies of pure $MgSiO_3$ and $(Mg_{0.9}Fe_{0.1})SiO_3$ samples, and several additional peaks were recognized in this study (Fig. 2). The frequencies of longitudinal (LO) modes were also deduced from the reflectivity data. WILLIAMS *et al.* (1987) observed Raman peaks at 251, 280, 378 and 499 cm^{-1}, which likely correspond to four of the seven expected A_g vibrations. Additional weak modes at 536 and 990 cm^{-1} were observed by HEMLEY *et al.* (1989). The best resolved spectrum obtained (at 78 K) to date is that of DURBEN and WOLF (1992), who observed eleven peaks between 249 and 666 cm^{-1} (Fig. 2). In neither the IR nor the Raman spectra have all the expected modes been observed, nor their symmetries reliably assigned experimentally (DURBEN and WOLF, 1992; LU *et al.*, 1994). This should be pursued in further studies, for the following reasons.

First, the vibrational data have proved useful in calculations of the specific heat and vibrational entropy of this phase (NAVROTSKY, 1989; AKAOGI and ITO, 1993b; GILLET *et al.*, 1993a; LU *et al.*, 1994). These calculated properties have helped to determine and rationalize the negative *P-T* slopes of reactions forming perovskite from lower pressure phases (NAVROTSKY, 1980; AKAOGI and ITO, 1993a). Of critical importance in determining the heat capacity and entropy are the positions of the lowest frequency vibrations of this phase. The lowest frequency Raman mode found by DURBEN and WOLF (1992) occurred at 249 cm^{-1}, and no obvious additional modes were found down to 30 cm^{-1} in the study by WILLIAMS *et al.* (1987) (Fig. 2). LU *et al.* (1994) found the lowest lying IR mode to occur at 180 cm^{-1}, although their data do not absolutely rule out the possibility of features at lower wavenumber. Further careful IR and Raman spectroscopy in the region below 200 cm^{-1} will be required to fully resolve this question.

Information on the low-lying vibrational modes, along with their dispersion across the Brillouin zone, is readily obtained from lattice dynamics calculations. HEMLEY *et al.* (1989) used a first principles method based on the modified electron gas approach, allowing for charge relaxation via spherical breathing of the O^{2-} anions (COHEN *et al.*, 1987), to calculate the structure and vibrational properties of $MgSiO_3$ perovskite. There was good agreement between some of the calculated mode frequencies and experimental values, but in general such approximate calculations are not expected to be able to reproduce the measured frequencies. For example, because such ionic model calculations underestimated the degree of distortion from the high-symmetry cubic structure, the frequencies of the lower octahedral librational modes were underestimated. Likewise, the frequencies of some of the higher modes are overestimated. The occurrence of the first-order Raman spectrum results from the descent in symmetry associated with the Brillouin zone folding arising from *R*- and *M*-point instabilities within the cubic aristotype, so that the Raman frequencies are particularly sensitive to details of the calculation (HEMLEY and COHEN, 1992; CHIZMESHYA *et al.*, 1994; BUKOWINSKI *et al.*, 1996). These modes are also particularly sensitive to potential symmetry changes in the perovskite structure with pressure and temperature (HEMLEY *et al.*, 1987b; MCMILLAN and ROSS, 1988; WOLF and BUKOWINSKI, 1987). In contrast to the approximate ionic model calculations, vibrational frequencies predicted from full electronic structure methods (*e.g.*, linearized augmented plane wave) give much better agreement with experiment as demonstrated for stishovite, as discussed below (KINGMA *et al.*, 1995). Although calculations at this level have been performed for $MgSiO_3$ perovskite (STIXRUDE and COHEN, 1993; WENTZCOVITCH *et al.*, 1993), the vibrational frequencies have not been reported.

The lowest lying modes in the Raman spectrum are associated with coupled rotations of the linked SiO_6 octrahedra, and these are expected to be extremely sensitive to the degree of orthorhombic distortion of the perovskite structure; in particular, they should soften if $MgSiO_3$ perovskite undergoes a displacive phase transition to tetragonal or cubic structures at high temperature (WOLF and JEANLOZ, 1985; WOLF and BUKOWINSKI, 1987; BUKOWINSKI and WOLF, 1988; HEMLEY *et al.*, 1989; SALJE, 1989). Accurate calculations indicate a wide stability field for the orthorhombic form, at least for the $MgSiO_3$ endmember (HEMLEY *et al.*, 1987b; D'ARCO *et al.*, 1993; STIXRUDE and COHEN, 1993; WENTZCOVITCH *et al.*, 1993; BUKOWINSKI *et al.*, 1996). *In situ* X-ray diffraction measurements show that the orthorhombic form remains stable over the *P-T* range of the lower mantle explored so far experimentally (MAO *et al.*, 1989; FUNAMORI and YAGI, 1993; WANG *et al.*, 1994; YAGI *et al.*, to be published). The possibility of such critical transitions still exists (WANG *et al.*, 1992; MEADE *et al.*, 1995; BUKOWINSKI *et al.*, 1996), because the entire *P-T* range of the lower mantle has not been explored experimentally with *in situ* techniques, nor have detailed experiments been carried out for Fe-bearing perovskite. This could be tested by carrying out Raman spectroscopy *in situ* under simultaneous high *P-T* conditions, via laser heating in the diamond anvil cell (GILLET *et al.*, 1993b).

In the IR spectrum, the bands between 770 and 950 cm^{-1} observed by LU *et al.* (1994) could be assigned to the ν_3 asymmetric stretching vibration of the SiO_6 octahedra (B_{1u} + $2B_{2u}$ + $2B_{3u}$ symmetries) from lattice dynamics calculations (HEMLEY *et al.*, 1989), and bands in the 600–725 cm^{-1} and 380–540 cm^{-1} regions to the ν_4 and ν_6 deformation vibrations (B_{1u} + $2B_{2u}$ + B_{3u}, and $2B_{1u}$ + B_{2u} + B_{3u}, respectively) (Fig. 2). Five modes (B_{1u} + $2B_{2u}$ + $2B_{3u}$) are expected in the low frequency region (~250–350 cm^{-1}) due to Mg^{2+} translation, and additional low frequency modes will arise from acoustic lattice modes, through folding of the Brillouin zone from the ideal cubic structure ($2B_{1u}$ + $2B_{2u}$ + $2B_{3u}$) (WILLIAMS *et al.*, 1987; WOLF and BUKOWINSKI, 1987; HEMLEY *et al.*, 1989). The behavior of these low frequency translational modes is of interest, because they determine the ferroic character of many perovskite materials (SCOTT, 1974). In their calculation of $CaSiO_3$ perovskite during decompression, HEMMATI *et al.* (1995) showed that a low frequency mode associated with off-centre Si^{4+} displacements within the SiO_6 octahedra became unstable, contributing to the amorphization phenomenon observed in this phase at or near ambient pressure. This type of low frequency mode would result in an increased far-IR reflectivity due to the large static dielectric constant, ϵ_o (SPITZER *et al.*, 1962; GRZECHNIK *et al.*, 1996). No such behaviour is indicated for $MgSiO_3$ perovskite: the far-IR reflectivity at low frequency is quite "normal" (~0.35; Fig. 2) (LU *et al.*, 1994). However, this should be investigated more fully in further experiments. In a recent theoretical study, BUKOWINSKI *et al.* (1996) have found that the lowest lying ferroic mode frequency for $MgSiO_3$ perovskite lies above 300 cm^{-1} at 50 GPa, and only softens near 3500 K, above the predicted instability temperatures for the soft modes associated with octahedral rotational motions.

Vibrational spectroscopy at high pressure and ambient temperature has been used to determine vibrational mode Grüneisen parameters ($\gamma_i = -d \ln \nu_i / d \ln V$), for comparison with the thermodynamic Grüneisen parameter γ_{th} ($=\alpha K_T V/C_v$). WILLIAMS *et al.* (1987) found frequency shifts ($d\nu_i/dP$) of 2.6 cm^{-1}/GPa for four IR-active modes between 500 and 850 cm^{-1} measured to 27 GPa, giving γ_i in the range 1.15–1.45. Using assumptions for non-observed vibrational modes, they estimated an average value of $\gamma \sim 1.9 \pm 0.2$, close to the experimental value γ_{th} = 1.77 obtained from the results of KNITTLE *et al.* (1986). Recent "best" estimates for this parameter at ambient density include γ_{th} = 1.7–1.9 (HEMLEY and COHEN, 1992); 1.3 ± 0.2 (WANG *et al.*, 1994), and 1.5 ± 0.2 (ANDERSON *et al.*, 1995). In the theoretical study by HEMLEY *et al.*, (1989), the value of γ_{th} was estimated to be 2.0 at ambient temperature, decreasing to 1.59 in the high temperature limit, by summing over all zone center vibrational modes ($\langle\gamma\rangle = \Sigma C_{vi}\gamma_i/\Sigma C_{vi}$, where C_{vi} is the contribution of each mode to C_v). Experimentally, HEMLEY *et al.* (1989) found frequency shifts ranging from 1.7 to 4.2 cm^{-1}/GPa, for seven Raman active modes measured to 26 GPa. CHOPELAS and BOEHLER (1992) found slightly smaller shifts (1.37–3.28 cm^{-1}/GPa) for the same modes observed over a greater pressure range (to 47 GPa). These values correspond to high mode Grüneisen parameters (HEMLEY and COHEN, 1992). Measurements of the anharmonicity of the material under lower mantle conditions could be obtained from analysis of the Raman linewidths and infrared reflectivity spectra under high pressure and high temperature conditions (GERVAIS, 1983; GILLET *et al.*, 1993c). *In situ* high *P-T* vibrational spectroscopic experiments could also yield information on melting (and possible pre-melting) behaviour of silicate perovskite (MATSUI and PRICE, 1991; WINKLER and DOVE, 1992; WASSERMAN *et al.*, 1993; BELONOSHKO, 1994; RICHET *et al.*, 1994), of considerable importance for mantle dynamics (VAN KEKEN *et al.*, 1994).

The temperature dependence of the Raman spectrum has been investigated at ambient pressure by DURBEN and WOLF (1992) (Fig. 3). These authors found no observable shift in the lowest frequency modes (the 249–254 cm^{-1} doublet), which might be expected to be extremely sensitive to changes in orthorhombic distortion, and only a small temperature dependence (~ −0.001 cm^{-1}/K) for the other modes. An obvious conclusion is that any critical behaviour in perovskite must lie at very high temperature, especially at high pressure (DURBEN *et al.*, 1991; STIXRUDE and COHEN, 1993; BUKOWINSKI *et al.*, 1996). DURBEN and WOLF (1992) also observed that the metastable perovskite phase transformed irreversibly to an amorphous material above approximately 300°C. During the transformation, the Raman peaks of the remaining perovskite material were found to be shifted irreversibly to higher wavenumber, as if the crystal had been placed under compression by the presence of the glass (DURBEN and WOLF, 1992). This illustrates the problems inherent in studying the material outside its stability field (HEMLEY and COHEN, 1992). The behavior of the vibrational spectra of silicate perovskites during amorphization has been examined theoretically (HEMMATI *et al.*, 1995; see also HEMLEY and COHEN, 1992).

To date, most vibrational studies have been focused on end-member $MgSiO_3$ perovskite. LU *et al.* (1994) and WANG *et al.* (1994) obtained the infrared and Raman spectra of $(Mg,Fe)SiO_3$ perovskites, and found little change in the spectra with

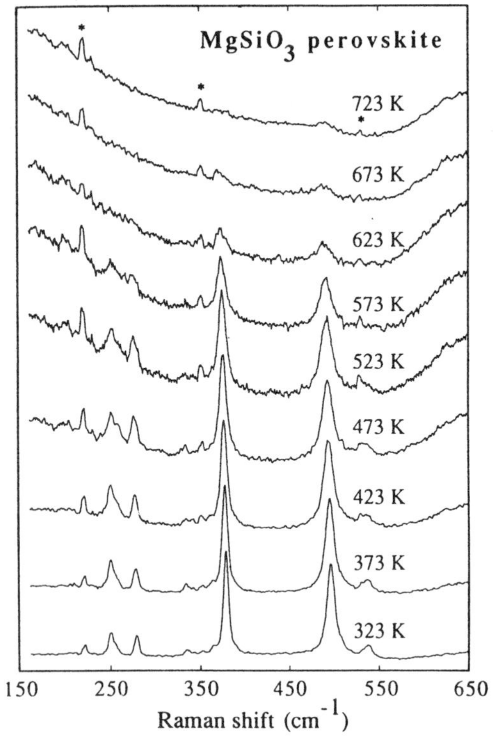

FIG. 3. Raman spectrum of $MgSiO_3$ perovskite as a function of temperature at ambient pressure, taken from DURBEN and WOLF (1992; Fig. 1, p. 891). At high temperature, the perovskite bands begin to weaken, and are replaced by broad bands due to $MgSiO_3$ glass (causing the rise in intensity to high and low frequencies), as the metastable perovskite reverts to an amorphous phase.

Fe^{2+}/Mg^{2+} substitution, except that the Raman peaks were broadened, indicating some structural disorder. Additionally, the Raman frequencies were shifted slightly to lower wavenumber, consistent with the small lattice expansion (KUDOH *et al.*, 1990; MAO *et al.*, 1991; WANG *et al.*, 1994).

$CaSiO_3$ perovskite

Phase equilibrium studies, together with constraints from cosmochemical constraints, indicate that $CaSiO_3$ perovskite is the next most abundant silicate in the lower mantle (HEMLEY and COHEN, 1992; IRIFUNE, 1994). It was first synthesized by LIU and RINGWOOD (1975). Unlike (Mg,Fe)SiO_3 perovskite, this phase appears to remain cubic over the entire *P-T* range of the lower mantle (MAO *et al.*, 1989; HEMLEY and COHEN, 1992; BUKOWINSKI *et al.*, 1996). To date, there has been no experimental study of the vibrational properties of $CaSiO_3$ perovskite, in part because this phase tends to amorphize upon decompression. *In situ* IR would be of particular interest to observe the behaviour of the far-IR reflectivity as a function of pressure, to give information on the low-lying ferroic mode as the sample is decompressed (HEMMATI *et al.*, 1995; BUKOWINSKI *et al.*, 1996). HEMMATI *et al.* (1995) have proposed that this vibration, which involves off-centre motions of the Si^{4+} ions within their octahedral sites, plays a role in the amorphization process. It is possible to recover at ambient pressure perovskites synthesized under pressure along the $CaSiO_3$-$CaTiO_3$ join (RINGWOOD and MAJOR, 1967, 1971). LEINENWEBER *et al.* (1994, 1996) have recently obtained Raman and infrared spectra for $CaSiO_3$—$CaTiO_3$ perovskites synthesized at high pressure. This system forms an orthorhombic or tetragonal solid solution between $CaSiO_3$ and 50 mol% $CaTiO_3$, and a new ordered compound at the Ca_2SiTiO_6 composition. Strong Raman bands appear near 750 and 450 cm^{-1} as the silicate content is increased, which can be assigned to Si—O stretching and OSiO bending vibrations associated with the SiO_6 octahedral groups.

(Mg,Fe)O magnesiowüstite

MgO and FeO have the cubic NaCl (B1) structure, which has no first order Raman spectrum. The zone centre optic mode has symmetry

$$\Gamma_{vib} = F_{1u}(IR) \qquad (3)$$

This triply degenerate vibration is actually split into transverse (2 TO) and longitudinal (LO) components by the macroscopic electric field associated with the vibration (FERRARO, 1984; MCMILLAN, 1985; MCMILLAN and HOFMEISTER, 1988). The infrared reflectance spectrum under ambient conditions shows a broad reflectance band extending from ~350 cm^{-1} to ~750 cm^{-1} with low- and high frequency limits determined by the TO and LO frequencies (PIRIOU, 1964, 1974; PIRIOU and CABANNES, 1968). The pure absorption frequency, which would be visible in thin film measurements, would give rise to a single peak near 380 cm^{-1}. However, additional features are often observed in powder measurements in the 550–650 cm^{-1} region, due to vibrational mode coupling, and particle size effects (LUXON *et al.*, 1969; PIRIOU, 1974; MCMILLAN, 1985).

The characterization of the thermophysical properties of (Mg,Fe)O solid solutions is crucial for understanding the lower mantle (ANDERSON *et al.*, 1992, 1993). There has been a large number of theoretical studies of the lattice dynamics of MgO under lower mantle conditions (COHEN *et al.*, 1987; AGNON and BUKOWINSKI, 1990; BURNHAM, 1990; BUKOWINSKI, 1994; CHIZMESHYA *et al.*, 1994; BUKOWINSKI *et al.*, 1996): however, the frequency of the LO modes is typically overestimated in these studies because of constraints imposed on the electronic relaxation associated with the vibration. The zone centre frequency fixes the upper limit of the vibrational spectrum. The IR spectra at ambient conditions are well understood (PIRIOU, 1964; BREHAT *et al.*, 1966; FARMER, 1974; FERRARO, 1984). The infrared reflectance spectrum of MgO has been studied at high temperature, giving information on the anharmonicities of the lattice modes (PIRIOU, 1964, 1974; PIRIOU and CABANNES, 1968). To date, there has been no direct study of the IR spectrum at high pressure. The far IR spectra of several halides with the B1 and B2 structures have been studied at high pressure in the diamond anvil cell (FERRARO, 1984), and this type of study could be extended to magnesiowüstite in future studies. CHOPELAS and NICOL (1982) noted that the side bands present on the V^{2+} or Cr^{3+} luminescence for these ions doped in MgO contained information on the vibrational density of states of the host oxide crystal, due to vibronic coupling between the electronic excitation of the transition metal ion and the lattice vibrations of MgO. They used these data to construct a density of states function $g(\omega)$ for MgO, and to calculate thermodynamic properties of this phase within the quasiharmonic approximation, up to 10 GPa. CHOPELAS (1990a, 1992) has revised and extended these studies to over 20 GPa.

Stishovite

Stishovite with the rutile structure is the high pressure phase of SiO_2, stable above 9 GPa (STISHOV and POPOVA, 1961). The natural occurrence of this mineral was first described as a product of impact metamorphism of quartz in sandstone (CHAO *et al.*, 1962). However, its high pressure stability means that it could be present throughout the mantle, being formed in disproportionation reactions between silicate minerals and oxides (HEMLEY *et al.*, 1994; KINGMA *et al.*, 1995). The expected vibrational modes for the stishovite structure are:

$$\Gamma_{vib} = A_{1g}(R) + A_{2g}(inactive) + B_{1g}(R) + B_{2g} + E_g(R) + 2B_{1u}(IR) + 2E_u(IR) \quad (4)$$

A powder transmission spectrum of stishovite separated from a natural sample was first recorded by (LYON, 1962). This spectrum showed many more peaks than expected from the symmetry analysis. However, the powder spectrum closely resembles the polycrystalline IR reflectance spectrum obtained by HOFMEISTER *et al.* (1990) (Fig. 4). Analysis of the reflectivity data gave the zone centre TO modes at ~650 cm^{-1} (A_{2u}), and 470, 580 and 820 cm^{-1} (E_u). These frequencies do not correspond simply to the positions of transmission minima in the powder spectrum, because of the variation in optical constants due to overlapping modes of different symmetry with large TO-LO splitting (HOFMEISTER *et al.*, 1990). The reflectivity data are noisy in the low-frequency region, and do not constrain the IR spectrum well in the region below 700 cm^{-1}.

WILLIAMS *et al.* (1993) measured the pressure dependence of peaks in the powder IR transmission spectrum of stishovite. Two bands were followed to approximately 36 GPa, with pressure shifts of 2.4 and 3.3 cm^{-1}/GPa, respectively. The two bands followed have absorbance maxima at 611 and 861 cm^{-1} at room pressure, and were assigned to two of the E_u modes observed by HOFMEISTER *et al.* (1990). As noted by WILLIAMS *et al.* (1993), these frequencies do lie close to (within 40 cm^{-1}) of the TO frequencies determined in the reflectivity study: however, this region of the stishovite spectrum is considerably complicated by overlap between E_u and the A_{2u} modes, and resulting interference effects (HOFMEISTER *et al.*, 1990) (Fig. 4). Based on the resemblance between the powder reflectance and the transmission data (HOFMEISTER *et al.*, 1990; WILLIAMS *et al.*, 1993), it is most likely that the transmission minima (inverted to give absorbance maxima, via $A(\nu) = -\log_{10} T(\nu)$) contain contributions from both TO and LO components, of both E_u and A_{2u} modes in the case of the lower frequency bands. However, the pressure shifts measured by WILLIAMS *et al.* (1993) give information on the average Grüneisen parameter for these modes. The high pressure behaviour of the IR spectrum should be subjected to further analysis, especially because of the impending transition to the $CaCl_2$-structured phase at ~50 GPa, as discussed below.

The Raman spectra of natural and synthetic stishovite have been reported by HEMLEY *et al.* (1986a) and HEMLEY (1987) (Fig. 4). The results of earlier studies were complicated by the introduction of impurities due to the extraction process of stishovite from natural samples and the presence of glassy material: this only became apparent once the spectrum of synthetic samples had been recorded (HEMLEY *et al.*, 1986a; HEMLEY, 1987). The Raman spectrum shows all four expected modes: A_{1g} at 753 cm^{-1}, B_{2g} at 967 cm^{-1}, E_g at 589 cm^{-1}, and B_{1g} at 231 cm^{-1}. HEMLEY (1987) measured the pressure

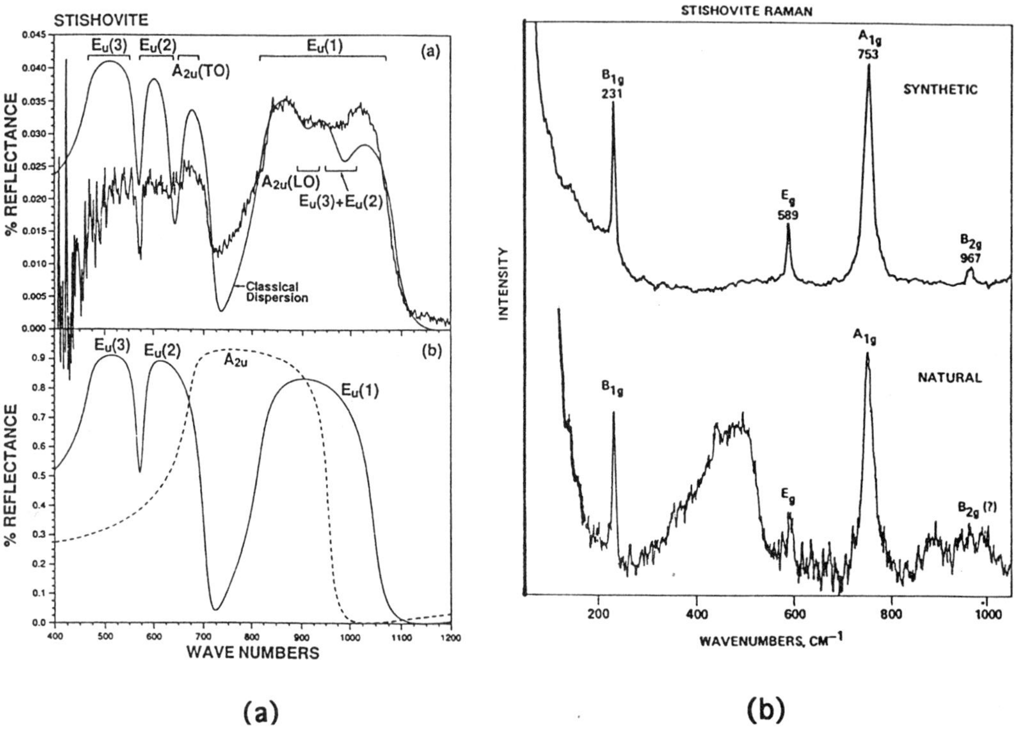

FIG. 4. (a) Infrared and (b) Raman spectra of SiO_2 stishovite. The infrared reflectance spectrum is taken from HOFMEISTER *et al.* (1990; Fig. 1, p. 952). The experimental (unpolarized) polycrystalline spectrum is shown at top, along with the TO and LO mode frequencies obtained from analysis of the reflectivity data. The synthetic spectrum at bottom was reconstructed using these parameters, for both mode symmetries (A_{2u} and E_u). The Raman spectrum at right is from HEMLEY *et al.* (1986a). The four expected modes are clearly visible in the spectrum of the synthetic sample at top. The natural sample (bottom) shows broad bands corresponding to amorphous SiO_2.

shifts of these Raman active vibrations. The behavior of the low-lying B_{1g} mode is particularly interesting, because its frequency decreases with increasing pressure (Fig. 5). The atomic displacements associated with this mode correspond to those required to transform stishovite from the rutile into the calcium chloride structure (HEMLEY, 1987; NAGEL and O'KEEFFE, 1971). Recent theoretical (COHEN, 1992; LACKS and GORDON, 1993; MATSUI and TSUNEYUKI, 1992) and experimental (TSUCHIDA and YAGI, 1989, 1990; KINGMA *et al.*, 1995, 1996) work has shown that the rutile structure transforms to a phase with the calcium chloride structure. From theoretical calculations of the vibrational mode frequency with increasing pressure, this transition was predicted to occur near 50 GPa (COHEN, 1992), in remarkable agreement with the recent experimental result of KINGMA *et al.* (1995) (Fig. 5). This work indicates that free SiO_2 would exist in the $CaCl_2$ structure below 1200–1500 km within the lower mantle (depending on the *P-T* slope of the transition, which has not yet been determined experimentally). A further transition to a modified fluorite or *Pa3* phase is predicted to occur at Earth core pressures, above ~150 GPa (COHEN, 1994; HAINES *et al.*, 1996).

GILLET *et al.* (1990) measured the temperature dependence of the Raman active modes of stishovite at ambient pressure, and calculated the heat capacity and entropy of this phase via vibrational modeling. In this study, the pressure- and temperature derivatives of the Raman peaks were used to estimate the intrinsic anharmonicity of the vibrational modes, and to obtain a high temperature heat capacity corrected for anharmonicity, beyond the quasiharmonic approximation. In the high-temperature study, broad bands were observed to appear due to back-transformation of the high pressure phase into glass dominated by tetrahedrally coordinated silicate groups (GILLET *et al.*, 1990; GRIMSDITCH *et al.*, 1994). These broad glass bands are also present in the Raman spectrum of natural

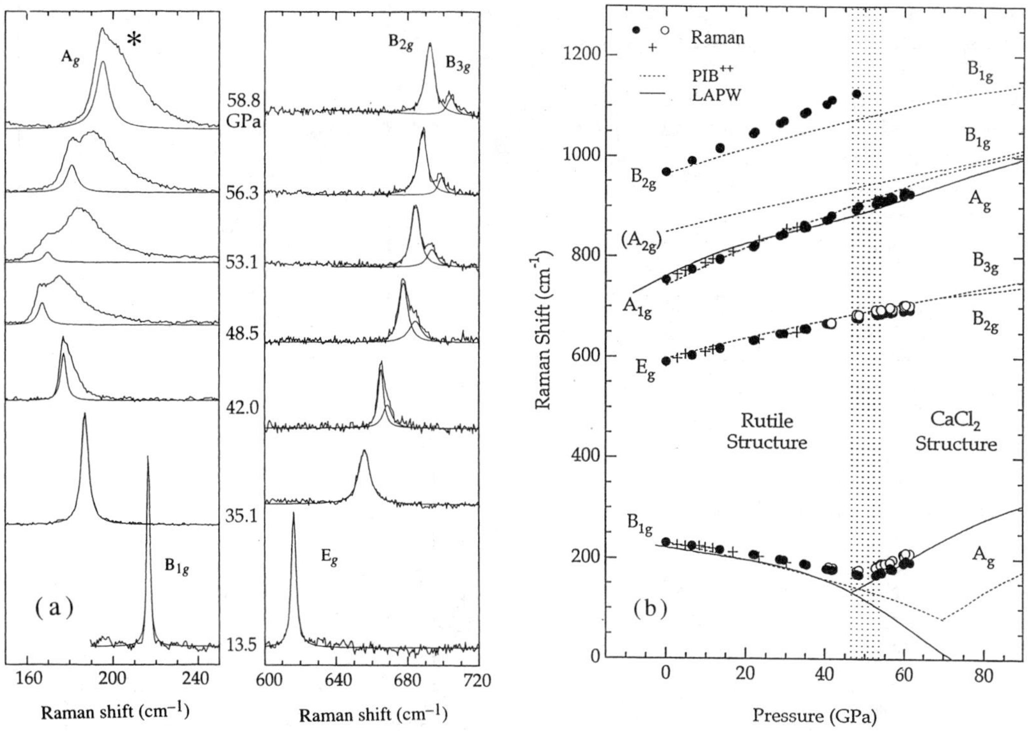

FIG. 5. (a) Raman spectra from KINGMA *et al.* (1995) showing evidence for the rutile-$CaCl_2$ transition in SiO_2 at high pressure. The smooth lines under the observed bands are Lorentzian fits for the expected fundamentals of both phases (KINGMA *et al.*, 1995). The broad band marked * is not predicted by theory, and is dependent upon the stress state of the sample. The low frequency B_{1g} peak for the rutile form softens with increasing pressure. Above the transition pressure (~50 GPa), this becomes the A_g mode of the $CaCl_2$ structure, and its frequency begins to increase with further compression. (b) Frequency shifts of Raman bands of stishovite and $CaCl_2$-structured SiO_2 with pressure at ambient temperature, during compression (filled circles) and decompression (open circles). The experimental data are compared with results of theoretical calculations, using first-principles LAPW (COHEN, 1992: full line) and parametrized PIB (potential induced breathing: dashed line) methods.

stishovite (HEMLEY *et al.*, 1986a). The phenomenon of pressure- and temperature-induced amorphization of SiO_2 phases has been studied extensively using vibrational spectroscopy (HEMLEY, 1987; GILLET *et al.*, 1990; WOLF *et al.*, 1992; WILLIAMS *et al.*, 1993; GRIMSDITCH *et al.*, 1994).

TRANSITION ZONE MINERALS

β- (Mg,Fe)$_2$SiO$_4$

Along with the spinel (γ) polymorph discussed below, a $(Mg,Fe)_2SiO_4$ polymorph with the wadsleyite, or "modified spinel", (β) structure is likely to form one of the dominant constituents of the mantle transition zone. The transition from the low pressure olivine (α-$(Mg,Fe)_2SiO_4$) to the β-phase occurs at approximately 13 GPa at ~1700 K, and marks the 410 km seismic discontinuity (BINA and WOOD, 1987; AKAOGI *et al.*, 1989; KATSURA and ITO, 1989; MENG *et al.*, 1993; MORISHIMA *et al.*, 1994). The Raman and infrared spectra of the β- and γ-polymorphs of Mg_2SiO_4 were first reported by AKAOGI *et al.* (1984), who used the vibrational data to construct model density of states functions $g(\omega)$ for specific heat and entropy calculations. This information was then combined with calorimetric data to establish a *P-T* phase diagram for the Mg_2SiO_4 polymorphs (AKAOGI *et al.*, 1984). This study has been superseded by more recent work (AKAOGI *et al.*, 1989; GASPARIK, 1990; CHOPELAS, 1991b), but it represented the first application of KIEFFER's (1979a,b,c) approach to mantle mineralogy.

The powder IR spectrum of β-Co_2SiO_4 was obtained by JEANLOZ (1980). Unlike the α-(olivine) and γ-(spinel) polymorphs of M_2SiO_4 phases, this spectrum showed a peak at 686 cm^{-1}, associated with a bending vibration of the SiOSi linkage between tetrahedral groups. Similar features at 700 and 675 cm^{-1} appear in the IR spectrum of β-Mg_2SiO_4 (AKAOGI *et al.,* 1984; WILLIAMS *et al.,* 1986; CYNN and HOFMEISTER, 1994) (Fig. 6). These weak but characteristic bands reveal the presence of the unusual Si_2O_7 units in the structure (JEANLOZ, 1980; HORIUCHI and SAWAMOTO, 1981), which require the presence of oxygen bound only to Mg atoms in the β-polymorph. This structural feature is important for the ability of this nominally anhydrous phase to accept large quantities of H into its structure (SMYTH, 1987; DOWNS, 1989), discussed further below.

The expected vibrational modes for β-Mg_2SiO_4 are:

$$\Gamma_{vib} = 11A_g(R) + 7B_{1g}(R) + 9B_{2g}(R) + 12B_{3g}(R) + 7A_u(inactive) + 14B_{1u}(IR) + 13B_{2u}(IR) + 11B_{3u}(IR) \quad (5)$$

The Raman spectrum of β-Mg_2SiO_4 originally reported by AKAOGI *et al.* (1984) was incorrect, because the sample partly back-transformed during heating by the laser in the Raman experiment. The Raman spectrum for this phase was later obtained by MCMILLAN and AKAOGI (1987) and by CHOPELAS (1991b). In both of these studies, a strong peak (most likely of A_g symmetry) was observed at 723 cm^{-1}, due to the SiOSi linkage vibrations, (Fig. 6). REYNARD *et al.* (1996) have recently found some differences with the previous spectra. They did not reproduce a peak found at 86 cm^{-1} by CHOPELAS (1991b), which is important for defining the low frequency limit of the density of states function for heat capacity and entropy calculations (AKAOGI *et al.,* 1984; CHOPELAS, 1991b), and they did not observe the weak peaks in the 1000–1100 cm^{-1} region reported by MCMILLAN and AKAOGI (1987). REYNARD *et al.* (1996) have suggested that these peaks could be due to Si_3O_{10} units, created locally within the structure associated with the presence of stacking faults in the structure (BREARLEY *et al.,* 1992). If this assignment is correct, Raman spectroscopy could provide a powerful *in situ* method for characterizing the defect structure in β-Mg_2SiO_4 at high pressure and temperature (*e.g.,* with a diamond-anvil cell).

The kinetics and mechanism of transformations between α-, β- and γ-polymorphs of $(Mg,Fe)_2SiO_4$ are important for understanding the seismicity

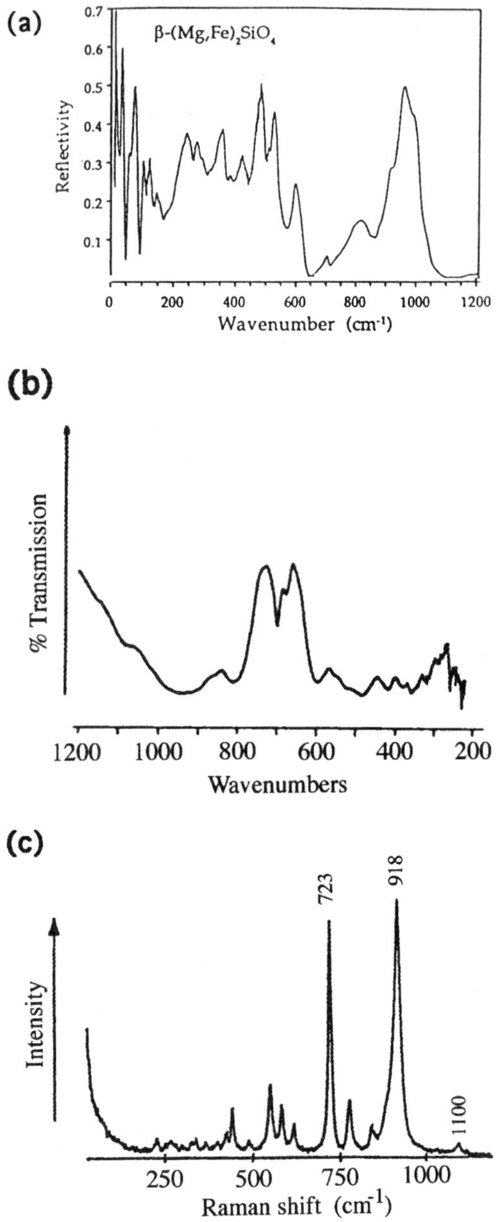

FIG. 6. Infrared (a,b) and Raman (c) spectra of β-$(Mg,Fe)_2SiO_4$. The infrared reflectance spectrum (a) is redrawn from CYNN and HOFMEISTER (1994; Fig. 1, p. 17,718). The weak peaks associated with the SiOSi linkages in the structure are more easily distinguished in the powder transmission spectrum, from AKAOGI *et al.* (1984) (b). The Raman spectrum (c) was obtained by MCMILLAN and AKAOGI (1987). The weak peak at 1100 cm^{-1} was not observed in the recent study by REYNARD *et al.* (1996), and may correspond to a structural defect in the sample (see text).

within subducting slabs (SUNG and BURNS, 1976; WU *et al.*, 1993; BREARLEY *et al.*, 1992; RUBIE and BREARLEY, 1994; REYNARD *et al.*, 1996). Raman spectroscopy provides a useful technique for characterizing these structural changes (CHOPELAS, 1991b; DURBEN *et al.*, 1993; REYNARD *et al.*, 1996). The accidentally back-transformed spectrum of AKAOGI *et al.* (1984) resembled that of Mg_2SiO_4 olivine, but with considerably broadened bands. In a high temperature single crystal X-ray diffraction study of β-Mg_2SiO_4 at ambient pressure, TSUKIMURA *et al.* (1988) had observed anomalous changes in the lattice parameters above 580°C which might be associated with a structural transformation. MCMILLAN *et al.* (1991) obtained a Raman spectrum for a sample that had been heated to 585°C at ambient pressure. This spectrum showed the broadened bands corresponding to a disordered olivine-like material found by AKAOGI *et al.* (1984), along with weak broad bands in the 650–700 cm^{-1} region. These suggested the presence of SiOSi linkages persisting within the structure, presumably as the β-Mg_2SiO_4 back-transformed to olivine via a series of spinelloid intermediate structures (HYDE *et al.*, 1982; MADON and POIRIER, 1983). REYNARD *et al.* (1996) have recently carried out a detailed kinetic study of this transformation at ambient pressure via Raman spectroscopy. DURBEN *et al.* (1993) have used Raman spectroscopy to investigate the compression mechanism in olivine (α-Mg_2SiO_4) at room temperature, and observed changes which might be associated with formation of SiOSi linkages within the structure, perhaps accompanied by Si coordination changes (GUYOT and REYNARD, 1992). Further compression of the olivine end-members at room temperature leads to amorphization, as shown by WILLIAMS *et al.* (1990) and RICHARD and RICHET (1990) for Fe_2SiO_4 and by ANDRAULT *et al.* (1995) for $(Mg,Fe)_2SiO_4$. Further experiments using both Raman and infrared spectroscopy under mantle *P-T* conditions are likely to yield new information on the nature of the forward- and back-transformations between the α-, β- and γ-phases of $(Mg,Fe)_2SiO_4$.

WILLIAMS *et al.* (1986) measured the powder IR spectrum of β-Mg_2SiO_4 to 27 GPa at room temperature, and CYNN and HOFMEISTER (1994) have obtained the mid- and far-IR spectrum of a sample with composition $(Mg_{1.80}Fe_{0.15})SiO_4$, to ~23 GPa. The spectrum of the Fe-containing sample is less well-resolved, presumably due to Fe/Mg disorder. The peaks in the powder absorption spectrum correspond quite closely with the TO modes determined from the IR reflectivity (CYNN and HOFMEISTER, 1994), in contrast to the case for stishovite mentioned above. It has been noted that, as the complexity of the mineral phase increases and more peaks are present in the IR spectrum, the powder transmission spectrum gives a better representation of the true TO mode frequencies, at least for modes which are not strongly polar (such as the Si—O stretching vibrations) (MCMILLAN and HOFMEISTER, 1988). The average mode Grüneisen parameter was determined to be $\langle\gamma\rangle = 1.28 \pm 0.02$, in good agreement with the thermodynamic value $\gamma_{th} = 1.05 \pm 0.2$ (WILLIAMS *et al.*, 1986). REYNARD *et al.* (1996) measured the temperature shifts of the Raman peaks of β-Mg_2SiO_4 up to 800–900 K (the back-transformation temperature), and combined the data with pressure shifts to determine intrinsic mode anharmonicity parameters (GILLET *et al.*, 1991). In her study of the Raman spectrum at high pressure, CHOPELAS (1991b) observed changes in the pressure shifts at 9.2 GPa and perhaps also at 17 GPa, which might indicate some structural transformation in β-Mg_2SiO_4. Similar changes in slope are apparent in the IR data taken at high pressure (CYNN and HOFMEISTER, 1994). Although the possibility of weak structural distortions deserves further study, no anomalies have been noted in X-ray diffraction measurements of β-$(Mg,Fe)_2SiO_4$ to 26 GPa and 900 K (FEI *et al.*, 1992). LIU *et al.* (1994) has also investigated the Raman spectrum of this phase to 18 GPa at ambient temperature.

Because the composition of β-Mg_2SiO_4 is that of an orthosilicate, formation of the SiOSi linkage in the high pressure phase results in one oxygen per asymmetric unit being bound to only Mg atoms, in an irregular five-fold coordinated site (HORIUCHI and SAMAMOTO, 1981; SAWAMOTO and HORIUCHI, 1990). SMYTH (1987) and DOWNS (1989) suggested that this might provide a potential site for protonation of the nominally anhydrous mineral, so that β-Mg_2SiO_4 could act as an important source or sink for water in the mantle (DOWNS, 1989; SMYTH, 1987). MCMILLAN *et al.* (1991) examined a synthetic sample of β-Mg_2SiO_4 using micro-Raman spectroscopy, and found that an asymmetric band was present at 3322 cm^{-1}, due to OH groups within the structure (Fig. 7). Measurement of the absorbance via micro-infrared spectroscopy allowed an estimate of the OH content as approximately 0.06 % by weight (as g OH per 100 g Mg_2SiO_4: this corresponds to ~10,300 H/10^6 Si) (MCMILLAN *et al.*, 1991). These authors considered that this value represented a maximum degree of hydration for the β-Mg_2SiO_4 phase, because other stable hydrous magnesium silicates were known to be pres-

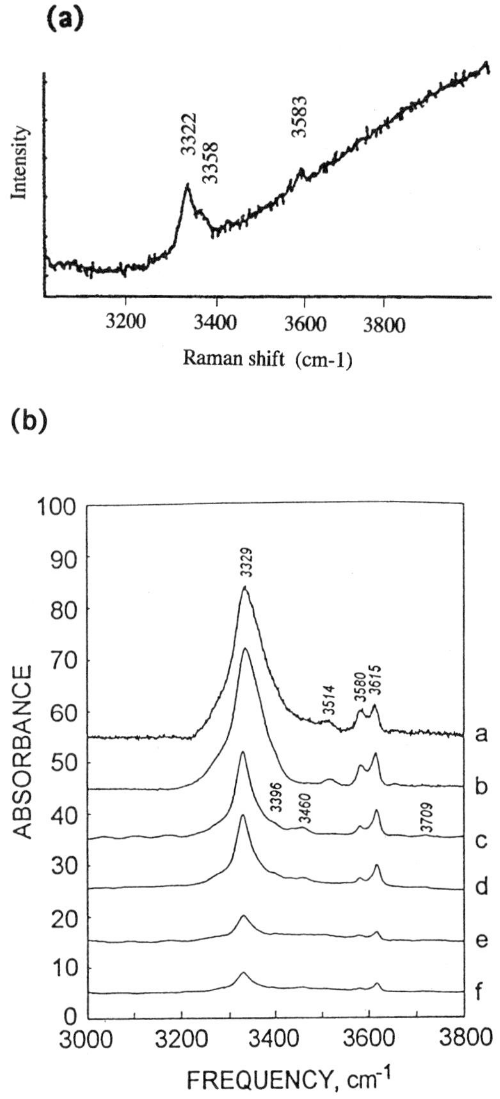

FIG. 7. (a) Raman and (b) infrared spectra of β-Mg_2SiO_4 in the 3000–3800 cm^{-1} region, showing evidence for OH groups within the structure (McMILLAN *et al.*, 1991; YOUNG *et al.*, 1993). Samples (a–f) on the right were determined to contain up to 65,000 H/10^6Si (YOUNG *et al.*, 1993). Spectra from YOUNG *et al.* (1993; Fig. 6, p. 415) used by permission of the editors of *Physics and Chemistry of Minerals* (Springer-Verlag GmbH).

ent. From quantitative IR absorption spectra on single crystal samples of $(Mg_{0.92},Fe_{0.08})_2SiO_4$, prepared at 14 GPa and 1550–1650 K, YOUNG *et al.* (1993) determined much higher OH concentrations, between 10,000 and 65,000 H/10^6 Si. It was suggested that these higher values could be due to the presence of Fe in the sample (YOUNG *et al.*, 1993). KOHLSTEDT *et al.* (1996) have recently prepared a suite of $(Fe,Mg)_2SiO_4$ samples from natural olivine at pressures and temperatures up to 15 GPa and 1100°C, and have found water contents ranging up to ~2.4 wt% (400,000 H/10^6 Si) via quantitative IR absorbance spectroscopy (Fig. 7). This is consistent with the recent determination of OH in β-Mg_2SiO_4 (to give an actual composition close to $Mg_{1.75}H_{0.5}SiO_4$) via SIMS (INOU *et al.*, 1995). KOHLSTEDT *et al.* (1996) also documented a similar quantity of hydrous component dissolved in γ-$(Fe,Mg)_2SiO_4$, prepared at 19.5 GPa and 1100°C. These results are extremely important for the mineralogy of the transition zone, and water storage within the Earth. CYNN and HOFMEISTER (1994) have investigated the behaviour of the OH stretching vibrations in the $(Mg,Fe)_2SiO_4$ sample prepared by YOUNG *et al.* (1993), and have documented an increase in hydrogen bonding in this phase with increasing pressure. Such behaviour could help in stabilizing hydrous phases at high pressure (WILLIAMS, 1992; KNITTLE *et al.*, 1992).

γ-$(Mg,Fe)_2SiO_4$

At pressures above 18–20 GPa for the temperature range encountered within the transition zone of the normal mantle, β-Mg_2SiO_4 transforms further into a phase with the spinel structure, γ-Mg_2SiO_4 (AKAOGI *et al.*, 1989; RIGDEN *et al.*, 1991). Unlike the magnesian end-member, Fe_2SiO_4 transforms directly from olivine into spinel at much lower pressure, ~6–7 GPa, over the same temperature range. For this reason, the Fe content of the $(Mg,Fe)_2SiO_4$ component is an important parameter in determining the depth and width of the α-(β)-γ transitions occurring within the mantle (AKAOGI *et al.*, 1989; BINA and WOOD, 1987; KATSURA and ITO, 1989; WU *et al.*, 1993; GILLET, 1995).

There have been several vibrational studies of the Mg_2SiO_4 and Fe_2SiO_4 spinel end-members, and their solid solutions (Fig. 8, 9). The expected vibrational modes for a cubic (*Fd3m*) spinel structure are:

$$\Gamma_{vib} = A_{1g}(R) + E_g(R) + T_{1g}(inactive) + 3T_{2g}(R) + 2A_{2u}(inactive) + 2E_u(inactive) + 4T_{1u}(IR) + 2T_{2u}(inactive) \quad (6)$$

JEANLOZ (1980) obtained a powder IR absorption spectrum of Fe_2SiO_4 spinel, and observed three infrared peaks at 848, 503 and 344 cm^{-1} (Fig. 8). The lower frequency peaks correspond quite well with the TO mode frequencies estimated from the reflectance spectrum obtained by HOFMEISTER (see

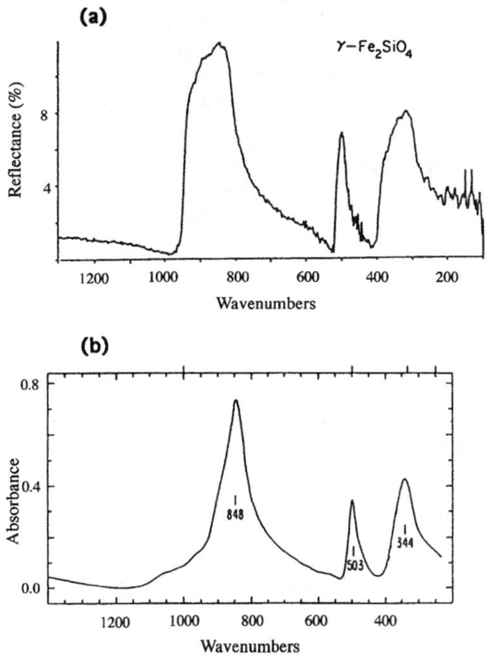

FIG. 8. (a) IR reflectance (obtained by A. HOFMEISTER, reported in MCMILLAN and HOFMEISTER, 1988) and (b) powder transmission (JEANLOZ, 1980) spectra for γ-Fe_2SiO_4. Three of the expected four resonances are clearly visible in these spectra. Spectra from JEANLOZ (1980; Fig. 2, p. 331) used by permission of the editors of *Physics and Chemistry of Minerals* (Springer-Verlag GmbH).

MCMILLAN and HOFMEISTER, 1988), and correspond to the deformation vibration (ν_4) of the SiO_4 tetrahedra, and an octahedral FeO_6 stretching mode, respectively. The highest frequency mode is attributed to the ν_3 asymmetric stretching vibration of the tetrahedral SiO_4 groups. This highly polar vibration is associated with a large TO—LO splitting, and gives rise to a broad reflectance band (Fig. 8). The upper limit of this band occurs near 960 cm^{-1}, which fixes the approximate upper limit in the density of states function, $g(\omega)$. The powder IR absorption frequency lies close to the inflexion on the low-frequency side of the reflectance band, which corresponds to the TO mode frequency (MCMILLAN and HOFMEISTER, 1988). The fourth mode expected in the IR spectrum has not yet been observed.

AKAOGI *et al.* (1984) observed broad bands at 830 and 445 cm^{-1} in their powder transmission IR spectrum for γ-Mg_2SiO_4 (Fig. 9). The remaining infrared modes for Mg_2SiO_4 have still not been characterized. There have been several lattice dynamical calculations for γ-Mg_2SiO_4 using empirical interatomic potentials, but the vibrational frequencies were not reported (MATSUI and BUSING, 1984; PRICE and PARKER, 1984; PRICE *et al.,* 1987).

AKAOGI *et al.* (1984) obtained a Raman spectrum for Mg_2SiO_4 spinel, which showed the strong peaks expected for the symmetric (ν_1) and asymmetric (ν_3) stretching vibrations of the SiO_4 tetrahedra, in the 780–900 cm^{-1} region. A similar spectrum was obtained by GUYOT *et al.* (1986) for a natural sample of γ-$(Mg_{0.74}Fe_{0.26})_2SiO_4$, although an additional weak peak was observed in that study at 844 cm^{-1}, which has not been assigned (and could be due to coating from the microscope objective). The remaining weak peaks expected in the spinel spectrum were recorded by MCMILLAN and AKAOGI

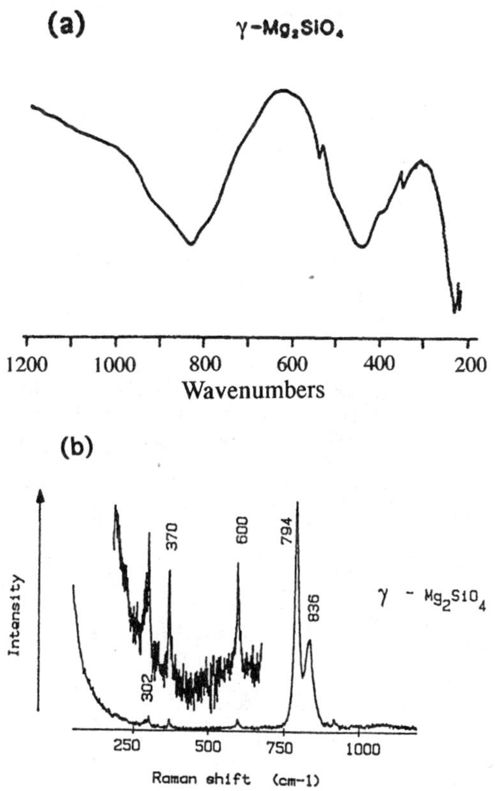

FIG. 9. (a) Infrared and (b) Raman spectra of γ-Mg_2SiO_4. The Raman spectrum shows all five peaks expected for the spinel structure. MCMILLAN and AKAOGI (1987) assigned the strong peak at 794 cm^{-1} to the A_g mode, and the 836 cm^{-1} peak to the T_{2g} mode, expected from the SiO_4 symmetric and asymmetric stretching vibrations. Based on results of a single crystal polarized study, CHOPELAS *et al.* (1994) have now reversed this assignment. The mid-IR transmission spectrum (a) was obtained by AKAOGI *et al.* (1984). This shows only two bands, instead of the expected four. Further work is required to locate the additional resonances.

(1987). In their unpolarized powder spectrum, these authors assigned the strong peak at 794 cm^{-1} peak to the ν_1 symmetric stretching vibration of the SiO_4 groups (A_{1g} symmetry), and the 836 cm−1 band to the ν_3 asymmetric stretch with T_{2g} symmetry, based on their relative intensities (Fig. 9). Based on oriented spectra of single crystals, CHOPELAS *et al.* (1994) re-assigned these peaks, suggesting that the symmetric stretching vibration corresponded to the *higher* frequency component (Fig. 10) (this has consequences for the Si—O stretch-stretch interaction force constant of the SiO_4 tetrahedron: MCMILLAN and HESS, 1990). These authors used the Raman data to determine a model vibrational density of states for quasiharmonic calculations of heat capacity and entropy. It is interesting to note that, although the calculated entropy differs by ~10% from the earlier model of AKAOGI *et al.* (1984) for the value calculated at room temperature, the difference at 1000 K is only ~2%. This serves to illustrate the relative insensitivity of the calculated high temperature heat capacity and entropy to details of the vibrational modeling. The pressure shifts in the Raman spectrum of γ-Mg_2SiO_4 were measured in the study by CHOPELAS *et al.* (1994) to 20 GPa.

$MgSiO_3$ ilmenite

The ilmenite phase of $MgSiO_3$ would be stable in a narrow *P-T* range within cold subducted regions at approximately 600–700 km depth (ITO and YAMADA, 1982). It enters into solid solution with Al_2O_3 component (LIU, 1977). There have been several studies of this phase by Raman and infrared spectroscopy, and by computer simulation

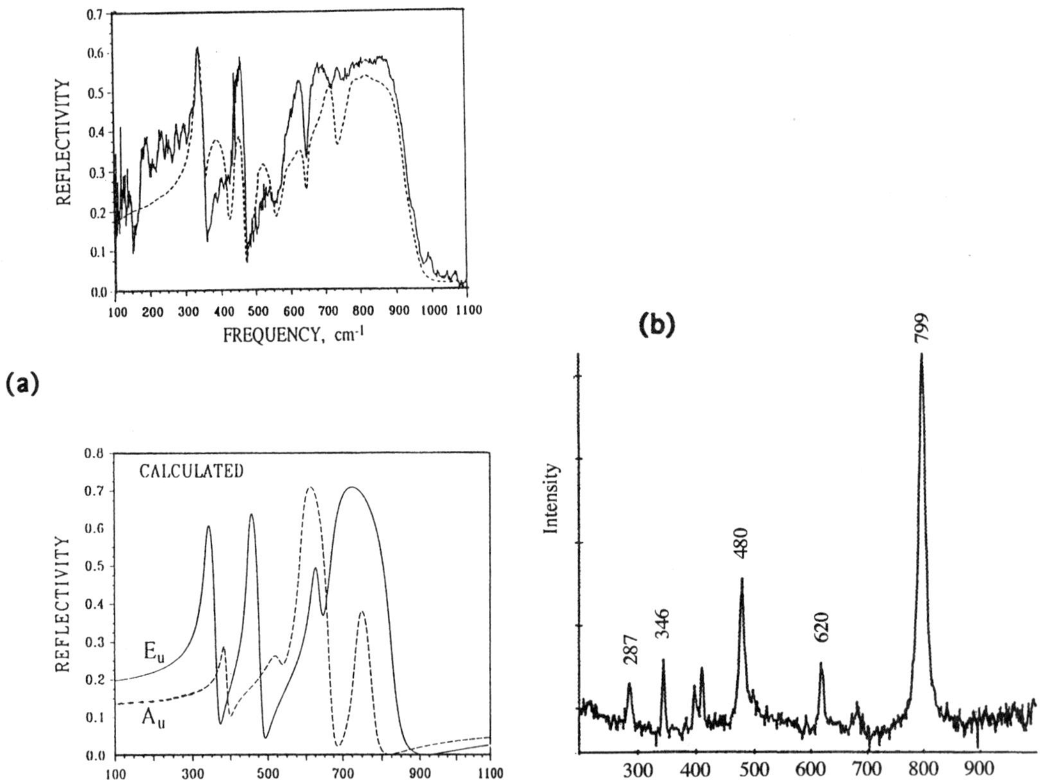

FIG. 10. (a) Infrared and (b) Raman spectra of $MgSiO_3$ ilmenite. The infrared reflectance spectrum is redrawn from HOFMEISTER and ITO (1992; Figs. 1 and 2, p. 426. Used by permission of the editors of *Physics and Chemistry of Minerals,* Springer-Verlag GmbH). The unpolarized experimental spectrum is shown at top. The synthetic spectrum at bottom was reconstructed for both mode symmetries (A_u and E_u). All expected modes are observed. The Raman spectrum at right is from MCMILLAN and ROSS (1987). Recent careful measurements (REYNARD and RUBIE, 1995) indicate that all indicated modes are real, including the weak shoulder at 499 cm^{-1}.

of its lattice dynamics. The expected vibrational mode symmetries are:

$$\Gamma_{vib} = 5A_g(R) + 5E_g(R) + 4A_u(IR) + 4E_u(IR) \quad (7)$$

The Raman spectrum of $MgSiO_3$ ilmenite was first recorded by ROSS and MCMILLAN (1984), who observed seven of the expected ten modes. A better quality spectrum was obtained by MCMILLAN and ROSS (1987), in which eight peaks were clearly resolved, and a further mode was apparent as a shoulder on the band at 480 cm^{-1} (Fig. 10). The spectrum is dominated by an intense peak at 799 cm^{-1}, assigned to an A_g mode associated with symmetric Si—O stretching of the SiO_6 groups (MCMILLAN and ROSS, 1987; HOFMEISTER and ITO, 1992). WALL and PRICE (1988) carried out a theoretical calculation of the lattice dynamics of $MgSiO_3$ ilmenite using a polarizable ion (shell) model with empirical potentials. These workers used their results to predict the positions of the missing Raman peaks not observed by MCMILLAN and ROSS (1987), and to reassign the suggested mode symmetries. However, HOFMEISTER and ITO (1992) noted that all of the calculated vibrational frequencies were overestimated, and they made a further re-assignment of the observed infrared and Raman active modes. They suggested that the "missing" Raman mode corresponded to an unresolved pair of peaks at 620 cm^{-1}, with A_g and E_g symmetry. However, REYNARD and RUBIE (1996) have recently obtained a high quality Raman spectrum for this phase, and find no evidence for splitting in the 620 cm^{-1} peak.

MADON and PRICE (1989) first obtained a powder IR transmission spectrum for $MgSiO_3$ ilmenite. Unfortunately, these authors observed *nine* peaks between 282 and 825 cm^{-1}, compared with the eight IR-active modes expected for the ilmenite structure. HOFMEISTER and ITO (1992) carried out infrared reflectance measurements for both polycrystalline and single samples of $MgSiO_3$ ilmenite, and observed all of the expected modes, although there may be some remaining uncertainty in the detailed analysis of the reflectance spectrum, because of noise in the data below 300 cm^{-1}, the occurrence of overlapping peaks, and the presence of weak additional oscillators, presumably due to combination bands (HOFMEISTER and ITO, 1992) (Fig. 10). There is generally good agreement between the spectra of MADON and PRICE (1989) and the reflectance data of HOFMEISTER and ITO (1992), except that the lowest frequency peak observed by Madon and Price (282 cm^{-1}) is not reproduced, and the transmission minima ("absorption" maxima) of the broad high frequency bands occur at considerably higher wavenumber than the true TO mode frequencies, as expected (MCMILLAN and HOFMEISTER, 1988; HOFMEISTER and ITO, 1992).

LIU *et al.* (1994) recorded the principal Raman peaks of $MgSiO_3$-ilmenite to 30 GPa. A more complete study has recently been carried out by REYNARD and RUBIE (1996), in which accurate pressure shifts were obtained for nine Raman modes, with $(\partial\nu_i/\partial P)_T$ ranging from 1.7–3.7 cm^{-1}/GPa. These investigators also measured the temperature shifts of the Raman peaks $(\partial\nu_i/\partial T)_{P,298} = -0.0148$ to -0.0285 cm^{-1}/K), which permitted evaluation of the intrinsic mode anharmonicity parameters. These are all very small, indicating that $MgSiO_3$-ilmenite behaves nearly quasiharmonically over the temperature range studied. Because the anharmonicity parameters for garnets are much larger (GILLET *et al.*, 1992), it has been proposed that intrinsic mode anharmonicity corrections might play a role in determining the slope of garnet-ilmenite phase boundaries (REYNARD and GUYOT, 1994; REYNARD and RUBIE, 1996). From their data, REYNARD and RUBIE (1996) calculated an average value of the Grüneisen parameter (defined as $\langle\gamma\rangle = \Sigma C_{vi}\gamma_{iT}/\Sigma C_{\nu i}$, where $\gamma_{iT} = K_T(\partial \ln \nu_i/\partial P)_T$) of 1.38 at room temperature, and 1.24 in the high temperature limit, compared with the value of $\gamma_{th} = 1.74$ (ambient *P,T*) which can be estimated from the available thermodynamic parameters (ASHIDA *et al.*, 1988; HOFMEISTER and ITO, 1992).

MCMILLAN and ROSS (1987) used the structural analogy between $MgSiO_3$ ilmenite and Al_2O_3 corundum, for which the full vibrational spectrum across the Brillouin zone was known from inelastic neutron scattering measurements, to construct a model vibrational density of states $g(\omega)$ for the high pressure silicate phase, and calculated its vibrational heat capacity and entropy within the quasiharmonic approximation. The calculated entropy was combined with calorimetric data by ASHIDA *et al.* (1988) to obtain phase boundaries between $MgSiO_3$ ilmenite, pyroxene, β-Mg_2SiO_4 + stishovite, and γ-Mg_2SiO_4 + stishovite in a *P-T* range appropriate for the transition zone. HOFMEISTER and ITO (1992) have constructed an alternative model for $g(\omega)$ based on their infrared spectra and the Raman data of MCMILLAN and ROSS (1987), combined with the lattice dynamics calculation of WALL and PRICE (1988). The calculated specific heat was in better agreement with the experimental data of ASHIDA *et al.* (1988).

REYNARD and RUBIE (1996) have recently obtained the Raman spectrum for a rhombohedral phase with composition $Mg_3Al_2Si_3O_{12}$ (corresponding to pyrope). This showed eight distinct peaks which matched up well with those observed for $MgSiO_3$ ilmenite, indicating that the $3MgSiO_3$-Al_2O_3 phase has an ilmenite structure, rather than the corundum structure expected for a fully-disordered material. A disordering transition between the ilmenite and averaged corundum structures might be expected at high temperature, as discussed below for $(Mg,Fe)SiO_3$ garnet (majorite), and this could be studied most easily by vibrational spectroscopy. However, the importance of such an ordering transition on phase relations or mineral properties in the transition zone has yet to be demonstrated.

$MgSiO_3$ garnet

Garnets in the $(Ca,Fe,Mg)_3(Al,Fe)_2Si_3O_{12}$ (pyrope-grossular-almandine-spessartine) system are important phases in the lower crust and upper mantle. In addition, a majorite garnet phase, containing octahedrally coordinated silicon, with composition in the system $(Mg,Fe,Ca)_3(Si,Al,Fe)_2Si_3O_{12}$, is likely to be present within the transition zone (Akaogi and Akimoto, 1977; Kato and Kumazawa, 1985; Akaogi *et al.,* 1987; Gasparik, 1990; Yusa *et al.,* 1993). A majorite garnet with composition close to $(Mg,Fe)SiO_3$ was first identified in natural samples from the Coorara and Tenham meteorites (Smith and Mason, 1970), following a laboratory synthesis from pyroxene (Ringwood, 1967). X-ray diffraction and vibrational spectroscopic studies of pure $MgSiO_3$ garnet indicate that this phase has a slight tetragonal distortion from cubic symmetry (Kato and Kumazawa, 1985; Akaogi *et al.,* 1987; McMillan et al., 1989). At high temperatures and pressures, there is extensive solid solution in a cubic garnet phase between pyrope ($Mg_3Al_2Si_3O_{12}$) and the majorite end member, extending to approximately 80 mol% $MgSiO_3$ component at 1000°C and 16 GPa (Akaogi *et al.,* 1987; Kanzaki, 1987).

The expected vibrational spectrum for an ideal cubic garnet structure, such as pyrope, almandine, grossular and the $Mg_3Al_2Si_3O_{12}$-$Mg_4Si_4O_{12}$ solid solutions, is already quite complex due to the large number of atoms in the unit cell:

$$\Gamma_{vib} = 3A_{1g}(R) + 5A_{2g}(inactive) + 8E_g(R) + 14T_{1g}(inactive) + 14T_{2g}(R) + 5A_{1u}(inactive) + 5A_{2u}(inactive) + 10E_u(inactive) + 17T_{1u}(IR) + 16T_{2u}(inactive) \quad (8)$$

There have been several systematic infrared and Raman spectra of garnet minerals (Moore *et al.,* 1971; Omori, 1971; Tarte and Deliens, 1973; Delany, 1981; Suwa and Naka, 1975; Kieffer, 1979b; Dietrich and Arndt, 1982; Geiger *et al.,* 1989; McMillan *et al.,* 1989; Hofmeister and Chopelas, 1991a; Pinet and Smith, 1993). In particular, Hofmeister and Chopelas (1991a) have assigned the symmetries for the expected Raman and infrared active modes for the principal garnet end members, through detailed polarized single crystal measurements (Fig. 11).

Hofmeister and Chopelas (1991b) used the vibrational spectra of pyrope and grossular to construct model density of states functions, g(ω), for calculation of the specific heat and entropy of these phases. The relative heat capacities of these garnets has been of some interest. From crystal chemical and mineral thermodynamics systematics, it was expected that the standard entropy for pyrope should be smaller than that for grossular, but instead, the reverse is observed (Kieffer, 1980; Hofmeister and Chopelas, 1991b). Kieffer (1979a, 1980) suggested that the "excess entropy" of pyrope could be attributed to low frequency vibrations of the small Mg^{2+} ion "rattling" in the large dodecahedral site in the garnet (Delany, 1981; Hofmeister and Chopelas, 1991a,b). Hofmeister and Chopelas (1991a) do find the lowest frequency (IR-active) vibrations of pyrope (140 cm^{-1}) to lie at lower wavenumber than the lowest frequency mode in grossular (159 cm^{-1}). However, they assign the low frequency mode of pyrope to a translation of the SiO_4 group, and that of grossular to a Ca^{2+} vibration, based on the relative cation masses. It seems that the alternative assignment proposed by Kieffer (1979a, 1980) could still hold, if the effect of the lengthened bond in pyrope on the Mg-O force constant were taken into account. This question seems to be an obvious candidate for study by $^{24}Mg/^{26}Mg$ isotopic substitution (Pâques-Ledent and Tarte, 1973), particularly since the result might have implications for the low temperature heat capacity and hence entropy of majorite ($(Mg,Fe)SiO_3$ garnet), which have not yet been determined (Yusa *et al.,* 1993).

Raman spectra for cubic garnets have been measured at high pressure and high temperature by several authors (Dietrich and Arndt, 1982; Mernagh and Liu, 1990; Gillet *et al.,* 1992; Liu *et al.,* 1995). Gillet *et al.* (1992) combined the high P and high T data for pyrope, grossular and almandine to calculate intrinsic mode anharmonicity parameters, showing that a relatively large (~3%) anharmonic correction to the heat capacity would be necessary. As already noted, this could affect the slope of the garnet-ilmenite phase boundary, in

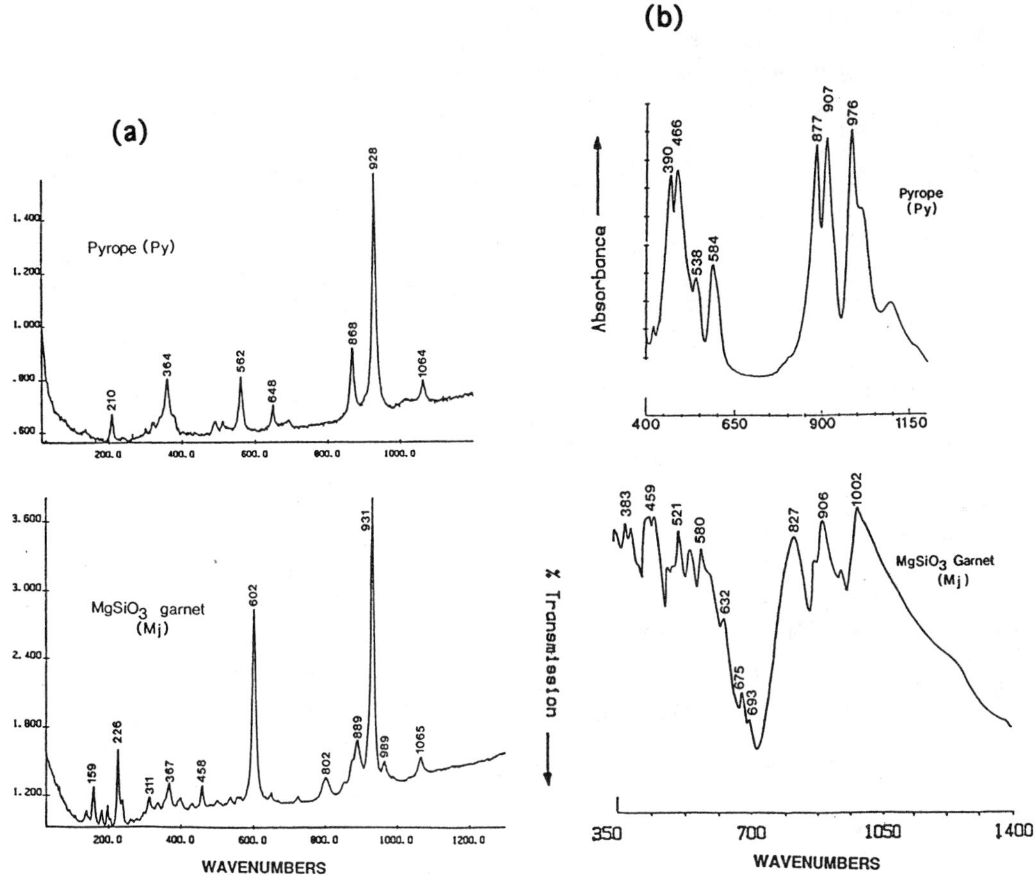

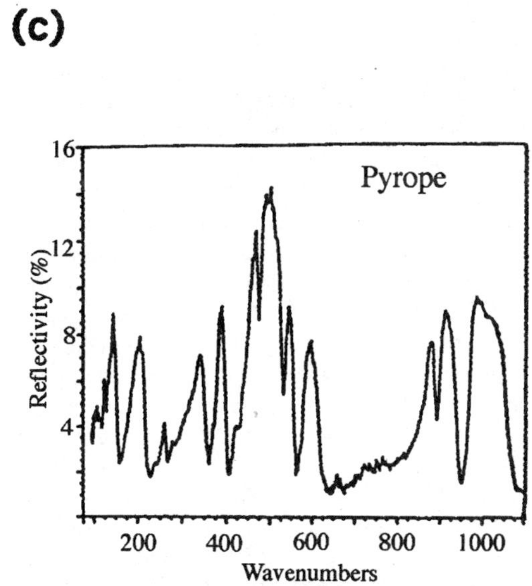

FIG. 11. Infrared and Raman spectra of pyrope and $MgSiO_3$ garnets. The Raman (a) and powder transmission IR data (b) are from McMILLAN *et al.* (1989), and show the additional peaks which appear due to the tetragonal distortion of $MgSiO_3$ garnet compared with cubic pyrope. The single crystal IR reflectance spectrum of pyrope (c) is taken from HOFMEISTER and CHOPELAS (1991a). Peaks are assigned to internal vibrations of the SiO_4 tetrahedra (ν_3 asymmetric stretching, ν_4 and ν_2 deformation) and external modes associated with the Mg and Al sites. Spectra from HOFMEISTER and CHOPELAS (1991a; Fig. 1, p. 507) used by permission of the editors of *Physics and Chemistry of Minerals* (Springer-Verlag GmbH).

hot regions of the transition zone (REYNARD and RUBIE, 1996). LIU *et al.* (1995) have recently recorded the high frequency (>800 cm^{-1}) Raman spectrum of a garnet with composition 0.9 $MgSiO_3$:0.1 Al_2O_3, to 13.6 GPa at room temperature, and 875 K at room pressure.

The pure $MgSiO_3$ end member garnet exhibits a small tetragonal distortion from cubic symmetry (KATO and KUMAZAWA, 1985; AKAOGI *et al.,* 1987; YAGI *et al.,* 1992), probably resulting from ordering Mg and Si atoms on the [6]-coordinated sites (HATCH and GHOSE, 1989). This ordering and tetragonal distortion causes the appearance of weak additional peaks and splittings in its infrared and Raman spectra, when compared with pyrope (KATO and KUMAZAWA, 1985; MCMILLAN *et al.,* 1989) (Fig. 11). It has already been noted that $Mg_3Al_2Si_3O_{12}$-$Mg_4Si_4O_{12}$ garnet solid solutions are cubic, as are high pressure $Fe_3Al_2Si_3O_{12}$-$Fe_4Si_4O_{12}$ garnets (AKAOGI and AKIMOTO, 1977). The X-ray, IR and Raman spectra of a natural majorite garnet ($(Mg_{0.79}Fe_{0.21})_4Si_4O_{12}$) from the Catherwood meteorite indicated that it was cubic also, consistent with some samples studied by KATO (1986). However, $(Mg,Fe)SiO_3$ majorites with Fe/(Fe+Mg) up to ~0.24 have been reported to exhibit a tetragonal distortion (KATO, 1986; MATSUBARA *et al.,* 1990; OHTANI *et al.,* 1991). It appears that the presence or absence of complete solid solution between tetragonal $(Mg,Fe)SiO_3$ and cubic majorite garnets may be associated with the energetics of cation ordering on the octahedral sites (AKAOGI *et al.,* 1987; MCMILLAN *et al.,* 1989; ANGEL *et al.,* 1989; HATCH and GHOSE, 1989; PHILLIPS *et al.,* 1992). On the basis of modulated microstructures observed by transmission electron microscopy, WANG *et al.* (1993) have suggested that a tetragonal-cubic transition associated with octahedral cation disordering occurs near 2400°C for pure $MgSiO_3$ garnet (HATCH and GHOSE, 1989), and that the transition temperature drops rapidly with increasing Fe content. This transition would have implications for garnet elastic properties within the transition zone (WANG *et al.,* 1993). The existence and nature of this possible phase transformation in (Mg,Fe,Al)-SiO_3 majorites merits further investigation, by *in situ* studies at combined high *P-T* conditions, and studies on phases equilibrated at high *P-T* and quenched to ambient. Although X-ray diffraction is probably a first choice for these studies, IR (or perhaps even Raman scattering, but cubic and tetragonal structures might be less easy to distinguish: see Fig. 11) spectroscopy combined with laser heating techniques in the diamond anvil cell (GILLET *et al.,* 1993b) might be useful in determining small deviations from local cubic symmetry, during development of the ordered domains (see, for example, analogous work on cordierite: MCMILLAN *et al.,* 1984).

UPPER MANTLE

$(Mg,Fe)_2SiO_4$ olivine

This phase, along with $(Mg,Fe)SiO_3$ pyroxenes and $(Mg,Fe,Ca)_3Al_2Si_3O_{12}$ garnets, dominates the mineralogy of the upper mantle (LIU and BASSETT, 1986). The garnet phases have been discussed in the previous section, and the chain silicates are described below. The lattice dynamics of synthetic and natural olivines have been thoroughly investigated, both experimentally and by theoretical calculation. The zone centre vibrational modes for the orthorhombic (*Pbnm*) structure are:

$$\Gamma_{vib} = 11A_g(R) + 11B_{1g}(R) + 7B_{2g}(R) + 7B_{3g}(R) + 10A_u(inactive) + 9B_{1u}(IR) + 13B_{2u}(IR) + 13B_{3u}(IR) \quad (9)$$

All of the infrared and Raman-active modes have been assigned via polarized measurements on single crystals (SERVOIN and PIRIOU, 1973; IISHI, 1978; PIRIOU and MCMILLAN, 1983; HOFMEISTER, 1987; CHOPELAS, 1991a; REYNARD, 1992). The lattice dynamics of forsterite have been investigated by single crystal and powder inelastic neutron scattering spectroscopy, coupled with lattice dynamical calculations within the rigid ion approximation (RAO *et al.,* 1988). The experimental data and the theoretical modeling have been used to construct vibrational density of states functions, for calculation of the specific heat, entropy, and other thermodynamic properties within the quasiharmonic approximation (AKAOGI *et al.,* 1984; CHOPELAS, 1990b; CHOUDHURY *et al.,* 1989; HOFMEISTER, 1987; HOFMEISITER *et al.,* 1989; PRICE *et al.,* 1987; RAO *et al.,* 1988). The pressure and temperature dependence of infrared and Raman modes in forsterite and fayalite has been measured in several studies (BESSON *et al.,* 1982; DIETRICH and ARNDT, 1982; XU *et al.,* 1983; GILLET *et al.,* 1988, 1991; HOFMEISTER *et al.,* 1989; CHOPELAS, 1990b; GILLET *et al.,* 1991; LIU and MERNAGH, 1992; SHARMA *et al.,* 1992; WANG *et al.,* 1993; DURBEN *et al.,* 1993). ANDERSON and SUZUKI (1983) have discussed the effects of anharmonicity on the thermodynamic properties of olivines at mantle temperatures, and GILLET *et al.* (1991) and REYNARD *et al.* (1992) have shown the importance of taking anharmonic effects into account for calculations of the vibrational heat capacity. More recently, GUYOT *et al.* (1996) have used available high-pressure and high-temperature IR and Raman data to calculate the equations of state of olivines at various *P-T* conditions. The agreement with experiment is excellent.

In a high-pressure Raman study of forsterite at room temperature, CHOPELAS (1990b) noted breaks in slope of the pressure shifts in several modes at 9.2 GPa, which was thought to be associated with a high order phase transformation occurring in olivine at high pressure and ambient temperature. WANG *et al.* (1993) noted similar breaks in slope, but concluded that no phase transition occurs, simply a change in the compression mechanism with pressure. Moreover, DOWNS *et al.* (1996) found no evidence for structural changes at these pressures in a single-crystal x-ray diffraction study to 17.2 GPa with a variety of pressure media. Such changes in the strength of pressure media used cause apparent changes in the slope of such frequency shifts with pressures. Compression of forsterite to higher pressure (50 GPa) at room temperature results in the appearance of new bands near 750 cm^{-1} and 960 cm^{-1} (DURBEN *et al.,* 1993). These are likely related to formation of SiOSi (dimer) linkages within the structure (perhaps related to the formation of spinelloid structures) or to the formation of high coordinate Si species (perhaps as a precursor to amorphization) (WILLIAMS *et al.,* 1990; GUYOT and REYNARD, 1992). In their infrared study of pressure-induced amorphization in fayalite, WILLIAMS *et al.* (1990) noted an increase in intensity of a band in the 600–800 cm^{-1} region above 35 GPa, which they attributed to the formation of six-coordinated silicon species. As already noted above, Raman and IR measurements *in situ* at high pressures and temperatures may be useful in phase identification and possibly for characterizing local and intermediate-range structures in kinetic studies of high-pressure transformations of olivine and other $(Mg,Fe)_2SiO_4$ polymorphs; such experiments are needed to further understand seismicity and rheology within the upper mantle and transition zone.

An area that deserves further attention is the exploration of pre-melting behavior of mantle minerals, through *in situ* high temperature vibrational spectroscopy, combined with diffraction and calorimetric experiments (RICHET *et al.,* 1994). The pre-melting regime is characterized by an anomalous rapid increase in heat capacity and entropy, below the thermodynamic melting point (RICHET and FIQUET, 1991). The origin of the phenomenon has been studied in detail using *in situ* Raman spectroscopy for $(Ca,Mg)GeO_4$ olivines, and has been related to M_1-M_2 disordering at high temperature, as well as to extreme anharmonicity (FIQUET *et al.,* 1992; RICHET *et al.,* 1994). This behavior has been observed to occur for silicate olivine, pyroxene and garnet, and has obvious implications for rheology of high temperature mineral assemblages in the mantle, as well as the kinetics of phase transformations. This phenomenon can now be studied with existing techniques by *in situ* high P-T IR and Raman spectroscopy to evaluate its importance in the mantle *P-T* regime.

Pyroxene

In comparison with the olivine group of minerals, there has been much less work done on the vibrational spectroscopy of pyroxenes, either at ambient or under mantle conditions (WHITE, 1975). The $(Mg,Fe)SiO_3$ rich enstatite phase shows complex polymorphism as a function of pressure and temperature (ANGEL *et al.,* 1992; LEE and HEUER, 1987; PREWITT, 1980; SCHRADER *et al.,* 1990; ANGEL and HUGH-JONES, 1994a,b). The orthorhombic phase (*Pbca*) is stable to approximately 8 GPa and 1300 K. Above this temperature, there is a displacive transformation to protoenstatite (*Pbcm*) at ambient pressure for pure $MgSiO_3$, but Fe-containing phases transform to *C2/c* structures (YANG and GHOSE, 1995). At pressures above 8 GPa, orthopyroxene transforms into a high density clinoenstatite structure (PACALO and GASPARIK, 1990; ANGEL *et al.,* 1992). CHOPELAS and BOEHLER (1992) prepared a sample of this phase by CO_2 laser heating single crystalline $MgSiO_3$ orthoenstatite in a diamond anvil cell, and investigated its Raman spectrum during decompression. The spectrum of the clinonenstatite polymorph is similar to that of diopside at the same pressure, consistent with the *C2/c* symmetry assigned by ANGEL *et al.* (1992), and it reverts to the orthorhombic polymorph on decompression below 5 GPa (CHOPELAS and BOEHLER, 1992). The pressure shifts of the Raman peaks in both phases are given by CHOPELAS and BOEHLER (1992). DIETRICH and ARNDT (1982) have investigated the infrared spectrum of natural orthopyroxene at lower pressures, to 5 GPa, at up to 250°C. SHARMA *et al.* (1987), SHARMA (1989, 1990) and GHOSE *et al.* (1994) have investigated the effect of temperature on the Raman spectra of $MgSiO_3$ pyroxene phases. In particular, GHOSE *et al.* (1994) have reported polarized single crystal Raman spectra of protoenstatite. The transitions between the ortho-, proto- and clino-enstatite phases are clearly observed by the appearance or disappearance of peaks in the spectra.

At lower pressures, a (Ca,Al)-rich pyroxene phase is found, forming a monoclinic (*C2/c*) solid solution in the system $CaMgSi_2O_6$ (diopside)-$NaAlSi_2O_6$ (jadeite) (GASPARIK, 1992). The Raman and infrared spectra of diopside have been characterized at ambient conditions via polarized single crystal measurements (ETCHEPARE, 1970, 1972; ZULUMYAN *et al.,* 1976), and an empirical lattice dynamics calculation has been carried out (TOMISAKA and IISHI, 1980). A Raman spectrum at 16 GPa is presented by CHOPELAS and BOEHLER (1992). SEKITA *et al.* (1988) have used Raman spectroscopy to study bands in the 600–700 cm^{-1} region of clinopyroxenes, and assigned peaks to SiOSi, SiOAl and AlOAl linkages. This type of calibration could prove useful in future high *P-T* studies of Ca-rich pyroxene phases. Raman and infrared studies of pyroxene phases, both at ambient conditions and *in situ* at high pressure and temperature, will be useful to complement diffraction studies in unraveling the detailed nature of the complex transformation mechanisms between the different polymorphs of $(Mg,Fe)SiO_3$ (YANG and GHOSE, 1994a,b; 1995; GHOSE *et al.,* 1994), as well as the onset of pre-melting behaviour as described above for olivine (RICHET *et al.,* 1994).

WATER AND CARBON IN THE MANTLE

Water in the mantle

The questions associated with ''water'' or the hydrogen content in the mantle (how much? where? in what form?) have been and still are a matter of intense interest and considerable debate. These are critical questions for mantle petrology, both to understand the volatile budget and recycling processes, and because of the large effects of even trace hydrous species on mineralogical phase relations, including melting, and rheology of mantle minerals and mineral assemblages (MACKWELL *et al.,* 1985; LIU, 1987; WYLLIE, 1988; MACKWELL and KOHLSTEDT, 1990; BELL and ROSSMAN, 1992a; THOMPSON, 1992; DUFFY and AHRENS, 1992; GASPARIK, 1993; GILLET, 1993a). Vibrational spectroscopy, particularly infrared absorbance measurements of natural mantle minerals and laboratory-synthesized samples, has played a role in helping establish our current estimates of the hydration state of the mantle.

It is well known that hydrous minerals such as phlogopitic mica and kaersutitic amphibole can be stable under the *P-T* conditions of the upper mantle. However, IR absorption studies of nominally anhydrous mineral phases have demonstrated that these can contain substantial amounts of OH in structural sites (BERAN and PUTNIS, 1983; AINES and ROSSMAN, 1984a; ROSSMAN, 1988; BELL and ROSSMAN, 1992a; BELL *et al.,* 1995). Measured OH contents for natural olivines, pyroxenes and garnets, which constitute >90% of peridotites sampled at the surface, range from 100–1500 H/10^6 Si (BERAN and PUTNIS, 1983; AINES and ROSSMAN, 1984b,c; MILLER *et al.,* 1987; BELL and ROSSMAN, 1992a,b, 1995; KUROSAWA *et al.,* 1992, 1993; BAI and KOHLSTEDT, 1993). Experiments are under way to determine the maximum OH concentration which can be incorporated in these phases (KOHLSTEDT *et al.,* 1996).

The case of garnet is particularly interesting, because grossular ($Ca_3Al_2Si_3O_{12}$) forms a solid solution with the hydrogarnet ($Ca_3Al_2(O_4H_4)_3$), in which each Si^{4+} site is occupied by a distorted tetrahedron of OH groups (COHEN-ADDAD *et al.,* 1967; KOBAYASHI and SHOJI, 1983; LAGER *et al.,* 1987, 1989). This substitution permits the OH concentration of Ca-rich garnets to become very high. From vibrational spectroscopy, the structural arrangement of OH groups is determined by bifurcated hydrogen bonds to adjacent oxygens forming the tetrahedral site (HARMON *et al.,* 1982). In a high pressure study of hibschite ($Ca_3Al_2(SiO_4)_{1.5}(O_4H_4)_{1.5}$), KNITTLE *et al.* (1992) found both Raman-active O—H stretching vibrations and the higher frequency IR mode to decrease in frequency with increasing pressure, as expected for increased hydrogen bonding, but the lower frequency IR band increased in frequency. It was suggested that this could result from H...H or H...Ca^{2+} repulsive interactions (SHEU and MCMILLAN, 1988; WILLIAMS, 1992).

The *P-T* stability and breakdown behaviour of hydrous minerals such as amphibole or mica play a critical role in volatile recycling. A few studies of the vibrational properties of these minerals at ambient conditions have been reported (see MCMILLAN and HOFMEISTER, 1988), and some high pressure work is now beginning to appear (HOLTZ *et al.,* 1993). Raman and IR studies under high *P-T* conditions will be extremely important in helping determine the thermodynamic properties of these phases (GILLET *et al.,* 1989b; ROBIE *et al.,* 1991), and will shed light on their formation or dehydration mechanisms within the upper mantle. A related field which is only beginning to receive attention is the study of hydrous silicate melt phases, under mantle *P-T* conditions (WILLIAMS, 1990; FARBER and WILLIAMS, 1992; WILLIAMS, 1992; CLOSMANN and WILLIAMS, 1995). Vibrational spectroscopy is expected to play a critical role in these studies, which can not be easily addressed by diffraction methods.

Considerable interest has been stimulated by the possibility that large amounts of water could be stored in the transition zone, contained in structural sites in β-$(Mg,Fe)_2SiO_4$, as discussed above (SMYTH, 1987; DOWNS, 1989; MCMILLAN *et al.*, 1991; YOUNG *et al.*, 1993; KOHLSTEDT *et al.*, 1996), or within high pressure hydrous minerals in the system $MgO—SiO_2—H_2O$ (LIU, 1987; KANZAKI, 1991; PREWITT and FINGER, 1992). These phases (chondrodite, clinohumite, phases A, B, C, D, E, etc.) have been partly characterized by infrared and Raman spectroscopy (AKAOGI and AKIMOTO, 1986; FINGER *et al.* 1989; MCMILLAN *et al.*, 1991), but much work remains to be done. This is an essential complement to X-ray diffraction, because it is much easier to study the H atom environment via vibrational spectroscopy. WILLIAMS (1992) has studied a natural chondrodite to ~9 GPa, showing the importance of H-bonding in one of the O—H stretching vibrations, and used IR spectroscopy to investigate phase changes following laser heating at pressures between 22 and 44 GPa.

Two other interesting cases of trace OH in nominally anhydrous mantle minerals have also recently been demonstrated by IR absorption spectroscopy. PAWLEY *et al.* (1993) have shown that stishovite, the high pressure phase of SiO_2, can accept hydrogen in defect sites within its structure, and the H content was shown to scale with Al substitution into the SiO_2 phase (PAWLEY *et al.*, 1993; SMYTH *et al.*, 1995). This is consistent with the observed dissolution of H into TiO_2 rutile (VLASSOPOULOS *et al.*, 1993; SWOPE *et al.*, 1995). LU *et al.* (1994) and MEADE *et al.* (1994) have also shown by IR spectroscopy that $(Mg,Fe)SiO_3$ perovskite, the dominant phase in the lower mantle, can accept trace OH within its structures, with considerable potential implications for the water content of the deep Earth. The study of MEADE *et al.* (1994) was carried out with the synchrotron infrared technique, a promising new method for micro-spectroscopy of materials at ambient and high-pressure conditions (REFFNER *et al.*, 1994).

Carbonates

A second problem of intense interest is the carbon budget and oxidation state of the deep Earth (KUSHIRO *et al.*, 1975; IRVING and WYLLIE, 1975; KATSURA and ITO, 1990; BLUNDY *et al.*, 1991; GILLET, 1993b; GILLET *et al.*, 1996a). Carbonate minerals are fundamental actors in the global carbon cycle because, on the one hand, they may store carbon, and on the other, they provide carriers for carbon when sediments are recycled in subduction zones. Experimental studies as well as observations of natural samples have shown that carbonates could provide major hosts for carbon in the Earth's upper and lower mantle. High-pressure experiments have shown that in the presence of pyroxene and olivine, or their high-pressure equivalents (β-$Mg,Fe)_2SiO_4$ and the assemblage $(Mg,Fe)SiO_3$ perovskite + $(Mg,Fe)O$ magnesiowüstite), $MgCO_3$ is the stable carbonate phase (BREY *et al.*, 1983; CANIL and SCARFE, 1990; KATSURA and ITO, 1990; BIELLMANN *et al.*, 1993b). Vibrational spectroscopy has been used to demonstrate the stability of carbonates under pressure and temperature conditions relevant to the mantle (GILLET, 1993b; GILLET *et al.*, 1996b) (Fig. 12).

The expected vibrational modes of calcite and magnesite with space group $R\bar{3}c$ are:

$$\Gamma_{vib} = A_{1g}(R) + 3A_{2g}(inactive) + 4E_g(R) + 2A_{1u}(inactive) + 4A_{2u}(IR) + 5E_u(IR) \quad (10)$$

The vibrational modes of dolomite $(CaMg(CO_3)_2)$ are related by the disappearance of the c-glide (space group $R\bar{3}$).

$$\Gamma_{vib} = 4A_g(R) + 4E_g(R) + 5A_u(IR) + 5E_u(IR) \quad (11)$$

The infrared and Raman spectra of these phases are well known, through extensive single crystal spectroscopy (WHITE, 1974; GILLET *et al.*, 1993a).

Several recent infrared and Raman studies of carbonate minerals have been carried out at ambient temperature, to pressures in excess of 30 GPa (GILLET *et al.*, 1988; 1993a; LIU and MERNAGH, 1990; KRAFT *et al.*, 1991; BIELLMANN and GILLET, 1992). Calcium carbonate has two stable polymorphs, calcite and aragonite, the high pressure form. Calcite also shows transitions to metastable forms of $CaCO_3$ (calcite-II and calcite-III) in the 1.4–2.0 GPa range (FONG and NICOL, 1971; HESS and GHOSE, 1988; GILLET *et al.*, 1988; LIU and MERNAGH, 1990; COLLERSON *et al.*, 1992; BIELLMANN and GILLET, 1992). BIELLMANN *et al.* (1993a) have shown that, after heating of calcite to 2000 K at 40 GPa, calcite II can be quenched to ambient conditions. In contrast to this behaviour, dolomite and magnesite show no evidence for any phase transitions to the highest pressures studied (34 GPa) (Fig. 12), nor does the stable high pressure aragonite phase of $CaCO_3$ (KRAFT *et al.*, 1991; COLLERSON *et al.*, 1992; BIELLMANN and GILLET, 1992; GILLET, 1993b; GILLET *et al.*, 1993a; GILLET *et al.*, 1996a).

The effect of temperature on the room pressure

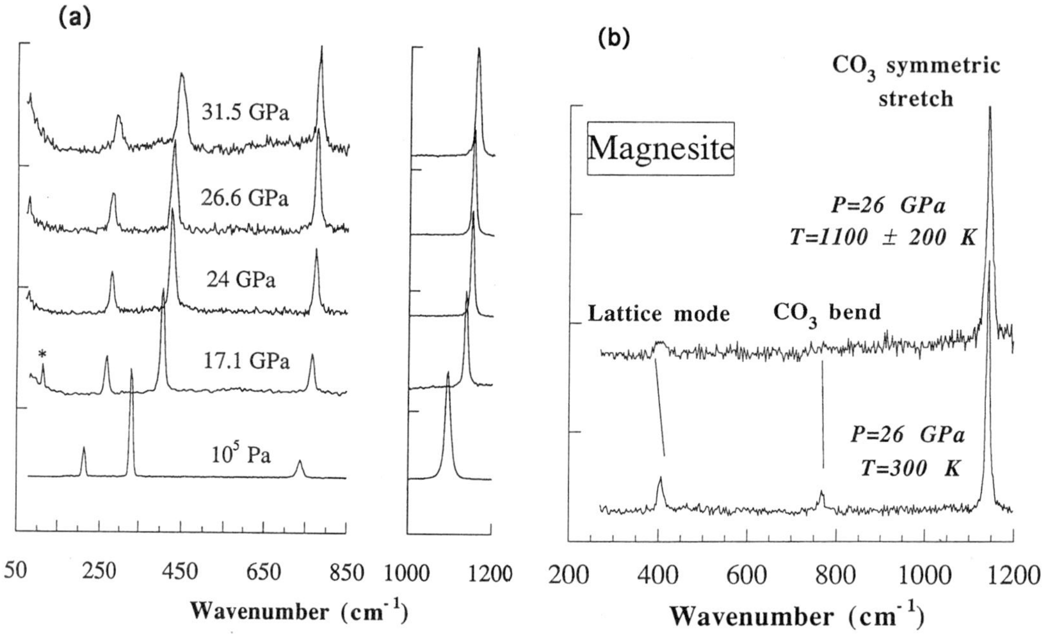

FIG. 12. (a) Effect of pressure on the Raman spectrum of magnesite, $MgCO_3$, to mantle pressures. (b) *In situ* high *P-T* spectrum of magnesite, at 26 GPa and 1100 K, showing the stability of this phase under mantle conditions (GILLET, 1993b).

infrared spectrum of calcite (SAKURAI and SATO, 1971), and the Raman spectra of calcite, dolomite and magnesite (GILLET *et al.*, 1993a), has been investigated up to 800–1200 K. The measured frequency shifts with temperature have been combined with the observed pressure-induced shifts to calculate mode anharmonic parameters (GILLET *et al.*, 1993a). These must be taken into account for a realistic calculation of $^{16}O/^{18}O$ isotopic fractionation factors at high pressure and temperature (GILLET *et al.*, 1996b). It is found that these anharmonic parameters are larger for the lattice modes than for the CO_3 internal modes. Moreover, they are larger in calcite than in dolomite and magnesite, consistent with increasing relaxational component to the libration of the CO_3 groups, premonitory to a rotational order-disorder phase transition in calcite at high temperatures (GILLET *et al.*, 1993a).

KRAFT *et al.* (1991) have investigated the vibrational spectrum of dolomite *in-situ* under high *P-T* conditions, at pressures up to 11.5 GPa and 550 K. The stability of magnesite ($MgCO_3$) has been studied by Raman spectroscopy at high pressure and high temperature, using a CO_2-laser-heated diamond anvil cell (GILLET, 1993b). Raman spectra recorded at simultaneous high pressure (26 GPa) and high-temperature (1200 K ± 200 K) show that under these conditions magnesite retains its ambient $R\bar{3}c$ structure (Fig. 12). Optical observations during heating as well as Raman spectra recorded at 30 GPa after laser heating up to 2000–2500 K showed that no decarbonation had occurred and that magnesite was stable. These results are in agreement with other experimental data that show that magnesite can act as a host for carbon storage down to at least 1000 km, and that this phase can also be the carrier for carbon in subducting plates (GILLET, 1993b; GILLET *et al.*, 1995).

The Raman data for magnesite, combined with compressibility measurements at high pressures and room temperature, have been used to compute the specific heat and vibrational entropy, and calculate the equation of state of magnesite up to 135 GPa and 3500 K (GILLET *et al.*, 1995), in good agreement with existing experimental data. The model was then used to calculate the decarbonation curve $MgCO_3 \Leftrightarrow MgO + CO_2$, under mantle conditions, which provides an upper limit for the stability of magnesite in the Earth's mantle.

Diamond

This high pressure form of carbon also provides an important source or sink for carbon in the mantle. Diamond is the window material of choice for *in situ* diffraction and spectroscopic experiments under high *P-T* conditions, in the diamond anvil cell, for mineral physics studies of the deep Earth. As discussed above, this same material also provides a ''window'' of a different type, acting as a host for mineral and fluid inclusions brought to the surface from great depths within the Earth (KESSON and FITZGERALD, 1992; HARTE and HARRIS, 1993; SAUTTER and GILLET, 1994). BELL (unpublished) has carried out measurements of the OH content in a suite of silicate inclusions in diamond with new synchrotron micro-IR techniques (REFFNER *et al.*, 1994). In contrast to studies on typical mantle minerals, OH was not detected in either olivines or orthopyroxene inclusions in diamonds, while its presence was clearly demonstrated in kyanite inclusions, with IR spectra similar to kyanites from mantle eclogite xenoliths. These results have important implications for the hydration history of the mantle and place constraints on processes of diamond formation. Notably, the measurements were not possible with conventional IR methods.

Vibrational studies of this phase have been carried out to help demonstrate the stability of diamond to high pressures and temperatures, and as a testing ground for first-principles theory. The vibrational spectrum of cubic diamond itself is very simple: the expected vibrational modes are

$$\Gamma_{vib} = T_{2g}(R) \qquad (12)$$

which gives rise to a single first order peak in the Raman spectrum at 1332 cm^{-1}. Weak features due to second order Raman scattering are observed in the 2200–2600 cm^{-1} range. The Raman spectrum of diamond has been studied to ~40 GPa (BOPPART *et al.*, 1985; HANFLAND *et al.*, 1985; HANFLAND and SYASSEN, 1985; SHARMA *et al.*, 1985; ALEKSANDROV *et al.*, 1986). These studies confirm both the high-pressure stability and the singular high strength of this mineral over a pressure range extending well into the lower mantle. MAO and HEMLEY (1991) have carried out micro-Raman measurements on diamond anvils at sample pressures up to 300 GPa, and found some evidence for structural transformations driven by the large nonhydrostatic stresses of the anvils under these conditions. VOHRA (1992) has found evidence for a transition in highly stressed diamond anvils in studies to 420 GPa. ZOUBOULIS and GRIMSDITCH (1991) have investigated the Raman scattering in diamond at high temperatures, to 1900 K. Recently, FRANTZ and MYSEN (1996) have exploited the frequency shift of the T_{2g} mode in ^{13}C relative to ^{12}C and its weak temperature shift for pressure calibration in a diamond-anvil cell at high temperature. An exciting possibility for future work is the use of micro-IR and Raman spectroscopy of mineral and fluid inclusions in deep mantle diamonds, as well as defect species in the diamond matrix itself (NAVON *et al.*, 1988; FIELD, 1992), to better understand the volatile budget of the deep Earth.

CONCLUSIONS

Vibrational studies via infrared and Raman spectroscopy provide a set of powerful techniques for the characterization and structural study of mantle minerals. The vibrational spectra can reveal subtle structural distortions which are not immediately obvious in diffraction experiments, and are sensitive to the presence of hydrogen within the mineral, even as a defect species. It is usually quite simple to obtain spectra *in situ* at high pressure or at high temperature, and combined *in situ* high *P-T* studies are beginning to appear, allowing minerals to be examined directly under mantle conditions. The spectra can be used to obtain important thermodynamic information on the high-pressure phases, including heat capacity, vibrational entropy, and Grüneisen parameters, and detailed information on phase transformation mechanisms. These Raman and infrared studies are particularly fruitful when they are combined with theoretical calculations of the lattice dynamics, which give information on all vibrational modes throughout the Brillouin zone, and permit a structural interpretation of the observed spectra. Because the vibrational spectrum provides a fingerprint for a mineral, micro-Raman and IR measurements are proving very useful for phase identification of natural and experimental polyphase aggregates with spatial resolution close to the diffraction limit of the radiation used (*e.g.*, ~1 μm for visible-light Raman; ~10 μm at 1000 cm^{-1} for IR methods).

Most of the principal mantle mineral phases have been identified, and their infrared and Raman spectra assigned, but data on relevant solid solution series are often lacking. There are obvious gaps in the data base: for example, the far infrared spectra of γ-Mg_2SiO_4, and infrared and Raman data on the polymorphs of $(Mg,Fe)SiO_3$. In addition, further experimental and theoretical work on $MgSiO_3$ perovskite is required to fully characterize the low-frequency vibrational spectrum, and determine the presence or absence of soft modes at simultaneous high *P-T* conditions. In general, *in situ* studies under combined high *P-T* conditions will be useful to gain a better understanding of anharmonic behaviour, and to investigate phenomena such as pre-melting, which may play an important role in determining transport properties at high temperature. Vibrational spectroscopy has played a central role in the characterization of OH-containing phases likely to be stable within the mantle, and thus helping to constrain the sources and sinks for a hydrous component at depth. But more work is required under high pressure conditions, to help determine the role that hydrogen bonding may play in stabilizing these mineral phases, and also to investigate the dehydration mechanisms of hydrous phases during processes such as subduction.

This review has focused on vibrational spectroscopic studies of crystalline minerals within the mantle. Much is now known about the structure and properties of aluminosilicate melts which yield magmas observed at or near the surface, generated by partial melting events at relatively shallow depths within the lower crust and upper mantle (STEBBINS *et al.,* 1995). However, melting may also occur deeper in the mantle. Of paramount interest is the role of structural changes (*e.g.,* changes in coordination) in determining the density contrast between melt and crystalline phases, and also in determining the rates of chemical diffusion and viscous flow in the high pressure melt phase (MCMILLAN and WOLF, 1995). *In situ* vibrational spectroscopy has been instrumental in demonstrating that large structural changes take place in silicate glasses at high pressures [*e.g.,* HEMLEY *et al.,* 1986b; WILLIAMS and JEANLOZ, 1988]. More recently, IR and Raman studies have begun to shed light on the structural nature of ultrabasic melts likely to present in the deep mantle (MCMILLAN and WOLF, 1995). Vibrational spectroscopy will play a key part in investigating the structural behaviour and relaxation dynamics of melts under deep mantle conditions, through *in situ* experiments at combined high pressure and temperature (FARBER and WILLIAMS, 1992; GILLET *et al.,* 1993b).

Finally, future progress is likely to come from continued development of experimental techniques. Synchrotron IR techniques now permit vibrational spectra to be easily measured in both absorption and reflectivity on 10 μm samples to 200 GPa (HEMLEY *et al.,* 1996). The development of such synchrotron as well as IR laser techniques is expected to continue. An exciting prospect is the adaptation of micro-Raman and IR techniques to three-dimensional sample imaging. Such methods have been developed for Raman scattering and are being extended to IR absorption (LEWIS *et al.,* 1995). Another prospect is the continued development of IR lasers and detectors for Raman scattering in the IR, which can alleviate problems associated with laser-induced sample fluorescence and degradation at visible wavelengths (CHASE *et al.,* 1986). Also, there may be important applications of time-resolved and non-linear optical techniques (stimulated Raman scattering) (*e.g.,* SHARMA, 1989). Notably, although the *P-T* regime of the entire mantle of the Earth can now be created in the laboratory (MAO and HEMLEY, 1996), there have been few measurements of material properties under simultaneous high *P-T* conditions. It should be possible to combine each of these new spectroscopic methods with high *P-T* techniques for the next generation of *in situ* measurements on mantle minerals.

Acknowledgements—The work of PFM and RJH on spectroscopy of mantle minerals has been supported by the National Science Foundation, Division of Earth Sciences. PG has benefited from financial support of the CNRS (INSU, programme Terre Profonde). This manuscript was prepared while PM was Professeur Associé at the Ecole Normale Supérieure, Lyon. We thank Q. Williams, P. Conrad, and an anonymous reviewer for their comments, and G. Wolf for many valuable discussions.

REFERENCES

AGNON A. and BUKOWINSKI M. S. T. (1990) Thermodynamic and elastic properties of a many-body model for simple oxides. *Phys. Rev. B* **41,** 7755–7766.

AHMED-ZAID I. and MADON M. (1995) Electron microscopy of high-pressure phases synthesized from natural garnets in a diamond anvil cell: Implications for the mineralogy of the lower mantle. *Earth Planet. Sci. Lett.* **129,** 233–244.

AINES R. D. and ROSSMAN G. R. (1984a) Water in minerals? A peak in the infrared. *J. Geophys. Res.* **89,** 4059–4071.

AINES R. D. and ROSSMAN G. R. (1984b) The hydrous component in garnets: Pyralspites. *Amer. Mineral.* **69,** 1116–1126.

AINES R. D. and ROSSMAN G. R. (1984c) Water content of mantle garnets. *Geology* **12,** 720–723.

AKAOGI M. and AKIMOTO S. (1977) Pyroxene-garnet solid solution equilibria in the systems $Mg_4Si_4O_{12}$-$Mg_3Al_2Si_3O_{12}$ and $Fe_4Si_4O_{12}$-$Fe_3Al_2Si_3O_{12}$ at high pressures and temperatures. *Phys. Earth Planet. Int.* **15,** 90–106.

AKAOGI M. and AKIMOTO S. (1986) Infrared spectra of high-pressure hydrous silicates in the system MgO—SiO_2—H_2O. *Phys. Chem. Minerals* **13,** 161–164.

AKAOGI M. and ITO E. (1993a) Refinement of enthalpy measurement of $MgSiO_3$ perovskite and negative pressure-temperature slopes for perovskite-forming reactions. *Geophys. Res. Lett.* **20,** 1839–1842.

AKAOGI M. and ITO E. (1993b) Heat capacity of $MgSiO_3$ perovskite. *Geophys. Res. Lett.* **20,** 105–108.

AKAOGI M., ROSS N. L., MCMILLAN P. and NAVROTSKY A. (1984) The Mg_2SiO_4 polymorphs (olivine, modified spinel, spinel)—thermodynamic properties from oxide melt solution calorimetry, phase relations, and models of lattice vibrations. *Am. Mineral.* **69,** 499–512.

AKAOGI M., ITO E. and NAVROTSKY A. (1989) Olivine-modified spinel-spinel transitions in the system Mg_2SiO_4-Fe_2SiO_4: calorimetric measurement, thermochemical calculation, and geophysical application. *J. Geophys. Res.* **94,** 15671–15685.

AKAOGI M., NAVROTSKY A., YAGI T. and AKIMOTO S.-I. (1987) Pyroxene-garnet transformations: thermochemistry and elasticity of garnet solid solutions, and application to a pyrolite mantle. In *High-Pressure Research in Mineral Physics* (eds. M. H. MANGHNANI and Y. SYONO), pp. 251–260. Terra Scientific Publishing Company, Tokyo/American Geophysical Union, Washington D.C.

ALEKSANDROV I. V., GONCHAROV A. F. and STISHOV S. M. (1986) Direct determination of the Gruneisen parameter of the LTO(G) mode of diamond at high pressures. *JETP Lett.* **44,** 611–614.

ANDERSON D. L. (1989) *Theory of the Earth.* Blackwell, Boston.

ANDERSON O. L. and SUZUKI I. (1983) Anharmonicity of three minerals at high temperature: forsterite, fayalite, and periclase. *J. Geophys. Res.* **88,** 3549–3556.

ANDERSON O. L., ODA H. and ISAAK D. (1992) A model for the computation of thermal expansivity at high compression and high temperatures: MgO as an example. *Geophys. Res. Letts.* **19,** 1987–1990.

ANDERSON O. L., ODA H., CHOPELAS A. and ISAAK D. (1993) A thermodynamic theory of the Gruneisen ratio at extreme conditions: MgO as an example. *Phys. Chem. Minerals* **19,** 369–380.

ANDERSON O. L., MASUDA K. and GUO D. (1995) Pure silicate perovskite and the PREM lower mantle model: a thermodynamic analysis. *Phys. Earth Planet. Int.* **89,** 35–49.

ANDRAULT D., BOUHIFD M. A., ITIÉ J. P. and RICHET P. (1995) Compression and amorphization of $(Mg,Fe)_2$-SiO_4 olivines: An X-ray diffraction study up to 70 GPa. *Phys. Chem. Minerals* **22,** 99–107.

ANGEL R. J. and HUGH-JONES. D. A. (1994a) A compressional study of $MgSiO_3$ orthoenstatite up to 8.5 GPa. *Am. Mineral.* **79,** 405–410.

ANGEL R. J. and HUGH-JONES. D. A. (1994b) Equations of state and thermodynamic properties of enstatite pyroxenes. *J. Geophys.Res.* **99,** 19777–19783.

ANGEL R. J., CHOPELAS A. and ROSS N. L. (1992) Stability of high-density clinoenstatite at upper-mentle pressures. *Nature* **358,** 322–324.

ANGEL R. J., FINGER L. W., HAZEN R. M., KANZAKI M., WEIDNER D. J., LIEBERMANN R. C. and VEBLEN D. R. (1989) Structure and twinning of single-crystal $MgSiO_3$ garnet synthesized at 17 GPa and 1800°C. *Amer. Mineral.* **74,** 509–512.

ASHIDA T., KUME S., ITO E. and NAVROTSKY A. (1988) $MgSiO_3$ ilmenite: heat capacity, thermal expansivity, and enthalpy of transformation. *Phys. Chem. Minerals* **16,** 239–245.

BAI Q. and KOHLSTEDT D. L. (1993) Effects of chemical environment on the solubility and incorporation mechanism for hydrogen in olivine. *Phys. Chem. Minerals* **19,** 460–471.

BELL D. R. and ROSSMAN G. R. (1992a) Water in Earth's mantle: the role of nominally anhydrous minerals. *Science* **255,** 1391–1396.

BELL D. R. and ROSSMAN G. R. (1992b) The distribution of hydroxyl in garnets from the subcontinental mantle of southern Africa. *Contrib. Mineral. Petrol.* **111,** 161–178.

BELL D. R., IHINGER P. D. and ROSSMAN G. R. (1995) Quantitative analysis of trace OH in garnet and pyroxenes. *Amer. Mineral.* **80,** 465–474.

BELONOSHKO A. B. (1994) Molecular dynamics of $MgSiO_3$ perovskite at high pressures: equation of state, structure, and melting transition. *Geochim. Cosmochim. Acta* **58,** 4039–4047.

BERAN A. and PUTNIS A. (1983) A model of the OH positions in olivine, derived from infrared-spectroscopic positions. *Phys. Chem. Minerals* **9,** 57–60.

BERTKA C. M., and FEI Y. (1996) Constraints on the mineralogy of an iron-rich martian mantle from high-pressure experiments. *Planet. Space Sci.*, in press.

BESSON J. M., PINCEAUX J. P., ANASTOPOULOS C. and VELDE B. (1982) Raman spectra of olivine up to 65 kbar. *J. Geophy.Res.* **87,** 10773–10775.

BIELLMANN C. and GILLET PH. (1992) High-pressure and high-temperature behaviour of calcite, aragonite and dolomite: a Raman spectroscopic study. *Eur. J. Mineral.* **4,** 389–393.

BIELLMANN C., GILLET PH., GUYOT F., PEYRONNEAU J. and REYNARD B. (1993a) Experimental evidence for carbonate stability in the Earth's lower mantle. *Earth Planet. Sci. Lett.* **118,** 31–41.

BIELLMANN C., GUYOT F., GILLET PH. and REYNARD B. (1993b) High-pressure stability of carbonates: quenching of calcite II, high-pressure polymorph of $CaCO_3$. *Eur. J. Mineral.* **5,** 503–510.

BINA C. R. and WOOD B. J. (1987) Olivine-spinel transitions: experimental and thermodynamic constraints and implications for the nature of the 400-km seismic discontinuity. *J. Geophys. Res.* **92,** 4835–4866.

BINA C. R. and HEMLEY R. J. (1996) Solid-state chemistry of the Earth's mantle. *Nature*, submitted.

BLUNDY J. D., BRODHOLT J. P. and WOOD B. J. (1991) Carbon-fluid equilibria and the oxidation state of the mantle. *Nature* **349,** 321–324.

BOPPART H., VAN STRAATEN J. and SILVERA I. F. (1985) Raman spectra of diamond at high pressures. *Phys. Rev. B* **32,** 1423–1425.

BOYER H., SMITH D. C., CHOPIN C. and LASNIER B. (1985) Raman microprobe (RMP) determinations of natural and synthetic coesite. *Phys. Chem. Minerals* **12,** 45–48.

BREARLEY A. J., RUBIE D. C. and ITO E. (1992) Mechanisms of the transformations between the α, β and γ polymorphs of Mg_2SiO_4 at 15 GPa. *Phys. Chem. Minerals* **18,** 343–359.

BREHAT F., EVRARD O., HADNI A. and LAMBERT J. (1966) Far infrared absorption spectra of Mg, Fe and Mn oxides and their mixed crystals (FeO—MgO and FeO—MnO). *C. R. Acad. Sci. Paris Sér. A.* **263B,** 1112–1116.

BREY G., BRICE W. R., ELLIS D. J., GREEN D. H., HARRIS K. L. and RYABCHIKOV I. D. (1983) Pyroxene-carbonate reactions in the upper mantle. *Earth Planet. Sci. Lett.* **62,** 64–74.

BUKOWINSKI M. S. T. (1994) Quantum geophysics. *Ann. Rev. Earth Planet. Sci.* **22,** 167–205.

BUKOWINSKI M. S. T. and WOLF G.H. (1988) Equation of state and possible critical phase transitions in $MgSiO_3$ perovskite at lower-mantle conditions. In *Advances in Physical Geochemistry, vol. 7, Structural and Magnetic Phase Transitions* (eds. S. GHOSE, J. M. D. COEY and E. SALJE), pp. 91–112. Springer-Verlag, New York.

BUKOWINSKI M. S. T., CHIZMESHYA A., WOLF G. H., and ZHANG H. (1996) Advances in electron-gas potential models: applications to some candidate lower mantle minerals. In *Modelling of Minerals and Silicated Structures* (eds. B. Silvi and Ph. d'Arco), Topics in Molecular Organization and Engineering, Kluwer Academic Publishers, Dordrecht, The Netherlands, in press.

Burnham C. W. (1990) The ionic model: perceptions and realities in mineralogy. *Amer. Mineral.* **75,** 443–463.

Canil D. and Scarfe C. M. (1990) Phase relations in peridotite + CO_2 systems to 12 GPa: implications for the origin of kimberlite and carbonate stability in the Earth Upper Mantle. *J. Geophys. Res.* **95,** 15,805–15,816.

Carroll M. and Holloway J. R. (1994), eds., *Volatiles in Magmas. Reviews in Mineralogy,* Vol. 30. Mineralogical Society of America, Washington D.C.

Catlow C. R. A and Price G. D. (1990) Computer modelling of solid-state inorganic materials. *Nature* **347,** 243–248.

Chao E. C. T., Fahey J. J., Littler J. and Milton D. J. (1962) Stishovite, a very high pressure new mineral from Meteor Crater, Arizona. *J. Geophys. Res.* **67,** 419–421.

Chase D. B. (1986) Fourier transform Raman spectroscopy. *J. Am. Chem. Soc.* **108,** 7485–7488.

Chizmeshya A., Laviolette R. and Wolf G. H. (1994) Variational charge relaxation in ionic crystals: an efficient treatment of statics and dynamics. *Phys. Rev. B.* **50,** 15559–15574.

Chopelas A. (1990a) Thermal expansion, heat capacity, and entropy of MgO at mantle pressures. *Phys. Chem. Minerals* **17,** 149–157.

Chopelas A. (1990b) Thermal properties of forsterite at mantle pressures derived from vibrational spectroscopy. *Phys. Chem. Minerals* **17,** 149–156.

Chopelas A. (1991a) Thermal properties of β-Mg_2SiO_4 at mantle pressures derived from vibrational spectroscopy: implications for the mantle at 400 km depth. *J. Geophys. Res.* **96,** 11817–11829.

Chopelas A. (1991b) Single crystal Raman spectra of forsterite, fayalite and monticellite. *Am. Mineral.* **76,** 1101–1109.

Chopelas A. (1992) Sound velocities of MgO to very high compression. *Earth Planet. Sci. Letts.* **114,** 185–192.

Chopelas A. and Nicol M. F. (1982) Pressure dependence to 100 kbar of the phonons of MgO at 90 and 295 K. *J.Geophys.Res.* **87,** 8591–8597.

Chopelas A. and Boehler R. (1992) Raman spectroscopy of high pressure $MgSiO_3$ phases synthesized in a CO_2 laser heated diamond anvil cell: perovskite and pyroxene. In *High-Pressure Research: Application to Earth and Planetary Sciences* (eds. Y. Syono and M. H. Manghnani), pp. 101–108. Terra Scientific Publishing Co., Tokyo/American Geophysical Union, Washington D.C.

Chopelas A., Boehler R. and Ko T. (1994) Thermodynamics and behavior of γ-Mg_2SiO_4 at high pressure: implications for Mg_2SiO_4 phase equilibria. *Phys. Chem. Minerals* **21,** 351–359.

Chopin C. (1984) Coesite and pure pyrope in high-grade blueschists of the Western Alps. *Contrib. Mineral. Petrol.* **86,** 107–118.

Choudhury N., Chaplot S. L. and Rao K. R. (1989) Equation of state and melting point studies of forsterite. *Phys. Chem. Minerals* **16,** 599–605.

Closmann C. and Williams Q. (1995) In-situ spectroscopic investigation of high-pressure hydrated (Mg,Fe)-SiO_3 glasses: OH vibrations as a probe of glass structure. *Amer. Mineral.* **80,** 201–211.

Cohen R. E. (1992) First-principles predictions of elasticity and phase transitions in high pressure SiO_2 and geophysical applications. In *High-Pressure Research: Application to Earth and Planetary Sciences* (ed. Y. Syono and M. H. Manghnani), pp. 425–431. Terra Scientific Publishing Co., Tokyo/American Geophysical Union, Washington D.C..

Cohen R. E. (1994) First-principles theory of crystalline SiO_2. In *Silica: Physica Behavior, Geochemistry, and Materials Applications,* (eds. P. J. Heaney, C. T. Prewitt and G. V. Gibbs), *Rev. Mineral.* **29** 369–402. Mineralogical Society of America, Washington D.C..

Cohen R. E., Boyer L. L. and Mehl M. J. (1987) Theoretical studies of charge relaxation effects on the statics and dynamics of oxides. *Phys. Chem. Minerals* **14,** 294–302.

Cohen-Addad C., Ducros P. and Bertaut E. F. (1967) Etude de la substitution du groupement SiO_4 par $(OH)_4$ dans les composés $Al_2Ca_3(OH)_{12}$ et $Al_2Ca_3(SiO_4)_{2.16}(OH)_{3.36}$ de type grenat. *Acta Cryst.* **23,** 220–230.

Collerson B., Williams Q. and Knittle E. (1991) Vibrational spectra of magnesite ($MgCO_3$) and calcite-III at high pressures. *Am. Mineral.* **77,** 1158–1165.

Cynn H. and Hofmeister A. M. (1994) High-pressure IR spectra of lattice modes and OH vibrations in Fe-bearing wadsleyite. *J. Geophys. Res.* **99,** 17717–17727.

d'Arco P., Sandrone G., Dovesi R., Orlando R. and Saunders V. R. (1993) A quantum mechanical study of the perovskite structure type of $MgSiO_3$. *Phys. Chem. Minerals* **20,** 407–414.

Delany J. M. (1981) A spectral and thermodynamic investigation of synthetic pyrope-grossular garnets. Ph. D. Dissertation, University of California, Los Angeles.

Dietrich P. and Arndt J. (1982) Effects of pressure and temperature on the physical behavior of mantle-relevant olivine, orthopyroxene, and garnet: 2. Infrared absorption and microscopic Gruneisen parameters. In *High Pressure Research in Geoscience* (eds. W. Schreyer), pp. 307–319. E. Schweizerbart'sche Verlagsbinchhandlung.

Downs J. W. (1989) Possible sites for protonation in β-Mg_2SiO_4 from an experimentally-derived electrostatic potential. *Amer. Mineral.* **74,** 1124–1129.

Downs R. T., Zha C. S., Duffy T. S. and Finger, L. W. (1996) The equation of state of forsterite to 17.2 GPa and effects of pressure media. *Amer. Mineral.* **81,** 51–55.

Duffy T. S. and Ahrens T. J. (1992) Lateral variations in lower mantle seismic velocity. In *High-Pressure Research: Application to Earth and Planetary Sciences* (eds. Y. Syono and M. H. Manghnani), pp. 197–205. Terra Scientific Publishing Co., Tokyo/American Geophysical Union, Washington D.C..

Duffy T. S., Zha C. S., Downs, R. T., Mao, H. K. and Hemley, R. J. (1995) Elasticity of forsterite to 16 GPa and the composition of the upper mantle, *Nature* **378,** 170–173.

Durben D. J. and Wolf G. H. (1992) High-temperature behavior of metastable $MgSiO_3$ perovskite: a Raman spectroscopic study. *Amer. Mineral.* **77,** 890–893.

Durben D. J., Wolf G. H. and McMillan P. F. (1991) Raman scattering study of the high-temperature vibrational properties and stability of $CaGeO_3$ perovskite. *Phys. Chem. Minerals* **18,** 215–223.

DURBEN D. J., MCMILLAN P. F. and WOLF G. H. (1993) Raman study of the high pressure behavior of forsterite (Mg_2SiO_4) crystal and glass. *Amer. Mineral.* **78,** 1143–1148.

ETCHEPARE J. (1970) Spectres Raman du diopside cristallisé et vitreux. *C. R. Acad. Sci. Paris Sér. II* **270,** 1339–1342.

ETCHEPARE J. (1972) Study by Raman spectroscopy of crystalline and glassy diopside. In *Amorphous Materials* (eds. R. W. DOUGLAS and B. ELLIS), pp. 337–346. Wiley, New York.

FARBER D. L. and WILLIAMS Q. (1992) Pressure-induced coordination changes in alkali-germanate melts: an in-situ spectroscopic investigation. *Science* **256,** 1427–1430.

FARMER V. C. (1974) The anhydrous oxide minerals. In *The Infrared Spectra of Minerals* (ed. V. C. FARMER), pp. 183–204. Mineralogical Society, London.

FEI Y., SAXENA S. K. and NAVROTSKY A. (1990) Internally consistent thermodynamic data and equilibrium phase relations for compounds in the system MgO-SiO_2 at high pressure and high temperature. *J. Geophys. Res.* **95,** 6915–6928.

FEI Y., MAO H. K. and MYSEN B. O. (1991) Experimental determination of element partitioning and calculation of phase equilibrium relations in the MgO-FeO-SiO_2 system at high pressure and high temperature. *J. Geophys. Res.* **96,** 2157–2169.

FEI Y., MAO H. K., SHU J., PARTHASARATHY G., BASSETT W. A. and KO J. (1992) Simultaneous high-P, high-T X ray diffraction study of β-$(Mg,Fe)_2SiO_4$ to 26 GPa and 900 K. *J. Geophys. Res.* **97,** 4489–4495.

FERRARO J. R. (1984) *Vibrational Spectroscopy at High External Pressures.* Academic Press, Orlando.

FIELD J. E. (1992) *The Properties of Natural and Synthetic Diamond.* Academic Press, Orlando.

FINGER L. W., KO J., HAZEN R. M., GASPARIK T., HEMLEY R. J., PREWITT C. T. and WEIDNER D. J. (1989) Crystal chemistry of phase B and a new anhydrous magnesium silicate: implications for water storage in the upper mantle. *Nature* **341,** 140–142.

FIQUET G., GILLET PH. and RICHET P. (1992) Anharmonicity and high-temperature heat capacity of crystals: the examples of Ca_2GeO_4, Mg_2GeO_4 and $CaMgGeO_4$ olivines. *Phys. Chem. Minerals* **18,** 469–479.

FONG M. Y. and NICOL M. (1971) Raman spectrum of calcium carbonate at high pressures. *J. Chem. Phys.* **54,** 579–585.

FRANTZ J. D. and MYSEN B. O. (1996) Spectral measurements of coexisting silicate melts and aqueous fluids in the system K_2Si4O_9-H_2O to 910°C and 8 kbar. *Geochim. Cosmochim. Acta.*, submitted.

FUNAMORI N. and YAGI T. (1993) High pressure and high temperature in situ X-ray observation of $MgSiO_3$ perovskite under lower mantle conditions. *Geophys. Res. Lett.*. **20,** 387–390.

GASPARIK T. (1990) Phase relations in the transition zone. *J. Geophys. Research B* **95,** 15751–15769.

GASPARIK T. (1992) Enstatite-jadeite join and its role in the Earth's mantle. *Contrib.Mineral. Petrol.* **111,** 283–298.

GASPARIK T. (1993) The role of volatiles in the transition zone. *J. Geophys. Res.* **98,** 4287–4299.

GEIGER C. A., WINKLER B. and LANGER K. (1989) Infrared spectra of synthetic almandine-grossular and almandine-pyrope garnet solid-solutions: evidence for equivalent site behavior. *Mineral. Mag.* **53,** 231–238.

GERVAIS F. (1983) High-temperature infrared reflectivity spectroscopy by scanning interferometry. In *Electromagnetic Waves in Matter, Part 1* (ed. K. J. BURTON), pp. 279–339. Academic Press, New York.

GERVAIS F. and PIRIOU B. (1975) Temperature dependence of transverse and longitudinal optic modes in the α and β phases of quartz. *Phys. Rev. B* **11,** 3944–3950.

GERVAIS F., PIRIOU B. and CABANNES F. (1972) Anharmonicity of infrared vibrational modes in beryl. *Phys. Status Solidi b* **51,** 701–712.

GERVAIS F., PIRIOU B. and CABANNES F. (1973) Anharmonicity in silicate crystals: temperature dependence of A_u type vibrational modes in $ZrSiO_4$ and $LiAlSi_2O_6$. *J. Phys. Chem. Solids* **34,** 1785–1796.

GHOSE S. (1985) Lattice dynamics, phase transitions and soft modes. In *Microscopic to Macroscopic. Atomic Environments to Mineral Thermodynamics* (eds. S. W. KIEFFER and A. NAVROTSKY), *Rev. Mineral.* **14,** 127–163. Mineralogical Society of America, Washington D.C.

GHOSE S. (1988) Inelastic neutron scattering. In *Spectroscopic Methods in Mineralogy and Geology* (ed. F. HAWTHORNE), *Rev. Mineral.* **18,** 161–192. Mineralogical Society of America, Washington D.C.

GHOSE S., CHOUDHURY N., PAL CHOWDHURY C., and SHARMA S. K. (1994) Lattice dynamics and Raman spectroscopy of protoenstatite $Mg_2Si_2O_6$. *Phys. Chem. Minerals* **20,** 469–471.

GILLET PH. (1993a) L'eau du manteau terrestre. *La Recherche* **24,** 676–685

GILLET PH. (1993b) Stability of magnesite ($MgCO_3$) at mantle pressure and temperature conditions. A Raman spectroscopic study. *Amer. Mineral.* **78,** 1328–1331.

GILLET PH. (1995) Mineral physics, mantle mineralogy and mantle dynamics. *C. R. Acad. Sci. Paris. Sér. II a* **320,** 341–356.

GILLET PH., INGRIN J. and CHOPIN C. (1984) Coesite in subducted continental crust: *P-T* history deduced from an elastic model. *Earth Planet. Sci. Lett.* **70,** 426–436.

GILLET PH., MALEZIEUX J. M. and DHAMELINCOURT M. C. (1988) Microraman multichannel spectroscopy up to 2.5 GPa using a sapphire-anvil cell: experimental set-up and some applications. *Bull. Minéral.* **111,** 1–15.

GILLET PH., GUYOT, F. and MALEZIEUX J. M. (1989a) High-pressure, high-temperature Raman spectroscopy of Ca_2GeO_4 (olivine form): some insights on anharmonicity. *Phys. Earth Planet. Int.* **58,** 141–154.

GILLET PH., REYNARD B. and TEQUI C. (1989b) Thermodynamic properties of glaucophane new data from calorimetric and spectroscopic measurements. *Phys. Chem. Minerals* **16,** 659–667.

GILLET PH., LE CLÉAC'H A. and MADON M. (1990) High-temperature Raman spectroscopy of SiO_2 and GeO_2 polymorphs: anharmonicity and thermodynamic properties at high-temperatures. *J. Geophys. Rese.B* **95,** 21635–21655.

GILLET PH., RICHET P., GUYOT F. and FIQUET G. (1991) High-temperature thermodynamic properties of forsterite. *J. Geophys. Res.B* **96,** 11805–11816.

GILLET PH., FIQUET G., MALEZIEUX J.-M. and GEIGER C. C. (1992) High-pressure and high temperature Raman spectroscopy of end-member garnets: pyrope, grossular and andradite. *Eur. J. Mineral.* **4,** 651–664.

GILLET PH., GUYOT F., PRICE G. D., TOURNERIE B. and LE CLÉAC'H A. (1993a) Phase changes and thermodynamic properties of $CaTiO_3$. Spectroscopic data, vibrational modelling and some insights on the properties of $MgSiO_3$ perovskite. *Phys. Chem. Minerals* **20,** 159–170.

GILLET PH., FIQUET G., DANIEL I. and REYNARD B. (1993b) Raman spectroscopy at mantle pressure and temperature conditions. Experimental set-up and the example of $CaTiO_3$ perovskite. *Geophys. Res. Lett.*. **20,** 1931–1934.

GILLET PH., BIELLMANN C., REYNARD B. and MCMILLAN P. F. (1993c) Raman spectroscopic studies of carbonates. Part I: High-pressure and high-temperature behaviour of calcite, magnesite, dolomite, aragonite. *Phys. Chem. Minerals* **20,** 1–18.

GILLET PH., GUYOT F., MARTINEZ I. and ZHANG Z. (1996a) Is magnesite ($MgCO_3$) stable throughout the Earth's mantle? Implications for carbon storage in the Earth. *J. Geophys. Res.* in press.

GILLET PH., MCMILLAN P. F., SCHOTT J., BADRO J. and GRZECHNIK A. (1996b) Thermodynamic properties and isotopic fractionation of calcite from vibrational spectroscopy of ^{18}O-substituted calcite. *Geochim. Cosmochim. Acta*, in press.

GRZECHNIK A., PETUSKEY W. T. and MCMILLAN P. F. (1996) High dielectric $SrGeO_3$ perovskite prepared by high pressure synthesis. *Proc. Mater. Res. Soc., Symp. Thermodynamics and Kinetics of Phase Transformations*, in press.

GRIMSDITCH M., POPOVA, S., BRAZHKIN, V. V. and VOLOSHIN, R. N. (1994) Temperature-induced amorphization of SiO_2 stishovite. *Phys. Rev. B.* **50,** 12984–12986

GUYOT F. and REYNARD B. (1992) Pressure-induced structural modifications and amorphization in olivine compounds. *Chem. Geol.* **96,** 411–420.

GUYOT F., BOYER H., MADON M., VELDE B. and POIRIER J. P. (1986) Comparison of the microprobe spectra of $(Mg,Fe)_2SiO_4$ and Mg_2GeO_4 with olivine and spinel structures. *Phys. Chem. Minerals* **13,** 91–95.

GUYOT F., WANG Y., GILLET, PH. and RICARD Y. (1996) Experimental measurements of high-pressure, high-temperature volumes in olivines by synchrotron X-ray diffraction. Comparison with quasi-harmonic calculations. *Phys. Earth Planet. Inter.*, in press.

HAINES J., LEGÉR J. M. and SCHULTE O. (1996) $Pa\bar{3}$ modified fluorite-type structures in metal dioxides at high pressure. *Science* **271,** 629–631.

HANFLAND M. and SYASSEN K. (1985) A Raman study of diamond anvils under stress. *J. Appl. Phys.* **57,** 2752–2756.

HANFLAND M., SYASSEN K., FAHY S., LOUIE S. G. and COHEN M. L. (1985) Pressure dependence of the first-order Raman mode in diamond. *Phys. Rev. B* **31,** 6896–6899.

HARMON K. M., GABRIELE J. M. and NUTTALL A. S. (1982) Hydrogen bonding. Part 14. Hydrogen bonding in the tetrahedral $O_4H_4^{4-}$ cluster in hydrogrossular. *J. Mol. Struct.* **82,** 213–219.

HARTE B. and HARRIS J. W. (1993) Lower mantle inclusions from diamonds. *Terra Nova.* **5** (suppl. 1), 101.

HATCH M. D. and GHOSE S. (1989) Symmetry analysis of the phase transition and twinning in $MgSiO_3$ garnet: Implications to mantle mineralogy. *Amer. Mineral.* **74,** 1221–1224.

HEMLEY R. J. (1987) Pressure dependence of Raman spectra of SiO_2 polymorphs: α-quartz, coesite and stishovite. In *High-Pressure Research in Mineral Physics* (eds. M. H. MANGHNANI and Y. SYONO), pp. 347–359. Terra Scientific Publishing Co., Tokyo/American Geophysical Union, Washington D.C.

HEMLEY R. J. (1991) Spectroscopic models for the thermodynamic properties of minerals at high pressure and high temperature. *Eos Trans. Am. Geophys. Union (Spring Meeting)* **72,** 282.

HEMLEY R. J. and GORDON R. G. (1985) Theoretical study of solid NaF and NaCl at high pressures and temperatures, *J. Geophys. Res.* **90,** 7803–7813.

HEMLEY R. J. and COHEN R. E. (1992) Silicate perovskite. *Ann. Rev. Earth Planet. Sci.* **20,** 553–600.

HEMLEY R. J., MAO H. K. and CHAO E. C. T. (1986a) Raman spectrum of natural and synthetic stishovite. *Phys. Chem. Minerals* **13,** 285–290.

HEMLEY R. J., MAO H. K., BELL P. M. and MYSEN B. O. (1986b) Raman spectroscopy of SiO_2 glass at high pressure. *Phys. Rev. Lett.* **57,** 747–750.

HEMLEY R. J., BELL P. M. and MAO M. K. (1987a) Laser techniques in high pressure geophysics. *Science* **237,** 605–612.

HEMLEY R. J., JACKSON M. D. and GORDON R. G. (1987b) Theoretical study of the structure, lattice dynamics and equations of state of perovskite-type $MgSiO_3$ and $CaSiO_3$. *Phys. Chem. Minerals* **14,** 2–12.

HEMLEY R. J., COHEN R. E., YEGANEH-HAERI A., MAO H. K., WEIDNER D. J. and ITO E. (1989) Raman spectroscopy and lattice dynamics of $MgSiO_3$-perovskite at high pressure. In *Perovskite: a Structure of Great Interest To Geophysics and Materials Science* (eds. A. NAVROTSKY and D. J. WEIDNER), pp. 35–44. American Geophysical Union, Washington D.C.

HEMLEY R. J., PREWITT C. T. and KINGMA K. J. (1994) High-pressure behavior of silica. In *Silica: Physica Behavior, Geochemistry and Materials Applications* (eds. P. J. HEANEY, C. T. PREWITT and G. V. GIBBS), *Rev. Mineral.* **29,** 41–81. Mineralogical Society of America, Washington D.C.

HEMLEY R. J., MAO H. K., GONCHAROV A. F., HANFLAND M. and STRUZHKIN V. V. (1996) Synchrotron infrared spectroscopy to 0.15 eV of H_2 and D_2 at megabar pressures, *Phys. Rev. Lett.* **76,** 1667–1670.

HEMMATI M., CHIZMESHYA A., WOLF G. H., POOLE P. H., SHAO J. and ANGELL C. A. (1995) Crystalline-amorphous transition in silicate perovskites, *Phys. Rev. B* **51,** 14841–14848.

HESS N. J. and GHOSE S. (1988) Raman spectra of the calcite-$CaCO_3$ (II) structural phase transition as a function of pressure. *Eos Trans. Am. Geophys. Union* **69,** 500.

HOFMEISTER A. M. (1987) Single-crystal absorption and reflection infrared spectroscopy of forsterite and fayalite. *Phys. Chem. Minerals* **14,** 499–513.

HOFMEISTER A. M. (1991) Calculation of bulk moduli and their pressure derivatives from vibrational frequencies and mode Grüneisen parameters: Solids with high symmetry or one nearest-neighbor distance. *J. Geophys. Res.* **96,** 16181–16203.

HOFMEISTER A. and CHOPELAS A. (1991a) Vibrational spectroscopy of end-member silicate garnets. *Phys. Chem. Minerals* **17,** 503–526.

HOFMEISTER A. M. and CHOPELAS A. (1991b) Thermodynamic properties of pyrope and grossular from vibrational spectra. *Amer. Mineral.* **76,** 880–891.

HOFMEISTER A. M. and ITO E. (1992) Thermodynamic properties of $MgSiO_3$ ilmenite from vibrational spectra. *Phys. Chem. Minerals* **18,** 423–432.

HOFMEISTER A. M., XU J., MAO H.-K., BELL P. M. and

HOERING T. C. (1989) Thermodynamics of Fe-Mg olivines at mantle pressures: mid- and far-infrared spectroscopy at high pressures. *Amer. Mineral.* **74,** 281–306.

HOFMEISTER A., XU J. and AKIMOTO S. (1990) Infrared spectroscopy of synthetic and natural stishovite. *Amer. Mineral.* **75,** 951–955.

HOLTZ M., SOLIN S. A. and PINNAVAIA T. J. (1993) Effect of pressure on the Raman vibrational modes of layered aluminosilicate compounds. *Phys. Rev. B* **48,** 13312–13317.

HORIUCHI H. and SAWAMOTO H. (1981) β-Mg_2SiO_4: single crystal X-ray diffraction study. *Amer. Mineral.* **66,** 568–575.

HYDE B. G., WHITE T. J., O'KEEFFE M. and JOHNSON A. W. S. (1982) Structures related to those of spinel and the β-phase and a possible mechanism for the transformation olivine-spinel. *Z. Kristall.* **160,** 53–62.

IISHI K. (1978) Lattice dynamics of forsterite. *Amer. Mineral.* **63,** 1198–1208.

INOU T., YURIMOTO H. and KUDOH Y. (1995) Hydrous modified spinel, $Mg_{1.75}SiH_{0.5}O_4$: a new water reservoir in the mantle transition region. *Gophys. Res. Lett.* **22,** 117–120.

IRIFUNE T. (1994) Absence of an aluminous phase in the upper part of the the Earth's lower mantle. *Nature* **370,** 131–133.

IRIFUNE T., FUJINO K. and OHTANI E. (1991) A new high-pressure form of $MgAl_2O_4$. *Nature* **349,** 409–411.

IRVING A. J. and WYLLIE P. J. (1975) Subsolidus and melting relationships for calcite, magnesite and the join $CaCO_3$-$MgCO_3$ to 36 kbar. *Geochim. Cosmochim. Acta* **39,** 35–53.

ITO E. and YAMADA H. (1982) Stability relations of silicate spinels, ilmenites and perovskites. *Adv. Earth Planet. Sci.* **12,** 405–419.

ITO E. and KATSURA T. (1992) Melting of ferromagnesian silicates under the lower mantle conditions. In *High-Pressure Research: Application to Earth and Planetary Sciences* (eds. Y. SYONO and M. H. MANGHNANI), pp. 315–322. Terra Scientific Publishing Co., Tokyo/ American Geophysical Union, Washington D.C..

JEANLOZ R. (1980) Infrared spectra of olivine polymorphs: α, β phase and spinel. *Phys. Chem. Minerals* **5,** 327–341.

JEANLOZ R. (1981) Majorite: Vibrational and compressional properties of a high-pressure phase. *J. Geophys. Research B.* **86,** 6171–6179.

JEANLOZ R. (1993) Chemical reactions at the Earth's core-mantle boundary: summary of evidence and geomagnetic implications. In *Relating Geophysical Structures and Processes: The Jeffreys Volume, Geophys. Monogr.* **76** (eds. K. AKI and R. DMOWSKA), pp. 121–127. American Geophysical Union, Washington D.C..

KANZAKI M. (1987) Ultrahigh-pressure phase relations in the system $Mg_4Si_4O_{12}$-$Mg_3Al_2Si_3O_{12}$. *Phys. Earth Planet. Int.* **49,** 168–175.

KANZAKI M. (1991) Stability of hydrous magnesium silicates in the mantle transition zone. *Phys. Earth Planet. Int.* **66,** 307–312.

KAPUSTA B. and GUILLOPÉ M. (1993) Molecular dynamics study of the perovskite $MgSiO_3$ at high temperature: structural, elastic and thermodynamical properties. *Phys. Earth Planet. Int.* **75,** 205–224.

KATO T. (1986) Stability relation of (Mg,Fe)SiO_3 garnets, major constituents in the Earth's interior. *Earth Planet. Sci. Lett.* **77,** 399–408.

KATO T. and KUMAZAWA M. (1985) Garnet phase of $MgSiO_3$ filling the pyroxene-ilmenite gap at very high temperature. *Nature* **316,** 803–804.

KATSURA T. and ITO E. (1989) The system Mg_2SiO_4-Fe_2SiO_4 at high pressures and temperatures: precise determination of stabilities of olivine, modified spinel and spinel. *J. Geophys. Res.* **94,** 15663–15670.

KATSURA T. and ITO E. (1990) Melting and subsolidus phase relations in the $MgSiO_3$-$MgCO_3$ system at high pressures: implications to evolution of the Earth's atmosphere. *Earth Planet. Sci. Lett.* **99,** 110–117.

VAN KEKEN P. E., YUEN D. A. and VAN DEN BERG A. P. (1994) Implications for mantle dynamics from the high melting temperature of perovskite. *Science* **264,** 1437–1439.

KESSON S. E., FITZGERALD J. D. and SHELLEY J. M. G. (1994) Mineral chemistry and density of subducted basaltic crust at lower-mantle pressures. *Nature* **372,** 767–769.

KIEFFER S. W. (1979a) Thermodynamics and lattice vibrations of minerals: 1. Mineral heat capacities and their relationships to simple lattice vibrational models. *Rev. Geophys. Space Phys.* **17,** 1–19.

KIEFFER S. W. (1979b) Thermodynamics and lattice vibrations of minerals: 2. Vibrational characteristics of silicates. *Rev. Geophys. Space Phys.* **17,** 20–34.

KIEFFER S. W. (1979c) Thermodynamics and lattice vibrations of minerals: 3. Lattice dynamics and an approximation for minerals with application to simple substances and framework silicates. *Rev. Geophys. Space Phys.* **17,** 35–59.

KIEFFER S. W. (1980) Thermodynamics and lattice vibrations of minerals: 4. Application to chain and sheet silicates and orthosilicates. *Rev. Geophys. Space Phys.* **18,** 862–886.

KIEFFER S. W. (1982) Thermodynamics and lattice vibrations of minerals: 5. Applications to phase equilibria, isotopic fractionation and high pressure thermodynamic properties. *Rev. Geophys. Space Phys.* **20,** 827–849.

KIEFFER S. W. (1985) Heat capacity and entropy: systematic relations to lattice vibrations. In *Microscopic to Macroscopic. Atomic Environments to Mineral Thermodynamics* (eds. S. W. KIEFFER and A. NAVROTSKY), *Rev. Mineral.* **14,** 65–126. Mineralogical Society of America, Washington D.C.

KINGMA K. J., COHEN R. E., HEMLEY R. J. and MAO H. K. (1995) Transformation of stishovite to a denser phase at lower mantle pressures. *Nature* **374,** 243–245.

KINGMA K. J., MAO H.-K. and HEMLEY R. J. (1996) Synchrotron X-ray diffraction of SiO_2 to multimegabar pressures. *High Press. Res.* in press.

KNITTLE E. and JEANLOZ R. (1987) Synthesis and equation of state of (Mg,Fe)SiO_3 perovskite to over 100 gigapascals. *Science* **235,** 668–670.

KNITTLE E. and JEANLOZ R. (1991) Earth's core-mantle boundary: Results of experiments at high pressures and temperatures. *Science* **251,** 1438–1443.

KNITTLE E., JEANLOZ R. and SMITH G. L. (1986) Thermal expansion of the silicate perovskite and stratification of the earth's mantle. *Nature* **319,** 214–216.

KNITTLE E., HATHORNE A., DAVIS M. and WILLIAMS Q. (1992) A spectroscopic study of the high-pressure behavior of the O_4H_4 substitution in garnet. In *High-Pressure Research: Application to Earth and Planetary Sciences* (eds. Y. SYONO and M. H. MANGHNANI), pp. 297–304. Terra Scientific Publishing Co., Tokyo/ American Geophysical Union, Washington D.C.

KOBAYASHI S. and SHOJI T. (1983) Infrared analysis of

the grossular-hydrogrossular series. *Mineral. J.* **11,** 331–343.

Kohlstedt D. L., Keppler H. and Rubie D. C. (1996) Solubility of water in the α, β and γ phases of $(Mg,Fe)_2SiO_4$. *Amer. Mineral.,* submitted.

Kraft S., Knittle E. and Williams Q. (1991) Carbonate stability in the Earth's mantle : a vibrational spectroscopic study of aragonite and dolomite at high pressures and temperatures. *J. Geophys. Res.* **96,** 17997–18010.

Kudoh Y., Prewitt C. T., Finger L. W., Darovskikh A. and Ito E. (1990) Effect of iron on the crystal structure of $(Mg,Fe)SiO_3$ perovskite. *Geophys. Res. Lett.* **17,** 1481–1484.

Kurosawa M., Yurimoto H., Matsumoto K. and Sueno S. (1992) Hydrogen analysis of mantle olivine by secondary ion mass spectrometry. In *High-Pressure Research: Application to Earth and Planetary Sciences* (eds. Y. Syono and M. H. Manghnani), pp. 283–287. Terra Scientific Publishing Co., Tokyo/American Geophysical Union, Washington D.C.

Kurosawa M., Yurimoto H. and Sueno S. (1993) Water in the Earth's mantle: Hydrogen analysis of mantle olivine, pyroxenes and garnet using SIMS. pp. 839–840. *24th Lunar Planet. Sci. Abstr.*, Houston, Texas.

Kushiro I., Satake H. and Akimoto S. (1975) Carbonate-silicate reactions at high pressures and possible presence of dolomite and magnesite in the upper mantle. *Earth Planet. Sci. Lett.* **28,** 116–120.

Lacks D. J. and Gordon R. G. (1993) Calculations of pressure-induced phase transitions in silica. *J. Geophys. Res.* **98,** 22147–22155.

Lager G. A., Armbruster T. and Faber J. (1987) Neutron and X-ray diffraction study of hydrogarnet $Ca_3Al_2(O_4H_4)_3$. *Amer. Mineral.* **72,** 756–765.

Lager G. A., Armbruster T., Rotella F. J. and Rossman G. R. (1989) OH substitution in garnets: X-ray and neutron diffraction, infrared and geometric-modelling studies. *Amer. Mineral.* **74,** 840–851.

Le Cléac'h A. and Gillet Ph. (1990) IR and Raman spectroscopic study of natural lawsonite. *Eur. J. Mineral.* **2,** 43–53.

Lee W. E. and Heuer A. H. (1987) On the polymorphism of enstatite. *J. Am. Ceram. Soc.* **70,** 349–360.

Leinenweber K., Grzechnik A. and McMillan P. (1994) Structure and spectroscopy of perovskite compounds on the $CaSiO_3$-$CaTiO_3$ join. *Eos Trans. Am. Geophys. Union, Washington D.C..* **75,** 662.

Leinenweber K., Grzechnik A., Voorhees M., Navrotsky A., Yao N. and McMillan P. F. (1996) Structural variation in $Ca(Ti_xSi_{1-x})O_3$ perovskites ($1 > x > 0.65$) and the ordered phase Ca_2TiSiO_6. *Phys. Chem. Minerals*, submitted.

Lewis E. N., Treado P. J., Reeder R. C., Story G. M., Dowrey A. E., Marcott C. and Levin I. W. (1995) Fourier transform spectroscopic imaging using an infrared focal-plane array detector. *Anal. Chem.* **67,** 3377–3381.

Liu L.-G. (1974) Silicate perovskite from phase transformations of pyrope-garnet at high pressure and temperature. *Geophys. Res. Lett.* **1,** 277–280.

Liu L.-G. (1975) Post-oxide phases of forsterite and enstatite. *Geophys. Res. Lett.* **2,** 417–419.

Liu L.-G. (1977) Ilmenite-type solid solutions between $MgSiO_3$ and Al_2O_3 and some structural systematics among ilmenite compounds. *Geochim. Cosmochim. Acta* **41,** 1355–1361.

Liu L.-G. (1987) Effects of H_2O on the phase behaviour of the forsterite-enstatite system at high pressures and temperatures and implications for the Earth. *Phys. Earth Planet. Int.* **49,** 142–167.

Liu L.-G. (1992) Bulk moduli of $MgSiO_3$ perovskite. *Phys. Earth Planet. Int.* **72,** 12–20.

Liu L.-G. (1993) Volume, pressure and temperature dependences of vibrational frequencies. *Phys. Lett. A* **176,** 448–453.

Liu L.-G. and Bassett W. A. (1986) *Elements, Oxides, Silicates*. Oxford University Press, New York.

Liu L.-G. and Ringwood A. E. (1975) Synthesis of a perovskite-type polymorph of $CaSiO_3$. *Earth Planet. Sci. Lett.* **28,** 209–211.

Liu L.-G. and Mernagh T. P. (1990) Phase transitions and Raman spectra of calcite at high pressures and room temperature. *Amer. Mineral.* **75,** 801–806.

Liu L. G., Mernagh T. P. and Jaques A. L. (1990) A mineralogical Raman spectroscopy study on eclogitic garnet inclusions in diamonds from Argyle. *Contrib. Mineral. Petrol.* **105,** 156–161.

Liu L.-G., Mernagh T. P. and Irifune T. (1994) High pressure Raman spectra of β-Mg_2SiO_4, γ-Mg_2SiO_4, $MgSiO_3$-ilmenite and $MgSiO_3$-perovskite. *J. Phys. Chem. Solids.* **55,** 185–193.

Liu L. G., Mernagh T. P. and Irifune T. (1995) Raman spectra of pyrope and $MgSiO_3$-$10Al_2O_3$ garnet at various pressures and temperatures. *High Temp.-High Press.* in press.

Lu R., Hofmeister A. and Wang Y. (1994) Thermodynamic properties of ferromagnesium silicate perovskites from vibrational spectroscopy. *J. Geophys. Res.* **99,** 11795–11804.

Luxon J. T., Montgomery D. J. and Summitt R. (1969) Effect of particle size and shape on the infrared absorption of magnesium oxide powders. *Phys. Rev.* **188,** 1345–1356.

Lyon R. J. P. (1962) Infra-red confirmation of 6-fold coordination of silicon in stishovite. *Nature* **196,** 266–267.

Mackwell S. J. and Kohlstedt D. L. (1990) Diffusion of hydrogen in olivine: implications for water in the mantle. *J. Geophys. Res.* **95,** 5079–5088.

Mackwell S. J., Kohlstedt D. L. and Paterson M. S. (1985) The role of water in the deformation of olivine single crystals. *J. Geophys. Res.* **90,** 11319–11333.

Madon M. and Poirier J. P. (1983) Transmission electron microscopy observation of α, β and γ $(Mg,Fe)_2SiO_4$ in shocked meteorites: planar defects and polymorphic transitions. *Phys. Earth Planet. Int.* **33,** 31–44.

Madon M. and Price G. D. (1989) Infrared spectroscopy of the polymorphic series (enstatite, ilmenite and perovskite) of $MgSiO_3$, $MgGeO_3$ and $MnSiO_3$. *J. Geophys. Res.* **94,** 15687–15701.

Malezieux J.-M. (1990) Contribution of Raman microspectrometry to mineral studies. In *Absorption Spectroscopy in Mineralogy* (eds. A. Mottana and F. Burragato), pp. 39–60. Elsevier Science Publishers.

Mammone J. F. and Sharma S. K. (1980) Pressure and temperature dependence of the Raman spectra of rutile-structured oxides. Carnegie Inst. Wash. *Yearbk.* **79,** 369–373.

Mao H. K. and Hemley R. J. (1991) New optical transitions in diamond at ultrahigh pressures. *Nature* **351,** 721–724.

Mao H. K. and Hemley R. J. (1996) Experimental studies of Earth's deep interior: accuracy and versatility of diamond-anvil cells, *Phil. Trans. R. Soc. Lond. A* **354,** 1–18.

MAO H. K., HEMLEY R. J. and CHAO E. C. T. (1987) The application of micro-Raman spectroscopy to analysis and identification of minerals in thin sections. *Scanning Microscopy* **1,** 495–501.

MAO H. K., CHEN L. C., HEMLEY R. J., JEPHCOAT A. P., WU Y. and BASSETT W. A. (1989) Stability and equation of state of $CaSiO_3$-perovskite to 134 GPa. *J. Geophys. Res.* **94,** 17889–17894.

MAO H. K., HEMLEY R. J., FEI Y., SHU J. F., CHEN L. C., JEPHCOAT A. P., WU Y. and BASSETT W. A. (1991) Effect of pressure, temperature and composition on lattice parameters and density of (Fe,Mg)SiO_3-perovskites to 30 GPa. *J. Geophys. Res.* **96,** 8069–8079.

MATSUBARA R., TOKAYA H. TANAKA S. and SAWAMOTO H. (1990) Precision lattice-parameter determination of (Mg,Fe)SiO_3 tetragonal garnets. *Science* **247,** 697–700.

MATSUI M. and BUSING W. R. (1984) Computational modelling of the structure and elastic constants of the olivine and spinel forms of Mg_2SiO_4. *Phys. Chem. Minerals* **11,** 55–59.

MATSUI M. and PRICE G. D. (1991) Simulation of the pre-melting behaviour of $MgSiO_3$ perovskite at high pressures and temperatures. *Nature* **351,** 735–737.

MATSUI Y. and TSUNEYUKI S. (1992) Molecular dynamics study of rutile-$CaCl_2$-type phase transition of SiO_2. In *High-Pressure Research: Application to Earth and Planetary Sciences* (eds. Y. SYONO and M. H. MANGHNANI), pp. 433–439. Terra Scientific Publishing Co., Tokyo/American Geophysical Union, Washington D.C.

MCMILLAN P. (1985) Vibrational spectroscopy in the mineral sciences. In *Microscopic to Macroscopic. Atomic Environments to Mineral Thermodynamics* (eds. S. W. KIEFFER and A. NAVROTSKY), *Rev. Mineral.* **14,** 9–63. Mineralogical Society of America, Washington D.C.

MCMILLAN P. (1989) Raman spectroscopy in mineralogy and geochemistry. *Ann. Rev. Earth Planet. Sci.,* **17,** 255–283.

MCMILLAN P. and AKAOGI M. (1987) The Raman spectra of β-Mg_2SiO_4 (modified spinel) and γ-Mg2SiO4 (spinel). *Amer. Mineral.* **72,** 361–364.

MCMILLAN P. and ROSS N. L. (1987) Heat capacity calculations for Al_2O_3 corundum and $MgSiO_3$ ilmenite. *Phys. Chem. Minerals* **14,** 225–234.

MCMILLAN P. F. and ROSS N. L. (1988) The Raman spectra of several orthorhombic calcium oxide perovskites. *Phys. Chem. Minerals* **16,** 21–28.

MCMILLAN P. and HOFMEISTER A. M. (1988) Infrared and Raman spectroscopy. In *Spectroscopic Methods in Mineralogy and Geology* (ed. F. HAWTHORNE), *Rev. Mineral.* **18,** 99–159. Mineralogical Society of America, Washington D.C.

MCMILLAN P. F. and HESS A. C. (1990) Ab initio force field calculations for quartz. *Phys. Chem. Minerals* **17,** 97–107.

MCMILLAN P. F. and WOLF G. H. (1995) Vibrational spectroscopy of silicate liquids. In *Structure, Dynamics and Properties of Silicate Melts* (eds. J. F. STEBBINS, P. F. MCMILLAN and D. B. DINGWELL) *Rev. Mineral.* **32,** 247–315. Mineralogical Society of America, Washington D.C.

MCMILLAN P., PUTNIS A. and CARPENTER M. A. (1984) A Raman spectroscopic study of Al-Si ordering in synthetic magnesium cordierite. *Phys. Chem. Minerals* **10**, 256–260.

MCMILLAN P., AKAOGI M., OHTANI E., WILLIAMS Q., NIEMAN R. and SATO R. (1989) Cation disorder in garnets along the $Mg_3Al_2Si_3O_{12}$-$Mg_4Si_4O_{12}$ join: an infrared, Raman and NMR study. *Phys. Chem. Minerals* **16**, 428–435.

MCMILLAN P., AKAOGI M., SATO R., POE B. and FOLEY J. (1991) Hydroxyl groups in β—Mg_2SiO_4. *Amer. Mineral.* **76**, 354–360.

MEADE C., REFFNER J. A. and ITO E. (1994) Synchrotron infrared absorbance of hydrogen in $MgSiO_3$ perovskite. *Science* **264**, 1558–1560.

MEADE C., MAO H.-K. and HU J. (1995) High-temperature phase transition and dissociation of (Mg,Fe)SiO_3 perovskite at lower mantle pressures. *Science* **268**, 1743–1745.

MENG Y., WEIDNER D. J., GWANMESIA G. D., LIEBERMANN R. C., VAUGHAN M. T., WANG Y., LEINENWEBER K., PACALO R. E., YEGANEH-HAERI A. and ZHAO Y. (1993) In situ high *P-T* X ray diffraction studies on three polymorphs (α, β, γ) of Mg_2SiO_4. *J. Geophys. Res.* **98**, 22199–22207.

MERNAGH T. P. and LIU L. (1990) Pressure dependence of Raman spectra of garnet end-members pyrope, grossularite and almandite. *J. Raman Spectroscopy* **21**, 305–309.

MILLER G. H., ROSSMAN G. R. and HARLOW G. E. (1987) The natural occurrence of hydroxide in olivine. *Phys. Chem. Minerals* **14**, 461–472.

MOORE R. K., WHITE W. B. and LONG T. V. (1971) Vibrational spectra of the common silicates: I. The garnets. *Amer. Mineral.* **56**, 54–71.

MORISHIMA H., KATO T., SUTO M., OHTANI E., URAKAWA S., UTSUMI W., SHIMOMURA O. and KIKEGAWA T. (1994) The phase boundary between α- and β-Mg_2SiO_4 determined by *in-situ* X-ray observation. *Science* **265,** 1202–1203.

NAGEL L. and O'KEEFFE M. (1971) Pressure and stress induced polymorphism of compounds with rutile structure. *Mat. Res. Bull.* **6**, 1317–1320.

NAVON O., HUTCHEON I. D., ROSSMAN G. R. and WASSERBURG G. J. (1988) Mantle-derived fluids in diamond micro-inclusions. *Nature* **335**, 784–789.

NAVROTSKY A. (1980) Lower mantle phase transitions may generally have negative pressure-temperature slopes. *Geophys. Res. Lett.* **7**, 709–711.

NAVROTSKY A. (1985) Crystal chemical constraints on the thermochemistry of minerals. In *Microscopic to Macroscopic. Atomic Environments to Mineral Thermodynamics* (ed. S. W. KIEFFER and A. NAVROTSKY) *Rev. Mineral.* **14**, 225–275. Mineralogical Society of America, Washington D.C.

NAVROTSKY A. (1989) Thermochemistry of perovskites. In *Perovskite: a Structure of Great Interest To Geophysics and Materials Science* (eds. A. NAVROTSKY and D. J. WEIDNER). American Geophysical Union, Washington D.C.

NAVROTSKY A. (1992) Silicate and germanate garnets, ilmenites and perovskites: thermochemistry, lattice vibrations and spectroscopy. In *High-Pressure Research: Application to Earth and Planetary Sciences* (eds. Y. SYONO and M. H. MANGHNANI), pp. 261–268. Terra Scientific Publishing Co., Tokyo/American Geophysical Union, Washington D.C.

NAVROTSKY A. and WEIDNER D. J. (1989) *Perovskite: a*

Structure of Great Interest To Geophysics and Materials Science, American Geophys. Union.

OHTANI E., KAGAWA N. and FUKINO K. (1991) Stability of majorite (Mg,Fe)SiO_3 at high pressures and 1800°C. *Earth Planet. Sci. Lett.* **102**, 158–166.

OMORI K. (1971) Analysis of the infrared absorption spectrum of almandine-pyrope garnet from Nijosan, Osaka Prefecture, Japan. *Amer. Mineral.* **56**, 841–849.

PACALO R. E. G. and GASPARIK T. (1990) Reversals of the orthoenstatite-clinoenstatite transition at high pressures and high temperatures. *J. Geophys. Res. B.* **95**, 15853–15858.

PÂQUES-LEDENT M. TH. and TARTE P. (1973) Vibrational studies of olivine-type compounds-I. The i.r. and Raman spectra of the isotopic species of Mg_2SiO_4. *Spectrochim. Acta* **29A**, 1007–1016.

PAWLEY A. R., MCMILLAN P. F. and HOLLOWAY J. R. (1993) Hydrogen in stishovite, with implications for mantle water content. *Science* **261**, 1024–1026.

PHILLIPS B. L., HOWELL D. A., KIRKPATRICK R. J. and GASPARIK T. (1992) Investigation of cation order in $MgSiO_3$-rich garnet using ^{29}Si and ^{27}Al MAS NMR spectroscopy. *Amer. Mineral.* **77**, 704–712.

PINET M. and SMITH D. C. (1993) La microspectrométrie Raman des grenats $X_3Y_2Z_3O_{12}$: I. La série calcique naturelle ouvarovite-grossulaire-andradite. *Schweiz. Mineral. Perogr. Mitt.* **73**, 21–40.

PIRIOU B. (1964) Etude des bandes restantes de la magnésie et du corindon. *Rev.Int. Hautes Températures Réfractaires* **3**, 109–114.

PIRIOU B. (1974) Etude des modes normaux par réflexion infrarouge. *Ann. Chim.* **9**, 9–17.

PIRIOU B. and CABANNES F. (1968) Validité de la méthode de Kramers-Kronig et application à la dispersion infrarouge de la magnésie. *Optica Acta* **15**, 271–286.

PIRIOU B. and MCMILLAN P. (1983) The high-frequency vibrational spectra of vitreous and crystalline orthosilicates. *Amer. Mineral.* **68**, 426–443.

POIRIER J. P. (1991) *Introduction to the Physics of the Earth's Interior*. Cambridge University Press, New York.

PREWITT C. T. (ed.) (1980) Pyroxenes. *Reviews in Mineralogy,* Vol. 7 Mineralogical Society of America, Washington D.C.

PREWITT C. T. and FINGER L. W. (1992) Crystal chemistry of high-pressure hydrous magnesium silicates. In *High-Pressure Research: Application to Earth and Planetary Sciences* (eds. Y. SYONO and M. H. MANGHNANI), pp. 269–274. Terra Scientific Publishing Co., Tokyo/ American Geophysical Union, Washington D.C..

PRICE G. D. and PARKER S. C. (1984) Computer simulations of the structural and physical properties of the olivine and spinel polymorphs of Mg_2SiO_4. *Phys. Chem. Minerals* **10**, 209–216.

PRICE G. D., PARKER S. C. and LESLIE M. (1987) The lattice dynamics and thermodynamics of the Mg_2SiO_4 polymorphs. *Phys. Chem. Minerals* **15**, 181–190.

RAO K. R., CHAPLOT S. L., CHOUDHURY N., GHOSE S., HASTINGS J. M., CORLISS L. M. and PRICE D. L. (1988) Lattice dynamics and inelastic neutron scattering from forsterite, Mg_2SiO_4: phonon dispersion relation, density of states and specific heat. *Phys. Chem. Minerals* **16**, 83–97.

REFFNER J., CARR G. L., SUTTON S., HEMLEY R. J. and WILLIAMS G. P. (1994) Infrared microspectroscopy at the NSLS. *Synchrotron Radiation News* **7,** 30–37.

REYNARD B. (1992) Single-crystal infrared reflectivity of pure Mg_2SiO_4 forsterite and $(Mg_{0.86}Fe_{0.14})_2SiO_4$ olivine. *Phys. Chem. Minerals* **18**, 19–25.

REYNARD B. and GUYOT F. (1994) High temperature properties of geikelite ($MgTiO_3$-ilmenite) from high-temperature high-pressure Raman spectroscopy—some implications for $MgSiO_3$ ilmenite. *Phys. Chem. Minerals* **21**, 441–450.

REYNARD B. and RUBIE D. C. (1996) High-pressure high-temperature Raman spectroscopic study of $MgSiO_3$-ilmenite. *Amer. Mineral.* in press.

REYNARD B., PRICE G. D. and GILLET PH. (1992) Thermodynamic and anharmonic properties of forsterite, α-Mg_2SiO_4: computer modelling versus high-pressure and high-temperature measurements. *J. Geophys. Res.* **97**, 19791–19801.

REYNARD B., TAKIR F., GUYOT F., GWANMESIA G. D., LIEBERMANN R. C. and GILLET PH. (1996) High temperature Raman spectroscopic and X-ray diffraction study of β-Mg_2SiO_4: some insights on its high temperature thermodynamic properties and on β- to α phase transformation. *Amer. Mineral.* in press.

RICHARD G. and RICHET P. (1990) Room temperature amorphization of fayalite and high pressure properties of Fe_2SiO_4 liquid. *Geophys. Res. Lett.* **17**, 2093–2096.

RICHET P. and FIQUET G. (1991) High-temperature heat capacity and pre-melting of minerals in the system MgO—CaO—Al2O3—SiO2. *J. Geophys. Res.* **96**, 445–456.

RICHET P., GILLET P. and FIQUET G. (1992) Thermodynamic properties of minerals: macroscopic and microscopic approaches. In *Thermodyamic Data—Systematics and Estimation* (ed. S. K. SAXENA), pp. 98–131. Springer-Verlag, New York.

RICHET P., INGRIN J., MYSEN B. O., COURTIAL P. and GILLET P. (1994) Premelting effects in minerals: an experimental study. *Earth Planet. Sci. Lett.* **121**, 589–600.

RIGDEN S. M., GWANMESIA G. D., FITZGERALD J. D., JACKSON I. and LIEBERMANN R. C. (1991) Spinel elasticity and seismic structure of the transition zone of the mantle. *Nature* **354**, 143–145.

RINGWOOD A. E. (1967) The pyroxene-garnet transformation in the Earth's mantle. *Earth Planet. Sci. Lett.* **2**, 255–263.

RINGWOOD A. E. and MAJOR A. (1970) The system Mg_2SiO_4-Fe_2SiO_4 at high pressures and temperatures. *Phys. Earth Planet Interiors* **3**, 89–108.

RINGWOOD A. E. and MAJOR A. (1971) Synthesis of majorite and other high-pressure garnets and perovskites. *Earth Planet. Sci. Lett.* **12**, 411–414.

ROBIE R. A., HEMINGWAY B. S., GILLET PH. and REYNARD B. (1991) On the entropy of glaucophane $Na_2Mg_3Al_2Si_8O_{22}(OH)_2$. *Contrib. Mineral. Petrol.* **107**, 484–486.

ROSS N. L. and MCMILLAN P. (1984) The Raman spectrum of $MgSi0_3$ ilmenite. *Amer. Mineral.* **69**, 719–721.

ROSSMAN G. R. (1988) Vibrational spectroscopy of hydrous components. In *Spectroscopic Methods in Mineralogy and Geology* (ed. F. HAWTHORNE) *Rev. Mineral.* **18**, 193–206. Mineralogical Society of America, Washington D.C.

RUBIE D. C. and BREARLEY A. J. (1994) Phase transitions between β and γ (Mg,Fe)$_2SiO_4$ in the Earth's mantle: mechanisms and rheological implications. *Science* **264**, 1445–1448.

SAKURAI T. and SATO T. S. (1971) Temperature dependence of vibrational spectra in calcite by means of emissivity measurement. *Phys. Rev. B* **4**, 583–591.

SALJE E. (1989) Characteristics of perovskite-relaed materials. *Phil. Trans. R. Soc. Lond. A* **328**, 409–416.

SAMARA G. A. and PEERCY P. S. (1973) Pressure and temperature dependence of the static dielectric constants and Raman spectra of TiO_2 (rutile). *Phys. Rev. B* **7**, 1131–1148.

SAUTTER V. and GILLET PH. (1994) Les diamants, messagers de la Terre profonde. *La Recherche* **271**, 1238–1245.

SAWAMOTO H. and HORIUCHI H. (1990) β $(Mg_{0.9},Fe_{0.1})_2$-SiO_4: single crystal structure, cation distribution and properties of coordination polyhedra. *Phys. Chem. Minerals* **17**, 293–300.

SAXENA S. K., CHATTERJEE N., FEI Y. and SHEN G. (1993) *Thermodynamic Data on Oxides and Silicates. An Assessed Data Set Based on Thermochemistry and High Pressure Phase Equilibria*. Springer-Verlag, Berlin.

SCHRADER H., BOYSEN H., FREY F. and CONVERT P. (1990) On the phase transformation proto- to clino-/orthoenstatite: neutron powder investigation. *Phys. Chem. Minerals* **17**, 409–415.

SCHRAUDER M. and NAVON O. (1993) Solid carbon dioxide in a natural diamond. *Nature* **365**, 42–44.

SCOTT J. F. (1974) Soft-mode spectroscopy: Experimental studies of structural phase transitions. *Rev. Mod. Phys.* **46**, 83–128.

SEKITA M., OHASHI H. and TERADA S. (1988) Raman spectroscopic study of clinopyroxenes in the system $CaScAlSiO_6$-$CaAl_2SiO_6$. *Phys. Chem. Minerals* **15**, 319–322.

SERVOIN J. L. and PIRIOU B. (1973) Infrared reflectivity and Raman scattering of Mg_2SiO_4 single crystal. *Physica Status Solidi b* **55**, 677–686.

SHARMA S. K. (1989) Applications of advanced Raman techniques in Earth sciences. In *Raman Spectroscopy: Sixty Years On, Vibrational Spectra and Structure*, Vol. 17B (eds. H. D. BIST, J. R. DURIG and J. F. SULLIVAN), pp. 513–568. Elsevier, Amsterdam.

SHARMA S. K. (1990) Applications of Raman spectroscopy in Earth and planetary sciences. In *From Mantle to Meteorites* (eds. K. GOPALAN, V. K. GAUR and J. D. MACDOUGALL), pp. 263–308. Indian Academy of Sciences, Bangalore.

SHARMA S. K., MAMMONE J. F. and NICOL M. (1981) Raman investigation of ring configurations in vitreous silica. *Nature* **292**, 140–141.

SHARMA S. K., MAO H. K., BELL P. M. and XU J. A. (1985) Measurement of stress in diamond anvils with micro-Raman spectroscopy. *J.Raman Spectroscopy* **16**, 350–352.

SHARMA S. K., GHOSE S. and URMOS J. P. (1987) Raman study of high temperature phase transition in ortho- and clinoenstatite. *Eos Trans. Am. Geophys. Union, Washington D.C.*. **68**, 433.

SHARMA S. K., COONEY T. F. and WANG, S.Y. (1992) Effect of high P and T on olivines: a Raman spectral study. *XIII AIRAPT International Conference on High Pressure Science and Technology*, 614–619.

SHEU J. L. and MCMILLAN P. (1988) An ab initio study of the O—H stretching vibration in tremolite. *Phys. Chem. Minerals* **16**, 114–119.

SHUTONG X., OKAY A. I., SHOUYAN J., SENGOR A. M. C., WEN S., YICAN L. and LAILI J. (1992) Diamond from the Dabie Shan metamorphic rocks and its implication for metamorphic setting. *Science* **256**, 80–82.

SMITH D. C. (1984) Coesite in clinopyroxene in the Caledonides and its implications for geodynamics. *Nature* **310**, 641–644.

SMITH J. V. and MASON B. (1970) Pyroxene-garnet transformation in Coorara meteorite. *Science* **168**, 832–833.

SMYTH J. R. (1987) β-Mg_2SiO_4- A potential host for water in the mantle? *Amer. Mineral.* **72**, 1051–1055.

SMYTH J. R., SWOPE R. J. and PAWLEY A. R. (1995) H in rutile-type compounds: II. Crystal chemistry of Al substitution in H-bearing stishovite. *Amer. Mineral.* **80**, 454–456.

SOBOLEV N. V. and SHATSKY V. S. (1990) Diamond inclusions in garnets from metamorphic rocks: a new environment for diamond formation. *Nature* **343**, 742–746.

SPITZER W. G., MILLER R. C., KLEINMAN A. and HOWARTH L. E. (1962) Far infrared dielectric dispersion in $BaTiO_3$, $SrTiO_3$ and TiO_2. *Phys. Rev.* **126**, 1710–1721.

STEBBINS J. F., MCMILLAN P. F. and DINGWELL D. B. (eds.) (1995) Structure, Dynamics and Properties of Silicate Melts. *Rev. Mineral.Vol. 32*, Mineralogical Society of America, Washington D.C.

STISHOV S. M. and POPOVA S. V. (1961) A new dense modification of silica. *Geochem. Int.* **10**, 923–926.

STIXRUDE L. and COHEN R. E. (1993) Stability of orthorhombic $MgSiO_3$ perovskite in the Earth's lower mantle. *Nature* **364**, 613–616.

SUNG C. M. and BURNS R. G. (1976) Kinetics of high-pressure phase transformations: implications to the evolution of the olivine-spinel transition in downgoing lithosphere and its consequences on mantle dynamics. *Tectonophysics.* **31**, 1–32.

SUWA Y. and NAKA S. (1975) Infrared spectra of the solid solution between uvarovite and spessartine. *Amer. Mineral.* **60**, 1125–1126.

SWOPE R. J., SMYTH J. R. and LARSON A. C. (1995) H in rutile-type compounds: I. Single-crystal neutron and X-ray diffraction study of H in rutile. *Amer. Mineral.* **80**, 448–453.

TARTE P. and RINGWOOD A. E. (1964) Infrared spectrum and germanium coordination in some high pressure meta-germanates. *Nature* **201**, 819.

TARTE P. and DELIENS M. (1973) Correlations between the infrared spectrum and the composition of garnets in the pyrope-almandine-spessartine series. *Contrib. Mineral. Petrol.* **40**, 25–37.

THOMPSON A. B. (1992) Water in the Earth's upper mantle. *Nature* **358**, 295–302.

TOMISAKA T. and IISHI K. (1980) Some aspects of the lattice dynamics of diopside. *Mineral.J.* **10**, 84–96.

TSUCHIDA Y. and YAGI T. (1989) A new, post-stishovite high-pressure polymorph of silica. *Nature* **340**, 217–220.

TSUCHIDA Y. and YAGI T. (1990) New pressure-induced transformations of silica at room temperature. *Nature* **347**, 267–269.

TSUKIMURA K., SATO-SORENSEN Y. GHOSE S. and SAWAMOTO H. (1988) High-temperature single-crystal study of β-Mg_2SiO_4. *Eos Trans. Amer. Geophys. Union* **69**, 498.

VLASSOPOULOS D., ROSSMAN G. R. and HAGGERTY S. E. (1993) Coupled substitution of H and minor elements in rutile and the implications of high OH contents in Nb- and Cr-rich rutile from the upper mantle. *Amer. Mineral.* **78**, 1181–1191.

VOHRA Y. K. (1992), Spectroscopic studies on diamond anvil under extreme static pressures, *Recent Trends in High-Pressure Research* (ed. A. K. SINGH) pp. 349–358, New York: International Science, New York.

WALL A. and PRICE G. D. (1988) Computer simulation of the structure, lattice dynamics and thermodynamics of ilmenite-type $MgSiO_3$. *Amer. Mineral.* **73**, 224–231.

WANG S. Y., SHARMA S. K. and COONEY T. F. (1993) Micro-Raman and infrared spectral study of forsterite under high pressure. *Amer. Mineral.* **78**, 469–476.

WANG X., LIOU L. G. and MAO H. K. (1989) Coesite-bearing eclogite from the Dabie Mountains in cnetral China. *Geology* **12**, 1085–1088.

WANG Y., GUYOT F. and LIEBERMANN R. C. (1992) Electron microscopy of $(Mg,Fe)SiO_3$ perovskite: evidence for structural phase transitions and implications for the lower mantle. *J. Geophys. Res.* **97**, 12327–12347.

WANG Y., WEIDNER D. J., LIEBERMANN R. C. and ZHAO Y. (1994) P-V-T equation of state of $(Mg,Fe)SiO_3$ perovskite: constraints on the composition of the lower mantle. *Phys. Earth Planet. Int.* **83**, 13–40.

WASSERMAN E. A., YUEN D. A. and RUSTAD J. R. (1993) Molecular dynamics study of the transport properties of perovskite melts under high temperature and pressure conditions. *Earth Planet. Sci. Lett.* **114**, 373–384.

WENG K., XU J., MAO H. K. and BELL P. M. (1983) Preliminary Fourier-transform infrared spectral data on the SiO_6^{8-} octahedral group in silicate perovskite. *Carnegie Inst. Wash. Yearb.* **82**, 355–356.

WENTZCOVITCH R. M, MARTINS J. L. and PRICE G. D. (1993) Ab initio molecular dynamics with variable cell shape: application to $MgSiO_3$. *Phys. Rev. Lett.* **70**, 3947–3950.

WHITE W. B. (1974) The carbonate minerals. In *The Infrared Spectra of Minerals* (ed. V. C. FARMER), pp. 227–284. Mineral Society.

WHITE W. B. (1975) Structural interpretation of lunar and terrestrial minerals by Raman spectroscopy. In *Infrared and Raman Spectra of Lunar and Terrestrial Minerals* (ed. C. KARR), pp. 325–358. Academic Press, New York.

WILLIAMS Q. (1990) Molten $(Mg_{0.88}Fe_{0.12})_2SiO_4$ at lower mantle conditions: melting products and structure of quenched glasses. *Geophys. Res. Lett.*. **17**, 635–638.

WILLIAMS Q. (1992) A vibrational spectroscopic study of hydrogen in high pressure mineral assemblages. In *High-Pressure Research: Application to Earth and Planetary Sciences* (eds. Y. SYONO and M. H. MANGHNANI), pp. 289–296. Terra Scientific Publishing Co., Tokyo/American Geophysical Union, Washington D.C..

WILLIAMS Q. and JEANLOZ R. (1988) Spectroscopic evidence for pressure-induced coordination changes in silicate glasses and melts. *Science* **239**, 902–905.

WILLIAMS Q., JEANLOZ R. and AKAOGI M. (1986) Infrared vibrational spectra of beta-phase Mg_2SiO_4 and Co_2SiO_4 to pressures of 27 GPa. *Phys. Chem. Minerals* **13**, 141–145.

WILLIAMS Q., JEANLOZ R. and MCMILLAN P. (1987) The vibrational spectrum of $MgSiO_3$ perovskite: zero-pressure and mid-infrared spectra to 27 GPa. *J. Geophys. Res.* **92**, 8116–8128.

WILLIAMS Q., KNITTLE E. and JEANLOZ R. (1989) Geophysical and crystal chemical significance of (Mg,Fe)-SiO_3 perovskite. In *Perovskite: A Structure of Great Interest To Geophysics and Materials Science* (eds. A. NAVROTSKY and D. J. WEIDNER), pp. 1–12. American Geophysical Union, Washington D.C.

WILLIAMS Q., KNITTLE E., REICHLIN R., MARTIN S. and JEANLOZ R. (1990) Structural and electronic properties of Fe_2SiO_4-fayalite at ultrahigh pressures: amorphization and gap closure. *J. Geophys. Res.* **95,** 21549–21563.

WILLIAMS Q., HEMLEY R. J., KRUGER M. B. and JEANLOZ R. (1993) High pressure vibrational spectra of quartz, coesite, stishovite and amorphous silica. *J. Geophys. Res.* **98,** 22157–22170.

WINKLER B. and DOVE M. T. (1992) Thermodynamic properties of $MgSiO_3$ perovskite derived from large scale molecular dynamics simulations. *Phys. Chem. Minerals* **18**, 407–415.

WOLF G. H. and JEANLOZ R. (1984) Lindemann melting law: anharmonic correction and test of its validity for minerals. *J. Geophys. Res.* **89**, 7821–7835.

WOLF G. H. and JEANLOZ R. (1985) Lattice dynamics and structural distortions of $CaSiO_3$ and $MgSiO_3$ perovskites. *Geophys. Res. Lett.*. **12**, 413–416.

WOLF G. and BUKOWINSKI M. S. T. (1987) Theoretical study of the structural properties and equations of state of $MgSiO_3$ and $CaSiO_3$ perovskites: implications for lower mantle composition. In *High Pressure Research in Mineral Physics* (eds. M. H. MANGHNANI and Y. SYONO), pp. 313–331. Terra Scientific Publishing Co., Tokyo/American Geophysical Union, Washington D.C.

WOLF G. H. and MCMILLAN P. F. (1995) Pressure effects on silicate melt structure and properties. In *Structure, Dynamics and Properties of Silicate Melts.* (eds. J. F. STEBBINS, P. F. MCMILLAN and D. B. DINGWELL) *Rev. Mineral.* **32,** 505–561. Mineralogical Society of America, Washington D.C.

WOLF G. H., WANG S., HERBST C. A., DURBEN D. J., OLIVER W. F., KANG Z. C., and HALVORSON K. (1992) Pressure induced collapse of the tetrahedral framework in crystalline and amorphous GeO_2. In *High-Pressure Research: Application to Earth and Planetary Sciences* (eds. Y. SYONO and M. H. MANGHNANI), pp. 503–517. Terra Scientific Publishing Co., Tokyo/American Geophysical Union, Washington D.C.

WU T. C., BASSETT W. A., BURNLEY P. C. and WEATHERS M. S. (1993) Shear-promoted phase transitions in Fe_2SiO_4 and Mg_2SiO_4 and the mechanism of deep earthquakes. *J. Geophys. Res.* **98,** 19767–19776.

WYLLIE P. J. (1988) Magma genesis, plate tectonics and chemical differentiation. *Rev. Geophys.* **26**, 370–404.

XU J., MAO H.-K., WENG K. and BELL P. M. (1983) High-pressure, Fourier transform infrared spectra of forsterite and fayalite. *Carnegie Inst. Wash. Yearb.* **82**, 350–354.

YAGI T., UCHIYIMA Y., AKAOGI M. and ITO E. (1992) Isothermal compression curve of MgSiO3 tetragonal garnet. *Phys. Earth Planet. Int.* **74**, 1–7.

YAGI T., SUZUKI T. and AKAOGI M. (1994) High pressure transitions in the system $KAlSi_3O_8$-$NaAlSi_3O_8$. *Phys. Chem. Minerals* **21**, 12–17.

YANG H. and GHOSE S. (1994a) In-situ Fe-Mg order-disorder studies and thermodynamic properties of orthopyroxene $(Mg,Fe)_2Si_2O_6$. *Amer. Mineral.* **79**, 633–643.

YANG H. and GHOSE S. (1994b) Thermal expansion, Debye temperature and Grüneisen parameter of synthetic $(Fe,Mg)SiO_3$ orthopyroxenes. *Phys. Chem. Minerals* **20**, 575–586.

YANG H. and GHOSE S. (1995) A transitional structural state and anomalous Fe-Mg order-disorder in Mg-rich orthopyroxene, $(Mg_{0.75}Fe_{0.25})Si_2O_6$. *Amer. Mineral.* **80**, 9–20.

YOUNG T. E., GREEN H. W. I., HOFMEISTER A. M. and WALKER D. (1993) Infrared spectroscopic investigation of hydroxyl in β-$(Mg,Fe)_2SiO_4$ and coexisting olivine: implications for mantle evolution and dynamics. *Phys. Chem. Minerals* **19**, 409–422.

YUSA H., AKAOGI M. and ITO E. (1993) Calorimetric study of $MgSiO_3$ garnet and pyroxene: heat capacities, transition enthalpies and equilibrium phase relations in $MgSiO_3$ at high pressures and temperatures. *J. Geophys. Res.* **98**, 6453–6460.

ZHANG J., LIEBERMANN R. C., GASPARIK T., HERZBERG C. T. and FEI Y. (1993) Melting and subsolidus relations of SiO_2 at 9 to 14 GPa, *J. Geophys. Res.* **98**, 19785–19793.

ZOUBOULIS E. S. and GRIMSDITCH M. (1991) Raman scattering in diamond up to 1900 K. *Phys.Rev.B* **43**, 12,490–12,493.

ZULUMYAN N. O., MIRGORODSKII A. P., PAVINICH V. F. and LAZAREV A. N. (1976) Study of calculation of the vibrational spectrum of a crystal with complex polyatomic anions. Diopside $CaMgSi_2O_6$. *Optics and Spectroscopy.* **41**, 622–627.

Mineral Spectroscopy: A Tribute to Roger G. Burns
© The Geochemical Society, Special Publication No. 5, 1996
Editors: M. D. Dyar, C. McCammon and M. W. Schaefer

Thermodynamic properties of stishovite at mantle conditions determined from pressure variations of vibrational modes

ANNE M. HOFMEISTER
Dept. of Earth and Planetary Sciences, Washington University, St. Louis, MO 63130, U.S.A.

Abstract—Thermodynamic properties were calculated from infrared (IR) spectra of synthetic stishovite that were measured at room temperature and at pressures up to 360 kbar generated by a diamond anvil cell. Two IR modes with 1 atm frequencies ν of 761 and 836 cm^{-1} have nearly constant $d\nu_i/dP$ and initial mode Grüneisen parameters $\gamma_{io} = (K_T/\nu_i)d\nu_i/dP|_o$ near 1, with $d\gamma_i/dP|_o$ less than −0.0005/kbar. Modes initially at 580 and 470 cm^{-1} both show moderate decreases of $d\nu_i/dP$ at high pressures, γ_{io} values near 2, and derivatives $d\gamma_i/dP|_o$ near −0.004/kbar. Averaging all the optic modes gives $\gamma_{ave} = [1.15 \pm 0.04] - [0.0014 \pm 0.0005]*P(kbar)$. The result compares well with the 1 atm high temperature value of $\gamma_{th} = 1.2 \pm 0.1$ and with the initial slope $d\gamma_{Th}dP|_o = -0.002$, which was calculated using the slope dC_V/dP obtained from Kieffer's model, below. The bulk modulus K_T of 3085 ± 25 kbar calculated from only V_o and $\nu_i(P)$ equals the acoustic value within the experimental uncertainties. The calculated dK/dP of 4.1 ± 0.3 is higher than previous determinations but reproduces available V(P) data, suggesting that this new value is correct. The calculated K″ of −0.0009 ± 0.0002/kbar is similar to those previously measured for dense oxides. The above quadratic representation of K(P) can be integrated to predict volumes up to the phase transition at 500 kbar. From previous data on $d\nu_i/dT$, dK/dT is predicted to be −0.185 ± 0.020 kbar/K which is similar to the 1 atm values for many silicates. Heat capacity and entropy calculated from Kieffer's model are in excellent agreement with calorimetric data from 150 to 660 K, suggesting that the calculations can be used to extrapolate C_P and S at temperatures higher than the region of calorimetric measurements. The calculated pressure dependence of C_V is weak at temperatures above 1000 K [*e.g.*, $C_{V,1000K}$(J/mol-K) = 73.69 − 0.00097P(kbar)], whereas the dS/dP is smallest at low temperatures (*e.g.*, $S_{V,298K}$ (J/mol-K) = 14.51 − 0.0173P + 1.873 × 10^{-6} P(kbar)^2). The excellent agreement of calculated macroscopic properties with all available thermodynamic and elastic measurements suggests that the predicted 1 atm quantities (dK/dT, K″, dC_v/dP, dS/dP and $d\gamma_{th}/dP$) are correct and that the above formulations can be extrapolated to mantle conditions with confidence.

INTRODUCTION

ROGER G. BURNS had long-standing interests in mantle minerals and in using spectra to understand the physical properties of materials (see *e.g.*, BURNS' 1970 monograph). He was particularly well-known for his applications of crystal field spectra to problems of chemical composition, phase transitions, and fractionation processes (*e.g.*, BURNS and FYFE, 1967; BURNS, 1985). Vibrational spectra can also be used to constrain thermodynamic properties of solids under conditions that cannot be attained during direct thermochemical measurements. This spectroscopic approach is particularly applicable to mantle petrology because the samples are small and metastable at ambient conditions, and because pressure/temperature conditions appropriate to the mantle are difficult to attain in the laboratory.

At high pressure, SiO_2 adopts the rutile structure (STISHOV and POPOVA, 1961; CHAO *et al.*, 1962). The physical response of stishovite to pressure is of interest because this phase has one of the simplest structures and chemistries of all compounds with Si cations that are octahedrally coordinated by O. Stishovite may be a important constituent of the Earth's mantle because free SiO_2 in this region could result either from transformation of quartz or disproportionation of a number of common silicate minerals. Therefore, knowledge of the thermodynamic properties of stishovite are needed for geophysical models.

Entropy and heat capacity can be accurately calculated through KIEFFER's (1979; 1980) model if the vibrational data are reasonably complete (*e.g.*, HOFMEISTER and CHOPELAS, 1991). In particular, spectroscopic modeling in conjunction with heat capacities measured over a limited temperature range serve to tightly constrain the entropy. This approach is useful because the volumes of material produced in high pressure syntheses are sufficient for calorimetric measurements at moderate temperatures (300 to 1000 K), whereas cryogenic measurements require dozens of synthesis runs.

Reliable knowledge of bulk moduli and their P and T derivatives for mantle minerals as functions of pressure and temperature are needed to construct a compositional model for the mantle. This may be accomplished through comparing the elasticity measurements to seismological observations. Uncertainties in the pressure derivatives of the bulk

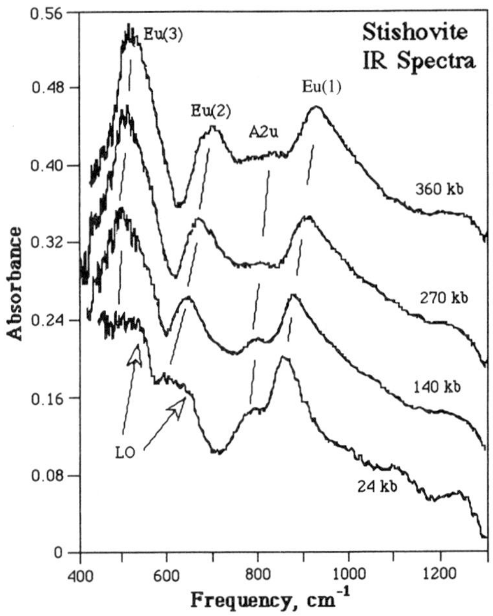

FIG. 1. Infrared spectra of stishovite at pressure. Pressures are labeled in kbar. Peaks are labled by short dashes. The series taken during decompression is shown. The spectra are shifted for clarity. The true absorbances are all nearly the same at 1300 cm^{-1}. Note: these spectra were scanned from hard copies. The resolution of the original spectra was degraded during the process.

modulus arise due to technical difficulties (*e.g.*, limited pressure range) and trade-offs between K and K′ that are made through the equation of state. The latter problems are severe for high-pressure, deep mantle phases with large bulk moduli, such as stishovite. Complementary approaches of deducing derivatives of the elastic moduli are desirable, *e.g.*, through spectroscopy.

This paper presents data on all four IR modes of stishovite from 1 atm to 360 kbar, which is an advance over previous IR data (WILLIAMS *et al.*, 1993) that depicted only two of the modes. The new IR data plus Raman data at pressure of KINGMA *et al.* (1994), and Raman data at temperature of GILLET *et al.* (1990) are used to calculate thermodynamic properties including heat capacity and bulk modulus as functions of temperature and pressure, using the semi-empirical models of KIEFFER (1979; 1980) and HOFMEISTER (1991). Because curvature in $d\nu/dP$ is observed, higher order pressure derivatives, *e.g.* d^2K/dP^2 and $d\gamma/dP$, are also determined.

EXPERIMENTAL

The sample of synthetic stishovite is described by HOFMEISTER *et al.* (1990a). Pressure was generated by a Mao-Bell diamond anvil cell (DAC) with type II stones. A thin film of stishovite was prepared by compressing powder with an ungasketed DAC. A preindented stainless steel gasket with a 200 μm aperture was mounted on the diamond with the sample. The aperture was filled with CsI for a pressure transmitting medium, and a dozen small (ca. 5 μm) ruby grains were sprinkled on top of the CsI. Pressure was determined by using the ruby fluorescence technique (*e.g.*, MAO *et al.*, 1986) and by averaging the pressure values for eight grains which were evenly scattered across the surface of the CsI. Mid-IR spectra were collected at a resolution of 1 cm^{-1} using a Nicolet 7199 optical bench, a Nicolet 1280 data processor, a KBr beam-splitter, and a HgCdTe detector. The number of scans varied from 5,000 to 44,000, such that longer collection times were used for the higher pressure runs in order to improve the signal-to-noise ratio. A detailed description of the procedures is given by HOFMEISTER *et al.* (1989).

RESULTS

Mid-infrared spectra of synthetic stishovite taken at quasi-hydrostatic pressures from 1 atm to 360 kbar posses three strong bands, one weak band, and a few shoulders whose frequencies ν all gradually increase with pressure (Fig. 1; Table 1). This intensity pattern (Fig. 1) is consistent with symmetry analysis (TRAYLOR *et al.*, 1971) and reflectivity measurements of various compounds with the rutile structure (HOFMEISTER *et al.*, 1990a,b). The three strong peaks represent the three modes with E_u symmetry and the weak band is associated with the single A_{2u} mode. The shoulders are the longitu-

Table 1. IR frequencies of stishovite vs. pressure

Pressure Kbar	Volume cm^3/mole	$E_u(1)$ cm^{-1}	A_{2u} cm^{-1}	$E_u(2)$ cm^{-1}	$E_u(3)$ cm^{-1}
0.001[1]	14.016	836		580	470
15±1	13.948	841.7	760	593	470
45±4	13.817	850.0	758	599	489
30±2	13.881	850.0	766	596	489
56±3	13.770	854	772	609	482±4
70±3	13.709	856	775	615	479±2
97±3	13.590	864±2	781	619	497
117±3	13.518	865±2	784	634±2	498
127±4	13.480	873	785	635±5	492±4
144±5	13.414	875	788	642	496
160±8	13.353	880	792±2	636±3	505±2
175±10	13.298	881	795	642±2	498±4
200±10	13.207	885	797	646	505±5
212±10	13.162	882	800	649	?
235±10	13.085	889	800	654±3	510±5
269±15	12.968	903	805	671±2	507±5
300±15	12.867	910	816±5	680±2	513±5
335±15	12.755	915	818±5	685±3	512±5
360±15	12.677	920	830±3	695±2	515±2
270±15	12.965	901±2	802	665±4	506±4
140±10	13.429	871	790±2	637±3	494±3
8±1	13.980	840±5	767	579±2	?
24±1	13.907	848	768±3	590	?

[1] Thin film frequencies from HOFMEISTER *et al.* (1990)

dinal optic (LO) components. These shoulders appear to follow the same trends as the strong peaks. However, the data are not presented here because the shoulders could not be traced over a pressure range that was sufficiently wide to accurately establish the slope $d\nu/dP$. The position of the lowest frequency E_u peak could not be determined from some of the low pressure runs (Table 1) due to a combination of two effects: interference fringes were present in a few runs, and this peak is located near the cut-off of the detector and beamsplitter.

The peaks broaden and decrease slightly in intensity such that the area is roughly constant during the decompression runs. Because only three spectra were acquired during decompression and because samples thin during compression, quantitative analyses of the pressure dependence of peak width, intensity, or area are not presented.

The pressure dependence of the two low frequency IR modes is clearly non-linear (Fig. 2), such that the slope $d\nu/dP$ decreases as pressure increases. This behavior has been previously observed (*e.g.*, HOFMEISTER *et al.*, 1989) and represents the increasing resistance of the structure to compression. Least squares analysis shows that the data are well described by a quadratic fit (Table 2). The initial frequencies were constrained from the previous thin film and reflectivity measurements (HOFMEISTER *et al.*, 1990a) in order to obtain the most accurate value for the initial slope. Similar fits were obtained by allowing ν_o to vary and the resultant parameters lie within the uncertainty of those reported in Table 2.

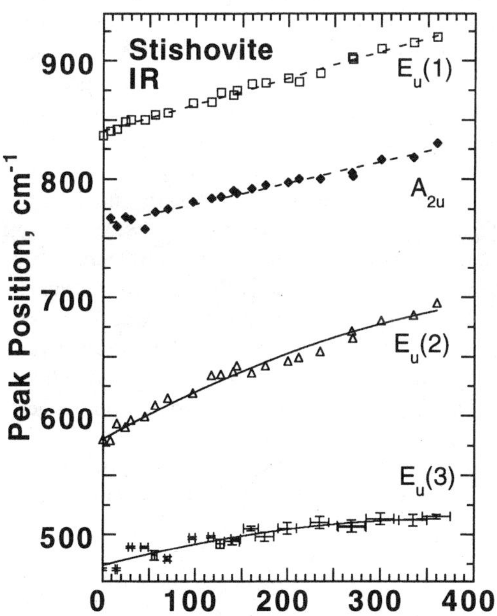

FIG. 2. IR peak positions of stishovite as a function of pressure. Dashed lines, linear fits to the data. Solid lines, quadratic fits to the data. Open square, $E_u(1)$. Filled diamond, A_{2u}. Open triangle, $E_u(2)$. Plus sign, $E_u(3)$. Error bars are shown for the lowest frequency peak. For the other modes, the uncertainty in frequency is generally less than or equal to the size of the symbol. The errors in pressure are less than the size of the symbol for P below 100 kbar, and as indicated for $E_u(3)$ at higher pressures.

Peak positions of the two high frequency peaks depend nearly linearly on pressure (Fig. 2). The $E_u(1)$ peak shows a slight, but measurable, curvature near 1 atm. Because the data at low pressures have the least experimental uncertainties, a quadratic fit and a fixed initial frequency (Table 2) are reported for the least squares analysis. The initial slope of 0.28 cm^{-1}/Kbar for the quadratic fit is slightly higher than the slope of 0.22 cm^{-1}/Kbar derived from the linear fit, which is consistent with negative curvature.

For the weak A_{2u} peak, the 1 atm position was not clearly discerned in the thin film experiment. Therefore, least squares analysis and a linear fit were used to constrain its 1 atm position. The resulting value of 761 cm^{-1} (Table 2) is higher that the TO position of roughly 670 cm^{-1} derived from reflectivity data (HOFMEISTER *et al.*, 1990a). Because the A_{2u} peak is very broad in rutile compounds (*e.g.* GeO_2: KAHAN *et al.*, 1971; ROESSLER and ALBERS, 1972), the absorption peak should occur at a considerably higher frequency than the TO position (WOOTEN, 1972). The slope derived from the linear fit represents a lower limit because this mode probably depends quadratically on pressure as did the three E_u modes.

COMPARISON WITH PREVIOUS SPECTROSCOPIC MEASUREMENTS

IR spectra of stishovite were previously collected from 0.5 to 2 μm size grains dispersed in CsI to pressures of 360 kbar with pressures determined from three to six grains of ruby (WILLIAMS *et al.*, 1993). The spectra presented of stishovite have the three E_u peaks and several shoulders (see Fig. 6 in WILLIAMS *et al.*, 1993). However, the intensity of $E_u(2)$ is high relative to the other peaks, the peaks are asymmetrically shaped, and the positions are about 30 cm^{-1} higher than ours at the same pressure. The differences are clearly due to sampling from a dispersion: (1) scattering contributes an LO component, causing the asymmetric profile and the shift of the maximum to higher wavenumbers than

Table 2. Pressure dependencies of vibrational modes in stishovite.

Mode	ν_o cm^{-1}	$d\nu/dP$ $cm^{-1}/kbar$	$d^2\nu/dP^2$ $cm^{-1}/kbar^2$	R	χ^2
IR					
$E_u(1)$ [1]	836.0[2]	0.28 ±0.01	-0.00015±0.00005	0.99	200
A_{2u}	761.4±1.5	0.176±0.008		0.98	318
$E_u(2)$	580.0[2]	0.44 ±0.02	-0.00038±0.00007	0.99	588
$E_u(3)$	470.0[2]	0.25 ±0.02	-0.00036±0.00007	0.93	443
Raman [3]					
B_{2g}	966.2[4]	0.383±0.006	-0.00009±0.00001	0.99	27
A_{1g}	754.1[4]	0.311±0.008	-0.000025±0.00002	0.99	75
E_g	589.0[4]	0.217±0.003	-0.000067±0.00006	0.99	11
B_{1g}	231.6[4]	-0.102±0.002	-0.000068±0.00005	0.99	5

[1] The linear fit (840.4±0.9)+(0.225±0.005)*P has R=0.99 and χ^2=148. [2] The initial frequencies were constrained from reflectance and thin film data of HOFMEISTER *et al.* (1990a). [3] Least squares analysis of data from KINGMA *et al.* (1994). [4] The initial frequencies were set to the 1 atm values of HEMLEY (1987) and KINGMA *et al.* (1994).

the TO position, (2) variations in crystal size cause weaker peaks to appear more intense than they should be, and (3) the grain sizes are larger than the sub-micron thickness needed to transmit light through a silicate mineral (for a detailed comparison of spectra produced from dispersions and thin films see HOFMEISTER, 1995). Furthermore, as the pressure changes, the change in intensity observed from the present study of thin films (Fig.1) means that a different proportion of LO and TO components are sampled as pressure increases. The degree to which this occurs depends on distribution of particulate sizes and the rate of change of intensity with pressure. There appears to be increased sampling of the LO component over the TO with pressure in the data of WILLIAMS *et al.* (1993) because their reported trends for E_u (1) and $E_u(2)$ are linear (trends for the other two peaks are not presented). Other reasons for curvature not being observed in the previous data set are (1) the curvature is most apparent at low pressure (Fig. 2) but WILLIAMS *et al.* data are sparse in that pressure region; and (2) determination of slight amounts of curvature requires accurate 1 atm values, whereas WILLIAMS *et al.* data is not as well constrained at 1 atm. Moreover, the pressures of WILLIAMS *et al.* study may not be as accurate as those reported here, because fewer ruby grains were used in computing the average pressure.

Raman modes of stishovite were acquired at pressures of up to 328 kbar (HEMLEY, 1987) and up to 500 kbar by KINGMA *et al.* (1994). The frequencies of the lower pressure data set appear to linearly depend on pressure, but the data set acquired over a larger pressure range shows curvature for all four modes, such that the lower frequency modes have more curvature (see Fig. 2 in KINGMA *et al.,* 1994). The values for $d^2\nu/dP^2$ in Table 2 were calculated from quadratic fits for pressures from 1 atm to 410 kbar to KINGMA *et al*'s. data (K. KINGMA, pers. commun. 1995). The data set was truncated because a phase transition occurs near 500 kbar. The frequency of the B_{1g} mode has a negative pressure shift (HEMLEY, 1987). This soft mode behavior is particular to the B_{1g} mode as its atomic motions (see TRAYLOR *et al.,* 1971) lead to a shear-type distortion of rutile to the $CaCl_2$ structure (NAGEL and O'KEEFE, 1971). Thus, the behavior of the Raman bands is entirely consistent with the response of the IR bands to pressure reported here.

MODE GRÜNEISEN PARAMETERS AND THEIR PRESSURE DERIVATIVES

Mode Grüneisen parameters were calculated from

$$\gamma_i = \frac{K_T}{\nu_i}\frac{\partial \nu_i}{\partial P} \qquad (1)$$

where K_T is the bulk modulus. We first discuss the initial values of γ_i, and then their pressure derivatives. An initial value K_{To} of 3060 Kbar established by Brillouin scattering techniques (WEIDNER *et al.,* 1982) was applied both to the IR data from this work and to previous measurements of the Raman modes (HEMLEY, 1987; KINGMA *et al.,* 1994). The initial mode gammas are calculated from eqn. (1), rather than from the volume dependence, because this approach gives an accurate initial value, minimizes manipulation of the spectroscopic data, and relies the least upon the equation of state.

The high frequency IR modes have low initial values ($\gamma_{io} \sim 1$, Table 3), which are similar to the behavior of the E_g bending mode that is active in the Raman (HEMLEY *et al.*, 1986; KINGMA *et al.*, 1994). Low values of γ_{io} for A_{2u} and E_g are consistent with their atomic motions lying along the c axis (TRAYLOR *et al.*, 1971) which is less compressible than the a axis (WEIDNER *et al.*, 1982). The A_{1g} and B_{2g} modes active in the Raman have similar stretching motions involving Si cations moving against a stationary O sublattice (TRAYLOR *et al.*, 1971) and similar intermediate values (HEMLEY *et al.*, 1986; KINGMA *et al.*, 1994). The $E_u(1)$ motions are closely related to those of the A_{1g} and B_{1g} Raman modes in that the Si and O sublattices move in opposite directions in $E_u(1)$. A value of 1 for γ_{io} is thus reasonable for $E_u(1)$. In contrast, the low frequency IR modes have high initial values (γ_{io} $\sim$2, Table 2). High values of γ_{io} for $E_u(2)$ and $E_u(3)$ are consistent with their atomic motions involving stretching along the a axis (TRAYLOR *et al.*, 1971). These motions differ significantly from that of $E_u(1)$, B_{2g} and A_{1g} in that pairs of O and Si move together in the former, whereas in the latter O and Si move in opposite directions. The B_{1g} mode in the Raman has a very different negative γ_{io} (see summary in Table 3), as expected for this soft mode.

That similar values of γ_{io} are found for modes with similar atomic motions suggests that mode Grüneisen parameters can be predicted for the inactive modes. By comparing the atomic motions of the modes as derived by TRAYLOR *et al.* (1971), we predict that the inactive $B_{1u}(1)$ and (2) modes which are derived from atomic motions along the c axis will have γ_{io} values like those of A_{2u} and E_g; respectively. A_{2g} could either have a have a γ_{io} about 0.4 larger than those of B_{2g} and A_{1g} (because γ_{io} of E_g is 0.4 larger than that of A_{2u}) or a γ_{io} that is 0.4 higher than that of E_g (because γ_{io} of $E_u(1)$ is 0.4 larger than that of A_{2u}). The value given is Table 3 is the average of these possibilities.

The pressure dependence of γ_i for each of the IR and Raman modes was calculated from the quadratic fits of Table 2 and by approximating the bulk modulus as:

$$K_T = K_{To} + K' P \qquad (2)$$

where the pressure derivative K′ of 2.8 ± 0.2 determined by ROSS *et al.* (1990) was used. ROSS *et al.* (1990) derived this value for K′ by combining their V(P) data with the ultrasonic measurement of K_{To} by WEIDNER *et al.* (1982). This combination of data sets constrains K′ value better than V(P) data alone (BASS *et al.*, 1981). The polynomial approach is used to calculate γ_i because it results in the best constraint of the initial value for $d\gamma/dP|_o$

Table 3. Mode Grüneisen parameters for stishovite

Mode	Description	ν_o cm^{-1}	γ_{io} [1]	$d\gamma_i/dP\,|_o$ [2] kbar^{-1}
IR				
$E_u(1)$	Stretch(Si vs. O) ‖a	836.0[3]	1.03±0.04	-0.0005±0.0001
A_{2u}	Stretch(Si vs. O) ‖c	761.4±1.5	0.71±0.03	~0
$E_u(2)$	Stretch ‖a	580.0[3]	2.3 ±0.1	-0.0036±0.0006
$E_u(3)$	Stretch ‖a	470.0[3]	1.6 ±0.1	-0.0040±0.0008
Raman				
B_{2g}	Stretch(Si vs. O) ‖a	966.2[4]	1.21±0.02[5]	~0
A_{1g}	Stretch(Si vs. O) ‖a	754.1[4]	1.26±0.03[5]	~0
E_g	Bend ‖c	589.0[4]	1.12±0.01[5]	-0.0001±0.0001
B_{1g}	Shear ‖a	231.6[4]	-1.34±0.03[5]	-0.0040±0.0008
Inactive				
A_{2g}	Bend ‖a	671[6]	1.6[7]	~0[7]
$B_{1u}(1)$	Bend ‖c	515[6]	1.1[7]	~0[7]
$B_{1u}(2)$	Stretch ‖c	143[6]	0.7[7]	~0[7]

[1] K_{To} of 3060 kbar was used (WEIDNER *et al.*, 1982). [2] K' of 2.8 was used (ROSS *et al.*, 1990). [3] The initial frequency is from HOFMEISTER *et al.* (1990a). [4] The initial frequency was constrained by data from KINGMA *et al.* (1994). [5] From a quadratic fit to the data of KINGMA *et al.* (1994). [6] Predicted from systematics (HOFMEISTER *et al.*, 1990b). [7] Predicted from modes with similar atomic motions: see text.

The calculated γ_i (Fig. 3) values for the IR modes clearly exhibiting curvature in $d\nu/dP$ decrease as pressure increases, as expected from the definition (eqn. 1). The decrease in $E_u(1)$ is an order of magnitude less than those for $E_u(2)$ and (3), consistent with the relative degrees of curvature. The calculation indicates that γ_i may not linearly depend on pressure; however, because the curvature in γ_i with P is slight (Fig. 3) and because eqn. (2) is a first order approximation in pressure, it is appropriate to model the slope $d\gamma_i/dP$ as a constant (Table 3). For A_{2u}, the increase in γ_i with pressure is an artifact of the linear dependence of frequency with pressure. The derivative $d\gamma_i/dP$ is therefore close to zero (Table 3): it can be no larger than that of $E_u(1)$.

The Raman modes behave similarly (Fig. 3, Table 3). The slope $d\gamma_i/dP$ is less and approaches zero for several of the modes, resulting in nearly constant mode gammas. The increase in γ_{io} for A_{1g} results because curvature of $\nu(P)$ equals 0 within experimental uncertainty. The soft mode B_{1g} has a large slope similar to those of $E_u(2)$ and $E_u(3)$.

The mode gammas of the inactive modes were assumed to be independent of pressure. This approximation is reasonable in that $d\gamma_i/dP$ is near 0 for most of the Raman and IR modes (Table 3).

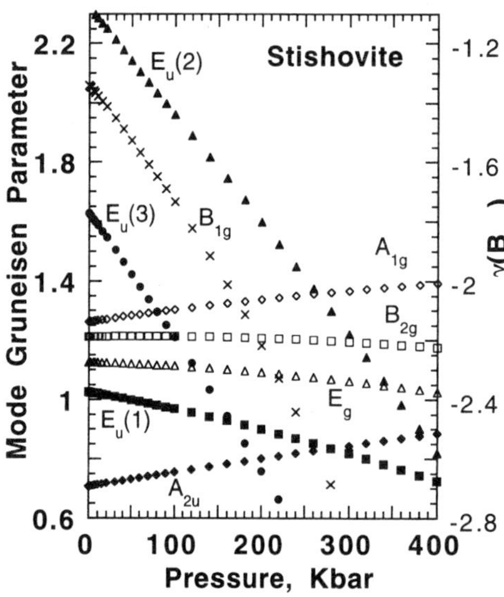

FIG. 3. Mode Grüneisen parameters of stishovite as a function of pressure. Filled triangle. $E_u(1)$. Filled square, $E_u(2)$. Filled circle, $E_u(3)$. Filled diamond, A_{2u}. Open diamond, A_{1g}. Open square, B_{2g}. Open triangle, E_g. X, the soft mode B_{1g}. The scale on the right side pertains only to B_{1g}. The increase in γ_{io} for A_{2u} is an artifact of using a linear fit for $\nu(P)$. The increase in γ_{io} for A_{1g} results because curvature of $\nu(P)$ equals 0 withing experimental uncertainty.

CALCULATION OF THERMODYNAMIC AND ELASTIC PROPERTIES

Bulk modulus and its pressure derivatives

Elastic properties can be calculated from vibrational frequencies (BROUT, 1959). The assumptions are electrostatic attractive forces, pair-wise central repulsive forces, rigid ions, and a structure which scales upon compression. The relation for compounds with the rutile structure is (HOFMEISTER, 1991):

$$K_T(P) = \frac{4P}{3} + \frac{0.0006554}{V} r_o^2 \sum_{i=1}^{18} \nu_i^2(\bar{k}) \quad (3)$$

where the numerical constant incorporates the reduced mass, coordination numbers, and physical constants and give K_T in kbar; P is pressure in kbar, V is the unit cell volume in Å^3; and r_o is the average Si-O distance in Å The sum can be taken at any point in the Brillouin zone: zone center is most convenient and thus the sum involves only the 15 optic modes. The LO and TO components of A_{2u} count as ½ mode each, whereas those of the E_u modes count as 1 mode each in order to properly account for degeneracy. In the calculation, we take advantage of the fact that r_o^3 maintains its 1 atm proportionality with V, that is the shape of the Si octahedron was not found to change with pressure (ROSS *et al.*, 1990). This observation allow elimination of r_o from eqn. (3).

The strong dependence of bulk modulus on frequency but weak dependence on volume (as $V^{1/3}$) or bond length, also allows $K_T(P)$ to be calculated independently of the equation of state through iteration. A trial function for $K_T(P)$ is obtained by substituting only $\nu_i(P)$ and one volume at one specific pressure (1 atm in this case) into eqn. (3); this result is integrated using:

$$V(P) = V_o \exp\left\{-\int_0^P \frac{dp}{K_T(p)}\right\} \quad (4)$$

to give V(P), which is then substituted into eqn. (3) with $\nu_i(P)$ to redetermine $K_T(P)$; these steps are repeated until iteration no longer changes V(P) or $K_T(P)$.

Implicit in equations (3) and (4) are that all quantities are measured at the same temperature. For determinations of dK/dP and d^2K/dP^2, room temperature is used. The spectroscopic data and estimates are given in Tables 2 and 3. The initial volume of 46.61 Å^3 of HILL *et al.* (1983) was used.

The calculated initial value K_{To} equals 3085 ± 25 Kbar, which lies midway between and within the experimental uncertainty of the acoustic determination of WIEDNER *et al.* (1982) and of the volumetric determination of ROSS *et al.*'s (1990). The present value is slightly lower than that of HOFMEISTER (1991) because the inactive modes in this work were derived from trends of all known rutile-type compounds (HOFMEISTER *et al.,* 1990b), whereas the previous estimates were made based on data for TiO_2 only. The agreement for compounds with the rutile structure is generally good (HOFMEISTER, 1991) because (1) the vibrational sums are well-constrained, and (2) the like atoms are further than second-nearest neighbors, suggesting that the assumed forms of the potentials are reasonable.

The accuracy of the pressure determination is clearly shown in the volume determination. Volume derived solely from V_o and $\nu_i(P)$ is in superb agreement with experiment at pressures below 130 kbar (Fig. 4). Above 130 kbar, the data points fall below the calculated V(P) curve, deviating 0.2% at 150 kbar. However, the calculation lies within the experimental uncertainty of all data points except the one at 150 Kbar. The small discrepancy for that data point could result if the uncertainty in the pressure determination were larger than ±1 kbar. A small systematic error could be present for this datum as it was the last measurement in one of the runs. In particular, a slight reduction in hydrostaticity is likely present as the methanol-ethanol mixture is known to become increasingly stiff above 100 kbar. The comparison suggests that the volume obtained by integrating the vibrational frequencies (eqns. 3 and 4) lies within 0.1% of the actual value. An equation for V(P) can be obtained by interating eqn. (5), below.

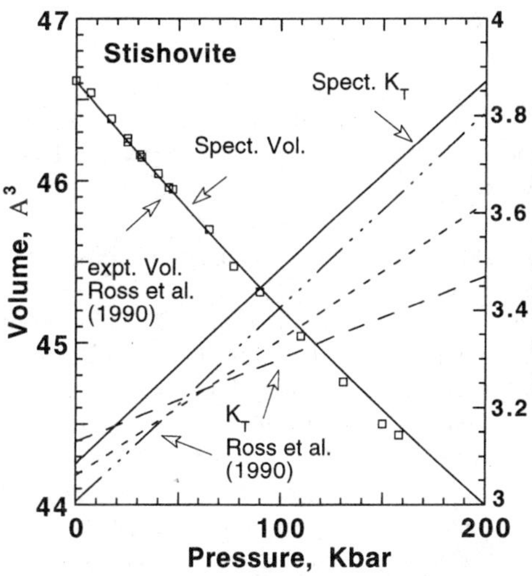

FIG. 4. Volume and bulk modulus of stishovite as a function of pressure. Solid lines, spectroscopic calculation based on V_o and $\nu_i(P)$. Open squares, data from ROSS *et al.* (1990). Short dashed line, K(P) from Birch-Murnaghan fit to the V(P) data with K_{To} set to the acoustic value. Dot-dashed line, K(P) from Birch-Murnaghan fit to the V(P) data with K′ set to 4. Long dashed line, K(P) from Birch-Murnaghan fit to the V(P) data allowing both parameters to float.

The pressure dependence of K_T which was derived solely from $\nu_i(P)$ and V_o (Fig. 4) is closely described by the quadratic equation:

$$K_T(P) = 3084.5 + (4.103 \pm 0.003)\,P - (0.000866 \pm 0.000008)\,P^2. \quad (5)$$

The low uncertainties in eqn. (5) reflect the goodness of the fit. The value of K′ calculated from spectroscopic measurements is 4.1 ± 0.3, considering the uncertainties in $d\nu_i/dP$. The spectroscopic determination for K′ is larger than the value of 2.8 that ROSS *et al.* (1990) derived from V(P) data assuming the acoustic value for K_{To}. However, V(P) data do not constrain both K and K′ (the quantities are interrelated through the chosen equation of state). As shown in Fig. 4, the V(P) data of ROSS *et al.* (1990) can be described by K′ = 4 in a Birch-Murnaghan equation if K_{To} = 3006 ± 18 kbar, which is within the uncertainty of the acoustic value. In contrast, the values of K, K′ and K″ derived from spectroscopic data are independent. It thus appears that stishovite has a K′ value similar to that observed for the majority of silicate phases. A more precise determination of K′ could be made by combining the spectroscopic calculation with one accurate determination of volume at high pressure (ca. 300 kbar) under hydrostatic conditions.

The calculated value for K″ of −0.0009 ± 0.0002/kbar is similar those measured for Mg_2AlO_4 with K″ of −0.065/kbar (CHANG and BARSCH, 1973; YONEDA, 1990); and for MgO with K″ of −0.006 ± 0.003/kbar (JACKSON and NIESLER, 1982) or −0.003 to +0.001 (YONEDA, 1990). To my knowledge, no other data on K″ exists for materials chemically or structurally similar to stishovite. Because K″ correlates roughly inversely with the bulk modulus (see discussion of alkali halides in HOFMEISTER, 1991) and because K_{To}'s are 3060 kbar for SiO_2, 1964 kbar for $MgAl_2O_4$, and 1603 kbar for MgO, the calculated K''_o of stishovite appears to be reasonable.

Temperature derivative of the bulk modulus and the Anderson-Grüneisen parameter

For determination of dK/dT, ambient pressure and the temperature variations of the volume measured by ENDO *et al.* (1986) were used in eqn. (3). Temperature variations of the Raman modes have been measured by GILLET *et al.* (1990). The peak positions depend linearly on temperature with slopes of $-0.022 \pm 0.002 cm^{-1}/K$ for A_{1g}, $-0.014 \pm 0.002 cm^{-1}/K$ for E_g, and 0 for B_{1g}. These slopes are roughly -0.068 times the numerical value of $d\nu_i/P$ (KINGMA *et al.*, 1994: see Table 3), allowing estimation of the temperature variations for the other modes from the data in Table 2.

The calculated $K_T(T)$ depends linearly on temperature (Fig. 5) with a slope of -0.185 ± 0.020 Kbar/K. The reported error stems from the uncertainty in the measured variations of the vibrational frequencies with temperature, but does not account for systematic differences between the measured Raman modes and estimated IR modes. The linear trend is due to the linear behavior of frequency with temperature, low slopes for $d\nu_i/T$, and the dominant contribution of ν_i in eqn. (3). No data on dK/dT for stishovite exist to my knowledge for comparison. Rutile has a considerably larger magnitude, -0.41 kbar/K (MANGHNANI, 1969) which is not common for solids. Our determination is similar to measured values commonly observed for dK/dT, *e.g.*, -0.27 kbar/K for MgO (ISAAK *et al.*, 1989) and -0.16 kbar/K α-Mg_2SiO_4 (SUMINO *et al.*, 1977).

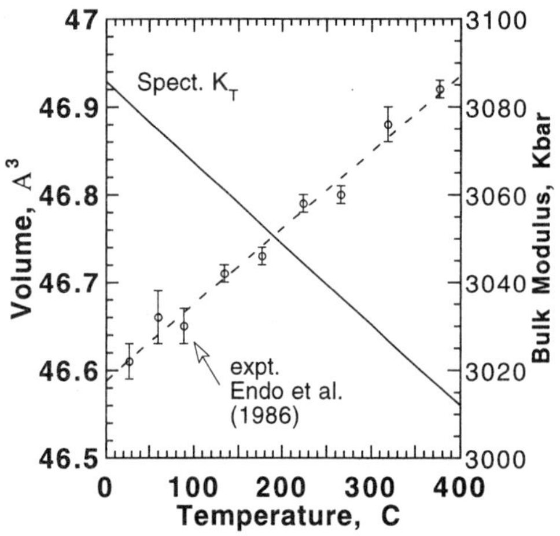

FIG. 5. Volume and bulk modulus of stishovite as a function of temperature. Open circles, volume data of ENDO *et al.* (1986). Dotted line, least squares fit to V(T). Solid line, $K_T(T)$ calculated from V_o and $\nu_i(T)$.

The Anderson-Grüneisen parameter is

$$\delta_T = \frac{-1}{\alpha K_T} \frac{\partial K_T}{\partial T}\bigg|_P \tag{6}$$

where α is thermal expansivity. The above result and V(T) data of ENDO *et al.* (1986) give δ_T equal to 3.3 ± 0.3 from 300 to 700 K. The predicted value for stishovite is significantly lower than the high temperature values near 5 for a variety of materials (ANDERSON *et al.*, 1990). One possibility is that dK/dT is low because $d\nu_i/dP$ of the IR modes are larger than that predicted from the Raman dependencies. Alternatively, stishovite may have a different value for δ_T because this material has ^{VIII}Si rather than ^{IV}Si and a bulk modulus that is 50% higher than those of the compounds previously investigated.

Construction of density of states for Kieffer's model

KIEFFER (1979; 1980) developed a model based on lattice dynamics that allows calculation of the lattice contribution to C_V and S from vibrational frequencies and acoustic velocities. A model for density of states $g(\nu)$ is constructed as follows. The primitive cell has 3N total degrees of freedom, where N is the number of atoms in a primitive cell. Three of the 3N vibrational modes are acoustic: their frequency values are constrained by the acoustic velocities of the one longitudinal and the two shear waves measured by WEIDNER *et al.* (1982). KIEFFER's (1979) model assumes that frequency ν of the acoustic modes varies sinusoidally with wavevector **k**, a phenomena termed dispersion). The contribution of the acoustic modes occurs at low temperature because the acoustic frequencies are low.

The remaining modes are optic. The density of states in KIEFFER's (1979) model is constructed by distributing the zone center vibrational modes into several continuua and Einstein oscillators. The three inactive modes were estimated (Table 3) from systematics of rutile-type structures (HOFMEISTER *et al.*, 1990b). Dispersion is ignored for the optic modes because it is unknown. For solids with a large number of optic modes this assumption contributes a minor error to the calculation because some modes disperse to higher frequencies, and others to lower. Twenty different models for $g(\nu)$ were investigated. Figure 6 shows the preferred model which separately enumerates most of the modes. Two of the inactive modes were estimated to fall within the TO and LO components of the

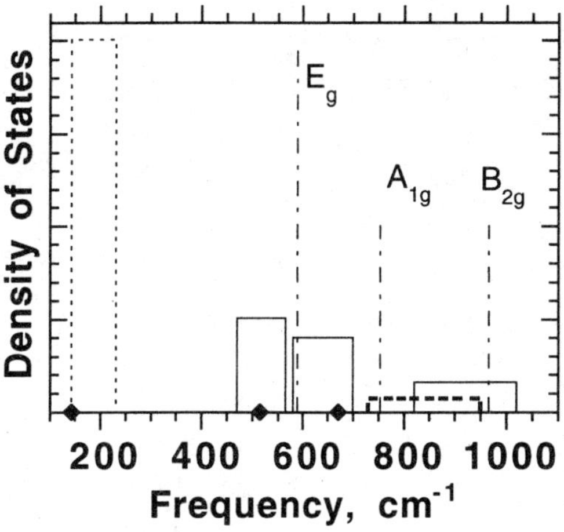

FIG. 6 Model density of states in arbitrary units. Dot-dashed lines with lables, Einstein oscillators representing three of the Raman modes. Solid lines, continuua representing the E_u modes. Thick dashed line, the continuum representing A_{2u}. Dotted lines, the continuum representing $B_{1u}(2)$ and B_{1g}. Solid triangles, positions of the three inactive modes. Relative heights of the three Einstein oscillators are correct, and relative heights of the continuua are correct, but the vertical scales differ for these two representations.

lower frequency E_u modes and thus their estimated frequencies will not affect the calculation. However, the lowest frequency inactive mode is estimated to lie below the lowest frequency Raman or IR modes, and thus its position influences the calculation, especially at low temperature.

If powder data are used, the model usually reproduces heat capacities C_P to within 5% at 298 K and 1% at 700 K. However, the accuracy considerably improves if single-crystal data and band assignments are used to construct the density of states. For example, the existence of a large number of modes over a small interval for olivines or garnets together with well-established band assignments, results in very low uncertainties (0.5%, *e.g.*, HOFMEISTER and CHOPELAS 1991). Considering (1) that number of modes (15 optical) is moderate, (2) that all but the $B_{1u}(2)$ mode are well constrained, and (3) that dispersion measured for rutile is low to moderate (TRAYLOR *et al.*, 1971), the uncertainty of the calculated values is estimated as <2% from 200 to 800 K.

Calculations can be compared to calorimetric measurements of heat capacity via the anharmonic correction:

$$C_P = C_V + TV\alpha^2K_T \qquad (7)$$

where V is volume, and α is the thermal expansivity. Single crystal volume is from SINCLAIR and RINGWOOD (1978) and HILL *et al.* (1983). K_T is from WEIDNER *et al.* (1982). Thermal expansivity of $1.863 \times 10^{-5}/K + (T - 273)^* \, 0.344 \times 10^{-10}/K^2$ was determined from a least squares fit to the data of ENDO *et al.* (1986). ENDO *et al.*'s (1986) result lies within the experimental uncertainty of the previous α by ITO *et al.* (1974ab). The temperature dependence of K_T calculated above was used. Given the uncertainties in the temperature derivatives, the entropy is also approximated from

$$S = S_V + TV\alpha^2K_T \qquad (8)$$

where S_V is the harmonic entropy calculated by the model. Because the anharmonic correction is about 3% of C_P or S, uncertainties in the parameters used contribute little (ca 0.3%). The cumulative uncertainties for the calculated S and C_P should be ~2% from 200 to 800 K.

Heat capacity, entropy, and Debye temperature

The calculated heat capacity (Fig. 7; Table 4) lies within the experimental uncertainty of the scanning calorimetric data of WATANABE (1982). Calculated C_P is equal to the cryogenic data of HOLM *et al.* (1967) within experimental uncertainties at temperatures greater than 150 K (Fig. 7). At low temperatures below 100 K, the discrepancies are greatest, which is typical of KIEFFER's model (*e.g.*, HOFMEISTER and CHOPELAS, 1991). For the case of stishovite, this discrepancy is partially due to estimation of the lowest frequency mode (Fig. 6). The

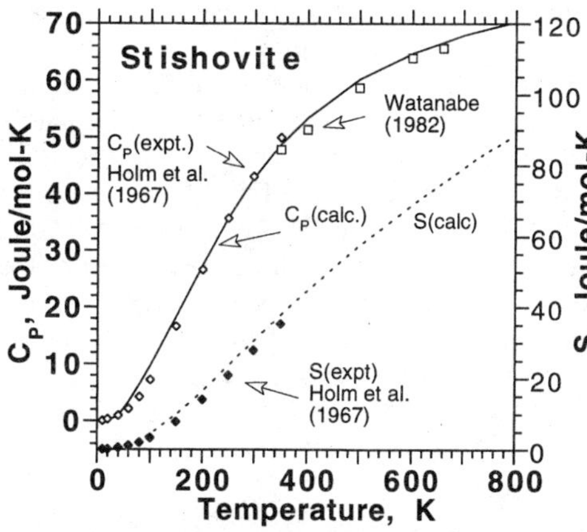

FIG. 7. Heat capacity and entropy of stishovite at 1 atm. Open diamonds, calorimetric data on C_P from HOLM *et al.* (1967). Open squares, calorimetric data on C_P from WATANABE (1982). Solid line, C_P calculated from spectroscopy. Dotted line, S calculated from spectroscopy. The axis on the right hand side refers to the entropy.

Table 4. Thermodynamic properties of stishovite at 1 atm

		C_P				S			γ_{Th}	Debye Temp.
Temp.	Anharmonic correction	Calculated	Expt.[1]	Expt.[2]	Del[3]	Calculated	Expt.[2]	Del[3]		Calculated
K	J/mol-K	J/mol-K	J/mol-K	J/mol-K	%	J/mol-K	J/mol-K	%		K
10	0.00126	0.0050		0.0511	90	0.0025	0.0163	85		1182
20	0.00352	0.0387		0.266	85	0.0136	0.112	88		1093
40	0.0766	0.945		0.928	-1.8	0.273	0.476	43		754
60	0.115	3.434		2.175	-58	1.086	1.065	-2		723
80	0.153	6.541		4.265	-53	2.495	1.954	-28		766
100	0.191	9.741		7.173	-36	4.295	3.206	-34		825
150	0.304	18.34		16.55	-11	9.849	7.814	-26		943
200	0.417	27.41		26.59	-3.1	16.38	13.96	-17		1014
250	0.542	35.74		35.66	-0.2	23.43	20.90	-12		1055
298	0.679	42.68		43.03	0.8	30.36	27.82	-9	1.89	1079
350	0.827	48.73	47.77	49.90	2.3[4]	37.72	35.28	-7	1.66	1099
400	0.974	53.4	51.33		-4.0	44.57			1.51	1108
500	1.30	60.18	58.66		-2.6	57.33			1.34	1125
600	1.68	64.72	63.90		-1.3	68.83			1.24	1134
660			65.62							1134
700	2.08	67.92				79.17			1.18	1134
800	2.55	70.32				88.55			1.14	1144
1000	3.59	73.78				111.7			1.09	1140
1500	7.08	79.82				137.5			1.02	1140
2000	12.0	85.67				163.5			0.97	1140

[1] WATANABE (1982). [2] HOLM *et al.* (1967). [3] Del. = (expt.-calc.)/expt. [4] The % difference for WATANABE'S (1982) data are +2.0.

match with experiment could be improved by using this lower cutoff as a fitting parameter.

The calculated entropy (Fig. 7) is slightly and uniformly higher than HOLM *et al.'s* (1967) data from temperatures of 100 K to their upper limit of measurement. This behavior is due to the slight mismatch of C_P near 100 K. As errors in S are additive due to the form of the integral defining S, the calculated values above 200 K in Table 4 should be reduced by 2.3 Joule/mol-K, representing the difference between experiment and calculation at 200 K.

The Debye temperature (Table 4; Fig. 8) varies slightly at temperatures above 150 K, increasing gradually form 1000 to 1140 K. Below 150 K, the calculation is not accurate because the cryogenic temperature dependences of the bulk modulus or sound velocities are not known.

Pressure dependence of heat capacity and entropy

Heat capacity and entropy were calculated as functions of pressure by utilizing the measured dependence of the optical frequencies on pressure and by applying the pressure dependence for bulk modulus to the sound velocities of WEIDNER *et al.* (1982). The latter approximation affects the low temperature values of S and C_V and becomes unimportant above 200 K. The calculated C_V values depend weakly on pressure (Fig. 9). The slope and the curvature in dC_V/dP decrease as temperature increases. Near 2000 K, heat capacity is essentially independent of pressure. Least squares fits are given in Table 5.

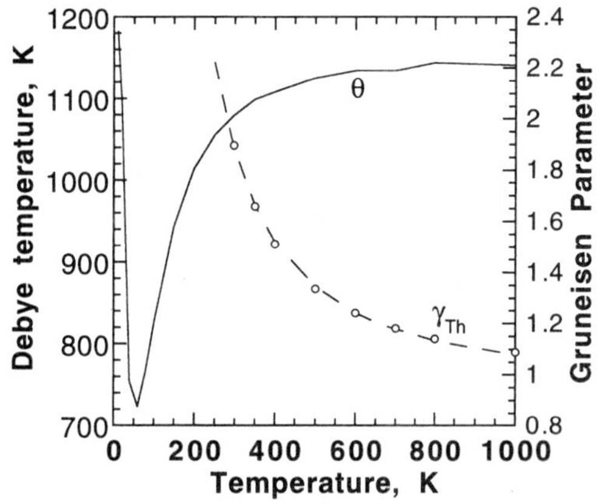

FIG. 8. Thermal Grüneisen parameter (dots) and Debye temperature (solid line) as a function of temperature. The dashed line is the least squares fit to γ_{Th}. The axis on the right hand side refers to γ only.

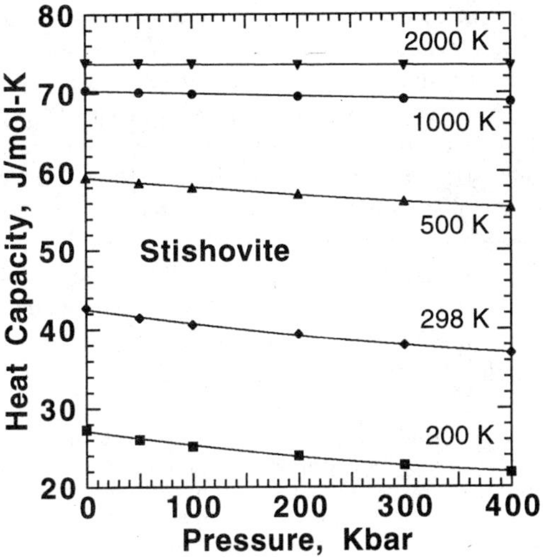

FIG. 9. C_V as a function of pressure. Solid lines are least squares fits to the calculation. Symbols are the calculated values at selected temperatures and pressures. The slope and the curvature decrease as temperature increases. By 2000 K, heat capacity is essentially independent of pressure.

Entropy (Fig. 10) behaves in the opposite manner due to its definition. Curvature is apparent at all temperatures except the lowest (Table 5).

The data in Table 3 and the equations in Table 5 allow extrapolation of S_V and C_V to mantle conditions. S and C_P can be obtained by adding the 1 atm anharmonic correction. The effect of pressure on $TV\alpha^2K_T$ is not significant and lies within the uncertainty of the calculated pressure dependence of C_V and S_V.

Grüneisen parameters and their pressure dependence

The thermal Grüneisen parameter is defined as

$$\gamma_{th} = \alpha K_T V/C_v. \quad (9)$$

Use of $\alpha = 18.5 \times 10^{-6}$/K from ENDO *et al.* (1986), V = 14.02 cm^3/mole from HILL *et al.* (1981), K_T = 3060 kbar from WEIDNER *et al.* (1982) and C_V ∓42.4 J/mol-K from HOLM *et al.* (1967) gives γ_{th} equal to 1.89 ± 0.07 at room temperature, decreasing to 1.2 ± 0.1 at 700 K (Fig. 8). The present calculations for dK/dT and C_V were used in eqn. (9) to obtain the temperature dependence. The value at 298 K reported here differs considerably from that of WATANABE (1982) because he used lower values for α, K and C_V. Least squares analysis gives the fit

$$\gamma_{Th} = 0.924 + 94.73/T + 57319/T^2, \quad (10)$$

which is valid between the 350 to 660 K range of the thermal expansivity determination. Extrapolation at temperatures below to 200 K and above to 1000 are probably less accurate than the uncertainty of ±10% for γ_{Th} at intermediate temperatures.

The initial pressure derivative $d\gamma_{Th}/dP|_o$ at 298 K is calculated to be −0.0020 ± 0.0005/kbar by using the thermodynamic identity

$$\left.\frac{\partial \alpha}{\partial P}\right|_{T=298K} = \left.\frac{1}{K_T^2}\frac{\partial K_T}{\partial T}\right|_{P=1atm} \quad (11)$$

Because these cross derivatives are not known at high temperature or pressure, the pressure dependence of γ_{Th} cannot be calculated except by assuming a linear dependence (*i.e.*, that $d\gamma_{Th}/dP$ is constant).

These above values derived from thermodynamic properties can be compared to the weighted sum of the microscopic properties

$$\gamma_{ave}(T,P) = \Sigma\sigma(\nu_i,T)\gamma_i(P)/\Sigma\sigma(\nu_i,T) \quad (12)$$

where the Einstein functions are

$$\sigma(\nu_i,T) = \left(\frac{h\nu_i}{kT}\right)^2 \frac{\exp\left(\frac{h\nu_i}{k_T}\right)}{\left[1 - \exp\left(\frac{h\nu_1}{k_T}\right)\right]^2} \quad (13)$$

Table 5. Thermodynamic properties of stishovite at pressure

Temp. K	C_V Joule/mole-K	S_V Joule/mole-K
200	27.107-0.0198 P+1.69x10^{-5} P^2	11.98-0.0033 P+2.73x10^{-6} P^2
298	42.491-0.0196 P+1.46x10^{-5} P^2	14.51-0.0173 P+1.87x10^{-6} P^2
500	59.198-0.0125 P+7.27x10^{-6} P^2	28.37-0.0249 P+2.35x10^{-5} P^2
1000	70.250-0.00349 P	54.95-0.0338 P+3.09x10^{-5} P^2
2000	73.691-0.000971 P	100.54-0.0393 P+3.37x10^{-5} P^2

P is in kbar.

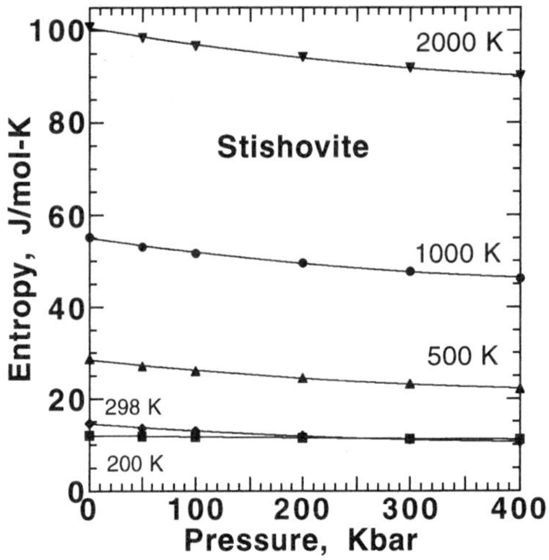

FIG. 10. Calculated entropy (S_V) as a function of pressure. Symbols are as in Fig. 9. At low temperature, S is essentially independent of pressure.

(*e.g.*, KRISHNAN *et al.*, 1979), k is Boltzman's constant and h is Planck's constant. The variation of γ_{ave} with temperature is not strong, such that above 700 K, γ_{ave} is roughly constant. Given this behavior and the need to estimate γ_i for the three inactive modes (Table 3), only the high temperature average of all optic modes of 1.15 ± 0.04 is reported here. Note that because E-type modes are doubly degenerate, these are counted twice in the average. Because estimates for mode gammas of the inactive modes are near the average values, the particular values chosen have a small effect on the computation. Averaging the slopes $d\gamma_i/dP$ gives $d\gamma_{ave}/dP = -0.0014$ with and estimated uncertainty of ±0.0005. Because most of the vibrational modes depend linearly on pressure, and because the measured curvature is small, γ_{ave} will depend linearly on pressure, within the uncertainty stated above.

The calculated values of γ_{ave} and $d\gamma_{ave}/dP$ are equal to γ_{Th} at 700 K and to $d\gamma_{Th}/dP$ at 298 K, given the stated uncertainties. At 298 K, the average is much lower than γ_{Th}. The difference is due to the strong temperature dependence of γ_{Th} which stems from the strong temperature dependence of C_V. Because the microscopic average is essentially linear with pressure, assumption of constant $d\gamma_{Th}/dP$ is reasonable, which allows calculation of γ_{Th} at mantle conditions from eqn. (10) and the average value of the two derivatives $d\gamma/dP = -0.0017$/kbar.

CONCLUSIONS

Thermodynamic and elastic properties and their temperature and pressure derivatives were calculated for stishovite from measurements of vibrational spectra at pressure. A new method of obtaining the cross-derivative $dK/dT|_P$ was presented. In all cases, agreement with independently measured physical properties was excellent (*i.e.*, equal within the stated uncertainties). The accuracy with which macroscopic properties can be determined from microscopic data suggests that the predicted 1 atm values for dK/dT, K″, dC_P/dP, dS/dP and $d\gamma_{Th}/dP$ are accurate and that formulations for all quantities presented here may be extrapolated to mantle conditions.

Acknowledgements—The sample of synthetic stishovite was graciously provided by S. Akimoto. I thank K. Kingma (Geophysical Laboratory) for providing the Raman peak positions. The IR data were collected at the Geophysical Laboratory during my postdoctoral research. I thank H. K. Mao and Thomas C. Hoering (Geophysical Laboratory) for their generous support and help. Data analysis was supported by NSF grant EAR-9496266 (original grant no. 9205991) and by a Fellowship from the David and Lucile Packard Foundation.

REFERENCES

ANDERSON O. L., CHOPELAS A. and BOEHLER R. (1990) Thermal expansivity versus pressure at constant temperature: a re-examination. *Geophys. Res. Let.* **17,** 685–688.

BASS J. D., LIEBERMANN R. C., WEIDNER D. J. and FINCH S. J. (1981) Elastic properties from acoustic and volume compression experiments. *Phys. Earth and Planet. Int.* **25,** 140–158.

BROUT R. (1959) Sum rule for lattice vibrations in ionic crystals. *Phys. Rev.* **113,** 43–44.

BURNS R. G. (1970) *Mineralogical Applications of Crystal Field Theory.* Cambridge University Press. Oxford, England.

BURNS R. G. (1985) Thermodynamic data from crystal field spectra. In *Microscopic to Macroscopic: Atomic Environments to Mineral Thermodynamics. Rev. Mineral., Vol 14.* (eds. S. W. KIEFFER and A. NAVROTSKY), p. 277–316. Mineralogical Society of America, Washington D.C.

BURNS R. G. and FYFE W. S. (1967) Crystal field theory and the geochemistry of transitions elements. In *Researches in Geochemistry, Vol. 2.* (ed. P. H. ABELSON) p. 259–285. J. Wiley and Son, New York.

CHANG Z. P. and BARSCH G. R. (1973) Pressure dependence of single-crystal elastic constants and anharmonic properties of spinel. *J. Geophys. Res.* **78,** 2418–2433.

CHAO E. C. T, FAHEY J. J., LITTLER J. and MILTON D. J. (1962) Stishovite, a very high pressure new mineral from Meteor Crater, Arizona. *J. Geophys. Res.* **67,** 419–421.

ENDO S., AKAI T., AKAHAMA Y., WAKATUSKI M., NAKAMURA T., TOMII Y., KOTO K. and ITO Y. (1986) High temperature x-ray study of single-crystal stishovite synthesized with Li_2WO_4 as flux. *Phys. Chem. Miner.* **13,** 146–151.

GILLET P., LE CLEAC'H A. and MADON M. (1990) High-temperature Raman spectroscopy of SiO_2 and GeO_2

polymorphs: anharmonicity and thermodynamic properties at high temperatures. *J. Geophys. Res.* **95,** 21635–21655.

HEMLEY R. J. (1987) Pressure dependence of Raman spectra of SiO_2 polymorphs: a-quartz, coesite, and stishovite. In *High Pressure Research in Mineral Physics* (ed. M. H. MANGHANI and Y. SYONO), p. 347–360. Terra Scientific Publishing Company, Tokyo.

HEMLEY R. J., MAO H. K. and CHAO E. C. T. (1986) Raman spectrum of natural and synthetic stishovite. *Phys. Chem. Miner.* **13,** 285–290.

HILL R. J, NEWTON M. D. and GIBBS G. V. (1983) A crystal chemical study of stishovite. *J. Solid State Chem.* **47,** 185–200.

HOFMEISTER A. M. (1991) Calculation of bulk moduli and their pressure derivatives from vibrational frequencies and mode Grüneisen parameters: solids with cubic symmetry or one-nearest-neighbor distance. *J. Geophys. Res.* **96,** 16181–16203.

HOFMEISTER A. M. (1995) Infrared microspectroscopy in the Earth Sciences. In *Practical Guide to Infrared Microspectroscopy* (ed. H. HUMICKI) p. 377–416. Marcel Dekker Inc., New York.

HOFMEISTER A. M. and CHOPELAS A. (1991) Thermodynamic properties of pyrope and grossular from vibrational spectroscopy. *Amer. Mineral.* **76,** 880–891.

HOFMEISTER A. M., XU J., MAO H. K., BELL P. M. and HOERING T. C. (1989) thermodynamics of fe-mg olivines at mantle pressures: Mid- and far-infrared spectroscopy at high pressure, *Amer. Mineral.* **14,** 281–306.

HOFMEISTER A. M., XU J. and AKIMOTO S. (1990a) Infrared spectroscopy of synthetic and natural stishovite. *Amer. Mineral.* **75,** 951–955.

HOFMEISTER A. M., HORIGAN J. and CAMPBELL J. A. (1990b) Infrared spectra of GeO_2 with the rutile structure and prediction of inactive modes for isostructural compounds. *Amer. Mineral.* **75,** 1238–1248.

HOLM J. L, KLEPPA O. J. and WESTRUM E. F. JR. (1967) Thermodynamics of polymorphic transformations in silica. Thermal properties from 5 to 1070 K and pressure-temperature stability fields for coesite and stishovite. *Geochim. Cosmochim. Acta* **31,** 2289–2307.

ISAAK D. G., ANDERSON O. L. and GOTO T. (1989) Measured elastic moduli of single-crystal MgO up to 1800 K. *Phys. Chem. Minerals* **16,** 704–713.

ITO H., KAWADA K. and AKIMOTO S. (1974a) Thermal expanstion of stishovite. *Phys. Earth and Planet. Int.* **8,** 277–281.

ITO H., KAWADA K. and AKIMOTO S. (1974b) *Erratum. Phys. Earth and Planet. Int.* **9,** 371.

JACKSON I. and NIESLER H. (1982) The elasticity of periclase to 3 GPa and some geophysical implications. In *High-Pressure Research: Applications in Geophysics* (ed. S. AKIMOTO and M. H. MANGHANI), p. 93–113. Academic Press, San Diego, CA.

KAHAN A., GOODRUM J. W., SINGH R. S. and MITRA S. S. (1971) Polarized reflectivity spectra of tetragonal GeO_2. *Jour. Appl. Phys.* **42,** 4444–4446.

KIEFFER S. W. (1979) Thermodynamics and lattice vibrations of minerals: 3. Lattice dynamics and an approximation for minerals with application to simple substances and framework silicates. *Rev. Geophys. Space Phys.* **17,** 20–34.

KIEFFER S. W. (1980) Thermodynamics and lattice vibrations of minerals: 4. Application to chain and sheet silicates and orthosilicates. *Rev. Geophys. Space Phys.* **18,** 862–886.

KINGMA K. J., COHEN R. E., HEMLEY R. J. and MAO H.-K. (1994) Novel signature of post-stishovite silica and implications for the Earth's mantle. *Nature* **374,** 243–245.

KRISHNAN R. S., SRINIVASAN R. and DEVANARAYANAN S. (1979) *Thermal Expansion of Crystals.* Permagon Press. Oxford, England.

MANGHNANI M. H. (1969) Elastic constants of single-crystal rutile under pressure to 7.5 kilobars. *J. Geophys. Res.* **74,** 4317–4328.

MAO H.-K., XU J. and BELL P. M. (1986) Calibration of the ruby pressure gauge to 800 kbar under quasi-hydrostatic conditions. *J. Geophys. Res.* **91,** 4673–4676.

NAGEL L. and O'KEEFFE M. (1971) Pressure and stress induced polymorphism of compounds with retile structure. *Mater. Res. Bull.* **6,** 1317–1320.

ROESSLER D. M. and ALBERS W. A. J. (1972) Infrared reflectance of single crystal tetragonal GeO2. *Jour. Phys. Chem. Solids* **33,** 293–296.

ROSS N. L., SHU J.-F., HAZEN R. M. and GASPARIK T. (1990) High-pressure crystal chemistry of stishovite. *Amer. Mineral.* **75,** 739–747.

SINCLAIR W. and RINGWOOD A. E. (1978) Single crystal analysis of the structure of stishovite. *Nature* **272,** 714–715.

STISHOV S. M. and POPOVA S. V. (1961) New dense polymorphic modification of silica. *Geokhimiya* **10,** 837–839.

SUMINO Y., NISHIZAWA O., GOTO T., OHNO I. and OZIMA M. (1977) Temperature variation of elastic constants of single-crystal forsterite between −190 and 400°C. *J. Phys. Earth* **25,** 377–392.

TRAYLOR J. G., SMITH H. G., NICKLOW R. M. and WILKINSON M. W. (1971) Lattice dynamics of rutile. *Phys. Rev. B* **10,** 3457–3472.

WATANABE H. (1982) Thermochemical properties of synthetic high-pressure compounds relevant to the Earth's mantle. In *High-Pressure Research: Applications in Geophysics* (ed. S. AKIMOTO and M. H. MANGHANI), p. 441–464. Academic Press, San Diego, CA.

WEIDNER D. J., BASS J. D., RINGWOOD A. E. and SINCLAIR W. (1982) The single-crystal elastic moduli of stishovite. *J. Geophys. Res.* **87,** 4740–4746.

WILLIAMS Q., HEMLEY R. J., KRUGER M. B. and JEANLOZ R. (1993) High-pressure infrared spectra of α-quartz, coesite, stishovite and silica glass. *J. Geophys. Res.* **98,** 22157–22170.

WOOTEN F. (1972) *Optical Properties of Solids.* Academic Press, Inc. San Diego.

YONEDA A. (1990) Pressure derivatives of elastic constants of single crystal MgO and $MgAl_2O_4$. *J. Phys. Earth* **38,** 19–55.

Mineral Spectroscopy: A Tribute to Roger G. Burns

Editors: M. D. Dyar, C. McCammon and M. W. Schaefer

Sound velocities of five minerals to mantle pressures determined by the sideband fluorescence method

A. CHOPELAS, H. J. REICHMANN,[1] and L. ZHANG[2]
Max Planck Institut für Chemie, Postfach 3060, 55020 Mainz, Germany

Abstract—Sound velocities determined from acoustic modes measured in the chromium sideband fluorescence of MgO to 37 GPa, $MgAl_2O_4$ to 11 GPa, Al_2O_3 to 62 GPa, YAG to 58 GPa, and pyrope to 52 GPa are presented. The pressure derivatives of the bulk and shear moduli obtained from our velocity measurements are in excellent agreement with those obtained from lower pressure ultrasonic measurements. Both shear and compressional velocities are linear with volume for all materials unless a phase change or change in compressional mechanism occurred. Such changes were observed for $MgAl_2O_4$ and pyrope at 11 and 28 GPa, respectively. The velocity-volume slope is different for all materials but is smaller for minerals containing significant proportions of alumina, particularly for the shear velocity. A linear velocity-volume relationship means that the elastic or anharmonic contribution to the geophysically important parameter $(\partial \ln \rho / \partial \ln v)_P$, where ρ is the density and v is a sound velocity, must increase with depth.

INTRODUCTION

IN 1992, A new method for measuring sound velocities from fluorescence spectra to very high pressures in the diamond anvil cell was introduced (CHOPELAS, 1992). In that paper, sound velocities of MgO were obtained to 22 GPa which were later extended to 36 GPa (CHOPELAS, 1996). This method has been successfully tested on other oxides, namely Al_2O_3 (alumina) to 63 GPa (ZHANG and CHOPELAS, 1994) and $MgAl_2O_4$ (spinel) to 11 GPa (CHOPELAS, 1996). The precision of this method approaches that of ultrasonic techniques but allows measurements to much higher pressures. Both bulk and shear moduli calculated from the sound velocities determined by the sideband fluorescence method were in excellent agreement with those obtained by ultrasonic measurements to lower pressures.

The method consists of monitoring the acoustic modes found in the vibrational sidebands of a fluorescing cation doped into the host lattice. The method requires that a fluorescing transition metal cation with 3 *d* electrons, *e.g.*, Mn^{4+}, Cr^{3+}, or V^{2+}, be doped into the lattice at an octahedral site; for all five minerals presented in this article, Cr^{3+} was used. The *d* to *d* transition that creates the very sharp fluorescence is the $^2E_g \rightarrow {}^4A_{2g}$ transition, which is spin, symmetry, and orbitally forbidden (see CHOPELAS, 1996, for a discussion). This very sharp electronic transition is vibrationally assisted by uneven or odd parity vibrations of the host lattice because it relaxes the symmetry forbidden selection rule. For Cr^{3+} cations, both T_{1u} and T_{2u} lattice phonons irrespective of their location in the Brillouin zone appear in the spectrum. These lattice phonons, which include the acoustic modes, appear as 'sidebands' on the low and high energy sides of the sharp electronic transition at an energy removed from the pure electronic or no phonon line corresponding to the energy of the phonon.

This sideband method has several advantages over the more conventional methods, such as ultrasonic techniques, Brillouin scattering, or the newer impulse stimulated scattering technique. The experimental requirements are very simple, requiring only a laser, diamond cell, and single monochromator with a photomultiplier tube. Since the orientation of the sample in the cell does not affect the results, an irregularly shaped chip in any quasi-hydrostatic pressure medium suffices. The signal is strong and does not degrade, even at high pressures, so that very small samples can be used, thus making it easily possible to measure velocities to lower mantle pressures. The measured acoustic modes are spherically averaged which provides a means to directly determine spherically averaged velocities (discussed later). High symmetry minerals are easiest to analyze. The velocity measurements by the fluorescence method are made typically with uncertainties of 0.3 to 0.5%, compared to the slightly smaller uncertainties of 0.1% for ultrasonic techniques. There are also some disadvantages: the inability of this method to determine individual elastic constants, the interpretation of the sidebands may be ambiguous especially in lower symmetry compounds, temperatures are lim-

[1]*Current address:* Bayerisches Geoinstitut, Universität Bayreuth, D-95440 Bayreuth, Germany
[2]*Current address:* Institut für Mineralogie, Universität Marburg, Lahnberge, D-35043 Marburg, Germany

ited to a few (three to four) hundred degrees above ambient, and the introduction of small quantities of iron (<1%) into a mineral suppresses the fluorescence.

In this article, we present the pressure dependence of the sound velocities of five of the eight minerals measured by the sideband method in our laboratory. We show that the sound velocities for these five minerals are linear versus volume to compressions as low as 0.8, as long as the mineral does not undergo any phase changes, even for the more complex garnet structure, which does not compress isotropically. We also show that the velocity-volume dependence is much smaller for those minerals containing alumina, particularly for the shear velocities and that the linear velocity-volume relationship requires that $(\partial \ln \rho/\partial \ln v)_T$ increase with increasing pressure. The geophysical implications of this result are discussed.

EXPERIMENTAL DETAILS

Fundamentally, the experimental techniques for all five minerals were similar, the mineral was confined in a gasketed diamond anvil cell with ruby for pressure calibration and a quasi-hydrostatic pressure medium, either a methanol:ethanol mixture (to 20 GPa) or argon. All minerals except the $MgAl_2O_4$ sample were synthesized with small quantities of Cr^{3+}; typically about 0.1%. The composition of the natural $MgAl_2O_4$ sample has been previously reported (CHOPELAS and HOFMEISTER, 1991). In the experiments over 20 GPa with MgO and Al_2O_3, the samples were heated with a CO_2 laser until the surrounding argon was molten, thus providing a truly hydrostatic environment for these samples to the highest pressures of these respective studies. The experiments on spinel and garnet could not be conducted this way because these materials undergo phase changes above 20 GPa. Pressures were measured before and after recording the sample spectra; maximum differences observed did not exceed 1% of the confining pressure. The fluorescence spectra of all minerals were excited with a few mW of either the 488.0- or 514.5-nm line of an argon laser, which was kept to the minimum power needed in order to avoid heating the sample due to light absorption. No peak broadening or shifting due to heating was observed. The laser light entered the diamond cell at an incident angle of approximately 25° and was focused to approximately 10-μm diameter in the sample. The signal was collected in a backscattering geometry with an f/2 lens and focused into the analyzing monochromator, either a double or single (used in the YAG experiments), which were equipped with photon counting detection systems. Typical spectral resolution was 1 cm^{-1}. All spectra were recorded in a climate controlled environment, with temperatures typically 293 ± 1 K, in order to prevent drift in monochromator calibration.

SPECTROSCOPY

In conventional vibrational spectroscopy, *i.e.*, Raman and infrared spectroscopy, the vibrational modes that appear in the spectrum must be very near the Brillouin zone center, where the acoustic modes are zero, because the process that allows the mode to appear in the spectrum depends on the direct interaction of light with the vibrational mode, requiring the wave vector of the phonon to match that of the light. In fluorescence spectroscopy, it is the interaction of light with the electronic transition and not the vibration that creates the spectrum, thus eliminating the Brillouin zone restriction. The selection rules for determining the presence of a vibrational mode in a sideband spectrum requires that the vibration not have inversion symmetry, *i.e.*, that it is of odd parity. For Cr^{3+}, both T_{1u} and T_{2u} vibrations, including the acoustic modes, appear in the sidebands irrespective of their location in the Brillouin zone. If the modes in the sidebands were restricted to one point in the Brillouin zone, sharp peaks as found in Raman spectra would be found instead of the broad and sometimes diffuse peaks found in the sidebands (illustrated for all five minerals in Fig. 1). The Cr^{3+} fluorescence spectrum in a host crystal at a six-fold coordinated site consists of a sharp central 'no-phonon' line onto which the vibrational modes are symmetrically superimposed (see Fig. 1). The intensity of the sidebands on the high energy or anti-Stokes side of the spectrum is dependent on temperature similar to the anti-Stokes Raman spectra; the number of modes excited at a given frequency is given by the Boltzmann distribution function. The vibrational energy of the phonon equals its energy separation from the central peak. There is usually more than one sharp peak near the central peak, which is either due to the lifting of the degeneracy of the d-orbitals because the six-fold site is not perfectly cubic (R_2 line) or the local lowering of site symmetry caused by the substitution of a trivalent cation impurity for a divalent host crystal cation (N-lines) requiring some nearby cation vacancies for charge compensation. The wavelengths of the electronic transition of interest in Cr^{3+} range from about 685 to 700 nm.

The sidebands are representative of the phonons of the unperturbed lattice because the frequencies of the sidebands are not measurably affected by the presence of the impurity by the comparison of the sideband frequencies in $MgO:Mn^{4+}$, Cr^{3+}, or V^{2+} (GLASS and SEARLE, 1967; CHOPELAS, 1996) to those measured by inelastic neutron scattering (SANGSTER *et al.*, 1970). Furthermore, comparison of the pressure dependence of the Raman and infrared modes to the pressure dependence of the sideband modes in $MgAl_2O_4$ revealed identical behavior for both types of phonon data (CHOPELAS and

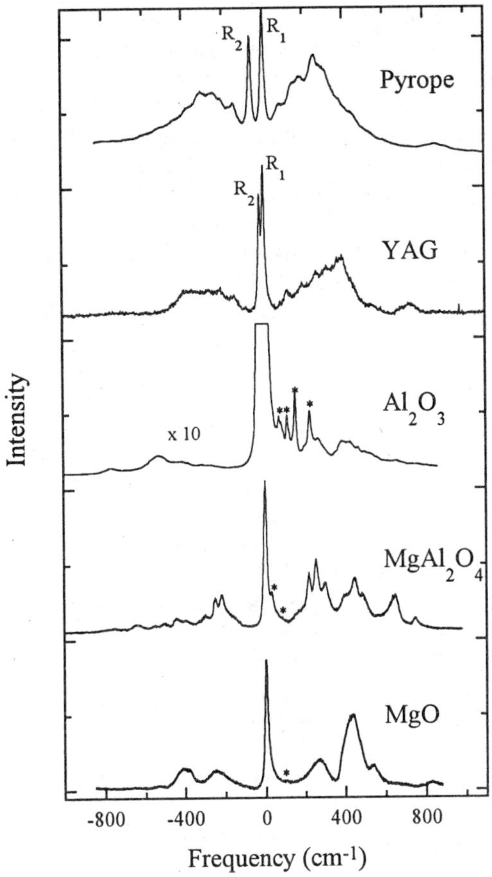

FIG. 1. The 1-atm fluorescence spectra of Cr^{3+} doped in the five minerals of this study. The R_1 lines are set to zero to facilitate the comparison of the phonons of the various materials. The central peaks for these minerals are as follows: MgO—699.2 nm; $MgAl_2O_4$—685.5 nm; Al_2O_3—694.3 nm; YAG—688.8 nm; and pyrope—698.9 nm. The R_2 lines for YAG and pyrope are well resolved and indicated in the figure. The sharp lines on the Stokes side of the central line in the $MgAl_2O_4$ and Al_2O_3 spectra are N-lines and are indicated by asterisks. The Al_2O_3 vibrational sidebands are significantly weaker relative to the R lines at one atm so the spectrum is magnified by a factor of 10 to enhance the sidebands.

HOFMEISTER, 1991). Substitution of different impurity species affect the symmetries of the vibrations appearing in the sideband and hence intensities of the modes may vary. For example, the substitution of V^{2+} in MgO changes the appearance of the sidebands dramatically because the T_{1u} symmetry modes are strongly preferred over the T_{2u} modes (CHOPELAS and NICOL, 1982) but the positions of the modes are not shifted from those found in Cr^{3+} doped MgO (CHOPELAS, 1996).

In order to obtain the sound velocities from the sideband spectra, the transverse and longitudinal acoustic modes need to be distinguished from the optical modes. Fortunately, the frequencies of the sideband modes closely correspond to those found in inelastic neutron scattering studies for MgO (SANGSTER *et al.*, 1970), Al_2O_3 (KAPPUS, 1975), and $MgAl_2O_4$ (THOMPSON and GRIMES, 1978) making assignment of the acoustic modes in the sidebands relatively straightforward. For the garnets, however, the large number of atoms in the unit cell means there are a large number of vibrations to be accounted for, complicating the acoustic mode assignment. Furthermore, two-phonon features may also appear in the sidebands although this doesn't seem to be the case for the other three compounds. There are other factors that can aid in distinguishing the acoustic modes from the optic aside from the fact that the acoustic modes are expected to be among the lowest in frequency in the spectrum. (1.) High pressure behavior can be used. In MgO, Al_2O_3 and $MgAl_2O_4$, the acoustic modes have much larger magnitude second pressure derivatives than the optic modes (CHOPELAS and HOFMEISTER, 1991; CHOPELAS, 1992; ZHANG and CHOPELAS, 1994; CHOPELAS, 1996). (2.) Process of elimination can be used. For example, in $MgAl_2O_4$, nearly all the optic modes in the sidebands could be accounted for by the infrared and Raman active modes, leaving the selection of the acoustic modes straightforward. (3.) Finally, the acoustic mode results can be verified with the low pressure sound velocity data from ultrasonic studies. In the MgO, Al_2O_3, and $MgAl_2O_4$ studies, using any of the optic modes for determining the velocities at high pressure did not yield results that even approached those found in ultrasonic studies.

SOUND VELOCITIES

The sound velocities of a mineral are related to the slope of the dispersion curves near the Brillouin zone center, $d\omega/dk$ where ω is the mode frequency and k is the wavevector ($= 2\pi/a$, where a is the lattice constant). However, the measured frequencies, or the peak in the fluorescence spectrum, are near values that one expects at the Brillouin zone edge. It is again emphasized that the appearance of a phonon in the sideband has little to do with Brillouin zone location but is only governed by its symmetry. Indeed, in examining the sideband spectrum of MgO (Fig. 1), we see that the transverse acoustic mode at 270 cm^{-1} is very broad with intensity down to 0 cm^{-1}, which is the frequency of the acoustic mode at the zone center. Therefore, this spectral feature obtains intensity from phonons throughout Brillouin zone, *i.e.*, from the center to the edge.

In order to extract sound velocity information from the frequencies in the spectra, we need to determine how this frequency relates to velocity. The simplest method to illustrate this is to use a simple diatomic 1-dimensional ''lattice''. The phonon frequencies of this ''lattice'' can be easily obtained in terms of the elastic constant C between atoms and the masses of the heavy and light atoms, m_1 and m_2, respectively, by solving the equation of motion. For small values of k (*e.g.*, Chapter 4 in KITTEL, 1976), the acoustic frequencies are

$$\omega = \sqrt{\frac{2C}{m_l + m_2}} \times ka \quad (1)$$

and at maximum k, the acoustic frequencies are

$$\omega = \sqrt{\frac{2C}{m_l}} \quad (2)$$

To determine the relationship between the slope of the acoustic mode dispersion curve at the zone center to an observed frequency at the zone edge, a linear extrapolation of the slope to the zone edge yields

$$\omega_D = \sqrt{\frac{2C}{m_l + m_2}} \times \pi \quad (3)$$

where the subscript D refers to the Debye frequency, which is the frequency at the zone edge with a linear dispersion curve. The ratio of the slope to the zone edge frequency is

$$\frac{slope}{\omega} = \sqrt{\frac{m_l}{m_l + m_2}} \times \frac{\pi}{2} \quad (4)$$

The ratio in Eqn. 4 above is an expression of constants and suggests that the velocity must be proportional to the measured frequency as expected for a continuous elastic medium.

$$v \propto \omega \times a \quad (5)$$

Aside from the measured frequencies ω, this method requires the lattice constant versus pressure and the zero pressure sound velocities, which have both been measured for a great number of materials. Equation 5 implies that it is immaterial what the exact function is for the acoustic dispersion curve except that it must not change general form during compression. Because all of the materials under consideration are oxides, it is not expected that the acoustic dispersion curves will be irregularly shaped. In fact, the available inelastic neutron scattering data for MgO (SANGSTER *et al.*, 1970), Al_2O_3 (KAPPUS, 1975), and $MgAl_2O_4$ (THOMPSON and GRIMES, 1978) show that for all three compounds the acoustic mode dispersion curves are nearly sinusoidal.

This simple one-dimensional model might suggest that the acoustic mode frequencies that we measured are directional and not spherically averaged. However, it has been recently shown for MgO and $MgAl_2O_4$ that the acoustic modes in these materials do, in fact, yield spherically average velocities by comparison of the directional velocity information provided by ultrasonic data to those from our experiments (see Fig. 2ab) (CHOPELAS, 1996). The sideband frequencies are compared to frequencies calculated from directional velocities using $\omega = \omega_0(v/v_0)(a_0/a)$, where the variables with the subscript zero represent values at one atm. The spherically averaged velocities from the ultrasonic studies compare very well with the sideband frequency measurements. This same test for the two garnets and Al_2O_3 yields similar results as for MgO and $MgAl_2O_4$, although not quite as decisively, because (1) the ultrasonic data for YAG (YOGURTCU *et al.*, 1980) and Al_2O_3 (GIESKE and BARSCH, 1968) are only to very low pressures, thus extrapolation to over 50 GPa for comparison have very large uncertainties, and (2) the ultrasonic (WEBB, 1989) and impulse stimulated scattering (CHAI *et al.*, 1993) data for pyrope are for substantially different compositions.

VELOCITY RESULTS

The primary plots of sideband spectra at various pressures and frequency versus pressure for MgO, $MgAl_2O_4$, and Al_2O_3 have been published (CHOPELAS, 1992; ZHANG and CHOPELAS, 1994; CHOPELAS, 1996) so only the highlights and results are shown here for those compounds. The velocity-pressure and velocity volume relationships for all materials are listed in Table 1 and the comparison of the sound velocity of all materials versus pressure to PREM values are illustrated in Fig. 3.

MgO and Al_2O_3

For MgO and Al_2O_3, no phase transitions or unusual phenomena occurred over the pressure ranges of these studies nor were they expected to occur (ZHANG and CHOPELAS, 1994; CHOPELAS, 1996). Above 20 GPa, the samples were heated until the surrounding argon melted providing a hydrostatic environment and eliminating the pressure uncertainties due to pressure gradients. The frequencies and thus the sound velocities calculated using Eqn.

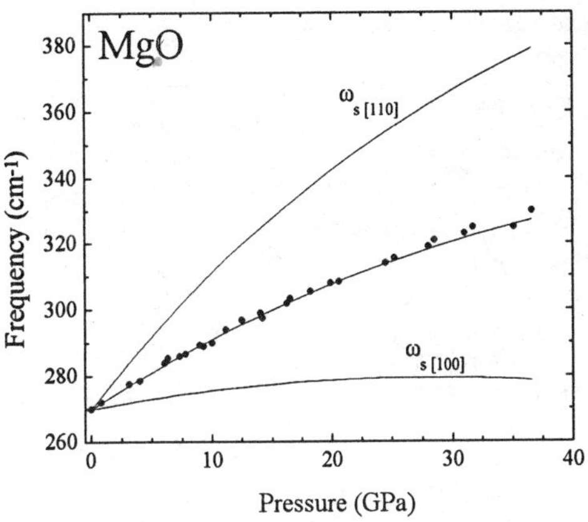

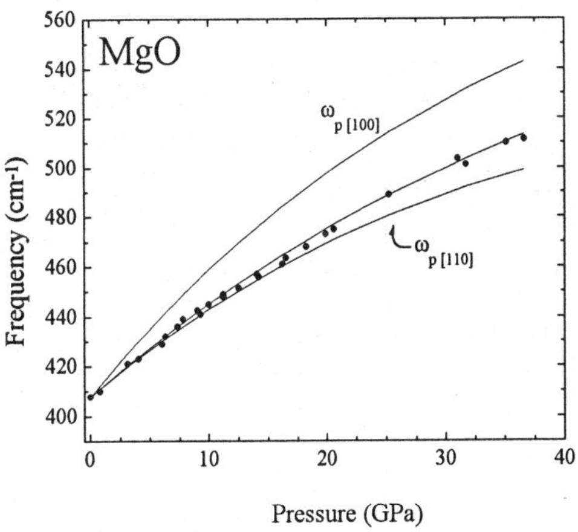

FIG. 2a. *Upper.* Comparison of the predicted effect that the directional shear velocities of MgO at high pressures would have on the transverse acoustic mode frequency to the actual frequencies measured at high pressure. The shear velocity in the 100 direction is calculated from $(C_{44}/\rho)^{1/2}$ and in the 110 direction is calculated from $\{(C_{11}-C_{22})/2\rho\}^{1/2}$. The "directional" frequencies ω at high pressure are then calculated from $\omega_0 \times (v/v_0) \times (a_0/a)$, where v are the respective directional velocities, a the lattice constant and the subscript zero represents the values at ambient pressure. The elastic constants used for these calculations are from ultrasonic studies (JACKSON and NIESLER, 1982; YONEDA, 1990). The intermediate curve that describes the data well is calculated using the averaged elastic moduli from those ultrasonic studies but with the second derivatives adjusted according to ultrahigh pressure compressional results (DUFFY and AHRENS, 1995).

FIG. 2b. *Lower.* Comparison of the predicted effect that the directional shear velocities of MgO at high pressures would have on the longitudinal acoustic mode frequency to the actual frequencies measured at high pressure. The compressional velocity in the 100 direction is calculated from $(C_{11}/\rho)^{1/2}$ and in the 110 direction from $\{(C_{11} + C_{12} + 2C_{44})/2\rho\}^{1/2}$. The remaining information from Fig. 2a applies to this figure.

5 increased monotonically with pressure (see Fig. 3) and followed the trends expected by extrapolating the low pressure ultrasonic data. Plots of the velocity versus volume for both materials yielded straight lines (Fig. 4ab). For Al_2O_3, the large number of optic modes surrounding the longitudinal acoustic mode rendered it impossible to measure this mode to high pressures with any confidence (ZHANG and CHOPELAS, 1994). We were only able to measure two of the transverse acoustic modes, which exhibited identical behavior, and obtained a good constraint on the shear velocity to over 60 GPa. The compressional velocity was calculated using the well-constrained bulk modulus and its pressure derivative, the density, and our shear velocity from

$$\nu_p = \left[\frac{Ks}{\rho} + \frac{4}{3} \times \nu_s^2\right]^{1/2} \quad (6)$$

This method yielded reasonable results (Fig. 3 and 4ab).

$MgAl_2O_4$

For $MgAl_2O_4$, data were taken to 21 GPa (CHOPELAS and HOFMEISTER, 1991) but a change in compressional mechanism was observed as a break in slope for some of the optical modes in the Raman, infrared, and sideband spectra and as decreases in frequencies of the acoustic modes. Significant changes in the sideband fluorescence spectra were not observed until 19 GPa, where the signal became significantly weaker and the shape of the sidebands changed dramatically. Therefore, we report the sound velocities of $MgAl_2O_4$ to only 11 GPa (CHOPELAS, 1996). It is noted here that the shear velocity does not change with pressure within the uncertainties of the data even though the compressional velocity behaves similarly to that of the other compounds.

YAG

Typical high pressure spectra of YAG {yttrium aluminum garnet: $Y_3Al_2(AlO_4)_3$} are shown in Fig. 5. It is clear that the signal quality or resolution does not degrade with pressure. It is also noted here that this sample was not heated at high pressures to relax the argon matrix but the resolution does not degrade substantially or suddenly with pressure. In these experiments, pressures were taken from ruby chips placed very close to the sample in order to minimize the uncertainties due to pressure gradients. The acoustic modes are indicated in Fig. 5 at

Table 1. Pressure and volume dependence of the sound velocities of MgO and $MgAl_2O_4$

Pressure Dependence in km/s	Volume Dependence in km/s
$v_s(P) = v_s(0) + (\partial v_s/\partial P)_T \cdot P + (\partial^2 v_s/\partial P^2) \cdot P^2$	$v_s = v_{s0} + \{\partial v_s/\partial(V/V_0)\} \cdot (1-V/V_0)$
MgO	
v_s 6.05(1) + 0.0381(13)·P - 3.6(4)·10^{-4} · P^2	6.05(1) + 6.37(10)·(1-V/V_0)
v_p 9.70(2) + 0.0704(20)·P - 5.6(6)·10^{-4} · P^2	9.69(1) + 12.5(2)·(1-V/V_0)
$MgAl_2O_4$	
v_s 5.49(5) + 0.001(1) · P	5.49(1) + 0.1(3)·(1-V/V_0)
v_p 9.785(11) + 0.047(5) · P - 0.0010(5) · P^2	9.79(1) + 8.0(4)·(1-V/V_0)
Al_2O_3	
v_s 6.41(1) + 0.0234(11)·P - 1.7(2)·10^{-4} ·P^2	6.40(3) + 5.70(13)·(1-V/V_0)
v_p 10.90(11) + 0.0532·P - 1.9·10^{-4}·P^2	10.87 + 17.2·(1-V/V_0)
YAG	
v_s 4.94(2) + 0.00484(46)·P	4.97(2) + 1.47(17)·(1-V/V_0)
v_p 8.52(3) + 0.0520(26)·P - 3.0(5)·10^{-4} · P^2	8.52(1) + 11.3(2)·(1-V/V_0)
Pyrope	
v_s 5.06(4) + 0.0271(22)·P	5.07(1) + 6.5(5)·(1-V/V_0)
v_p 9.16(4) + 0.0572(14)·P	9.06(1) + 12.2(6)†·(1-V/V_0)

†this value for pyrope is to 28 GPa ; above here the slope changes to 20.1(11) (see fig. 8b)
Note: densities ρ_0 in g cm^{-3} and bulk moduli K in GPa of the minerals above are as follows- MgO: ρ_0=3.583, K=160.5 (JACKSON and NIESLER, 1982); Al_2O_3: ρ_0=3.982, K=253.0 (GIESKE and BARSCH, 1968); $MgAl_2O_4$: ρ_0=3.578, K=195.7 (CHANG and BARSCH, 1973); YAG: ρ_0=4.550, K=180.3 (YOGURTCU *et al.*, 1980); Pyrope: ρ_0=3.601, K=172.8 (O'NEILL *et al.*, 1991); $MgSiO_3$ perovskite: ρ_0=4.096; K=264.0(WEIDNER *et al.*, 1993)

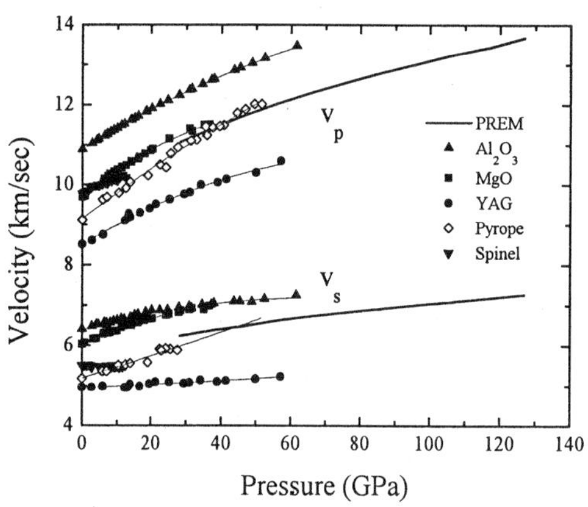

FIG. 3. Plot of compressional and shear velocities of five minerals versus pressure at room temperature. The data are indicated by the symbols and the fitted functions are listed in Table 1. The lower mantle PREM (DZIEWONSKI and ANDERSON, 1981) compressional and shear velocities at mantle pressure and temperatures are shown for comparison. Velocities for perovskite should lie slightly above those for Al_2O_3 at v_{s0} = 6.58 and v_{p0} = 11.1 km/s (WEIDNER *et al.*, 1993).

194 and 319 cm^{-1}. Justifications for these identifications are: (1) These are the only two modes that did not correlate to either an infrared or Raman mode (HURRELL *et al.*, 1968; HOFMEISTER and CAMPBELL, 1992), (2) the LA mode at 319 cm^{-1} exhibited the largest magnitude second derivative, and (3) these frequencies yielded high pressure velocities consistent with the available ultrasonic data (YOGURTCU *et al.*, 1980). Moreover, these frequencies for YAG are very near those found for $MgAl_2O_4$ and are therefore not unreasonable. The pressure dependencies are shown in Fig. 6a and their velocities calculated from Eqn. 5 are shown in Fig. 6b, with the uncertainties indicated in both of these figures. As in $MgAl_2O_4$, the shear velocity of YAG changes very little with pressure. The velocities of YAG are proportional to volume (Fig. 4ab) to the highest pressure of this study.

Pyrope

The velocity results for pyrope ($Mg_3Al_2Si_3O_{12}$) are preliminary but are worth mentioning here because they are of significant geophysical interest.

The sideband spectrum of pyrope (Fig. 1) strongly resembles that of YAG and the frequencies of most of the modes correspond well to the positions of the available infrared and Raman data (HOFMEISTER and CHOPELAS, 1991). As in the case of spinel, there appears to be a change in compressional mechanism or phase change at about 28 GPa in pyrope. First, the slope of the optical mode sideband at 852 cm^{-1} decreases by about 45% at this pressure. Second, the resolution of the sideband spectra decreased suddenly above this pressure (Fig. 7). Since no Raman measurements exist to 28 GPa, we attempted to measure the Raman spectrum of this sample above this pressure but the inherently weak spectrum allowed us to monitor only four modes with any confidence. No discernible changes could be detected to 35 GPa. A similar decrease in resolution of sideband fluorescence spectra was observed for $MgSiO_3$ majorite (CHOPELAS, 1991) at 38 GPa where the Raman spectrum exhibited dramatic changes, with major new peaks appearing which are characteristic of cubic garnets and peaks characteristic of the tetragonal structure disappeared.

A plot of velocity versus pressure from our results (Fig. 8a) shows that there is excellent agreement with impulse stimulated scattering data on a different composition pyropic garnet (with 16% iron in the Mg^{2+} site) to 22 GPa (CHAI *et al.*, 1993) and, from this, it appears that these compressional velocity data can be simply extrapolated to higher pressures. However, a plot of velocity-versus volume reveals a break in slope for the velocity volume function at approximately 28 GPa (Fig. 8b), the same pressure the resolution of the spectra degraded. The transverse acoustic mode could not be followed with any confidence above this pressure due to this loss of resolution. The data scatter for pyrope is significantly higher than for the other materials so pyrope warrants another set of experiments before these results can be called final.

Pressure derivatives of the elastic moduli

The first and second pressure derivatives of the bulk and shear moduli can be estimated from the velocity results here from

$$Ks = \rho(v_p^2 - 4/3 \times v_s^2) \tag{7}$$

for the adiabatic bulk modulus and

$$G = \rho v_s^2 \tag{8}$$

for the shear modulus. The results for the elastic moduli versus pressure are fitted to

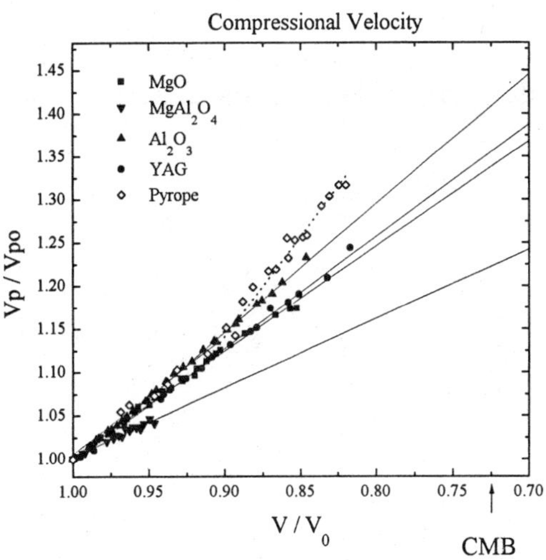

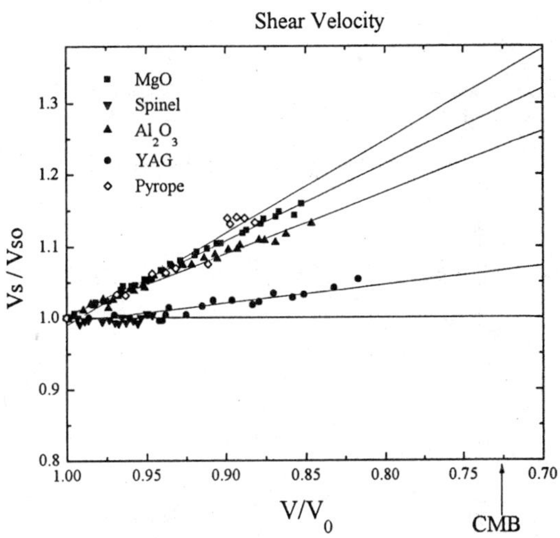

FIG. 4a. *Left.* Plot of reduced compressional velocity versus compression. The compression at the core mantle boundary (CMB) is indicated in the figure. The data are shown as symbols and the fitted lines are listed in Table 1 as velocity. The results for pyrope (open diamond) show a increase in slope at about $V/V_0 = 0.9$ corresponding to 28 GPa (see Fig. 8b). Reduced volumes were calculated with a third-order finite strain equation of state (BIRCH, 1978) and bulk moduli listed in the footnote of Table 1 and pressure derivatives listed in Table 2.

FIG. 4b. *Right.* Plot of reduced shear velocity versus compression. The remaining information pertaining to Fig. 4a applies to this figure.

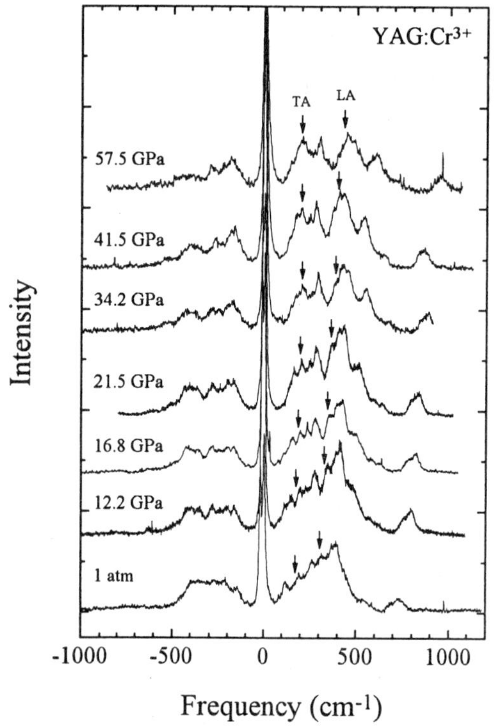

FIG. 5. Fluorescence spectrum of chromium-doped YAG at various pressures. The R_1 line is set to 0 cm^{-1} in order to emphasize the pressure effect on the sidebands. The R_1 line at 688.8 nm shifts at a rate of 0.0385 nm/GPa. The acoustic modes are indicated at each pressure. Note that the quality, intensity and resolution of the sidebands does not deteriorate even to the highest pressures.

$$M(P) = M_0 + M_{0'} \cdot P + \tfrac{1}{2} \times M_0'' \times P^2 \quad (9)$$

where M is the respective modulus. Since the density is required in the calculation of the moduli, the density is first calculated from a third order finite strain equation of state (BIRCH, 1978) with the zero pressure values of the bulk modulus obtained in ultrasonic studies or volume compressional studies (see Table 1). Then if the pressure derivative of the bulk modulus obtained here differed from that used to calculate the density, the volumes were recalculated with the new pressure derivative. This process was continued iteratively until the two results converged. Table 2 shows the excellent agreement between the low pressure ultrasonic results and our results for all five minerals.

DISCUSSION

Comparison of the sound velocities of candidate minerals at high pressures and temperatures is the most definitive way to constrain compositional models of the mantle. Using density alone does not distinguish between the two most discussed lower mantle compositional models, a perovskitic model composed of almost entirely of $Mg_{0.88}Fe_{0.12}SiO_3$ with minor percentages of other phases or a pyrolitic model composed primarily of magnesiowüstite and $MgSiO_3$ perovskite where the iron is preferentially partitioned into the magnesiowüstite phase. Extrapolation of the densities of each of these three phases, the two different perovskite compositions and magnesio-

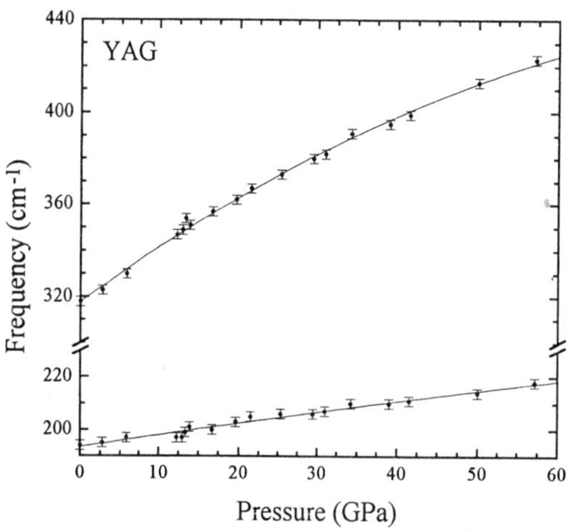

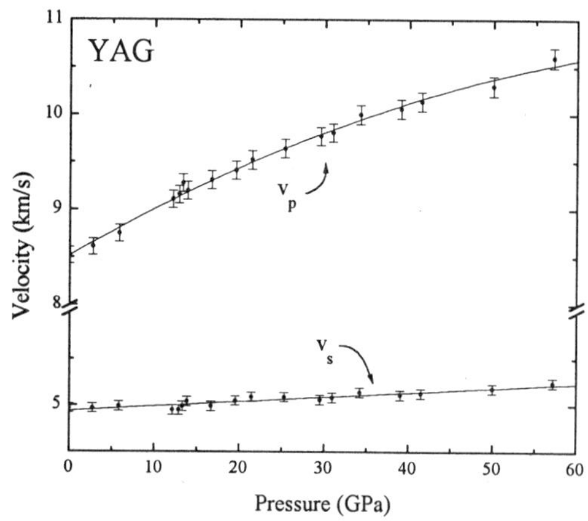

FIG. 6a. *Left.* The transverse and longitudinal acoustic modes of YAG at 190 and 319 cm^{-1}, respectively, versus pressure. The uncertainties indicated by the error bars are about 0.5%.

FIG. 6b. *Right.* The shear and compressional velocities of YAG calculated using Eqn. 5 and the frequencies shown in Fig. 6a. The lattice constant at each pressure was calculated with the third-order finite strain equation of state (BIRCH, 1978) and a bulk modulus of 180.3 GPa and its pressure derivative of 5.2.

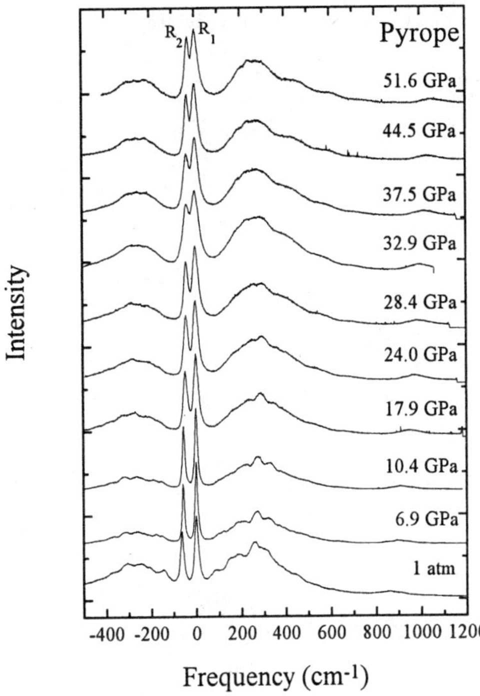

FIG. 7. Fluorescence spectra of pyrope ($Mg_3Al_2Si_3O_{12}$) doped with chromium at various pressures. The R_1 peak is plotted as 0 cm^{-1} in order to emphasize pressure induced changes in the sidebands. Note that above 28 GPa, the R_1 and R_2 lines and the features in the sidebands broaden substantially, decreasing the resolution markedly. Above this pressure, the transverse acoustic mode became impossible to resolve.

wüstite, to lower mantle conditions brings them to within 1–2 % of each other (CHOPELAS and BOEHLER, 1992) and to PREM densities for the lower mantle (DZIEWONSKI and ANDERSON, 1981). Using the bulk (hydrodynamic) sound velocity v_ϕ from equations of state leads to less ambiguous results. However, for an adiabatic extrapolation to lower mantle conditions, the required high temperature values of the bulk modulus or its pressure derivative for perovskite are not very accurately known (and are current subject for debate) nor can they be measured at 1 atm or low pressures because of metastability problems. Bulk sound velocity can not yet be used as a much more rigorous constraint than the density.

To illustrate the difference between density and velocity comparison, the density of the mantle minerals and alumina versus pressure are plotted in Fig. 9, where we see the density these minerals converge at high pressures. The differences between MgO and $MgSiO_3$ perovskite become smaller at high temperatures because of their respective thermal expansion coefficients. However, in examining Fig. 3 where the velocities measured here are shown relative to PREM, there are differences of at least 10% in velocities instead of differences less than 5% (for density) at room temperature. The perovskite velocities are expected to lie slightly above those for alumina with the 1 atm values at 6.58 and 11.1 km/s for v_s and v_p, respectively (WEIDNER *et al.*, 1993). Distinguishing the

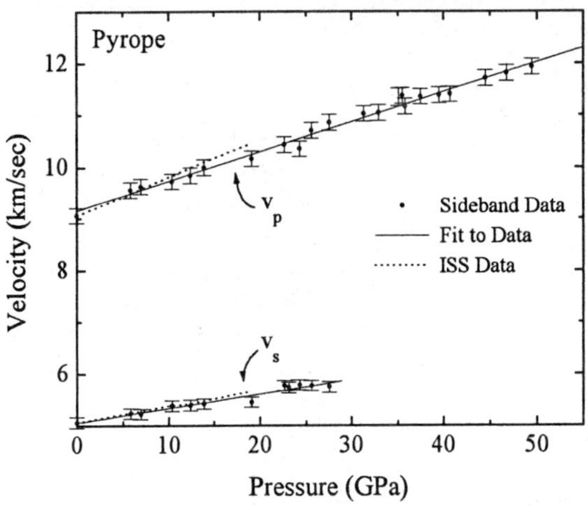

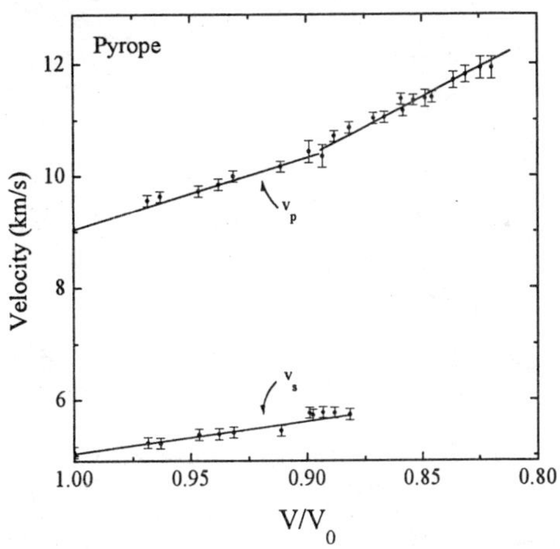

FIG. 8a. *Left.* Preliminary velocity results for pyrope for the compressional and shear velocity versus pressure. The pressure dependence of impulse stimulated scattering data for a slightly different composition garnet $\{(Mg_{0.84},Fe_{0.16})_3Al_2Si_3O_{12}\}$ to 22 GPa (CHAI *et al.*, 1993) are shown for comparison.

FIG. 8b. *Right.* Plot of the pyrope shear and compressional velocities from Fig. 8a versus compression. The increase in slope of v_p versus V/V_0 by over 60% at 28 GPa is clearly shown (see Table 1).

Table 2. Comparison of pressure derivatives of the elastic moduli obtained by the sideband method to those obtained by ultrasonic techniques.

		Sideband	Ultrasonics#	Reference
MgO	P_{max} (GPa)	36.5	3.2	JACKSON and NIESLER, 1982
	K'	4.08	4.13	YONEDA, 1990
	K" (GPa^{-1})	-0.036	-0.022	DUFFY and AHRENS, 1995
	G'	2.56	2.53	
	G" (GPa^{-1})	-0.030	-0.027	
YAG	P_{max} (GPa)	57.5	0.3	YOGURTCU *et al.*, 1980
	K'	5.16	4.44	
	K" (GPa^{-1})	-0.0036	--	
	G'	0.68	0.62	
$MgAl_2O_4$	P_{max} (GPa)	11.0†	1.0	CHANG and BARSCH, 1973
	K'	5.05	5.15	YONEDA, 1990
	K" (GPa^{-1})	-0.13	-0.48	
	G'	0.072	0.51	
Al_2O_3	@P_{max} (GPa)	53.0	1.0	GIESKE and BARSCH, 1968
	G'	1.94	1.98	
	G" (GPa^{-1})	-0.02	-0.02	
Pyrope	P_{max} (GPa)	28.0†	3.2	WEBB, 1989
	K'	4.5	4.7*	
	G'	1.73	1.81*	

#See Fig. 2a for explanation of values for MgO.
@Only shear velocities were measured.
†phase change or change in compressional mechanism at this pressure
*pyrope with 36% iron at the Mg^{2+} site

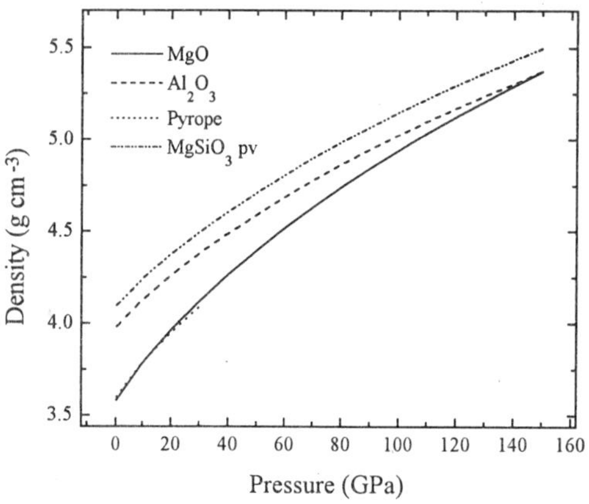

FIG. 9. The extrapolated densities of MgO, Al_2O_3, pyrope, and $MgSiO_3$ perovskite versus pressure at room temperature to the core-mantle boundary pressure. The MgO and $MgSiO_3$ perovskite curves converge even further at high temperatures because of the lesser value of α for perovskite and the greater value of α for MgO at high pressures. The densities at high pressure were calculated with the third-order finite strain equation of state (BIRCH, 1978) and the bulk modulus data listed in Tables 1 and 2. The density and bulk modulus values used for perovskite are listed in the Table 1 footnote with $K_{0T}' = 4$ assumed.

compositional models of the mantle will have to wait, however, until high pressure measurements for perovskite are completed.

$(\partial \ln \rho/\partial \ln v)$

Aside from distinguishing between mantle compositional models, measurements of sound velocities to these high pressures allows us to examine more closely the velocity-volume relationship for oxide minerals. This relationship in its logarithmic form is useful for interpreting lateral velocity anomalies in terms of lateral temperature anomalies from

$$\delta T = \frac{(\partial \ln \rho/\partial \ln v)_P}{\alpha} \times \delta v \qquad (10)$$

where δv is the velocity anomaly, α the thermal expansion coefficient, and ρ the density. {Note: $\ln(\rho/\rho_0) = -\ln(V/V_0)$.} Eqn. 10 assumes that the velocity variation is due to a temperature variation. Furthermore, note here that Eqn. 10 requires the isobaric value of $(\partial \ln \rho/\partial \ln v)$ and our measurements yield the isothermal value. The following

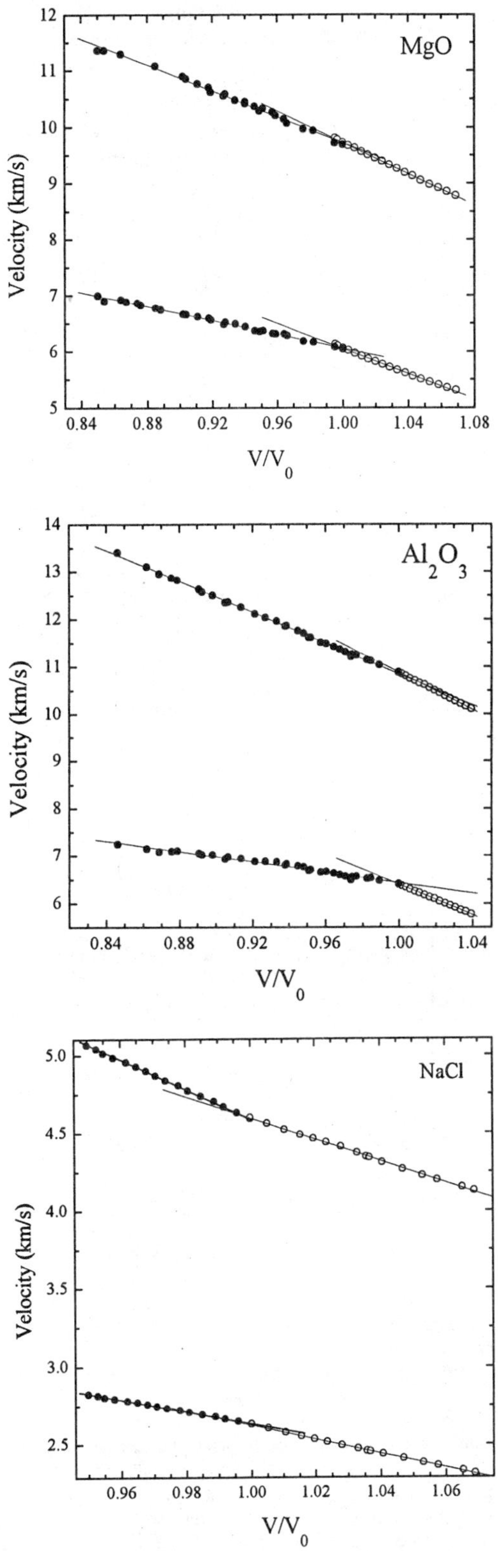

discussion provides constraints for the high pressure isobaric $(\partial \ln \rho/\partial \ln v)_P$.

As seen in Fig. 4ab, all the sound velocities depend linearly on volume as long as there is no change in compressional mechanism. It is also noted that the minerals with high alumina content have the smallest relative changes in shear velocity with compression. In fact, $MgAl_2O_4$ exhibited no measurable change in shear velocity over the pressure range of that study. From examination of this figure, it is further clear that the velocity-volume relationship is quite different for all the materials particularly for the shear velocity. For compressional velocity, the variation of slope between minerals is much less but nonetheless, extrapolation of velocity to high pressures using this method requires at least some lower pressure data in order to establish the velocity-volume function.

The linear relationship between velocity and volume requires that the value of $(\partial \ln \rho/\partial \ln v)_T$ increases with pressure. Likewise, in the available data on sound velocities versus temperature {summarized in (ANDERSON *et al.*, 1992)}, there is a linear velocity-volume dependency for MgO and Al_2O_3 (see Fig. 10a,b) although the slope is of different magnitude than that versus pressure at con-

FIG. 10a. *Top.* Plot of shear and compressional velocities of MgO (Θ_D = 945 K) versus reduced volume from high pressure data at room temperature (solid points are data from this study) and from high temperature data at 1 atm {open points are fitted values from (ISAAK *et al.*, 1989)}. The slopes of the high temperature data for v_s and v_p are 11.1 and 14.0 km/s, respectively, compared to 6.4 and 12.5 km/s for the high pressure data (Table 1).

FIG. 10b. *Center.* Plot of shear and compressional velocities of Al_2O_3 (Θ_D = 1034 K) versus reduced volume from high pressure data at room temperature (solid points are data from this study) and from high temperature data at 1 atm {open points are fitted values from (GOTO *et al.*, 1989)}. The slopes of the high temperature data for v_s and v_p are 15.8 and 19.6 km/s, respectively, compared to 5.5 and 16.3 km/s for the high pressure data (Table 1).

FIG. 10c. *Bottom.* Plot of shear and compressional velocities of NaCl (Θ_D = 321 K) versus reduced volume from high pressure data at room temperature {solid points are fitted values from (SPETZLER *et al.*, 1972)} and from high temperature data at 1 atm {open points are data from (YAMAMOTO *et al.*, 1987)}. The slopes of the high temperature data for v_s and v_p are 4.6 and 6.9 km/s, respectively, compared to 3.8 and 9.4 km/s for the high pressure data. Recent very high pressure data obtained by Brillouin scattering (CAMPBELL and HEINZ, 1992) yielded approximately 6.7 km/s for v_p versus reduced volume at room temperature. Compressional velocity versus reduced volume for KCl (Θ_D = 235 K) is 6.3 km/s at 1 atm and high temperature (YAMAMOTO and ANDERSON, 1987) and is 6.2 km/s at room temperature and high pressure (CAMPBELL and HEINZ, 1992).

stant temperature. This linearity versus temperature requires that $(\partial \ln \rho/\partial \ln v)_P$ *decreases* with increasing temperature. Note that for both of these compounds the velocity-volume slope for the compressional velocity is very nearly the same whereas there are substantial differences for the shear velocity. In order to draw some conclusions on the required isobaric value of $(\partial \ln \rho/\partial \ln v)$ at mantle conditions, its relationship to the isothermal value needs to be clarified.

The relationship between the isothermal and isobaric values of $(\partial \ln \rho/\partial \ln v)$ can be obtained from straightforward mathematics

$$\left(\frac{\partial \ln \omega}{\partial T}\right)_v = \alpha\left[\left(\frac{\partial \ln \omega}{\partial \ln \rho}\right)_T - \left(\frac{\partial \ln \omega}{\partial \ln \rho}\right)_P\right] \quad (11)$$

The first term the right hand side of the equation can be related to $(\partial \ln v/\partial \ln \rho)_T$ by taking the natural log on both sides of Eqn. 5 and differentiating with respect to $\ln \rho$ at constant temperature. This results in $(\partial \ln v/\partial \ln \rho)_T = (\partial \ln \omega/\partial \ln \rho)_T - 1/3$ and when the same is done at constant pressure, it results in $(\partial \ln v/\partial \ln \rho)_P = (\partial \ln \omega/\partial \ln \rho)_P - 1/3$. Therefore, $(\partial \ln v/\partial \ln \rho)_{T,P}$ may be directly substituted in Eqn. 11 above for $(\partial \ln \omega/\partial \ln \rho)_{T,P}$ which then shows that the difference between the isobaric and the isothermal derivatives is directly related to the intrinsic anharmonicity $(\partial \ln \omega/\partial T)_V$. We do have evidence that the intrinsic harmonicity vanishes at lower mantle conditions (see following discussion), and thus, the differences between the isobaric and isothermal derivatives also vanishes. Note that the values of $(\partial \ln v/\partial \ln \rho)$ lie between $1/2$ and 2 therefore taking the inverse of these also gives us values that lie between $1/2$ and 2. Thus, small differences between $(\partial \ln v/\partial \ln \rho)$ values mean that $(\partial \ln \rho/\partial \ln v)$ must approach each other as their inverses do.

The evidence that the intrinsic anharmonicity diminishes substantially at temperatures above the Debye temperatures is theoretically (HARDY, 1980) and experimentally based (ANDERSON *et al.*, 1992). In high temperature experiments on several minerals including Al_2O_3, MgO, and pyrope, ANDERSON *et al.* (1992) showed that thermal pressure is linear with temperature above the Debye temperature Θ_D, meaning intrinsic anharmonicity could not be detected. Furthermore, the velocity-volume functions of two alkali halides, KCl and NaCl, are nearly the same at room pressure and temperature (which is near or above the Debye temperatures of these materials, Fig. 10c). This means that the isobaric and isothermal derivatives of $(\partial \ln \rho/\partial \ln v)$ are nearly the same as well.

We have clues as to what $(\partial \ln \rho/\partial \ln v)_T$ should be at the high temperatures and pressures of the lower mantle by examining the PREM adiabatic values, which should be very near the isothermal values (shown in Fig. 11). These values are very similar to the values found for the mantle materials at room temperature and high pressure of about 1.0 for v_p and 1.4 for v_s, which are approximately double those found for $(\partial \ln \rho/\partial \ln v)_P$ at room temperature and 1 atm (ANDERSON *et al.*, 1992) for a large number of minerals.

In summary, the value of the elastic (or anharmonic) contribution to $(\partial \ln \rho/\partial \ln v)_P$ appears to increase from the reported 1-atm values as pressure increases. In examining Eqn. 10, the combined effect of $(\partial \ln \rho/\partial \ln v)_P$ increasing and thermal expansivity decreasing with depth implies that the same magnitude velocity anomalies at greater depths yields larger temperature anomalies. Thus, the structural features in terms of velocity anomalies found in three-dimensional tomographic models significantly change when put in terms of temperature anomalies, reducing the magnitude of the features at shallower depths and increasing their magnitude at greater depths. The structural features in lower mantle tomographic models now in terms of temperature appear markedly similar to com-

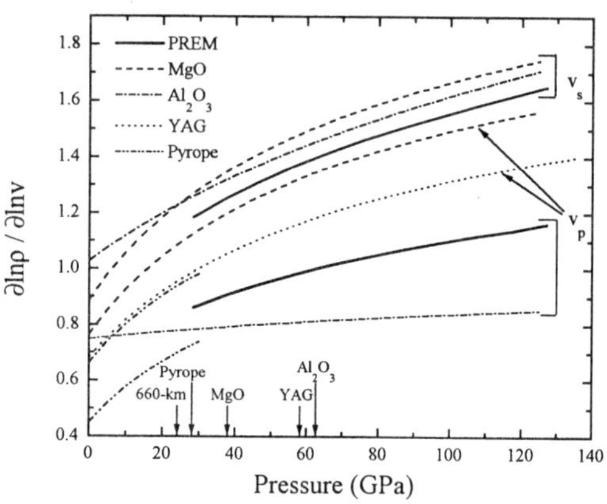

FIG. 11. Plot of $(\partial \ln \rho/\partial \ln v)_T$ versus pressure at room temperature for four of the five minerals of this study. The YAG shear velocity values are off the scale (see Table 3) since v_s changes very little with pressure. The adiabatic PREM values at lower mantle pressures and temperatures (DZIEWONSKI and ANDERSON, 1981) are included in the figure for comparison. The highest pressure attained in each of the experiments for the four minerals are indicated on the pressure axis. The extrapolated values for three of the four minerals were obtained by assuming a linear velocity-volume relationship (see Table 1). The values for pyrope are shown only to 28 GPa since a change in compressional mechanism is observed here (see Fig. 8b).

Table 3. Values of the density dependence of $(\partial \ln\rho/\partial \ln v)_T$ for five minerals.

	$(\partial \ln\rho/\partial \ln v_s)_T$	$(\partial \ln\rho/\partial \ln v_p)_T$
MgO	$0.85 + 3.7 \cdot \ln(\rho/\rho_0)$	$0.76 + 1.9 \cdot \ln(\rho/\rho_0)$
$MgAl_2O_4$*	$\approx\infty$	$0.97 + 11 \cdot \ln(\rho/\rho_0)$
Al_2O_3	$1.03 + 2.5 \cdot \ln(\rho/\rho_0)$	$0.75 + 0.39 \cdot \ln(\rho/\rho_0)$
Pyrope*	$0.67 + 2.7 \cdot \ln(\rho/\rho_0)$	$0.46 + 2.4 \cdot \ln(\rho/\rho_0)$
YAG	$3.3 + 5.5 \cdot \ln(\rho/\rho_0)$	$0.69 + 2.7 \cdot \ln(\rho/\rho_0)$

*pyrope values to 28 GPa and $MgAl_2O_4$ values to 11 GPa

puter simulations of plumes (YUEN *et al.*, 1993), which supports the hypothesis that $(\partial \ln \rho/\partial \ln v)_P$ increases with depth.

The results obtained here are insufficient to determine the exact value of the thermal anomalies because the anelastic contribution to $(\partial \ln \rho/\partial \ln v)_P$ is not known. There is evidence that the anelastic contribution may be as significant as the elastic (anharmonic) contribution, and thus will reduce the value of $(\partial \ln \rho/\partial \ln v)_P$ (KARATO, 1993) and which, in turn, reduces the calculated thermal anomaly.

CONCLUSIONS

Sound velocities of five minerals measured using the new sideband method have been presented and are shown to be in excellent agreement with the lower pressure results obtained by other methods. This means that the acoustic modes measured here yield spherically averaged information. We find that the velocity is linear with volume as long as no phase change or change in compressional mechanism occurs. However, this velocity-volume function varies from mineral to mineral and is smaller for minerals with significant alumina contents, particularly for the shear velocity. The linear velocity-volume relationship requires that $(\partial \ln \rho/\partial \ln v)_T$ increase with pressure or decreasing volume. Since the intrinsic anharmonicity decreases with depth, the isobaric $(\partial \ln \rho/\partial \ln v)$ must also increase with depth. This in turn increases the magnitude of the calculated thermal anomalies from the seismic tomographic velocity anomalies with depth.

Acknowledgements—Partial support for L. Zhang and H. R. Reichmann was provided by the Bundes Ministerium für Forschung und Technik (BMFT). A. Zerr assisted in the CO_2 laser heating of MgO and Al_2O_3 samples. Critical comments by G. C. Serghiou, R. Boehler, A. M. Hofmeister, and O. L. Anderson improved the manuscript.

REFERENCES

ANDERSON O. L., ISAAK D. and ODA H. (1992) High-temperature elastic constant data on minerals relevant to geophysics. *Rev. Geophys.* **30,** 57–92.

BIRCH F. (1978) Finite strain isotherm and velocities for single crystal and polycrystalline NaCl at high pressures and 300 K. *J. Geophys. Res.* **83,** 1257–1268.

CAMPBELL A. J. and HEINZ D. L. (1992) A high-pressure test of Birch's law. *Science* **257,** 66–68.

CHAI M., BROWN J. M. and SLUTSKY L. J. (1993) The elasticity of pyrope garnet to 22 GPa (abstract). *EOS, Trans. Am. Geophys. U.* **74,** 676.

CHANG Z. P. and BARSCH G. R. (1973) Pressure dependence of single-crystal elastic constants and anharmonic properties of spinel. *J. Geophys. Res.* **78,** 2418–2433.

CHOPELAS A. (1991) Evidence of a pressure-induced phase transition in tetragonal majorite: implications for the mantle (abstract). *EOS, Trans. Am. Geophys. U.* **72,** 478.

CHOPELAS A. (1992) Sound velocities of MgO to very high compression. *Earth Planet. Sci. Lett.* **114,** 195–192.

CHOPELAS A. (1996) The fluorescence sideband method for obtaining acoustic velocities at high compressions: Application to MgO and $MgAl_2O_3$. *Phys. Chem. Minerals,* **23,** 25–37.

CHOPELAS A. and BOEHLER R. (1992) Thermal expansivity in the lower mantle. *Geophys. Res. Lett.* **19,** 1983–1986.

CHOPELAS A. and HOFMEISTER A. M. (1991) Vibrational spectroscopy of aluminate spinels at 1 atm and of $MgAl_2O_4$ to over 200 kbar. *Phys. Chem. Minerals* **18,** 279–293.

CHOPELAS A. and NICOL M. F. (1982) Pressure dependence to 100 kbar of the phonons of MgO at 90 and 295 K. *J. Geophys. Res.* **87,** 8591–8597.

DUFFY T. S. and AHRENS T. J. (1995) Compressional sound velocity, equation of state, and constitutive response of shock-compressed magnesium oxide. *J. Geophys. Res.* **100,** 529–542.

DZIEWONSKI A. M. and ANDERSON D. L. (1981) Preliminary reference Earth model. *Phys. Earth Planet. Inter.* **25,** 297–356.

GIESKE J. H. and BARSCH G. R. (1968) Pressure dependence of the elastic constants of single crystal aluminum oxide. *Phys. Stat. Solidi* **29,** 121–131.

GLASS A. M. and SEARLE T. M. (1967) Reactions between vacancies and impurities in magnesium oxide. II. Mn^{4+}-

ion and OH^--ion impurities. *J. Chem. Phys.* **46,** 2092–2101.

GOTO T., YAMAMOTO S., OHNO I. and ANDERSON O. L. (1989) Elastic constants of corundum up to 1825 K. *J. Geophys. Res.* **94,** 7588–7602.

HARDY R. J. (1980) Temperature and pressure dependence of intrinsic anharmonic and quantum corrections to the equation of state. *J. Geophys. Res.* **85,** 7011–7015.

HOFMEISTER A. M. and CAMPBELL K. R. (1992) Infrared spectroscopy of yttrium aluminum, yttrium gallium, and yttrium iron garnets. *J. Appl. Phys.* **72,** 638–646.

HOFMEISTER A. M. and CHOPELAS A. (1991) Vibrational spectroscopy of end-member silicate garnets. *Phys. Chem. Minerals* **17,** 503–526.

HURRELL J. P., PORTO S. P. S., CHANG I. F., MITRA S. S. and BAUMAN R. P. (1968) Optical phonons of yttrium aluminum garnet. *Phys. Rev.* **173,** 851–858.

ISAAK D. G., ANDERSON O. L. and GOTO T. (1989) Measure elastic modulus of single-crystal MgO up to 1800 K. *Phys. Chem. Minerals* **16,** 703–714.

JACKSON I. and NIESLER H. (1982) The elasticity of periclase to 3 GPa and some geophysical implications. *High Pressure Research in Geophysics*. (ed. S. AKIMOTO and M. H. MANGHNANI) Tokyo. pp. 98–113. Center of Academic Publications of Japan.

KAPPUS W. (1975) Lattice dynamics of sapphire (corundum), part II: calculation of the phonon dispersion. *Z. Physik* **B21,** 325–331.

KARATO S. I. (1993) Importance of anelasticity in the interpretation of seismic tomography. *Geophys. Res. Lett.* **20,** 1623–1626.

KITTEL C. (1976) *Introduction to Solid State Physics*. John Wiley & Sons, Inc.

O'NEILL B., BASS J. D., ROSSMAN G. R., GEIGER C. A. and LANGER K. (1991) Elastic Properties of Pyrope. *Phys. Chem. Minerals* **17,** 617–621.

SANGSTER M. J. L., PECKHAM G. and SAUNDERSON D. H. (1970) Lattice dynamics of magnesium oxide. *J. Phys. C* **3,** 1026–1036.

SPETZLER H., SAMMIS C. G. and O'CONNELL R. J. (1972) Equation of state of NaCl: Ultrasonic measurements to 8 kbar and 800°C and static lattice theory. *J. Phys. Chem. Solids* **33,** 1727–1750.

THOMPSON P. and GRIMES N. W. (1978) Observation of low energy phonons in spinel. *Solid St. Comm.* **25,** 609–611.

WEBB S. L. (1989) The elasticity of the upper mantle orthosilicates olivine and garnet to 3 GPa. *Phys. Chem. Minerals* **16,** 684–692.

WEIDNER D. J., WANG Y. and YEGANEH-HAERI A. (1993) Equation of state properties of mantle perovskite (abstract). *EOS, Trans. Am. Geophys. U.* **74,** 571.

YAMAMOTO S. and ANDERSON O. L. (1987) Elasticity and anharmonicity of potassium chloride at high temperature. *Phys. Chem. Minerals* **14,** 332–340.

YAMAMOTO S., OHNO I. and ANDERSON O. L. (1987) High temperature elasticity of sodium chloride. *J. Phys. Chem. Solids* **48,** 143–151.

YOGURTCU Y. K., MILLER A. J. and SAUNDERS G. A. (1980) Elastic behavior of YAG under pressure. *J. Phys. C: Sol. St. Phys.* **13,** 6585–6597.

YONEDA A. (1990) Pressure derivatives of elastic constants of single crystal MgO and $MgAl_2O_4$. *J. Phys. Earth* **38,** 19–55.

YUEN D. A., CADEK O., CHOPELAS A. and MATYSKA C. (1993) Geophysical evidences for thermal chemical plumes in the lower mantle. *Geophys. Res. Lett.* **20,** 899–903.

ZHANG L. and CHOPELAS A. (1994) Sound velocity of Al_2O_3 to 616 kbar. *Phys. Earth Planet. Inter.* **98,** 77–83.

Mineral Spectroscopy: A Tribute to Roger G. Burns
© The Geochemical Society, Special Publication No. 5, 1996
Editors: M. D. Dyar, C. McCammon and M. W. Schaefer

Crystal chemistry of FeO at high pressure and temperature

YINGWEI FEI
Center for High Pressure Research and Geophysical Laboratory, Carnegie Institution of Washington, 5251 Broad Branch Road, NW, Washington, DC 20015, U.S.A.

Abstract—The phase relations and equations of state of Fe_xO have been studied up to a pressure of 100 GPa and a temperature of 1100 K, using a high-temperature diamond-anvil cell combined with synchrotron x-ray diffraction. Under a hydrostatic environment, Fe_xO transforms from the NaCl-type cubic structure (B1) to a rhombohedral structure at 16 GPa and 300 K. The transition pressures increase as a function of temperature with a pressure-temperature slope of 0.070(±0.003) GPa/K. No significant effect of nonstoichiometry on the transition boundary was observed. The transition occurs at lower pressure under nonhydrostatic conditions. The rhombohedral phase increases its distortion with increasing pressure. Further distortions from the rhombohedral cell to a lower symmetry phase were observed at pressures above 40 GPa at room temperature. Upon heating at high pressure, the NiAs-type hexagonal phase (B8) forms at 90 GPa and 600 K. Least-squares fits of the hydrostatic compression data yielded the room-temperature bulk modulus K_{T0} = 146(±2) GPa with its pressure derivative $(\partial K_{T0}/\partial P)_T = 4$ for the NaCl-type cubic phase (B1). The effect of degree of nonstoichiometry on the bulk moduli is small, within the uncertainty of the experiments. Although the bulk modulus for the rhombohedral phase cannot be determined accurately, by comparing the structurally related interplanar *d* spacings it is evident that the rhombohedral phase is less compressible than the cubic phase. The bulk modulus and its pressure derivative for the NiAs-type hexagonal phase (B8), derived from the static compression data, are 172 GPa and 4.3 at 900 K, respectively.

INTRODUCTION

EXPERIMENTAL STUDIES in the system MgO-FeO-SiO_2 at high pressure and temperature have provided tight constraints on mantle mineralogy. The crystal chemistry of the end-member FeO at high pressure and temperature is of particular importance in understanding the role of iron in the Earth's lower mantle. Furthermore, FeO has been suggested as a possible light-alloying component in the Earth's core (RINGWOOD, 1977). There has been increasing experimental evidence that high pressure and temperature would enhance the solubility of oxygen in molten iron (KATO and RINGWOOD, 1989; OHTANI and RINGWOOD, 1984; OHTANI *et al.*, 1984; RINGWOOD and HIBBERSON, 1990, 1991; and ITO *et al.*, 1995). In order to understand the fundamental physiochemical basis for those observations, it is essential to determine the crystal and electronic structure transitions and the compression behavior of FeO at high pressure and temperature.

Ferrous oxide is invariably nonstoichiometric with compositions from $Fe_{0.88}O$ to $Fe_{0.98}O$ (HAZEN and JEANLOZ, 1984). Three polymorphs in Fe_xO, the NaCl-type cubic (B1), rhombohedral, and NiAs-type hexagonal (B8) phases, have been identified (ZOU *et al.*, 1980; JEANLOZ and AHRENS, 1980; YAGI *et al.*, 1985; and FEI and MAO, 1994). The rhombohedral and NiAs-type hexagonal phases are not quenchable. The cubic to rhombohedral transition accompanies a magnetic transition from paramagnetic to antiferromagnetic state (TOMBS and ROOKSBY, 1950; WILLIS and ROOKSBY, 1953; ROTH, 1960; VAUGHAN and DRICKAMER, 1967; ZOU *et al.*, 1980). The transition to the NiAs-type hexagonal phase may be associated with the metallization of Fe_xO (KNITTLE and JEANLOZ, 1986, 1991; KNITTLE *et al.*, 1986; and FEI and MAO, 1994). The equation of state for the cubic phase has been studied extensively by static compression method (MAO *et al.*, 1969; WILL *et al.*, 1980; HAZEN, 1981; YAGI *et al.*, 1985; JEANLOZ and SATO-SORENSEN, 1986; LIU and LIU, 1987; and FEI and MAO, 1991) and by ultrasonic and shockwave experiments (MIZUTANI, 1971; SUMINO *et al.*, 1980; BERGER *et al.*, 1981; BONCZAR and GRAHAM, 1982; JACKSON *et al.*, 1990; and JEANLOZ and AHRENS, 1980). Despite considerable study by both static and dynamic compression methods, many questions still remain unanswered regarding the effect of the degree of nonstoichiometry, the deviatoric stress, and temperatures on the phase boundaries and the equations of state. In order to address these issues, we have conducted a series of static compression experiments on Fe_xO over a wide composition range under hydrostatic and nonhydrostatic conditions.

EXPERIMENTAL PROCEDURE

Sample synthesis and characterization

The physical properties and structural states of Fe_xO are complicated by its nonstoichiometric nature. The synthesized Fe_xO samples often retain a memory of their thermal and pressure history. Careful characterization of the starting materials is very important for interpreting the high-pressure data of Fe_xO. Three synthetic Fe_xO samples ($Fe_{0.92}O$, $Fe_{0.94}O$, and $Fe_{0.98}O$) were used in the experiments. The $Fe_{0.92}O$ and $Fe_{0.94}O$ samples were prepared by heating Fe_2O_3 to 1473 K under controlled oxygen fugacity in a CO-CO_2 gas-mixing furnace. The compositions of the samples were determined by x-ray diffraction measurements on the basis of the composition-lattice parameter relationship of Fe_xO (x = $[a_0$ - 3.900]/0.433, SIMONS, 1980). The lattice parameters of the synthesized samples are a_0 = 4.296(1) Å and a_0 = 4.307(1) Å, corresponding to compositions $Fe_{0.92}O$ and $Fe_{0.94}O$, respectively. The $Fe_{0.98}O$ sample was synthesized from a mixture of $Fe_{0.94}O$ and metallic Fe at 9 GPa and 1273 K in a multi-anvil apparatus. Its lattice parameter is a_0 = 4.323(1) Å.

High-pressure and high-temperature techniques

High pressures were generated between two gem-quality single-crystal diamonds in a diamond-anvil cell, similar to that described by MAO *et al.* (1979). In order to achieve simultaneous high pressure and temperature, an external resistance heating system was incorporated in the diamond-anvil cell. The cell was accordingly modified to optimize pressure generation at high temperature (MAO *et al.*, 1991). The major modifications include (1) inconel piston-cylinder, (2) thermally insulating the piston-cylinder from the lever arm body, and (3) a double-ring alignment system on the piston to freely retract and advance the piston at high temperature.

Two types of external resistance heating designs were used to achieve high temperature in the diamond-anvil cell in this study. A large sleeve-shaped platinum-wire heater fitted around the protruding portion of the piston-cylinder was used to heat the sample to temperatures up to 900 K (MAO *et al.*, 1991; FEI *et al.*, 1992). This design is not advantageous for generating high pressures >40 GPa) at temperatures above 600 K because of stress relaxation through plastic flow at high temperature. To overcome the problem, we designed a double-heater configuration in which an additional small molybdenum-wire heater was positioned around the diamond anvils (Fig. 1). This double-heater high-temperature diamond-anvil cell is capable of achieving pressures greater than 125 GPa at temperatures up to 1100 K in a mildly reducing atmosphere (Ar with 1% H_2) (FEI and MAO, 1994).

Sample configurations

Experiments on Fe_xO were conducted under both hydrostatic and nonhydrostatic environments. In the hydro-

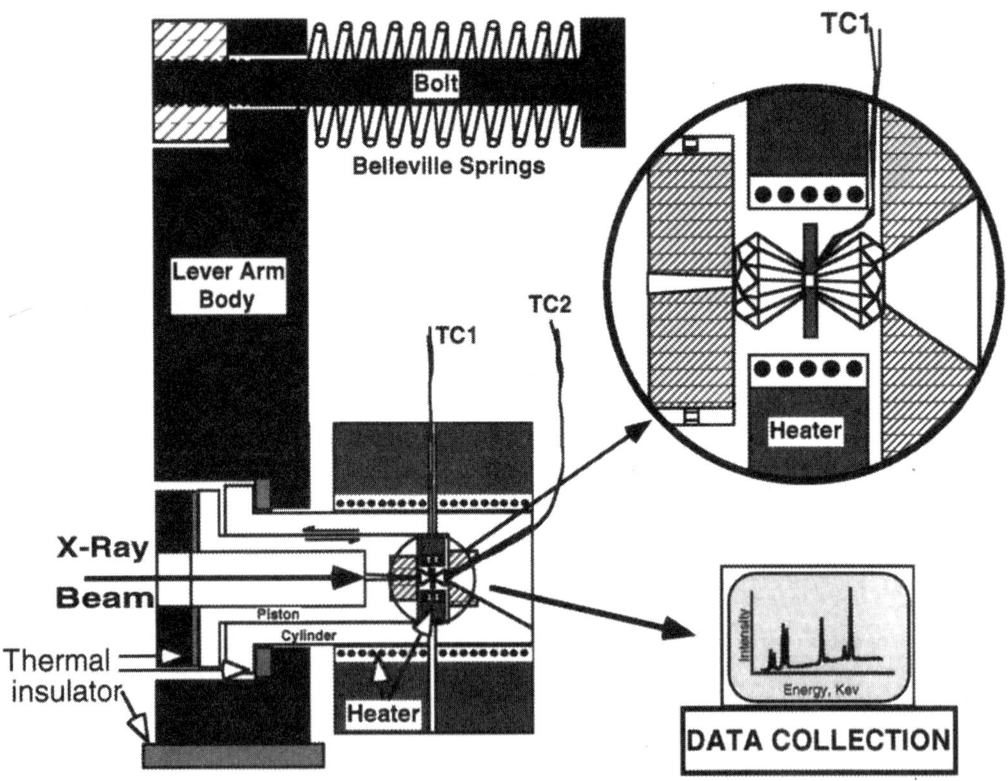

FIG. 1. Experimental configuration. The design of the diamond-anvil cell is similar to that of MAO *et al.* (1991). High temperatures were achieved by using a double-heater system. Temperatures were measured with thermocouples (TC1 and TC2). Energy-dispersive x-ray diffraction techniques were used for data acquisition.

static compression experiments, neon was used as a pressure-transmitting medium. The polycrystalline sample was loaded into a sample chamber, 200 μm in diameter by 70 μm in thickness, drilled from a preindented rhenium gasket. Only one-third of the chamber volume was filled with sample. Small ruby grains and gold foil were placed in the sample chamber as pressure calibrants. The sample chamber was then filled with neon gas at 200 MPa in a high-pressure gas-loading device (JEPHCOAT *et al.*, 1987) and subsequently sealed at pressures above 1 GPa. All the hydrostatic compression experiments were carried out at pressures below 36 GPa, using 600-μm flat diamond anvils. The nonhydrostatic compression experiments were designed to generate higher pressures, using 300-μm culet or beveled diamond anvils. Similar to the hydrostatic compression experiments, the rhenium gasket was preindented to the desired thickness. The powder sample with thin gold foil was compacted in a sample chamber, typically 75 μm in diameter by 25 μm in thickness. The rhenium gasket works very well at high temperature because of its high strength. A simultaneous pressure of 100 GPa and temperature of 1100 K was achieved in this sample configuration.

Synchrotron X-ray diffraction techniques

The experiments were conducted at x17c beamline, the National Synchrotron Light Source, Brookhaven National Laboratory. Polychromatic (white) wiggler synchrotron x-radiation was used for energy-dispersive x-ray diffraction measurements. A highly collimated x-ray beam, regulated by two mutually perpendicular slits, was aligned with the detection system and the center of the sample chamber in the diamond-anvil cell. The sample can be scanned in vertical and horizontal directions by a motor system installed in the sample stage. A 25-μm beam spot was used in diffraction measurements. The diffraction data were collected with an intrinsic germanium 4096-channel solid-state detector at a fixed 2θ angle. The energy-channel number relationship was determined by measuring the energies of well-determined x-ray emission lines (K_α and K_β) of Mn, Cu, Rb, Mo, Ag, Ba, and Tb. The 2θ angle was calibrated by measuring the energies of diffraction peaks, corresponding to the known interplanar spacings d_{hkl}, of gold at ambient conditions.

The energy-dispersive x-ray diffraction technique has the advantage of quick acquisition of diffraction data. Although it is not a problem to maintain simultaneous high pressure and temperature in the high-temperature diamond-anvil cell for extended period of time, quick data acquisition helps to obtain sufficient data over a wide *P-T* range within the allocated short synchrotron beam time. The diffraction of Fe_xO is relatively strong. A complete diffraction pattern of Fe_xO with reasonable signal-to-noise ratio can be obtained in less than 10 minutes. The diffraction data were fitted to Voight functions using a peak-fitting program. The interplanar spacings d_{hkl} (Å) were calculated from the measured peak energies E_{hkl} (kev) by $d_{hkl} = (6.199/\sin\theta)/E_{hkl}$.

Temperature and pressure measurements

It is essential to establish a pressure calibration standard at high temperature for the high *P-T* experiments. At room temperature, pressures in the diamond-anvil cell were determined by measuring the specific volumes of metals such as Cu, Mo, Pd, Ag, Au, and Pt by x-ray diffraction, based on their volume-pressure relationships derived from shock compression data. The ruby pressure scale was established by measuring simultaneously the shift of the ruby R_1 luminescent line and the specific volume of those metals (MAO *et al.*, 1978; 1986). In this study, pressures at high temperature were determined from the measured lattice parameter of gold and sample temperature, based on the *P-V-T* equation of state of gold proposed by ANDERSON *et al.* (1989). The sample temperatures were measured with a Pt-Pt13%Rh thermocouple placed near the sample chamber (Figure 1). The temperature difference between the sample and the thermocouple junction is calibrated by observing the melting point of NaCl at ambient pressure. The accuracy and precision of measurements are within 10 K.

RESULTS AND DISCUSSION

Phase relations

The phase transitions in Fe_xO were determined by in situ synchrotron x-ray diffraction techniques. In the experiment where $Fe_{0.92}O$ powder was used as starting material, we confirmed the transition from the NaCl-type cubic (B1) to a rhombohedral phase at 16 GPa and 300 K under hydrostatic conditions (in a neon pressure transmitting medium). The transition was characterized by the splitting of the 111, 220, and 222 diffraction peaks of the cubic phase (*cf.* Fig. 2). The 200 peak remained sharp and strong. The observed new diffraction pattern

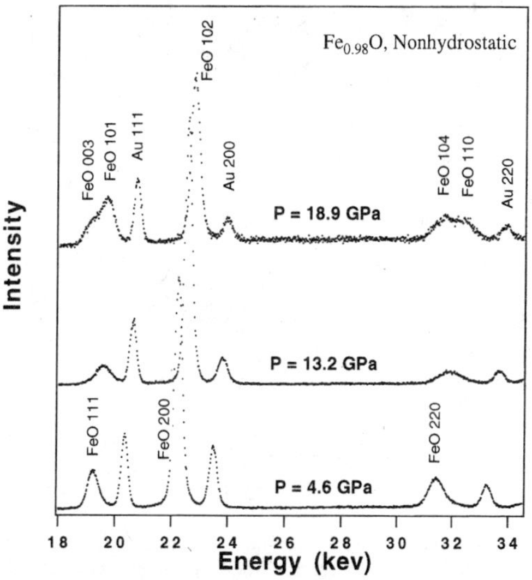

FIG. 2. Energy-dispersive x-ray diffraction spectra of Fe_xO at pressures across the cubic-rhombohedral transition under nonhydrostatic conditions. Under hydrostatic conditions, the splitting of the 111 and 220 peaks of the cubic phase occurred in a small pressure interval (~1 GPa).

can be indexed by using a rhombohedral cell at pressures near the transition, as shown by YAGI *et al.* (1985). The cubic-rhombohedral transition was also supported by the single-crystal compression experiment where helium was used as pressure medium (MAO *et al.*, 1993). The rhombohedral cell is closely related to the NaCl-type cubic cell and can be derived by stretching along the body diagonal direction 111 of the cubic cell, as suggested by ZOU *et al.* (1980). However, with increasing pressure, deviation from the rhombohedral cell becomes evident, even when neon was used as pressure transmitting medium.

After observing the rhombohedral phase at room temperature, we started to heat the sample at 22 GPa. The rhombohedral phase transformed back to the cubic phase at 21 GPa and 400 K (Fig. 3). The small pressure drop was due to material relaxation at high temperature. With further increase of pressure at 400 K, the transition from the cubic to the rhombohedral phase was observed again at 23 GPa. Similarly, the transition was observed at 31 GPa and 500 K. At the maximum simultaneous pressure and temperature of 37 GPa and 600 K, the cubic phase was observed. The *P-T* path of the experiment is shown by the arrows in Fig. 3. The transition boundary has a positive pressure-temperature slope with $P(\text{GPa}) = -5.0 + 0.070(\pm 0.003)T(\text{K})$.

To evaluate the effect of degree of nonstoichiometry on the transition pressures, we also carried out an experiment where $Fe_{0.98}O$ powder was used as starting material, using neon as pressure transmitting medium. The observed transition pressures at 300 K, 400 K, and 500 K were identical to those of $Fe_{0.92}O$ described above, within experimental uncertainty (±0.4 GPa). The observation at 300 K is also consistent with that of YAGI *et al.* (1985) who used a $Fe_{0.98}O$ sample in their experiment.

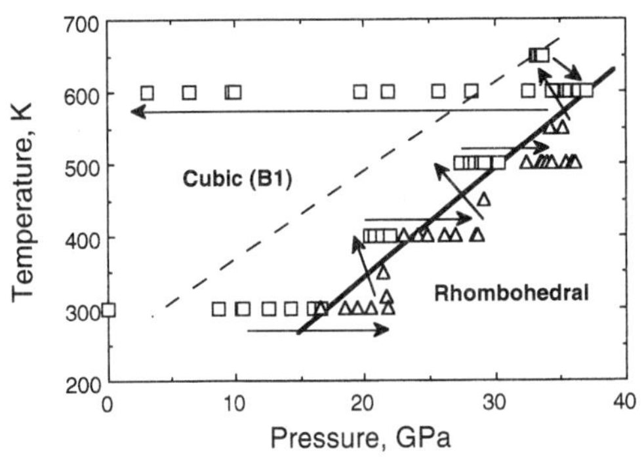

FIG. 3. The cubic-rhombohedral transition determined by in situ x-ray diffraction measurements. The arrows indicate the *P-T* path of the experiment. The solid and dashed lines represent phase boundaries under hydrostatic and nonhydrostatic conditions, respectively. Experimental data (squares, cubic phase; and triangles, rhombohedral phase) are shown for the $Fe_{0.92}O$ sample. Similar results were obtained when $Fe_{0.98}O$ was used as the starting material.

From the experimental results of ZOU *et al.* (1980) and YAGI *et al.* (1985), it is evident that the transition from the cubic to the rhombohedral phase is affected by the stress environments of the sample. ZOU *et al.* reported the transition at 9 GPa and 298 K when no pressure medium was used in the experiment. In contrast to the sharp transition at 16 GPa under hydrostatic conditions (YAGI *et al.*, 1985), the transition pressures cannot be clearly defined under nonhydrostatic conditions. We found that the broadening of the 111 and 220 diffraction peaks occurred at pressures below 8 GPa. The lattice parameters calculated from the 111, 200, and 220 lines indicate significant deviation from the cubic cell at pressures above 6 GPa and room temperature. If we define the transition by the mismatches of the observed and calculated interplanar d_{hkl} spacings, the transition pressure is about 6 GPa at room temperature, 10 GPa lower than that under hydrostatic conditions. The measured pressure-temperature slope (dT/dP) of the transition is lower than the hydrostatic value, resulting from the change of the stress environment with increasing temperature (Fig. 3).

At ambient pressure, Fe_xO undergoes a magnetic transition from paramagnetic to antiferromagnetic state at 198 K (TOMBS and ROOKSBY, 1950; WILLIS and ROOKSBY, 1953; and ROTH, 1960). The magnetic transition was also observed at high pressure and room temperature by high-pressure Mössbauer studies (VAUGHAN and DRICKAMER, 1967; and ZOU *et al.*, 1980). The antiferromagnetic phase has a rhombohedral cell, derived from distortion of the NaCl-type cubic cell, with the rhombohedral angle $a = 59.4°$ at 90 K (WILLIS and ROOKSBY, 1953). ZOU *et al.* (1980) and YAGI *et al.* (1985) suggested that the high-pressure transition from the cubic to the rhombohedral phase was the extension of the low-temperature transition caused by the increase of the Néel temperature with pressure. Linear extrapolation of our measured cubic-rhombohedral boundary to ambient pressure yielded the Néel temperatures of 71 K and 250 K for hydrostatic and nonhydrostatic compression, respectively. The discrepancy on the Néel temperature at ambient pressure may result from the complex effect of the stress conditions on the transition. The rhombohedral angle α is about 57.7° (±0.3°) for the high-pressure rhombohedral phase at 20 GPa and 300

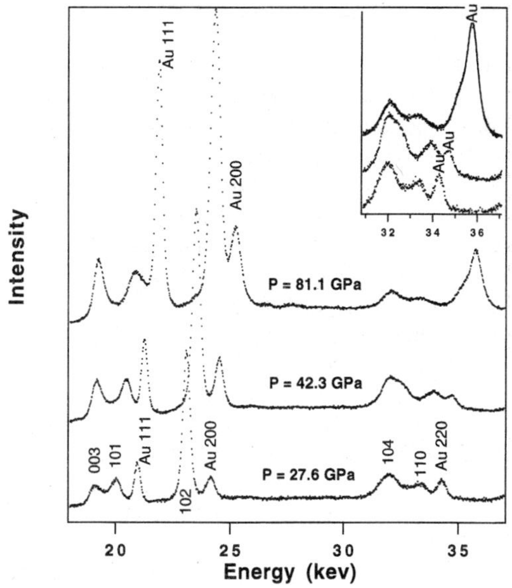

FIG. 4. Energy-dispersive x-ray diffraction spectra of Fe_xO at pressures above 27 GPa. At pressures below 40 GPa, the diffraction pattern can be qualitatively indexed with a rhombohedral cell. With increasing pressure, the splitting of the 104 peak was observed, as shown in the inset.

K, which indicates that the degree of distortion at high pressure is higher than that at low temperature.

Although the 111 and 220 peaks begin to broaden at pressures as low as 5 GPa under nonhydrostatic conditions, clear splitting of the 111 and 220 peaks was resolved at pressures above 16 GPa (Fig. 2). The two peaks resulting from the splitting of 111, corresponding to 003 and 101 in the rhombohedral cell assignment, were significantly broader than the 102 peak, corresponding to 200 in the cubic cell. Similarly, the 104 and 110 peaks, of the split 220 in the cubic cell, were very broad. Having examined the diffraction data from three nonhydrostatic compression experiments of $Fe_{0.94}O$ and $Fe_{0.98}O$ up to 58 GPa, 71 GPa and 82 GPa, respectively, at room temperature, we observed changes in the relative intensities of the 003 and 101 diffractions, indicating strong preferred orientation. Furthermore, the deviation from the rhombohedral cell increases with increasing pressure. At pressures above 40 GPa, further splitting of the 104 peak was observed (Fig. 4), indicating further distortion from the rhombohedral cell to a lower symmetry phase. YAGI *et al.* (1985) reported evidence of further splitting of the 104 peak at about 40 GPa, but the observed peak was very weak. From our diffraction data, the existence of the lower symmetry phase is evident, but the nature of this lower symmetry phase is unclear. Our data also indicate that degree of the distortion increases with increasing degree of nonstoichiometry, reflected by the separation of the 003 and 101 peaks in $Fe_{0.94}O$ and $Fe_{0.98}O$ at any given pressure.

Shock wave studies revealed the existence of a denser phase of Fe_xO at pressures above 70 GPa along the Hugoniot (JEANLOZ and AHRENS, 1980; and YAGI *et al.*, 1988). Transition to this denser phase was not observed in static compression experiments at room temperature (YAGI *et al.*, 1985; this study). The structure of this denser phase was the subject of speculation (*e.g.*, JEANLOZ and AHRENS, 1980; JACKSON and RINGWOOD, 1981; NAVROTSKY and DAVIES, 1981; and JACKSON *et al.*, 1990). JACKSON and RINGWOOD (1981) and NAVROTSKY and DAVIES (1981) argued that the denser phase may have the NiAs structure based on thermochemical systematics. Very recently, FEI and MAO (1994) discovered a NiAs-type hexagonal phase of Fe_xO at high pressure and temperature by *in situ* x-ray diffraction measurements. Under nonhydrostatic conditions, the distorted rhombohedral phase transformed to the NiAs phase with transition pressure $P(\text{GPa}) = 118(\pm 2) - 0.051 \times (\pm 0.004) T(\text{K})$ (Fig. 5) (FEI and MAO, 1994). On the basis of this transition and the transition from the cubic to the rhombohedral phase discussed above, the cubic phase (B1) would directly trans-

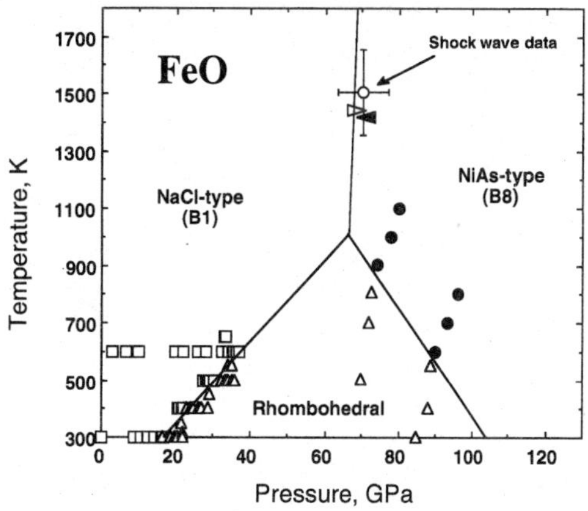

FIG. 5. Phase diagram of Fe_xO. Data are shown by symbols: open squares, NaCl structure (B1); open triangles, rhombohedral phase; and solid circles, NiAs-type structure (B8) (FEI and MAO, 1994). Shock-induced transition point (JEANLOZ and AHRENS, 1980) is represented by open circle. The open and solid arrows indicate metallization of FeO inferred from electrical resistance measurements in the laser-heated diamond-anvil cell (KNITTLE and JEANLOZ, 1991).

form to the NiAs phase (B8) at temperatures above the triple point temperature (~1020 K). The transition from the cubic to the NiAs phase at high pressure and temperature is consistent with the observation of the shock-wave compression experiments.

Isothermal bulk modulus

The compression behavior of Fe_xO has been studied by ultrasonic and shock-wave compression methods and by static compression in the diamond-anvil cell, as reviewed by JEANLOZ and HAZEN (1983) and JACKSON *et al.* (1990). The bulk modulus for the cubic phase (B1) of Fe_xO has a large uncertainty, derived from the static compression data (WILL *et al.*, 1980; HAZEN, 1981; YAGI *et al.*, 1985; JEANLOZ and SATO-SORENSEN, 1986; and LIU and LIU, 1987). The scattering of the compression data, especially at pressures above 6 GPa, is largely due to the deviatoric stress effect under nonhydrostatic environments. We have carried out compression experiments of Fe_xO under hydrostatic and nonhydrostatic conditions and observed similar data scattering when no pressure medium was used in the experiments. Under nonhydrostatic conditions, the diffraction peaks broadened with increasing pressure. The mismatches of d_{111}, d_{200}, and d_{220} were about 0.3 % at 6 GPa, increasing to 1 % at 8 GPa. The lattice parameter of the B1 phase is very sensitive to the deviatoric stress. Even when a soft pressure transmitting medium is used in the experiments, it is important to have sufficiently large medium to sample volume ratio to prevent premature bridging of the diamond anvils by the sample at high pressure. To evaluate the effect of the degree of nonstoichiometry on the bulk modulus, we obtained compression data on $Fe_{0.98}O$ and $Fe_{0.92}O$, using neon as pressure transmitting medium (Table 1, Fig. 6). The volumes were calculated by fitting the d_{111}, d_{200}, and d_{220} spacings, while the pressures were determined using gold as an internal pressure calibrant. In the $Fe_{0.98}O$ experiment, pressures were also determined by using the ruby pressure scale to check the consistency in pressure determination. Least-squares fits of the compression data to the Birch-Murnaghan equation of state yielded the bulk moduli K_{T0} of 147(±2) GPa and 144(±2) GPa for $Fe_{0.92}O$ and $Fe_{0.98}O$, respectively, assuming the pressure derivative $(\partial K_{T0}/\partial P)_T = 4$. The result does not show any significant

Table 1. Hydrostatic compression data of $Fe_{0.92}O$ and $Fe_{0.98}O$ at room temperature

$Fe_{0.92}O$		$Fe_{0.98}O$		
P, GPa[a]	*V*, cm³/mol	*P*, GPa[a]	*P*(ruby), GPa[b]	*V*, cm³/mol
0.0001	11.93(1)	0.0001	-	12.16(1)
0.81(10)	11.86(1)	1.90(5)	1.80	12.04(1)
1.55(12)	11.79(1)	3.24(6)	-	11.91(1)
2.63(8)	11.73(2)	4.96(16)	5.20	11.76(1)
4.58(15)	11.58(1)	7.61(16)	7.70	11.60(2)
8.65(29)	11.31(1)	8.91(19)	9.00	11.51(1)
10.51(27)	11.20(1)	10.50(10)	10.43	11.40(1)
12.47(30)	11.09(2)	11.49(12)	11.34	11.32(2)
14.19(56)	11.01(2)	12.12(40)	12.34	11.34(2)
16.02(12)	10.90(2)	13.14(31)	13.19	11.27(2)
		13.94(26)	-	11.19(2)
		14.50(30)	14.80	11.17(2)
		15.60(23)	15.90	11.10(2)

[a]Calculated from the lattice parameter of gold, based on its *P-V-T* equation of state proposed by ANDERSON *et al.* (1989). The uncertainties in pressure were propagated from the errors in lattice parameter measurements.

[b]Determined by using the ruby pressure scale (MAO *et al.*, 1986)

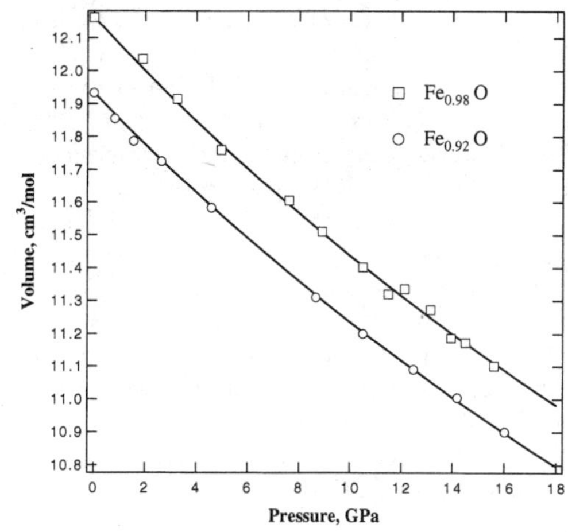

FIG. 6. Hydrostatic compression data of $Fe_{0.92}O$ (open circles) and $Fe_{0.98}O$ (open squares) at room temperature. The solid curves are calculated isotherms.

effect of the degree of nonstoichiometry on the bulk modulus over the composition range from x = 0.92 to 0.98, contrary to the speculation on the compositional dependence of Fe_xO bulk modulus (MCCAMMON, 1993). A similar conclusion was also reached by HAZEN (1981) who determined the bulk moduli of $Fe_{0.90}O$, $Fe_{0.93}O$, and $Fe_{0.947}O$ using methanol-ethanol (4:1) as pressure medium by the single-crystal diffraction method.

There is some confusion in the comparison of the bulk moduli of Fe_xO derived from static compression data, because the comparisons were made without specifying the pressure derivatives. It is important to evaluate the correlation between the bulk modulus and its pressure derivative, especially when the compression range is limited (JEANLOZ, 1981). To compare the compression data with different compositions, we plotted the relative volume, V/V_0, as a function of pressure. Fig. 7 shows all the hydrostatic or quasi-hydrostatic compression data of the cubic

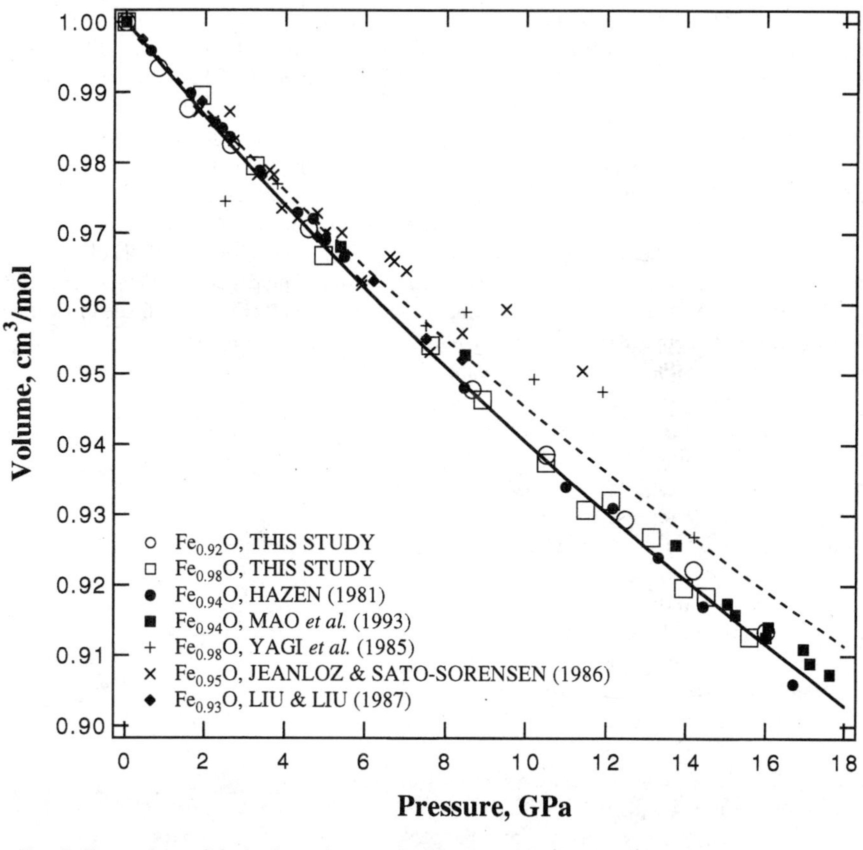

FIG. 7. Comparison of the hydrostatic compression data of Fe_xO. The solid curve is a least-squares fit of all the data except those of YAGI *et al.* (1985) and JEANLOZ and SATO-SORENSEN (1986) to the Birch-Murnaghan equation of state. The isentrope based on K_{S0} = 154.9 GPa and $(\partial K_{S0}/\partial P)_S$ = 4.84 derived from the ultrasonic data (JACKSON *et al.*, 1990) are also shown for comparison (the dashed curve).

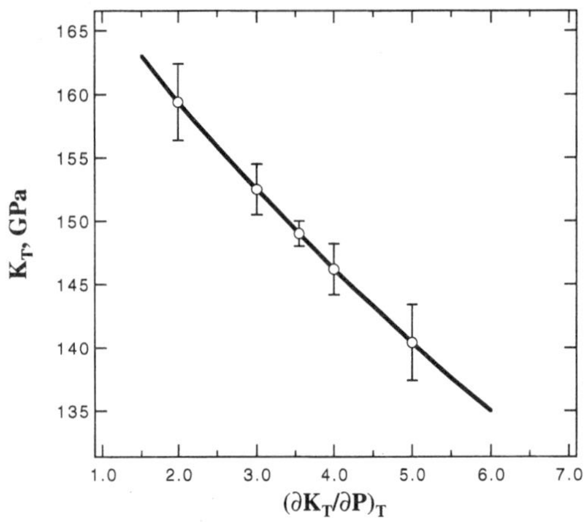

FIG. 8. The trade-off between the bulk modulus K_{T0} and its pressure derivative $(\partial K_{T0}/\partial P)_T$. The best-fit results are K_{T0} = 149 GPa and $(\partial K_{T0}/\partial P)_T$ = 3.5.

Fe_xO phase. The data obtained in this study, using neon as pressure medium, are consistent with the single-crystal studies of HAZEN (1981) who used methanol-ethanol (4:1) as pressure medium and of MAO *et al.* (1993) who used helium as pressure medium. Our result is also in a good agreement with that of LIU and LIU (1987) who obtained the compression data up to 8.3 GPa using methanol-ethanol (4:1) as pressure medium, by the powder x-ray diffraction method. In comparison with the data of YAGI *et al.* (1985) and JEANLOZ and SATO-SORENSEN (1986) who used methanol-ethanol (4:1) as pressure medium, there is general agreement between our data and their data at pressures below 6 GPa. However, their data at pressures above 6 GPa showed relatively large scatter which may result from the increase of nonhydrostatic stress with increasing pressure.

The Fe_xO samples used to obtain the compression data plotted in Fig. 7 represent a wide range of compositions ($0.90 \leq x \leq 0.98$). There is no systematic effect of the degree of nonstoichiometry on the bulk modulus within the uncertainties of the data. Least-squares fits of all the data except those of YAGI *et al.* (1985) and JEANLOZ and SATO-SORENSEN (1986) to the Birch-Murnaghan equation of state yielded the best-fit results, K_{T0} = 149 GPa and $(\partial K_{T0}/\partial P)_T$ = 3.5. However, The K_{T0} and $(\partial K_{T0}/\partial P)_T$ are strongly correlated. Figure 8 shows the covariance between K_{T0} and $(\partial K_{T0}/\partial P)_T$.

JEANLOZ and HAZEN (1983) pointed out that the bulk moduli determined by dynamic methods, including ultrasonic and shock-wave experiments, were significantly higher than those obtained by static compression. Recent ultrasonic measurements indicate that the bulk moduli of single-crystal Fe_xO determined by ultrasonic wave propagation have very small variation, ranging from 151 GPa to 154 GPa (BERGER *et al.*, 1981; and JACKSON *et al.*, 1990). The bulk moduli determined by resonance techniques (SUMINO *et al.*, 1980), by polycrystalline ultrasonic measurements (MIZUTANI, 1971; and BONCZAR and GRAHAM, 1982), and by shock-wave compression (JEANLOZ and AHRENS, 1980) range from 172 GPa to 182 GPa. JACKSON *et al.* (1990) showed that essentially all of the previous ultrasonic, static and shock compression data could be reconciled with their K_{S0} and $(\partial K_{S0}/\partial P)_S$ values of 155 GPa and 4.9 respectively. However, assessment of the hydrostatic compression data of Fe_xO (HAZEN, 1981; MAO *et al.*, 1993; this study) shows a lower bulk modulus, K_{T0} = 149 GPa, for a $(\partial K_{T0}/\partial P)_T$ value of 3.5. The ultrasonic and static compression data could be reconciled with the ultrasonic value (K_{T0} = 153 GPa), if a $(\partial K_{T0}/\partial P)_T$ value of 3.1 is assumed.

The bulk modulus for the rhombohedral phase of Fe_xO was not well constrained because the powder diffraction data for the rhombohedral phase were limited between 16 GPa and 23 GPa. The data collected above 23 GPa under nonhydrostatic conditions indicated significant deviation from the rhombohedral cell, and also indicated further distortion to a lower symmetry phase. Despite the splitting and broadening of the 111, 220, and 311 peaks, the 200 of the cubic phase, corresponding to the 102 in the rhombohedral cell, remained sharp and strong. The pressure dependence of the *d* spacing of this peak (Fig. 9) indicates that the rhombo-

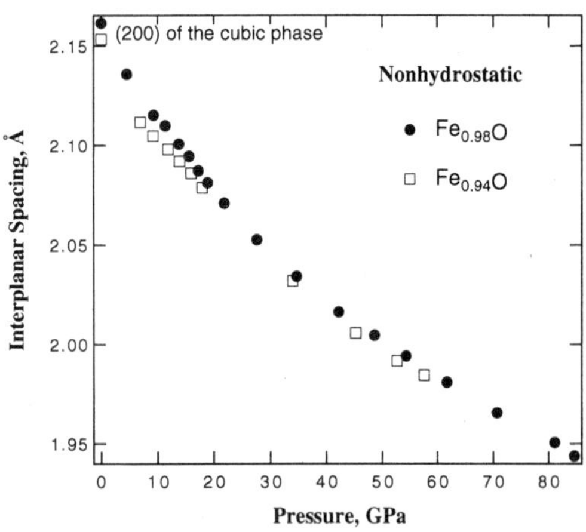

FIG. 9. Pressure dependence of the interplanar *d* spacings related to the 200 peak of the cubic cell. Experimental data are shown for two compositions, $Fe_{0.98}O$ (solid circles) and $Fe_{0.94}O$ (open squares).

hedral phase (or the distorted rhombohedral phase) is less compressible than the cubic phase. The density of the rhombohedral phase is only slightly larger than that of the cubic phase at the transition pressure. The difference in density is about 0.5%.

FEI and MAO (1994) reported the bulk modulus $K_T = 172(\pm 14)$ GPa and its pressure derivative $(\partial K_T/\partial P)_T = 4.3(\pm 0.6)$ for the NiAs hexagonal phase of Fe_xO at 900 K. Their results are in good agreement with the shock compression data (JEANLOZ and AHRENS, 1980; and YAGI *et al.*, 1988). A comparison of the densities of the NiAs hexagonal phase between the static and shock-wave data was shown by FEI and MAO (1994).

It is very difficult to determine the temperature effect on the bulk modulus even for the cubic phase of Fe_xO because the composition of Fe_xO changes as a function of temperature and pressure. Figure 10 shows the results from two high *P-T* experiments in which neon was used as pressure medium. In the first experiment, a $Fe_{0.92}O$ sample was used as the starting material. Its room temperature compression data were discussed above. Upon heating, the cubic phase is stable to higher pressure (cf. Fig. 3). The lattice parameters for the cubic phase at 400 K and 500 K plotted on the extrapolation of the 300-K compression curve indicate very small thermal expansivity at high pressure (Fig. 10). With further increasing temperature to 600 K, the lattice parameters increased substantially. The change cannot be simply attributed to thermal expansion. After we conducted the experiment with the sample close to the stoichiometric composition, we realized that the sudden change in lattice parameter at 600 K was due to the compositional change. Figure 10 also shows the lattice parameters as a function of pressure from the experiment where $Fe_{0.98}O$ sample was used as the starting material. The lattice

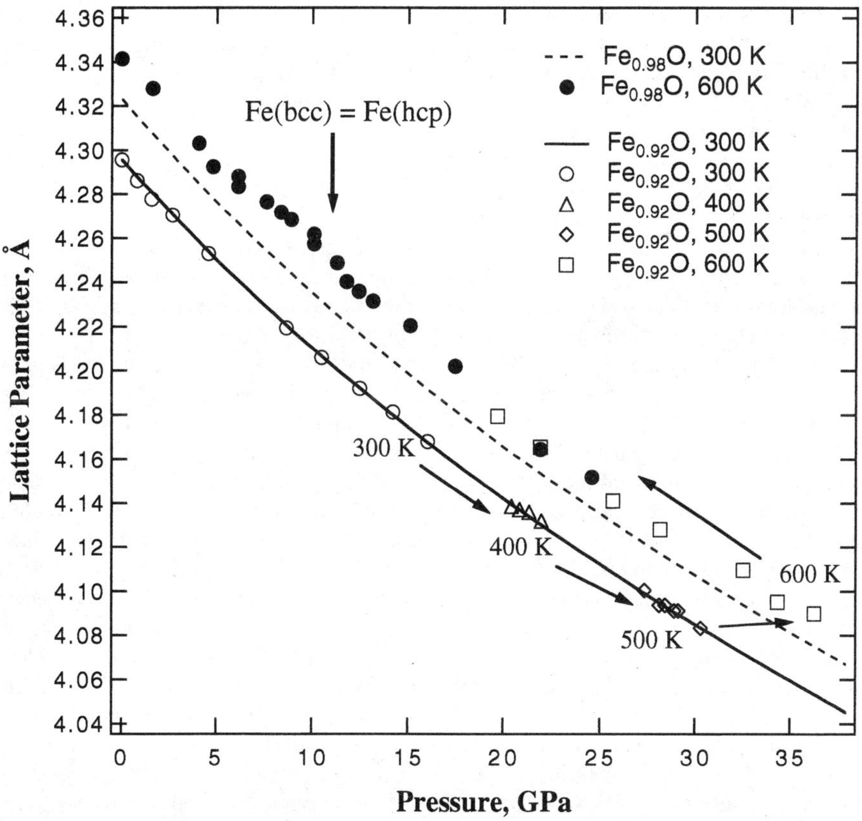

FIG. 10. Lattice parameters of the cubic phase as a function of pressure and temperature. The open symbols are data from the experiment where $Fe_{0.92}O$ was used as the starting material. The solid circles represent compression data of $Fe_{0.98}O$ at 600 K. The solid and dashed curves are the calculated 300-K isotherms of $Fe_{0.92}O$ and $Fe_{0.98}O$, respectively. The arrows indicate the *P-T* path of the experiment.

parameters at 600 K are consistent with those from the first experiment where the $Fe_{0.92}O$ sample was used as starting material, implying that the Fe_xO samples may have similar equilibrium composition at high pressure and temperature regardless of the compositional differences in the starting materials. Upon decompression at 600 K, a small kink in the lattice parameter was observed at about 11 GPa, coincident with the $\alpha-\epsilon$ transition in Fe. The kink is reproducible upon compression or decompression. If this kink in the lattice parameter reflects the compositional change of Fe_xO, it may explain why the Fe_xO sample with maximum Fe content (close to the stoichiometric composition) can be synthesized at about 10 GPa.

Crystal chemistry

HAZEN and JEANLOZ (1984) reviewed the crystal chemistry and defect structure of Fe_xO at ambient pressure. Our *in situ* x-ray diffraction measurements of Fe_xO provide new data for understanding the crystal chemistry and phase relations of Fe_xO at high pressure and temperature. On the basis of the x-ray diffraction studies (ZOU *et al.*, 1980; YAGI *et al.*, 1985; MAO *et al.*, 1993; and this study), it is quite clear that the 16-GPa transition at room temperature is a cubic to rhombohedral transition caused by distortion of the cubic cell, an elongation along a body diagonal direction (111) of the NaCl-type cubic cell. Our determination of the temperature and stress dependence of the transition supports that this transition is closely related to the observed paramagnetic to antiferromagnetic transition at low temperature, but the degree of the distortion may be a complex function of pressure and stress. The rhombohedral distortion is very sensitive to the sample stress environments. Further distortion to a low symmetry phase at room temperature is evident, but it is not clear if this low symmetry phase is metastable.

At high pressure and temperature, Fe_xO has the NiAs-type hexagonal structure (B8) (FEI and MAO, 1994). The NiAs structure consists of hexagonally close-packed layers of oxygen and iron alternately stacked along the *c* axis. It can be derived from the NaCl-type cubic cell by varying the stacking along the body diagonal direction (111) of the cubic cell, corresponding to the *c* axis in the NiAs structure. The Fe-Fe distance in the NiAs structure is about 8% shorter than that in the NaCl structure. The shorter Fe-Fe distance across shared FeO_6 octahedral faces in the NiAs structure could lead to metallization of Fe_xO (JACKSON *et al.*, 1990), supporting the metallic behavior of this phase observed by electrical resistivity measurements (KNITTLE and JEANLOZ, 1986, 1991; and KNITTLE *et al.*, 1986). The increase of metallicity of Fe_xO at high pressure would enhance oxygen solubility (in the form of FeO) in molten iron, which has important implications for the incorporation of lighter elements into the Earth's core (FEI and MAO, 1994).

It is well documented that Fe_xO exsolves iron or magnetite lamellae on the scale of several tens of unit cells (HAZEN and JEANLOZ, 1984). The exsolution in Fe_xO is the result of equilibrium processes in the system Fe-O. Attempts to model the Fe-Fe_xO equilibria at high pressure and temperature were made previously by MAO, 1974; FEI and SAXENA, 1986; and MCCAMMON, 1993. Those models were based on the results from quenched experiments. Many measurements and observations were difficult to interpret as a result of complexities in quenching thermal and pressure history. In this study, we provided new *in situ* measurements on lattice parameters of Fe_xO as a function of pressure and temperature (Fig. 10). To completely understand equilibrium processes in the Fe-Fe_xO system at high pressure and temperature, more *in situ* measurements in pressure-temperature-composition space are required.

CONCLUSIONS

Despite the complexities in physical properties and structure states of Fe_xO at high pressure and temperature, a clear picture of its phase relations and compression behavior is begining to emerge. On the basis of our experimental results, we can conclude that the cubic-rhombohedral phase transition boundary is not affected by the degree of nonstoichiometry of Fe_xO, but is strongly dependent on the deviatoric stress environment. The bulk modulus for the cubic phase, derived from hydrostatic compression data, is K_{T0} = 146 GPa with its pressure derivative $(\partial K_{T0}/\partial P)_T$ = 4, independent of the degree of nonstoichiometry of Fe_xO. Under nonhydrostatic conditions, the rhombohedral phase increases its distortion with increasing pressure. A lower symmetry phase, resulting from the rhombohedral distortion, was observed at pressures above 40 GPa and room temperature. The structure and nature of this lower symmetry phase are still not clear. Transition from the rhombohedral distortion to a NiAs-type hexagonal phase was observed upon heating, consistent with shock compression experiments. The transition has been studied only for composition $Fe_{0.98}O$. Therefore, the effect of the degree of nonstoichiometry still needs to be addressed.

The effect of Mg content on the transitions observed in Fe_xO has important implications for lower mantle mineralogy. We carried out one experiment on $(Fe_{0.95}, Mg_{0.05})O$. The transition from the cubic to rhombohedral phase was observed. However, no NiAs-type hexagonal phase was observed up to 101 GPa and 600 K. Evidently, further study is required to understand the role of Mg in the transitions at high pressure and temperature.

Acknowledgements—I am grateful to Jingzu Hu for assistance with the experiments and to Constance Bertka, Hokwang Mao, Robert Hazen, and Charles Prewitt for helpful discussions. This work was supported by the National Science Foundation Center for High-Pressure Research, the Norton Company, and the Carnegie Institution of Washington.

REFERENCES

ANDERSON O. L., ISAAK D. G. and YAMAMOTO S. (1989) Anharmonicity and the equation of state for gold. *J. Appl. Phys.* **65**, 1534–1543.

BERGER J., BERTHON J., REVCOLEVSCHI A. and JOLLES E. (1981) Elastic constants of $Fe_{1-x}O$ single crystals. *J. Am. Ceram. Soc.* **64**, C153–C154.

BONCZAR L. J. and GRAHAM E. K. (1982) The pressure and temperature dependence of the elastic properties of polycrystal magnesiowüstite. *J. Geophys. Res.* **87**, 1061–1078.

FEI Y. and MAO H. K. (1991) In-situ x-ray diffraction study of wüstite ($Fe_{0.947}O$) up to 70 GPa and 600 K (abstr.). *EOS* **72**, 465.

FEI Y. and MAO H. K. (1994) In-situ determination of the NiAs phase of FeO at high pressure and temperature. *Science* **266**, 1678–1680.

FEI Y. and SAXENA S. K. (1986) A thermochemical data base for phase equilibria in the system Fe-Mg-Si-O at high pressure and temperature. *Phys. Chem. Mineral.* **13**, 311–324.

FEI Y., MAO H. K., SHU J. and HU J. (1992) *P-V-T* equation of state of magnesiowüstite $(Mg_{0.6}Fe_{0.4})O$. *Phys. Chem. Miner.* **18**, 416–422.

HAZEN R. M. (1981) Systematic variation of bulk modulus of wüstite with stoichiometry. *Year Book Carnegie Inst. Washington* **80**, 277–280.

HAZEN R. M. and JEANLOZ R. (1984) Wüstite ($Fe_{1-x}O$): A review of its defect structure and physical properties. *Rev. Geophys. Space Phys.* **22**, 37–46.

ITO E., MOROOKA K., UJIKE O. and KATSURA T. (1995) Reactions between molten iron and silicate melts at high pressure: Implications for the chemical evolution of the Earth's core. *J. Geophys. Res.* **100**, 5901–5910.

JACKSON I. and RINGWOOD A. E. (1981) High-pressure polymorphism of the iron oxides. *Geophys. J. R. Astron. Soc.* **64**, 767–783.

JACKSON I., KHANNA S. K., REVCOLEVSCHI A. and BERTHON J. (1990) Elasticity, shear-mode softening and high-pressure polymorphism of wüstite ($Fe_{1-x}O$). *J. Geophys. Res.* **95**, 21,671–21,685.

JEANLOZ R. (1981) Finite-strain equation of state for high-pressure phases. *Geophys. Res. Lett.* **8**, 1219–1222.

JEANLOZ R. and AHRENS T. J. (1980) Equations of state of FeO and CaO. *Geophys. J. R. Astron. Soc.* **62**, 505–528.

JEANLOZ R. and HAZEN R. M. (1983) Compression, nonstoichiometry and bulk viscosity of wüstite. *Nature* **304**, 620–622.

JEANLOZ R. and SATO-SORENSEN Y. (1986) Hydrostatic compression of $Fe_{1-x}O$ wüstite. *J. Geophys. Res.* **91**, 4665–4672.

JEPHCOAT A. P., MAO H. K. and BELL P. M. (1987) Operation of the megabar diamond-anvil cell. In *Hydrothermal Experimental Techniques* (ed. G. C. ULMER and H. L. BARNES), pp. 469–506. Wiley-Interscience, New York.

KATO T. and RINGWOOD A. E. (1989) Melting relationships in the system Fe-FeO at high pressure: Implications for the composition and formation of the Earth's core. *Phys. Chem. Miner.* **16**, 524–538.

KNITTLE E. and JEANLOZ R. (1986) High-pressure metallization of FeO and implications for the Earth's core. *Geophys. Res. Lett.* **13**, 1541–1544.

KNITTLE E. and JEANLOZ R. (1991) The high-pressure phase diagram of $Fe_{0.94}O$: A possible constituent of the Earth's core. *J. Geophys. Res.* **96**, 16,169–16,180.

KNITTLE E., JEANLOZ R., MITCHELL A. C. and NELLIS W. J. (1986) Metallization of $Fe_{0.94}O$ at elevated pressures and temperatures observed by shock-wave electrical resistivity measurements. *Solid State Commun.* **59**, 513–515.

LIU M. and LIU L. (1987) Bulk moduli of wüstite and periclase: a comparative study. *Phys. Earth Planet. Inter.* **45**, 273–279.

MAO H. K. (1974) A discussion of the iron oxides at high pressure with implications for the chemical and thermal evolution of the Earth. *Year Book Carnegie Inst. Washington* **73**, 510–518.

MAO H. K., TAKAHASHI T., BASSETT W. A., WEAVER J. S. and AKIMOTO S. (1969) Effect of pressure and temperature on the molar volumes of wüstite and of three $(Fe, Mg)_2SiO_4$ spinel solid solutions. *J. Geophys. Res.* **74**, 1061–1069.

MAO H. K., BELL P. M., SHANER J. W. and STEINBERG D. J. (1978) Specific volume measurements of Cu, Mo, Pd, and Ag and calibration of the ruby R1 fluorescence pressure gauge from 0.06 to 1 Mbar. *J. Appl. Phys.* **49**, 3276–3283.

MAO H. K., BELL P. M., DUNN K. J., CHRENKO R. M. and DEVRIES R. C. (1979) Absolute pressure measurements and analysis of diamonds subjected to maximum static pressures of 1.3–1.7 Mbar. *Rev. Sci. Instrum.* **50**, 1002–1009.

MAO H. K., XU J. and BELL P. M. (1986) Calibration of the ruby pressure gauge to 800 kbar under quasihydrostatic conditions *J. Geophys. Res.* **91**, 4673–4676.

MAO H. K., HEMLEY R. J., FEI Y., SHU J. F., CHEN L. C., JEPHCOAT A. P., WU Y. and BASSETT W. A. (1991) Effect of pressure, temperature, and composition on lattice parameters and density of $(Fe,Mg)SiO_3$-perovskites to 30 GPa, *J. Geophys. Res.* **96,** 8069–8079.

MAO H. K., SHU J., FEI Y., HU J. and HEMLEY R. J. (1993) High pressure phase transition in $Fe_{1-x}O$ wüstite (abstr.). *EOS* **74**, 169.

MCCAMMON C. (1993) Effect of pressure on the composition of the lower mantle end member Fe_xO. *Science* **259**, 66–68.

MIZUTANI H. (1971) Elasticity of mantle minerals and composition of the upper mantle. Ph.D. dissertation, Univ. Tokyo.

NAVROTSKY A. and DAVIES P. K. (1981) Cesium chloride versus nickel arsenide as possible structures for (Mg,

Fe)O in the lower mantle. *J. Geophys. Res.* **86**, 3689–3694.

OHTANI E. and RINGWOOD A. E. (1984) Composition of the core, I. Solubility of oxygen in molten iron at high temperatures. *Earth Planet. Sci. Lett.* **71**, 85–93.

OHTANI E., RINGWOOD A. E. and HIBBERSON W. (1984) Composition of the core, II. Effect of high pressure on solubility of FeO in molten iron. *Earth Planet. Sci. Lett.* **71**, 94–103.

RINGWOOD A. E. (1977) Composition of the core and implications for origin of the earth. *Geochem. J.* **11**, 111–135.

RINGWOOD A. E. and HIBBERSON W. (1990) The system Fe-FeO revised. *Phys. Chem. Miner.* **17**, 313–319.

RINGWOOD A. E. and HIBBERSON W. (1991) Solubilities of mantle oxides in molten iron at high pressures and temperatures: Implications for the composition and formation of Earth's core. *Earth Planet. Sci. Lett.* **102**, 235–251.

ROTH W. L. (1960) Defects in the crystal and magnetic structures of ferrous oxide. *Acta Cryst.* **13**, 140–149.

SIMONS B. (1980) Composition-lattice parameter relationship of magnesiowüstite solid solution series. *Year Book Carnegie Inst. Washington* **79**, 376–380.

SUMINO Y., KUMAZAWA M., NISHIZAWA O. and PLUSCHKELL W. (1980) The elastic constants of single crystal $Fe_{1-x}O$, MnO and CoO, and the elasticity of stoichiometric magnesiowüstite. *J. Phys. Earth* **28**, 475–495.

TOMBS N. C. and ROOKSBY H. P. (1950) Structure of monoxides of some transition elements at low temperatures. *Nature* **165**, 442–443.

VAUGHAN R. W. and DRICKAMER H. G. (1967) High-pressure Mössbauer studies on a-Fe_2O_3, $FeTiO_3$, and FeO. *J. Chem. Phys.* **47**, 1530–1536.

WILL G., HINZE E. and NUDING W. (1980) The compressibility of FeO measured by energy dispersive x-ray diffraction in a diamond anvil squeezer up to 200 kbar. *Phys. Chem. Miner.* **6,** 157–167.

WILLIS B. T. M. and ROOKSBY H. P. (1953) Change of structure of ferrous oxide at low temperature. *Acta Cryst.* **6**, 827–831.

YAGI T., SUZUKI T. and AKIMOTO S. (1985) Static compression of wüstite ($Fe_{0.98}O$) to 120 GPa. *J. Geophys. Res.* **90**, 8784–8788.

YAGI T., FUKUOKA K., TAKEI H. and SYONO Y. (1988) Shock compression of wüstite. *Geophys. Res. Lett.* **15**, 816–819.

ZOU G., MAO H. K., BELL P. M. and VIRGO D. (1980) High pressure experiments on the iron oxide wüstite ($Fe_{1-x}O$). *Year Book Carnegie Inst. Washington* **79**, 374–376.

Mineral Spectroscopy: A Tribute to Roger G. Burns
© The Geochemical Society, Special Publication No. 5, 1996
Editors: M. D. Dyar, C. McCammon and M. W. Schaefer

In situ ^{57}Fe Mössbauer spectroscopy of iron and olivine at high pressure and temperature

E. HÅLENIUS, H. ANNERSTEN, and S. JÖNSSON
Institute of Earth Sciences, Uppsala University, Norbyvägen 18 B, S-752 36 Uppsala, Sweden

Abstract—*In situ* Mössbauer measurements have been performed at high pressure and temperature by use of a diamond anvil cell equipped with a resistivity furnace. Iron-57 enriched iron metal and olivine are investigated in this study to set up the routine conditions for this type of Mössbauer spectroscopy. Sample temperatures in the DAC are calibrated using the second order Doppler shift of α-iron and pressures are determined from the shift of the fluorescence line from Sm:YAG. Methods are developed to prevent the point source from being heated during the experiment. The pressure dependent centre shift of α-iron is observed to be $-4.0.10^{-3}$ mms^{-1}GPa^{-1} at temperatures up to 600 K.

INTRODUCTION

THERE IS A growing need for *in situ* measurements of minerals and rocks at elevated temperatures and pressures to simulate conditions during rock formation. The kinetics of transformation reactions at the high temperatures prevailing in the Earth's mantel and core, are so rapid that physical-chemical properties of solids at high temperatures usually cannot be preserved by quenching. The application of different spectroscopic methods *e.g.* such as Raman spectroscopy, optical spectroscopy, X-ray diffraction, neutron diffraction and Mössbauer spectroscopy (HAWTHORNE, 1988), all in combination with the diamond anvil cell allow for determining properties of earth materials under conditions resembling their natural origin. Because of the simple alignment of a Mössbauer spectrometer with a diamond anvil cell device, this method has the potential to become an important method for such types of measurements.

Early use of a high-pressure device for obtaining high pressure Mössbauer spectra was developed by DEBRUNNER *et al.* (1966) to study the valence electron behaviour in metal organic complexes. Later several groups have developed diamond anvil cells for ultrahigh pressure Mössbauer spectroscopy (see STERER and PASTERNAK, 1990). The advantage of using diamond as anvils in high pressure Mössbauer spectroscopy is the low absorption of gamma rays by the diamonds. AMTHAUER *et al.* (1979) used B_4C as anvils, in conjunction with a hydraulic system, to apply pressure on their investigated samples. The use of B_4C was favoured because it allows large area samples to be examined, but limits the pressure to moderate values compared to conditions in the Earth's interior. In the present paper we have performed high pressure Mössbauer spectroscopy using the diamond anvil cell fitted with an external Pt-wired furnace to obtain high temperatures in the sample simultaneously. We discuss the difficulties in using this technique before it can be applied as a routine method for measuring minerals under these conditions.

EXPERIMENTAL SET UP

We have used a modified type of diamond anvil cell (DAC) described by MING *et al.* (1987) for performing X-ray diffraction at high pressure and temperature. The main parts of the DAC are depicted in Fig 1. There are several precautions one has to take before Mössbauer measurements can be performed using the DAC.

(1) the small sample size necessitates the use of ^{57}Fe in the sample which results in thickness effects of the absorption lines.

(2) reasonable counting rates need the source close to the sample and a strong point source, in our case 50 mCi. At the same time heating of the source has to be avoided.

(3) the pressure and temperature calibration is very sensitive to the mounting of the sample in the DAC.

(4) the measurements must be performed in an inert or reduced atmosphere to avoid oxidation of diamonds, gasket, rockers and sample which may limit the maximum temperature used in the experiment.

Our exploratory experiments have resulted in a new Mössbauer technique. We have investigated powdered samples of metallic iron produced by reducing Fe_2O_3 in hydrogen, and a medium iron-rich olivine $Fa_{41}Fo_{59}$ obtained from hydrothermal experiments (see below). All synthetic samples contained ^{57}Fe enriched to 95%. Samples, approximately 0.1–0.2 mg, and chips of Sm:YAG crystals were placed together with a 1:4 mixture of ethanol/methanol in a 350 μm diameter hole of a Ta-gasket on the diamond anvil face. The closed cell was then wrapped tightly with tape and the inner part of the pressure cell was slowly flushed with argon gas. The upper and lower part of the cell was connected to a water cooling system. Note the copper shield (Fig. 1), in contact with the cooled pressure cell, preventing the radioactive point source, ^{57}Co in Rh, from becoming heated during the experiments.

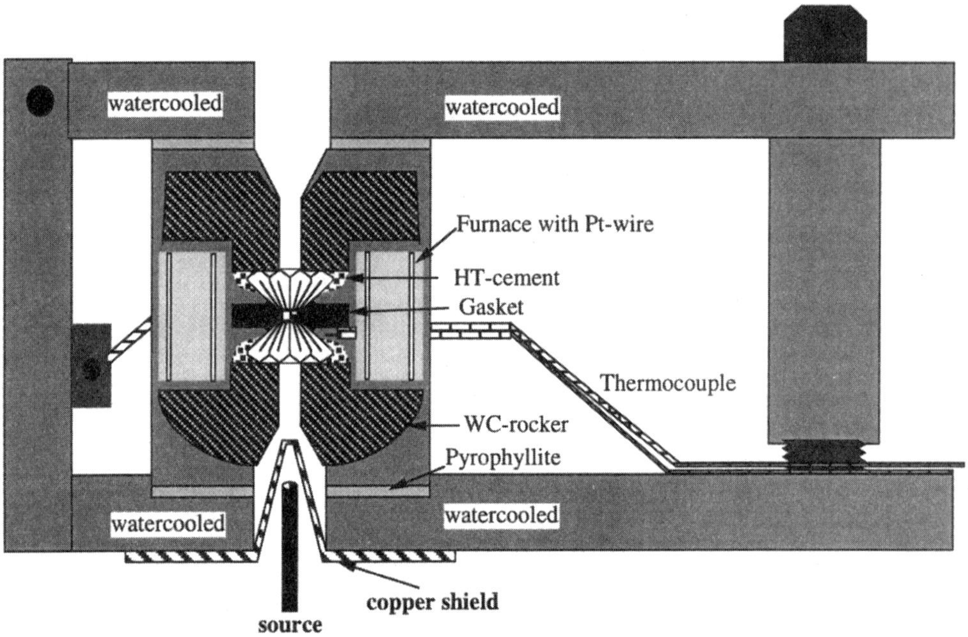

FIG. 1. Schematic cross section of the diamond anvil cell (not in scale).

Temperature calibration

A circular Pt wired furnace around the diamonds was temperature regulated using a thermocouple $Pt_{100}/Pt_{90}Rh_{10}$ attached close to the gasket (see Fig. 1). We believe that there is no larger temperature gradient in the sample due to the shape of the furnace and the good thermal conductivity of diamonds. However there are several positions in the DAC where the temperature of the system is of vital concern.

The temperature of the sample between the diamonds was calibrated using two different procedures:

(1) comparing melting temperatures observed under the microscope of congruently melting oxides placed in the sample chamber with well known melting points;

(2) measuring the temperature dependent centre shift of metallic iron in the sample chamber under ambient pressure due to the second order Doppler shift (SOD).

The result of the temperature calibration is shown in Fig. 2. Similar to MING *et al.* (1987) we observed a somewhat lower temperature of the sample (real temperature) compared to the set value of the regulator using calibration method 1. At a set value of 473 K a real temperature of 448 K was recorded, and at a set value of 823 K the real temperature in the sample chamber was observed to be 723 K; a similar difference was observed by MING *et al.* (1987) in their melting calibration. However when using the well known SOD (-7.3×10^{-4} $mms^{-1}deg^{-1}$) INGALLS (1967), much larger deviations are observed from the set value (see Fig. 2). At 873 K a sample temperature of only 613 K is estimated from the SOD value. The SOD measurements were most similar to the actual Mössbauer spectroscopy set up with the DAC in a horizontal position, the source cooling shield in place and a steady flow of argon gas. A plausible explanation for the observed large temperature difference in the diamond anvil cell could be the high thermal conductivity of diamonds, in combination with the cooling media brought close to the lower rocker and the diamond anvils (see Fig. 1). It is possible that the horizontally situated diamond anvil cell will enhance thermal convection in the system. The obtained temperature calculated from the SOD values are further supported by the observed changes in the hyperfine fields of α-iron with temperature (PRESTON *et al.*, 1962).

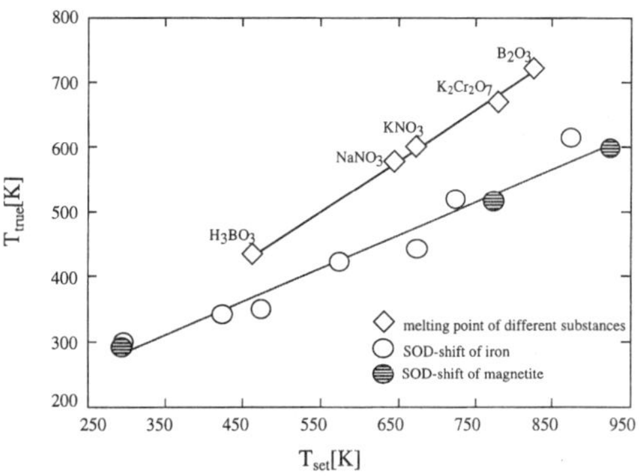

FIG. 2. Results of calibration of the sample chamber temperature. Diamonds: from salt oxides, Circles: from ^{57}Fe second order Doppler shifts. The size of the symbols indicates the estimated errors.

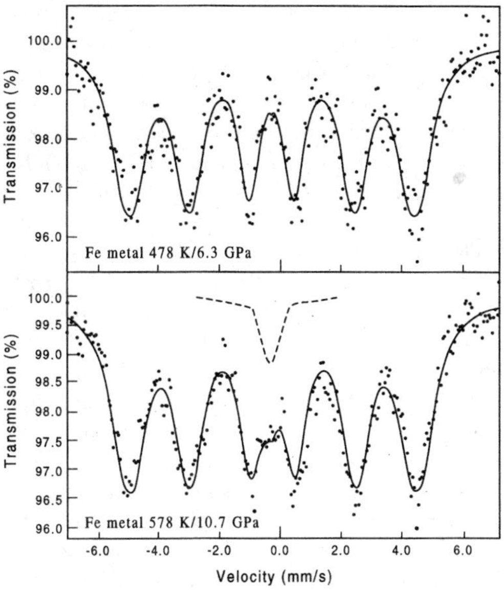

FIG. 3. Mössbauer spectra of ^{57}Fe at a) 478 K/6.3 GPa, b) 578 K/10.7 GPa. The dashed line represents the fitted single absorption line due to the ϵ-phase. Note the intensity ratio of individual lines indicating substantial thickness effect of the resonance.

The use of the SOD as a temperature calibration would be incorrect if the source were heated, which would result in too high centre shift values. In order to establish the temperature of the source during heating experiments, centre shifts of a thin iron foil above the cell, ventilated at room temperature, were measured while operating the empty pressure cell over a range of temperatures with the source cooled in the usual manner. At a DAC temperature of 873 K the first signs of an elevated temperature of the source was noticed. The SOD indicated that the source was approximately 10 K above room temperature. It is therefore justified to assume that heating of the source is negligible when performing experiments below 873 K.

Pressure calibration

The pressure calibration of the sample chamber was carried out at each set temperature in situ before and after each Mössbauer measurement. A laser optical system, situated close to the Mössbauer source, was used to determine the pressure dependent shift of the fluorescence line at 617.7 nm (ambient pressure) of a Sm doped YAG, excited by a 20 mW Ar-laser operating at 488 nm (BI *et al.*, 1990). Pressure calibration measurements were performed with the DAC in a vertical position after which it was put back over the Mössbauer source for Mössbauer spectra collection in a horizontal position without disconnecting the power and cooling lines. Use of ruby was avoided since the shift of the fluorescence lines depends on both pressure and temperature. In the laser optical system an ORIEL Multispec spectrograph was used to measure the shift of the fluorescence line. Precision of pressure measurements are estimated to be within ± 0.02 GPa. The pressure dependence of the centre shift is too small ($\approx -8.0 \times 10^{-3}$ mms^{-1}GPa^{-1}) to allow for a practical calibration of the pressure.

RESULTS

In our aim to explore the routine performance of *in situ* high P and T Mössbauer spectroscopy we

Table 1. Hyperfine parameters of iron at different temperatures and pressures

Pressure [GPa]	Temperature [K]	CS [mm/s]	B [Tesla]	FWHM [mm/s]	
6.3	478	-0.15	31.0	0.39	
	555	-0.21	29.8	0.38	
	573	-0.24	29.5	0.39	
	590	-0.24	29.5	0.39	
	605	-0.26	29.13	0.30	
10.7	471	-0.17	30.7	0.39	
	522	-0.21	30.1	0.39	
	522	-0.39		0.77	ε-phase
	522	-0.22	30.1	0.33	
	522	-0.35		0.83	ε-phase
	574	-0.26	29.3	0.32	
	574	-0.35		0.53	ε-phase

The uncertainty in pressure is ± 0.2 GPa, and in temperature ± 20 K. The uncertainty in the centre shift (CS) and the full width at half maximum (FWHM) are both ± 0.02 mm/s, and in the magnetic field (B) approximately ± 0.2 Tesla. The uncertainty in the parameters for ε-iron are larger because the singlet is mainly hidden under the sextet.

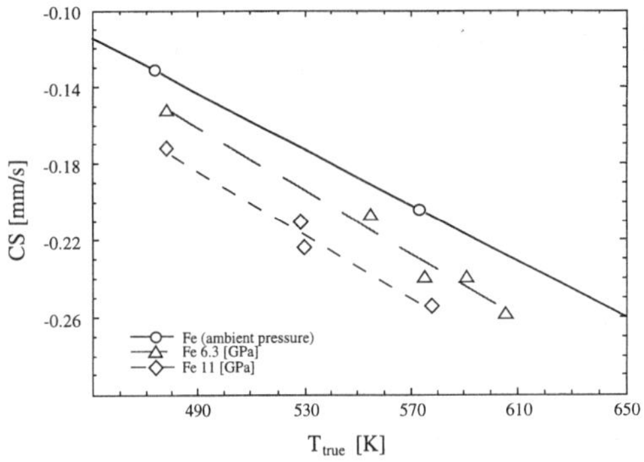

FIG. 4. Change in centre shift of ^{57}Fe with temperature and pressure.

have examined iron rich phases with well known properties. Metallic iron was chosen because it is the most iron rich phase, but with the present method it is also possible to investigate phases less rich in iron. For example, olivine with only 41% fayalite in solid solution can be measured within reasonable times, 20–24 hours.

Measurements on metallic iron

Our first spectra of metallic iron are shown in Fig 3. and the hyperfine parameters are given in Table 1. The spectra are broad and deviate considerably from the usual 3:2:1 peak intensity ratio due to the strong thickness effects (effective thickness calculated to 150 compared to 2.0 for a thin absorber). At high temperatures, however, the broadening of the spectra decreases as a result of the decreasing recoil free fraction and accordingly give a more favourable situation for site population estimates, *e.g.* in Fe-Mg silicates.

The increase in pressure and temperature are clearly demonstrated in the decrease of the centre shift values, partly due to the SOD caused by temperature but also from the change in electron density around the nucleus, a pressure effect discussed by INGALLS (1967). This behaviour explains the parallel lines representing data at 6.3 and 10.7 GPa, respectively, below the SOD lines (see Fig. 4). The results indicate a pressure shift in α-iron in the order of $-4.0 \times 10^{-3} mms^{-1}GPa^{-1}$, somewhat less compared to the tabulated observations for α-iron, -7×10^{-3} to -9×10^{-3} $mms^{-1}GPa^{-1}$, at room temperature in WILLIAMSON (1978).

Although we were not able to reach the $\alpha-\gamma$ transition, ϵ-iron appears in all spectra at 10.7 GPa in agreement with the phase diagram produced by AKIMOTO *et al.* (1987) for iron. The appearance of a single line demonstrates its paramagnetic state. The ϵ-phase seems always to occur simultaneously with the α-phase due to a sluggish phase transformation (MILLET and DECKER, 1969) and/or pressure gradients in the cell.

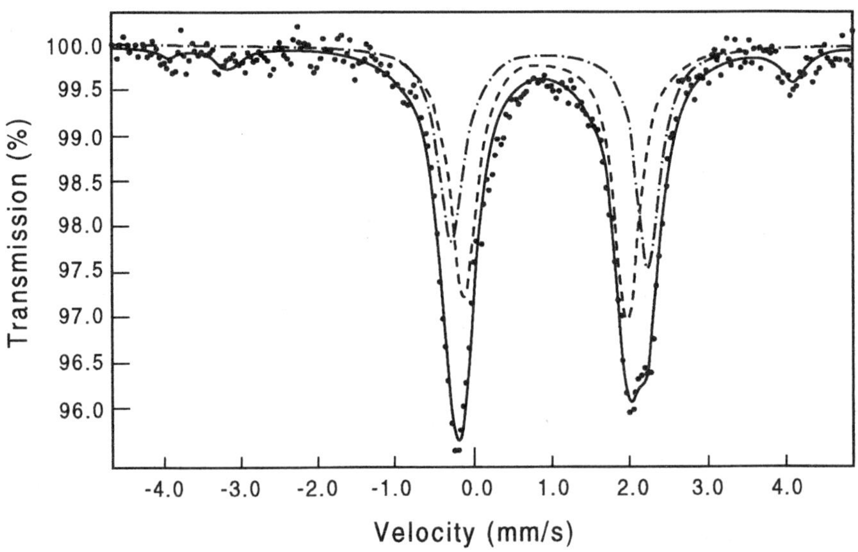

FIG. 5. Mössbauer spectrum of olivine $Fa_{41}Fo_{59}$ at 595 K/2.4 GPa. The week magnetic pattern is due to trace amounts of magnetite.

Table 2. Mössbauer parameters of synthetic olivine $Fa_{41}Fo_{59}$ at 573 K

Pressure [GPa]	Intensity M1 [%]	Intensity M2 [%]	CS M1 [mm/s]	CS M2 [mm/s]	QS M1 [mm/s]	QS M2 [mm/s]	FWHM [mm/s]
0.4	55	45	0.90	0.97	2.08	2.43	0.35*
2.4	56	44	0.92	0.98	2.12	2.49	0.36*
3.0	59	41	0.95	0.96	1.97	2.41	0.40*

*Full width at half maximum (FWHM) for M1 and M2 doublets was constrained to be equal.

All velocities are given relative to α-iron at room temperature. The estimated error for CS, QS and FWHM is approximately ± 0.02 mm/s and for the intensity ± 3%. The uncertainty in pressure is approximately ± 0.2 GPa.

Measurements on olivine

The Mössbauer spectra of olivine have the advantage of showing an increased resolution when measured at elevated temperature reflecting the distribution of Fe among the available M1 and M2 sites in the olivine structure (ANNERSTTEN *et al.*, 1982).

A synthetic olivine was produced from a reduced oxide mix (MgO, $^{57}Fe_2O_3$ and SiO_2) in cold sealed bombs at 0.2 GPa/975 K using graphite as an oxygen buffer. Small amounts of magnetite were present after the run and could clearly be seen in the Mössbauer spectra. From X-ray diffraction analysis of the run products the composition of the olivine was calculated according to the formula: $X_{Fa} = 7.522 - 14.9071(3.0199 - d_{130})^{1/2}$ (SCHWAB and KUSTNER, 1977) to be $Fa_{41}Fo_{59}$ instead of the original $Fa_{50}Fo_{50}$ oxide mixture and in agreement with the presence of small amounts of magnetite. Microprobe analysis could not be performed due to the small crystalline size of the olivine.

Figure 5 shows the spectra of olivine $Fa_{41}Fo_{59}$ at the temperature 575 K and the pressure 2.4 GPa. The resulting Mössbauer parameters obtained from the computer fit of the spectra are listed in Table 2. Over the small pressure range measured the centre shifts are constant within the error limit. The room temperature spectra are not resolved, as expected, that will allow a separation of the overlapping doublets.

It is interesting to observe that our data suggest a slight increase of iron preferring the M1 site with increasing pressure. Similar observations were made by AKAMATSU *et al.* (1993) on olivine samples quenched from high pressure. Our present data on iron site population in olivine are too sparse to confirm these data, but they are in qualitative agreement with the results of AKAMATSU *et al.* (1993).

CONCLUSIONS

Our work, so far, on high pressure-high temperature Mössbauer spectroscopy has shown the possibility of the DAC technique to study iron containing minerals in situ under those conditions under which rock forming processes prevail. Very high pressures can be reached with DAC but the temperature is limited to around 1200 K when the cell is put under vacuum. Studies at higher pressure and temperature are underway to further study the pressure influence on cation distribution in silicate minerals. Local heating, using laser radiation, may increase the temperature to higher values but will probably result in unpractically long measuring times due to the extreme low recoil free fractions of the absorber at such high temperatures.

Acknowledgement—Helpful discussions with Drs G. Shen, P. Lazor and T. Ericsson are acknowledged. Without the technical assistance from E. Meland, this project would not have been able to be performed. We thank for the constructive criticism from G. Amthauer, R. Wäppling and C. McCammon. This project was made possible through a grant from the Swedish Natural Research Council grant G-AA/GU 085431-307 to HA.

REFERENCES

AKAMATSU T., KUMAZAVA M., AIKIWA N. and TAKEI N. (1993) Pressure effect on the divalent cation distribution in non ideal solid solution of forsterite and fayalite. *Phys. Chem. Minerals* **19**, 431–444.

AKIMOTO S., SUZUKI T., YAGRI T. and SHIMOMURA O. (1987) Phase diagram of iron determined by high-pressure/temperature X-ray diffraction using synchrotron radiation. In *High Pressure Research in Mineral Physics* (eds. M. H. MANGHNANI and Y. SYONO) pp.149–154. Am. Geophys. Union, Washington D.C.

AMTHAUER G., FENNER J., HAFNER S. S., HOLZAPEL W. B. and KELLER R. (1979) Effect of pressure on resistivity and Mössbauer spectra of the mixed valence compounds Sn_2S_3. *J. Chem. Phys.* **70**, 4837–4842.

ANNERSTEN H., ERICSSON T. and FILIPPIDIS A. (1982) Cation ordering in Ni-Fe olivines. *Amer. Mineral* **67**, 1212–1217.

BI Q., BROWN M. and SATO-SORENSEN Y. (1990) Calibration of Sm:YAG as an alternate high pressure scale. *J. Appl. Phys.* **68**, 5357–5359.

DEBRUNNER P., VAUGHAN R. W., CHAMPION A. R., COHEN J., MOYZIS J. A. and DRICKAMER H. G. (1966) Versatile high pressure Mössbauer apparatus. *Rev. Sci. Instrum.* **37**, 1310–1315.

HAWTHORNE F. C. (1988) Spectroscopic methods in mineralogy and geology. In *Review in Mineralogy* (ed. F. C. HAWTHORNE) Vol. 18, 696 p. Mineralogical Soc. of America.

INGALLS R. (1967) Isomer shift of Fe57 in iron. *Phys. Rev.* **155**, 155–165.

MILLET L. E. and DECKER D. L (1969) Mössbauer measurements on iron at high pressure and elevated temperatures. *Phys. Lett.* **A 29**, 7–8.

MING L. C., MANGHNANI M. H. and BALOGH (1987) Restive heating in the diamond- anvil cell under vacuum conditions. In *High pressure research in mineral physics* (eds. M. H. MANGHNANI and Y. SYONO) pp. 69–74. Am. Geophys. Union, Washington D.C.

PRESTON R. S., HANNA S. S. and HEBERIE J. (1962) Mössbauer effect in metallic iron. *Phys. Rev.* **128**, 2207–2218.

SCHWAB R. B. and KUSTNER D. (1977) Präzisionsgitterkonstantenbestimmung zur Festlegung röntgenographischer Bestimmungkurven für synthetische Olivine der Mischkristallreihe Forsterit-Fayalit. *N. Jahrb. Mineral. Mh.* 205–215.

STERER E. and PASTERNAK M. P. (1990) A multipurpose miniature diamond anvil cell. *Rev. Sci. Instrum.* **61**, 1117–1119.

WILLIAMSON D. L. (1978) Influence of pressure on isomer shifts. In *Mössbauer isomer shifts* (eds. G. K. SHANOY and F. E. WAGNER) pp. 317–360. North-Holland Publishing Company.

Mineral Spectroscopy: A Tribute to Roger G. Burns
© The Geochemical Society, Special Publication No. 5, 1996
Editors: M. D. Dyar, C. McCammon and M. W. Schaefer

The hydrothermal diamond anvil cell (HDAC) and its applications

WILLIAM A. BASSETT[1], TZY-CHUNG WU[1], I-MING CHOU[2], H. T. HASELTON[2], JR., JOHN FRANTZ[3], BJORN O. MYSEN[3], WUU-LIANG HUANG[4], SHIV K. SHARMA[5], DAVID SCHIFERL[6]

[1]Department of Geological Sciences, Snee Hall, Cornell University, Ithaca, NY 14853, U.S.A.
[2]MS 959 U.S. Geological Survey, Reston, VA 22092, U.S.A.
[3]Geophysical Laboratory, 5251 Broad Branch Rd. NW, Washington, DC 20015, U.S.A.
[4]Exxon Production Research Company, 3319 Mercer St. Houston, TX 77027, U.S.A.
[5]Hawaii Institute of Geophysics, University of Hawaii, 2525 Correa Rd., Honolulu, HI 96822, U.S.A.
[6]Materials Research, Los Alamos National Laboratory, Los Alamos, NM 87545, U.S.A.

Abstract—A new style of diamond anvil cell, which is called the hydrothermal diamond anvil cell (HDAC), has been developed to make accurate measurements on samples under pressure and at temperatures between −190°C and +1200°C. It is especially suitable for fluid samples and for observing reactions between solids and fluids. The techniques for making observations and determinations used to date include optical microscopy, x-ray diffraction, Raman spectroscopy, luminescence, and infrared absorption. Three studies described in this paper deal with pressure-temperature calibration. They are (1) H_2O equation of state, (2) rapidly reversible second order phase transitions, and (3) shift of the Raman spectrum of diamond. Three applications of the HDAC described here include (1) luminescence and infrared absorption analyses of kerogen, lignite, and crude oil, (2) Raman spectroscopy of silicate melts, and (3) x-ray diffraction studies of dehydration and rehydration of montmorillonites using synchrotron radiation.

INTRODUCTION

THE HYDROTHERMAL diamond anvil cell (HDAC) is a new style of diamond anvil cell that was designed and built to meet a specific set of requirements (BASSETT *et al.*, 1993). It has been used to subject samples to a very uniform (±0.5°C) temperature over the range from −190 to +1200°C, hold the temperature constant within ±2°C, and measure the temperature by thermocouple with an accuracy of ±1.5°C. A rhenium gasket properly prepared is capable of holding a constant volume within ±2% or better over most of the temperature range. If a fluid such as water is used as the sample or pressure medium, its equation of state can be used to determine the pressure from the measured temperature. When the equation of state is well known and the density of the fluid is accurately measured by homogenization temperature or freezing temperature, the accuracy of such pressure measurements is significantly greater than by other pressure measuring methods formerly used with diamond anvil cells. This and two other new pressure measuring techniques are described in this paper.

Optical microscopy can be used to make direct *in-situ* observations of changes in the solid, liquid, and vapor phases of a sample as a function of pressure and temperature. Raman spectroscopy, luminescence, infrared absorption, x-ray diffraction, as well as other techniques can also be used for making *in-situ* observations in the new diamond anvil cell. In this paper we describe some new results from these techniques applied to a variety of samples held at high pressures and temperatures in the HDAC.

THE HYDROTHERMAL DIAMOND ANVIL CELL (HDAC)

Until recently, very few studies of hydrothermal reactions or of equations of state of fluids had been made in the diamond anvil cell (*e.g.*, VAN VALKENBURG *et al.*, 1987). The hydrothermal diamond anvil cell (HDAC), which is described in detail by BASSETT *et al.* (1993), consists of two platens with diamond anvils and heaters mounted at their centers (Fig. 1). The two platens are drawn together by tightening three nuts, thus applying pressure to a sample held between the diamond anvils. Samples are typically placed in a rhenium gasket consisting of a hole 200 to 500 μm in diameter and 75 to 500 μm thick. A rocker under one anvil allows for orientational alignment and a sliding disk under the other allows for translational alignment. Holes through the centers of the platens allow access to the sample for viewing and for various forms of electromagnetic radiation. Molybdenum wires wrapped around the tungsten carbide seats which support the diamond anvils serve as heaters (Fig. 2). These can heat the anvils and the sample very uniformly and very constantly to temperatures as high as 1200°C. By introducing liquid nitrogen the temperature can be reduced to −190°C. Electrical leads for the heaters and the thermocouples are fed through the platens. The vol-

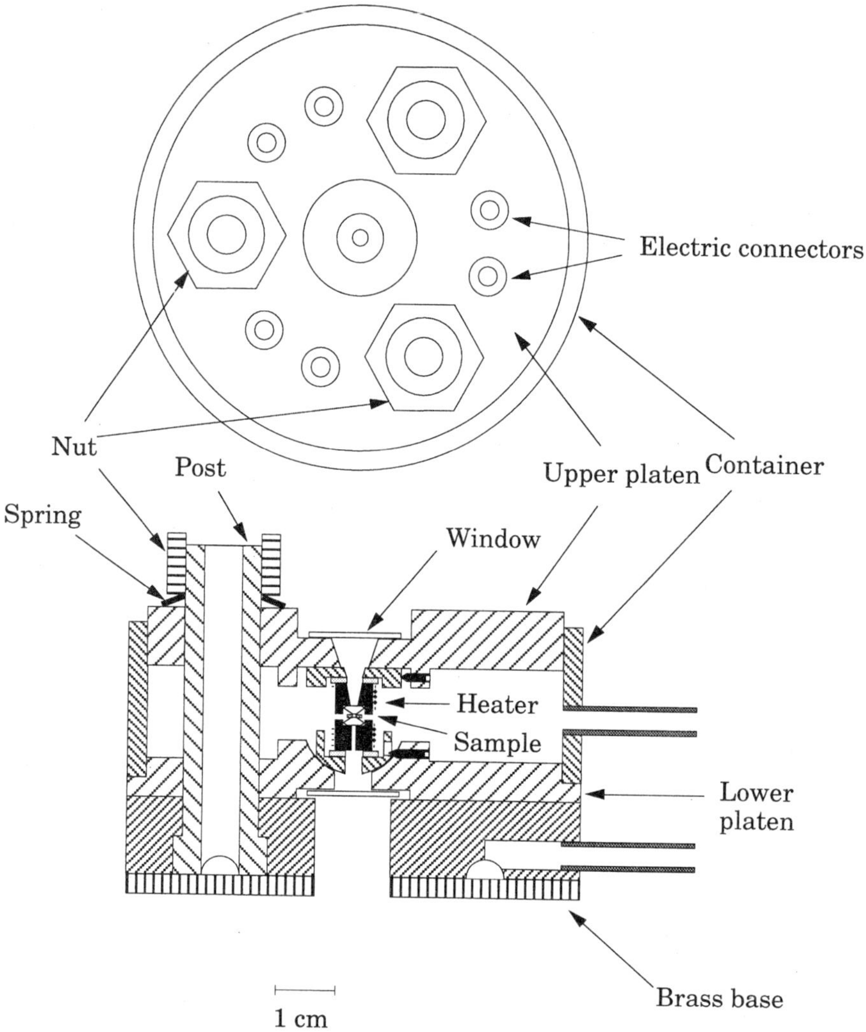

FIG. 1. Plan and elevation of the new diamond anvil cell. The diameter of the cell is 7.62 cm. The height is 5.72 cm. The base is constructed of brass; the platens, posts, and cylinder are constructed of stainless steel.

ume containing the heaters, the anvils, and the sample can be enclosed and completely surrounded with a gas to prevent oxidation. We have found that a constant flow of a mixture of Ar with 1% H_2 is very satisfactory for this purpose.

Laser light with any wavelength in the visible spectrum can be introduced into the top of the diamond cell to produce interference fringes in the sample chamber as well as the solid samples. Interference fringes produced by laser light reflected from top and bottom anvil faces provide a means for monitoring the volume of sample chamber. This is made possible by assuming that the refractive index of the homogeneous fluid portion of the sample remains constant during isochoric experiments. Fringes produced by laser light reflected from the top and bottom of a doubly polished solid sample provide a very sensitive means for observing volume and refractive index changes in samples due to transitions and reactions.

PRESSURE MEASUREMENT BY EQUATION OF STATE OF A FLUID AND HYDROTHERMAL STUDIES USING VISUAL OBSERVATION

When a sample is loaded in a gasketed diamond anvil cell surrounded by a fluid of known equation of state such as water, the pressure at which an

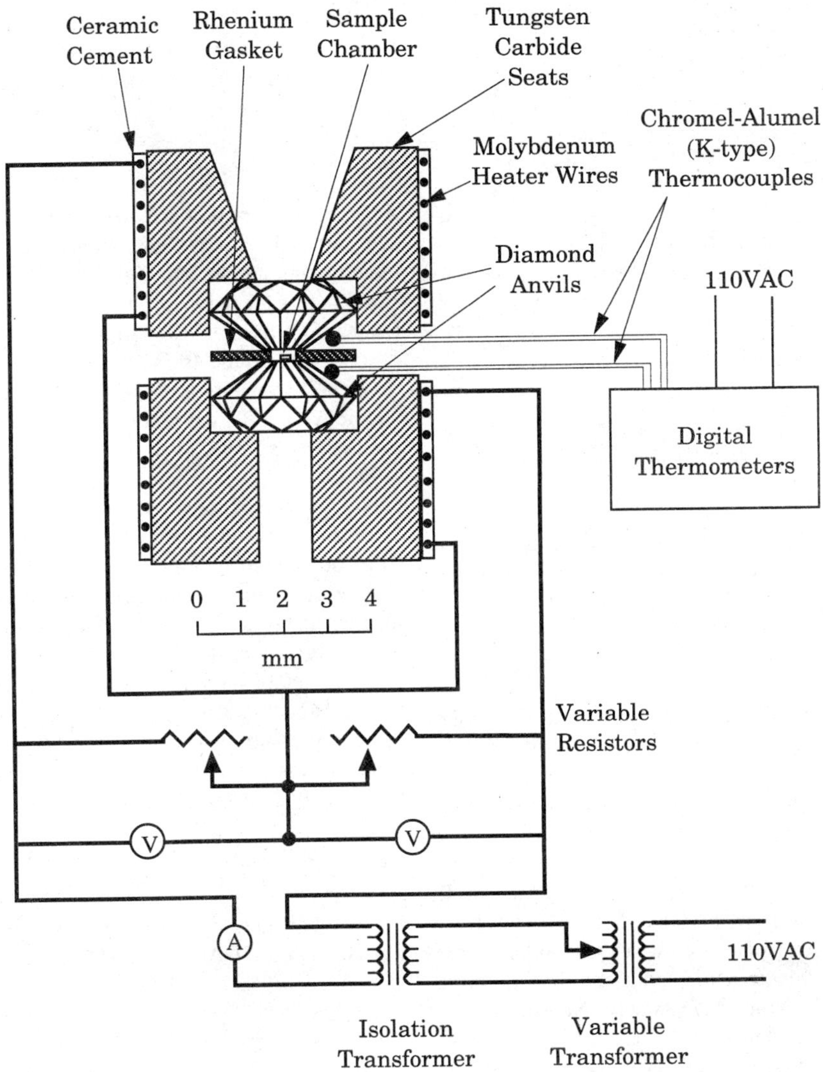

FIG. 2. Schematic diagram showing details of the sample, diamond anvils, heaters, and the electric circuits for heating, controlling, and measuring the temperature of the sample.

observation is made can be ascertained by the following steps: (1) determine the density of the fluid by measuring the temperature of a suitable transition such as homogenization or melting using the equation of state or P-T-ρ diagram of the fluid (Fig. 3); (2) monitor the sample chamber size to see that it does not change significantly by observing interference fringes caused by reflected laser light from the upper and lower anvil faces; (3) measure the temperature at which you make an observation on your sample; (4) using the equation of state or P-T-ρ diagram of the fluid, solve for the pressure. In practice the most accurate results are obtained during the cooling part of the cycle during which there is much less change of the gasket thickness. The path that is typically followed through the P-T-ρ diagram is illustrated in Fig. 4.

Independent Check of the Equation of State of H_2O

At Cornell University we have used the HDAC to compare the equation of state of H_2O with the P-T boundary of the $\alpha-\beta$ transition in quartz. The sample for these experiments consisted of a doubly-polished quartz platelet, distilled, deionized water, and an air

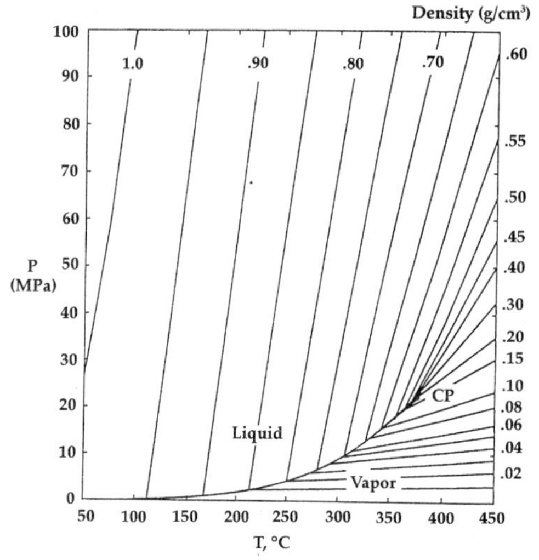

FIG. 3. Pressure-temperature plot of the equation of state of H_2O. LV represents the liquid-vapor coexistence curve and CP represents the critical point. The isochores are labeled by density (ρ) in g/cm^3.

bubble (Fig. 5). The sample was then heated until the bubble disappeared (homogenization). As the sample was further heated, both the temperature and pressure of the homogeneous fluid increased. This was continued until sudden motion of interference fringes was observed indicating the $\alpha-\beta$ transition. The sample was then cooled until a vapor bubble reappeared. If the interference fringes produced by reflection from the upper and lower anvil faces and the outline of the chamber hole indicated insignificant change during cooling, the volume of the sample chamber was assumed to have remained constant and, therefore, to have been isochoric. The sample was then slowly reheated until the vapor bubble disappeared, and the temperature at the moment of bubble disappearance was measured. If there was no loss of fluid during the experiment, this new homogenization temperature is typically found to be lower than the starting homogenization temperature indicating an increase in density due to relaxation of the gasket during the heating. The isochore density was determined from this homogenization temperature and the transition pressure was determined from the temperature along the isochore at the time fringe movement in the quartz was observed. Our results (SHEN *et al.*, 1992) show that the equation of state of H_2O of HAAR *et al.* (1984) is in good agreement with the $\alpha-\beta$ quartz boundary of MIRWALD and MASSONNE (1980) for several different isochores.

Isochores of increasing density were studied this way until temperatures required were so high that rapid solution of the quartz crystal took place before the $\alpha-\beta$ transition in the quartz could be observed. These techniques are described in more detail in BASSETT *et al.* (1993) and SHEN *et al.* (1992).

Precision and Accuracy

The thermocouples are not subject to any pressure-induced error because they are external and therefore not pressurized. Temperature can be measured with a precision of ±0.1°C over the range of temperature from −110°C to 380°C. The accuracy of our measurements is ±0.5°C over the same range. At temperatures from −190°C to −1101°C and above 380°C, the precision is ±0.5°C and the accuracy is ±1.5°C.

The accuracy of a pressure determination depends on the accuracies of the temperature measurements and the equation of state of H_2O (Fig. 3) or other fluid being used as pressure medium. The accuracy for our homogenization temperature measurements in H_2O is believed to be ±0.5°C, which corresponds to a maximum density uncertainty of ±0.004 (g/cm^3). When all of the uncertainties are collected, this method is considered to provide pressure determinations with an accuracy of 1% in the range of temperatures and pressures up to 1000°C and 0.5 GPa respectively.

NEW SOLID PRESSURE CALIBRANTS USING VISUAL OBSERVATION

Three new pressure calibrants have been characterized in the HDAC using H_2O as the pressure medium. They are $BaTiO_3$ (CHOU *et al.*, 1993), $Pb_3(PO_4)_2$ (CHOU and NORD, 1994) and $PbTiO_3$ (CHOU and HASELTON, 1994), and their applicable temperature ranges are below about 120, 180, and 490°C, respectively (Fig. 6). The common features for these calibrants include: (1) they are all based on ferroelastic phase transitions that are rapid and can be easily and precisely determined visually; (2) the transition temperatures decrease as pressure increases, such that the phase boundaries in P-T space intersect isochores of common geologic fluids at high angles (Fig. 6) and, thus, facilitate their application; and (3) that they are inert in most pressure media in a wide P-T range.

To apply these calibrants in the HDAC, a doubly polished platelet or a cleavage fragment (typically 200 x 200 x 40 μm) is loaded with a fluid pressure medium in the HDAC. The tetragonal-cubic phase transition for $BaTiO_3$ and the monoclinic-trigonal transition for $Pb_3(PO_4)_2$ can both be detected by the disappearance and reappearance of twin boundaries during heating and cooling, respectively. The hysteresis is less than

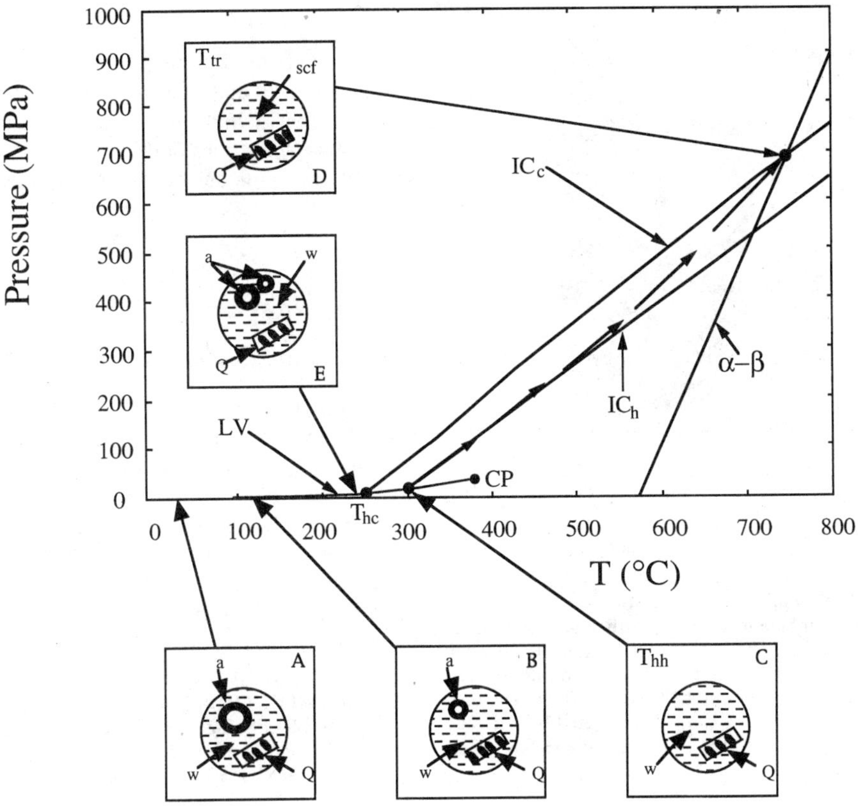

FIG. 4. This diagram shows a typical pressure-temperature path of an experiment to examine the $\alpha-\beta$ transition in quartz. Insets show the appearances of the sample chamber at various steps along the path. CP is the critical point, LV is the liquid-vapor coexistence curve, IC_h is the isochore followed during heating, IC_c is the isochore followed during cooling, T_{hh} is the homogenization temperature during heating, T_{hc} is the homogenization temperature during cooling, T_{tr} is the transition temperature, a is an air bubble, w is water, scf is supercritical fluid, Q is a sample of quartz. The steep straight line represents the $\alpha-\beta$ transition in quartz.

1°C and 0.2°C for $BaTiO_3$ and $Pb_3(PO_4)_2$, respectively. We can determine the transition temperatures (T_{tr}) with a precision of ±0.2°C and an accuracy of ±0.5°C. The tetragonal-cubic phase transition in $PbTiO_3$, which is indicated by the movements of phase fronts, was detected optically at a heating or cooling rate of about 1°C/min. The temperature interval in which the phase transition occurs ranges from 0.8°C to 2.8°C, depending on pressure. The temperature at which the movement of phase fronts starts (or ends) on heating (or cooling) is taken as the transition temperature [$T_{tr,h}$ (or $T_{tr,c}$)]; these temperatures can be determined with a precision of ±0.4°C and an accuracy of ±1.5°C. Results of our calibration are shown in Fig. 6 and described in the following sections.

$BaTiO_3$

Different regions of our $BaTiO_3$ sample, synthesized by MATTHIAS (1948) and characterized by EVANS (1961), have different T_{tr} values (113-120°C at 1 atm.); these differences may be due to variations in impurity content or local strain. For a specific region of the crystal in our heating experiments, we observed T_{tr} = 115.3°C at 0.17 MPa (1.7 bar) along the liquid-vapor curve of pure H_2O (SAUL and WAGNER, 1989) and 49.2°C at 1404.8 MPa (14048 bar) along the ice VI-water boundary (WAGNER *et al.*, 1994). On the basis of these measurements, we obtained the relation

$$P_{tr}(\text{MPa}) = 0.17 - 21.25\,[(T_{tr})_P-(T_{tr})_{LV}] \quad (1)$$

where LV represents the liquid-vapor pressure of pure H_2O. Observations at intermediate pressures, when compared with H_2O equation of state calculations, are consistent with a linear relationship. The slope of −21.25 MPa/°C (−212.5 bar/°C) agrees very well with the value reported by SAMARA

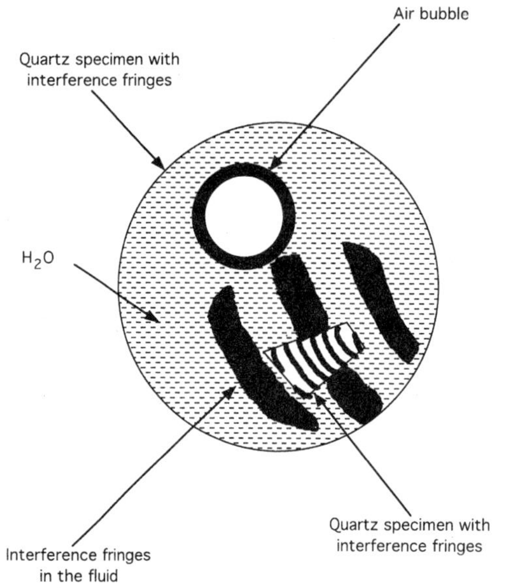

FIG. 5. Typical appearance of a solid sample in liquid water with an air bubble in a circular rhenium gasket hole 500 μm in diameter. The portion of the sample illuminated with laser light shows coarse fringes due to interference between light reflected from the two anvil faces. The fine fringes in the solid quartz sample are caused by interference between light reflected from the top and bottom of the platelet.

(-21.0 MPa/°C with $(T_{tr})_{0.1\ MPa}$ = 122°C (-210 bar/°C with $(T_{tr})_{1bar}$ = 122°C) cited in RAMIREZ *et al.* (1990). The data of DECKER and ZHAO (1989) indicate that the equation given above is probably valid to the tricritical pressure for $BaTiO_3$ of about 3.5 GPa (35 kbar).

$Pb_3(PO_4)_2$

In our heating experiments, the monoclinic-trigonal transition occurs at 180.2°C and 1.01 MPa (10.1 bar) along the L-V curve of H_2O and at 49.1°C and 1402.7 MPa (14027 bar) along the ice VI and water boundary. These two P-T points define a linear phase boundary having a slope of -10.62 MPa/°C (-106.2 bar/°C). Observations at intermediate pressures (along several isochores of H_2O), when combined with H_2O equation of state calculations, are consistent with this linear relation. Our measured T_{tr} value agrees very well with the value of 180.4°C at 1 atm. reported by BISMAYER and SALJE (1981). This calibrant has a pressure resolution of ±2.1 MPa (21 bar), which is twice the sensitivity of $BaTiO_3$ described above.

$PbTiO_3$

The sample used in our calibration experiments was synthesized and characterized by BING-NAN SUN (personal communication, 1994). The tetragonal-cubic phase transition temperatures were determined at 1 atm. and along six isochores of H_2O in our heating and cooling experiments (CHOU and HASELTON, 1994). Least squares regression of our data give

$$P_{tr,h}(\mathrm{MPa}) = 7021.7\text{-}14.235\ T_{tr,h}(°\mathrm{C}), \quad (2)$$

$$P_{tr,c}(\mathrm{MPa}) = 6831.3\text{-}14.001\ T_{tr,c}(°\mathrm{C}), \quad (3)$$

with the R-squared values of 0.9966 and 0.9987, respectively. For clarity, only the first equation is plotted in Fig. 6. The slopes for the phase boundary are -14.235 and -14.001 MPa/°C (-142.35 and -140.01 bar/°C), respectively. They are more negative than the -11.91 MPa/°C (-119.1 bar/°C) reported by SAMARA (1971) for uranium-doped $PbTiO_3$; the variation may result from the compositional difference.

RAMAN SPECTROSCOPY OF ^{13}C DIAMOND AS PRESSURE SENSOR AT HIGH TEMPERATURES: MELT STUDIES OF HYDRATED SILICATES

Raman spectroscopy of ^{13}C diamond has been developed as a new method for measuring pressure

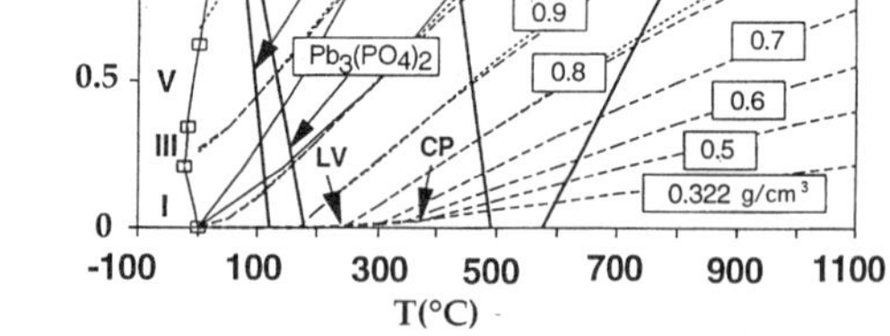

FIG. 6. A P-T plot showing the phase relations in pure H_2O and four pressure calibrants (quartz, $PbTiO_3$, $Pb_3(PO_4)_2$, and $BaTiO_3$). I, III, V, and VI are ice polymorphs; LV is the liquid-vapor curve of H_2O; CP is the critical point of H_2O; the long dashed and dotted curves are isochores of H_2O derived from equations of state formulated by HAAR *et al.* (1984) and SAUL and WAGNER (1989), respectively. The numbers in the rectangles give the densities of the isochores in g/cm³.

at high temperatures in diamond anvil cells (SCHIFERL *et al.*, in preparation). When combined with Raman spectroscopy of ^{12}C diamond, the method becomes a powerful means for measuring pressure at high temperatures, especially in samples that are corrosive to oxide pressure sensors such as ruby and Sm:YAG.

At the University of Hawaii, experiments are being conducted at simultaneous high P and T conditions in diamond anvil cells for studying minerals and hydrated silicate melts. The HDAC style of cell as well as an externally heated Merrill-Bassett DAC have been adapted for use with a multichannel micro-Raman spectrometer with a fiber-optics based technique to acquire pressure sensitive fluorescence spectra of Sm:YAG simultaneously with Raman spectra of the sample. High-frequency Raman spectra of forsterite have been measured to 650°C (923 K) and pressures above 8 GPa (80 kbar) (SCHIFERL *et al.*, 1993).

The HDAC is easily mounted on the stage of the Raman microscope for optical observation and Raman spectral measurements of samples pressurized to 10 GPa (100 kbar) and simultaneously heated to 900°C (1173 K). With the HDAC we measured Raman spectra of hydrous potassium silicate and hydrous albite to 10 GPa (100 kbar) at room temperature. We have also measured spectra of hydrous potassium silicate melts to 550°C (823 K) and at pressures of about 0.7 GPa (7 kbar) but encountered technical difficulties. One of the major problems with hydrous silicate systems is that these are extremely corrosive, even at moderate temperature, and would invariably react with either ruby or Sm:YAG pressure calibrants. Furthermore, silicate glasses and melts are very weak Raman scatterers. In order to overcome these technical difficulties, we have refined the confocal micro-Raman technique so that it can be used with diamond anvil cells for recording spectra of very weak Raman scatterers (SHARMA *et al.*, 1994), and developed the $^{13}C/^{12}C$ diamond Raman sensor system for measuring pressure for all samples under any simultaneous P-T conditions that can be achieved in externally heated diamond-anvil cells (SCHIFERL *et al.*, in preparation).

The use of the Raman spectra of natural diamond (99% ^{12}C and 1% ^{13}C) for measuring pressure has been suggested by a number of authors (SHERMAN, 1985; HANDFLAND *et al.*, 1985; SHARMA *et al.*, 1985; ALEKSANDROV *et al.*, 1987; TARDIEU *et al.*, 1990). However, even with the micro-Raman technique it is very difficult to separate the Raman spectrum of the sensor chip from the strong Raman signal of diamond anvils both of which appear at 1332.5 cm^{-1} under ambient conditions. This problem persists to pressures above 14 GPa (140 kbar), where the Raman signal from the sensor chip under hydrostatic pressure moves (at 2.90 ± 0.05 cm^{-1}/GPa) higher and becomes well separated from the Raman signals of the anvils.

We have used the HDAC as well as the Merrill-Bassett and Mao-Bell cells for measuring pressure dependence of the first-order Raman line of synthetic ^{13}C-enriched (91 atomic %) diamond film. The pressure dependence of the spectra of small (20 μm × 20 μm × 20 μm) fragments of the polycrystalline diamond film was measured with ruby R_1 and R_2 fluorescence, using either methanol-ethanol (4:1) mixture or water as pressure transmitting medium. At room temperature, 23°C (296 K), the Raman line of the ^{13}C-enriched film appears at 1287.79 cm^{-1}, and a fit of our data and that of MUNIOV *et al.* (1994) for the same ^{13}C diamond film yields a shift of 2.8294 ± 0.047 cm^{-1}/GPa (SCHIFERL *et al.*, in preparation). The temperature dependence of the Raman line of the ^{13}C diamond has been measured to 1056°C (1329 K) and is given by:

$$V(T) = 1294.5 - 0.022\,T - 3.8 \times 10^{-6}\,T^2, \tag{4}$$

where V is the frequency in cm^{-1} and T is the temperature in Kelvin (K = °C + 273.13). Using these calibrations, the ^{13}C-diamond chip provides the first means of measuring pressure up to 16 GPa (160 kbar) accurately (±0.3 GPa or ±3 kbar) at temperatures as high as 927°C (1200 K) for samples in corrosive environments. Above 14 GPa (140 kbar), the Raman spectrum from the ^{12}C diamond chip under hydrostatic pressure comes out from under the diamond-anvil Raman line. Recently, the temperature dependence of the Raman line of natural diamonds has been measured to 1627°C (1900 K) (HERCHEN and CAPPELLI, 1991; ZOUBOULIS and GRIMSDITCH, 1991). The $^{13}C/^{12}C$ Raman pressure sensor system is now ready for measuring simultaneous high temperature and high pressure samples in diamond anvil cells.

IN SITU MONITORING OF ORGANIC TRANSFORMATION USING LUMINESCENCE AND FTIR SPECTROSCOPY

In the research lab of the Exxon Production Research Company we have developed techniques to monitor *in situ* organic transformation using visual observation, fluorescence, and infrared spectroscopy. The responses of organic materials to heating were observed in real-time using a microscope equipped with a charge couple device (CCD) video monitoring system, and fluorescence was monitored using a luminescence spectrometer. Progressive changes of functional groups were monitored during transformation by micro-fourier transform infrared spectroscopy (micro-FTIR). Each spectrum accumulated from 50 scans was recorded with a spectral resolution of 4 cm^{-1} in the 700–4000 cm^{-1} range. Experiments were conducted in open and closed systems at anhydrous or hydrous conditions for kerogen, lignite, and crude oil samples. The merit of the HDAC over conventional hot-stage methods is its capability of conducting experiments at hydrous and high-pressure conditions. Preliminary results illustrate the capabilities of the technique to examine reaction pathways and the timing and rates of organic transformations. The visual experiments provide direct observations and estimations of the temperature (or timing) and yields of liquid and gas generation at a variety of experimental conditions.

Pyrolysis of kerogen

A concentrated kerogen from shale of the Green River Formation (Eocene) was heated at a rate of 25°C/min to 380°C. The temperature was then held at 380°C for 118 hours. The results show a successive change in spectrum distribution in the range of 1700 to 3100 cm^{-1}. Deconvolution of the spectrum in this range using curve fitting has been performed to calculate five aliphatic C-H_x stretching bands ($CH_{3(asym)}$, $CH_{2(asym)}$, CH, $CH_{3(sym)}$ and $CH_{2(sym)}$) and one aromatic C-H stretching band. The dramatic increase in the $CH_{3(asym)}/CH_{2(asym)}$ intensity ratio of asymmetric aliphatic stretching bands at the late stage of pyrolysis indicates that the cleavage of aliphatic chains occurs during the kerogen transformation (LIN and RITZ, 1993). The trend of the aromatic/aliphatic stretching band ratio, which first increases then dramatically decreases, is probably attributable, respectively, to the formation of the alkyl-substituted aromatic hydrocarbons and the rapid cracking of aliphatic chains in the aromatic compounds at the late stage of pyrolysis (BENKHEDDA *et al.*, 1992). This preliminary study shows that in situ FTIR spectroscopic analysis during HDAC experiments has a potential to monitor the reaction pathways or derive kinetic information during kerogen transformation.

Cracking of crude oil

A crude oil sample from Gwydr Bay (API gravity = 31) was heated to 450°C at a rate of 25°C/min. The temperature was then held at 450°C for 45 minutes. The results show that (1) a progressive decrease in the absorption band ratio of carbonyl or carboxyl C=O stretching (~1710 cm^{-1}) to aromatic C=C stretching (~1630 cm^{-1}); (2) the peak intensities for both aliphatic and aromatic groups progressively decrease with maturity, suggesting the cracking of oil to gas; (3) the aliphatic/aromatic band ratio (1460 cm^{-1}/1630 cm^{-1}) increases significantly (about two-fold) during the early and middle stages of maturation then decreases slightly with increasing maturity; (4) the methane absorption band at ~3020 cm^{-1} was detected by probing the bubble space of the sample during cooling. The progressive variation of the functional groups in the samples during pyrolysis suggests that FTIR spectra can be used to follow the kinetics of oil cracking.

Monitoring timing of gas generation during pyrolysis of coals

A low maturity coal (lignite with vitrinite reflectance (Ro) = 0.31 and hydrogen index (HI) = 130) was heated at a rate of 25°C/min. to 380°C and the temperature was then held for 22 hours before cooling. The vapor generated within the sample chamber has been probed. The vapor species were identified from their designated functional groups. The results (Fig. 7) show that gas generation occurring from early to late follows the sequence: CO_2, carbonyl-carboxylate vapors, hydroxyl-vapor, hydrocarbon wet gases and carbon monoxide. This technique can be an effective tool for studying the timing of generation of gases from source rocks. This technique is particularly useful for identifying the type of coals or kerogens which can thermally generate hydrocarbon gases at low maturity levels or for quantifying the effect of geological catalysts on generating premature gases from coals or kerogens.

ION-PAIRING IN $MgSO_4$ SOLUTIONS BY RAMAN SCATTERING

Gaskets lined with platinum were used to contain chemically reactive samples of $MgSO_4$ (FRANTZ *et al.*, 1994). A 0.5 m solution of $MgSO_4$ was analyzed using a micro-Raman spectrometer while conditions were changed along an isochore from room temperature to 500°C and 1.1 GPa (11 kbar). Relative abundances of species and the extent of ion pairing of $MgSO_4$ were measured along an isochore of ~1g/cm^3. Ion pairing was observed to increase with increasing pressure and the mass-action constants for the reaction Mg^{+2} + SO_4^{2-}

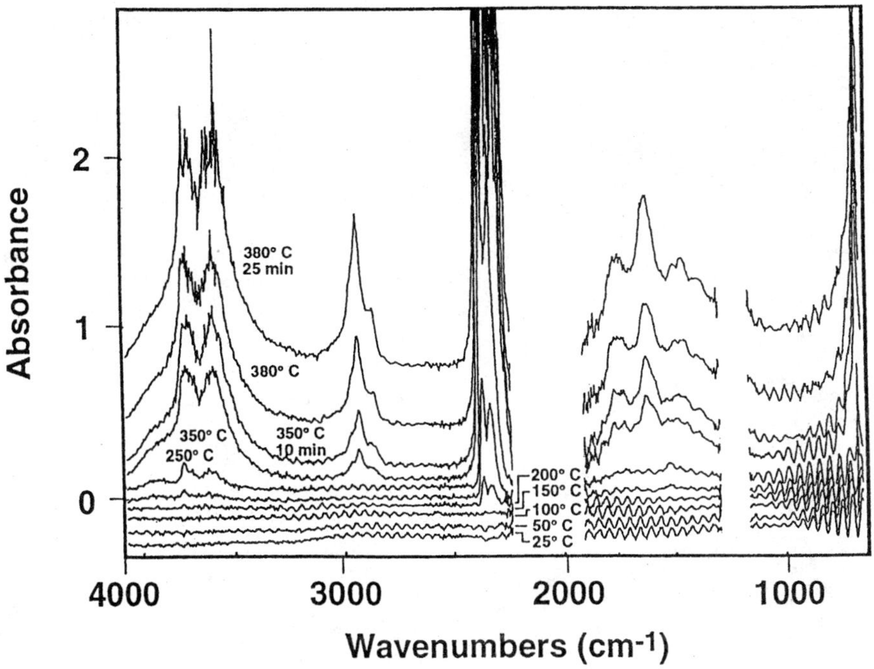

FIG. 7. Infrared spectra showing the generation of CO_2, carbonyl-carboxylate vapors, hydroxyl-vapor, hydrocarbon wet gases and CO during heating of a lignite sample.

= $MgSO_4$ were found to range from 35 at 20°C and 0.1 MPa (1 bar) to 1.25 x 10^4 at 500°C and 1.14 GPa (11.4 kbar). These results are in good agreement with those obtained by RITZERT and FRANCK (1968) using electrical conductance measurements.

This trend of $MgSO_4$ to become a weaker electrolyte under hydrothermal conditions indicates a greater disruption of the relatively weak ion-water dipole bonds thus allowing a higher frequency of ion-ion interactions that result in ion pairing. This study of the effect of pressure and temperature on ion-water dipole bonds provides information that helps us to better understand viscosity as a function of pressure and temperature, a relationship that plays an important role in heat and mass transfer in the Earths crust and mantle by flow of aqueous fluids.

HYDRATION AND DEHYDRATION OF CLAYS USING SYNCHROTRON RADIATION

The montmorillonite members of the clay family of minerals contain large quantities of water and are abundant in sediments. Therefore, they are among the most important hydrous phases to understand. These are sheet silicates with low negative charges on the sheets, charges that are balanced by cations filling the spaces between the sheets. Because of the low charges only a small number of cations fill the spaces and when they fill the spaces, they carry with them envelopes of H_2O molecules. These H_2O molecules reside between the sheets forming layers of water resulting in d-spacings along the c-axis of 12.5Å, 15Å, and 19Å. These cations and their accompanying water molecules are readily exchanged when other cations are made available in solution.

Combining the HDAC with the very intense synchrotron radiation available on the white x-ray beamline at the Cornell High Energy Synchrotron Source (CHESS) has enabled us to make real-time measurements of changes in the hydration states of montmorillonites in response to changes in pressure and temperature. The sample consists of a grain of montmorillonite, distilled H_2O, and an air bubble. Pressure can be determined from the equation of state of the water as described in an earlier section. Energy dispersive x-ray diffraction patterns of our sample were obtained by passing a white beam of x-rays through the diamond anvils as the sample was heated and compressed. X-rays diffracted by our sample were collected by an intrinsic germanium detector placed at a 2θ angle of

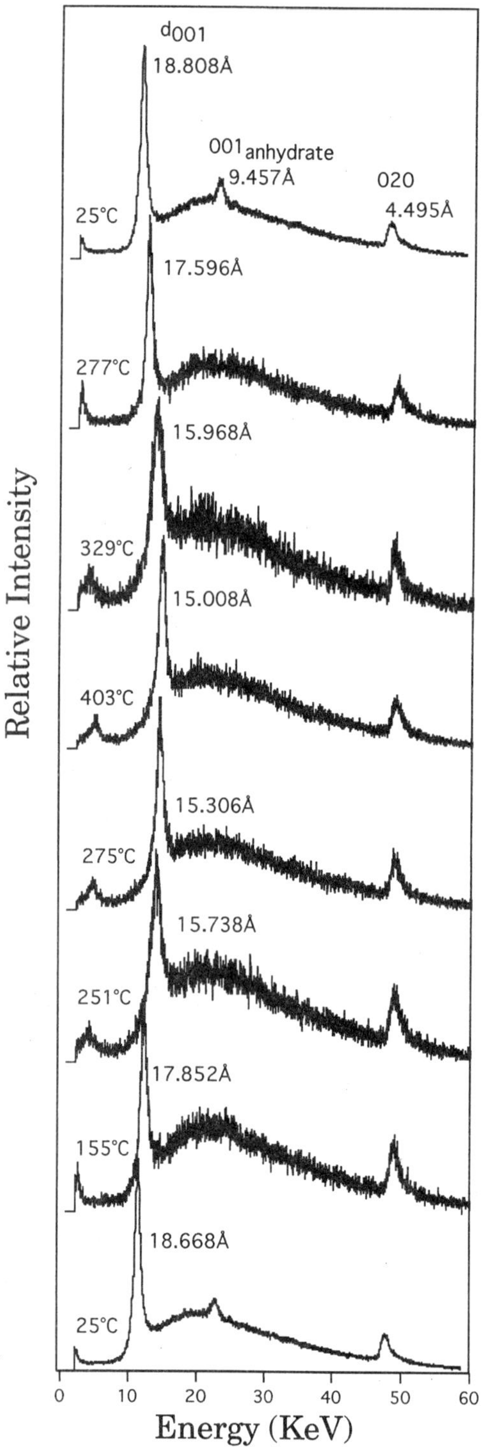

FIG. 8. A series of diffraction patterns of Ca-montmorillonite as the sample is heated and then cooled along the 1.053 g/cm^3 isochore.

3° (Fig. 8). Less than a minute was required to collect enough information to measure d-spacings. This allows us not only to observe the changes in the hydration states of water but also the response rate when temperature and pressure are changed. The three prominent hydration states of montmorillonite are easily recognized from the c-axis d-spacings.

Dehydration and rehydration experiments have been performed on montmorillonite with Ca, Mg, and Na cations in the exchange positions. Our results show that the 19Å hydrate is stable up to 200–380°C when the water pressure is greater than that along the H_2O liquid-vapor (LV) curve. The 15Å hydrate is stable up to 450–600°C. At pressures greater than the LV curve the dehydration temperatures are not very sensitive to pressure but are sensitive to the interlayer cation species (Fig. 9). Increasing ionic potential (charge/radius) is correlated with a decrease in the temperature of the 19Å-15Å transition and an increase in the temperature of the 15Å-12.5Å transition.

The effect of pressure and temperature on the hydration states of montmorillonite based on our measurements are compared in Fig. 10 with a recently published thermodynamic model proposed by RANSOM and HELGESON (1995). In their model Ransom and Helgeson propose a starting hydration state of 15Å and a 50% dehydration by approximately 180°C during burial. In contrast, our measurements indicate that the starting hydration state

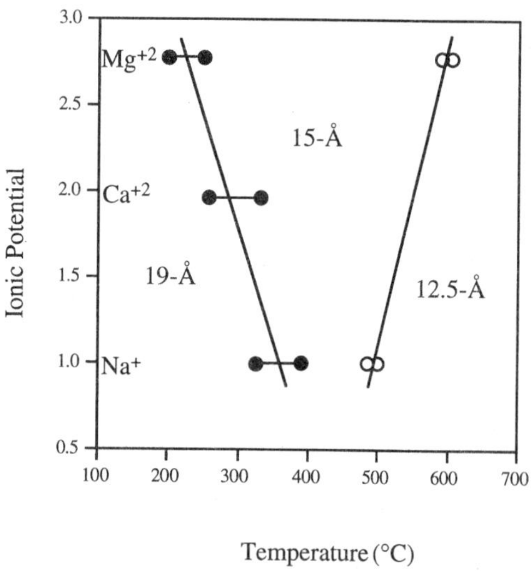

FIG. 9. Plot of the dependence of the 19Å-15Å and 15Å-12.5Å transitions in montmorillonite as a function of ionic potential (ionic charge/radius).

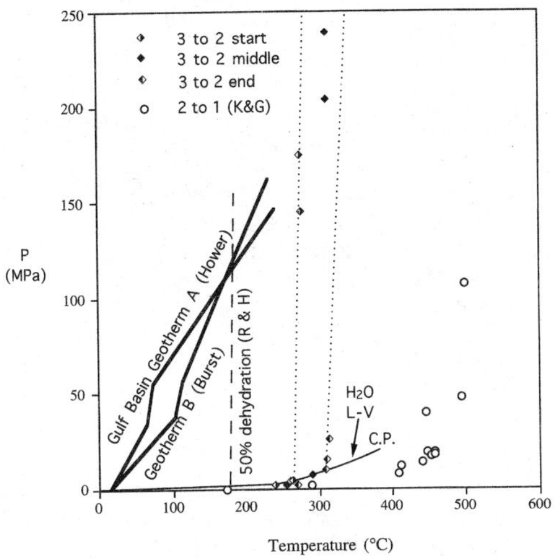

FIG. 10. The two dotted lines bracket our measured dehydration from 19Å to 15Å hydrates and the open circles indicate dehydration from 15Å to 12.5Å hydrates in montmorillonites (KOSTER VANGROOS and GUGGENHEIM, 1987). These data are compared with the recently proposed 50% dehydration (dashed line) assuming an initial 15Å hydrate (RANSOM and HELGESON, 1995). Heavy solid lines show two versions of geotherms from BURST (1969) and HOWER *et al.* (1976).

is 19Å and that dehydration to 15Å does not occur until approximately 200–380°C and the dehydration to the 12.5Å hydrate does not occur until 450–600°C. Thus, our observations indicate that montmorillonite retains more water to higher temperatures than previously thought. These observations are important for sedimentary basin and petroleum generation modeling. Our results indicate that for any water-releasing reaction shallower than 5 km, the 19Å hydrate probably dominates and may be responsible for the release of 44% more water than previously thought. Another consequence of our observations concerns the mechanical properties of montmorillonite clay. Our results show that those clays, if they can escape chemical alteration to illite by potassium, are expected to remain ductile down to considerable depths where they may be the best candidate to form a good seal for deep oil reservoirs.

Our observation that the 15Å hydrate of montmorillonite is stable up to 450–600°C at pressures higher than the H_2O LV curve suggest that clays might be transported down subducting slabs before the remaining water is released. The dehydration in subducting slabs is considered to be an important cause of shallow earthquakes as well as the source of volatiles that create volcanic activity along some of the boundaries of tectonic plates.

CONCLUSIONS

The hydrothermal diamond anvil cell (HDAC) has found a wide variety of applications in the study of geological samples employing a variety of observational techniques where there is need to make observations and measurements at simultaneous high pressure and high temperature. A new paper by CHOU *et al.* (1995), reports the use of the HDAC to study eutectic melting of the assemblage $Ca(OH)_2$ + $CaCO_3$ with excess H_2O and lack of evidence for "portlandite II" phase. Another study is under way at the Bayerisches Geoinstitut in Bayreuth, Germany by Spetzler, Shen, and Angel to develop techniques for making ultrasonic velocity measurements as a function of pressure and temperature in the HDAC with gigahertz sound waves. Judging from the interest in the HDAC, many more novel applications will be developed in the years ahead.

Acknowledgements—We wish to thank the Exxon Production Research Company, Houston, TX for supporting the research on clays and organic samples. Some of the research was conducted at the Geophysical Laboratory of the Carnegie Institution of Washington and in US Geological Survey (USGS) research labs. Most of the rest of the technique development and research reported in this paper was supported by diverse grants from the National Science Foundation over a period of several years. We wish to thank the staff of the Cornell High Energy Synchrotron Source for their help in carrying out synchrotron radiation experiments.

REFERENCES

ALEKSANDROV I. V., GONCHAROV A. F., ZISMAN A. N. and STISHOV S. M. (1987) Diamond at high pressures: Raman scattering of light, equation of state and high pressure scale. *Soviet Phys. JETP* **66**, 384–390.

BASSETT W. A., SHEN A. H., BUCKNUM M. and CHOU I.-M. (1993) A new diamond anvil cell for hydrothermal studies to 2.5 GPa and from −190 to 1200°C. *Rev. Sci. Instrum.* **64**, 2340–2345.

BENKHEDDA Z., LANDAIS P., KISTER J., DEREPPE J.-M. and MONTHIOUS M. (1992) Spectroscopic analysis of aromatic hydrocarbons extracted from naturally and artificially matured coals. *Energy and Fuels* **6**, 166–172.

BISMAYER U. and SALJE E. (1981) Ferroelastic phases in $Pb_3(PO_4)^2$-$Pb_3(AsO_4)_2$; X-ray and optical experiments. *Acta Cryst.* **A37**, 145.

BURST J. F. (1969) Diagenesis of Gulf Coast clayey sediments and its possible relation to petroleum migration. *Amer. Assoc. Petrol. Geol. Bull.* **53**, 73–93.

CHOU I.-M., HASELTON H. T., JR., NORD G. L., JR., SHEN A. H. and BASSETT W. A. (1993) Barium titanate as a pressure calibrant in the diamond anvil cell. *Amer. Geophys. Union Trans.* **74**, 170.

CHOU I.-M. and HASELTON H. T., JR. (1994) Lead titanate as a pressure calibrant in the diamond anvil cell. *Abstr., 16th Gen. Mtng., IMA*, 74–75.

CHOU, I.-M. and NORD G. L., JR. (1994) Lead phosphate

as a pressure calibrant in the hydrothermal diamond anvil cell. *GSA Abstr. Progr.* **26**, A-291.

Chou I.-M., Bassett W. A. and Bai T. B. (1995) Hydrothermal diamond anvil cell study of melts: eutectic melting of the assemblage $Ca(OH)_2 + CaCO_3$ with excess H_2O and lack of evidence for "portlandite II" phase, *Amer. Mineral.* **80**, 852–853.

Decker D. L. and Zhou Y. X. (1989) Dielectric and polarization measurements on $BaTiO_3$ at high pressures to the tricritical point. *Phys. Rev. B* **39**, 2432–2438.

Evans H. T. Jr. (1961) An X-ray diffraction study of tetragonal barium titanate. *Acta Cryst.* **14**, 1019–1026.

Frantz J. D., Dubessy J. and Mysen B. O. (1994) Ion-pairing in aqueous $MgSO_4$ solutions along an isochore to 500°C and 11 kbar using Raman spectroscopy in conjunction with the diamond-anvil cell. *Chem. Geol.* **116**, 181–188.

Haar L., Gallagher J. S. and Kell G. S. (1984) *NBS/NRC Steam Tables: Thermodynamic and Transport Properties and Computer Programs for Vapor and Liquid States of Water in SI Units.* Hemisphere Publ. Corp., Washington D. C., 320 pp.

Handfland M., Syassen K., Fahy S., Louis S.G. and Cohen M. L. (1985) Pressure dependence of first-order Raman mode in diamond. *Phys. Rev. B* **31**, 6896–6899.

Herchen H. and Cappelli M. A. (1991) First-order Raman spectrum of diamond at high temperatures. *Phys. Rev. B* **43**, 11740–11744.

Hower J. E., Hower M. E. and Perry E. A. (1976) Mechanism of burial metamorphism of agilaceous sediments I. mineralogical and chemical evidence, *Geol. Soc. Amer. Bull.* **87**, 725–737.

Koster van Groos, A. F. and Guggenheim, S. (1987) Dehydration of Ca- and Mg-exchanged montmorillonite (SWy-1) at elevated pressure. *Amer. Mineral.* **72**, 292–298.

Lin R. and Ritz G. P. (1993) Studying individual macerals using I.R. microspectroscopy and implications on oil versus gas/condensate proness and "low-rank" generation: *Org. Geochem.* **20**, 695–706.

Matthias B. (1948) The growth of barium titanate crystals. *Phys. Rev.* **73**, 808–809.

Mirwald P. W. and Massonne H.-J. (1980) The low-high quartz and quartz-coesite transition to 40 kbar between 600°C and 1600°C and some reconnaissance data on the effect of $NaAlO_2$ component on the low quartz-coesite transition. *J. Geophys. Research* **85**, 6983–6990.

Muinov M. H., Kanda H. and Stishov S. M. (1994) Raman scattering in diamond at high pressure: isotope effects. *Phys. Rev. B* **50**, 13860–13862.

Ramirez R., Lapena M. F. and Gonzalo J. A. (1990) Pressure dependence of free-energy expansion coefficients in $PbTiO_3$ and $BaTiO_3$ and tricritical-point behavior. *Phys. Rev. B* **42**, 2604–2606.

Ransom B. and Helgeson H. (1995) A chemical and thermodynamic model of dioctahedral 2:1 layer clay minerals in diagenetic processes: dehydration of dioctahedral aluminous smectite as a function of temperature and depth in sedimentary basins. *Amer. J. Sci.* **295**, 245–281.

Ritzert V. G. and Franck E. U. (1968) Elektrische Leitfähigkeit wässeriger Lösungen bei hohen Temperaturen und Drücken, I. KCl, $BaCl_2$, $Ba(OH)_2$, und $MgSO_4$ bis 750°C und 6 kbar. *Ber. Bunsenges. Phys. Chem.* **72**, 798–808.

Samara G. A. (1971) Pressure and temperature dependence of the dielectric properties and phase transitions of the ferroelectric perovskites: $PbTiO_3$ and $BaTiO_3$. *Ferroelectrics.* **2**, 277–289.

Saul A. and Wagner W. (1989) A fundamental equation of water covering the range from the melting line to 1273 K at pressures up to 25000 MPa. *J. Phys. Chem. Ref. Data* **18**, 1537–1565.

Schiferl D., Sharma S. K., Cooney T. F., Wang S. Y. and Mohanan K. (1993) Multichannel Raman spectroscopy system for weakly scattering materials at simultaneous high pressures and high temperatures. *Rev. Sci. Instrum.* **64**, 2821–2827.

Schiferl D., Sharma S. K., Cooney T. F., Wang S. Y., Anthony T. R. and Fleischer J. F. (in preparation). The use of the diamond C^{13}/C^{12} isotope Raman sensor system at high temperatures and pressures for aqueous and other chemically reactive environments.

Sharma S. K., Mao H. K., Bell P. M. and Xu J. A. (1985) Measurement of stress distribution in diamond anvils with micro-Raman spectroscopy. *J. Raman Spectrosc.* **16**, 350–352.

Sharma S. K., Cooney T. F. and Wang S.Y. (1994) Structure of alkali-silicate and germania glasses at high temperature and pressure. *J. Non-Cryst. Solids* **179**, 125–134.

Shen A. H., Bassett W. A. and Chou I.-M. (1992) Hydrothermal studies in a diamond anvil cell: Pressure determination using the equation of state of H_2O. In *High-Pressure Research: Application to Earth and Planetary Sciences* (eds. Y. Syono and M. H. Manghnani), 61–68. Terra Sci. Pub. Co./ AGU, Tokyo/ Washington, D.C.

Sherman W. F. (1985) The diamond Raman band as a high pressure calibrant. *J. Phys. C* **18**, L973-L978.

Tardieu A., Cansell F. and Petitet J. P. (1990) Pressure and temperature dependence of first-order Raman mode of diamond. *J. Appl. Phys.* **68**, 3243–3245.

van Valkenburg A., Bell P. M. and Mao H. K. (1987) High-pressure mineral solubility experiments in the diamond-window cell In *Hydrothermal Experimental Techniques* (eds. G. C. Ulmer and H. L. Barnes), pp. 458–468. J. Wiley and sons, New York

Wagner W., Saul A. and Pruss A. (1994) International equations for the pressure along the melting and along the sublimation curve of ordinary water substance. *J. Phys. Chem. Ref. Data* **23**, 515–527.

Zouboulis E. S. and Grimsditch M. (1991) Raman scattering in diamond up to 1900 K. *Phys. Rev. B* **43**, 12490–12493.

Mineral Spectroscopy: A Tribute to Roger G. Burns
© The Geochemical Society, Special Publication No. 5, 1996
Editors: M. D. Dyar, C. McCammon and M. W. Schaefer

Crystal field spectrum of synthetic clinoferrosilite

NANCY L. ROSS and JOHN R. SOWERBY
Department of Geological Sciences, University College London, London WC1E 6BT, U.K.

Abstract—The optical and near-infrared absorption spectrum of synthetic clinoferrosilite, $FeSiO_3$, was measured from 4,000 to 20,000 cm^{-1} at ambient conditions and at high pressure. Polarised and unpolarised spectra obtained from single crystals at ambient conditions display four absorption bands: (a) a very strong absorption band between 11,000–11,300 cm^{-1} that is attributed to Fe^{2+} in the M2 site ($^5A_1 \rightarrow {}^5A_1$); (b) a strong peak near 10,100 cm^{-1} assigned to Fe^{2+} in M1 ($^5B_{2g} \rightarrow {}^5A_{1g}$); (c) a weaker band near 8,200 cm^{-1} attributed to Fe^{2+} in M1 ($^5B_{2g} \rightarrow {}^5B_{1g}$), and (d) a peak at 5,100 cm^{-1} due to Fe^{2+} in M2 ($^5A_1 \rightarrow {}^5B_1$). The latter two bands are not observed in the spectra obtained from polycrystalline samples of clinoferrosilite.The estimated crystal-field splitting parameter (Δ_0) and crystal-field stabilisation energy (CFSE) of Fe^{2+} in the octahedral sites are: (i) M1: $\Delta_0 \simeq 8{,}640$ cm^{-1} and CFSE $\simeq 3{,}956$ cm^{-1} ; and (ii) M2: $\Delta_0 \simeq 7{,}325$ cm^{-1} and CFSE $\simeq 3{,}830$ cm^{-1}.

High-pressure spectra were collected between 0 and 5 GPa from a polycrystalline sample of clinoferrosilite. Between 0 and 1.85 GPa, the absorption bands near 11,000 cm^{-1} and 10,300 cm^{-1} increase at rates of 104 cm^{-1}/GPa and 121 cm^{-1}/GPa, respectively. At 1.85 GPa, the absorption bands show a discontinuity corresponding to the transition from the structure with $P2_1/c$ symmetry to a high-density structure with C2/c symmetry. Above the transition, the peak near 11,000 cm^{-1} increases to higher energy at a rate of 163 cm^{-1}/GPa, and the band near 10,300 cm^{-1} increases to higher energy at a rate of approximately 149 cm^{-1}/GPa. The transition is fully reversible and displays no detectable hysteresis.

INTRODUCTION

PYROXENES WITH composition $(Mg,Fe)SiO_3$ are important constituents of the Earth's upper mantle (*e.g.*, RINGWOOD, 1975). Until recently, the stable structure of $(Mg,Fe)SiO_3$ pyroxene throughout the Earth's mantle has always been considered to be an orthopyroxene. However, high-pressure single-crystal X-ray diffraction experiments on low-clinoenstatite (ANGEL *et al.*, 1992) and clinoferrosilite (HUGH-JONES *et al.*, 1994), which both have $P2_1/c$ symmetry, combined with phase equilibria data for $MgSiO_3$ (PACALO and GASPARIK, 1990; KANZAKI, 1991) and $FeSiO_3$ (LINDSLEY, 1965; AKIMOTO *et al.*, 1965) have confirmed the stability of a high-pressure clinopyroxene phase with C2/c symmetry (Fig. 1).

Orthoenstatite of composition $MgSiO_3$ transforms to this high-pressure clinoenstatite structure at pressures corresponding to a depth of 200–300 km within the Earth's upper mantle. ANGEL *et al.* (1992), for example, found that the transition from low-clinoenstatite to the high-density C2/c structure occurs at approximately 8 GPa at room temperature, and that the C2/c phase persists with release of pressure to approximately 5 GPa where it back-transforms to the $P2_1/c$ structure. The addition of iron in the structure has a dramatic effect on the $P2_1/c \rightleftharpoons C2/c$ transition. The transition pressure is lowered and little hysteresis is observed. In $FeSiO_3$, HUGH-JONES *et al.* (1994) have bracketed the transition between 1.48 and 1.75 GPa at room temperature (Fig. 1).

Electronic absorption spectra provide information about the role of transition metals such as iron in mineral structures not obtainable through other methods (*e.g.*, BURNS, 1993). Although many optical and near-infrared absorption spectra of natural and synthetic orthoferrosilites have been reported (WHITE and KEESTER, 1966; BANCROFT and BURNS, 1967; MAO and BELL, 1971; RUNCIMAN *et al.*, 1973; GOLDMAN and ROSSMAN, 1976, 1977; and 1979; LANGER and ABU-EID, 1977; ROSSMAN, 1980 and 1988; STEFFEN *et al.*, 1988; BURNS, 1993), none have been described for clinoferrosilite. We report here the first electronic absorption spectra of clinoferrosilite and compare it with spectra of natural and synthetic orthoferrosilites. In addition, spectra collected *in situ* at high pressure are described and the effect of the $P2_1/c \rightleftharpoons C2/c$ transition on the crystal field spectrum of $FeSiO_3$.

EXPERIMENTAL METHOD

Clinoferrosilite was synthesised at 8 GPa and 1200°C in a multianvil press at the Bayerisches Geoinstitut. Further details of the sample assembly and experimental run are described in WOODLAND and O' NEILL (1993) and HUGH-JONES *et al.* (1994). Compositional analyses using energy-dispersive X-ray analysis with a Phillips CM20 FEG transmission electron microscope at the Bayerisches Geoinstitut indicated that the sample was pure endmember $FeSiO_3$ and contained no impurities. Mössbauer experiments performed by C. McCammon on the clinofer-

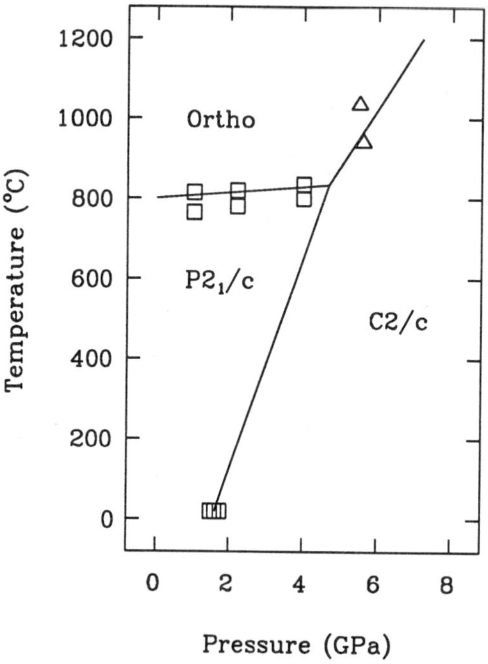

FIG 1. Phase diagram of $FeSiO_3$ as postulated by HUGH-JONES *et al.* (1994) with their reversal of $P2_1/c \rightleftharpoons C2/c$ at 298K (squares). Also shown are results from LINDSLEY's (1965) reversals (squares), and AKIMOTO *et al.*'s (1965) data (triangles).

rosilite crystals displayed no peaks that could be attributed to Fe^{3+}, implying that the concentration of Fe^{3+} in the sample was less than 0.8% (HUGH-JONES *et al.*, 1994).

Optical and near-infrared absorption spectra were collected at room pressure from single crystals of clinoferrosilite in the range from 4,000 to 20,000 cm^{-1} using a Bruker IFS HR high-resolution FTIR spectrometer equipped with a Bruker IR microscope at the Bayerisches Geoinstitut. Typically, two hundred scans with 4 cm^{-1} resolution were accumulated between 4,000 to 10,000 cm^{-1} with a CaF_2 beamsplitter, tungsten source and MCT detector. Between 9,000 and 15,000 cm^{-1}, two hundred scans were collected with 8 cm^{-1} resolution using CaF_2 and quartz beamsplitters, tungsten source and Si diode detector, and, in the region from 12,500 to 20,000 cm^{-1}, a quartz beamsplitter, xenon arc source and Si diode detector were used. Spectra were merged and baseline-corrected to produce the final spectrum. Unpolarised spectra collected on different crystals in different orientations yielded similar spectra. Polarised spectra were collected from the (100) face of a doubly-polished euhedral single crystal with dimensions of $100 \times 80 \times 30$ μm using a 60 μm diameter beam.

Spectra were collected between 0 and 5 GPa from a polycrystalline sample of clinoferrosilite at University College London using a Bruker IFS88 FTIR spectrometer. A thin film of powdered sample was loaded into a Diacell Products lever-arm diamond anvil cell (DAC) with a 4:1 mixture of methanol-ethanol as the pressure-transmitting medium. A small ruby chip was also included in the cell for pressure determination by measuring the shift of R_1 fluorescence peak with pressure (MAO *et al.*, 1986). For each spectrum, five hundred scans with 4 cm^{-1} resolution were accumulated in the spectral ranges 4000 to 9000 cm^{-1} (CaF_2 beamsplitter, tungsten source, MCT detector), and from 9000 to 20,000 cm^{-1} (quartz beamsplitter, tungsten source, Si diode detector).

RESULTS AND DISCUSSION

Single crystal spectra at room pressure

The merged, unpolarised spectrum from the doubly-polished single crystal of clinoferrosilite is shown in Fig. 2. The spectrum is dominated by a strong absorption band near 10,500 cm^{-1} with a shoulder at approximately 8,000 cm^{-1} and a moderately intense band near 5,000 cm^{-1}. In order to investigate the fine structure of the crystal field, peaks with Gaussian components were fit to the spectra. A minimum of four peaks is required to obtain a reasonable fit to the spectrum. The refined peak positions are 10,926 cm^{-1}, 10,026 cm^{-1}, 8,166 cm^{-1}, and 5,081 cm^{-1} with widths of 2,366 cm^{-1}, 1,006 cm^{-1}, 1,445 cm^{-1}, and 1,149 cm^{-1}, respectively (Fig. 2). No evidence for any charge-transfer bands was observed in any of the spectra collected which is consistent with the lack of evidence for any Fe^{3+} from the Mössbauer spectra.

The observed absorption bands of clinoferrosilite correspond well with spin-allowed Fe^{2+} bands observed in the electronic of natural and synthetic orthoferrosilites (MAO and BELL, 1971; GOLDMAN and ROSSMAN, 1977 and 1979; LANGER and ABU-EID, 1977; ROSSMAN, 1980 and 1988; STEFFEN *et al.*,1988; BURNS, 1993). The crystal field spectra arise from electronic transitions among the partially-filled d orbitals of the Fe^{2+} cations occupying six-coordinated sites in the orthopyroxene structure. The orthopyroxene structure contains single-chains of corner-sharing $[SiO_4]$ tetrahedra running parallel to the c-axis which are cross-linked by cations located in sites designated as M1 and M2. Although the M1 site has site symmetry C_1, it is only slightly distorted from octahedral symmetry and is usually approximated with O_h symmetry. STEFFEN *et al.* (1988), however, have shown that the M1 octahedron is better approximated with D_{4h} symmetry because of the observed quadrupole splitting in the Mössbauer spectra of orthoferrosilite. The M2 site also has site symmetry C_1 , but it is much more distorted than the M1 site. It is generally agreed that the effective symmetry of the M2 site is C_{2v} (*e.g.*, ROSSMAN, 1988).

In order to aid in the band assignments, polarised spectra from the (100) face of the doubly-polished single crystal of clinoferrosilite were collected parallel and perpendicular to its c-axis (Fig. 3). The

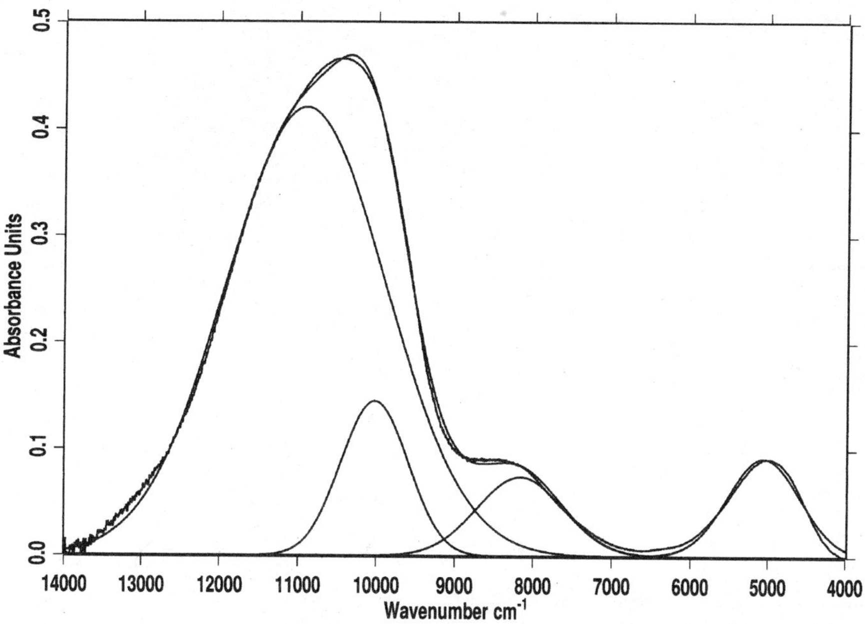

FIG. 2. Unpolarised optical and near infrared absorption spectrum from the (100) face of a doubly-polished (30 μm thick) single-crystal of clinoferrosilite after subtraction of an almost linear background. Deconvolution into four Gaussian components and the resulting simulated profile are also shown.

polarised spectrum collected parallel to the c-axis is approximately 50% weaker than the one collected perpendicular to the c-axis. Three peaks with Gaussian components were fit to the spectrum collected parallel to the c-axis as shown in Fig. 3a. The refined peak positions are 10,290 cm^{-1}, 8,200 cm^{-1} and 5,050 cm^{-1}. This spectrum is similar to the γ spectrum of natural and synthetic orthoferrosilites. MAO and BELL (1971), for example, observed bands at 10,549 cm^{-1}, 8333 cm^{-1}, and 4878 cm^{-1} in the γ spectrum of a synthetic crystal of pure endmember orthoferrosilite. BURNS (1993) reported bands in the γ spectra of two orthopyroxenes, enstatite $Fs_{14.5}$ and ferrosilite $Fs_{86.4}$, at approximately 11,000 cm^{-1}, 8,500 cm^{-1} and 5,300 cm^{-1} with the absorption band at 8,500 cm^{-1} becoming more prominent in the Fe^{2+}-rich sample. The spectrum of clinoferrosilite collected perpendicular to the c-axis was also deconvoluted into three Gaussian components shown in Fig. 3b. The spectrum is dominated by an intense absorption band at 11,310 cm^{-1} with less intense bands at 10,080 cm^{-1} and 5,140 cm^{-1}. MAO and BELL (1971) measured a combined α and β spectrum that has features similar to this spectrum. The $\alpha + \beta$ spectrum is much stronger than the γ spectrum and is dominated by an intense absorption band at 10,627 cm^{-1} and a less intense band at 4,878 cm^{-1}. In the polarised spectra of natural ferrosilites (*e.g.,* BURNS 1993), the α spectrum is dominated by an intense band at approximately 10,700 cm^{-1} and the β spectrum generally contains moderately intense bands near 11,000 cm^{-1} and 5,000 cm^{-1}.

The strong correspondence between the electronic spectra of clinoferrosilite and orthoferrosilite is due to their similar structures. The γ spectrum of clinoferrosilite resembles that of orthoferrosilite because the chains of SiO_4 tetrahedra cross-linked by strips of edge-sharing M1 and M2 octahedra run parallel to the c-axis in both structures (*e.g.,* CAMERON and PAPIKE, 1980). In addition, the average bond distances and distortions in the M1 and M2 sites of the two structures are almost identical (Table 1). The average Fe—O bond lengths for the M1 and M2 sites in clinoferrosilite are 2.137 Å and 2.224 Å, respectively, compared with 2.135 Å and 2.228 Å in orthoferrosilite, respectively (BURNHAM, 1966; SUENO *et al.,* 1976). The M1 sites in both structures are close to regular octahedra and have similar quadratic elongation (QE) parameters (Table 1). The M2 sites are much more distorted with QE values of 1.0641 for clinoferrosilite and 1.0706 for orthoferrosilite (BURNHAM, 1966; SUENO *et al.,* 1976).

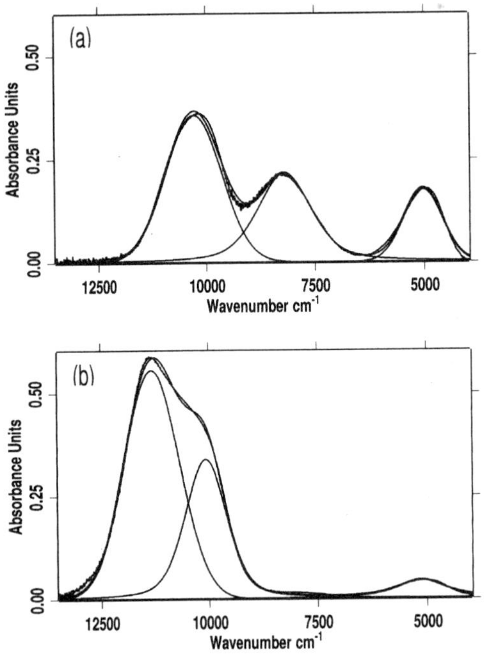

FIG. 3. Polarised optical and near infrared absorption spectra of the same crystal in Fig. 2 collected (a) parallel to the *c*-axis and (b) perpendicular to the *c*-axis. Deconvolution of both spectra into three Gaussian components and the resulting simulated profile are also shown.

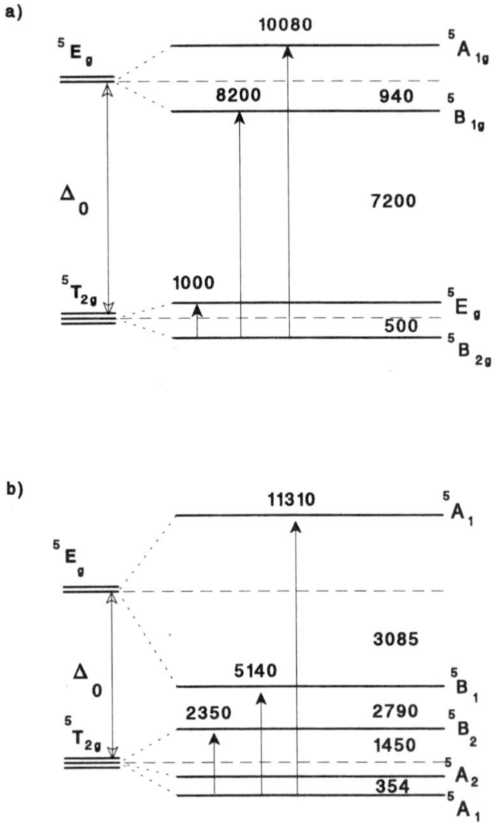

FIG. 4. Schematic 3*d* orbital energy level diagrams for Fe^{2+} ions in clinoferrosilite on (a) M1 site and (b) M2 site. Observed transitions in clinoferrosilite (Fig. 3) are indicated.

Given the similarity between the crystal field spectra clinoferrosilite spectra and orthoferrosilite, tentative band assignments for the Fe^{2+} crystal field spectra of clinoferrosilite can be made using the band assignments of orthoferrosilite as a guide. Figure 4 shows schematic 3d orbital energy level diagrams for Fe^{2+} ions in the M1 and M2 sites of clinoferrosilite constructed following BURNS' (1993) and STEFFEN *et al.*'s (1988) analyses of orthoferrosilite. We assign the strong 11,310 cm^{-1} band (Fig. 3b) to $^5A_1 \rightarrow {}^5A_1$ (Fe^{2+} in M2) and the less intense band at 10,080 cm^{-1} (Fig. 3a) to $^5B_{2g} \rightarrow {}^5A_{1g}$ (Fe^{2+} in M1) in agreement with MAO

Table 1. Crystal chemical data determined at room pressure for clinoferrosilite and orthoferrosilite and data for high-density clinoferrosilite determined at 1.87 GPa

COMPOUND:	Clinoferrosilite	Orthoferrosilite	High-density Clinoferrosilite
Space Group:	$P2_1/c$	Pbca	C2/c
M1 Site:			
⟨Fe-O⟩ (Å)	2.137	2.135	2.14
QE[1]	1.0091	1.0087	1.0083
M2 Site:			
⟨Fe-O⟩ (Å)	2.224	2.228	2.18
QE[1]	1.0641	1.0706	1.0182
REFERENCES:	BURNHAM (1966)	SUENO et al. (1976)	HUGH-JONES et al. (1994)

[1] QE is the quadratic elongation parameter as defined by ROBINSON et al. (1971).

and BELL (1971) and STEFFEN *et al.* (1988). The band at 5,140 cm^{-1} (Fig. 3b) is assigned to the $^5A_1 \rightarrow {}^5B_1$ transition of Fe^{2+} in M2 octahedra, and the band at 8,200 cm^{-1} (Fig. 3a) is attributed to the $^5B_{2g} \rightarrow {}^5B_{1g}$ transition of Fe^{2+} in M1 octahedra. In order to estimate the crystal field splitting parameter (Δ_0) and crystal field stabilisation energy (CFSE) for Fe^{2+} in the M1 and M2 sites, the splitting of the lower level t_{2g} orbitals is needed. Since the splittings of the lower t_{2g} orbitals for the M1 and M2 sites are unknown, we follow BURNS' (1993) approximation. For the M1 site, a small value of 500 cm^{-1} is assumed since the M1 coordination polyhedron is not significantly distorted from octahedral symmetry. For the M2 site, lower-level splittings of 2,350 cm^{-1} and 354 cm^{-1} determined for enstatite (GOLDMAN and ROSSMAN, 1977 and 1979; ROSSMAN, 1988) are assumed and applied to clinoferrosilite. Values of Δ_0 and CFSE for Fe^{2+} in the two sites are therefore estimated to be:

M1 site: $\Delta_0 = 8640\ cm^{-1}$;

$$CFSE = 0.4\Delta_0 + 500 = 3956\ cm^{-1}$$

M2 site: $\Delta_0 = 7325\ cm^{-1}$;

$$CFSE = 0.4\Delta_0 + 900 = 3830\ cm^{-1}$$

These values are compared with natural and synthetic orthoferrosilites in Table 2. Both Δ_0 and CFSE of Fe^{2+} in the M1 site of clinoferrosilite are less than those of orthoferrosilite whereas Δ_0 and CFSE of Fe^{2+} in the M2 site are greater. Thus our approximations suggest that the difference in CFSE between the M1 and M2 sites in clinoferrosilite is less than those found in orthoferrosilites. These results also suggest that the splitting of the e_g orbitals of the M1 and M2 sites in clinoferrosilite, 1,880 cm^{-1} and 6,170 cm^{-1}, respectively, differ from those observed in orthoferrosilites which typically have splittings of 2,200 cm^{-1} and 5,800 cm^{-1}, respectively (MAO and BELL, 1971; BURNS, 1993). This suggests that the distortion of the M2 site is greater in clinoferrosilite than orthoferrosilite and the M1 site is less distorted although the site distortions determined from X-ray structure refinements are virtually identical (Table 1). X-ray refinements, however, obtain an average of all the M1 and M2 octahedra whereas absorption spectroscopy sees more local detail of the octahedral sites.

Polycrystalline spectra at high pressure

Representative spectra collected from a polycrystalline sample in the diamond anvil cell are shown in Fig. 5. At 0.34 GPa (Fig. 5a), the spectrum contains a single, broad absorption feature at approximately 10,500 cm^{-1} which corresponds to the strong absorption band observed at approximately the same position in the unpolarised single crystal spectrum (Fig. 2). Two peaks with Gaussian components were fit to the spectrum yielding refined positions of 10,876 cm^{-1} and 10,245 cm^{-1} and widths of 1,584 cm^{-1} and 1,193 cm^{-1}, respectively. These compare well with the peak positions at 10,837 cm^{-1} (width 1,523 cm^{-1}) and 10,110 cm^{-1} (width 1,144 cm^{-1}) obtained from a polycrystalline spectrum at 0 GPa outside the DAC. Based on the discussion above, the peak at approximately 11,000 cm^{-1} is most likely due to the $^5A_1 \rightarrow {}^5A_1$ transition of Fe^{2+} in M2, and the peak near 10,300 cm^{-1} can be attributed to the $^5B_{2g} \rightarrow {}^5A_{1g}$ transition of Fe^{2+} in M1. Refined peak positions for the other pressures are given in Table 3.

Figure 6 shows the shift of the peaks between 0 and 5 GPa. In general, both peaks show a shift to higher energy with increasing pressure. However, the shift of the peaks with pressure is not linear, but displays a discontinuity at approximately 1.85 GPa. In a high-pressure single-crystal X-ray diffraction study, HUGH-JONES *et al.* (1994) observed a reversible transition in $FeSiO_3$ between 1.48 and 1.75 GPa at room temperature from a structure with

Table 2. Comparison of Δ_0 and CFSE for clinoferrosilite ($P2_1/c$) with a synthetic orthoferrosilite (MAO and BELL 1971) and natural orthoferrosilite (BURNS 1993) calculated using the splittings of the lower level t_{2g} orbitals for the M1 and M2 sites described in the text

COMPOUND:	Synthetic $FeSiO_3$: Clinoferrosilite	Synthetic $FeSiO_3$: Orthoferrosilite	Natural Fs86: Orthoferrosilite
M1 site:			
Δ_0	8640 cm^{-1}	8941 cm^{-1}	9150 cm^{-1}
CFSE	3956 cm^{-1}	4076 cm^{-1}	4160 cm^{-1}
M2 site:			
Δ_0	7325 cm^{-1}	6852 cm^{-1}	6900 cm^{-1}
CFSE	3830 cm^{-1}	3641 cm^{-1}	3660 cm^{-1}

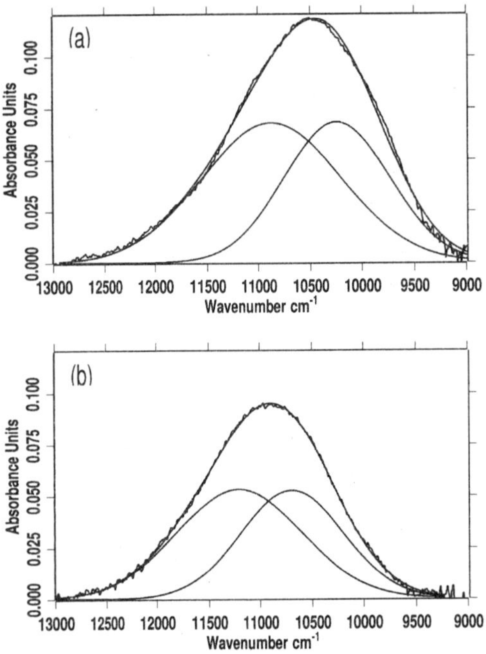

FIG. 5. Electronic absorption spectrum of polycrystalline $FeSiO_3$ clinoferrosilite collected at (a) 0.34 GPa and (b) 2.28 GPa after subtraction of a linear background. The two Gaussian components of both spectra and the resulting simulated profile are also shown.

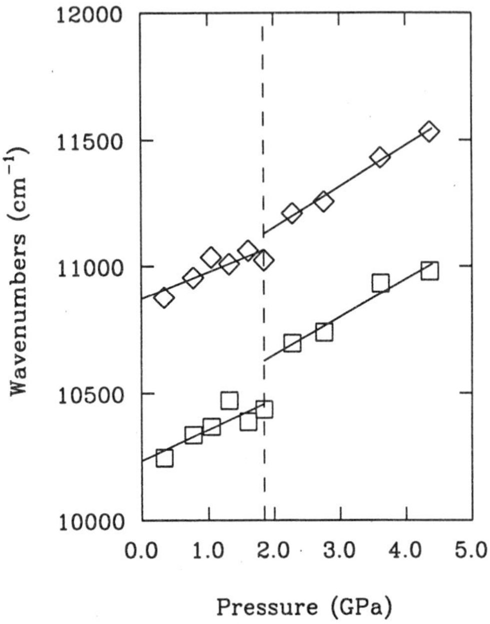

FIG. 6. Shift of the crystal-field bands of a polycrystalline sample of clinoferrosilite with pressure.

$P2_1/c$ symmetry to one with $C2/c$ symmetry. This corresponds well with the results shown in Fig. 6. The slightly higher transition pressure observed in this study may be due to obtaining spectra from a polycrystalline sample rather than a single crystal. A representative spectrum of the high-density C2/c phase collected at 2.28 GPa is shown in Fig. 5b.

HUGH-JONES *et al.* (1994) determined that the transition is first-order in character accompanied by a 3% decrease in volume of the unit cell. The change in symmetry is related to the rotation of the chains of silicate tetrahedra. The low-pressure $P2_1/c$ structure consists of two distinct chains of tetrahedra, the A chain which is S-rotated and more extended than the B chain which is O-rotated and considerably kinked (BURNHAM, 1966). The edge sharing of the tetrahedra and the M2 site, resulting

Table 3. Optical absorption bands of clinoferrosilite as a function of pressure obtained by fitting two Gaussian components to the spectra. Peak #1 is attributed to the $^5B_{2g} \rightarrow {}^5A_{1g}$ transition of Fe^{2+} in M1 and Peak #2 is assigned to the $^5A_1 \rightarrow {}^5A_1$ transition of Fe^{2+} in M2. All peak positions and widths are given in cm^{-1}

	Peak #1		Peak #2	
Pressure (GPa)	Position	Width	Position	Width
0.34	10,245	1,193	10,876	1,584
0.78	10,335	890	10,954	1,170
1.05	10,367	975	11,034	1,188
1.32	10,472	915	11,008	1,298
1.61	10,388	1,010	11,060	1,194
1.85	10,436	890	11,023	1,148
2.28	10,696	1,175	11,208	1,429
2.76	10,739	945	11,255	1,358
3.62	10,932	979	11,429	1,249
4.36	10,980	910	11,531	930

from the S-rotation of the A chain, makes the M2 site extremely distorted. At the transition pressure, the A chain rotates through a straight chain configuration, becoming O-rotated and geometrically equivalent to the B chain above the transition pressure (HUGH-JONES *et al.*, 1994). In the high-pressure C2/c structure, the tetrahedra of the A chain no longer share edges with the M2 sites and therefore the M2 octahedra are significantly smaller and less distorted than in the $P2_1/c$ structure. The QE parameter of the M2 site changes from 1.0641 in the $P2_1/c$ structure to 1.0182 in the C2/c structure, and the M1 site shows a slight decrease from 1.0091 to 1.0083 (Table 1). Corresponding changes are observed in the high pressure electronic spectra. Between 0 and 1.85 GPa, the peak near 11,000 cm^{-1} attributed to the $^5A_1 \rightarrow {}^5A_1$ transition of Fe^{2+} in M2, increases at a rate of 104 cm^{-1}/GPa. Above 1.85 GPa, however, the band increases at a rate of approximately 163 cm^{-1}/GPa (Fig. 6). The peak near 10,200 cm^{-1} associated with $^5B_{2g} \rightarrow {}^5A_{1g}$ transition of Fe^{2+} in M1 also shows a change across the transition. Between 0 and 1.85 GPa, this band shifts to higher energy at a rate of approximately 121 cm^{-1}/GPa, and above the transition pressure, the band increases a greater rate of approximately 149 cm^{-1}/GPa. Both bands show a distinct break at 1.85 GPa (Fig. 6).

MAO and BELL (1971) studied the effects of pressure up to approximately 9.0 GPa on a polycrystalline sample of orthoferrosilite. The spectrum is similar to the spectra of the $P2_1/c$ and C2/c polymorphs of $FeSiO_3$ (Fig. 5) in that it is dominated by a broad, absorption band near 10,700 cm^{-1}. All of these spectra resemble the α spectrum obtained from polarised studies of single crystals and therefore may indicate a preferred orientation of both polymorphs when loaded as polycrystalline aggregates in the diamond anvil cell. MAO and BELL (1971) also performed experiments on single crystals to 2.5 GPa in order to obtain better resolution of the crystal field bands. The absorption band at 10,627 cm^{-1} corresponding to the $^5A_1 \rightarrow {}^5A_1$ transition of Fe^{2+} in M2 shifts to higher energy at a rate of approximately 130 cm^{-1}/GPa which is intermediate between what we observed in the $P2_1/c$ structure and the C2/c structure. The absorption band at 10,549 cm^{-1} attributed to the $^5B_{2g} \rightarrow {}^5A_{1g}$ transition of Fe^{2+} in M1 shifts to higher energy at a rate of approximately 120 cm^{-1}/GPa which again is similar to what we observed in the $P2_1/c$ structure but slightly less than what we observed in the C2/c structure. MAO and BELL (1971) also determined that the bands at 8333 cm^{-1} (α spectrum) and 4878 cm^{-1} (β spectrum) shift more rapidly with pressure at rates of approximately 210 cm^{-1}/GPa and 280 cm^{-1}/GPa. As a result, the crystal field splitting of the e_g orbital group decreases with pressure for both M1 and M2, suggesting less distortion in the octahedral field at high pressure in orthoferrosilite. MAO and BELL (1971) also showed that Δ_o of M1 and M2 increases with pressure.

Single-crystal measurements at high pressure of clinoferrosilite are needed in order to determine the Δ_o and CFSE for the M1 and M2 sites of the high-density C2/c structure. However, we can make some predictions based on the available structural data of the C2/c structure from HUGH-JONES *et al.* (1994). In the C2/c structure at 1.87 GPa, the average Fe—O distance for the M2 site is 2.18 Å with a QE parameter of 1.0182 which compares with an average Fe—O distance of 2.224 Å and a QE of 1.0641 in the $P2_1/c$ structure at ambient conditions (Table 1). Assuming a simple point charge model, Δ_0 is proportional to the inverse fifth power of the interatomic metal-oxygen distance (*e.g.*, BURNS, 1993). Therefore Δ_0 of the M2 site should increase from the $P2_1/c$ structure to the C2/c structure. Since the average bond length and distortion of the M2 site in the C2/c structure are more similar to the M1 site than observed in either the $P2_1/c$ or orthoferrosilite structures (Table 1), we expect that it will have crystal-field parameters closer to M1 than observed above. In addition, the decrease in the distortion of the M2 site will lead to a decrease in the splitting of the e_g orbitals as detected by MAO and BELL (1971) in their high pressure experiments of orthoferrosilite. A discontinuity should therefore be observed in the shift of the absorption band near 5,100 cm^{-1} with pressure. In contrast, we predict smaller change in Δ_0 and CFSE of the M1 site across the transition since there is little change in the average bond length and distortion of the octahedra between the two polymorphs (Table 1).

CONCLUSIONS

We have obtained the first crystal field spectra of clinoferrosilite from single crystals at ambient conditions and from a polycrystalline sample at high pressure. In general, the spectra are similar to orthoferrosilite with absorption bands near 11,000 cm^{-1}, 10,000 cm^{-1}, 8,200 cm^{-1}, and 5,100 cm^{-1} that correspond to the spin-allowed transitions of Fe^{2+} on the M1 and M2 sites. Band assignments obtained from comparison of polarised single-crystal spectra of clinoferrosilite with orthoferrosilite suggest, however, that are subtle differences between the two polymorphs: (a) a greater splitting of the e_g orbitals of the M2 site in clinoferrosilite, and (b) a smaller splitting of the e_g orbitals of the M1 site with smaller values of Δ_0 and CFSE than observed in natural and synthetic orthoferrosilites.

High-pressure spectra collected between 0 and 5 GPa from a polycrystalline sample of clinoferrosilite show a broad absorption band that can be fit with two Gaussian components corresponding to spin-allowed transitions of Fe^{2+} on the M1 and M2 sites. The shift of the peaks with pressure is not linear and displays a discontinuity at approximately 1.85 GPa where the structure transforms from $P2_1/c$ to $C2/c$ symmetry. The transition is reversible with no detectable hysteresis. Work is in progress to obtain more detailed crystal field spectra from single crystals of clinoferrosilite at high pressure. From these, we shall be able to determine Δ_0 and estimate the CFSE of the M1 and M2 sites as a function of pressure. The CFSE may be a contributing factor to why the $P2_1/c \rightleftarrows C2/c$ transition in $FeSiO_3$ has little hysteresis but the analogous transition in $MgSiO_3$ has a hysteresis of approximately 3 GPa (ANGEL *et al.*, 1992).

Acknowledgments—We would like to thank A. B. Woodland for providing the powder clinoferrosilite sample used in the high-pressure study. We would also like to thank R. J. Angel, D. A. Hugh-Jones, H. Keppler, C. McCammon, and S. Webb for their assistance and helpful discussions. N.L.R. thanks H. Schultz for polishing the single crystal specimen, and J.R.S. acknowledges P. A. Woods for technical assistance. This work was supported by an NERC studentship (GT4/93/267/G) to J.R.S., a Nuffield Foundation Research Fellowship to N.L.R. and an NERC research grant GR9/1141 to N.L.R.. Multi-anvil and spectroscopic experiments were performed at the Bayerisches Geoinstitut under the EC "Human Capital and Mobility-Access to Large Facilities" programme (Contract No. ERBCHGECT940053 to D.C. Rubie).

REFERENCES

AKIMOTO S., KATSURA T., SYONO Y., FUJISAWA H. and KOMADA E. (1965) Polymorphic transition of pyroxenes $FeSiO_3$ and $CoSiO_3$ at high pressures and temperatures. *J. Geophys. Res.* **70,** 5269–5278.

ANGEL R. J., CHOPELAS A. and ROSS N. L. (1992) Stability of high-density clinoenstatite at upper mantle pressures. *Nature* **358,** 322–324.

BANCROFT G. M. and BURNS R. G. (1967) Interpretation of the electronic spectra of iron in pyroxenes. *Amer. Mineral.* **52,** 1278–1287.

BURNHAM C. W. (1966) Ferrosilite. *Carnegie Inst. Washington Year Book* **65,** 285–290.

BURNS R. G. (1993) *Mineralogical Applications of Crystal Field Theory* (2nd Ed.). Cambridge University Press.

CAMERON M. and PAPIKE J. J. (1980) Crystal chemistry of silicate pyroxenes. In *Pyroxenes* (ed. C. T. PREWITT), *Rev. Mineral.***7,** 5–92. Mineralogical Society of America.

GOLDMAN D. S. and ROSSMAN G. R. (1976) Identification of a mid-infrared electronic absorption band of Fe^{2+} in the distorted M(2) site of orthopyroxene $(Mg,Fe)SiO_3$. *J. Phys.Chem. Lett.* **41,** 474–475.

GOLDMAN D. S. and ROSSMAN G. R. (1977) The spectra of iron in orthopyroxenes revisited: the splitting of the ground state. *Amer. Mineral.* **62,** 151–157.

GOLDMAN D. S. and ROSSMAN G. R. (1979) Determination of quantitative cation distribution in orthopyroxenes from electronic absorption spectra. *Phys. Chem. Minerals* **4,** 43–53.

HUGH-JONES D. A., WOODLAND A. B. and ANGEL R. J. (1994) The structure of high-pressure C2/c ferrosilite and crystal chemistry of high-pressure C2/c pyroxenes. *Amer. Mineral.* **79,** 1032–1041.

KANZAKI M. (1991) Ortho/clinoenstatite transition. *Phys. Chem. Minerals* **17,** 726–730.

LANGER K. and ABU-EID R. M. (1977) Measurement of the polarized electronic spectra of synthetic transition metal-bearing silicate microcrystals in the spectral range 44000–4000 cm^{-1}. *Phys. Chem. Minerals* **1,** 273–299.

LINDSLEY D. H. (1965) Ferrosilite. *Carnegie Inst. Washington Year Book* **65,** 148–149.

MAO H. K. and BELL P. M. (1971) Crystal field spectra. *Carnegie Inst. Washington Year Book* **70,** 207–215.

MAO H. K., XU J. and BELL P. M. (1986) Calibration of the ruby pressure gauge to 800kbar under quasi-hydrostatic conditions. *J. Geophys. Res.* **91,** 4673–4676.

PACALO R. E. G. and GASPARIK T. (1990) Reversals of the orthoenstatite-clinoenstatite transition at high pressures and high temperatures. *J. Geophys. Res.* **95,** 15853–15858.

RINGWOOD A. E. (1975) *Composition and Petrology of the Earth's Mantle.* McGraw-Hill, New York.

ROBINSON K., GIBBS G. V. and RIBBE P. H. (1971) Quadratic elongation: A quantitative measure of distortion in coordination polyhedra. *Science* **172,** 567–570.

ROSSMAN G. R. (1980) Pyroxene Spectroscopy. In *Pyroxenes* (ed. C. T. PREWITT), *Rev. Mineral.* **7,** 93–115. Mineralogical Society of America.

ROSSMAN G. R. (1988) Optical Spectroscopy. In *Spectroscopic Methods in Mineralogy and Geology* (ed. F. C. HAWTHORNE), *Rev. Mineral.* **18,** 207–254. Mineralogical Society of America.

RUNCIMAN W. A., SENGUPTA D. and MARSHALL M. (1973) The polarized spectra of iron in silicates. I. Enstatite. *Amer. Mineral.* **79,** 1032–1041.

STEFFEN G., LANGER K. and SEIFERT F. (1988) Polarized electronic absorption spectra of synthetic (Mg,Fe)-orthopyroxenes, ferrosilite and Fe^{3+}-bearing ferrosilite. *Phys. Chem. Minerals* **16,** 120–129.

SUENO S., CAMERON M. and PREWITT, C. T. (1976) Orthoferrosilite: high-temperature crystal chemistry. *Amer. Mineral.* **61,** 38–53.

WHITE E. B. and KEESTER K. L. (1966) Optical absorption spectra of iron in rock-forming silicates. *Amer. Mineral.* **51,** 444–450.

WOODLAND A. B. and O'NEILL, H. St. C. (1993) Synthesis and stability of $Fe_2^{2+}Fe_2^{3+}Si_3O_{12}$ garnet and phase transitions with $Fe_3Al_2Si_3O_{12}$—$Fe_3^{2+}Fe_2^{3+}Si_3O_{12}$ solutions. *Amer. Mineral.* **78,** 1000–1013.

Mineral Spectroscopy: A Tribute to Roger G. Burns
© The Geochemical Society, Special Publication No. 5, 1996
Editors: M. D. Dyar, C. McCammon and M. W. Schaefer

High-pressure Mössbauer study of synthetic clinoferrosilite, $FeSiO_3$

CATHERINE A. MCCAMMON[1] and CRAIG TENNANT[1,2]
[1]Bayerisches Geoinstitut, Universität Bayreuth, D-95440 Bayreuth, Germany
[2]Department of Chemistry, University of Canterbury, Christchurch, New Zealand

Abstract—Mössbauer spectra were recorded of a polycrystalline sample of clinoferrosilite ($FeSiO_3$) with $P2_1/c$ space group that was synthesised at 8 GPa and 1150°C using a multianvil press. In situ experiments were performed at 293 K using a diamond anvil cell at pressures up to 3.8 GPa. The transition to the C2/c structure was observed between 1.33 and 1.74 GPa, and is characterised by a large increase in quadrupole splitting (ΔE_Q) for the M1 site, and a small decrease in ΔE_Q for the M2 site. The centre shift (δ) data for the M1 site show no change within experimental error through the transition, while δ increases slightly for the M2 site. The values of $\partial\delta_o/\partial P$ (δ_o = isomer shift) were determined by correcting centre shift data for the pressure dependence of the second-order Doppler shift. For the $P2_1/c$ structure $\partial\delta_0/\partial P = -6(4) \times 10^{-3}$ and $+8(4) \times 10^{-3}$ mm s^{-1} GPa^{-1}, and for the C2/c structure $\partial\delta_o/\partial P = +1(3) \times 10^{-3}$ and $+1(3) \times 10^{-3}$ mm s^{-1} GPa^{-1} for the M1 and M2 sites, respectively. The positive slopes imply a contribution to the electron density from changes in electron orbital occupation. For the $P2_1/c$ structure $\partial\Delta E_Q/\partial P = +6.3(9) \times 10^{-2}$ and $+4.0(9) \times 10^{-2}$ mm s^{-1} GPa^{-1}, and for the C2/c structure $\partial\Delta E_Q/\partial P = +6(6) \times 10^{-3}$ and $+3.6(6) \times 10^{-2}$ mm s^{-1} GPa^{-1} for the M1 and M2 sites, respectively. The results are used to deduce the relative changes in the splitting of the Fe^{2+} ground state ($^5T_{2g}$) as a function of pressure.

INTRODUCTION

PYROXENES CONSTITUTE a significant proportion of the Earth's upper mantle. Recent experiments have established that a high-pressure phase with space group C2/c has a large stability region at high P,T for the endmember phases $MgSiO_3$ (ANGEL *et al.*, 1992) and $FeSiO_3$ (HUGH-JONES *et al.*, 1994). The transition from the low pressure form with space group $P2_1/c$ to the C2/c modification can be observed at room temperature and high pressures and is reversible. *In situ* single crystal X-ray diffraction experiments on $FeSiO_3$ by HUGH-JONES *et al.* (1994) have shown that the transition involves a 3% decrease in unit cell volume with considerable structural rearrangement of the tetrahedral chains. The two sites occupied by Fe^{2+}, M1 and M2, behave differently through the transition: the size and distortion of the M2 site changes significantly, while the M1 site remains relatively unaltered. Recent optical absorption spectra measurements of $FeSiO_3$ at high pressure provide information on the crystal-field parameters of the two polymorphs (ROSS and SOWERBY, 1996). Complementary data can be obtained from ^{57}Fe Mössbauer spectroscopy, which is highly sensitive to changes in the electronic environment of the iron atoms. There are relatively few studies of the effects of pressure on the hyperfine parameters of silicate minerals, and the system $FeSiO_3$ provides an excellent opportunity to combine results from structural and electronic studies with Mössbauer data to characterise the effect of pressure on the $P2_1/c$ and C2/c structural modifications, and the changes that occur through the phase transition.

EXPERIMENTAL DETAILS

Preparation of clinoferrosilite

Clinoferrosilite was synthesised from a starting mixture of Fe powder, Fe_2O_3 and SiO_2, with a net enrichment in ^{57}Fe of 21%, roughly ten times the natural abundance. A mixture of the starting material was loaded into a Ag capsule and run at 8 GPa and 1150°C for 20 h using a multianvil press at the Bayerisches Geoinstitut. The run product was checked by X-ray powder diffraction for crystalline impurities and by Mössbauer spectroscopy for iron-containing impurities. Nearly all diffraction peaks corresponded to published reflections for $P2_1/c$ clinoferrosilite, with only a few extremely weak peaks (<5% relative intensity) that could not be identified. The room temperature Mössbauer spectrum contains two doublets, which are assigned to Fe^{2+} in the M1 and M2 sites. The two doublets account for at least 99% of spectrum absorption, with the possibility of a small impurity corresponding to less than 1% of the total area. Addition of absorption corresponding to the impurity changed the hyperfine parameters of the M1 and M2 doublets by less than 0.002 mm/s; therefore the effect of the impurity is neglected in subsequent data analysis.

High-pressure Mössbauer measurements

Mössbauer data were collected at high pressure using a modified Merrill-Bassett diamond anvil cell (DAC) (HAZEN and FINGER, 1982) mounted with 0.125 carat diamonds with culet diameters of 1.0 mm. A 250 μm thick $Ta_{90}W_{10}$ gasket (pre-indented to 150 μm) with a 500 μm diameter hole provided collimation for the gamma radiation. Powdered clinoferrosilite was compressed between two glass slides and loaded as a thin plate into the DAC to ensure uniform absorber distribution. A 4:1 methanol:ethanol liquid was added as a pressure medium along with small ruby grains (~10 μm diameter) for pressure calibration. The DAC was mounted in a custom cryostat with the base of the bottom diamond at a distance of approximately 1 mm from the radioactive source. Further description of the apparatus is given in McCAMMON (1992). Mössbauer spectra were collected at each pressure over time periods ranging from 2 to 4 days. Experiments were made during several cycles of increasing and decreasing pressure; the numerical sequence of the measurements is given in Table 1.

Mössbauer spectra were recorded in transmission mode on a constant acceleration Mössbauer spectrometer with a nominal 50 mCi ^{57}Co source in a 6 μm Rh matrix (room pressure) and a nominal 20 mCi high specific activity source (2 Ci/cm^2) in a 12 μm Rh matrix (high pressure). The velocity scale was calibrated relative to 25 μm α-Fe foil using the positions certified for National Bureau of Standards standard reference material no. 1541. Room temperature line widths for the outer lines of α-Fe were 0.28 mm/s for the conventional source and 0.42 mm/s for the high specific activity source. The density of iron in the room pressure absorber was 5 mg Fe/cm^2, and in the high-pressure experiments the density was estimated to correspond to 10 mg of natural Fe/cm^2. Mössbauer spectra were fit using the commercially available fitting program NORMOS written by R. A. Brand (distributed by Wissenschaftliche Electronik GmbH, Germany).

Table 1. Mössbauer parameters of clinoferrosilite at room temperature.

	M1		M2		
P GPa	δ^a mm s^{-1}	ΔE_Q mm s^{-1}	δ^a mm s^{-1}	ΔE_Q mm s^{-1}	N[b]
clinoferrosilite in air					
0.00	1.174 ±0.005	2.609 ±0.005	1.124 ±0.005	1.959 ±0.005	
clinoferrosilite in cell[c]					
P2₁/c phase					
0.00	1.177	2.638	1.126	1.933	3
0.31	1.181	2.650	1.129	1.958	8
0.59	1.171	2.672	1.128	1.953	4
1.00	1.170	2.700	1.137	1.987	11
1.14	1.170	2.700	1.130	1.985	9
1.33	1.169	2.722	1.133	1.983	14
mixed phase					
1.40	1.169	2.800	1.143	1.915	13
1.54	1.168	2.829	1.146	1.907	12
C2/c phase					
1.74	1.166	2.925	1.145	1.937	7
1.95	1.172	2.908	1.158	1.933	10
2.02	1.165	2.919	1.145	1.946	2
2.24	1.149	2.909	1.154	1.945	5
2.90	1.159	2.928	1.150	1.985	1
3.84	1.166	2.928	1.147	2.005	6
0.0[d]	1.165 ±0.007	2.904 ±0.014	1.152 ±0.007	1.870 ±0.014	
orthoferrosilite in air (DOWTY and LINDSLEY, 1973)					
0.00	1.182 ±0.003	2.490 ±0.005	1.130 ±0.002	1.906 ±0.004	

[a]relative to α-Fe
[b]numerical sequence of experiment
[c]estimated standard errors for high-pressure determinations of δ and ΔE_Q are 0.005 and 0.01 mm s^{-1}, respectively
[d]extrapolated from high-pressure data

The Mössbauer spectrum of clinoferrosilite recorded in air at atmospheric pressure is illustrated in Fig. 1a, and was fitted to two Lorentzian lines using the conventional constraint of equal component linewidths. The areas of component doublets are not exactly 1:1 because of slight preferred orientation due to the method used for sample preparation.

Figure 1b shows the spectrum of the same sample mounted in the diamond anvil cell. Increased source linewidth and the relatively high effective thickness of the absorber cause broader linewidths and non-Lorentzian lineshapes; therefore all high-pressure spectra were fit using the Voigt lineshape. As well as the conventional constraints of equal component Lorentzian linewidths and Gaussian standard deviations, the following additional constraints were applied to high-pressure spectra of the $P2_1/c$ phase: the Lorentzian linewidth, Gaussian standard deviation and spectrum area of the M1 doublet were constrained to be equal to the corresponding parameters of the M2 doublet. Such constraints were justified on the basis of fits to the spectrum of the absorber recorded in air, where the lines are well resolved (Fig. 1a). Application of these constraints during the fitting process to the spectrum recorded in air changed the hyperfine parameters by less than 0.0005 mm/s compared to an unconstrained fit. Component areas were allowed to vary in the high-pressure spectra to account for slight preferred orientation effects. The similar component area ratios for all spectra are consistent with the similar method of sample preparation, where crystallites were pressed at right angles to the plane of the absorber in both cases, either by the sample holder (sample in air) or by a glass slide (diamond anvil cell sample).

Mössbauer spectra from the C2/c phase are well resolved (Fig. 1c) and robust hyperfine parameters could be determined using only the conventional constraints of equal component Lorentzian linewidths and Gaussian standard deviations.

RESULTS AND DISCUSSION

The hyperfine parameters determined for clinoferrosilite are listed in Table 1. By analogy with orthoferrosilite, the doublet with large quadrupole splitting is assigned to the M1 site and the doublet with small quadrupole splitting to the M2 site. The Mössbauer parameters of orthoferrosilite (DOWTY and LINDSLEY, 1973) are given in Table 1 for comparison. X-ray diffraction experiments have shown that the transition $P2_1/c \rightarrow C2/c$ occurs between

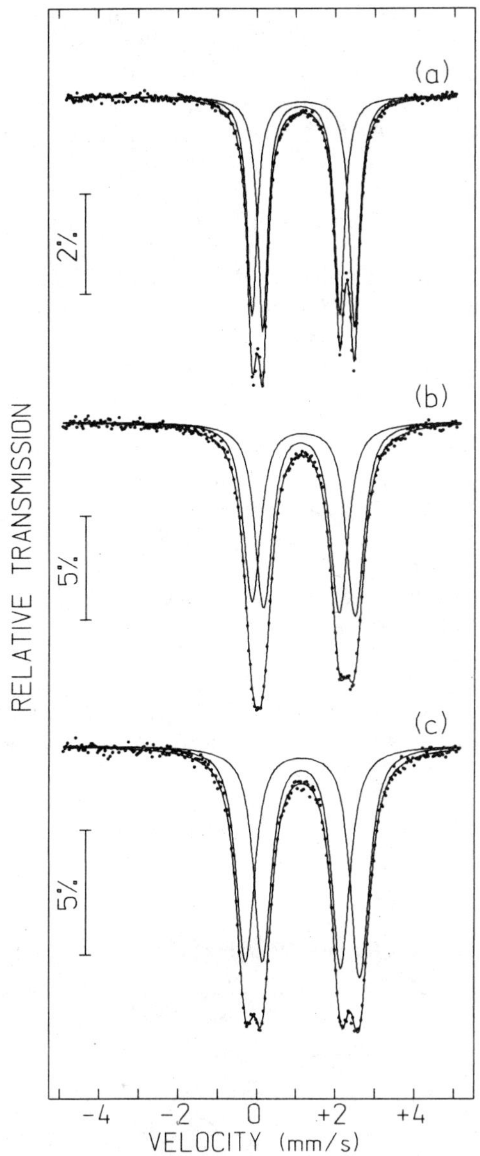

FIG 1. Room temperature Mössbauer spectra of clinoferrosilite (a) in air, and in a diamond anvil cell at (b) 0.1 MPa and (b) 3.84 GPa. The velocity scales are relative to Fe in Rh (−0.114 mm/s relative to α-Fe).

1.48 and 1.75 GPa (HUGH-JONES *et al.*, 1994), and measurements using optical absorption spectroscopy have found the transition to occur between 1.61 to 1.85 GPa (ROSS and SOWERBY, 1996). Our observation of the transition between 1.33 and 1.74 GPa (Fig. 2) is in good agreement with these previous observations. In the Mössbauer data there is a region between 1.44 and 1.74 GPa with apparently intermediate values of the hyperfine parameters. Since the transition is first order with a 3% decrease in volume (HUGH-JONES *et al.*, 1994), the intermediate values represent a mixture of the $P2_1/c$ and $C2/c$ phases rather than an intermediate structure. The hysteresis might be due in part to microstructure in the sample caused by excessive cycling through the transition (the spectra with intermediate hyperfine parameters were recorded on the sixth cycle; see Table 1). HUGH-JONES (1995) has noted significant broadening of X-ray diffraction peaks after several cycles through the transition, which could arise from strain within the lattice. Such strain could broaden the pressure interval over which the transition occurred. The presence of strain does not affect the hyperfine parameters, however; results from the final spectrum recorded (1.33 GPa) are fully consistent with all other data from the $P2_1/c$ phase.

Effect of pressure on the centre shift

The centre shift (δ) is composed of the isomer shift (δ_o) plus the second-order Doppler shift (δ_{SOD}). The centre shift and isomer shift are often taken to

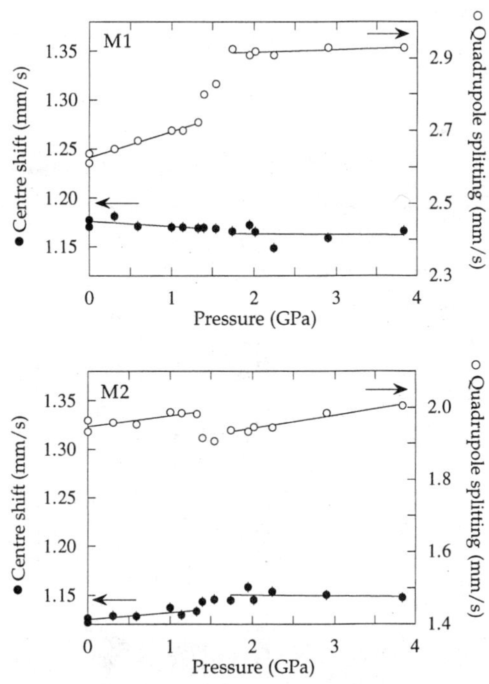

FIG. 2. Variation of hyperfine parameters for the M1 and M2 sites in clinoferrosilite as a function of pressure. Solid lines indicate linear least-square fits to the data corresponding to the $P2_1/c$ and $C2/c$ phases. Arrows indicate the relevant axis for the data.

be equal, since δ_{SOD} is similar for most materials and is relatively small. However both δ_o and δ_{SOD} are pressure dependent, so the pressure dependence of δ at constant temperature is approximated by

$$\partial\delta/\partial P = \partial\delta_o/\partial P + \partial\delta_{SOD}/\partial P. \quad (1)$$

We can estimate the volume dependence of δ_{SOD} following the method of WILLIAMSON (1978):

$$\frac{\partial\delta_{SOD}}{\partial \ln V} = \gamma\frac{9k\Theta}{16mc}F(T/\Theta) \quad (2)$$

where γ is the lattice Grüneisen parameter, Θ is the Debye temperature, m is the mass of the Mössbauer isotope, T is the temperature, k is the Boltzmann constant, c is the speed of light, and the function F is given by:

$$F(T/\Theta) = 1 + \frac{8}{e^{T/\Theta}-1} - 24\left(\frac{T}{\Theta}\right)^{4T/\Theta}\int_0 \frac{x^3}{e^x - 1}\,dx. \quad (3)$$

The Grüneisen parameter can be calculated from the thermal expansion, volume, bulk modulus and heat capacity data according to $\gamma = \alpha V K_T/C_V$. Since thermodynamic data are not available for clinoferrosilite, we used the Grüneisen parameter ($\gamma = 2.3$) calculated from enstatite data taken from AHRENS (1995). Using the value $\Theta = 719$ K for enstatite (KIEFFER, 1985), we calculated ($\partial\,\delta_{SOD}/\partial \ln V$) to be 0.25 mm/s, which is comparable to the values 0.15–0.57 mm/s determined by ZHANG and HAFNER (1992) for hedenbergite. The pressure dependence of δ_{SOD} can be calculated using the elastic data reported by HUGH-JONES (1995) for the $P2_1/c$ phase of clinoferrosilite ($K_T = 100 \pm 4$ GPa for K_T' fixed at 4.3) and is reported in Table 2. The C2/c phase of clinoferrosilite is less compressible than the $P2_1/c$ phase ($K_T = 114 \pm 7$ GPa for K_T' fixed at 4.3; HUGH-JONES, 1995) and if we assume a 10% increase in Θ, this results in an increase of only 5% in the value of $\partial\delta_{SOD}/\partial P$. The uncertainties for the calculated quantities in Table 2 were estimated by propagating the likely uncertainties in the input data through the calculation.

The pressure dependence of the centre shifts ($\partial\delta/\partial P$) were determined from the experimental data using linear regression, and the results are listed in Table 2. It is evident that while $\partial\delta_{SOD}/\partial P$ can be ignored to a first approximation in calculating $\partial\delta_o/\partial P$ for the $P2_1/c$ phase, it cannot be ignored for the C2/c phase. For consistency we used equation (1) to calculate $\partial\delta_o/\partial P$ for all phases.

The isomer shift (δ_o) is directly proportional to the difference in s electron density at the nucleus between the source and the absorber, where the constant of proportionality is negative for ^{57}Fe. Two effects have been suggested for changing the s electron density at the nucleus with pressure: (1)

Table 2. Pressure dependence of δ and ΔE_Q for clinoferrosilite.

	phase	site	$P2_1/c$	C2/c	ref
$\partial\delta/\partial P$	cfs	M1	$-8(4) \times 10^{-3}$	$-7(28) \times 10^{-4}$	1
mm s^{-1} GPa^{-1}	cfs	M2	$6(4) \times 10^{-3}$	$-1(3) \times 10^{-3}$	1
	Hd	M1		-8×10^{-3}	2
$\partial\delta_{SOD}/\partial P^*$	cfs		$-2(1) \times 10^{-3}$	$-2(1) \times 10^{-3}$	1
mm s^{-1} GPa^{-1}	Hd			-1×10^{-3}	2
$\partial\delta_o/\partial P^*$	cfs	M1	$-6(4) \times 10^{-3}$	$1(3) \times 10^{-3}$	1
mm s^{-1} GPa^{-1}	cfs	M2	$8(4) \times 10^{-3}$	$1(3) \times 10^{-3}$	1
	Hd	M1		-7×10^{-3}	2
$\partial\Delta E_Q/\partial P$	cfs	M1	$6.3(9) \times 10^{-2}$	$6(6) \times 10^{-3}$	1
mm s^{-1} GPa^{-1}	cfs	M2	$4.0(9) \times 10^{-2}$	$3.6(6) \times 10^{-2}$	1
	Hd	M1		-3×10^{-2}	2

*calculated
cfs: clinoferrosilite; Hd: hedenbergite
numbers in parentheses represent the standard deviation of the last digit
References: [1]this work; [2]ZHANG and HAFNER (1992)

changes in orbital occupation; and (2) distortion of the wavefunctions—either compression of the s electrons or spreading out of the 3d electrons (increase in covalency) (DRICKAMER *et al.*, 1969). The second factor, an increase in covalency, causes an increase in s electron density and hence a decrease in δ_o. The first factor includes transfer of electrons between the iron 3d and 4s levels, between different iron 3d levels, and between iron orbitals and orbitals of the co-ordinating anions. It can either increase or decrease δ_o depending on the nature and direction of transfer. The similarity of pressure coefficients for ionic compounds despite a variety of ligands suggests that metal—ligand transfer plays a minor role for these compounds (WILLIAMSON, 1978).

For the $P2_1/c$ phase, $\partial\delta_o/\partial P$ is negative for the M1 site while it is positive for the M2 site (Table 2). This cannot be attributed to differential compression of the sites: excluding changes in orbital occupation, the effect of polyhedral compression is to decrease δ_o, and differential compression would change the magnitude of the decrease, but not the sign. One possibility is electron transfer from 4s to 3d iron orbitals as a function of pressure at the M2 site, which would decrease s electron density and hence increase δ_o. Such a bonding change might be expected to accompany changes in the distortion of the coordination octahedron with pressure. As discussed by HUGH-JONES *et al.* (1994), increasing pressure tends to rotate the tetrahedral chains in the $P2_1/c$ structure to maintain stability, hence changing the configuration of the M2 site. It should be emphasised, however, that the overall change in δ_o with pressure for the M2 site is small, implying only minor changes in the configuration.

There is no observed change in δ_o for the M1 site between the $P2_1/c$ and C2/c phases, which is consistent with single crystal structure refinements which show negligible change in the M1 site bond lengths and distortion through the transition (HUGH-JONES *et al.*, 1994). The slight increase in δ_o for the M2 site through the transition to the C2/c phase implies a decrease in s electron density. This suggests that the increase in s electron density which accompanies the shorter Fe—O bond distances in the C2/c phase (HUGH-JONES *et al.*, 1994) must be offset by a greater decrease in s electron density caused by changes in orbital occupation.

For the C2/c phase, both M1 and M2 sites show similar behaviour: no change in δ_o with pressure in this pressure range within experimental error (Table 2). HUGH-JONES (1995) has shown that compression of the M1 and M2 octahedra occurs in this pressure range, so decreasing bond distances must be offset by slight changes in orbital occupation that are similar for both sites.

Finally, we note that $\partial\delta_o/\partial P$ for the M1 site in hedenbergite is identical within experimental error to that observed for the M1 site in the $P2_1/c$ phase of clinoferrosilite. This suggests a similar compression of the M1 octahedra in both structures.

Effect of pressure on the quadrupole splitting

The change in quadrupole splitting (ΔE_Q) during the transition from the $P2_1/c$ to the C2/c phase is significantly higher for the M1 site compared to the M2 site (Fig. 2). Yet high-pressure structure refinements by HUGH-JONES *et al.* (1994) show that the M1 octahedron in the C2/c phase at high pressure is similar to the M1 octahedron in the $P2_1/c$ phase at room pressure. DOWTY and LINDSLEY (1973) used the model of INGALLS (1964) to account for differences in ΔE_Q between the M1 and M2 sites in clinopyroxenes. The same model can be used to account for the above discrepancy, and to explain qualitatively the changes in ΔE_Q in clinoferrosilite with pressure.

In the model of INGALLS (1964) ΔE_Q for Fe^{2+} in octahedral coordination is expressed in terms of a reduction function F:

$$\Delta E_Q = C\,\alpha^2\,F(\Delta_1, \Delta_2, \lambda_o, \alpha^2, T) \qquad (4)$$

where C is a constant ($= 2/7e^2\,Q\,(1 - R_o)\,\langle r^{-3}\rangle_o$), Δ_1 and Δ_2 are the ground state splittings of the crystal field levels, α^2 is a covalency factor which decreases with increasing covalency, λ_o is the spin-orbit coupling constant for the free ion and T is the temperature. The reduction function, F, has two contributions, one from the valence electrons (valence term) and one from the influence of the crystal lattice (lattice term). The values of Δ_1 and Δ_2 are related to the distortion of the coordination octahedron from cubic symmetry. The valence term is highly temperature dependant and increases with increasing distortion, eventually reaching a maximum value. The lattice term, which is largely independent of temperature, also increases with increasing distortion, but is of opposite sign. The reduction function F therefore increases initially with increasing distortion to a maximum value, but then decreases with further increase in distortion as the contribution from the lattice term becomes greater.

We have calculated the reduction function according to the INGALLS (1964) model for the M1 and M2 sites of $P2_1/c$ clinoferrosilite. Since crystal field parameters are not available for this phase, we used values for orthopyroxene determined by

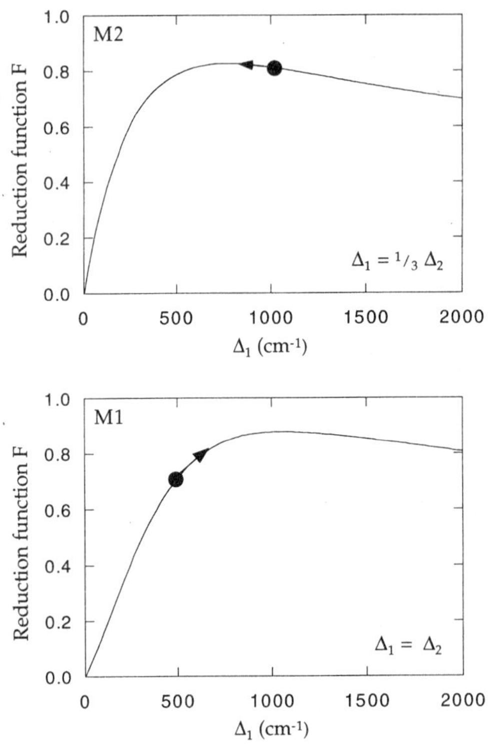

FIG. 3. Calculated curves for the reduction function as a function of ground-state splitting (Δ_1 and Δ_2) for the M1 and M2 sites in P2_1/c clinoferrosilite according the the model of INGALLS (1964). The solid circles indicate the approximate position corresponding to the site at ambient pressure, and the arrows indicate the direction of movement during compression.

LIN *et al.* (1993) and VAN ALBOOM *et al.* (1993). Assuming the M1 site to have tetragonal distortion ($\Delta_1 = \Delta_2$), and the M2 site to have rhombic distortion (assuming $\Delta_1 = \frac{1}{3}\Delta_2$, which approximates the situation for orthopyroxene), we calculated the reduction functions illustrated in Fig. 3. The curves should not be regarded as a quantitative determination of the quadrupole splitting, being based on an approximate model with a number of simplifying assumptions, but rather as a qualitative guide to factors that influence the quadrupole splitting.

Using the values for ground-state splitting for orthopyroxene of $\Delta_1 = \Delta_2 = 500$ cm^{-1} for the M1 site and $\Delta_1 \approx \frac{1}{3}\Delta_2 = 1000$ cm^{-1} for the M2 site (LIN *et al.*, 1993; VAN ALBOOM *et al.*, 1993), the approximate position of clinoferrosilite on the reduction function curves can be located (indicated by solid circles on Fig. 3). As postulated by DOWTY and LINDSLEY (1973) based on values of ΔE_Q for M1 and M2 sites, we see that the two sites indeed fall on opposite sides of the reduction function maximum.

The increase in ΔE_Q for the M1 site implies an increase in distortion with pressure. The increase is likely to be small since $\partial\delta_o/\partial P$ is also small (see above), but the steep slope in that region of the reduction curve produces a relatively large increase in ΔE_Q with small distortion increase for the M1 site. In contrast, the position of the M2 site on the reduction function curve implies that increasing pressure produces a decrease in Δ_1. This is consistent with $\partial\delta_o/\partial P$ results above which also suggest a change in distortion as well as compression of the M2 site in the P2_1/c phase with increasing pressure.

There is a large change in ΔE_Q for the M1 site between the P2_1/c and C2/c phases of clinoferrosilite. This is surprising considering the results of structure refinements that show negligible change in bond lengths and site distortion of the M1 site between the two polymorphs (HUGH-JONES *et al.*, 1994). But ΔE_Q is sensitive not only to the nearest neighbour coordination shell, but also to the next-nearest neighbour shell, which includes the iron atoms in M2 sites (the M1 and M2 octahedra share common edges). It is therefore likely that the significant change in M2 site geometry in the C2/c phase compared to the P2_1/c phase is reflected in ΔE_Q for the M1 site. The decrease in ΔE_Q for the M2 site between the P2_1/c and C2/c phases would appear to suggest an increase in Δ_1 according to Fig. 3a, but we know from single crystal structure refinements that the distortion of the M2 site is lower in the C2/c structure than in the P2_1/c structure (HUGH-JONES *et al.*, 1994). One possibility is that the change in Δ_1 is so large that the value for the P2_1/c phase moves to the left side of the reduction function maximum in Fig. 3a, where ΔE_Q decreases with decreasing Δ_1. Another explanation, which is probably more likely, is that the change in bonding of the M2 site is so large between the P2_1/c and C2/c structures that Fig. 3a cannot be used for both phases. This is consistent with structure refinements that show that the P2_1/c → C2/c transition involves breaking of one of the M2—O bonds in the P2_1/c structure and reconstruction at a different location in the C2/c structure (ANGEL *et al.*, 1992; HUGH-JONES *et al.*, 1994).

For the C2/c phase, ΔE_Q increases with increasing pressure for both the M1 and M2 sites, where the slope for the M2 site is significantly greater than the slope for the M1 site (Fig. 2 and Table 2). Since $\partial\Delta E_Q/\partial P$ for the M2 site in the C2/c structure is identical within experimental error to the value for the P2_1/c structure (Table 2), this would suggest a similar change in Δ_1 for the M2 site in both structures as a function of pressure.

The hedenbergite structure provides an interesting comparison for the clinoferrosilite data. In hedenbergite Ca occupies the M2 site, which is eight-coordinated and very distorted compared to the M2 sites in both clinoferrosilite polymorphs. The signs of $\partial\Delta E_Q/\partial P$ are opposite for the M1 site in the hedenbergite structure compared to the M1 site in the $P2_1/c$ phase (Table 2). According to the INGALLS (1964) model and Fig. 3a, this would imply that the distortion of the M1 site in hedenbergite decreases with increasing pressure. ZHANG and HAFNER (1992) reached the same conclusion based on a similar analysis. The values of $\partial\delta_0/\partial P$ are identical for the two phases, implying that compression of the octahedra with pressure is similar (see above). One explanation for the opposite slopes might be the effect of next-nearest neighbours (i.e. the different M2 sites) on $\partial\Delta E_Q/\partial P$.

Comparison of clinoferrosilite with orthoferrosilite

The Mössbauer parameters of orthoferrosilite (DOWTY and LINDSLEY, 1973) are given in Table 1 for comparison with the $P2_1/c$ and $C2/c$ phases of clinoferrosilite. The centre shifts are nearly identical: differences are close to the experimental error and are too small to interpret with the qualitative models used in this paper. In contrast, there are significant differences in ΔE_Q. The greatest difference is in values for the M1 site, which vary from a maximum of 2.904 mm/s for the $C2/c$ phase (extrapolated to zero pressure) to a minimum of 2.490 mm/s for orthoferrosilite (Table 1). As discussed above, according to the INGALLS (1964) model the value of ΔE_Q is highly sensitive to the distortion due to the steep slope of the reduction function in that region. Differences in the distortion may reflect variation in next-nearest neighbour coordination (M2 sites) at least as much as they reflect variation in M1 site coordination between structures. Values for ΔE_Q of M2 sites vary less between the different structures (Table 1), which is consistent with the position of the M2 site on the reduction function plot in the INGALLS (1964) model (Fig. 3b). The nearly identical values for ΔE_Q corresponding to the M2 site in the $C2/c$ and orthoferrosilite polymorphs is consistent with structure refinements of HUGH-JONES *et al.* (1994) that show identical size and only small differences in distortion for the M2 sites in these two polymorphs.

CONCLUSIONS

Mössbauer spectra have been recorded of clinoferrosilite as a function of pressure. The relative change in hyperfine parameters has been interpreted in terms of qualitative models that provide insight into changes in the electronic and structural environment of the iron atoms, both as a function of pressure and through the phase transition from the $P2_1/c$ to the $C2/c$ structure.

From the perspective of Mössbauer spectroscopy, the M1 and M2 sites behave quite differently under pressure in the $P2_1/c$ structure. For the M1 site there is only a small change in electronic structure: a slight increase in covalency of the Fe—O bonds and a small increase in the ground-state splitting. The latter is probably due to changes in the next-nearest neighbour environment (i.e. the M2 atoms). For the M2 site the changes are greater: the orbital occupation changes (e.g. transfer of electron density from the iron 4s to 3d orbitals) in addition to a probable increase in covalency of Fe—O bonds, and the ground-state splitting likely decreases. Due to the difficulty of synthesising untwinned single crystals of the $P2_1/c$ phase, single-crystal refinements as a function of pressure are not available. If refinements could be performed, however, one would not expect to observe changes in the M1 site as a function of pressure beyond perhaps a slight shortening of Fe—O bond distances, but changes in the M2 site might be large enough to be observed as a change in site distortion as a function of pressure.

The M1 and M2 sites behave in a more similar manner as a function of pressure in the $C2/c$ structure. In addition to a probable increase in covalency of Fe—O bonds in both sites, there is a slight change in orbital occupation that decreases the s electron density. The ground-state splitting changes for both sites, and the magnitude of the change for the M2 site in the $C2/c$ structure appears to be similar to the change for the M2 site in the $P2_1/c$ structure. Recent refinements of the $C2/c$ phase show that the distortion of the M2 site decreases with increasing pressure to 6.5 GPa (HUGH-JONES, 1995), which is consistent with the above observation.

The M1 and M2 sites behave quite differently through the phase transition from the $P2_1/c$ to the $C2/c$ structure. The M1 site shows no change in covalency or orbital occupation, and a small increase in ground-state splitting; whereas the M2 site undergoes a change in bonding as well as a change in ground-state splitting. These results are consistent with single-crystal refinements of the

$P2_1/c$ phase at room pressure compared to the C2/c phase just above the transition—there is little change in bond lengths or site distortion of the M1 site, but an observable change in the M2 site (HUGH-JONES *et al.*, 1994).

The model used in the present work to interpret relative changes in quadrupole splittings gives a qualitative indication of the trends, but cannot provide quantitative information on observables such as ground-state splittings. Such data would be useful in the interpretation of optical spectra of clinoferrosilite (*e.g.* ROSS and SOWERBY, 1996) and in the prediction of properties such as Fe-Mg partitioning in solid solutions. In view of the large amount of information available on the crystal structure (HUGH-JONES *et al.*, 1994) and electronic structure (ROSS and SOWERBY, 1996), a valuable extension of this work would be a crystal field analysis of the high-pressure spectra similar to the work of LIN *et al.* (1993) or VAN ALBOOM *et al.* (1993), which would provide quantitative data on the detailed electronic structure of these phases.

Acknowledgements—Helpful comments on the manuscript were provided by R. Angel, T. Ericsson, D. Hugh-Jones, and F. Seifert.

REFERENCES

AHRENS T. J. (1995) *Mineral Physics and Crystallography: A Handbook of Physical Constants.* AGU Reference Shelf 2, Amer. Geophys. Union.

ANGEL R. J., CHOPELAS A. and ROSS N. L. (1992) Stability of high-density clinoenstatite at upper mantle pressures. *Nature* **358**, 322–324.

DOWTY E. and LINDSLEY D. H. (1973) Mössbauer spectra of synthetic hedenbergite-ferrosilite pyroxenes. *Amer. Mineral.* **58,** 850–868.

DRICKAMER H. G., VAUGHAN R. W. and CHAMPION A. R. (1969) High-pressure Mössbauer resonance studies with iron-57. *Acc. Chem. Res.* **2,** 40–47.

HAZEN R. M. and FINGER L. W. (1982) *Comparative Crystal Chemistry.* John Wiley and Sons.

HUGH-JONES D. A. (1995) High pressure behaviour of pyroxenes. Ph.D. dissertation, Univ. London.

HUGH-JONES D. A., WOODLAND A. B. and ANGEL R. J. (1994) The structure of high-pressure C2/c ferrosilite and crystal chemistry of high-pressure C2/c pyroxenes. *Amer. Mineral.* **79,** 1032–1041.

INGALLS R. (1964) Electric-field gradient tensor in ferrous compounds. *Phys. Rev.* **133A,** 787–795.

KIEFFER S. W. (1985) Heat capacity and entropy: Systematic relations to lattice vibrations. In *Rev. Mineral. Vol. 14* (eds. S. W. KIEFFER and A. NAVROTSKY), pp. 65–126. Mineral. Soc. America.

LIN C., ZHANG L. and HAFNER S. S. (1993) Local electronic states of Fe^{2+} ions in orthopyroxene. *Amer. Mineral.* **78,** 8–15.

MCCAMMON C. A. (1992) A new apparatus to conduct variable temperature high pressure Mössbauer spectroscopy. In *Recent Trends in High-Pressure Research: Proceedings of XIII AIRAPT International Conference on High Pressure Science and Technology* (ed. A. K. SINGH), pp. 824–826. Oxford & IBH Publishing Co. Pvt. Ltd.

ROSS N. L. and SOWERBY J. R. (1996) Electronic absorption of synthetic clinoferrosilite. this volume.

VAN ALBOOM A., DE GRAVE E. and VANDENBERGHE R. E. (1993) Study of the temperature dependence of the hyperfine parameters in two orthopyroxenes by ^{57}Fe Mössbauer spectroscopy. *Phys. Chem. Minerals* **20,** 263–275.

WILLIAMSON D. L. (1978) Influence of pressure on isomer shifts. In *Mössbauer Isomer Shifts* (eds. G. K. SHENOY and F. E. WAGNER), pp. 317–360. North-Holland Publishing Co.

ZHANG L. and HAFNER S. S. (1992) High-pressure ^{57}Fe γ resonance and compressibility of $Ca(Fe,Mg)Si_2O_6$ clinopyroxenes. *Amer. Mineral.* **77,** 462–473.

Mineral Spectroscopy: A Tribute to Roger G. Burns
Editors: M. D. Dyar, C. McCammon and M. W. Schaefer

Crystal chemistry of Fe^{3+}, H^+, and D/H in mantle-derived augite from Dish Hill: Implications for alteration during transport

M. DARBY DYAR,[1] SUZETTE V. MARTIN,[2] STEPHEN J. MACKWELL,[3] SUSAN CARPENTER,[3] CHRISTOPHER A. GRANT[4] and ANNE V. MCGUIRE[5]

[1]Department of Geology and Astronomy, West Chester University, West Chester, PA 19383, U.S.A.
[2]Mathematics and Science Department, Navajo Community College, P.O. Box 580, Shiprock, Navajo Nation, NM 87420, U.S.A.
[3]Department of Geosciences, The Pennsylvania State University, University Park, PA 16802, U.S.A.
[4]Department of Chemistry, University of Oregon, Eugene, OR 97403, U.S.A.
[5]Department of Geosciences, University of Houston, Houston, TX 77204, U.S.A.

Abstract—Chemical and crystal chemical analyses have been performed on a suite of sub-continental, mantle-derived augite samples and their accompanying magmas from a single cinder cone at Dish Hill, California. Mössbauer techniques have been used to investigate Fe oxidation state, uranium-extraction techniques have been used to determine bulk wt%H_2O contents and D/H ratios, wet chemistry has been used to corroborate $Fe^{3+}/\Sigma Fe$, and electron microprobe techniques coupled with the above measurements have been used to determine major element contents of pyroxene. Comparison of Mössbauer and wet chemical results on Fe^{3+}/Fe^{2+} permitted determination of empirical correction factors that can be used to bring Mössbauer results of various fitting models into agreement with true Fe^{3+}/Fe^{2+} ratios, which range from 0.16 to 1.53. D/H ratios of augite samples range from −93.0‰ to −154.3‰, possibly suggesting exchange with meteoric waters, while the entraining basalts have somewhat heavier D/H ratios in the range −87.0‰ to −102.2‰, also indicating the likelihood of some exchange with surface waters. Results of experimental work to determine hydrogen diffusivity in high-Fe clinopyroxene, when applied to the Dish Hill augite, suggest that the augite megacrysts have exchanged hydrogen with the basaltic melts during ascent. However, as the time scale for transport is short, it is possible that equilibration of deuterium in the augite with that of the surrounding basalts may not have occurred. Differences in D/H ratios in the augite probably result from hydrogen exchange with the basalt during transport, and therefore do not reflect conditions in the mantle source region. The lack of a correlation between hydrogen remaining in the augite samples and Fe^{3+}/Fe^{2+} may therefore either result from the large uncertainties in the bulk hydrogen analysis or reflect significant variability in H^+ or Fe^{3+} in the source region.

INTRODUCTION

PIONEERING STUDIES by Roger Burns and Michael Bancroft on the electrochemical states of iron in pyroxene minerals were among the first investigations using the techniques of optical and Mössbauer spectroscopy in the study of minerals (BURNS, 1966; BANCROFT and BURNS, 1967; BANCROFT *et al.*, 1967a). These early studies are still cited frequently, particularly in the context of studies on Fe in pyroxenes with high pressure and temperature (upper mantle) compositions. Although his later interests expanded to include a wide variety of theoretical and experimental problems dealing with transition metal-bearing minerals, Roger Burns' interest in common rock-forming minerals and his respect for the geologic problems they could solve remained strong throughout his long career. He was never too busy to explain crystal chemistry to his geologic colleagues who came bearing questions. It seems appropriate, therefore, to dedicate this study of augite from the upper mantle to the broad-thinking genius and memory of Roger Burns.

Until very recently, research programs assessing the water contents, intrinsic oxygen fugacity, and hydrogen transport in the minerals of the Earth's upper mantle have proceeded largely independently, with minimal interaction between groups. For example, studies by Rossman and coworkers (MILLER *et al.*, 1987; SKOGBY and ROSSMAN, 1989, SKOGBY *et al.*, 1990 and 1992; SMYTH and ROSSMAN, 1991; BELL and ROSSMAN, 1992) pioneered standards and techniques for quantitative measurement of water contents in mantle phases, with emphasis on olivine, pyroxene, garnet and amphibole. Their work has shown that incorporation of small amounts of structurally-bound hydroxyl is ubiquitous in nominally anhydrous minerals from mantle xenoliths. Their conclusions have profound implications for the accommodation of water in the upper mantle. However, their studies were focussed primarily on characterization of hydroxyl contents, and have largely neglected experimental determination of the mechanisms by which hydroxyl contents of mantle phases may have been altered during transport.

Studies of mantle f_{O_2} have also utilized measurements of as-erupted Fe^{3+}/Fe^{2+} contents in xenoliths to infer mantle conditions. Research by this group (DYAR *et al.*, 1989 and 1992; MCGUIRE *et al.*, 1989 and 1991), Wood and coworkers (WOOD and VIRGO, 1989; BRYNDZIA and WOOD, 1989; LUTH *et al.*, 1990; LUTH and CANIL, 1993), and others has characterized the Fe^{3+}/Fe^{2+} contents of olivine, pyroxene, spinel, garnet, and amphibole. This body of research has been widely used to derive estimates of upper mantle f_{O_2}, again with little consideration of the effects of hydrogen diffusion during transport on the Fe^{3+}/Fe^{2+} contents of as-erupted samples.

Meanwhile, a largely separate body of work within the geophysics community addresses the kinetics of hydrogen diffusion. MACKWELL and KOHLSTEDT (1990) showed that hydrogen diffusion in olivine is extremely rapid, facilitating the loss of H^+ via the mechanism

$$H^+_{ol} + Fe^{2+}_{ol} \rightleftarrows \tfrac{1}{2}[H_2] + Fe^{3+}_{ol}. \qquad (1)$$

Further work by BAI and KOHLSTEDT (1992) suggested that H^+ contents of as-erupted olivine may therefore be under-representative of actual mantle conditions. Although these studies postulated a correlation between H^+ and Fe^{3+} via the dehydrogenation reaction (eqn. 1 above), Fe^{3+} contents were not actually measured. This was also true of a recent study of hydrogen diffusion in a diopside single crystal (INGRIN *et al.*, 1995), where diffusivities fast enough to account for significant modification of Fe^{3+} (not measured) and H^+ (measured) contents were observed. However, in an earlier study on synthetic magnesian clinopyroxene, SKOGBY (1994) did observe a near 1:1 correlation between H^+ incorporated in clinopyroxene and Fe^{2+} increase. For amphibole, measured Fe^{3+} and H^+ data were reported by DYAR *et al.* (1993), who used results from GRAHAM *et al.* (1984) to demonstrate that dehydrogenation of hornblende occurs more slowly than in olivine or pyroxene, leading to only minor modification of mantle H^+ and Fe^{3+} during transport. This suggests that as-erupted H^+ and Fe^{3+} contents of amphibole may directly reflect mantle equilibrium conditions, a conclusion that was also reached by BELL and HOERING (1994). Data on H^+ diffusion in orthopyroxene, spinel, and garnet are currently being investigated (S. MACKWELL, personal communication; H. SKOGBY, personal communication). In summary, as-erupted Fe^{3+} and H^+ contents of olivine and clinopyroxene may be substantially modified during transport, while amphibole is much less affected by ascent.

Accordingly, in the present study we seek to measure and understand the interrelationships between H_2O and Fe^{3+} in augite and Fe-rich diopside. We apply new data on hydrogen diffusivity in Fe-rich diopside compositions to our results on Fe^{3+}, H^+, and D/H ratios in a suite of mantle-derived augite samples collected from a single cinder cone at Dish Hill, California. Hydrogen isotope data on augite are then compared with results on the host magmas at this locality, and with previous work on Dish Hill amphibole megacrysts (DYAR *et al.*, 1993; BELL and HOERING, 1994). Through this detailed study of naturally-occurring augite samples and applicable experimental results on Fe-rich diopside, we can assess 1) the isotopic and crystal chemical variation among samples from a single cinder cone, 2) the relationship(s) among H^+, Fe^{3+}, and other major elements within the augite structure, and 3) the extent to which the compositions of the natural samples are in any way representative of mantle equilibria.

BACKGROUND

Geology of Dish Hill

The Dish Hill locality was chosen for this study for three important reasons. First, in order to rule out the effects of tectonic setting, this locality was chosen to provide an example of a single eruptive event. Second, the cone is extremely rich in megacrysts (as well as xenoliths), facilitating sampling of a number of large augite single crystals. The area was also the site of hornblende studies by BELL and HOERING (1994). Finally, the chemistry of the magmas that coexist with the augite megacrysts has been relatively well studied in the context of the southern California basaltic volcanoes (WILSHIRE *et al.*, 1986 and 1988).

Dish Hill is a cinder cone of basanite. Fission-track age dating shows the cone to be approximately 2 Ma old (WILSHIRE and TRASK, 1971). The present study used samples of the Dish Hill basanite and single crystals of augite (megacrysts), ranging from 1 cm to several cms in diameter (Table 1), found as inclusions within the host basalt. The megacrysts possess rounded glassy margins implying disequilibrium with the host basalt at the time of eruption. These megacrysts are likely to be either xenocrysts, fragments of augite from aluminous-augite pyroxene mantle rocks, or phenocrysts crystallized from the host magmas at mantle pressures. Aluminous augite pyroxenite xenoliths commonly form dikes or veins in mantle peridotite, which are generally inferred to originate by crystallization of mafic alkalic magmas within the mantle (*e.g.*, FREY and PRINZ, 1978). Whether they are xenocrysts or phenocrysts, the megacrysts in this

Table 1. Physical Characteristics of Augite Samples.*

Sample Number	Weight in g** as collected	Maximum Dimensions of Megacryst, in cm		
DH-201	7.10	2.70	0.95	1.20
DH-202	47.39	3.70	2.10	3.40
DH-203	10.45	2.70	0.80	1.90
DH-204	9.35	3.40	1.20	2.20
DH-205	19.35	3.90	1.20	2.00
DH-206	18.70	2.50	1.50	2.40
DH-207	5.45	2.20	1.00	1.90
DH-209	38.90	3.40	2.20	2.80
DH-210	40.40	3.30	2.30	2.50
DH-211	204.10	2.40	0.60	1.40
DH-212	20.95	2.30	1.50	0.40
DH-213	299.80	3.10	1.80	1.20
DH-217	n.d.	1.30	1.10	n.d.
DH-218	n.d.	3.00	1.80	n.d.
DH-220	n.d.	2.20	1.84	n.d.
DH-222	n.d.	1.55	1.10	n.d.
DH-227	n.d.	1.60	1.12	n.d.
DH-231	n.d.	1.90	1.10	n.d.

*All samples showed evidence of slight melting on some edges.
**Includes sample and its attached lava matrix.
n.d. ≡ Some samples were crushed before complete measurements were obtained.

study are presumed to be the products of crystallization of alkalic basalts at mantle pressures, and thus their measured Fe^{3+}/Fe^{2+} and D/H ratios should provide information about f_{O_2} and D/H of the mafic magma within the mantle, unless these ratios are altered during transport and eruption of the host magma.

H^+ Contents of Pyroxene

This subject is well reviewed in the work by Rossman and coworkers (see, especially, SKOGBY *et al.*, 1990; SMYTH *et al.*, 1991, and BELL, 1993). Only two aspects of these studies bear upon the present work. First, the existing IR studies show that mantle derived diopside, omphacite, and augite are all significant reservoirs for OH^-. Reported hydrogen contents for as-erupted clinopyroxene samples range up to 620 ppm H_2O (BELL, 1993), and suggest that clinopyroxene has the highest OH^- contents of the volumetrically significant "anhydrous" minerals in the mantle. Second, the determination of clinopyroxene OH^- contents is precise and relatively straightforward by the FTIR technique, BUT accurate quantitative measurements have been hampered by the lack of mineral-specific calibration standards for OH^-. Even when such well-characterized standards are used, determination of OH^- contents in upper mantle pyroxenes may have uncertainties ranging from ±10% to ±50% depending on composition (BELL, 1993). For this reason, and because of our interest in D/H as well as H_2O contents of our samples, we have chosen in this study to use manometric methods for hydrogen extraction from augite instead of FTIR. Although the extraction technique brings its own inherent errors to our work (see Methods section below), it does avoid the calibration problems associated with FTIR.

Fe^{3+} Contents of Pyroxene

The technique of Mössbauer spectroscopy has been used in numerous studies of the Fe^{3+} contents of mantle clinopyroxene, as reviewed in DYAR *et al.* (1989). More recently, the methodology of the Mössbauer experiment has been examined in detail and refinements have been made that improve its accuracy. LUTH and CANIL (1993) tested the effects of absorber thickness on a single clinopyroxene sample from British Columbia and found constant values within experimental error of $Fe^{3+}/\Sigma Fe$ over a concentration range of 2–5 mg Fe/cm^2. Because our samples have nearly identical bulk compositions to the sample studied by LUTH and CANIL (1993), we might assume that for samples prepared within that concentration range, and having similar Fe^{3+} and total Fe contents, absorber thickness should have little effect on the measured $Fe^{3+}/\Sigma Fe$ values. However, this presumes that the relative amounts and distributions of all species on sites are also similar.

A more significant problem in the determination of $Fe^{3+}/\Sigma Fe$ values by Mössbauer spectroscopy is the differential recoil-free fractions (f) of Fe atoms having different oxidation states (WHIPPLE, 1972; AMTHAUER *et al.*, 1976; DEGRAVE and VAN ALBOOM, 1991; DYAR *et al.*, 1993). This effect arises because Fe^{3+} atoms are more tightly bound into their sites than are less highly charged Fe^{2+} atoms, resulting in $f_{Fe^{3+}}$ being generally higher than $f_{Fe^{2+}}$. Differences have also been observed in the recoil-free fractions of Fe atoms having the same oxidation state but located in different crystallographic sites (BANCROFT *et al.*, 1967b; DEGRAVE and VAN ALBOOM, 1991).

The area of a Mössbauer doublet depends explicitly on the absorber recoil-free fraction (PRESTON *et al.*, 1962; BANCROFT, 1967; BANCROFT, 1969). In conventional analysis, ratios of peak areas are used to determine site population or redox ratios.

More completely, the areal ratio is proportional to the true redox ratio:

$$\frac{A_{Fe^{3+}}}{A_{Fe^{2+}}} = C\,\frac{N(Fe^{3+})}{N(Fe^{2+})}, \qquad (2)$$

where

$$C = \frac{\Gamma_{Fe^{3+}} f_{Fe^{3+}} G(x)_{Fe^{3+}}}{\Gamma_{Fe^{2+}} f_{Fe^{2+}} G(x)_{Fe^{2+}}}, \qquad (3)$$

and N is the amount of each species actually present, f is the site-specific recoil-free fraction, Γ is peak width, and $G(x)$ is a saturation term (a similar expression may be constructed for site-population ratios). Commonly, it is assumed that $\Gamma_{Fe^{2+}} = \Gamma_{Fe^{3+}}$, $f_{Fe^{2+}} = f_{Fe^{3+}}$, and $G(x)_{Fe^{2+}} = G(x)_{Fe^{3+}}$, so that $C = 1.0$. In general, $f_{Fe^{2+}} \neq f_{Fe^{3+}}$ (WHIPPLE, 1972; DEGRAVE and VAN ALBOOM, 1991; KARFUNKEL and POLLAK, 1993).

To the extent that f values are known (and caution must be used here, because f is sensitive to mineral composition) in principle it is possible to calculate "true" $Fe^{3+}/\Sigma Fe$ values for given minerals by determining an appropriate value for C. However, complications arise when a mineral contains Fe in more than one type of crystallographic site AND more than one oxidation state. In the case of clinopyroxene, both Fe^{3+} and Fe^{2+} are present, and may be distributed between the M2 and M1 sites (and/or in the tetrahedral site if the composition is Si- and Al-deficient, but such samples will not be considered here). Ideally this distribution of species should be treated as:

$$\frac{A^{M1}_{Fe^{3+}}}{A^{M1}_{Fe^{2+}} + A^{M2}_{Fe^{2+}}} = \frac{f^{M1}_{Fe^{3+}} N(Fe^{3+})^{M1}}{f^{M1}_{Fe^{2+}} N(Fe^{2+})^{M1} + f^{M2}_{Fe^{2+}} N(Fe^{2+})^{M2}}. \qquad (4)$$

For diopside, f^{M1}_{Fe3+} and f^{M1}_{Fe2+} have been experimentally determined to be 0.862 and 0.708, respectively, at room temperature (DEGRAVE and VAN ALBOOM, 1991), but there has been no comparable determination of f^{M2}_{Fe2+}. Also, SKOGBY *et al.* (1992) demonstrated that for orthopyroxene, f^{M2}_{Fe2+} and f^{M1}_{Fe2+} are different. In our samples, the site-specific recoil-free fractions are all different, and there is no way to formulate a simple constant, C, that depends on recoil-free fractions and that can be used to extract "true" redox ratios from Mössbauer peak-area ratios.

A final complication in the derivation of "true" $Fe^{3+}/\Sigma Fe$ values from Mössbauer peak areas arises from the use of different models for interpreting the Mössbauer data. Historically, mineralogical studies have employed Lorentzian line shapes, using the assumptions inherent in the thin absorber approximation. Although it has been generally recognized that other line shapes such as Voigt lines (LANG, 1963; EVANS and BLACK, 1970) may be more appropriate, use of these improved line shapes has until recently been computationally intractable, particularly when processing large numbers of different samples. In this study we have calculated Fe^{3+}/Fe^{2+} values for spectra fit with a number of different models including eight and six Lorentzian lines and four Voigt lines. Using this strategy, we have derived appropriate correction terms that bring the results of the various fitting models into agreement with true Fe^{3+}/Fe^{2} ratios as determined by wet chemistry.

SAMPLE SELECTION, PREPARATION, AND ANALYSES

Augite megacrysts were collected for this study by M.D.D. and A.V.M., with the assistance of co-participants in a U.S.G.S.-sponsored workshop on mantle composition, structure, and processes. Samples were collected from the south flank of the Dish Hill cinder cone adjacent to the north side of the railroad bridge. Specimens were weighed and measured before removing small pieces for thin sections. Additional pieces were broken off and ground by hand with constant wetting using acetone (to prevent oxidation) into fragments <1 mm to facilitate hand-picking of grains. Only pristine augite grains were used for this study; any grains containing visible alteration, inclusions, or impurities were rejected. In particular, we sought to use only the interior portions of megacrysts to guard against contamination by surface carbonate and to avoid the need to wash the samples in acid and risk modification of D/H. Approximately 530 mg of separates were prepared from each sample: 150 mg for Mössbauer analyses, 50 mg for wet chemistry, 300 mg for H^+ extraction, and 30 mg for future oxygen isotope study. Basalt samples were collected for us by Howard Wilshire; clean splits were crushed and prepared for isotopic analysis by the U.S. Geological Survey.

Electron microprobe analyses were performed at the University of Houston using the JEOL 8600 Superprobe with Noran (Tracor Northern) automation. Natural and synthetic mineral standards were used and matrix correction was done by the Noran ZAF routine. Routine analytical conditions were used: 15 keV acceleration voltage, 20 nA beam current measured on a Faraday cup, 40 s count times (20 s count time on sodium), and a 10 μm beam diameter. At least 5 points were analyzed on each megacryst. Probe scans were used to screen for zoning, and to eliminate zoned samples from this study. Analytical errors are ±0.5–2% for major elements and ±10–20% for minor elements.

Mössbauer analyses were performed in the Mineral Spectroscopy Laboratory at West Chester University. A source of 50–20 mCi ^{57}Co in Pd was used on an Austin Science Associates constant acceleration spectrometer. Results were calibrated against an α-Fe foil of 6 μm thickness and 99.99% purity. Spectra were folded at a channel number selected by use of autocorrelation to ensure the proper match of peak positions after folding. Spectra were fitted using a version of the program STONE modified to run on IBM and compatible personal computers. The program uses a nonlinear regression procedure with a facility for constraining any set of parameters or linear combination of parameters. Lorentzian and Voigt line shapes were used for resolving peaks as discussed above. Fitting procedures in general followed those described by DYAR *et al.* (1989) and MCGUIRE *et al.* (1989), with modifications described by GRANT (1995). A statistical best fit was obtained using each model for each spectrum using the χ^2 and/or Misfit parameters; practical application of these parameters is discussed elsewhere (DYAR, 1984; GRANT, 1995). Errors are estimated at $\pm1\%$ for doublet areas, and ±0.02 mm/s for peak width, isomer shift, and quadrupole splitting.

Hydrogen contents were determined by means of a procedure for collecting and measuring all the structural hydrogen using a volumetric measurement of water vapor extracted from silicates (see BIGELEISEN *et al.*, 1952 and HOLDAWAY *et al.*, 1986 for details of the technique). For this study, 300 mg of each sample were weighed into molybdenum crucibles, then degassed under vacuum for at least 8 h at 50–85°C to drive off adsorbed atmospheric moisture. Samples were then fused in an induction furnace to liberate structural water. Distillation processes involving transfer of evolved gases through a series of liquid nitrogen and methanol-dry ice slush traps were used to separate water molecules effectively from other condensible and non-condensible gases. Water vapor was then passed over a hot (>750 °C) uranium furnace to liberate free H^+. A mercury-piston Toepler pump was used to collect hydrogen vapor in a volumetrically calibrated reservoir for yield measurement. Approximately half the samples were run at least twice. Distilled H_2O was used for calibration. All data are corrected for the blank of 2.5 torr that arises from degassing of the crucibles. Errors are estimated at ≤0.1 wt% H_2O based on replications.

Wet chemical determinations of $Fe^{3+}/\Sigma Fe$ were determined by analyst Jun Wu in the laboratory of Joseph Stucki at the University of Illinois, using methods developed by Stucki for clay minerals. His colorimetric method (STUCKI, 1981; STUCKI and ANDERSON, 1981; KOMANDEL and STUCKI, 1988) includes an organic chromophore during the sample digestion which complexes the iron as it is released into solution, thereby protecting against aerial oxidation. This method has a greater specificity for iron than do oxidimetric methods. In the analyses, 1,10-phenanthroline (phen) is used, which forms a red complex with Fe^{2+} and a colorless complex with Fe^{3+}. Several aspects of this procedure are significant. First, Stucki's group has done an extensive analysis of the sources of variability in the method. Second, they have demonstrated that the instability of the Fe^{3+}-phen complex, which bedeviled earlier colorimetric analyses using phen, is due to photoreduction and can be eliminated by excluding light. Third, the photoreduction of the Fe^{3+}-phen complex to the Fe^{2+}-phen complex can be carried out quantitatively. Samples are digested while excluding light, Fe^{2+} is measured colorimetrically, the sample is exposed to light to reduce all Fe^{3+}, and finally Fe^{2+} is measured again, this time to obtain total iron. Thus, both Fe^{2+} and total iron are obtained from a single dissolution. Errors based on 10 replicate analyses of geochemical reference standards are ±0.01–0.03% of the % of Fe species (*e.g.*, Fe^{2+}or Fe^{3+}) present (J. STUCKI, personal communication, 1995). Errors on the wet chemical analyses of our samples are at least as good as $\pm0.1\%$ of the $\%Fe^{2+}$ present, or approximately ±0.07 of the %Fe for each species.

Single crystal diopside samples ($Ca_{0.969}Mg_{0.946}Fe_{0.065}$-$Si_2O_6$) from Jaipur in India (CARPENTER, 1996) were used for the hydrogen diffusivity experiments, which are somewhat more iron-rich than those in the study reported by INGRIN *et al.* (1995). The experiments were performed in a 1-bar controlled-atmosphere furnace at 750° and 810°C, with the oxygen fugacity controlled by using CO/CO_2 gases at $10^{-10-14.0\pm0.5}$ atm (near the Ni/NiO buffer), for time lengths from 1 to 100 hours. The hydrogen contents were measured from unpolarized infrared spectra obtained using a Fourier transform infrared spectrometer in the wavelength range from 4000 to 2000 cm^{-1}. Using the integral absorbance for a series of bands near 3500 cm^{-1}, resulting from O-H stretching vibrations, and an integral molar absorption coefficient of 1.0 cm^{-2} per $H/10^6Si$ (PATERSON, 1982), we were able to calculate water contents as a function of the length of time of heat treatment. Although the extinction coefficients have not been determined explicitly for O-H absorption bands in diopside, we have assumed a value typical of O-H bands in other silicates (PATERSON, 1982); we note that our diffusion analysis is insensitive to the absolute values of water content and only depends on the relative results from a sequence of measurements on an individual sample.

RESULTS

Chemical formulae

Chemical compositions for all samples and for the average of all samples from Dish Hill determined by electron probe, corrected Mössbauer, and H-extraction techniques as described above are given in Table 2. The FeO and Fe_2O_3 contents were calculated from the electron probe data using Mössbauer results corrected to "true" Fe^{3+}/Fe^{2+} using the empirical correction factors discussed below. Data were normalized to formula units with 6 oxygens, and iterated to ensure charge balance.

Cation data from the analyses show that $\Sigma(Ca+Na+K)$ is always considerably less than one formula unit, requiring the substitution of some combination of divalent cations to fill the site. In most cases, even the addition of the all the Fe^{2+} determined for the sample cannot completely fill the M2 site; thus it is apparent that some Mg must occupy M2. The tetrahedral sites are systematically low in Si, requiring the substitution of Al or Fe^{3+} to fill them. These compositional characteristics place constraints on the interpretation of the site occupancy data provided by the Mössbauer work, because the presence of tetrahedral Fe^{3+} will change the shape of the spectral absorption profile. However, without corroborating X-ray structure refinement data, it is unclear whether Al or Fe^{3+} fills the tetrahedral sites. It is also possible that sufficient concentrations of vacant lattice sites may be present to affect stoichiometry in the minerals, although we have no evidence to support or refute this possibility.

Table 2. Augite Compositions

	DH-201	DH-202	DH-203	DH-204	DH-205	DH-206	DH-207	DH-209	DH-210	DH-211	DH-212	DH-213	DH-217	DH-218	DH-220	DH-222	DH-227	DH-231
SiO_2	48.25	47.87	48.07	47.45	47.74	47.29	45.91	50.89	50.08	47.71	47.22	47.74	47.17	47.34	47.97	47.55	47.56	47.52
Al_2O_3	9.11	9.44	9.27	9.15	9.42	9.35	8.86	6.64	7.80	9.01	9.39	9.41	9.26	9.30	9.25	9.33	9.31	9.15
TiO2	1.42	1.53	1.35	1.79	1.54	1.85	2.04	0.70	0.67	1.50	2.25	1.63	1.78	1.72	1.34	1.61	1.78	1.70
FeO	4.73	4.92	4.96	5.33	5.09	5.30	7.50	3.98	3.79	5.00	5.39	4.64	5.18	5.08	5.19	5.32	5.31	5.31
Fe_2O_3	2.33	2.27	2.21	2.76	2.53	2.78	4.26	1.66	1.87	2.87	2.52	2.80	2.72	2.96	2.51	2.62	2.92	3.09
MgO	14.23	14.17	14.27	13.36	14.12	13.28	9.49	17.11	16.51	13.94	13.83	13.77	13.39	13.56	14.23	13.61	13.22	13.35
MnO	0.15	0.15	0.16	0.15	0.16	0.15	0.34	0.15	0.13	0.18	0.15	0.16	0.15	0.16	0.16	0.16	0.15	0.16
Cr_2O_3	0.00	0.01	0.01	0.00	0.00	0.00	0.00	0.57	0.45	0.00	0.12	0.01	0.00	0.01	0.01	0.00	0.00	0.01
CaO	19.54	19.60	19.64	19.54	19.13	19.74	20.96	18.44	18.46	19.30	18.48	19.48	19.76	19.26	18.92	19.55	19.45	19.39
Na_2O	1.11	1.09	1.04	1.25	1.17	1.20	1.79	0.95	1.03	1.21	1.32	1.18	1.23	1.24	1.19	1.16	1.30	1.24
F	0.07	0.04	0.50	0.01	0.00	0.02	0.03	0.03	0.02	0.08	0.04	0.03	0.01	0.00	0.02	0.07	0.03	0.03
H_20	0.18	0.07	0.86	0.06	0.06	0.06	0.13	0.16	0.07	0.05	0.13	0.09	0.07	0.11	0.12	0.05	0.07	0.10
Total	101.11	101.16	102.33	100.85	100.95	101.02	101.30	101.29	100.88	100.85	100.84	100.94	100.72	100.74	100.91	101.03	101.10	101.05
$Fe^{3+}/\Sigma Fe$	30.7	29.3	28.6	31.8	30.9	32.1	33.8	27.3	30.7	34.1	29.6	35.2	32.1	34.4	30.3	30.7	33.1	34.4
Si	1.731	1.730	1.578	1.736	1.740	1.725	1.706	1.816	1.799	1.722	1.714	1.731	1.728	1.733	1.742	1.718	1.730	1.729
Al	0.385	0.402	0.359	0.395	0.405	0.402	0.388	0.279	0.330	0.383	0.402	0.402	0.400	0.401	0.396	0.397	0.399	0.392
Ti	0.038	0.042	0.033	0.049	0.042	0.051	0.057	0.019	0.018	0.041	0.061	0.044	0.049	0.047	0.037	0.044	0.049	0.047
Fe^{2+}	0.142	0.149	0.136	0.163	0.155	0.162	0.233	0.119	0.114	0.151	0.164	0.141	0.159	0.155	0.158	0.161	0.162	0.162
Fe^{3+}	0.063	0.062	0.055	0.076	0.069	0.076	0.119	0.045	0.050	0.078	0.069	0.076	0.075	0.082	0.069	0.071	0.080	0.085
Mg	0.761	0.763	0.698	0.729	0.767	0.722	0.526	0.910	0.884	0.750	0.748	0.744	0.731	0.740	0.770	0.733	0.717	0.724
Mn	0.005	0.005	0.004	0.005	0.005	0.005	0.011	0.005	0.004	0.006	0.005	0.005	0.005	0.005	0.005	0.005	0.005	0.005
Cr	0.000	0.000	0.000	0.000	0.000	0.000	0.000	0.016	0.013	0.000	0.003	0.000	0.000	0.000	0.000	0.000	0.000	0.000
Ca	0.751	0.759	0.691	0.766	0.747	0.772	0.834	0.705	0.710	0.747	0.719	0.757	0.776	0.756	0.736	0.757	0.758	0.756
Na	0.077	0.076	0.066	0.089	0.083	0.085	0.129	0.066	0.072	0.085	0.093	0.083	0.087	0.088	0.084	0.081	0.092	0.087
F	0.008	0.005	0.052	0.001	0.000	0.002	0.004	0.003	0.002	0.009	0.005	0.003	0.001	0.000	0.002	0.008	0.003	0.003
H	0.042	0.017	0.188	0.015	0.015	0.015	0.031	0.038	0.017	0.012	0.031	0.022	0.017	0.027	0.029	0.012	0.017	0.024

H^+ contents

Hydrogen data collected in this study show a range of water contents from 0.05–0.21 wt% H_2O for 17 of the 18 samples. These values are consistent with the ranges observed in previous studies. The interesting outlier is sample DH-203, which has a very unusual H_2O content of 0.86 wt%. IR study of this sample is underway to confirm this unusually high value. It is tempting to discard this analysis except for the fact that this sample is also distinct by virtue of low Ca, Si, and Mg accompanied by high F and H. Perhaps this sample originated in a different compositional regime of the upper mantle from the other samples in the suite. Alternatively, it may have reacted with a metasomatic fluid with unusually high F and H contents.

Hydrogen isotope values for the augite samples range from −93.0‰ up to −154.3‰, strongly suggesting contact and exchange with meteoric waters either during ascent or cooling (Table 3). The entraining basalts have heavier D/H ratios, falling in a narrower range between −87.0‰ and −102.2‰. It must be noted that contamination from meteoric waters (locally about −46 to −99‰, according to data from INTERNATIONAL ATOMIC ENERGY AGENCY, 1979) is not the only plausible explanation for the observed deuterium depletions. Such compositions are known from mantle sources, having been measured in mantle-derived amphiboles and interpreted using similar arguments to those presented here (DELOULE *et al.*, 1991). However, both basalts and augite samples are distinctly different from hornblende megacrysts previously reported from the Dish Hill locality, which have D/H ratios of −46.8 ± 7.7‰ (BELL and HOERING, 1994) that can be interpreted as mantle values. Therefore, the difference between the D/H of the augites and basalts vs. the Dish Hill hornblendes, combined with the slower H diffusion in hornblende, does argue strongly for the meteoric contamination interpretation.

Table 3. D/H Data on Dish Hill Samples

	Wt%H_2O	D/H,‰
Hornblende Ba-5[1]	1.03	-47.9
Assorted hornblendes[2]	0.93-1.03	-46.8±7.1
Cpx DH-201	0.18	-102.9
Cpx DH-202	0.11	-111.0
Cpx DH-203	0.86	-121.9
Cpx DH-204	0.06	-115.4
Cpx DH-205	0.06	-123.5
Cpx DH-206	0.06	-124.7
Cpx DH-207	0.13	-93.0
Cpx DH-209	0.15	-114.6
Cpx DH-210	0.07	-134.3
Cpx DH-211	0.05	-154.3
Cpx DH-212	0.13	-118.2
Cpx DH-213	0.09	-130.7
Cpx DH-217	0.06	-139.1
Cpx DH-218	0.11	-128.3
Cpx DH-220	0.12	-145.2
Cpx DH-222	0.07	-129.8
Cpx DH-227	0.07	-140.5
Cpx DH-231	0.10	-113.3
Basalt Ba-1-101	0.06	-87.0
Basalt Ba-2-101	2.28	-98.1
Basalt Ba-2-102	1.19	-99.6
Basalt Ba-3-101	1.74	-102.2
Basalt Ba-3-102	1.30	-95.1
Basalt Ba-5-101	0.87	-100.2

[1]Data from DYAR *et al.*, 1992
[2]Data from BELL and HOERING, 1994

Fe^{2+} and Fe^{3+} data

Results of the wet chemical study of ten augite specimens from this suite show a range of oxidation from 19.3% up to 35.1% of the total Fe as Fe^{3+} (Table 4). These values are within the ranges for Fe^{3+} proposed by previous Mössbauer studies. Wet chemical data were then used as "true" Fe^{3+} contents to provide a check on the interpretation of various models of our Mössbauer data ("M.E." on Table 4) and, ultimately, as part of the calculation of improved C values.

Mössbauer data collected for all 18 samples in our suite were processed to derive several different models for each spectrum, as shown in Table 5. Eight peak fits modeled two (octahedral) Fe^{2+} and two (one octahedral and one tetrahedral) Fe^{3+} doublets, six peak fits used two Fe^{2+} and one Fe^{3+}(octahedral) doublet, and four peak fits used one doublet for each oxidation state of Fe (Fig. 1). A comparison of results of these models vs. the wet chemical data is shown in Table 4. These data can be used to illustrate the large differences between fitting models, and to assess the need for correction factors to provide agreement between Mössbauer and wet chemical results. For reasons discussed earlier, a true value of C as defined by BANCROFT (1973) cannot be defined for these augites, so we have defined an empirical correction factor (ECF):

$$ECF = \frac{(Fe^{3+}/Fe^{2+})\ by\ Mössbauer}{(Fe^{3+}/Fe^{2+})\ by\ wet\ chemistry}. \quad (5)$$

Table 4. %Fe^{3+} Contents and ECF Values from Different Fitting Models as Compared to Wet Chemistry

Sample	Thickness mg Fe /cm²	Wet Chemistry %Fe^{3+}*	L6_14 %Fe^{3+} M.E.†	L6_11 %Fe^{3+} M.E.	L6_11 %Fe^{3+} "true"	L6_08** %Fe^{3+} M.E.	L6_08** %Fe^{3+} "true"	L8_18** %Fe^{3+} M.E.	L8_18** %Fe^{3+} "true"	V4_U %Fe^{3+} M.E.	V4_U %Fe^{3+} "true"
DH-201	3.74	29.3	29.1	28.9	30.7	34.8	31.0	31.0	30.1	36.6	31.7
DH-202	3.80	29.6	31.6	24.6	29.3	26.8	27.8	30.2	29.4	30.2	29.7
DH-203	3.63	30.1	32.5	22.8	28.6	27.3	28.1	32.0	30.9	23.9	27.1
DH-204	4.30	n.a.	34.0	32.9	31.8	n.f.		32.2	31.1	n.f.	
DH-205	4.06	29.2	31.1	29.5	30.9	29.9	29.2	29.6	28.8	30.1	29.7
DH-206	4.28	33.3	35.4	34.6	32.1	36.0	31.4	33.8	32.5	45.6	33.9
DH-207	6.18	n.a.	45.8	43.3	33.8	n.f.		42.9	39.9	n.f.	
DH-209	3.01	19.3	20.5	19.9	27.3	30.3	29.3	19.0	19.2	17.5	23.5
DH-210	3.01	21.0	28.6	28.9	30.7	21.8	25.3	31.1	30.1	19.1	24.5
DH-211	4.15	n.a.	35.8	45.3	34.1	n.f.		35.7	33.6	n.f.	
DH-212	4.21	n.a.	30.1	25.6	29.6	n.f.		28.7	28.0	n.f.	
DH-213	3.93	35.1	28.3	53.8	35.2	53.8	35.9	36.3	34.6	19.6	24.8
DH-217	4.19	n.a.	40.3	34.6	32.1	n.f.		39.3	37.0	n.f.	
DH-218	4.26	n.a.	35.1	47.7	34.4	n.f.		34.2	32.8	n.f.	
DH-220	4.09	n.a.	31.2	27.7	30.3	n.f.		29.9	29.1	n.f.	
DH-222	4.20	31.3	33.0	28.9	30.7	32.9	30.3	31.7	30.7	34.2	31.0
DH-227	4.34	n.a.	34.0	39.1	33.1	n.f.		32.4	31.3	n.f.	
DH-231	4.45	34.7	38.1	47.2	34.4	45.3	34.0	37.0	35.1	38.0	32.1

*Errors on the wet chemical analyses of our samples are at least as good as ±0.1% of the %Fe^{2+} present, or approximately ±0.07 of the %Fe for each species.

†"True" %Fe^{3+} was not calculated for L6_14 fits because the regression line fit to the data was not of significant variance. M.E. represents results obtained using the Mössbauer effect.

**Convergent fits for 6 Lorentzian peaks, 8 constraints and 4 Voigt peaks were not obtained.

Figure 2 presents a comparison between ECF and the Mössbauer Fe^{3+}/Fe^{2+} for the samples on which wet chemistry was performed. Analysis of variance shows that all of the fits except L6_14 are significant when outliers are omitted. The regression data on these models are given in Table 5. Figure 2 illustrates the major differences between fitting models. Regression lines derived from these figures can be used to convert Mössbauer Fe^{3+}/Fe^{2+} to agree with wet chemical Fe^{3+}/Fe^{2+} for augites in the compositional range studied here. Of course, this makes the assumption that wet chemistry gives the ''true'' value for Fe^{3+}/Fe^{2+}; certainly wet chemistry yields Fe^{3+}/Fe^{2+} ratios with the smallest errors.

For the purposes of this study, our goal was to determine the Fe^{3+}/Fe^{2+} ratios of all our samples; therefore, we were forced to choose one consistent model for fitting. Unfortunately, the statistical parameters commonly used to evaluate Mössbauer fits were not helpful in this case. Although χ^2 values for the fits uniformly declined when additional

Table 5. Models Used to Interpret Mössbauer Results and Regression Results

Model	# of Peaks	# of Doublets	Widths* Constrained?	Areas* Constrained?	Line Shape Used	Intercept	Slope	r
L6_14	6	3	yes	yes	Lorentzian			
L6_11	6	3	yes	no	Lorentzian	0.26	1.62	0.99
L6_8	6	3	no	no	Lorentzian	0.43	1.42	0.96
L8_18	8	4	yes	yes	Lorentzian	0.92	0.28	0.69
V4_U	4	2	no	no	Voigt	0.37	1.51	0.89

*Widths and areas are constrained to vary in pairs. For example, when the width of one Fe^{3+} peak increases, the width of its pair will also be increased by a similar amount.

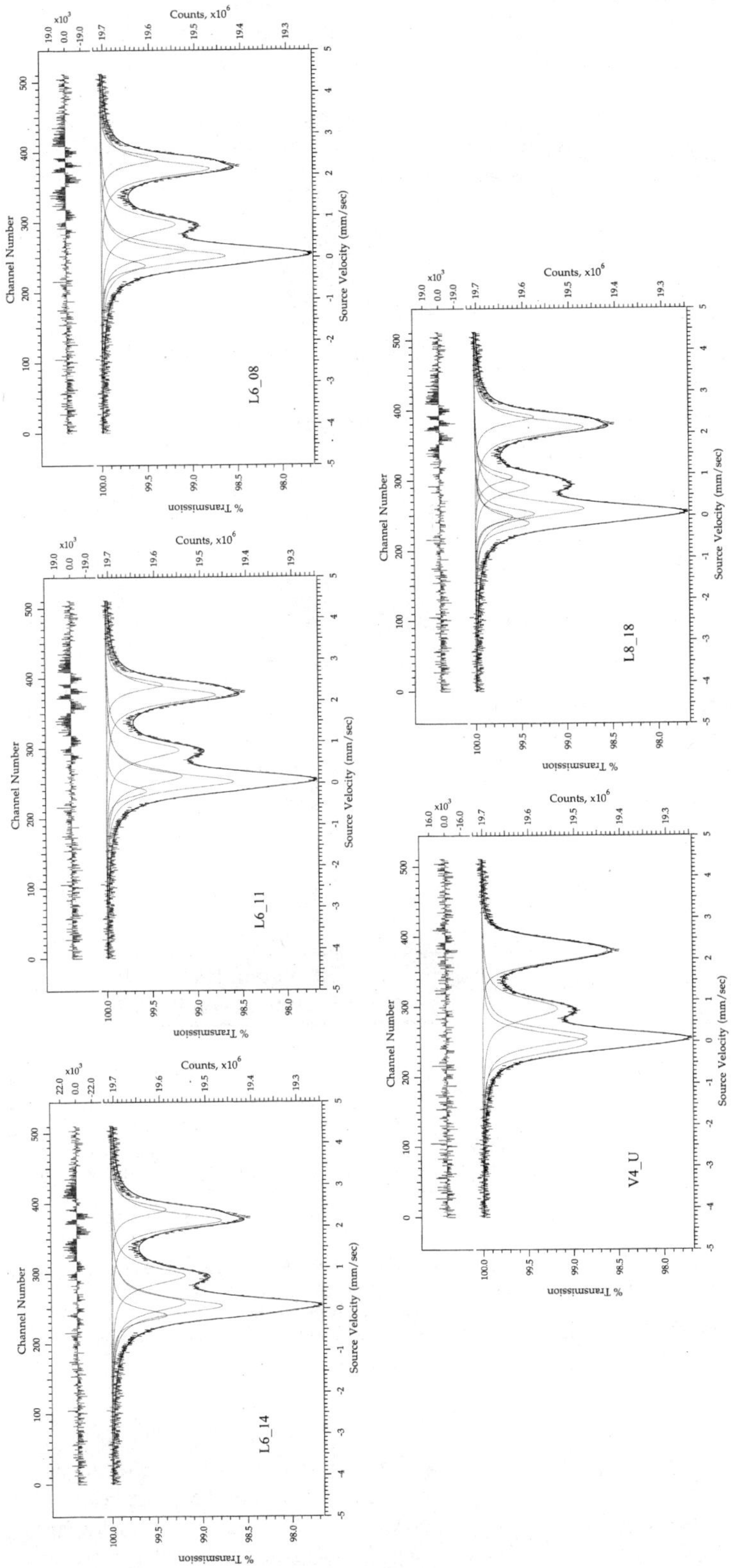

FIG. 1. Mössbauer spectra of augite sample DH-206, fit with the five different models described in Table 5. The total areas of the Fe3+ doublets in each case are 35.4, 34.6, 36.0, 33.8, and 45.6 for L6_14, L6_11, L6_08, L8_18, and V4_U respectively; the total Fe^{3+} as determined by wet chemistry is 33.3% of the total iron.

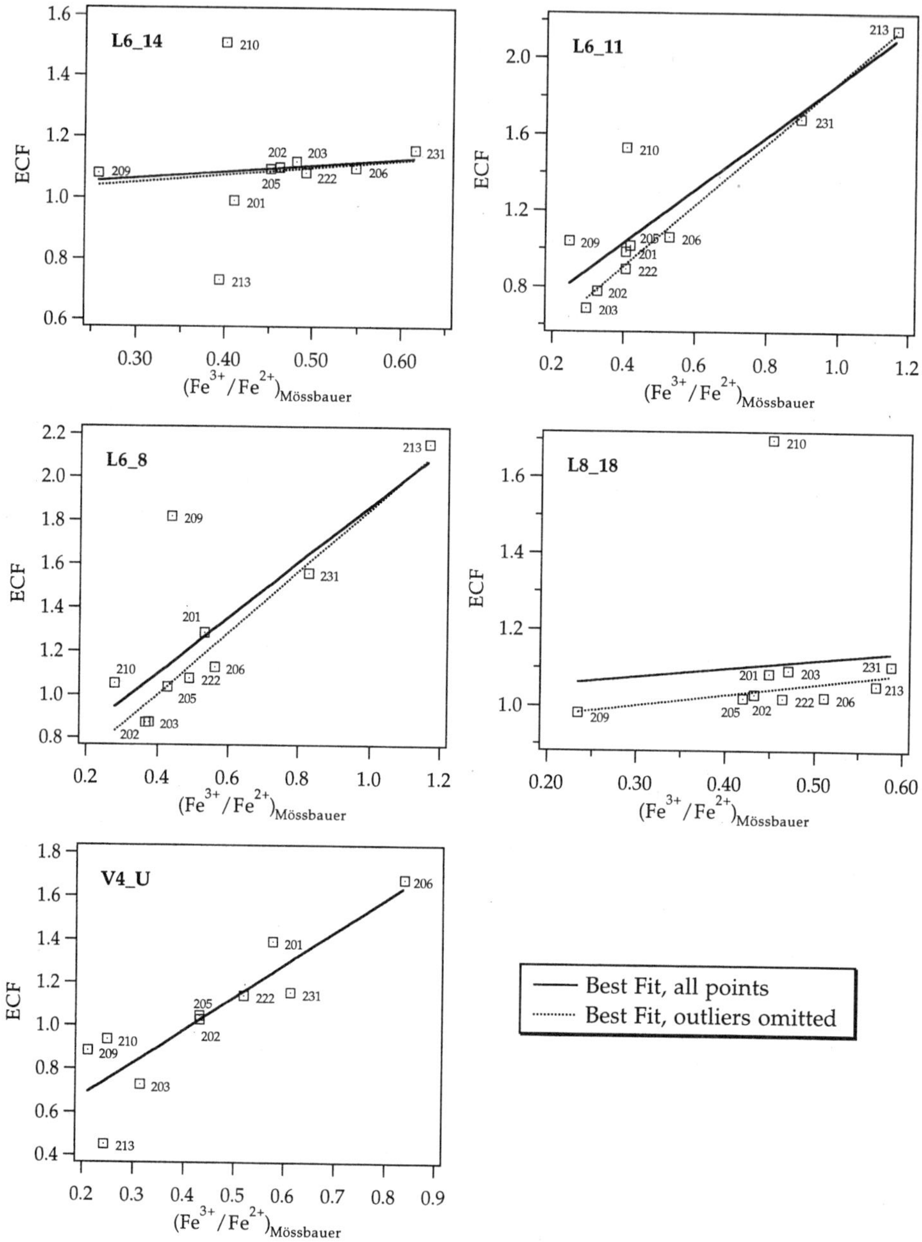

FIG. 2. The relationship between ECF and Mössbauer Fe^{3+}/Fe^{2+} is plotted for each of the five models presented in Table 5, with regression lines shown as before. Analysis of variance shows that all of the dashed line fits except L6_14 are significant. The equations of these regression lines may be used to calculate "true" Fe^{3+}/Fe^{2+} from Mössbauer results when the fitting model is known.

doublets were added to the fits (*i.e.*, going from four to six to eight peak fits), the improvement was not dramatic, and may be completely due to improvement in the fit to the spectral noise (as suggested by RUBY, 1973). Either Al or Fe^{3+} (or possibly some combination of the two) might occupy the tetrahedral sites in the structures, so that crystal chemistry does not distinguish between fitting models. No X-ray diffraction data were available to assist in distinguishing the six peak models

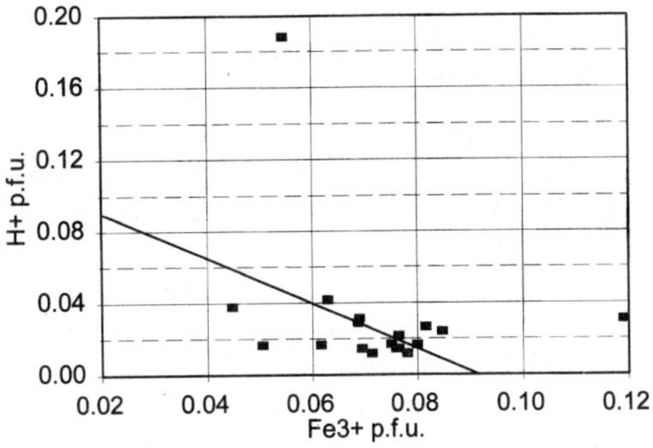

FIG. 3. Plot of Fe^{3+} vs. H^+ contents of samples, with a line of slope = −1.

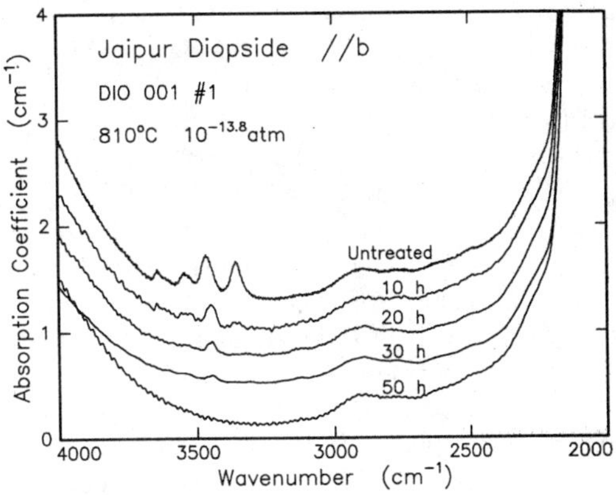

FIG. 4. Sequence of infrared absorption spectra of a sample of Jaipur diopside after treatment for cumulative times of 0, 10, 20, 30 and 50 hours at 810°C, with the oxygen activity controlled at $10^{-13.8}$ atm. The absorption bands near 3500 cm^{-1} correspond to O-H stretching vibrations.

from the eight peak models. Accordingly, in this study we have made an arbitrary choice to use the Mössbauer Fe^{3+}/Fe^{2+} results from L6_11 fits to determine the "true" Fe^{3+}/Fe^{2+} values used to recalculate the formulas given in Table 2. The L6_11 model was chosen because it yields the best fit (r = 0.99) to the wet chemical data.

Comparison of H^+ and Fe^{3+}

It might be tempting to use the data collected here to look for dependent relationships between H^+ and Fe^{3+} as predicted by the dehydrogenation mechanism discussed above. However, Fig. 3 demonstrates that it would be risky to do so. Error bars on the H^+ determinations are extremely large, making any correlation between H^+ and Fe^{3+} extremely tenuous. Unless the unusually high H^+ content sample is included, points on this plot do not define a real trend. In a qualitative fashion, this plot suggests that an upper limit for a hydrogenated augite in the mantle might be on the order of 0.05 or 0.20 H^+ ions per formula unit (p.f.u.) if all the Fe^{3+} in the augite is assumed to be the result of dehydrogenation. Conversely, a maximum amount of Fe^{3+} that might occur in a fully dehydrated augite following eruption might be about 0.20 Fe^{3+} atoms p.f.u. However, these values are questionable due to the lack of precision in the respective analyses.

Results of dehydration experiments

After each heat treatment experiment on a diopside crystal, we measured the infrared spectrum of the sample. Figure 4 shows a sequence of infrared spectra taken after a series of heat treatments of a sample of Jaipur diopside at 810°C, indicating the progressive loss of water at high temperature. Because our samples were prepared as thin slabs, and assuming isotropic diffusion of hydrogen (INGRIN *et al.*, 1995), we were able to use a theoretical analysis for diffusion in an infinite slab to calculate hydrogen diffusivities as a function of temperature for our samples. Figure 5 shows values of the water con-

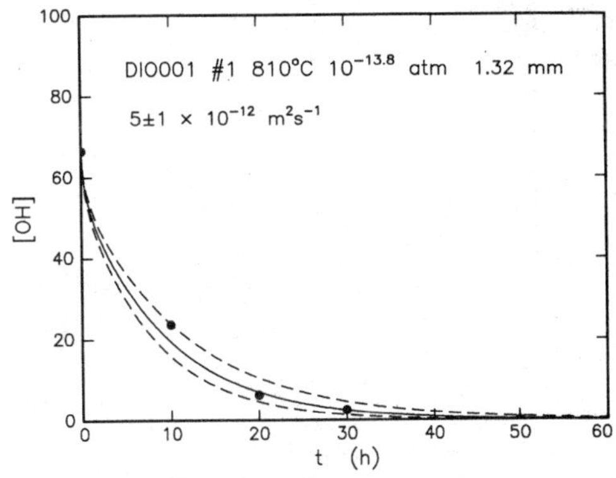

FIG. 5. Dehydration kinetics for the sample of Jaipur diopside illustrated in Fig. 5 showing dehydration after a range of heat treatment times at 810°C. The concentrations of OH were determined by integrating the area beneath the O-H bands in the spectra shown in Fig. 4. The solid line is a fit to the data using a solution of Fick's law for diffusion in an infinite slab of finite thickness, yielding a diffusivity for hydrogen of 5×10^{-12} m^2s^{-1}. The dashed lines correspond to 4 and 6×10^{-12} m^2s^{-1}. Note that essentially all the H is diffused out of these samples by 50 hours at this temperature.

tent of the Jaipir diopside sample from Fig. 4 as a function of time. A fit to the data, assuming isotropic diffusion in a slab, is superimposed on the figure and indicates a diffusivity of $D_{H+} = (5 \pm 1) \times 10^{-12}\ m^2s^{-1}$. We have performed similar analyses on samples treated at 750°C. Fig. 6 shows a comparison of these results with those of INGRIN *et al.* (1995) for a diopside with ~50% lower Fe content. Also shown are values for hydrogen diffusion in diopside and augite, calculated from the results of SKOGBY and ROSSMAN (1989) using the infinite slab model for diffusion. We have also plotted in Fig. 6 the hydrogen diffusivities determined in olivine (MACKWELL and KOHLSTEDT, 1990), and hornblende (GRAHAM *et al.*, 1984), and water diffusivities in rhyolite glass (ZHANG *et al.*, 1991). The hornblende curve was determined by a linear regression fit to the data for both hornblende samples reported by

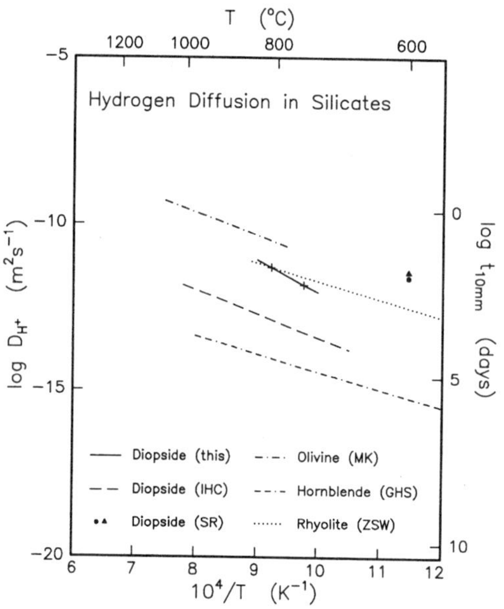

FIG. 6. Hydrogen diffusion in silicates, based on data for hornblende from GRAHAM *et al.* (1984) GHS, for diopside and augite from SKOGBY and ROSSMAN (1989) SR, for olivine from MACKWELL and KOHLSTEDT (1990) MK, for diopside by INGRIN *et al.* (1995) IHC, and for diopside in the present study (this). Also shown are data for water diffusion in rhyolite glass by ZHANG *et al.* (1992) ZSW. Horizontal axes show temperature (top) and inverse temperature (bottom axis); the left vertical axis shows the log of hydrogen diffusivity, while the right vertical axis shows the log of time expressed in days for diffusive loss of 90% of the hydrogen from 10 mm thick slabs. Of the three diopside data sets, the Ingrin data are for the most iron-poor composition, and the Skogby and Rossman data are for the most iron-rich compositions.

GRAHAM *et al.*, assuming a plate geometry. The clinopyroxene diffusion data from the various studies span a wide range in hydrogen diffusivity at any temperature. Because the only apparent compositional difference between the clinopyroxene is the Fe content, a positive correlation appears to exist between the Fe content and the hydrogen diffusivity (Fig. 6). Thus, we would expect that if we were to include the Dish Hill augites in Fig. 6, they would fall in the upper range of hydrogen diffusivities measured for clinopyroxene.

DISCUSSION

Implications for mantle f_{O_2}

Traditionally, petrologic studies of mantle xenoliths have been used to gain information about mantle lithologies, mineral compositions, P-T conditions, and major and trace element effects of fluid-rock interactions in the mantle. These studies have made the assumption that major, trace, and rare earth element (REE) compositions of mantle xenoliths are not significantly changed during transport to the surface. The relatively fast rate of transport (hours to weeks) compared to the slow rates of diffusion of large major and trace element cations in mantle minerals makes this assumption fairly safe. In contrast, many recent studies measure Fe^{3+}/Fe^{2+}, H^+ contents, and D/H ratios in mantle samples and use these data to make inferences about mantle f_{O_2} conditions and mantle fluid/rock interactions. Such inferences are risky because D and H diffusion rates are generally orders of magnitude faster in mantle minerals than major element diffusion rates, and dehydrogenation may change the Fe^{3+}/Fe^{2+} ratio in the process. It is important to know if measured Fe^{3+}/Fe^{2+}, D/H and H^+ contents actually represent original mantle values, or if they may have been altered during transport.

Dehydrogenation of minerals

During transport of basalt and included megacrysts from their sources deep in the Earth to the surface, hydrogen and deuterium dissolved in the minerals and melt will partially exchange, and some interactions with meteoric water in the country rock may occur. In particular, minerals containing dissolved water in a subsolidus environment may dehydrate significantly during entrainment and transport in an undersaturated melt phase. The degree of dehydration will depend on the partitioning of water among the various phases, the kinetics of water diffusion within the crystals and melt, and the temperature/time history of entrainment and transport.

Some constraints on water exchange are provided by experimental investigations of the kinetics of hydration and dehydration of the appropriate minerals and melts. DYAR *et al.* (1993) used hydrogen diffusion data in hornblende (GRAHAM *et al.*, 1984) to infer the history of H^+ loss (and a parallel increase in Fe^{3+}/Fe^{2+}) in a suite of mantle-derived hornblendes, with inferences for mantle metasomatic processes. In this way, they were able to show that significant H^+ loss from hornblende must have occurred due to a metasomatic event in the mantle prior to entrainment. Other studies have investigated the rates of hydration/dehydration of minerals such as olivine (MACKWELL and KOHLSTEDT, 1990), and diopside (SKOGBY and ROSSMAN, 1989; SKOGBY, 1994; INGRIN *et al.*, 1995). There are also indications that equilibration of point defect concentrations in mineral grains will take significantly longer than H^+ transport rates, so that changes in H^+ content will be accompanied by changes in Fe^{3+}/Fe^{2+}, according to the reaction

$$H^+_{aug} + Fe^{2+}_{aug} \rightleftharpoons \tfrac{1}{2}[H_2] + Fe^{3+}_{aug}. \qquad (6)$$

For example, at temperatures of about 1200°C, dehydration of olivine with a 10 mm grain size will take only 4 hours, while point defect equilibration (and hence equilibration of Fe^{3+}/Fe^{2+}) will take about 4 days (MACKWELL *et al.*, 1988; MACKWELL and KOHLSTEDT, 1990).

As discussed above, we have extended the data on H^+ diffusion in diopside compositions (INGRIN *et al.*, 1995) to Fe-rich clinopyroxene compositions (augite). In the absence of diffusion measurements for deuterium in clinopyroxene, we speculate that the kinetics of deuterium diffusion in these minerals are likely to be equal to or slower than that for hydrogen, depending on the nature of the diffusing species (H_2O/D_2O or H^+/D^+). In addition, although no water diffusion measurements have been reported in molten basalt or basalt glass, a number of studies have investigated water transport for relatively silica-rich magma and glass compositions (KARSTEN *et al.*, 1982; LAPHAM *et al.*, 1984; ZHANG *et al.*, 1991). We use these results and the observations of H_2O contents, D/H ratios, and Fe^{3+}/Fe^{2+} ratios in Dish Hill, California basalts and megacrysts to infer the history of H and D exchange during the transport of the magma and included minerals from their sources to the surface at Dish Hill.

Mantle sources and alteration of Dish Hill basalt and megacrysts

The compositional data compiled from Dish Hill basalts and included megacrysts, coupled with the diffusion information described above provide insight into the origins of the various rock and mineral compositions and the effects of their entrainment and transport history. Clearly the basalt was generated at depth in the Earth, and it entrained the xenoliths and megacrysts during transport to the Earth's surface. It is noteworthy that a basaltic magma will be initially undersaturated with water, and therefore water dissolved in entrained minerals might be expected to partition strongly into the melt at a rate dependent on either the diffusion of hydrogen in the mineral (KARATO, 1986) or of water through the melt (ZHANG *et al.*, 1991), whichever is slower.

The right hand axis of Fig. 6 provides an indication of the rates of hydrogen loss at any particular temperature from 10 mm grains of natural olivine, hornblende, and clinopyroxene of various Fe contents. Thus, in a dehydrating environment, a grain of forsteritic (mantle composition) olivine 10 mm across will lose 90% of its water in about 4 hours at 1200°C. By contrast, a 10 mm grain of hornblende will take several years for 90% dehydration under the same conditions. The diopside grains used in this study will give dehydration rates similar to olivine under these conditions; grains with higher Fe contents are likely to yield even faster transport rates for hydrogen. Also shown on Fig. 6 is the diffusivity for water in rhyolite compositions determined by ZHANG *et al.* (1991), who demonstrated that the mobile hydrogen-bearing species in rhyolitic glasses and melts is molecular water. Centimeter-scale transport of water in these materials might be expected to occur in about 2 days at 1200°C; however, transport rates in less polymerized glass and melts of higher Fe (and Si-poorer) compositions, such as basalt, may be faster.

The observed partial breakdown of the Dish Hill hornblende megacryst samples provides an indication of the temperatures (probably 1100°–1250°C) during transport, and of the time scales for entrainment and ascent that are likely to take from several days to a month (*e.g.*, SPERA, 1984). While it is not clear that the basalt and megacrysts have a common depth of origin, oxygen isotope studies by FARMER *et al.* (1995) indicate that both are derived from a mantle source. The D/H ratios for the basalts are very uniform and show that some mixing with meteoric water has occurred during transport, presumably due to contact metamorphism of the country rock by the melt and to interactions between the melt and shallow crustal hydrothermal fluids. Although the diffusion rate for water in silicate melts and glasses suggests relatively short diffusion distances over the time required to transport magma to the surface (Fig. 6), turbulent mixing could provide a more effective mechanism for homogenization of D/H ratios within the magma during ascent. The D/H ratios for the hornblende megacrysts are consistent with a mantle source region and show little evidence of exchange with the surrounding basalt, as might be expected given the relatively slow diffusion rates for hydrogen in hornblende. By contrast, the fine-grained hornblende selvages on composite xenoliths show some exchange with the magma; because 1 mm grains equilibrate approximately 100 times faster than 10 mm grains, this observation does not suggest a different origin for the xenoliths and the megacrysts.

Based on the kinetics of hydrogen diffusion in clinopyroxene, augite megacrysts would be expected to exchange hydrogen with a surrounding basaltic melt on time scales much less than those required for transport to the surface. Thus, rapid hydrogen loss might be anticipated during mobilization, resulting in an increase in the Fe^{3+}/Fe^{2+} in the augite, following eqn. (6). The diffusion of D^+ in clinopyroxene is expected, on steric grounds, to be slower than the diffusion of H^+. Thus, loss of D^+ from the clinopyroxene may lag behind the loss of H^+, resulting in a transient increase in D/H. While there are no data to constrain this supposition, it might provide a reason for the observed higher D/H in the augite than in the basalt. Such an effect would produce augites that were zoned in D/H. Any variation in D/H ratios and in Fe^{3+}/Fe^{2+} may thus result from hydrogen exchange with the basalt during transport, and would not necessarily reflect mantle compositions.

In contrast, diffusion rates are likely to be similar for H_2O and D_2O in the basalt. Due to convective mixing in the magma, no equivalent compositional zoning of D/H or Fe^{3+}/Fe^{2+} is likely to be observed in the basalt. If D^+ and H^+ can diffuse so rapidly in large megacrysts (*i.e.,* on the time scales of transport), then dehydrogenation must also occur within the smaller grains in polycrystalline xenoliths, due to greater rates of grain boundary diffusion compared to rates of lattice transport.

CONCLUSIONS

(1) Mössbauer spectra measured on a suite of 18 mantle aluminous augite samples reveal a range of Fe^{3+} doublet areas from 14% Fe^{3+} to 60% Fe^{3+}. Spectra are interpreted with simple three doublet models constrained by holding peak widths constant in doublet pairs; more complicated fits are also mathematically possible.

(2) H^+ contents as determined by uranium extraction techniques range from 0.06 to 0.18 wt% H_2O for augite (with an outlier at 0.86 wt% H_2O), and 0.06 to 2.28 wt% H_2O for their entraining basalts.

(3) Hydrogen isotope data range from −93.0‰ to −154.3‰, for augite, and −87.0‰ to −102.2‰ for related basalts. Both basalt and augite samples show evidence of contact and exchange with meteoric waters either during ascent or cooling, and they are distinctly different from hornblendes previously reported from the Dish Hill locality.

(4) Predictions of hydrogen transport based on diffusion data for hydrogen and water in augites and basalt are consistent with observations of significant interactions with meteoric water during transport to the surface of the Earth, while similar data for amphibole indicate that hornblende megacrysts will exchange little on the timescales of transport, as observed by BELL and HOERING (1994).

(5) The lack of a clear systematic trend in H^+ versus Fe^{3+} data could be masked by the uncertainties in the hydrogen analysis; it could also suggest significant variability in H^+ or Fe^{3+} contents of augite in the mantle source region.

Acknowledgments—Supported by National Science Foundation grants EAR-9304304 and EAR-9405310 (M.D.D.), EAR-9305187 (S.J.M.), EAR-9005414 and EAR-9315762 (A.V.M.), with travel assistance provided by Navajo Community College. We thank Howard Wilshire and Jane Nielson for their guidance and assistance with sample collection, Sanjay Monie at Penn State for assistance with IR spectroscopy, Joseph W. Stucki and analyst Jun Wu at the University of Illinois for wet chemical data, and Kipp Bajaj, Heather Wilcox, and Marjorie Taylor for assistance with sample preparation and Mössbauer spectral data reduction. Thorough and constructive reviews of this manuscript were generously provided by Jane Nielson, Colin Graham, and Catherine McCammon.

REFERENCES

AMTHAUER G., ANNERSTEN H. and HAFNER S. S. (1976) The Mössbauer spectrum of ^{57}Fe in silicate garnets. *Zeits. Krist.* **143,** 14-55.

BAI Q. and KOHLSTEDT D. L. (1992) Substantial hydrogen solubility in olivine and implications from water storage in the mantle. *Nature* **357,** 672–674.

BANCROFT G. M. (1967) Quantitative estimates of site populations in an amphibole by the Mössbauer effect. *Phys. Lett.* **26A**(1), 17–18.

BANCROFT G. M. (1969) Quantitative site population in silicate minerals by the Mössbauer effect. *Chem. Geol.* **5,** 1969, 255–258.

BANCROFT G. M. (1973) *Mössbauer Spectroscopy. An Introduction for Inorganic Chemists and Geochemists.* McGraw-Hill.

BANCROFT G. M. and BURNS R. G. (1967) Interpretation of the electronic structure of iron in pyroxenes. *Amer. Mineral.* **52,** 1278–1282.

BANCROFT G. M., BURNS R. G. and HOWIE R. A. (1967a) Determination of the cation distribution in the orthopyroxene series by the Mössbauer effect. *Nature* **213,** 1221–1223.

BANCROFT G. M., BURNS R. G. and MADDOCK A. G. (1967b) Determination of cation distribution in the cummingtonite-grunerite series by Mössbauer spectroscopy. *Amer. Mineral.* **52,** 1009–1026.

BANCROFT G. M. and BROWN J. R. (1975) A Mössbauer study of coexisting hornblendes and biotites: Quantitative Fe^{3+}/Fe^{2+} ratios. *Amer. Mineral.* **60,** 265–272.

BELL D. R. (1993) Hydroxyl in mantle minerals. Ph.D. Thesis, California Institute of Technology.

BELL D. R. and HOERING T. C. (1994) D/H ratios and H_2O contents of mantle-derived amphibole megacrysts from Dish Hill, California. *Lunar Planet. Inst. Conf. on Deep Earth and Planetary Volatiles.*

BELL D. R. and ROSSMAN G. R. (1992) Water in earth's mantle: The role of nominally anhydrous minerals. *Science* **255,** 1391–1397.

BIGELEISEN J., PERLMAN M. L. and PROSSER H. C. (1952) Conversion of hydrogenic materials to hydrogen for isotopic analysis. *Anal. Chem.* **24,** 1356–1357.

BRYNDZIA L. T. and WOOD B. J. (1990) Oxygen thermobarometry of abyssal spinel peridotites: The redox state and C-O-H volatile composition of the Earth's suboceanic upper mantle. *Amer. J. Sci.* **290,** 1093–1116.

BURNS R. G. (1966) Origin of optical pleochroism in orthopyroxenes. *Mineral. Mag.* **35,** 715–719.

CARPENTER S. (1996) The kinetics of hydrogen diffusion in single crystal clinopyroxene. M.S. Thesis, Pennsylvania State University.

DELOULE, E., ALBARÈDE, F. and SHEPPARD, S. M. F. (1991) Hydrogen isotope heterogeneities in the mantle from ion probe analysis of amphiboles from ultramafic rocks. *Earth Planet. Sci. Letts.* **105,** 543–553.

DEGRAVE E. and VAN ALBOOM A. (1991) Evaluation of ferrous and ferric Mössbauer fractions. *Phys. Chem. Mins.* **18,** 337–342.

DYAR M. D. (1984) Precision and interlaboratory reproducibility of measurements of the Mössbauer effect in minerals. *Amer. Mineral.* **69,** 1127–1144.

DYAR M.D., MCGUIRE A.V. and ZIEGLER R.D. (1989) Redox equilibria and crystal chemistry of coexisting minerals from spinel lherzolite mantle xenoliths. *Amer. Mineral.* **74,** 969–980.

DYAR M. D., MACKWELL S. M. and MCGUIRE A. V. (1992) Fe^{3+}/H^+ and D/H in mantle kaersutites - Misleading indicators of mantle fugacities. *Geology* **20,** 565–568.

DYAR M. D., MACKWELL S. J., MCGUIRE A. V., CROSS L. R. and ROBERTSON J. D. (1993) Crystal chemistry of Fe^{3+} and H^+ in mantle kaersutite: Implications for mantle metasomatism. *Amer. Mineral.* **78,** 968–979.

EVANS M. J. and BLACK P. J. (1970) Voigt profile of Mössbauer transmission spectra. *J. Phys. C* **3,** 2167–2177.

FARMER G. L., GLAZNER A. F., WILSHIRE H. G., WOODEN J. L., PICKTHORN W. J. and KATZ M. (1995) Origin of the later Cenozoic basalts at the Cima volcanic field, Mojave Desert, California. *J. Geophys. Res.* **100,** 8399–8415.

FREY F. A. and PRINZ M. (1978) Ultramafic inclusions from San Carlos, Arizona: Petrologic and geochemical data bearing on their petrogenesis. *Earth Planet. Sci. Lett.* **38,** 129–176.

GRAHAM C. M., HARMON R. M. and SHEPPARD S. M. F. (1984) Experimental hydrogen isotope studies: Hydrogen isotope exchange between amphibole and water. *Amer. Mineral.* **69,** 128–138.

GRANT C. A. (1995) Sources of experimental and analytical error in measurements of the Mössbauer effect in amphibole. Ph.D. Thesis, University of Oregon.

HOLDAWAY M. J., DUTROW B. L., BORTHWICK J., SHORE P., HARMON R. S., and HINTON R. W. (1986) H content of staurolite as determined by H extraction line and ion microprobe. *Amer. Mineral.* **71,** 1135–1141.

INGRIN J., HERCULE S. and CHARTON T. (1995) Diffusion of hydrogen in diopside: Results of dehydration experiments. *J. Geophys. Res.* **100,** 15489–15499.

INTERNATIONAL ATOMIC ENERGY AGENCY (1979) Environmental isotope data No. 6, World survey of isotope concentration in precipitation (1972–1975). *I.A.E.A. Techn. Rep. Ser.* **192,** I.A.E.A. Vienna.

KARATO S. (1986) Does partial melting reduce the creep strength of the upper mantle? *Nature* **319,** 309–310.

KARFUNKEL U. and POLLAK H. (1993) A new approach to the determination of the Fe(3+)/Fe(2+) ratio and recoil-free factors in two garnets by Mössbauer spectroscopy. *Phys. Stat. Solidi* **180B,** 6772.

KARSTEN J. L., HOLLOWAY J. R. and DELANEY J. R. (1982) Ion microprobe studies of water in silicate melts: Temperature-dependent water diffusion in obsidian. *Earth Planet. Sci. Lett.* **59,** 420–428.

KOMANDEL P. and STUCKI J. W. (1988) The quantitative assay of minerals for Fe^{2+} and Fe^{3+} using 1,10-phenanthroline: III. A rapid photochemical method. *Clays Clay Mins.* **36,** 379–381.

LAPHAM K. E., HOLLOWAY J. R., HERVIG R. L. and STOLPER E. M. (1984) Diffusion of H_2O and D_2O in obsidian at elevated temperatures and pressures. *J. Non-Cryst. Solids* **67,** 179–191.

LONG G. (1963) Interpretation of experimental Mössbauer spectrum areas. *Nucl. Instr. Methods* **24,** 425-428.

LONG G. J., CRANSHAW T. E. and LONGWORTH G. (1984) The ideal Mössbauer absorber thickness. *Mössbauer Effect Ref. Data J.* **6**(2), 42–49.

LUTH R. W., VIRGO D., BOYD F. R. and WOOD B. J. (1990) Ferric iron in mantle-derived garnets. *Contrib. Mineral. Petrol.* **104,** 56–72.

LUTH R. W. and CANIL D. (1993) Ferric iron in mantle-derived pyroxenes and a new oxybarometer for the mantle. *Contrib. Mineral. Petrol.* **113,** 236–248.

MACKWELL S. J., DIMOS D. and KOHLSTEDT D. L. (1988) Transient creep of olivine: Point defect relaxation times. *Philos. Mag.* **57,** 779–789.

MACKWELL S. J. and KOHLSTEDT D. L. (1990) Diffusion of hydrogen in olivine: Implications for water in the mantle. *J. Geophys. Res.* **95,** 5079–5088.

MCGUIRE A. V., DYAR M. D. and WARD K. W. (1989) Neglected Fe^{3+}/Fe^{2+} ratios--a study of Fe^{3+} content of megacrysts from alkali basalts. *Geology* **17,** 687–690.

MCGUIRE A. V., DYAR M. D. and NIELSON J. E. (1991) Metasomatic oxidation of upper mantle peridotite. *Contrib. Mineral. Petrol.* **107,** 252–264.

MILLER G. H., ROSSMAN G. R. and HARLOW G. E. (1987) The natural occurrence of hydroxide in olivine. *Phys. Chem. Minerals* **14,** 461–472.

PATERSON M. S. (1982) The determination of hydroxyl by infrared absorption in quartz, silicate glasses and similar materials. *Bull. Mineral.* **105,** 20–29.

PRESTON R. S., HANNA S. S. and HEBERLE J. (1962) Mössbauer effect in metallic iron. *Phys. Rev.* **128,** 2207-2218.

RUBY S. L. (1973) Why Misfit when you already have χ^2? In I.J. Gruverman, C.W. Seidel, and D.K. Dieterly, eds., *Mössbauer Effect Methodology* **9,** 277–305.

SKOGBY H. (1994) OH incorporation in synthetic clinopyroxene. *Amer. Mineral.* **79,** 240–249.

SKOGBY H. and ROSSMAN G. R. (1989) OH^- in pyroxene: An experimental study of incorporation mechanisms and stability. *Amer. Mineral.* **74,** 1059–1069.

SKOGBY H., BELL D. R. and ROSSMAN G. R. (1990) Hydroxide in pyroxene: Variations in the natural environment. *Amer. Mineral.* **75,** 769–774.

SKOGBY H., ANNERSTEN H., DOMENEGHETTI M. C., MOLIN G. M. and TAZZOLI V. (1992) Iron distribution in orthopyroxene: A comparison of Mössbauer spectroscopy and X-ray refinement results. *Eur. J. Mineral.* **4,** 441–452.

SMYTH J. R., BELL D. R. and ROSSMAN G. R. (1991) Incorporation of hydroxyl in upper-mantle clinopyroxenes. *Nature* **351,** 732–635.

SPERA F. J. (1984) Carbon dioxide in petrogenesis III: Role of volatiles in the ascent of alkaline magmas with special reference to xenolith-bearing mafic lavas. *Contrib. Mineral. Petrol.* **88,** 217–232.

STUCKI J. W. (1981) The quantitative assay of minerals for Fe^{2+} and Fe^{3+} using 1,10-phenanthroline: II. A photochemical method. *Soil Sci. Soc. Amer. J.* **45,** 638–641.

STUCKI J. W. and ANDERSON W. L. (1981) The quantitative assay of minerals for Fe^{2+} and Fe^{3+} using 1,10-phenanthroline: I. Sources of variability. *Soil Sci. Soc. Amer. J.* **45,** 633–637.

WHIPPLE E. R. (1972) Quantitative Mössbauer spectra and chemistry of iron. Ph.D. Thesis, Massachusetts Institute of Technology.

WILSHIRE H. W. and TRASK N. J. (1971) Structural and textural relationships of amphibole and phlogopite in peridotite inclusions, Dish Hill, California. *Amer. Mineral.* **56,** 240–255.

WILSHIRE H. W. and NIELSON-PIKE J. E. (1986) Upper-mantle xenoliths in alkaline basalt, Dish Hill, California. *Guidebook to S. Calif. Field Trips, G.S.A. Cordilleran Section, 82nd Ann. Mtng.*, Los Angeles, CA, 9–11.

WILSHIRE H. W., MEYER C. E., NAKATA J. K., CALK L. C., SHERVAIS J. W., NILESON J. E. and SCHWARZMAN E. C. (1988) Mafic and ultramafic xenoliths from volcanic rocks of the western United States. *U.S. Geol. Survey Prof. Paper* 1443.

WOOD B. J. and VIRGO D. (1989) Upper mantle oxidation state: Ferric iron contents of lherzolite spinels by ^{57}Fe Mössbauer spectroscopy and resultant oxygen fugacities. *Geochim. Cosmochim. Acta* **53,** 1277–1291.

ZHANG Y., STOLPER E. M. and WASSERBURG G. J. (1991) Diffusion of water in rhyolitic glasses. *Geochim. Cosmochim. Acta* **55,** 441–456.

Part C.
Remote Sensing of Planetary Surfaces

INTRODUCTION

ROGER BURNS had an active interest in planetary spectroscopy and mineralogy for most of his professional career. Beginning with his studies of lunar and meteorite samples in the early 1970s, he added to this the application of spectroscopy to remote sensing of planetary surfaces in the late 1970s and 1980s. He had a particular fascination with Martian mineralogy, and published most heavily in this field beginning in the mid to late 1980s. In the early 1990s he began adding to this range of interests the mineralogy of Venus as well. Roger was gifted not only with the ability to perform meaningful laboratory measurements of extraterrestrial materials, but also with a unique insight into what sorts of processes might be important in the mineralogy of other planetary bodies. His papers in this field are classics of their kind.

This section contains a variety of papers concerning planetary spectroscopy and mineralogy. The first paper is on remote sensing and the others all concern the mineralogy of Mars. Remote sensing of planetary surfaces and spectroscopic analyses of extraterrestrial analogs in the laboratory are dependent on the crystal field transitions that cause mineral absorption bands. The application of crystal field theory to mineralogy, and subsequently to remote sensing, was pioneered by ROGER BURNS, and summarized in his monograph, *Mineralogical Applications of Crystal Field Theory.*

We start this section with a paper by PIETERS, MUSTARD, and SUNSHINE, describing quantitative mineral analyses of planetary surfaces by means of reflectance spectroscopy. This paper comprises a short review of the present state of reflectance spectroscopy as it is applied to remote compositional analyses. Three different methods of identifying and quantifying mineralogy are described, with examples. These methods include empirical approaches (comparing spectra to a library of spectra of known samples), spectral mixture analyses (comparing spectra with linear or non-linear combinations of spectra of known samples), and deconvolution of individual absorption bands (a more theoretical approach, allowing quantitative assessment of the diagnostic properties of individual electronic transitions).

The next two papers describe terrestrial ferric minerals that also may be present on Mars. MORRIS, MING, GOLDEN, and BELL discuss the occurrence of jarositic tephra on Mauna Kea, Hawaii, which is formed by the hydrothermal dissolution of silicate glass and minerals by sulfuric acid solutions derived from volcanic gases. Jarosite is more commonly associated with sulfide deposits. They note the similarity of the spectra of these samples to spectra from some Martian regions, and make the point that jarosite could have formed similarly in these regions in conjunction with sulfur dioxide release during Martian volcanism. BISHOP and MURAD propose schwertmannite, a hydroxylated ferric sulfate mineral, as a possible component of the surface material of Mars. Spectra of terrestrial schwertmannites are compared to other ferric minerals and Mars soil analogs, the formation processes of schwertmannite are discussed in relation to possible Martian chemical weathering processes, and the implications for the surface chemistry of Mars should schwertmannite exist there are discussed.

The paper by BELL discusses ferric oxides/hydroxides, sulfates, carbonates, and hydrates as diagnostic minerals that can be used to assess past climatic changes on Mars. The paper begins with a review of current telescopic and spacecraft observational data and interpretations about the occurrence of these phases on Mars, and then presents new results and interpretations from telescopic observations in 1990. New models for the weathering and alteration of surface materials are outlined, and a set of observational measurements are described that can be used to discriminate against possible weathering and alteration scenarios, and provide new constraints on the composition and mineralogy of the Martian surface.

The last paper, by SCHAEFER, is a detailed discussion based on an idea of ROGER BURNS, that there are good geochemical reasons to think that there may be iron formations on Mars, similar to the Archean banded iron-formations (BIFs) found in many locations on Earth. Several abiotic mechanisms proposed for producing terrestrial BIFs are discussed, along with their possible relevance to Martian geochemistry. The implications of the existence of iron-formations on Mars are also discussed.

Mineral Spectroscopy: A Tribute to Roger G. Burns
© The Geochemical Society, Special Publication No. 5, 1996
Editors: M. D. Dyar, C. McCammon and M. W. Schaefer

Quantitative mineral analyses of planetary surfaces using reflectance spectroscopy

CARLÉ M. PIETERS,[1] JOHN F. MUSTARD[1] and JESSICA M. SUNSHINE[2,3]

[1]Department of Geological Sciences, Brown University, Providence RI 02912 U.S.A.
[2]SETS Technology Inc., 300 Kahelu Ave., Mililani, HI 96769 U.S.A.
[3]Department of Earth, Atmospheric, and Planetary Science, MIT, Cambridge, MA 02139 U.S.A.

Abstract—There are several uses of reflectance spectroscopy for remote compositional analyses: mapping of lithologic units, identifying minerals present on a surface, quantifying the abundance of minerals present, and characterizing the composition of minerals present. Three very different approaches currently used to identify and quantify mineralogy are summarized with examples. (1) Empirical approaches. These include comparing or 'matching' spectra with a library of spectra for natural and prepared materials with known physical and chemical properties, documenting trends of spectra for known materials and correlating with observations, developing statistical frameworks through which to evaluate spectra of unknown materials, developing spectral parameters which measure or are correlated with spectral trends that follow composition. (2) Spectral mixture analyses. These approaches are based on the premise that natural surfaces consist of mixtures of a small number of mineralogical or lithologic components (endmembers), and the reflectance spectrum of the surface is a systematic combination of the reflectance spectra of the endmembers. Models for both linear (spatially distinct) and non-linear (intimate) mixtures have been compared, evaluated, and applied to planetary science problems that require information on the relative proportion of materials in a naturally mixed surface. (3) Deconvolution of individual mineral absorption bands allowing quantitative assessment of the diagnostic properties of individual electronic transitions or groups of such diagnostic absorptions. Examples to date include quantification of properties of pyroxenes and their mixtures leading to abundance determination for mixtures and determination of olivine composition from spectral variations of deconvolved electronic transitions. Extrapolating from the rapid progress of the past, our ability to contribute to the exploration and understanding of various solar system bodies using reflectance spectroscopy will continue to expand productively as quantitative approaches mature.

INTRODUCTION

REFLECTANCE SPECTROSCOPY, or spectrophotometry, is the quantitative measurement and study of light reflected from a surface as a function of wavelength. The use of reflectance spectroscopy to evaluate the mineralogy of planetary surfaces came of age in the 1970's with the fortunate convergence of several research activities. First was publication of a book that clearly describes the fundamental physics and crystal field theory that give rise to the diagnostic absorption bands of many rock forming minerals in the visible to near-infrared (BURNS, 1970a). Second was the development of detectors and telescopic instruments that could acquire high precision spectroscopic data for planetary objects (*e.g.*, ADAMS and MCCORD, 1969; MCCORD *et al.*, 1970, 1972, MCCORD and JOHNSON, 1970). Third, was parallel efforts to identify and characterize diagnostic features in laboratory reflectance spectra of rocks and minerals in order to develop the interpretive base for the remote observations (HUNT and SALISBURY, 1970; ADAMS, 1974, 1975). Over the last few decades, each of these areas of research has matured, and with the publication of the revised and expanded second edition of Burns' book (BURNS, 1993) it is appropriate to also assess the progress, accomplishments, and future challenges for the use of reflectance spectroscopy in remote compositional analysis. In the discussion below it is assumed the reader is familiar with crystal field theory and the origin and nature of diagnostic electronic transition absorptions that occur in minerals containing transition element ions (BURNS, 1993).

Before discussing recent advances in the use of reflectance spectroscopy, a clarification of important distinctions between several different types and scientific applications of remote spectroscopic measurements is warranted. There is no single approach to spectroscopic analysis of a planetary surfaces and the different objectives, as well as specific instruments and analytical tools available, can be confusing. In all cases, variations in the spectral properties of natural materials provide valuable compositional information. In general, however, increasing complexity of spectral measurement and sophistication of analytical capabilities provides increased understanding and thus increased utility for various scientific applications.

Mapping lithologic units

Color photography is the oldest form of remote spectral measurements. This naturally evolved to multi-spectral imaging when digital detectors became available. When spectral information is obtained in two spatial dimensions (image format), spectral differences can be examined and their spatial relations mapped. An example of a digital image for a region on the Moon is shown in Fig. 1 along with a spectral ratio image (the reflectance at 0.41 μm divided by the reflectance 0.66 μm) of the same area. Typically, geometrically coregistered images of the surface are obtained in a small number of spectral channels. Although such multispectral images cannot identify minerals, if the wavelengths of the spectral channels span parts of the spectrum containing absorption features and if the detectors achieve appropriate precision and accuracy, then different lithologies can usually be distinguished from each other and their spatial relations mapped. Obviously, a large the number of spectral channels usually increases the ability to distinguish between different lithologies. Most multispectral imaging systems contain between 3–12 bandpasses. Multispectral cameras have continued to improve along with development of efficient and accurate algorithms to distinguish and map surface materials. The relation between multispectral imaging and spectroscopy is largely complementary. As separate approaches, the former is principally a tool to *map* the spatial distribution of distinct materials, the latter a tool to *characterize* the composition of surface materials (*e.g.,* see concept discussion in HEAD *et al.,* 1978).

Identifying minerals present

When the number of spectral channels of an instrument is sufficiently high (often loosely referred to as high spectral resolution) and spectral coverage spans continuously across individual absorption bands to fully determine their properties (shape, strength, etc.), then the measurement approach is called spectroscopy, or hyperspectral, rather than multispectral. Reflectance spectra of several common materials of interest to planetary scientists are shown in Fig. 2. An up-to-date survey and review of the reflectance properties of most rock forming minerals can be found in GAFFEY *et al.* (1993c). Remote measurements with an appropriate spectrometer (high spectral resolution, spectral coverage, precision, accuracy) can measure the presence or absence of a particular diagnostic absorption band, which in turn usually indicates the presence or absence of a mineral. Such use of spectroscopy blossomed in the 1980's and has been highly successful with the advent of near infrared spectrometers used on earth-based telescopes. These instruments allowed spectra (hundreds of spectral channels) to be obtained for individual areas, or spots, on the surface (typically 5–15 km on the Moon, hundreds of km on Mars, and the whole body for asteroids). Most work has focused on mafic minerals (orthopyroxene, clinopyroxene, olivine), ferric oxides, hydrous minerals, and plagioclase. A variety of analyses have accumulated since

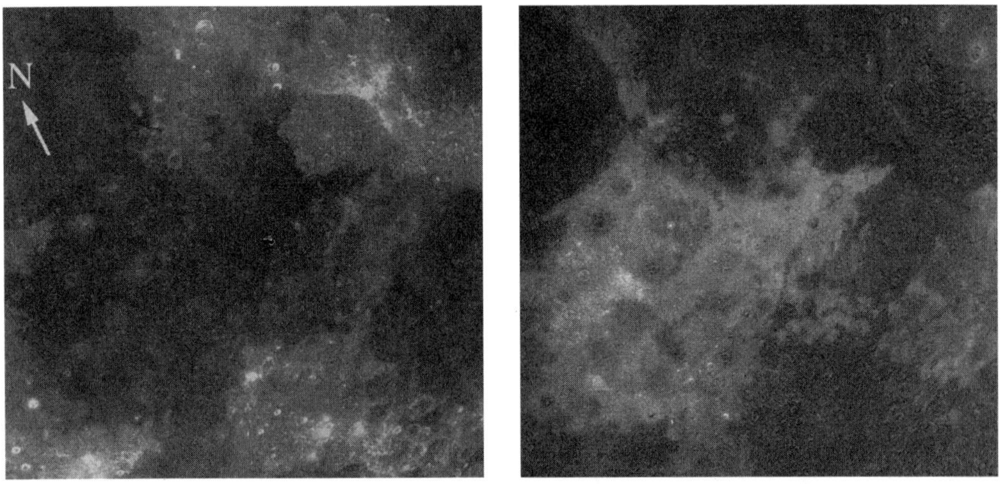

FIG. 1. Galileo digital image of the Mare Tranquillitatis to Mare Serenitatis region of the Moon. The image on the left is a reflectance image obtained at 0.66 μm and the image on the right is a UV/VIS spectral ratio image (0.41/0.66 μm) for the same area. This spectral ratio is particularly sensitive to variations in basalt type. The Galileo data are from the second Earth-Moon encounter (BELTON *et al.,* 1994).

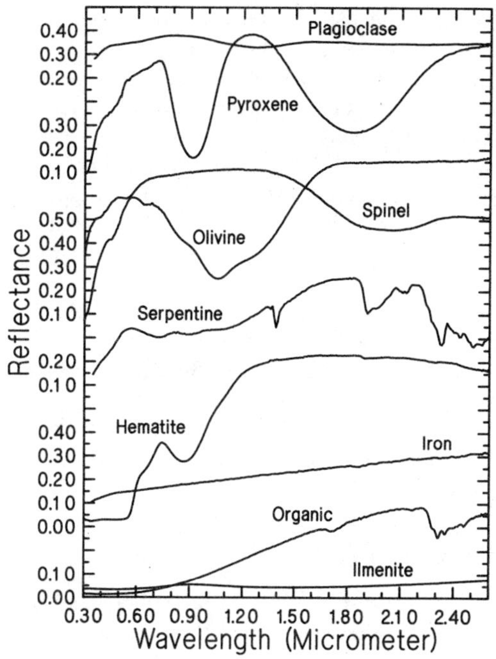

FIG. 2. Reflectance spectra of various rock forming minerals (after PIETERS and MCFADDEN, 1994)

the early observations and there are several recent reviews that summarize the type of minerals or mineral assemblages identified for solar system silicate bodies: Mars (ADAMS and MCCORD, 1969; SODERBLOM, 1992; ROUSH *et al.* 1993); Asteroids (MCCORD *et al.,* 1970; GAFFEY *et al.,* 1993a; PIETERS and MCFADDEN, 1994); and the Moon (MCCORD and JOHNSON, 1970; PIETERS, 1986; 1993).

Quantifying the abundance of minerals present

Identifying the presence of minerals requires accurate spectrometers, and sometimes a bit of luck. Nature tends to produce minerals in mixtures, sometimes simple, sometimes complex. In many applications, further complications arise due to alteration, both physical and chemical, by the local environment. The more interesting scientific issue, and more difficult task for spectroscopy, is to go beyond simply identifying the presence of a mineral, and to quantify the actual abundance of minerals present in a natural mixture. The ultimate goal is to describe the first order petrology of the material. As summarized below, several approaches to this problem have been developed for spectroscopy including empirical measurements of a range of materials, theoretical modeling of spectra in a mixture, and separating the individual components of a composite spectrum.

Characterizing the composition of minerals present

Several important rock forming minerals are part of a solid solution series and can exist over a range of compositions (*e.g.,* olivine, pyroxene). The composition of the mineral is often as important scientifically as its presence on the surface of a silicate body. Various empirical approaches that document regular variations of absorption bands with composition have been used to obtain good approximations of mineral composition. More recently, deconvolution of individual absorptions in reflectance spectra has been successful in quantifying compositional trends for some observed absorptions and linking them directly with crystal field theory predictions.

The instrument technology available for remote mineralogical analyses has advanced considerably over the last two decades along with enormous expansion of computing capabilities and the development of very sophisticated analytical capabilities for information extraction. Although the technical capabilities will not be reviewed here, the concepts are simple. Spectroscopy and imaging have readily merged in new instruments called imaging spectrometers which produce an "image cube" of coregistered data: two dimensions of spatial information and one dimension of spectral information, with each dimension containing hundreds of elements. Such hyperspectral imaging data is the preferred type of data to address compositional issues in a geologic context and it is anticipated such data will gradually replace the more limited single-point spectroscopy and multi-spectral imaging.

The discussion below focuses in more detail on quantitative approaches to spectroscopic analysis. An excellent earlier discussion can be found in CLARK and ROUSH (1984) who describe appropriate methods for quantifying features observed in reflectance spectra. The discussion here concentrates on applications to electronic transition absorptions, although these more recent methods can be generalized to other forms of absorption as well. Results from three very different approaches currently used to identify and quantify mineralogy are summarized here. The first covers empirical approaches for which laboratory analogs of anticipated surface components are prepared and systematic variations of spectral properties are noted. The second relies heavily on theoretical frameworks which describe the interaction of light with particulate media and specifically how the spectral properties of individual components combine together to form a spectrum of a mixture. The third approach models individual absorption features in a reflectance spectrum and provides a direct link to the physical understanding of absorptions provided by crystal field theory.

EMPIRICAL APPROACH TO COMPOSITION AND MIXTURES

From the onset, empirical methods of comparing laboratory and remote reflectance spectra served as the primary tool for identifying the composition of planetary surfaces (*e.g.,* ADAMS and MCCORD, 1968; MCCORD *et al.,* 1970; MCCORD and JOHNSON, 1970). This approach has been quite successful in providing an initial, but usually qualitative, assessment of the mineralogy of planetary surfaces. The assumptions are simple. If the spectral properties of minerals, rocks, and soil are well documented, a spectrum of an unknown material needs only be compared (or "matched") to the library of known materials. Since spectral properties are controlled by composition, when a good match is found in well known laboratory materials, the composition of the unknown material is thought to be comparable to the known material. Similarly, trends observed in a series of known materials are often extrapolated and applied to unknown surfaces. Many of the early applications relied on visual inspection and simple qualitative correlation. The basic analytical procedures have recently been implemented into more sophisticated and rigorous computer programs that use numerical algorithms to determine goodness of fit. These new methods are thus more objective and can often accommodate noisy data, while allowing rapid searches over large spectral libraries (*e.g.,* CLARK *et al.,* 1993).

The greatest utility of these approaches is in the identification of minerals present on a surface, which, through the use of computers and spectral libraries, can now be used to search through large data sets. There are, however, some fundamental limitations. The identification of minerals present is necessarily limited by the completeness of the spectral library upon which comparisons are made. Results are constrained to the presence or absence of comparable materials in the library, and though mineral assemblages may be identified if present in the library, modal abundance is not easily determined. A further complication is the frequent presence of overlapping absorptions, which require libraries to accommodate various combinations of mixtures and their textures. In addition, this approach is generally unable to reproduce processes such as weathering and alteration, which are at the heart of some of the more important problems in studies of planetary surface composition.

In several commonly used forms, empirical approaches document systematic trends in spectral variation through a series of spectral measurements made for well defined material with a controlled variable such as particle size or proportion of mineral in a mixture. Although this procedure often provides numerical values (see examples below), it is important to continually recognize the basic assumptions and scope of the data inherent in empirical comparisons. In short, the *uniqueness* of a given interpretation or "match" in empirical approaches is often difficult to assess.

Mineral mixtures

Several laboratory studies have addressed spectral trends observed in carefully prepared mixtures of common rock forming minerals. Some of the earliest (NASH and CONEL, 1974; PIETERS, 1974) documented effects of particle size and opaques on diagnostic absorptions of mafic minerals in mixtures (*e.g.* Fe^{+2} band near 1 μm). It was noted that spectral properties of individual components in a reflectance spectrum combine in a very non-linear manner, with dark absorbing materials having a disproportionately large influence. The nonlinear effects of opaque components on a reflectance spectrum was later expanded substantially by CLARK (1983) who analyzed abundance and, more importantly, particle size of the absorbing species in a mixture with more translucent materials. Fine grained opaque particles (a few micrometers in size) dispersed in a translucent (or silicate) matrix are *exceptionally* efficient at absorbing radiation and darkening the mixture.

The properties of a variety of additional two component mixtures involving one or more mafic minerals were documented by SINGER (1981), PIETERS (1983), CLOUTIS (CLOUTIS *et al.*, 1986; 1990a,b), and CROWN and PIETERS (1987). These studies were principally used to evaluate trends in mixtures that might be useful in interpreting remotely acquired spectra. For example, Singer's data allowed bounds to be set on the relative abundance of orthopyroxene to clinopyroxene in a two pyroxene mixture based on the band center of the combined absorption near 1 μm. Such a trend can easily be noted in the spectra of pyroxene mixtures shown in Fig. 3. Similarly, the detectability in a mixture of crystalline plagioclase and pyroxene by

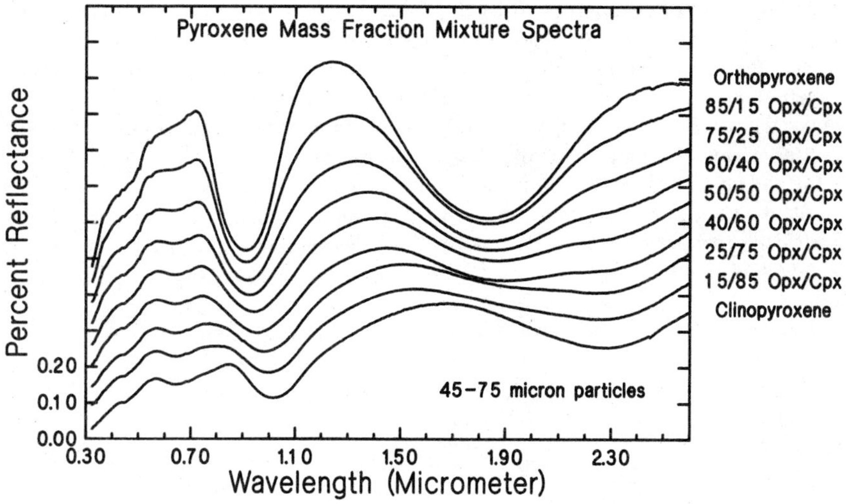

FIG. 3. Reflectance spectra for a sequence of prepared mineral mixtures of 45–75 μm particles which contain various proportions of low-calcium orthopyroxene and high-calcium clinopyroxene (after SUNSHINE *et al.*, 1990). The spectra are displayed from bottom to top with increasing orthopyroxene fraction and a 5% offset in reflectance between each successive spectrum.

their ferrous absorptions (CROWN and PIETERS, 1987) was shown to be much easier for pyroxene ($\geq$5%) than for plagioclase ($\geq$50%). Parameters derived from the trends for mixtures of olivine and orthopyroxene observed by CLOUTIS *et al.* (1986) are some of the most widely used in asteroid studies and are discussed further below.

Most of these mineral mixture studies also examined the effects of particle size with ideal size separates. Since it is radiation transmitted through grains that produces the observed diagnostic absorption bands, the spectrum of a sample of large particles generally has stronger absorption bands than spectra of smaller particles. This is true until the band is so strong or the particles so large that the band saturates. The effects of well defined particle size can be modeled for reflectance, but most natural soils have a range of particle sizes leading to a non-unique solution (see next section). As an example, three particle size separates for an orthopyroxene are shown in Fig. 4 along with synthetic soils prepared with various particle size distributions. The small particles of a soil dominate the optical properties, even when present at only 10%.

Correlation and monotonic trends

More commonly, an empirical relation is observed between a particular type of reflectance measurement and composition, although the details of what causes the relation may not be well known. Once established, this relation can nevertheless be used to estimate the composition of unknown surfaces. The Moon has been a principal beneficiary of this approach since ground truth information is available to define compositional relationships with measured spectra.

One of the most widely used empirical relations for lunar spectra is that between the red-sloped

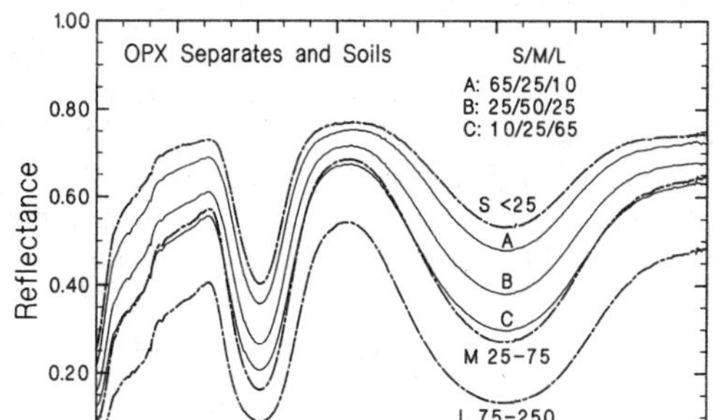

FIG. 4. Reflectance spectra of prepared size separates for an orthopyroxene. Individual size separates (<25 μm, 25–75 μm, and 75–250 μm) are shown in dashed lines. Synthetic monomineralogic soils are shown in solid lines prepared with the following proportions of the small, medium and large particles: 65/25/10, 25/50/25, and 10/25/65. Note that the smaller particles tend to dominate the soil spectra and that the spectral contrast is reduced.

lunar visible continuum, often referred to as the UV/VIS ratio, and the TiO_2 content of mature mare soils, in which flatter (or bluer) spectra are associated with higher abundances of TiO_2 (CHARETTE *et al.,* 1974; PIETERS, 1978). It was originally suspected the observed relation may have something to do with an electronic transition of Ti^{+3} since lunar materials are very reduced. No appropriate carrier has yet been found, however, and reduced Ti-rich glass exhibits an effect opposite to that observed (BELL *et al.,* 1976). After much debate within the community, this relation between TiO_2 content and the UV/VIS continuum slope of basaltic soils is now believed to be due largely to spectral contrast issues rather than any specific absorption feature. It is important to note that although the precision of reflectance data can readily be measured to better than 1% (JOHNSON *et al.,* 1991; MELENDREZ *et al.,* 1994), the accuracy of the compositional estimation is nevertheless limited to several percent because the uniqueness issues cannot be resolved (see discussion in PIETERS, 1993 and PIETERS *et al.,* 1993).

A more comprehensive empirical approach uses sophisticated statistical analyses, such as principal components, to identify the primary spectral variations of a collection of spectra. These statistically derived principal components of spectral variation can then be associated with quantified information on compositional variations. Such an approach to compositional analyses has been developed using a suite of lunar sample spectra to define compositional associations of spectral variations within the suite (JAUMANN, 1991). As would be expected, some compositions are highly correlated with spectral variations, particularly those associated with albedo. Spectral variations of rock types with abundant minerals exhibiting diagnostic absorptions are also correlated with compositional trends. This type of probability analysis can be very useful for estimating expected bulk compositions when there is some assurance that the original data suite used to define the correlation are representative of the materials being studied remotely or in the laboratory. It is, however, not suitable for evaluating complex mineral assemblages or for assessing the importance of subtle spectral variations.

Paramaterization

Bridging pure empirical approaches and approaches that are solidly based on a physical and chemical understanding of cause and effect are a variety of paramaterized approaches that contain elements of a physical basis but fall short of being able to quantify the detailed interaction of radiation with specific solid surfaces. These approaches nevertheless have been and will continue to be enormously useful, some more so than others, depending on the closeness of their link with physical reality.

One of the most solidly based relations is that between composition and the band center of pyroxene absorptions established by ADAMS (1974) using reflectance spectra. This was carried further by HAZEN *et al.* (1978) using transmission spectra and expanded with additional reflectance spectra by CLOUTIS and GAFFEY (1991). In these studies, the band centers observed for pyroxene samples were selected as a simple minimum relative to a general continuum. Adams showed that the band centers of pyroxene, both those near 1 μm and those near 2 μm, move to longer wavelengths as Fe and Ca substitute for Mg in the pyroxene structure. He also noted that additional ions (such as Al or Ti) in the pyroxene structure disrupt the smooth trend. Nevertheless, if the band center of an unknown pyroxene can be accurately measured, then its composition can typically be estimated to about 10%. Due to the abundance of pyroxene across many solar system bodies, this has proven to be one of the most widely used relationship in remote mineralogical analyses.

Because the spectral properties of olivine and low-Ca pyroxene are very different from each other (see Fig. 2), mixtures of the two follow very systematic trends that are well suited to parameter measurement. Low-Ca pyroxenes have two very well defined absorptions centered near 0.9 and 1.8 μm; olivines have a broad composite feature near 1.05 μm but no feature at 2.0 μm. The relative strength of the feature near 2 μm in a mixture should thus indicate the relative proportion of low-Ca pyroxene present. CLOUTIS *et al.* (1986) devised a parameter to approximately cancel the effect of particle size on band strength measurements by measuring the relative strength of absorptions through a band area ratio. For olivine and low-Ca pyroxene mixtures, the ratio (BII/BI) of the absorption observed near 2 μm to the combined absorption observed near 1 μm (both measured by the band area) varied systematically with the relative abundance of the two minerals. This band area parameter was shown to be able to estimate olivine/low-Ca pyroxene abundances of mixtures to about a 10% accuracy (CLOUTIS *et al.,* 1986). It should be noted, of course, that this band area parameter is only suitable for mixtures of these two specific minerals and is demonstrably inappropriate if other minerals are present with absorptions in this part of the spectrum (*e.g.,* clinopyroxene, spinel, etc.).

Olivine and low-Ca pyroxene are some of the most common mafic minerals found in stony meteorites and mixtures of these two minerals are thus expected to also be common on asteroid surfaces. GAFFEY *et al.* (1993b) later used the band area ratio of CLOUTIS *et al.* (1986) and a measure of the center of the composite band near 1 μm (BI) to derive a taxonomy for S-type asteroids based on near-infrared spectra. A comparison of these two spectral parameters derived for the S-type asteroids and for several meteorite groups is shown in Fig. 5. GAFFEY *et al.* (1993b) concluded from such relationships that only their SIV subset of the S-asteroids might be consistent with ordinary chondritic mineral assemblages. It was later shown that the proposed zones in Fig. 5 are certainly diffuse, that different sides of the same asteroid stretch the limits of the SIV boundaries (MURCHIE and PIETERS, 1996), and that alteration effects can move an ordinary chondrite out of the SIV zone completely (MOROZ *et al.,* 1996).

Recently, several approaches have been developed to estimate the abundance of iron for the surface of the Moon using optical parameters. Each approach must separate optical effects due to composition from those due to surface exposure or maturity (space weathering). The first method, derived by FISCHER and PIETERS (1994), uses a spectral ratio parameter (950/750 nm) to identify mture soils and the measured strength of the ferrous absorption band near 1 μm to estimate the abundance of iron. The 950/750 nm ratio is a composite measure of both continuum slope and band strength; it is empirically related to the soil maturity parameter I_s/FeO of MORRIS (1978) through measurements of lunar samples. Similarly, band strength of a mature soil is empirically related to FeO content of lunar highland soil samples. This empirical relation between FeO and band strength essentially mirrors the abundance of Fe-bearing minerals in lunar anorthositic soils. The second and third optical method for estimating iron abundance use the 950/750 nm ratio and a measure of albedo (*e.g.,* reflectance at 750 nm). LUCEY *et al.* (1995) uses measurements of these two parameters for lunar samples to empirically derive a coordinate transformation from which a parameter Θ is derived and FeO estimated. FISCHER and PIETERS (1996) start with the 950/750 nm spectral parameter to derive soil Is/FeO and then use this soil maturity index and the albedo (750 nm reflectance) to estimate iron and titanium abundance, again using lunar samples for calibration. These latter two methods are essentially measuring the correlation of composition with the abundance of optically bright, but iron-poor, anorthosite. Since all three approaches use limited empirical data for calibration, the accuracy of the extrapolated composition estimates is difficult to determine. Results from the LUCEY *et al.* (1995) method suggest that large regions of the northern lunar farside may contain vast areas of extremely low abundance of iron, an observation with very important implications for the early crustal evolution of the Moon.

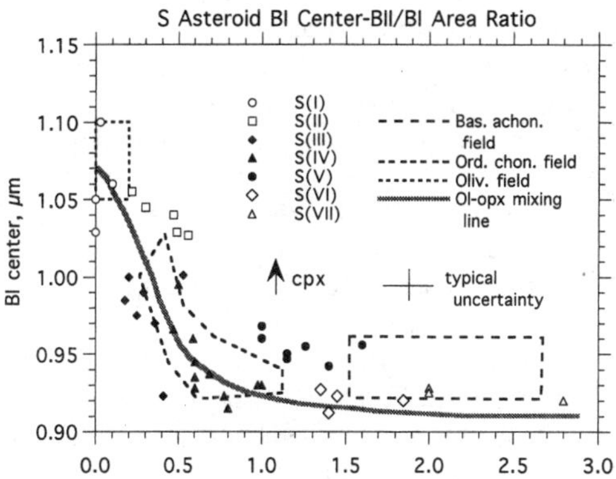

FIG. 5. Wavelength of the Band I band center versus the Band II/Band I area ratio for S-type asteroids, selected meteorite types, and laboratory mixtures of olivine and orthopyroxene (after GAFFEY *et al.,* 1993b; modified by MURCHIE and PIETERS, 1996). Band I refers to the 1 μm band of pyroxenes and olivines; Band II refers to the 2 μm band of pyroxenes.

SPECTRAL MIXTURE ANALYSIS

The empirical approaches discussed above are based on specific absorption features or broad scale trends in spectral properties that are readily characterized by simple parameters or statistical techniques. A suite of parallel approaches have be pursued that are based on physical modeling from first principles, using fundamental properties such as spatial relation of components in a mixture, optical constants, particle size and shape, internal and external scattering efficiencies, and physical properties of a surface (*e.g.,* HAPKE, 1981; 1993; HIROI and PIETERS, 1992a,b). These models, particularly the radiative transfer model of Hapke, have been used widely in studies ranging from photometric studies of planetary bodies and the determination of modal abundances from reflectance spectra of mineral mixtures to the mapping of relative abundances across the surface.

The empirical trends in the spectral properties of mineral mixtures noted in the previous section are explicitly accommodated in reflectance models for particulate surfaces. Over the last two decades, there has been much activity in the testing and application of models that derive surface mineral or lithologic abundance. The basic premise of mixture modeling is that natural surfaces consist of mixtures of a small number of mineralogical or lithologic components (endmembers), and the reflectance spectrum of the surface is a systematic combination of the reflectance spectra of the endmembers. Thus if the spectral properties of the endmembers can be defined, either locally from within a remotely sensed data set, or globally through a library of reflectance spectra or optical properties, then the endmember abundances within the surface mixture can be derived through an appropriate model that combines them to form the spectrum of a mixture. Endmember spectra are usually treated as a whole (all wavelengths) and some statistical measure of goodness of fit is used to assess the degree to which the spectrum computed from endmember spectra models the actual or natural spectrum. Numerical values are provided for the fractional abundances of each endmember required for the best fit.

The basic equation for spectral mixture analysis is:

$$r_b = \sum_{i=1}^{N} F_i R_{ib} + E_b \quad \text{and} \quad \sum_{i=1}^{N} F_i = 1.0$$

where r_b is the reflectance of the observed spectrum in band b, N is the number of endmembers, F_i is the fractional abundance of endmember i, R_{ib} is the reflectance of endmember i in band b, and E_b is the error of the fit of the spectrum calculated from endmembers to the observed spectrum in band b. The second equation constrains the sum of the fractions to be unity. If M is the number of bands, then there will be M equations in N unknowns. As long as the number of bands is equal to or greater than the number of endmembers, a solution can be found and the fractional abundance derived.

Linear mixing (spatial mixtures)

The theoretical development and application of mixing models fall into two broad categories: linear and nonlinear mixing. Linear mixing models are the simple case where it is assumed that the components of interest are located in physically discrete areas on the surface (i.e. a checkerboard). In this case, the reflectance of the mixture is a linear combination of the reflectance of the endmembers, and the abundance is the aerial fraction of each endmember within the field of view. A basic assumption of the linear spectral mixture model is that reflected radiation encounters only one component of the mixture, and thus there is no multiple scattering of radiation between components. This approach was initially developed for planetary applications by SINGER and MCCORD (1979), but has since evolved to an increased level of sophistication largely through the efforts of Adams and coworkers (*e.g.* ADAMS *et al.*, 1986; SMITH *et al.*, 1990; ADAMS *et al.*, 1993). The approach has been developed and tested on multispectral and hyperspectral data using lithologic and mineralogic endmembers derived from spectra acquired remotely as well as from library spectra.

This model has enjoyed wide application in recent years, fueled to some extent by the ease of use and versatility, by the recognition that it is based on a physically realistic model of many surfaces, and, more importantly, by the new insights into surface composition and processes it can provide. A linear mixing approach is particularly appropriate when distinguishing and mapping different lithologies in a spatial context when the size of individual components is below the measurement spatial resolution (*e.g.*, trees in a field). It is especially useful for data with low spatial resolution when the measured signal is a combination of signals from different neighboring lithologies. This approach was used, for example, with telescopic spectra of asteroids (effectively a point source of radiation) which were assumed to contain spatially coherent different lithologies (HIROI *et al.*, 1993). In this case, the spectra of a few (but not all) S-type asteroids were successfully fit with various proportions of different laboratory spectra obtained from primitive achondrites and iron meteorites. If the assumption of spatially coherent lithologies is correct, and if the laboratory meteorite spectra accurately represent these lithologies, then the proportion of each meteorite spectrum used to produce the best fit is an indication of the spatial extent of that material on the asteroid.

A more common approach is to model spectral variations in image data, where the result of the mixture model is maps (fraction images) of the abundance of the spectral endmembers on the surface (ADAMS *et al.*, 1993). Fraction images then become fundamental new tools for understanding surface and physical processes. Strictly speaking, linear spectral mixture models treat each endmember as a spatially coherent material (see below). An example of such a spectral mixture analysis performed with Galileo multi-spectral images (six channels from 0.41 to 1.00 μm) is shown is Fig. 6. This particular example (from STAID *et al.*, 1996) uses a linear mixture analysis with four endmembers chosen to represent the properties of two types of basalt, one relatively blue (a) and one relatively red (b), freshly excavated mare crater (c), and highland soil (d). These endmember fraction images separate highland contamination from the basalt units, allow the spatial and stratigraphic relations between several mare units to be distinguished, and provide the basis for a more direct link between the basalt units of Mare Tranquillitatis and samples returned from Apollo 11 (STAID *et al.*, 1996).

Given the high quality data often provided by CCD multispectral imaging, the Moon has been a common target for scientific investigations using image mixture analysis. Additional recent examples include an examination of mixing relationships between mare and highland lithologies in order to investigate modes of emplacement of basin ejecta and to explore for evidence of buried mare deposits in highland terrain (MUSTARD *et al.*, 1992; 1994; HEAD *et al.*, 1993; MUSTARD and HEAD, 1995; BLEWETT *et al.*, 1995), an evaluation of the spatial extent of feldspathic surface units across the crater Copernicus (PINET *et al.*, 1993), and an evaluation of the distribution and maturity of basaltic units in the region between Mare Serenitatis and Mare Tranquillitatis (BELL and HAWKE, 1995).

Nonlinear mixing (intimate mixtures)

A fundamental limitation of the linear model is that it is only strictly valid for the situation where

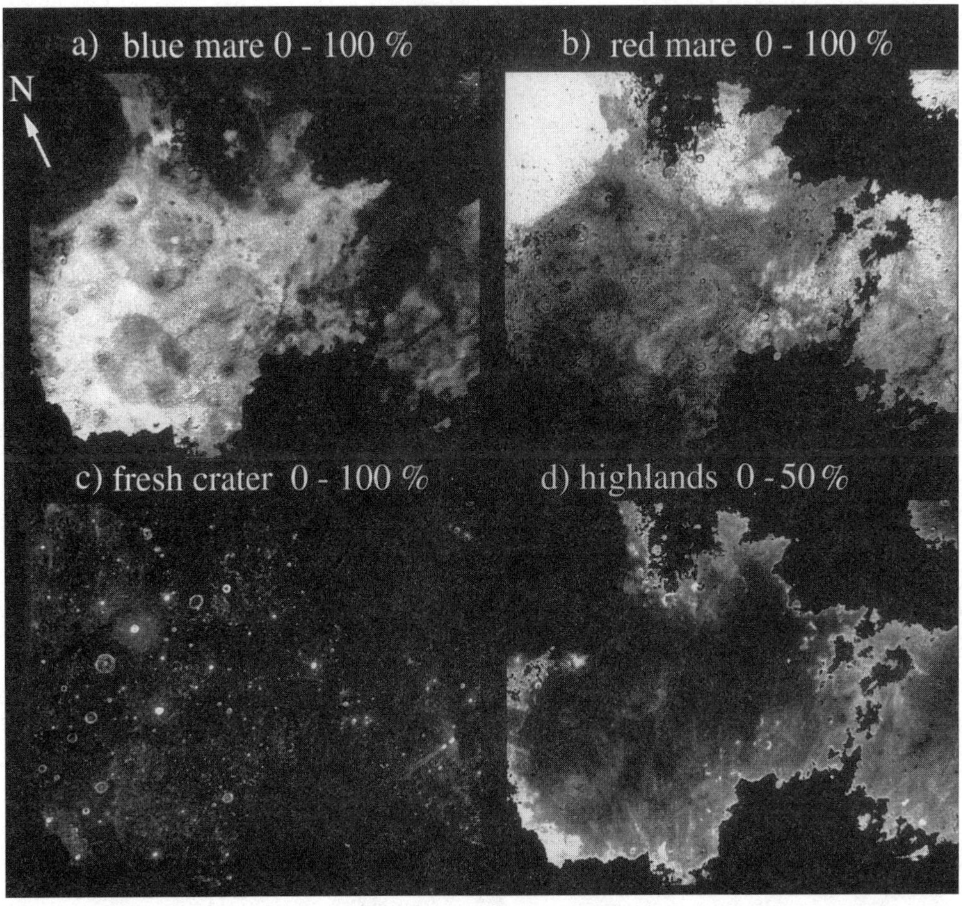

FIG. 6. Spectral mixture analysis for the same lunar area as in Fig. 1 centered on Mare Tranquillitatis (after STAID *et al.*, 1996). The Galileo data contain six spectral bands from 0.41 to 1.00 μm (BELTON *et al.*, 1994). Shown are fractional abundance images for the four endmembers used in the linear mixing analysis: blue mare, red mare, fresh mare crater, and highlands.

the endmembers are arranged in discrete patches on the surface. The scale of components can be from centimeters up to several kilometers. Many science issues of interest in planetary settings, however, are associated with particulate regoliths, where the endmember minerals that make up the soil are intimately mixed on spatial scales smaller than the path length of photons through the medium (typically—1 mm). In these settings, light usually interacts with more than one component, and the measured spectrum is a complex convolution of the endmember spectra rather than a simple additive (linear) mixture.

The nonlinear effects in spectra of particulate mixtures has been recognized for many years (*e.g.*, NASH and CONEL, 1974; PIETERS, 1974) and a variety of methods have been developed to treat this situation. The photometric model of HAPKE (1981; 1993) has been shown to be a powerful and useful model for application to nonlinear spectral mixing. The validity of the model for linearizing the mixture systematics has been demonstrated in laboratory studies of directional-hemispherical reflectance (JOHNSON *et al.*, 1983; 1992) and bidirectional reflectance (MUSTARD and PIETERS, 1987a; 1989). For controlled laboratory samples, the Hapke model has been shown to accurately predict mineral abundance to approximately 5% absolute abundance. The technique has also been successfully applied to imaging spectrometer data for desert soils in Utah (MUSTARD and PIETERS, 1987b) allowing the spatial extent and abundance of serpentized ultramafic microbreccia to be mapped. Reflectance spectra of the icy satellites of Jupiter have also been successfully modeled using a nonlinear mixing model (CALVIN and CLARK, 1991; ROUSH *et al.*, 1990) that includes methods to model grain sizes. A second model for nonlinear intimate mixing using different assumptions for bidirectional reflectance is that of Hiroi. In modified form, both the Hapke and Hiroi models can model reflectance spectra of mixtures within 4% accuracy (HIROI and PIETERS, 1992a,b).

Comparison of linear and nonlinear mixing

The effects of nonlinear mixing on reflectance spectra can be quite dramatic, as illustrated in Fig. 7. This 2-dimensional figure presents a reflectance data cloud generated using five endmembers, labeled A to E, in a mixing scenario. Reflectance in TM band 3 is plotted against the reflectance in TM band 4. In Fig. 7a, the 75 mixture points that constitute the cloud were calculated using a linear mixing model with prescribed fractions of the endmembers. Along the planes joining the endmembers (*e.g.* line A-C), the mixtures are in 25% increments. Visually, one recognizes this instantly as a linear relationship, with the points spread equidistant along the lines joining the endmembers in this reflectance space. The systematics are very different in Fig. 7b where the same 5

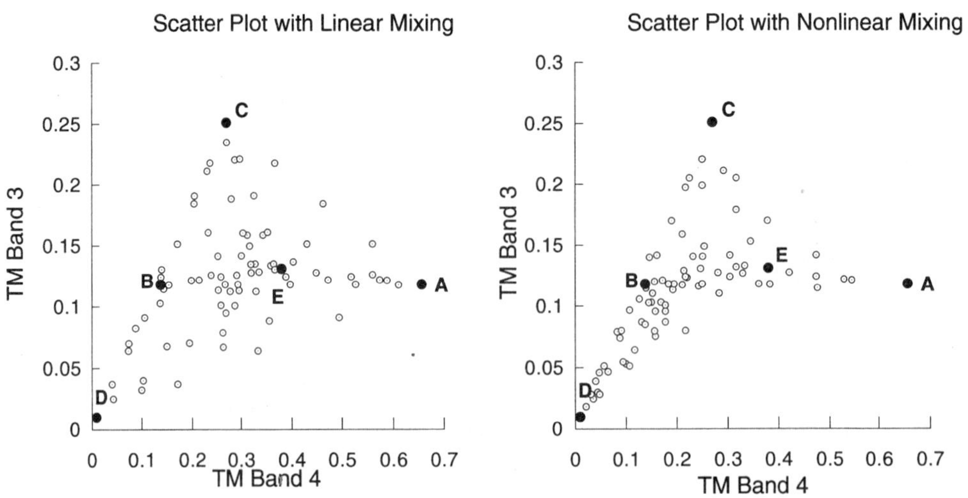

FIG. 7. Data clouds modeled for mixtures of five endmembers (A, B, C, D, and E). Individual points are reflectance values of each sample for TM bands 3 and 4 (near 0.68 μm and 0.88 μm, respectively). (a) Mixtures modeled with a linear mixing model. (b) Mixtures modeled with a nonlinear (Hapke) mixing model.

endmembers and the 75 mixture points with the same fractional abundances are shown, but here the mixture spectra were calculated using the nonlinear mixing model of MUSTARD and PIETERS (1989) adapted from HAPKE (1981). The nonlinear effects on reflectance values for mixtures are clearly indicated by the curvilinear segments joining endmembers (*e.g.* A-C, A-D). In addition, the entire data cloud is shifted to the left against the segment C-D, and towards the low albedo endmember (D). What drives this shift is the predominance of low albedo endmembers in nonlinear mixing situations.

There are some important implications of these differences for spectral mixture analysis. If a linear mixing model is used on data where the relation between components are nonlinear, the calculated fractions will be significantly in error. In tests of linear vs. nonlinear mixing on laboratory data, the calculated fractions can show differences by as much as 30% (MUSTARD, unpublished data). A linear model can cause considerable ambiguity and incorrect fractions when used on intimate mixtures. For example, a least square fitting approach will minimize errors using any of the endmembers in the equations. Thus in the fitting process endmembers not present in a mixture may be calculated to be present simply to minimize the error. This is illustrated in Figs. 8 and 9. In Fig. 8 are shown reflectance spectra of intimate mineral mixtures prepared in the laboratory with known abundances of particulate samples of three minerals, an anorthite, an olivine, and an enstatite. In Fig. 9, the abundances used to prepare the mixtures (open circles) are shown on a ternary diagram together with the best fit calculated abundances using a nonlinear (solid circles) and linear (crosses) mixture model. It is evident that the nonlinear model most accurately predicts the modal abundances of the mixtures from the reflectance spectra. In general, the accuracy for such intimate mixtures is about 5%. The linear model fractions, on the other hand, are significantly in error, and for the enstatite-anorthite mixture series a component of olivine is predicted which is not present in the actual mixtures. Although a nonlinear model is clearly most appropriate for intimate mixtures, it is nevertheless important to note that for any given endmember, the abundances predicted using the linear model increase with an increase in that endmember. That is, if a linear model is used for intimate mixtures, the predicted fractional abundances *do* increase monotonically with the abundance of the real component, but the *rate* of increase is not accurate.

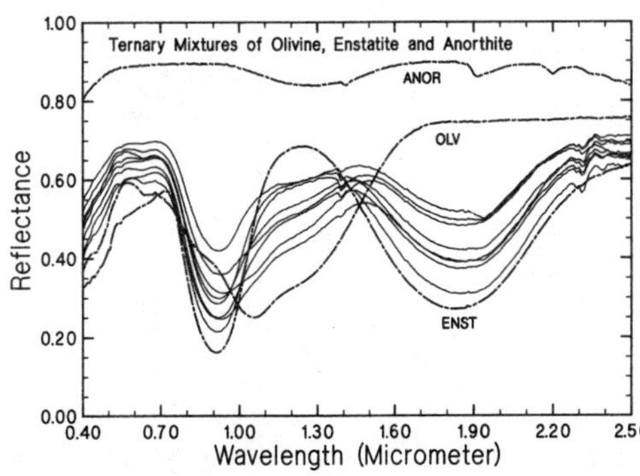

FIG. 8. Reflectance spectra of particulate samples of three mineral separates (Anorthite, Olivine, and Enstatite) shown in dashed lines, and intimate mixtures of these minerals with known abundances shown in solid lines (samples discussed in MUSTARD and PIETERS, 1989).

Despite the obvious advantages of using a nonlinear approach for intimate mixtures, this has not been widely applied to remotely acquired data of particulate planetary surfaces principally because such models are much more complex to implement than a linear model and are thus impractical. In their complete formulation nonlinear mixing models require detailed information on viewing geometry (incidence and emergence) and physical properties such as particle size and scattering properties of each endmember. One comparison of the two models undertaken by ADAMS and co-workers in which the mixing of primary material with local material was followed along a crater ray (PIETERS *et al.,* 1985) showed that the surface material was indeed better modeled with an intimate mixture than a linear mixture and the observed relative abundances of local to foreign material using the intimate mixing model was quite comparable to that predicted by crater experiments. Similarly, more thorough consideration of the nature of the remote measurements may allow many parameters in the nonlinear models to be reasonably approximated, and as spectral libraries become more well endowed with data, many of the obstacles may be sufficiently removed for nonlinear modeling to replace linear modeling as the approach of choice in remote sensing applications.

Spectral mixture analysis nevertheless is, of course, a tool and not a panacea. As described above, it is highly productive for some applications. A critical issue in the use of spectral mixture analysis is the accuracy of the choice of endmembers and the degree to which they characterize a scene uniquely (*e.g.,* ADAMS *et al.,* 1993; TOMPKINS *et al.,* 1995). There can be scientific issues, however, where neither linear nor non-linear mix-

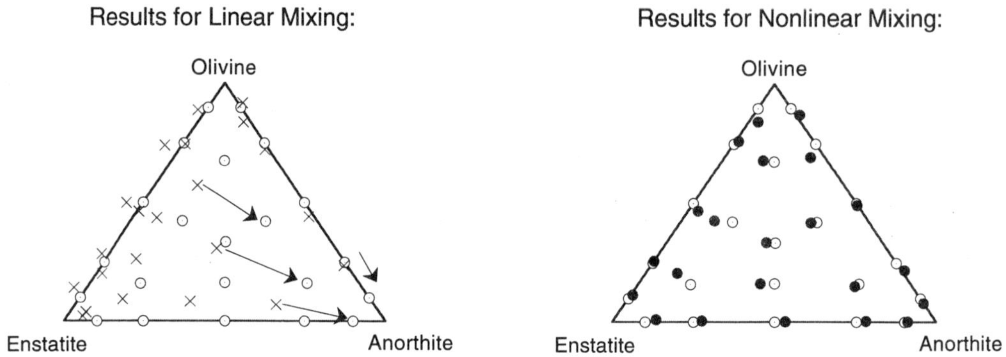

FIG. 9. Ternary diagrams of mineral abundances. Spectra of the endmember minerals were used to model the mixture spectra of Fig. 8. The open circles indicate the known abundances of the prepared mixture. The filled circles represent abundances predicted from a fit to the mixture spectra using a nonlinear mixing model. The crosses indicate the predicted abundances derived using a linear mixing model of the same mixture spectra.

ing approaches appear to converge on a viable solution, and assumptions about endmember characteristics and/or physical and compositional processes active on a surface need to be reassessed. A classic example is the continuing discussion about the character of S-type asteroids and the relation, if any, to ordinary chondrites. Although spectral mixture analyses as applied to this issue certainly accommodate a range of hypotheses (cited in several above sections), the approach can also be used to demonstrate that some endmembers logically assumed to be present on the asteroid as we currently understand them (such as various forms of Fe, Ni), simply cannot be modeled as a component for a large fraction of these asteroids using any current mixing model (CLARK, 1995).

DECONVOLUTION OF INDIVIDUAL ABSORPTION BANDS (MGM)

It is the individual absorption bands, such as the diagnostic crystal field electronic transitions (BURNS, 1993), that contain the most compositional information in reflectance spectra. However, in many, if not most spectra of natural surfaces, absorptions occur not as single isolated bands, but instead superimposed on other spectral features. The modified Gaussian model (MGM) of SUNSHINE *et al.* (1990) was specifically developed to address this problem and to provide a method to isolate individual electronic transition absorption bands in reflectance spectra. At the core of the MGM is a physically based mathematical expression, a modified Gaussian distribution, which (in sharp contrast to the often used Gaussian model) has been demonstrated to accurately describe the shape of electronic transition absorptions. As an example, a comparison of modeling a single electronic absorption with a MGM distribution and with a Gaussian distribution is shown in Fig. 10. Having successfully described the shape of isolated bands, it is now possible to use the MGM to deconvolve reflectance spectra into their constituent absorptions. To facilitate this goal, the original development of the MGM includes implementation of fast, reliable, and flexible inversion techniques which together provide a powerful tool for quantitative mineralogic analysis (see SUNSHINE and PIETERS, 1993a,b and SUNSHINE *et al.,* 1993).

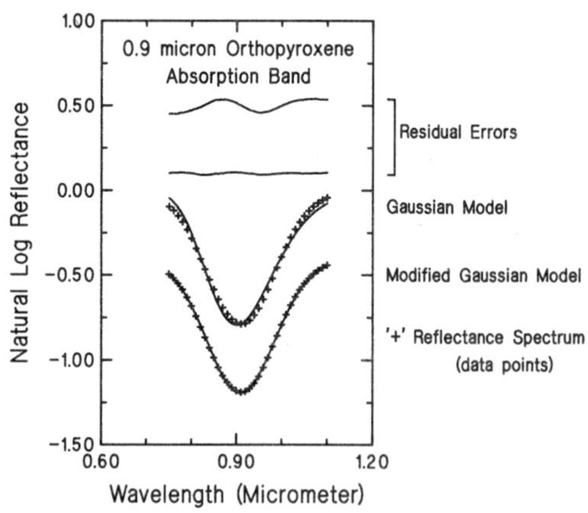

FIG. 10. Comparison of a Gaussian and a Modified Gaussian fit to an electronic transition absorption band of Fe^{+2} in an orthopyroxene. The reflectance spectrum of orthopyroxene is for a <45 μm particulate sample.

In the MGM, spectra are modeled simultaneously as a sum of overlapping absorption bands superimposed on a continuum (see SUNSHINE *et al.,* 1990 for model details). Each absorption is represented by a modified Gaussian distribution, $m(x)$, and described by three model parameters: a band center (μ), a band width (s), and a band strength (σ) where the MGM distribution is expressed as

$$m(x) = s \cdot \exp\left\{\frac{-(x^{-1} - \mu^{-1})^2}{2\sigma^2}\right\}$$

The continuum is normally characterized by a slope (straight line in energy) and an offset. Modeling is carried out in natural log reflectance and energy. Examples of MGM fits for particulate samples of two pyroxenes and their mixtures are shown in Fig. 11. It should be noted that in contrast to other methods described above, the MGM neither relies on a library of spectra that may or may not reproduce all permutations and combinations of absorption bands, nor requires knowledge of endmember spectra, as is needed for both linear and non-linear mixing. Instead, the MGM derives compositional information directly from a measured spectrum. This is particularly useful in planetary remote sensing applications, where the exact petrology, major and minor element chemistry, and physical state (particle size, alteration state, etc.) of materials is unknown and often irreproducible from either spectra of terrestrial samples or samples that have been simulated in the laboratory.

The successful validation of the MGM approach for reflectance spectroscopy opens the way for quantitative assessment of the diagnostic properties of individual electronic transitions or groups of

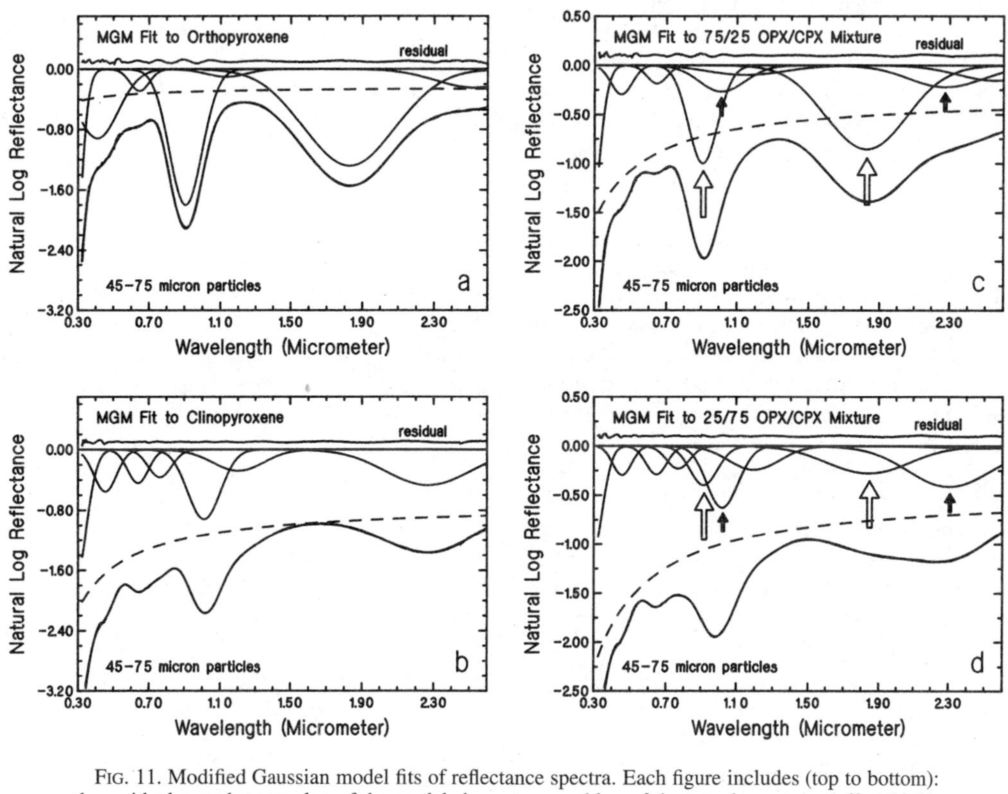

FIG. 11. Modified Gaussian model fits of reflectance spectra. Each figure includes (top to bottom): the residual error between log of the modeled spectrum and log of the actual spectrum (offset 10%), the individual modified Gaussian distributions representing absorption bands, the continuum (dashed line), and the modeled spectrum superimposed on the actual spectrum. a) MGM fit to orthopyroxene size separate. The symmetric non-random residual error near 0.9 μm is a characteristic of minor band saturation (Sunshine and Pieters, 1993). b) MGM fit to clinopyroxene size separate. c) MGM fit to 75% orthopyroxene-25% clinopyroxene mass fraction mixture. d) MGM fit to 25% orthopyroxene-75% clinopyroxene mixture. In (c) and (d) open arrows correspond to absorptions from the low-Ca orthopyroxene phase. Filled arrows correspond to absorptions from the high-Ca clinopyroxene phase.

such diagnostic absorptions. The process of building a solid understanding of the properties of individual minerals and mixtures of minerals is in its early stages. While the MGM is a powerful tool which holds great promise, it currently needs to be used cautiously, particularly with remote data that have substantial noise and often lower spectral resolution, while the foundation of experience with well controlled laboratory samples is properly expanded. Discussed below are some of the benchmark results and a taste of the applications that have emerged in the last few years since the MGM was developed.

Pyroxenes and their mixtures

A thorough analysis of two different types of pyroxene and their mixtures (a low-calcium orthopyroxene [OPX] and a high-calcium clinopyroxene [CPX]) was undertaken to determine whether discrete absorptions retained their well defined properties when mixed with similar materials (SUNSHINE and PIETERS, 1993a). MGM analyses were performed on three different particle sizes (<45 μm, 45–75 μm, and 75–125 μm) of the pure minerals and seven mass fraction mixtures. Mass fraction spectra for one particle size are shown in Fig. 3 and example MGM analyses are shown in Fig. 11. The results are remarkably consistent: the band centers and band widths are invariant for the primary diagnostic bands near 1 and 2 μm which are due to ferrous iron in the M2 site. That is, the MGM fitting routine always converged on the same value for the orthopyroxene component and a separate value for the clinopyroxene component, regardless of the particle size or mass fraction. Such a task is virtually impossible with visual inspection. Furthermore, the absorptions bands determined from the mixture spectra were found to occur at the same wavelength as those of the endmember OPX and CPX pure mineral spectra. Thus for the first time, with the MGM, it is now possible to determine endmember absorption bands directly from a mixture spectra.

What is equally important for remote sensing applications is that the relative strengths of the diagnostic pyroxene absorptions in a mixture (the strength of the primary OPX band relative to the CPX band) were seen to vary in a very systematic manner with modal abundance. Such a band strength ratio is shown in Fig. 12 as a function of the clinopyroxene mass fraction in the mixture. The same relationship is seen for the absorptions near 1 μm and for the absorptions near 2 μm, consistent with the crystal field theory that shows the absorption in both regions arise from asymmetries in the M2 site. It is readily apparent that this relation is independent of particle size, making it an extremely valuable tool for estimating relative mineral abundances. Requirements for use are currently that pyroxenes be a dominant mafic component in a spectrum and that whatever the particle size is, it is more or less comparable for all components in the mixture. For further detail see discussion in SUNSHINE and PIETERS (1993a).

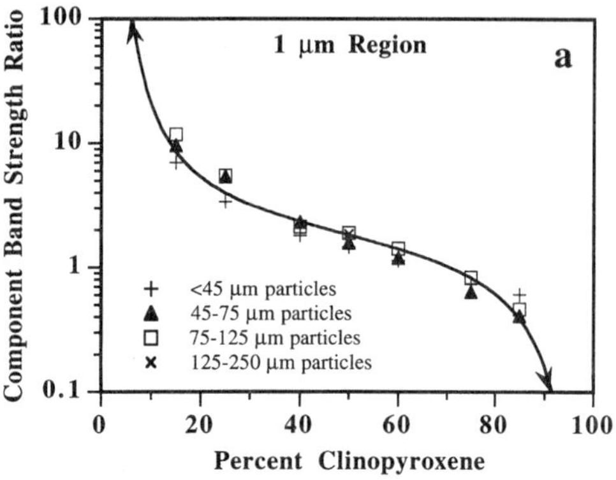

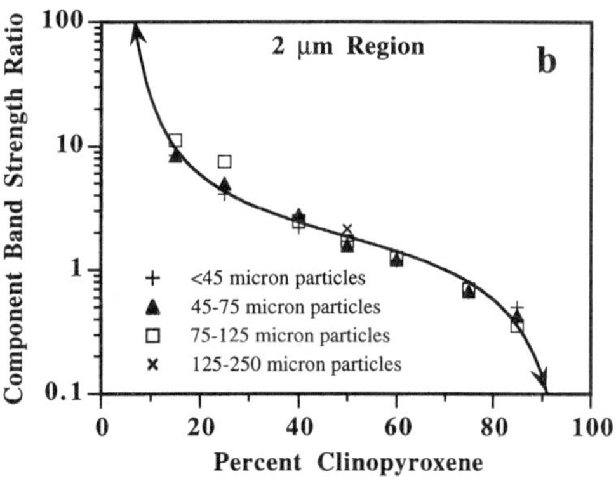

FIG. 12. Relation between modal abundance (wt %) in a mixture of pyroxenes and the 1 and 2 μm band strength ratios for the primary orthopyroxene and clinopyroxene absorption bands (after SUNSHINE and PIETERS, 1993). a) ratio of OPX/CPX absorptions in the 1 μm region, b) ratio of OPX/CPX absorptions in the 2 μm region. The curves are the best fits to the data from all particle sizes. Note the particle size independence and the consistency between the 1 and 2 μm regions.

One of the first applications of the MGM pyroxene systematics was an evaluation of the two different basaltic lithologies that occurs in a shergottite meteorite, EETA 79001 (SUNSHINE *et al.*, 1993), believed to have come from Mars. MGM analyses of reflectance spectra of the two lithologies present

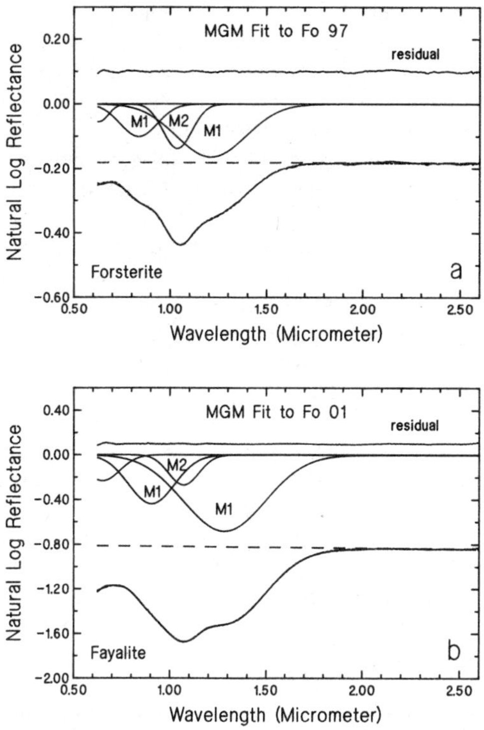

FIG. 13. MGM fit for two compositions of olivine. Note the difference in relative band strengths (M2 vs M1 bands) between the Mg-rich forsterite and the Fe-rich fayalite.

in this meteorite were used to determine that two pyroxenes are present in each of the lithologies, to estimate the composition of the two pyroxenes (based on band center), and to estimate the relative abundances of the two pyroxenes using the relationship of Fig. 12. These results compare very favorably (5–10%) to those of more traditional geochemical and petrographic methods to determine mineral abundance and composition. This first application not only provides a quantitative comparison of compositional information derived from spectroscopy with actual "ground truth", but also serves as a model for how the MGM could be used in remote planetary applications.

One of the significant results of the MGM pyroxene analysis was that it also clearly documented an absorption band near 1.2 μm in both pyroxenes (more prominent in clinopyroxenes) attributed to ferrous iron in the M1 site (*e.g.,* BURNS, 1993). Since such a feature can easily be confused with a similar band in plagioclase (see Fig. 2), its identification is of great importance to remote mineral identification (see discussion in SUNSHINE *et al.,* 1993).

Olivine

Olivine exhibits a composite absorption band near 1.05 μm consisting of three electronic transition absorptions. Transmission spectra have documented systematic trends with composition for the band centers of individual absorption bands in oriented olivine crystals (BURNS, 1970b). Previous work with olivine reflectance spectra (*e.g.,* KING and RIDLEY, 1987) was able to document small shifts of the composite band minimum. Because the olivine absorption distributions overlap so much, almost any mathematical model can produce a "good" fit to a measured spectrum. However, since the MGM has been demonstrated to be an accurate characterization of electronic transitions, the MGM parameters derived for each of the three bands of olivine can be used with some confidence in compositional analyses across the olivine solid solution series.

Reflectance spectra for a suite of olivines of different compositions were obtained and analyzed with the MGM, examples of which are shown in Fig. 13 (from SUNSHINE, 1994). As might be expected from crystal field theory, the relative band strengths (probability) of the two absorptions derived from iron in the M1 site remained the same as iron composition varies. In contrast, the relative strengths of the M2 and M1 bands vary with composition. The widths of the olivine absorptions are also seen to be controlled by absorption site and to be largely compositionally invariant (SUNSHINE, 1994).

Results for the full suite of reflectance spectra for olivine samples are shown in Fig. 14 (SUNSHINE

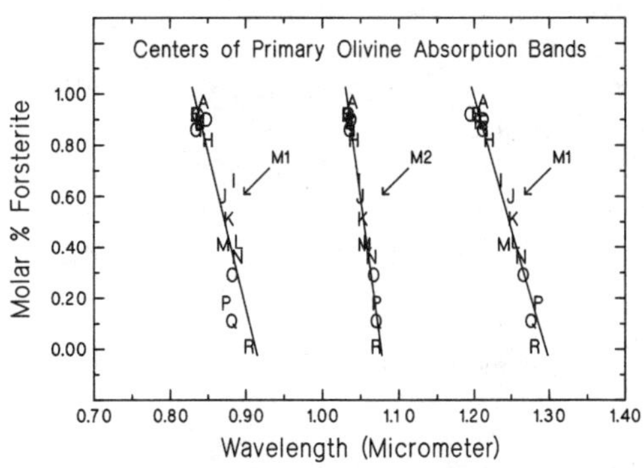

FIG. 14. MGM derived band centers for a suite of olivines of different compositions. Although all absorptions move to longer wavelengths with increasing iron content, both absorptions associated with the M1 site show a stronger change of energy with composition than the absorption from the M2 site.

and PIETERS, 1990, revised by SUNSHINE, 1994). The MGM derived band center for the three overlapping absorptions are seen to vary systematically with composition. This duplication of equivalent compositional results as those observed for transmission spectra of oriented crystals (BURNS, 1970b) is a substantial breakthrough for reflectance spectroscopy which effectively measures the combined signatures of randomly oriented crystals. The relations shown in Fig. 14 can now be used to estimate the composition of an unknown olivine from reflectance spectra.

Remote sensing applications

Given high quality remote spectra, MGM analysis can be straightforward. More commonly, however, remotely acquired data have lower signal to noise and/or lower spectral resolution than laboratory data. With such uncertainties, simple application of the MGM to produce to best mathematical model can result in physically unrealistic and uninterpretable solutions. Under these circumstances, additional constraints or assumptions may need to be imposed on the modeling approach. Constraints such as restricting solutions to be consistent with laboratory trends (*e.g.,* coupling of the relative strengths of the 1 and 2 μm pyroxene bands or the M1 olivine bands) and only allowing parameters to range within natural compositions (*e.g.,* olivine band center that lie within 0–100% fosterite) are quite reasonable. The mathematical framework of the MGM explicitly allows for such constraints to be including in modeling (see SUNSHINE and PIETERS, 1993a,b; SUNSHINE, 1994).

The importance of applying physically reasonable constraints was demonstrated by MGM analysis of the telescopic spectrum of the surface of the asteroid 243 Asporina, an "olivine-rich" asteroid (SUNSHINE, 1994; SUNSHINE and PIETERS, 1993b). When the mathematically best fit for all parameters was determined with somewhat noisy data, the resulting absorptions bands parameters (centers, widths, and relative strengths) were found to be inconsistent with those derived for any olivines in the laboratory. However, when a series of model variations were tested, which were partially constrained by band parameter coupling (*e.g.,* coupling of the relative strength of the M1 bands observed for olivine compositions), the results were quite dramatic. The surface of Asporina was shown to be totally inconsistent with fayalitic (Fe-rich) olivine, yet was successfully characterized to be distinctly fosteritic (Mg-rich).

Another elegant application of the MGM approach used high spatial resolution ISM imaging spectrometer data for Mars (MUSTARD and SUNSHINE, 1995) to make important compositional distinctions between different volcanic terrain. In this case, pyroxene mixtures and ferric alteration were considered as possible components contributing to the remotely acquired spectra. Solutions were sought which satisfied a set of criteria based on laboratory experience with pyroxene mixtures. These included such constraints as the band centers and band widths must be consistent with known properties of pyroxenes and the relative strengths of absorptions for two pyroxenes must be comparable for the 1 and the 2 μm regions. Using this constrained MGM it was possible to model varying amounts of ferrous components in the martian spectra and to place compositional bounds on the ferric constituents. Successful solutions showed that volcanic terrain on Mars included two pyroxenes, that low-Ca pyroxene dominated the abundances, that the relative proportions of pyroxenes varied on the order of 20%, and that various amounts of a ferric component was present. This inferred basaltic composition is relatively uncommon on Earth, but is typical of the SNC meteorites which are believed to come from Mars. Through this combination of ISM data and MGM analysis it is now possible to conclude that the basaltic compositions present in the SNC meteorites are in fact representative of large regions of Mars.

EXPECTATIONS AND HOPES FOR THE FUTURE

Imaging spectrometers are high on the list of desirable instruments for most planetary exploration missions that are under consideration. There is no question that abundant information currently exists to clearly specify the type of spectroscopic measurements desired for quantitative remote compositional analysis. The above review of techniques and recent progress in quantitative approaches to remote spectral analysis has emphasized applications that involve electronic transitions of transition element ions, which typically occur in mafic rock forming minerals. However many of the same approaches have been or can be used for almost any type of diagnostic absorptions, including absorptions due to charge transfer electronic transitions and vibrational, bending, and stretching modes of molecular excitation that occur in spectra of minerals important to solar system bodies.

Many of the approaches described here, although highly productive, are still in the developmental stage. Due to the newness of these research efforts, it is not surprising that little work has been done to combine these various methods. However, given that many approaches are complimentary, a natural direction would be to combine or link several in an integrated fashion to enhance interpretive capabilities and strengthen confidence in results. Technology and instrumentation is clearly heading in that direction with the implementation of imaging spectrometers. It can be envisioned that future analytical approaches may, for example, utilize MGM deconvolution methods to identify and couple critical diagnostic absorptions and then synthesize appropriate spectra to be used in various mixing models which will derive abundance information. Fortunately, computing capabilities appear to be keeping ahead of, or at least in line with, instrumentation and analytical approaches.

The last few decades of activity in spectroscopic applications for remote sensing have been very exciting. The future of remote reflectance spectroscopy is likely to include larger and more complex data sets with full spectral resolution that cover large, perhaps global, areas of planetary surfaces at higher spatial resolution. This expansion in data alone will provide significant new breakthroughs. Extrapolating from the past, parallel advances in advanced analysis techniques and computing capabilities should similarly add new dimensions to our ability to contribute to the exploration and understanding of various solar system bodies. As we progress to new horizons, the small but vigorous spectroscopy community is expected to continually expand upon the solid foundation of theoretical understanding provided by Roger Burns and as such provide a living legacy to his life's work.

Acknowledgments—Research support to the authors from NASA grants [NAGW-28, NAGW-3379, NASW-5026] is gratefully acknowledged. Spectra were acquired with RELAB, a multi-user facility supported under NASA grant NAGW-748.

REFERENCES

ADAMS J.B. (1974) Visible and near-infrared diffuse reflectance spectra of pyroxenes as applied to remote sensing of solid objects in the solar system. *J. Geophys. Res.* **79,** 4829–4836.

ADAMS J. B. (1975) Interpretation of visible and near-infrared diffuse reflectance spectra of pyroxenes and other rock-forming minerals, in *Infrared and Raman Spectroscopy of Lunar and Terrestrial Materials* (ed. C. KARR Jr.), Academic, New York, 91–116.

ADAMS J. B. and MCCORD T. B. (1969) Mars: Interpretation of spectral reflectivity of light and dark regions. *J. Geophys. Res.* **74,** 4851–4856.

ADAMS J. B., SMITH M. O. and JOHNSON P. E. (1986) Spectral mixture modeling: a new analysis of rock and soil types at Viking Lander 1. *J. Geophys. Res.* **91,** 8113–8125.

ADAMS J. B., SMITH M. O. and GILLESPIE A. R. (1993) Imaging spectroscopy: Interpretation based on specrtral mixture analysis. In *Remote Geochemical Analysis: Elemental and Mineralogical Composition.* (eds. C. M. PIETERS and P. A. J. ENGLERT), Cambridge University Press, New York, 145–166.

BELL J. and HAWKE B. R. (1995) Compositional variability of the Serenitatis/Tranquillitatis region of the Moon from telescopic multispectral imaging and spectroscopy. *Icarus* **118,** 51–68.

BELL P. M., MAO H. K. and WEEKS R. A. (1976) Optical spectra and electron paramagnetic resonance of lunar and synthetic glasses: A study of the effects of controlled atmosphere, composition, and temperature. *Proc. Lunar Sci. Conf. 7th,* 2543–2559.

BELTON M. J. S., GREELEY R., GREENBERG R., MCEWEN A., KLAASEN K. P., HEAD J. W., PIETERS C. M., NEUKUM G., CHAPMAN C. R., GEISSLER P., HEFFERNAN C., BRENEMAN H., ANGER C., CARR M. H., DAVIES M. E., FANALE F. P., GIERASCH P. J., INGERSOLL A. P., JOHNSON T. V., PILCHER C. B., THOMPSON R., VEVERKA J. and SAGAN C. (1994) Galileo multispectral imaging of the North Polar and Eastern Limb regions of the Moon. *Science* **264,** 1112–1115.

BLEWETT D. T., HAWKE B. R., LUCEY P. G. and TAYLOR G. J. (1995) Remote sensing and geologic studies of the Shiller-Schickard region of the Moon. *J. Geophys. Res.* **100,** E8, 16959–16977.

BURNS R. G. (1970a) *Mineralogical Application of Crystal Field Theory,* Cambridge Univ. Press, Cambridge, 234 pp.

BURNS R. G. (1970b) Crystal field spectra and evidence of cation ordering in olivine minerals. *Amer. Mineral.* **55,** 1608–1632.

BURNS R. G. (1993) *Mineralogical Application of Crystal Field Theory,* Second Ed., Cambridge Univ. Press, Cambridge, 551 pp.

CALVIN W. M. and CLARK R. N. (1991) Modeling the reflectance of Calisto 0.25–4.1 mm. *Icarus* **89,** 305–317.

CHARETTE M. P., MCCORD T. B., PIETERS C. M. and ADAMS J. B. (1974) Application of remote spectral reflectance measurements to Lunar geology classification and determination of titanium contents of Lunar soils. *J. Geophys. Res.* **79,** 1605–1613.

CLARK B. E. (1995) Spectral mixing models of S-Type asteroids. *J. Geophys. Res.,* **100,** E7, 14443–14456

CLARK R. N. (1983) Spectral properties of mixtures of montmorillonite and dark carbon grains: Implications for remote sensing minerals containing chemically and physically adsorbed water. *J. Geophys Res.* **88,** B12, 10635–10644.

CLARK R. N. and ROUSH T. L. (1984) Reflectance spectroscopy: Quantitative analysis techniques for remote sensing applications. *J. Geophys. Res.* **89,** 6329–6340.

CLOUTIS E. A. and GAFFEY M. J. (1991) Pyroxene spectroscopy revisited: Spectral-compositional correlations and relationship to geothermometry. *J. Geophys. Res.* **96,** 22809–22826.

CLOUTIS E. A., GAFFEY M. J., JACKOWSKI T. and REED K. (1986) Calibration of phase abundance, composition, and particle size distribution for olivine-orthopyroxene mixtures from reflectance spectra, *J. Geophys. Res.* **91,** 11641–11653.

CLOUTIS E. M., GAFFEY M. J., SMITH D. G. W. and LAM-

BERT R. S. J. (1990a) Reflectance spectra of mafic silicate-opaque assemblages with applications to meteorite spectra. *Icarus* **84,** 315–333.

CLOUTIS E. M., GAFFEY M. J., SMITH D. G. W. and LAMBERT R. S. J. (1990b) Metal silicate mixtures: Spectral properties and applications to asteroid taxonomy. *J. Geophys. Res.* **95,** 8323–8338.

CROWN D. A. and PIETERS C. M. (1987) Spectral properties of plagioclase and pyroxene mixtures and the interpretation of Lunar soil spectra. *Icarus* **72,** 3, 492–506.

FISCHER E. and PIETERS C. M. (1994) Remote determination of exposure degree and iron concentration of Lunar soils using VIS-NIR spectroscopic methods. *Icarus* **111,** 475–488.

FISCHER E. M. and PIETERS C. M. (1996) Composition and exposure age of the Apollo 16 Cayley and Descartes regions from Clementine data: Normalizing the optical effects of space weathering. *J. Geophys. Res.* **101,** E1, 2225–2234.

GAFFEY M. J., BURBINE T. H. and BINZEL R. P. (1993a) Asteroid spectroscopy: Progress and perspectives. *Meteoritics* **28,** 161–187.

GAFFEY M. J., BELL J. F., BROWN R. H., BURBINE T. H., PIATEK J. L., REED K. L. and CHAKY D. A. (1993b) Mineralogical variations within the S-type asteroid class. *Icarus* **106,** 573–602.

GAFFEY S. J., MCFADDEN L. A., NASH D. and PIETERS C. M. (1993c) Ultraviolet, visible, and near-infrared reflectance spectroscopy: Laboratory spectra of geologic materials. In *Remote Geochemical Analysis: Elemental and Mineralogical Composition* (eds. C. M. PIETERS and P. A. J. ENGLERT), Cambridge University Press, New York, 43–77.

HAPKE B. (1981) Bidirectional reflectance spectroscopy 1: Theory. *J. Geophys. Res.* **86,** 3039–3054.

HAPKE B. (1993) *Theory of Reflectance and Emittance Spectroscopy,* Cambridge University Press, New York, 455 pp.

HAZEN R. M., BELL P. M. and MAO H. K. (1978) Effects of compositional variation on absorption spectra of Lunar pyroxenes. *Proc. Lunar and Planet. Sci. Conf. 9th.,* 2919–2934.

HEAD J., PIETERS C., MCCORD T. B., ADAMS J. B. and ZISK S. (1978a) Definition and detailed characterization of Lunar surface units using remote observations. *Icarus,* 145–172.

HEAD J. W., MURCHIE S., MUSTARD J. F., PIETERS C. M., NEUKUM G., MCEWEN A., GREELEY R. and BELTON M. J. S. (1993) Lunar Impact Basins: New Data for the Western Limb and Farside (Orientale and South Pole-Aitkin Basins) from the First Galileo Flyby, *J. Geophys. Res., 98,* No. E9, 17,149–17,181.

HIROI T. and PIETERS C. M. (1992a) Effects of grain size and shape in modeling reflectance spectra of mineral mixtures. *Proc. Lunar Planet. Sci.* **22,** 313–25.

HIROI T. and PIETERS C. M. (1992b) Estimation of grain sizes and mixing ratios of fine powder mixtures of common geologic minerals. *J. Geophys Res.* **99,** E5, 10867–10879.

HIROI T., BELL J. F., TAKEDA H. and PIETERS C. M. (1993) Modeling of S-type asteroid spectra using primitive achondrites and iron meteorites. *Icarus* **102,** 107–116.

HUNT G. R. and SALISBURY W. (1970) Visible and near-infrared spectra of minerals and rocks: I. Silicate minerals. *Modern Geol.* **1,** 283–300.

JAUMANN R. (1991) Spectral-chemical analysis of Lunar surface materials. *J. Geophys. Res.* **96,** E5, 22793–22807.

JOHNSON J., LARSON S. M. and SINGER R. (1991) Remote sensing of potential lunar resources 1. Near side compositional properties. *J. Geophys Res.* **96,** 18861–18882.

JOHNSON P. E., SMITH M. O., TAYLOR-GEORGE S. and ADAMS J. B. (1983) A semiempirical method for analysis of the reflectance spectra of binary mineral mixtures. *J. Geophys. Res.* **88,** 3357–3361.

JOHNSON P. E., SMITH M. O. and ADAMS J. B. (1992) Simple algorithms for remote determination of mineral abundances and particle sizes from reflectance spectra. *J. Geophys. Res.* **97,** 2649- 2658.

KING T. V. V. and RIDLEY W. I. (1987) Relation of spectroscopic reflectance of olivines to mineral chemistry and some remote sensing implications. *J. Geophys. Res.* **92,** 11457–11469.

LUCEY P. G., TAYLOR G. J. and MALARET E. (1995) Abundance and distribution of iron on the Moon. *Science* **268,** 1150–1153.

MCCORD T. B. and JOHNSON T. V. (1970) Lunar spectral reflectivity (0.3 to 2.50 μm) and implications for remote mineralogical analysis. *Science* **169,** 855–858.

MCCORD T. B., ADAMS J. B. and JOHNSON T. V. (1970) Asteroid Vesta: Spectral reflectivity and compositional implications. *Science* **168,** 1445–1447.

MCCORD T. B., CHARETTE M., JOHNSON T. V., LEBOFSKY L. A., PIETERS C. and ADAMS J. B. (1972) Lunar spectral types. *J. Geophys. Res.* **77,** 1349–1359.

MELENDREZ D. E., JOHNSON J., LARSON S. and SINGER R. (1994) Remote sensing of potential lunar resources 2. High spatial resolution mapping of spectral reflectance ratios and implications for near side mare TiO2 content. *J. Geophys. Res.* **99,** 5601–5620.

MOROZ L. V., FISENKO A. V., SEMJONOVA L. F. and PIETERS C. M. (1996) Optical effects of regolith processes on S-asteroids as simulated by laser shots on ordinary chondrite and other mafic materials. *Icarus,* in press.

MORRIS R. V. (1978) The surface exposure (maturity) of lunar soils: some concepts and Is/FeO compilation. *Proc. Lunar Planet. Sci. Conf. 9th,* 2287–2297.

MURCHIE S. L. and PIETERS C. M. (1996) Spectral properties and rotational spectral heterogeneity of 433 Eros. *J. Geophys. Res.* **101,** E1, 2201–2214.

MUSTARD J. F. and HEAD J. W. (1995) Buried stratigraphic relationships along the southwestern shores of Oceanus Procellarum. *J. Geophys. Res.,* submitted.

MUSTARD J. F. and PIETERS C. M. (1987a) Quantitative abundance estimates from bidirectional reflectance measurements. *Proc. Lunar and Planet. Sci. Conf. 17th, J. Geophys. Res.* **92,** E617-E626.

MUSTARD J. F. and PIETERS C. M. (1987b) Abundance and distribution of ultramafic microbreccia in Moses Rock Dike: Quantitative application of mapping spectroscopy. *J. Geophys. Res.* **92,** B10, 10376–10390.

MUSTARD J. F. and PIETERS C. M. (1989) Photometric phase functions of common geologic minerals and application to quantitative analysis of mineral mixture reflectance spectra. *J. Geophys. Res.* **94,** 13619–13634.

MUSTARD J. F. and SUNSHINE J. M. (1995) Seeing through the dust: Martian crustal heterogeneity and links to the SNC meteorites. *Science* **267,** 1623–1626.

MUSTARD J. F., HEAD J. W., MURCHIE S. M., PIETERS C. M., BELTON M. S. and MCEWEN A. S. (1992) Schickard cyrptomare: Interaction between Orientale ejecta and pre-basin mare from spectral mixture analysis of Galileo SSI data. *Lunar and Planet. Sci. XXIII,* 957–958.

MUSTARD J. F., HEAD J. W. and ANTONENKO I. (1994) Mare-highland mixing relationships along the southwestern shores of Oceanus Procellarum. *Lunar and Planet. Sci. XXV,* 693–694.

NASH D. B. and CONEL J. E. (1974) Spectral reflectance systematics for mixtures of powdered hypersthene, labradorite, and ilmenite. *J. Geophys. Res.* **79,** 1615–1621.

PIETERS C. (1974) Polarization in a mineral absorption band. In *Planets, Stars, and Nebulae: Studied with Photopolarimetry,* (ed. T. GEHRELS), Univ. of Arizona Press, 405–418.

PIETERS C. M. (1978) Mare basalt types on the front side of the Moon: A summary of spectral reflectance data. *Proc. Lunar and Planet. Sci. Conf. 9th,* 2825–2849.

PIETERS C. M. (1983) Strength of Mineralogical Absorption Features in the Transmitted Component of Reflection: First Results from RELAB, *J. Geophys. Res. 88,* No. B11, 9534–9544.

PIETERS C. M. (1986) Composition of the Lunar highland crust from near-infrared spectroscopy. *Reviews of Geophysics* **24,** 3, 557–578.

PIETERS C. M. (1993) Compositional diversity and stratigraphy of the Lunar crust derived from reflectance spectroscopy, in *Remote Geochemical Analysis: Elemental and Mineralogical Composition.* (eds. C. M. PIETERS and P. A. J. ENGLERT), Cambridge University Press, New York, 309–339.

PIETERS C. M. and MCFADDEN L. A. (1994) Meteorite and asteroid reflectance spectroscopy: Clues to early solar system processes. *Ann. Rev. Earth Planet. Sci.* **22,** 457–497.

PIETERS C. M., ADAMS J. B., MOUGINIS-MARK P., ZISK S. H., HEAD J. W., MCCORD T. B. and SMITH M. (1985) The nature of crater rays: The Copernicus example. *J. Geophys. Res.* **90,** B14, 12393–12413.

PIETERS C. M., HEAD J. W., SUNSHINE J. M., FISCHER E. M., MURCHIE S. L., BELTON M., MCEWEN A., GADDIS L., GREELEY R., NEUKUM G., JAUMANN R. and HOFFMANN H. (1993) Crustal diversity of the Moon: Compositional analyses of Galileo SSI data. *J. Geophys. Res.* **98,** E9, 17,127–17,148.

PINET P. C., CHEVREL S. D. and MARTIN P. (1993) Copernicus: A regional probe to the Lunar interior. *Science* **260,** 797–801.

ROUSH T. L., BLANEY D. L. and SINGER R. B. (1993) The surface composition of Mars as inferred from spectroscopic observations, in *Remote Geochemical Analysis: Elemental and Mineralogical Composition.* (eds. C. M. PIETERS and P. A. J. ENGLERT), Cambridge University Press, New York, 367–393.

ROUSH T. L., POLLACK J. B., WITTEBORN F. C., BERGMAN J. D. and SIMPSON J. P. (1990) Ice and minerals on Callisto: A reassessment of the reflectance spectra. *Icarus* **86,** 355–382.

SINGER R. B. (1981) Near-infrared spectral reflectance of mineral mixtures: Systematic combinations of pyroxenes, olivine, and iron oxides. *J. Geophys. Res.* **86,** 7967–7982.

SINGER R. B. and MCCORD T. B. (1979) Mars: Large scale mixing of bright and dark surface materials and implications for analysis of spectral reflectance. *Proc. Lunar Planet. Sci. Conf. 10th,* 1835–1848.

SMITH M. O., USTIN S. L., ADAMS J. B. and GILESPIE A. R. (1990) Vegetation in deserts: I. Estimating abundance from remotely sensed images. *Remote Sensing Environ.* **31,** 1–26.

SODERBLOM L. A. (1992) The Composition and mineralogy of the Martian surface from spectroscopic observations: 0.3 μm to 50 μm, in *Mars* (eds. H. H. KIEFFER *et al.,*), Univ. Arizona Press, 557–593.

STAID M. I., PIETERS C. M. and HEAD J. W. (1996) Mare Tranquillitatis: Basalt emplacement history and relation to lunar samples. *J. Geophys. Res.,* submitted.

SUNSHINE J. M. (1994) Inferring the composition of mafic lithologies from multi- and hyper-spectral data sets: Implications for remote sensing of terrestrial bodies, Ph.D. dissertation, Brown University.

SUNSHINE J. M. and PIETERS C. M. (1990) Extraction of compositional information from olivine reflectance spectra: A new capability for lunar exploration. *Lunar and Planet. Sci. XXI,* Houston, 1223–1224.

SUNSHINE J. M. and PIETERS C. M. (1993a) Estimating modal abundances from the spectra of natural and laboratory pyroxene mixtures using the modified Gaussian Model. *J. Geophys. Res.* **98,** 9,075–9,087.

SUNSHINE J. M. and PIETERS C. M. (1993b) Determining the composition of olivine on asteroidal surfaces. *Lunar & Planet. Sci. XXIV,* LPI, Houston, TX, 1379–1380.

SUNSHINE J. M., PIETERS C. M. and PRATT S. F. (1990) Deconvolution of mineral absorption bands: An improved approach. *J. Geophys. Res.* **95,** 6955–6966.

SUNSHINE J. M., MCFADDEN L. A. and PIETERS C. M. (1993) Reflectance spectra of the Elephant Moraine A79001 Meteorite: Implications for remote sensing of planetary bodies. *Icarus* **105,** 79–91.

TOMPKINS S., MUSTARD J. F., PIETERS C. M. and FORSYTH D. W. (1995) Optimization of end-members for spectral mixture analysis. *Remote Sensing of Environ.,* submitted.

Mineral Spectroscopy: A Tribute to Roger G. Burns
© The Geochemical Society, Special Publication No. 5, 1996
Editors: M. D. Dyar, C. McCammon and M. W. Schaefer

An occurrence of jarositic tephra on Mauna Kea, Hawaii: Implications for the ferric mineralogy of the Martian surface

RICHARD V. MORRIS,[1] DOUGLAS W. MING,[1] D. C. GOLDEN[2] and JAMES F. BELL, III[3]
[1]NASA Johnson Space Center, Houston, TX 77058, U.S.A.
[2]Dual Inc., Houston, TX 77058, U.S.A.
[3]Center for Radiophysics and Space Research, Cornell University, Ithaca, NY 14853, U.S.A.

Abstract—A sample of yellow-colored tephra (HWMK24) from a cone on the summit of Mauna Kea, Hawaii, was examined by a variety of analytical techniques, including X-ray diffraction, scanning electron microscopy, Mössbauer spectroscopy, and visible and near-IR reflectance spectroscopy. The yellow pigment was determined to be the ferric hydroxysulfate jarosite ($(H_3O,Na,K)Fe_3\times(SO_4)_2(OH)_6$) by positions of the (101), (003), (021), (033), and (220) reflections in the XRD powder pattern, 0.35 mm/s isomer shift and 1.19 mm/s quadrupole splitting for the Mössbauer doublet, and well-defined band minima at 430, 910, 1480, and 1840 nm in reflectivity spectra. The XRD, Mössbauer, and reflectivity data also indicate that phyllosilicates are not present and that jarosite is the predominant iron-bearing phase. As determined by SEM imaging, the jarosite has a cubic morphology and occurs as aggregates of individual crystals on exterior surfaces and in vesicles. Jarosite formation in HWMK24 likely proceeded by hydrothermal dissolution of silicate glass and minerals by sulfuric acid solutions (from volcanic SO_2 and H_2O gases) and subsequent precipitation, under oxidizing conditions, of relatively insoluble jarosite (K-rich) during or shortly after cone formation. The occurrence of band minima near 910 nm in spectral data for certain Martian regions (*e.g.*, East and West Oxia and Lunae Planum) might indicate the presence of jarosite formed during acid-sulfate weathering, possibly in conjunction with SO_2 gases release during Martian volcanism.

INTRODUCTION AND BACKGROUND

THE MINERALOGY of iron-bearing weathering products on the Martian surface is largely constrained by the Viking Lander geochemical and magnetic properties experiments and by earth-based telescopic and spacecraft spectral observations at visible and near-IR wavelengths. Elemental abundances provided by the Viking X-ray fluorescence experiment (CLARK *et al.*, 1982) constrain mineralogy to the extent that mineral assemblages must be consistent with bulk composition. The magnetic properties experiment implies that a 1–7 wt. % strongly magnetic component is present in Martian fines (HARGRAVES *et al.*, 1977, 1979). Mineralogies that have been suggested for this component include Fe-Ti spinels (magnetites and maghemites), pyrrhotite, feroxyhyte, and nanophase hematite (HARGRAVES *et al.*, 1979; BURNS, 1980, 1988; MOSKOWITZ and HARGRAVES, 1982; POSEY-DOWTY *et al.*, 1986; MORRIS *et al.*, 1989, 1990; COEY *et al.*, 1990). All these mineralogies are consistent with the elemental data for 1–7 wt. % magnetic component. Earth-based telescopic data of Martian bright regions (weathered Martian surface material) are characterized by an absorption edge extending from approximately 400 to 750 nm with inflections near 520 and 600 nm, a relative reflectivity maximum near 750 nm, and a shallow absorption band centered near 860 nm (*e.g.*, SINGER *et al.*, 1979; MCCORD *et al.*, 1982; SINGER, 1985; BELL *et al.*, 1990). Spectral data from the Phobos-2 spacecraft, which have significantly better spatial resolution than the telescopic data, confirm the presence of surface regions with a ~860 nm band minimum and also show other regions where the shallow band minimum is located as high as ~910 nm (MURCHIE *et al.*, 1993). It is not known whether the other spectral features indicated above also change position because the Phobos-2 data do not extend to wavelengths shorter than ~800 nm. Composite spectra for Martian bright regions with the 860 nm band have been published by SINGER *et al.* (1979) and MUSTARD and BELL (1994). What mineralogies are implied by the positions of these spectral features?

Analyses of both synthetic and naturally-occurring spectral analogues of Mars have been important for making specific mineralogical assignments for Martian spectral features. For synthetic samples, MORRIS *et al.* (1989) and MORRIS and LAUER (1990) showed that nanophase hematite particles (hematite particles having diameters less than ~10 nm) and subordinate amounts of well-crystalline hematite dispersed in spectrally-neutral matrix materials can reproduce spectral data for Martian bright regions where the shallow ~860 nm band is present. The ferric absorption edge results predominantly from absorption by nanophase hematite and the shallow 860 nm band, 750 nm relative reflectivity maximum, and 600 and 520 nm inflections are all features traceable to well-crystalline hematite. Identification of well-crystalline hematite on Mars is unequivocal to the extent that hematite spectral features are unique. Nanophase hematite, however, does not have diagnostic spectral features, so its presence is consistent with (but not required by) Martian spectral data. Other nanophase ferric oxides may be present instead.

For naturally-occurring material, samples that contain palagonite (weathering product of mafic volcanic glass) have received considerable attention, in part because palagonitic samples are generally reasonable Martian spectral analogues (EVANS and ADAMS, 1979, 1980; BERKLEY and DRAKE, 1981; ALLEN *et al.* 1981; SINGER, 1982; MORRIS *et al.*, 1990, 1993; BELL *et al.*, 1993; GOLDEN *et al.*, 1993) and in part because processes favorable to palagonite formation are considered to be active on Mars, both now and in times past. These processes include hydrothermal alteration induced by hot, oxidizing fluids and/or gases mobilized by impact events, volcanism, or geothermal gradients (NEWSOM, 1980; ALLEN *et al.*, 1982; BELL *et al.*, 1993; MORRIS *et al.*, 1995), subpermafrost magmatic intrusion (SODERBLOM and WENNER, 1978; ALLEN *et al.*, 1981), subaerial extrusion above the permafrost layer (BERKLEY and DRAKE, 1981), and static gas-glass weathering (GOODING and KEIL, 1978). Samples of palagonitic tephra collected from the same location on the Puu Nene cinder cone on Mauna Kea Volcano, Hawaii, are among the best spectral analogues and have been studied in detail as sample Hawaii-34 (EVANS and ADAMS, 1979, 1980), VOLO2A (SINGER, 1982), and PN-9 (MORRIS *et al.*, 1993).

EVANS and ADAMS (1979, 1980) reported spectral data for Hawaii-34 and showed that its nearly constant reflectivity in the near-IR and strong absorption edge through the visible were very similar to Martian bright regions spectra. Based on compositional data and the absence of coherent scattering in powder X-ray diffraction patterns, they concluded that the palagonitic tephra was primarily iron allophane (an iron-rich amorphous hydrated aluminosilicate gel) with iron present as Fe^{3+}. The absence of well-defined Fe^{3+} crystal-field bands was attributed to the lack of long-range crystalline ordering as observed in the XRD data. SINGER (1982), who obtained similar spectral, compositional, and XRD data for sample VOLO2A, also argued that the low intensity of Fe^{3+} crystal-field bands apparently results from the lack of crystalline order. Although not explicitly stated, the discussion of the XRD data in these studies implies a model where iron is present as isolated paramagnetic Fe^{3+} ions at sites having highly-variable size and distortion. In addition to reproducing the spectral and XRD data for Hawaii-34 and VOLO2A for sample PN-9, MORRIS *et al.* (1993) reported observations made with transmission electron microscopy and data from spectral and Mössbauer measurements of size and magnetic separates which were subjected to chemical extraction procedures to determine the siting of Fe^{3+}. These data showed that the Fe^{3+} is present as discrete particles having dimensions less than ~20 nm (*i.e.*, nanophase ferric oxide) embedded in a hydrated aluminosilicate matrix and that these particles are responsible for the strong ferric absorption edge. The nanophase ferric oxide particles were not detected in XRD data because they are too small to coherently scatter X-rays. The XRD mineralogy of these particles is uncertain, but electron diffraction indicates the mineralogy of these nanophase particles is nanophase hematite and/or ferrihydrite (MORRIS *et al.*, 1993).

The iron mineralogy of both synthetic and palagonitic samples provides a strong basis for inferring the presence of well-crystalline hematite for Martian bright regions which are characterized by spectral data with a ~860 nm band minimum. Other ferric mineralogies are implied for regions with band minima near 910 nm. As noted by MURCHIE *et al.* (1993), potential mineralogies include the ferric hydroxysulfate jarosite, which has the general composition $((H_3O,Na,K)Fe_3(SO_4)_2(OH)_6)$. Iron-bearing sulfates like jarosite are possible mineralogies because sulfur is a chemically important component of Martian surface materials (~7% SO_3 (CLARK *et al.*, 1982)) and because the phase will probably precipitate under Martian surface conditions (BURNS, 1988, 1993; BURNS and FISHER, 1990a,b). Are jarosites reasonable assignments when all spectral features are considered; for example, are the jarosite 1480 and 1840 nm bands present? Are there spectral analogues among altered tephra samples from Mauna Kea that contain sulfates and have shallow band minima near 910 nm? The only ferric-bearing alteration products observed to date are nanophase ferric oxide, hematite, and titanohematite (*e.g.*, BELL *et al.*, 1993; GOLDEN *et al.*, 1993; MORRIS *et al.*, 1993).

In this paper, we report an occurrence of oxidatively-altered tephra (sample HWMK24) from the summit of Mauna Kea in which the dominant ferric mineralogy is jarosite. The implication of its spectral data and formation conditions are discussed with respect to the potential occurrence of the phase on Mars. In addition to spectral data, the jarosite-bearing tephra was characterized by powder X-ray diffraction analysis, Mössbauer spectroscopy, and photomicrographs obtained by scanning electron microscopy. For purposes of comparison, corresponding data were also obtained for four other terrestrial occurrences of jarosite.

SAMPLES AND METHODS

Samples

HWMK24. Sample is yellow-colored tephra collected from a road cut on the northwest flanks of the unnamed summit cone of Mauna Kea Volcano upon which the Japan National Subaru Telescope is located. The cone is part of the Laupahoehoe Volcanics (hawaiitic composition) which form the magmatically evolved cap of Mauna Kea (WOLFE *et al.,* 1996). The tephra occurs as a layer ~50 cm wide and ~5 cm high. The sample was later fractionated in the laboratory by dry sieving with a 1 mm sieve. A <5 μm separate was obtained from part of the <1 mm fraction by ultrasonic dispersion and water sedimentation. Several tephra particles 1–4 cm in diameter were ground to fine powders (<90 μm).

LNVJAR1. Sample is a fine powder (<90 μm) of a hand sample of massive jarosite from Luning, NV (obtained from Wards Natural Science Establishment).

GCJAR1. Sample is a fine powder (<90 μm) of a hand sample of jarositic material from Grant County, NM. Location and description of the sample has been documented by NORTHROP and WHITNEY (1987).

TT#12 and TT#35. Samples are <1 mm sieve fractions of jarositic soil that have been previously examined by TOWNSEND (1987). They are his samples 12F and 35F and were collected near Goldfield, NV.

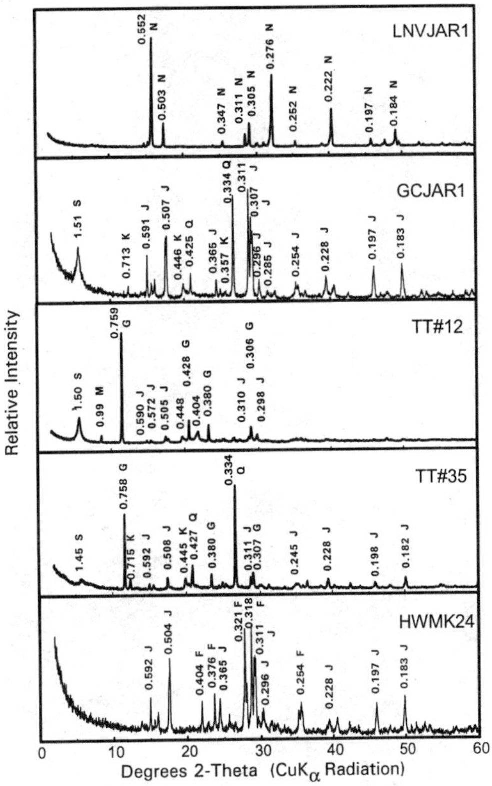

FIG. 1. X-ray diffraction patterns for jarosite-containing samples LNVJAR1 (Luning, NV), GCJAR1 (Grant Co., NM), TT#12 and TT#35 (Goldfield, NV), and HWMK24 (<1 mm; Mauna Kea, HW). All **d**-spacings are in nm. K = kaolinite, S = smectite, Q = quartz, G = gypsum, M = mica, J = jarosite, N = natrojarosite, F = plagioclase feldspar.

Methods

A Cary-14 spectrophotometer configured with a 14 cm diameter integrating sphere was used to obtain visible and near-IR reflectance spectra. Ranger Mössbauer spectrometers using $^{57}Co(Rh)$ sources were used to obtain iron Mössbauer spectra. Mössbauer spectra were fit to theoretical line shapes by an in-house computer program. A Scintag XDS 2000 X-ray diffractometer using CuK radiation was employed to obtain powder X-ray diffraction patterns. A JEOL JSM-35CF scanning electron microscope equipped with a PGT EDS was employed for SEM imaging and chemical analysis. More detailed information on analytical procedures is given by MORRIS *et al.* (1989, 1990), GOLDEN *et al.* (1993), and BELL *et al.* (1993).

RESULTS AND DISCUSSION

X-ray diffraction

Throughout this paper, we will use the name jarosite to refer to the general composition $(H_3O,Na,K)Fe_3(SO_4)_2(OH)_6$, although the name actually refers to the K endmember. The H_3O and Na endmembers are hydronium jarosite and natrojarosite, respectively. Jarosite was identified in all five samples (Fig. 1), although the apparent quantities in the samples varies significantly as suggested by XRD peak intensities. LNVJAR1 was monomineralic natrojarosite (Fig. 1). Primary peaks used to identify natrojarosite were 0.552 nm (003) and 0.276 nm (006) peaks. The high-intensity diffraction peaks for the (003) and (006) indicate a highly-oriented sample along the **c**-axis.

In addition to jarosite (K-rich as indicated by SEM EDS analysis), GCJAR1 contained smectite, quartz, and minor amounts of kaolinite. The primary XRD **d**-spacings used to identify (K-rich) jarosite were 0.592 nm (101), 0.572 nm (003), 0.311 nm (021), 0.197 nm (033), and 0.183 nm (220) peaks. The sharp, narrow diffraction peaks of jarosite in GCJAR1 indicate a high degree of crystallinity. TT#12 and TT#35 contained minor quantities of jarosite as indicated by weak diffraction peaks (Fig. 1). TT#12 sample contained pri-

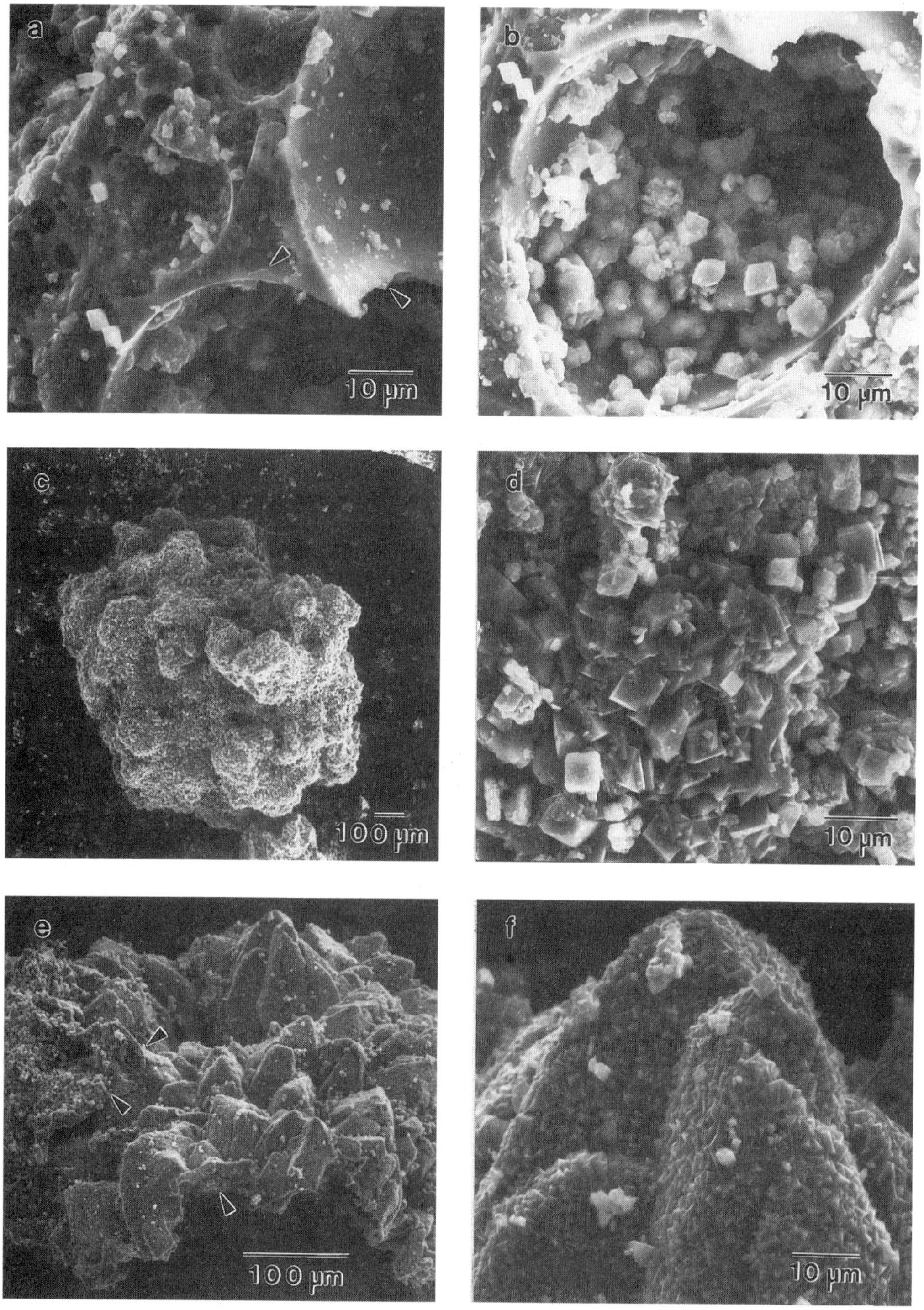

FIG. 2. Scanning electron micrographs of jarosite from the Mauna Kea tephra. (a) dissolution of intervesicular wall and dissolution pits in glass (arrows); (b) jarosite crystals precipitated in glass vesicles; (c) jarosite replica of an unknown cluster of cubic crystals; (d) jarosite cubes enlarged from (c); (e) jarosite replica of an unknown cluster of prismatic crystals (arrows show hollow interior of replica); (f) jarosite cubes enlarged from (e).

marily gypsum and lesser amounts of smectite and mica. Gypsum and quartz dominated the mineralogy of the TT#35. This sample also contained minor quantities of smectite and kaolinite. Stronger peak intensities for jarosite in TT#35 suggested more jarosite in this sample as compared to TT#12.

Jarosite (K-rich as indicated by SEM EDS analysis) and plagioclase feldspar (Ca-rich) were the major phases identified in the Mauna Kea tephra (sample HWMK24, <1 mm). No phyllosilicates were detected. The XRD pattern for jarosite in HWMK24 is very similar to the pattern obtained for jarosite in GCJAR1. The sharp, narrow diffraction peaks of jarosite in sample HWMK24 indicate a high degree of crystallinity, which is supported by SEM morphological analysis discussed next.

Scanning electron microscopy

Morphological characteristics of jarosite in the Mauna Kea tephra are photogenic and are nearly identical with those reported in soils (*e.g.,* DIXON *et al.,* 1982; DONER and LYNN, 1989). The jarosite from GCJAR1 has a similar morphology to the jarosite in the tephra. Scanning electron photomicrographs of jarosite from HWMK24 are shown in Fig. 2. The jarosite occurs ubiquitously as aggregates of individual cubic crystals that are 2–5 μm in size. The aggregates normally are present as small clusters on exterior surfaces or within glass vesicles (Fig. 2a,b). Two unrepresentative aggregates of jarosite crystals were noted (Fig. 2c-f). Thin layers of cubic jarosite crystals appear to have replicated the form of larger crystals. In Fig. 2e, the broken edge indicates the jarosite replication has separated from the larger crystals, and the overturned rind (arrows) shows the hollow interior of the replica.

The occurrence of jarosite crystals in glass vesicles (Fig. 2a,b) suggests precipitation from acid sulfate solutions. Acid sulfate weathering conditions which produce sulfuric acid have apparently facilitated dissolution of intervesicular walls (Fig. 2a) and formation of dissolution pits in glass and plagioclase feldspar (not shown). Energy dispersive spectroscopy showed that the jarosite is K-rich (data not shown).

Mössbauer spectroscopy

The Mössbauer spectra of all five samples are shown in Fig. 3, and the jarosite Mössbauer parameters are compiled in Table 1. The average values of IS and QS for the jarosite Fe^{3+} doublet, which are 0.35 $\pm$ 0.02 mm/s and 1.19 $\pm$ 0.02 mm/s at room temperature, are within the range published in the literature (STEVENS *et al.,* 1983). Jarosite is the only iron-bearing phases detected for GCJAR1, TT#12, and TT#35. For HWMK24, jarosite is virtually the only iron-bearing phase detected for the <5 μm and <1 mm size fractions. Minor amounts of other iron-bearing mineralogies (especially for the <1 mm fraction) are indicated by low-intensity sextets (probably hematite and (titano)magnetite) and low-intensity ferrous doublets (probably glass and olivine).

Mössbauer spectra of powders of 1–4 cm tephra particles from HWMK24 were highly variable. The spectrum of HWMK24-R3 (Fig. 3), which is a powder of a coherent, dense tephra particle, is dominated by ferrous mineralogies. Its spectrum is very similar to that published by MORRIS *et al.* (1993) for hawaiitic basalt HAW-16, for which the major iron-bearing phases are olivine and titanomagnetite. This type of material is probably the ferrous-bearing component in the spectrum of HWMK24 (<1 mm). Mössbauer spectra for powders of friable, vesicular (porous) tephra particles were inter-

Table 1. Phase identification based on XRD, values of Mössbauer parameters (IS and QS) for jarosite doublets, and positions of band minima for ferric ${}^6A_1 \rightarrow {}^4T_{1g}$ and ${}^6A_1 \rightarrow ({}^4E,{}^4A)$ electronic transitions. Phases are listed in order of apparent decreasing abundance based on XRD peak intensities.

Sample	Phases Based on XRD	IS (mm/s)	QS (mm/s)	${}^4T_{1g}$ (nm)	${}^4E,{}^4A$ (nm)
LNVJAR1	jarosite (natrojarosite)	0.36	1.19	920	434
GCJAR1	jarosite, smectite, quartz, kaolinite	0.33	1.20	920	432
TT#12	gypsum, smectite, mica, jarosite	0.33	1.22	910	432
TT#35	quartz, gypsum, jarosite, kaolinite, smectite	0.33	1.20	918	432
HWMK24	<1 mm: jarosite, plagioclase	0.38	1.19	910	432
HWMK24	<5 μm: jarosite	0.36	1.15	910	432
Uncertainty		0.01	0.01	8	5

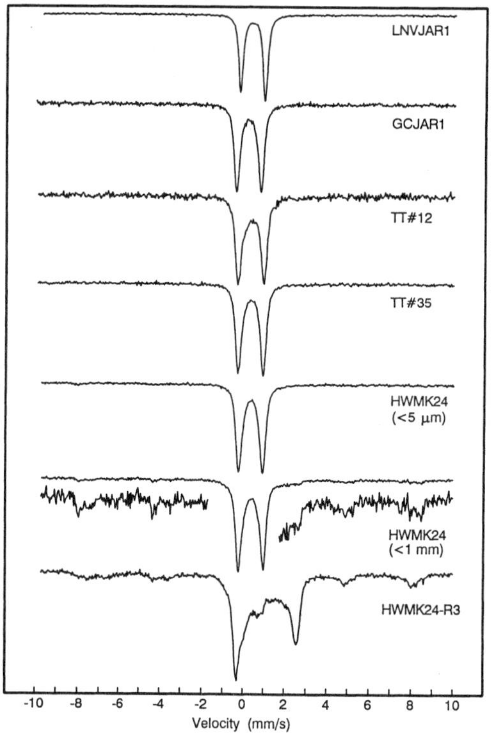

FIG. 3. Mössbauer spectra at 298 K. The spectra of LNVJAR1, GCJAR1, TT#12, TT#35, and HWMK24 (<1 mm and <5 μm) are characterized by a doublet resulting from Fe^{3+} in jarosite. The spectrum of HWMK24-R3, which is a powder of a dense, coherent 1–4 cm diameter tephra particle, is complex but contains a doublet resulting from Fe^{2+} in olivine and sextets resulting from Fe^{2+} and Fe^{3+} in (titano)magnetite.

mediate to those for HWMK24 (<1 mm) and HWMK24-R3. In keeping with SEM data, these results imply that the jarosite is formed by penetration of sulfur-bearing gases and/or fluids or gases through porous and glassy tephra under oxidizing conditions. Tephra particles like HWMK24-R3 are relatively impervious and are not altered to jarosite. This view is supported by observations made during pulverization of the large tephra particles. Coherent particles like HWMK24-R3 had a yellow (jarosite) rind or coating and black (hawaiitic) interiors, and friable particles were yellow with scattered black islands throughout their volume.

Diffuse reflectance spectroscopy

Reflectivity spectra are shown in Fig. 4; the spectrum of LNVJAR1 is equivalent to jarosite spectra published previously (*e.g.*, HUNT *et al.*, 1971; HUNT and ASHLEY, 1979; CLARK *et al.*, 1990). All spectral features at wavelengths shorter than 1100 nm result from ferric iron, which has four electronic transitions in this spectral region. Three are single-electron transitions from the $^6A_1$ ground state of ferric iron to the (^4E,^4A), $^4T_{2g}$, and $^4T_{1g}$ levels of its first (quartet) excited state and one is a pair transition of two electrons from two magnetically coupled ferric ions from $^6A_1$ to $^4T_{1g}$ (SHERMAN *et al.*, 1982; MORRIS *et al.*, 1985; SHERMAN and WAITE, 1985). The well-defined band minima near 430 and 920 nm are assigned to the (^4E,^4A) and $^4T_{1g}$ transitions, respectively. The region between the inflection near 600 nm and the relative reflectivity maximum near 720 nm is a manifestation of the unresolved $^4T_{2g}$ transition, and the region between the two inflections near 480 and 600 nm is a manifestation of the pair transition.

As shown best by the spectrum of LNVJAR1, jarosite also has two sharp bands near 1480 and

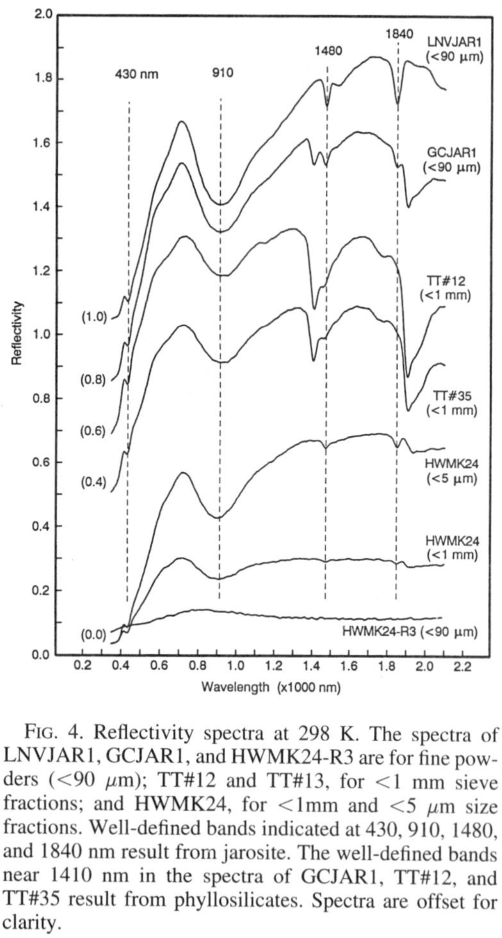

FIG. 4. Reflectivity spectra at 298 K. The spectra of LNVJAR1, GCJAR1, and HWMK24-R3 are for fine powders (<90 μm); TT#12 and TT#13, for <1 mm sieve fractions; and HWMK24, for <1mm and <5 μm size fractions. Well-defined bands indicated at 430, 910, 1480, and 1840 nm result from jarosite. The well-defined bands near 1410 nm in the spectra of GCJAR1, TT#12, and TT#35 result from phyllosilicates. Spectra are offset for clarity.

1840 nm. These two features are also reported and discussed by HUNT and ASHLEY (1979) and CLARK *et al.* (1990), who attribute the 1480 nm band to first overtones of OH stretching fundamental vibrations ($2\upsilon_{OH}$). Assignment of the 1840 nm band to $\upsilon_{FeOH} + \upsilon_{OH}$ is consistent with its calculated position (1830 nm) and with assignment of the alunite ($KAl_3(SO_4)_2(OH)_6$) 1770 nm band to $\upsilon_{AlOH} + \upsilon_{OH}$ (HUNT and ASHLEY, 1979). In any event, these two bands are at distinctly different positions than those for common phyllosilicate minerals, including nontronite and montmorillonite (*e.g.*, CLARK *et al.*, 1990). A good example of this difference is the spectrum of GCJAR1 (Fig. 4), which has OH spectral features associated with both jarosite (1480 nm) and phyllosilicate (1410 nm). The spectra of HWMK24, like LNVJAR1, have weak bands near 1480 nm but no detectable feature near 1410 nm, which implies the presence of jarosite and the absence of phyllosilicates.

The reflectivity spectrum of the powder (<90 μm) of tephra particle HWMK24-R3 is relatively featureless and has generally low reflectivity (Fig. 4). This spectrum can be attributed to opaque (titano)magnetites observed in the Mössbauer spectra. The reduced spectral contrast of the absorption bands in HWMK24 (<1 mm) as compared to HWMK24 (<5 μm) likely results from a higher proportion of dark particles (like HWMK24-R3) in the former compared to the latter.

Jarosite formation processes on Mauna Kea

WOLFE *et al.* (1996) report an occurrence of the hyroxysulfate mineral alunite in a sample from another Mauna Kea cinder cone (Puu Poliahu) and attribute its formation to alteration of glass and phenocrysts by hot, sulfur-bearing volcanic gases percolating up through the cone during or soon after its formation. Because its last eruption was ~4 ka before present (WOLFE *et al.*, 1996), there are no volcanic gas measurements available for Mauna Kea. However, volcanic gases from nearby Kilauea Volcano are rich in sulfur. GERLACH (1993) reports that gas compositions from the east rift zone eruption (January, 1983) average 14 and 80 mole % SO_2 and H_2O, respectively. Thus, jarosite formation in HWMK24 likely proceeded by hydrothermal dissolution of silicate glass and minerals by sulfuric acid solutions and subsequent precipitation, under oxidizing conditions, of relatively insoluble jarosite (K-rich). K and Na, whose concentrations are typically ~2 and ~4% as K_2O and Na_2O in hawaiite (*e.g.*, WEST *et al.*, 1988; WOLFE *et al.*, 1996), are derived from the glass and minerals. The formation of jarosite is generally expected at pH <3.5 (VAN BREEMAN, 1982; CARSON *et al.*, 1982; DIXON *et al.*, 1982). This process is similar to sulfuricization as described by CARSON *et al.* (1982) except that the source of sulfur is volcanic gases and not sulfide minerals. Sulfides are not reported in petrographic studies of hawaiitic lavas (*e.g.*, WEST *et al.*, 1988; WOLFE *et al.*, 1996). We know of no other occurrences of jarosite in tephra, and the Mauna Kea occurrence may be the first reported occurrence of jarosite from a sulfur source other than sulfides.

IMPLICATIONS FOR MARTIAN SURFACE MINERALOGY AND PROCESSES

Our results show that jarosite (in addition to nanophase ferric oxide and well-crystalline hematite and titanohematite) occurs as an iron-bearing, oxidative alteration product of Mauna Kea tephra, although it does not appear to be widely distributed. The alteration is pervasive and occurs throughout the volume of porus tephra particles and as a rind on coherent, dense tephra particles. The cubic crystal morphology suggests the jarosite (K-rich) precipitated directly from solution; XRD data show the phase is well-crystalline. The jarosite likely formed as a hydrothermal alteration product of glassy tephra in the cinder cone when hot, sulfur-bearing volcanic gases percolated up through the tephra during or soon after cone formation. What are the implications of these results for Mars?

MURCHIE *et al.* (1993) discuss the ferric mineralogy of Martian bright soils obtained from the ISM near-infrared imaging spectrometer on the Phobos-2 spacecraft. Some of the spectra (*e.g.*, Amazonis Planitia, Pavonis Base, and Juventae Dorsa) have band minima near 850 nm (Fig. 5) and are consistent with a mineral assemblage containing well-crystalline hematite (*e.g.*, MORRIS *et al.*, 1989; BELL *et al.*, 1990; MORRIS and LAUER, 1990; MORRIS *et al.*, 1993). Other spectra (*e.g.*, East and West Oxia and Lunae Planum) have band minima near 910 nm and imply a ferric mineralogy other than hematite. Jarosite is one of the phases suggested by MURCHIE *et al.* (1993) as an explanation for the 910 nm band. As shown in Fig. 5, this suggestion is not inconsistent with the spectrum of HWMK24 (Fig. 5) because the jarosite 910 nm band minimum in HWMK24 corresponds to the feature in East and West Oxia and Lunae Planum.

If the Martian feature at 910 nm results from jarosite, then there are potentially jarosite features at 430, 1480, and 1840 nm. None are apparent in available reflectance spectra of Martian bright

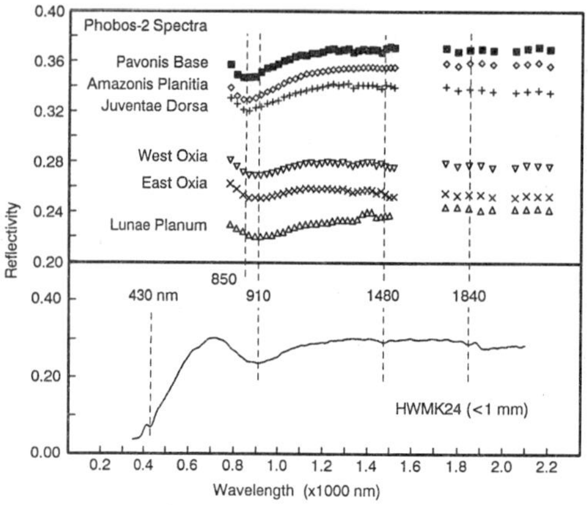

FIG. 5. Comparison of reflectivity spectra for HWMK24 (<1 mm jarositic tephra) and Martian bright regions acquired by the Phobos-2 spacecraft (MURCHIE *et al.,* 1993; MURCHIE, unpublished spectra). Band minima near 850 nm in spectra of Amazonis Planitia, Pavonis Base, and Juventae Dorsa are consistent with the presence of well-crystalline hematite. Band minima near 910 in the spectra of East and West Oxia and Lunae Planum are consistent with the presence of jarosite. The spectrum of HWMK24 results from K-rich jarosite.

regions. This implies (1) that jarosite is not an optically important component in Martian surface materials and some other, presumably ferric, mineralogy is present, or (2) that jarosite is present but these spectral features are not observed for some reason. In the first case, other ferric-mineralogies that have band minima at ~910 nm include ferrihydrite (~$Fe_5HO_8 \cdot 4H_2O$), schwertmannite (~$Fe_8O_8(OH)_6SO_4$), goethite (α-FeOOH), lepidocrocite (γ-FeOOH), nontronite ($Na_{0.33}Fe_2(Si,Al)_4O_{10} \cdot nH_2O$), and maghemite ($\gamma$-$Fe_2O_3$) (*e.g.,* SHERMAN *et al.,* 1982; SINGER, 1982; MORRIS *et al.,* 1985; MORRIS, unpublished results).

In the second case, the 1480 and 1840 nm jarosite bands might not be apparent in the spectra of East and West Oxia and Lunae Planum because they are too weak to be detected at the spectral resolution and precision of the Phobos-2 data (MURCHIE *et al.,* 1993). That this might be the case is shown by the spectrum of HWMK24 for which the 1480 and 1840 nm features are significantly less intense than the 910 nm feature. The 430 nm jarosite band is more likely to be detected, but the Phobos-2 data do not extend to wavelengths below ~750 nm, and ground-based observations near this wavelength are frequently contaminated by strong solar and stellar emission features (*e.g.,* BELL, 1992). An additional complication is that, even if 430 and 920 nm bands are present in Martian spectral data, assignment to jarosite is equivocal because nontronite (an Fe-smectite) has ferric bands at nearly the same positions (*e.g.,* SINGER, 1982).

In summary, assignment of the Martian 910 nm band to jarosite on the basis of spectral data is equivocal until the presence of bands near 430, 1480, and 1840 nm can be demonstrated. If the results of elemental analyses for Martian soil and SNC meteorites, which indicate low abundances of K (CLARK *et al.,* 1982; MCSWEEN, 1985), are valid for surface regions with the band at 910 nm, hydronium- and/or sodium-rich jarosite ($(H_3O,Na)Fe_3(SO_4)_2(OH)_6$) would be present (BURNS, 1988). Alternatively, the presence of jarosite could imply surface regions of Mars that are rich in K and Na.

If jarosite is present on Mars, what are its formation processes? Burns and coworkers (BURNS, 1988, 1993; BURNS and FISHER, 1990a,b) have suggested that massive and disseminated iron sulfide mineralization occurs near the Martian surface in association with komatiitic volcanism. Jarosite could result from oxidative weathering of the sulfides (pyrrhotite and pentlandite) disseminated in komatiitic basalts (BURNS, 1988, 1993). Oxidative weathering of iron sulfides produced the jarosite in our samples GCJAR1, TT#12, and TT#35 (NORTHROP and WHITNEY, 1987; TOWNSEND, 1987). This study documents an additional formation process for jarosite on Mars which does not involve oxidative alteration of iron sulfides (BURNS, 1987, 1988; BURNS and FISHER, 1990a,b). Namely, acid sulfate weathering of iron-bearing glass and minerals under oxidizing conditions and for which the source of sulfur is volcanic gases and/or fluids.

Acknowledgements—We dedicate this manuscript on rogerburnsite (jarosite) to the late Roger G. Burns who was a strong advocate for its presence on the Martian surface. Roger is greatly missed, both as an exemplary person and as a distinguished scientist. This paper has benefited from discussions with Ed Wolfe, Terry Gerlach, and Del Fanning and the reviews of Scott Murchie and Phil Ihinger. We thank Scott Murchie for revised Phobos-2 IMS spectra. This work was supported by NASA's Planetary Materials and Geochemistry Program.

REFERENCES

ALLEN C. C., GOODING J. L., JERCINOVIC M. and KEIL K. (1981) Altered basaltic glasses: A terrestrial analog to the soil of Mars. *Icarus* **45,** 347–369.

ALLEN C. C., GOODING J. L. and KEIL K. (1982) Hydrothermally altered impact melt rock and breccia: Contributions to the soil of Mars. *J. Geophys. Res.* **87,** 10083–10101.

BELL III J. F. (1992) Charge-coupled device imaging

spectroscopy of Mars. 2. Results and implications for Martian ferric mineralogy. *Icarus* **100,** 575–597.

BELL III J. F., MCCORD T. B. and OWENSBY P. D. (1990) Observational evidence of crystalline iron oxides on Mars. *J. Geophys. Res.* **95,** 14,447–14,461.

BELL III J. F., MORRIS R. V. and ADAMS J. B. (1993) Thermally altered palagonitic tephra: A spectral and process analog to the soil and dust of Mars. *J. Geophys. Res.* **98,** 3373–3385.

BERKLEY J. L. and DRAKE M. J. (1981) Weathering on Mars: Antarctic analogue studies. *Icarus* **45,** 231–249.

BURNS R. G. (1980) Does feroxyhyte occur on the surface of Mars? *Nature* **185,** 647.

BURNS R. G. (1987) Ferric sulfates on Mars. *Proc. Lunar Planet. Sci. Conf. 17th,* E570-E574.

BURNS R. G. (1988) Gossans on Mars. *Proc. Lunar Planet. Sci. Conf. 18th,* 713–721.

BURNS R. G. (1993) Rates and mechanisms of chemical weathering of ferromagnesium silicate minerals on Mars. *Geochim. Cosmochim. Acta* **57,** 4555–4574.

BURNS R. G. and FISHER D. S. (1990a) Iron-sulfur mineralogy of Mars: Magmatic evolution and chemical weathering products. *J. Geophys. Res.* **95,** 14415–14421.

BURNS R. G. and FISHER D. S. (1990b) Evolution of sulfide mineralization on Mars. *J. Geophys. Res.* **95,** 14169–14173.

CARSON C. D., FANNING D. S. and DIXON J. B. (1982) Alfisols and Utisols with acid sulfate weathering features in Texas, p. 127–146. In *Acid Sulfate Weathering* (eds. J. A. KITTRICK et al.), Spec. Pub. 10, SSSA, Madison, WI.

CLARK B. C., BAIRD A. K., WELDON R. J., TSUSAKI D. M., SCHNABEL L. and CANDELARIA M. P. (1982) Chemical composition of martian fines. *J. Geophys. Res.* **87,** 10059–10067.

CLARK R. N., KING T. V. V., KEJWA M. and SWAYZE G. A. (1990) High spectral resolution reflectance spectroscopy of minerals. *J. Geophys. Res.* **95,** 12,653–12,680.

COEY J. M. D., MORUP S., MADSEN M. B. and KNUDSEN J. M. (1990) Titanomaghemite in magnetic soils on Earth and Mars. *J. Geophys. Res.* **95,** 14423–14425.

DIXON J. B., HOSSNER L R., SENKAYI A. L. and EGASHIRA K. (1982) Mineralogical properties of lignite overburden as they relate to mine spoil reclamation, p. 169–191. In *Acid Sulfate Weathering* (eds. J. A. KITTRICK *et al.*), Spec. Pub. 10, SSSA, Madison, WI.

DONER H. E. and LYNN W. C. (1989) Carbonate, halide, sulfate, and sulfide minerals, p. 279–330. In *Minerals in Soil Environments,* 2nd edition, (eds. J. B. DIXON and S. B. WEED) SSSA, Madison, WI.

EVANS D. L. and ADAMS J. B. (1979) Comparison of Viking Lander multispectral images and laboratory reflectance spectra of terrestrial samples. *Proc. Lunar Planet. Sci. Conf. 10th,* 1829–1834.

EVANS D. L. and ADAMS J. B. (1980) Amorphous gels as possible analogs to martian weathering products. *Proc. Lunar Planet. Sci. Conf. 11th,* 757–763.

GERLACH T. M. (1993) Oxygen buffering of Kilauea volcanic gases and the oxygen fugacity of Kilauea basalt. *Geochim. Cosmochim. Acta* **57,** 795–814.

GOLDEN D. C., MORRIS R. V., MING D. W., LAUER JR. H. V. and YANG S. R. (1993) Mineralogy of three slightly palagonitized tephra samples from the summit of Mauna Kea, Hawaii. *J. Geophys. Res.* **98,** 3401–3411.

GOODING J. L. and K. KEIL (1978) Alteration of glass as a possible source of clay minerals on Mars. *Geophys. Res. Lett.* **5,** 727–730.

HARGRAVES R. B., COLLINSON D. W., ARVIDSON R. V. and SPITZER C. R. (1977) The Viking magnetic properties experiment: Primary mission results. *J. Geophys. Res.* **82,** 4547–4558.

HARGRAVES R. B., COLLINSON D. W., ARVIDSON R. E. and CATES P. M. (1979) Viking magnetic properties experiment: Extended mission results. *J. Geophys. Res.* **84,** 8379–8384.

HUNT G. R., SALISBURY J. W. and LENHOFF C. J. (1971) Visible and near-infrared spectra of minerals and rocks: IV. Sulfides and sulfates. *Modern Geology* **3,** 1–14.

HUNT G. R. and ASHLEY R. P. (1979) Spectra of altered rocks in the visible and near infrared. *Econ. Geol.* **74,** 1613–1629.

MCSWEEN H. Y., Jr. (1985) SNC meteorites: Clues to Martian petrolgic evolution? *Rev. Geophys.* **23,** 391–416.

MCCORD T. B., SINGER R. B., HAWKE B. R., ADAMS J. B., EVANS D. L., HEAD J. W., MOUGINIS-MARK P. J., PIETERS C. M., HUGENIN R. L. and ZISK S. H. (1982) Mars: Definition and characterization of global surface units with emphasis on composition. *J. Geophys. Res.* **87,** 10129–10148.

MORRIS R. V., AGRESTI D. G. LAUER Jr. H. V., NEWCOMB J. A., SHELFER T. D. and MURALI A. V. (1989) Evidence for pigmentary hematite on Mars based on optical, magnetic, and Mössbauer studies of superparamagnetic (nanocrystalline) hematite. *J. Geophys. Res.* **94,** 2760–2778.

MORRIS R. V., GOLDEN D. C., BELL III J. F., LAUER JR. H. V. and ADAMS J. B. (1993) Pigmenting agents in Martian soils: Inferences from spectral, Mössbauer, and magnetic properties of nanophase and other iron oxides in Hawaiian palagonitic soil PN-9. *Geochim. Cosmochim. Acta* **57,** 4597–4609.

MORRIS R. V., GOLDEN D. C., BELL III J. F. and LAUER JR. H. V. (1995) Hematite, pyroxene, and phyllosilicates on Mars: Implications from oxidized impact melt rocks from Manicouagan Crater, Quebec, Canada. *J. Geophys. Res.* **100,** 5319–5328.

MORRIS R. V., GOODING J. L., LAUER JR. H. V. and SINGER R. B. (1990) Origins of Marslike spectral and magnetic properties of a Hawaiian palagonitic soil. *J. Geophys. Res.* **95,** 14427–14434.

MORRIS R. V. and LAUER JR. H. V. (1990) Matrix effects for reflectivity spectra of dispersed nanophase (superparamagnetic) hematite with application to Martian spectral data. *J. Geophys. Res.* **95,** 5101–5109.

MORRIS R. V., LAUER JR. H. V., LAWSON C. A., GIBSON JR. E. K., NACE G. A. and STEWART C. (1985) Spectral and other physicochemical properties of submicron powders of hematite (-Fe_2O_3), maghemite (-Fe_2O_3), magnetite (Fe_3O_4), goethite (-FeOOH), and lepidocrocite (-FeOOH). *J. Geophys. Res.* **90,** 3126–3144.

MOSKOWITZ B. M. and HARGRAVES R. B. (1982) Magnetic changes accompanying the thermal decomposition of nontronite (in air) and its relevance to martian mineralogy. *J. Geophys. Res.* **87,** 10,115–10,128.

MURCHIE S., MUSTARD J., BISHOP J., HEAD J. and PIETERS C. (1993) Spatial variations in the spectral properties of bright regions on Mars. *Icarus* **105,** 454–468.

MUSTARD J. F. and BELL III J. F. (1994) New composite reflectance spectra of Mars from 0.4 to 3.14 m. *Geophys. Res. Lett.* **21,** 3353–3356.

NEWSOM H. E. (1980) Hydrothermal alteration of impact melt sheets with implications for Mars. *Icarus* **44,** 207–216.

NORTHROP H. R. and WHITNEY G. (1987) Clays and associated minerals in the alum mountain fossil geothermal system near Gila Hot Springs, Grant County, New Mexico, p. 1–12. In MUMPTON F. A. (ed.), Zeo-Trip '87, International Committee on Natural Zeolites, Brockport, NY.

POSEY-DOWTY J., MOSKOWITZ B., CRERAR D., HARGRAVES R., TANENBAUM L. and DOWTY E. (1986) Iron oxide and hydroxide precipitation from ferrous solutions and its relevance to martian surface mineralogy. *Icarus* **66,** 105–116.

SHERMAN D. M., BURNS R. G. and BURNS V. M. (1982) Spectral characterization of the iron oxides with application to the martian bright region mineralogy. *J. Geophys. Res.* **87,** 10169–10180.

SHERMAN D. M. and WAITE T. D.(1985) Electronic spectra of Fe^{3+} oxides and oxide hydroxides in the near IR to near UV. *Am. Mineral.* **70,** 1262–1269.

SINGER R. B. (1982) Spectral evidence for the mineralogy of high-albedo soils and dust on Mars. *J. Geophys. Res.* **87,** 10159–10168.

SINGER R. B. (1985) Spectroscopic observation of Mars. *Adv. Space Res.* **5,** 59–68.

SINGER R. B., MCCORD T. B., CLARK R. N., ADAMS J. B. and HUGUENIN R. L. (1979) Mars surface composition from reflectance spectroscopy: A summary. *J. Geophys. Res.* **84,** 8415–8425.

SODERBLOM L. A. and WENNER D. B. (1978) Possible fossil H_2O liquid-ice interfaces in the Martian crust. *Icarus* **34,** 622–637.

STEVENS J. H., POLLAK H., ZHE L., STEVENS V. E., WHITE R. M. and GIBSON J. L. (1983) *Mössbauer Handbook. Mineral: Data Handbook.* Mössbauer Effect Data Center, University of North Carolina, Ashville, NC.

TOWNSEND T. E. (1987) Discrimination of iron alteration minerals in visible and near-infrared reflectance data. *J. Geophys. Res.* **92,** 1441–1454.

VAN BREEMAN N. (1982) Genesis, morphology and classification of acid sulfate soils in coastal plains. p. 95–108. In *Acid Sulfate Weathering,* (eds. J. A. KITTRICK *et al.*) Spec. Pub. 10, SSSA, Madison, WI.

WEST H. B., GARCIA M. O., FREY F. A. and KENNEDY A. (1988) Nature and cause of compositional variation among the alkalic cap lavas of Mauna Kea Volcano, Hawaii. *Contrib. Mineral. Petro.* **100,** 383–397.

WOLFE E. W., WISE W. S. and DALRYMPLE G. B. (1996) *The Geology and Petrology of Mauna Kea Volcano, Hawaii: A Study of Postshield Volcanism,* U. S. Geological Survey Prof. Paper 1557, in press.

Mineral Spectroscopy: A Tribute to Roger G. Burns

Editors: M. D. Dyar, C. McCammon and M. W. Schaefer

Schwertmannite on Mars? Spectroscopic analyses of schwertmannite, its relationship to other ferric minerals, and its possible presence in the surface material on Mars.

JANICE L. BISHOP[1] and ENVER MURAD[2]

[1]DLR, Institute for Planetary Exploration, Rudower Chaussee 5, D-12489 Berlin, Germany
[2]Bayerisches Geologisches Landesamt, D-96033 Bamberg, Germany

Abstract—Schwertmannite, a hydroxylated ferric sulfate mineral, is presented as a possible component of the surface material on Mars. Schwertmannite could have formed on Mars through chemical weathering processes as acidic groundwater came in contact with the atmosphere. Visible and infrared reflectance spectra of schwertmannite are presented here and analyzed in the context of spectra of other ferric minerals and Mars soil analogs. Infrared transmittance spectra and Mössbauer spectra of schwertmannite are studied as well to facilitate interpretation of the reflectance spectra. The visible region spectra of schwertmannite are similar to those of jarosite and akaganéite in that they are dominated by a broad Fe^{3+} transition near 0.91 μm and a local reflectance maximum near 0.73 μm. Spectra of schwertmannite are relatively bright in the near infrared and exhibit broad bound and adsorbed water features near 1.45, 1.95 and 3 μm. The mid-infrared region spectra are significantly darker and include a water feature near 6.1 μm and several weak sulfate and OH features at longer wavelengths. The formation conditions and spectral features of schwertmannite are consistent with those of the ferric sulfate phase present in chemically-treated ferric sulfate-bearing montmorillonites, which have been examined as Mars soil analogs. Methods suitable for detecting the possible presence of schwertmannite in the martian surface soil are discussed, as well as implications for the surface geochemistry of Mars should schwertmannite occur there.

INTRODUCTION

Ferric sulfates and gossans have been proposed as components of the martian surface material due to their occurrence on Earth as weathering products under aqueous conditions that could have arisen on Mars under an earlier warm, wet climate (BURNS, 1987; BURNS, 1988). More recently BURNS (1994) suggested the formation of the hydroxylated ferric sulfate mineral schwertmannite in equatorial regions of Mars, where acidic permafrost melts and is oxidized by the martian atmosphere. Schwertmannite is often associated with ferrihydrite (BIGHAM *et al.*, 1992; SCHWERTMANN *et al.*, 1995), a ferric oxyhydroxide that has been proposed as a component of the soil on Mars (BISHOP *et al.*, 1993; BISHOP and PIETERS, 1995). BURNS (1994) described schwertmannite formation under conditions of laterization and ferric oxide deposition in a warm and humid environment depleted in dissolved K and Al, followed by a period of aridity. This pattern has been observed in regions of Australia and was proposed by BURNS (1994) for the equatorial regions on Mars. It is consistent with the development of groundwater under ancient martian climatic conditions, from martian crustal materials thought to be low in K and Al.

Many areas in the martian bright regions, such as the Tharsis and Amazonis plateau plains, exhibit weak visible ferric absorption features consistent with those of hematite (*e.g.*, MCCORD *et al.*, 1982; BELL *et al.*, 1990; MURCHIE *et al.*, 1993). Spectral analyses of hematites with variable particle sizes indicate that Mars-like spectral signatures can be achieved from combinations of crystalline and nanophase hematite, but neither alone (MORRIS *et al.*, 1989). Initial spectral mixing experiments indicate that a maximum of 1 wt.% bulk, crystalline hematite should be present in the martian soil in order to give the observed spectral properties for many martian bright regions (SINGER, 1982). Additional mixing experiments by MORRIS *et al.* (1989) and MORRIS and LAUER (1990) using fine-grained ("type S + D", approximately <2.0 μm particle size) hematite that exhibits characteristic bulk hematite extended visible region spectral features, showed that mixtures containing ~1, ~3, and ~5 wt.% Fe_2O_3 of this fine-grained, crystalline hematite produced a spectrum in the extended visible region consistent with a Mars bright region spectrum (SINGER *et al.*, 1979).

There is ~18–20 wt.% Fe_2O_3 present in the martian soils measured by Viking (TOULMIN *et al.*, 1977). Based on the work of SINGER (1982) and MORRIS *et al.* (1989), ~1–5 wt.% Fe_2O_3 in the martian bright region soil can be "crystalline" hematite, i.e. hematite exhibiting the spectral properties characteristic of bulk hematite. The remaining ~13–19 wt.% Fe_2O_3 in the martian surface material is not present as "crystalline" hematite and is probably present as (i) one or more nanophase ferric minerals, (ii) incorporated into the silicate matrix material (*e.g.,* spectrally neutral basalt or smectites, where some of the structural cation sites are filled by Fe) and/or (iii) amorphous ferric species. Nanophase hematite grown in silica gel (MORRIS *et al.,* 1989; MORRIS and LAUER, 1990), nanophase hematite found in palagonitic soils (MORRIS *et al.,* 1990, 1993; GOLDEN *et al.,* 1993), ferrihydrite (BISHOP *et al.,* 1993), and nanophase ferric oxides/oxyhydroxides/oxyhydroxysulfates grown in chemically-treated montmorillonites (BANIN *et al.,* 1993; BISHOP *et al.,* 1993, 1995) have been studied as potential candidates for this nanophase component of the ferric iron in Mars surface material.

In addition to the bright region Mars soils having a weak visible feature near 0.86 μm, which is consistent with hematite, the less common bright region Mars soils having weak visible features at longer wavelengths must be considered. Lunae Planum and the Oxia region of western Arabia exhibit ferric absorption features near 0.88 and 0.92 μm, respectively (MCCORD *et al.,* 1977, 1982; MURCHIE *et al.,* 1993; S. MURCHIE, pers. comm., 1995). These absorptions are not consistent with the component of bulk hematite inferred to be present in the more common bright regions, such as Tharsis and Amazonis, and may indicate a different crystalline ferric phase. However, nanophase and non-pigmentary components of the surface material in these uncommon regions might still be similar to those in the regions where an 0.86 μm hematite band is observed. A small amount of crystalline hematite (probably less than 1 wt.% based on the work of MORRIS *et al.,* 1989) may also be present in these uncommon regions; however, it would no longer be spectrally dominant.

BURNS (1993a) outlined detailed mechanisms for chemical weathering on Mars that could lead to the formation of smectites in combination with ferric oxides, oxyhydroxides, and hydroxysulfates. POLLACK *et al.* (1990) detected spectral evidence for sulfates in the airborne dust, which have been associated with the surface material on Mars. If schwertmannite were present pervasively in bright martian soil, it could account for some of the S measured by Viking (CLARK *et al.,* 1977) and for the spectral observations of POLLACK *et al.* (1990), and it would be consistent with the weathering scenarios described by BURNS (1993a). Local concentrations of schwertmannite may also be related to the uncommon bright martian regions where there is a longer-wavelength ferric absorption, although other minerals could also be responsible for this feature and there is no direct evidence for schwertmannite on Mars.

Reflectance spectra of a natural and a synthetic schwertmannite are presented here in the visible to infrared regions (0.3 to 25 μm). The spectral features in schwertmannite due to ferric species in the extended visible region and due to H_2O, OH, and sulfate complexes in the infrared region are analyzed in comparison with the spectral features of other ferric minerals, Mars soil analogs, and selected regions on Mars. As the mid-infrared spectral features are very weak in reflectance spectra of schwertmannite, transmittance spectra of schwertmannite are discussed here as well. Mössbauer spectra have been measured to provide additional information about the iron sites for interpretation of the optical spectra. Based on the visible to infrared reflectance spectra and other properties, schwertmannite is considered as a possible form of the nanophase ferric component of the bright region surface material on Mars. Remote sensing and sample analysis techniques are discussed for detecting schwertmannite, if present, in the martian surface material. Given the presence of schwertmannite or mixtures of schwertmannite and other ferric minerals, implications for the geochemistry of the surface of Mars are discussed.

BACKGROUND

Discovery of schwertmannite

Schwertmannite is an iron oxyhydroxysulfate with the empirical unit cell formula $Fe_{16}O_{16}(OH)_y(SO_4)_z \cdot nH_2O$, where $16-\mathbf{y} = 2\mathbf{z}$ and $2.0 \leq \mathbf{z} \leq 3.5$ (BIGHAM *et al.,* 1994). It was approved as an independent mineral species by the Commission on New Minerals and Mineral Names of the International Mineralogical Association in September, 1992. The first description of schwertmannite as a new mineral was published by BIGHAM *et al.* (1994). Prior to the formal approval of schwertmannite, selected data were published under the designation "mine drainage mineral" abbreviated as MDM (MURAD *et al.,* 1994).

Early studies of schwertmannite were carried out by BRADY *et al.* (1986) and MURAD (1988) before it was recognized as a discrete mineral. Lacking chemical analyses, the described phase was tentatively identified as ferrihydrite and "well"-crystallized ferrihydrite, respectively. The subsequent observation that this material always has sulfate contents from 12 to 16 wt.% led to the realization that sulfate must in some way be incorporated into the structure (BIGHAM *et al.,* 1990; MURAD *et al.,* 1990) and that it is not an iron oxyhydroxide, such as ferrihydrite. The X-ray diffraction diagram furthermore has two peaks at higher d-spacings (0.486 and 0.339 nm) than ferrihydrite; this information was not taken into consideration in the work mentioned earlier.

Schwertmannite differs from the majority of "new" minerals in that it is not an exotic phase that exists in only a few, or even just a single locality. On the contrary, it is a mineral of relatively common occurrence in acidic environments rich in sulfate and iron. Thus, it is a typical mineral of acid mine drainage precipitates, but has also been observed to occur in totally natural environments (SCHWERTMANN *et al.*, 1995). Major complications in the identification of schwertmannite arise from its small particle size, leading to significant broadening of X-ray diffraction lines and its frequent association with other, related minerals (typically jarosite, goethite, and/or ferrihydrite). Previous studies involving "amorphous ferric gels" or similar materials from environments such as acid mine drainage may, therefore, in fact refer to schwertmannite or mineral assemblages of which schwertmannite is a constituent.

In laboratory experiments, schwertmannite has been synthesized both abiotically and under the participation of *Thiobacillus ferrooxidans*. The possibility of bacterial activity on Mars has been presented recently by FARMER (1995). Although not essential for the formation of schwertmannite, the bacterium decreases the reaction time for oxidation of sulfides and, thus, facilitates the formation of secondary precipitates. If schwertmannite is found in the martian surface material, then evidence for bacterial activity on Mars should be investigated as well (FARMER, 1995).

Characteristic properties of schwertmannite

The X-ray diffraction diagram of schwertmannite consists of eight peaks between 0.486 and 0.146 nm. The X-ray diffraction data show schwertmannite to have a structure similar to that of akaganéite (nominally ß-FeOOH), except that the *c*-dimension is doubled (BIGHAM *et al.*, 1990). In schwertmannite the sulfate is located within the approximately 0.5 × 0.5 nm wide, square tunnels that run parallel to the *c*-axis. The sulfate is, therefore, relatively readily accessible, and can be exchanged with other anions. Schwertmannite is quite stable in a dry environment, but becomes unstable when stored in an aqueous environment. A schwertmannite stored in distilled water was transformed nearly completely to goethite within a period of two months (MURAD *et al.*, 1994; BIGHAM *et al.*, 1996). Thermal transformation of schwertmannite to hematite occurs by 560°C (BIGHAM *et al.*, 1990).

Schwertmannites exhibit some variability in chemical compositions, due primarily to the extent of additional sulfate species adsorbed on the particle surfaces (BIGHAM *et al.*, 1990). The Fe/S ratio ranges from ~5–8 (BIGHAM *et al.*, 1990; SCHWERTMANN *et al.*, 1995). The purest forms of schwertmannite contain ~62–65 wt.% Fe_2O_3, ~10–13 wt.% SO_3, ~9–10 wt.% adsorbed H_2O (lost by drying at 100°C for 12–18 hours), and ~14 wt.% bound H_2O plus structural OH (lost by drying at 800°C for 12–18 hours) (BIGHAM *et al.*, 1990). As schwertmannite contains much more Fe than S, schwertmannite could not account for all of the S present in the martian soil, even if it were a major source of Fe in some regions. Other sources of S may be sulfate salts, such as $MgSO_4$ (CLARK, 1993).

Natural schwertmannites are orange to yellowish brown in color. Quantitative color measurements indicate a broad distribution of color values for natural and synthetic schwertmannite. The Munsell color hues for schwertmannite range from ~8 YR to ~2.5 Y (BIGHAM *et al.*, 1992; BIGHAM *et al.*, 1994). Schwertmannite is typically associated with other genetically related phases, which is common for many microcrystalline minerals. This may influence the color of schwertmannite, thus creating the appearance of more color variation than really exists among pure schwertmannites.

Infrared spectra taken in the wavenumber range 1800–400 cm^{-1} show a number of bands that can be assigned to SO_4 (BIGHAM *et al.*, 1990). The position of these IR bands, combined with the XRD data, provide information about the structure of schwertmannite. BIGHAM *et al.* (1990) showed that about half of the $FeO_3(OH)_3$ octahedra in the akaganéite structure are replaced by $FeO_3(OH)_2O{-}SO_3$ in schwertmannite. Infrared spectra have also been used to demonstrate the exchange of SO_4 by other anions, *e.g.*, SeO_4 or $HAsO_4$ (MURAD *et al.*, 1994).

Electron micrographs confirm the small particle size (0.2–0.4 × 6–9 μm) indicated by X-ray diffraction line broadening, and show schwertmannite to have a fibrous morphology that has been termed "pincushion"-like by BIGHAM *et al.* (1994).

Mössbauer spectra taken at room temperature consist of broadened, somewhat asymmetric doublets. The line widths in schwertmannite spectra are high, requiring that the spectra be fitted with distributions of doublets instead of simple Lorentzian lines. Because of this asymmetry at least two distributions of quadrupole-split doublets were necessary. The distributions proved to have similar quadrupole splittings of maximum probability (0.65 mm/s on average), but different isomer shifts (0.39 and 0.33 mm/s). These parameters are typical for trivalent iron in moderately distorted octahedral coordination in paramagnetic minerals (*e.g.*, ferric oxides or oxyhydroxides above their respective Néel temperatures) or in superparamagnetic minerals (*e.g.*, ferric oxides or oxyhydroxides of small particle size). Thus, they do not differ sufficiently from those of numerous other minerals with which schwertmannite may be associated (*e.g.*, jarosite and ferric oxides) to allow the identification of schwertmannite in assemblages of such minerals on the basis of Mössbauer spectra taken in the paramagnetic state (see Table 1).

Mössbauer spectra taken at low temperatures have shown schwertmannite to order magnetically at 75 ± 5 K. Although the small particle size suggests that superparamagnetism could play a role in determining the magnetic properties of schwertmannite, Mössbauer spectra taken under external magnetic fields show the above temperature to correspond to a genuine Néel point. At 4.2 K the Mössbauer spectra of schwertmannite are again asymmetric and require fits entailing two distributions of magnetic hyperfine fields. Both distributions, however, have similar hyperfine fields of maximum probability of about 45.4 T. This is lower than that observed for any iron oxide or oxyhydroxide (MURAD and JOHNSTON, 1987; MURAD, 1996), and will usually allow distinction between schwertmannite and other minerals at 4.2 K (MURAD *et al.*, 1994). A summary of the Möss-

Table 1a. Average Mössbauer parameters of schwertmannite and other ferric minerals

Mineral	Room temperature			4.2 K			Néel or Curie Temperature (K)
	IS (mm/s)	QS (mm/s)	B_{hf} (T)	IS (mm/s)	QS (mm/s)	B_{hf} (T)	
Goethite[1,2]	0.37	-0.26	38.0	0.48	-0.25	50.6	400
Akaganéite[1,2]	0.37 0.38	0.55 0.95	— —	0.50 0.48 0.49	-0.02 -0.24 -0.81	48.9 47.8 47.3	299
Lepidocrocite[1,2]	0.37	0.53	—	0.47	0.02	45.8	77
Feroxyhite[1,2]	0.37	-0.06	41	0.48	~0	52 53	450
Ferrihydrite[1,2]	0.35	0.62 to 0.78	—	0.48	-0.02 to -0.07	47 to 50	25 - 115*
Hematite[1,2](wfm) (afm)	0.37	-0.20	51.8	0.49 0.49	-0.20 or 0.41	53.5 or 54.2	955
Maghemite[1,2]	0.23 0.35	≤|0.02| ≤|0.02|	50.0 50.0	0.36 0.48	≤|0.02| ≤|0.02|	53.0 52.0	~950
Jarosite[3,4]	0.37	1.16	—	0.49	0.11	49.2	54
Schwertmannite[5,6]	0.39 0.33	0.65	—	0.49	-0.35 -0.04	45.4	75

Table 1b. Average Mössbauer parameters of chemically-treated montmorillonites containing nanophase ferric minerals

Mineral	Room temperature			4.2 K		
	IS (mm/s)	QS (mm/s)	B_{hf} (T)	IS (mm/s)	QS (mm/s)	B_{hf} (T)
Ferrihydrite-bearing montmorillonite †,3	0.35	0.6 to 0.8	—	0.47 0.48	0.57 -0.1	 48
Ferric sulfate-bearing montmorillonite †,3	0.37	0.7	—	0.47 0.54 0.48	0.57 -0.1 -0.2	 48 45

[1] MURAD and JOHNSTON (1987), [2] MURAD (1996) and references therein. [3] BISHOP *et al.* (1995), [4] AFANASEV *et al.* (1974), [5] MURAD *et al.* (1990), [6] BIGHAM *et al.* (1994).

* Superparamagnetic blocking temperature. This varies as a function of particle size.

† Low-temperature Mössbauer spectra of chemically-treated smectites have a doublet due to structural iron (shown on the first line) and one or more sextets due to interlayer iron (shown on the following line(s)); the values given are averages for several samples.

bauer parameters of schwertmannite and other ferric minerals is provided in Table 1a, and of Mars soil analogs in Table 1b. The Mössbauer parameters listed in Table 1a are average values for crystalline materials. It should be noted that the magnetic hyperfine fields decrease with decreasing crystallinity and particle size (*e.g.*, MURAD and JOHNSTON, 1987), and that the values given in Table 1a are those of pure, bulk minerals. Schwertmannite and ferrihydrite are nanophase minerals, which should be taken into consideration when comparing their Mössbauer parameters with those of other materials.

Martian climate and surface material

The runoff and outflow channels on Mars provide evidence for abundant water there in the past (CARR, 1981). Climatic models describing a warm and wet early Mars were developed in order to explain these features (*e.g.*, POLLACK *et al.*, 1987; FANALE *et al.*, 1992). Current temperatures on Mars range from 130 to 290 K, varying both diurnally and regionally across the surface (KIEFFER *et al.*, 1977). The hypothesis of a warm and wet early Mars has been challenged recently due to a lack of spectral evidence for carbonates in the martian surface material, re-calculation of greenhouse models, proposed new methods of valley network formation, and the possibility of recent network formation or glaciation events (CARR, 1995 and references therein). Recent analyses of the valley networks and channels on Mars suggest that these features formed primarily by mass wasting and groundwater seepage (CARR, 1995). Seasonal variations in martian atmospheric water vapor levels also support the presence of near-surface reservoirs of water or ice in the regolith (JAKOSKY and HABERLE, 1992). Each of these models includes water in some form that would enable the kinds of weathering processes necessary for the formation of schwertmannite and other ferric minerals.

Geochemical results of the Viking missions led CLARK *et al.* (1977) and TOULMIN *et al.* (1977) to propose a mixture of smectites, sulfates, carbonates, and oxides as the fine-grained surface material covering Mars. Given liquid water on early Mars, chemical weathering processes and thermodynamic considerations indicate that smectites and sulfate minerals could have formed at that time, and that they would be stable under the arid conditions found today on Mars (GOODING, 1978; GOODING and KEIL, 1978; GOODING *et al.*, 1992; BURNS and FISHER, 1993). Current models of the surface material on Mars include smectite clays and clay-like mineraloids, such as palagonitic soils (BANIN *et al.*, 1992; GOODING *et al.*, 1992). The abundance of Fe and S in the martian soil, the oxidizing conditions on the surface of the planet, and the chemical reactivity of ferric species and sulfates imply that hydrated ferric sulfate minerals may have formed on Mars (BURNS, 1987, 1988, 1994; BURNS and FISHER, 1993).

Reflectance spectroscopy

Mars—The spectra of the bright regions on Mars are characterized by gently sloping ferric bands in the extended visible region (ADAMS and MCCORD, 1969; SINGER *et al.*, 1979; BELL *et al.*, 1990), weak features near 2.2–2.3 μm (MCCORD *et al.*, 1978; CLARK *et al.*, 1990; ARNOLD, 1992; MURCHIE *et al.*, 1993; BELL and CRISP, 1993; BELL *et al.*, 1994; BEINROTH and ARNOLD, 1996), a strong and broad bound water band at ~3 μm (HOUCK *et al.*, 1973; PIMENTEL *et al.*, 1974; BLANEY and MCCORD, 1989; BIBRING *et al.*, 1990; ERARD *et al.*, 1991), the possibility of a weak sulfate feature near 4.5 μm (BLANEY and MCCORD, 1995), and broad silicate emittance maxima near 10 and 20 μm (HANEL *et al.*, 1972; HUNT *et al.*, 1973). Recent reviews of the visible to infrared spectral features of Mars can be found in SODERBLOM (1992) and ROUSH *et al.* (1993).

Variations in the spectral character of Mars have been observed spatially across bright regions of the surface in the extended visible region (MCCORD *et al.*, 1977; SINGER, 1982; BELL, 1992). Specifically, Fe^{3+} transition bands are observed near 0.84–0.86 μm in the Tharsis region (BELL *et al.*, 1990), near 0.88 μm in Lunae Planum (MCCORD *et al.*, 1977; MURCHIE *et al.*, 1993), and near 0.92 μm in some of the bright soils throughout Arabia (MCCORD *et al.*, 1982; MURCHIE *et al.*, 1993).

Ferric Minerals—Reflectance spectroscopy has been measured in the visible and infrared regions of a number of ferric oxides, oxyhydroxides, and hydrated sulfates. Extended visible region reflectance spectra have been analyzed for numerous ferric minerals that have been characterized with Mössbauer spectra and other techniques (SHERMAN *et al.*, 1982; SHERMAN and WAITE, 1985; MORRIS *et al.*, 1985; BISHOP *et al.*, 1993). SHERMAN *et al.* (1982) assigned Fe^{3+} crystal field theory bands and Fe—O charge transfer bands to these visible reflectance spectra and to spectra from Mars. Ferric oxides composed of face-sharing [FeO_6] octahedra also exhibit the paired Fe^{3+} transition: $^6A_1 + {}^6A_1 \rightarrow {}^4T_1 + {}^4T_1$ near 0.5 μm, which is present, but much weaker, in ferric oxyhydroxides (SHERMAN and

WAITE, 1985; BURNS, 1993b). The intensity of this paired transition decreases, as well, with decreasing particle size and substitutions of Al^{3+} for Fe^{3+} (MORRIS *et al.,* 1989, 1992; MORRIS and LAUER, 1990). Table 2a includes a summary of the extended visible region spectral properties of several ferric minerals. Detailed analyses of the crystal field theory absorptions in mineral spectra are available in BURNS (1993b).

Chemically-treated montmorillonites—Reflectance spectra of cation-exchanged montmorillonites and montmorillonites bearing hydrated ferric oxides and sulfates, such as ferrihydrite and schwertmannite, have been studied in detail by BISHOP *et al.* (1993, 1994, 1995). The ferrihydrite-bearing and ferric sulfate-bearing montmorillonites were prepared in the laboratory as analog materials for the soils in the bright regions on Mars. Reflectance spectra of these samples include gently sloping, extended visible region ferric features, bound H_2O features near 1.4, 1.9, 3, and 6 μm, cation—OH bands near 1.41, 2.75, and 10–12 μm, and broad silicate features near 10 and 20 μm. The extended visible region spectral properties of these Mars soil analog materials are summarized in Table 2b.

EXPERIMENTAL PROCEDURES

Sample preparation or location

Schwertmannite—Two schwertmannite samples are included in these analyses—a natural sample collected near Glenn's run, Belmont Co., Ohio, and a synthetic sample. Sample Bt-4 is an almost pure schwertmannite, formed at pH 3.2 in a creek that is contaminated with acid mine

Table 2a. Spectroscopic properties of schwertmannite and other ferric minerals

	Band minimum	Band maximum	Band minimum	Reference
Schwertmannite (Bt-4)		0.740	0.915	this work
Schwertmannite (Z510b)		0.730	0.918	this work
Lepidocrocite	0.65		0.96	1
	0.70	0.82	0.98	2
		0.79	~0.94	3
Maghemite			0.93	1
		~0.78	0.92-0.94	2
Goethite	~0.64		0.91-0.93	1
	~0.65	~0.76	0.92-0.94	2
		0.765	0.91	3
Ferrihydrite			0.91	1
		0.80	0.93	3
Jarosite		0.74	0.905	this work
Akaganéite			0.908	1
		0.73	0.90	3
Feroxyhite			0.88	1
		0.76	0.88	3
Hematite			0.86-0.88	1
	0.63-0.64	0.74-0.76	0.85-0.87	2
		~0.75	0.86	3

Table 2b. Spectroscopic properties of chemically-treated montmorillonites containing nanophase ferric minerals

	Band maximum	Band minimum	Reference
Ferrihydrite bearing-montmorillonite	0.77-0.80	0.89-0.92	3,4
Ferric sulfate bearing-montmorillonite	0.74-0.75	0.88-0.89	4

[1] SHERMAN *et al.* (1982). [2] MORRIS *et al.* (1985). [3] BISHOP *et al.* (1993). [4] BISHOP *et al.* (1995).

drainage precipitates (BIGHAM *et al.,* 1990). Sample Z510b is a synthetic schwertmannite formed in the laboratory at pH 2.9 under the participation of *Thiobacillus ferrooxidans* (BIGHAM *et al.,* 1990). Chemical analyses of these samples are available in BIGHAM *et al.* (1990). A portion of the natural sample was dry sieved to less than 45 μm; reflectance spectra were measured of both the bulk and <45 μm particle size samples for Bt-4. Sample Z510b was only measured as a bulk sample. These particle size separates indicate the size of aggregates of individual grains, which are much smaller.

Ferric oxides—Goethite (α—FeOOH), akaganéite (β—FeOOH), feroxyhite (δ'—FeOOH), ferrihydrite ($Fe_5HO_8 \cdot 4H_2O$), and hematite (α-Fe_2O_3) samples were collected or synthesized for an earlier study (SHERMAN *et al.,* 1982). A previously studied natural specimen of magnetite (Fe_3O_4) from Grangeburg, NY (BISHOP *et al.,* 1993) and a previously studied jarosite ($KFe_3(SO_4)_2(OH)_6$) sample (BURNS, 1987) are included here. Insufficient quantity of these samples was available to separate them by particle size. The aggregated particle sizes were estimated by visual inspection at approximately less than 125 μm.

Goethite, lepidocrocite (γ—FeOOH), and ferrihydrite were synthesized following procedures described in SCHWERTMANN and CORNELL (1991) for a previous study (BISHOP *et al.,* 1993). These samples were dry sieved to less than 45 μm.

Chemically-treated montmorillonites—SWy-1 montmorillonite (The Clay Minerals Society, Source Clay Minerals Repository) normally contains Na and Ca cations in the interlayer sites. This natural montmorillonite has been chemically treated in the laboratory to produce montmorillonites containing a variety of interlayer ferric complexes. Ferric-doped montmorillonites have been produced and studied in many laboratories (BISHOP *et al.,* 1993, 1995; BANIN and RISHPON, 1979; BANIN *et al.* 1988; ORENBERG and HANDY, 1992). Low-temperature Mössbauer spectra of the samples prepared by BISHOP *et al.* (1993) have strong similarities with the mineral ferrihydrite and have been therefore named ferrihydrite-montmorillonite assemblages or ferrihydrite-bearing montmorillonites. The ferric sulfate-bearing montmorillonites were prepared by adding ferric sulfate solution to a protonated montmorillonite suspension under low pH (1.6–1.8), then raising the pH to 3.0–4.0, depending on the sample (BISHOP *et al.,* 1995). These samples were dry sieved to less than 45 μm; the actual grain sizes of the natural and chemically-treated montmorillonites are on the order of 1–2 μm.

Reflectance spectroscopy

Spectra were measured relative to Halon from 0.3 to 3.6 μm under ambient conditions with the RELAB (reflectance experiment laboratory) spectrometer at Brown University. The Halon spectrum is multiplied by wavelength- and angle-dependent correction factors, giving absolute reflectance values for the samples. A photomultiplier tube is used for the visible to near-infrared region and an InSb detector is used for the infrared region. This instrument has been described in detail in PIETERS (1983). The RELAB spectrometer allows for variable illumination (i) and emergence (e) angles. Spectra included here were measured at i = 30° and e = 0°. The sample dish is rotated during the measurement to eliminate orientation effects.

Infrared reflectance spectra were measured relative to a rough gold surface, using a Nicolet 740 Fourier transform interferometer (FTIR), in an H_2O- and CO_2(g)-purged environment. A PbSe detector was used in the range 0.9 μm to 3.2 μm and a DTGS (deuterated triglycine sulfate) detector between 1.6 μm and 25 μm. The sample chamber, including a remotely controlled sample train, was manufactured by SpectraTech and uses a biconical arrangement. Composite spectra were prepared by splicing RELAB data at 1.2 μm with Nicolet data (1.2 μm to 25 μm). The Nicolet spectra were scaled to those measured using the RELAB spectrometer.

Sample preparation for the spectral measurements involved pouring the particulate material into a sample dish 12 mm in diameter and 3 mm in depth. The dish was tapped gently on a hard surface to settle the particles, then re-filled. This procedure was continued until the dish remained full. The sample surface was not smoothed with a spatula. For measurement of some of the ferric minerals, where very small amounts of material were available, it was necessary to use smaller sample dishes.

Additional infrared reflectance spectra were measured using a Nicolet Magna 550 FTIR spectrometer with a DTGS detector and a Harrick "Drift" sample measurement system at the Bayerisches Geologisches Landesamt in Bamberg. In order to give absolute reflectance, these spectra were scaled at 1390 cm^{-1} to the reflectance spectra obtained at Brown University.

Transmittance spectroscopy

Infrared transmittance spectra were measured in the range 400 to 4000 cm^{-1} (2.5 to 25 μm) using a Nicolet Magna 550 FTIR spectrometer with a DTGS detector. One mg of sample was pressed between two KBr disks (but not mixed with KBr, thus avoiding reaction between the schwertmannite and KBr). A KBr pressed pellet was prepared as a blank standard. Additional spectra were obtained by attenuated total reflectance using an ATR kit equipped with a ZnSe plate. Peak positions of the spectral features were determined either directly or from first derivatives using the Nicolet OMNIC software.

Mössbauer spectroscopy

Mössbauer spectra of sample Bt-4 were taken at 295 and 4.2 K using a ^{57}Co/Rh source. In order to obtain a homogeneous iron distribution, the sample was mixed with 2 parts sugar and placed in a sample holder with an area of 2 cm^2. The transmitted radiation was recorded with a proportional counter and stored in a 1024 channel analyzer. A metallic Fe foil was used for velocity calibration and as an isomer shift reference. The room-temperature spectrum was taken in the velocity range ± 2 mm/s on a sample containing 20 mg schwertmannite until 3.6×10^6 counts per channel had been accumulated. The spectrum at 4.2 K was taken in the velocity range ± 10.6 mm/s on a sample containing 36 mg schwertmannite, cooling both source and absorber in a bath cryostat, until 4.7×10^6 counts per channel had been accumulated. The mirror halves of the spectra were folded and fitted with distributions of Lorentzian lines by a computer procedure, constraining corresponding lines of every doublet and sextet to have equal widths and intensities. This fitting procedure has been described in more detail by MURAD (1988).

RESULTS

The spectral properties of schwertmannite are presented here, with a focus on reflectance spectroscopy because that is the primary means at this time of gaining spectral information from Mars. As schwertmannite is a fine-grained mineral, many of the mid-range infrared spectral features are weak and difficult to interpret. In order to understand the origin of the spectral features and gain additional information about the mineral structure, transmittance spectra are evaluated in this region as well. The character of the Fe^{3+} sites in schwertmannite and the nature of the Fe—O bonds are responsible for the dominant extended visible region features in the reflectance spectra of schwertmannite. Mössbauer spectra also provide information about these Fe^{3+} sites, and are, therefore, analyzed here in combination with the optical spectra.

Reflectance spectroscopy—extended visible region

The reflectance spectrum of a natural schwertmannite (sample Bt-4) is shown in Fig. 1 from 0.3–1.3 μm (~30,000–8000 cm^{-1}) along with spectra of several other ferric minerals. The spectrum of schwertmannite exhibits a steep absorption edge beginning at about 0.5 μm, a gently sloping shoulder near 0.6 μm, a local reflectance maximum near 0.74 μm and a broad absorption band with a minimum near 0.91 μm. Included in Fig. 1 are spectra of akaganéite and jarosite that exhibit a local reflectance maximum near 0.73–0.74 μm and an absorption feature with a minimum near 0.90 μm. The positions of these reflectance maxima and minima in the schwertmannite spectrum in this region resemble more closely the spectral character of akaganéite and jarosite than those of other ferric minerals shown in Fig. 1. The broadness of the ferric transition near 0.91 μm in the spectrum of schwertmannite is similar to that of ferrihydrite, although for ferrihydrite this band occurs at a longer wavelength.

In Fig. 2 are shown extended visible region reflectance spectra of two schwertmannite samples, and chemically-treated montmorillonites containing ferric sulfate species and ferrihydrite. The spectral properties of the natural (Bt-4) and synthetic (Z510b) schwertmannites shown in Fig. 2 are in general similar, although subtle differences are present in the reflectance spectra of these two samples in this region. For example, the ferric band near 0.91 μm in the spectrum of the synthetic schwertmannite is deeper than the band in the natural schwertmannite spectrum. First derivatives of the spectra give local reflectance maxima of 0.740 μm for the natural schwertmannite Bt-4 and 0.730 μm for the synthetic schwertmannite Z510b and

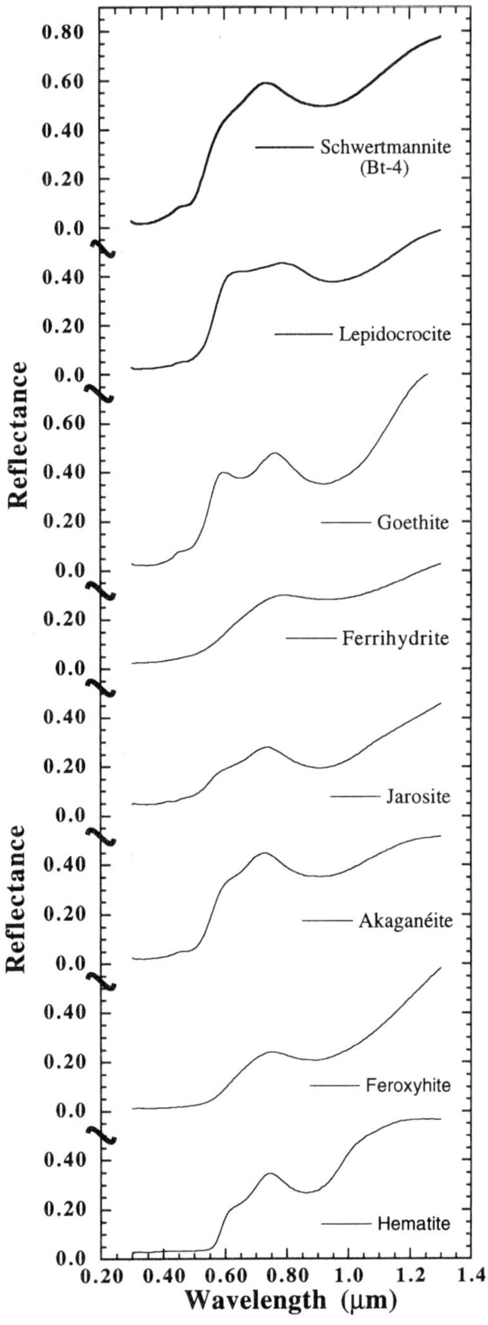

FIG. 1 Reflectance spectra of schwertmannite and ferric minerals in the extended visible region (0.3–1.3 μm). The ferric oxides and oxyhydroxides are arranged in approximate order of the position of their ferric band near 0.9 μm.

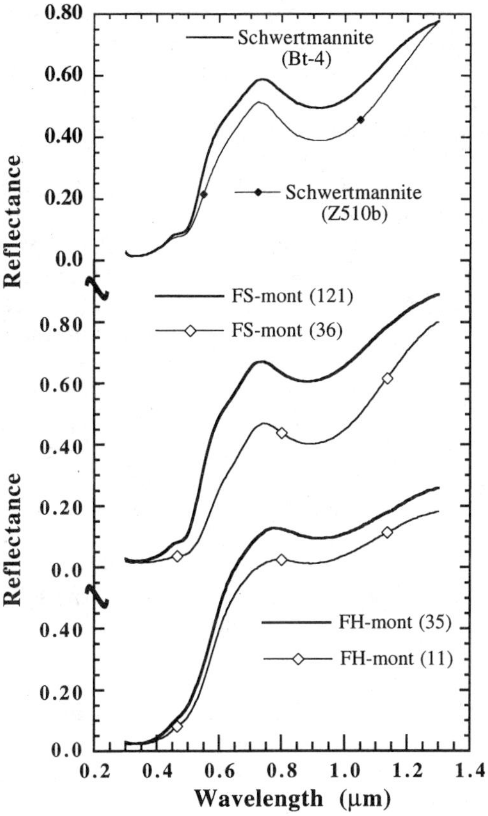

FIG. 2 Reflectance spectra of natural and synthetic schwertmannites and Mars soils analogs in the extended visible region (0.3–1.3 μm). Spectra of two ferrihydrite-bearing montmorillonites (FH-mont) and two ferric sulfate-bearing montmorillonites (FS-mont) are also shown here for comparison.

local reflectance minima at 0.915 μm for sample Bt-4 and 0.918 μm for sample Z510b. The ferric absorption occurs near 0.91–0.92 μm in spectra of the ferrihydrite-bearing montmorillonites and near 0.88–0.89 μm in spectra of the ferric sulfate-bearing montmorillonites shown in Fig. 2.

Mössbauer spectroscopy

Mössbauer spectra of sample Bt-4 measured at room temperature and at 4.2 K are shown in Figs. 3a and b, respectively. These spectra exhibit the characteristic features of all schwertmannites: broad asymmetric doublets at room temperature (Fig. 3a) and broad, asymmetric sextets at 4.2 K (Fig. 3b). The room-temperature spectrum was fitted with two distributions of quadrupole-split doublets of different isomer shifts (0.33 and 0.39 mm/s); both distributions have a quadrupole splitting value with a maximum probability of 0.68 mm/s. The spectrum taken at 4.2 K was fitted with two distributions of sextets with different quadrupole interactions (−0.03 and −0.33 mm/s); the magnetic hyperfine fields of maximum probability were 44.9 and 45.0 T, which are somewhat lower than the averge values for schwertmannite given in Table 1a. These parameters for sample Bt-4 lie within the range of values reported for other, natural and synthetic, schwertmannites (MURAD *et al.*, 1990; BIGHAM *et al.*, 1990). Average values of IS, QS, and B_{hf} for several schwertmannites are shown in Table 1. No Mössbauer data are available for sample Z510b; however, substantial Mössbauer data have been published on sample Z500a, a synthetic sample prepared by the same procedure as for Z510b (MURAD, 1988; MURAD *et al.*, 1990; BIGHAM *et al.*, 1990).

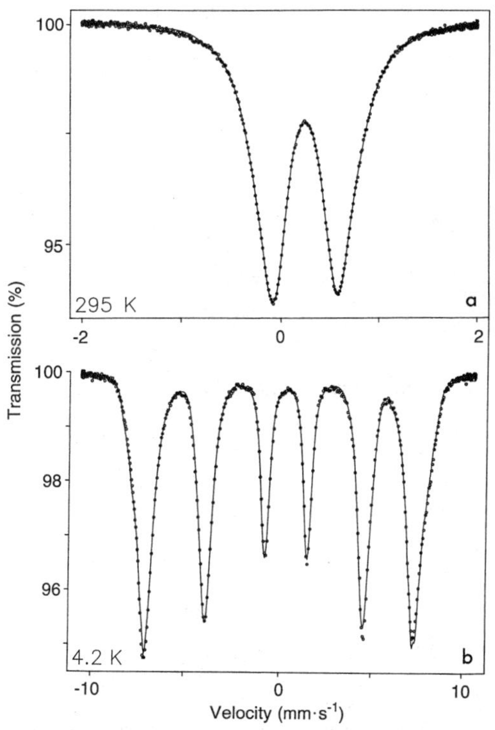

FIG. 3 Mössbauer spectra of the natural schwertmannite sample Bt-4 measured at (a) room temperature and (b) 4.2 K. Fits of the room-temperature spectrum resulted in two distributions of Lorentzian-split doublets that differ in their isomer shifts. For the spectrum taken at 4.2 K the best fit includes two distributions of magnetic hyperfine fields with different quadrupole interactions.

Reflectance spectroscopy—near infrared region

Near-infrared (NIR) reflectance spectra of the natural schwertmannite Bt-4 and the synthetic schwertmannite Z510b are shown in Fig. 4a. The synthetic schwertmannite sample exhibits darker reflectance values than the natural sample in the visible to near-infrared region, except for the range 1.3–1.4 μm. The spectra of each of these schwertmannites contain broad features due to bound H_2O near 1.45 μm, 1.95 μm, and near 3 μm. The schwertmannite spectra were measured in a dry environment (the air in the sample chamber was purged of water and CO_2 for more than 12 hours), where all or most of the adsorbed water was removed from the sample.

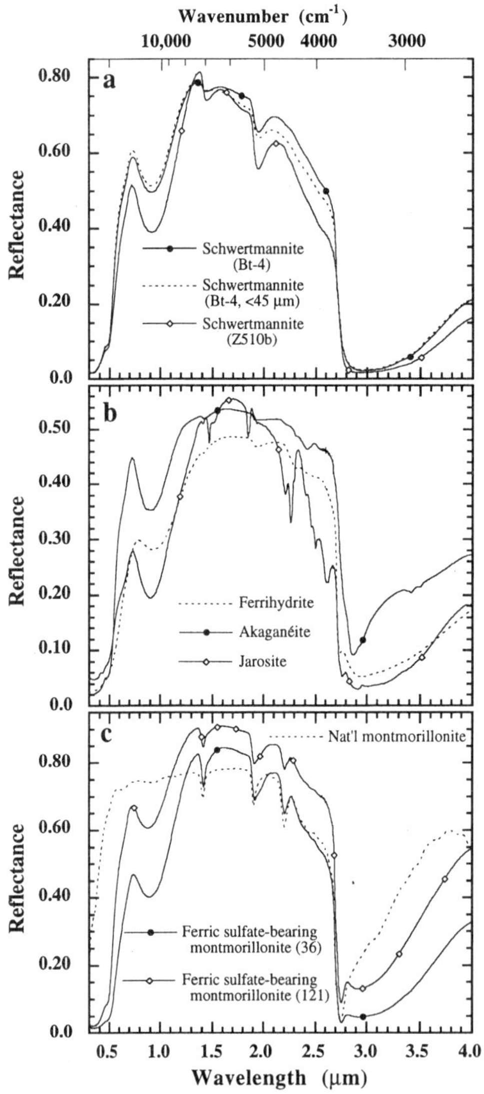

FIG. 4 Reflectance spectra in the visible to near-infrared region (0.3–4.0 μm): a) natural (Bt-4) and synthetic (Z510b) schwertmannite, b) akaganéite, jarosite, and ferrihydrite, and c) natural SWy-1 montmorillonite and two ferric sulfate-bearing montmorillonites.

Near-infrared reflectance spectra of ferrihydrite, akaganéite, and jarosite are shown in Fig. 4b for comparison to schwertmannite. Although the spectral properties of these three ferric minerals are similar to those of schwertmannite in the extended visible region, they exhibit significant differences in the near-infrared region. The reflectance spectra of schwertmannite in this region are much brighter and have a stronger negative continuum slope from ~1.4 to 2.5 μm. The jarosite spectrum has narrow features near 1.4, 1.9, 2.2–2.3, and 2.75 μm due to structural OH species and additional features near 2.6 μm due to combinations of OH or O—H—O vibrations (GAFFEY *et al.,* 1993). Weak absorption features are observed in the spectrum of akaganéite near 1.45 and 1.95 μm due to bound water, and near 2.4 μm probably due to structural OH. The broad, bound H_2O features observed in the spectrum of ferrihydrite are similar to those seen in the spectra of schwertmannite (Fig. 4a).

Reflectance spectra of natural SWy-1 montmorillonite and two ferric sulfate-bearing montmorillonites are shown in Fig. 4c. Several spectral changes can be seen as a result of the chemical treatment of the natural montmorillonite. The spectra of the ferric sulfate-bearing samples exhibit intense ferric features near 0.9 μm and a broad, bound water feature near 3 μm. The spectrum of the natural montmorillonite exhibits narrow features near 1.41 and 1.91 μm, while broad shoulders near 1.45 and 1.97 μm are observed in the spectra of the ferric sulfate-bearing montmorillonites. These broad, bound water features are similar to those observed in spectra of ferrihydrite and schwertmannite.

Reflectance spectroscopy—mid-infrared region

Reflectance spectra are shown as a function of wavenumber from 3330 to 400 cm^{-1} (~3–25 μm) in Fig. 5a for natural and synthetic schwertmannite. The primary absorption features in the reflectance spectrum of schwertmannite in this region occur near 3000 cm^{-1} (~3–3.5 μm) and 1640 cm^{-1} (6.1 μm) and are due to bound water. A reflectance minimum (also called Christiansen feature, see below) occurs near 1200 cm^{-1} (8.3 μm) and weaker features are observed near 1150 cm^{-1} (8.7 μm), 1060 cm^{-1} (9.4 μm), 985 cm^{-1} (10.2 μm), ~730

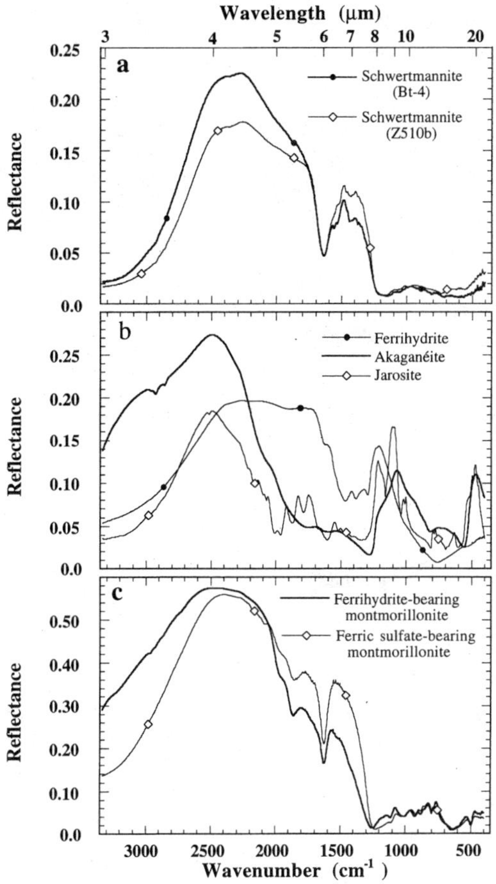

FIG. 5 Reflectance spectra in the mid-infrared region (3330–400 cm^{-1} or ~3–25 μm): a) natural (Bt-4) and synthetic (Z510b) schwertmannite, b) akaganéite, jarosite, and ferrihydrite, and c) ferrihydrite-bearing montmorillonite and ferric sulfate-bearing montmorillonite.

cm^{-1} (13.7 μm), 630 cm^{-1} (15.9 μm), and 565 cm^{-1} (17.7 μm) in the spectra of schwertmannite and differ somewhat for the two samples measured here. These bands are summarized in Table 3. A few weak and poorly resolved features in the range 1350–1550 cm^{-1} (~6.4–7.4 μm) are attributed to a minor amount of jarosite. Weak features, observed near 915 cm^{-1} (10.9 μm) and 800 cm^{-1} (12.5 μm) in Bt-4, are also assigned to impurities in this sample.

Mid-infrared reflectance spectra are shown in Fig. 5b of ferrihydrite, akaganéite, and jarosite. The spectrum of schwertmannite in this region (Fig. 5a) is much darker than the spectra of ferrihydrite, akaganéite, and jarosite. The Christiansen feature is a strong reflectance minimum that is associated with the principal molecular vibration in silicates (SALISBURY, 1993) and occurs where the real part of the index of refraction, n, approaches that of the surrounding medium, generally unity (CONEL, 1969). This has been measured for a number of geological samples and may be useful in remote characterization of planetary surfaces (SALISBURY, 1993). This reflectance minimum occurs near 1400 cm^{-1} (~7.1 μm) for jarosite, near 1280 cm^{-1} (~7.8 μm) for akaganéite; in both cases at shorter wavelengths than for schwertmannite.

Reflectance spectra of a ferric sulfate-bearing montmorillonite and a ferrihydrite-bearing montmorillonite are displayed in Fig. 5c. The feature near 1635 cm^{-1} in these spectra is due to bound H_2O. The Christiansen feature occurs as a broad reflectance minimum at 1240–1255 cm^{-1} (~8 μm) for ferrihydrite-bearing montmorillonites and at 1210–1250 cm^{-1} (~8–8.3 μm) for ferric sulfate-bearing montmorillonites. In the scattering regime, at wavelengths longer than the Christiansen feature, these spectra are very dark and exhibit weak features, much like the spectra of schwertmannite in Fig. 5a.

Transmittance and ATR spectroscopy

Transmittance spectra of samples Bt-4 and Z510b were measured from 4000 to 400 cm^{-1} (2.5–25 μm) and ATR spectra from 4000 to 630 cm^{-1} (2.5–16 μm). These are shown in Fig. 6 from 1300 to 400 cm^{-1} (7.7–25 μm). A strong, broad spectral band near 3300 cm^{-1} (~3 μm) and another feature at ~1630 cm^{-1} (6.14 μm) due to water are observed in these spectra, but are not shown in Fig. 6. Additional spectral features are observed at approximately 1180 cm^{-1} (8.5 μm), 1100 cm^{-1} (9.1 μm), 1045 cm^{-1} (9.6 μm), 980 cm^{-1} (10.2 μm), 800–850 cm^{-1} (12.5–12 μm), 700 cm^{-1} (14.3 μm), 610 cm^{-1} (16.4 μm), 520 cm^{-1} (19.2 μm), 465 cm^{-1} (21.5 μm), and 415 cm^{-1} (24.1 μm) (Table 3). Subtle differences in the strengths and positions of these features can be seen in Fig. 6 for the two schwertmannite samples studied here. Band assignments for the mid-IR transmittance features of schwertmannite are listed in Table 3. The features due to impurities found in the reflectance spectra are either absent or much weaker in the transmittance spectra of schwertmannite.

DISCUSSION

Spectral features due to ferric iron

The spectral character of ferric minerals in the extended visible region is dominated by ferric crystal field theory bands (SHERMAN *et al.*, 1982; SHER-

Table 3. Infrared absorptions in spectra of natural (Bt-4) and synthetic (Z510b) schwertmannites

Transmittance **Bt-4** (cm^{-1})	Reflectance **Bt-4** (cm^{-1})	Transmittance **Z510b** (cm^{-1})	Reflectance **Z510b** (cm^{-1})	Band Assignment
~1635	~1640	~1630	~1640	bound H_2O bend[1]
~1180†	~1200†	~1190†	~1210†	ν_3 sulfate vib. or ν_4 acid sulfate vib.[2]
~1100	~1140	~1120	~1150	ν_3 sulfate asymmetric stretch[1,2]
~1045	~1060	~1070	~1060	ν_3 sulfate asymmetric stretch[1,2] or ν_2 acid sulfate vib.[2]
~980	~985	~980	~985	ν_1 sulfate symmetric stretch[1]
~800-850*		~800-850		OH bend[1,2]
~700	~720	~700	~735	OH bend[2]
~610	(620)†	~610	~630	ν_4 sulfate asymmetric bend[1,2]
~520	(565)†	(520)†	~565	ν_2 or ν_4 sulfate bend or acid sulfate bend[2]
~465		(465)†		ν_2 sulfate symmetric bend[2]
~415	~400-420	~420	~400-420	Fe-O stretch[1,2]

[1] BIGHAM *et al.* (1990), BIGHAM *et al.* (1994).
[2] Based on assignments for other ferric sulfate minerals in Farmer (1974).
* This band is partially masked by weak features due to a small amount of goethite in the sample.
† Probable location of band centers; bands are broad or weak.
Note: All transmission features above 650 cm^{-1} were confirmed by ATR.

MAN and WAITE, 1985; BURNS, 1993b). The absorption edge near 0.5 μm, due to the $^6A_1 + {}^6A_1 \rightarrow {}^4T_1 + {}^4T_1$ paired transition, is evident in the spectra of schwertmannite, as well as in spectra of the other ferric minerals (Figs. 1 and 2). A poorly-defined shoulder near 0.6 μm is due to a weak $^6A_1 \rightarrow {}^4T_2$ transition that is angular in the spectra of akaganéite, goethite, and hematite shown in Fig. 1. The broadness of the $^6A_1 \rightarrow {}^4T_1$ band near 0.91 μm is consistent with multiple Fe^{3+} sites in the structure of schwertmannite or vibronic coupling (BURNS, 1993b). Broad $^6A_1 \rightarrow {}^4T_1$ bands are also observed for akaganéite and ferrihydrite (Fig. 1). As the mineral structures of akaganéite and schwertmannite are similar, the observed reduction in the intensity of the $^6A_1 \rightarrow {}^4T_2$ absorption feature and broadening of the $^6A_1 \rightarrow {}^4T_1$ band imply that schwertmannite is less crystalline than akaganéite or that there is a broader distribution of Fe^{3+} sites in the mineral structure.

Some of the Fe^{3+} cations in schwertmannite occur in octahedral coordination with O, OH, and O—SO_3, while others are coordinated only with O and OH (BIGHAM *et al.*, 1990, 1994). Mössbauer spectra of schwertmannite measured at room temperature exhibit two isomer shift values and at 4.2 K two quadrupole splitting values. The electron density on the Fe^{3+} nuclei is sufficiently different in these two sites that two Fe^{3+} sites can be detected in the paramagnetic state. In akaganéite the Fe^{3+} cations are octahedrally coordinated to both O and OH, which gives a broad ferric $^6A_1 \rightarrow {}^4T_1$ transition located near 0.90 μm. Therefore, the O—SO_3 species in schwertmannite are probably responsible for the longer wavelength (~0.91 μm) portion of this broad band, while the $FeO_3(OH)_3$ sites would be expected to give a transition near 0.90 μm, as seen in spectra of akaganéite.

BURNS (1993b) discussed the influence of covalent bonding on the metal-ligand bond length and crystal field splitting energy (CFSE). The ligand's ability to delocalize electronic charge bears directly on the electronic environment of the metal cation and, hence, its CFSE. BURNS (1993b) described the situation for hematite and goethite as examples. The FeO_6 bonding configuration in hematite requires a higher electron density on O and lower electron density on Fe^{3+} than does the $FeO_4(OH)_2$ bonding configuration in goethite. The CFSE is higher for hematite than for goethite, and correspondingly, the ferric $6A_1 \rightarrow 4T_1$ transition occurs at a shorter wavelength for hematite than for goethite.

These results from BURNS (1993b) can be extrapolated to the structures of akaganéite and schwertmannite. The replacement of OH by O—SO_3 in some of the $FeO_3(OH)_3$ sites causes a change in the electron density distribution and degree of covalent bonding. Based on the polarizing power of the sulfate ion (*e.g.*, HUHEEY *et al.*, 1993), the SO_3 groups are better able to delocalize the electronic charge on O than H, thus, the electron density on O will be lower for the $FeO_3(OH)_2O—SO_3$ sites than for the $FeO_3(OH)_3$ sites. Therefore, a lower CFSE and longer wavelength ferric absorption feature for the sulfate-bearing sites compared to the akaganéite-like sites in schwertmannite are predicted. This interpretation of the nature of the chemical bonds in schwertmannite is consistent with the observed optical spectra in Figs. 1 and 2.

Additional features in the mid-infrared spectral region result from ferric iron in the schwertmannite structure. These include absorption features due to Fe—O stretching vibrations, Fe—OH bending vibrations, and $Fe—O—SO_3$ vibrations. Strong absorption features at 704 and 483 cm^{-1} were tentatively assigned to Fe—O stretching modes in schwertmannite (BIGHAM *et al.,* 1994). Although Fe—O stretching vibrations do have energies in this region, OH bending vibrations and sulfate ν_2 vibrations also occur in this region for hydrated ferric sulfates (ROSS, 1974). Based on the spectra discussed by Ross (1974), the absorption features observed in the transmittance spectra of schwertmannite near 700 and 465 cm^{-1} in Fig. 6 are assigned here to an OH bending mode and a ν_2 symmetric sulfate bending mode, respectively, and are discussed in more detail in the following sections. However, the lower energy absorption feature near 425–430 cm^{-1} is assigned to an Fe—O stretching mode. This Fe—O absorption feature is centered at ~415 cm^{-1} for sample Bt-4 and at ~420 cm^{-1} for sample Z510b in the transmittance spectra shown in Fig. 6.

Spectral features due to adsorbed and bound H_2O

The broad, bound H_2O features observed near 1.45, 1.95, 2.75–3.5, and 6.1 μm in the spectra of natural and synthetic schwertmannite are much stronger than these features in the akaganéite spectrum. In each case spectra were measured after H_2O and CO_2 were purged overnight from the air in the sample chamber. For smectite clays, dehydrating the samples by placing them overnight in a H_2O- and CO_2-purged sample chamber or heating them at ~100°C overnight proved effective in removing

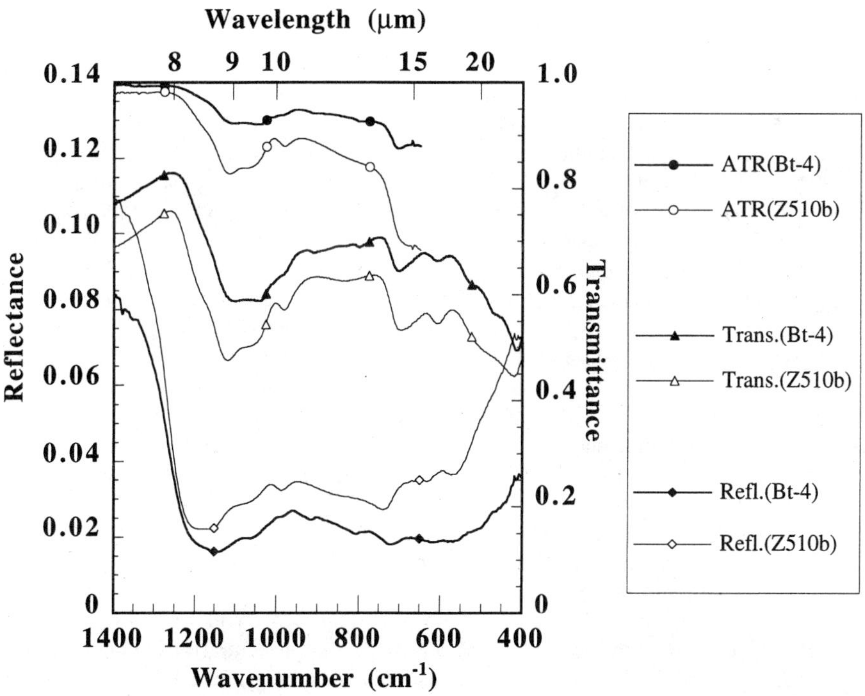

FIG. 6 Reflectance, transmittance, and attenuated total reflectance (ATR) spectra of natural and synthetic schwertmannite in the mid-infrared region (1400–400 cm^{-1} or ~7.1–25 μm). The natural schwertmannite (Bt-4) spectra are labeled with filled symbols and the synthetic schwertmannite (Z510b) spectra are labeled with open symbols.

adsorbed H_2O (BISHOP *et al.*, 1994). Assuming that the adsorbed H_2O molecules in schwertmannite behave similarly to those in smectites, then the spectral features due to water in Fig. 4a are primarily due to the presence of bound water, rather than adsorbed water, in schwertmannite. The highly polarizing nature of the sulfate species causes an increase in the intensity of the spectral water features in two ways: (i) by holding additional bound water molecules in the sample and (ii) because the polarization enhances the band strength for a given amount of water molecules.

The bound water features in spectra of the ferric sulfate-bearing montmorillonites occur at longer wavelengths and are broader than the bound water features in spectra of natural montmorillonites (Fig. 4c). The broadening of these spectral features near 1.45 and 1.97 μm in the ferric sulfate-bearing montmorillonite relative to the natural montmorillonite could be due to the presence of schwertmannite. Based on low temperature Mössbauer spectroscopy and other analyses, the ferric sulfate-bearing montmorillonite samples are thought to contain both schwertmannite and ferrihydrite (BISHOP *et al.*, 1995). The ferric sulfate-bearing montmorillonites, ferrihydrite, and schwertmannite all contain adsorbed and bound water under ambient conditions.

Band depths of the near-infrared features in schwertmannite are dependent on the environmental conditions and were measured here according to the method of CLARK and ROUSH (1984). Reflectance spectra measured under ambient conditions resulted in band depths of 6 % and 19% at 1.45 and 1.95 μm, respectively, for sample Bt-4. Similarly, sample Z510b exhibited band depths of 14% and 31% for the absorption features at 1.45 and 1.95 μm, respectively. Reflectance spectra of these samples under controlled, dry conditions gave band depths of 3% and 10% at 1.45 and 1.95 μm, respectively, for sample Bt-4 and band depths of 8% and 18% at 1.45 and 1.95 μm, respectively, for sample Z510b. By assuming that the spectra measured under dehydrated conditions do not include any adsorbed water, the portion of these spectral features attributed to the adsorbed water can be roughly determined by subtracting the band depths of the ''dry'' spectra from those of the ambient spectra. This results in 3% and 9% band depths at 1.45 and 1.95 μm, respectively, due to adsorbed water in Bt-4, and 6% and 13% band depths at 1.45 and 1.95 μm, respectively, due to adsorbed water in Z510b. The differences in the near-infrared band depths of the bound water features between the two samples imply that the amount of bound water contained in the mineral schwertmannite can vary from sample to sample. This is consistent with the variability in OH and SO_4 in the schwertmannite structure (BIGHAM *et al.*, 1994).

The amount of water lost by heating schwertmannite samples to 100°C and 800°C was measured for several samples (BIGHAM *et al.*, 1990). This study determined that 8.8 wt.% H_2O was lost on heating to 100°C and an additional 11.3 wt.% H_2O was lost on heating to 800°C for sample Bt-4. Similarly, for sample Z510b, 9.8 and 14.1 wt.% H_2O were lost on heating to 100°C and 800°C, respectively. These results are consistent with the greater magnitude of the bound water absorption features observed in the near-infrared reflectance spectra of Z510b than in spectra of Bt-4 (Figure 4a). The formula for schwertmannite indicates that H_2O molecules are associated with this mineral, but that H_2O molecules are not included in the structure (BIGHAM *et al.*, 1994). Water must be chemisorbed or bound to the grain surfaces or in the tunnels because it is retained in the sample at elevated temperatures as shown by measurements in BIGHAM *et al.* (1990) and under dehydrated conditions as shown by the near-infrared spectra in Fig. 4a.

The wt.% H_2O lost upon heating to 800°C is due to a combination of bound water and structural OH. If the amount due to structural OH is assumed to be about the same in each sample, then the differences measured in wt.% H_2O lost upon heating to 800°C would be due to differences in the amount of bound water in each sample. The band depths due to bound water in smectites were found to be linearly related to the wt.% H_2O lost upon heating to 500°C (BISHOP *et al.*, 1994). It can be assumed that the NIR band depths due to bound water are also linearly related to the wt.% H_2O lost upon heating of schwertmannite. The NIR bound water band depths (at 1.45 and 1.95 μm) for Z510b are roughly twice as strong as those for Bt-4. It can then be assumed that the amount of bound water lost upon heating to 800°C (as measured by BIGHAM *et al.*, 1990) is about twice as large for sample Z510b as for sample Bt-4. This gives approximately 8 wt.% H_2O as structural OH in both samples, ~3 wt.% H_2O as bound water in Bt-4, and ~6 wt.% H_2O as bound water in Z510b. Additional thermal and spectroscopic analyses could be used to more precisely estimate the structural OH, bound H_2O, and adsorbed H_2O components of schwertmannite.

Spectral features due to OH

A broad spectral shoulder near 2.5 μm in the schwertmannite spectra may be related to the features from 2.4–2.6 μm in akaganéite and jarosite in Fig. 4. Spectral features in this region for sulfate minerals are assigned to combinations of the OH stretching and bending modes (*e.g.*, GAFFEY *et al.*, 1993). The broad water band near 3 μm in the spectra of schwertmannite has a steep slope and angular character near 2.75 μm that may be due to structural OH, as observed in the spectrum of jarosite (Fig. 4).

Bending modes of OH species attached to Fe^{3+} cations occur in the range ~700–900 cm^{-1} for most hydroxylated ferric minerals, but they also occur at higher energies for jarosite and a few other hydroxylated ferric sulfate minerals (FARMER, 1974). In the transmittance spectrum of sample Bt-4, Fe—OH bending modes are observed near 700, 800, and 900 cm^{-1}. The features at ~800 and ~915 cm^{-1} are very weak in comparison to the feature near 700 cm^{-1} and are not observed in the spectrum of Z510b (Fig. 6). Strong bands are observed near 800 and 900 cm^{-1} in transmittance spectra of goethite (not shown). Jarosite exhibits a spectral feature near 790 cm^{-1} and plumbojarosite has a feature at 915 cm^{-1} due to OH bending modes (ROSS, 1974). X-ray powder diffraction patterns indicate that sample Bt-4 contains a small amount (<1wt.%) of jarosite and goethite, which are not found in sample Z510b. Therefore, weak features observed near 800 and 900 cm^{-1}, are assigned to small amounts of goethite, jarosite and/or plumbojarosite in the Bt-4 schwertmannite sample. Weak features in the range 1350–1550 cm^{-1} are also assigned to small amounts of jarosite.

BIGHAM *et al.* (1994) assigned a broad, weak band from 800–880 cm^{-1} to a bending mode of OH bound to Fe^{3+} cations. A broad band centered near 800–850 cm^{-1} is observed in the transmittance spectra, but only very weakly, if at all, in the reflectance spectra of our samples. An additional feature is observed at ~700 cm^{-1} in the transmittance spectra and from 725–735 cm^{-1} in the reflectance spectra. Small differences such as these in the position of fundamental vibrations between transmittance and reflectance spectra of minerals are common (*e.g.*, SALISBURY *et al.*, 1991). We have assigned these features to an additional Fe—OH bending mode in schwertmannite. Two distinct Fe—OH bending modes are expected for schwertmannite because of two ferric sites, $FeO_3(OH)_3$ and $FeO_3(OH)_2O—SO_3$, brought on by interactions (or bridges) between the SO_4^{2-} anions and the schwertmannite structure.

Spectral features due to sulfate species

Detailed spectral analyses of hydroxylated ferric sulfates are provided by ROSS (1974). Four sulfate vibrational modes exist in infrared spectra and are defined (i) for isolated, tetrahedral (Td) SO_4^{2-} at 983 cm^{-1} as ν_1 (symmetric stretch), at 450 cm^{-1} as ν_2 (symmetric bend), at 1105 cm^{-1} as ν_3 (asymmetric stretch), and at 611 cm^{-1} as ν_4 (asymmetric bend), and (ii) for acid sulfates at 887 cm^{-1} as ν_1 (S—OH stretch), at 1051 cm^{-1} as ν_2 (SO_3 symmetric stretch), at 594 cm^{-1} as ν_3 (SO_3 symmetric bend), at 1200 cm^{-1} as ν_4 (SO_3 asymmetric stretch), at 594 cm^{-1} as ν_5 (degenerate OSO bend), and at 429 cm^{-1} as ν_6 (degenerate OH—S—O bend). The form of the sulfate species in schwertmannite is probably somewhere in between these two forms. Evidence for these spectral modes is observed in schwertmannite spectra (Fig. 6) and summarized in Table 3 for the samples measured here.

Numerous spectral bands due to SO_4^{2-} species have been observed in several schwertmannite and related samples in the mid-infrared region (BIGHAM *et al.*, 1990, 1994). The strength and presence of these features was found to be dependent on the crystallinity of the sample (BIGHAM *et al.*, 1990) and band assignments were made for an absorption feature at 976 cm^{-1} to the ν_1 mode, for three bands at 1186, 1124, and 1038 cm^{-1} to a splitting of the ν_3 mode, and for a feature at 608 cm^{-1} to the ν_4 mode (BIGHAM *et al.*, 1994).

The spectra of samples Bt-4 and Z510b shown in Fig. 6 exhibit a strong ν_3 sulfate vibration near 1100 cm^{-1}, a medium-strength ν_4 vibration at 610 cm^{-1}, and a weak ν_1 vibration at 980 cm^{-1}, which are similar to those previously assigned by BIGHAM *et al.* (1990, 1994). Spectral features in the range 445–480 cm^{-1} have been assigned to a ν_2 sulfate mode for hydrated ferric sulfates (ROSS, 1974). Therefore, the absorption feature observed at 465 cm^{-1} in the transmittance spectra of samples Bt-4 and Z510b, is assigned here to a ν_2 symmetric sulfate bending vibration. When the sulfate group exhibits lower than Td symmetry, the ν_2, ν_3, and ν_4 sulfate bending vibrations are often split into multiple vibrations (FARMER, 1974). A strong spectral feature is observed near 1045 and 1070 cm^{-1} in transmittance spectra of samples Bt-4 and Z510b, respectively (Fig. 6, Table 3). A similar feature has been assigned as a degenerate ν_3 vibration in schwertmannite (BIGHAM *et al.*, 1994), but may also be partially due to a ν_2 SO_3 symmetric stretching vibration of acid sulfate species, forming in the schwertmannite tunnels or along the grain surfaces. The ν_1 vibrational feature near 980 cm^{-1} is often not observed in sulfates unless the structure is slightly distorted (ROSS, 1974). This deviation from Td structure is accompanied by the appearance of a band near 530 cm^{-1} and a shifting of the ν_4 feature near 610 cm^{-1} towards lower wavenumbers. As the ν_2 and ν_4 vibrations are often degenerate for lower symmetry sites, the feature observed near 520 cm^{-1} in the schwertmannite transmittance spectra could be due to either a ν_2 or ν_4 vibration; alternatively, this feature could be due to the presence of acid sulfate groups.

In addition, a weak spectral feature is observed near 1200 cm^{-1} in the reflectance, transmittance and ATR spectra of both samples (Fig. 6), that has been assigned to a degenerate ν_3 vibration in schwertmannite (BIGHAM *et al.* (1994), but would also be consistent with an acid sulfate vibrational mode (ROSS, 1974). Transmittance spectra of schwertmannite mixed with KBr (not shown) exhibited a much stronger absorption near 1200 cm^{-1} indicating that this feature is related to reaction of the sample. KBr reacting with water in the schwertmannite could be protonating some of the sulfate species. Protonation or hydration of the $O—SO_3^-$ ligand is likely if water is present and probably occurs instead of formation of bridging ligands in some cases. BIGHAM *et al.* (1990) pointed out that stretching vibrations occurring at wavenumbers greater than 1200 cm^{-1} indicate the presence of non-bridging sulfate bonds, whereas stretching vibrations occurring at wavenumbers less than 1200 cm^{-1} indicate the presence of sulfates as bidentate bridging complexes (NAKAMOTO, 1986). The non-bridging sulfates would be more likely to form S—OH acid sulfate bonds than the sulfate species acting as bidentate (or doubly) bridging ligands. This is consistent with the spectral feature observed near 1220 cm^{-1} in reflectance spectra (or near 1190 cm^{-1} in transmittance spectra) as resulting from an acid sulfate stretching vibration. The fact that this feature is observed weakly in all spectra indicates that some of the sulfate species in schwertmannite exist in the form of $SO_3(OH)$, as well as SO_4.

Weak spectral features near 1130 and 1030 cm^{-1} were observed in spectra of ferric sulfate-bearing montmorillonites and the intensity of these features was found to increase with the sulfate content of the samples (BISHOP *et al.*, 1995). These are similar to weak features observed in reflectance spectra of schwertmannite and strong features observed in transmittance spectra of schwertmannite.

Recent telescopic observations of Mars have shown a weak feature near 4.5 μm that has been attributed to sulfates in the surface material or atmospheric dust (BLANEY and MCCORD, 1995). Comparison of spectra from 4.4–5.1 μm across several regions of Mars (BLANEY and MCCORD, 1995) shows variations in the spectral slope in the range 4.5–4.8 μm. The reflectance spectra of schwertmannite (Fig. 5a) exhibit a relative reflectance maximum near 4.4 μm and a broad absorption feature from 4.5–5.0 μm. A decrease in reflectance of greater that 10% was observed in our spectra from 4.5–4.8 μm. Perhaps this broad schwertmannite feature is related to the subtle slope changes observed in the martian spectra. Other sulfate minerals, such as jarosite (Fig. 5b), as well as gypsum and anhydrite (BLANEY and MCCORD, 1995), exhibit sharper features in this region. Given mixing of multiple materials in the Mars soil and the large spatial regions observed, the specific sulfate mineral(s) responsible for the spectral observations near 4.5 μm cannot be determined at this time.

Variations in the spectral character of schwertmannites

The trace amounts of goethite and jarosite present in sample Bt-4 (as evidenced by XRD data and weak mid-IR spectral features) do not appear to influence the extended visible region spectral properties as the Bt-4 spectrum is brighter than the Z510b spectrum, whereas the impurities all give darker spectra in this region.

Subtle differences in the character of the extended visible region spectral features between the natural and synthetic schwertmannites (Fig. 2) imply minor structural variations between these two samples. The reflectance maximum in sample Bt-4 occurs at 0.740 μm for both the bulk and <45 μm samples, while it occurs at 0.730 μm for sample Z510b. The $6A_1 \rightarrow 4T_1$ band is broad and nearly flat from 0.900 to 0.915 μm for both samples, although the band minimum lies at 0.915 μm for sample Bt-4 and at 0.918 μm for sample Z510b. Based on colorimetric measurements, samples Bt-4 and Z510b are fairly similar to each other compared to the range of schwertmannites examined so far (*e.g.*, BIGHAM *et al.*, 1992). This is consistent with the similarities in the extended visible region reflectance spectra of samples Bt-4 and Z510b. Optical spectra of other schwertmannites have not yet been measured in this region.

Reflectance spectra of particulate samples have been observed to exhibit brighter near-infrared reflectance and slightly shallower band depths in the visible and near-infrared regions for smaller particle size separates than for larger size separates of the same mineral (*e.g.*, CROWN and PIETERS, 1987). No significant spectral differences were observed in the visible region as a function of particle size for sample Bt-4. Schwertmannite is generally composed of ultra fine-grained particles that cluster together forming larger aggregates, much like what is observed for smectites. Individual grain sizes of natural and synthetic schwertmannites are on the order of 0.2–0.5 μm (BIGHAM *et al.,* 1994). Therefore, the Bt-4 schwertmannite that was dry sieved in this study to <45 μm particle size may not be very different from the bulk schwertmannite sample. Near-infrared reflectance spectra of the bulk and <45 μm Bt-4 sample, measured under H_2O- and CO_2-purged conditions, showed no significant differences; however, spectra (not shown) of these samples measured under ambient conditions showed increased water absorption in the <45 μm sieved sample.

The reflectance spectra of the two schwertmannite samples measured here are very dark in the surface-scattering region. This is also consistent with small (few μm) grain size. A reflectance minimum or Christiansen feature is observed near 1160 cm^{-1} (~8.6 μm) for both particle size separates of Bt-4 and near 1220 cm^{-1} (~8.2 μm) with a shoulder near 1160 cm^{-1} (~8.6 μm) for sample Z510b. Some subtle variations in band position and shape exist, as well, for several of the mid-IR features shown in Fig. 6 and listed in Table 3. This is consistent with small variations in chemical composition and XRD analyses for several schwertmannites (BIGHAM *et al.,* 1990).

APPLICATIONS FOR MARS

Numerous materials are currently under study as potential analogs for the surface soil on Mars; however, few of these contain substantial amounts of S. The martian soils measured by Viking contained 6.5 to 9.5 wt.% SO_3 (TOULMIN *et al.,* 1977). Because of these large variations in S contents at different sites along the surface (TOULMIN *et al.,* 1977; BAIRD *et al.,* 1977), it is important and interesting to find out what sulfate minerals or salts are present in the martian soil. Perhaps knowing what these sulfate-bearing phases are will provide clues to the chemical weathering processes that occurred on Mars. How much of this fine-grained surface material is present on Mars is not well known, but it may be as deep as several meters in some areas, such as the Tharsis region (CHRISTENSEN and MOORE, 1992).

The extended visible region spectral character of schwertmannite includes a gently sloping shoulder near 0.6 μm, a reflectance maximum at 0.730–0.740 μm, and a broad absorption band having a minimum near 0.915 μm. This is consistent with the spectra of Mars in many regions (*e.g.,* MCCORD *et al.,* 1982; BELL *et al.,* 1990; MURCHIE *et al.,* 1993). The minerals schwertmannite and ferrihydrite often form together (BIGHAM *et al.,* 1992; BIGHAM *et al.,* 1994; SCHWERTMANN *et al.,* 1995); ferrihydrite has been found in at least one meteorite thought to have come from Mars (TREIMAN *et al.,* 1993) and exhibits spectral features similar to those observed for some regions on Mars (BISHOP *et al.,* 1993). BURNS (1994) and BIGHAM *et al.* (1996) have described the geochemical formation patterns of schwertmannite and ferrihydrite under warm and humid conditions, followed by long-term stability under dry conditions. Geochemical and thermodynamic analyses of possible weathering products on Mars indicate that all of the "essential ingredients" for formation of ferric oxyhydroxy sulfate minerals such as schwertmannite existed on Mars (BURNS, 1987, 1988). Although none of these findings are direct evidence for the presence of schwertmannite on Mars, they imply that schwertmannite could be a component of the martian soil.

Ferric sulfate-bearing montmorillonites exhibit similar spectral properties in the extended visible region, except that the broad band minimum occurs at slightly shorter wavelengths (~0.89 μm) than in the spectra of schwertmannite. As this band position in spectra of ferrihydrite, another mineral thought to be present in the ferric sulfate-bearing montmorillonites, is at still longer wavelengths (0.91–0.93 μm), a third ferric component with a shorter wavelength band here must be present as well in the ferric sulfate-bearing montmorillonites. A combination of schwertmannite, ferrihydrite, and smectites is a possible model for the surface material on Mars in the uncommon bright regions, where a longer-wavelength ferric band is observed. A combination of schwertmannite, ferrihydrite, hematite, and smectites is a possible model for the surface material on Mars in the more common bright regions, where the primary ferric absorption occurs near 0.86 μm. Other plausible analogs include terrestrial palagonitic soils and crystalline hematite—nanophase hematite—silicate mixtures. The fine-grained nature of these materials is consistent with the texture observed for the martian surface soil. Schwertmannite, ferrihydrite, and smectites are weathering products that could have formed during an earlier, wetter time on Mars, and would have persisted during the more recent period of aridity.

The concept of a salt-cemented soil, or "duricrust", on Mars was first presented by BINDER *et al.* (1977). Chemical analyses of this martian duricrust showed higher S and Cl abundances, but otherwise little chemical difference from the surface fines (CLARK *et al.,* 1982). Statistical analyses of these elemental trends indicate that sulfates on Mars are correlated with either Mg, Ca, Al or Fe cations (CLARK and VAN HART, 1981). GREELEY *et al.* (1992) suggested that this duricrust is composed of fine-grained particles held together by soluble salts. CLARK *et al.* (1982) interpreted duricrust on Mars as "sulfate salt-enriched and cemented variants of the fines material". The chemically-treated montmorillonites included in this study contain ferric sulfate- and ferric oxide-bearing species that were allowed to form in and around the smectite clay grains in an aqueous environment. Based on the spectral properties of these chemically-treated montmorillonites and other factors, clusters of schwertmannite, ferrihydrite and amorphous ferric iron and sulfate complexes coating the surfaces of smectite grains or aggregates have been proposed as a possible model for duricrust (BISHOP *et al.,* 1995). Were schwertmannite present on Mars, it could act as a binding agent for silicate particles and nanophase ferric oxides, thus resulting in a cemented soil or duricrust having an elevated S content, but otherwise similar chemical composition to the more dust-like surface material on Mars.

Schwertmannite forms under aqueous conditions at low pH in the presence of high concentrations of iron and sulfur and an oxidant (BIGHAM *et al.,* 1992). Schwertmannite often forms together with ferrihydrite, goethite, or jarosite. Which of these minerals form is highly dependent on the concentrations of Fe and S and the pH. SO_4^{2-} concentrations between 1000 and 3000 ppm and pH values between 3 and 4 favor schwertmannite formation (BIGHAM *et al.,* 1992). Future analyses of the martian surface material may provide evidence of schwertmannite or other ferric minerals, and thus provide information on the environmental conditions during the formation of these minerals.

Wind ablation and other physical factors have also acted on the martian surface material. Ferric minerals, such as schwertmannite or hematite, may have formed in mineral deposits, then with time become intermixed with the silicate-bearing phases. Alternatively, nanophase ferric minerals such as ferrihydrite or schwertmannite may have formed along smectite grain surfaces or in the smectite interlayer regions. The presence of silica could inhibit the transformation of schwertmannite to hematite as observed previously for ferrihydrite (CARLSON AND SCHWERTMANN, 1981). Nanophase ferric minerals, such as schwertmannite and ferrihydrite, could be precursors to hematite formed during thermal weathering. If localities on Mars are identified where the relative abundance of the nanophase minerals schwertmannite or ferrihydrite vs. the relative abundance of crystalline hematite are different, then those areas having reduced amounts of crystalline hematite could be more sheltered from the processes that produce hematite. Examination of the physical properties of the matrix materials (*i.e.* silicate phase) could lead to insights, as well, into the geochemical conditions during formation and the subsequent weathering of the martian soil.

Given the possibility of schwertmannite on Mars, a pertinent question is: how can it be detected there? The optical spectra of schwertmannite are distinct from those of many other iron oxides and sulfates; however, high spatial-resolution and high spectral-resolution spectra would be required to differentiate schwertmannite from all other ferric minerals. The dominant features in infrared reflectance spectra of schwertmannite are due to OH and H_2O, and are found in the spectra of other hydroxylated minerals such as ferrihydrite. Transmittance spectra of schwertmannite exhibit several strong sulfate and Fe—OH absorption features that are observed only weakly in reflectance spectra. Because of the fine-grained nature of schwertmannite, it is likely that these features would also be weak in emittance spectra. Therefore, high spectral-resolution spectroscopic observation of Mars might provide spectra consistent with that of schwertmannite, but uniquely identifying schwertmannite from visible and infrared spectra alone would be difficult.

KNUDSEN *et al.* (1991) and AGRESTI *et al.* (1992) have designed Mössbauer experiments for Mars that should be useful for characterizing the iron-rich phases on that planet. Because of the asymmetry of its Mössbauer spectra, and because the magnetic hyperfine field of schwertmannite at 4.2 K is lower than that of any iron oxide and jarosite, Mössbauer spectra taken at 4.2 K would be suitable for the identification of this mineral, and its quantification in assemblages of complex iron mineralogy (MURAD *et al.,* 1994). Mössbauer spectra taken in the paramagnetic state are also unique and would allow the identification of schwertmannite in monomineralic samples. The Mössbauer spectra of paramagnetic schwertmannite, however, are not distinctive enough to discern the presence of this mineral in complex mineral mixtures containing substantial amounts of other paramagnetic ferric components.

The best procedure for detecting schwertmannite, as in the case of other minerals, would be to combine the results of as many techniques as possible. Optical, infrared and Mössbauer spectroscopy, X-ray fluorescence, and X-ray diffraction would be useful. A sample return mission might facilitate a number of these analyses; however, as mentioned by KNUDSEN *et al.* (1991), it is possible that the oxidation state of the iron component might be altered during transit. For this reason, *in situ* measurements of the martian surface materials would be advantageous.

CONCLUSIONS

The reflectance spectrum of schwertmannite is dominated by a broad ferric band near 0.91 μm and a broad water band near 3 μm. These features are consistent with spectra of some areas on Mars. The optical and Mössbauer spectra show evidence of the ferric sites in the schwertmannite structure, namely Fe(III) octahedrally coordinated as $FeO_3(OH)_3$ and $FeO_3(OH)_2O—SO_3$. The extended visible region reflectance spectra of natural and synthetic schwertmannite are characterized by a steep incline near 0.5 μm, a gently sloping shoulder near 0.6 μm, a reflectance maximum at 0.73–74 μm, and a broad band with a minimum at 0.915–0.918 μm. Additional infrared spectral features due to bound and adsorbed water are observed at 1.45, 1.95, and 6.1 μm. Numerous spectral features from ~8–25 μm (~1200–400 cm^{-1}) are due to sulfate and OH species.

The results of optical, infrared, and Mössbauer spectroscopy indicate that ferric sulfate-bearing montmorillonites, prepared as Mars soil analog materials, contain schwertmannite or a schwertmannite-like phase. Schwertmannite could be present in the surface material on Mars, intermixed with other ferric minerals, such as ferrihydrite, and with silicates, such as smectites. High-resolution visible to infrared spectroscopy and low-temperature Mössbauer spectroscopy of the martian soil would be useful for detection of schwertmannite if it is present on Mars. Identifying the ferric oxide minerals and silicate components of the martian soil would provide information about the chemical weathering processes on Mars.

Acknowledgements—We are indebted to U. SCHWERTMANN for placing schwertmannite samples at our disposal and providing access to room-temperature Mössbauer facilities and to F. E. WAGNER for providing access to liquid-helium Mössbauer facilities. We would like to express gratitude to A. DUMMEL, T. HIROI and S. PRATT for technical assistance and to J. F. BELL, V. C. FARMER, R. V. MORRIS, S. MURCHIE, and U. SCHWERTMANN for helpful editorial comments. Financial support from the Alexander von Humboldt Foundation and use of the facilities at the DLR-Berlin are much appreciated by J. BISHOP. The RELAB and Nicolet spectrometers at Brown University are part of a multi-user system supported by NASA grant NAGW-748.

REFERENCES

ADAMS J. and MCCORD T. B. (1969) Mars: Interpretation of spectral reflectivity of light and dark regions. *J. Geophys. Res.* **74**, 4851–4856.

AFANASEV A. M., GOROBCHENKO V. D., KULGAWCZUK D. S. and LUKASHEVICH I. I. (1974) Nuclear γ-resonance in iron sulphates of the jarosite group. *Phys. Stat. Sol.(a)* **26**, 697–701.

AGRESTI D. G., MORRIS R. V., WILLS E. L., SHELFER T. D., PIMPERL M. M., SHEN M.-H., CLARK B. C. and RAMSEY B. D. (1992) Extraterrestrial Mössbauer spectrometry. *Hyperfine Interactions* **72**, 285–298.

ARNOLD G. (1992) Infrarotsondierung planetarer Oberflächen. *Dtsch. Luft Raumfahrt-Nachrichten* **69**, 25–31.

BAIRD A. K., CASTRO A. J., CLARK B. C. III, TOULMIN P. III, ROSE H. J. JR., KEIL K. and GOODING J. L. (1977) The Viking X ray fluorescence experiment: Sampling strategies and laboratory simulations. *J. Geophys. Res.* **82**, 4595–4624.

BANIN A. and RISHPON J. (1979) Smectite clays in Mars soil: Evidence for their presence and role in viking biology experimental results. *J. Molecular Evolution* **14,** 133–52.

BANIN A., CARLE G., CHANG S., COYNE L., ORENBERG J. and SCATTERGOOD T. (1988) Laboratory investigations of Mars: Chemical and spectroscopic characteristics of a suite of clays as Mars soil analogs. *Origins of Life and Evolution of the Biosphere* **18,** 239–265.

BANIN A., CLARK B. C. and WÄNKE H. (1992) Surface chemistry and mineralogy. In *Mars* (eds. H. KIEFFER *et al.*), pp. 594–625. Univ. Arizona Press, Tucson.

BANIN A., BEN-SHLOMO T., MARGULIES L., BLAKE D. F., MANCINELLI R. L. and GEHRING A. U. (1993) The nanophase iron minerals in Mars soil. *J. Geophys. Res.* **98,** 20831–20853.

BEINROTH A. and ARNOLD G. (1996) Analysis of weak surface absorption bands in the ISM (Phobos-2) near-infrared spectra of Mars. *Vibrational Spectroscopy,* in press. **11,** in press.

BELL J. F. III (1992) Charge-coupled device imaging spectroscopy of Mars. 2. Results and implications for martian ferric mineralogy. *Icarus* **100,** 575–597.

BELL J. F. III and CRISP D. (1993) Groundbased imaging spectroscopy of Mars in the near-infrared: Preliminary results. *Icarus* **104,** 2–19.

BELL J. F. III, MCCORD T. B. and OWENSBY P.D. (1990) Observational evidence of crystalline iron oxides on Mars. *J. Geophys. Res.* **95,** 14447–14461.

BELL J. F. III, POLLACK J. B., GEBALLE T. R., CRUIKSHANK D. P. and FREEDMAN R. (1994) Spectroscopy of Mars from 2.04 to 2.44 μm during the 1993 opposition: Absolute calibration and atmospheric vs mineralogic origin of narrow absorption features. *Icarus* **111,** 106–123.

BIBRING J.-P., COMBES M., LANGEVIN Y., CARA C., DROSSART P., ENCRENAZ T., ERARD S., FORNI O., GONDET B., KSANFOMALITI L., LELLOUCH E., MASSON P., MOROZ V., ROCARD F., ROSENQVIST J., SOTIN C. and SOUFFLOT A. (1990) ISM observations of Mars and Phobos: First results. *Proceedings 20th Lunar Planet. Science Conf.,* LPI, Houston, pp. 461–471.

BIGHAM J. M., SCHWERTMANN U., CARLSON L. and MURAD E. (1990) A poorly crystallized oxyhydroxysulfate of iron formed by bacterial oxidation of Fe(II) in acid mine waters. *Geochim. Cosmochim. Acta* **54,** 2743–2758.

BIGHAM J. M., SCHWERTMANN U. and CARLSON L. (1992) Mineralogy of precipitates formed by the biogeochemical oxidation of Fe(II) in mine drainage. In *Biomineralization Processes of Iron and Manganese—Modern and Ancient Environments* (eds. H. C. W. SKINNER and R. W. FITZPATRICK), pp. 219–232, Catena Verlag.

BIGHAM J. M., CARLSON L. and MURAD E. (1994) Schwertmannite, a new iron oxyhydroxysulfate from Pyhäsalmi, Finland, and other localities. *Miner. Mag.* **58,** 641–648.

BIGHAM J. M., SCHWERTMANN U., TRAINA S. J., WINLAND R. L. and WOLF M. (1996) Schwertmannite and the chemical modeling of iron in acid sulfate waters. *Geochim. Cosmochim. Acta* **60,** 2111–2121.

BINDER A. B., ARVIDSON R. E., GUINNESS E. A., JONES K. L., MORRIS E. C., MUTCH T. A., PIERI D. C. and SAGAN C. (1977). The geology of the Viking Lander 1 site. *J. Geophys. Res.* **82,** 4439–4451.

BISHOP J. L. and PIETERS C. M. (1995) Low-temperature and low atmospheric pressure infrared reflectance spectroscopy of Mars soil analog materials. *J. Geophys. Res.* **100,** 5369–5379.

BISHOP J. L., PIETERS C. M. and BURNS R. G. (1993) Reflectance and Mössbauer spectroscopy of ferrihydrite-montmorillonite assemblages as Mars soil analog materials. *Geochim. Cosmochim. Acta* **57,** 4583–4595.

BISHOP J. L., PIETERS C. M. and EDWARDS J. O. (1994) Spectroscopic analyses on the nature of water in montmorillonite. *Clays Clay Minerals* **42,** 702–716.

BISHOP J. L., PIETERS C. M., BURNS R. G., EDWARDS J. O., MANCINELLI R. L. and FRÖSCHL H. (1995) Reflectance spectroscopy of ferric sulfate-bearing montmorillonites as Mars soil analog materials. *Icarus* **117,** 101–119.

BLANEY D. L. and MCCORD T. B. (1989) An observational search for carbonates on Mars. *J. Geophys. Res.* **94,** 10159–10166.

BLANEY D. L. and MCCORD T. B. (1995) Indications of sulfate minerals in the martian soil from Earth-based spectroscopy. *J. Geophys. Res.* **100,** 14433–14441.

BRADY K. S., BIGHAM J. M., JAYNES W. F. and LOGAN T. J. (1986) Influence of sulfate on Fe-oxide formation: Comparisons with a stream receiving acid mine drainage. *Clays Clay Miner.* **34,** 226–274

BURNS R. G. (1987) Ferric sulfates on Mars. *Proc. 17th Lunar Planet. Sci. Conf.* LPI, Houston, E570–574.

BURNS R. G. (1988) Gossans on Mars. *Proc. 18th Lunar Planet. Sci. Conf.* LPI, Houston, 713–721.

BURNS R. G. (1993a) Rates and mechanism of chemical weathering of ferromagnesian silicate minerals on Mars. *Geochim. Cosmochim. Acta* **57,** 4555–4574.

BURNS R. G. (1993b) *Mineralogical Applications of Crystal Field Theory.* 2nd Ed. Cambridge Univ. Press.

BURNS R. G. (1994) Schwertmannite on Mars: Deposition of this ferric oxyhydroxysulfate mineral in acidic saline meltwaters (abstr.). *Lunar and Planet. Sci.* **XXV,** 203–204.

BURNS R. G. and FISHER D. S. (1993) Rates of oxidative weathering on the surface of Mars. *J. Geophys. Res.* **98,** 3365–3372.

CARLSON L. and SCHWERTMANN U. (1981) Natural ferrihydrites in surface deposits from Finland and their association with silica. *Geochim. Cosmochim. Acta* **45,** 421–429.

CARR M. H. (1981) *The Surface of Mars.* Yale Univ. Press, New Haven.

CARR M. H. (1995) The martian drainage system and the origin of valley networks and fretted channels. *J. Geophys. Res.* **100,** 7479–7507.

CHRISTENSEN P. R. and MOORE H. J. (1992) The martian surface layer. In *Mars* (eds. H. KIEFFER *et al.*), pp 686–729. Univ. Arizona Press, Tucson.

CLARK B. C. (1993) Geochemical components in martian soil. *Geochim. Cosmochim. Acta* **57,** 4575–4581.

CLARK B. C. and VAN HART D. C. (1981) The salts of Mars. *Icarus* **45,** 370–378.

CLARK B. C. III, BAIRD A. K., ROSE H. J. JR., TOULMIN P. III, CHRISTIAN R. P., KELLIHER W. C., CASTRO A. J., ROWE C. D., KEIL K. and HUSS G. R. (1977) The Viking X ray fluorescence experiment: Analytical methods and early results. *J. Geophys. Res.* **82,** 4577–4594.

CLARK B. C., BAIRD A. K., WELDON R. J., TSUSAKI D. M., SCHNABEL L. and CANDELARIA M. P. (1982) Chemical composition of martian fines. *J. Geophys. Res.* **87,** 10059–10067.

CLARK R. N. and ROUSH T. L. (1984) Reflectance spectroscopy: Quantitative analysis techniques for remote sensing applications. *J. Geophys. Res.* **89,** 6329–6340.

CLARK R. N., SWAYZE G. A., SINGER R. B. and POLLACK J. B. (1990) High-resolution reflectance spectra of Mars in the 2.3 μm region: Evidence for the mineral scapolite. *J. Geophys. Res.* **95,** 14463–14480.

CONEL J. E. (1969) Infrared emissivities of silicates: Experimental results and a cloudy atmosphere model of spectral emission from condensed particulate mediums. *J. Geophys. Res.* **74,** 1614–1634.

CROWN D. A. and PIETERS C. M. (1987) Spectral properties of plagioclase and pyroxene mixtures and the interpretation of Lunar soil spectra. *Icarus* **72,** 492–506.

ERARD S., BIBRING J.-P., MUSTARD J., FORNI O., HEAD J. W., HURTREZ S., LANGEVIN Y., PIETERS C. M., ROSENQVIST J. and SOTIN C. (1991) Spatial variations in composition of the Valles Marineris and Isidis Planitia regions of Mars derived from ISM data. *Proc. 21st Lunar Planet. Sci. Conf.,* LPI, Houston, 437–455.

FANALE F. P., POSTAWKO S. E., POLLACK J. B., CARR M. H. and PEPIN R. O. (1992) Mars: Epochal climate change and volatile history. In *Mars* (eds. H. KIEFFER *et al.*), pp. 1135–1179. Univ. Arizona Press, Tucson.

FARMER J. (1995) Did Mars once harbor life? *EOS* **76,** 442.

FARMER V. C. (1974) The layer silicates. In *The Infrared Spectra of Minerals* (ed. V. C. FARMER), pp. 331–363. Bartholomew Press.

GAFFEY S. J., MCFADDEN L. A., NASH D. and PIETERS C. M. (1993) Ultraviolet, visible, and near-infrared reflectance spectrsocopy: Laboratory spectra of geologic materials. In *Remote Geochemical Analyses: Elemental and Mineralogical Composition* (eds. C. M. PIETERS and P. ENGLERT), pp. 367–393. Cambridge Univ. Press.

GOODEN D. C., MORRIS R. V., MING D. W., LAUER H. V. JR. and YANG S. R. (1993) Mineralogy of three slightly palagonitized basaltic tephra samples from the summit of Mauna Kea, Hawaii. *J. Geophys Res.* **98,** 3401–3411.

GOODING J. L. (1978) Chemical weathering on Mars. *Icarus* **33,** 483–513.

GOODING J. L. and KEIL K. (1978) Alteration of glass as a possible source of clay minerals on Mars. *Geophys. Res. Let.* **5,** 727–730.

GOODING J. L., ARVIDSON R.E. and ZOLOTOV M. Y. (1992) Physical and chemical weathering. In *Mars* (eds. H. KIEFFER *et al.*), pp. 626–651. Univ. Arizona Press, Tucson.

GREELEY R., LANCASTER N., LEE S. and THOMAS P. (1992) Martian aeolian processes, sediments, and features. In *Mars* (eds. H. KIEFFER *et al.*), pp.730–766. Univ. Arizona Press, Tucson.

HANEL R. B., CONRATH B., HOVIS W., KUNDE V., LOWMAN P., MAGUIRE W., PEARL J., PIRRAGLIA J., PRABHAKARA C. and SCHLACHMAN B. (1972) Investigation of the martian environment by infrared spectroscopy on Mariner 9. *Icarus* **17,** 423–442.

HOUCK J. R., POLLACK J. B., SAGAN C., SCHAACK D. and DECKER J. A. JR. (1973) High altitude infrared spectroscopic evidence for bound water on Mars. *Icarus* **18,** 470–480.

HUHEEY J. E., KEITER E. A. and KEITER R. L. (1993) *Inorganic Chemistry*. 4th Ed. HarperCollins, New York.

HUNT G. R., LOGAN L. M. and SALISBURY J. W. (1973) Mars: Components of infrared spectra and the composition of the dust cloud. *Icarus* **18,** 459–469.

JAKOSKY B. M. and HABERLE R. M. (1992) The seasonal behavior of water on Mars. in *Mars*, eds. H. KIEFFER *et al.*, Univ. Arizona Press, Tucson, 969–1016.

KIEFFER H. H., MARTIN T. Z., PETERFREUND A. R., JAKOSKY B. M., MINER E. D. and PALLUCONI F. D. (1977) Thermal and albedo mapping of Mars during the Viking primary mission. *J. Geophys. Res.* **82,** 4249–4291.

KNUDSEN J. M., MADSEN M. B., OLSEN M., VISTISEN L., KOCH C. B., MØRUP S., KANKELEIT E., KLINGELHÖFER G., EVLANOV E. N., KHROMOV V. N., MUKHIN L. M., PRILUTSKI O. F., ZUBKOV B., SMIRNOV G. V. and JUCHNIEWICZ J. (1991) Mössbauer spectroscopy on the surface of Mars. Why? *Hyperfine Interact.* **68,** 83–94.

MCCORD T. B., HUGUENIN R. L., MINK D. and PIETERS C. (1977) Spectral reflectance of martian areas during the 1973 opposition: Photoelectric filter photometry. *Icarus* **31,** 25–39.

MCCORD T. B., CLARK R. N. and HUGUENIN R. L. (1978) Mars: Near-infrared spectral reflectance and compositional implications. *J. Geophys. Res.* **83,** 5433–5441.

MCCORD T. B., CLARK R. N. and SINGER R. B. (1982) Mars: Near-infrared spectral reflectance of surface regions and compositional implications. *J. Geophys. Res.* **87,** 3021–3032.

MORRIS R. V. and LAUER H. V. JR. (1990) Matrix effects for reflectivity spectra of dispersed nanophase (superparamagnetic) hematite with application to martian spectral data. *J. Geophys Res.* **95,** 5101–5109.

MORRIS R. V., LAUER H. V. JR., LAWSON C. A., GIBSON E. K. JR., NACE G. A. and STEWART C. (1985) Spectral and other physicochemical properties of submicron powders of hematite (α-Fe_2O_3), maghemite (γ-Fe_2O_3), magnetite (Fe_3O_4), goethite (α-FeOOH), and lepidocrocite (γ-FeOOH). *J. Geophys. Res.* **90,** 3126–3144.

MORRIS R. V., AGRESTI D. G., LAUER H. V. JR., NEWCOMB J. A., SHELFER T. D. and MURALI A. V. (1989) Evidence for pigmentary hematite on Mars based on optical, magnetic and Mössbauer studies of superparamagnetic (nanocrystalline) hematite. *J. Geophys Res.* **94,** 2760–2778.

MORRIS R. V., GOODING J. L., LAUER H. V. JR. and SINGER R. B. (1990) Origins of marslike spectral and magnetic properties of a Hawaiian palagonitic soil. *J. Geophys Res.* **95,** 14427–14434.

MORRIS R. V., SCHULZE D. G., LAUER H. V. JR., AGRESTI D. G. and SHELFER T. D. (1992) Reflectivity (visible and near IR), Mössbauer, static magnetic, and X ray diffraction properties of aluminum-substituted hematites. *J. Geophys Res.* **97,** 10257–10266

MORRIS R. V., GOLDEN D. C., BELL J. F., LAUER H. V. and ADAMS J. B. (1993) Pigmenting agents in martian soils: Inferences from spectral, Mössbauer, and magnetic properties of nanophase and other iron oxides in Hawaiian palagonitic soil PN-9. *Geochim. Cosmochim. Acta* **57,** 4597–4609.

MURAD E. (1988) The Mössbauer spectrum of ''well''-crystallized ferrihydrite. *J. Magnetism Magnetic Mater.* **74,** 153–157.

MURAD E. (1996) Magnetic properties of microcrystalline iron(III) oxides and related materials as reflected in their Mössbauer spectra. *Phys. Chem. Miner.* **23,** in press.

MURAD E. and JOHNSTON J. H. (1987) Iron oxides and oxyhydroxides. In *Mössbauer Spectroscopy Applied to Inorganic Chemistry* (ed. G. J. Long), Vol.2, pp. 507–582. Plenum Pub.

MURAD E., BIGHAM J. M., BOWEN L. H. and SCHWERTMANN U. (1990) Magnetic properties of iron oxides produced by bacterial oxidation of Fe^{2+} under acid conditions. *Hyperfine Interact.* **58,** 2373–2376.

MURAD E., SCHWERTMANN U., BIGHAM J. M. and CARLSON L. (1994) The mineralogical characteristics of poorly crystalline precipitates formed by oxidation of Fe^{2+} in acid sulfate waters. In *Environmental Geochemistry of Sulfide Oxidation* (eds. C. N. ALPERS and D. W. BLOWES), pp. 191–200, American Chemical Society.

MURCHIE S., MUSTARD J., BISHOP J., HEAD J. and PIETERS C. (1993) Spatial variations in the spectral properties of bright regions on Mars. *Icarus* **105,** 454–468.

NAKAMOTO K. (1986) *Infrared and Raman Spectra of Inorganic and Coordination Compounds,* 4th ed. J. Wiley & Sons.

ORENBERG J. and HANDY J. (1992) Reflectance spectroscopy of palagonite and iron-rich montmorillonite clay mixtures: Implications for the surface composition of Mars. *Icarus* **96,** 219–225.

PIETERS C. M. (1983) Strength of mineral absorption features in the transmitted component of near-infrared reflected light: First results from RELAB. *J. Geophys Res.* **88,** 9534–9544.

PIMENTEL G. C., FORNEY P. B. and HERR K. C. (1974) Evidence about hydrate and solid water in the martian surface from the 1969 Mariner infrared spectrometer. *J. Geophys. Res.* **79,** 1623–1634.

POLLACK J. B., KASTING J. F., RICHARDSON S. M. and POLIAKOFF K. (1987) The case for a warm, wet climate on early Mars. *Icarus* **71,** 203–224.

POLLACK J. B., ROUSH T., WITTEBORN F., BREGMAN J., WOODEN D., STOKER C., TOON O., RANK D., DALTON B. and FREEDMAN R. (1990) Thermal emission spectra of Mars (5.4–10.5 μm): Evidence for sulfates, carbonates, and hydrates. *J. Geophys. Res.* **95,** 14595–14627.

ROSS S. D. (1974) Sulphates and other oxy-anions of Group VI. In *The Infrared Spectra of Minerals* (ed. V. C. FARMER), pp. 423–444. Bartholomew Press.

ROUSH T. L., BLANEY D. L. and SINGER R. B. (1993) The surface composition of Mars as inferred from spectroscopic observations. In *Remote Geochemical Analyses: Elemental and Mineralogical Composition* (eds. C. M. PIETERS and P. ENGLERT), pp. 367–393. Cambridge Univ. Press.

SALISBURY J. W. (1993) Mid-infrared spectroscopy: Laboratory data. In *Remote Geochemical Analyses: Elemental and Mineralogical Composition* (eds. C. M. PIETERS and P. ENGLERT), pp. 367–393. Cambridge Univ. Press.

SALISBURY J. W., WALTER L. S., VERGO N. and D'ARIA D. M. (1991) *Infrared (2.1–25 μm) Spectra of Minerals,* Johns Hopkins Univ. Press.

SCHWERTMANN U. and CORNELL R. M. (1991) *Iron Oxides in the Laboratory,* VCH Publishers.

SCHWERTMANN U., BIGHAM J. and MURAD E. (1995) The first occurrence of schwertmannite in a natural stream environment. *Eur. J. Miner.* **7,** 547–552.

SHERMAN D. M. and WAITE T. D. (1985) Electronic spectra of Fe^{3+} oxides and oxide hydroxides in the near-IR to near-UV. *Amer. Miner.* **70,** 1262–1269.

SHERMAN D. M., BURNS R. G. and BURNS V. M. (1982) Spectral characterization of the iron oxides with application to the martian bright region mineralogy. *J. Geophys Res.* **87,** 10,169–10,180.

SINGER R. B. (1982) Spectral evidence for the mineralogy of high-albedo soils and dust on Mars. *J. Geophys. Res.* **87,** 10,159–10,168.

SINGER R. B., MCCORD T. B., CLARK R. N., ADAMS J. B. and HUGUENIN R. L. (1979) Mars surface composition from reflectance spectroscopy: A summary. *J. Geophys. Res.* **84,** 8415–8426.

SODERBLOM L. A. (1992) The composition and mineralogy of the martian surface from spectroscopic observations: 0.3 μm to 50 μm. In *Mars* (eds. H. KIEFFER *et al.*), pp. 557–593. Univ. Arizona Press, Tucson.

TOULMIN P. III, BAIRD A. K., CLARK B. C., KEIL K., ROSE H. J. JR., CHRISTIAN R. P., EVANS P. H. and KELLIHER W. C. (1977) Geochemical and mineralogical interpretation of the Viking inorganic chemical results. *J. Geophys. Res.* **82,** 4625–4634.

TREIMAN A. H., BARRETT R. A. and GOODING J. L. (1993) Preterrestrial aqueous alteration of the Lafayette (SNC) meteorite. *Meteoritics* **28,** 86–97.

Mineral Spectroscopy: A Tribute to Roger G. Burns
© The Geochemical Society, Special Publication No. 5, 1996
Editors: M. D. Dyar, C. McCammon and M. W. Schaefer

Iron, sulfate, carbonate, and hydrated minerals on Mars

JAMES F. BELL III

Cornell University, Department of Astronomy, Center for Radiophysics and Space Research, Ithaca, NY 14853–6801 U.S.A.

Abstract—The Martian surface is a mixture of highly oxidized secondary minerals and relatively pristine basalts. The currently-known and inferred mineralogic inventory provides evidence for possible large-scale climatic variability during past epochs. Because they are often formed under unique environmental conditions, the most diagnostic minerals that can be used to assess the duration and magnitude of past climatic changes are ferric oxides/oxyhydroxides, sulfates, carbonates, and hydrates. This paper (a) reviews the current telescopic and spacecraft observational data and interpretations regarding the occurrence of these phases on Mars; (b) presents some new results and interpretations from visible to near-IR telescopic observations of Mars obtained in 1990; (c) outlines some new models for the weathering and alteration of surface materials that are consistent with the available data; and (d) defines a set of observational measurements (telescopic, spacecraft, and in situ) that can be used to discriminate between possible weathering and alteration scenarios, as well as to provide important new constraints on the composition and mineralogy of the Martian surface.

PREFACE

TO THOSE WHO use spectroscopy to study the mineralogy of planets, satellites, and asteroids, Roger Burns was the person who literally wrote the book that defines our field (BURNS, 1970, 1993a). This extraordinary book, which was revised and updated by Roger just prior to his death, is required reading for every graduate student entering the field, and is an essential element of the reference libraries of the many professionals who have gone on to pursue careers in planetary spectroscopy over the past several decades. In particular, Roger's detailed theoretical treatment of absorption features in visible to near-infrared spectra of Fe-bearing phases like pyroxene, olivine, and iron oxides spurred a generation of observational astronomers to (successfully) search for the occurrence of these minerals on other solar system bodies. Beyond this initial impetus, Roger also provided consistent and valuable theoretical and laboratory guidance throughout his prolific career on the interpretation of the spectral features that were detected.

Mars was a particular favorite of Roger's, and was the focus of much of his non-terrestrial research. Specifically, Roger realized the potential importance that a small number of spectrally-active minerals could have towards elucidating both the current and the past chemical weathering environment of the red planet (*e.g.*, BURNS, 1980, 1987, 1988, 1993b). To Roger, the question "Why is Mars red?" demanded an answer not only dealing with which specific minerals are responsible for the color of the surface, but also with which specific surface weathering or alteration processes can be inferred based on the current mineral inventory (*e.g.*, BURNS, 1993; BURNS and FISHER, 1990, 1993).

Roger was an enthusiastic supporter of efforts to obtain new spectroscopic observations of Mars at key wavelengths, and his excitement with new or controversial discoveries was especially inspiring to young researchers. His probing comments, gentle criticisms, and insightful suggestions for follow-up investigations are keenly missed among the planetary science community. While he would certainly have blushed over such praise and the attention that his research is receiving in this volume and elsewhere, there is perhaps no other researcher who has studied the details of Mars surface mineralogy whose impact was so large, knowledge so broad, and opinions so well respected.

INTRODUCTION

The answer to the question "Why is Mars red?" is that sunlight incident on the surface is preferentially absorbed from the ultraviolet to about 0.6 μm via a series of ligand field and cation pair transitions in ferric iron-bearing minerals (*e.g.*, SHERMAN *et al.*, 1982; BELL *et al.*, 1990a). This is well understood, although important debates continue about what specific ferric minerals are responsible for these absorption features. The statement "Mars is red" applies to the planet as a whole, and usually viewed from a large distance. Upon closer examination, a perhaps more interesting question arises: "Why isn't Mars all red?" Or, more specifically, why are some regions of Mars covered by bright, heavily oxidized ferric-rich minerals (SINGER, 1982; BELL *et al.*, 1990a) while many adjacent regions appear to be much less altered mineralogically, containing high abundances of dark, relatively unoxidized ferrous-rich minerals (SINGER *et al.*, 1979; MUSTARD *et al.*, 1993)? I refer to this as the "Mars Weathering Paradox", the solution of which has important implications for understanding the physical and chemical weathering and alteration processes that have operated on Mars.

The abundance, form, and distribution of sulfates and carbonates on the Martian surface also have important implications for past weathering environments, the most important of which may have been acidic groundwaters that chemically altered pristine mafic crustal rocks. The detection of specific hydroxo ferric sulfate and bisulfate ions, Fe-, Al- and Mg-bearing sulfate minerals, bicarbonate ions, carbonate minerals, and clay silicates could provide key indicators of the past and present pH of Martian soils and of the efficiency of chemical weathering under current Martian climatic conditions (BURNS, 1987).

Finally, it has been known since the discovery of the strong 3-μm absorption feature in Mars spectra that the surface minerals contain OH and/or bound H_2O (*e.g.*, HOUCK *et al.*, 1973). The specific abundance and mineralogy of the OH/H_2O-bearing phase(s) has remained uncertain, however. The importance of this issue in terms of surface weathering and alteration is similar to the importance of water in terrestrial weathering processes: water acts as a catalyst to substantially speed chemical reactions on mineral surfaces because reactions involving aqueous phases proceed orders of magnitude faster than gas-solid phase reactions (GOODING and KEIL, 1978; BURNS and FISHER, 1993). Other factors like temperature and pH also strongly influence chemical reaction rates on Mars (BURNS, 1993b), as they do on Earth.

In this paper, I review the currently-available spectroscopic data on the iron, sulfate, carbonate, and OH-bearing mineralogy of the Martian surface, present some new observational results, and discuss some new models of surface weathering processes that are consistent with the available data. Particular emphasis is placed on current and past attempts to identify the types of minerals that Roger Burns first realized were so important for untangling complex and often non-unique surface weathering processes. For clarity, in Table 1 I define many of the specific weathering and mineralogical terms used throughout this paper.

MARS FE MINERALOGY

Background

The most spectrally-active cation with relevance to Mars is iron. Fortunately, the Martian crust and mantle have relatively high iron contents (*e.g.*, TOULMIN *et al.*, 1977; DREIBUS and WÄNKE, 1985), and so Fe-bearing minerals and alteration products exist in enough abundance to be detectable using remote sensing techniques. Almost all of the absorption features that have been definitively detected on Mars and identified with specific mineralogies are due to electronic transitions within Fe^{3+} cations and charge transfers between these cations and O^{2-} anions (*e.g.*, SHERMAN *et al.*, 1982; SHERMAN and WAITE, 1985; MORRIS *et al.*, 1985), or to electronic transitions within Fe^{2+} cations and charge transfers between these cations and Fe^{3+} (*e.g.*, BURNS, 1970; ADAMS, 1974).

Spectroscopic observations

Many telescopic observations of Mars have been performed during this century with the explicit goal

Table 1. Definition of terms as used in this paper

Term	Definition
Weathering	Steady-state oxidation and/or hydration of exposed surface materials, leading to, for example, the conversion of $Fe^{2+} \rightarrow Fe^{3+}$.
Alteration	Forced, non-steady-state oxidation and/or hydration of exposed surface materials. Forcing may come from impact, hydrothermal circulation, volcanism, etc.
Bedrock	Unweathered lava flows, tephra deposits, or Fe^{2+}-rich glass.
Nanophase Iron Oxide	Aggregates of Fe^{3+}-rich particles having crystallite sizes in the 5-20 nm range. X-ray amorphous. Also called superparamagnetic.
Palagonite	Secondary weathering product of basaltic glass. Composition and mineralogy vary with environment, but usually includes nanophase iron oxide. Degree of crystallinity can range from well-crystalline to X-ray amorphous.

of identifying and quantifying the surface and airborne dust Fe mineralogy (for reviews of the earliest observations, see DE VAUCOULEURS, 1954; MCCORD and ADAMS, 1969; BELL, 1992). To within a few percent, the overall shape of the mean visible to near-IR spectral reflectance curve of Mars has been known since the late 1960's.

Continued advances in instrumentation and detector design have allowed observers to obtain detailed, higher spectral resolution and spectral sampling measurements of many small, discrete regions of the Martian surface in the visible to near-IR. Parallel advances in laboratory spectroscopic instruments and other sample characterization techniques have also occurred during this time. As the technology has improved, so has the ability to detect subtle spectral absorption features and to assign them to specific Fe-bearing minerals. For example, MCCORD and WESTPHAL (1971) observed 200 km-sized bright and dark regions on Mars during the 1969 opposition and found that both bright and dark regions are quite red, with the strong 0.4–0.7 μm absorption edge having a steeper slope for brighter regions. This has been interpreted as indicating an enrichment in ferric (Fe^{3+}) minerals in the bright areas (*e.g.,* ADAMS and MCCORD, 1969; BINDER and JONES, 1972). Also, a broad and weak absorption band from 0.9–1.1 μm was observed in dark region spectra. This band was ascribed to ferrous (Fe^{2+}) cations occurring in silicate

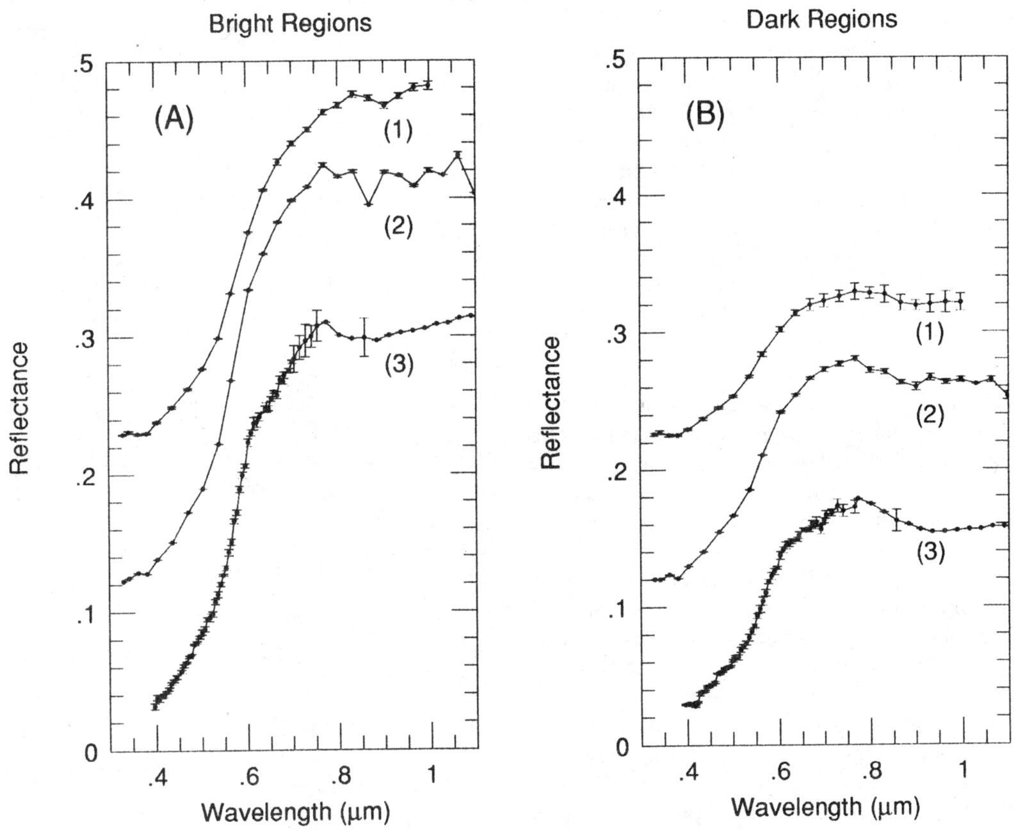

FIG. 1. Comparison among published Mars visible to near-IR spectra obtained between 1969 and 1989. (A) Bright region spectra. Spectrum (1) is Spot 69–1 (Arabia) from MCCORD and WESTPHAL (1971), spectrum (2) is Spot 73–9 (Dust Cloud) from MCCORD *et al.* (1977), and spectrum (3) is the composite telescopic-ISM spectrum of spot 88–41 (Olympus) from MUSTARD and BELL (1994). (B) Dark region spectra. Spectrum (1) is Spot 69–6 (Iapygia) from MCCORD and WESTPHAL (1971), spectrum (2) is Spot 73–18 (Margaritifer) from MCCORD *et al.* (1977), and spectrum (3) is the composite telescopic-ISM spectrum of spot 88–22 (Oxia Palus) from MUSTARD and BELL (1994). Reflectances for the 1969 and 1973 data are estimated by scaling to the value of the 1988 composite spectrum at 0.4 μm. For clarity, the 1969 and 1973 spectra are offset by 0.2 and 0.1 reflectance units respectively from the 1988 spectra.

minerals such as olivines and pyroxenes (*e.g.*, ADAMS, 1968; ADAMS and MCCORD, 1969; BINDER and JONES, 1972), and the correlation of the strength of this feature with low albedo areas led to the general impression that the dark regions on Mars are more "mafic" in character (with a greater exposure of basalts or other bedrock materials) than the more weathered and altered bright regions. Figure 1 presents some examples of typical Mars bright and dark region spectra obtained by MCCORD and WESTPHAL (1971) and others.

More recent telescopic and spacecraft observations and re-interpretations of previous data sets have concentrated on providing more specific information on band centers, band depths, and the spatial distribution of observed spectral features, in order to constrain further the surface mineralogy (detailed reviews can be found in SODERBLOM, 1992 and ROUSH *et al.*, 1993). For example, MORRIS *et al.* (1989) re-examined spectra obtained by MCCORD *et al.* (1982a) and found that the 0.84–0.90 μm band seen in bright region spectra is most likely caused by crystalline hematite (α-Fe_2O_3) on the Martian surface, mixed with extremely fine-grained or nanophase ferric oxide particles. BELL *et al.* (1990a) confirmed the existence of hematite in a large range of particle sizes on Mars and placed an upper limit ($\approx$4%) on the abundance of well-crystalline hematite in the bright region spectra, although the existence of other ferric-bearing minerals with only subtly differing spectral properties could not be ruled out. Differences in the hematite abundance and distribution between and among bright and dark regions have been reported by SINGER *et al.* (1990), BELL *et al.* (1990b), and BELL (1992) in high resolution imaging spectroscopic observations of Mars.

Other researchers have interpreted several of these previous data sets as providing evidence of different Fe^{3+}-bearing phases. For example, BISHOP *et al.* (1993) noted 0.4 to 1.0 μm spectral similarities between Mars spectra and a suite of ferrihydrite-bearing smectite clays. BANIN *et al.* (1993) have presented and reviewed evidence based on 0.4 to 1.0 μm spectroscopy and geochemical weathering constraints that lepidocrocite-bearing smectite clays are consistent with the available Mars spectra.

The more recent data sets also confirmed the existence of the weak 0.9–1.0 μm Fe^{2+} absorption feature reported by MCCORD and WESTPHAL (1971) and others. This feature consistently appears deeper in spectra of low albedo regions, and comparisons with laboratory reflectance studies have suggested that this band is due predominantly to low-Ca pyroxene (*e.g.*, SINGER *et al.*, 1979). No firm evidence has yet been found to support the claim by HUGUENIN (1987) of more olivine-rich spectral units on the Martian surface.

An important source of new spectroscopic data was provided by the Phobos-2 ISM instrument, which was an imaging spectrometer that obtained 0.8 to 3.2 μm data in 1989 for a part of the Mars equatorial region at $\approx$22 km/pixel spatial resolution. MURCHIE *et al.* (1993) have concentrated on the analysis of bright region mineralogy from the ISM spectra, and have found broad general agreement with previous groundbased studies. These relatively high-spatial-resolution data show clear evidence for crystalline hematite in the bright regions examined, and also show distinct absorptions in a few intermediate-albedo regions that resemble those due to ferric oxyhydroxide phases such as goethite (α-FeOOH) (Fig. 2). This interpre-

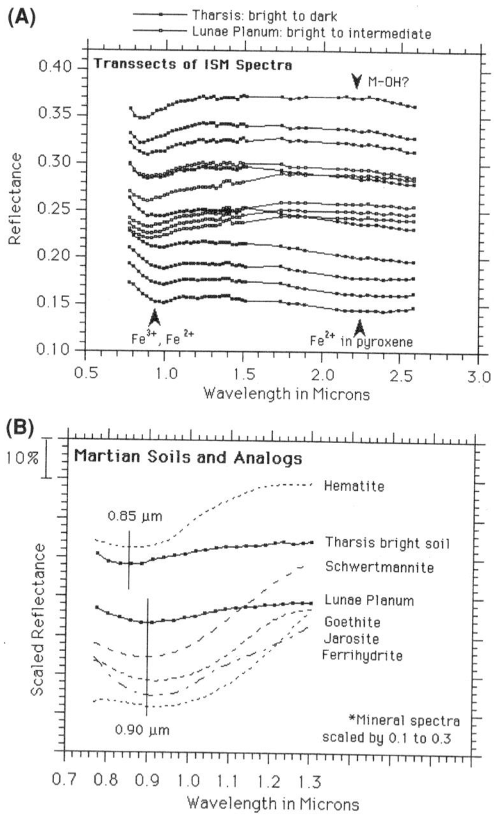

FIG. 2. (A) Representative spectra of Mars bright and intermediate regions measured by the Phobos-2 ISM spectrometer. (B) Comparison of ISM spectra of Tharsis and Lunae Planum with laboratory spectra of ferric oxides/oxyhydroxides and ferric sulfates. Both figures based on MURCHIE *et al.* (1993).

tation is also supported by the GEISSLER *et al.* (1993) analysis of ISM spectra in the Valles Marineris region. They found that spectra of isolated regions in the canyon system show greatly enhanced ferric absorptions suggesting a combination of hematite possibly with a hydrous ferric oxide such as goethite, and this observation along with the presence of associated canyon structural features may imply the existence of a past hydrothermal alteration zone on this part of Mars. MUSTARD *et al.* (1993) analyzed the ISM spectra with an emphasis on the ferrous mineralogy of Mars and found conclusive evidence for calcic pyroxene in dark region spectra, as well as for subtle variations in pyroxene chemistry from region to region (Fig. 3a). MUSTARD and SUNSHINE (1995) have expanded on this analysis and found that many of the volcanics are two-pyroxene basalts (high- and low-Ca, likely pigeonite and augite) analogous to basaltic compositions found in several of the SNC meteorites (see below), and that the modal mineralogy of these pyroxenes varies across the surface (Fig. 3b).

Newer Mars imaging spectroscopic observations were obtained from the Mauna Kea Observatory 2.24-m telescope in November 1990. These spectra cover the 0.50 to 0.95 μm region at a spectral resolution of R = 200 to 350, and were obtained in order to address the issue of Mars surface ferric and ferrous mineralogy for regions of the planet not observed during the 1986 and 1988 oppositions. Poor atmospheric

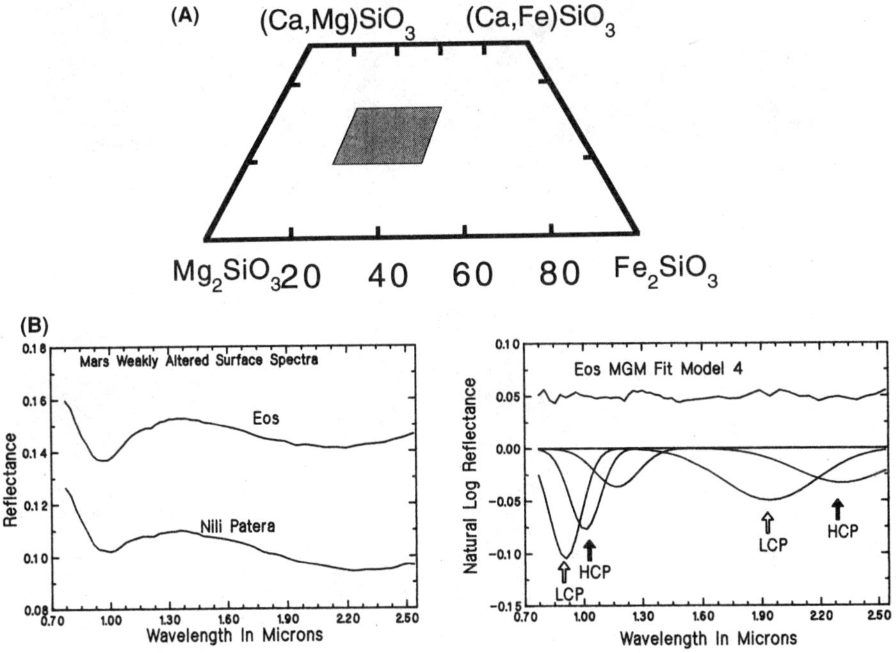

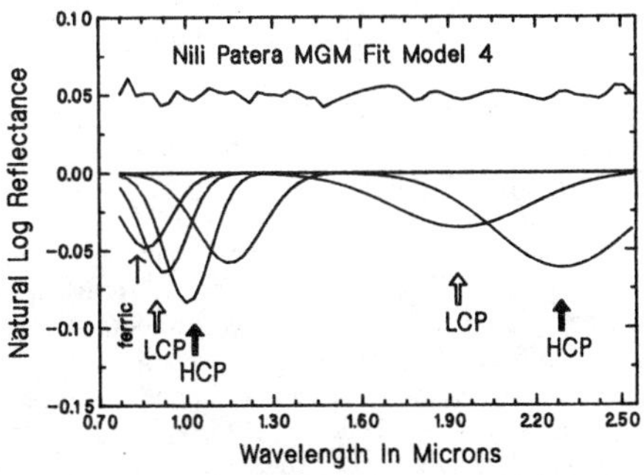

FIG. 3. (A) Plot of the pyroxene quadrilateral with the range of pyroxene compositions consistent with Phobos-2 ISM spectra of Mars dark regions shaded (MUSTARD *et al.*, 1993). (B) Modified Gaussian Model (MGM) analysis of ISM spectra of the dark regions Eos in Valles Marineris and Nili Patera in Syrtis Major (top left frame) from MUSTARD and SUNSHINE (1995). The Eos spectrum was modeled (top right frame) as a combination of low-calcium pyroxene (LCP) and high-calcium pyroxene (HCP), and the Nili spectrum was modeled (bottom right frame) as a combination of LCP, HCP, and a ferric-bearing component. The relative strengths of the modeled pyroxene absorptions indicate a ≈20% increase in HCP in Nili relative to Eos.

seeing conditions during the observing run severely limited the resolution of the images to worse than 750 km/pixel (*cf.* Fig. 4a), but useful mineralogic information can still be extracted by treating the image cubes as a traditional point-spectroscopy data set. However, the effective spatial resolution of these spectra is much lower than was obtained in 1986 and 1988 because the poor atmospheric seeing conditions tend to blur regions together and reduce the ability of detecting subtle spectral differences. Details of the characteristics, reduction, and calibration of these data are presented in the Appendix.

Figures 4b and c show some representative spectra from a wide range of regions of different albedo, topography, latitude, and geologic setting that were observed. Several interesting features are seen in these new spectra. First, both bright and dark regions exhibit the characteristic red visible spectral slope typical of Mars. Second, inflections or ab-

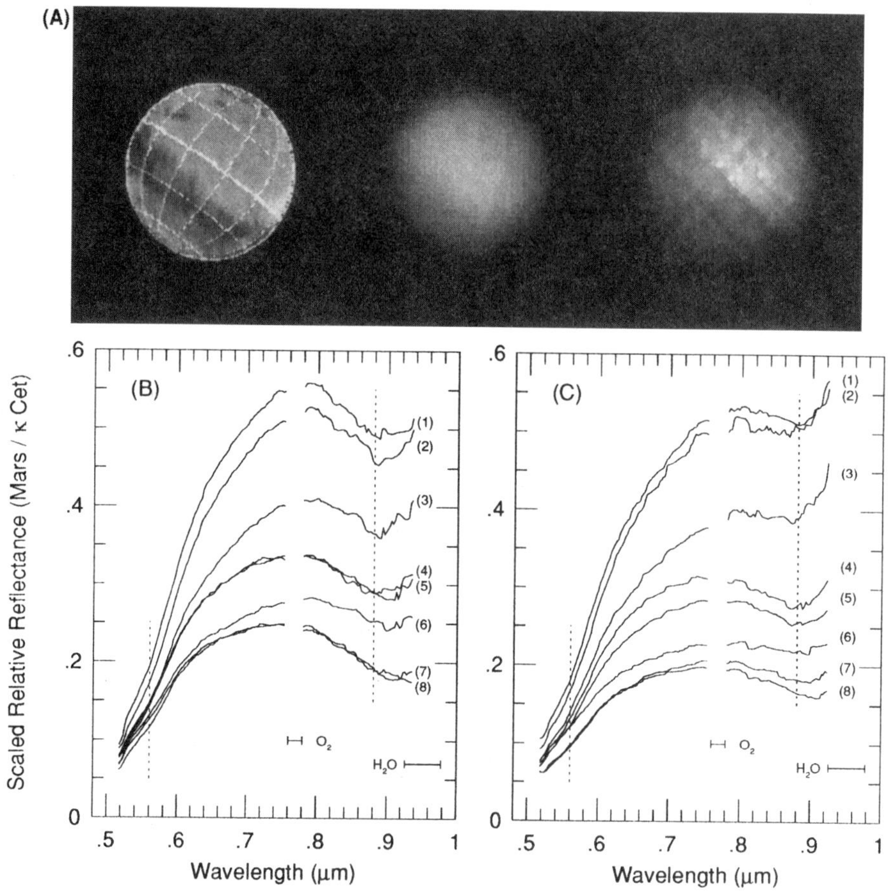

FIG. 4. (A) Example of image quality obtained during November 9–10, 1990 Mauna Kea Mars imaging spectroscopy observations. On the left is a schematic representation of the Viking IRTM albedo projected to the same viewing geometry as these data. In the middle is a red-filter CCD image of Mars obtained just before the image cubes were acquired, and on the right is shown one image cube plane at 0.8 μm. (B) Spectra of regions spanning a wide range of albedo, topography, latitute, and geologic setting extracted from one of the November 9, 1990 image cubes. The original spectra at 2.5 nm/pixel have been smoothed using a Gaussian function to an effective resolution of 7.5 nm/pixel. Data near the region of residual telluric O_2 absorption around 0.76 μm have been removed. Spectral regions associated with possible telluric H_2O contamination are noted. Spot locations are: (1) Elysium; (2) Center of Disk (S. Amazonis); (3) Olympus Mons; (4) Gusev crater region; (5) Southern Highlands; (6) N. Polar region; (7) Daedalia Planitia; (8) S. Polar region. (C) Same regions as in (B) except extracted from one of the November 10, 1990 image cubes having nearly identical viewing geometry as the November 9 cube.

sorption features are evident near 0.6 to 0.7 μm and 0.8 to 0.9 μm in both types of spectra. The longer wavelength band has a minimum near 0.867 μm (derived using a 2nd order polynomial fit) in the bright regions and near 0.893 μm in the dark regions. And finally, the dark region spectra exhibit a more negative spectral slope in the 0.75 to 0.94 μm region than the bright regions.

These spectra and others among the many thousands in this new data set show that absorption band positions in most bright regions indicate the presence of a small amount (less than 5%) of well-crystalline hematite that must occur in a matrix of much more poorly crystalline ferric-bearing materials which account for the overall color of Mars (MORRIS *et al.,* 1989; BELL *et al.,* 1990a). The slight shifts in band position observed among and between bright and dark regions and the negative 0.75 to 0.94 μm spectral slope are most likely caused by the presence of pyroxene-bearing materials in the dark regions that have ferrous absorption band positions longward of 0.9 μm. Alternately, the band shift may be partly due to the presence of hydrous ferric oxides like goethite. Unfortunately, the spectra do not extend to long enough wavelength to unambiguously resolve this question. The position of the near-UV absorption edge, position of the 0.86 μm band minimum, and the decreasing near-IR spectral slope argue that band shifts between regions may be indicating slight differences in mixing between hematite-rich dust and soils and pyroxene-rich volcanics at spatial scales smaller than the resolution of the data (*e.g.,* MUSTARD and BELL, 1994; MORRIS *et al.,* 1995).

Because these data cover a hemisphere dominated by bright, windblown materials and their spatial resolution was poor, there is certain to be a high amount of spatial mixing occurring in the data, such that the spectral differences between regions are muted and even the darkest spectra measured have a substantial "bright" component to them. The only solution to this common problem is increased spatial resolution, utilizing improvements in groundbased techniques, spaceborne observatories, or Mars orbiting or lander spacecraft. Initial tests of this hypothesis that compared Viking Orbiter and Lander color data and Phobos-2 ISM and groundbased spectroscopy have confirmed the importance of spatial resolution in understanding the mineralogy of the Martian surface (*e.g.,* ARVIDSON *et al.*, 1989; MUSTARD and BELL, 1994). All of these approaches are currently being attempted by many different groups, greatly improving our understanding of Mars spectral heterogeneity.

It has become generally apparent that increases in spectral and spatial resolution and wavelength coverage of Mars observations are necessary in order to identify the surface mineralogy. Higher spectral resolution is needed in order to accurately quantify and map the subtle band shifts associated with the rather weak (typically 1 to 10%) ferric and ferrous bands seen in the data. Higher spatial resolution is needed in order to increase the detectability of minerals that have only a limited spatial extent and whose signature is muted in spectra of regions hundreds to thousands of km in size (*e.g.,* MUSTARD and BELL, 1994). Greater wavelength coverage, encompassing both visible and NIR wavelengths, is critical for discriminating ferric and ferrous minerals having only subtle differences in absorption band shape and position from each other. Thus, while there is no debating the value of high spatial resolution Mariner and Viking Orbiter/Lander imaging of Mars for geologic unit mapping and photogeologic studies, these spacecraft observations have not been optimal for determining the surface mineralogy because of their low spectral resolution, coarse spectral sampling, and limited wavelength coverage.

Geochemical, magnetic, and other data

Much of what we have learned about the Fe mineralogy of Mars comes from telescopic and spacecraft reflectance spectroscopic observations. However, the Viking Landers obtained additional diagnostic information on the bulk chemical composition of the Martian soil and on its optical and magnetic properties. Table 2 presents a summary of the average chemical composition of the Martian soil at the two landing sites from the X-Ray Fluorescence (XRF) analyses of CLARK *et al.* (1982). All of the soils sampled exhibited high iron contents (ranging from 16% to 19% Fe as Fe_2O_3). The soil composition was interpreted using normative calculations and compositional mixing models as secondary weathering products of mafic igneous rocks, possibly resulting from palagonitization. In particular, the models yielding the best fits to the data included iron-bearing smectite (nontronite) and/or iron oxides (hematite, maghemite, magnetite) as the major iron-bearing minerals (TOULMIN *et al.,* 1977). The Viking Landers also revealed that the soils are highly magnetic, and HARGRAVES *et al.* (1977) interpreted the magnetism as due to 1% to 7% maghemite (γ-Fe_2O_3) dispersed as a pigment throughout the surface particles. Estimates of the light scattering and other properties of the Martian airborne dust were obtained by POLLACK *et al.* (1977) using Viking Lander sky images. They found that the best match to their derived imaginary index values for the dust came from fine-grained magnetite ($Fe^{2+}Fe_2^{3+}O_4$), which could also satisfy the magnetic properties results. It is possible that both of these results are correct, and that both of these iron oxides occur in the Martian soils. The surface magnetic interpretations incorporate not only the bulk magnetic behavior results but also the observed surface color, which is more consistent with maghemite than magnetite; the airborne dust optical properties results are more sensitive to materials with uniformly high absorption coefficients like magnetite, rather than generally less-absorbing minerals like maghemite. Finally, ADAMS *et al.* (1986) and GUINNESS *et al.* (1987) analyzed Viking Lander multicolor reflectance data and noted that most of the soils are redder than most palagonites, suggesting that the soils may have a greater degree of ferric iron crystallinity. Dark, gray blocks were found to exhibit colors similar to mafic rocks covered by varying amounts of ferric dust. These authors concluded that the ferric-rich soils observed by the Viking Landers have not been derived from local weathering of the mafic blocks, but more likely have been deposited by aeolian processes such as global dust storms.

Summary: Fe mineralogy

Only two iron-bearing mineral phases have been definitively identified spectroscopically on Mars: hematite and pyroxene (low- and high-Ca). Pyroxene-bearing volcanics are an abundant component of many dark regions, and hematite-rich soil/dust is found in varying levels of abundance and crystallinity in regions of all albedos. Tentative and potentially important spectroscopic evidence for the existence of ferric oxyhydroxide minerals, such as goethite, has been presented for a few isolated intermediate-albedo regions. No firm spectroscopic evidence for other ferric oxide phases exists, although magnetic studies of the Mars soil by the Viking Landers indicate the possible presence of maghemite, and studies of the optical properties of Mars atmospheric dust reveal many similarities with magnetite. Only indirect evidence exists for iron-bearing smectite clay (nontronite), based on modeling of the Viking compositional data. Evidence for other ferrous-bearing phases, such as olivine, has been only circumstantial to date.

EVIDENCE FOR SULFATES AND CARBONATES ON MARS

Theoretical and terrestrial analog considerations

There is abundant geomorphic evidence that water once flowed across the surface of Mars, suggesting that perhaps the climate was substantially different at some time in the past. In particular, many researchers believe that early Mars went through a warmer and wetter climatic period that may have included a thicker greenhouse atmosphere and/or a higher geothermal heat flux (*e.g.*, POLLACK *et al.*, 1987; FANALE *et al.*, 1992; SQUYRES and KASTING, 1994). If greenhouse warming was important, then many of the atmospheric constituents responsible (CO_2, SO_2, NH_3, CH_4, or H_2O; see FANALE *et al.*, 1992) have since been lost. One possible sink for the most important and relevant greenhouse gases for Mars (CO_2 and volcanically-outgassed SO_2 and H_2O) is the regolith. For example, on Earth, CO_2 is continually being recycled through the crust-atmosphere system in the form of carbonate rocks. This process occurs when CO_2 is dissolved in solution, forming an acid which leaches cations like Mg and Ca from silicate rocks. Bicarbonate anions (HCO_3^-) are also formed in this process, and the recombination of the Mg and Ca with the HCO_3^- complexes in aqueous solution leads to the formation of carbonate minerals like dolomite and calcite (*e.g.*, FANALE *et al.*, 1992; SCHAEFER, 1993). A similar process could also occur if atmospheric SO_2 was dissolved in solution, forming bisulfate anionic complexes (HSO_4^-) and possibly leading to the formation of sulfates. Yet another possible method of sulfate formation is the weathering of surface minerals by sulfuric acid contained in volcanic sulfate aerosol particles (SETTLE, 1979).

Additional theoretical support for the existence of carbonates and sulfates on Mars comes from thermodynamic calculations of the most stable gas-solid decomposition products of major mafic and ultramafic minerals under Marslike environmental conditions. O'CONNOR (1968), GOODING (1978), and GOODING *et al.* (1992) found that even if Mars did not have a warmer and wetter past climate, gas-solid weathering of olivine, pyroxene, and plagioclase could lead to the formation of stable or metastable carbonate phases like magnesite or siderite. These thermodynamic calculations also indicated that the weathering of igneous sulfide minerals by either gas-solid or liquid-solid mechanisms should lead to the formation of sulfates like $FeSO_4$ and possibly $CaSO_4$ and $MgSO_4$.

On Earth, most chemical weathering processes are cyclical because of the extensive hydrosphere and the role of plate tectonics in recycling crustal materials. On Mars, Earth-like plate tectonics is absent, and the extent and duration of the Martian hydrosphere is highly uncertain, though almost certainly less pervasive than the Earth's. It is also clear that if the Martian climate was warmer and wetter, this period occurred long ago and the climate has been more similar to its present state for at least several billion years (*e.g.*, SQUYRES and KASTING, 1994). These factors argue that for Mars the formation of carbonates or sulfates from an early dense atmosphere or by gas-solid weathering may have been a one-way trip, and that the current surface and crustal mineral inventory may still sequester a substantial abundance of CO_2 and SO_2. This possibility has spurred substantial interest in searching for quantitative evidence of carbonate and sulfate minerals on Mars.

A recent complication has arisen from the calculations of KASTING (1991), who has found that condensation of CO_2 clouds in a dense early Mars atmosphere (and assuming a fainter young Sun) would substantially reduce the magnitude of the greenhouse effect. Thus, attention has returned to the need for alternate trace greenhouse gases as sources of heat for early Mars, with SO_2 being perhaps the most plausible gas, as it is a common outgassing product of terrestrial volcanism (*e.g.*, POSTAWKO and KUHN, 1986). CH_4 and NH_3 are also good greenhouse gases, but they are photochemically unstable in the current high-UV flux environment of the Martian atmosphere (KUHN and ATREYA, 1979), so they would not be able to influence atmospheric heating on long timescales unless there were a constant and abundant resupply of these gases over geologic time.

If SO_2 was more abundant in the early Mars atmosphere, then its aqueous dissolution and the formation of HSO_4^- could have played an important role in chemical weathering of surface igneous silicates and sulfides by leaching out Fe, Mg, and Ca and forming sulfates and other silicates from solution. Alternately, sulfur could have segregated from magma at depth to form massive iron sulfide deposits, and the interaction of groundwater with these deposits would yield acidic solutions capable of weathering the surrounding rock. BURNS (1987, 1988) invoked such a weathering process to explain the high S abundance observed by the Viking Landers, and also showed that spectra of the Martian surface in the visible and near-IR are consistent with spectra of ferric-bearing sulfate phases (like jarosite). According to BURNS (1987), a byproduct of this aqueous weathering process, dependent on pH, would be the formation of poorly crystalline iron oxides and clay silicates which, after desiccation, could also yield fine-grained and highly magnetic phases like maghemite (γ-Fe_2O_3) in the soils. Most recently, MORRIS *et al.* (1996) have provided an example of the formation of sulfates (jarosite in this case) by a different process with relevance to Mars, involving the action of S-rich volcanic gases and the subsequent hydrothermal alteration of tephra by acidic solutions.

Viking Lander and SNC meteorite measurements

One of the biggest surprises from the Viking XRF measurements was the discovery of a much higher than expected abundance of sulfur in the soils at both landing sites (Table 2). TOULMIN *et al.* (1977) interpreted the high abundance of sulfur

Table 2. Average percent chemical composition of the Martian soil , mantle, and SNC meteorites

	VL-1[a]	VL-2[a]	VL-1[b]	Shergotty[c]	Nakhla[d]	EETA79001 Lith-C[e]	Mars Mantle[f]
SiO_2	44	43	55.1	51.36	48.24	51.0-54.9	40.0-45.0
Fe_2O_3	17.5	17.3	–	–	–	–	–
FeO	–	–	19.7	19.41	20.64	9.2-13.0	15.1-18.5
SO_3	6.7	7.9	–	–	–	0.23-0.88	–
Al_2O_3	7.3	7	9.1	7.06	1.45	2.3-6.4	3.0-6.4
MgO	6	6	7.5	9.28	12.47	17.6-25.4	29.8-33.2
CaO	5.7	5.7	7.1	10.00	15.08	3.4-10.4	2.45-5.16
TiO_2	0.62	0.54	0.78	0.87	0.29	0.14-0.40	0.14-0.63
K_2O	<0.5	<0.5	<0.6	0.16	0.10	<0.1-0.15	0.01-0.12
MnO	–	–	–	0.53	0.54	0.25-1.3	0.12-0.46
Na_2O	–	–	–	1.29	0.42	<0.1-0.61	0.1-0.5
P_2O_5	–	–	–	0.80	0.12	0.45-0.80	0.05-0.16
Cr_2O_3	–	–	–	0.20	0.42	0.52-1.60	0.38-0.76
Cl	0.8	0.4	–				
Other	2.0	2.0	–				
TOTAL:	91	90					

[a]Viking Lander XRF data from CLARK *et al.* 1982; "Other" includes unmeasured P, Mn, and Na.
[b]VL-1 corrected for SO_3 and Cl, Fe_2O_3 converted to FeO, normalized to 100% (LAUL *et al.* 1986).
[c]Bulk composition data obtained from literature values summarized in Table 1 of LAUL *et al.* 1986.
[d]MCCARTHY *et al.* 1974.
[e]Lithology-C (glassy inclusions) data from range of values reported by GOODING and MUENOW (1986).
[f]Average Mars bulk mantle compositional model ranges from literature values summarized in Table 2 of BOUSKA and BELL 1993.

as sulfates occurring in a surface duricrust, responsible for the hardpan or cemented appearance of some regions around the landers (Fig. 5a). The apparently oxidizing nature of the soils argued that the sulfur occurs in sulfates rather than in reduced forms like pyrrhotite ($Fe_{1-x}S$), thus ruling out the possibility that the sulfur-bearing phases could be responsible for the observed soil magnetism.

Evidence for carbonates in the Martian soils from the Viking Lander data has only been indirectly provided. BIEMANN *et al.* (1977) reported that CO_2 was evolved from the surface soils in the Viking GCMS experiment, and TOULMIN *et al.* (1977) use this result and thermodynamic stability arguments to conclude that CO_2 may represent a significant portion, second only to H_2O, of the "missing" (undetectable) chemical components in the soil (Table 2). In their normative mineralogy and mixing calculations, TOULMIN *et al.* (1977) favored a model having from 7% to 9% calcite in the surface soils as consistent with the Viking XRF data. However, simulations of the Viking Pyrolitic Release (PR) and Labeled Release (LR) biology experiments (*e.g.*, HUBBARD, 1979; BANIN and RISHPON, 1979) demonstrated that inclusion of more than 0.5 to 1.0% calcite, magnesite, or dolomite results in a substantial reduction of the release of $^{14}CO_2$ and is not consistent with the Viking data. These authors found that a much higher concentration of siderite could exist and still be consistent with the Viking data, though, so the interpretation of the PR and LR results (as well as the XRF data) in terms of carbonates is still somewhat uncertain.

Recent detailed petrographic analyses of the SNC (Shergottites, Nakhlites, and Chassigny) meteorites, thought to be samples of the Martian crust and mantle delivered to Earth by impact(s) on Mars, have revealed traces of water-precipitated minerals, including carbonates and sulfates (*e.g.*, GOODING and MUENOW, 1986; GOODING, 1992; MCSWEEN, 1994). These minerals provide evidence for regolith weathering products similar to what have been theoretically predicted. Isotopic analysis of the SNC carbonates has been used to show that the weathering took place at low temperature (CLAYTON and MAYEDA, 1988). The nonuniform distribution of these low-temperature weathering products within the meteorites and the chemical similarities between some of the weathered minerals and their precursor phases argues that the weathering occurred during episodic infiltration of small volume of saline water (TREIMAN *et al.*, 1993; MCSWEEN, 1994), rather than during a steady-state aqueous environment.

Spectroscopic observations

There have been many spectroscopic searches for sulfate and carbonate minerals on Mars in order to try to quantitatively substantiate the theoretical and indirect arguments discussed above for the occurrence of these phases. Accurate detection and identification of these minerals requires observations in the near-IR to mid-IR, where the strongest vibrational fundamental and overtone bands from CO_3^{2-}, HCO_3^-, SO_4^{2-}, and HSO_4^- anions occur (*e.g.*,

FIG. 5. (A) *Left.* Image of hardpan or duricrust surface region at the Viking Lander 1 site. This platy and fractured surface texture was exposed by the action of the Lander descent engines, and may represent a region where the surface soils have been cemented by the precipitation of evaporites. Long rock at the top of the image is 23 cm long. Segment of Viking Lander 1 image 12C158. (B) *Right.* Image of 1-m high boulder dubbed "Big Joe" with sprinkling of fine, presumably genetically-unrelated dust. Segment of Viking Lander 1 image 11C162.

BURNS, 1970; HUNT and SALISBURY, 1971; SWAYZE and CLARK, 1990; GAFFEY *et al.,* 1993). Thus, many of the highest-quality data sets with which to assess the presence or absence of sulfate and carbonate minerals have been obtained only recently and have coincided with advances in modern infrared detector technologies. Important early exceptions are: (1) the Mariner 6 and 7 Infrared Spectrometer (IRS) data set, which, although more than 25 years old, continues to remain unique in terms of spatial and spectral coverage among telescopic and spacecraft investigations of Mars. ROUSH *et al.* (1986) and MCKAY and NEDELL (1988) searched through the IRS data set for evidence of anhydrous carbonate absorption features, but found nothing conclusive. And (2) the Mariner 9 infrared interferometer-spectrometer (IRIS) data set, obtained in 1971 and which continues to be the primary source of high-quality thermal emission spectra of Mars. HUNT *et al.* (1973) and TOON *et al.* (1977) searched through the IRIS data set for evidence of carbonates, but were only able to place broad upper limits on the abundance of carbonate minerals in the global dust (a few to 10%).

More recent spectroscopic searches for evaporite minerals on Mars have been conducted by several groups. BLANEY and MCCORD (1989a) observed large regions of the planet in the 2.5 to 4.2 μm region and provided upper limits on the possible presence of carbonates based on the detection of weak spectral features near 4 μm. Subsequent observations reported by BLANEY and MCCORD (1989b) also detected these weak spectral features, and they were interpreted at the time as possibly originating from H–S stretch absorption features in hydroxy-acid salts or other sulfates. However, modeling by ENCRENAZ and LELLOUCH (1990) demonstrated that much of the spectral structure detected near 4 μm could be attributed to isotopically heavy CO_2 in the Martian atmosphere. BLANEY and MCCORD (1995) extended their observations into the 4 to 5 μm region and detected an absorption feature near 4.5 μm that was interpreted as an SO_4^{2-} overtone band in sulfates in the Martian surface and/or airborne dust. Careful atmospheric modeling revealed that this feature is likely not due to any known species in the Martian atmosphere; however, the specific mineralogy responsible was not identified because of the inherent difficulties of analyzing a spectral feature that occurs on the wing of the strong 4.2–4.4 μm Mars atmospheric CO_2 band. POLLACK *et al.* (1990) used the Kuiper Airborne Observatory (KAO) to obtain high-quality spectra in important wavelength regions that cannot be observed from groundbased telescopes. These data showed evidence for spectral features due to carbonate (near 6.7 μm) and sulfate (near 8.7 and 9.8 μm) anionic complexes in the airborne dust, but specific mineralogies could not be identified. CLARK *et al.* (1990) observed Mars in the 2.0 to 2.5 μm region and reported bicarbonate and/or bisulfate absorption features with positions and relative intensities that they claimed were diagnostic of the mineral scapolite. However, the occurrence on Mars of large quantities of this rather rare metamorphic mineral was met with substantial skepticism by geochemists and others. BELL *et al.* (1994a) obtained new spectra in this wavelength region and showed that several of the spectral features reported by CLARK *et al.* (1990) and others were probably due to residual telluric, Martian, or solar atmospheric bands. The remaining mineralogic features could still be caused by (bi)carbonate or (bi)sulfate complexes, although assignment of a specific mineralogy was not possible using the available data. Finally, CALVIN *et al.* (1994) have used a combination of groundbased spectra and a re-analysis of the Mariner IRS data to infer the existence of hydrous carbonates on Mars. Hydrated carbonates such as hydromagnesite $[Mg_5(CO_3)_4(OH)_2 \times 4H_2O]$ show much weaker spectral features than their anhydrous counterparts, and the positions of some of the weak absorption bands seen in terrestrial samples was found to match the positions of certain features seen in telescopic and Mariner IRS spectra of Mars. Hydrous carbonates represent a double-edged sword of sorts because they may occur on Mars in enough quantity to store a substantial reservoir of CO_2, but they may not be detectable spectroscopically even at the tens of percent level.

Summary: Sulfates and carbonates

Compelling theoretical arguments suggest that Mars once had a denser atmosphere, and strong evidence for this theory would be provided by the discovery of a global inventory of carbonate or sulfate minerals that have sequestered a large fraction of important greenhouse gases. A high sulfur abundance was measured in the Martian soils by the Viking Landers, but the specific sulfate mineralogy was not identified by Viking nor has it been well-constrained by groundbased or spacecraft spectroscopic observations since then. Carbonate minerals were also inferred based on the Viking Lander measurements. The SNC meteorites, thought to be samples of the Mars crust and/or mantle, have been found to contain small amounts of pre-terrestrial carbonates and sulfates. Groundbased and spacecraft searches for anhydrous carbonates have not conclusively identified any specific minerals in the uppermost few microns of the Martian surface, but evidence for (bi)carbonate and (bi)sulfate anionic complexes has been found in several spectroscopic data sets.

EVIDENCE FOR OH-BEARING MINERALS ON MARS

Background

The distribution and abundance of water in the surface rocks and soils of Mars has important implications for the style and rate of chemical weathering, for surface/atmosphere volatile transport, for the history of volatile outgassing, and for exobiology. The low temperatures and water contents of the current Martian environment place severe constraints on weathering rates, to the point that mineral phases predicted to be thermodynamically unstable on Mars (*e.g.*, O'CONNOR, 1968; GOODING, 1978) may actually persist over geologic timescales. For example, BURNS (1993b) postulated that (partially) dehydroxylated clays may have formed earlier in Martian history and remained metastable under current atmospheric conditions. Thus, observations of relict water- or hydroxyl-bearing minerals which are currently thermodynamically unstable at the surface could provide a way to quantify previous climatic regimes, assuming that the detected phases formed under equilibrium conditions with an earlier atmosphere.

Spectroscopic observations

Spectroscopic detection of adsorbed or structurally-bound water or OH in minerals is achieved by observations of rather broad features at many of the same wavelengths as the fundamental and overtone/combination bands of water vapor (*e.g.*, HUNT *et al.*, 1971; GAFFEY *et al.*, 1993; BISHOP and PIETERS, 1995). Thus, it is difficult to detect unambiguously these features from groundbased telescopes observing Mars through the Earth's water-rich atmosphere. Nonetheless, SINTON (1967) provided the first indication of bound water on Mars by observing indications of a depression in the spectrum near 3 μm (the O–H stretching vibrational fundamental). Additional indications of continuum absorption ascribed to bound water were provided by the high-resolution spectra of BEER *et al.* (1971). However, the most convincing measurements were those obtained by HOUCK *et al.* (1973) using an airborne telescope at 12 km altitude, above most of the Earth's water vapor. These authors observed a broad, $\approx$80% deep absorption band centered near 2.85 μm, and their modeling results indicated that the fraction of bound water and/or OH^- in the surface materials is $\approx$1%.

These groundbased and airborne results were verified by analysis of the Mariner 6 and 7 IRS spectra of Mars (PIMENTEL *et al.*, 1974). The IRS data show a strong 2.85 to 3.30 μm absorption band that varies in strength from region to region on Mars. Various spectral ratios and correlations with other parameters (albedo, latitude, etc.) were used to discriminate between hydrated minerals and water ice as the cause of the 3-μm absorption. Water (and CO_2) ice absorptions were found in the IRS data over the polar regions, but at other latitudes the 3-μm band is entirely consistent with the O–H stretching vibrational overtone observed in many hydrated silicate minerals.

The strong 3-μm hydrated mineral feature has been observed more recently from groundbased telescopes and from the Phobos-2 ISM spectrometer, which have revealed more details on the spatial distribution of the 3-μm absorption. BELL and CRISP (1993) found spatial variations in the strength of the 3-μm absorption in equatorial and mid-latitudes that is correlated with albedo (Fig. 6). MURCHIE *et al.* (1993) also found that the 3-μm band strength is generally greater in high albedo regions than in low albedo regions. However, there are discrete areas exhibiting enhanced 3-μm absorptions, including layered deposits in Valles Marineris and (perhaps not coincidentally) the regions having Fe absorptions resembling those of ferric oxyhydroxides. Their ISM data revealed two classes of soils based on 3-μm band strength: "normal soils" with a band typically 50 to 60% deep, and less common "hydrated bright soils" with a band several percent deeper than in other materials of comparable albedo. MURCHIE *et al.* (1993) inferred that variations in soil hydrated mineral lithology offered the most likely explanation for these 3-μm band variations.

The 3-μm hydrated mineral feature indicates that a substantial amount of water is tied up in the rocks and minerals on the Martian surface (20 pr μm in the upper few mm as estimated by HOUCK *et al.*, 1973). However, the specific minerals responsible for this spectral feature have not been unambiguously identified because the 3-μm feature does not exhibit very sensitive spectral variations with changes in mineralogy. Other features that involve both adsorbed and structurally-bound water do exhibit diagnostic variations, however. Most notable

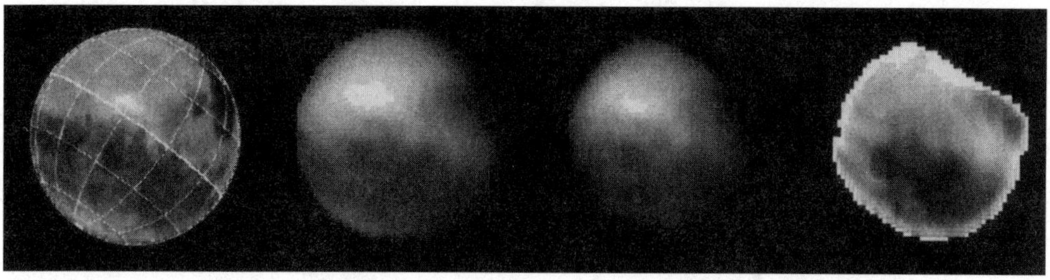

FIG. 6. Near-IR Mars images obtained from the Mauna Kea Observatory Infrared Telescope Facility on November 15, 1990. The left image is a schematic representation of the Viking IRTM albedo projected to the same viewing geometry as these observations. The adjacent two images show the albedo at 2.42 μm and 3.02 μm respectively. The right image shows the 3.02 μm relative band depth image constructed from the 2.42 μm and 3.02 μm albedo images using the technique of BELL and CRISP (1993). The most intense 3.02 μm absorption, associated with limb darkening and polar clouds and ices, has been masked off in order to show the more subtle variations in the strength of the surface 3-μm band. This absorption band is likely due to the presence of bound water in minerals, and the strength of the absorption correlates well with albedo.

are absorption features caused by cation–OH stretching vibrational fundamentals, overtones, and combinations in clay silicate minerals. Many of these features occur in the 2.2 to 2.3 μm region, which is a transparent window in both the terrestrial and Martian atmospheres. Examples include the 2.2 μm Al–OH feature in montmorillonite and the 2.3 μm Fe–OH feature in nontronite. However, even in this wavelength range the Martian atmosphere contains absorptions due to CO_2 and CO that complicate the identification and assignment of absorptions from surface minerals.

Spectroscopic searches for cation–OH and OH^- overtone and combination features have been conducted by many researchers, often with conflicting results. Data obtained and analyzed by MCCORD *et al.* (1978, 1982a) showed a complex absorption feature near 1.95 μm that was interpreted as being composed of several lattice vibrational modes of clay minerals like montmorillonite and kaolinite. Small, distinct absorption bands in the 2.2–2.5 μm region can also be seen in their spectra. These authors and SINGER *et al.* (1979) attribute weak features near 2.3 μm and possibly 2.4 μm to Mg-OH in sheet silicates (serpentine, talc, Mg-smectites) or amphiboles (anthophyllite, actinolite). They interpret the lack of a 2.2 μm feature (characteristic of Al-OH) as evidence that montmorillonite or other dioctohedral clays are not a major component of the Martian regolith. Further study by SINGER (1982) concluded that nontronite is not a major component of the Martian soil since its visible reflectance signature is unlike that of the available data. However, the presence of other iron-poor clays (such as montmorillonite) could not be totally excluded.

As discussed previously, narrow features in the 2.1 to 2.5 μm spectra of CLARK *et al.* (1990) have been interpreted as due to carbonate- or sulfate-bearing minerals, as opposed to clay or other hydrated silicates. BELL and CRISP (1993) noted evidence for absorption features near 2.25 μm in their imaging spectroscopic observations of Mars, and speculated on their possible origin as Fe–OH overtones. MURCHIE *et al.* (1993) also observed a weak absorption from 2.20 to 2.25 μm in ISM spectra covering bright regions, and interpreted this feature as consistent with a mixture of poorly-crystalline phyllosilicates (Fig. 2). As with the 3-μm H_2O absorption there is a general correlation of band strength with albedo, but discrete high-albedo features (layered deposits in Candor Chasma) lack this absorption. This spatial coherence of variations in band strength with geology is perhaps the strongest evidence from ISM data for association of this feature with surface mineralogy. BELL *et al.* (1994a) were able to confirm that at least some of the narrow, weak features seen in Mars spectra in the 2.0 to 2.5 μm region are due to surface mineralogy (with the most likely origin of these features being either (bi)carbonate or (bi)sulfate anions in framework silicates or (Fe, Mg)–OH bonds in sheet silicates), while other features appear to originate from telluric, Martian, or solar atmospheric absorptions.

The KAO mid-IR spectra of POLLACK *et al.* (1990) discussed above showed evidence for an emission maximum near 6.2 μm that was attributed to hydrated surface minerals, but the specific mineralogy could not be identified. Spectroscopic observations by the Mariner 9 IRIS instrument detected spectral features near 9.0 μm in the atmospheric dust that were interpreted to be most consistent with montmorillonite (HUNT *et al.*, 1973) or a basalt-montmorillonite mixture (TOON *et al.*, 1977). More recent re-analyses of the IRIS measurements have shown that poorly-crystalline weathering products like palagonite may in fact be more consistent with the data than smectite clay, however, based on improved knowledge of the optical constants of palagonites and on the scattering properties of atmospheric dust particles (ROUSH *et al.*, 1993; CLANCY *et al.*, 1995).

A perplexing observation, noted by BELL *et al.* (1994a), is that many of the observed spectral features on Mars that occur at or near wavelengths characteristic of OH-bearing minerals are much narrower than their counterpart bands in terrestrial samples. A possible explanation was proposed, based on the dehydroxylated clays study of BURNS (1993b): one effect of dehydroxylation could be to reduce the strengths and possibly, under certain circumstances, the widths of both OH^- and cation–OH absorption features in the near-infrared. However, evidence for changes in the nature of structural OH absorption features with variations in pressure and temperature appropriate for the Martian environment were not found by BISHOP and PIETERS (1995) in their analysis of Mars soil analog materials. These authors did find evidence for systematic changes in the depth and width of H_2O stretching and bending combination bands in the near-IR with changes in environmental conditions and with variations in the nature of interlayer smectite anionic complexes. An alternate explanation for the narrow band widths, frustrating for Earthbound mineralogists, is that a continuum of crystalline and amorphous salts, clays, and/or clay-like phases may exist on Mars that simply do not have good terrestrial analogs.

Other sources of information

As mentioned above, although the Viking Landers did not directly determine surface mineralogy, the chemical composition results of the XRF experiment have been used to construct mixing models of possible surface mineralogies. The best-fit models included nontronite and montmorillonite as prominent surface minerals, in abundances around 50% and 20%, respectively (TOULMIN *et al.*, 1977). However, these models are based on normative mineralogy calculations, and make certain geochemical assumptions that may not necessarily be unique or appropriate for the Martian surface. BANIN and RISHPON (1979) and BANIN and MARGULIES (1983) in their simulations of the Viking Lander LR experiment concluded that smectites, but not the smectite-poor palagonites that they examined, could reproduce the LR data. In particular, Fe-bearing montmorillonite and nontronite were found as the best-fit smectites, while Na- or Ca-bearing smectites did not fit as well. Thus, it is clear that the available data are consistent with the presence of a significant iron-bearing clay silicate component in the Martian soils; however, to date such minerals have not been directly detected or measured either by in situ analyses or remote sensing.

Analyses of SNC meteorites have also detected evidence of presumably Martian smectites, most notably the deposits of "iddingsite" (ferrihydrite + smectite clay) detected in the Lafayette meteorite by TREIMAN *et al.* (1993), and the possible identification of illite and smectites in EETA79001 and Nakhla (GOODING and MUENOW, 1986; GOODING, 1992). GOODING (1992) found that a mixture of these two phyllosilicate assemblages provides a close match to the elemental abundances measured by the Viking XRF experiment, and MURCHIE *et al.* (1993) noted that such a mixture would also yield a metal–OH overtone having a wavelength position similar to that found by them and BELL and CRISP (1993).

Finally, observations of water vapor in the Martian atmosphere as a function of time by the Viking Orbiter Mars Atmospheric Water Vapor Detection (MAWD) experiment obtained results relevant to the study of hydrated minerals on and in the surface (FARMER *et al.*, 1977). These data, acquired over the course of more than a Mars year, showed evidence for the transport of water vapor from the northern to the southern hemisphere during the northern spring and summer, but they did not show any evidence for the return circulation in the northern fall and winter (JAKOSKY and FARMER, 1982). Because MAWD was sensitive only to water vapor and not to hydrated minerals or surface/atmospheric water ice, the current view of what maintains the seasonal hemispheric asymmetry in water vapor or what processes recharge the water reservoirs (regolith and polar cap) each year in the north is obviously incomplete. However, it is likely that the exchange of water between the atmosphere and the surface, through adsorbed and possibly even structural water in minerals, plays an important role. Unfortunately, detailed spectroscopic studies of hydrated minerals as a function of latitude and season on Mars are lacking at the present time.

Summary: Hydrated minerals

A large body of evidence indicates that substantial amounts of adsorbed and/or structurally-bound water exist in the surface minerals of Mars. Identification of the specific hydrated mineral(s) occurring on the surface is equivocal, however, because of the limited diagnostic ability of the O–H and H_2O stretching vibrational features in minerals and because of limitations on the quantity and quality of appropriate Mars spectral data. The fact that a large inventory of water appears to be sequestered in the surface minerals is intriguing, and argues that conditions more favorable for aqueous alteration and weathering may have been more common during past epochs when these minerals were originally formed. Although many H_2O and OH bearing minerals should not be thermodynamically stable in the current desiccating Martian environment, convincing arguments have been presented indicating that such minerals may in fact persist metastably because current chemical weathering rates are exceedingly slow.

WEATHERING AND ALTERATION SCENARIOS CONSISTENT WITH THE AVAILABLE DATA

The data discussed above allow for the development of a set of consistent and reasonable models of the Mars surface chemical weathering and alteration environment over geologic time. Such models have been discussed in detail for carbonates and sulfates (*e.g.*, FANALE *et al.*, 1982, 1992; SCHAEFER, 1993; BURNS, 1993b), as well as hydrated minerals (*e.g.*, SCHAEFER, 1990; GOODING, 1992). Here, I discuss possible scenarios for the weathering and alteration of ferric and ferrous minerals that are consistent with available Mars observations.

Mars, as a whole, appears quite red as viewed by naked eye or by other low spatial resolution techniques. Upon closer examination, such as from high resolution telescopic or spacecraft views, it becomes clear that all regions of Mars are not equally red and oxidized, but that there are substantial variations from region to region (*e.g.*, BELL *et al.*, 1990a,b, 1995; MURCHIE *et al.*, 1993; MUSTARD *et al.*, 1993). These variations generally correlate with albedo (although there are exceptions), with bright regions typically showing the most oxidized spectral character and the dark regions, including many adjacent to the bright regions, typically showing the least oxidized spectral behavior. This is the Mars weathering paradox discussed above, the solution to which has important implications for understanding the physical and chemical weathering and alteration processes that have operated on Mars.

Two possible scenarios for the weathering and/or alteration of Mars surface Fe-bearing minerals are proposed (BELL *et al.*, 1994b):

Decoupled soil and bedrock

In this model the dark, less-red, and relatively pristine surface material such as that observed in Syrtis Major is unweathered pyroxene-bearing bedrock that is covered by or simply mechanically mixed with unrelated ferric materials. Thus, the unweathered rocks are not the source of the bright, ferric-bearing global dust, which must have been formed under different past conditions.

An important question that must be addressed if this model is valid is: Where do the oxidized (ferric) minerals come from? There are at least two possible answers. First, there may have been different climatic conditions in the distant past that allowed much higher weathering rates than at present. Second, individual alteration event(s), such as impacts or volcanism, may have catalyzed the $Fe^{2+} \rightarrow Fe^{3+}$ conversion, regardless of the climatic regime. The implications are that the present inventory of ferric phases on Mars may be a fossil remnant from a past epoch or past event(s), and that the current distribution of ferric minerals may have little or no genetic relationship to the underlying bedrock. In addition, this model offers the intriguing possibility that the degree of oxidation of dark basaltic materials can be used as a relative age-dating tool on Mars, with dark oxidized regions being older and dark unoxidized regions being younger.

Coupled soil and bedrock

In this model the dark material is weathered or weathering (slowly), creating ferric-bearing surface materials that are mobilized through spallation and erosion. Thus, the bright, ferric-bearing global dust is a direct weathering product of the dark bedrock units, raising the possibility of localized genetic links between weathered and un-weathered materials.

If this model is valid, an important question that must be answered is: Why are the ferrous-bearing surface materials apparently so pristine? One solution is the extremely slow weathering reaction rates in the current environment, which effectively prevent oxidative weathering from occurring at the surface over even geologic timescales (BURNS,

1993b). Another possible solution is that the environment conducive to oxidative weathering occurred in association with episodic or sporadic events, none of which have occurred in recent geologic time. The implications of this model are that the basaltic units on Mars are currently weathering, albeit slowly, and that the ferric-bearing dust and soils are a direct product of in situ weathering by surface reactions.

There is evidence supporting both the coupled and decoupled models in currently-available remote sensing data. For example, the presence of well-defined and rather deep Fe^{2+} (pyroxene) absorption features in the ISM spectra (*e.g.*, MUSTARD *et al.*, 1993) and evidence from spacecraft and HST imaging for temporal variations in the distribution of bright materials in and around dark regions (*e.g.*, JAMES *et al.*, 1994) argues for the decoupled soil/bedrock model. The alternative, that the basalts are either very coarse-grained or extremely young, are not consistent with other available data, such as thermal inertia and relative age dating from crater statistics. Also, Viking Lander images showing dark, basaltic rocks covered by an apparent sprinkling of bright but presumably genetically unrelated red dust (Fig. 5b; *e.g.*, ADAMS *et al.*, 1986) argue that the soils and bedrock, at least in those regions, are decoupled. The presence of weak or intermediate-position Fe^{2+} absorption features in some ISM and telescopic dark region spectra provides possible evidence for the coupled model, in that such spectral signatures are consistent with partial weathering of initially more pristine ferrous-bearing materials (*e.g.*, MORRIS *et al.*, 1996). This evidence is not unique, though, as it has been shown that spatial averaging of bright and dark regions as often occurs in low spatial resolution observations of Mars can lead to apparent weakening of Fe^{2+} band positions (MUSTARD and BELL, 1994). Additional evidence for the coupled soil/bedrock model comes from telescopic, Viking, and ISM observations of ferric-bearing dark regions (*e.g.*, ARVIDSON *et al.*, 1982; MCCORD *et al.*, 1982b; SINGER and ROUSH, 1983; SINGER *et al.*, 1990; FISCHER and PIETERS, 1993; MUSTARD and SUNSHINE, 1995; MURCHIE *et al.*, 1996). Negative visible to near-IR spectral slope and correlations with higher thermal inertia values indicate that the spectra of some of these areas may not represent simple mixtures of ferric-bearing dust and pristine basalt, but instead may represent basaltic regions coated by a ferric stain, somewhat analogous to terrestrial desert varnish.

Discriminating between models

Clearly, the quality and quantity of currently-available remote sensing evidence do not strongly favor one weathering model over the other. In fact, both models may be valid independent of each other in localized regions and/or epochs. The key to solving this ambiguity is to accurately measure the spectrum of the rock "endmember" on a region-by-region basis. This will allow an unmixing of the soil and bedrock phases to be performed and an analysis of whether the bedrock materials are truly unoxidized or whether they are in fact actively oxidizing. Specific observational tests that could be used in this analysis are discussed in the next section.

OBSERVATIONAL TESTS

Continued advances in the study of Mars surface mineralogy are dependent on the ability to obtain higher-quality observations at increased spatial resolution in order to reduce the spectral mixing effects introduced by averaging many regions into a single measurement. It is particularly important to capitalize on the opportunity to observe Mars from groundbased and spaceborne platforms during the upcoming series of favorable oppositions over the next decade and during the current extremely low average atmospheric dust opacity conditions (*e.g.*, Fig. 7). The presence of any amount of dust in the Martian atmosphere acts to effectively degrade the spatial resolution of surface observations and to reduce the detectability of many materials that exhibit less spectral contrast than minerals in the airborne dust. Three specific types of continuing observations are needed:

Visible to near-IR (VNIR) imaging and imaging spectroscopy

VNIR spectra continue to provide the best remote means of identifying the mineralogy of specific alteration products on Mars. As discussed above, spectra at the wavelengths where reflected solar flux is primarily being measured (0.3 to $\approx$3.5 μm) are sensitive to variations in ferric iron mineralogy, hydrated minerals including clays, and certain salts and evaporites. In addition, diagnostic information on the pyroxene mineralogy of primary basalts can be obtained from spectra in the 1- and 2-μm regions. The key to using VNIR techniques to extend the current knowledge is to apply them in an imaging context, so that regional variations in spectral properties at fine scales can be correlated with similar-scale variations in surface

FIG. 7. Hubble Space Telescope 0.673 μm image of the Syrtis Major hemisphere, acquired on Feb. 24, 1995. This image demonstrates the level of spatial detail that can be examined spectroscopically using the best available telescopic techniques. The spatial resolution in this image (obtained during one of the worst oppositions of the planet in terms of Earth-Mars distance) is 25 km/pixel near the center of the disk, comparable to the 22 km/pixel resolution obtained by the Phobos-2 ISM instrument from Mars orbit. This image is part of a 5-filter multispectral imaging set obtained during HST Cycle 4 (1994–95). In Cycle 5 (1995–96) the spectral coverage was increased to 9 filters and in Cycle 6 (1996–97) the coverage from multiple observing programs will include 19 filters at key mineralogic wavelengths from 0.25 to 1.04 μm.

morphology, geology, topography, thermal inertia, etc. For example, it would be extremely interesting to determine whether spectral slope and/or band depth (diagnostic of the degree of surface weathering and alteration) correlates with surface age as determined by other techniques (craters, superposition). The result could be used to assess whether the coupled or decoupled weathering scenarios discussed above are operating in each region. The Phobos-2 ISM data set has effectively demonstrated the value of VNIR imaging spectroscopy for Mars; however, such data need to be obtained over all solar reflectance wavelengths at higher spectral resolution (in order to resolve better narrow absorption features) and over the entire planet at scales of meters to kilometers.

Infrared (IR) thermal emission spectroscopy

IR spectroscopy ($\approx$3.5 to 25 μm) is much more sensitive to variations in primary silicate mineralogy than VNIR techniques (*e.g.*, FARMER, 1974; SALISBURY *et al.*, 1991). Also, many of the strongest and most diagnostic fundamental spectral features of evaporite minerals occur at thermal wavelengths, making detection and identification of these materials potentially easier in the IR than in the VNIR. The Mars Global Surveyor Thermal Emission Spectrometer (TES) instrument (CHRISTENSEN *et al.*, 1992) will have a much greater ability than previous IR investigations to spectrally characterize the Martian surface, and will do this globally from Mars orbit at several km resolution. In particular, TES will be able to search for any evidence of carbonate or sulfate spectral features in the IR, and to map the distribution of these features. Recent groundbased spectral observations in the IR have also shown dramatic improvements in sensitivity and spatial resolution (*e.g.*, MOERSCH *et al.*, 1993; ROWLAND *et al.*, 1995). Such observations have the potential of substantially enhancing the TES data by observing surface regions at different times of day and different photometric angles. It must also be remembered that groundbased and orbital IR measurements are also observing the highly altered soils and dust, and so a more complete understanding of the spectral properties of poorly-crystalline or highly-oxidized materials in the IR (*e.g.*, ROUSH and BELL, 1995) will be critical for interpreting Mars thermal IR data sets.

In-Situ analyses

Finally, there are a host of measurements that can be conducted from surface landers or (eventually) trained human explorers that will be able to answer key questions on the iron, sulfur, carbonate, and hydrate mineralogy of Mars. Examples include (1) surface lander VNIR and IR imaging and spectroscopy, able to address the mineralogic issues discussed above but at meter to centimeter spatial resolution; (2) chemical composition experiments such as XRF or DSC/TGA which can determine the abundances of key rock-forming elements or constrain the mineralogy of volatile-bearing phases in the soils at different places; and (3) Mössbauer spectroscopy, which can uniquely and definitively identify many iron-bearing minerals like iron oxides. Many such experiments are currently being proposed by NASA and others as part of planned Mars landing missions over the next decade.

SUMMARY

The Martian surface is composed of a complex mixture of rocks and minerals that tell a story about the past history of the atmosphere, hydrosphere, and crust. Fundamental to untangling this story is a more complete knowledge of the current mineralogic inventory of Mars and of formation pathways for secondary minerals, specifically concentrating on the kinds of climatically-diagnostic materials (ferric oxides/oxyhydroxides, carbonates, sulfates, and hydrates) that Roger Burns frequently advocated as the key to understanding the past and current Martian environment.

Our current knowledge is limited to a few specific items:

(a) Most bright surface regions and the airborne dust are highly oxidized and oxidizing, as evidenced by the abundance of Fe^{3+}-bearing minerals like hematite and the results of several of the Viking Lander wet chemistry experiments. Evidence suggesting other (unidentified) iron oxide/oxyhydroxide phases exists from groundbased and spacecraft imaging, spectroscopy, and magnetic experiments, and efforts are being made at determining the spatial distribution of these highly altered phases.

(b) Many dark surface regions are composed of relatively unaltered pyroxene-bearing (Fe^{2+}) basalts. It is unclear why or how these rather pristine, unweathered regions coexist with the highly oxidized areas. Detailed observations and models are beginning to be able to detect variability in the composition of these basalts and to map their spatial distribution.

(c) Abundant geomorphic evidence suggests that water once flowed on Mars. The possible implications for a past warmer, wetter Mars climate have led to intense searches for hydrated mineral phases and evaporites, as these minerals could have sequestered much of the water and greenhouse gases of a putative past thicker atmosphere. To date, tentative evidence of carbonate minerals has been provided by groundbased, airborne, and spacecraft observations, but definitive mineralogic identifications and detailed spatial mapping of these materials is lacking. Evidence for sulfur bearing minerals comes primarily from Viking XRF measurements of high S in the soils and tentative groundbased detection of sulfate mineral absorption features. Again, however, detailed mineralogic identifications and spatial distributions have not been determined.

There are a number of groundbased, spacecraft, and in situ measurements that can be performed in order to try to remove some of the ambiguity in mineralogic identification and in order to try to correlate the observed mineralogic variations with variations in other physical properties like geology and topography. Collection and interpretation of such new measurements represents a fundamental goal in the continued exploration of Mars.

Acknowledgments—Much of this research has benefited greatly from ongoing collaborations with Richard Morris (NASA/JSC) and John Adams (Univ. Washington). I thank Thomas Ozoroski, Karl Bornhoeft, and Paul Lucey for assistance with the collection and reduction of the 1990 Mauna Kea Mars telescopic data. I am also sincerely grateful to Diana Blaney, Scott Murchie, John Mustard, and Ted Roush for providing detailed reviews of earlier versions of this paper. This research was supported by grants from the NASA Planetary Astronomy and Planetary Geology Programs.

REFERENCES

ADAMS J. B. (1968) Lunar and Martian surfaces: Petrologic significance of absorption bands in the near-infrared. *Science* **159,** 1453–1455.

ADAMS J. B. (1974) Visible and near-infrared diffuse reflectance spectra of pyroxenes as applied to remote sensing of solid objects in the solar system. *J. Geophys. Res.* **79,** 4829–4836.

ADAMS J. B. and MCCORD T. B. (1969) Mars: Interpretation of spectral reflectivity of light and dark regions. *J. Geophys. Res.* **74,** 4851–4856.

ADAMS J. B., SMITH M. O. and JOHNSON P. E. (1986) Spectral mixture modeling: A new analysis of rock and soil types at the Viking Lander I site. *J. Geophys. Res.* **91,** 8098–8112.

ARVIDSON R. E., GUINNESS E. A. and ZENT A. P. (1982) Classification of surface units in the equatorial region of Mars based on Viking Orbiter color, albedo, and thermal data. *J. Geophys. Res.* **87,** 10149–10157.

ARVIDSON R. E., GUINNESS E. A., DALE-BANNISTER M. A., ADAMS J. B., SMITH M. O., CHRISTENSEN P. R. and SINGER R. B. (1989) Nature and distribution of surficial deposits in Chryse Planitia and vicinity, Mars. *J. Geophys. Res.* **94,** 1573–1587.

BANIN A. and RISHPON J. (1979) Smectite clays in Mars soil: Evidence for their presence and role in Viking biology experimental results. *J. Mol. Evol.* **14,** 133–152.

BANIN A. and MARGULIES L. (1983) Simulation of Viking biology experiments suggests smectites not palagonites, as martian soil analogs. *Nature* **305,** 523–525.

BANIN A., BEN SHLOMO T., MARGULIES L., BLAKE D. F., MANCINELLI R. L. and GEHRING A. U. (1993) The nanophase iron mineral(s) in Mars soil. *J. Geophys. Res.* **98,** 20831–20853.

BEER R., NORTON R. H. and MARTONCHIK S. V. (1971) Astronomical infrared spectroscopy with a Connes-type interferometer, 2, Mars, 2500–3500 cm^{-1}. *Icarus* **15,** 1–10.

BELL J. F., III (1992) Charge-Coupled Device Imaging Spectroscopy of Mars. 2. Results and Implications for Martian Ferric Mineralogy. *Icarus* **100,** 575–597.

BELL J.F., III and CRISP D. (1993) Groundbased Imaging Spectroscopy of Mars in the Near-Infrared: Preliminary Results. *Icarus* **104,** 2–19.

BELL J. F., III, MCCORD, T. B. and OWENSBY P. D. (1990a) Observational Evidence of crystalline iron oxides on Mars. *J. Geophys. Res.* **95,** 14447–14461.

BELL J. F., III, MCCORD T. B. and LUCEY P. G. (1990b) Imaging Spectroscopy of Mars (0.4–1.1 μm) During the 1988 Opposition. *Proc. Lunar Planet. Sci. Conf. XX,* 479–486.

BELL J. F., III, POLLACK J. B., GEBALLE T. R., CRUIKSHANK D. P. and FREEDMAN R. (1994a) Spectroscopy of Mars from 2.04 to 2.44 μm during the 1993 opposition: Absolute calibration and atmospheric vs. mineralogic origin of narrow absorption features. *Icarus* **111,** 106–123.

BELL J. F., III, MORRIS R. V. and ADAMS J. B. (1994b) Why Mars is red: A process-oriented viewpoint. *EOS, Trans. A.G.U.* **75,** 406.

BELL J. F., III, JAMES P. B., MARTIN L. J., CLANCY R. T., LEE S. W. and CRISP D. (1995) Mars surface mineralogy from Hubble Space Telescope multispectral imaging: 1994 pre-opposition data. *Lunar Planet. Sci. Conf. XXVI,* 95–96.

BIEMANN K., ORO J., TOULMIN P., III, ORGEL L. E., NIER A. O., ANDERSON D. M., SIMMONDS P. G., FLORY D., DIAZ A. V., RUSHNECK D. R., BILLER J. E. and LAFLEUR A. L. (1977) The search for organic substances and inorganic volatile compounds in the surface of Mars. *J. Geophys. Res.* **82,** 4641–4658.

BINDER A. B. and JONES J. C. (1972) Spectrophotometric studies of the photometric function, composition, and distribution of the surface materials of Mars. *J. Geophys. Res.* **77,** 3005–3020.

BISHOP J. L. and PIETERS C. M. (1995) Low-temperature and low atmospheric pressure infrared reflectance spectroscopy of Mars soil analog materials. *J. Geophys. Res.* **100,** 5369–5379.

BISHOP J. L., PIETERS C. M. and BURNS R. G. (1993) Reflectance and Mössbauer spectroscopy of ferrihydrite-montmorillonite assemblages as Mars soil analog materials. *Geochim. Cosmochim. Acta* **57,** 4583–4595.

BLANEY D. L. and MCCORD T. B. (1989a) An observational search for carbonates on Mars. *J. Geophys. Res.* **94,** 10159–10166.

BLANEY D. L. and MCCORD T. B. (1989b) Telescopic spectrometric observations of Mars between 3.5 μm and 4.2 μm: Evidence for compounds containing H–S. *Fourth International Conference on Mars,* pp. 71–72, Tucson, AZ, January 10–13, 1989.

BLANEY D. L. and MCCORD T. B. (1995) Indications of sulfate minerals in the Martian soil from Earth-based spectroscopy. *J. Geophys. Res.,* **100,** 14433–14442.

BOUSKA V. and BELL J. F., III (1993) Assumptions about the presence of natural glasses on Mars. *J. Geophys. Res.* **98,** 18719–18726.

BURNS R. G. (1970) *Mineralogical Applications of Crystal Field Theory.* Cambridge Univ. Press, New York, 551 pp. First edition.

BURNS R. G. (1980) Does feroxyhyte occur on the surface of Mars? *Nature* **285,** 647.

BURNS R. G. (1987) Ferric sulfates on Mars. *J. Geophys. Res.* **92,** E570-E574.

BURNS R. G. (1988) Gossans on Mars. *Proc. Lunar Planet. Sci. Conf. 18th,* 713–721.

BURNS R. G. (1993a) *Mineralogical Applications of Crystal Field Theory.* Cambridge Univ. Press, New York, 551 pp. Second edition.

BURNS R. G. (1993b) Rates and mechanisms of chemical weathering of ferromagnesian silicate minerals on Mars. *Geochim. Cosmochim. Acta* **57,** 4555–4574.

BURNS R. G. and FISHER D. S. (1990) Iron-sulfur mineralogy of Mars: Magmatic evolution and chemical weathering products. *J. Geophys. Res.* **95,** 14415–14421.

BURNS R. G. and FISHER D. S. (1993) Rates of oxidative weathering on the surface of Mars. *J. Geophys. Res.* **98,** 3365–3372 1993.

CALVIN W. M., KING T. V. V. and CLARK R. N. (1994) Hydrous carbonates on Mars?: Evidence from Mariner 6/7 infrared spectrometer and groundbased telescopic spectra. *J. Geophys. Res.* **99,** 14,659–14,676 1994.

CHRISTENSEN P. R., ANDERSON D. L., CHASE S. C., CLARK R. N., KIEFFER H. H., MALIN M. C., PEARL J. C., CARPENTER J., BANDIERA N., BROWN F. G. and SILVERMAN S. (1992), Thermal emission spectrometer experiment: Mars Observer mission. *J. Geophys. Res.* **97,** 7719–7734.

CLANCY R. T., LEE S. W., GLADSTONE G. R., MCMILLAN W. W. and ROUSH T. (1995) A new model for Mars atmospheric dust based upon analysis of ultraviolet through infrared observations from Mariner 9, Viking, and Phobos. *J. Geophys. Res.* **100,** 5251–5264.

CLARK B. C., BAIRD A. K., WELDON R. J., TSUSAKI D. M., SCHNABEL L. and CANDELARIA M. P. (1982) Chemical composition of Martian fines. *J. Geophys. Res.* **87,** 10059–10067.

CLARK R. N., SWAYZE G. A., SINGER R. B. and POLLACK J. B. (1990) High-resolution reflectance spectra of Mars in the 2.3 μm region: Evidence for the mineral scapolite. *J. Geophys. Res.* **95,** 14463–14480.

CLAYTON R. N. and MAYEDA T. K. (1983) Oxygen isotopes in eucrites, shergottites, nakhlites and chassignites. *Earth Planet. Sci. Lett.* **62,** 1–6.

DE VAUCOULEURS G. (1954) *Physics of the Planet Mars.* Faber and Faber, London, 365 pp.

DREIBUS G. and WÄNKE H. Mars: a volatile rich planet. *Meteoritics* **20,** 367–382 1985.

ENCRENAZ T. and LELLOUCH E. (1990) On the atmospheric origin of weak absorption features in the infrared spectrum of Mars. *J. Geophys. Res.* **95,** 14589–14593.

FANALE F. P., SALVAIL J. R., BANERDT W. B. and SAUNDERS R. S. (1982) Mars: The regolith-atmosphere-cap system and climate change. *Icarus* **50,** 381.

FANALE F. P., POSTAWKO S. E., POLLACK J. B., CARR M. H. and PEPIN R. O. (1992) Mars: Epochal climate change and volatile history. In *Mars* (ed. H. H. KIEFFER, B. M. JAKOSKY, C. W. SNYDER and M. S. MATTHEWS), pp. 1135–1179, Univ. Arizona Press, Tucson.

FARMER V. C. (1974) The anhydrous oxide minerals. In *The Infrared Spectra of Minerals* (ed. V. C. FARMER), pp. 183–204, Mineral. Soc. Monograph 4, Mineral. Soc., London.

FARMER C. B., DAVIES D. W., HOLLAND A. L., LA PORTE D. D. and DOMS P. E. (1977) Mars: Water vapor observations from the Viking orbiters. *J. Geophys. Res.* **82,** 4225–4248.

FISCHER E. M. and PIETERS C. M. (1993) The continuum slope of Mars: Bidirectional reflectance investigations and applications to Olympus Mons. *Icarus* **102,** 185–202.

GAFFEY S. J., MCFADDEN L. A. and NASH D. B. (1993) Ultraviolet, visible, and near-infrared reflectance spectroscopy: Laboratory spectra of geologic materials. In *Remote Geochemical Analysis: Elemental and Mineralogical Composition* (ed. C. PIETERS and P. ENGLERT), pp. 43–71. Cambridge Univ. Press.

GEISSLER P. E., SINGER R. B., KOMATSU G., MURCHIE S. and MUSTARD J. (1993) An unusual spectral unit in West Candor Chasma: Evidence for aqueous or hydrothermal alteration in the Martian canyons. *Icarus* **106,** 380–391.

GOODING J. L. (1978) Chemical weathering on Mars. *Icarus* **33,** 483–513.

GOODING J. L. (1992) Soil mineralogy and chemistry on Mars: Possible clues from salts and clays in SNC meteorites. *Icarus* **99,** 28–41.

GOODING J. L. and KEIL K. (1978) Alteration of glass as a possible source of clay minerals on Mars. *Geophys. Res. Lett.* **5,** 727–730.

GOODING J. L. and MUENOW D. W. (1986) Martian volatiles in shergottite EETA 79001: New evidence from oxidized sulfur and sulfur-rich aluminosilicates. *Geochim. Cosmochim Acta* **50,** 1049–1059.

GOODING J. L., ARVIDSON R. E. and ZOLOTOV M. Y. (1992) Physical and Chemical Weathering. In Mars (ed. H. H. KIEFFER, B. M. JAKOSKY, C. W. SNYDER and M. S. MATTHEWS), pp. 626–651, Univ. Arizona Press, Tucson.

GUINNESS E. A., ARVIDSON R. E., DALE-BANNISTER M. A., SINGER R. B. and BRUCKENTHAL E. A. (1987) On the spectral reflectance properties of materials exposed at the Viking landing sites. *J. Geophys. Res.,* **92,** E575-E587.

HARDORP J. (1978) The Sun among the stars. I. A search for solar spectral analogs. *Astron. Astrophys.* **63,** 383–390.

HARGRAVES R. B., COLLINSON D. W., ARVIDSON R. E. and SPITZER C. R. (1977) The Viking magnetic properties experiment: Primary mission results. *J. Geophys. Res.* **82,** 4547–4558.

HOUCK J. R., POLLACK J. B., SAGAN C., SCHAACK D. and DECKER J. (1973) High altitude infrared spectroscopic evidence for bound water on Mars. *Icarus* **18,** 470–480.

HUBBARD J. S. (1979) Laboratory simulation of the Pyrolitic Release experiment. *J. Molec. Evol.* **14,** 211–222.

HUGUENIN R. L. (1987) The silicate component of martian dust. *Icarus* **70,** 162–168.

HUNT G. R. and SALISBURY J. W. (1971) Visible and near-infrared spectra of minerals and rocks: II. Carbonates. *Mod. Geol.* **2,** 23–30.

HUNT G. R., SALISBURY J. W. and LENHOFF C. J. (1971) Visible and near-infrared spectra of minerals and rocks: III. Oxides and Hydroxides. *Mod. Geol.* **2,** 195–205.

HUNT G. R., LOGAN L. M. and SALISBURY J. W. (1973) Mars: Components of infrared spectra and the composition of the dust cloud. *Icarus* **18,** 459–469.

JAKOSKY B. M. and FARMER C. B. (1982) The seasonal and global behavior of water vapor in the Mars atmosphere: Complete global results of the Viking atmospheric water detector experiment. *J. Geophys. Res.* **87,** 2999–3019.

JAMES P. B., CLANCY R. T., LEE S., MARTIN L. J., SINGER R. B., SMITH E., KAHN R. and ZUREK R. (1994) Monitoring Mars with the Hubble Space Telescope: 1990–1991 observations. *Icarus* **109,** 79–101.

KASTING J. F. (1991) CO_2 condensation and the climate of early Mars. *Icarus* **94,** 1–13.

KUHN W. R. and ATREYA S. K. (1979) Ammonia photolysis and the greenhouse effect in the primordial atmosphere of the Earth. *Icarus* **37,** 207–213.

LAUL J. C., SMITH M. R., WÄNKE H., JAGOUTZ E., DREIBUS G., PALME H., SPETTEL B., BURGHELE A., LIPSCHITZ M.E. and VERKOUTERAN R. M. (1986) Chemical systematics of the Shergotty meteorite and the composition of its parent body (Mars). *Geochim. Cosmochim. Act,* **50,** 909– 926.

MCCARTHY T. S., ERLANK A. J., WILLIS J. P. and AHRENS L. H. (1974) New chemical analyses of six achondrites and one chondrite. *Meteoritics* **9,** 215–222.

MCCORD T. B. and ADAMS J. B. (1969) Spectral reflectivity of Mars. *Science* **163,** 1058–1060.

MCCORD T. B. and WESTPHAL J. A. (1971) Mars: Narrowband photometry, from 0.3 to 2.5 microns, of surface regions during the 1969 apparition. *Astrophys. J.* **168,** 141–153.

MCCORD T. B., HUGUENIN R. L., MINK D. and PIETERS C. (1977) Spectral reflectance of Martian areas during the 1973 opposition: Photoelectric filter photometry, 0.33–1.10 μm. *Icarus* **31,** 25–39.

MCCORD T. B., CLARK R. N. and HUGUENIN R. L. (1978) Mars: Near-infrared spectral reflectance and compositional implications. *J. Geophys. Res.* **83,** 5433–5441.

MCCORD T. B., CLARK R. N. and SINGER R. B. (1982a) Mars: Near-infrared spectral reflectance of surface regions and compositional implications. *J. Geophys. Res.* **87,** 3021–3032.

MCCORD T. B., SINGER R. B., HAWKE B. R., ADAMS J. B., EVANS D. L., HEAD J. W., MOUGINIS-MARK P. J., PIETERS C. M., HUGUENIN R. L. and ZISK S. H. (1982b) Mars: Definition and characterization of global surface units with emphasis on composition. *J. Geophys. Res.* **87,** 10129–10148.

McKay C. P. and Nedell S. S. (1988) Are there carbonate deposits in the Valles Marineris. Mars? *Icarus* **73,** 142–148.

McSween H. Y., Jr. (1994) What we have learned about Mars from SNC meteorites. *Meteoritics* **29,** 757–779.

Moersch J., Nicholson P., Squyres S., Van Cleve J., Lee P., Hayward T., Houck J. and Miles J. (1993) Thermal infrared observations of Mars during the 1993 opposition. *Bull. Amer. Astron. Soc.* **25,** 1033.

Morris R. V., Agresti D. G., Laver H. V. Jr., Newcomb J. A., Shelfer T. D. and Murali A. V. (1989) Evidence for pigmentary hematite on Mars based on optical, magnetic, and Mössbauer studies of superparamagnetic (nanocrystalline) hematite. *J. Geophys. Res.* **94,** 2760–2778.

Morris R. V., Lauer H. V., Jr., Lawson C. A., Gibson E. K., Jr., Nace G. A. and Stewart C. (1985) Spectral and other physicochemical properties of submicron powders of hematite (α-Fe_2O_3), maghemite (γ-Fe_2O_3), magnetite (Fe_3O_4), goethite (α-FeOOH), and lepidocrocite (γ-FeOOH). *J. Geophys. Res.* **90,** 3126–3144.

Morris R. V., Golden D. C., Bell J. F., III and Lauer H. V., Jr. (1995) Hematite, pyroxene, and phyllosilicates on Mars: Implications from oxidized impact melt rocks from Manicouagan Crater, Quebec, Canada. *J. Geophys. Res.* **100,** 5319–5329.

Morris R. V., Ming D. W., Golden D. C. and Bell J. F., III (1996) An occurrence of jarositic tephra on Mauna Kea, Hawaii: Implications for the ferric mineralogy of the Martian surface. *This volume.*

Murchie S., Mustard J., Bishop J., Head J., Pieters C. and Erard S. (1993) Spatial variations in the spectral properties of bright regions on Mars. *Icarus* **105,** 454–468.

Murchie S., Merényi E., Singer R. and Kirkland L. (1996) Visible-NIR spectroscopic evidence for the composition of low-albedo altered soils on Mars. *Lunar and Planet. Sci. XXVI,* in press.

Mustard J. F. and Bell J.F., III (1994) New composite reflectance spectra of Mars from 0.4 to 3.14 μm. *Geophys. Res. Lett.* **21,** 353–356.

Mustard J. F. and Sunshine J. M. (1995) Seeing through the dust: Martian crustal heterogeneity and links to the SNC meteorites. *Science* **267,** 1623–1626.

Mustard J. F., Erard S., Bibring J.-P., Head J. W., Hurtrez S., Langevin Y., Pieters C. M. and Sotin C. J. (1993) The surface of Syrtis Major: Composition of the volcanic substrate and mixing with altered dust and soil. *J. Geophys. Res.* **98,** 3387–3400.

O'Connor J. T. (1968) Mineral stability at the Martian surface. *J. Geophys. Res.* **73,** 5301–5311.

Pimentel G. C., Forney P. B. and Herr K. C. (1974) Evidence about hydrate and solid water in the Martian surface from the 1969 Mariner infrared spectrometer. *J. Geophys. Res.* **79,** 1623–1634.

Pollack J. B., Colburn D., Kahn R., Hunter J., Van Camp W. and Baldwin B. (1977) Properties of aerosols in the Martian atmosphere, as inferred from Viking Lander data. *J. Geophys. Res.* **82,** 4479–4496.

Pollack J. B., Kasting J. F., Richardson S. M. and Poliakoff K. (1987) The case for a wet, warm climate on Mars. *Icarus* **71,** 203–224.

Pollack J. B., Roush T. L., Witteborn F., Bregman J., Wooden D., Stoker C., Toon O. B., Rank D., Dalton B. and Freedman R. (1990) Thermal emission spectra of Mars (5.4–10.5 μm): Evidence for sulfates, carbonates, and hydrates. *J. Geophys. Res.* **95,** 14,595–14,627.

Postawko S. E. and Kuhn W. R. (1986) Effect of the greenhouse gases (CO_2, H_2O, SO_2) on Martian paleoclimate. *J. Geophys. Res.* **91,** D431-D438.

Roush T. L. and Bell J. F., III (1995) Thermal emission measurements (5–25 μm, 2000–400 cm^{-1}) of Hawaiian palagonitic soils and their implications for Mars. *J. Geophys. Res.* **100,** 5309–5318.

Roush T. L., Blaney D., McCord T. B. and Singer R. B. (1986) Carbonates on Mars: Searching the Mariner 6 and 7 IRS measurements. *Lunar Planet. Sci. XVII,* 732–733.

Roush T. L., Blaney D. L. and Singer R. B. (1993) The surface composition of Mars as inferred from spectroscopic observations. In *Remote Geochemical Analysis: Elemental and Mineralogical Composition,* (ed. C. Pieters and P. Englert), pp. 367–393. Cambridge Univ. Press.

Rowland C. M., Roush T. L., Sloan G. C. and Bell J. F., III (1995) Thermal infrared (7–14 μm) spectral imaging of Mars. *Lunar Planet. Sci. Conf. XXVI,* 1195–1196.

Salisbury J. W., Walter L. S., Vergo N. and D'Aria D. M. (1991) *Infrared (2.1–25 μm) Spectra of Minerals.* Johns Hopkins Univ. Press, Baltimore, 267 pp.

Schaefer M. W. (1990) Geochemical evolution of the northern plains of Mars: Early hydrosphere, carbonate development, and present morphology. *J. Geophys. Res.* **95,** 14,291–14,300.

Schaefer M. W. (1993) Aqueous geochemistry on early Mars. *Geochim. Cosmochim. Acta* **57,** 4619–4625.

Settle M. (1979) Formation and deposition of volcanic sulfate aerosols on Mars. *J. Geophys. Res.* **84,** 8343–8354.

Sherman D. M. and Waite T. D. (1985) Electronic spectra of Fe^{3+} oxides and oxide hydroxides in the near-IR to near-UV. *Amer. Mineral.* **70,** 1262–1269.

Sherman D. M., Burns R. G. and Burns V. M. (1982) Spectral characteristics of the iron oxides with application to the martian bright region mineralogy. *J. Geophys. Res.* **87,** 10169–10180.

Singer R. B. (1982) Spectral evidence for the mineralogy of high-albedo soils and dust on Mars. *J. Geophys. Res.* **87,** 10159–10168.

Singer R. B. and Roush T. L. (1983) Spectral reflectance properties of particulate weathered coatings on rocks: Laboratory modeling and applicability to Mars. *Lunar Planet. Sci. XIV,* 708–709.

Singer R. B., McCord T. B., Clark R. N., Adams J. B. and Huguenin R. L. (1979) Mars surface composition from reflectance spectroscopy: A summary. *J. Geophys. Res.* **84,** 8415–8426.

Singer R. B., Miller J. S. and Wells W. K. (1990) Observed variations in martian crustal composition (abstract). *Bull. Amer. Astron. Soc.* **22,** 1061.

Sinton W. M. (1967) On the composition of martian surface materials. *Icarus* **6,** 222–228.

Soderblom L. A. (1992) The composition and mineralogy of the martian surface from spectroscopic observations: 0.3–50 μm. In Mars (ed. H. H. Kieffer, B. M. Jakosky, C. W. Snyder and M. S. Matthews), pp. 557–593. Tucson: Univ. Arizona Press.

Squyres S. W. and Kasting J. F. (1994) Early Mars: How warm and how wet? *Science,* **265,** 744–749.

Swayze G. A. and Clark R. N. (1990) Infrared spectra

and crystal chemistry of scapolites: Implications for Martian mineralogy. *J. Geophys. Res.* **95,** 14481–14495.
TOON O. B., POLLACK J. B. and SAGAN C. (1977) Physical properties of the particles composing the Martian dust storm of 1971–1972. *Icarus* **30,** 663–696.
TOULMIN P., III, BAIRD A. K., CLARK B. C., KEIL K., ROSE H. J., Jr., CHRISTIAN R. P., EVANS P. H. and KELLIHER W. C. (1977) Geochemical and Mineralogical Interpretation of the Viking inorganic chemical results. *J. Geophys. Res.* **82,** 4625–4634.
TREIMAN A. H., BARRETT R. A. and GOODING J. L. (1993) Preterrestrial aqueous alteration of the Lafayette (SNC) meteorite. *Meteoritics* **28,** 86–97.

APPENDIX: COLLECTION, REDUCTION, AND CALIBRATION OF 1990 MARS SPECTRAL DATA

Imaging spectroscopic data of Mars were obtained from the University of Hawaii 2.24 m telescope at Mauna Kea Observatory on 1990 November 9 and 10 UT (Table 3). The data were obtained with the Wide Field Grism Spectrograph (WFGS), a facility instrument operated by the Institute for Astronomy at the University of Hawaii. WFGS uses an 800 × 800 CCD and a transmission grating ruled on a prism to disperse a spectrum along one axis of the CCD array. We used a grating blazed at 4800 Å in first order to obtain data from 0.50 to 0.95 μm at a spectral resolution of R = 200 to 350. WFGS is a field widened spectrograph that can accommodate a variety of user-designed slits. Our Moon/Mars slit design had projected dimensions of 0.29″ × 153″, allowing for high spectral resolution and adequate cross-slit spatial sampling of the Martian disk. Individual images at a number of slit positions on the sky were obtained by sequentially scanning the slit across the disk of Mars. In this way, image cubes were built up by using telescope motion to construct one spatial axis (across-slit) and obtaining individual images in a spatial × spectral format (along slit × dispersed spectrum). A typical Mars image cube consisted of approximately 20 × 30 spatial pixels at ≈160 useful wavelengths. Due to rather poor atmospheric seeing conditions during the observing run, the actual image quality corresponded to only 1″ to 2″, equivalent to regions from 750 to 1500 km in diameter, and thus the imaging aspect of these measurements did not yield much success (Fig 4a).

Eleven image cubes were obtained centered near 100–140° W longitude. Mars subtended 17.8″ during this time, was at a phase angle of 17°, and the season was late northern winter (L_s = 330°). We also obtained 10 image cubes of the solar analog star κ Cetus (Type G5V, MV = +4.82; HARDORP, 1978) and numerous bias and flatfield calibration frames. As an aid in reconstruction of the slit-scan images, we also obtained direct red filter CCD images of Mars. Wavelength calibration was performed using Hg-Cd-Zn and Ne calibration lamps.

Data reduction procedures are similar to those used for 1988 WFGS data described previously by BELL *et al.* (1990b) and BELL (1992) for different regions of the planet. These procedures include bias and dark current removal, flatfield (pixel-to-pixel non-uniformity) corrections, assembly of the individual slit positions into image cubes, and spatial reconstruction of the imaging data so that the geographic locations of spectra could be determined. We were not able to completely remove the residual effects of telluric O_2 absorption near 0.76 μm, so the spurious data near that wavelength have been removed.

Table 3. 1990 Imaging spectroscopic observations of Mars from Mauna Kea

Date (UT)	Ls	Phase	Mars Cubes	Spectra	κ Cet Cubes	Spectra
9 Nov. 1990	330°	17°	5	≈4000	4	≈30
10 Nov. 1990	330°	16°	6	≈6000	5	≈35

Mineral Spectroscopy: A Tribute to Roger G. Burns
© The Geochemical Society, Special Publication No. 5, 1996
Editors: M. D. Dyar, C. McCammon and M. W. Schaefer

Are there abiotically-precipitated iron-formations on Mars?

M. W. SCHAEFER
Yale University, Department of Geology and Geophysics, P. O. Box 208109, New Haven, CT 06520-8109, U.S.A.

Abstract—In early times, surface conditions on Mars and Earth were probably quite similar, however, the two planets took different paths in their evolution. Banded iron-formations (BIFs) were prevalent on Earth in the Archean, and the likelihood of their formation on Mars at the same time is investigated. There are several abiotic mechanisms for producing banded iron-formations on Earth. These are: (1) oxidation of ferrous iron by free oxygen, (2) photooxidation of ferrous iron, (3) oxidation of ferrous iron by hydrogen peroxide, and (4) anoxic precipitation of ferrous iron by changes in pH. Any of these mechanisms could have been active on Mars in its distant past, to produce BIFs there as well, which should be observable today. The minerals one would expect to observe on the surface of a BIF on Mars are indeed thought to exist on Mars, but unfortunately BIFs are not unique in containing these minerals, and confirmation of the existence of BIFs on Mars will have to wait for more data. Proof of the existence of BIFs on Mars would give us valuable information about the early history of the planet Mars as well as Earth. It would act to confirm models of Mars that include a warm, wet period, and the existence of bodies of standing water.

INTRODUCTION

IN THE SURFACE of Mars we see a relict of the early evolution of the planet. During the early history of the solar system, conditions were probably not dissimilar on the surface of Earth and Mars, but as time passed, the two planets diverged. Earth, larger and with a greater source of internal heating, stayed active and its surface kept evolving. Mars, on the other hand, smaller and with less internal heating, froze and its surface almost completely stopped evolving. Therefore, when we look at the surface of Mars, we have the opportunity not only to observe the early history of Mars itself, but also to observe a landscape perhaps more similar to that of early Earth than is the present terrestrial landscape.

The surface of Mars seems to have changed relatively little over the past several billion years. Processes that have acted on its surface in the past include meteorite bombardment, which was at its peak over 3.5 Gy ago; volcanism, which is thought to have largely ceased by about 1 Gy ago; and fluvial and glacial processes, about which there is active speculation (BAKER *et al.,* 1991; BAKER *et al.,* 1992; KARGEL *et al.,* 1995), but which neither appear to have been active within the last 1 Gy (although periglacial processes may be active today (ROSSBACHER and JUDSON, 1981). The role of hydrothermal processes is as yet unknown, however they could be very important, both to the chemical evolution of the regolith and to the isotopic evolution of the atmosphere (JAKOSKY and JONES, 1996). Physical and chemical weathering, which is still active today, acts only to alter the characteristics of the uppermost surface layers (GOODING, 1978; GOODING *et al.,* 1992).

It is likely that Earth and Mars were not dissimilar when they were first formed. By studying processes that occurred on the early Earth, we may gain insight into what processes might have been important on early Mars. And conversely, as we obtain more detailed information about the present surface of Mars, we will learn more about the early history of both planets. BURNS (1993) originally suggested that banded iron-formations (BIFs) could have occurred on Mars. The present paper discusses the likelihood of the occurrence of BIFs on Mars, and what their existence might imply about the early environment of Mars.

TERRESTRIAL BANDED IRON-FORMATIONS

Some of the oldest surviving sedimentary features on Earth are the banded iron-formations, chemically-precipitated sedimentary deposits containing layers of silica-rich and iron-rich bands. GROSS (1980) classifies banded iron-formations into two groups, based on associated types of rock, lithological features, and interpretation of their depositional environment. These two groups are Algoma-type and Lake-Superior-type. Both of these types exhibit cherty layers interbedded with iron in four facies: oxides, carbonates, silicates, and sulfides (JAMES, 1954). In addition to the banded iron-formations, there is another type of iron-formation called an oolitic ironstone. Oolitic ironstones are higher in aluminum than BIFs, and have a more clastic association. They are found in rocks of a variety of ages from Proterozoic to Pliocene (MAYNARD, 1983).

Although BIFs are agreed to be chemical precipitates of iron, the mechanism by which the precipitation occurs is a controversial topic. Both biotic (CLOUD, 1973; GARRELS *et al.,* 1973; LABERGE, 1973; LABERGE *et al.,* 1987; ROBBINS *et al.,* 1987) and abiotic (BRATERMAN and CAIRNS-SMITH, 1987; CAIRNS-SMITH, 1978; EUGSTER and CHOU, 1973; FRANÇOIS, 1986; HOLLAND, 1973) mechanisms have been proposed.

The major idea behind both of these classes of theories is that the atmosphere of the early Earth was depleted in O_2, compared to the present atmospheric level (PAL). Under conditions of low oxygen fugacity, especially in an acidic environment, iron exists in the ferrous form, and is highly soluble in water. Thus any iron input into a lacustrine or oceanic system by weathering or other process will stay in solution in the water until it reaches saturation with respect some iron silicate or carbonate, or is oxidized to form ferric iron, which is relatively insoluble in water. Then, for any of a variety of reasons, depending on the model, the iron is oxidized, and precipitates out, forming the extensive deposits that still exist today.

In the classic biotic models the factor that causes this precipitation of iron is the increase in atmospheric O_2 caused by the development of photosynthetic organisms. However, it is not necessary to invoke the development of life to achieve oxidation of iron in Precambrian waters. Also, there is evidence that very little oxygen was present in the atmosphere during the formation of many of the deposits. The presence of detrital siderite in Lake-Superior-type deposits in Labrador and the occurrence of detrital grains of uraninite and pebbles of pyrite in the 2.5-Gy-old Witwatersrand conglomerate indicate a vanishingly low p_{O_2} content in the atmosphere at that time (GARRELS *et al.,* 1973). However, biotic models explain this by postulating procaryotic microorganisms that reduced CO_2 to CH_2O in the hydrosphere by taking electrons from ferrous iron (CLOUD, 1972, 1973).

ABIOTIC MECHANISMS OF MAKING IRON-FORMATIONS

There are four abiotic processes that are capable of forming BIFs. (1) *Oxidation of ferrous iron by dissolved oxygen.* Although this is also descriptive of what happens when photosynthesis produces oxygen, there are abiotic reactions that may produce free oxygen. (2) *Photooxidation of ferrous iron by ultraviolet irradiation.* Ultraviolet (UV) radiation is capable of directly oxidizing ferrous iron in solution. (3) *Oxidation of ferrous iron by hydrogen peroxide.* Hydrogen peroxide (H_2O_2) is a strong oxidant, and is also able to directly oxidize ferrous iron in solution. (4) *Anoxic precipitation of ferrous iron.* Under anoxic conditions, it is possible to precipitate ferrous iron carbonates and silicates under the right conditions of pH and p_{CO_2}. These mechanisms are shown schematically in Fig. 1 and will be discussed in detail below.

The question of whether banded iron-formations on Earth formed primarily by a biotic or an abiotic mechanism is still an open question. Another open question is the likelihood of life ever having existed on Mars. The main thrust of the present paper is the discussion of abiotic mechanisms of making BIFs. Such mechanisms require fewer assumptions, when translated to Martian conditions, than do the biotic models. However, if life did exist on Mars and was instrumental in causing iron precipitation in bodies of water there, then the iron formations developed would be a likely place to look for fossil evidence of such life.

Dissolved oxygen

There is geological evidence that free oxygen was present in the early atmosphere (even as early as 3.8 Gy ago), although not in quantities approaching PAL (TOWE, 1983). HOLLAND (1973) demonstrated that neither subaerial weathering or volcanic emanations were likely to have been the immediate sources of soluble iron to form the Hamersley Basin BIF of Australia. He then proposed that upwelling of bottom waters in the ocean could

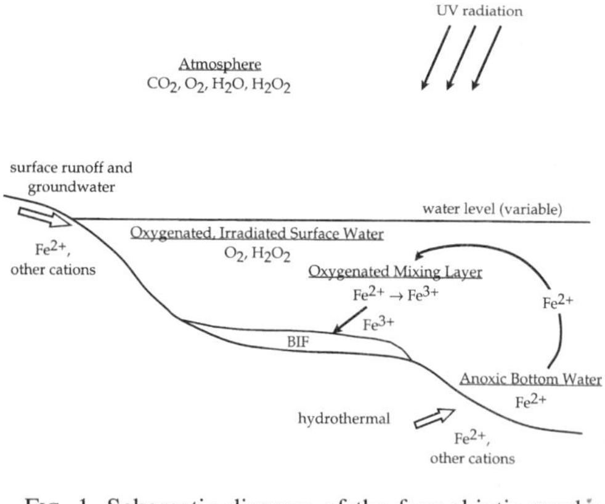

FIG. 1. Schematic diagram of the four abiotic mechanisms of BIF formation. Ferrous iron may be oxidized to ferric iron in the mixing layer by means of dissolved oxygen or hydrogen peroxide, or nearer the surface by UV radiation. Anoxic water (not necessarily bottom water, as in the figure, if the oxygen content of the atmosphere is low) may precipitate ferrous iron if the pH is very high (caused by evaporation) or very low (caused by an input of acidic ground or surface water).

have provided the necessary quantities of dissolved iron to produce this deposit, and generalized this mechanism to explain other Lake Superior-type deposits where there is little clastic debris (HOLLAND, 1973). TOWE (1983) then finds a corollary to Holland's proposition, namely, that free oxygen does not have to be absent from the atmosphere to maintain the anoxic conditions necessary to transport large quantities of ferrous iron in solution, instead, that the deeper basins must remain stagnant and anoxygenic for reasonable periods of time.

In this model, anoxygenic bottom waters, carrying dissolved ferrous iron, circulate upwards to oxygen-rich surface layers in shallower regions of the ocean, where the iron is oxidized and deposited. It is not necessary to invoke photosynthesis here to produce the relatively oxygen-rich surface waters. In fact, there is evidence against photosynthesis being an important process here. BIF oxide facies are remarkably free of organic carbon (TOWE, 1983). Anoxic atmosphere models of BIF formation require that photosynthesis in the basin waters produce the dissolved oxygen necessary to oxidize the iron. However, if there were large algal communities living in these basins, one would expect to find evidence of them in the fossil record, which is not seen (TOWE, 1983). This observation is also supported by the very low contents of phosphorus in BIFs (BEUKES and KLEIN, 1992).

In the absence of photosynthesis to produce free oxygen in the atmosphere and dissolved in the surface waters, UV photolysis of water and CO_2 can become important processes:

$$H_2O \xrightarrow{h\nu} H_2 + \frac{1}{2}O_2 \qquad (1)$$

$$CO_2 \xrightarrow{h\nu} CO + \frac{1}{2}O_2 \qquad (2)$$

When this ferrous iron in solution is exposed to free oxygen, it will oxidize according to the following reactions (BURNS, 1993)

$$Fe^{2+}(aq) + \frac{1}{4}O_2(aq) + H^+ \rightleftharpoons Fe^{3+}(aq) + \frac{1}{2}H_2O \qquad (3)$$

$$Fe^{3+} + 2H_2O \rightleftharpoons FeO(OH)(s) \downarrow + 3H^+. \qquad (4)$$

Measurements of UV radiation from T-Tauri stars (thought representative of the Sun at an early stage in its evolution) by the International Ultraviolet Explorer (IUE) spacecraft indicate that such stars produce far higher UV fluxes than originally thought (CANUTO *et al.*, 1983). Although the Sun's total luminosity is thought to have been lower in the past, the UV flux was likely much higher (Table 1).

CANUTO *et al.* (1983) performed a detailed photochemical model of the Earth's early atmosphere to determine the abundance of free oxygen. They calculated that whereas the surface O_2 mixing ratio for an atmosphere containing 1 PAL CO_2 (2.8×10^{-4} bars), irradiated by the present solar UV flux, is about 10^{-15}, if an atmospheric abundance of 100 PAL is assumed, irradiated by a solar UV flux 4 times the present value (as would be appropriate at a solar age of about 1 Gy), the surface O_2 mixing ratio is a factor of over 10^3 greater (Fig. 2). This surface O_2 mixing ratio of 3×10^{-12} is within the range of O_2 values (10^{-13} to 10^{-3}) required to explain the simultaneous existence of oxidized iron and reduced uranium noted earlier in the oldest rocks (CANUTO *et al.*, 1983). Even assuming the present solar UV flux, and an atmospheric partial pressure of 100 PAL (0.028 atm) CO_2, the CANUTO *et al.* (1983) model calculates a surface O_2 mixing ratio of about 10^{-12}, still within this range. The model of KASTING *et al.* (1984), on the other hand, calculated that the O_2 surface mixing ratio in an atmosphere containing 1000 PAL CO_2 under the present solar UV flux would be 10^{-11}, and that an increase in solar UV flux would act to reduce, not increase, the O_2 surface mixing ratio, due to the higher H_2 densities produced by more rapid H_2O photolysis. Clearly this is a complicated problem, but we seem close to understanding how surface O_2 levels in the atmosphere before the development of life could have been in the range

Table 1: *Stellar UV flux as a function of age, from* CANUTO *et al. (1983).*

Age (years)	UV enhancement
10^6	10^4
10^7	500
5×10^7	100
10^8	32
5×10^8	8
10^9	4
5×10^9	1

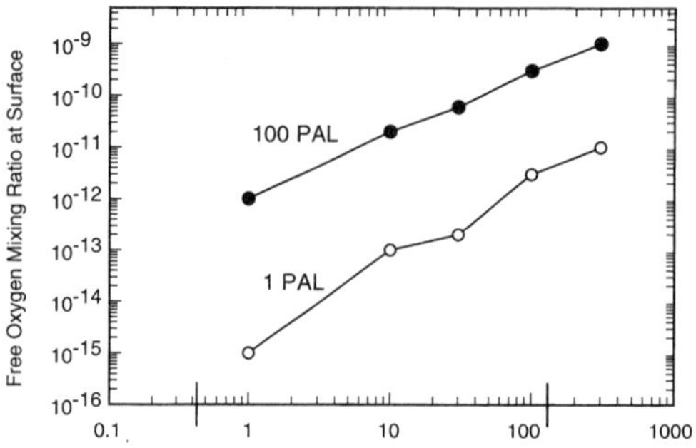

FIG. 2. Free oxygen mixing ratio at the surface, *vs.* solar UV flux, from data in CANUTO *et al.* (1983). Two cases are given, an assumed p_{CO_2} of 1 PAL and one of 100 PAL. A range of solar UV fluxes from 1 to 300 times the present Earth value are shown. In addition, the two bars at about 0.4 and 130 present UV flux represent the corresponding fluxes at Mars, for this range of fluxes at Earth.

necessary to explain the mineralogy of the sediments.

Ultraviolet irradiation

Another method of oxidizing iron in solution is by means of direct UV irradiation (WEISS, 1935; JORTNER and STEIN, 1962a and b; PAPP and VINCZE, 1980).

CAIRNS-SMITH (1978) originally proposed that BIFs were formed by the photochemical oxidation of iron, according to the reaction

$$Fe^{2+}(aq) + H^+(aq) \xrightarrow{h\nu} Fe^{3+} + \frac{1}{2}H_2\uparrow. \quad (5)$$

Further studies showed that photo-oxidation of iron is an important reaction in ferrous solutions with a wide range of pH values, 7.5 and less (BRATERMAN *et al.,* 1983 and 1984). At pH values less than about 3.5, hydrolysis and precipitation are not important, but in the pH range 6.2–4, goethite is precipitated according to the following reactions:

$$Fe^{3+}(aq) \rightleftharpoons [Fe(OH)]^{2+}(aq) + H^+(aq) \quad (6)$$

$$[Fe(OH)]^{2+}(aq) \rightleftharpoons \frac{1}{2}[Fe_2(OH)_4]^{2+}(aq) + H^+(aq) \quad (7)$$

$$\frac{1}{2}[Fe_2(OH)_4]^{2+}(aq) \rightleftharpoons Fe(OH)_3(aq) + H^+(aq) \quad (8)$$

$$Fe(OH)_3(aq) \rightarrow FeO(OH)(s)\downarrow. \quad (9)$$

The above scenario occurs in both saline and non-saline solutions, with UV irradiation at $\lambda < 300$ nm (BRATERMAN *et al.,* 1983 and 1984).

At pH values between 6.5 and 7.5, a different set of reactions becomes important (BRATERMAN *et al.,* 1983 and 1984; BRATERMAN and CAIRNS-SMITH, 1987):

$$Fe^{2+}(aq) \rightleftharpoons [Fe(OH)]^+(aq) + H^+(aq) \quad (10)$$

$$[Fe(OH)]^+(aq) \xrightarrow{h\nu} Fe(OH)_3 + H^+(aq). \quad (11)$$

The $Fe(OH)_3$ then forms aggregates and eventually precipitates out as $FeO(OH)$, and the system evolves gaseous hydrogen. The reaction rate for this process is about 10 times greater than at lower pH, and longer wavelengths ($\lambda > 400$ nm) are still effective.

Using these reactions and assuming the present solar UV flux (with 50% loss through scattering and clouds), in terrestrial Fe^{2+}-containing waters having a pH > 6.5, a rate of photoprecipitation of 150–750 mg cm^{-2} yr^{-1} at the tropics or 50–250 mg cm^{-2} yr^{-1} at the poles can be calculated (BRATERMAN *et al.,* 1983). These are somewhat larger than estimates of iron deposition in the Hamersley BIFs of 9–43 mg cm^{-2} yr^{-1} (TRENDALL and BLOCKLEY, 1970), indicating that photo-oxidation of iron is a sufficient explanation for their formation. The Precambrian ocean model of FRANÇOIS (1986, 1987), which also calculates deposition of iron formations based on the photo-oxidation of $[Fe(OH)]^+$, agrees with these results.

The work of ANBAR and HOLLAND (1992) on the photochemistry of manganese is also consistent with a photochemical model for the formation of BIFs. However, DES MARAIS (1994) points out that the absence of BIF deposition on shallow-water Archean platforms, even at water depths some distance below the photic zone, indicates that not only was the surface seawater oxidizing, but that a mobile oxidant such as dissolved O_2 must have penetrated below the photic zone to prevent the deposition of BIFs at those depths.

Hydrogen peroxide

KASTING *et al.* (1984, 1985) suggested that hydrogen peroxide (H_2O_2) could be formed in the early, carbon dioxide-rich atmosphere of Earth and then rain out to oxidize and precipitate the iron in the ocean. They determined that hydrogen peroxide is probably the dominant oxidant in rainwater at free oxygen levels less than 10^{-2}–10^{-3} PAL (about 2×10^{-3} to 2×10^{-4} bars), and was probably an important oxidant even in O_2-free atmospheres, provided that the CO_2 levels were significantly higher than today. These results indicate that hydrogen peroxide is also a viable candidate for an inorganic oxidizer in the Precambrian ocean, that could have acted to precipitate the BIFs according to the reactions

$$2Fe^{2+}(aq) + H_2O_2(aq) + 4H_2O \rightarrow 2Fe(OH)_3 \downarrow + 4H^+(aq) \quad (12)$$

$$Fe(OH)_3(aq) \rightarrow FeO(OH)(s) \downarrow. \quad (13)$$

This oxidant, like dissolved oxygen gas, would not have had a problem penetrating below the photic zone to prevent the deposition of BIFs at such depths.

However, subsequent work by KASTING (1990) indicates that when sulfur chemistry is included in a low-O_2 atmosphere, the dominant photochemically-produced oxidant is sulfuric acid, and the hydrogen peroxide abundances are greatly reduced. In this case, the sulfuric acid would react with water in the atmosphere to form sulfate, which would then rain out into the ocean (KASTING, private communication). Sulfate is not nearly as powerful an oxidant as hydrogen peroxide, so this mechanism may not be sufficient to oxidize the ferrous iron to form iron formations.

In summary, there appear to be an abundance of adequate abiotic mechanisms to precipitate out iron in a largely anoxic Precambrian ocean, under a carbon dioxide-rich atmosphere. At our current level of knowledge, it is difficult to determine unequivocally which must have been the dominant process. Although certainly the development of living organisms radically changed the oxidation state of the early terrestrial environment, however, it is not necessary to invoke the origin of photosynthesis to begin the deposition of the banded iron formations.

Anoxic deposition and the banded nature of the iron-formations

In addition, the banded nature of the iron formations requires explanation, and the most common idea is that the iron supply is periodically variable (BUTTON *et al.*, 1982; DREVER, 1974; EUGSTER and CHOU, 1973; HOLLAND, 1973; MORRIS, 1993; TOWE, 1983).

EUGSTER and CHOU (1973) proposed that banded iron-formations were deposited in a playa-lake complex. They assumed that the atmosphere was free of oxygen, and that precipitation of iron occurred in response to changes in pH and f_{CO_2} at an f_{O_2} level between that of the hematite + magnetite and magnetite + quartz + fayalite buffers. In their model, evaporative precipitation of carbonates occurs at the margins of a largely spring-fed lake, or barred lagoon, with magadiite ($NaSi_7O_{13}(OH)_3 \times 4H_2O$) and iron-silica gel precipitation occurring in the deeper parts of the lagoon. Occasional floods, which bring in fresh, low-pH waters, are invoked to precipitate the magadiite or sodium silica gel, and ferrous hydroxide ($Fe(OH)_2$), iron-silicate and sodium-iron-silicate gels and to wash the carbonates into the lake, producing bands. In addition, ferric hydroxide ($Fe(OH)_3$) precipitates if free oxygen should become available.

This model represents not a mechanism for oxidizing the iron, but a means of depositing it in the ferrous state. The presently-observed chert and ferric iron deposits are caused by the subsequent diagenesis of the primary precipitates mentioned above (EUGSTER and CHOU, 1973). This scenario is significantly different than the other models discussed here. The other models assume that the depths of the ocean were low in oxygen, but that ferric iron was oxidized and precipitated in an upper layer, either by dissolved oxygen or by some other means. The arguments for oxidization of the iron in solution are strong, and seem to preclude the likelihood of the EUGSTER and CHOU (1973) mechanism [developed over ten years earlier, before the work of Braterman, Cairns-Smith, Canuto, Kasting, Towe, and their co-workers (BRATERMAN *et al.*, 1983; BRATERMAN *et al.*, 1984; BRATERMAN and CAIRNS-SMITH, 1987; CAIRNS-SMITH, 1978; CANUTO *et al.*, 1983; KASTING *et al.*, 1984; KASTING *et al.*, 1985)] having been dominant. In addition, there is evidence from rare earth element (REE) analysis that Precambrian BIFs are the result of chemical precipitation from solutions that are mixtures of seawater and hydrothermal input (KLEIN and BEUKES, 1992). However, the thermodynamic calculations and predominance diagrams that they developed are still useful in determining what species are stable under given conditions of f_{O_2}, pH, and f_{CO_2}, particularly in the anoxic conditions prevalent deep in the ocean.

A more recent model for explaining the banded nature of the BIFs was developed by MORRIS (1993). This detailed model of the Hamersley BIF in Australia relies on the interaction of two major oceanic supply systems to produce bands of iron and silica: (1) surface currents and (2) convective upwelling from mid-ocean ridges or hot spots (MORRIS, 1993). The surface currents were saturated in silica, but relatively free of iron, either because the system was oxidizing or because of photoprecipitation. Precipitation from these surface waters, possibly driven by evaporative concentration, formed the chert-rich layers. Periodically, upwelling from iron-rich deeper waters, driven by convective overturn, overwhelmed the deposition of silica to produce the iron-rich layers. This model includes a low-oxygen to anoxic atmosphere, higher heat flow under the oceans than at present, and a modest level of photosynthetic activity, but none of these conditions are essential to the model (MORRIS, 1993). Another factor considered in this model of the formation of the Hamersley deposits was the effects of surface volcanic activity.

The bulk of both ferrous and ferric iron presently in the Hamersley BIF is therefore attributed to oxidation in the surface zone of the ocean. The FeO(OH) precipitate eventually is transformed to (1) hematite, if oxidizing conditions are maintained in the sediments, or (2) magnetite, if partial reduction occurred due to the action of organic matter, or (3) siderite/silicate, if substantial reduction occurred in the photic zone (increasing its iron content) or in the sediment itself (MORRIS, 1993).

EARLY GEOCHEMISTRY OF EARTH AND MARS

When Mars and Earth were in their infancy, they were far more alike than they are at present. Both were rocky bodies, heated by their accretion, and with atmospheres dominated by carbon dioxide, nitrogen, and water vapor. Volcanic activity produced vast basalt plains, as well as volcanic edifices. The bombardment of the crust with meteorites also kept hydrothermal circulation systems active. In topographically low areas, lakes and oceans formed. Active weathering of the silicate crust by carbonic acid-rich waters provided a rich source of iron, silica, and other cations to supply the early oceans, and left behind it a clay and oxide-rich weathering profile, which might still be seen today on Mars (TOWE, 1983).

The existence of lakes or oceans on early Mars is still controversial. Values of total outgassed CO_2 from several to about 10 bars are consistent with present knowledge (POLLACK *et al.,* 1987), and this amount of CO_2 implies an amount of water outgassed at least equal to an equivalent depth of 500 to 1000 meters (CARR, 1986). POLLACK *et al.* (1987) have made extensive calculations deriving the amount of carbon dioxide necessary to achieve a strong enough greenhouse effect to raise the surface temperature of Mars above the freezing point of water, and have determined that, for different values of surface albedo, latitude, and orbital position, pressures of from 0.75 to 5 bars are necessary. Work by KASTING, however (KASTING, 1991), indicates that CO_2 alone may not be sufficient to warm the surface of Mars above the freezing point of water. He finds that under Martian conditions carbon dioxide cannot be present in the atmosphere at such high pressures, but instead will condense out. There is recent work from FANALE and POSTAWKO (1995) and from YUNG *et al.* (1995) indicating that adding SO_2 to the models may eliminate the problem of CO_2 condensation. Even if a carbon dioxide atmosphere alone was not enough to cause a warm early Mars, there is still sufficient evidence that Mars outgassed large amounts of carbon dioxide (POLLACK *et al.,* 1987), and there is much geological evidence of flowing water upon the surface, implying warm temperatures. In addition, PARKER and coworkers (PARKER *et al.,* 1987; PARKER and GORSLINE, 1992; PARKER *et al.,* 1993) have seen geomorphic evidence for ancient shorelines in the northern plains of Mars.

The existence of lakes or oceans on early Mars will therefore be assumed as a working hypothesis for this paper, but I will show that if BIFs do exist on Mars, their existence offers proof of the existence of large bodies of water on Mars at some time.

Mars is farther from the Sun, which means that it receives less radiation than Earth, and would have been cooler. More importantly, it is smaller than Earth, so that its internal heat was not enough to support plate tectonics, or to keep the crust from thickening and cutting off volcanic activity to the surface. As it lost volatiles from the atmosphere, both to the surface and to space, the surface cooled past the freezing point of water, and its geochemical evolution slowed and nearly ceased.

But before about 3500 million years ago, the two planets were relatively similar, and geochemical processes, such as the formation of BIFs, that we observe on Earth may have occurred on Mars as well. The second part of this paper investigates how one must modify the mechanisms discussed above to apply to Martian conditions, and discusses what implications the existence of BIFs on Mars would have for the history of the planet (such as confirmation of the warm, wet, early Mars hypothesis), and how we might determine their existence there today.

MECHANISMS OF MAKING IRON-FORMATIONS ON MARS

Dissolved oxygen

At present, the Martian atmosphere contains, on average, 0.13% O_2, in an atmosphere of about 7.5

$\times 10^{-3}$ bars CO_2, or about 10^{-5} bars of O_2 (OWEN, 1992). This oxygen is produced by the photolysis of water and CO_2, as it was described above for the prebiotic Earth. If there could be liquid water on the surface of Mars today, the upper layers, at least, would be oxygenated, and ferrous iron in solution could not exist. In the past, when there was more carbon dioxide and water vapor in the atmosphere of Mars, there might have been correspondingly more free oxygen as well. However, the oxygen produced in the atmosphere of Mars today is in contact with a highly-oxidized surface, whereas in the distant past it would not have been, and the oxygen in the atmosphere would have been decreased by its reactions with the surface materials and the cations in solution in water. The amount of oxygen in the atmosphere of Mars today is thought to be controlled by redox balance, such that the H escape rate is equal to twice the O escape rate (MCELROY and DONAHUE, 1972) plus surface oxidation. Active volcanism in the past would have released H_2 and other reduced gases, which also would have lowered the p_{O_2} relative to present levels.

If an early Earth-type atmospheric model, such as that of CANUTO *et al.* (1983) or KASTING *et al.* (1984), is used to calculate the amount of oxygen in the atmosphere of Mars, it produces a much lower mixing ratio of oxygen than is presently seen. The CANUTO *et al.* (1983) model produces a surface oxygen mixing ratio for a solar UV flux of 1.7 times the present solar flux at Earth (appropriate for about 3.5 Gy ago, when the solar UV flux at Earth would have been about 4 times the present flux) and an atmosphere of 100 terrestrial PAL (2.8 $\times 10^{-2}$ bars) of approximately 10^{-13} (Fig. 2). If this model is further extrapolated to 10,000 terrestrial PAL, or 2.8 bars (a value that may have been necessary to produce enough of a greenhouse effect on Mars to raise the temperature above the freezing point of water), a mixing ratio of about 10^{-10} may be produced.

The KASTING *et al.* (1984) model, extrapolated in a similar fashion, also produces a surface oxygen mixing ratio of about 10^{-10} to 10^{-11}. It is not really a valid approximation to extrapolate either of these two models in this fashion, because there is a good deal of Earth-specific atmospheric chemistry in each of them, but it is useful to give us a vague approximation of the surface oxygen mixing ratio that might have been expected on early Mars.

Therefore, let us assume that the surface oxygen mixing ratio in the warm, wet period of early Mars may have been approximately 10^{-10} to 10^{-13}, based on these two terrestrial models. This is more than enough oxygen to oxidize ferrous iron in solution so that it will precipitate out as ferric iron. [Note: There does appear to be a small stability field for siderite at f_{O_2} equal to 10^{-13} and below, and pH ≈ 6 (see Fig. 3, discussed in more detail later).] The problem with forming BIFs under these conditions, conversely, it that there must have been some mechanism for maintaining the ferrous iron in solution so that it could be transported. Possibly, any bodies of water on early Mars could have been deep enough and still enough (Mars, after all, can have had no tides) so that conditions near the bottom could have been anoxic.

Ultraviolet irradiation

Because Mars is only about 1.5 times as far from the Sun as is the Earth, the flux of UV (and other solar) radiation at its surface is only about a factor of 2.3 times less than that at the surface of the Earth. The yield of a photochemical process is di-

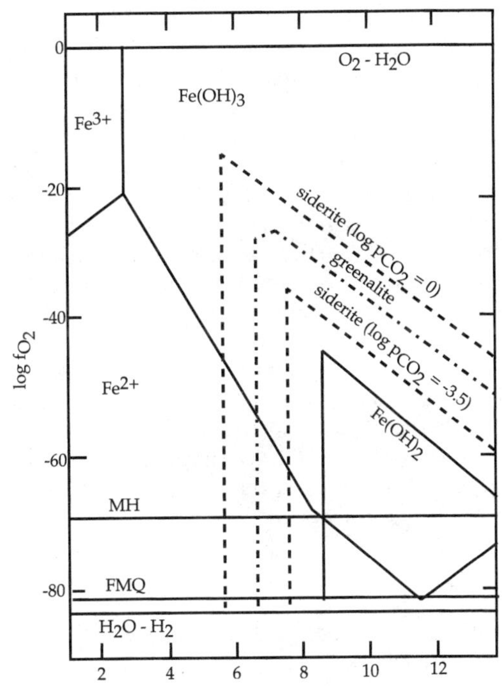

FIG. 3. Solubilities of $Fe(OH)_3$, $Fe(OH)_2$, greenalite (dot-dashes, in equilibrium with amorphous silica), and siderite (dashes, with a log p_{CO_2} of 0 and -3.5), at 25°C and a log total iron content of -4. The magnetite-hematite (MH) and fayalite-magnetite-quartz (FMQ) boundaries are also shown. Adapted from EUGSTER and CHOU (1973).

rectly proportional to the number of photons absorbed, so the efficiency of iron photoprecipitation on Mars at a given time should be about 2.3 times less than at the same time on Earth, if scattering and loss of UV flux through the atmosphere is disregarded. The range of iron photoprecipitation rates calculated by BRATERMAN *et al.* (1983) for Earth may be thus scaled to Mars, to get 65–320 mg cm^{-2} yr^{-1} at the tropics or 22–110 mg cm^{-2} yr^{-1} at the poles. These rates are still larger than the rate estimates for the terrestrial Hamersley BIF deposition (9–43 mg cm^{-2} yr^{-1} (TRENDALL and BLOCKLEY, 1979)), indicating that photoprecipitation of iron is a sufficient mechanism to produce BIFs on Mars as well.

Hydrogen peroxide

The same photochemical reactions that could have formed hydrogen peroxide in the early atmosphere of Earth could have acted in the early atmosphere of Mars. As a first approximation, one may look at the terrestrial model of KASTING *et al.* (1985), which determined that hydrogen peroxide is probably the dominant oxidant in rainwater at free oxygen levels less than 10^{-2}–10^{-3} PAL (about 2×10^{-3} to 2×10^{-4} bars). This model was also run with a mixing ratio of CO_2 of 100 PAL, and the H_2O_2 rain-out rate was found to decrease by only 5–30%. This implies that the crossover point between free oxygen or hydrogen peroxide as the major oxidant was probably the same, within an order of magnitude or so, on Mars as it was on Earth. The calculations for Mars above indicate that there may have been between 10^{-11} and 10^{-3} O_2 surface mixing ratio, or between 10^{-11} and 10^{-3} bars O_2, assuming a total atmospheric pressure of 1 bar. This range is from far less than the amount required to make free oxygen the dominant oxidant, to roughly the borderline between oxygen and hydrogen peroxide at the dominant oxidant. Based on this, hydrogen peroxide also was capable of precipitating BIFs on Mars.

Anoxic deposition

The anoxic model of terrestrial BIF formation developed by EUGSTER and CHOU (1973) attempts to explain not only iron-rich bands, but also the silica-rich chert bands. Their explanation of the formation of the chert bands is that these were originally deposited in the form of magadiite ($NaSi_7O_{13}(OH)_3 \times 4H_2O$). Magadiite, which is observed as the precursor to chert beds in Lake Magadi, in Africa, has a minimum in its solubility curve at pH = 8.6 (EUGSTER and CHOU, 1973). This implies that magadiite can be precipitated by either evaporative concentration (raising the pH), or by flooding of silica-rich, high-pH brines by fresher, more acid waters. This mechanism is not dependent on any factors that are different on Mars than on Earth, and would therefore work equally well on Mars as it does on Earth.

The deposition of the iron-rich layers, in this anoxic model, is more complicated. The stabilities of various iron oxides, hydroxides, and iron carbonate relative to one another must be determined, to derive which species are stable under likely Martian conditions. The solubility of iron hydroxides and hydroxides depends on the f_{O_2} and the pH, and that of siderite ($FeCO_3$) depends as well on p_{CO_2}.

Figure 4 shows a predominance diagram (from EUGSTER and CHOU, 1973) for iron species in solution at 25°C. At low f_{O_2} the diagram is dominated by ferrous species, with Fe^{2+} being the most important at pH below 8. The pH of an early Martian ocean is likely to have varied between 5.4 (initially) and 10 (as the carbon dioxide in the atmosphere is reduced), over a time period of about 0.5 Gy, according to the model of SCHAEFER (1993). Therefore, at least at earlier times, ferrous iron should

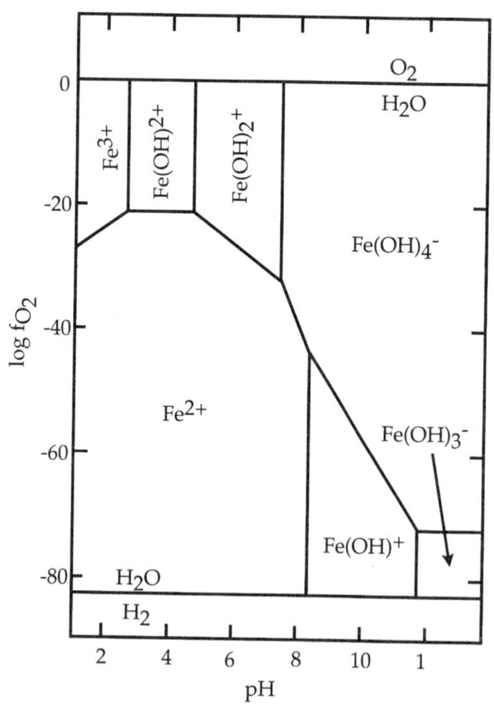

FIG. 4. Predominance diagram (adapted from EUGSTER and CHOU (1973) for iron species in solution at 25°C.

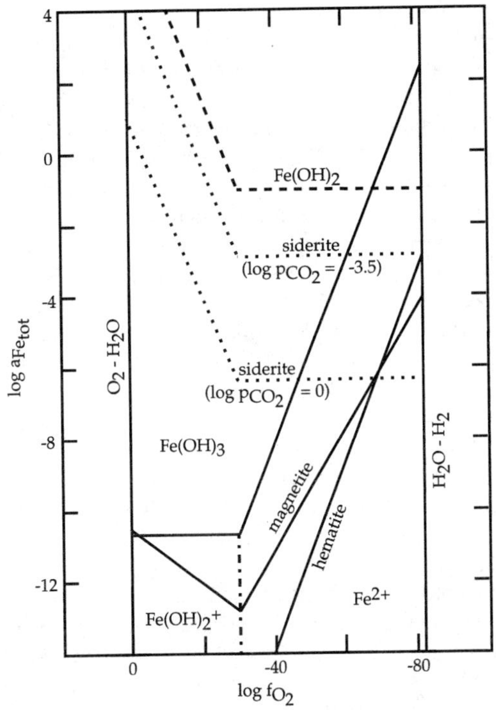

FIG. 5. The solubility of magnetite, hematite, siderite, and some iron hydroxides at a pH of 7 and 25°C, adapted from EUGSTER and CHOU (1973). The solubilities of $Fe(OH)_3$, magnetite, and hematite are shown as solid lines, $Fe(OH)_2$ as dashes, and siderite (at log f_{CO_2} of -3.5 and 0) as dots. The boundary between Fe^{2+} and $Fe(OH)_2^+$ is shown as dash-dot.

have been highly mobile in relatively anoxic waters (log $f_{O_2} < \sim -20$) on Mars. As was mentioned above, the atmosphere of Mars is now, and may always have been, oxidizing, but still waters may have been anoxic at depth. Also, groundwater that is not in direct contact with the atmosphere (*i.e.*, it has a long residence time in the aquifer) can be extremely anoxic, particularly if it is buffered by a redox buffer (such as Fe^{2+}–$Fe(OH)_3$ or Fe^{2+}–Fe_2O_3) (DREVER, 1988).

Figure 5 (after EUGSTER and CHOU, 1973) shows the solubility of magnetite, hematite, siderite, and some iron hydroxides at a pH of 7 and 25°C. The chief feature of interest here is the change in solubility of siderite with change in p_{CO_2}, it being far more stable under conditions of high p_{CO_2}, for any value of oxygen fugacity. At log f_{O_2} values of less than about -50, and a p_{CO_2} of 1 bar, siderite is preferentially precipitated over $Fe(OH)_3$.

The solubility of siderite relative to calcite for low f_{O_2} (after EUGSTER and CHOU (1973)) is seen in Fig. 6. Under conditions of high p_{CO_2}, siderite will precipitate out at a much lower pH than calcite ($CaCO_3$). Siderite begins to precipitate at an iron concentration of only 0.5–0.05 ppm (log $a_{Fe^{2+}_{tot}}$ of 10^{-5}–10^{-6}) at low pH, and even lower at high pH.

Figure 3 is a log f_{O_2}-pH plot showing the comparative solubilities of $Fe(OH)_3$, $Fe(OH)_2$, greenalite ($Fe_3Si_2O_5(OH)_4$), and siderite, at 25°C and a log total iron content of -4, corresponding to about 5 ppm iron (after EUGSTER and CHOU (1973)). Under present terrestrial p_{CO_2} conditions, greenalite will precipitate preferentially to siderite at conditions of low f_{O_2} and moderate pH. However, in the higher pH regime thought to exist on early Mars, siderite will precipitate preferentially. If the log f_{O_2} is higher, above about -15 at a pH of 6, then $Fe(OH)_3$ is the preferential phase of iron in this system.

In summary, while the terrestrial anoxic model of EUGSTER and CHOU (1973) produces iron-rich layers that are dominated by calcite + greenalite, with some siderite present, the Martian version of this model strongly favors the precipitation of siderite. Both models produce chert bands, originally deposited as magadiite.

DISCUSSION

The application of abiotic models of producing terrestrial BIFs to the conditions likely on early

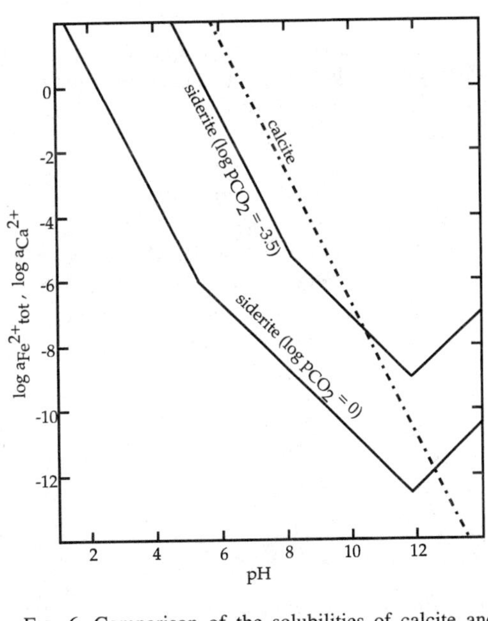

FIG. 6. Comparison of the solubilities of calcite and siderite, at low f_{O_2} and 25°C, adapted from EUGSTER and CHOU (1973). Siderite curves are drawn for log f_{CO_2} of -3.5 and 0.

Mars indicates that these deposits could have formed there, as they did on Earth. How can we determine if BIFs do indeed exist on Mars? And if they do, what implications does the existence of BIFs on Mars have for the evolution of the Martian atmosphere and hydrosphere?

Observations

The three mechanisms of BIF formation involving free oxygen, photoprecipitation, or hydrogen peroxide all produce deposits of FeO(OH). Under different formation conditions, FeO(OH) will precipitate in different crystal phases: α-FeO(OH) (goethite), β-FeO(OH) (akaganeite), γ-FeO(OH) (lepidocrocite), δ-FeO(OH), or δ'-FeO(OH) (feroxyhyte) (BURNS, 1993). Akaganeite may subsequently transform to either goethite or hematite (α-Fe_2O_3); lepidocrocite may revert to maghemite (γ-Fe_2O_3); feroxyhyte may transform to goethite, hematite, or mixtures of the two (BURNS, 1993). Many of these phases have been suggested as existing on the surface of Mars (*e.g.*, BURNS, 1980; HARGRAVES *et al.*, 1979; MORRIS *et al.*, 1989; MORRIS and LAUER, 1990; SINGER, 1982), but in the present Martian environment, hematite may be the only stable iron oxide (GOODING, 1978).

Under Martian conditions, the anoxic mechanism of BIF formation produces deposits of siderite. Siderite is unstable under present Martian conditions, and is converted via gas-solid decomposition into hematite (GOODING, 1978). Nanocrystalline hematite is a strong candidate to exist on the Martian surface (MORRIS *et al.*, 1989; MORRIS and LAUER, 1990), and a nanophase product could be expected as a result of gas-solid decomposition of siderite.

BURNS (1993) suggested that the bright deposits littering the Argyre and Hellas basins could be composed of dust derived from dessicated hydrous iron oxide-silica deposits that remained there after the water had evaporated from these deep depressions.

There is evidence, therefore, that one source for the iron oxides that are observed on Mars may be iron-formations. However, this is not the only possible source, because hematite and other iron oxides could have formed as weathering rinds on the rocks of the surface, without having been transported in solution, or could have formed by hydrothermal processes, in the presence of groundwater alone.

Spectroscopic data alone, whether ground-based or from a Mars-orbiting spacecraft, are not enough to determine the existence of BIFs on Mars, because the only surface expression of BIFs we expect is hematite.

As a result, in order to determine the existence of BIFs on Mars, we must see below the surface. A lander capable of penetrating past the uppermost weathered layer of the surface could, perhaps even by visual inspection alone, determine the existence of BIFs on Mars. Terrestrial BIFs form deposits of iron oxides and silica, with other associated minerals, which are quite distinctive in hand specimens. Adding to this simple visual inspection any spectoscopic measurements would enable easier identification, especially if the proposed Martian iron-formations do not have the same banded structure as the terrestrial ones. On Mars, one would wish to examine shallowly-buried impact debris in the northern plains or in large basins, such as Hellas or Argyre. It is in these localities that one might expect to find evidence of BIFs on Mars, should they exist. By examining impact debris we increase greatly the depth of sampling into the solid bedrock of the planet over what any feasible lander might be able to sample on its own. And by looking at shallowly-buried debris, such as might still be accessible to a lander, we allow the possiblilty of finding some material that has not entirely been altered to hematite.

The Mars Pathfinder, to be launched in 1996, is not likely to find evidence for iron-formations, because it will not be looking beneath the surface. The Mars Volatiles and Climate Surveyor, proposed for launch in 1998, has the necessary instumentation to search for iron-formations, but will be visiting the south polar cap region, which is not a likely place to find iron-formations on Mars. We will have to wait for one of the later Mars Surveyor missions to search for iron-formations on Mars.

Implications

If iron-formations are found on Mars, this will be confirmation of those models of the early environment of Mars that include the existence of a warm, wet atmosphere and bodies of liquid water. Their existence implies the existence of liquid water basins, probably of significant depth, and of a active weathering profile on the land, supported by a groundwater system and possibly rain as well. If it is possible to determine which of the different formation mechanisms produced the BIFs, at least between the oxidizing mechanisms (producing FeO(OH)) and the anoxic mechanism (producing siderite), then we can place limits on the oxygen fugacity of the early Martian atmosphere, or on the necessary depth of ocean basins.

In all the abiotic mechanisms of BIF formation discussed above, gaseous hydrogen is produced.

On Mars, this hydrogen will be lost from the upper atmosphere, contributing to the oxidation of the planet, and to the loss of its water to space. There is not likely to be a large isotopic effect due to banded iron formation, other than that resulting from preferential escape of D to space relative to H (Bruce Jakosky, personal communication).

Hydrothermal systems. The existence of hydrothermal systems could have a strong impact on the geochemical evolution of the surface and atmosphere of Mars (JAKOSKY and JONES, 1986). Hydrothermal systems could have relevance to the formation of BIFs on Mars as well. On Earth, hydrothermal plumes in the ocean have been suggested as a means of supplying the iron-rich fluids necessary to precipitate iron-formations (ISLEY, 1995; MORRIS, 1993). In the more volcanically-active environment of early Mars (compared to present-day Mars), such hydrothermal circulation systems could have existed as well, driven either by volcanism or by large impacts (NEWSOM, 1980). Such circulation cells below the lakes or oceans of Mars could have provided iron-rich fluids to these bodies of water. Martian hydrothermal systems in the crust could also provide additional cations to the groundwater, which would then flow into bodies of standing water.

Hydrothermal systems could have existed on Mars for a much longer time than the warm, wet period is likely to have existed. As these systems transported cations through the crust, they may also have deposited iron minerals there. If the groundwater was relatively anoxic, then ferrous iron would have been freely transported in it. If the water erupted onto the surface, as might have happened in the formation of the catastrophic outflow channels, and the atmosphere was relatively oxygen-rich, then ferric iron minerals could have been precipitated. One would expect, then, to find FeO(OH) (now converted to hematite?) in the outwash plains of these channels. If the hydrothermal circulation systems remained below the surface, but gradually cooled and stopped circulating, the increase in pH of the fluid would drive the precipitation of siderite. Neither of these types of deposits are banded iron-formations, but the chemistry involved in their formation is the same.

CONCLUSIONS

There are several abiotic mechanisms for producing banded iron-formations on Earth. These are: (1) oxidation of ferrous iron by free oxygen, (2) photooxidation of ferrous iron, (3) oxidation of ferrous iron by hydrogen peroxide, and (4) anoxic precipitation of ferrous iron by changes in pH. Any of these mechanisms could have been active on Mars in its distant past, to produce BIFs there as well, which should be observable today.

The minerals one would expect to observe on the surface of a BIF on Mars are indeed thought to exist on Mars, but unfortunately BIFs are not unique in containing these minerals, and confirmation of the existence of BIFs on Mars will have to wait for more data. Proof of the existence of BIFs on Mars would give us valuable information about the early history of the planet, and about Earth. It would act to confirm models of Mars that include a warm, wet period, and the existence of bodies of standing water.

Acknowledgements—This paper is inspired by and dedicated to Roger G. Burns. It was improved by comments from James Kasting and Bruce Jakosky. This work was supported through the University of Maryland under NASA grant NAGW-3588.

REFERENCES

ANBAR A. D. and HOLLAND H. D. (1992) The photochemistry of manganese and the origin of banded iron formations. *Geochim. Cosmochim. Acta* **56,** 2595–2603.

BAKER V. R., CARR M. H., GULICK V. C., WILLIAMS C. R. and MARLEY M. S. (1992) Channels and valley networks. In *Mars,* chapter 15, pages 493–522. University of Arizona Press.

BAKER V. R., STROM R. G., GULICK V. C., KARGEL J. S., KOMATSU G. and KALE V. S. (1991) Ancient oceans, ice sheets and the hydrological cycle on Mars. *Nature* **352,** 589–594.

BEUKES N. J. and KLEIN C. (1992) Models for iron-formation deposition In *The Proterozoic Biosphere: A Multidisciplinary Study.,* pages 147–152. Cambridge University Press, New York.

BRATERMAN P. S. and CAIRNS-SMITH A. G. (1987) Iron photoprecipitation and the genesis of the banded iron-formations. In *Precambrian Iron Formations,* p. 215–245. Theophrastus Publications, Athens.

BRATERMAN P. S., CAIRNS-SMITH A. G. and SLOPER R. W. (1983) Photo-oxidation of hydrated Fe^{2+}—Significance for banded iron formations. *Nature* **303,** 163–164.

BRATERMAN P. S., CAIRNS-SMITH A. G., SLOPER R. W., TRUSCOTT T. G. and CRAW M. (1984) Photo-oxidation of iron (II) in water between pH 7.5 and 4.0. *J. Chem. Soc., Dalton Trans.,* p. 1441–1445.

BURNS R. G. (1980) Does feroxyhyte occur on the surface of Mars? *Nature* **285,** 647.

BURNS R. G. (1993) Rates and mechanism of chemical weathering of ferromagnesian silicate minerals on Mars. *Geochim. Cosmochim. Acta* **57,** 4555–4574.

BUTTON A., BROCK T. D., COOK P. J., EUGSTER H. P., GOODWIN A. M., JAMES H. L., MARGULIS L., NEALSON K. H., NRIAGU J. O., TRENDALL A. F. and WALTER M. R. (1982) Sedimentary iron deposits, evaporites, and phosphorites: State of the art report. In *Mineral Deposits and the Evolution of the Biosphere,* (eds. H. D.

HOLLAND and M. SCHIDLOWSKI), p. 259–273. Springer-Verlag, Berlin.

CAIRNS-SMITH A. G. (1978) Precambrian solution photochemistry, inverse segregation, and banded iron formations. *Nature* **276,** 807–808.

CANUTO V. M., LEVINE J. S., AUGUSTSSON T. R. and IMHOFF C. L. (1983) Oxygen and ozone in the early Earth's atmosphere. In *Developments and Interactions of the Precambrian Atmosphere, Lithosphere, and Biosphere,* (eds. B. NAGU, R. WEBER and J. C. GUERRERO), p. 1–12. Elsevier.

CARR M. H. (1986) Mars: A water-rich planet? *Icarus* **68,** 187.

CLOUD P. (1972) A working model of the primitive Earth. *Am. J. Sci.* **272,** 537–548.

CLOUD P. (1973) Paleoecological significance of the banded iron-formation. *Econ. Geol.* **68,** 1135–1143.

DES MARAIS D. J. (1994) The Archean atmosphere: Its composition and fate. In *Archean Crustal Evolution,* (ed. K. C. CONDIE), p. 505–523. Elsevier.

DREVER J. I. (1974) Geochemical model for the origin of Precambrian banded iron formations. *Geol. Soc. Am., Bull.* **85,** 1099–1106.

DREVER J. I. (1988) *The Geochemistry of Natural Waters.* Prentice Hall.

EUGSTER H. P. and CHOU I.-M. (1973) The depositional environment of Precambrian banded iron-formations. *Econ. Geol.* **68,** 1144–1168.

FANALE F. P. and POSTAWKO S. E. (1995) Mass loading and the early Mars greenhouse: When a pollutant becomes the environment. *Lunar Planet. Sci. XXVI,* 391–392.

FRANÇOIS L. M. (1986) Extensive deposition of banded iron formations was possible without photosynthesis. *Nature* **320,** 352–354.

FRANÇOIS L. M. (1987) Reducing power of ferrous iron in the Archean ocean, 2. Role of $FeOH^+$ photooxidation. *Paleoceanography* **2,** 395–408

GARRELS R. M., PERRY E. A. J. and MACKENZIE F. T. (1973) Genesis of Precambrian iron-formations and the development of atmospheric oxygen. *Econ. Geol.* **68,** 1173–1179.

GOODING J. L. (1978) Chemical weathering on Mars—Thermodynamic stability of primary minerals (and their alteration products) from mafic igneous rocks. *Icarus* **33,** 483–513.

GOODING J. L., ARVIDSON R. E. and ZOLOTOV M. Y. (1992) Physical and chemical weathering. In *Mars,* (eds. H. H. KIEFFER *et al.*), p. 626–651. University of Arizona Press.

GROSS G. A. (1980) A classification of iron formations based on depositional environments. *Cana. Mineral.,* **18,** 215–222.

HARGRAVES R. B., COLLINSON D. W., ARVIDSON R. E. and GATES P. M. (1979) Viking magnetic properties experiment: Extended mission results. *J. Geophys. Res.,* **84,** 8379–8384.

HOLLAND H. D. (1973) The oceans: A possible source of iron in iron-formations. *Econ. Geol.* **68,** 1169–1172.

ISLEY A. E. (1995) Hydrothermal plumes and the delivery of iron to banded iron formations. *J. Geol.* **103,** 169–185.

JAKOSKY B. M. and JONES J. H. (1996) Mars volatile evolution from stable isotope abundances. In *Workshop on Evolution of Martian Volatiles, LPI Tech. Rpt. 96-01, Pt. 1,* p. 20. Lunar and Planetary Institute.

JAMES H. L. (1954) Sedimentary facies of iron-formation. *Econ. Geol.* **49,** 235–293.

JORTNER J. and STEIN G. (1962a) The photochemical evolution of hydrogen from aqueous solutions of ferrous ions. part I. the reaction mechanism at low pH. *J. Phys. Chem.* **66,** 1258–1264.

JORTNER J. and STEIN G. (1962b) The photochemical evolution of hydrogen from aqueous solutions of ferrous ions. part II. effect of changing pH. *J. Phys. Chem.* **66,** 1264–1271.

KARGEL J. S., BAKER V. R., BEGÉT J. E., LOCKWOOD J. F., PÉWÉ T. L., SHAW J. S. STROM R. G. (1995) Evidence of ancient continental glaciation in the Martian northern plains. *J. Geophys. Res.* **100,** 5351–5368.

KASTING J. F. (1990) Bolide impacts and the oxidations state of carbon in the Earth's early atmosphere *Origins of Life and Evolution of the Biosphere* **20,** 199–231.

KASTING J. F. (1991) CO_2 condensation and the climate of early Mars. *Icarus* **94,** 1–13.

KASTING J. F., HOLLAND H. D. and PINTO J. P. (1985) Oxidant abundances in rainwater and the evolution of atmospheric oxygen. *J. Geophys. Res.* **90,** 10497–10510.

KASTING J. F., POLLACK J. B. and CRISP D. (1984) Effects of high CO_2 levels on surface temperature and atmospheric oxidation state of the early Earth. *J. Atmos. Chem.* **1,** 403–428.

KLEIN C. and BEUKES N. J. (1992) Time distribution, stratigraphy, and sedimentologic setting, and geochemistry of Precambrian iron-formations In *The Proterozoic Biosphere: A Multidisciplinary Study.,* p. 139–146. Cambridge University Press, New York.

LABERGE G. L. (1973) Possible biological origin of Precambrian iron-formations. *Econ. Geol.* **68,** 1098–1109.

LABERGE G. L., ROBBINS E. I. and HAN T.-M. (1987) A model for the biological precipitation of Precambrian iron-formations-A: Geological evidence. In *Precambrian Iron Formations,* p. 69–96. Theophrastus Publications, Athens.

MAYNARD J. B. *Geochemistry of Sedimentary Ore Deposits.* Springer-Verlag, New York.

MCELROY M. B. and DONAHUE T. M. (1972) Stability of the Martian atmosphere *Science* **177,** 986–988.

MORRIS R. C. (1993) Genetic modelling for banded iron-formation of the Hamersley Group, Pilbara Craton, Western Australia. *Precambrian Res.* **60,** 243–246.

MORRIS R. V., AGRESTI D. G., LAUER, H. V. J., NEWCOMB J. A., SHELFER T. D. and MURALI A. V. (1989) Evidence for pigmentary hematite on Mars based on optical magnetic, and Mössbauer studies of superparamagnetic (nanocrystalline) hematite. *J. Geophys. Res.* **94,** 2760–2778.

MORRIS R. V. and LAUER, H. V. J. (1990) Matrix effects for reflectivity spectra of dispersed nanophase (superparamagnetic) hematite with application to Martian spectral data. *J. Geophys. Res.* **95,** 5101–5109.

NEWSOM H. E. (1980) Hydrothermal alteration of impact melt sheets with implications for Mars. *Icarus* **44,** 207.

OWEN T. (1992) The composition and early history of the atmosphere of Mars. In *Mars,* (eds. H. H. KIEFFER *et al.*), p. 818–834. Arizona University Press, Tucson, Arizona.

PAPP S. and VINCZE L. (1980) Photooxidation of iron (II) evolving hydrogen. *Inorg. Chim. Acta* **44,** L241–L243.

PARKER T. J. and GORSLINE D. S. (1992) Distribution of coastal morphology in the Martian northern lowlands In *Workshop on the Martian Surface and Atmosphere*

Through Time, LPI Tech. Rpt. 92-02, p. 113–114. Lunar and Planetary Institute.
PARKER T. J., GORSLINE, DONN S., SAUNDERS R. S., PIERI, DAVID C. and SCHNEEBERGER D. M. (1993) Coastal geomorphology of the Martian northern plains *J. Geophys. Res.* **98,** 11,061–11,078.
PARKER T. J., SCHNEEBERGER D. M., PIERI D. C. and SAUNDERS R. S. (1987) Geomorphic evidence for ancient seas on Mars. In *MECA Symposium on Mars: Evolution of its Climate and Atmosphere, LPI Tech. Rpt. 87-01,* p. 96–98. Lunar and Planetary Institute.
POLLACK J. B., KASTING J. F., RICHARDSON S. M. and POLIAKOFF K. (1987) The case for a wet, warm climate on early Mars. *Icarus* **71,** 203.
ROBBINS E. I., LABERGE G. L. and SCHMIDT R. G. (1987) A model for the biological precipitation of Precambrian iron-formations-B: Morphological evidence and modern analogs. In *Precambrian Iron Formations,* p. 97–139. Theophrastus Publications, Athens.
ROSSBACHER L. A. and JUDSON S. (1981) Ground ice on Mars: Inventory, distribution, and resulting landforms. *Icarus* **45,** 39.
SCHAEFER M. W. (1993) Aqueous geochemistry on early Mars. *Geochim. Cosmochim. Acta* **57,** 4619–4625.
SINGER R. B. (1982) Spectral evidence for the mineralogy of high albedo soils and dust on Mars. *J. Geophys. Res.* **87,** 10159–10168.
TOWE K. M. (1983) Precambrian atmospheric oxygen and banded iron formations: A delayed ocean model. In *Developments and Interactions of the Precambrian Atmosphere, Lithosphere, and Biosphere,* (eds. B. NAGU, R. WEBER and J. C. GUERRERO), p. 53–62. Elsevier.
TRENDAL A. F. and BLOCKLEY J. G. (1970) The iron formation of the Precambrian Hamersley group Western Australia. *West. Aust., Geol. Surv., Bull.* **119,** 366 p.
WEISS J. (1935) Photochemical reactions connected with the quenching of fluorescence of dyestuffs by ferrous ions in solution. *Nature* **136,** 794–795
YUNG Y. L., NAIR H., GERSTELL M. F., ZHOU X. and CRISP D. (1995) Sulfur dioxide and the early Mars greenhouse. *Bull. American Astron. Soc.* **27,** 1060.

Subject Index

Temporary Measures

Migrant Workers and the Developmental State in the Philippines and South Korea

SUZY K. LEE

OXFORD
UNIVERSITY PRESS

OXFORD
UNIVERSITY PRESS

Oxford University Press is a department of the University of Oxford. It furthers the University's objective of excellence in research, scholarship, and education by publishing worldwide. Oxford is a registered trademark of Oxford University Press in the UK and in certain other countries.

Published in the United States of America by Oxford University Press
198 Madison Avenue, New York, NY 10016, United States of America.

CIP data is on file at the Library of Congress

ISBN 9780197789971

DOI: 10.1093/9780197790007.001.0001

Printed by Marquis, Canada

MIX
Paper | Supporting responsible forestry
FSC® C103567
FSC
www.fsc.org

Acknowledgments

I have incurred many debts in the writing of this book. It is hard to pinpoint a date at which I can say I started—there are some questions in here I have wondered about, consciously or not, for most of my life—but by even the shortest accounting, I have spent more than a decade and a half researching and writing about the forces that compel people to leave their hometowns and families and make their lives somewhere new. In the process, I have had to make many moves of my own, some of them for this project and others for personal reasons. In each new place, I found people who were welcoming and generous with their time and talents, without whom this book would not exist. The list is too long, and my debt too great, to express adequately here, but I will try to name at least a few.

I should begin at New York University's Department of Sociology, where the first recognizable version of the project began as a dissertation. The most important person I met there was Vivek Chibber, my teacher and adviser, who spent countless hours helping develop the project and problem-solve, not only offering encouragement and insightful critique but making sure I had the time and space to complete it. More valuable than this direct support, perhaps, has been the intellectual community that Vivek and Jeff Goodwin—another important mentor to me—have created at NYU Sociology. When I arrived at NYU, I was intellectually unformed, with a handful of questions and some vague notions about social-science research but no real framework for thinking about and analyzing the phenomena I was observing. It was my good fortune to join Vivek's and Jeff's Economic and Political Sociology workshop, a sometimes terrifying but always also nurturing place, where I got to practice articulating and taking apart my ideas, see serious work as it was being done, and eventually hammer out the details of this project. EPS is also where I met so many of the friends and colleagues who have been influential in my intellectual development, and who, over the years, have read and given feedback on hundreds of pages of drafts. Thanks especially to Nada Matta, Rene Rojas, Adaner Usmani, Erez Maggor, Ercan Sadi, Jonah Birch, Jeremy Cohan, Paul Heideman, John Clegg,

Mark Cohen, Virgilio Urbina Lazard, Patrick Inglis, David Calnitsky, and Ahmad Al Sholi. Writing a book has been described as a lonely and difficult endeavor; I did sometimes find it difficult, but because of these wonderful people, I never felt lonely.

I also need to thank the scholars beyond the EPS bubble who helped shape this project and offered support at crucial moments. Robyn Rodriguez's work on the Philippine migration policy was an important early inspiration, and it gave me confidence that the phenomena I thought I was seeing were real. Robyn, in person, gave time and encouragement in the early years of the project and helped shepherd it through as a member of the dissertation committee. Thank you also to Natasha Iskander, another scholar whose work was influential in setting the direction of this project, who I later came to know in person, and whose advice and encouragement I have relied on over the years. I have also found support during this process through the community of scholars working on international migration and Asian studies, many of whom read draft chapters and gave me opportunities to present and work out my ideas. My thanks especially to Roger Waldinger, Cate Bowman, David Cook-Martin, Rubén Hernández-Léon, David Trouille, Kathryn Babineau, Lieba Faier, Yasmin Ortiga, Min Zhou, and Sungdai Cho.

In the Philippines, I thank the Third World Studies Center at the University of the Philippines, Dilliman, which offered me an institutional home while I was in the field there, as well as the Salabrini Migration Center, which has been such an important center for migration research in the Philippines. A special thanks also to Erika Rey, who worked as my research assistant over the years, helping to scour archives and gather documents in the time between my research visits. In South Korea, I need to thank Hye Young Min, who helped me gather data during the COVID pandemic, when it became impossible for me to travel there in person.

In the final stage of this project, I received critical support from my writing coach, K. Anne Amienne, who helped me set deadlines and then helped me set new ones when I missed them, all the while keeping me working steadily toward my goal.

I also received generous financial support during the research and writing of this project from the National Science Foundation (#1435946) and the Ford Foundation.

Portions of Chapters 4 and 6 were previously published in my articles "Migration beyond networks: The mechanisms of sending state intervention" (*Migration and Development*, vol 10, no. 3) and "Sustaining inequality: The incorporation of migrant remittances in the Philippine political economy" (in *Race, gender and contemporary international labor migration regimes: 21st-century coolies?*, edited by Leticia Saucedo and Robyn Magalit Rodriguez). I am grateful to Sage Publications and Edward Elgar Publishing, respectively, for their kind permission to use them here.

During the years that I have been working on this book, the other big project in my life has been my three children, who were born in the time between field visits and dissertation defenses. Given the conflicting demands of raising young children and completing a book-length project, I would have had to make some painful compromises between these two commitments if not for the generous support of a small army of grandmothers, aunts, uncles, babysitters, teachers, and friends. This book exists in no small part because they cared for our small family while I pursued my dream. There are no words to express the depth of my gratitude to them. And to my husband and partner in all projects, who has supported me in my impractical career choice, let me drag our children around the world in its pursuit, and shared my joy in it: This is your book as much as it is mine.

When I started this project, I thought of it as an intellectual endeavor, believing that, by answering the research question I had set myself, I would complete my training to become a professional scholar. Of course, as a child of immigrants I always knew that the project was not unconnected to my own life story. But I did not understand the extent to which my curiosity was driven by a need to know the histories that shaped my family and the places from which they came, and through these histories to better know myself. In the years that I have worked on this project, as it ballooned from a single case study of modern bureaucracy to a history of how migrant workers have shaped the destinies of nations, I have come to realize how much it had always been a labor of love. For my mother and father, my aunts and uncles, whose lives - beyond the domestic spaces and the everyday struggles through which I knew them - I was trying to comprehend. And for all the workers who leave their homes in search of better lives. Through their journeys and sacrifices, they build new worlds. This book is dedicated to them.

Contents

Chapter 1
Introduction

Crisis and Migration

When Ferdinand Marcos's regime initiated a labor-export policy in the Philippines in the early 1970s, the country was in profound political crisis. The decade began with mass protests, led by students but supported by a wide swath of society—workers, peasants, the urban poor—against the failures of the state to address long-standing problems of corruption and poverty, and the acute threat to democracy posed by Marcos's reluctance to leave executive office. The protests, which ran from January 26 to March 17 of 1970, involved violent confrontations between the Philippine military and thousands of barely armed protestors, six of whom were killed in what came to be known as the "First Quarter Storm." These protests, and others that continued into 1971, seemed to have little effect in preventing Marcos's seizure of power and at times seemed to hasten it. That year, the rally of an opposition political party was bombed, resulting in over a hundred casualties. Despite unresolved questions about the involvement of the Marcos administration itself in the bombing, Marcos presented the bombing as evidence of a conspiracy against the state. He suspended the writ of habeas corpus, and began rounding up opposition activists. A year later, he declared outright martial law.

The collapse of democracy occurred against the backdrop of an intensifying economic crisis. A decade of sluggish growth and the failure of the country's import-substitution industrialization (ISI) strategy had ended in an acute balance-of-payments crisis in 1969. The International Monetary Fund (IMF), as a condition of bailout, forced the Philippines to devalue the peso by more than 60 percent, spiking the inflation rate to 14.8 percent in 1970 (from 1.3 percent in 1969) and throwing the already impoverished population into even greater suffering (Bello et al., 1982, p. 136). The Marcos regime, whose profligate spending in the 1969 reelection cycle had been at least somewhat responsible for the economic crisis, justified its seizure of power with the promise of returning the country to economic stability and

Temporary Measures. Suzy K. Lee, Oxford University Press. © Oxford University Press (2025).
DOI: 10.1093/9780197790007.003.0001

growth. Promising an authoritarian route to a "New Society," the Marcos regime introduced a series of reforms intended to restructure the state and the economy; included among them was a new approach to labor migration.

Blas Ople, then Minister of Labor, who would eventually come to be known as the father of the Philippine labor-migration system, announced in 1975: "We no longer apologize for the outflow of Filipino labor abroad.... We have decided it in such a manner that it will redound to the national interest. We are scouting aggressively for job markets for excess Filipino skills in many countries of the world."[1] This statement was remarkable, not because governments in crisis never relied on emigration to release social and political pressures (that was common) but because it is rare for officials to admit to it. Mass migration betrays the state's failure to secure some necessary condition of life for its populace, and as such can never *not* be a rebuke to the regime in power. Moreover, the labor-export strategy would seem inconsistent with the thrust of the New Society initiatives that were promising solutions to the country's problems, to deliver prosperity and stability at home. At the time, the Marcos regime presented the move as a short-term, "stop-gap"[2] solution to the acute crises. But if this was all it was, why not do it quietly? The answer to this question lies in Ople's invocation of the *national interest.* Labor export was intended not just to relieve the economic and political pressure of the immediate moment; it was also seen as a significant tool through which the New Society would be built. By migrating out, Filipino workers would not be fleeing a desperate situation but would instead be participating in a national project to collectively pull the country not only out of crisis but up the ladder of economic development.

Though this was not a commonplace strategy, one very compelling contemporary example gave the regime's plan some plausibility: South Korea. In 1961, only a year after a student-led revolution against the autocratic and corrupt rule of South Korea's first president, Syngman Rhee, General Park Chung Hee seized power in a coup d'état, promising to deliver stability, security, and economic growth. The Park regime soon began promoting its citizens as a migrant labor force to advanced industrial economies in Western Europe and North America (Hong, 2015, p. 260). By

[1] Speech at the opening session of the National Tripartite Conference on the Labor Code, Development Academy of the Philippines, Tagaytay City, October 23, 1975. (published in Ople, 1979), p. 223.

[2] Speech at the presentation of the Ten Outstanding Workers of the Republic (TOWER) Awards for 1977, sponsored by the Rotary Club of Manila, Manila Hilton, April 28, 1977 (published in Ople, 1979), p. 145.

the mid-1960s, Park's regime had also negotiated a deal to send hundreds of thousands of soldiers and other workers to the Vietnam War, in exchange for billions of dollars in development aid and loans, as well as lucrative military-procurement contracts for Korean businesses from the United States. This arrangement helped fund the country's push into heavy and chemical industries and transformed companies like Hyundai and Samsung from domestically oriented businesses into international players. By the 1970s, South Korean firms were consolidating their position in international markets by pushing into the petrodollar-fueled construction boom in the Middle East, successfully competing against European and American firms on the basis of their ability to import reliable, cheap Korean workers to the labor-poor region. This state-led labor-export policy served as a handmaiden to industrial development, helping South Korea to leapfrog over other countries—including the Philippines—on the development ladder. By the time the Philippines initiated its own labor-export program in the mid-1970s, the Korean economy had been so dramatically transformed that it seemed probable that in the very near future the labor-export policy itself would be rendered obsolete, as Korea's newly established international conglomerates moved into different export lines—such as automobiles, electronics, and shipbuilding—that could be produced at home. If the Korean strategy could be reproduced in the Philippine context, labor export would represent more than a temporary measure to weather the immediate crisis; It could help deliver the "economic miracle"[3] that was ostensibly the long-term goal of the Marcos regime.

It is now over half a century since these temporary measures were introduced, and nearly four decades since the authoritarian dictatorships that imposed them were replaced by formal democracies. In this time, the two countries have followed very different paths. South Korea continued on the trajectory described above. Though there was no dramatic closing of the labor-export program, by the 1980s domestic wage rates were so high that Korea's international construction firms were abandoning the labor-export model to instead staff their worksites with cheaper labor from lower-wage countries, the Philippines chief among them. South Korea soon became a country of net *in-migration*, which in recent years has experimented with its own temporary contract migration schemes to *import* workers into what has become one of the largest economies in the world. In contrast, labor

[3] Ibid.

migration from the Philippines grew exponentially in the 1980s and 1990s, and eventually evolved into a fact of Filipino life. Today over 2 million temporary workers leave the country each year. The labor-export apparatus that facilitates these departures is now one of the most enduring institutions of the Philippine state, persisting through dramatic political disruptions and significant economic developments, including reforms that have brought an end to the cycles of debt and crisis that were the labor-export policy's initial reason for being.

The obvious conclusion to draw from this comparison is to confirm the importance of economic development to migration flows. Korea's labor-export program waned as industrialization and economic growth rendered it unnecessary, while the Philippines' economic instability and more sluggish growth seem to explain the country's continued reliance on out-migration. Yet the facts about the Korean case sketched out above suggest that the directions of the causal arrows are not quite this straightforward. I will lay out an argument in this book that South Korea's economic development might not have proceeded as it did without the openings created by the country's willingness and ability to export hundreds of thousands of temporary workers. The Philippine case, when examined more closely, also resists a straightforward interpretation. Though the Philippine economy did fall significantly behind its "tiger" neighbors in the first two decades of the labor-export policy, in more recent years it has managed a sustained turnaround, first stabilizing and then growing at a rate—6–7 percent in the past decade—that regularly outperforms the rest of the East Asian and Pacific region by nearly 2 percent each year. While there are many explanations for the Philippines' changing fortunes, all of them acknowledge that migrant remittances play a key role. These funds, the vast majority of which go directly to contract workers' families, buttress an economic model that depends on high levels of spending on real estate and consumption goods, associated with sectors of the Philippine economy where growth has been fastest.

Looking closely at the histories of labor export in these two countries raises a host of questions about the relationship between labor migration and economic development. Why was South Korea able to leverage labor export for industrialization in the mid-20th century so effectively? Why was the Philippines unable to reproduce this effect? And how do we explain the effect of remittances on the modern Philippine economy? Is this an example of migrant remittances successfully working together with neoliberal development policies? If so, why are they working in this case when they so rarely do in other contexts? Why are they working now when they failed to do so for three decades? And what do the answers to these questions tell us about labor-export policy in general, about the states that engage in it, and about the conditions under which it can become entrenched?

The Argument

This book approaches these questions by examining the relationship of labor export and remittances to capital and the economy's dominant sectors. The current literature on migration and economic development tends to target its analysis at two scales: the microstructural scale of individuals and households, where migration creates opportunities for skills development and remittances affect savings and consumption, and the macrostructural scale, where remittances and human capital are aggregated to affect the national economy. This study pitches its analysis somewhere between these two extremes, at the point where labor migration affects various sectors of capital, political and economic elites, and their accumulation strategies.

Moving away from the microstructural view of migration allows us to distinguish *labor migration*, which highlights the economic motivations of individual migrants, from the institutional processes that generate different forms of migration. Here, I identify the ways in which sending-state policy can catalyze migration and channel it toward temporary, contract-based migration in order to serve the accumulation goals of different sectors of domestic capital—a process I call *labor export.*

Examined from this viewpoint, we can see two different models for the incorporation of labor export into a national economic-development strategy, only one of which is captured by the existing literature on remittance, consumption, and investment. In this first model, labor export supports economic growth through the inflow of remittances, which, through the multiplier effect of consumption spending, has spurred the expansion of economic sectors that capture this consumption, including real estate, retail trade, and services. This process can be considered "state-led" to the extent that state policy intervenes to promote labor migration, high levels of remittance, and/or spending or investment of those remittances in specific sectors of the economy. I refer to this model as a *remittance-driven growth* (RDG) strategy. Some version of this has been occurring in the Philippines since the inception of the labor-export program, and it is currently the most important way in which labor export affects the Philippine economy. However, for the decade between the mid-1970s and mid-1980s, the Philippine state pursued another strategy, in which remittances were secondary or tangential. If an RDG strategy can be considered *indirect*, in the sense that growth is fostered not immediately through the activity of labor export but through the circulation of migrant workers' earnings back into the national economy, the second model, which I call the *labor-export-mediated industrialization* (LEMI) strategy, should be considered a more direct yoking of labor export in service of national development. LEMI strategies seek to leverage the labor migration itself—specifically the willingness of a cheap domestic labor force to work abroad—to buttress the competitiveness of domestic businesses in international markets. LEMI strategies are a variation of the development policies that capitalize on comparative advantages in low-cost labor that most late-developing countries enjoy, applied to the subset of economic activities in which at least some portion of the work must be performed on-site. Opportunity for this kind of labor export/business partnership existed in the 1970s and early 1980s in the construction boom of the oil-rich, labor-poor countries of the Middle East. In those years, South Korea and the Philippines both developed their labor-export apparatuses in order to support their domestic construction industry in their efforts to break into the international construction market.

Labor-export mediated industrialization (LEMI)

Only a decade previous, the South Korean construction industry had demonstrated that an LEMI strategy could effectively complement national industrialization efforts. Beginning in the early 1960s, the South Korean state carried out an extensive program of labor export that would become the model for the Philippines and other labor-sending countries. The South Korean state not only provided for the recruitment and training of migrant workers but paired these policies with an extensive system of support for construction exports, including giving domestic construction firms protected access to low-wage and highly productive Korean labor. With this support, the Korean construction industry managed to disrupt a European and North American oligopoly in international construction and gain significant market share, first in Southeast Asia and then in the Middle East. The contracts Korean firms were able to win through this arrangement were orders of magnitude larger than any they had won up to that point. These revenues not only helped Korean capital to survive the oil shocks and ensuing debt crises of the 1970s and early 1980s, but also provided markets for the products of the country's fledgling heavy industries, allowing them the time and space required for the trial-and-error learning through which the Korean economy industrialized (Amsden, 1989).

The Philippine foray under the LEMI strategy, while modeled on the Korean experience, did not produce the same outcomes. In attempting to reproduce the successes of Korean international construction, the Philippine state transplanted many of the policies used by the Korean state, including protected access to Filipino workers for national firms and export incentives and subsidies for the construction industry. The labor-export aspect of the Philippine LEMI strategy was highly successful, to the point that by the late 1970s Filipino workers were replacing Korean workers even in Korean firms. On the basis of this workforce, Philippine construction firms were able to gain a foothold in the Middle Eastern market, beginning as subcontractors in the low-tech, labor-intensive sector of the market that had hitherto been dominated by South Korea (Linder, 1994, p. 156). Where the LEMI strategy was less successful, however, was in linking the construction industry's opportunities into a larger national industrialization project. Unlike in South Korea, where top construction exporters like Hyundai were able to use windfall revenues from their labor-export-related opportunities to support moves into heavy industries and seed investments in new technologies and product lines, the activities of Philippine construction firms in the Middle East failed to generate structural changes in the industrial capacity of individual firms or the Philippine economy in general.

Many contributing factors may explain why LEMI strategies produced such different outcomes in these two cases. The Middle East construction boom slowed in the early 1980s, giving the Philippine LEMI program only a handful of years before the conditions for its own success were eroded. Still, problems had begun to appear even before the slowdown. For example, the Philippines' top construction exporter at the time, the Construction and Development Corporation of the Philippines (CDCP), struggled to deliver on its contracts and by 1982 had ceased competing for Middle East contracts entirely. The failure of the CDCP, in which the Marcos family and its cronies were reported to have major ownership stakes, suggests that the LEMI programs were hindered by issues of governance and rent-seeking that affected other parts of the Philippine development project. In the Korean case, LEMI policies were embedded within a centralized and coordinated push into heavy and chemical industries, run by a developmentalist state that, while not without clientelist structures and rent-seeking practices, had demonstrated the capacity to design and direct a national industrialization policy. Firms that benefited from LEMI policies were subject to the same system of discipline that administered the rest of the country's export-oriented subsidy and incentive program. The background to the labor-export-aided expansion in international construction were massive investments in shipbuilding and steel (often with funding secured through leveraging labor export), products of which could be used as inputs in the Middle East construction projects. At the same time, the companies that were making windfall profits were directed by state managers to plow those profits into new sectors, such as automobile production, that were being nurtured by the state through subsidies and coordination.

In the Philippines, not only did the entanglements between the LEMI-beneficiary firms and the state prevent the kind of direction and discipline evidenced in the Korean case, but the problems of governance also prevented the establishment of a robust industrialization program that could make use of the opportunities LEMI policies *did* manage to secure. While the Philippine state's development programs also emphasized the need for industrialization, the strategy for reaching its goals was built around a much less centralized model. This decentralization reflects the greater impact of patrimonial clientelism on the Philippine case, where oligarchic kin-networks representing different regional power bases have long dominated politics and economics. The consensus view in the economic-development literature argues that the specific form of rent-seeking found in the Philippines—highly unpredictable and central to the accumulation strategy of capital—was particularly disruptive of efforts to implement an industrialization plan (Hutchcroft, 1998). The Philippine experience demonstrates that, while LEMI programs might crucially rely on migration as a powering mechanism, they are, in essence, industrialization projects vulnerable to the problems of state capacity that hinder all such efforts.

On the issue of state capacity, the early years of the Philippine labor-export program tell us something about the distinction between bureaucratic capacity and autonomy. At least in this one area, the Philippine state has proven itself to have a significant level of bureaucratic capacity—sufficient to build an internationally competitive labor-export program, one that developed and continues to administer a program that facilitates and coordinates the movement of millions of people under conditions that maximize the returns to the national economy. I argue that the labor-export program was able to operate because the activities associated with labor export—training, recruitment, documentation, coordination, information dissemination—did not generate the kind of large-scale rents that could be vulnerable to the "looting" practices found in other parts of the Philippine economy. Where we do see this kind of corrosive rent-seeking is in the other half of the LEMI policy toolbox, where the protections and supports offered to business capital created opportunities for "crony capitalists" to profit.

Remittance-based growth (RDG)

In the second half of the 1980s, the Middle East construction boom slowed and the LEMI strategy withered in both countries. In South Korea, after industrial transformation had created other directions for economic expansion, the labor-export policy was largely abandoned. In the Philippine case, however, the labor-export apparatus was not dismantled but was instead repurposed to find new markets for Filipino/a workers. The Philippine state moved aggressively to open migration flows to East Asia and Europe, where social and economic changes were driving up demand for care workers. This meant not only lobbying and negotiating with receiving-country governments but also changing the domestic recruitment and training strategy to produce a migrant workforce that employers of care workers wanted—namely women. Rates of contract-labor migration continued to grow even as the demographics of the migrant population changed dramatically, as did the structure of contract migration. Throughout the 1990s, the migration of construction workers, recruited and managed in large groups, into enclave-style industrial worksites in the Middle East was replaced by a flow of care workers—overwhelmingly women—who migrated and worked in relative isolation, often in the Middle East but also in East Asia and Europe. What did not change was the temporary nature of these migrations. The new markets that the Philippine state opened up were primarily in countries that were seeking to import short-term guest workers, who were permitted entry only without their families and with limited visas that required them to return to the Philippines at the end of the contract term. In the book's second part I argue that the Philippine state continued to operate the labor-export program because it had become the linchpin for an accumulation strategy that required a continuing and growing flow of migrant remittances.

My account of the RDG strategy, at its core, is not at odds with the existing theories that analyze the relationship between remittances and economic development. Remittance inflows, as they get spent or saved and invested, have the potential to promote growth in the recipient economy—as a source of foreign exchange, investment capital, and stimulating consumer demand. In explaining the long-term involvement of the Philippine state in labor export, we could just stop at this point, observing that these potential effects motivated policymakers to continue with the migration policy. What is unsatisfying about stopping the analysis here, however, is that the Philippines is such an outlier among developing countries that have participated in labor export. Most similarly situated countries never developed a labor-export apparatus comparable to that of the Philippines, and while the Philippines has often been criticized for the failures of its economic-development policy, its actual problems with foreign debt and the balance of payments were not, after the 1980s, so egregious as to justify aggressively pursuing such a socially costly strategy. A version of the RDG strategy that can explain this persistence must go beyond the broad macroeconomic analysis of remittances' impact on the economy and, again, focus on the point at which those remittances come in contact with the various sectors of Philippine capital and their accumulation strategies. If we do this, we can see that, in the past fifty years of the Philippine labor-export policy, the economy of the country has transformed in ways that have given migrant remittances an outsize impact.

Though it cannot be said that this economic transformation was entirely a product of the labor migration, it did grow alongside—and was aided by—these remittances. Even as the LEMI strategies to boost the construction-export sector were faltering, the growing influx of migrant remittances was creating opportunities in other sectors of the Philippine economy. Migrant workers' spending on housing drove a massive expansion in mid- to low-end real estate development in metropolitan areas and their surroundings. Within short distances of these new housing developments, or often even built together with them, were retail operations that served the consumption needs of these communities. These two sectors, retail and real estate, grew together during the 1980s and 1990s. This growth occurred not only through the emergence of a new economic elite that originated in these sectors but through the reorientation of the oligarchic family conglomerates, which increasingly moved their investments toward these remittance-dependent areas.

Thus, over the fifty years of its state-led labor-export program, the Philippines has transitioned from an economy dominated by agricultural activity, with some manufacturing supported by import-substitution industrialization policies, to an economy in which growth is driven by real estate, services, and retail trade (Ofreneo, 2015). The transition itself was not unusual—many developing countries experienced this kind of "premature deindustrialization" during the neoliberal era—but in most of these cases it has been associated with a slowdown in growth and prospects for development (Rodrik, 2016). In the Philippines, by contrast, premature deindustrialization has not had this effect but rather has been associated with an increase in average growth rates, because of the synergies of this economic structure with labor export and migrant remittances. On its own, the latter represent a significant share of the Philippine economy; at 10 percent of gross domestic product, labor would be the country's top export product if remittances were counted as export revenue. But the impact on industries like retail and real estate is even greater than this number would indicate, because remittance income is more likely to be spent in the domestic economy than revenue earned by the oligarchic conglomerates, who have no obligation to invest or spend their profits locally. In the real estate sector, remittance spending also drives the rise in property values, which assume that a continuing stream of migrant workers will sustain demand for housing. While the rate of growth under this model has been somewhat less spectacular than the industrialization-driven growth of the Asian miracle, I argue in this book that this path to growth was viable in a way that industrialization was not, given the entrenchment of the oligarchic family networks in the Philippine political economy.

This is because labor export is different from other forms of economic activity that generate export earnings, as a result of the way in which the associated revenue streams enter and impact the economy. In most other export industries—including manufacturing, agriculture, even services—the revenue generated by the industry flows through the businesses that control the activity itself. In labor export, while the businesses that facilitate the export (the recruiters and manpower suppliers) do make profits, the vast preponderance of the economic benefit to the Philippines enters through another route: in the form of remittances, in millions of relatively small installments, to the households of migrant workers.

The relatively small size of the actual "business" of labor export has discouraged the entrance of the oligarchic family conglomerates. This does not mean, however, that Philippine capital has no interest in labor export. As I said above, in the wake of the debt crises and the structural-adjustment programs that ravaged the country's economy, the Philippine economy was rebuilt on the basis of remittance-supported consumption, with the major family conglomerates making heavy investments in the remittance-dependent sectors. But because most of the profit to be made from remittances is *indirect*, through the aggregation of remittances as they are invested, saved, or consumed, the activity of labor export itself is relatively insulated from the "booty capitalism" practices that hindered the country's industrialization efforts. In the Philippines, as was true almost everywhere else, import-substitution industrialization policies were undermined by the very industrial capitalists who are their beneficiaries because individual firms—which are agnostic about whether their profits come from state largesse or market competition—tend to undermine the capacity of states to ensure that subsidies are actually going toward the targeted economic activities. In labor-export and remittance-dependent sectors, that negative feedback loop could not be established; thus, the parts of the Philippine state associated with promoting and managing the labor export, while not entirely free of clientelist rent-seeking, has remained relatively unhindered in their attempts to develop and implement policy.

Not only does the RDG model make the Philippine state's long-term intervention *possible*, by separating the sources of profit (remittances) from the economic activity that is ultimately responsible for it (labor export); the model makes state intervention *more likely* to occur. The economic sectors most closely associated with migrant remittances have driven the relatively high growth rates that the Philippines has experienced; in the first two decades of the 21st century, retail and real estate grew an average of 10.2 and 16.8 percent per year, respectively, doubling and tripling the rate of growth of the entire economy. The oligarchic families, whose capture of the state had proven so detrimental to industrialization in the past, as well as the new entrepreneurial class, which increasingly wields influence in the political realm, all have significant interests in sustaining the flow of remittances on which this new economy has been built. Sustaining remittances, in turn, requires a continuing and growing flow of the kind of

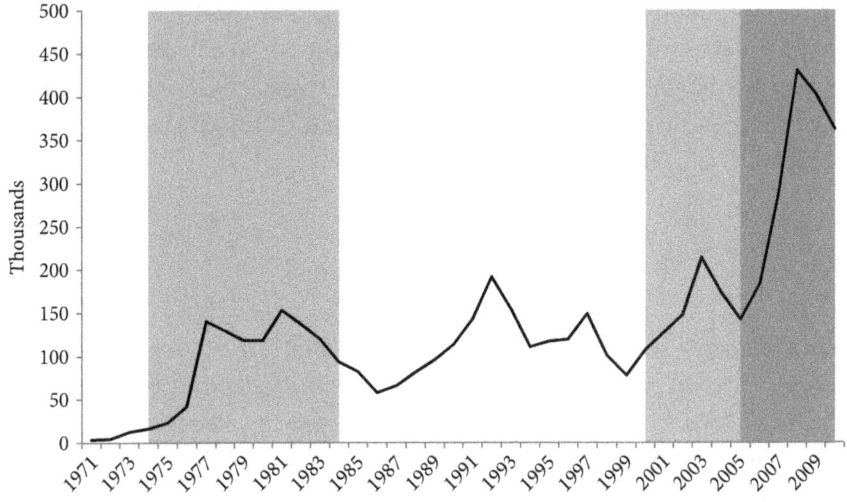

Figure 1.1 Number of Annual Emigrants, Pakistan, 1971–2010.

Note: The light grey bars highlight years in which the average price of crude oil exceeded $25 per barrel, the dark grey bar, $50 per barrel. The only other major spike in emigration corresponds to the first Gulf War in the early 1990s.

Data: Pakistan Bureau of Emigration and Overseas Employment (BEOE, http://www.beoe.gov.pk).

migration that produces high rates of remittances—namely, migration that is temporary and based on contract employment, in which migrants maximize their remittances to save for an expected return or to support families left behind.

While there are some examples of high rates of migration and remittances without significant state intervention, these typically involve some extraordinary circumstance, such as Mexico's geographic proximity to the United States. Most other countries experience much more variability in both migration and remittance flows. Take, for example, Pakistan, a country that also sent migrant labor to the Middle East during its construction boom in the 1970s and 1980s. When this boom ended, migration from Pakistan declined by two-thirds and remittances fell by one-half, recovering only when rising oil prices catalyzed another infrastructure boom in the Middle East (see Figure 1.1). Philippine migration and remittance rates have been much less vulnerable to these types of fluctuations because the country's migration flow is highly diversified, never dependent on any single market, and pivots quickly whenever labor demand or other conditions in any one market changes. It has these characteristics, in large part, because of state policy that constantly monitors and responds to conditions that might affect demand for migrant labor in receiving countries, and that works to facilitate Filipino/a workers' employment abroad.

Methods and Evidence

The arguments of this book are built on a comparative historical analysis of the labor-export programs of South Korea and the Philippines, beginning in the early 1960s, when these programs were first implemented in South Korea and contemplated in the Philippines. The end points of the analysis differ between the two countries because of the different durations of their labor-export programs. South Korea's program essentially ended in the mid-1980s with the end of the Middle East construction boom and the concomitant decline in the demand for migrant labor. The Philippine labor-export program has extended, though in changed form, all the way to the present day. I have divided this longer arc into two periods. The first, from approximately 1973 to 1983, corresponds to the period during which the Philippine state utilized labor export to support the construction exports of the Philippine national firms, similar to South Korea's LEMI strategy. There was a break in the early to mid-1980s, when the Philippine labor-export program was reorganized away from the construction industry and toward a growing demand for service workers, not only in the Middle East but also in East Asia, Europe, and North America.

I begin with a direct comparison between the South Korean labor-export policy and that of the Philippines during the first period. This part of the analysis proceeds through a rough method of difference that first identifies the many ways in which the two countries reflected each other during the relevant times, then analyzes the ways in which they diverged. The labor-export policies they pursued were not similar merely by coincidence; there is evidence to suggest that the Marcos regime had studied the Korean model and was attempting to reproduce it. In addition, the political and economic characteristics of the two countries at the time—including general regional location, levels of economic development and industrialization, the close and dependent relationship of their state elites to the United States, and even, broadly speaking, their political systems for most of the 1970s and 1980s—are similar enough that we can account for how these may have affected the ways in which migrant labor affected the national economy. The key difference that remains involves the overall industrial policies pursued by the two countries, which in turn require us to examine the reasons why these policies differed so significantly: issues of state capacity, corruption, and the balance of power between the state and economic elites. These explanations are carried over in the second part of the book, which presents the second, modern period of the Philippine labor-export policy. Here, I trace the process through which the Philippines' failed LEMI strategies paved the way for a policy regime that enables the country's current RDG. Comparisons with the LEMI model pursued by both the Philippines and South Korea in the earlier period are used to illustrate how and why this model appears to be more successful in generating economic growth, despite the persistent limits on state capacity imposed by the Philippines' oligarchic political-economic structure.

The data I present in this book are drawn from a mix of primary and secondary sources. Primary sources include legislation and legislative history, policy documents, and internal memoranda collected at executive and legislative archives, including material published by relevant executive agencies. Key among the latter were the Philippine Department of Labor and Employment (DOLE), Philippine Overseas Employment Administration (POEA), and Philippine National Economic Development Authority (NEDA), as well as South Korea's Ministry of Employment and Labor and Ministry of Land, Infrastructure and Transport. These materials were supplemented by archival research in contemporary corporate reports, newspapers, trade publications – in particular the *Far Eastern Economic Review* and the *Engineering News Record*. For the more recent years of the Philippine case, I was also able to interview legislators, policymakers, industry representatives, civil-society leaders, and migration activists. The statistical data here have primarily been drawn from the Philippine Statistics Authority and the World Bank Development Indicators. The secondary sources were particularly important for my account of the industrialization and economic policies of these two countries, which relies heavily on the existing historical and public-policy scholarship.

Organization of the Book

This book presents the LEMI and RDG strategies, and the conditions under which these strategies can work, by following the trajectory of labor-export policy regimes in South Korea and the Philippines. Chapter 2 lays the conceptual foundation for the analysis to follow by examining existing approaches to migration, economic development, and the developmental sending state. It also considers how the Philippines and South Korea have been treated in these accounts. The remaining empirical chapters follow the development of the labor-export policies of the two countries, beginning with the early period—the mid-20th century—when both countries pursued LEMI strategies. Chapter 3 focuses on the Korean experience, which could be seen as a paradigmatic case in which LEMI strategies buttress the country's development policy during a key period of industrial transformation. Chapter 4 describes the Philippine experience with LEMI, which diverges from the Korean case in important ways. Operating in the same labor markets in a similar time frame, the Philippine policy effectively utilizes labor export to support domestic construction exports, but fails to translate those successes into a more fundamental economic transformation. In Chapter 5, I argue that the two countries' diverging experiences with LEMI demonstrate that the ability of migration-related strategies to produce economic growth is dependent on, and limited by, the larger industrial and development policy framework into which these policies are incorporated, as well as the capacities of the state to implement that framework. In the Philippine case, the gains delivered by LEMI strategies fed into a development policy that was fractured and subverted by the oligarchic political-economic elite, whose accumulation strategies relied on a form of rent-seeking that, because it depended on appropriation of state-controlled funds, was highly disruptive of state-led industrialization and development efforts. Chapter 6 follows the Philippine labor-export program as it continues into the 21st century, converging again with the country's economic-development policy. I outline the surprising performance of the Philippine economy under premature deindustrialization and the expansion of those economic sectors particularly associated with deindustrialization, such as real estate, retail trade, services, and finance. I trace how labor export enabled the growth of these sectors, and particularly the role of remittances in sustaining the relatively high levels of domestic consumption that these sectors require. Chapter 7 addresses the question of how an RDG model was able to produce sustained growth in the Philippines where LEMI strategies had failed. The explanation lies in the different routes through which economic gains associated with labor export have entered the Philippine economy under LEMI and RDG systems. I argue that a remittance-focused strategy is less vulnerable to the types of rent-seeking and appropriation that undermined the LEMI strategy, because remittances—unlike profits to capital—enter the economy in relatively small amounts, dispersed across millions of receiving households. Manipulation of state policy cannot serve as an effective means of accessing these remittance earnings because the state never aggregates or exercises any control over the bulk of these funds. While RDG strategies might be a "work-around" for leveraging migration for economic growth in countries where an industrialization strategy is unviable, it tends to produce an economy that is dependent, in the long term, on labor export—an outcome at odds with the promise of temporariness that justifies most labor-export programs, as well as the welfare of migrant workers and their families.

Chapter 2
The State, Economic Development, and Labor Export

Migration and Economic Development

Migration and capitalist development have always gone together. The early theorists of migration, writing in the 19th century, built their analyses on the churn of population created by the industrial revolution (see, e.g., Ravenstein, 1889). It was assumed and understood that the disruptions of capitalist development—both in the opportunities created by emerging manufacturing centers and the changes to rural life resulting from new property relations—would impel people to move (Marx, 1992). Seminal works in early economic-development theory also acknowledged this relationship, going so far as to equate capitalist development with the movement of people—surplus labor—out of subsistence agriculture into the modern manufacturing sectors (Gollin, 2014, citing Lewis, 1954). Though this sectoral mobility does not require *physical* relocation, a certain level of physical movement was inevitably associated with capitalist development, given the typical geographic distribution of these types of production.

Beginning in the latter part of the 20th century, this once commonsense association began to weaken as the countries on which the early analyses were built matured into advanced industrial economies and later-developing countries struggled to catch up. Industrialization and capitalist development proved not only to have a geographically uneven distribution, but this unevenness was shown to be structured and enforced through national boundaries (Wallerstein, 1974; Smith, 1984; Chang, 2002). The consequences of migration across those boundaries were different from migration—rural-to-urban or otherwise—that occurred entirely within the borders of a single nation's economy. The modern scholarship on migration reflects this understanding, treating international migration as a phenomenon conceptually distinct from internal migration (King & Skeldon, 2010). Capitalist development is not irrelevant to international migration;

Temporary Measures. Suzy K. Lee, Oxford University Press. © Oxford University Press (2025).
DOI: 10.1093/9780197790007.003.0002

it continues to be true that people are pushed into migration by economic transformations (see, e.g., Massey, 1988). But in the context of international migration, those migrants are pushed into economies that are separated from the sending country by political borders, and whatever gains that are made through migrant workers' participation in modern industries cannot be automatically claimed by that sending country. Instead, scholars interested in understanding the impact of international migration on sending-country economies must analyze what, if anything, flows back across the border.

This results in an analysis that is focused not on how domestic industries *directly* interact with workers on the move but on the *indirect* impact of migration on the sending economy. We can see this orientation toward migration and economic development in lines of research that seek to measure the effects of migration on human capital. In these accounts, what might be lost as people migrate out—"brain drain"—can potentially be outweighed by what migrants acquire and bring back upon their return, whether that be skills ("brain gain") or an entrepreneurial ethos that can catalyze new investment (Bhagwati & Hamada, 1974; Stark et al., 1997; Beine et al., 2001; McCormick & Wahba, 2001; Black & Castaldo, 2009; Blouin & Debnath, 2011). Even when migrants fail to return, however, many regularly send portions of their earnings back to the country of origin, and these migrant remittances have become the most important mechanism through which migration is understood to impact economic development. In the past half-century, the earnings migrants have remitted to their countries of origin have mushroomed, from less than two billion dollars in 1970 to over half a trillion dollars by the mid-2010s. At over $750 billion today, remittances dwarf official development aid and are also significantly higher than the total amount of foreign direct investment into the low- and middle-income countries of the world. Because certain countries experience greater rates of migration than others, the impact of migrant remittances on individual countries ranges, but for more than twenty-five countries remittances are equivalent to at least 10 percent of gross national product.[1] Given the scale of financial flows, it is not surprising that intergovernmental development institutions, including the World Bank and the United Nations, have increasingly promoted migrant remittances as a tool for developing economies (Bakker, 2015; Agarwala, 2022).

[1] World Bank, at http://data.worldbank.org/indicator/BX.TRF.PWKR.DT.GD.ZS?order=wbapi_data_value_2013+wbapi_data_value&sort=desc.

However, the track record of remittance-dependent economies does not support a great deal of optimism about the potential of labor migration to generate economic growth. Because remittance flows are not actually investment capital, but instead small amounts channeled through thousands or millions of individual households, their impact on the economy is mediated through the consumption and savings patterns of migrants and their families, which do not always align with the requirements of economic transformation or sustainable growth. Research conducted in the 1970s and 1980s in migrant-sending communities in Mexico showed that, while remittance income undoubtedly ameliorated the poverty of individual households, the macrostructural impact of these flows was not industrial development, but rather a long-term dependence on migration (Binford, 2003). Researchers found that remittances were primarily spent on consumption goods or nonproductive investments, such as housing, rather than on investments that might contribute to employment-generating economic transformations. Even when remittances were used to invest in productive assets—the key example in this literature was agricultural land—the flow of income unconnected to local production distorted local economies in ways that were often harmful. In the case of agricultural land, because migrants rarely used this land as a primary source of income and because long absences made intensive cultivation difficult, migrants were less likely to farm in ways that yielded productivity increases. At the same time, migrants' participation in local land markets drove inflation, pricing local producers out and consequently exacerbating unemployment and inequality (Oberai & Singh, 1980; Reichert, 1981; Wiest, 1984).

In the ensuing decades, the scholarship has become more hopeful about the potential impact of remittances. However, this new optimism is notably *not* based on any change in the consumption and investment patterns of remittance-receiving households; most modern research on the use of remittances confirms the earlier findings (see, e.g., Acosta et al., 2009; Catrinescu et al., 2009; Chami et al., 2003; Durand et al., 1996). Rather, the shift in thinking is based on a reinterpretation of the implications of these enduring remittance-spending patterns. For example, the tendency for remittances to be spent on consumption and non-tradable goods is now weighed not only as investment opportunity lost, but also in terms of the growth-inducing "multiplier effect" that *any* spending can have on an economy (Stahl & Habib, 1989; Adelman & Taylor, 1990; Glytsos, 1993;

Taylor, 1999). Other research along these lines has focused on identifying the institutional conditions that can transform the consumption tendencies of remittance-recipients away from consumption and toward saving and investment (Catrinescu et al., 2009; Mundaca, 2009; Bettin & Zazzaro, 2012). Critics of this approach have noted how neatly remittance-based growth strategies fit into the market-fundamentalist, financialized approach of the neoliberal ideology that currently dominates global discourses on economic development (Bakker, 2015).

These debates remain unresolved because both sides are correct in their observations: remittances do tend to get spent on consumption and nonproductive investment, but that spending can still have growth-generating effects in some contexts. The Philippines is proof of this, and in looking closely at this case I do find evidence to support the view that the Philippine model was shaped by the prevailing neoliberalism of global development policy. I also find, however, that this focus on indirect effects, like remittance-driven growth, misses more direct routes through which labor migration can serve a sending state's development agenda.

Labor Export as Development Policy

It is the argument of this book that the type of migration that occurred in the periods under consideration should be considered a *state-led labor export.* What, exactly, do we mean when we say that the Philippines or South Korea pursues a policy of labor export? While many governments of countries with high rates of out-migration develop policies responding to that migration, most of these policies treat migration as a process determined primarily by forces exogenous to sending-state policy—such as migrant social networks or labor-market demand in receiving countries. For example, diaspora engagement and remittance-promotion policies presuppose an existing migrant community, but are agnostic as to the *flow* of migration that determines the location, size, and occupational characteristics of that community. Countries that utilize these types of policies may do so, in part, to maximize the remittances migrants send back, but because their governments do not intervene in the flow itself, they can do little to address volatility in remittance flows that result from changes in receiving labor markets or the settlement patterns of migrants. The Philippines and South Korea are among a small subset of sending countries that have sought not only to encourage remittances from existing migrant communities but also to *direct* migration into new labor markets and to *shape* that flow into one that is dominated by contract-based laborers who have little hope or expectation of settlement in the receiving country. These policies have produced a migration that generated gains for the sending economy: first, in the 1970s and 1980s, by providing cheap labor for Korean and Philippine construction companies competing in international markets, and later, in the Philippines, by sustaining high and regular rates of remittances through diversification of the migrant flow.

Because Korea's labor-export policy was relatively short-lived, we rarely see it studied or discussed in the migration literature. However, in research on the longer-lasting Philippine case, the term "labor export" often appears in reference to the country's contract-based migration flow, though it is rarely precisely defined. Scholars and policy makers often use the term interchangeably with "labor migration," or at least with reference to large-scale labor migration accompanied by significant remittance flows (see, e.g., Hugo & Stahl, 2004; Rodriguez, 2010). The treatment of these terms as synonyms is, in many ways, justifiable. The phenomenon that "labor export" is commonly used to indicate—the large-scale, employment-driven movement of a country's nationals beyond its borders—is literally also labor migration. The term "export," though rarely explicitly defined even in economics textbooks, typically refers to the sale of some good or service in a foreign country. Thus, when a person born and raised in one country moves to another country in order to sell their labor in that new place, there is a way in which this approximates the idea of export.

Of course, there are also many ways in which a migration resulting in the sale of labor cannot be equated with export, whether of goods or services. One important distinction between labor migration and export is in the nature of the exchange. When a good or service is exported, not only is the thing being exported delivered to the destination country, but payment is remitted to a seller, who receives it (if not literally, then legally) in the receiving country's economy. To the extent that export trade is considered desirable, it is because of this return flow of currency. With labor migration, if we understand the sellers of the labor to be the migrants themselves, then the fact of their migrant status means that the payment is *not* directed to the country of origin. Not unless and until the migrants sends some portion of their earnings back home can an international transaction be said to have occurred—and though this often occurs in instances of labor or economically driven migration, it is not a given.

One way in which we might distinguish labor export from labor migration—or at least define labor export as a distinct subtype of labor migration—might then be in the extent to which a particular migration flow is associated with an economic return. For example, we could use the term "labor export" to distinguish between a migration that is associated with high rates of remittance and one characterized by lower rates. Or we could go further, to define labor export not only by relative rates of remittance but by structural characteristics of a migration flow that might condition the likelihood of remitting. Whether and how much migrants remit depend on a number of factors, but among the most important of these is the connection that migrants retain to their countries of origin. Permanent or settlement-oriented migrants may tend to have higher incomes and thus a higher capacity to remit, but because their attachment to the country of origin is increasingly eroded by competing commitments in the new country, they tend, over time, to remit at lower rates—not only relative to their income, but also in absolute numbers—than temporary migrants (Merkle & Zimmerman, 1992; Lucas, 2005). Migrants who have less expectation of settlement, who plan to return—whether because of short-term employment contracts, spouses and children left behind in the country of origin, prohibitive legal or political obstacles to settlement, simply a strong preference, or some combination thereof—maintain a much stronger connection to their country of origin. As a result, temporary migrants are more likely to send their money back, not only because their families rely on remittances for basic consumption, but because they see the country of origin as the appropriate destination for their saving and investment. While permanent migrants tend to treat their remittances as gifts, sending them occasionally or in an ad hoc fashion, temporary migrants are more likely to treat remittances as an obligation, reliably and regularly sending fixed portions of their income (Glytsos, 1997; Dustmann & Mestres, 2010). Given this association between the temporal orientation of migrants and remittance rates, sending states that intervene in labor migration in order to direct migration flows toward temporary forms and away from permanent settlement could be understood to be enacting a labor-export policy. Other policies developed to secure migrants' long-term economic engagement, such as the diaspora engagement policies that promote real-estate investment in the country of origin, could be categorized as labor-export policies, even when directed at migrants who have permanent settlement plans in the destination country. In the empirical portion of this book, I will explore the ways in which South Korea and the Philippines pursued such strategies in part by encouraging migration into receiving countries whose immigration policy reflected a corresponding preference for temporary, contract-based, or seasonal migration.[2]

Thus far, everything I have proposed about labor export has focused on our understanding of the word "export" and the extent to which labor migrations share characteristics with other types of export activity. But what of the term "labor" in "labor export"? Is there a difference between what "labor" signifies when used in reference to export rather than migration? I argue that there is: that when it is paired with "export" it more clearly refers to the *work* that migrant workers do in the receiving country (rather than the *worker* who embodies and carries that work across borders) than when paired with "migration." Delgado-Wise and Márquez Covarrubias, in their analysis of NAFTA and Mexican development (2007), have suggested that it could be both, proposing that we use the term "*direct* labor export" when referencing migration and "*indirect* labor export" for the export of goods produced in maquiladoras, where the value added in Mexico consists entirely of the work Mexican laborers do in assembling and packaging imported components. In this view, it seems that any sort of export-oriented economic activity in which the chief competitive advantage a country has is in the availability and cost of labor—for example, in the complete knock-down assembly that occurs in export-processing zones, or in the telecommunication-enabled services exports of business-process outsourcing (BPO)—should be considered a form of labor export. Their proposal makes visible a distinction between migration and labor export: If "labor export" can be used to denote instances where the laborer *does not* migrate, then we can say that what is being exported in labor export is primarily the *work* itself rather than the *worker*. What Delgado-Wise and Márquez Covarrubias call direct labor export, involving the physical migration of the laborer, happens in the Mexican-U.S. case because of long histories that tie the two countries' labor markets together. But in other cases—particularly the newer migration flows that emerged in the neoliberal period—direct labor export occurs because the movement of the worker is necessary for some forms of export activity, in those sectors where production or service-provision must occur on-site.

[2] We could call these "policies of *labor import*."

The movement of the person is necessary but subordinate to the economic phenomenon.

This definition of labor export gives us further tools to make distinctions between types of labor migration policy. Policy regimes whose goals are primarily to export—or, from the perspective of the receiving country, to import *labor*—will look different from those regimes that treat migration flows as the movement of *people* (Castles, 2006). We can see this most clearly in the context of receiving-country policies, where countries oriented toward labor import will seek to minimize the impact of those aspects of the migrant that are not directly related to their status as worker. Guest-worker or temporary-employment-based migration programs often have this character. Limitations on settlement are combined with restrictions on political participation and establishing a social life: While it is common for the vote to be withheld from migrants, temporary migrants are very rarely offered any sort of pathway to enfranchisement, something permanent immigrants can expect in most liberal democracies. Similarly, while permanent immigrants are often permitted to migrate with their families, or are offered an opportunity to sponsor their families' migration at some future point, temporary migrants—particularly the ones working in the lowest-paid occupations—must often migrate on their own, with little hope of reunifying with their families unless they return to their countries of origin (Castles, 2006).[3] Limitations on the social life of the migrant can be exacerbated by the relative isolation that migrant workers might experience, not only because of cultural and linguistic barriers or the physical isolation that workers housed in encampments might experience, but also because of the social stigma associated with their migrant status. These policies and practices function to carve away, to the extent that this is possible, the economic person from the social and the political, to extract the labor from the worker while limiting the engagement with the worker as mother, father, friend, or political stakeholder, and abdicating responsibility for any care they may need when they are no longer capable of work (Bird, 2018; Chin, 2007; Tseng & Wang, 2013).

[3] For example, many importing countries, including the United States and the oil-rich countries of the Middle East, impose minimum income or salary requirements on migrants who want to sponsor the migration of their families. In the United States, any applicant for a visa can be denied admission if they are deemed likely to become a "public charge." Many factors can go into making this determination, including a requirement that the household demonstrate income and assets greater than 250 percent of the federal poverty line—nearly $70,000 for a family of 4 in 2022. The average hourly wage for field and livestock workers in 2021 was $15.56; thus, an H-2A agricultural worker working at that wage year-round, full-time, makes less than half of what would be required to sponsor a family. (See, e.g., Shah & Menon, 1999, on Kuwait's minimum income requirement.)

Labor-exporting policies serve a complementary function, embracing the social and political person even as they facilitate the departure of their physical body to work and earn wages elsewhere (Basch et al., 1994; Levitt & de la Dehesa, 2003; Rodriguez, 2002).

This policy of dissociation suggests another distinction that we can make between labor migration and labor export, on the question of agency and where it lies in the migration process. This distinction is best illustrated by the debates in Philippine politics about the appropriate term to use for the state's labor migration/export policies. "Migration" is the term preferred by the Philippine state, used to emphasize the voluntariness of individual migration and to characterize the state's interventions in the migrant flow as regulatory or protective in nature. In contrast, Philippine migrant-rights activists—most prominently Migrante International—have insisted on "export" as the appropriate term to describe the migrant flow. In utilizing this term, activists underscore migrant workers' lack of agency; although the migration might be voluntary in a legal sense, the economic and social circumstances of that migration render it something less than a free choice. They argue that the Philippine state's reliance on remittances, and the policies it has implemented to ensure the source of those remittances—namely a large outflow of temporary workers—amounts to a *forced* migration. As such, these policies are more precisely characterized as a "labor export policy" rather than "migration management" (Connie Bragas Regalado, Migrante official, Interview, July 21, 2014).

Put another way, "export" implies an orientation toward migrant workers that views them as *objects*—commodities to be exchanged—rather than as *agentive subjects* who choose migration to serve their own life goals. To be sure, both things can be true at the same time; even the most explicit labor-export policy, if it is not to be outright slavery or human trafficking, has to appeal to the person and motivate them to participate in the labor-export process. Still, we can make a broad distinction between, on the one hand, policies that seek to *create* migration flows and *shape* them in ways that serve some economic and political objective, and, on the other, policies that respond to the evolving interests of migrant communities as the economic and social conditions contributing to migration change (Massey, 2002; Lee, 2017). A labor-export-oriented policy will be more likely to intervene early in the migration process to shape workers into potential migrants (both in their internationally marketable skills and in their psychological orientation to working abroad) and respond to declines in demand in a particular foreign labor market by redirecting that flow to new destinations.

If the agents of migration in a labor-export situation are not the migrants themselves, then who is? If we want to find a precise match for the export metaphor, then the agents of labor export—the sellers of migrant labor—could be identified as those businesses that earn revenue through the provision of labor for international markets. We can divide these businesses into two broad categories: (1) those that function primarily as recruiters for foreign employers, earning revenue both through labor-procurement contracts and through fees charged to migrant workers,[4] and (2) those that export labor more directly, transporting migrants to work in their operations in foreign countries, and that compete in international markets on the basis of their access to low-cost labor.[5] Where the first type of labor exporter predominates, the actual business of labor export represents only a small part of the economic gains associated with migration, and labor exporters are, relatively speaking, less able to wield power in the economy at large or in relation to the state. Instead, the greater share of the revenue associated with migration enters the country as remittances, and it is those sectors of the economy that are able to capture remittance spending that make the greatest gains from the labor-export system. Where labor export occurs through national businesses that are operating in foreign countries, however, the potential returns to the labor-exporting firms are much higher, as the latter are paid not as labor coordinators but instead as producers who claim the surplus generated by the migrant laborers. As I will argue in Chapters 6 and 7 of this book, the way the business of labor export is structured will affect its impact on the national economic-development project, as well as the ensuing politics.

[4] These are primarily what is contemplated in the literature on the commercial migration industry (Hernández-León, 2013). It is possible to make a distinction within this first category and say that only those businesses whose revenue is derived primarily from fees charged to employers for procuring migrant labor should be considered labor exporters proper, while those businesses whose revenue is derived primarily from fees charged to workers for their administering and coordinating the process of out-migration should be considered service providers to the migrants themselves. However, in practice, those service fees play a significant role in creating the relations that tie migrant workers to the sending economy: these fees are rarely paid up front, but are credited to migrants as debt, secured against the wages they are expected to earn in the foreign country.

[5] For both types of business, there are sanctioned and unsanctioned (criminal) forms (Hernández-León, 2013). Sanctioned labor exporters would include licensed manpower companies or construction companies bringing their own labor to international worksites. Unsanctioned forms could include *coyotes* or smugglers who transport migrants across closed borders, or human traffickers who engage in the forced labor of migrants in both legal and illegal industries.

Of course, the other actor in the labor-export system, one that may not always qualify as a labor exporter but can shape the legal and institutional backdrop against which labor exporters operate, is the state. While sending states can act directly as labor exporters in some instances—providing labor-recruiting services to foreign employers through state agencies or through businesses in which the state has significant ownership shares—states can impact labor export more significantly not as labor exporters but *as states*, deploying their legal, diplomatic, and bureaucratic powers to facilitate and shape international migration. A sending state's orientation toward labor export can range from (1) a relatively laissez-faire approach toward the migrant flow, which might include accommodation of and engagement with migrant communities in an acknowledgment of their continued economic and political relevance to the sending country, to (2) higher levels of intervention that attempt to regulate employers and labor exporters and facilitate migration through existing channels, to (3) efforts to induce migration, influence its direction, and shape the kinds of workers who participate. The first two types of policy orientations could be interpreted as labor-*migration* policy, in the sense that these policies could be understood to be treating the migrant as a subject, whose choice to migrate is being served, accommodated, or facilitated in some way. However, in the third type of policy orientation, the state treats migrant workers more like a product, and migration policy becomes more akin to economic development or industrial policy, with the state setting targets for how many migrants they want to send out and to which labor markets, and coordinating with receiving countries, employers, and labor-exporting businesses to meet those targets. This third orientation can be regarded as a type of export policy.[6]

Migration and the Developmental Sending State

It is only in recent years that the migration literature has developed the conceptual tools to analyze the kinds of states that design and implement labor-export policies. For most of the history of migration scholarship, the

[6] Labor-export policy, as the term is used in this book, could be interchangeable with Robyn Rodriguez's term "labor brokerage," which she applied to the suite of sending-state policies—institutional and discursive practices through which the Philippine state mobilizes its citizens and sends them abroad to work for employers throughout the world while generating a "profit" from the remittances that migrants send back to their families and loved ones remaining in the Philippines (Rodriguez, 2010, p. x). In my analysis, I find "labor export policy" to be the more effective term for highlighting the connections to national economic development and industrial policy.

dominant trends pushed toward analyses in which there was very little room for the state at all. The classical theoretical framework, which treated international migration as of a piece with domestic migration, understood large movements of people to be driven largely by economic factors, mediated by the feasibility of relocation, measured in terms of both geographic and sociocultural distance (Ravenstein, 1889). And to some degree this basic logic—the push and pull of need and opportunity that get people from places where there are no jobs or only low-paying jobs to places where there are abundant and well-paying jobs, combined with the migrant networks that make such movement feasible—is still reflected in much of the analysis that we see today (Lee, 1966; Massey et al., 2005). Some variations on the basic formulation have arisen over the years, most prominent among them a "new economics of labor migration" (NELM) that has argued for an analytical shift away from individuals—as opposed to households and other kinship groupings—as the unit on which migration decisions are made. NELM proponents argue that individuals rarely think exclusively about their own well-being when they migrate; rather, the migration decision is made within a web of social relationships in which the migrant plays a particular role. That role is rarely simply one of income maximization, but instead addresses a complex of household needs, such as the need to overcome local market failures in the provision of credit or insurance (Massey et al., 2005; Stark & Bloom, 1985; Stark, 1991). The logic of NELM better explains the persistence of remittances and circulatory migration, which is more difficult to account for under the simple push/pull analysis. Similarly, the migrant-networks theory that explains how social capital and migrant institutions facilitate migration flows is a more sophisticated elaboration of the logic of Ravenstein's laws: people go to places that they can reach, whether because those places are literally close or because networks bridge the distance by providing information and material support for the transition (Massey et al., 2005; Massey et al., 1990; MacDonald & MacDonald, 1974; Choldin, 1973; Tilly & Brown, 1967). Whether in its classical or more modern form, there is very little room in this economic and social framework for an analysis of the state's role.

When attention turned to states and migration policy in the late 20th century, it began as a critique of the microsocial focus of the classical theory. Scholars like Saskia Sassen, Alejandro Portes, and Aristide Zolberg argued that microsocial theories presented an oversimplified account of labor migration as a series of individual choices. The simplification elided the role of state policy and capital-investment strategies in setting the conditions that *delineate the range of choices* available to any individual migrant at a given time. As a corrective, they analyzed migration through the theories of dependency and world systems that were utilized in analyses of capitalism: industrialized countries of the "core" invade, invest, and make other interventions into the poorer, less developed countries of the "periphery," disrupting their economic structures and producing labor surpluses for which migration into the core countries becomes a solution. This potential migrant labor finds its way to international labor markets through channels fundamentally similar to those described by the micro-structuralists: pre-existing networks. For these scholars, however, it is not (at least at first) a matter of social capital, a network of migrants who share information and facilitate new migration; rather, it is networks or linkages created by *actual* capital—as well as the state that supports capital—as it penetrates new markets and connects them to the markets of the core (Sassen, 1988; Portes, 1978). In this literature, Aristide Zolberg was among the most prominent to argue for an examination of policy and politics around migration, at least with regard to receiving states. He excavated the history of U.S. immigration policy, demonstrating how its invitations and exclusions helped to shape the immigrant population, and thereby the nation (Zolberg, 1989, p. 406; Zolberg, 2006).

Zolberg's impact on migration scholarship was powerful, and research on immigration policy, its causes and consequences, is now commonplace. The literature is filled with empirically rich and rigorous studies about how receiving countries have tried to shape the migration flows that cross their borders—sometimes failing, sometimes succeeding, and often generating unintended impacts (see, e.g., Massey et al., 2002; Cornelius, 2001; Jasso, 2005). Combined with these empirical studies are efforts to build robust theoretical frameworks for understanding the role of the state and policy on migration. These include frameworks to systematize the various types of state intervention, as well as analyses of the political and economic processes through which immigration policy is shaped (Meyers, 2000; Sciortino, 2000; Teitelbaum, 2002; Castles, 2004). Many of these theorists posited that migration policy was not incidental to the state but central—that the management of population flows across borders was as important a state function as economic or national security policy (Torpey, 1998; Hollifield, 2004).

Sending states, however, were excluded from this opening of migration theory. This exclusion was explicit in Zolberg's arguments. Like Ravenstein, he found sending-state policy to be largely irrelevant for understanding migration flows, though for different reasons. Unlike Ravenstein, he argued that states were crucial for understanding international migration. For Zolberg, the border crossing, and the political and jurisdictional transfer it implicated, was essential to understanding *international* migration as a distinct phenomenon. Moreover, this crossing, and the management of the crossing, was not merely social or economic but political in nature, because it was fundamental to determining who belongs to the national community and who does not. Zolberg's understanding of the national community was essentially territorial—political membership transferring with geographical movement—and for this reason he focused on the *control over entry and exit* as the primary issue over which states could intervene in the realm of international migration (Zolberg, 1981, p. 7). As Zolberg observed, because most countries with populations likely to emigrate do not restrict exit, sending-state emigration policy could not be causally relevant to migration flows (Zolberg, 1989).

This view of the politics of migration began to change with the intervention of scholars of transnationalism, whose work demonstrated that national membership is fluid, multiple, and not always tethered to territory (Basch et al., 1994; Glick Schiller et al., 1995; Rodriguez, 2002). This literature, which began with a focus on how nation-states might be transformed or displaced by the cross-border practices of transmigrants—practices that often operated below or beyond the level of the state (Basch et al., 1994; Heisler, 1986)—evolved toward a view of states, particularly sending states, as more than just passive recipients or reactors to the changes wrought by transmigrants, but as crucial participants in the creation and maintenance of a national identity that transcends territory (Smith, 2003a, 2003b; Levitt & de la Dehesa, 2003).

This new orientation catalyzed an explosion of research on the sending state, demonstrating how sending states engage with emigrant and diaspora communities, creating opportunities for the political, social, and economic participation of transmigrants, and in turn are reshaped by their participation (Brand, 2006; Fitzgerald, 2006, 2008; Brinkerhoff, 2012; Mügge, 2012). Some of these works engage in basic theory-building, offering different approaches to organizing sending states' policies based on different frameworks for understanding the relationship between states and migration (Gamlen, 2006; Levitt & Glick Schiller, 2004). Moreover, researchers have used these theories to ask deeper questions about how these policies are formed and implemented, and interrogating the role of domestic politics, state capacity, international politics, and engagement of migrant communities (see, e.g., Levitt & de la Dehesa, 2003; Brand, 2006; Cook-Martin, 2008; Mügge, 2012; Pearlman, 2014; Waldinger, 2014).

Perhaps because transnational migration studies originated as a critique of assimilationist approaches to immigration, one bias in the research has been to focus on what happens *after* the initial act of migration—both the practices of existing emigrant or diaspora communities and the policies that states direct toward them. For our understanding of sending states, this has meant the exclusion of one section of potential policy: that is, those *ex ante* interventions in which sending states engage to facilitate, manage, or foster the movements on which transnational communities are built. The consequence of this "blind spot" has been to keep research on the sending state from being deployed to address some key questions asked by the general migration theory: specifically, the questions regarding causation, why and how migration flows look the way they do. On these questions, migration studies remain dominated by theories of economic forces and social networks, micro- and meso-level cumulative causation (Massey et al., 2005; Fussell, 2012). Studies of policy do exist, but research is focused on the immigration policies of receiving states (Calavita, 1992; Massey et al., 2005; Cornelius, 2001; Zolberg, 2006).

The transnationalism scholarship's turn toward the sending state has thus far had little impact on the migration theories dealing with causation because their data and analytical framework, focused as they are on diaspora and *ex post* migration policies, *cannot* engage with those types of questions. Yet the sending state's potential impact on migration causation is a crucial implication of transnationalism's insights on the sending state. If national identification survives the migration to a new political territory, then it becomes possible to conceive of policies other than exit restrictions being available to sending states. After all, the political community that entry/exit policies define is not being defined for the sake of its own existence; a nation-state defines its members as a preliminary step required for *taking actions on behalf of those members* to "maximize [their] collective goals" (Zolberg, 1981, p. 7). States limit entry and full political participation of foreigners in order to reserve the "wealth of the nation"—public goods, welfare-state benefits, the products of its industries, etc.—for its members, even as it opens its doors to immigrant labor where such labor is needed to further such national goals as industrialization and economic expansion. However, sending countries, which have less wealth to hoard and less industry that could benefit from increased supplies of labor (Lewis, 1954), are not necessarily interested in the *restriction* of migration. If, as transnational scholars have shown, migrants maintain their identification with their countries of origin, migration can be a means of gaining access to the wealth of the richer receiving countries, such that the question becomes how to *utilize* migration to serve national goals.

Some of the most recent research on sending states has begun to address these issues. Not only has there been interest in these interventions in the development and policy-oriented literature (Hugo, 2009; Rajan & Mishra, 2010; United Nations, 2013); the space for sending states that the transnationalism literature has opened has made possible a number of new studies which view sending-state policy more broadly (see, e.g., Biao, 2003; Siracusa & Acacio 2004; Fitzgerald, 2008; Iskander, 2010; Rodriguez, 2010). Some of these works include discussions of *ex ante* policies, particularly those promoting temporary labor migration, in a more general analysis of sending-state policy that also includes diaspora-management programs (Biao, 2003; Fitzgerald, 2008; Iskander, 2010). It is here that we have begun to see a shift toward analyzing these *ex ante* migration policies as something more explicitly economic in orientation—as policies that not only promote or support labor migration but constitute an export activity (Rodriguez, 2010; Siracusa & Acacio, 2004). This literature also makes explicit the ways in which sending states utilize migration policies to further national economic goals, particularly the domestic accumulation of capital. Rina Agarwala, writing in the Indian context, has shown how the benefits to national economic policy are not incidental to migration policy but can actually be its central driver. As such, the suite of policies, as well as the practices they reflect and shape, should be considered a "migration-development regime" (Agarwala, 2022).

Comparing the Philippines and South Korea

In the scholarship on sending-state policy, there is perhaps no country more influential than the Philippines. It drew the attention of early scholars on sending states because of the depth and breadth of state intervention in international migration. It operates in the labor-export industry as a principal organizer and coordinator, providing services to the workers, employers, private recruiting agencies, and educational institutions that occupy this economic niche (Rodriguez, 2010). Almost every aspect of the modern Philippine migration policy has been carefully documented and analyzed—from the system of training and education (Ortiga, 2018, 2021) to the regulation of private labor brokers (Guevarra, 2010) and methods through which the sending state disciplines and protects migrant workers (Parreñas, 2021). The only aspect of the Philippine migration policy that has not been thoroughly explored is its history and longevity. While there are studies of specific subsectors of the migration policy—for example, of Filipino seafarers (Galam, 2022)—there have been few accounts of how the program as a whole arose and developed over time.

This omission is particularly notable because of the unusual longevity of the Philippine state program, as well as its role in pioneering policies on which other sending states have modeled their own. Among countries with high rates of out-migration, there are very few that engage with the migration flow as directly or as comprehensively as the Philippines. From the perspective of a nation-state—a political community defined by its connection to and control over clearly demarcated, bounded territories—a significant out-migration carries an existential risk. The mass dispersal of its people beyond its borders erodes the state's claim to represent that political community, or, worse, threatens the continued existence of that community altogether. Thus, when significant out-migrations occur, we understand them to be accidents of history caused by some catastrophe or failure of governance, where the state's response is either to abdicate completely its claims on the people who leave or to attempt to salvage some part of the political community through diaspora management and the promotion of transnational identities. Yet the notion that mass migrations are political accidents is only partially correct. It *is* true that most mass migrations ultimately originate in economic or political hardships, which can sometimes be attributed to intentional state policy but more often than not are caused by factors beyond explicit state control. Still, these hardships can only be considered necessary, not sufficient, causes of mass migration, which is driven not only by the level of popular desperation but by the availability of viable paths of escape. These pathways are often extra-political, established through social and economic networks in which state involvement is minor and peripheral, but there are a significant number of cases where the state's role is much more central. In these cases, the pathways were created or maintained through the explicit policy intervention of sending-country governments, which sought to produce or expand mass migration in service of some vision of the national interest. Here, policymakers measured the risks to nationhood against the potential gains and determined that a policy of mass migration was worth pursuing.

Among these cases, we can identify at least two types. The first are those states for which the mass migrations are intended to be *permanent* departures, expulsions in the service of some new vision of the nation that excludes those being forced to leave. The second type of mass migration in which sending states play a leading role, what I will refer to in this book as *state-led labor export*, differs from the first in that the national project that the migration is intended to serve is not one of political redefinition or exclusion, but rather one of economic transformation or renewal in which the migrant, as worker, plays an integral part. Migrant workers in these cases may also be motivated by a desire to escape difficult conditions in the country of origin, but that individual goal is harnessed and directed to serve a more collective escape from the circumstances that make migration necessary: endemic poverty, underdevelopment, the country's peripheral status in the world economy. As such, these migrations are understood to be *temporary* for the migrant, who, after doing their part for the nation—working under alienated and difficult conditions, enduring months or years of separation from family and community, in order to maximize the remittances they send back home—can expect to return to enjoy the fruits of their sacrifice. Those fruits include not only the individual wealth the migrant might have accumulated during their sojourn abroad, but a society transformed by and, importantly, *through* mass migration, where the problems that made migration necessary in the first place will have been overcome. Thus, the policies of state-led labor export are themselves also understood to be *temporary*, measures reluctantly imposed that would be repealed in some happier future that is brought nearer with each migrant's sacrifice.

In reality, the actual duration of these migrations, as well as the policies that produce them, vary from case to case. We know that some migrations that were initially designed to be temporary, such as the flow of *Gastarbeiter* migrants into Europe in the mid-20th century, ended up establishing permanent African and Asian communities in countries like France, Germany, and Belgium. Still other migrations, such as the contract-based flows into the Middle East and East Asia, have produced lower rates of settlement, continuing for decades as circuits of short-term migration. Many of these variations can be attributed to the policies of receiving countries, some of which are more willing to tolerate settlement than others. Another important contributor to the production and maintenance of these temporary flows are the sending states themselves, which can direct migrants toward temporary contract labor and prevent defection from contract terms by facilitating the maintenance of social connections, coordinating with receiving-country governments and employers to ease return migration and, should these efforts fail, imposing penalties for those who violate their contract terms. Their short-term orientation—sending migrant workers abroad under conditions that tether them to the country of origin—combined with efforts to maximize the national economic gain from these migrations, are what establish these policies as a form of labor export, distinguishing them from other types of migration- and migrant-supporting policies in which sending states might engage. And like the migrations themselves, the actual duration of labor-export policy varies depending on the case. Most cases of active labor-export promotion by sending states have been ad hoc and short-lived, developed to exploit a particular labor-market opportunity or to respond to an acute crisis, and expiring as these specific circumstances change. In more recent years, we have perhaps seen labor-export programs become more entrenched and comprehensive in some sending countries, particularly in Southeast Asia. The Philippines is not merely one among these more long-lived programs but was actually the original prototype on which many of the subsequent programs were modeled.

In contrast to the Philippine case, the South Korean labor-export program is almost entirely absent from the current scholarly debates on sending states. This absence is interesting because, in many ways, the Korean program was the prototype on which the Marcos regime modeled the Philippine program in its early years. This is not to say that Korea has been ignored in the migration literature; Korean migration in general has been well studied. But most of this research—at least in the Western scholarship—has focused on the more permanent migrations that established the "Koreatown" ethnic enclaves in U.S. metropolises like Los Angeles and New York, or the inverted transnationalism of contemporary "wild geese" families that raise their children in English-speaking countries while breadwinners remain in South Korea (Logan et al., 2002; Yuh, 2005; Yoon, 2012; Lee, 2021). This is partly a result of timing. The Korean labor-export program essentially ended in the 1980s, decades before migration scholars and policymakers began seriously engaging with questions of sending-state policy, by which time South Korea had become a country of net *immigration*. Consequently, there is much more written about Korea's immigration policy and efforts to administer and manage its foreign-born population than there is on its relatively short-lived labor-export program, even in the context of migration's role in economic development (Dewind et al., 2012; Chung et al., 2024). It is also partly a result of the Korean state's active efforts to downplay the role of migration and labor export during its industrial transformation.

In the story of South Korea's economic miracle, the export of South Korean labor to the Middle East is often treated as a footnote to the more dramatic developments in the second half of the twentieth century. Given everything that happened—the creation of heavy industries out of whole cloth, the *chaebol* conglomerates' rapid transition from technological imitators to industry leaders, the aggressive role of state intervention—it seems unremarkable that, for a decade or two during the early period of development, a few hundred thousand people were sent abroad to work as contract labor. However, for understanding the relationship between migration and economic development, the period of South Korea's Middle East migration should be considered a paradigmatic example—as important as the Philippines' more long-running migration. The export of South Korean labor was a necessary condition for the participation of the *chaebol* in international construction markets, which not only helped these companies survive the turbulence of the international economy in this period, but, more importantly, generated enough revenue to subsidize the learning, experimentation, and initial failures in those industries, such as automobile production and electronics, that would go on to become the flagships of the Korean economy. Moreover, Korean construction export created downstream markets for Korean steel and other products of Korea's Heavy-Chemical Industry (HCI) drive, providing an outlet at a time when there were few others, domestic or international, for Korea's nascent heavy industry. Thus, South Korea's Middle East migration is a key example of how migration, specifically the export of contract labor, can be deployed to serve national development goals.

If South Korea's Middle East migration has been relatively ignored, the general comparison between South Korea and the Philippines has not (see, e.g., Kang, 2004; Hutchcroft, 2011). Particularly in the field of economic development, the two countries have come to represent two ends of the spectrum of possibility for countries attempting industrialization and economic transformation. South Korea and the Philippines entered the 1970s at very similar levels of development. Throughout the 1960s, both countries had attempted to transition away from a reliance on agricultural activity and the production of primary goods toward manufacturing, through the import-substitution strategies that were conventional at the time. Both countries had met with mixed success, establishing some new industries but also confronting the limits of these strategies, as well as instabilities caused by volatility in the global economy. They were both client states of the United States, that were being pressured to reorient their economies toward exports. In both countries, the labor-export programs were initiated by states that were undergoing authoritarian transitions. Within weeks of each other in the fall of 1972, both Ferdinand Marcos in the Philippines and Park Chung Hee in South Korea imposed martial law in their respective countries, in response to challenges to their leadership amidst economic instability. There were, of course, significant differences, including the impact of colonization on local elites and their relationship to the state, but there was nothing, from the standpoint of the mid-20th century, to indicate that the two countries would develop in such dramatically different ways, much less that South Korea would end the century as a major industrial power.[7] From this very similar starting point, in the same period and experiencing the same competition in the global market, with comparable policy tools available to them, the two countries took nearly opposite trajectories. South Korea experienced twenty-five years of stratospheric growth, while the Philippines struggled through crisis and stagnation. Throughout the 1960s and early 1970s, the two countries' per capita GDPs were comparable, in the low hundreds of U.S. dollars; by 1995, South Korea's per capita GDP had grown to $12,564, nearly *twelve times* that of the Philippines.

This combination of similar initial endowments and divergent outcomes provides a particularly rich field for examining the sources of economic development, and these countries have often been treated as comparative cases. The findings from these comparisons are diverse, but the consensus view places the state at the center of most explanations. Something about the Korean state—its developmental character, its autonomy relative to social actors, its bureaucratic capacity, the levels and forms of corruption that could be incorporated into the development project—differentiated it from

[7] If anything, development economists and policymakers of the time were more optimistic about the Philippines' prospects: The Philippines was rich in natural resources and had an industrial sector that had been expanding gradually for decades (de Dios & Williamson, 2015), while South Korea, having lost its access to the resource-rich North after a devastating—and technically unresolved—civil war, had embarked on a program of debt-fueled heavy industrialization that was considered by international observers to be infeasible and ill-advised (Woo, 1991).

the vast majority of states in developing countries (Amsden, 1989; Evans, 1995; Chibber, 2003; Kang, 2004). For the Philippine state, that "something" was not only lacking but often inverted—predatory and patrimonial in character, weak relative to society and in its bureaucratic capacity, with a form of corruption that was particularly harmful to the country's developmental prospects (Evans, 1995; Hutchcroft, 1998; Kang 2004).

Labor export, however, is one aspect of the two states' projects in which they were not polar opposites but actually quite similar—similar in their strategies as well as in the actual levels of out-migration and wage remittance they managed to produce. For understanding the relationship between migration and economic development, this similarity amidst difference raises questions about the interaction between these two phenomena. One possible conclusion to draw from the comparison is that migration plays no role in economic development or growth; the stagnation of the Philippine economy through the 1970s and 1980s, despite a labor-export program very similar to that of Korea, could be read this way. However, this interpretation requires us to ignore the pivotal role that labor export played for South Korean development between 1964 and 1984, two decades in which the foundations for the country's modern dynamism were laid. Moreover, the Philippines' more recent economic growth—and the dependence for that growth on migrant remittances—suggest that the differences between these two countries constitute two different models through which migration can interact with economic development. It can also tell us something about how countries move into and out of labor-export policies, and how those programs might get entrenched.

Chapter 3

South Korea and Labor-Export-Mediated Industrialization, 1964–1984

Introduction

Beginning in the 1960s, under the regime of Park Chung Hee, South Korea embarked on an aggressive program of industrialization. This program included a project to export Korean workers to foreign markets, initially as a means of earning foreign exchange. Within a few years, however, the Korean state found that migrant workers could be much more beneficial to the national development project than the wages they remitted: they could be bartered for development aid and market access. This strategy, pursued by the South Korean state roughly between 1964 and 1984, I call a strategy of *labor-export-mediated industrialization* (LEMI).

We know that most developing countries, because of their relative poverty, have comparative advantages in low-cost labor, and that the effective deployment of this labor is a key to most export-oriented development projects. LEMI is a variation on this labor-dependent strategy, in which the nature of the industries involved require that production, or the provision of some service, occur abroad. In the Korean case, the "industries" in question were, first, the military, and second, construction. It is self-evident why military service—in this case, to support the US military intervention in Vietnam—requires the movement of the participating servicemen. Construction similarly requires migration because the products of construction—that is, buildings, roads, and other large-scale infrastructure—cannot entirely be fabricated in one location and shipped to a buyer. Some significant portion of the labor must occur in the importing country. Thus, to the extent that international construction can be called an "export" activity, it is so because of the labor that is deployed abroad. Construction firms from advanced industrial countries, which compete on the basis of technological innovation, limit that labor export to a small

Temporary Measures. Suzy K. Lee, Oxford University Press. © Oxford University Press (2025).
DOI: 10.1093/9780197790007.003.0003

number of skilled engineers and managers, but for firms from developing countries that hope to compete on the basis of cheap labor, access to a pool of workers that are cheaper and unavailable to their competitors is a necessity. An LEMI strategy, then, is a policy regime in which support for labor export is provided by states in order to buttress the export competitiveness and profitability of domestic capital. While opportunities for LEMI are not always abundant, when it works—as I will argue it did in the case of South Korea—these strategies can play a pivotal role in economic transformation.

In this chapter, I will outline the model demonstrated by South Korea between the mid-1960s and the mid-1980s: labor-export-mediated industrialization (LEMI). Unlike the policies promoted by international development institutions today, which focus on the potential impact of migrant remittances—*remittance-dependent growth*, or RDG, strategies—what drives growth under LEMI strategies is not the flow of remitted wages but the competitive advantage that migrant workers provide to the companies of the labor-sending country. In the Korean case, the state's willingness and ability to mobilize hundreds of thousands of workers, who could be made to work longer hours and at lower cost than workers from competing countries, gave Korean construction firms extraordinary and, in some cases, protected access to international markets—at first in Vietnam War-related contracts in Southeast Asia, then in the petroleum-driven infrastructure boom in the Middle East. It is an uncontroversial view that this period was key to South Korea's economic development; South Korea entered a period of rapid and sustained growth in the 1960s that transformed the country's economy, launching it into the ranks of the advanced industrial economies within a few decades (Amsden, 1989; Harvie & Lee, 2003). The current consensus on South Korea's economic development attributes this "miracle" to a developmental state that, through a combination of support and discipline, curated the process of learning and catch-up required for Korean industry to compete in international export markets (Amsden, 1989; Evans, 1995; Chibber, 2003). Here I will argue that LEMI strategies were critical to this overall developmental project, securing a number of factors that the Korean state needed to pursue its agenda, including opportunities for firms to practice, as well as sources of capital and foreign exchange that the Korean state and capital needed to stay afloat during the early years of "learning," when technological progress was achieved through costly experimentation and short-term failures.

The Beginnings of the Korean Labor-Export Program: The Vietnam War (1964–1973)

Like most labor-sending countries, Korea had experience with migration that preceded the South Korean state's labor-export projects of the 1970s and 1980s. Significant Korean diaspora communities were established in China and Russia through large-scale migrations that occurred in the 19th century, and in Japan through colonial-era (1910–1945) migrations. There was also a flow of Korean migrants to Western countries, particularly the United States, beginning in the early years of the 20th century, with Korean laborers recruited to work on the plantations of the territory of Hawaii. While some of these flows may have involved return migration, most of them were not designed to be temporary or tied explicitly to employment.

South Korea's modern experiences with temporary and contract-based migration began in the late 1950s, as Western European countries, rapidly reconstructing after World War II, recruited migrants from other parts of Europe, Africa, and Asia to address their acute labor needs. Many of these migrants would eventually end up settling in Europe, but in their initial design these migrations were intended to be temporary guest-worker schemes, granting migrants limited rights to bring their families or to remain at the end of the employment contract period (Castles, 1986). And while recruitment occurred most heavily from countries in relative geographic proximity—most migrants in this period were from Northern Africa and Southern Europe—migrant workers from farther afield were hired for occupations for which an adequate labor force could not be found from the abovementioned sources. South Korea, connected to Western European through political linkages and religious missions, became a source country for such workers (Jung, 2018).

The first of these contract migrations occurred with very little participation from the Korean state, and at a relatively small scale. Migrant workers—primarily in the nursing field—were recruited by priests and other church officials on mission trips to Asia to staff church-run hospitals (Na, 2019). This began to change in the early 1960s, with the military government of Park Chung Hee, who, from very early in his regime, included labor export as an element in the country's overall development efforts. The military took power in South Korea on May 16, 1961, deposing the civilian government of Chang Myon, who had been elected prime minister only ten months prior, following a period of protest and political unrest that had forced

South Korea's longtime authoritarian leader, Syngman Rhee, out of office. The military junta had legitimized its coup through a critique of the corruption and mismanagement of the civilian leadership[1] and the promise of accelerated economic growth (Han, 2011). Plans for generating this growth involved developing the country's capacities in heavy industries, requiring massive capital investments that Korea's domestic economy—largely agricultural even before the devastation of the Korean War—did not have the resources to make.

The early years of the Park regime were marked by international outreach to raise funding and other support for the industrialization drive, and these overtures often included offers to send Koreans abroad. Less than a year after taking power, Park was approaching the Kennedy administration about sending Korean soldiers to the Vietnam War and trying to negotiate a traineeship program with West Germany that would circumvent the country's restrictions on Asian migration and permit the entry of up to 2,400 Korean workers (Cho, 2013; Hong, 2015). Though these initial offers were rejected, Park continued to pursue the labor-export strategy, building Korea's domestic infrastructure to support out-migration. In 1962 the regime enacted the Overseas Migration Act, which created the legal structure to regulate and promote migration, and in 1965 it formed the Korean Overseas Development Corporation (KODC), a state-connected agency to coordinate and facilitate labor export.

For the first half of the 1960s, the Park regime's efforts to raise funds for the heavy industrialization effort met with limited success. Ambitious industrialization plans, including the development of a world-class steel sector, for a country that had no comparative advantage in heavy industry—neither natural resources nor technical expertise—or even much of a domestic market for capital goods, seemed like a fool's errand, and US aid officials, international lenders, and investors initially declined to support the effort (Yoon, 2011). However, the regime was more successful in negotiating for labor export. By the mid-1960s, Park had negotiated for the entry of a small but significant contingent of Korean miners and nurses to West Germany. More importantly, the rapid escalation of US involvement in Vietnam meant that by 1964 the United States was looking much more favorably on

[1] Chang, along with all the other members of the civilian government elected in 1960, had been closely associated with the Rhee regime. Chang himself served as prime minister and vice president under Rhee for most of the 1950s. Thus, Chang's government was largely considered a continuation of Rhee's.

South Korea's offer of soldiers. The Johnson administration was now actively seeking allies in the region to participate in military action and, importantly, finding it difficult enough that it was willing to offer generous terms for this participation. South Korea was one of only a handful of countries willing to assist the US efforts in Vietnam, and was the only country other than Thailand willing to send troops in significant numbers. Over 300,000 South Korean troops would end up serving in Vietnam, the largest foreign contingent after the United States itself. And while troops from other countries often served in noncombat and support roles, Korean soldiers served under some of the most dangerous conditions, not only engaging in heavy fighting but clearing brush in areas where the defoliant chemical Agent Orange was being sprayed. In all, 15,462 Korean soldiers were reported killed or injured, and another 74,230 were poisoned by defoliant (Park, 2003).

At the time, South Korea's participation in the war was presented as a matter of political alliance. For the Johnson administration, this framing was key to selling the American intervention in Vietnam to a skeptical domestic public (Baldwin, 1975). This framing largely persists today, with South Korea's role in the Vietnam War—to the extent that it is considered at all[2]—examined for its human-rights implications or for the complex imperial relationship between the United States and South Korea rather than as a question of migration and economic development (see, e.g., Armstrong, 2001; Lee, 2009). While not gainsaying the relevance of these factors to understanding South Korea's participation in Vietnam, this focus obscures the extent to which Korea's participation in the Vietnam War was an economic arrangement—an explicit and deliberate exchange of Korean labor for US support for Korea's development agenda, support that would be crucial to Korea's industrial transformation.

Though Korea had offered troops as early as 1961, the actual dispatch of Korean combat troops required a protracted negotiation, in which the Park regime was able to secure not only billions of dollars in military aid but economic support for its development project. In terms of military aid, the United States not only agreed to cover all the costs of South Korea's military, including equipment and combat pay at rates more than twenty times the standard Korean military salary of the time, but a major expansion of general military aid (Baldwin, 1975). Between 1961 and 1965, Korea received $820

[2] South Korea's participation in the Vietnam War tends to be overlooked in accounts of Korean history, so much so that it has been called Korea's "forgotten war" (Han, 2006).

million in US military aid, a number that grew to $1.608 billion between 1966 and 1970 (Park, 2003). The total of additional military aid, combat pay, and expenses for Korea's participation in the Vietnam War eventually surpassed $4 billion (Cho, 2013). As part of the economic incentive for sending combat troops to Vietnam, the Park regime negotiated for Korean firms to bid for US military procurements and construction contracts in Vietnam, an extraordinary opportunity given that bidding for procurement contracts was typically limited to firms that had been vetted through high levels of security clearance and prior experience (Baldwin, 1975; Glassman & Choi, 2014). Not only did the Korean state secure the right to bid on US military procurements, it lobbied for and won *protected* access to the procurements market: that is, the United States gave Korea priority in the bidding process against all non-US firms, making it possible for Korean firms to win contracts that would otherwise have gone to Japanese firms with more technical expertise and long-standing relationships with the US military (Glassman & Choi, 2014). To ensure that Korean businesses would be able to make the most of this opportunity, the concessions included US Agency for International Development (AID) loans, as well as technical assistance in export promotion. Through the procurement process, the South Korean economy received another $1.22 billion between 1965 and 1978 (Park, 2003). In addition to these direct payments, the United States helped Park raise funds for the industrialization drive, pressuring Japan to normalize relations with Korea and organizing a consortium of international-development agencies and wealthy countries to provide an additional $4 billion in loans and investments, half of which was funded by the United States (Park, 2003).

It is not surprising that the incentives South Korea received for sending its troops to Vietnam contributed to the country's economic growth. Relative to the country's economy, whose annual GDP hovered around $3.3 billion in the years immediately prior to the Vietnam deployment,[3] the funds injected into the Korean economy were astronomical. Thus, we can say that this period represents a case in which labor export was leveraged to support a nation's economic development project. While it may be controversial to identify military deployment as a form of labor export, in this case there was very little to distinguish it from other forms of contract labor apart from the violent nature of the work.[4] Korean soldiers were paid by the United

[3] Korea's average annual GDP between 1960 and 1965.
[4] On "violence work," see, e.g., Tahir, 2019.

States and recruited specifically to deploy to Vietnam; the United States even increased its financial incentives when the pay rates initially offered failed to attract sufficient numbers of recruits (Baldwin, 1975). Moreover, in addition to soldiers, 23,000 civilian Korean workers deployed to Vietnam to work for the various firms that were hired to build infrastructure and to supply and service American and Korean troops (Steers, 1999).

For my analysis, however, what is more important than the sheer scale of the financial contribution is the way in which those funds entered the Korean economy. While remittances from the deployed soldiers and civilian workers represented a significant portion of Korea's Vietnam earnings—combat pay alone totaled $2.3 billion—a larger share went directly to the Korean state (in the form of military and economic aid) and Korean capital (in the form of procurement and construction contracts). Research on the impact of military aid has found that it is very fungible, being used to cover defense costs to which the state is already committed, and thus freeing up funds to be used in other endeavors (Khilji & Zampelli, 1994). This, combined with the direct economic aid/loans from the United States and its allies, gave the Park regime the financial leeway to pursue the industrial agenda it had struggled to implement during its first five years. For example, after nearly a decade of trying and failing to secure funding to build a domestic steel plant, the Park regime was finally able to break ground on an integrated steel mill, using war-reparations payments the United States had pressured Japan to make (Yoon, 2011).[5] These funds not only allowed the Korean state to make direct investments in heavy industry, but also enabled the policies of subsidy and discipline with which the state directed Korean private investment. While some of the subsidies the Korean state provided came in the relatively cheap forms of exclusive licenses and tax concessions, others, such as the foreign loan guarantee program or subsidized loans from state banks (at negative interest rates) for capital investments required reserves of international currency (Park & Weber, 2006). In the beginning of the 1960s, the Korean state had almost no independent capacity to provide that kind of support to capital; it actually relied on US aid for half of its annual budget. The mushrooming of that aid with the deployment to Vietnam made it possible for the Korean state to pursue a more aggressive development policy, giving it the leverage to force Korean businesses into sectors they might otherwise have been too risk-averse to move toward.

[5] By 1992, Pohang Iron & Steel Co., Ltd (POSCO), the state-owned steel company funded by Japanese colonial reparations, was the second-largest steel producer in the world (Hogan, 2001).

Perhaps as significant as the funds directly paid to the Korean state were the gains Korean capital made through Vietnam War-related opportunities, which neatly complemented the state's developmental goals. The United States had essentially promised to provide an export market for Korean businesses, not only for a broad range of goods related to the war effort but for war-related construction and reconstruction projects as well. Vivek Chibber (2003), in explaining the efficacy of Korea's developmentalist state, has argued that the Korean state's capacity to direct the economy came from the acquiescence of Korean capital, which in turn resulted from the availability of export markets in which they had *reasonable expectations* of success. As a source of this expectation, Chibber's account focuses on Japan's "flying geese" effect—the low-end markets that Japan was leaving behind as it climbed up the industrial ladder. In some ways, South Korea *was* following in Japan's footsteps with the US military-procurements contracts. A decade earlier, Japan's economic development had been aided by the participation of Japanese companies in procurements for the Korean War. However, at the time, Japan was not a major participant in the construction export market and thus was not in a position to mentor or provide market access to South Korea. In fact, in the mid-1960s, Japanese construction firms were seeking opportunities to expand their business abroad as the domestic building boom began to slow, making Japan a *competitor* to Korea in this sector (Linder, 1994). While the country's overall export-orientation may have originated in the opportunities made available through Japanese assistance, in this particular sector, the Korean state secured market access on its own by offering the country's workers as migrant soldiers.

Several of the companies that would become Korea's flagship international conglomerates won their first major international contracts in Vietnam: Hyundai (construction), Daewoo (construction), Dong-A (construction), and Hanjin (shipping). And they won these contracts, many of which were initially beyond these companies' experience or existing technical capacity, on terms that included the time and space to practice and learn (what Amsden has called "learning-by-doing"), as well as instruction and guidance on meeting international standards. For example, Hyundai's first foray in international construction—the Pattani-Narathiwat Highway in Thailand, built to support the US war effort in Vietnam—was, by all accounts, a fiasco. Hyundai was so unprepared, and the build was so inadequate, that portions of completed roadway had to be torn up and rebuilt. Yet this experience is not counted as a failure in Hyundai's and Korea's development history because the US military responded not by withdrawing the contract but by *teaching* the firm how to meet the US requirements and giving it time to upgrade its equipment and experiment with management techniques.

One key management technique that Korean firms like Hyundai experimented with during this time was an "enclave model" of project management, in which all labor for the project was imported from Korea. The model emerged in response to the labor struggles Hyundai faced in the Pattani-Narathiwat Highway build. Initially, Hyundai managers followed the typical practice in international construction of procuring labor from the local or regional labor markets, hiring the majority of its workforce from among the Thai and Malay population of the region. Among the many problems Hyundai encountered in this project, the difficulty of managing the local labor force was a significant one: there was high turnover, resistance to the duration and intensity of work Hyundai expected, language and cultural barriers, and violent disruptions over pay equity. An all-migrant Korean labor force solved many (though not all) of these problems, not only because of the cultural affinity between managers and workers but also because of the different expectations of contract-based migrant workers, who, having traveled far from family and friends for the express purpose of earning the high wages associated with overseas work, were more willing to submit to the difficult conditions and extreme labor demands. While it required that Hyundai pay wages much higher than those paid to local labor, as well as additional costs for enclave facilities, the gains in productivity were such that Hyundai thereafter increasingly opted for Korean-only workforces (DiMoia, 2018).

Though Hyundai lost money on the Pattani-Narathiwat Highway, it did manage to successfully complete the project. And as the Vietnam War progressed, the company was given the opportunity to absorb that loss as well as to practice its newly acquired technical and managerial knowledge through further US contracts (Glassman & Choi, 2014). Between 1966 and 1972, Hyundai averaged an annual profit of over $1.5 million, more than what it had been making in annual *total sales* in prior years (Park, 2003). The profits were pivotal to the company's growth, helping to fund the *chaebol's* diversification into heavy industry and consumer products, in addition to the expansion of its construction business. More important, however, was the enclave model of project management that Hyundai developed, which would prove particularly well suited to a new, much bigger opportunity that was appearing on the horizon.

Middle East Construction Boom (1973–1984)

By the early 1970s, the US military had begun to draw down its operations in Vietnam, and in 1973 it withdrew from the country altogether. In 1969, at the height of Korean involvement in the war, US military expenditures on and exports to Vietnam accounted for 5.65 percent of South Korea's GDP, and the loss of that revenue stream came at an already precarious moment for the Korean economy and Park's regime. The economy had started to show the strain of the regime's breakneck growth strategies, including a slowdown in growth and the beginnings of a foreign debt crisis. Korean firms, particularly those at the forefront of the country's industrialization efforts, had been encouraged by the Park regime to fund their growth through foreign loans and enabled to do so through the Foreign Capital Inducement Act (1962), which made the Korean state the guarantor for these loans. Though this system fueled the country's double-digit growth rates through the 1960s, by the end of that decade major companies were struggling to make their interest payments and entering bankruptcy. Korea was forced to accept significant structural adjustments in exchange for support from the International Monetary Fund, which had the short-term effect of exacerbating the economic contraction (Woo, 1991; Im, 2011). The downturn was short-lived, with the economy returning to double-digit growth by 1973, but it had lasting effects on the country's politics. After a decade in which the Park regime had bought acquiescence, even popular support, by delivering rapid growth, the slowdown galvanized political mobilization against the regime, with protests erupting around the country and an opposition party, the New Democratic Party (NDP), gaining seats in the National Assembly in the 1971 elections. By the fall of 1972, Park had dissolved the National Assembly and imposed martial law. The Vietnam War's close exacerbated all of these problems. Both remittance rates and newly awarded construction and procurement contracts fell sharply, resulting in not only the loss of hundreds of millions of dollars in foreign exchange but the idling of capacity that had been built up during this period (Kim, 1982, 1988; Cho, 2013). At the same time, tens of thousands of soldiers and migrant workers, who had been earning premium wages and trained to work in the difficult conditions of wartime Southeast Asia, had to be provided with alternative sources of livelihood.

In October of 1973, less than a year after Park had imposed martial law, and just as the Korean economy was beginning to regain its footing, the Organization of Petroleum Exporting Countries (OPEC) initiated an oil embargo against the United States and other industrialized nations, causing the price of crude oil to triple within a few short months. The price rise itself had immediate impacts on the South Korean economy. As in other nations, Korea's oil bill more than tripled between 1972 and 1974, from $300 million to $1 billion. It caused an immediate foreign exchange crisis and drove inflation from 3.2 percent in 1973 to over 24 percent in 1974 (Azad, 2015). The return of economic instability so shortly after the debt crisis was, at the very least, a short-term threat to the Park regime. For Korea's economy, however, it was a more long-term problem, potentially fatal to the country's development hopes. Having lost access to the Korean Peninsula's only indigenous fossil-fuel reserves in its partition from the North, South Korea was entirely reliant on fuel imports. Any attempt to enlarge the country's industrial structure—such as the heavy and chemical industrialization drive Park had just introduced[6]—would require massive increases in those imports. The resetting of oil prices tipped the already perilous accounting balance that the Park regime had maintained throughout the 1960s, in which rapid export growth had been only just able to keep ahead of the foreign-exchange requirements for servicing the foreign debt that was used to finance that growth.

Within months of the onset of the crisis, however, the Park regime had begun to consider the oil price rise as not only a crisis but a "once in a thousand years" opportunity (Cho, 2013). In December 1973, Samwhan Construction Company won a contract worth nearly $25 million to build a highway in Saudi Arabia, becoming the first Korean construction firm to win a bid in the Middle East (Moon, 1986). The Samwhan contract suggested that South Korea could earn back the dollars it was spending on petroleum, and more, by sending its construction firms to build the infrastructure on which those petrodollars were being invested. By January of 1974, Oh Won Chol, head of the Ministry of Commerce and Industry and responsible for implementing the Heavy and Chemical Industry Promotion drive, was reporting to Park about how Korean construction could compete in the Middle East construction market through what would be called the "Middle East Expansion Strategy" (Cho, 2013). It was a strategy that relied almost entirely on the export of Korean workers.

[6] The Heavy and Chemical Industry Promotion Act (HCIPA) was enacted in 1973. The Pohang Iron and Steel Company (POSCO), the Korean state's steel-making company that had been created with an investment of $3.6 billion, had only just begun producing steel that year.

If the petroleum-exporting countries of the Middle East were rich in capital, they were relatively poor in labor. In Saudi Arabia, for example, the total population was just over 7 million in 1974. Oil exports raised its GDP to over $45 billion that year, giving it a GDP per capita that exceeded those of many developed countries.[7] At the same time, its labor force reflected the pre-industrial economy that prevailed in the region prior to the discovery of oil reserves. Labor-force participation rates were extremely low, due to the relative youth of the population and gender dynamics that kept women out of the workforce; the active labor force was estimated to be around 1 million, less than 15 percent of the total population. Those 1 million workers were a poor match for the development needs of the country, with low rates of education and little industrial experience. Moreover, because Saudi Arabian citizens had access to petrodollar-funded public-sector jobs, few sought blue-collar employment in manufacturing or construction. Already in the early 1970s, Saudi Arabia was heavily dependent on foreign labor, importing nearly half of its labor force (Knauerhase, 1974). Any infrastructure drive, let alone one on the scale made possible by the petroleum boom, would require not only the services of foreign architects, engineers, and general contractors but a massive influx of construction workers and supporting staff.

Oh argued that South Korean construction firms were uniquely situated to compete in the Middle East's infrastructure boom, reasoning that Korea's access to cheap and reliable manpower was an exceptional advantage in this context. He predicted that construction firms from developed countries would be reluctant to enter the Middle East market due to the difficulty of securing labor and the harsh living and working conditions they could expect. These conditions included not only the extreme heat but also the isolation resulting from many Middle Eastern states' efforts to minimize the social and political impact of migrant workers and to prevent the conversion of temporary migrants into permanent settlers (AlShehabi, 2015). The Vietnam War experience had already demonstrated that Korean workers could be recruited to labor under extremely difficult and dangerous conditions. More importantly, Korean construction firms were now experienced in operating under such conditions, knowing how to deliver projects on time and to international standards, and keeping rates of labor unrest

[7] In 1974, GDP per capita in the United Kingdom was $3,665, compared to $6,437 in Saudi Arabia. GDP per capita in the United States was $7,225 that year.

and turnover under tolerable levels. Korean workers might be less skilled and productive than workers from developed countries, but Oh argued that these differences could be made up by the savings on the wage bill (Cho, 2013).

The implementation of the Middle East Expansion Strategy was immediate and comprehensive. The Park regime transformed its diplomatic approach to the countries of the Middle East: 1974 marked the beginning of a more serious engagement with the region, particularly the oil-rich Persian Gulf countries. While South Korea had already established diplomatic relations with most of the largest countries in the region, including Saudi Arabia, Iran, and Turkey, until the 1970s these relations had been so perfunctory that the news media's reporting at the time referred to the Middle East as a "dark corner" of Korea's foreign policy (Azad, 2015, p. 50). The oil shock precipitated a flurry of diplomatic activity. Top-level Korean officials began to frequent the region in outreach missions (culminating in a first-ever presidential-level visit to Saudi Arabia in 1980), attempting to establish formal diplomatic relations with the smaller oil-rich states of the Gulf and negotiate thicker economic relations with the major economic powers of Iran and Saudi Arabia. Beginning in July of 1974, Park's regime and the Saudi government executed a series of agreements to facilitate the export of Korean goods and construction services while creating a framework for mutual investment and joint ventures. Through similar agreements with the government of Iran, the Park regime secured long-term access to Iranian crude oil, contracts for Korean firms to build housing and infrastructure, and significant increases in the numbers of Korean workers who could be imported to work on those projects (Azad, 2015).

On the domestic front, the Park regime ordered Korean construction firms to pursue contracts in the Middle East and instituted a package of policies to support them (Steers, 1999). Under the 1975 Overseas Construction Promotion Law, construction firms operating overseas were categorized as export companies, and were thus given access to the subsidies and incentives that manufacturing exporters enjoyed. These included preferential access to loans, tax deductions on income earned abroad, and subsidies for exploration of new markets. In some instances, construction firms received extraordinary treatment, even in comparison to other exporting businesses. Until 1981, construction firms were given a 50 percent reduction in corporate income tax on overseas construction income, while other exporters were permitted a much smaller 1 percent deduction of foreign-exchange revenues from their taxable income. The state was also critical in providing the credit and financing necessary for Korean firms to compete for Middle East contracts. In the first years of construction exporting to the Middle East, a problem arose because the scale of the projects were so large that Korean construction firms could not independently finance the initial operating costs, lacking the capital outright or the collateral sufficient to guarantee advance-payments bonds. The Park regime responded by creating a system of collective guaranties among Korea's major banks to provide insurance for these giant projects and thereby enable the necessary loans (Kim, 1988).

Finally, the Korean state provided a number of supports to ensure the flow of Korean migrant workers, whose cheap and reliable labor was the basis for Korean construction firms' ability to compete with the North American and European firms that had historically dominated the international construction market. A system for coordinating and administering the recruitment of overseas workers had been put in place in the Vietnam War era, but it needed to be expanded to accommodate the export of labor to the Middle East, where out-migration numbers were an order of magnitude greater than in the Vietnam years. The Korean Overseas Development Corporation continued to coordinate the migration of Korean workers employed by foreign businesses, but the preponderance of the labor export occurred through Korean construction firms, which were heavily regulated by the state. This regulation imposed a standard employment contract that set minimum employment standards and pay, and required employers to provide for airfare to and from the worksite, food, shelter, recreational facilities, medical care, and compensation for workplace injury or death. It also required firms to make investments in skills development. Overseas construction firms were required to provide training to workers; the number of workers trained and the types of skills taught were prescribed by the Korean state (Kim, 1982).[8] The regulatory framework also built in incentives and subsidies for Korean construction. For example, the state essentially took over the costs of the required training programs by allowing firms to deduct these costs from their taxes. Korean regulations also gave Korean firms preferential treatment in the market for Korean overseas workers, imposing different minimum wage rates for Korean and foreign employers[9] and prohibiting

[8] In 1979, 35,000 workers were trained under these programs (Kim, 1982).

[9] Until 1981, foreign employers hiring for the Middle Eastern market were required to pay a minimum wage that was 60 percent higher than the minimum wage required of Korean labor-exporting firms (Kim, 1982).

foreign employers from hiring in those occupations where labor supply was limited (Kim, 1982).

The Korean state became directly involved in labor-force development as well. State intervention began even before workers were directly engaged in the labor market, with high-school curricula amended to include skills development relevant to the construction export labor market (Oh, 2006). Recruitment efforts were aided by policies allowing for early discharge and exemptions from military service—which was mandatory for all men—for workers who took contracts in the Middle East. The military was also directly involved in training the migrant labor force. During the Vietnam War period, the Korean state, in partnership with labor-exporting construction companies, had converted a number of military bases into construction training camps. The training regimes here were intense, with the ability to turn out skilled workers on short timelines, and the training not only included education and skills development but imposed militaristic standards of workplace discipline (Disney, 1977). These camps continued to train workers through the 1970s, though now for the Middle East labor market, where this orientation to the workplace was especially useful. The difficulties and dangers of the work environment, the pressures imposed by tight timelines, and the isolation of the work camps resulted in a work experience that was in many ways more akin to a military campaign than to domestic employment. Many observers at the time noted this militaristic quality of the Korean construction sites, which was often cited as a source of Korean firms' ability to compete despite the technological advantages of the more established European and North American firms.[10]

The state's intervention in the labor force was not limited to training and regulation of labor contracts but also extended to the management of labor relations. Given how pivotal the performance of Korean workers was to the Middle Eastern expansion, it was imperative that labor unrest be kept to a minimum. This was in part accomplished through the "enclave model" developed during the Vietnam years, which gave employers almost total control over the lives of their workforce. Korean migrant workers were segregated from the surrounding population in project enclaves that

[10] "These guys are something else. They go in there like a tornado. They bring their own workers from engineers on down, their own clerks, cooks, food, doctors, clothing, movies. They build their own housing. They do the job faster than anyone else, working 14-hour shifts, seven-day weeks.... They only mix with one another, and once they are through with a job they're moved on to another or sent back home. They're like a disciplined, paramilitary, single-minded, self-sufficient machine—simply amazing" (Ibrahim, 1978).

operated as "self-contained units," with all activities, including rest and leisure, occurring at company-controlled project sites. Construction firms not only provided room and board (including Korean cuisine) in on-site facilities but also arranged for a slew of services—including recreation, counseling, and health care—that made it unnecessary for Korean workers to ever leave the project enclave (Laporte, 1984). These enclaves were essentially mandated by Korean state regulations, which not only required employers to provide housing and other accommodations but detailed *how* those accommodations should be provided, down to the square footage allotted per worker, the allocation of toilet facilities, and the number and type of sports facilities (Kim, 1982). Workers who were unhappy with the labor or living conditions had few options other than a return to Korea. State regulations forbade union organizing and collective bargaining. Workers were permitted to send elected representatives to conferences with management to deal with labor disputes, but the Ministry of Labor's attaché reserved the right to overrule the worker's election process and directly select labor representatives (Kim, 1982; Azad, 2015). While it was not impossible to find alternative employment with another international contractor, finding such work required workers to make connections beyond the enclave—a difficult task given the language and cultural barriers. Remaining in the Middle East for even a short time without employment was prohibited by law, and moreover practically infeasible. A worker outside of the enclave would not only struggle with the language and cultural barriers but would lack the basic funds to survive for long; though salaries were high, the 80 percent of earnings that were required by law to be remitted to Korea was paid in *won* and often deposited directly into banks accessible only in Korea. With not much else to do but work, and under short-term contracts that incentivized maximizing earnings, Korean workers could be pushed to a grueling pace, enabling firms like Hyundai to operate twenty-four-hour worksites and establish a reputation for rapid, reliable construction (Steers, 1999; Azad, 2015; Cho, 2013).

Labor unrest, when it did occur, was treated as a state matter, with labor attachés directly overseeing meetings of companies involved in serious labor disputes. In the most serious instances, labor unrest could warrant the attention of top-level state officials. The most well-known of these occurred in the construction of the Jubail Harbor Oil Terminal Facility in Saudi Arabia. The contract was won by Hyundai in 1976, in the early years of the Middle East Expansion Strategy. Worth nearly $1 billion, it was not only the largest bid won by a Korean firm but the largest construction project in the world at the time; if successful, it was expected to put Hyundai, and the Korean construction industry with it, on the map as a serious player in international construction. Success, however, was not a given: the technology required for the harbor build was not something Hyundai had in-house, and it was unclear whether Hyundai could actually deliver the project at the cost (which was 50 percent lower than the bid offered by its nearest competitor) and on the timelines it had promised the Saudi government. To meet these targets, Hyundai managers pushed the enclave labor model to its limits, demanding fifteen-hour days in "hundred-degree-plus" heat. By March of 1977, Hyundai was facing wildcat strikes and labor violence, including the killing of two managers and the summary execution of three workers by Saudi security forces (Clifford, 1998; Linder, 1994). The Park regime intervened directly, creating a task force to resolve the issues and sending Prime Minister Kyu-ha Choi to Saudi Arabia to make a formal apology to the Saudi government (Azad, 2015). The workers, whose return to work was critical to Korea's fortunes in the Middle East, also received a formal apology, along with the concessions—higher wages, a slowdown in the pace of work, improvements in accommodations and social supports, and the removal of problematic supervisors—necessary to prevent future disruption.

The Impact of the LEMI Policy

With this labor-force advantage, Korean construction firms moved rapidly into the Middle Eastern market. While Korean firms did not have the capacity to compete for projects requiring advanced technology, much of what the Middle East required during the early years of the petrodollar boom was basic infrastructure—roads, housing, and harbors—that could be readily constructed with the labor-intensive, low-technology methods that Korean industry had mastered during the Vietnam War (Kim, 1982; Moon, 1986). By 1978, only five years after they entered the market, Korean construction firms accounted for 25 percent of the Middle Eastern market, making the country, for a time, a leader in construction exports—an unprecedented achievement in an industry that had been dominated by a North American and European oligarchy since the 19th century (Kim, 1988; Linder, 1994). Korean bids for projects regularly came in at 20–30 percent below their international competitors', enabled by state subsidies that made it possible to accept lower profit rates and, importantly, a wage bill that was a fraction of what other international construction firms paid (Linder, 1994). South

Korea became a top labor exporter to the region; by 1977 the annual flow of Korean workers into the Middle East exceeded 50,000, having grown from less than 400 in 1974, and by 1982 that number had tripled to over 150,000 (Figure 3.1). In 1977, wages for Korean manual laborers in the Middle East, the overwhelming majority of whom worked for Korean companies,[11] were one-fourth of those commanded by Western laborers and one-half of those paid to Arab workers; the wage differentials for skilled workers and managers were even larger. In the early years of the Middle East construction boom, labor constituted 30 percent of total costs for North American- and European-led projects, compared to the 12.4 percent it represented for Korean-led projects (Moon, 1986).

The centrality of this labor-force advantage to South Korea's performance in the Middle East can be seen in what happened to Korean construction as that advantage dissipated. Wage rates in Korea grew rapidly in the 1970s, more than doubling in that decade (Amsden, 1989). By the early 1980s, Korean workers' wages were two or three times those of workers from other developing countries that were sending labor to the region, such as the Philippines, Thailand, and Turkey. These countries started to supply labor

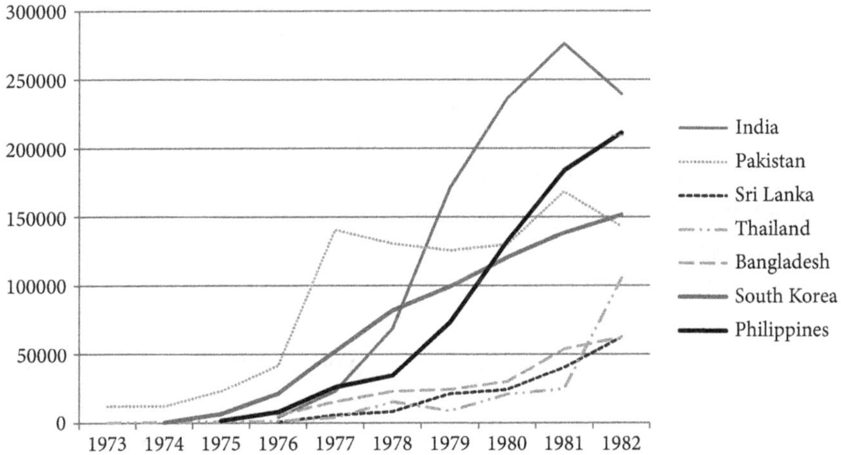

Figure 3.1 Gross Flows of International Contract Labor to the Middle East (from Asia), 1975–1982

Note: The numbers for Pakistan represent total national outflows, rather than outflows specifically to the Middle East. Because 94% of Pakistani migration during this period went to the Middle East, in the absence of exact statistics, I have included these numbers as an approximation that gives a sense of the comparative scale of migration into the region.

Source: Demery 1986.

[11] 98.7 percent in 1980 (Kim, 1982).

to Korean firms, who began abandoning their national labor-recruitment program and transitioning to these cheaper workforces. More importantly, construction firms from these countries began to emulate the successful Korean strategy, leveraging their national labor force to compete in the low-tech end of the Middle Eastern market, often with the help of states that had implemented their own LEMI programs. The pressure on the Korean model was exacerbated by the crash in oil prices in the mid-1980s, with most petroleum-dependent states paring down their construction budgets by shifting their investment priorities away from labor-intensive basic infrastructure and toward technology-intensive industrial infrastructure (Moon, 1986). The overall decline in the international construction market that followed the oil price decline affected all construction-exporting countries,[12] and Korea's construction sector was hit particularly hard, losing not only in overall earning but in international *market position*[13] during this period (Kim, 1988). The international construction industry would recover, but there would be few other places in the world with the kinds of labor shortages that would favor Korea's migrant enclave model to such a high degree (Linder, 1994, pp. 149–150). Many Korean construction firms did not survive the market restructuring. Those that did, such as Hyundai, succeeded because of a long-term diversification strategy that had spun off profits from Middle Eastern construction into various non-construction-related industries (discussed later in this chapter).

The impact of the Middle Eastern construction boom, short as it was, on the Korean economy was enormous. Between 1974 and 1981, Korean firms won over $40 billion in contracts in the Middle East, equivalent to 11.2 percent of Korea's GDP in those years. South Korea's average annual GDP growth rate in those years was 8.5 percent. In the decade of global economic turbulence that was the 1970s, this revenue saved the Park regime's developmental project. The regime's strategy of favoring debt financing over foreign direct investment, while effective for keeping profits and control over development policy in Korean hands, was feasible only when revenues were high enough to stay ahead of debt-service requirements. The debt crisis of the late

[12] Western international construction companies experienced steep declines in revenue. At Bechtel, for example, revenue declined by 24 percent *per year* between 1984 and 1988 before recovering growth again in 1989. ("Bechtel Business Grows to Six-Year High," United Press International (April 3, 1990), available at https://www.upi.com/Archives/1990/04/03/Bechtelbusiness-grows-to-six-year-high/2656639115200/).

[13] From 11 percent of the international construction market in 1981 to less than 6 percent in 1985 (Kim, 1988).

1960s and early 1970s demonstrated that even when growth rates were—relative to the rest of the world—quite high, climbing debt-service ratios could destabilize the economy enough to trigger international intervention and domestic political consequences.

If the Middle Eastern revenues were critical during the decade of the petroleum boom, the structural impact of construction exports on the Korean economy was even greater for the country's long-term developmental prospects. Remember that, throughout the late 1960s and early 1970s, the Park regime had been making investments in heavy industry, creating state-owned enterprises in steel-making and pushing *chaebols* to move into automobile manufacturing and shipbuilding. Some of these projects were just coming online—POSCO began steel production in 1973, and Hyundai's first very large crude carrier (VLCC) ship was completed in 1974—when the oil shocks collapsed demand for these products. At this crucial moment, the Middle East construction exports provided markets for these upstream industrial products. Hyundai's shipyards, which struggled to find markets, were repurposed to produce the steel components for the Jubail Harbor project, which required over 2 million tons of steel and concrete, all sourced in Korea and shipped 7,000 miles for installation (Amsden, 1989; Steers, 1999). At the same time, the remittance consumption of Korean overseas workers was used to buttress the market for domestically produced goods. Migrant workers often used remittances to buy durable consumer goods abroad. In 1980, the Korean state created the "export-price coupons" program, through which migrant workers could purchase Korean-made products while in the Middle East, at an "export" price significantly lower than the domestic retail price. The purchases would be delivered in Korea, which also saved the costs of transport for both the migrant worker and the exporting Korean firm. In this way, Korean migrant workers were leveraged to support Korean capital: Korean overseas workers helped Korean steel and Korean ships find buyers for industrial products and services in international markets while buttressing domestic demand for Korean consumer products (Kim, 1982).

On the firm level, the Middle East period offered opportunities that were at least as transformative as those of the Vietnam War era. If the export of soldiers in the 1960s enabled companies like Hyundai to enter international construction markets, the export of construction workers in the 1970s and early 1980s laid the foundation on which these companies would become globally dominant conglomerates. Hyundai, the leading Korean company in the Middle East, had an annual growth rate of 38 percent throughout the 1970s, becoming not only Korea's biggest business conglomerate but the fourth-largest construction company in the world by the early 1980s (Steers, 1999; Lim, 2003). Such windfalls would have had significant impacts on any company, but in the Korean case the *chaebols'* expansive but tightly coordinated diversified structure, combined with the Park regime's directed subsidies, meant that these funds were funneled into learning trials in product lines far afield from the core Middle Eastern business, in sectors where opportunities for market growth were much greater than in construction and civil engineering (Amsden, 1989).

One such important sector was the automobile industry, which had long been a priority for the Park regime. The industry figured in the regime's first five-year plan in 1962, which included provisions to protect the domestic auto market and promote Korean car production. While a nascent automobile industry did appear in the 1960s, for most of that decade Korean firms had not been able to progress beyond the very bottom of the production ladder—that is, junior partnerships with North American and Japanese companies in which Koreans assembled complete knockdown kits of components imported from abroad (Lansbury et al., 2007). The Korean government made repeated attempts to climb above the bottom rungs, announcing plans to increase the proportion of parts sourced in Korea and eventually to develop Korea's own models for export.[14] Hyundai—whose automobile subsidiary was established in 1967 and by the 1970s had become one of the country's top three auto manufacturers—made promises to the Park regime to deliver on these goals, and began developing its own low-price model in 1973. The first car produced, the original Pony, was good enough to win Hyundai nearly 60 percent of the protected domestic market by 1977. However, the car was rejected in international markets, where there was less patience for the Pony's myriad quality-control issues, including problems with steering and brakes, peeling roofs, and fading paint (Steers, 1999). To develop a car that could compete in export markets, Hyundai made heavy and long-term investments in product development and mass-production capacity. Its motor subsidiary lost so much money throughout the 1970s and into the early 1980s that one Hyundai executive called it "essentially a bankrupt company supported by other Hyundai companies" (Clifford, 1998, p. 257). But this strategy of learning-by-failing did eventually

[14] Long-Term Plan for Development of the Automobile Industry (1973).

pay off, and by the mid-1980s, as the Middle Eastern construction business started to slow, Hyundai had succeeded in gaining a foothold in the international automobile market. In the decade between 1976 and 1985, its exports grew over a hundredfold, from 1,042 units to 120,000, then doubled again to over 260,000 by 1987. Hyundai was selling not only in the developing countries where the first Pony models had been sent but in the major consumer markets of Europe and North America.[15] It would take another decade for Hyundai to consolidate its position in the automobile market; between 1987 and 1991, its auto exports declined by 40 percent, forcing the company to reorganize its production systems and introduce new product lines (Lansbury et al., 2007). Though the company could no longer rely on its Middle Eastern construction profits to shore up its losses in the automobile sector, other parts of its business—seeded and subsidized during its Middle East period—had by this time grown to fill the gap. For example, Hyundai's shipbuilding arm, which had relied on backward linkages to its Middle Eastern construction business to survive the 1970s, had become, by the 1980s, the world's largest shipbuilder, with profits sufficient to support Hyundai's investments in auto and other sectors (Amsden, 1989). With these supports, Hyundai Motor would survive its period of learning-by-failing to become, by 1994, the thirteenth-largest automobile producer in the world (Steers, 1999; Kim, 1997; Lansbury et al., 2007). By the early 2000s, when the Hyundai Group was broken up and the major subsidiary businesses split off into independent companies, Hyundai Motor had grown large enough to hold the title of Korea's fifth-largest *chaebol* on its own (Feenstra & Hamilton, 2006).

What I have just described is a fairly textbook illustration of how the *chaebol's* diversification strategy enabled Korea's industrialization-by-learning. This process has been well examined in the economic-development literature, and it is not the purpose of this book to revise the core of that account, which I find largely convincing (Amsden, 1989). However, what is often elided in these accounts—referred to in passing but rarely given serious consideration—is the role labor export can have in facilitating this process. In the 1960s and 1970s, the most dynamic parts of Hyundai's business, the engine of its growth, came from sectors that relied on exported Korean labor for its market access and competitiveness. Moreover, in its efforts to promote

[15] In 1986 and 1987, Hyundai had the top-selling import car in the US market (Lansbury et al., 2007; Steers, 1999; Kim, 1997).

industrialization, and to support companies like Hyundai in particular, the Korean state implemented policies that promoted contract-based migration and harnessed it to the internationalization goals of domestic industry. This connection between migration and the industrializing firms of the sending country is the central characteristic of the LEMI model, distinguishing it from strategies that focus on the behavior of migrants alone.

From the perspective of the economic-development literature, LEMI should be understand as a variation of late-developing countries' comparative advantage in low-wage labor, applied in circumstances where production must occur offshore. Because capital is generally easier to move than people, production moves to the sites where it can access laborers, and late developers are able to take advantage of their cheap labor without adopting migration-promoting policies. There are certain industries, however, where the sites of production are less interchangeable. Construction is one of those industries: while some prefabrication is possible, some significant portion of the work that goes into "producing" buildings and physical infrastructure must occur on location. In order to compete on the basis of cheap labor in international construction, late developers *must* have a strategy for moving that labor to the sites of production—that is, a labor-export policy. And unlike out-migration policies motivated by efforts to alleviate unemployment or political discontent—which can be agnostic to the specific forms of migration they produce—LEMI strategies are designed to maximize the productivity of workers, in arrangements where migrants are tightly bound to their employers and where settlement is discouraged or altogether blocked.

The claim here is not that labor export is *necessary* for every late-developing country attempting to catch up, but that in certain circumstances it can be a critical policy tool for the midwifing of export industry. There are many examples of successful late developers that have not relied so heavily on a labor-export strategy. A very specific confluence of circumstances—the United States' need for regional partners in the Vietnam War effort, the oil shocks, and the consequent infrastructure boom in a region with relatively low human-capital endowments—made the labor-export strategy viable for South Korea in the mid- to late 20th century. Still, the conditions of that period are not entirely singular: the oil-rich countries of the Middle East continue to import most of their workforce, and in recent decades we have seen yet another infrastructure boom in this region, fueled by yet another rise in the price of crude oil. And where such opportunities exist, it is to be expected that some late developers will seek to take advantage of them. In Amsden's analysis of South Korea's economic development, she finds that the extreme diversification pursued by firms like Hyundai—and championed by the Korean state—was a function of the country's industrialization-by-learning strategy. Because firms from these countries were competing not on the basis of technological invention or innovation but on the cost advantages of cheap labor and managerial technique, there was no reason for them to stay within the relatively narrow product areas prescribed by technological expertise. Attempting to push higher up the ladder in existing industries would require direct competition with firms from advanced industrial economies, in which late developers, particularly in the early stages of their industrialization, were not likely to fare well. The more viable path to growth was through expansion into the labor-intensive end of many different markets (Amsden, 1989, p. 320). Given the opportunities available in the lower technology end of the international construction market in the 1970s, it is unsurprising that a country like Korea would pursue that market and, in so doing, develop complementary LEMI strategies.

Chapter 4
Labor-Export-Mediated Industrialization in the Philippines, 1974–1984

Introduction

Having laid out the basic parameters of the labor-export-mediated industrialization (LEMI) strategy, we now turn to the Philippine case. The current Philippine labor-export system, dominated by flows of service workers, looks very different from the heavy-industry-focused enclave model described in the previous chapter. Yet observers of sending-state migration policy in the 1970s and 1980s would not have seen much difference between the two programs in those early decades. In those years, the Philippine labor-export program was conceived of in LEMI terms, with state migration policy designed to help the construction industry bootstrap its way into international markets, just as South Korean policy was doing. The labor-export aspect of the project was successful on its own terms, resulting in migration into the Middle East at a rate comparable to that of Korea. The Philippine construction industry was even able to convert some of those labor advantages into gains in the Middle Eastern construction market; though they never approached the market dominance of the South Korean firms, for a short window of time Philippine builders ranked among the world's top construction exporters. The problems lay in translating those gains into a basis for sustained national economic growth, as the Korean state had been able to do through construction's linkages to domestic heavy industry.

Beginnings of an LEMI Strategy

In Chapter 2, I noted the many political and economic resemblances between South Korea and the Philippines in the 1960s and 1970s, the period in which the LEMI strategy was most energetically implemented by the South Korean state. These similarities also carried over into their experiences with international labor migration. As in the South Korean case, by the

Temporary Measures. Suzy K. Lee, Oxford University Press. © Oxford University Press (2025).
DOI: 10.1093/9780197790007.003.0004

1960s, before the onset of the state-led labor-export program, the Philippines had already achieved an established flow of migration. In both cases, these older flows were based on links forged through their historical relationships to larger colonial powers, but were much smaller in scale and different in distribution than the migrations that grew after labor-export policies were introduced. In 1972, the year before the Philippine labor-export policy was established, 14,366 Filipino/as lived abroad on state-approved foreign employment contracts (Philippine Department of Labor, 1974). Most of this labor migration was associated with the United States, either directly, or indirectly through its colonies in the Pacific or its areas of military intervention in Southeast Asia. A similar number of individuals left with permanent immigrant visas—again, primarily to the United States. In 1975, 14,492 emigrants registered with the Philippine Department of Labor (Philippine Department of Labor, 1981, Table 12.6). The picture these statistics draw is of a migration typical of countries with very little sending-state intervention, dominated by flows to a receiving country or region with long-established migration-facilitating networks, where permanent settlement is at least as likely as temporary or circular migration.

Both South Korea and the Philippines began to diverge from this pattern in the latter half of the 20th century, though they differed somewhat in the timing of the first major temporary outflows. This difference reflects differences in state policy, with the Philippine state reluctant to engage in an overt labor-export strategy until nearly a decade after the Korean program had demonstrated the gains such a strategy could bring. As I described in previous chapters, South Korea's labor-export program should be dated to the early years of the Park Chung Hee regime, when Park was negotiating with European countries to export Korean workers in the early 1960s and, crucially, leveraging Korea's soldiers during the Vietnam War to gain access to the military procurements and reconstruction markets for Korean industrialists. The Philippine labor-export program began in 1973 with the promotion of construction workers to the Middle East.

It is not impossible that the Philippine state could have similarly employed a labor-export policy as early as the 1960s, as it had access to many of the same opportunities—most importantly, an existing military relationship with the United States—that were exploited by the Park regime. In fact, the Philippine government had been involved in Vietnam since the 1950s, working with the United States on humanitarian and clandestine operations. It was also one of the few Asian countries, along with South Korea and Thailand, that responded to the Johnson administration's call for "more flags" to join the Vietnam intervention in May of 1964. Within months, the Philippines had sent a forty-man team of medical personnel to the region and were engaged in negotiations to send a much larger contingent. However, the Philippine government—under Diosdado Macapagal in 1964–1965 and Ferdinand Marcos from 1966 on—limited its participation in the war to noncombat personnel, primarily engineers and medical teams, and in numbers much smaller than those sent by South Korea. At the height of Philippine involvement, between 1966 and 1968, there were never more than 2,070 Filipino troops in any given year, none of them cleared for combat duty. Compare this to over 44,500 Korean combat troops in the same period (Blackburn, 1994).

There are many factors that might explain why the two countries responded differently to the opportunities presented by the United States' military action in Vietnam, and it is beyond the scope of this book to examine all of the possibilities. We do know that the Vietnam War was unpopular throughout Asia, and accusations that state leaders in both Korea and the Philippines were selling the national military as a mercenary force arose almost as soon as negotiations with the United States were initiated. The crucial difference may have been that Park, though by 1964 the formally elected president of South Korea, was the head of a military junta that had seized control of the country in 1961, whereas the Philippine leadership in the 1960s was still subject to a democratic process, with public debate and meaningful competition at least among the country's economic and political elite. Marcos ran for and won the presidency in 1965 on an explicit critique of Philippine involvement in Vietnam. Once he took office, Marcos took care to insulate the Philippines' military involvement in Vietnam against accusations of mercenarism, laundering the US payments through the Congress-controlled treasury, even when this meant smaller troop deployments and less generous development aid (Blackburn, 1994). And when the extent of US financing became publicized in American Senate hearings, Marcos abruptly announced the withdrawal of all Philippine troops save for a small medical contingent.

While criticism of the Vietnam War in Korea was significant, Park, facing a much-diminished opposition, was better able to control the public discourse. The Park regime kept the details of the Korean involvement in Vietnam, including the number of casualties and the atrocities of which Korean soldiers were accused, away from the Korean public for most of the war; details of the Korean deployment were not released until 1970, and then only because they had been publicized by the US Congress (Baldwin, 1975). Even without such details, and under conditions of significant repression, public opposition to the deployment was so widespread that Park was not able to send any more battalions to Vietnam after the initial contingent. Though the troops that were sent in 1966 remained until the US withdrawal, no *new* regiments were committed after 1966, despite continuing pressure from the Johnson administration (Blackburn, 1994).

Whatever the cause, the Philippines' unwillingness to commit combat troops to the Vietnam War effort made it impossible for the Marcos administration to negotiate for the kinds of development aid and access that the Park regime had won. Philippine officials were well aware that, despite the interests of all parties in denying it, this was essentially an exchange of economic support for Filipino bodies, and they made demands that were nearly identical to the ones made by the Park regime. They negotiated with the United States to finance the entire cost of the deployment, including generous per diems and overseas allowances as well as additional monies for replacement units and military equipment. They also secured support for Philippine agricultural exports, technical assistance, and tens of millions of dollars in loans and grants. However, with such a small number of personnel committed, and none of them actual combat troops, they were not able to bargain for the more impactful industrialization support that Korea received. Citing the Philippines' low level of domestic savings and state managerial capacity, the United States limited support for major industrialization projects to consultation and continued talks, and rebuffed specific requests by Marcos for the kind of access Korea was being given to the special procurements process (Blackburn, 1994; White House, 1966). Though the United States offered to share information about opportunities, it would not agree to give Philippine businesses protection from competition (Glassman & Choi, 2014, p. 1176).

It was not until the early 1970s, as the Vietnam War was drawing to a close, that we can say the Philippine state demonstrated its willingness to export Filipino workers for development gains. During the years of the Vietnam War, changes on a number of fronts made this policy shift possible. For one, the Philippines' economic situation was more desperate than it had been in the decade previous. During the 1960s, the Philippines lost the economic advantage it had had against its regional neighbors, struggling with stagnating growth and debt crises while countries like South Korea and Taiwan caught up with and then surpassed it in per capita income and industrial capacity. Intertwined with the economic difficulties was a political instability that culminated in September of 1972 in Marcos's declaration of martial law. He moved quickly to consolidate his power, dissolving Congress, taking repressive measures against his opposition and the press, and imposing a flurry of major structural reforms that he dubbed the "New Society" program. These political changes significantly reduced the political constraints on a labor-export program, even as it produced a flow of people seeking to flee the repression and instability. Finally, against this backdrop, the Organization of Petroleum Exporting Countries initiated an embargo in October of 1973, causing the price of oil to spike. These shifts, described in detail in Chapter 3's discussion of the expansion of South Korean construction into the Middle East, presented the Philippine economy with a similar array of exigencies and opportunities: the sudden escalation in the need for foreign exchange combined with an equally dramatic increase in the demand for construction expertise and migrant labor.[1]

The specific influence of the oil price shock of 1973—and the acute intensification of demand for foreign exchange that it caused—in pushing the Philippine state toward a labor-export policy is suggested in the timing of the policy's appearance. Development plans promulgated by the Philippine state throughout the 1960s, and even up to the four-year plan adopted in 1973 for the fiscal years 1974 to 1977, made no mention of migration, whether as a tool for economic development or a stopgap solution for unemployment and poverty (Republic of the Philippines, 1966, 1969, 1970, 1971, 1973). Less than a year later, and half a year after the OPEC oil embargo began, labor export was introduced as a major innovation of the new labor code promulgated under Marcos's "New Society" reform program (Presidential Decree No. 442, 1974). Having publicly rejected labor export in the context of the Vietnam War less than a decade prior, the Marcos regime now spoke openly about its intent to pursue a migration strategy. Blas Ople, who served as Marcos's chief labor minister for most of his presidency and the entirety of

[1] It should also be noted that sending construction laborers to the Middle East was, politically speaking, a less contentious undertaking than the deployment of military troops to an American intervention in one of the Philippines' neighboring states, in that it lacked the mercenary, imperialist charge that tainted the Vietnam deployment. And while the increase in the price of oil had made countries like Saudi Arabia suddenly and dramatically wealthier, the region was, by all other measures, still in the developing periphery of the world economy, making it possible to represent the arrangement as something more like an equal partnership. This was how Marcos introduced the idea, a kind of "interdependent development" in which the Philippines provided to the Middle East the human capital crucial for its infrastructure projects, even as it eased the economic crisis caused by the oil shocks (Gonzalez, 1998, p. 34).

the authoritarian period, characterized the regime's new approach to labor export in this way:

> We no longer apologize for the outflow of Filipino labor abroad under such labels as the brain drain. We have decided it in such a manner that it will redound to the national interest. We are scouting aggressively for job markets for excess Filipino skills in many countries of the world. (Ople, speech to the opening session of National Tripartite Conference on the Labor Code, October 23, 1975, reprinted in Ofreneo, 2010)

Despite a not-insignificant history with labor migration—both permanent and contract-based—state involvement in that migration had been negligible up to this point. Emigrants had to interact with the state for the basic documentation necessary for international travel and employment, but there was no bureaucratic apparatus to support or promote migration. The recruiting companies that coordinated much of this migration operated largely without state regulatory oversight. The Labor Code of 1974 changed this dramatically, creating two new government agencies, the Overseas Employment Development Board (OEDB) and the National Seamen Board (NSB), that were intended to take over the recruitment and coordination of overseas employment entirely. Private recruitment was to be prohibited after a short phaseout period of four years. The OEDB was tasked with implementing a "systematic program for overseas employment" of all Filipino workers excepting maritime workers, who were covered by the NSB. The OEDB's mandate included both the improvement of the terms and conditions of overseas employment for Filipino workers, generating foreign exchange through these workers, and promoting skills development that matched Filipino workers with international demand (Labor Code Article 16). The Labor Code also established a system for gathering and disseminating information on conditions in foreign labor markets. It allowed the Secretary of Labor to designate labor attachés working in consular offices, whose duties included—in addition to providing support and counseling for Filipino workers overseas—making "continuous and special researches/studies and recommendations on the various aspects of the employment market within their jurisdictional area" and making such information available to the Department of Labor and the Department of Foreign Affairs for "dissemination to the public" (Presidential Decree No. 442, 1974).

This initial iteration of the labor-export policy does not seem to have been drafted along LEMI terms, but rather represented an attempt by the state to organize and capture existing labor-migration flows and, in so doing, to rationalize and encourage their growth. The explicit abolition of private recruitment, and the absence of any language referencing coordination with the construction industry or any private capital, support this view. However, the development reports from this period indicate that Philippine economic managers were actively studying the development programs of the emerging Asian tiger economies, South Korea among them (see, e.g., Republic of the Philippines, National Economic Council, 1971). They would not have missed the fact that Korea's revenue from construction exports grew tenfold between 1974 and 1976, to $2.5 billion, almost entirely on the explosion of contracts in the Middle East made competitive by Korean migrant labor. The role of the Korean state in supporting and facilitating this expansion was also not a secret; major publications in the construction industry were offering detailed reports by early 1977 (Engineering News-Record, 1977). At the same time, Philippine builders began to pressure the state for support in entering the Middle East construction market.

Beginning in 1977, the Marcos regime reoriented the labor-export program to duplicate some of the strategies successfully implemented in Korea. In June of that year, Marcos signed the Overseas Construction Incentives Decree (Presidential Decree No. 1167), which set up a structure for promoting and supporting Philippine construction exports. The Decree offered a series of tax incentives and explicit priority in hiring Filipino workers. It also created the Philippine Overseas Construction Board (POCB), a new agency tasked with regulating export construction, coordinating among exporting firms, and managing state incentives. All builders operating internationally would henceforth be required to register with the POCB and to obtain POCB approval before making international bids or exporting construction materials and equipment. POCB was also authorized to engage in negotiations with foreign governments on behalf of the construction industry, and—in coordination with other economic and labor agencies including the OEDB, the National Economic Development Authority (NEDA), and the Philippine Export and Foreign Loan Guaranty Corporation—to design and implement additional policy for promoting construction exports. In addition, the Labor Code was amended the following year (Presidential Decree No. 1412, 1978) to again allow private-sector participation in labor export, making it possible for construction firms to directly recruit and hire workers for their Middle East projects. The OEDB's apparatus for marketing and training Filipino workers for overseas employment would also continue to operate. The link between labor exports and the growth of domestic industry was made explicit in the five-year development plan drafted by NEDA and adopted by presidential decree in September of 1977:

> Service export contracts on a package deal basis will be promoted to include exports of manpower, raw materials, and other producer goods. This is expected to further expand foreign exchange earnings and create direct linkage between overseas jobs and domestic industries. (Republic of the Philippines, National Economic and Development Authority, 1977, p. 138)

The Philippine LEMI Strategy in Practice

Disassembled into its component parts, LEMI strategies seek to achieve three discrete objectives: (1) export labor, (2) leverage this labor export to promote and support a critical national industrial sector (here, construction) in international competition, and (3) transfer the gains made by that sector throughout the domestic economy, particularly through backward linkages to heavy and intermediate industry (Figure 4.1). With regard to objectives 1 and 2, the Philippine LEMI program was quite successful. Contract-based migration from the Philippines grew rapidly after the initiation of the labor-export program, and veritably exploded once domestic construction companies began to participate in the labor-export process. Access to Filipino laborers helped Philippine national firms to gain entry into international markets, especially in the Middle East, where working with Philippine builders became the primary path through which foreign firms could hire cheap and skilled Filipino construction workers. Where the program fell short, however, was in the translation of those gains into a larger national industrialization project. Whereas, in the Korean case, construction exports' demand for steel, cement, and the ships to transport them supported the country's push into heavy industry during a period of turbulence in global markets, even as the windfall profits enabled investments in more advanced sectors, the gains from the Philippines' construction exports proved to be short-lived, profiting individual industrialists without setting a foundation for long-term international competitiveness.

Figure 4.1 The Labor-Export Mediated Industrialization Process

The Labor Code of 1974's promotion of contract migration, and the participation of construction exporters in that labor migration beginning in 1977, had a dramatic impact on temporary migration flows from the Philippines. The Philippines did have a history of providing migrant labor prior to 1974, but at a scale much smaller than what it became; in 1971, for example, 14,366 contract migrant workers, or less than 0.03 percent of the population, were deployed (Philippine Department of Labor, 1982), primarily to its neighbors in Southeast Asia and the islands of the Pacific (Philippine Department of Labor, 1974, Table 13.4). This was combined with another 30,000 to 40,000 individuals who left every year with permanent immigration visas (Philippine Department of Labor, 1994, Table 13.15), the vast majority to the United States, the country of which the Philippines had only recently been a formal colony and on which it remained politically, militarily, and economically dependent. The implementation of the labor-export policy precipitated a drastic transformation in the country's migration patterns, from a flow characterized by permanent, settlement-oriented migration toward North America (primarily the United States) to one dominated by temporary, contract-based migrations, which would grow explosively after the mid-1970s (Figure 4.2). Importantly, this growth was driven by diversification, a significant shift away from the United States as a receiver of Filipino/a migrant and toward new markets for international labor, first in the Middle East, then expanding to East Asia in the 1980s, Europe in the 1990s, and Australia/New Zealand and Canada in the first decade of the 2000s. The temporary migration itself also changed its character in this period. In 1975, the first year in which the data were collected, less than 35 percent of overseas workers were placed in land-based occupations, with the vast majority of workers instead finding employment on ships. By 1981, that balance had flipped completely, with over 80 percent

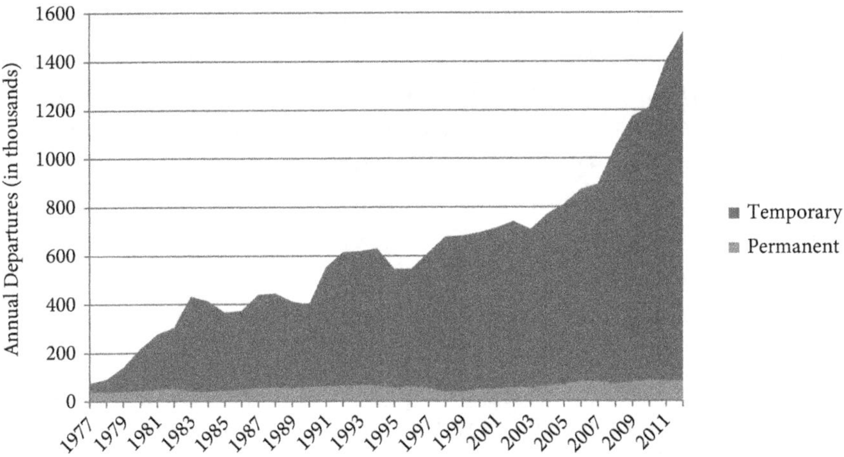

Figure 4.2 Philippine migration, Permanent v. Temporary—1977–2012
Source: Philippine Overseas Employment Administration; Commission on Filipinos Overseas.

of workers finding employment in land-based occupations. It was also during this time that the Middle East became the dominant destination for land-based workers (Philippine Department of Labor, 1981).

While some of this expansion can be attributed to the exploding demand for migrant labor in the Middle East and elsewhere, the data also suggest that the Philippine state's labor-export policies were helping Filipino workers gain an outsize share of these new labor markets. Within five years of entering the labor market of the Middle East, the Philippines had become one of the top Asian labor-sending countries to the region—sending 132,044 workers—surpassed only by India, which, in addition to having a much larger source population, had a history of labor migration to the Middle East that went back to the beginning of the 20th century. It had overtaken Pakistan, a country that, like India, had long provided labor to the Middle East (see Figure 3.1), and was sending more workers than three other top Asian labor-sending countries—Thailand, Sri Lanka, and Bangladesh—combined, even though these countries had initiated labor export to the Middle East at around the same time. By 1981, the Philippines was sending nearly 300,000 workers to the Middle East each year. In 1986, Filipino/as accounted for nearly 5 percent of all foreigners documented entering Saudi Arabia.[2] The only sending country that showed comparable growth was South Korea (Arnold & Shah, 1986).

[2] From the Saudi Arabia Statistical Yearbook 1985–1989. This total includes the 851,761 pilgrims who traveled to Saudi Arabia that year to perform the *hajj*.

Though the policy to promote international construction exports was not initiated until three years after the Labor Code's promulgation, this would become so symbiotically intertwined with labor export that it is difficult to determine which was the driving force in their mutual growth. The changes that came with the 1977 Overseas Construction Incentives Decree set up Philippine construction firms as necessary middlemen for any international firms seeking Filipino workers, who were restricted, on pain of losing their passports, to working for registered Philippine national firms. American and European firms trying to regain market share from Korean firms during the Middle East's infrastructure boom had begun to treat the Philippines as a "happy hunting ground" of migrant workers who had developed a reputation for skill and reliability as well as low cost. The new policies forced international firms to take on Filipino construction companies as partners and subcontractors, giving them crucial entrée into this lucrative market (Engineering News-Record, 1978, pp. 22–26).

At the same time, the coordination and facilitation that Philippine construction firms provided helped drive the demand for Filipino labor. Philippine firms became specialists in solving labor-supply problems, which constituted the value they brought as subcontractors and partners in joint ventures. For example, Rudy Cuenca, president of the Construction and Development Corporation of the Philippines (CDCP), the Philippines' flagship international construction company in the 1970s, won bids for projects by finding niche populations of workers for the specific requirements of the Muslim countries of the Persian Gulf—in one prominent case, by recruiting and training workers from the Muslim Mindanao region of the Philippines for infrastructure projects in Mecca, where non-Muslims are forbidden (Dalisay & Reyes, 2010, p. 81; Engineering News-Record, 1979).

Construction firms were pivotal to ensuring that, once recruited and transported to the Middle East, Filipino workers stayed through the project period and delivered the labor for which they had been contracted. As described in the previous chapter, conditions for migrant workers were so difficult as to make labor recruitment and retention a significant challenge for all construction firms operating in the area. These conditions included desolate worksites lacking basic comforts, where workers often had to build their own living quarters before anything else; physical isolation from the surrounding society; prohibitions against socializing with the local population; and round-the-clock surveillance. Philippine firms, like their Korean counterparts, specialized in the management of workers under these conditions, ensuring that a critical mass of workers stayed through a project and maintained their productivity. Project managers not only oversaw the work during the day but also policed the activities of off-duty workers, confiscating alcohol, breaking up gambling in workers' quarters, and acting as informal counselors (Consunji, 2004, pp. 220–222). The active management of these companies helped to deliver the two seemingly contradictory characteristics that were desirable to the labor-importing countries of the Middle East: that they were able to adapt to the culture shock and the harsh labor conditions, but at the same time were unlikely to try to settle in the Middle East. These qualities, combined with their relatively high skill and education level (facilitated by the state) and low wages[3], made Filipino workers among the most in-demand in the Middle East. The participation of the construction firms also helped the Filipino workforce secure employment at higher rungs on the employment ladder. Not only did the Philippines become the second-largest source of Asian laborers in the Middle East by 1981, it was one of the few countries from which a substantial share of migrant workers worked in the more highly remunerated (and more highly remitting) professional and technical occupations (Arnold & Shah, 1986).[4]

Whatever the workers might have gained from this arrangement, it could be argued that the Philippine construction industry gained more. Labor export boosted the revenue of the construction industry in the Philippines by billions of dollars, and this, combined with ever-increasing remittance flows—which more than tripled from $339 million in 1977 to $1.1 billion in 1983—had countercyclical effects against the economic problems caused by the spikes in oil prices that were financing the construction boom in the Middle East. The construction industry's gross value added (GVA) *quadrupled* between 1976 and 1984, a period in which almost every other major sector of the economy, due to domestic political instability as well as the oil price spikes, experienced decline or stagnation.[5] The link between the construction industry's fortunes in the 1970s and 1980s and its labor export to the Middle East is demonstrated by how dramatically the industry's profits dropped when contract migrations to the region began to fall: the industry's

[3] In 1979, Filipino wage rates in the Middle East for construction trades were 32–66 percent lower than those of Korean workers (Engineering News-Record, 1979, p. 107).

[4] The other country exporting labor with this skills profile to the Middle East was South Korea.

[5] David Consunji, CEO of the conglomerate DCMI, writes of this period: "Business was so bad and credit so tight that there was almost no construction going on. [The Middle East projects] tided us over somehow, providing jobs for our people when there were no projects here" (Consunji, 2004, p. 230).

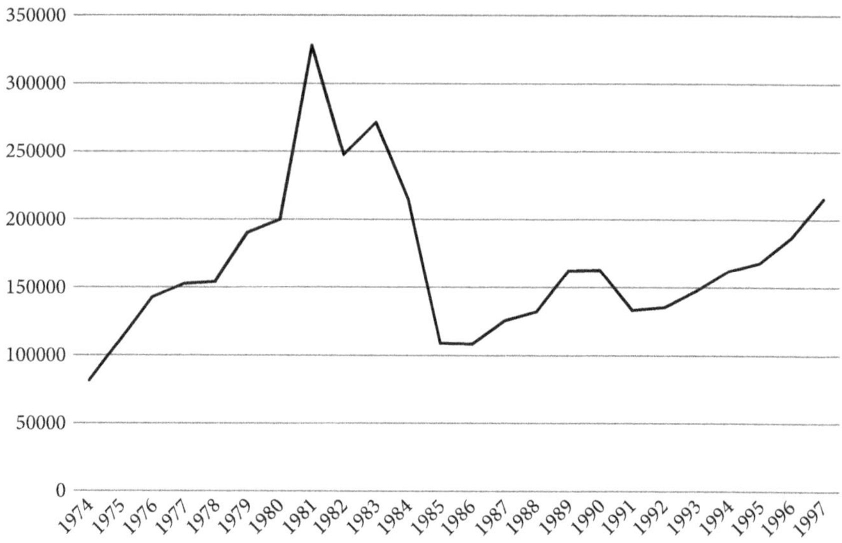

Figure 4.3 Construction Sector Gross Value Added, 1974–1997—in Philippine pesos, millions (at 2000 prices)
Source: Philippine Statistics Authority.

GVA declined by 40 percent in a single year between 1984 and 1985, and would not recover for a decade (Figure 4.3).

Beginning nearly a decade later, the Philippines' LEMI-supported construction industry never dominated in the Middle East the way that the Korean firms did. Still, for having entered the Middle East market several years after Korea, and without the near-decade of experience and expansion that Korean firms had gained during the Vietnam War, the Philippines did relatively well. Though the Philippine construction industry did not win its first contract of any kind in the Middle East until 1976, by the early 1980s only Japan and South Korea, among Asian construction industries, were more successful in the international markets. At least seven Philippine companies did well enough to rank among the leading international construction firms, with at least one—the CDCP—making revenues comparable to those of the major Korean firms (Rimmer, 1984; Engineering News-Record, 1979b, 1980, 1981). Most of their work in the Middle East was as subcontractors and junior partners to larger Western or Middle Eastern firms, but this gave Philippine companies the opportunity to undergo the same kind of learning process that had nurtured Korea's construction industry in the previous decade. David M. Consunji, founder and former CEO of

DMCI, a major Philippine conglomerate with a core construction business, explicitly attributed DMCI's ability to compete in the international market to the learning that was made possible in its period as a labor provider in the Middle East. He writes:

> Working with international construction companies made me see the inadequacies of our company and realize how far off our local standards were from international standards. The international contractors had certain requirements that were totally new to us. . . . We learned how to deal with international clients, and how to negotiate with them. We become more aware of the finer points of drafting contracts. We would submit our work to foreign inspectors and consultants and prove that our work compared favorably to that of international builders. We were also able to test some of our engineering ideas and we gained new technological knowledge from European firms. . . . The lessons we learned from the Middle East became our competitive advantage in our ensuing engagements. (Consunji, 2004, p. 222)

And very quickly Philippine contractors were beginning to show signs of moving beyond labor provision in the international markets. By the late 1970s, Philippine companies were already the leading construction firms in the Southeast Asian market; and even in the Middle East they were beginning to compete for, and sometimes win, contracts as primary contractors or managing partners in joint ventures with Middle Eastern firms. CDCP won its first general contract in the Middle East, worth $300 million, in 1980 (Dalisay & Reyes, 2010, p. 84). Others joined soon after, including DMCI, which formed a joint venture in 1983 with a Saudi Arabian firm, Saulex, and was able to successfully bid for villas, schools, and eventually larger infrastructure projects including irrigation dams and motorways (Consunji, 2004).

We do not know how the Philippine LEMI strategy might have unfolded had these companies continued on this trajectory, because soon after Philippine firms began to make these moves, a confluence of political and economic changes brought the period of intense opportunity in the Middle East to an end. The price of oil began a two-decade decline in 1980, forcing oil-rich countries to adjust their infrastructure development plans. While continuing to invest, they shifted their focus to higher-technology projects in which the labor cost advantage of countries like the Philippines were less impactful (Moon, 1986). In the same year, Iran and Iraq—representing two of the biggest markets for construction in the region—began a war that would last for the better part of a decade. Though Philippine firms would continue to operate in those countries, the danger and chaos of the war raised the costs of production and made it more difficult to secure payment on contracts (Dalisay & Reyes, 2010, pp. 89–90; Steers, 1999, p. 117). Given these exogenous shocks to the LEMI model, it would not be unreasonable to conclude that the Philippines' inability to transfer the gains from international construction to other parts of the domestic economy was an accident of history. However, a closer look at the process through which the LEMI strategy fell apart suggests that, even if the market for labor-dependent construction exports had persisted, there were problems with the Philippines' LEMI strategy that would have made it difficult to deliver the kinds of long-term gains that Korea achieved.

The best illustration of the limits of the LEMI strategy in the Philippines may be the trajectory of the Construction and Development Corporation of the Philippines (CDCP)—the closest that the Philippines came to a flagship international construction export firm like Korea's Hyundai or Samwhan—in the waning years of the Middle East infrastructure boom. After a number of years when the CDCP grew into one of the top construction exporters in the world, the company suddenly collapsed over a two-year period. Between 1980 and 1981, the value of its contracts shrank from $707.4 million to $168.6 million. Unable to meet its debt obligations, the CDCP was bailed out by the Philippine government in 1981, and then fully nationalized in the spring of 1983. Most of its subsidiary businesses and assets were sold off, and the core business was reconstituted as the Philippine National Construction Corporation, a company that operated at a much smaller, domestic scale and existed primarily to maintain and collect revenue from the toll roads it had built in the 1970s, over which it had a long-term franchise.

Given all the changes in the Middle East that I have described above, it is not surprising that a company dependent on that region for most of its revenue, like the CDCP in the late 1970s and early 1980s, would founder. Many Korean firms reported similar troubles. Over half of them had withdrawn from the Middle East entirely by the mid-1980s, often forced out by severe liquidity crises and bankruptcy. The difference was that in the Korean case, enough of the windfall of the boom years had been used to nurture alternative sources of export revenue—ships, automobiles, steel, electronics—that the decline of the Middle East market had little impact on the survival of the country's major conglomerates, or, consequently, the country's overall economic prospects. Like the Korean firms, the CDCP also operated under a conglomerate structure and channeled significant portions of its revenues from the Middle East into other ventures. It likewise spun off subsidiaries in areas directly related to its construction export business—steel prefabrication, cement, heavy-equipment rehabilitation, shipping—along with a number of other activities much further afield, including insurance, hostelry, mining, agriculture, and even automobile engine manufacturing (Dalisay & Reyes, 2010, pp. 75–77; Gonzaga, 1983). However, none of these ventures ever generated the kind of revenue that could replace the Middle East construction business, and some of them—including its forays into copper mining and shipping—proved to be costly mistakes that further weakened the CDCP's position when its construction revenue began to fall (Dalisay & Reyes, 2010, 92–99).

What accounts for this difference? On paper, the CDCP's strategy does not look very different from that of the major Korean conglomerates. The CDCP's non-construction-related investments, while eclectic, were no more so than those of a company like Hyundai, whose wide-ranging diversification has been identified by scholars as a key factor in the success of the Asian late-development model (Amsden, 1989). And even if individual firms like the CDCP failed in their individual diversification strategies, why did their construction exports not contribute to the growth of upstream industries like steel production or shipbuilding, as they had in the Korean case? The different outcomes between these two countries points to the limits of the LEMI strategy: the labor-export-mediated expansion of a sector does not, on its own, induce an industrial transformation or sustained economic growth. When labor-dependent international opportunities exist, they can deliver enormous advantages to a national economic-development program. In the Korean case, it was crucial to the state's ability to maintain its relative autonomy from foreign capital. This autonomy allowed it to pursue its much-vaunted industrialization-by-learning strategy, developing domestic capacity in industries in which it had little experience or competitive advantage. However, in order for a LEMI strategy to have this effect, a country requires a coherent domestic industrial policy that can accommodate the opportunities created by labor export. The Philippine case diverges from the Korean on this point.

Chapter 5
Labor Export, Industrial Policy, and State Capacity

Introduction

This chapter explores how and why the Philippine labor-export-mediated industrialization (LEMI) effort failed to generate the kinds of linkages to the domestic economy that were observed in the Korean case. I argue that the key difference between the two cases was the presence (or absence) of a coherent industrialization policy. In pointing out the differences between the two countries in this area, I am following a long literature on economic development in which these countries have often stood as exemplars for the two poles of the state's potential role in late development. South Korea is seen as a near-ideal developmental state whose interventions in the market helped propel the country into the ranks of the advanced industrial economies within a few decades, whereas the Philippines is viewed as a predatory state that cannibalized its economy and trapped its population in stagnation and poverty (see, e.g., Evans, 1995). While both countries were able to deploy hundreds of thousands of workers in their labor-export drives, and to leverage those workers in order to win opportunities for domestic firms to participate in international markets, only in the Korean case did those opportunities help support industrial transformation and sustained growth, under a coordinated national industrial policy. In the Philippines, without the state's overall coordination and direction, these opportunities were squandered on diversification efforts that not only floundered without adequate support but were sometimes actively undermined by factions within the state. This chapter also further reviews the explanations for differences in these countries' industrial policy, reviewing the economic-development literature on development-state capacity. I explore the two prevailing explanations—bureaucratic capacity and state/society relations—in the context of the LEMI program, and find that the evidence bolsters the argument that the Philippines was unable to reproduce the LEMI strategy of

Temporary Measures. Suzy K. Lee, Oxford University Press. © Oxford University Press (2025).
DOI: 10.1093/9780197790007.003.0005

South Korea because of an unfavorable balance of power between the state and society (in the form of the oligarchic family networks).

In Chapter 3, I described how, in the Korean case, the construction exports interacted with a heavy industrialization drive that began in the 1960s and continued through the labor-export period. As most of the economic-development literature on this period makes clear, that drive was a *state* project, in which the state set the goals and pushed economic elites to participate through a mix of incentives and discipline. Where private industry failed or refused, the Korean state also engaged in direct investment and production. All of Hyundai's diversifications, in those early years, were in response to opportunities created by the Korean state policy. Some accounts, such as the stories told of Hyundai's first efforts to develop its own car for export (rather than serving as the assembler of knockdown kits for American or European companies), suggest that at least some of these moves were not just supported or facilitated but *directly ordered* by the Park regime (Kirk, 1994, p. 136). This state-directed industrialization push created a fertile soil in which the opportunities created by labor export could grow. Backward linkages could be made because the difficult work of establishing heavy industries like steel production was already being done. And once that heavy industry was established, it was within the realm of possibility that Korean firms could, given enough time to fail and learn, produce a national car. Korean firms were able to buy that time with the windfall profits they earned in the Middle East, enabled by Korea's productive, cheap, and mobile labor force.

The Philippines industrial policy during the period of the LEMI strategy, in contrast, was much less primed to produce the kinds of synergies Korea experienced. The starting points for Korea and the Philippines were not very different; in fact, it could even be said that the Philippines in the 1960s, having benefited from over a decade of relative stability and an import substitution industrialization program, had a stronger base from which to develop heavy industries and manufacturing exports. As early as the 1970–1974 Five-Year Development Plan, the Marcos administration was writing about its plans for an integrated steel mill and developing the country's shipbuilding capacity, and similar plans were included in all the development plans published by the Marcos regime throughout the authoritarian period. Most of the ballooning public debt incurred by the Philippines during the Marcos presidency and dictatorship—which grew from $600 million to over $25 billion between 1965 and 1983—was secured on promises of the country's development plans, which included these heavy industrialization projects. However, in actual practice, the state demonstrated a much greater internal ambivalence about both the direction of industrialization and the feasibility of the more capital-intensive projects. There were reports of disagreements among ministers and Marcos, some of which resulted in conflicting public pronouncements about major industrial projects. Very little progress was ever made on these projects, mired down as they were in negotiations with a private sector that was hesitant to make what were considered risky investments (see, e.g., Far Eastern Economic Review, 1980; Gonzaga, 1982). The Marcos development strategy, which did incorporate policies to promote Filipino control of the economy, particularly in those sectors identified as critical to the national project, relied much more heavily on multinational corporations to implement its major industrial projects than Korea had. Thus, the regime was not positioned to make the kinds of demands on its partners in private industry that Park regularly made on Korea's industrialists. For example, while South Korean firms were developing a national car on the orders of Park Chung Hee, the Philippines was struggling to enforce minimal national sourcing requirements in its automobile assembly partnerships with foreign car companies (Gonzaga, 1981). Nor did the state demonstrate any willingness to enter directly into production to meet its industrialization goals when the private sector failed.[1]

The question of why these two states differed so much in their pursuit of industrialization in the later 20th century has been extensively explored in the economic-development literature. The consensus view on this question can essentially be understood as a difference in state "strength." The concept of "strength" can be understood to have two different senses. First,

[1] In multiple development plans, the Marcos government repeated the mantra that the industrialization plan would be led by the private sector, with the public sector taking a supporting role. For example, from the 1971–1974 Four-Year Development Plan:

> The industrial program is merely an indicative plan. As such, it confines itself to spelling out the magnitude and direction that private investment should take to enable the industrial sector to achieve its development objectives, and defining the policy measures the government must initiate to influence the flow of investments into areas which will permit rational and orderly development of the sector . . . the private sector is expected to assume the task of industrial development. The government's role will be limited to improving the investment atmosphere and in helping the promotion of projects which are commercially viable and essential to the industrial targets of the plan. (Republic of the Philippines, 1970, p. 107)

And again in the summary of the 1978–1982 Five-Year Development Plan: "the private sector will continue to be the prime mover of development activities" (xxi).

the state can be "weak" or "strong" in the *relative* sense, depending on its ability to resist the demands made upon it by some powerful social group (e.g., industrial capital or a landed oligarchy). This relative strength is often termed *autonomy*, which theorists of development and the state define as both independence from society—the "ability to formulate and pursue goals that are not simply reflective of the demands or interests of social groups, classes, or society" (Skocpol, 1985, p. 9)[2]—as well as a corporate or institutional coherence, meaning that those society-autonomous goals that the state is able to formulate are *collective* goals rather than goals based on the personal self-interest of officials acting as individuals (Evans, 1995, p. 45). Second, a state can also be "strong" or "weak" in a *substantive* sense, in that it has or lacks the *capacity* to do things that it sets out to do. The issue of capacity is causally linked to autonomy, but the direction of causation is not unambiguous. In one sense, the issue of autonomy *must* precede the capacity. If autonomy, as Skocpol defines it, has to do with the formulation of goals and capacity with the ability to achieve those goals (1985, p. 9), then without autonomy the question of capacity becomes moot.[3] However, a lack of capacity may be not only a result but also a cause of the lack of autonomy: for example, a state without the capacity sufficient to implement tax reform to secure revenue would not, in turn, be able to pay civil servants enough that most would opt out of clientelistic arrangements with extra-state actors. Thus, for the state to establish some autonomy against a powerful sector of society, it requires some measure of capacity. For our purposes, the underlying kernel of capacity, when it is distinguishable from autonomy, is that the term refers specifically to what the state *does* rather than what it *is*. Peter Evans, in his description of an ideal developmental state, adds that this autonomy/capacity must be leavened with a measure of embeddedness—or sufficient connection to allow for communication and repeated negotiations—with social groups that support the project of economic transformation (1995).

However "strong"/"weak" may be defined, most scholars of developmental states agree that South Korea, during the period of its industrialization, had the characteristics of a "strong" state while the Philippines largely lacked them.

[2] Evans uses a similar definition (Evans, 1995, p. 45).

[3] If not moot, it becomes less compelling for most state theorists. If the state's goals are to further enrich the oligarchy, then the Philippine state has demonstrated some capacity, but there seems to be little curiosity about the sources of this type of capacity.

Bureaucratic Capacity

If we define state capacity as a measure of what the state can do, there is danger that analyses of capacity in developmental states may become teleological. We know that the Korean state had the capacity to oversee an industrial transformation because it did so. All of the industrialization policies Park implemented are not only the things that the state did to achieve its goals but also the evidence that it was capable of doing those things. Development scholars have worked around this by focusing their analysis on the structure of the state itself, and attributing successful policy-making to characteristics of the state that can be reasoned to contribute to effectiveness. Perhaps the most influential of this type of analysis is Evans's argument about the origins of autonomy in Weberian "legal-rational bureaucracies." Evans specifically used the Korean case as an example of a state with this kind of bureaucratic structure—a cohesive, coherent, reliable staff created through a history of meritocratic hiring and high-status, long-term career tracks—and argued that this was the key reason that it was able to formulate goals and act independently of the industrialists with which they worked (1995, p. 50). In contrast, the Philippine state, which Evans termed "predatory," was characterized by an absence of Weberian bureaucratic norms. Bureaucratic and administrative positions were often distributed through personal and patron-client networks, and were thus more beholden to those networks than to the bureaucratic hierarchy. Corruption—whether it be petty graft or large-scale looting—was common, and resulted in legal and administrative systems that operated on the basis of personal connections and money politics rather than on the rule of law (Hutchcroft, 1998, pp. 8–9; Abinales & Amoroso, 2005, p. 216).

There has been some pushback against this account of the Korean bureaucracy. Shortly after Evans's analysis was published, the newly industrialized countries of Asia experienced a financial crisis that revealed the vulnerabilities of the region's economic miracle. South Korea was one of the countries that required a bailout from the International Monetary Fund (IMF) to survive the crisis, and this experience sparked a cohort of research that indicated that Korea's vaunted bureaucracy was not quite what Evans had described. The researchers documented levels of rent-seeking and cronyism in South Korea that were not so different from those observed in the Philippines (Khan, 2000; Kang, 2004). They pointed out that South Korean political history had been marked by corruption scandals, beginning with the post-independence regime of Syngman Rhee but continuing throughout the country's economic "miracle." These scandals were glimpses into the political economy of the Korean developmental state, where *chaebols* secured their access to the state's industrial subsidies and market protection by paying for election campaigns and contributing to the slush funds of the military leadership (Kang, 2004; Nam, 1995). Moreover, the bureaucracy through which these exchanges occurred, though it may have been filled through a meritocratic process, was woven through with educational, military, and regional networks which often cut across official channels of promotion or decision-making (Yee, 2000; Kim, 2007).

State/Society Relations

This reassessment of Korean bureaucratic capacity and corruption led subsequent researchers to put much more weight on the question of the state's *relative* strength: not whether the bureaucracy had innate legal-rational characteristics, but whether its relationships with other power centers in society—particularly capital—were such that it could exercise independent control over development policy. There is disagreement about the precise extent of the Korean state's power over its capitalists, but most scholars agree that, in the Korean case, capitalists could not ignore the state or run roughshod over development policy. Kang argues that the Korean state's power came from its control over the financial system. The Korean business model at the time was one in which most firms carried high levels of debt, and the state's ability to withhold or distribute credit gave it leverage over capitalists. This prevented the state from being entirely dependent on the country's flagship *chaebol* conglomerates, whose sheer size made them critical to the national economy. Rent-seeking occurred, but the balance of power created a situation of "mutual hostages" that prevented either side from completely plundering the other and forced a kind of mutually beneficial détente (Kang, 2004). The economist Mushtaq Khan also describes the mechanisms through which rent-seeking in Korea was made to serve the developmental agenda as one that relied on the state's power relative to capital. Unlike Kang, Khan does not describe a situation of mutual hostages, but rather one in which a state with a developmental agenda had the power to both incentivize and discipline capital. Khan also proposes a different explanation for that power—not control of finance, but the historical contingency of Japanese imperialism, which destroyed the landed elite in Korea. Groups of this kind, in many other contexts—such as India and, as we'll see, the Philippines—have acted as an alternative source of power *not* based on economic production, which interrupted and mediated the rent-seeking process between state and capital. Without this noneconomic power base with which they could politically triangulate, Korean capitalists could *only* protect their access to rents through economic performance (Khan, 2000). Vivek Chibber offers yet another reason for the Korean state's ability to direct its capitalist class, based in terms not of power but of interests. Japan is also important for Chibber's explanation, not through its historical imperialism but because of its "flying geese" development model, wherein it handed off the production of consumer goods to its regional neighbors as it moved up the industrial ladder toward the production of capital goods. Korean capitalists required state support to access these opportunities and were thus willing to submit to the state's coordination and discipline (2003).

Chibber's explanation circles back to the question of bureaucratic capacity, which he argues can be built relatively quickly, even absent any sort of bureaucratic tradition, when a capitalist class acquiesces to the leadership of the state. What matters most for this capacity is not a meritocratic structure that prevents corruption and graft at the level of individual bureaucrats, but instead a centralized meso-level structure, at the level of state agencies and ministries, that ensures an economic policy with a coherent and consistent direction. Such centralization of authority has been shown to be critical to successful late industrialization (Johnson, 1982). In the Korean case, the meso-level structures of the developmental state were clearly organized and centralized around a single agency, the Economic Planning Board (EPB), which led the state's industrialization efforts. In the area of economic policy, the EPB served as a "nodal" agency that not only coordinated the activities of other agencies but had the authority to command their compliance with its policy decisions (Chibber, 2003). This type of coherence made it possible to establish a policy direction—such as the pursuit of heavy and chemical industries—and follow through on it.

Whether balance of power or convergence of interests, what is clear is that, in the Philippine case, the state lacked the leverage to prevent economic elites from manipulating development policy for their short-term gains, to the detriment of the country's long-term economic prospects. The links between the political and economic elites in the Philippines are extraordinarily tight, and together they compose a small power elite or oligarchy that controls both the economic and political spheres. This oligarchy has been described as "predatory" for the way in which it uses its control of the state not only to carve out monopolies and make windfall profits but to engage in the active looting of the state's resources (Hutchcroft, 1998). The influence of this oligarchy over the state has meant that, in the economic sectors where they operate, including banking and agriculture, the state can do little to regulate or discipline their behavior, much less direct or coordinate their activities for industrialization or development programs.

The Philippine oligarchy is organized not as corporations or organizations with discrete areas of influence in economic sectors or the state but as competing, cooperating, and intermarrying extended families. The origins of the system lie in the Spanish colonial period, when a number of families still prominent today were established and developed their wealth through their control over land and agricultural exports. The Zóbel de Ayala/Roxas family, founders and controlling shareholders of the Ayala Corporation, are one such family, but others dating from the colonial period remain prominent; these include the Aranetas, who, through the Araneta Group, control interests in real estate and retail; the Aquinos, whose economic base was in sugar and rice; and the Cojuangcos, who built their wealth on sugar plantations and now control the San Miguel Corporation, the country's largest food and beverage company. It is hardly accurate to list these families and their companies as separate, distinct entities, as most of them are connected to each other through both interlocking directorates and the crisscrossing ownership of shares (Krinks, 2002, p. 9), as well as ties of marriage.[4] To anyone with a passing familiarity with Philippine politics, several of these names will be recognizable: for example, Manuel Roxas was the first president of the independent Philippines, and Benigno Aquino, son of former president Corazon Aquino (née Cojuangco), only left office in 2016. The legislature is likewise dominated by members of the landed oligarchy; this is particularly true of the House of Representatives, where the geographic distribution of representation has translated their local economic power as landlords and patrons into political office (Hedman & Sidel, 2000). In 1987, *after* the People Power Revolution that unseated the Marcos dictatorship, 169 of the 200 Congressional Representatives (85 percent) were members, by birth or

[4] For example, the Cojuangco family can be connected by marriage to the Aquinos, and, through the San Miguel Corporation (SMC), with the Zóbel de Ayala/Roxas family, whose members own significant shares in SMC. The Zóbel de Ayala/Roxas family is, in turn, connected by marriage to the Araneta family, as well as the Arroyo family (of former President Gloria Macapagal-Arroyo).

marriage, of the oligarchic families (Krinks, 2002; Gutierrez, 1994). By 2013, that number had only declined slightly to 74 percent (Purdey et al., 2016).

The newer wealth created since the country's independence has largely followed this model—organizing itself into family-run conglomerates with strong links to the state (Rivera, 2003). A part of this newer cohort is composed of Marcos-era cronies; the most well-known among them is Lucio Tan, whose LT Group has interests in tobacco, banking, real estate, and airlines, and who is believed to have used his influence over President Estrada not only to obtain favorable treatment for his businesses but to avoid conviction in a multi-billion-peso tax-evasion case. Another significant section of this new elite, whose wealth has been built in the consumption industries (which rely on migrant remittance spending) has participated directly in the Philippine state. David Consunji, of DMCI Holdings, served as Secretary of the Department of Public Works, Transportation and Communications from 1971 to 1975, during the second term of Marcos's presidency and the early years of martial law, designing much of the regime's early infrastructure investments (Consunji, 2004). Manuel Villar, Jr., of Vista Land and Lifescapes, a major real estate conglomerate, has had a political career running parallel to his career in business. He entered national government in 1992, winning the congressional seat for Las Piñas-Muntinlupa, the area where he had built his first subdivision and where his family sits atop the patronage networks. After rising to the post of House Speaker, Villar was elected to the Senate in 2001, leaving the congressional seat for the Las Piñas district to his wife, Cynthia Villar. In 2010, when Villar left to run for the presidency as the candidate of the Nacionalista Party, the center-right party that held the presidency for most of the 20th century, Cynthia Villar was elected to her husband's Senate seat. The congressional seat was then taken by their son, Mark Villar, who held it until 2016, when he was appointed Secretary of Public Works and Highways by President Rodrigo Duterte. Mark Villar's wife, Emmeline Villar, replaced him as interim congressional representative. This level of dynasticism is not unusual in the Philippines, where it is fairly common for a legislative office to pass among members of the same family over decades. Political parties, apart from some minority parties formed around specific issues, cannot be differentiated on the basis of ideology, but rather represent regional family networks.

Some families of the economic elite, particularly among the ethnic Chinese who built their fortunes on commerce, manufacturing, and trade rather than agriculture, have fewer direct links to the state. John Gokongwei, Henry

Sy, Jr., and George S. K. Ty are some prominent examples. This does not mean that they exert no influence over the state, but only that their influence is transmitted through channels with which liberal democracies are more familiar: money, including campaign contributions, sinecures, and graft (Hedman & Sidel, 2000; Krinks, 2002). The Federation of Chambers of Commerce, the organization of the ethnic Chinese traders, was the biggest source of campaign funds for the 1992 election (Krinks, 2002; de Castro, 1992), and these conglomerates regularly employ or partner with former public officials, whose value is in their access to state power.[5]

Scholars of the Philippine state have described this manipulation of the state by oligarchic networks for their own benefit as a kind of large-scale political looting, or "booty capitalism" (Hutchcroft, 1998). Rents were extracted through the state via multiple channels, including monopolies protected by the state's licensing and regulatory powers, low-cost loans, access to lucrative government contracts, and control over—and sometimes outright expropriation of—large pools of public funds. The profitability of these oligarchic businesses has relied not on their productivity but on their success in capturing these state rents. This contest, in turn, has played out in the electoral arena, where the major parties reflect the regional patronage systems and alliances among powerful families rather than any ideology or vision for the country (McCoy, 2009; de Dios & Williamson, 2015; Hutchcroft, 1991, 1998). This patrimonial contest has been highly destabilizing for the country's economy, producing corruption that has been not only massive in scale but so unpredictable as to hinder the ability of businesses to account for and work around it (Hutchcroft, 2000).

It is beyond the scope of this book to examine why the Philippine state and its capital class have never arrived at the kind of agreement or détente that, in the Korean case, made it possible to effectively implement an industrial policy despite clientelism and rent-seeking in the state apparatus. It suffices to say that *none* of the factors that were identified as possible explanations for the Korean state's relative strength existed in the Philippine case. Temario C. Rivera (2003) and other proponents of the "continuity of elites"

[5] George S. K. Ty has partnered with or employed the following former government officials in various enterprises: former finance secretary Pio Pedrosa, former secretary of agriculture Placido Mapa, Sr., former ambassador Emilio Abello, former Central Bank governor Andres Castillo, and two former presidents of the Philippine National Bank, Edgardo Espiritu and Placido Mapa, Jr. Two additional former high-ranking government officials sit on the board of his bank, Metrobank: former chief justice of the Supreme Court Querube C. Makalintal and former prime minister Cesar E. A. Virata (Rivera, 2003).

thesis have argued that Philippine elite formation never experienced the kind of radical break that occurred in Korea during the colonial period and the Korean War. Instead, efforts at industrialization were controlled by the landlord classes, whose conglomerates diversified into manufacturing even as they retained significant interests in export agriculture. Because of this continuity, there were few cleavages within the economic elite that the Philippine state could exploit to build a developmental alliance. Kang's account, which attributes the Korean state's relative strength to its control over credit and the banking system, finds that the Philippine state had no similar point of leverage against its own economic elites. There is also evidence to suggest that Chibber's explanation for capital's cooperation in the Korean case may apply here. The Philippines was not offered the same export opportunities Japan gave to Korea. The first iteration of the "flying geese" strategy excluded most economies in Southeast Asia (except Singapore and Hong Kong). Japan would not invest heavily in Southeast Asia until the 1980s, by which time the debt crises and instability of the Marcos years had so depressed economic conditions that the Philippines was largely bypassed for countries, like Thailand and Indonesia, that offered better prospects (Bello et al., 2005, pp. 19–22). Without the guarantee of Japan-bestowed export markets, an export-led industrialization strategy was unlikely to be successful. And without a viable alternative source of profit, Philippine capital had no reason to abandon its "booty capitalism" accumulation model and acquiesce to a risky state-directed endeavor at economic transformation.

LEMI's Demise in the Philippines

Whatever the reason, the Philippine state was never able to obtain the cooperation of the country's economic elite for a coherent industrialization project. As a result, there was never any meaningful reform to address the state's bureaucratic weaknesses, despite widespread acknowledgment of corruption and repeated efforts to spark such reform. Marcos himself, who would eventually come to symbolize political looting, came to power with the promise to break the oligarchies' hold on the country's political economy. The martial law-era "New Society" program was initially seen by many to be a credible effort at reform, and his first acts under martial law were to announce a reorganization of the state which sought to address the bureaucratic issues on both the micro- and meso-levels. To address the issues of bureaucratic culture, he streamlined the process for enforcing the state's anti-corruption provisions and purged nearly 8,000 government personnel (Presidential Decree No. 6, 1972a; Presidential Decree No. 46, 1972b; Moratalla, 1999). On the meso-level, Marcos announced a plan to reorganize the economic-development apparatus, creating the National Economic Development Authority (NEDA), in which planning authority would be centralized (Presidential Decree No. 1, 1971; Republic of the Philippines, 1972). Prior to this, planning had been dispersed among a number of organs, including the National Economic Council and the Presidential Economic Staff, which published their own separate multi-year development plans. The Marcos regime's declared intent was to emulate South Korea and the other newly industrialized countries (NICs) of Asia: industrialization through a "labor-intensive export-oriented" (LEIO) strategy, with heavy investments in infrastructure and the creation of special export processing zones (EPZs) that would attract foreign investment through a combination of monetary incentives and labor controls (Ofreneo, 2015; McKay, 2006, pp. 55–56).

These reforms, however, never fully succeeded in restructuring the Philippine state around a nodal developmental agency. Even as NEDA was given the authority to plan for the national economy, this authority was undermined by a regionalized structure that dispersed that authority among twelve administrative regions, each overseen by a Regional Development Council (RDC). These RDCs were the agencies that did the actual work of research, planning, and implementation. Even though, after 1972, it was understood that the regions would operate under guidelines set by NEDA, NEDA had little authority to direct or discipline the work of the RDCs, but instead operated largely as an advisory body (Republic of the Philippines, National Economic and Development Authority, 1973; Krinks, 2002, p. 41). This regionalized structure gave an outsize influence over development policy to the landed oligarchy, who populated or otherwise controlled local governments through their patronage networks. Within this regionalized system, then, development policy was just another arena in which the oligarchic competition played out. Though Marcos did seize the properties of some prominent landed families, these seizures were used to punish Marcos's rivals and reward his close supporters, not to restructure the system of oligarchic competition entirely or to force the RDCs to follow a national developmental plan (Hutchcroft, 1991; Krinks, 2002, p. 41). Without a centralized, coordinated system, the opportunities that Philippine firms were finding in the Middle East fell into a void. Philippine firms would have had to negotiate this complex of regional and national agencies to find the linkages to domestic industry and state subsidies that were more or less directly handed to Korean firms like Hyundai.

This is not to say that the Marcos regime had no role in the direction of domestic investments. Marcos had very close relationships with a cohort of Philippine industrialists and financiers, the "cronies" who were dependent on his regime and the economic opportunities it dispersed (Hutchcroft, 1991).[6] Reports of these relationships suggest that investment decisions were made in response to direction and incentives from the Marcos regime, but given the chaos of the overall industrial policy, these moves rarely produced the synergies that were seen in the Korean case. For example, Rudy Cuenca, head of the CDCP (Construction and Development Corporation of the Philippines) and well-known as a member of the Marcos crony group, reported that the company's moves in the Middle East, including its initial forays into subcontracting and its later independent bids, were made at the behest of the Marcos regime (Dalisay & Reyes, 2010). Even before the Middle East expansion, the CDCP had pushed into new types of construction—from hotels to bridges and other major infrastructural projects—as a result of the Marcos regime's domestic infrastructure program, for which it served as the government's preferred private partner. These domestic projects in the 1960s and early 1970s helped the company to achieve the scale necessary to later compete with Korean firms in the Middle East; even throughout the late 1970s and early 1980s, nearly half of the CDCP's revenue was generated from domestic contracts.

However, its very closeness to the state's erratic and ineffective industrial policy would eventually become the CDCP's undoing. Having entered, at Marcos's behest, into a major contract in Iraq on the eve of that country's war with Iran, the company found itself unable to weather the volatility—particularly the rapid rise in the costs of operations—caused by the military conflict. The CDCP's contracts with the state, rather than serving as an alternative source of revenue to buttress the company's losses in the Middle East,

[6] Marcos famously had a weekly golf date at the presidential palace complex with a group that constituted the core of what were labeled in the press as "Marcos cronies."

proved to be an additional drain. As Cuenca began to decline relative to his rivals in Marcos's circle, the company increasingly experienced problems on the domestic front, not only in winning construction contracts but also in securing payment on existing contracts. One particularly costly example was a 3,000-hectare land-reclamation project in Manila Bay. For this project, the CDCP had agreed to provide the upfront financing in exchange for the right to develop and sell half of the land that was reclaimed—approximately 1,500 hectares of what was expected to become prime real estate in the country's political and economic capital (Dalisay & Reyes, 2010). Once the project was completed, however, the land became entangled in the Marcos regime's internecine conflicts, with the result that the Ministry of Human Settlements—helmed by Imelda Marcos—withheld authorization for the CDCP's sale of its stake and she herself instead took significant portions of the reclaimed real estate for her own projects. The company also faced pressure from the Minister of Commerce and Industry, Roberto Ongpin, whose brother led the CDCP's competitor, the Benguet Corporation. Cuenca reports Ongpin's continued refusal to order state agencies to fulfill its payment obligations to the CDCP for multiple state projects. Ongpin would only act on these payments after the CDCP had been taken over by the state, with Cuenca removed as chairman and Ongpin himself named as his replacement (Dalisay & Reyes, 2010).

The CDCP's collapse in 1983 marked the beginning of the end of the LEMI strategy. No other Philippine company was large enough to make a significant play in the Middle East market, which, in any case, had begun to move away from the labor-intensive low-tech construction in which developing countries like Korea and the Philippines could compete. By 1984, the problems with the international construction industry were so severe that the Philippine state shifted its approach, essentially abandoning the LEMI strategy and encouraging contractors to focus their efforts in the Middle East on labor export exclusively (International Labour Organization, 1984, p. 154). Though it did not deliver the industrialization that South Korea experienced, the LEMI period was not a total loss for the Philippines. Enormous amounts of revenue were generated, and some of the companies that first went to the Middle East under LEMI programs, such as EEI Corporation, have continued to have a presence there. The Philippine state's infrastructure programs built roads and bridges that changed the face of the country and continue to be used today. It was also through LEMI strategies that the modern Philippine patterns of temporary contract migration and the state institutions that facilitate them were established. What did not occur in this period, however, was the kind of industrial transformation that, in Korea, would make labor export unnecessary within two decades. What the comparison between the two countries makes clear is that, absent a functioning, robust industrial policy, even a labor-export policy designed to support domestic firms cannot generate this kind of transformation. Where such a policy exists, however, LEMI policies can play a crucial role.

Chapter 6

Remittance-Driven Growth and the Continuation of Labor Export in the Philippines, 1984–Present

Introduction

If we were to close our analysis in the 1980s, as the Philippines and South Korea were pulling back from their labor-export-mediated industrialization (LEMI) strategies, the Philippines would remain simply the negative case in which migration was not associated with economic development. It demonstrates that even if a labor-export policy is able to generate international opportunities for a sending country's capitalists, these will be, at best, short-term gains absent an industrial policy that connects these opportunities to a coherent national development strategy. With the failure of LEMI, the Philippines looked like many of the other countries that had become dependent on migration and remittances in the late 20th century, where remittances might benefit individual recipient families but were found to distort local economies by raising property prices, reducing local employment, and enabling consumption patterns that could not be sustained without a continuing stream of migration and remittances (Oberai & Singh, 1980; Reichert, 1981; Wiest, 1984).

The Philippine case has more to tell us about the relationship between migration and development, because its program of labor export did not end in the 1980s but instead has continued to the present day. Moreover, changes in the country's economic performance, most evident in the 21st century but visible even in the 1990s, have made the Philippines a positive case for a different kind of relationship between migration, remittances, and economic development. It happened that the Philippine transformation dovetailed with, and provided support for, a new optimism about how remittance spending could catalyze economic growth. This move reflected the shift in international development policy away from its focus on industrialization

Temporary Measures. Suzy K. Lee, Oxford University Press. © Oxford University Press (2025).
DOI: 10.1093/9780197790007.003.0006

with high levels of state intervention and toward what became known as the Washington Consensus or neoliberal model, which repudiated state-led industrial transformation for the growth promised by free-market competition. Within this context, where the goal was not necessarily structural transformation (Atolia et al., 2018) but rather growth through trade, the effects of remittances on local communities were evaluated on the basis of their association with marginal growth, regardless of their impacts on factors such as productive capacity, employment, or long-run sustainability.[1] In practice, it has been difficult to prove that this theory of remittance-driven growth actually works, some modern research finding positive (though small) effects and others finding no impact at all (Fayissa & Nsiah, 2010; Meyer & Shera, 2017; Barajas et al., 2009). Certainly, remittances are not generating the kinds of "miraculous" growth that transformed the late-developing countries of the 20th century. Instead, typical remittance-dependent countries tend to resemble El Salvador, Haiti, or Nepal, where remittances account for 15–25 percent of GDP, and growth rates, hovering around 1–2 percent, can almost entirely be accounted for in the growth of remittances themselves.[2]

Within this context, the Philippines has emerged as a surprising, positive outlier. While its labor-export and remittance rates remain some of the highest in the world, the Philippines has also become one of the fastest-growing economies in the Asian region, with numbers comparable to China and Vietnam. More importantly, the modern Philippine economy has managed to avoid the wild swings it experienced in the late 20th century, managing growth rates that have been steady enough to begin to repair its reputation as Asia's economic "stray cat."[3] The Philippines has managed this without the kinds of industrial transformation that propelled growth in China, Vietnam, or any of the previous "Asian tigers." In fact, the modern Philippine economy

[1] It should also be noted that the disciplines taking up this question had changed—from anthropologists and sociologist to development economists, which, in turn changed methodological approaches.

[2] Based on World Bank data from 2010–2019. Note that remittances are not included in GDP calculations, but would, ostensibly, contribute to future GDP as households receiving remittances spent them in the economy. In Nepal, remittances, which account for around 24 percent of GDP, grew at a rate of 18 percent per year from 2010 to 2019. If we assume that most of these remittances went to households to support consumption, this represents an additional 4 percent in overall household income each year. Overall GDP in Nepal did not grow at anywhere near this rate, but was closer to 1 percent per year, suggesting that remittance spending had very little positive effect on growth. The numbers in El Salvador and Haiti are less extreme but similar.

[3] See, e.g., "Japan Inc. sees PH among most promising investment destinations" (https://www.theworldfolio.com/news/japan-inc-sees-ph-am/4297/).

looks more similar to the prematurely deindustrialized[4] countries of Latin America, where decades of neoliberal policies have produced low-income economies dominated by services and other low-growth sectors (Rodrik, 2016; Ravago et al., 2019; Vu, 2022).

In this chapter and the next, I will examine this new turn in the Philippine economy and demonstrate how the labor-export policies that failed to generate growth during the LEMI era have been crucial to the country's economy in recent years. I will argue that, in the modern Philippines, migrant remittances have been the engine of growth that scholars and policymakers posited they could be, but only because historical contingencies, combined with state policy, have channeled remittance consumption into the dominant sectors of the deindustrialized economy. This new approach to connecting labor export to economic growth I call *remittance-driven growth,* or RDG, distinguishing it from LEMI because of the way in which it orientates labor-export policy (unlike labor-export-dependent businesses like construction) toward individual migrants, as well as its relative agnosticism toward industrialization as the growth strategy. Chapter 6 describes how the RDG model operates in the Philippines, mapping out how remittances are pulled into sectors like residential real estate, retail trade, and banking, and in turn help draw consumers and investors into these sectors. Chapter 7 discusses the reasons why the RDG model has fared better than the LEMI model of the 20th century. Here I argue that RDG has been able to avoid the most corrosive effects of the Philippines' "booty capitalism" because of the way in which remittances enter the economy, in the form of millions of small sums directed toward households rather than as windfall profits collected via monopolized markets or large pools of funds controlled by the state. Because the actual "business" of labor export—the recruitment and supplying of manpower—constitutes so little of the revenue generated by labor export, the incentive to intervene in labor-export policy to capture this revenue does not exist. Instead, oligarchic conglomerates must compete for remittance consumption by producing goods and services.

[4] The term "premature deindustrialization" describes low- and middle-income countries in which the industrial sector's share of the economy, relative to the service sector, is at levels similar to those of the post-industrial advanced economies of the world. Dani Rodrik, the development economist who identified the phenomenon, argues that deindustrialization should be considered premature in these cases, because these poorer countries never experienced the period of manufacturing-driven growth that enriched the advanced economies before their pivot away from manufacturing (Rodrik, 2016).

A Neoliberal Transformation

When scholars and policymakers discuss the Philippine economy's evolution in the second half of the 20th century, it is to describe what is largely seen as a failure. While its neighbors in East and Southeast Asia, one after another, underwent industrial transformation and "liftoff" into sustained growth, the Philippines lagged behind, hobbled by corruption scandals, political crises, and external debt. This account is undoubtedly correct; however, in most discussion of the Philippine case, the focus is on what types of change *did not* occur—namely, the development of an export-oriented manufacturing and industrial sector, capable of making backward and forward linkages within the economy. Much of the focus has been on the cronyism of the Marcos era and the long-term consequences of the debt incurred during this period. In order to understand how labor export has impacted the Philippine economy, however, it is necessary to delineate what kinds of change *did* occur in this period. While the Philippines may not have managed the transformations of the Asian Tiger economies, it is not the case that the failed development projects resulted in stagnation. These changes, delimited by the problems above, resulted in strategies of capital accumulation that focused on the consumption sector—retail, real-estate, and construction in particular—and produced an economy that has grown, while not at the fantastic rates experienced historically by some of its neighbors, an average of 5.5 percent a year since 2003.

The Philippine Economy at the Beginning of Labor Export

This modern economy is quite different from what existed when the labor-export program began in the 1970s. At the time, agriculture and the export of natural resources still played a dominant role in the economy, though their importance had been reduced by two decades of import-substitution industrialization (ISI). The ISI period saw a significant restructuring of the Philippine economy, with the emergence of a manufacturing sector—composed largely of the established agrarian elite families diversifying their holding—that quickly grew to service the protected domestic market. After the first decade (1950s), the Philippine ISI program struggled to deepen the nation's industrial base beyond the finishing and assembly of consumer products, with little to no vertical development in the production of intermediate goods or machinery. The 1960s were marked by economic crises typical of ISI regimes in the mid-20th century—primarily in the form of balance-of-payments crises caused by the continual need to import capital goods—as well as an instability in food prices as shifts in investment toward manufacturing affected productivity in the agricultural sectors. By the 1970s, the agricultural sector was experiencing a resurgence—a devaluation of the peso and the relaxation of trade controls had led to a return of investment in export crops—while the manufacturing sector, still larger than it had been before the ISI period, had stagnated, propped up by tariff protection but unable to claim a greater share of GDP or the labor force (Bello et al., 1982; Hutchcroft, 1989; Krinks, 2002). When Ferdinand Marcos imposed martial law in 1972, the Philippines had been moving away from import-substitution toward an export-oriented model for several years, though most of what was being exported were agricultural products and raw materials rather than products of manufacturing or industry (Gonzalez, 1998, p. 60). Export of manufactured goods did increase in the early years of martial law, becoming the biggest single source of export earnings by 1975 and propelling the country to annual growth rates of 6 percent. However, as in the ISI period, the manufacturing that was occurring in the Philippines was light, not much more than final assembly, and the state was, again, unable to leverage its partnerships with the transnational corporations operating the export processing zones (EPZs) to gain the technologies and experience that could lead to industrial deepening. The patrimonialism of the state never fundamentally changed. Though Marcos did seize the properties of some prominent families, this did little to restructure the relationship between state and economic elites, as the seized properties were primarily used to reward Marcos's close supporters (Krinks, 2002, p. 41), and thus was never able to impose the kind of discipline necessary to spur domestic industry to become competitive at an international level. Moreover, because most of the investment, both in public infrastructure and private manufacturing, was funded by "petro-dollar" debt, the export-oriented program eventually collapsed in another debt crisis. The economic instability eventually induced the Philippines' economic elites to collaborate with the mass movement to remove Marcos and restore formal democracy to the country in 1986.

The fallout from the Marcos period would continue to hamstring subsequent governments, which committed to servicing the debt and accepted the structural-adjustment conditions imposed by the World Bank and the International Monetary Fund (IMF). These conditions included trade liberalization and privatization, essentially cutting off the state-led industrialization path of the East Asian newly industrializing countries (NICs). More importantly, debt payments bled the country of the capital it needed to invest in industrialization; in the first years after democratization, the Philippines paid $3.5 billion a year, or approximately 10 percent of the country's GDP, to service its foreign debt (Bello et al., 2005, p. 14). The costs to the Philippine economy were compounded by loss in foreign direct investment from Japan and the East Asian tiger economies, which were heavily investing in Southeast Asia in the late 1980s and early 1990s. These investments triggered an economic boom in Southeast Asia that helped countries like Thailand, Malaysia, and Indonesia move onto a path of industrialization and growth. Until 1984, for example, Thailand's GDP per capita had been smaller than that of the Philippines, but it would grow to double the Philippines' GDP per capita in less than a decade. Bello, Docena, Guzman, and Malig argue that, while the economic dogma of the time attributed these differences to better management on the part of the "tiger cub" economies, the Asian crisis in the late 1990s demonstrated that corruption was no less pervasive in these countries than in the Philippines. The crucial difference, they argue, was in the scale of foreign direct investment: while Thailand received $24 billion in investments from Japan, Taiwan, and Hong Kong between 1987 and 1991, the Philippines received only $1.6 billion in the same period. This difference, in turn, was the result of the Philippines commitment to debt repayment, which depressed growth and rendered the country unattractive to foreign investors (Bello et al., 2005, pp. 19–22).

The Road to a Consumption-Driven Economy

With the industrialization path to accumulation essentially cut off by international creditors, Philippine capital found an alternative path to growth, a path that led through deagriculturalization and industrial stagnation, toward an economy built on real estate, retail, and services, sustained by high rates of household consumption. Household consumption as a share of GDP in the Philippines was at 75 percent in 2015, over 20 percentage points higher than in Thailand or Indonesia, its neighbors in Southeast Asia (47 percent and 55 percent, respectively), and over twice as high as the consumption share of GDP in China (37 percent) (Figure 6.1). This number is also higher than in the United States or the United Kingdom, wealthy, advanced industrialized countries whose consumption levels are facilitated by high wage levels and expansive consumer credit (Figure 6.2). The role of consumption in the Philippine economy becomes even more apparent if we compare the Philippine trajectory of household consumption since the 1970s with these countries. These Asian NICs, even those that began the 1970s with higher rates of consumption than the Philippines, such as India, Indonesia, and Korea, have seen a steep and generally steady decline in the consumption share of the economy. These declines reflect the industrialization and growth that was occurring in these countries, both the diversion of national income to investment and the return on those investments in the form of higher incomes, which made possible higher rates of savings without sacrificing absolute consumption. That Philippine consumption does not follow this trajectory simply reflects, to some degree, the reality we know—that the Philippines failed to industrialize along with its neighbors. Under this interpretation, the relative share of consumption in the economy could be high, even when, in absolute terms, Filipino households consume less, because of stagnation or contraction in the economy.

However, this is not how observers of the Philippines interpret these consumption rates. They describe a country whose economy is "consumer-driven" (see, e.g., Lopez, 2016; Cahiles-Magkilat, 2017), in which the metropolitan centers—where population is becoming increasingly concentrated—are dominated by megamalls that have become public spaces and thoroughfares, surrounded by new housing developments

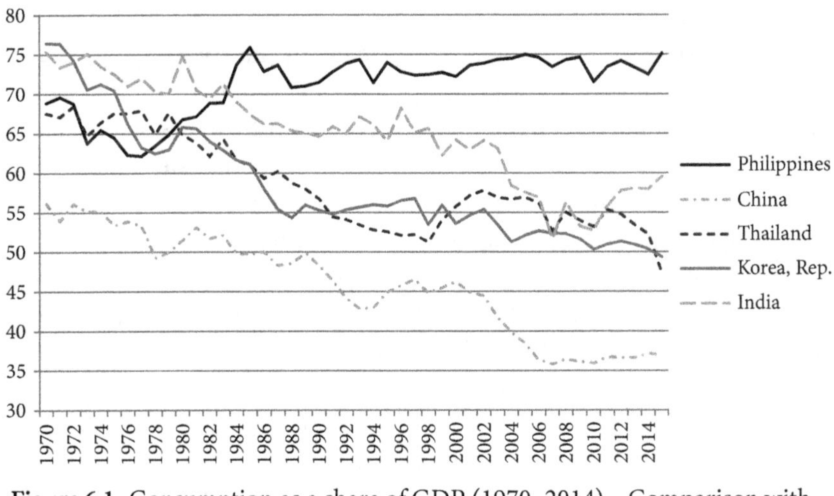

Figure 6.1 Consumption as a share of GDP (1970–2014)—Comparison with selected Asian economies

Data: World Bank World Development Indicators.

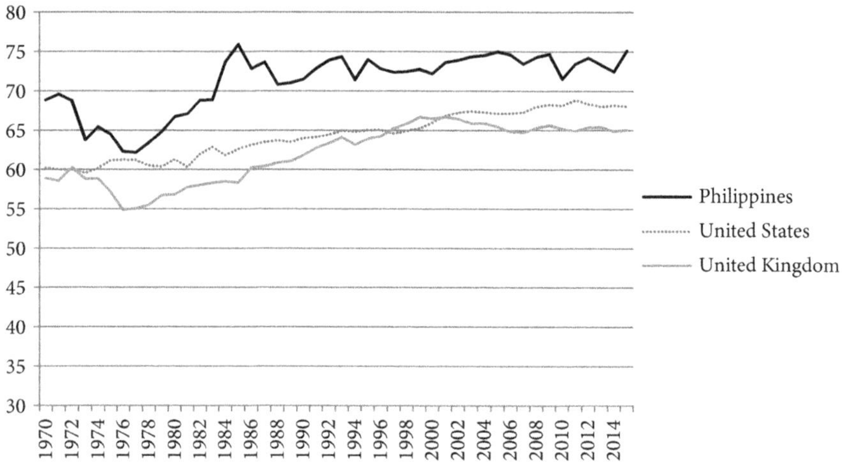

Figure 6.2 Consumption as a share of GDP (1970–2012)—Comparison with United States and the United Kingdom
Data: World Bank World Development Indicators.

targeting a middle class with tastes and expectations comparable to those in Europe or the United States. They point to the country's most recently minted billionaires, most of whom have built their fortunes in retail and real estate, not industrial production or export. The economic statistics likewise support this interpretation. As mentioned above, the Philippine economy has grown at a fairly high rate—over 5 percent a year—since 2003, but this growth occurred without a significant decline in the consumption share of GDP, either in the years leading up to the period of growth or after it had been established (Figure 6.1). A comparison of the gross-value-added share of GDP among the major economic sectors demonstrates gains in the sectors associated with consumption—real estate, retail, and services—at the expense of manufacturing and agriculture (Figure 6.3).

While most of the focus in the literature on policy and economic development has been on deindustrialization, the decline in agriculture was actually much steeper. In the late 20th century, the Philippines transitioned from being a country whose economy was largely built on agricultural export to being a net agricultural importer, needing to import staples like rice even after having adopted Green Revolution technologies (Ofreneo, 2015). While most countries in Southeast Asia experienced some decline in the importance of agriculture to their economies, most of these countries experienced

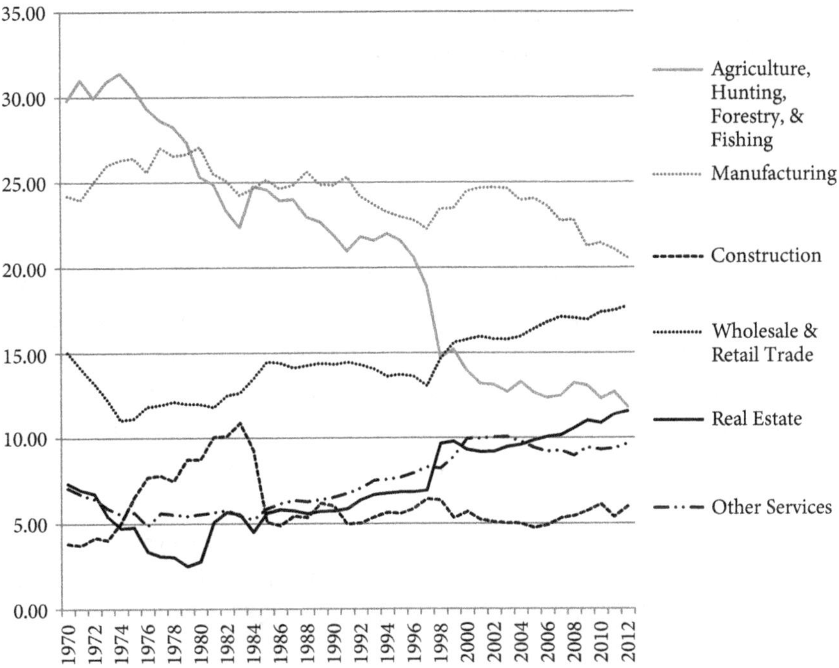

Figure 6.3 Gross Value Added of Major Economic Sectors, as Share of GDP—1970–2012
Data: Philippine Statistics Authority.

growth in the manufacturing sector, which suggests some degree of industrialization. The same is true for the Philippines throughout the 1970s, the end of the ISI and the beginning of the LIEO period; most other sectors of the economy, including retail, real estate, and services, in addition to agriculture, gave ground to the growing manufacturing sector. A similar trend can be seen in household consumption, whose share of the economy declined by nearly 8 percent between 1971 and 1977 (Figure 6.1). All of these trends would reverse in the late 1970s and early 1980s as the impact of the debt crisis and structural-adjustment policies forced an end to the industrialization path. The de-agriculturalization, however, continued apace, furthered by an agrarian-reform program that excepted agricultural lands converted to commercial or residential use (Republic Act No. 6657, 1988; Bello et al., 2005, pp. 52–53). Particularly for those agrarian elites whose holdings are near major metropolitan centers like Manila or Cebu, conversions meant transformation of agricultural land into suburban centers, with residential developments and retail spaces to serve the inhabitants.

Migration and Consumption

The conversion of these lands for residential developments, and the "consumption-driven" nature of the modern Philippine economy more generally, could not be possible without a sustained high level of domestic consumption. Higher levels of consumption are usually associated with countries that have achieved some measure of industrialization, as the level of wages required to support such consumption are usually impossible to achieve on an economy-wide scale without a significant manufacturing base (Usui, 2012, 2; Rodrik, 2007). In the Philippines, however, manufacturing had never been a major source of employment (Figure 6.4); even during the period of "labor-intensive, export-oriented" industrialization of the early 1970s, when manufacturing output represented 25 percent of the economy, the sector never employed more than 10 percent of the labor force. Unemployment and underemployment, more generally, remained high throughout this period and even through the period of stable GDP growth beginning in the early 2000s (Figure 6.5).

Yet household consumption expenditure did grow over this period by over $400 in real terms between 1983 and 2003, even as per capita GDP remained essentially unchanged (Table 6.1). Between 1984 and 1985, the country experienced a nearly 20 percent drop in GDP per capita, which it would not recover for two decades. During this time, however, household consumption grew steadily, only declining slightly by 4 percent during the most extreme downswing, and appearing largely impervious to smaller declines in GDP per capita over this time (Figure 6.6). This suggests that, by the 1980s, consumption had become uncoupled from the performance of the domestic economy to some degree. Because consumer credit was largely unavailable in the Philippines at that time, and the country's extreme inequality—49 percent of the population lived in poverty in 1985—made it unlikely that average households were able to draw on savings to support consumption during economic downturns, we must consider migrant remittances—which began to grow at precisely this period—as a significant source of income to sustain consumption.

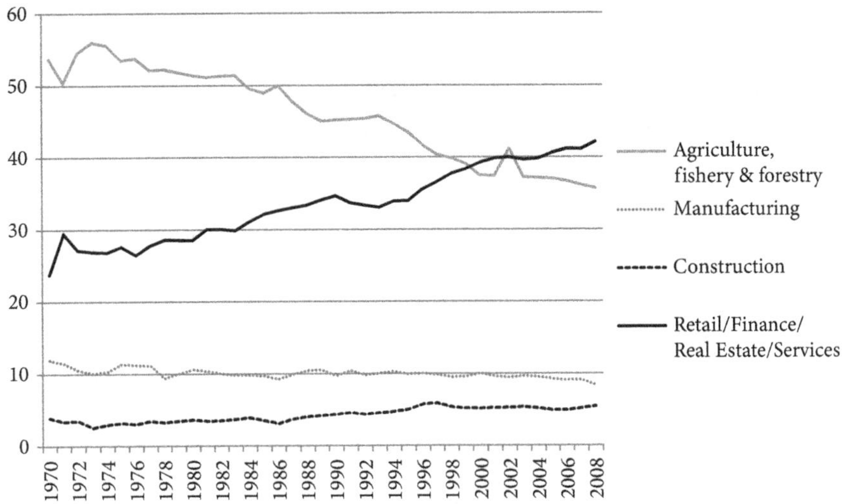

Figure 6.4 Employment share of key sectors, 1970–2008
Data: Philippine Statistics Authority.

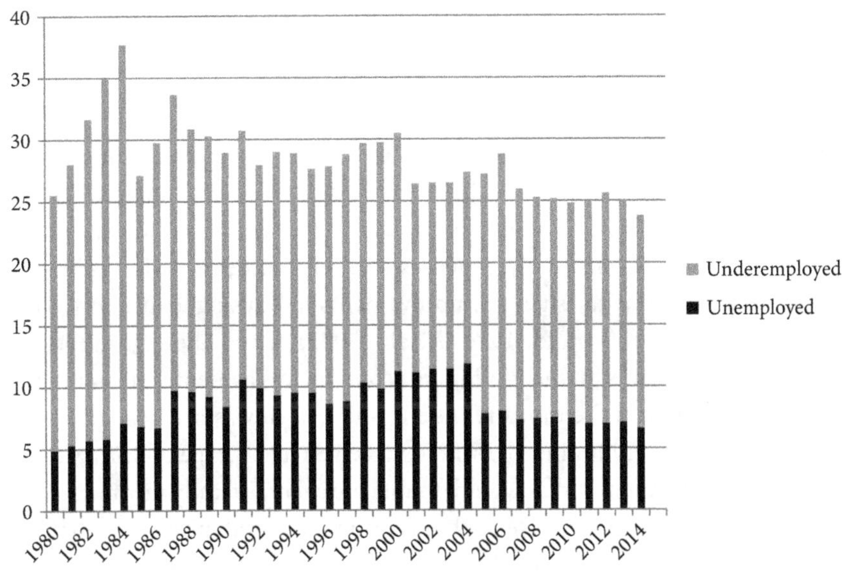

Figure 6.5 Unemployment and Underemployment, 1980–2014
Data: Philippine Yearbook of Labor Statistics.

The research on migrant remittances and economic development has shown that, in most sending countries, remittances serve precisely this function: the maintenance of current consumption. Labor migrants are typically

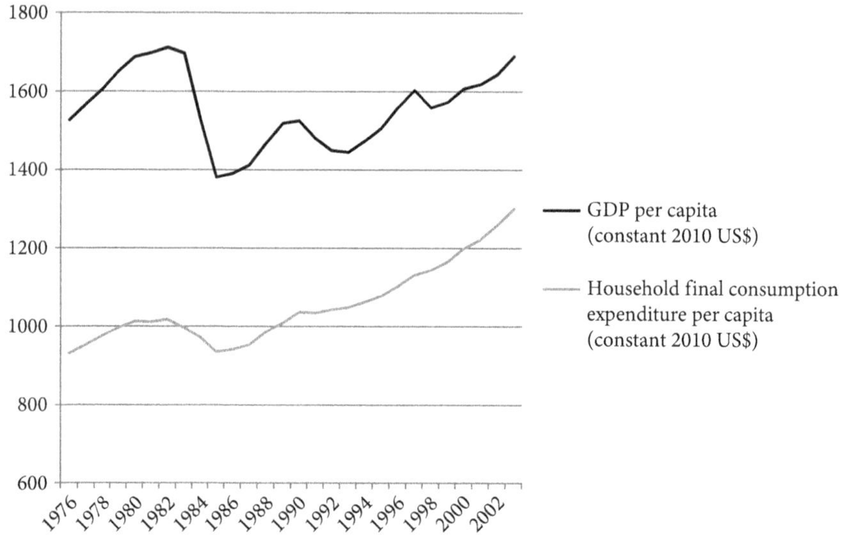

Figure 6.6 Per capita Household Consumption Expenditure v. GDP (1983–2003)
Data: World Bank Development Indicators.

drawn from, if not the most impoverished classes, nevertheless the lower and middle classes, who rarely have the luxury of saving migrant earnings or investing them in industrial projects but instead need those funds to pay for their families' daily expenses and tend to devote what remains to personal investments, such as improvements in housing, education, and healthcare (Stahl & Arnold, 1986; Airola, 2007). Thus, the consumption impact of remittances on the Philippine economy was evidently greater than the absolute value of the funds remitted, which itself was not negligible. In 1985, 1.4 million Philippine households received an average of $575 a year from abroad, an amount that would double the average family income at the time. By 2006, the number of households receiving remittances would triple, to over 4.2 million (Ducane, 2015). Moreover, the effect of those remittances would have been, to some extent, countercyclical. When economic crises are localized to the domestic economy migrant workers are motivated, and—because of the tendency of economic crises to depreciate national currency—are able, to send more to sustain their families' consumption (Frankel, 2011). Yang (2005) specifically documents this in the Philippine case.

Moreover, unlike in Morocco or Mexico, where labor migrants tend to come from primarily agricultural or peripheral regions such as the Rif mountains or the state of Oaxaca, respectively, the modern Philippine migration has clustered around urban/metropolitan and industrial centers (Table 6.1). In 1988, 65 percent of all contract migrants came from the country's two major metropolitan areas, Manila and Cebu, or the provinces immediately surrounding them; 39 percent came from the National Capital Region (Metropolitan Manila) alone. Though this number had declined somewhat, to 55 percent, by 2008, the Philippine migration continues to draw heavily on the two cities and their immediate surroundings.[5] The tendency of migrants to settle in urban centers made it easier for the consumption-related sectors, such as real estate and retail, to build an accumulation model on migrant remittances. In countries like Mexico, where migration tends to draw from rural areas, migrants' investments in property are typically in undeveloped agricultural lands for which stable markets rarely exist—unfavorable conditions for the expansion of a real estate sector.[6] By contrast, Philippine migrants were sending money back to families living in urban or suburban areas, where remittances fueled a demand for housing, both as investment and to shelter children and spouses left behind, creating opportunities for the real estate industry.

For the Philippines' "deindustrialized" economy, then, the consumption patterns that development economists often argue is an obstacle to the deployment of remittances for industrialization purposes has become an advantage, the basis of accumulation and growth. And while policymakers and economists debate whether the consumption-driven economy is sustainable or can deliver improvements in the lives of the general population, the absence of alternatives, particularly in the 1980s and 1990s, has meant that the Philippine state had little choice but to promote labor export and Philippine capital had little reason to oppose it. While the social costs of large-scale contract migration make an explicit labor-export promotion

[5] A National Migration Survey, conducted in 2018, found that migration from urban areas remains significant. 49.3 percent of households reporting an OFW member were from urban areas (Tabuga et al., 2021, 4).

[6] Mexican immigrants often face difficulties when applying for benefits from the American welfare state, as many of them own small parcels of land in their home regions. Such properties are taken into account when assessing an applicant's income and wealth for benefits like Supplemental Security Income, and applicants are required to disclose all such properties. However, Mexican immigrants from rural regions cannot provide a "fair market value" for their properties, as no real market for these lands exist.

Table 6.1 Comparative Distribution of Overseas Workers by Region (in thousands)

	1988	1998	2008
TOTAL	477	904	2002
National Capital Region	140	161	280
Cordillera Administrative Region	9	22	38
Region I: Ilocos Region	64	92	156
Region II: Cagayan Valley	16	50	110
Region III: Central Luzon	80	132	290
Region IV: Southern Tagalog*	80	157	412
Region V: Bicol Region	6	30	62
Region VI: Western Visayas	38	89	156
Region VII: Central Visayas	11	39	112
Region VIII: Eastern Visayas	2	19	58
Region IX: Zamboanga Peninsula**	8	24	36
Region X: Northern Mindanao	5	14	56
Region XI: Davao Region***	8	26	56
Region XII: SOCCSKSARGEN****	10	18	90
Caraga	-	14	24
Autonomous Region in Muslim Mindanao	-	16	66

Data: *Philippine Yearbook of Labor Statistics, various years*
The regions in grey represent metropolitan or metropolitan-adjacent areas.
*subdivided in 2003 into Region IV-A: CALABARZON and Region IV-B: MIMAROPA
**formerly Western Mindanao
***formerly Southern Mindanao
****formerly Central Mindanao

policy problematic, the state has found it difficult to dismantle a program upon which the most important sectors of the economy now rely.

A Remittance-Dependent Economy

It is likely, given the importance of the real estate and service sectors in other Asian economies and the constraints imposed on the Philippines by its economic circumstances and its social and political structure, that these sectors would have gained in importance in the last twenty years of the 20th century regardless of the labor migration. Even before the onset of the migration to the Middle East, some of the landed elite families had begun to convert their properties into real estate developments, the most well-known and lucrative of which was the development of the Makati financial sector, beginning in the 1960s, by the Ayala family. The deagriculturalization of the economy has produced a rural-to-urban migration, which, combined with a general population growth,[7] has created a growing demand for commercial and residential space, as well as urban transportation and other infrastructure. Thus, it is unsurprising that the real estate and retail sectors did well during the periods of recovery and growth that have occurred in the first part of the 1990s and in the new millennium (Ortega, 2016; Krinks, 2002). However, some of the biggest fortunes in these sectors were built not during these periods but in the late 1970s and 1980s, when the country was in the midst of its political and economic crises, and were made possible by the countercyclical spending of the migrants and their families.

The most well-known of the real estate companies to emerge in this period was Vista Land and Lifescapes, founded by Manny Villar, a man who was famously born in the Tondo slum area of Manila. Villar started building housing developments in 1977, just as the export in labor to the Middle East was beginning to boom. He began by taking over irregularly shaped, difficult-to-sell lots in areas that were already being developed near Manila, subdividing them into smaller sections for a lower-priced mass market. His company offered a number of innovations not then prevalent in the Philippine real estate market: American-style subdivisions in which lots and structure were sold together, community based amenities, easy to access financing (Asiaweek, 2006). Though the business tapped into a general need for low-cost housing in the Philippines, what it offered was also particularly attractive to overseas Filipino/a workers (OFWs). Of the first 100,000 units that Villar built and sold, over a third of them went to OFWs.

The company grew throughout the 1980s, and its reliance on OFW remittances meant that its phenomenal growth occurred not just in spite of but *because* of the economic crises of the Marcos years. Villar describes the plummeting of the peso in the early 1980s as a "business opportunity" that no one else saw, since the devaluation suddenly increased the value of the earnings OFWs were making overseas, making it increasingly possible for them to purchase real estate. Vista marketed its properties to OFWs, promising them the fulfillment of a "dream" of homeownership (Asiaweek, 2006). Villar's company grew while the rest of the economy was in freefall (Lee Flores, 1997, pp. 5–6), and he then turned those earnings around to invest further in land, whose value had collapsed during the political and economic crises of the early 1980s. Villar started the company with a small loan from the World Bank, and when it went public in 1995 it was valued at nearly

[7] The population has more than doubled since 1980, from 47.4 million to 100.7 million in 2015.

$2 billion. By 2015, Vista Land, the conglomerate now holding all of Villar's ventures, which include expansions into higher-cost and commercial markets, attributed 60 percent of total sales to overseas Filipinos (Vista Land, 2007; Vista Land, 2015).

Villar was able to capture this OFW market because it offered a number of things that temporary migrant workers particularly desired: safe and comfortable homes for the children and elderly parents left behind, easy and rapid transactions and financing for those who were not in the Philippines and were dealing with large sums for the first time, and the symbolic value of a home as something physical that would be there when they returned to the Philippines. As the remittance-based demand in the real estate market proved itself durable, these properties offered something as important: the promise of a wise investment, a place to put their remittance dollars that had the potential to increase in value (see, e.g., Nelson, 2016). In some areas, such as Metro Cebu, the value of residential real estate lots rose as much as 52 percent *per year* during the late 1980s and early 1990s (Sajor, 2001). While the Philippine real estate market has not sustained such fantastic growth rates since, they remain high; between 2014 and 2016, prices on all types of housing increased by nearly 14 percent, with condominium units—a type of housing particularly marketed to overseas Filipinos—increasing by nearly 20 percent (Bangko Sentral ng Pilipinas, 2017). The high prices fueled by and fueling high demand represent a kind of virtuous circle that would not exist without OFW investments. It is not only that OFWs constitute 60 percent of the buyers for companies like Vista Land, but that the characteristics of these OFW buyers have helped to reassure future investors and buyers that the market is stable and safe. Overseas Filipino buyers, apart from those at the very top of the market, are primarily buying residences for occupancy by their families left behind or upon their return, rather than for speculative purposes, and this fact has been publicized widely to reassure investors that the real estate boom is not an unsustainable bubble (see, e.g., Kritz, 2016).

The impact of labor migration on the real estate market is greater than that of remittances alone, because purchases are not only funded through the personal remittances sent by OFWs who have left family in the Philippines. A significant sector of migrant buyers, particularly at the high end, consists of overseas Filipinos/as who have relocated permanently to countries like the United States and Australia. While they may continue to send funds to support family in the Philippines, they are also wealthy enough to make investments in second homes or rental properties, and often do so in the Philippines. In 2016, Philippine nationals living abroad accounted for 71.3 percent of all approved foreign investments, approximately $10.4 billion out of $15.2 billion. While the data are not sufficiently disaggregated to show where the investments of Philippine nationals go, real estate investments constitute a significant portion of foreign investments in general, representing 21 percent of such investments in 2016 (Philippine Statistics Authority, 2017).

The importance of the overseas Filipino market can be seen in the marketing drives of residential developers. Materials specifically targeting OFWs include explicit mention of the labor migration itself, with emotional appeals invoking the family left behind. Marketing is not limited to the Philippines. Since the 1980s, property developers have promoted their developments abroad, with sales agents placed in cities with significant Filipino populations. These cities are also serviced regularly by international promotional caravans, complete with Filipino celebrities, glossy brochures, and all the paperwork for processing sales contracts and financing (Ortega, 2016).

The model that Villar spearheaded has now become the industry norm, and because Philippine capital is primarily structured around highly diversified family-owned conglomerates rather than clearly sectorally differentiated firms (Kang, 2004), Villar's model spread quickly throughout the economy. Most of the major conglomerates operated by the economic elites of the old landed oligarchy include subsidiaries building homes for a residential housing market. The biggest of these is the Ayala Corporation, the corporate form of the Zóbel de Ayala family, whose wealth originates in lands from the Spanish colonial era. The Ayala Corporation has interests in utilities (including water, power, and telecommunications), electronics, and banking, but real estate is the basis of its modern wealth, Ayala being the country's leading real estate developer (Ayala Corporation, 2015). Ayala Land, its real estate subsidiary, represents about 30 percent of Ayala's total earnings, and overseas Filipino buyers account for 18–25 percent of its residential sales (Delmendo, 2013; Ayala Corporation, 2015). For a conglomerate like Ayala, the impact of the OFW market cannot be measured in residential sales alone, since residential developments are typically developed in conjunction with, and clustered around, gigantic malls and industrial parks, or, in the cities, combined with malls and parks into mixed-use developments. The residences drive foot traffic into the malls, which also serve as hubs for public transportation and are linked by walkways to office buildings. The resultant sales help to support the commercial real estate investments, which are often held by the same company.

The expansion of the retail sector, which, like real estate, also expanded in the rest of Asia from the 1970s on, suggests that its share of the Philippine economy would have grown to some degree even without the influx of remittances from migrants. However, as in real estate, the remittance flows have helped to sustain a level of consumption that would not otherwise have been possible. Like real estate, retail trade was one of the few sectors that not only survived but managed to expand during the economic downturns precipitated by the Marcos debt crises. Some of the biggest corporations in this sector, such as SM Investments Corporation (formerly Shoemart) and Robinsons Retail Holdings, either were established or experienced breakthrough growth during this period. For example, SM, founded by Henry Sy, Jr., began as a couple of shoe stores in Manila in 1958. In the first fifteen years, growth was relatively slow, with only two new storefronts being established. Beginning in 1972, however, the company relaunched itself as a less specialized department store and began to grow more quickly, opening five additional department stores in Manila in the next twelve years, then opening a 125,000-square-meter shopping mall (including an SM department store and an SM-branded supermarket) in the fall of 1985, at the height of the country's political and economic crisis, when GDP had declined by nearly 10 percent for two straight years. The company continued its expansion through the 1990s, opening three new malls between 1990 and 1993, and has opened nearly two malls a year since then, now operating nearly sixty malls in the Philippines alone.

It is difficult to track how much of SM's sales revenues can be attributed to OFWs and their families. However, SM itself considers OFWs to be an important market. Not only does it report on the remittances and their impact on overall consumption in their annual report to shareholders, particularly tracking remittance flows following the global economic crisis in 2008 (see, e.g., SM Investment Corporation, 2010–2015), it caters to OFWs in its business operations in a number of ways. To some degree, the phenomenon of "malling"—spending one's leisure time in shopping malls—is partly facilitated by the cosmopolitan tastes and practices that OFWs develop through their time abroad, to which malls like those operated by SM cater by franchising and partnering with transnational restaurants and retailers—Starbucks, H&M, and Lego—and offering international films in their cinemas. More explicitly, SM operates a number of programs targeting

OFWs, including special lounges called "Global Pinoy Centers" (GPCs) reserved for visiting OFWs and their families, as well as specialized banking services offered through SM's subsidiary bank, Banco de Oro, branches of which are embedded throughout the malls and in the GPCs, that offer currency exchange and remittance processing (SM Investment Corporation, 2014). The SM and other malls have also partnered with the Philippine state to offer "one-stop shops," some of them combined with GPCs, where OFWs can access a myriad of state services. Though these services do not include those involving the labor-migration program directly—such as the POEA, OWWA, or DFA—almost any other state service that an OFW might need to access during a short visit home can be accessed at these locations, including the Social Security System, PhilHealth, the Bureau of Internal Revenue, and the Pag-IBIG Fund, one of the state's home-mortgage programs.

Business Process Outsourcing (BPO)

The final prong of the Philippines consumption-driven economy, the services industry, cannot claim the same dependence on remittances as real estate or retail. However, the sector itself, particularly in its most dynamic, export-oriented business process outsourcing (BPO) arm, reflects a growth strategy that parallels the labor export. BPO is the term applied to the outsourcing of a broad range of services—such as information technology, customer service, and accounting—to a third-party provider. The Philippines is most well-known for its call centers, which constitute approximately 50 percent of the industry worldwide, but Filipino/a BPO workers also provide software management services, data processing, and financial and accounting services. BPOs represent essentially another form of labor export, though communication technology has obviated the need to actually move bodies across borders. Call-center workers serving the American market start their workday as the sun sets, interacting with customers who might be receiving nursing, domestic, or other physical services from OFWs.

The BPO industry began operating in the Philippines in the early 2000s, drawn to the characteristics that have helped the Philippines export its workers to foreign labor markets, including the relatively low cost of labor and the high rate of English proficiency in the population, as well as a readily available population of skilled, educated workers. It has grown exponentially: from 2001 to 2008, BPO revenues jumped from $100 million to $6 billion, and by 2015 had reached approximately $21 billion. BPO revenues are now comparable with migrant remittances and are expected to start exceeding them in the next few years (Mitra, 2011). Successfully drawing market share from early industry leaders like India, the Philippines is now one of the top global BPO destinations in the world. Combined, BPO revenues and remittances and other revenues associated with labor export constitute 20 percent of the Philippine economy.

The revenue from the BPO industry works itself into the Philippine economy much like the remittances of labor migrants. Many of the major BPOs operating in the country, such as Accenture, are foreign multinational corporations (Kleibert, 2013), and even those owned by Philippine nationals benefit from tax holidays and exclusions, so the state sees little of the revenue generated by the BPO industry. In terms of its potential to stimulate growth in other parts of the economy, the BPO industry has so few backward or forward linkages that policymakers doubt that it could replace manufacturing as a basis for sustainable, inclusive growth (Usui, 2012, pp. 42–43). Where BPO does produce backward linkages is in real estate. The BPO boom has created a demand for commercial real estate, which companies like SM have leveraged to further buttress the consumption economy. Commercial spaces for BPO processing are rolled into the shopping mall/residential real estate development projects described above, which offer convenience and proximity for BPO workers in metropolises where transportation and traffic represent huge costs. For the consumption sector, the workers in the BPO industry represent a coveted market. The industry's direct impact on unemployment is not actually huge: In 2016, it employed 1.3 million workers, only 3 percent of the total work force (Errighi et al., 2016). However, because most BPO jobs require college degrees, the impact on unemployment and underemployment among workers with tertiary education has been quite large.[8] The salaries associated with these jobs start at $400 a month for the least skilled positions in call centers; while this is less than the minimum monthly wage of an OFW working as a domestic helper in Singapore, it is nevertheless more than double the country's average monthly wage. BPO workers in more specialized fields, like information technology, finance, and medicine, can make much more. In the past decade, BPO employment has become as important as migrant remittances for consumption sectors such as real estate and retail (SM Investment Corporation, 2013, 2015); BPO workers

[8] The numbers of unemployed and underemployed among workers with tertiary education, in 2008, were 1.2 million and 1.4 million, respectively.

constitute another category of workers with incomes sufficient to support the country's consumption-driven economy. The impact of migrant remittances or BPO salaries cannot technically be called a forward linkage, but the consumption this income makes possible does have important downstream effects.

In the years during which the BPO sector has developed, the Philippine state has helped to nurture the growth of the industry, offering incentives and subsidies to attract foreign BPO firms. These have access to the subsidies and privileges of the special economic zones (SEZs) operated under the Philippine Economic Zone Authority (PEZA). SEZs were initially created, during the Marcos regime, to be geographically separate industrial areas for export manufacturing. The PEZA has expanded the definition of SEZs to accommodate BPO operations, making it possible to declare a single building a "special economic zone" for the purposes of PEZA incentives. By 2012, there were 175 such "zones" specified for the IT-BPO sector (Kleibert, 2013).

For the Philippine state, the BPO industry has become an alternative to the export of labor, serving very similar purposes while eliminating the biggest of labor export's social costs—namely, the migration itself and the social dislocation it causes. The state's rhetoric regarding labor migration has long been ambivalent, lamenting the unemployment and poverty that have made migration a necessity while affirming that it remains necessary. The promise has never been to retreat from the migration promotion regime, which, since the 1990s, has always been presented as a means of protecting workers who would leave regardless of the state's position. The promise has been to promote, on a parallel track, an economic transformation great enough to reduce the level of domestic unemployment and underemployment.[9] In the last decade, the BPO sector has represented a partial delivery—at least to the highly educated portion of the population—on that promise. And, importantly, the BPO sector has made it possible

[9] President Benigno S. Aquino III: "For the longest time, it seemed that the fate of the Filipino was to be an economic refugee. This phenomenon has even been given a name: the Philippine diaspora. And as the word implies, it was a dispersal of some of our best and brightest to the four corners of the world, in search of prosperity and a future . . . our administration has been on a mission to reverse this trend. Working abroad should not be a necessity, but a choice. And our policies have always been guided by our desire to give Filipinos that choice . . . This is precisely why our reform efforts are focused on attracting more investments, creating more jobs, and giving our countrymen reason to believe that a good life awaits them here, at home. In this regard, we have been quite successful. Investors are regaining their confidence and putting up more and more businesses here; and if we build on our momentum, I think we can achieve our dream of keeping more of our countrymen here" (Speech at the Second Ministerial Meeting on the Abu Dhabi Dialogue, April 19, 2012).

to deliver on that promise without drastically transforming the structure of Philippine political economy. This is, essentially, the fruit of a neoliberal development agenda that encourages poorer, peripheral countries to lean into their comparative advantage without attempting to nurture or protect domestic industry. In the Philippines, those advantages were natural resources and the low cost of labor—and migrant labor, being tied to wage levels in wealthy economies, has turned out to be a more reliable source of income than agricultural exports (Prebisch, 1959).

The Role of State Policy

While there are some elements of serendipity in the emergence of RDG—specifically the buildup of the labor-export apparatus under the LEMI regime that established the flows of migration and remittance on which RDG could build—a significant element of policy intervention was required to enable its consolidation. Despite the country's economic challenges in the late 20th century, significant investments continued to be made, particularly in regions targeted for industrialization and urban development. These projects have transformed formerly agricultural regions into economic powerhouses where markets—for real estate as well as for consumer goods—now flourish. One prominent example of this is the Calabarzon region immediately south and east of Metropolitan Manila. As the country oriented itself around an export-led industrialization model of development, beginning under the Marcos regime but accelerating in the late 1980s, this hitherto largely agricultural region received heavy public investments in infrastructure in order to attract foreign investment for industrial production. In 1995, the Philippine state established the Philippine Economic Zone Authority (PEZA) to formalize and oversee a system of export-processing zones that had been built up since the Marcos era. These Special Economic Zones (SEZs), offering some of the most generous incentives in Asia and lax enforcement of labor protections, helped make the Philippines a major electronics exporter (McKay, 2006, pp. 59–74). In the first decade of its existence, PEZA established over 200 SEZs, most of which were concentrated in the regions surrounding Metro Manila, including Calabarzon (Sanders & Brown, 2012). Though there have been many criticisms of the SEZ model, citing illegal land conversions, labor abuses, and the failure to deepen the country's overall industrial capacity, SEZs have succeeded in transforming the Calabarzon region. In the decade following 1988, 44.7 percent of the region's land was converted from agricultural to industrial, commercial, and residential use (Magno-Ballesteros, 2000, p. 9). Since 1988 its population has nearly tripled, double the national rate. Enormous residential communities—of both suburban-style tract homes and more urban condominiums—dot the landscape, particularly along the expressways that have been built since the 1980s to connect the region to Manila. This transformation has dovetailed with the expansion of labor export from this region. Between 1988 and 2008, the region's share of OFWs surpassed that of Metro Manila-National Capital Region (NCR). It remains the top labor-sending region, representing about 16 percent of all contract-based migrant workers, which in 2010 represented about 16 percent of the region's population. These workers remitted approximately $2.7 billion in that year, or approximately 15 percent of the region's GDP, and this was spent almost entirely on housing, consumer goods, and services (Philippine Statistics Authority, 2023; National Economic and Development Authority, 2011, 2013).

In addition to this regional industrial planning, the state supports the RDG-related sectors in other ways. In the real estate sector, it does this through support for mortgage financing. In the private sector, the state encourages a private market through tax incentives for mortgage issuers and guaranteeing private loans against defaults. The state also acts directly as a source of financing, offering mortgages through a number of institutions, the most important of which is the Home Development Mutual Fund (also known as the Pag-IBIG Fund) (Ortega, 2016). Since 2009, OFWs have been required to participate in the Fund, which gives them access to mortgage rates that are among the lowest in the country, as well as mortgage periods that are longer than those offered by most private banks. In 2009, the Philippine state further buttressed real estate markets by passing the Real Estate Investment Trust (REIT) Act, which was intended to enable small investors to participate in the rental real estate market. Though it has yet to significantly impact the real estate sector, the law was written to encourage investment by OFWs and overseas Filipinos, with tax incentives on income generated through REITs (Republic Act No. 9856, 2009; Securities and Exchange Commission, 2020).

The other important policy support that the state provides for RDG is, of course, its labor-export program, which ensures a continuing out-migration of temporary contract workers with its associated remittance flow. The next chapter will explore the program's evolution since the 1980s, describe how it shapes and maintains the migrant flow, and explore why this particular feature of the Philippine developmental state has managed to avoid the effects of oligarchic rent-seeking. As was the case with LEMI in South Korea, the labor-export program did not, on its own, generate the economic transformations of RDG. It is undoubtedly true that, given the neoliberal turn in international development policy and the Philippines' vulnerability to pressure from the United States and foreign lenders, the country's deindustrialization and its concurrent push into real estate and services would have occurred regardless of remittance flows. However, these flows did make it possible for the Philippine economy to fare better than other developing countries have under similar conditions, and to achieve even better and more stable rates of growth than in its own LEMI period.

Chapter 7
The Remittance Workaround—Explaining the Relative Success of RDG

Introduction

In order to operate consistently, as it has in the Philippine case, the remittance-driven growth described in the previous chapter requires households to have a continuing source of funds to support the consumption that drives growth. It is crucial not only that this flow of funds exist but, in order to generate the investment, that the flow be perceived to be *stable*: investors must believe that there will be consumers to shop in the malls and buy the apartments they are building, and that home prices will continue to rise because people will continue to have money to sustain demand. The labor-export policy plays a crucial role in securing this steady and growing flow of remittances. It does this both by directing labor migration toward a specific type of migration—predominantly temporary and employment-based—associated with higher and more reliable rates of remittances, and by actively diversifying the countries to which overseas Filipino workers (OFWs) deploy. Migration flows unsupported by a labor-export policy like that of the Philippines rely primarily on the networks of social and economic relations that are generated by the migration itself. Such flows tend to adhere to well-established routes, and consequently are highly vulnerable to changes in economic or political conditions in the destination countries. For example, migration from a country like Pakistan, which is directed primarily toward the oil-rich Middle Eastern countries, fluctuates with the price of oil. Similar fluctuations in the Philippine economy would wreak havoc on the remittance-dependent sectors that drive the country's growth. But in fact migration and remittance growth from the Philippines have both been remarkably steady in the past fifty years, enabling companies dependent on remittance spending to sustain high prices and attract investments based on expected remittance spending.

Temporary Measures. Suzy K. Lee, Oxford University Press. © Oxford University Press (2025).
DOI: 10.1093/9780197790007.003.0007

In this chapter, I will first describe how the Philippine state's labor-export policy achieves this outcome, and then take up the question of how the Philippine state developed and sustained this policy over half a century. There are two reasons why we should be surprised by the long-term performance of the Philippine state's labor-export apparatus. First, any economic policy that relies on the out-migration of a significant portion of its population will be controversial—particularly where, as in this case, those migrants are being sent to work in some of the lowest-status sectors of the international labor market, with high instances of labor-rights violations, verbal, physical, and sexual abuse, and death. Many labor-sending countries have been unwilling to promote, or even permit, certain types of contract migration, such as the migration of women into domestic labor, in which the Philippines dominates. Even if we were to grant that, in the Philippine case, popular resistance to a labor-export policy has been overridden by the overwhelming dependence of the economy on remittance spending, we are still left with the question of how the Philippine state was able to develop the bureaucratic capacity to implement this policy.

In Chapter 5, I offered a description of the Philippine state that would be familiar to people with even a passing interest in recent Philippine history or economic development: a state that has often been labeled "failed" for the ways in which patrimonial clientelism and cronyism have undermined the state's efforts not only to catalyze industrial transformation and economic growth but even to achieve the most basic requirements of statehood, such as monopolizing violence or imposing rule of law. Ferdinand Marcos may be the most prominent symbol of the failures of the Philippine state, but many of these problems existed before Marcos imposed martial law and persisted after his removal in 1986. Only one area of policy escapes this generalized critique of the Philippine state: labor migration. Here, the Philippines is often studied in international development and migration policy circles as a model to be emulated by other states. And it has maintained this reputation for decades, surviving major economic and political transformations, including the end of the authoritarian regime that created it.

How can this be explained? Why have the forces that eroded the Philippines' efforts at LEMI not undermined the labor-export apparatus? Has the labor-export bureaucracy been immune to the clientelism and corruption that characterize most other parts of the Philippine state? If so, why? If not, how does it continue to operate effectively?

The answer I propose to these questions rests fundamentally on the different targets of LEMI (labor-export-mediated industrialization) and RDG (remittance-driven growth) policies, which in turn reflect the different ways in which the economic gains from migration enter the Philippine economy. While LEMI policies included those oriented toward migrants themselves, their more important targets were the construction companies that were doing the labor exporting. These companies were also the greatest beneficiaries of labor export, profiting directly from the opportunities—created by their access to low-wage, productive Filipino workers—to participate in the Middle East construction boom. RDG differs from LEMI in the relationship between the *business* of labor export and the economic gains that labor export brings to the Philippine economy. Under RDG, the labor exporters are no longer industrialists transporting migrant workers to engage in production abroad; instead, they are manpower recruitment companies whose sole role is to coordinate the hiring of Filipino/a workers by foreign employers. Though migration produces upwards of $35 billion for the Philippine economy every year, almost all of this is represented by remittances. Recruitment companies win only a small fraction of this income, and the rest enters the country in small amounts sent directly to millions of households.[1]

If, as I've argued, the weakness of the Philippine state is a function of its relationship with an oligarchy which, while having a power base outside of the state, relies on access to state-controlled rents (monopolies, licenses, subsidies, foreign aid and loans, etc.) for its profit, this separation between the business of labor export and remittances interrupts the typical process of rent-seeking through political influence. Because remittance income is widely dispersed, unlike easily monopolized pools of funds or revenue streams, oligarchic conglomerates cannot capture the gains from labor export by manipulating the state's labor-export policy. Capturing those gains requires Philippine capital to operate downstream from where migrant families spend their remittances, primarily in housing and household consumption. This creates a different set of incentives than during the LEMI period, when labor-export and industrial policy acted directly on the most influential sectors of Philippine capital. Under RSM, labor-export policy acts directly only on the relatively small-scale operators of the manpower-recruitment industry. A state that effectively directs and

[1] In 2018, over 3 million households reported having at least one OFW member (Tabuga et al., 2021, p. 4).

disciplines economic activity in this sector poses little threat to oligarchic conglomerates, while the latter benefit from the flow of remittances. This shared interest in labor export has resulted in the development of the state's bureaucracy around labor migration, which, while not impervious to the quotidian micro-level corruption that is endemic to the Philippine bureaucracy, notably lacks the chaos and incoherence that characterizes the meso-level structures of the Philippine state. Rather, the labor-export apparatus has, since the 1980s, demonstrated a high level of centralization, with a nodal agency, the POEA, serving as an organizational and informational hub through which policy is designed, vetted, and implemented—much like the centralized structures that have been described in the developmental states of East Asia.

The Role of Labor-Export Policy in Sustaining Remittances

In its most basic logic, migrant remittances as a basis for economic growth do not *require* sending-state intervention. If migration and remittances could be relied upon to grow on their own over a sufficiently long time horizon, then it is possible that a remittance-based economy could emerge independent of state support. There are reasons—whether from the perspective of countries like the United States, to which there never seems to be a shortage of workers willing to migrate, or from theories of migration built on migrant networks and the logic of cumulative causation—to believe that migration can be self-perpetuating in certain respects (Massey et al., 2005). While this may be true, it is also true that, year on year, migration and remittance flows within any migration channel can fluctuate significantly, making it difficult to establish a long-term development program based on remittance flows, absent the kind of sending-state management that we see in the Philippines.

Because migration, in the absence of state intervention, is a process enabled by networks and social capital, migration flows tend to move in established circuits, formed by geographic proximity or by historical political and economic linkages. For example, over 85 percent of Moroccan migrants live in a handful of Western and Southern European countries;[2]

[2] These countries include Belgium, France, Germany, the Netherlands, Italy, and Spain (UN Population Division, Department of Economic and Social Affairs, Trends in International Migrant Stock: Migrants by Destination and Origin, 2015).

93 percent of Haitian migrants live in North America, primarily the United States and the Dominican Republic; 98 percent of Mexican migrants live in the United States.[3] This means that migration from most sending countries (and consequently the remittance flow) is highly vulnerable to economic and political conditions in that primary receiving region (de Haas, 2010). Migration from a country like Pakistan, which is heavily dependent on a single market—the oil-rich Middle East—fluctuates wildly, with the price of crude oil, which dictates the level of infrastructure investment in the region and thus the demand for construction labor (Figure 1.1). Even in the migration between Mexico and the United States, which is often considered a paradigmatic case of a self-perpetuating, networks-driven migration, we can see that flows have been transformed repeatedly by changes in both immigration policy and economic conditions in the United States. For example, Mexico's emigrant stock grew explosively in the 1990s, then largely stagnated in the 2010s (see Figure 7.1), with net migration falling to zero (Passel et al., 2012). The growth in the 1990s reflects shifts in US immigration and border policy, namely the passage of a series of immigration laws in the late 1980s and 1990s, including the Immigration Reform and Control Act (IRCA) of

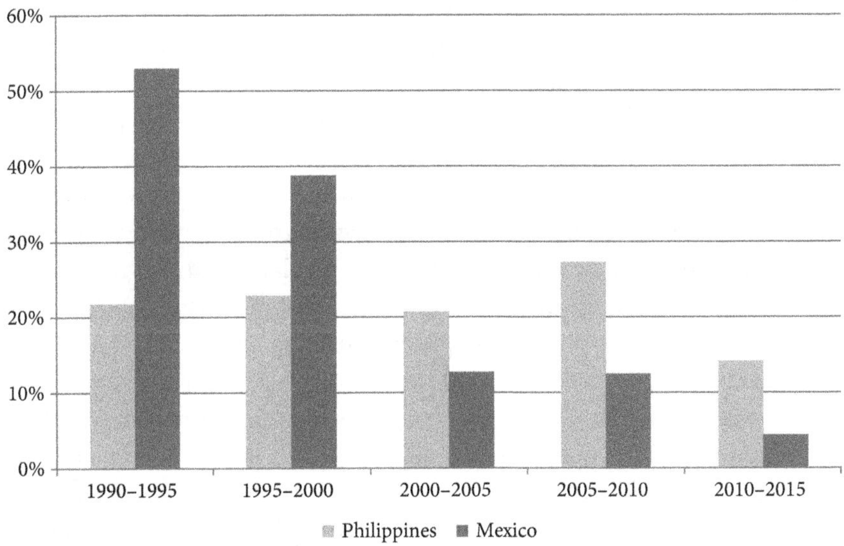

Figure 7.1 Rate of growth in emigrant stock, 1990–2015

Data: United Nations Population Division; https://www.un.org/development/desa/pd/content/international-migrant-stock.

[3] UN Population Division, ibid.

1986, the Immigration Act of 1990, and the Illegal Immigration Reform and Immigrant Responsibility Act (IIRIRA) of 1996. Together, these acts combined a pathway to legalization and citizenship for migrants present in the United States prior to 1986 with a militarization of the border between the two countries. Up to this point, Mexican-US migration had been characterized by temporary, seasonal, and circular migrations in which high numbers of undocumented border crossings were tacitly accepted by both states. The militarization of the border had a negligible impact on rates of migration, but it transformed temporary migrations into permanent resettlement as the costs and risks associated with border-crossing now rendered circular, seasonal migrations infeasible. In addition, the amnesty of 1986 created a pool of Mexican immigrants, over 2 million strong, who were now able to sponsor legal immigration visas for their family members,[4] again increasing the likelihood of permanent settlement (Massey et al., 2002). The steep decline in migration in the 2010s was precipitated by the financial crisis of 2007–2008 and the ensuing recession, as well as Mexico's own demographic transition (Durand & Massey, 2019). These swings in migration flows are reflected in remittance rates, which grew steadily—even exponentially— from the 1980s to the early 2000s, then dropped steeply beginning in 2008 (see Figure 7.2).

Even without the 2008 financial crisis in the United States, it is possible that we would have seen declines in migrant remittances to Mexico over time. As described above, the primary reason for the growth in Mexico's emigrant stock was a change in the form of migration precipitated by US migration policy, from temporary and circular seasonal migrations to more permanent migrations in which women and children joined working-age men. The research has shown that temporary migrants tend to remit a greater portion of their earnings, and more regularly, than migrants who plan to settle long-term in the destination country (Glytsos, 1997; Dustmann & Mestres, 2010). For example, in a sending country like Morocco, whose emigrant population evolved from largely temporary workers entering through European *Gastarbeiter* programs to more permanently settled communities, remittance rates have been relatively stagnant compared to the Philippines (Figure 7.2).

In comparison, the growth in migration from the Philippines has been remarkably consistent (Figure 7.1). Similarly, remittance rates have grown steadily; despite some volatility during the economic crisis in Asia in the

[4] While employment-based migration to the United States was quite restricted in this period, migration for family reunification was much larger.

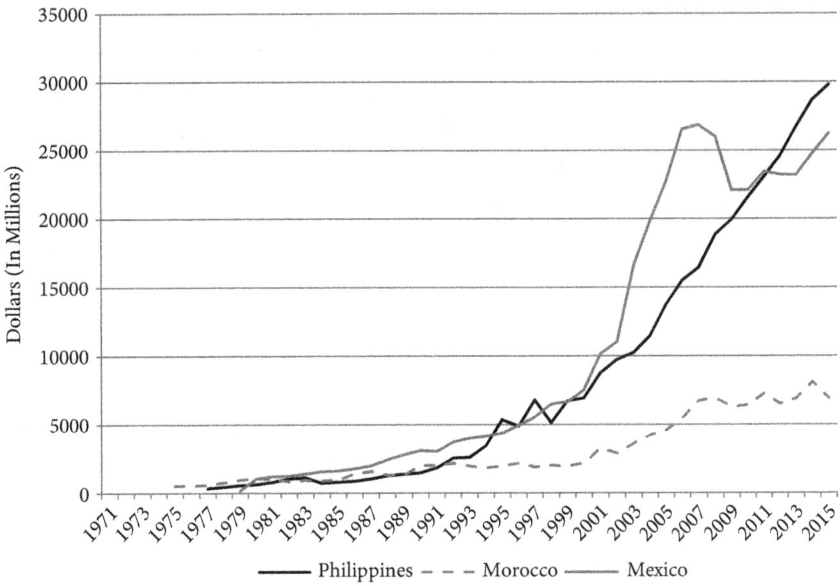

Figure 7.2 Yearly Remittances—Philippines, Mexico, Morocco
Data: World Bank World Development Indicators.

late 1990s, rates returned quickly to a steady growth that remained largely unaffected by the economic crisis of the late 2000s. While it is difficult, in a two-country comparison, to determine what factors contributed to these different migration trajectories, it is clear that the stability in Philippine migration cannot be attributed to a greater stability in the politics and economics of the regions receiving its migrant workers. Not only did Philippine migration also flow to the United States in the period in question, but the other major receiving regions for Filipino/a migrants—the Middle East and East Asia—were not without their own disruptions, including the steep decline in oil prices in the mid-1980s that ended the Middle East's construction boom and the Asian financial crisis of the late 1990s. If anything, the Philippines faced greater instability in terms of receiving-country conditions,[5] though that instability, crucially, was distributed across a multiplicity of receiving countries and regions. Filipino/a migrants live in over 160 countries and territories, with significant populations in all of the major

[5] Despite the policy shifts and economic crises described above, the United States has been open to immigrant labor relative to most other labor-importing countries and has sustained a consistently high demand for it.

migrant-receiving regions: Asia (38 percent), Europe (10 percent), North America (46 percent), and Australia and New Zealand (5 percent).[6] The dispersal of Filipino/a migrants across most of the major labor-importing regions of the world helped to sustain growth rates in migration, compensating for declines in any individual market with increases in others (Figure 7.3). Similarly, growth in remittance rates was maintained despite stagnation or decline from individual countries by sustained remittance income from other regions[7] (Figure 7.4).

At the initiation of its labor-export program in the 1970s, Philippine migration did not look very different from that of the countries mentioned above; like them, it was limited in its destinations, primarily the United States and its own neighbors in Southeast Asia. Moreover, it was a migration that was as likely to be permanent as temporary (Figure 4.2) (Philippine Department of Labor, 1974, Table 13.4, 13.15). Though created initially to

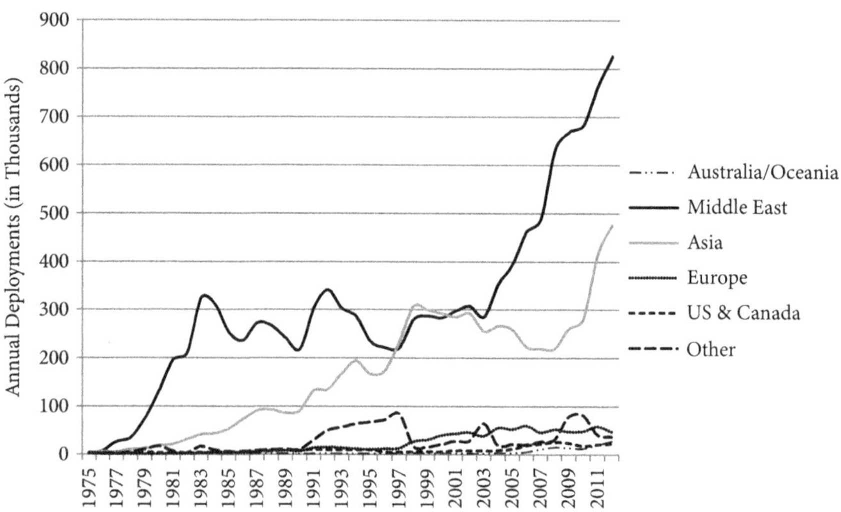

Figure 7.3 Deployments of land-based overseas workers by key destinations—Philippines, 1973–2012
Source: Philippine Bureau of Labor and Employment Statistics.

[6] These percentages are derived from the estimates of the UN Population Division (2015). We get similar results when we use the statistics gathered by the Philippine Commission of Filipinos Overseas: Asia (41 percent), Europe (8 percent), the Americas (42 percent), Oceania (5 percent).

[7] While the share of migrant remittances coming from the United States has declined steadily since the 1970s, it remains the major source of remittance income for the Philippines. This is due in part to the large settled immigrant population of Filipinos/as in the United States relative to other destination countries, as well as the higher wages paid; the latter reflects (1) the higher US wage rate and (2) the fact that Filipino/a migrants to the United States tend to enter higher-skilled occupations than in the Middle East, Europe, or East Asia.

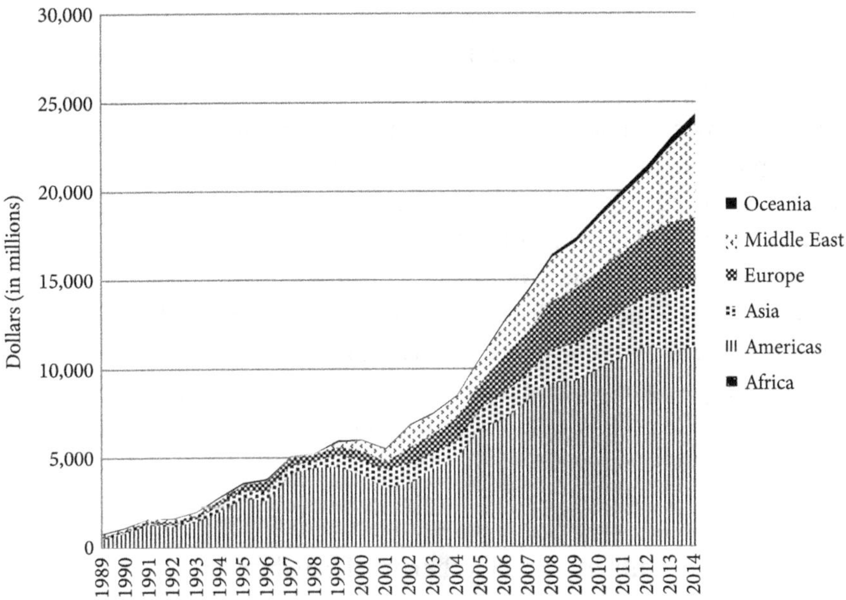

Figure 7.4 Migrant Remittances to the Philippines, by Region of Origin—1989–2014
Source: Bangko Sentral ng Pilipinas.

take advantage of the opportunities for construction labor in the Middle East, the Philippine state's labor-export program drove the diversification of migration, helping workers push into new market for international labor: East Asia in the 1980s, Europe in the 1990s, and Australia/New Zealand and Canada in the first decade of the 2000s. In so doing, it also transformed the dominant form of migration to one that was contract-based and temporary in character.

The Philippine Labor-Export Policy under RDG

How did the Philippines induce these changes to its migration flow? In chapter 4, I described the labor-export policies associated with LEMI, which involved supporting national construction-exporting companies in their recruitment of migrant workers. LEMI policies could also include training and education, as well as bilateral diplomacy with receiving states, though both remained closely tied to the needs of construction exporters. In the Philippines, many of these policies carried over into the RDG era, particularly the training, education, and diplomatic efforts. However, because the goal of RDG is to ensure a steady flow of worker remittances rather than to support a particular national industry, modern Philippine labor-export policy puts much more emphasis on the cultivation and preparation of potential migrants, helping to foster a "culture of migration" and developing new markets and occupational niches to which Filipino workers can be deployed. Moreover, it has taken on much of the task of migrant-worker support, which under LEMI had been provided by the managers of labor-exporting Philippine construction firms.

Because RDG requires that migration continue, many of the policies of this era in the Philippines mirror the mechanisms of migrations that are self-perpetuating—namely, social capital and material support that flow through migrant networks. These policies can supplement, or sometimes replace, the operation of these networks.

Information Gathering/Distributing

One important way in which migrant networks facilitate migration is through gathering and disseminating information about job opportunities and conditions in the receiving country, as well as the procedures, risks, and costs associated with getting there. The Philippine state enables diversification by gathering and distributing information on countries where migrant networks are weak. It also ensures a larger flow of migration by making that information widely available, reaching those potential migrants who have less access to migrant networks. Moreover, the data gathered by the state, which tends to be more oriented toward macro-policy, law, and national trends than the ethnographic and personal information that flows through social networks, can be a useful source even for those individuals who have access to migrant networks.

This information system is coordinated by a central agency. Until 2022, that agency was the Philippine Overseas Employment Administration (POEA); the POEA was absorbed by the newly created Department of Migrant Workers (DMW), which continues to carry out most of the POEA's mandate. The POEA/DMW remit is very broad, covering every aspect of labor-migration governance, from administration and documentation of labor migrants (OFWs), to regulating and monitoring recruiting agencies, to the marketing of OFWs to foreign governments and employers. As OFWs, recruiting agencies, and employers move through its administrative processes, basic data is generated through its operations. Other, more fine-grained information is collected through the system of the Philippine

Overseas Labor Offices (POLOs) in key destination countries, which, as providers of material support to OFWs abroad (see below), deal directly with employers, help resolve problems that OFWs face, and report on background conditions in destination countries. The labor attachés who head each POLO submit regular, formal reports that include both labor-market updates and details of the support work in which the POLO is engaged.

The information gathered through this system informs the POEA/DMW's public migration information and education program. Migration information is disseminated through a number of different platforms. At the most general level, the POEA/DMW disseminates information on migration through broadcast, print, and internet media; the POEA/DMW produces its own materials, including reports, press releases, brochures, posters, cartoons, radio programs, and websites, which cover a broad range of topics. These include announcements of job opportunities; the economic, social, and political conditions in destination countries; foreign immigration and employment laws; potential risks and safety tips; Philippine government policies; and key phone numbers and agency websites. In addition to its own production, the POEA/DMW is the source for much of the data on migration published by local journalists (Siddiqui et al., 2008). The result is a media environment in which information on migration is timely, abundant, and accessible. The Philippine state also oversees a system of seminars at multiple stages in the migration process, which ensures that migrants have access to detailed information specific to the destination country and occupation. These seminars provide, through formal channels, the kind of information on local conditions and practices—social, cultural, and religious practices and mores, criminal-law provisions that may differ from those in the Philippines, climate differences, and logistically useful information like airport and flight procedures—that might otherwise be obtained through migrant networks.

Material Support

The other type of support that migrant networks provide, support that helps create a social context in which migration becomes feasible, is physical and emotional. Workers who migrate to a region where migration is well established, with a robust transnational community, can access a panoply of material aid and services through the migrant network, which can act as a kind of safety net. In the context of LEMI, much of this support was provided directly by the employer, which was also the recruiter and labor exporter. In the RDG context, where most labor migrations are not mediated by domestic employers, the Philippine state has become the provider. The state provides support primarily through the Overseas Workers Welfare Administration (OWWA), which develops most of the services offered to migrants once they have left the Philippines. OWWA officers staff the POLOs, and together they provide a host of social, medical, legal, and financial services in the destination country. These include airport assistance, psychosocial counseling, catastrophic medical coverage, repatriation assistance, mediation with employers, financial assistance, legal assistance, and hospital/prison/work-camp visitations, as well as short-term shelter to OFWs in crisis.[8]

Regulation and Certification

Unlike information or material support, regulation and certification are functions that the sending state is uniquely positioned to provide. While it is possible that non-state institutions could provide some aspects of regulation, ultimately the state, *as a state*, has the authority to demand information and enforce rules that other institutions do not. The Philippine state licenses, certifies, and regulates all the various participants in the labor-export process. For migrant workers, the Philippine state provides the certification and documentation necessary to demonstrate education, skill, health, and other requirements that potential migrants might need to prove in order to secure employment contracts and visas for travel. Prospective employers must likewise go through an approval process, though at present the process seems relatively cursory, requiring only documentary review rather than on-site visits. Most important, however, is the state's regulation of the recruiting and manpower companies that facilitate the labor export itself. In order to recruit labor for the international market, companies must obtain a license through the POEA/DMW, a process that requires them to prove sufficient capitalization and capacity to deploy migrant workers. In order to keep their licenses, companies must comply with a host of regulations regarding

[8] Some of these services, such as shelter, are provided to all Filipinos/as living abroad. Others, such as catastrophic medical assistance and financial assistance to families, are provided exclusively to OFWs who have paid the US$25 fee for membership in OWWA.

rules on fees charged, the provision of documents, and the treatment of workers throughout the process. Violations can lead to the revocation of licenses and future denial of licenses (Republic Act No. 8042, 1995; Philippine Overseas Employment Administration, 2016). The Philippine state, via its information-distribution arm, also regularly publishes information on recruiting agencies, including their license status and their verified contracts with foreign employers. While there are many critiques of its porousness, the licensing and renewal process and public dissemination of licensing information provide a minimal ground for assessment by potential migrants, particularly for those without firsthand experience or access to migrant networks.

Diplomacy

A strategy that has been crucial for the diversification of Philippine migration is the opening of new markets and jurisdictions, and here the Philippine state plays a unique and crucial role. By virtue of its statehood, it is the only actor involved in the migration process that has the standing and authority to engage in formal diplomacy and negotiate the kinds of formal memoranda of understanding, bilateral agreements, and treaties through which guest-worker programs are often created. While such programs do not *require* sending state participation—the bracero and H-1 programs in the United States are examples of unilateral action—many receiving states seek sending-state partners, particularly when attempting to initiate new migrations where migrant networks and recruiting infrastructure do not yet exist (Lee, 2017). The Philippine state has gone further than responding to the entreaties of receiving countries implementing guest-worker programs, and has often been instrumental in *initiating* some of these programs, proactively lobbying receiving countries with labor shortages to consider importing temporary Filipino labor.

In an example that echoes the policies of the Vietnam era, then-Philippine President Gloria Macapagal-Arroyo pledged Filipino soldiers to the US President's 2003 invasion of Iraq. It is an indication of how far from an LEMI strategy the Philippines had come that Macapagal-Arroyo did not seek anything like the industrialization support that Park and Marcos had sought in the 1960s. Instead, she negotiated for the hiring of thousands of Filipino workers as civilian employees of the US Army and civilian contractors (Simon, 2004). (In this case, however, soldiers and workers were quickly withdrawn amidst the public outcry over the kidnapping of a Filipino worker, Angelo de la Cruz.)

An example of diplomacy that led to a longer-lasting migration flow was the campaign to formally open the Taiwanese labor market to Filipino workers. Like many of the other newly industrialized countries of Asia, Taiwan was struggling with a growing demand for labor throughout the 1980s. At the time, this demand was being met informally through workers, 40,000–70,000 strong, who were overstaying their short-term visitor visas. The POEA sent missions to lobby the Taiwanese government for a formal labor agreement, which led to negotiations between Taiwan's Chairman of the Commission of Labor Affairs and the Philippines' Chief of Immigration Affairs (Philippine Overseas Employment Administration, 1991, 1993; Liu, 1996). This diplomacy ensured that the Philippines would be included, along with only two other countries, as Taiwan cautiously opened its labor market to foreign workers. The Philippines remains a major source country for Taiwan's foreign labor to the present day (Tierney, 2007; Tseng & Wang, 2013).

Training and Education

Another tool that has been important for diversifying and sustaining Philippine labor migration is the shaping of the country's labor pool to meet the needs of foreign markets. The sending state's potential role here is obvious. Education and vocational training is a policy arena in which most states are already heavily involved, for purposes of their own economic development and as a basic component of their obligation to their citizenry. In most countries, the state is the authority that issues licenses and certification; when private institutions serve this purpose, it is often with the state's explicit permission and oversight. The Philippine state has shown itself to be both willing and able to modify its educational and vocational-training policies to support labor export. The agencies responsible for education and vocational training, including the Commission of Higher Education (CHED), the Department of Education (DepEd), and the Technical Education and Skills Development Authority (TESDA), develop programming in collaboration with the POEA/DMW, using the international labor-market research that it collects and distributes. The relationship between TESDA and international contract migration is particularly close. TESDA's mandate includes the promotion of the Philippines' "international competitiveness" (Republic Act No. 7796, 1994) and the creation of a labor supply that "matches demand" on an international scale (Technical Education and Skills Development Authority, 2005). There are thousands of TESDA-certified courses throughout the Philippines, and job categories—such as maritime, domestic-service, health-care, entertainment, and construction workers—that tend to be highly correlated with overseas contract labor constitute a large and very visible portion of available programs.

While the agencies managing Philippine education—the Commission on Higher Education (CHED), which sets standards for the country's universities and oversees the system of accreditation, and the Department of Education (DepEd), which is responsible for primary and secondary education—may not be as closely linked to the contract migration system as TESDA, both agencies have adopted policies to promote the international competitiveness of Philippine labor where their work overlaps with the needs of labor migrants. For example, in 2016 the DepEd implemented a plan to extend the years of primary and secondary schooling from ten to twelve. Key to this change was the need to align the Philippine educational system with the requirements of foreign certification bodies for professions such as engineering, which placed Philippine degree-holders at a lower status because of the country's short primary and secondary education schedule (Mateo, 2016).[9] Similarly, the CHED works to coordinate educational requirements in disciplines that have high rates of contract migration with the certification and employment conditions that workers will have to meet in major destination countries. The CHED's 2011–2016 Strategic Plan explicitly states an objective to support the country's "global competitiveness" and to offer programs that meet the needs of international industry (Commission on Higher Education, 2012). In the case of nursing, for example, this translates into a required curriculum that emphasizes the technology-intensive medical care offered in those countries that import

[9] The English-language proficiency of the Philippine population is often cited as an explanation for the country's high rates of labor migration. This comparative advantage, however, cannot be separated from state policy, because English is not a language that is spoken in the home except among the elite and professional classes. Filipino/a workers entering low-skilled professions like domestic labor and construction acquire English primarily through the educational system, a system that is unique in its promotion of bilingualism through English immersion beginning in secondary school. Even other sending countries with significant colonial histories under English-speaking countries, such as Pakistan or India, do not offer English-only public education as a national policy.

health-care workers from the Philippines—such as the United States, Japan, Germany, Saudi Arabia, and Canada—over medical services adapted to the resource-poor conditions in much of the Philippines (Ortiga, 2014).

Direct Recruitment

The Philippine state's role in direct recruitment is small relative to the overall migration numbers,[10] but it is critical to the labor-export strategy. The POEA/DMW's direct recruitment arm, the Government Placement Branch (GPB), steps in to correct situations where private recruitment efforts are failing, and works preemptively with government clients or in sectors that are valuable in shaping overall migration policy. For example, the GPB was active in the recruitment of labor to Taiwan, when the problem of prohibitively high fees charged by private recruiters for placement to that country threatened the Philippines' ability to establish itself in a potentially important market[11] (R. Casco, former POEA official, personal communication, August 5, 2015). The GPB's participation not only ensured the supply of Filipino workers at a pivotal time, it also succeeded in forcing private firms' compliance, as OFWs now had a cheap, reliable alternative to private recruiters. In more recent years, the GPB's hiring has been focused on the export of health-care workers to countries with labor shortages in these occupations, including the UAE, Canada, Germany, and Japan (Philippine Overseas Employment Administration, 2015), in support of the Philippine state's goal of increasing the proportion of OFWs who migrate to higher-paid, skilled, and professional occupations.

The Philippines in Comparative Context

The description of the modern Philippines' labor-export policies above is just a rough sketch, intended to point out both the continuities and divergences from the LEMI period, and summarize the more in-depth analyses that many migration scholars, including myself, have conducted(see, e.g., Rodriguez, 2010; Lee, 2021; Tyner, 2009; Ortiga, 2018; Parreñas, 2021).

[10] In 2014, approximately 0.5 percent of land-based OFWs, or 8,234 migrants, were deployed through the GPB (Philippine Overseas Employment Administration, 2015).

[11] Between 1995 and 2010, OFWs going to Taiwan represented 13 percent of yearly new hires (POEA 2016b, Deployment per Country per Skill per Sex, 1993–2010).

Even from this description, however, one can get a sense of how different this part of the Philippine state is from the predatory, incoherent state that economic development and state capacity scholars have found in most other policy arenas. Before I begin my analysis of the bureaucratic apparatus that oversees the labor-export policy, I want to take a moment to point out how much the Philippine state stands out among other sending states in both the extent and the longevity of its interventions in migration.

It is true that most sending states, even those with ostensibly laissez-faire emigration policies, are involved in some way in the migration process. Particularly for labor-migration flows enabled by formal guest-worker programs, some minimal level of administrative, documentary, and coordinating support from sending states is required for basic functioning. Thus, even Mexico, which has been famous for its "policy of no policy," has a history of participation in the management and coordination of its labor migration during the years of the bracero program. The Mexican state intervened to facilitate and regulate the northward migration, and helped to determine the original migration flows through which the future migrant network was built. Though this facilitation at the federal level ended shortly after the end of the bracero program, governments continue to provide, at the state level, the services needed for Mexican migrant workers' participation in the H-2 temporary worker visa program of the United States (Fitzgerald, 2006).[12] And as rich economies increasingly turned to guest-worker programs to solve their labor-shortage problems—from the European *Gastarbeiter* programs in the mid-20th century, the import of construction and domestic workers in the Middle East, and the "maid trade" in East Asia to the rise of guest-worker programs even in traditional countries of immigration like the United States and Australia[13]—sending states have become important

[12] The agricultural section of the program, called H-2a, is a smaller-scale continuation of the bracero program, facilitating the import of Mexican farmworkers under similar conditions. In 2014, more than 90 percent of H-2a visas, or 83,674 out of a total 89,274 visas granted, were to Mexican nationals (US Department of State, 2024a).

[13] In the United States, the H-2a and H-2b visa programs in the United States, which cover the admission of workers for agricultural and other temporary or seasonal employment, and which provide no pathway to permanent residence, have grown exponentially since the 1980s. In 1987, a total of 106 workers were admitted under the two programs; by 2016, that number had ballooned to 219,995 (US Department of State, 2024b,). Moreover, revision of the immigration law to expand the H-2 program or implement a similar guest-worker program remains a significant part of the debate on immigration, drawing support from members of both major parties (see, e.g., Porter, 2016).

In the late 1990s, Australia introduced a temporary, employment-based visa, Temporary Work (Skilled) Subclass 457. While the 457 visa does provide a pathway to permanent residency and citizenship, access to that pathway has become increasingly restricted, essentially converting the partners. Particularly at the initiation phase of the migration flow, states have typically played a crucial role, implementing policies and building institutions to facilitate and regulate these migrations. The Philippines is but one example in a list that includes most of the countries that have participated in the temporary migration booms in the Middle East and Asia, including Turkey, South Korea, Indonesia, Pakistan, Thailand, Sri Lanka, Bangladesh, and India (Paine, 1974; Shah & Arnold, 1986).

In many of these countries, state support for labor migration was not limited to facilitation and regulation, but in fact extended to diversifying and directing interventions that worked to support competition in international labor markets and to gain access to new markets, including negotiating bilateral agreements to supply labor and providing skills training to meet foreign demand. In the 1970s, countries like Bangladesh created manpower-training agencies, like TESDA in the Philippines, which were explicitly tasked with labor-force development for foreign labor markets. Morocco, for a short time, attempted to compete in the Middle Eastern labor market and negotiated a series of bilateral agreements with receiving countries. Turkey even pursued its own version of an LEMI policy, competing with the Philippines and South Korea for a portion of the Middle East construction market through the export of Turkish workers.

However, the development of labor-export policy in these countries followed different trajectories. Some, like South Korea, Morocco, and Mexico, would eventually withdraw from *ex ante* interventions in labor migration. Others, like Bangladesh, Turkey, and the Philippines, have maintained their support for the export of temporary contract labor, though the extent of their market-diversification strategies has varied. For Bangladesh, labor-export policies have remained focused on the oil-rich Middle Eastern countries, while Turkey and the Philippines have succeeded in the diversification of their labor migration (United Nations Economic and Social Commission for Asia and the Pacific, 2011; İçduygu & Karaçay, 2012). Still other countries would not shift toward diversifying and directing strategies until much later, as Indonesia did in the late 1990s.

457 program into a guest-worker program for those occupations not strategically important to the Australian economy. (Gothe-Snape & Elton Pym, 2017)

Some of this variation is easy to explain; in particular, we can see how the South Korean state's withdrawal from labor export reflected the country's economic growth, which altered employment patterns and wage levels, changing both the push factors faced by individual migrant workers and the economic value of migration to state managers. Yet the Korean case is unusual: most sending states have withdrawn from *ex ante* interventions in labor export not because economic growth renders such policies unnecessary[14] but because changes in the foreign labor market or immigration policy have interrupted their participation in a particular migration channel. This was the case in both Mexico and Morocco, where, for political and economic reasons, guest-worker programs collapsed in the receiving countries. In both cases, the sending governments initially responded to these changes by attempting to maintain their participation in temporary contract migration—in the Mexican case, through a decade-long effort to renegotiate a migration agreement with the United States, in the Moroccan case through the Middle East negotiations described above. And in both cases, the failures of these efforts eventually led to the countries' retreat from *ex ante* labor-export interventions altogether and to a focus on interventions that sought to maximize remittances through diaspora management.

Looking only at these two cases, the apparent explanation is consistent with the one offered by migration theory: that state migration institutions, like other migration institutions, are products of the migration itself, arising more or less automatically to service migration flows that are largely shaped and driven by other factors (Goss & Lindquist, 1995; Massey et al., 2002). Thus, we would expect them to change forms to reflect the changes in the conditions of migration, as in the Mexican and Moroccan cases. Yet the actual universe of labor-sending states includes states that fail to react to the decline of their primary migration channels, maintaining regimes of facilitation for decades despite steep declines or stagnation in migration and remittance rates (an example of a long-term facilitative regime is Pakistan), as well as states that reorganize their labor-export regimes to access and

[14] While labor-export programs are often built on the premise that they will be temporary and will become unnecessary at some future time when the domestic economy has the capacity to employ the surplus population, that kind of economic development is so rare that there are few cases other than the Korean in which this has occurred; modern Spain and Ireland might qualify as two other examples.

redirect migration flow into new markets. The Philippines exemplifies this second type of response.[15]

Finally, even among sending states that have pursued directing or diversifying strategies toward labor export, the Philippine case stands out. Among those states that have pursued policies intended to redirect or diversify their migration streams, very few—South Korea, and perhaps Turkey—have done so with a level of effectiveness comparable to that of the Philippines. More typical are cases like that of Morocco (described above), Bangladesh, and Indonesia, where efforts to promote existing migration flows and create new flows through policies formally similar to those of the Philippines have had less impressive results. In Bangladesh, the promotion of labor export through skills development has been a policy goal that the state has failed to meet since the creation of the Bureau of Manpower Employment and Training (BMET) in 1976. The country still lacks a nationwide skills-development program; it has only thirty-eight skills-training institutions, public or private, and lacks any infrastructure for conducting research on the labor-market conditions of labor-receiving countries, even in the regions where Bangladeshi migrants already go. Even on the level of regulation and facilitation, the state's migration institutions perform unreliably; foreign governments have periodically halted migration from Bangladesh because of unreliable skills certification and unscrupulous practices by poorly regulated recruiting companies. These failures do not reflect a lack of effort; significant modifications to the labor-migration program have been made every decade since the initiation of the program in the 1970s. But the result of these changes has been to produce multiple institutions with crisscrossing mandates and responsibilities, while the problems with the program have continued (United Nations Economic and Social Commission for Asia and the Pacific, 2011, pp. 127–128, 134).

All this is to say that the kind of labor-export policy that we have seen in the Philippines over the past fifty years does not spontaneously arise out of existing migrant networks, nor does it necessarily develop in response to labor-exporting opportunities. Even when such a policy does develop, it may well fail to persist over generations as the conditions in foreign labor markets change. Success of this kind requires a state with the capacity to make it happen.

[15] In the Turkish case, the state responded to the termination of the European guest-worker programs in the 1970s by turning to the Middle East, where it helped the construction sector to leverage low-wage Turkish workers to compete for building contracts during the petroleum boom.

State Capacity in Labor Export

The Philippine labor-export apparatus is not only an outlier among sending states; it is an outlier *within* the Philippine state in terms of its ability to design and deliver policy. This apparatus has been able to effectively administer and regulate a system that, in coordinating Filipino workers with employers, must operate sufficiently well to synchronize with the demands of immigration law and procedure in not just one or a handful of receiving countries but almost all the countries of the world (Tyner, 2009, p. 177), and do so with a speed sufficient to satisfy the institutions of the developed world that hire many of these workers. If the state did only this much, it would still be an interesting deviation from the economic-development literature's account of the Philippine state. But the Philippine labor-export apparatus does much more: It operates in the labor-export industry as a principal organizer and coordinator, providing services to the workers, employers, private recruiting agencies, and educational institutions that occupy this economic niche. Moreover, it has been able to shape the industry through the manipulation of private labor brokers through goals-based support and discipline. This suggests that the reason why the RDG strategy has worked to generate growth in the Philippines is because this apparatus has demonstrated something like the "transformative capacity" that scholars of economic development have found so absent in the rest of the Philippine state.

So where does the Philippine state's apparent capacity come from? Does the Philippine labor-export policy apparatus look so different from other parts of the Philippine state? In some ways, no. As described in Chapter 5, theorists of state capacity generally agree that the bureaucratic structure of the state—its cohesiveness and autonomy with regard to social and economic actors—is a key requirement for the development of the kinds of capacity required to intervene successfully in national economies.

The question of the state's bureaucratic structure has been analyzed in the literature on two different levels: At the microstructural level, state capacity is attributed to Weberian bureaucratic cultures and practices that protect the integrity of the state against the pull of patrimonial and clientelistic forces. Even within patrimonial states, scholars have found that "pockets of efficiency" can be created, in which Weberian bureaucratic forms are insulated from the rest of the state (see, e.g., Geddes, 1990, 1996; Evans, 1995; Dumitrescu, 2021). Other theorists have argued that the microstructural coherence of the state is less important than organization at the mesostructural level, where a clear, centralized structure of authority that limits disruptive competition between agencies can facilitate the implementation of policy (Johnson, 1982; Chibber, 2003).

In the Philippine case, the consensus is that the country struggles with bureaucratic performance at both levels: While the biggest culprits in the story are the oligarchic elites engage in large-scale rent-seeking and looting of the state's resources, the patrimonialism that facilitates this elite behavior operates through the corruption of state bureaucracies at every level. At the microstructural level, the bureaucracy is staffed on the basis of personal contacts rather than merit; laws go unenforced and services undelivered because of incompetence or the willingness of state employees to overlook such lapses in exchange for direct payments or patrimonial favors. At the mesostructural level, the bureaucratic structure is often so convoluted that agencies compete with each other over decision-making power, and individuals and businesses must pay "fixers" with extraordinary access to officials to complete even basic transactions. The Philippine labor-export apparatus is not immune from the clientelism that characterizes state's bureaucracy at the level of individual staff. However, in the meso-level structures that organize decision-making among agencies, the labor-export system looks very different from the chaotic and dispersed authority structures seen in the development apparatus.

Microstructural coherence and corruption

There is some evidence to suggest that the agencies involved in labor export— the Philippine Overseas Employment Agency (POEA) and the Department of Labor and Employment (DOLE, which housed the POEA until 2022), in particular—are among the less corrupt government institutions in the Philippines. An annual survey of businesses on their perceptions of and experiences with corruption in the Philippines, which has been run by the Philippine nonprofit Social Weather Stations since 2000, has consistently found the DOLE to be among the top 30 to 50 percent of government agencies in terms of its efforts to combat corruption (Social Weather Stations, 2016), ranking well above those considered the worst offenders: the Bureau of Customs (BOC), the Bureau of Internal Revenue (BIR), and the Philippine National Police (PNP) (Hutchcroft, 1998, p. 40). Relying on the state's own assessment of the performance of its bureaucracy—which, admittedly, can be problematic, as the reported cases are likely to represent only a small subset of questionable interactions that businesses and individuals might have with state officials—very broadly confirms the SWS assessment, though it also reveals that a significant portion of the microstructural corruption occurs at the local level. The Office of the Ombudsman reports that, of the 4,577 cases filed in 2016, 61 percent involved local and municipal government officials. Of the 1,778 cases filed against national government agencies, more than 86 percent, or 1,022, were made against the PNP; none were made against the DOLE or POEA. The case docket of the Philippine Civil Service Commission (CSC) reveals a similar distribution—thirty-one of the eighty-six cases involving nepotism, dishonesty, and graft/corruption in the first quarter of 2015 were against local or provincial government officials; the PNP was responsible for another eleven cases, and only one complaint was brought against an employee of DOLE.[16]

Still, these statistics cannot stand as an explanation for the Philippine labor-export system's effectiveness. While agencies like the DOLE and the POEA may perform better with regard to nepotism, graft, and petty corruption than the most problematic agencies, they do not rank any higher than agencies involved in policy areas in which the Philippines has struggled, such as the Bureau of Trade and Industry and the Securities and Exchange Commission (Social Weather Stations, 2016). Moreover, among the observers of Philippine labor migration, the problem of petty corruption is widely acknowledged; the use of "fixers" to expedite processing, or even to circumvent regulations, is enough of a problem that the agency regularly warns potential migrants against hiring them (Philippine Overseas Employment Administration, 2013), and incoming POEA administrators cite the curtailment of the practice as a policy goal (Global Media Arts News Online, 2012).

While most of the former and current labor-migration officials have been circumspect in their discussion of corruption, Jose Sarmiento, who ran the Overseas Workers Welfare Administration (OWWA) and the POEA under Corazon Aquino, heading each agency for three years, has been more direct in talking about corruption in the agencies in which he worked. Though he

[16] These statistics are drawn from my own analysis of Case Decisions and Resolutions reports of the CSC. I included in my analysis the following categories: Grave Misconduct; Dishonesty; Examination Irregularity; Failure to Declare Truth in a Personal Data Sheet; Failure to File Statement of Assets, Liabilities, and Net Worth; Failure to Meet Education and Training Requirements; Failure to Meet Eligibility Requirements; Falsification of Official Documents; Graft and Corruption; and Grave Abuse of Authority.

prefaced his statements by affirming that he believed most bureaucrats in the POEA to be honest and capable, and that the culpable actor in the corruption issue was the private sector, he also reported a system in which corruption was commonplace and prevalent. He stated first that there is probably a good deal of petty graft—twenty pesos here and there to jump the line—but that he did not really consider this corruption. On a larger scale, he said that POEA administrators were known for making money during their tenure, and that his refusal to accept money from promoters was something of an anomaly.

Sarmiento also reported that he knew that recruitment agencies had "friends" in the POEA who would alert them of inspections, and that he made a policy of joining the inspection teams and adding agencies to the inspection list at the last minute in order to ensure the "surprise" nature of the inspections. This practice would turn up irregularities—agencies charging up to six times the legal rate for job placement, rural applicants being forced to sell off productive farm assets in order to pay the exorbitant fees—that did not appear on reports from "unannounced" inspections that were planned well in advance. Sarmiento also reported being offered millions of pesos (the exchange rate in 1990 was forty pesos per US dollar, so the scale of the bribes offered was in the tens or hundreds of thousands of dollars) for the reinstatement of suspended recruiting licenses. He also reported bribes being offered indirectly, through close personal friends and subordinates in the POEA, and confirmed that his refusal to participate in the graft led to physical threats against his person (Sarmiento, J., personal communication, August 19, 2015).

Even if what Sarmiento described was anomalous, or the level of corruption he experienced was still sufficiently low relative to other Philippine government agencies that it did not hinder the ability of the POEA to deliver on its labor-export policy, the process of preparing a worker for overseas employment implicates so many other government agencies that it is unlikely that labor export could be insulated from the microstructural corruption that plagues the rest of the state. The process of labor export requires potential migrants to provide a host of certifications and documents, including proof of identity, birth and marriage certificates, medical certifications, skills certifications, and criminal background checks. While some of this processing occurs through the POEA and other subsidiary agencies of the Department of Labor and Employment (DOLE), some requires interaction with two of the most problematic parts of the Philippine state: local governments and the national police. Some positions require a "Barangay Clearance," a certification by a local government official that the applicant is a resident of that *barangay* and of good moral character. Even obtaining a copy of a birth certificate can become a complicated, timely, and expensive process. If a child's birth was not properly recorded at the time of birth (not unusual in rural settings), the process of getting a birth certificate, which is required not only for some overseas job applications but also minimally in order to get a passport, requires multiple steps. In some municipalities, the process is so labyrinthine that it cannot be done in a timely manner without the assistance of a "fixer," who typically charged 25,000 pesos in the early 1990s, a time when the average monthly wages for skilled workers hovered at around 3,000 pesos (Philippine Department of Labor, 1991) for the service (Felix Fariñas, University of the Philippines employee, personal communication, January 14, 2015).

Finally, and as José Sarmiento suggested in his statement that petty graft should not be considered corruption, the real reason why the microstructural weakness of a bureaucracy matters for state capacity is not fundamentally about whether individual officials charge unauthorized fees for the discharge of their duties. What has most hampered the operation of the Philippine state is its cronyism at the highest levels: the way in which the state distributes privileges and rents on the basis of social ties, the way high-level officials use the treasury for their personal gain, and the extent to which Philippine capital is dependent on the state for its reproduction. Some, though not all, of these problems are facilitated by the weakness of the bureaucracy, particularly the way in which nepotism undermines the chain of command of the state's institutions (Hutchcroft, 2000). Here, too, the evidence suggests that the labor-export agencies were not immune to these problems. The Overseas Workers Welfare Administration Fund, which is funded through a mandatory $25 contribution from every OFW going abroad, has been a constant source of controversy. With nearly 2 million OFWs departing every year, it represents a significant pool of resources to be managed by the state. Like many such pools in the Philippine state, however, it has been used regularly for the political and economic benefit of elected officials and their supporters. President Fidel Ramos ordered an investment by the OWWA Fund, without the approval of the Fund trustees, for a development project. The Fund also extended a billion-peso loan that was never repaid (Connie Bragas Regalado, chairperson of the Migrante Sectoral Party, personal communication, July 21, 2014). In another instance,

President Gloria Macapagal-Arroyo used 530 million pesos from the Fund to pay for an expansion of the PhilHealth national medical insurance during her reelection bid in 2004; the PhilHealth expansion was explicitly used in her campaign rallies, where she distributed PhilHealth membership cards that were printed with her name and image (Philippine Daily Inquirer, 2012).

If such misappropriations had little impact on the operation of labor export itself, it may be because, as already mentioned, the sums associated with labor export were relatively small. The OWWA Fund is one of the few parts of the labor-export program that generate a large pool of money that could potentially be looted, and it is not surprising that the major financial scandals associated with the labor-export program have involved the OWWA Fund. While the state's initial policies on labor export did attempt to create a state-controlled monopoly, the failure of that approach has meant that profit in the industry does not rely as heavily on the state's favor as it does in other subsidized or protected industries. While it is true that the POEA engages in direct negotiations with employers and governments and is in a position to favor certain companies for the contracts that result from such negotiations, the share of the labor migration that flows through such channels is relatively small.

The other channel through which favors and rents might be extracted is in the granting of licenses to recruitment agencies. The reports from José Sarmiento suggest that bribery in relation to such licenses was not uncommon. The POEA's current policy on agency licensing is called "Hard to enter—easy to go," which, while its goal is to ensure the quality of recruiting agencies, could be used to provide protection for a few state-favored agencies. However, in practice, the licensing system does not appear to have created a small group of agencies with monopoly power. The entry requirements are not onerous, requiring only a capitalization of about $11,000 (500,000 pesos), some proof of the market demand, and no prior record of recruiting violations by the business proprietors (Philippine Overseas Employment Administration, 2016). In 2017, there were 828 licensed land-based recruiting agencies and another 403 engaged in recruiting for sea-based employment (Philippine Overseas Employment Administration, 2017b). In an industry with revenues that range from 1 to 2 billion dollars, these numbers suggest a relatively competitive market.

Taken all together, the discussion above suggests the following: First, that the agencies of the labor-export program have not been, on the level of microstructural bureaucratic capacity, immune from the petty graft and rent-seeking that characterizes the rest of the Philippine government. This supports the arguments of scholars such as Kang (2004) and Khan & Jomo (2004) who argue that the presence of rent-seeking alone does not predict poor outcomes. Institutional and political variables play key roles in the impact of rent-seeking behaviors. In the Philippine case, Hutchcroft has argued that corruption hinders economic development because it is unpredictable, arbitrary, and variable (1998, p. 41). What makes Philippine corruption arbitrary is not the petty graft in the processing of day-to-day documents, but rather the exercise of power and influence by a small oligarchic elite who misappropriate state funds, distributing subsidized loans to cronies and making payments to clients for infrastructure projects that are never built. Apart from the OWWA Fund,[17] most of the activity of labor export has simply occurred at a scale too small to be drawn into this system of looting. Unlike banking or real estate, participation in which is *de rigueur* for the Philippine family conglomerates, the recruiting business is primarily a sphere of small businesses. If anything insulates the labor-export agencies from this large-scale rent-seeking, it is this division between the "industry" of migration and the remittances that make up the bulk of migration's impact on the domestic economy.

The performance of the labor-export system suggests, then, that the petty graft and microstructural corruption characteristic of the Philippine state do not impede the basic and even efficient operation of the state, as long as certain other factors are present. One such factor is the relative absence of looting by economic elites.

Mesostructural bureaucratic coherence

The other factor, related to the first, is the interagency organizational structure of the country's labor-export apparatus. Since the early 1980s, the labor-export system has been organized around a single, nodal agency—initially the POEA, and now the Department of Migrants Workers (DMW)—with explicit authority over the management of the program. While other agencies, such as the Department of Foreign Affairs (DFA) and TESDA, are

[17] Because the purview of the Overseas Workers Welfare Administration (OWWA) is the protection of migrant workers *after* they have been deployed, the looting of the OWWA Fund has had few direct consequences for labor export itself.

closely involved in various aspects of labor export, the policymaking and implementation centers around this nodal agency, which acts as a hub through which information flows and which coordinates the support work of these other agencies, in addition to administering the recruiting agencies and processing and documenting the migrant flow. This structure is particularly remarkable when we compare it to the organizational structure around other policy areas, such as agrarian reform and economic development, which are characterized by balkanized agencies with overlapping mandates and central agencies that lack any authority to implement or enforce its decisions.

As it was originally designed in the 1974 Labor Code, the labor-export program was not an outlier in the Philippine state. Its agency-level structure was disorganized and confusing, scattered among three agencies: the preexisting Bureau of Employment Services (BES), the Overseas Employment Development Board (OEDB), and the National Seamen Board (NSB). The OEDB and the NSB were tasked with market development, direct recruitment, and the oversight of contract conditions for land-based and sea-based workers, respectively. The BES, which had previously served as a government-run temporary-employment agency for the domestic labor market, primarily providing labor to multinational corporations in the Philippines, would now become responsible for "developing and monitoring a comprehensive employment program"; this would include regulating and licensing the recruiting agencies, developing a labor-market information system that would inform manpower development and planning, and maintaining a central registry of skills (Presidential Decree No. 442, 1974). The program planned to phase out private recruitment altogether within five years, creating a state monopoly of labor export.

This system quickly proved unworkable. The state's attempt to take over the recruitment industry translated in practice into restrictions on the issuing of licenses to private agencies, so that licensees—nineteen licensed agencies existed in 1974—shared with the state a monopoly over a contract migration that had grown explosively during the decade, from fewer than 2,000 migrants in 1970 to over 36,000 in 1976 (Abella, 1979, p. 9). The value of licenses created opportunities for favoritism and graft, and labor-export agencies quickly developed a reputation for corruption. More problematic was the three-pronged institutional structure. Patricia Santo Tomas, the first administrator of the POEA, writes:

> There were organizational conflicts between and among agencies involved in the overseas employment program. Each of the three agencies—BES, OEDB, and NSB—had its own administrative support services, marketing and placement departments, licensing and regulation units, and welfare and adjudication staffs. However, because of the differentials in resources and orientations, there was also wide varieties of procedures, approaches, capabilities, and priorities. Without meaning to, the three agencies seemed to be pulling in different directions. (1984)

The disorganization produced a number of problems. Employers complained about the screening process, which failed to properly find workers with the required skills, leading to high rejection rates at the worksite and costly delays caused by the slow mobilization times once workers had been matched. For the state itself, it was clear that the systems were failing; increasing rates of deployment were not being reflected in revenue, and contributions to the Welfare Fund for Overseas Workers (the predecessor to OWWA) remained stagnant even as rates of deployment doubled and tripled. Since the three agencies did not coordinate their data collection, working with different conceptual and operational definitions, it was impossible to obtain accurate totals of contract migration flows (Santo Tomas, 1984). The problems immediately impacted the contract migration flow; though the statistics from this period vary, there is general agreement among the available data that contract labor migration experienced a stagnation in the first years of the labor-export program.[18]

While similar problems in the agrarian-reform apparatus went unaddressed for decades, in the case of labor export the response was reasonably swift. Within the first five years, the state had abandoned its policy of phasing out the private recruiting agencies and begun actively supporting certain types of labor exporters, specifically the construction firms competing for contracts in the Middle East. Rates of contract migration not only recovered but grew explosively, from 36,676 in 1977 to 380,263 in 1980. To some extent, this rapid growth in contract migrations was an effect of an inefficient state getting out of the way. However, the end of the Middle East

[18] Abella, whose data is drawn from the selected labor indicators provided by the Ministry of Labor, finds that between 1973 and 1975 contract migration rates declined slightly, from 36,418 to 36,022 (1979). The statistics published in the *Yearbook of Labor Statistics* report a larger decline, from 36,418 in 1973 to 19,222 in 1976 (Philippine Department of Labor, 1976; Philippine Ministry of Labor and Employment, 1982).

petroleum boom in the early 1980s, with the concomitant drop in demand for construction labor, demonstrated that the private recruiting agencies on their own were vulnerable to volatility in labor demand in specific markets and could not sustain the high rate of labor export without effective support from the state.[19]

This time, the state's response was quite swift. In 1982, the labor-export apparatus was completely reorganized. The BES, OEDB, and NSB were dissolved. Recruitment for domestic employment was separated from overseas employment, and another agency, the Bureau of Local Employment, took over the functions of the BES and other agencies involved in domestic employment. All the overseas employment functions were centralized under a newly created agency, the POEA. Not only did the reorganization centralize labor-export activities, it also transformed the leadership and decision-making structure of the labor-export apparatus. Under the old structure, responsibility was divided between three agencies, and the governing bodies of the NSB and OEDB were composed of eleven members, each representing different interests or areas of expertise, including not only representatives of capital and labor but also the Department of National Defense and the Department of Education, Culture, and Sports (Presidential Decree No. 442, 1974). This structure, as well as the relatively low rank of the executive directors of the NSB and the OEDB, had contributed to the problems of the labor-export apparatus by hindering decision-making. The POEA's governing board was slashed to three members: the Secretary of Labor,[20] the POEA administrator, and a private-sector representative. The POEA administrator thus gained larger influence on the smaller governing board, and the position itself was upgraded to the rank of Deputy Minister. The expansion of the administrator's influence on the governing board helped to forge a closer link between policy-making and implementation. And while the smaller size of the board, and the removal of representatives from labor in particular, meant that there were fewer checks and balances, it also made the agency much more nimble, able to respond quickly to changes and opportunities in the overseas labor market (Santo Tomas, 1984).

The internal structure of the POEA was organized around functions rather than the client-based divisions that characterized the old system.

[19] Santo Tomas writes: "[I]t became clear that the Philippine competitive edge relative to overseas employment may be preserved only by improving on its previous performance. Among others, this meant more responsive structures, streamlined procedures and in some cases, policy reorientation." (1984, p. 67)

[20] At the time, the Department of Labor and Employment (DOLE) was called the Ministry of Labor and Employment and was led by the Minister, not the Secretary, of Labor and Employment. The institution was renamed by the Aquino administration after the end of the Marcos regime.

All interactions with recruiting agencies, including licensing and regulation enforcement, whether of sea- or land-based recruiters, was centralized under one department administrator for licensing and adjudication. Similarly, all worker and employer services were centralized, as was marketing. Importantly, the POEA also created a centralized information-management system to gather data on the contract migration flow (Santo Tomas, 1984). The POEA became the central organization through which all (legal) contract migration flows were processed; workers, employers, and recruiters were all documented and certified through the agency. It also served as the information hub for all matters related to labor export, not only through its own data-gathering services but through the Philippine Overseas Labor Offices located in the major destination countries for Filipino/a workers. Though POLOs are generally placed inside of embassies, POLO labor attachés are employees of the DOLE rather than the DFA. In addition to providing services to OFWs while abroad and assisting in the evaluation of employers, a key responsibility of the POLOs is to gather data on labor-market conditions, to be used by the POEA's marketing branch in developing its plans for training and skills development and outreach. POLOs are expected to report every semester not only on changes in the labor law but on anything that could signal a shift in labor demand in a country, and specifically which skills should be given more attention by TESDA programs. The POEA coordinates with the country's education agencies, in particular TESDA, to coordinate vocational and technical education with the international market. All the agencies involved in labor export form what is called the "Migration Cluster," which can meet as often as once a week if necessary. While each agency may be responsible for its discrete function, the POEA administrator remains at the head of the apparatus, chairing key meetings and heading any negotiation with international counterparts (Saul de Vries, Director, International Labor Affairs Bureau (ILAB), personal communication, August 18, 2015).[21]

The Stability of the Labor-Export Program

This structure, established in the early 1980s, has been remarkably stable. Though some related agencies, such as the OWWA and TESDA, are newer, reflecting reorganization at the margins, the essential structure, with

[21] ILAB is the subagency within DOLE that oversees the work of the POLOs.

the POEA at the center, has remained in place for over forty years. It has survived despite institutional competition from other agencies—what POEA staffers called "turfing"—particularly by the Commission of Filipinos Overseas (CFO), which has sought to bring some aspects of contract migration under its purview, including management of remittances. Because of its participation in the absentee voting program, which began in 2004, the CFO has had political influence (Ricardo Casco, personal communication August 5, 2015). Thus far, however, competition from the CFO has not produced a significant change in the structures around labor export. Some major changes to the structure were enacted at the end of 2021 with the signing of the Department of Migrant Workers Act (Republic Act No. 11641, 2021). Though the actual operation of the DMW remains to be seen, on paper the changes seem oriented toward an even greater centralization and prioritization. Labor migration has now been raised to the level of an executive department (the POEA was previously a subagency of the Department of Labor and Employment), with most of the agencies working on OFW issues now being folded into the DMW. The operation of the program seems to have changed little, with the POEA, now operating under the DMW, remaining the "backbone" of the labor-migration program (Jaymalin, 2021).

This stability is more remarkable given that, ever since labor migration became a major project for the country, the state's involvement in labor export has been a contentious issue in the public discourse. Almost a thousand OFWs are killed every year, and another 3,000 are incarcerated (del Callar, 2014). These numbers do not include the thousands of cases of discrimination and of physical and sexual abuse, nor does it take into account the isolation that OFWs experience even under the best employment conditions, or the costs of long separation from children and family. The state defends its labor-export policy as economic necessity, but that very necessity reveals the state's failure with regard to domestic employment and economic growth. Yet it is that continuing failure that generates the labor-export program's stability. Even organizations like Migrante International, which has been openly critical of the Philippine state's labor-migration program—explicitly naming it labor export and "forced migration" (Connie Bragas Regalado, chairperson of the Migrante Sectoral Party, personal communication, July 21, 2014;Migrante International 2018—do not call for an immediate end to migration. The grassroots opposition to the program instead takes the form of a generalized condemnation of the state for its inability to resolve the fundamental economic and social issues that make migration a necessity for many workers, as well as its exploitation of that desperation for its own ends. The immediate demands, then, are for more and *better* state intervention, specifically in the regulation of private recruiters, the protection of workers, and the management of the funds allotted to the migration program.Within such a context, the aspects of the migration program that could be more easily characterized as "export" rather than "protection"—such as the state's attempts to diversify labor export through active marketing—are the most controversial and the ripest for reform. When governments were under pressure, as during the period before the fall of the Marcos regime, the marketing aspect of the migration program was a target of criticism. POEA bureaucrats working in the marketing department in the 1980s remember a shift, around 1985, in the organization's attitude toward marketing, which up to that time had been considered a priority. As the mobilizations against Marcos intensified, the regime began to withdraw support from a publicly acknowledged marketing strategy.[22] Yet such retreats from an explicit labor-export strategy were only rhetorical; the same activities that had been criticized as labor export—marketing, training and skills development, licensing of private agencies—were relabeled as a form of worker protection. Properly trained and skilled workers, with many options about where they could find employment, are workers that are less likely to be abused or exploited. The state participates in the management of private recruiters to prevent unscrupulous practices. While these claims are not necessarily false, POEA officials have also made clear that they have been repackaging a set of policies to be more acceptable to particular audiences.

> The NGOs, it is just that their lingo, marketing, so sometimes when I do speeches and face them, I adopt different language so that they would understand that what we're doing, what are of marketing significance, are according to welfare, no? (R. Casco, former POEA official, personal communication, August 5, 2015)

The private recruiting firms, on the other hand, have called for reforms in the other direction. They resisted the efforts of the Marcos regime to phase them out of labor export altogether. Particularly after the first Aquino presidency placed a new emphasis on worker protection and welfare, the recruiting

[22] Ricardo Casco, a former POEA official who worked in the marketing department at the time, reports being "scolded" in 1985 for a report on marketing strategies that he had been assigned to write only the year before.

firms have had a contentious relationship with the POEA, arguing that state regulation was "killing the market" (R. Casco, former POEA official, personal communication, August 5, 2015). Though there have been complaints about the POEA's lax enforcement of recruitment regulations, the state policies can and do impact the industry. The state has banned the issuance of new licenses for up to five years at a time, and regularly imposes bans on labor export to certain markets in response to problems caused not only by conditions in the receiving country but also by widespread complaints against recruiters operating in that region. The state has also used its own recruitment arm—the Government Placement Branch, or GPB—to discipline private recruiters, entering markets that private firms are failing to serve in some capacity.[23] While the scale of the GPB's operation is small, involving only a few thousand OFWs per year, it tends to enter where the POEA has had some conflicts with private recruiters, and so is considered a threat to these businesses. Private recruiters have pushed for a return to Blas Ople's original position embracing labor export and strengthening its marketing and promotion efforts while removing government agencies from recruitment altogether, closing the GPB, and deregulating the private sector (Guevarra, 2010). Their position is essentially the typical demand of all industries with regard to the state: that is, subsidies and support without discipline.

Had the private recruiters been a more influential or important sector of capital, it is possible that the labor-export regime might have been revised to reflect their demands. The oligarchic structure of the Philippine political economy has meant that very few policy programs can be designed or implemented over the protests of the economic elite. However, apart from the labor recruiters, Philippine business has not been particularly critical of the labor-export program. Certainly, the export of skilled workers does have some impact on domestic firms. Nieves Confesor, the Secretary of Labor and Employment during the mid-1990s, when major reforms of the labor-export program were debated, reported that those industries that relied on engineering and skilled technical labor—including energy, the airlines, software engineering, and construction—had lodged complaints with the DOLE about the loss of skilled workers to foreign markets (Joint Committee on Foreign Relations/Labor, Employment

[23] For example, the GPB engaged in direct recruitment for the Taiwanese market, where the POEA struggled for years to curtail the private recruiters' practice of charging prohibitively and illegally high fees (R. Casco, former POEA official, personal communication, August 5, 2015).

and Human Resources Development testimony, November 3, 1994). Yet the tone of their complaints has been notably muted. For example, some of the companies specifically named by Confesor in her testimony, such as Philippine Airlines and DMCI, were owned by prominent businesses with ties to politicians at the highest level. Lucio Tan, the majority owner of Philippine Airlines, had built his wealth through his political connections, and it seems unlikely that he would not have used it to protect his investment in the airline if labor shortage had been a significant problem. The construction industry, on the other hand, made contradictory and inconsistent statements, with some major companies lodging complaints while the president of the Philippine Contractors Association argued against limiting labor migration for construction workers, who were crucial to the construction industry's international projects (Republic of the Philippines, 1995a, b).

This wavering, even by sectors experiencing negative impacts from the labor export, reflects the diversified conglomerate form of Philippine capital. By the mid-1990s, the remittance-driven consumption sector had produced a number of new billionaires, and the more established family conglomerates were now making investments in this area. In the case of Lucio Tan, his holdings in Philippine Airlines were only a small component of a larger portfolio that included tobacco and alcohol, banking, and property development. Similarly, by the mid-1990s major construction firms like DMCI had begun to expand into residential real estate development. As an organized group, employers chose not to mobilize against it, opting instead to mitigate their losses through other means, such as "lock-in" provision in employment contracts, which limit the mobility of workers who benefit from company-sponsored training program for a period of time, and pressuring the state to increase vocational skills training (Romeo Garcia, Employers Confederation of the Philippines (ECOP), personal communication, August 12, 2015). And those new billionaires whose wealth had been built on remittances were more explicit in their support for continued migration. Manny Villar, the founder of Vista Land, has built his political career on migration issues. Though he rarely explicitly supports the export of labor, he has developed a reputation as a supporter of OFW causes, particularly those that facilitate remittances and OFW investments. One of his first sponsored pieces of legislation was an act to create an investment-fund portfolio for OFWs and reserve 20 percent of all privatizing state-owned corporations for OFW investment (Villar, 1994).

The only times that significant reform of the labor-export program has been considered have been during periods of mass mobilization—immediately after the People Power Revolution in 1986, and again in the mid-1990s when the conviction and execution of a Filipina domestic worker in Singapore sparked widespread protests against the labor-migration program. In both cases, while some changes were instituted, particularly extensions of the political rights of workers and their access to services and protection from the Philippine state while abroad, the essential thrust of the labor-export program has remained unchanged.

In 1986, among the many mandates that Corazon Aquino received through her election was one to transform the labor-migration system. Marcos had long been criticized for the labor-export program, blamed both for economic conditions that made labor export necessary and for the state's lackluster protection of workers while abroad. OFW groups and NGOs felt they had a better rapport with the Aquino administration and hoped that Aquino would oversee a shift in policy direction. Aquino did issue an executive order in July of 1987 to reorganize the POEA, but the measure was not radical, leaving most of the structure in place; more importantly, it explicitly reiterated the state's commitment to labor export: "[POEA reorganization was necessary] in order to enhance its effectiveness in responding to changing market and economic conditions" (Executive Order No. 247). Initially, there seemed to be more activity in Congress, where a total of 142 bills and resolutions were filed between 1987 and 1991. However, the only bill that passed into law dealt with remittances: Republic Act 7111, "An Act Establishing the Overseas Workers Investment Fund to Provide Incentives to Overseas Workers, Reduce the Foreign Debt Burden and for Other Purposes," enacted in 1991.

By the time of the Ramos administration (1992–1998), the labor-export program had become a much bigger political problem. A number of high-profile OFW deaths—including the execution in Singapore of Flor Contemplacion, a domestic worker who many believed had been falsely accused of murder—had incited mass mobilizations against the program. The primary demand was for better protection for OFWs, but the underlying critique was of the country's reliance on labor export to address its unresolved economic problems. While public pressure forced Congress to discuss a number of bills relating to OFWs and labor export in the session year 1994–1995, Congress adjourned in February of 1995 without having passed a comprehensive migration-reform bill. In March, when Flor Contemplacion was executed despite intervention from President Ramos, her death and the plight of OFWs became the key issue of the general election that occurred in May. Ramos was forced to appoint a Presidential Fact-Finding and Policy Advisory Commission on the Protection of Overseas Filipinos (called the Gancayco Commission), then reconvened Congress for a special session immediately after the election to address the crisis. Prior to the special session, the Gancayco Commission published a report that recommended, in addition to a number of provisions for legal assistance and services to OFWs, the phasing-out of certain types of labor export (specifically, export of domestic workers and entertainers), as well as stronger regulations on and better oversight of the activities of private recruiters.

The law that was enacted, the Migrant Workers and Overseas Filipinos Act (Republic Act No. 8042, 1995), included a host of new legal and social services for OFWs, but it notably did not include any restrictions on labor export itself. The Congressional debates on R.A. 8042 and its predecessor bills reveal an explicit intent to preserve the labor export; only one speech on the House floor supported the Gancayco Commission's recommendation on the phaseout, made by a sectoral representative of labor, Andres L. Dinglasan, Jr. All the other arguments focused on the importance of migration to the Philippine economy and interpreted a protection measure in that context. Increased regulations were challenged on the grounds that they would affect the competitiveness of Filipino/a workers in the international labor market:

> Today there are 260 million unemployed in China. Our labor officials should prepare for the worst and assume that this untapped work force might be eventually utilized in Hong Kong. . . . The government's primary purpose is to assist the plight of our overseas workers and not to place any more hindrances on fellow Filipinos trying to better their lives by daring to seek employment abroad. . . . The delay alone in ensuring compliance is indicative of the folly of this endeavor. I wonder if we are prepared to accept the bitter consequences if thousands of Filipino entertainers suddenly find themselves out of a job. More frightening is the possible scenario whereby work of OCWs [overseas contract workers] is taken over by Romanians, Russians, and South Americans who have less strict administrative policies. Nigata [Japan] is only an hour and 50 minutes from Vladivostok. (Representative Roque R. Ablan, Jr., February 27, 1995, in Republic of the Philippines, Record of the House of Representatives 1995)

Even rather tepid language, expressing less than full support for a labor-export policy, was challenged on the floor until assurances were given that it would not be interpreted as a retreat from labor-export promotion. The law's "Declaration of Policies" states first that "the State does not promote overseas employment as a means to sustain economic growth and achieve national development" and "as soon as practicable, the government shall deploy and/or allow the deployment only to skilled Filipino workers" (Republic Act No. 8042, 1995, Section 2(c, g)). Both were carefully worded concessions to the national mood, attempting to replace the rhetoric attributing labor export to state policy with new language merely stating that the state had tried to manage the flow,[24] as well as promising a retreat from the most controversial aspects of the policy in some unspecified future. Yet Representative Roilo Golez, from a district in Metro Manila, demanded an explicit statement from the bill's sponsors that those provisions would not be interpreted to discourage the deployment of OFWs.[25]

One major change that R.A. 8042 did include, however, was a provision to deregulate the private recruitment agencies over the next decade. This was a reform that private recruiters had long pursued (Tigno, 2004), arguing that the state was incapable of effectively implementing regulations without damaging the international competitiveness of Filipino/a workers. OFW groups and POEA officials, who considered the private sector to be a main source of the problems associated with labor export, opposed absolute deregulation, though there was support for streamlined processes and more effective governance. Private recruiters were able to win this concession because of a confluence of factors. Congress was looking for a way to claim that it was withdrawing from its active role in labor-export promotion. Abolishing the POEA and the labor-export program altogether was proposed

[24] A White Paper on the Overseas Employment Program, published by DOLE at around the same time as the Gancayco Commission Report, suggests that the state's position should move away from an explicit promotion of labor export toward "managing effectively the natural process of labor migration—which will continue even if we ban the outflow of our workers." (DOLE, 1995, cited in Tigno, 2004)

[25] "Golez: In other words, Mr. Speaker, we take it that the sponsoring committee has no intention of ceasing to promote the export of overseas workers.

"Lopez: It is not the intention of the House Foreign Affairs Committee to discourage the deployment of migrant workers. . . .

"Golez: Now, may we have the assurance from the chairman that in the formulation of the rules and regulations, it will contain provisions that will clearly state that it is not the intention of this particular provision to discourage the deployment of overseas workers?

"Lopez: I will do that. That is the intention; we will not discourage the deployment of migrant workers." (Republic of the Philippines, 1995)

but rejected, and deregulation became acceptable as a "soft compromise" (Republic of the Philippines, 1995b). The state could claim to have withdrawn from labor-export promotion in a way that would not hinder the activity of the private sector and moreover was consistent with the general liberalization strategy of the Ramos administration. The market-promotion and skills-development aspects of the program would remain, relabeled as "protection." The OFW organizations and labor NGOs that might have objected to the reform were largely excluded from the discussion over this provision, which appeared at the very end of a years-long debate during the special session for which these groups had not been given time to prepare or participate.

While the deregulation was supported by Nieves Confesor, the Secretary of Labor and Employment, the lower-level bureaucrats who would implement the deregulation were skeptical of deregulation. Some of this skepticism had to do with a concern for OFW welfare, but another key issue was the ability of private recruiters, on their own, to maintain the policy of diversification that had helped to expand the Philippines' migration. Without the state's pressure, private recruiters would not push their way into new markets, choosing to remain in the territories with which they were familiar and where they incurred fewer risks (Guevarra, 2010). In addition, without the state's regulations on fees, even if they were not always strictly enforced, migration into the more highly paid receiving countries and occupations would become prohibitively expensive, harming the country's overall migration rates. There seems to be some understanding in the legislature that deregulation might have this effect, and so R.A. 8042 delayed the implementation of deregulation for five years, with the possibility of further delay, which legislators understood would be tied to the country's economic condition (Republic of the Philippines, 1995b).[26] Given the unlikelihood of an economic transformation on a scale required to negate the need for labor export occurring in the subsequent five years, the deregulation was probably never intended as more than a public-relations delay tactic. Practically speaking, this turned out to be the case: the five-year period passed without

[26] "Sen. Lopez: Can we be so sure that within that period it would be justified to phase out this particular agency of the government considering the changes going in the world right now?

"Sen. Maceda: . . . Why should you doubt that in five years you can be so sure under order of representation of President Ramos that we are on the route to economic NIChood? So if you have any confidence with the President's statement, this five-years period is really a long time."

deregulation, and in 2009 the deregulation provisions of R.A. 8042 were quietly repealed (Republic Act No. 10022, 2010).

Oligarchic Looting and the Remittance Work-Around

To fully understand the role of labor migration in the Philippine political climate, it is necessary to discuss briefly the two other sectors in which labor migration and remittances affect the country's economy: the recruitment/employment placement agencies and the banking sector. Some analysts of the Philippine state's policies on labor migration have argued that it is the influence of these two industries that has pushed the state to support and promote the labor-migration program. Banking and finance, as a sector, is deeply connected to the oligarchic family conglomerates that dominate the Philippine economy (Krinks, 2002, pp. 193–196; Hutchcroft, 1998) and plays a critical role in the economy. In 2004, the Asian Development Bank estimated that revenues from remittance processing were approximately $250–$500 million (Table 7.1). Considering that migrant remittances have nearly doubled since 2004, those revenues would now approach $1 billion. For the Philippines' major banks, this represents a significant share of revenue. The Bank of the Philippine Islands (BPI, a subsidiary of the Ayala Corporation) captures approximately 30 percent of the remittance market; at $300 million, remittance processing constitutes approximately one-fourth of the bank's yearly revenue (Bank of the Philippine Islands, 2014).

Labor recruitment, as an industry, is similar in scale to that of the remittance-processing section of banking. In 2016, the Annual Survey of Philippine Business and Industry (PASBI) reported the income of

Table 7.1 Estimates of Banking Revenue from Remittances

Estimates of Remittances	$7.6–$14 billion
Average Transaction Size	$350
Transaction Volume	21.7–40 million
Average Price	$8
Average Foreign Exchange (FX) Spread	1 percent
Transaction Revenue	$174–$320 million
FX Revenue	$76–$140 million
Total Revenue	$250–$460 million

Source: Asian Development Bank (2004)

"employment placement agencies" to be 43.6 billion pesos, or roughly $1 billion (Philippine Statistics Authority, 2018). While this category of the survey also includes employment placement for the domestic market, it serves as a very rough estimate of the scale of the market.[27] As a share of an economy whose GDP was $318.6 billion in 2016, this is significant but not so exceptional that we would expect the industry to be influential in the direction of state policy. Moreover, the major conglomerates of the Philippine economy, concentrated in banking, real estate, and retail, have not made investments or set up operations in this sector.

This is a peculiar fact about labor export as an "industry" or economic phenomenon. While it contributes over $35 billion to the Philippine economy every year—equivalent to 10 percent of the country's GDP—the "business" of exporting labor represents less than 5 percent of that figure. The money enters the country not in the form of the sales and profits of large firms but instead as millions of small remittances to individuals and households. Philippine capital has figured out how to capture and profit from these flows through the banking system and, as described in the previous chapter, through those sectors in which remittance dollars are spent—real estate and retail consumption. But while direct labor export by producers, such as international construction firms, survived the LEMI period, it still constitutes only a small fraction of the total labor export (less than 10 percent of the new land-based hires in 2015 were in occupations related to construction). The vast majority of businesses that engage in labor-exporting activity are manpower-recruitment firms, whose earnings are limited to fees charged to employers and/or the migrants themselves. Staffhouse International Resources, a top recruitment firm, reports an annual revenue of approximately $20 million—less than 5 percent of the value of international contracts of the Construction and Development Corporation of the Philippines (CDCP) in 1980.

[27] Another rough estimate of the recruitment industry could be made by extrapolating from the fee rate that private recruiters are permitted to charge (equivalent to one month's salary), the average OFW salary (40,000–50,000 pesos, or approximately $1,000), and the number of OFWs deployed in 2014 (1.8 million). This method would overestimate the number of OFWs charged this fee, as some classes of OFWs, including construction workers hired through Philippine contractors and seafarers, are typically (or by law in the case of seafarers) not charged recruitment fees, which are covered by the employer. It also underestimates the fees charged; overcharging of fees for certain occupations, particularly those associated with lower salaries such as domestic workers, is common practice. However, as a very rough estimate, the total, $1.8 billion, is higher than the $1 billion from the PASBI. Again, this suggests that the scale of the industry size is only a small fraction of the value of the annual remittances.

This is not to say that private recruiters are irrelevant. They arrange 95 percent of the contract-labor migrations from the Philippines (Philippine Overseas Employment Administration, 2008), a share that has grown from 72.65 percent in 1987 (Philippine Overseas Employment Administration, 1988). The failure of the Marcos regime's initial attempts to phase out the private recruiters and restrict labor recruitment to state agencies demonstrates how critical the private sector is to the system of contract labor migration. Most of the labor migration from Southeast Asia operates through private recruiters (Massey et al., 2005), and the quality of those recruiters matters. The willingness of receiving countries to open their borders to contract workers from the Philippines has to do, in part, with the reliability of private recruiters in providing workers with the skills and other characteristics that employers require. Problems with private recruiters have sometimes led receiving states to impose bans on labor migration from other sending countries. Private recruiters thus hold an unusual place in the country's political economy: they are both crucial and insignificant, small businesses with little influence whose work nevertheless impacts 10 percent of the country's GDP.

The relatively coherent organization of the labor-export apparatus and the relative absence of state looting are both functions of the small size of the "business" of labor export relative to its impact on the Philippine economy. At 10 percent of GDP, remittances have a significant impact on the domestic economy, amplified by how that economy has oriented itself around those economic sectors that capture remittance spending. This gives state actors an interest in subsidizing and promoting this industry, an industry that the country's politically powerful elite also support. At the same time, because most of the economic elite are not directly involved in the "business" of labor export, any discipline that the state must impose on the industry to ensure its proper operation is not likely to be opposed by capital. Chibber (2003) has argued that the installation of nodal agencies and generally the promotion of interagency coherence was crucial to the success of developmental states like Korea. This, in turn, was facilitated by the acquiescence of industrial capital, which was willing to submit to the state's discipline because the latter was necessary to compete in the international export market. Where no such incentive for submission exists, as in the case of India, capitalists are more likely to resist the installation of a state with the capacity to effectively oversee a policy. In the Philippines, the indirect way in which labor export affects the most powerful sectors of capital creates another route through which capital's acquiescence is achieved.

Chapter 8
Conclusion

Nations and Migration

Moving across an international border is an act loaded with meaning. For tourists traveling for pleasure, it can mean adventure, leisure, escape. For an asylee, it also means escape—not from the tedium of everyday life but from fear and danger, persecution and violence. Migrant workers cross borders seeking the opposite of leisure, even if the conditions are difficult and the hours are long, because this too promises a kind of escape, in this case from poverty and economic exclusion. There are probably as many different meanings to international movement as there are people who move, but what makes them similar is this sense of escape, of pulling away from the geographically bound worlds that shape our experience and life-chances—whether it be for a brief sojourn or for the remainder of the migrant's life.

And this uprooting is what makes migration such a supremely individualizing act. Even if we are expecting friends or kin to meet and help us at our destination, even if we are moving for the sake of our families, to support their lives back home, the process of departure, separation, and reconfiguration implies that the bonds tying us to the communities with which we identify are mutable and loose, that we can survive their severance. At the level of the individual, this process could be seen as positive: whatever is lost is balanced by the discovery of a new freedom, the warmth of new bonds. From the perspective of a nation-state—a political community defined by its connection to and control over a clearly demarcated, bounded territory—this same process carries an existential risk. The mass dispersal of its people beyond its borders erodes the state's claim to represent that political community, or, worse, threatens the continued existence of that community altogether. Thus, when significant out-migrations occur, we understand them to be accidents of history caused by some catastrophe or failure of governance, where the state's response may be either to abdicate completely its claims on the people who leave or to attempt to salvage

Temporary Measures. Suzy K. Lee, Oxford University Press. © Oxford University Press (2025).
DOI: 10.1093/9780197790007.003.0008

some part of the political community through diaspora management and the promotion of transnational identities.

The view of mass migrations as political accidents, however, is only partially correct. It is true that most mass migrations are ultimately driven by economic or political hardships, some of which can be attributed to intentional state policy, but more often than not they are caused by factors beyond explicit state control. However, mass migration only *sometimes* emerges as a by-product of these hardships—a function not only of the level of popular desperation but of the availability of viable paths of migration. These pathways are often extra-political, established through social and economic networks in which state involvement is minor and peripheral, but in a significant number of cases the state's role is much more central. In these cases, the pathways were created or maintained through the explicit policy intervention of sending-country governments, seeking to produce or expand mass migration in service of some vision of the national interest. Here, policymakers measured the risks to nationhood against the potential gains and determined that a policy of mass migration was worth pursuing.

This book has presented two such cases: South Korea, in the mid-20th century, and the Philippines, beginning at a similar time but continuing through to the present day. It has traced the development of both countries' labor-export policy regimes, from their origins in economic and political crises to their deployment by authoritarian regimes to support national industrialization programs, and eventually (in the Korean case) to their irrelevance, or (in the Philippine case) to their incorporation into a more neoliberal growth strategy. This analysis offers a definition of labor export—distinct from labor migration—which emphasizes the ways in which state policy can both produce migration and channel it into forms that maximize returns to the national economy. In so doing, the state effectively separates the working person who goes abroad—the labor—from the political and social person who remains tethered to the country of origin through policies that promote temporary, contract-based migration over more permanent forms.

I have argued that states that pursue labor-export policies often do so to support national economic goals, but that this relationship can take different forms. I identify two such forms, labor-export-mediated industrialization (LEMI) and remittance-driven growth (RDG). Both South Korea and the Philippines pursued LEMI strategies at the beginning of their interventions in labor migration, though their experiences diverged in important ways. In South Korea, migrant workers were successfully leveraged to create opportunities for domestic industry to access international markets in sectors requiring migration such as construction. The Philippines was not able to reproduce the industrial transformation that South Korea underwent during its years of LEMI. However, the labor-export aspect of its own policy turned out to be quite successful, producing a rapidly expanding labor migration and with it a significant flow of remittances. That remittance flow has, over time, proven to be the engine of growth that the construction-exporting firms of the 1970s were not. I argue that the different experiences of these two countries can be attributed ultimately to state structure and political economy. Because LEMI requires a coherent industrial policy that can make use of the opportunities created through labor export, the same processes of oligarchic state capture and rent-seeking that hindered overall Philippine economic development and industrialization hobbled the effectiveness of LEMI in its case. In contrast, the RDG that we have seen in more recent decades is made possible because it works around these obstacles. Remittances, the engine of growth under RDG, enter the country in small amounts, widely dispersed among the population, and as such cannot easily be appropriated through the patrimonial rent-seeking processes that hampered the country's previous efforts at economic development.

I conclude here by exploring the implications of this analysis, as well as directions in which it could be expanded.

Challenging Cultures of Migration

In making its arguments, this book challenges the "culture of migration" explanation for the Philippines' current high rates of migration. In this view, contemporary Philippine migration is a self-perpetuating phenomenon, driven by a historically contingent sociocultural process through which the population has come to see migration as an ordinary part of the life course. The propensity to migrate internationally is now seen as a social characteristic of the Filipino people who *choose* migration. While this explanation does not gainsay the political and economic challenges for which migration serves as a solution, the *choice* of the migration solution over other options is understood to be culturally determined. This is not only a commonplace view in the migration scholarship and political commentary on the Philippines, but one that is propounded by the Philippine state itself,[1] which often describes its interventions in migration as the "management" of a phenomenon, the growth of which it has little power to halt or even slow (Battistella, 1999; Asis, 2006; Rodriguez, 2010; Estrada, 2015). This understanding of labor migration as culturally driven has made it possible for the state to resist public demands to dismantle its labor-export apparatus. Scholars of Philippine migration have already pointed out the extent to which this culture of migration is itself shaped by the state, which not only initiated the 1970s-era migrations that enabled the proliferation of this culture but perpetuates it through a myriad of institutions that normalize contract migration and make it a viable choice for those in search of opportunity (Asis, 2006; Guevarra, 2010). This book contributes to these debates by describing the mechanisms through which the Philippine state and capital directly contribute to and benefit from the formation of the modern migration flow, and by outlining the political-economic processes through which labor export can transform from a temporary measure to a way of life. Migration can be recognized as a way of life without being encoded as culture. It can simply represent acceptance of what, to the Philippine working class, must now feel obvious: that the promise of an end to the state's labor-export policy is and has always been illusory, as illusory as the promise of an inclusive, egalitarian economic development has proven to be.

Developing a Typology

The book has framed its argument in terms of the relationship between migration policy and economic development. LEMI and RDG are two ways in which international migration can be made to serve the capitalist transformations of the countries in which they are utilized. We should consider them to be what Rina Agarwala has called migration-development regimes: sets of "emigration practices and policies that . . . enable sending countries to ensure domestic capital accumulation as well as their own political legitimacy" (Agarwala, 2022). One way in which to expand on this approach is to

[1] See, e.g., the pledge of the administration of Benigno Aquino III (President, 2010–2016) to transition the Philippines "[f]rom a government that treats its people as an export commodity and a means to earn foreign exchange, disregarding the social cost to Filipino families, to a government that creates jobs at home, so that working abroad will be a choice rather than a necessity; and when its citizens do choose to become OFWs, their welfare and protection will still be the government's priority" (*A social contract with the Filipino people*, Article 10).

view LEMI and RDG as the beginnings of a typology of labor-export-based development regimes. This makes it possible to consider the development of other regime types, but also to ask how broadly the typology can be used to understand sending-state policy. For RDG, which is essentially the paradigm promoted by international development agencies, the broader applicability is easy to establish. But how often does LEMI emerge as a strategy? In the 1970s, we can identify at least one other state—Turkey—that paired construction exports with migrant labor to the Middle East, with some success. However, there are reasons to suspect that LEMI strategies are unlikely to reemerge, because the opportunities from the 1970s were so specific to the conditions of the petroleum-driven infrastructure boom. The Middle East, at that time, presented ideal conditions for an LEMI approach, not just because heavy investments in infrastructure were being made but because of three additional factors: (1) the infrastructure that was being built was of the type—mostly roads and housing—in which advanced technology did not confer an insurmountable advantage against more labor-intensive techniques, (2) domestic industry lacked the capacity to meet new demand, and (3) the local labor market could not be expanded quickly enough to match the speed of investment. Without all three factors, which existed in the Middle East by historical and geological accident—migrant-labor-based construction exports would not have been competitive.

It is possible to conceive of other conditions in which LEMI could work if we understand LEMI, from the economic-development perspective, to be a type of export-led industrialization policy. It is a variant that applies for industries in which some part of production needs to occur on-site, and thus requires the mobility of a workforce if that workforce is to be the basis of one's competitive advantage. I identified construction as one such industry. The military might be another, to the extent that war is connected to economic activity—though in the Korean case the advantage was less in the availability of the workforce than in the Park regime's willingness to negotiate with it. While there are other economic sectors—health services, for example—where *export* would require some level of human migration, there are few that can be linked to industrialization in the way that construction was.

Another possible way into this analysis might be to examine the developmental states that experienced industrial transitions like that of South Korea. While the story of RDG in the Philippines is very much a story of the internal dynamics of the country's political economy, it cannot be denied that it happened in the context of a neoliberal international regime in which the kind of developmentalism that South Korea engaged in was discouraged. It is difficult to imagine the Philippines, as debt-ridden as it was in the 1980s, being able to disregard the Washington Consensus and pursue anything other than the deindustrialized, remittance-dependent path it ended up following. The experience of China, a country that *did* experience an industrial transformation in this period, and did so using export-led developmentalist policies like those used by South Korea a generation earlier, provides an intriguing contrast. A comparative analysis here would be tricky, given the huge differences in size—of both their economies and their populations—and global political influence. For example, while China is, in total numbers, among the top labor-sending countries of the world, remittances have never represented more than 0.5 percent of the country's annual domestic product. However, the way in which it uses that labor migration suggests strategies that are reminiscent of the LEMI period. Its construction firms, which build infrastructure around the developing world through the country's One Belt One Road initiative, are responsible for the migration of millions of Chinese workers, who they import to worksites in much the same way that Korean and Philippine firms did in the 1970s and 1980s (Hlongwa, 2023). The economic and political relationships between the labor-exporting and -receiving countries are very different in this case, as is the domestic economic context. Still, China provides an example of state policy in which labor export has been paired with construction exports to provide opportunities to domestic capital and further broader national goals.

Future of Labor Export Under RDG

Another avenue for further study is in the possible futures for RDG, in the Philippines and elsewhere. While the Korean example suggests that LEMI may indeed contribute to economic transformations that can eventually render migration unnecessary, it is unlikely that remittance-driven growth can do the same. Thus far, it has not proven to generate the kind of industrial transformation that the Asian tiger economies, including Korea, underwent at the end of the 20th century. In fact, because growth occurs through the multiplier effect of remittance spending, which is concentrated in consumption, the sectors that drive growth require a steadily growing level of remittance spending. These sectors, such as real estate and retail trade, are not known for their ability to lay the foundation for sustained economic growth, but rather are associated with speculative bubbles and vulnerability to swings in consumer demand. Growth under the RDG model has survived this long only because the Philippine state has proven itself adept at generating a steady and growing flow of remittances, which in turn requires an ever-expanding flow of labor migration.

This book has not focused on the social costs of that expansion, which has been well documented both by migrant and labor activists and by scholars writing about the Philippine migration. The vast majority of OFWs are sent into low-wage and vulnerable occupations such as domestic labor. Tens of thousands of abuse cases are reported to OWWA each year. During the COVID-19 pandemic, Filipino seafarers not only were among the first exposed to the virus but were stranded on ships for months because of pandemic-related entry restrictions and the disruptions to shipping traffic. And then there are the disruptions to family life—mothers and fathers separated from their children, couples separated, sometimes for years. Millions of OFWs and their families have borne these costs because of the absence of a viable alternative, and because the state has always promised that the costs would be temporary, necessary only until the nation overcame the conditions that made a labor-export policy necessary. What the Philippine case suggests is that a remittance-based growth strategy, on its own, can never deliver on this promise.

What, then, should our stance on migration and remittances be? Given the array of forces operating in the Philippine political economy, it may be that, short of revolution, there were no better options than RDG at the end of the 20th century. The Philippines has, in fact, undergone at least one political revolution in the fifty years since the beginning of the labor-export program, but it failed to reorganize the country's underlying social structure. As of this writing, another Ferdinand Marcos, albeit a junior, holds the presidency. His election reflects a justifiable popular frustration with the failure of governments, since the restoration of democracy, to meaningfully change the conditions of the country's political economy. But neither has industrial transformation proven to be a panacea for South Korea. State-led developmentalism has produced a system of *chaebol* families who dominate the country's political economy in ways that feel very similar to Philippine oligarchs. On average, Koreans may be richer than Filipinos, but they increasingly face the same inequality and stagnant politics.

List of Figures

List of Tables

References

Abella, M. I. (1979). *Export of Filipino manpower.* Manila: Institute of Labor and Manpower Studies.

Abinales, P. N., & Amoroso, D. J. (2005). *State and society in the Philippines.* Rowman & Littlefield.

Acosta, P. A., Lartey, E. K. K., & Mandelman, F. S. (2009). Remittances and the Dutch disease. *Journal of International Economics, 79*(1), 102–116.

Adelman, I., & Taylor, J. E. (1990). Is structural adjustment with a human face possible? The case of Mexico. *Journal of Development Studies, 26*(3), 387–407.

Agarwala, R. (2022). *The migration-development regime: How class shapes Indian emigration.* Oxford University Press.

Airola, J. (2007). The use of remittance income in Mexico. *International Migration Review, 41*(4), 850–859.

AlShehabi, O. (2015). Histories of migration to the Gulf. In A. Khalaf, O. AlShehabi, & A. Hanieh (Eds.), *Transit states: Labour, migration and citizenship in the Gulf.* Pluto Press.

Amsden, A. H. (1989). *Asia's next giant: South Korea and late industrialization.* Oxford University Press.

Aquino, B., Jr. (2012, April 19). Speech during the 2nd ministerial meeting of the Abu Dhabi dialogue. Presidential Broadcast Staff – Radio Television Malacañang.

Armstrong, C. K. (2001). America's Korea, Korea's Vietnam. *Critical Asian Studies, 33*(4), 527–540.

Arnold, F., & Shah, N. M. (1986). *Asian labor migration: Pipeline to the Middle East.* Westview Press.

Asiaweek. (October 18, 2006) Mr. Billion: How cheap homes made a Filipino rich. https://web.archive.org/web/20061018050250/http://www.asiaweek.com/asiaweek/95/0818/biz1.html

Asis, M. M. B. (2006). The Philippines' culture of migration. *Migration Information Source.* Migration Policy Institute. www.migrationinformation.org/article/philippines-culture-migration

Atolia, M., Loungani, P., Marquis, M., & Papageorgiou, C. (2018). Rethinking development policy: Deindustrialization, servicification and structural transformation. *International Monetary Fund Working Paper.*

Ayala Corporation. (2015). *Life: What we make of it: 2015 annual report.*

Azad, S. (2015). *Koreans in the Persian Gulf: Policies and international relations.* Routledge.

Bakker, M. (2015). *Migrating into financial markets: How remittances became a development tool.* University of California Press.

Baldwin, F. (1975). America's rented troops: South Koreans in Vietnam. *Bulletin of Concerned Asian Scholars, 7*(4), 33–40.

Bangko Sentral ng Pilipinas. (2017). *Residential real estate price index.* https://www.bsp.gov.ph/Pages/MediaAndResearch/PublicationsAndReports/regular_RREPI.aspx

Bank of the Philippine Islands. (2014). *Let's make the best happen: 2014 annual report.*

Barajas, A., Chami, R., Fullenkamp, C., Gapen, M., & Montiel, P. (2009). Do workers' remittances promote economic growth? *International Monetary Fund Working Paper.*

Basch, L., Glick Schiller, N., & Szanton Blanc, C. (1994). *Nations unbound: Transnational projects, postcolonial predicaments, and deterritorialized nation-states.* Gordon & Breach.

Battistella, G. (1999). Philippine migration policy: Dilemmas of a crisis. *Sojourn, 14*(1), 229–248.

Beine, M., Docquier, F., & Rapoport, H. (2001). Brain drain and economic growth: Theory and evidence. *Journal of Development Economics, 64*(1), 275–289. https://doi.org/10.1016/s0304-3878(00)00133-4

Bello, W., Docena, H., de Guzman, M., & Malig, M. (2005). *The anti-development state: The political economy of permanent crisis in the Philippines.* Zed Books.

Bello, W., Kinley, D., & Elinson, E. (1982). *Development debacle: The World Bank in the Philippines.* Institute for Food and Development Policy.

Bettin, G., & Zazzaro, A. (2012). Remittances and financial development: Substitutes or complements in economic growth? *Bulletin of Economic Research, 64*(4), 509–536.

Bhagwati, J., & Hamada, K. (1974). The brain drain, international integration of markets for professionals and unemployment: A theoretical analysis. *Journal of Development Economics, 1*(1), 19–42. https://doi.org/10.1016/0304-3878(74)90020-0

Biao, X. (2003). Emigration from China: A sending country perspective. *International Migration, 41*(3), 21–48.

Binford, L. (2003). Migrant remittances and (under) development in Mexico. *Critique of Anthropology, 23*(3), 305–336.

Bird, G. (2018). The biopolitical economy of 'guest' worker programmes. In H. Richter (Ed.), *Biopolitical Governance: Race, Gender and Economy* (pp. 99–120). Rowman & Littlefield.

Black, R., & Castaldo, A. (2009). Return migration and entrepreneurship in Ghana and Côte d'Ivoire: The role of capital transfers. *Tijdschrift voor Economische en Sociale Geografie, 100*(1), 44–58. https://doi.org/10.1111/j.1467-9663.2009.00504.x

Blackburn, R. M. (1994). *Mercenaries and Lyndon Johnson's "More Flags": The hiring of Korean, Filipino, and Thai soldiers in the Vietnam War.* McFarland.

Blouin, C., & Debnath, P. (2011). CARICOM return migration and brain circulation: Case study of Caribbean-born nurses. *Canadian Foreign Policy Journal, 17*(2), 101–114. https://doi.org/10.1080/11926422.2011.602842

Brand, L. A. (2006). *Citizens abroad: Emigration and the state in the Middle East and North Africa.* Cambridge University Press.

Brinkerhoff, J. M. (2012). Creating an enabling environment for diasporas' participation in homeland development. *International Migration 50*(1), 75–95.

Cahiles-Magkilat, B. (2017). PH economic growth will continue to be consumer-driven. *Manila Bulletin*. (February 3).

Calavita, K. 1992. Inside the state: The bracero program, immigration, and the I.N.S. New York: Routledge.

Castles, S. (1986). The guest-worker in Western Europe—an obituary. *International Migration Review, 20*(4), 761–778.

Castles, S. (2004). The factors that make and unmake migration policies. *International Migration Review, 38*(3), 852–884.

Castles, S. (2006). Guestworkers in Europe: A resurrection? *International Migration Review, 40*(4), 741–766.

Catrinescu, N., Leon-Ledesma, M., Piracha, M., & Quillin, B. (2009). Remittances, institutions, and economic growth. *World Development, 37*(1), 81–92.

Chami, R., Fullenkamp, C., & Jahjah, S. (2003). Are immigrant remittance flows a source of capital for development? *IMF Working Paper*.

Chang, H.-J. (2002). *Kicking away the ladder: Development strategy in historical perspective*. Anthem Press.

Chibber, V. (2003). *Locked in place: State-building and late industrialization in India*. Princeton University Press.

Chin, R. (2007). *The guest worker question in postwar Germany*. Cambridge University Press.

Cho, Su-Jong. (2013). 70 nyeondae jungdongjinchul-e daehan jaejonmyeong [Revisiting the 1970s Era Middle East Expansion]. In S. Chung (Ed.), *Park Jung Hee sidaewa jungdong-geonseol [Park Jung Hee Regime and the Middle East Construction]: Vol. 1* (pp. 47–83). Son-In Press.

Choldin, H. M. (1973). Kinship networks in the migration process. *International Migration Review, 7*(2), 163–176.

Chung, E. A., Draudt-Véjares, D., & Tian, Y. (2024). The developmental migration state. *Journal of Ethnic and Migration Studies, 50*(3), 637–656.

Clifford, M. L. (1998). *Troubled tiger: Businessmen, bureaucrats, and generals in South Korea* (2nd ed.). M. E. Sharpe.

Commission on Higher Education. (2012). *Strategic plan for 2011–2016*. Quezon City: Commission on Higher Education.

Consunji, D. M. (2004). *A passion to build: A memoir of David M. Consunji*. Philippine Construction Association.

Cook Martin, D. (2008). Rules, red tape, and paperwork: The archeology of state control over migrants. *Journal of Historical Sociology, 21*(1), 82–119.

Cornelius, W. A. (2001). Death at the border: Efficacy and unintended consequences of US immigration control policy. *Population and Development Review, 27*(4), 661–685.

Dalisay, J. Y., Jr., & Reyes, A. (2010). *Builder of bridges: The Rudy Cuenca story*. Anvil.

de Castro, I. (1992). Money and moguls: Oiling the campaign machinery. In L. Kalaw-Tirol & S. S. Coronel (Eds.), *1992 and beyond: Forces and issues in Philippine elections* (pp. 36–77). Philippine Center for Investigative Journalism and Ateneo Center for Social Policy and Public Affairs.

de Dios, E. S., & Williamson, J. G. (2015). Deviant behavior: A century of Philippine industrialization. In A. Balisacan, U. Chakravorty, & M. Ravago (Eds.), *Sustainable economic development: Resources, environment, and institutions* (pp. 371–400). Elsevier Science and Technology.

de Haas, H. (2010). The internal dynamics of migration processes: A theoretical inquiry. *Journal of Ethnic and Migration Studies, 36*(10), 1587–1617.

Delgado-Wise, R., & Márquez Covarrubias, H. (2007). The reshaping of Mexican labor exports under NAFTA: Paradoxes and challenges. *International Migration Review, 41*(3), 656–679.

del Callar, M. (2014, February 12). Over 3,000 OFWs jailed, 800 others arrive home dead—DFA report. *GMA News Online.* http://www.gmanetwork.com/news/story/348132/news/pinoyabroad/over-3-000-ofws-jailed-800-others-arrived-home-dead-dfa-report

Delmendo, L. C. (2013). House price rises accelerating in the Philippines. *Global Property Guide.* https://www.globalpropertyguide.com/Asia/Philippines/Price-History-Archive/House-price-rises-accelerating-in-the-Philippines-127206

DeWind, J., Kim, E. M., Skeldon, R., & Yoon, I.-J. (2012). Korean development and migration. *Journal of Ethnic and Migration Studies, 38*(3), 371–388.

DiMoia, J. (2018). In Pursuit of "Peace and construction": Hyundai construction and infrastructure in Southeast Asia, 1965–1973. In H. Mizuno, A. S. Moore, & J. DiMoia (Eds.), *Engineering Asia: Technology, colonial development, and the Cold War order* (pp. 209–239). Bloomsbury Academic.

Ducanes, G. (2015). The welfare impact of overseas migration on Philippine households: Analysis using panel data. *Asian and Pacific Migration Journal*, 24(1): 79–106.

Disney, N. (1977). South Korean workers in the Middle East. *MERIP Reports*, 61, 22–26.

Dumitrescu, L. (2021). Pockets of efficiency in a low capacity state: Dealing with the COVID-19 pandemic in Romania. *Southeast European and Black Sea Studies, 21*(3), 373–391.

Durand, J., W. Kandel, E. A. Parrado, & D. S. Massey. (1996). International migration and development in Mexican communities. *Demography, 33*(2), 249–264.

Durand, J., & Massey, D. S. (2019). Evolution of the Mexico-U.S. migration system: Insights from the Mexican Migration Project. *Annals of the American Academy of Political and Social Science, 684*, 21–42.

Dustmann, C., & Mestres, J. (2010). Remittances and temporary migration. *Journal of Development Economics, 92*, 62–70.

Engineering News-Record. (1977). "The Koreans are coming! The Koreans are coming!" *Engineering News-Record* (March 31), 16.

Engineering News-Record. (1978). The Asian invasion of the Middle East . . . competitors from developing countries. *Engineering News-Record.* (November 23), 22–26.

Engineering News-Record. (1979). Manila-based firm becomes Southeast Asia's biggest. *Engineering News-Record* (April 26), 21.

Engineering News-Record. (1979a). Cost of Middle East Labor: Monthly wage rates for expatriate personnel, U.S. dollars. *Engineering News-Record* (December 20), 107.

Engineering News-Record. (1979b). The world's top international contractors. *Engineering News-Record* (December 6), 28–35.

Engineering News-Record. (1980). The world's top international contractors. *Engineering News-Record* (July 17), 48–54.

Engineering News-Record. (1981). The world's top international contractors. *Engineering News-Record* (July 16), 74–88.

Errighi, L., Khatiwada, S., & Bodwell, C. (2016). *Business process outsourcing in the Philippines: Challenges for decent work*. ILO Asia-Pacific Working Paper Series.

Estrada, E. S. (2015). *Mi figue, mi raisin*: The Filipinos' culture of migration in the age of globalization. *CIRSS Commentaries*, *2*(15). https://fsi.gov.ph/wp-content/uploads/2015/06/2015-0515-Vol-2-No-15-Mi-Figue-Mi-Raisin-The-Filipinos-Culture-of-Migration-in-the-Age-of-Globalization-Estrada.pdf

Evans, P. (1995). *Embedded autonomy: States and industrial transformation*. Princeton University Press.

Executive Order No. 247. (1987). Reorganizing the Philippine Overseas Employment Administration and for Other Purposes. Republic of the Philippines.

Far Eastern Economic Review. (1980). Setbacks for ambition: Philippine industrial projects are hampered by an array of problems and may have to be revised. *Far Eastern Economic Review* (December 12), p. 55.

Fayissa, B., & Nsiah, C. (2010). The impact of remittances on economic growth and development in Africa. *American Economist*, *55*(2), 92–103.

Feenstra, R. C., & Hamilton, G. G. (2006). Economic organization in South Korea and Taiwan: A first test of the model. In R. C. Feenstra & G. G. Hamilton (Eds.), *Emergent economies, divergent paths: Economic organization and international trade in South Korea and Taiwan* (pp. 120–166). Cambridge University Press.

Fitzgerald, D. (2006). Inside the sending state: The politics of Mexican emigration control. *International Migration Review*, *40*(2), 259–293.

Fitzgerald, D. (2008). *A nation of emigrants: How Mexico manages its migration*. University of California Press.

Frankel, J. (2011). Are bilateral remittances countercyclical? *Open Economies Review*, *22*(1), 1–16.

Fussell, E. (2012). Space, time, and volition: Dimensions of migration theory. In R. Rosenblum & D. J. Tichenor (Eds.), *Oxford handbook of the politics of international migration* (pp. 25–52). Oxford University Press.

Galam, R. (2022). The Philippines and seafaring labour export: State, non-state and international actors in the assembly and employability of Filipino seafarers. *International Migration*, 1–15.

Gamlen, A. (2006). Diaspora engagement policies: What are they, and what kinds of states use them? COMPAS Working Paper No. 32, University of Oxford.

Geddes, B. 1990. Building "state" autonomy in Brazil, 1930–1964. *Comparative Politics*, *22*(2), 217–235.

Geddes, B. 1996. *Politician's dilemma: Building state capacity in Latin America*. University of California Press.

Glassman, J., & Choi, Y.-J. (2014). The *chaebol* and the US military-industrial complex: Cold War geopolitical economy and South Korean industrialization. *Environment and Planning A, 46*, 1160–1180.

Glick Schiller, N., Basch, L., & Szanton Blanc, C. (1995). From immigrant to transmigrant: Theorizing transnational migration. *Anthropological Quarterly, 68*(1), 48–63.

Global Media Arts News Online. (2012). New POEA chief assumes post. http://www.gmanetwork.com/news/news/content/243308/new-poea-chief-assumes-post/story/

Glytsos, N. P. (1993). Measuring the income effects of migrant remittances: A methodological approach applied to Greece. *Economic Development and Cultural Change, 42*(1), 131–168.

Glytsos, N. P. (1997). Remitting behaviour of "temporary" and "permanent" migrants: The case of Greeks in Germany and Australia. *Labour: Review of Labour Economics and Industrial Relations, 11*(3), 409–435.

Gollin, D. (2014). The Lewis model: A 60-year retrospective. *Journal of Economic Perspectives, 28*(3), 71–88.

Gonzaga, L. (1981). Knock-down knockout: The Philippines orders three companies to stop assembling cars after this year. *Far Eastern Economic Review* (May 15), 89–90.

Gonzaga, L. (1982). The hesitation waltz: The tempo is still in doubt as the president says go, the prime minister says slow and the industry minister says quick. *Far Eastern Economic Review* (January 15), 50–51.

Gonzaga, L. (1983). The reluctant victor: A Japanese firm wins a second diesel-engine contract in the Philippines, but possibly wishes it had never entered the race. *Far Eastern Economic Review* (May 19), 56–57.

Gonzalez, J. L., III. (1998). *Philippine labour migration: Critical dimensions of public policy.* Institute of Southeast Asian Studies.

Goss, J., & Lindquist, B. (1995). Conceptualizing international labor migration: A structuration perspective. *International Migration Review, 29*(2), 317–351.

Gothe-Snape, J., and Elton-Pym, J. (2017, April 19). Concerns 457 visa changes will create 'guest worker' system. *SBS News.* http://www.sbs.com.au/news/article/2017/04/19/concerns-457-visa-changes-will-create-guest-worker-system.

Guevarra, A. R. (2010). *Marketing dreams, manufacturing heroes: The transnational labor brokering of Filipino workers.* Rutgers University Press.

Gutierrez, E. U. (1994). *The ties that bind: A guide to family, business, and other interests in the Ninth House of Representatives.* Philippine Center for Investigative Journalism.

Han, H. (2006). South Korea and the Vietnam War. In B.-c. Lee (Ed.), *Developmental dictatorship and the Park Chung-hee era: The shaping of modernity in the Republic of Korea* (pp. 248–270). Homa & Sekey.

Han, Y. (2013). The May Sixteenth Military Coup. In B. Kim & E. Vogel (Eds.), *The Park Chung Hee era: The transformation of South Korea* (pp. 35–57). Harvard University Press. https:doi.org/10.4159/harvard.9780674061064.c1

Harvie, C. & Lee, H.-H. (2003). Export-led industrialisation and growth: Korea's economic miracle, 1962–1989. *Australian Economic History Review, 43*(3), 256–286. https://doi.org/10.1046/j.1467-8446.2003.00054.x

Hedman, E.-L. E., & Sidel, J. T. (2000). *Philippine politics and society in the twentieth century: Colonial legacies, post-colonial trajectories*. Routledge.

Heisler, M. O. (1986). Transnational migration as a small window on the diminished autonomy of the modern democratic state. *Annals of the American Academy of Political and Social Science, 485*, 153–166.

Hernández-León, R. (2013). Conceptualizing the migration industry. In T. Gammeltoft-Hansen & N. N. Sørensen (Eds.), *The migration industry and the commercialization of international migration* (pp. 24–44). Routledge.

Hlongwa, L. (2023). Trapped in no man's land: Chinese migrant workers along the Belt and Road Initiative. *Conflict, Justice, Decolonization: Critical Studies of Inter-Asian Societies*. https://cjdproject.web.nycu.edu.tw/2023/01/23/trapped-in-no-mans-land-chinese-migrant-workers-along-the-belt-and-road-initiative/

Hogan, W.T. (2001). *The Posco Strategy*. Lexington Books.

Hollifield, J. F. (2004). The emerging migration state. *International Migration Review, 38*(3), 885–912.

Hong, Y.-S. (2015). *Cold War Germany, the Third World, and the global humanitarian regime*. Cambridge University Press.

Hugo, G. (2009). Best practice in temporary labour migration for development: A perspective from Asia and the Pacific. *International Migration, 47*(5), 23–74.

Hugo, G., & Stahl, C. (2004). Labor export strategies in Asia. In D. S. Massey & J. E. Taylor (Eds.), *International migration: Prospects and policies in a global market* (pp. 174–200). Oxford University Press.

Hutchcroft, P. (1989). Situating the Philippine political economy: Import-substitution industrialization in comparative perspective. *Kasarinlan: Philippine Journal of Third World Studies, 4*(4), 39–56.

Hutchcroft, P. D. (1991). Oligarchs and cronies in the Philippine state: The politics of patrimonial plunder. *World Politics, 43*(3), 414–450.

Hutchcroft, P. D. (1998). *Booty capitalism: The politics of banking in the Philippines*. Cornell University Press.

Hutchcroft, P. (2000). Obstructive corruption: The politics of privilege in the Philippines. In M. H. Khan & Jomo K. S. (Eds.), *Rents, rent-seeking and economic development: Theory and evidence in Asia* (pp. 207–247). Cambridge University Press.

Hutchcroft, P. (2011). Reflections on a reverse image: South Korea under Park Chung Hee and the Philippines under Ferdinand Marcos. In B. Kim & E. F. Vogel (Eds.), *The Park Chung Hee era: The transformation of South Korea* (pp. 542–572). Harvard University Press.

Ibrahim, Y. M. (1978, June 19). How Koreans built Saudi success. *New York Times*, D5.

İçduygu, A., & Karaçay, A. B. (2012). The international migration system between Turkey and Russia: Project-tied migrant workers in Moscow. *International Migration, 50*(1), 55–75.

Im, H. B. (2011) The origins of the Yushin regime: Machiavelli unveiled. In B.-K. Kim & E. F. Vogel (Eds.), *The Park Chung Hee era: The transformation of South Korea* (pp. 233–262). Harvard University Press.

International Labour Organization. (1984). Overseas employment administration in selected Asian countries: Proceedings of the ILO/ARPLA Inter-Country Symposium on Overseas Employment Administration. Pattaya, Thailand (May 21–25, 1984).

Iskander, N. (2010). *Creative state: Forty years of migration and development policy in Morocco and Mexico*. Cornell University Press.

Jasso, G., Massey, D. S., Rosenzweig, M. R., & Smith, J. P. (2005). Immigration, health, and New York City: Early results based on the U.S. new immigrant cohort of 2003. *Economic Policy Review*, *11*(2), 127–151.

Jaymalin, M. (2021). Passage of migrant workers department law lauded. *The Philippine Star* (December 17).

Johnson, C. (1982). *MITI and the Japanese miracle: The growth of industrial policy, 1925–1975*. Stanford University Press.

Jung, Y.-S. (2018). Beyond the bifurcated myth: The medical migration of female Korean nurses to West Germany in the 1970s. *Korean Journal of Medical History*, *27*(2), 225–266.

Kang, D. C. (2004). *Crony capitalism: Corruption and development in South Korea and the Philippines*. Cambridge University Press.

Khan, M. H. (2000). Rent seeking as a process. In M. H. Khan & Jomo K. S. (Eds.), *Rents, rent-seeking and economic development: Theory and evidence in Asia* (pp. 70–144). Cambridge University Press.

Khan, M. H. & Jomo K. S. (2004). Introduction. In M. H. Khan & Jomo K. S. (Eds.), *Rents, rent-seeking and economic development: Theory and evidence in Asia* (pp. 1–20). Cambridge University Press.

Khilji, N. M., & Zampelli, E. M. (1994). The fungibility of U.S. military and non-military assistance and the impacts on expenditures of major aid recipients. *Journal of Development Economics*, *43*(2), 345–362. https://doi.org/10.1016/0304-3878(94)90012-4

Kim, L. (1997). *Imitation to innovation: The dynamics of Korea's technological learning*. Harvard Business School Press.

Kim, S. (1982) Contract migration in the Republic of Korea. International Working Paper, International Labour Organisation.

Kim, S. (1986). Labor migration from Korea to the Middle East: Its trends and impact on the Korean economy. In F. Arnold & N. M. Shah (Eds.), *Asian labor migration: Pipeline to the Middle East* (pp. 163–174). Westview.

Kim, S. (1988). The Korean construction industry as an exporter of services. *World Bank Economic Review*, *2*(2), 225–238.

Kim, Y. T. (2007). Korean elites: Social networks and power. *Journal of Contemporary Asia*, *37*(1), 19–37.

King, R., & Skeldon, R. (2010). "Mind the gap!" Integrating approaches to internal and international migration. *Journal of Ethnic and Migration Studies*, *36*(10), 1619–1646.

Kingdom of Saudi Arabia Ministry of Economy and Planning, General Authority for Statistics. (2016). *Saudi Arabia Statistical Yearbook 1985–1989*.

Kirk, D. (1994). *Korean dynasty: Hyundai and Chung Ju Yung*. Routledge.

Kleibert, J. M. (2013). Strategic coupling in "next wave cities": Local institutional actors and the offshore service sector in the Philippines. *Singapore Journal of Tropical Geography*, *35*(2), 245–260.

Knauerhase, R. (1974). Saudi Arabia's economy at the beginning of the 1970s. *Middle East Journal*, *28*(2), 126–140.

Krinks, P. (2002). *The economy of the Philippines: Elites, inequalities and economic restructuring*. Routledge.

Kritz, B. D. (2016). "Bubble or Not?" *Manila Times* (July 8).

Lansbury, R. D., Suh, C.-S., & Kwon, S.-H. (2007). *The global Korean motor industry: The Hyundai Motor Company's global strategy*. Routledge.

LaPorte, R. J., Jr. (1984). The ability of South and East Asia to meet the labor demands of the Middle East and North Africa. *Middle East Journal*, *38*(4), 699–711.

Lee, E. S. (1966). A theory of migration. *Demography*, *3*(1), 47–57.

Lee, J.-k. (2009). Surrogate military, subimperialism, and masculinity: South Korea in the Vietnam War, 1965–73. *Positions*, *17*(3), 655–682.

Lee, S. (2017). The three worlds of emigration policy: Towards a theory of sending state regimes. *Journal of Ethnic and Migration Studies*, *43*(9), 1453–1471.

Lee. S. (2021). Migrating beyond networks: The mechanisms of sending state intervention. *Migration and Development*, *10*(3), 342–358. https://journals.sagepub.com/doi/10.1080/21632324.2019.1637616

Lee, S. H. (2021). "I am still close to my child": Middle-class Korean wild geese fathers' responsible and intimate fatherhood in a transnational context. *Journal of Ethnic and Migration Studies*, *47*(9), 2161–2178.

Lee Flores, W. Y. (1997). *Manuel B. Villar Jr.* De La Salle University Press.

Levitt, P., & de la Dehesa, R. (2003). Transnational migration and the redefinition of the state: Variations and explanations. *Ethnic and Racial Studies*, *26*(4), 587–611.

Levitt, P., & Glick Schiller, N. (2004). Conceptualizing simultaneity: A transnational and social field perspective on society. *International Migration Review*, *38*(3), 1002–1039.

Lewis, W. A. (1954). Economic development with unlimited supplies of labour. *Manchester School*, *22*(2), 139–191.

Lim, W. (2003). The emergence of the *chaebol* and the origins of the *chaebol* problem. In S. Haggard, W. Lim, & E. Kim (Eds.), *Economic crisis and corporate restructuring in Korea: Reforming the chaebol* (pp. 35–52). Cambridge University Press.

Linder, M. (1994). *Projecting capitalism: A history of the internationalization of the construction industry*. Greenwood.

Liu, D. S. (1996). The 1992 Employment Service Act and the Influx of Foreign Workers in Taiwan and Translation of the 1994 Implementary Provisions. *Pacific Rim Law and Policy Journal*, *5*(3), 599–637.

Logan, J. R., Alba, R. D., & Zhang, W. (2002). Immigrant enclaves and ethnic communities in New York and Los Angeles. *American Sociological Review*, *67*(2), 299–322.

Lopez, E. J. (2016). The Philippines' consumer economy and Singapore's crisis. *The Philippine Star*. http://www.philstar.com/business/2016/02/03/1549276/philippines-consumer-economy-and-singapores-crisis

Lucas, R. E. B. (2005). Migration internationale vers les pays à haut revenu: quelles conséquences pour le développement économique des pays d'origine. *Revue d'économie du développement, 13*(4), 123–171.

MacDonald, J. S., & MacDonald, L. D. (1974). Chain migration, ethnic neighborhood formation, and social networks. In C. Tilly (Ed.), *An urban world* (pp. 226–236). Little, Brown.

Magno-Ballesteros, M. (2000). *Land use planning in Metro Manila and the urban fringe: Implications on the land and real estate market.* Philippine Institute for Development Studies Discussion Paper Series No. 2000-20.

Marx, K. (1992). *Capital: A critique of political economy, volume I.* Penguin Books. (Original work published 1876)

Massey, D. S. (1988). Economic development and international migration in comparative perspective. *Population and Development Review, 14*(3), 383–413.

Massey, D. S. (2002). A synthetic theory of international migration. *World in the mirror of international migration*, pp. 143–153. MAX Press.

Massey, D. S., Alarcón, R., Durand, J., & González, H. (1990). *Return to Aztlan: The social process of international migration from western Mexico.* University of California Press.

Massey, D. S., Arango, J., Hugo, G., Kouaouci, A., Pellegrino, A., & Taylor, J. E. (2005). *Worlds in motion: Understanding international migration at the end of the millennium.* Oxford University Press. (Original work published 1998)

Massey, D. S., Durand, J., & Malone, N. J. (2002). *Beyond smoke and mirrors: Mexican migration in an era of economic integration.* Russell Sage Foundation.

Mateo, J. (2016). CHED: Filipino engineers, architects not in danger of losing jobs in Qatar. *The Philippine Star.* https://www.philstar.com/headlines/2016/02/03/1549647/ched-filipino-engineers-architects-not-danger-losing-jobs-qatar

McCormick, B., & Wahba, J. (2001). Overseas work experience, savings and entrepreneurship amongst return migrants to LDCs. *Scottish Journal of Political Economy, 48*(2), 164–178. https://doi.org/10.1111/1467-9485.00192

McCoy, A. W. (2009). "An anarchy of families": The historiography of state and family in the Philippines. In A. W. McCoy (Ed.), *An anarchy of families: State and family in the Philippines* (pp. 1–28). University of Wisconsin Press.

McKay, S. C. (2006). *Satanic mills or silicon islands? The politics of high-tech production in the Philippines.* ILR Press/Cornell University Press.

Merkle, L., & Zimmerman, K. F. (1992). Savings, remittances, and return migration. *Economic Letters, 38*(1), 77–81.

Meyer, D., & Shera, A. (2017). The impact of remittances on economic growth: An econometric model. *EconomiA, 18*(2), 147–155.

Meyers, E. (2000). Theories of international immigration policy–A comparative analysis. *International Migration Review, 34*(4), 1245–1282.

Migrante International. (2018). Migrante International Orientation: The Progressive Movement of Overseas Compatriots. https://migranteinternational.org/wp-content/uploads/2021/05/Final-Migrante-Orientation-approved-8th-Congress-2018-1.pdf https://migranteinternational.org/resources

Mitra, R. M. (2011). BPO Sector Growth and Inclusive Development in the Philippines. World Bank Group. http://documents.worldbank.org/curated/en/715341468295535070/pdf/660930WP0P122100B0BPO0Sector0Growth.pdf

Moon, C.-i. (1986). Korean contractors in Saudi Arabia: Their rise and fall. *Middle East Journal*, *40*(4), 614–633.

Moratalla, N. N. (1999). Graft and corruption: The Philippine experience. *Resource Material Series No. 56*, 501–523.

Mügge, L. (2012). Managing transnationalism: Continuity and change in Turkish state policy. *International Migration*, *50*(1), 20–38.

Mundaca, B. G. (2009). Remittances, financial market development, and economic growth: The case of Latin America and the Caribbean. *Review of Development Economics*, *13*(2), 288–303.

Na, H.-S. (2019). South Korean nurses to West Germany? Reassessing the role of the state in postwar global labour migration. *Histoire sociale/Social History*, *52*(105), 193–207.

Nam, C.-H. (1995). South Korea's big business clientelism in democratic reform. *Asian Survey*, *35*(4), 357–366.

National Economic and Development Authority of the Philippines. (2011). *Calabarzon Regional Development Plan 2011–2016*.

National Economic and Development Authority of the Philippines. (2013). *Calabarzon Updated Regional Development Plan 2011–2016*.

Nelson, Chris. (2016, September 14). Booming Philippine property market inspires confidence among expat investors. The National: *UAE Edition*. https://www.thenationalnews.com/business/property/booming-philippine-property-market-inspires-confidence-among-expat-investors-1.232220

Oberai, A. S., & Singh, H. K. (1980). Migration, remittances and rural development: Findings of a case study in the Indian Punjab. *International Labour Review*, *119*(2), 229–241.

Ofreneo, R. E. (Ed.). (2010). *Balik-Tanaw Bagong Pananaw: Essays and writings on the 75 years of the Department of Labor and Employment*. Department of Labor and Employment.

Ofreneo, R. E. (2015). Growth and employment in de-industrializing Philippines. *Journal of the Asia Pacific Economy*, *20*(1), 111–129.

Oh, W. C. (2006). *Park Chung Hee neun bagjeonghuineun eotteohge gyeongjegang-gugeul mandeul-eossna [Park Chung Hee: How Park Chung-Hee created an economic powerhouse]*. Dongeo Culture Publishing.

Ople, B. F. (1979). *Frontiers of Labor and Social Policy: Selected Speeches*. Institute of Labor and Manpower Studies.

Ortega, A. A. (2016). *Neoliberalizing spaces in the Philippines: Suburbanization, transnational migration, and dispossession*. Lexington Books.

Ortiga, Y. Y. (2014). Professional problems: The burden of producing the "global" Filipino nurse. *Social Science & Medicine*, *115*, 64–71.

Ortiga, Y. Y. (2018). *Emigration, employability and higher education in the Philippines*. Routledge.

Ortiga, Y. Y. (2021). Shifting employabilities: Skilling migrants in the nation of emigration. *Journal of Ethnic and Migration Studies, 47*(10), 2270–2287.

Paine, S. (1974). *Exporting workers: The Turkish case.* Cambridge University Press.

Park, K. (2003). The Vietnam War and the "Miracle of East Asia." (H. K. Clayton, Trans.). *Inter-Asia Cultural Studies, 4*(3), 372–400.

Park, K. H., & Weber, W. L. (2006). A note on efficiency and productivity growth in the Korean Banking Industry, 1992–2002. *Journal of Banking and Finance, 30*(8), 2371–2386.

Parreñas, R. S. (2021). Discipline and empower: The state governance of migrant domestic workers. *American Sociological Review 86*(6), 1043–1065.

Passel, J. S., Cohn, D., & Gonzalez-Barrera, A. (2012). Net migration from Mexico falls to zero—and perhaps less. Pew Research Center. https://www.pewresearch.org/race-and-ethnicity/2012/04/23/net-migration-from-mexico-falls-to-zero-and-perhaps-less/

Pearlman, W. (2014). Competing for Lebanon's diaspora: Transnationalism and domestic struggles in a weak state. *International Migration Review, 48*(1), 34–75.

Philippine Daily Inquirer. (2012, January 12). What went before: The OWWA fund transfer. http://newsinfo.inquirer.net/126665/what-went-before-the-owwa-fund-transfer

Philippine Department of Labor. (1974). *Labor Statistics Service,'73 year book of labor statistics.*

Philippine Department of Labor. (1994). *Labor Statistics Service, 1993 year book of labor statistics.*

Philippine Department of Labor. (1991). *Labor Statistics Service, 1990 year book of labor statistics.*

Philippine Ministry of Labor and Employment. (1982). *Labor Statistics Service, 1981 year book of labor statistics.*

Philippine Overseas Employment Administration. (1988). Annual report 1987.

Philippine Overseas Employment Administration. (1991). Annual report 1990.

Philippine Overseas Employment Administration. (1993). Annual report 1991–1992.

Philippine Overseas Employment Administration. (2008). Annual report 2007.

Philippine Overseas Employment Administration. (2013). News advisory. Facebook post (July 5).

Philippine Overseas Employment Administration. (2015). Annual report 2014.

Philippine Overseas Employment Administration. (2016). Revised POEA rules and regulations governing the recruitment and employment of land-based overseas Filipino workers of 2016.

Philippine Overseas Employment Administration. (2016b). Deployment per Country per Skill per Sex, 1993–2010. Website: OFW Statistics. https://dmw.gov.ph/archives/ofwstat/ofwstat.html

Philippine Overseas Employment Administration. (2017b). Website: Status of recruitment agencies as of July 27, 2017. http://www.poea.gov.ph/cgi-bin/agList.asp?mode=act

Philippine Statistics Authority. (2017). Total approved foreign investments down by 12.8 percent in Q1 2017.

Philippine Statistics Authority. (2018). 2016 annual survey of Philippine business and industry (ASPBI)–Administrative and support service activities sector for establishments with total employment of 20 and over: Preliminary results. https://psa.gov.ph/content/2016-annual-survey-philippine-business-and-industry-aspbi-administrative-and-support

Philippine Statistics Authority. (2023). Special release: 2022 overseas Filipino workers (final results).

Porter, E. (2016, October 25). A framework for managing immigration. *New York Times*, p. B1.

Portes, A. (1978). Migration and underdevelopment. *Politics and Society*, *8*(1), 1–48.

Prebisch, R. (1959). Commercial policy in the underdeveloped countries. *The American Economic Review*, *49*(2), 251–273.

Presidential Decree No. 1. (1971). Reorganizing the executive branch. Republic of the Philippines.

Presidential Decree No. 6. (1972a). Amending certain rules on discipline of government officials and employees. Republic of the Philippines.

Presidential Decree No. 46. (1972b). Making it punishable for public officials and employees to receive, and for private persons to give gifts on any occasion, including Christmas. Republic of the Philippines.

Presidential Decree No. 442. (1974). Labor code of the Philippines. Republic of the Philippines.

Presidential Decree No. 1167. (1977). Developing and regulating the overseas construction industry, providing incentives therefor, and for other purposes. Republic of the Philippines.

Presidential Decree No. 1412. (1978). Further amending certain provisions of Book I, Presidential Decree No. 442, otherwise known as the Labor Code of the Philippines. Republic of the Philippines.

Purdey, J., Tadem, T. S. E., & Tadem, E. C. (2016). Political dynasties in the Philippines: Persistent patterns, perennial problems. *South East Asia Research*, *24*(3), 328–340.

Rajan, S. I., & Mishra, U. S. (2010). Managing migration from India: Lessons from the Philippines. In S. I. Rajan (Ed.), *Governance and labour migration: India migration report 2010*, 288–305. Routledge.

Ravago, M.-L. V., Zandro Brucal, A. Z., Roumasset, J., & Punongbayan, J. C. (2019). The role of power prices in structural transformation: Evidence from the Philippines. *Journal of Asian Economics*, *61(C)*, 20–33.

Ravenstein, E. G. (1889). The laws of migration. *Journal of the Royal Statistical Society*, *52*(2), 241–305.

Reichert, J. (1981). The migration syndrome: Seasonal U.S. wage labor and rural development in Central Mexico. *Human Organization*, *40*(1), 56–66.

Republic Act No. 6657. (1988). Comprehensive agrarian reform law of 1988. Republic of the Philippines.

Republic Act No. 7111. (1991). An act establishing the overseas workers investment fund to provide incentives to overseas workers, reduce the foreign debt burden and for other purposes. Republic of the Philippines.

Republic Act No. 8042. (1995). An act to institute the policies of overseas employment and establish a higher standard of protection and promotion of the welfare of migrant workers, their families and overseas Filipinos in distress, and for other purposes. Republic of the Philippines.

Republic Act No. 9856. (2009). The real estate investment trust (REIT) act of 2009. Republic of the Philippines.

Republic Act No. 10022. (2010). An act amending Republic Act No. 8042, otherwise known as the Migrant Workers and Overseas Filipinos Act of 1995, as amended, further improving the standard of protection and promotion of the welfare of migrant workers, their families and overseas Filipinos in distress, and for other purposes. Republic of the Philippines.

Republic Act No. 11641. (2021). Department of Migrant Workers act. Republic of the Philippines.

Republic of the Philippines, Congress of the Philippines, House of Representatives. (1995.) Record of the House of Representatives: Third Regular Session, 1994–1995.

Republic of the Philippines, Congress of the Philippines, Senate. (1994.) Joint Committee Hearings on Foreign Relations and Committee on Labor, Employment and Human Resources Development, November 3, 1994.Republic of the Philippines, Congress of the Philippines, Senate. (1995a.) Record of the Senate: Third regular session, February 13 to 28, 1995 and May 22 to June 5, 1995.

Republic of the Philippines, Congress of the Philippines, Senate. (1995b). Conference committee meeting on the disagreeing provisions of the S. No. 2077 and H. No. 14314. May 30, 1995.

Republic of the Philippines, Congress of the Philippines, Senate. (2009, November). A social contract with the Filipino People: The platform of Senator Benigno "Noynoy" S. Aquino III. Republic of the Philippines, National Economic and Development Authority. (1973). Four-year development plan FY 1974–77.

Republic of the Philippines, National Economic and Development Authority. (1977) Five-Year Philippine Development Plan, 1978–1982.

Republic of the Philippines, National Economic Council with the Presidential Economic Staff. (1969). Five-year development program for the Philippines: Fiscal years 1970–1974.

Republic of the Philippines, National Economic Council. (1970). Four year development plan, FY 1971–74. (Adopted by the National Economic Council 26 June 1970)

Republic of the Philippines, National Economic Council. (1971). Four year development plan, FY 1972–75. (Adopted by the National Economic Council 23 June 1971)

Republic of the Philippines, Presidential Economic Staff. (1966). Four year economic program for the Philippines: Fiscal years 1967–1970.

Republic of the Philippines. (1972). Reorganization of the executive branch of the national government: Integrated reorganization plan.

Rimmer, P. J. (1984). Japanese, Korean and Filipino international construction contractors: Structure and performance. Australian National University Development Studies Centre. Working Paper *No. 40.*

Rivera, T. C. (2003). The leading Chinese-Filipino business families in post-Marcos Philippines. In Jomo K. S. & B. C. Folk (Eds.), *Ethnic business: Chinese capitalism in Southeast Asia* (pp. 106–118). RoutledgeCurzon.

Rodriguez, R. M. (2002). Migrant heroes: Nationalism, citizenship and the politics of Filipino migrant labour. *Citizenship Studies, 6*(3), 341–356.

Rodriguez, R. M. (2010). *Migrants for export: How the Philippine state brokers labor to the world.* University of Minnesota Press.

Rodrik, D. (2007). Industrial development: Stylized facts and policies. *Industrial development for the 21st century: Sustainable development perspectives* (pp. 7–28). United Nations, Department of Economic and Social Affairs.

Rodrik, D. (2016). Premature deindustrialization. *Journal of Economic Growth, 21*(1), 1–33.

Sajor, E. E. (2001). *The real estate boom in the 1990s and land use allocation for socialized housing in Metro Cebu, Philippines.* UMP-Asia Occasional Paper No. 52, United National Urban Management.

Sanders, S. R., & Brown, D. L. (2012). The migratory response of labor to special economic zones in the Philippines, 1995–2005. *Population Research and Policy Review, 31*(1), 141–164.

Santo Tomas, P. (1984). Organizational change and innovation in overseas employment administration. Overseas employment administration in selected Asian countries: Proceedings of the ILO/ARPLA Inter-Country Symposium on Overseas Employment Administration, Pattaya, Thailand, May 21–25, 1984, 65–73.

Sassen, S. (1988). *The mobility of labor and capital: A study in international investment and labor flow.* Cambridge University Press.

Sciortino, G. (2000). Toward a political sociology of entry policies: Conceptual problems and theoretical proposals. *Journal of Ethnic and Migration Studies, 26*(2), 213–228.

Securities and Exchange Commission of the Philippines. (2020). Launching of the Real Estate Investment Trust (REIT) Regulations.

Shah, N. M., & Arnold, F. (1986). Government policies and programs regulating labor migration. In F. Arnold & N. M. Shah (Eds.), *Asian labor migration: Pipeline to the Middle East* (pp. 65–77). Westview.

Shah, N. M., & Menon, I. (1999). Chain migration through the social network: Experience of labour migrants in Kuwait. *International Migration, 37*(2), 361–382.

Siddiqui, T., Rashid, R., & Zeitlyn, B. (2008). *Information campaigns on safe migration and pre-departure training.* Development Research Centre on Migration, Globalisation & Poverty.

Simon, S. (2004). U.S.-Southeast Asian relations: Philippines withdraws from Iraq and JI strikes again. *Comparative Connections, 6*(3), 65–76.

Siracusa, C., & Acacio, K. (2004). State migrant-exporting schemes and their implications for the rise of illicit migration: A comparison of Spain and the Philippines. *Journal of International Migration and Integration*, 5, 321–342.

Skocpol, T. (1985). Bringing the state back in: Strategies of analysis in current research. In P. B. Evans, D. Rueschemeyer, & T. Skocpol (Eds.), *Bringing the state back* (pp. 3–43). Cambridge University Press.

SM Investment Corporation. (2010). *Scaling new heights: 2010 annual report.*

SM Investment Corporation. (2011). *Creating value at all times: 2011 annual report.*

SM Investment Corporation. (2012). *We build to serve millions: 2012 annual report.*

SM Investment Corporation. (2013). *The dynamic world of SM: 2013 annual report.*

SM Investment Corporation. (2014). *Pursuing new opportunities in growth: 2014 annual report.*

SM Investment Corporation. (2015). *Sustaining growth, delivering value: 2015 annual report.*

Smith, N. (1984). *Uneven development: Nature, capital, and the production of space.* Basil Blackwell.

Smith, R. C. (2003a). Migrant membership as an instituted process: Transnationalization, the state and the extra-territorial conduct of Mexican politics. *International Migration Review, 37*(2), 297–343.

Smith, R. C. (2003b). Diasporic memberships in historical perspective: Comparative insights from the Mexican, Italian and Polish cases. *International Migration Review, 37*(3), 724–759.

Social Weather Stations [SWS]. (2016). The 2016 surveys of enterprises on corruption.

Stahl, C. W., & Arnold, F. (1986). Overseas workers' remittances and Asian development. *International Migration Review, 20*(4), 899–925.

Stahl, C. W., & Habib, A. (1989) The impact of overseas workers' remittances on indigenous industries: Evidence from Bangladesh. *The Developing Economies, 27*(3), 269–285.

Stark, O. (1991). *The migration of labor.* Basil Blackwell.

Stark, O., & Bloom, D. E. (1985). The new economics of labor migration. *American Economic Review, 75*(2), 173–178.

Stark, O., Helmenstein, C., & Prskawetz, A. (1997). A brain gain with a brain drain. *Economics Letters, 55*(2), 227–234. https://doi.org/10.1016/s0165-1765(97)00085-2

Steers, R. M. (1999). *Made in Korea: Chung Ju Yung and the rise of Hyundai.* Routledge.

Tabuga, A., Baiño, M. L., & Vargas, A. R. P. (2021). *Analyzing the characteristics of migration in the Philippines using the 2018 national migration survey.* Philippine Institute for Development Studies.

Tahir, M. (2019). Violence work and the police order. *Public Culture, 31*(3): 409–418.

Taylor, J. E. (1999). The new economics of labour migration and the role of remittances in the migration process. *International Migration, 37*(1), 63–88.

Technical Education and Skills Development Authority. (2005). *The national technical education and skills development plan, 2005–2009.* Manila: Technical Education and Skills Development Authority.

Teitelbaum, M. S. (2002). The role of the state in international migration. *The Brown Journal of World Affairs, 8*(2), 157–167.

Tierney, R. (2007). The guest labor system in Taiwan. *Critical Asian Studies, 39*(2), 205–228.

Tigno, J. V. (2004). Governance and public policy making in the Philippines: RA 8042 and deregulating the overseas employment sector. *Philippine Political Science Journal, 25*(48), 1–24.

Tilly, C., & Brown, C. H. (1967). On uprooting, kinship, and the auspices of migration. *International Journal of Comparative Sociology, 8(2)*, 139–164.

Torpey, J. (1998). Coming and going: On the state monopolization of the legitimate "means of movement. *Sociological Theory, 16*(3), 239–259.

Tseng, Y.-f., & Wang, H.-z. (2013). Governing migrant workers at a distance: Managing the temporary status of guestworkers in Taiwan. *International Migration, 51*(4), 1–19.

Tyner, J. A. (2009). *The Philippines: Mobilities, identities, globalization.* Routledge.

United Nations Department of Economic and Social Affairs. (2013). *International migration policies: Government views and priorities.* United Nations Publications.

United Nations Economic and Social Commission for Asia and the Pacific. (2011). *Situation report on international migration in South and South-west Asia.*

United Nations Population Division, Department of Economic and Social Affairs. (2015) Trends in International Migrant Stock: Migrants by Destination and Origin.

United Press International. (1990). Bechtel business grows to six-year high.https://www.upi.com/Archives/1990/04/03/Bechtelbusiness-grows-to-six-year-high/2656639115200/

United States Department of State. (retrieved 2024a). Nonimmigrant Visa Issuances by Visa Class and by Nationality, FY2014 NIV Detail Table. Website: Nonimmigrant Visa Statistics. https://travel.state.gov/content/travel/en/legal/visa-law0/visa-statistics/nonimmigrant-visa-statistics.html

United States Department of State. (retrieved 2024b). Nonimmigrant Visas by Individual Class of Admission. Website: Nonimmigrant Visa Statistics. https://travel.state.gov/content/travel/en/legal/visa-law0/visa-statistics/nonimmigrant-visa-statistics.html

Usui, N. (2012). *Taking the right road to inclusive growth: Industrial upgrading and diversification in the Philippines.* Asian Development Bank.

Villar, M., Jr. (1994). House Bill No. 13287. An act requiring government-owned or controlled corporations (GOCCs) on the privatization program to reserve at least twenty percent (20%) of their initial public offering (IPO) of shares to overseas workers, authorizing the Overseas Workers Investment (OWI) fund under R.A. 7111 to manage an investment fund portfolio for overseas workers, and for other purposes.

Vista Land & Lifescapes, Inc. (2007). *Annual report.*

Vista Land & Lifescapes, Inc. (2015). *Annual report: Unfolding possibilities.*

Vu, T. K. (2022). Remittances, the Dutch disease, and premature deindustrialization in the Philippines. *The Hosei University Economic Review, 90*, 147–164. http://doi.org/10.15002/00025878

Waldinger, R. (2014). The politics of cross-border engagement: Mexican emigrants and the Mexican state. *Theory and Society, 43*(5), 483–511.

Wallerstein, I. (1974). *The modern world-system, vol. 1: Capitalist agriculture and the origins of the European world-economy in the sixteenth century.* Academic Press.

White House. 1966. Joint Communique, September 15. *Department of State Bulletin, 55*, 531–534.

Wiest, R. E. (1984). External dependency and the perpetuation of temporary migration to the United States. In R. C. Jones (Ed.), *Patterns of undocumented migration: Mexico and the United States* (pp. 110–135). Rowman & Littlefield.

Woo, J.-e. (1991). *Race to the swift: State and finance in Korean industrialization.* Columbia University Press.

Yang, D. (2005). International migration, human capital, and entrepreneurship: Evidence from Philippine migrants' exchange rate shocks. *Policy Research Working Paper* No. 3578. World Bank Group.

Yee, J. (2000). The social networks of Koreans. *Korea Journal, 40*(1), 325–352.

Yoon, I.-J. (2012). Migration and the Korean diaspora: A comparative description of five cases. *Journal of Ethnic and Migration Studies, 38*(3), 413–435.

Yoon, S. (2011). POSCO: Building an institution. In H. Kim & C. W. Sorensen (Eds.), *Reassessing the Park Chung Hee Era, 1961–1979: Development, political thought, democracy, and cultural influence* (pp. 43–65). University of Washington Press.

Yuh, J.-Y. (2005). Moved by war: Migration, diaspora, and the Korean War. *Journal of Asian American Studies, 8*(3), 277–291.

Zolberg, A. R. (1981). International migrations in political perspective. In M. M. Kritz, C. B. Keely, & S. M. Tomasi (Eds.), *Global trends in migration: Theory and research on international population movements* (pp. 3–27). Center for Migration Studies.

Zolberg, A. R. (1989). The next waves: Migration theory for a changing world. *International Migration Review, 23*(3), 403–430.

Zolberg, A. R. (2006). *A nation by design: Immigration policy in the fashioning of America.* Harvard University Press.

Index

For the benefit of digital users, indexed terms that span two pages (e.g., 52–53) may, on occasion, appear on only one of those pages.

Tables and figures are indicated by an italic *t* and *f* following the page/paragraph number.